Clinical and Translational Science

Clinical and Translational Science

Principles of Human Research

Second Edition

Edited by

David Robertson MD

Director, Clinical & Translational Research Center, Elton Yates Professor of Medicine,
Pharmacology and Neurology, Vanderbilt University,
Nashville, TN, United States

Gordon H. Williams MD

Chief, Hormonal Mechanisms of Cardiovascular Injury Laboratory,
Brigham and Women's Hospital, and Professor of Medicine,
Harvard Medical School, Boston, MA, United States

AMSTERDAM • BOSTON • HEIDELBERG • LONDON • NEW YORK • OXFORD • PARIS
SAN DIEGO • SAN FRANCISCO • SINGAPORE • SYDNEY • TOKYO

Academic Press is an imprint of Elsevier

Library of Congress Cataloging-in-Publication Data
A catalog record for this book is available from the Library of Congress

British Library Cataloguing-in-Publication Data
A catalogue record for this book is available from the British Library

ISBN: 978-0-12-802101-9

For information on all Academic Press publications
visit our website at https://www.elsevier.com/

  Working together
to grow libraries in
developing countries

www.elsevier.com • www.bookaid.org

Publisher: Mica Haley
Acquisition Editor: Mica Haley
Editorial Project Manager: Lisa Eppich
Production Project Manager: Laura Jackson
Designer: Victoria Pearson Esser

Typeset by TNQ Books and Journals

Contents

4. Introduction to Epidemiology

Donna K. Arnett and Steven A. Claas

5. The Patient-Centered Outcomes Research Institute: Current Approach to Funding Clinical Research and Future Directions

Joe V. Selby and Danielle M. Whicher

Section II
Approaches

14. Principles of Biostatistics

Kush Kapur

15. Good Clinical Practice and Good Laboratory Practice

Nathalie K. Zgheib, Stephanie L. Tomasic and Robert A. Branch

23. Regulatory Environment

*Christine Nguyen, Audrey Gassman and
Hylton V. Joffe*

24. Ethical Issues in Translational Research and Clinical Investigation

Greg Koski

25. Clinical Research in the Public Eye

Mary Woolley

Section VI
Research in Special Populations

26. Research in Special Populations: Acute Illnesses; Critical Care; and Surgical Patients

Todd W. Rice and Gordon R. Bernard

27. Research in the Emergency Care Environment

James Quinn and Daniel J. Pallin

36. A Stepwise Approach to a Career in Translational Research

William F. Crowley Jr.

37. Physician Careers in the Pharmaceutical Industry

Ronald L. Krall

Section IX
Research in Academia

38. Industry-Sponsored Clinical Research in Academia

Italo Biaggioni

39. Governmental Support of Research

Sten H. Vermund and Salim Abdool Karim

Section X
Prospectus

List of Contributors

Salim Abdool Karim University of KwaZulu–Natal, Durban, South Africa; Columbia University, New York, NY, United States

Donna K. Arnett University of Kentucky, Lexington, KY, United States

James R. Baker, Jr. University of Michigan, Ann Arbor, MI, United States

Seema Basu Partners HealthCare Innovation, Cambridge, MA, United States

Stacey Berg Texas Children's Hospital, Houston, TX, United States

Gordon R. Bernard Vanderbilt University School of Medicine, Nashville, TN, United States

Italo Biaggioni Vanderbilt University, Nashville, TN, United States

Lisa Bomgaars Texas Children's Hospital, Houston, TX, United States

Robert A. Branch University of Pittsburgh, Pittsburgh, PA, United States

Nancy J. Brown Vanderbilt University School of Medicine, Nashville, TN, United States

Robert M. Califf U.S. Food and Drug Administration, Silver Spring, MD, United States

Henry C. Chueh Massachusetts General Hospital, Boston, MA, United States

Steven A. Claas University of Alabama at Birmingham, Birmingham, AL, United States

William F. Crowley, Jr. Massachusetts General Hospital, Boston, MA, United States

Joann Data Data Consulting, Sparta, TN, United States

George D. Demetri Dana-Farber Cancer Institute and Ludwig Center at Harvard, Harvard Medical School, Boston, MA, United States

Zeruesenay Desta Indiana University, Indianapolis, IN, United States

Ruth M. Dunne Harvard Medical School, Boston, MA, United States

Luigi Ferrucci NIA, Baltimore, MD, United States

David A. Flockhart Indiana University, Indianapolis, IN, United States

Audrey Gassman U.S. Food and Drug Administration, Silver Spring, MD, United States

Rashmi Gopal-Srivastava National Institutes of Health, Bethesda, MD, United States

Glenn Gormley Daiichi Sankyo Inc., Edison, NJ, United States

Steven Grinspoon Harvard Medical School, Boston, MA, United States

Stephen C. Groft National Institutes of Health, Bethesda, MD, United States

Katherine E. Hartmann Vanderbilt University School of Medicine, Nashville, TN, United States

Elizabeth Heitman Vanderbilt University School of Medicine, Nashville, TN, United States

Christopher D. Herrick Massachusetts General Hospital, Boston, MA, United States

Hylton V. Joffe U.S. Food and Drug Administration, Silver Spring, MD, United States

Kush Kapur Harvard Medical School, Boston, MA, United States

Mark D. Kellogg Boston Children's Hospital, Boston, MA, United States

Richard B. Kim The University of Western Ontario, London, ON, Canada

Bruce R. Korf University of Alabama at Birmingham, Birmingham, AL, United States

Greg Koski Harvard Medical School, Boston, MA, United States; Massachusetts General Hospital, Boston, MA, United States

Ronald L. Krall University of Pittsburgh, Pittsburgh, PA, United States; GlaxoSmithKline

Jessica Lasky-Su Harvard Medical School and Brigham and Women's Hospital, Boston, MA, United States

Shawn N. Murphy Massachusetts General Hospital, Boston, MA, United States

Christine Nguyen U.S. Food and Drug Administration, Silver Spring, MD, United States

Ailbhe C. O'Neill Harvard Medical School, Boston, MA, United States

Daniel J. Pallin Harvard Medical School, Boston, MA, United States

James Quinn Stanford University, Stanford, CA, United States

Keren Regev Harvard Medical School, Boston, MA, United States

Uwe E. Reinhardt Princeton University, Princeton, NJ, United States

Todd W. Rice Vanderbilt University School of Medicine, Nashville, TN, United States

Rose Marie Robertson American Heart Association, Dallas, TX, United States

David Robertson Clinical & Translational Research Center, Elton Yates Professor of Medicine, Pharmacology and Neurology, Vanderbilt University, Nashville, TN, United States

Dan M. Roden Vanderbilt University Medical Center, Nashville, TN, United States

Angela J. Rogers Stanford University, Stanford, CA, United States

Daniel E. Salazar EMS Pharma, Hortolàndia, SP, Brazil

J. Sanford Schwartz University of Pennsylvania, Philadelphia, PA, United States

Alan F. Schatzberg Stanford University, Stanford, CA, United States

Ellen W. Seely Harvard Medical School, Boston, MA, United States

Joe V. Selby Patient-Centered Outcomes Research Institute, Washington, DC, United States

César Serrano Vall d'Hebron University Hospital, Barcelona, Spain

Donald C. Simonson Brigham and Women's Hospital, Harvard Medical School, Boston, MA, United States

Ann R. Stark Monroe Carell Jr. Children's Hospital, Nashville, TN, United States

Stephanie Studenski NIA, Baltimore, MD, United States

Clare M. Tempany Harvard Medical School, Boston, MA, United States

Marcia A. Testa Harvard T. H. Chan School of Public Health, Boston, MA, United States

Thommey P. Thomas University of Michigan, Ann Arbor, MI, United States

Rommel G. Tirona The University of Western Ontario, London, ON, Canada

Stephanie L. Tomasic The Pitt-Bridge: Gateway to Success, Pittsburgh, PA, United States

Suzie Upton American Heart Association, Dallas, TX, United States

Sten H. Vermund Vanderbilt University, Nashville, TN, United States

Brent B. Ward University of Michigan, Ann Arbor, MI, United States

Howard L. Weiner Harvard Medical School, Boston, MA, United States

Scott T. Weiss Harvard Medical School, Boston, MA, United States; Partners HealthCare Personalized Medicine, Boston, MA, United States; Channing Division of Network Medicine, Boston, MA, United States

Danielle M. Whicher Patient-Centered Outcomes Research Institute, Washington, DC, United States

Gordon H. Williams Hormonal Mechanisms of Cardiovascular Injury Laboratory, Brigham and Women's Hospital, and Professor of Medicine, Harvard Medical School, Boston, MA, United States

Mary Woolley Research!America, Alexandria, VA, United States

Nathalie K. Zgheib American University of Beirut, Beirut, Lebanon

Section I

Fundamental Principles

Chapter 1

Introduction to Clinical Research

Gordon H. Williams[1] and David Robertson[2]

[1]Hormonal Mechanisms of Cardiovascular Injury Laboratory, Brigham and Women's Hospital, and Professor of Medicine, Harvard Medical School, Boston, MA, United States; [2]Clinical & Translational Research Center, Elton Yates Professor of Medicine, Pharmacology and Neurology, Vanderbilt University, Nashville, TN, United States

Chapter Outline

During the past quarter century, the term "translational and clinical science" has come into popular use. Originally it was used to describe the activity of translating results from animal studies to humans. Shortly thereafter translational research was used to define activities on the proverbial two-way street between animal and human studies. However, more recently the term has been applied to a variety of activities ranging from knowledge gained by translating cell-based experiments to whole organs to information provided by translating epidemiologic data to the delivery of health services. While many of these "translating" disciplines use similar tools and resources, in the process the definition of "translational" has become obscured. This book will focus on the tools, techniques, and infrastructure available to assist clinical researchers including human translational investigators—sometimes termed "patient-oriented" investigators—accomplish their research goals. While the material contained herein is comprehensive, it is not encyclopedic. Some of the resources specifically addressed may be applicable to other types of clinical researchers; these individuals are not the primary audience to whom this textbook is directed. There are several excellent textbooks that cover some topics in considerably more depth, e.g., on clinical epidemiology, health services, statistics, outcomes research, genetics, and pharmacology. In the chapters related to these areas we refer the reader to some of these textbooks. However, whether population- or patient-oriented, clinical investigators should find value, for example, in the following topics: genetics, statistics, study design, imaging, careers, ethics, regulatory issues, funding, epidemiology, and studies in special populations.

HISTORICAL BACKGROUND

The resources that are the focus of this book have their origin in the span of 6 years in the middle of the 19th century. Three pivotal works were published that initiated the modern era of clinical and translational research, although their significance was not clearly recognized until some years later. The three authors came from diverse backgrounds: a Czech priest, a British naturalist, and a French physician and physiologist.

Charles Darwin was born in 1809 and, after dropping out of medical school at the University of Edinburgh, completed training for the clergy at Christ's College, Cambridge. He became interested in botany and natural history and as a result joined the crew of *HMS Beagle* for 5 years of exploration. His experiences on this ship dramatically shaped his future and ultimately resulted in his famous 1859 treatise *On the Origin of Species by Means of Natural Selection, or the Preservation of Favoured Races in the Struggle for Life.*

As reported in this work, Darwin's seminal hypothesis was:

> This is the doctrine of Malthus, applied to the whole animal and vegetable kingdoms. As many more individuals of each species are born than can possibly survive; and as, consequently, there is a frequently recurring struggle for existence, it follows that any being, if it vary however slightly in any manner profitable to itself, under the complex and sometimes varying conditions of life,

will have a better chance of surviving, and thus be naturally selected. *From the strong principle of inheritance, any selected variety will tend to propagate its new and modified form.*

Darwin (1889).

Darwin carefully avoided using the term "evolution" in his *On the Origin of Species* likely because of concerns of rejection of his theory based on a charged word. As a result, within a decade of its publication, his theories had gained substantial popularity, particularly in England, although critics abounded in the rest of Europe and in the United States. While dissenters still exist in the 21st century and some of his theories have been modified by others, his 1859 publication remains a cornerstone of modern-day clinical and translational research.

Claude Bernard was also born in the early 19th century (1813), in France, and was educated both as a physician and physiologist. He has often been referred to as the "Father of modern physiology and of the experimental method." Following completion of his training, he began to challenge the traditional dogma concerning how scientific discoveries should be made. In contrast to most scientists of his time, he was a champion of hypothesis testing and the experimental protocol to answer scientific questions—not "expert" opinions and/or case studies. He stated that carefully crafted experimental design with objectively evaluated results would lead to true advances in medical and biological sciences. In 1865, 6 years after *On the Origin of Species* was published, Bernard published his seminal work *An Introduction to the Study of Experimental Medicine*. This book detailed most of Bernard's ideas as to how advances in medical science would be accomplished. Many of the principles that he proposed are still fundamental to human research in the 21st century, including cause and effect, hypothesis testing and experimental fact, having positive and negative controls in an experiment, and never accepting a hypothesis as truth until one has tried and failed to disprove it. In contrast to Darwin, Bernard used a combative approach to put forth his controversial ideas. Illustrative of this was that *An Introduction to the Study of Experimental Medicine* was written in the first person singular. Bernard also framed his discussions in a somewhat acrimonious manner. For example, he stated:

In these researches I followed the principles of the experimental method that we have established, i.e., that, in presence of a new fact which contradicts a theory, instead of keeping the theory and abandoning the fact, I should keep and study the fact, … When we meet a fact which contradicts a prevailing theory, we must accept the fact and abandon the theory, even when the theory is supported by great names and generally accepted.

Bernard (1927 [1865], p. 164).

And:

To sum up, theories are only hypotheses, verified by more or less numerous facts. Those verified by the most facts are the best; but even then they are never final, never to be absolutely believed.

Bernard (1927 [1865], p. 165).

However, in contrast to the 21st-century scientist, he was not enthusiastic about the statistical method but cautioned about narrowly designing experiments where the only positive outcome will be to support the preconceived hypothesis. Finally, he had an inspiring view of the true medical scientist:

Yet truth itself is surely what concerns us and, if we are still in search of it, that is because the part which we have so far found cannot satisfy us. In our investigations, we should else be performing the useless and endless labor pictured in the fable of Sisyphus, ever rolling up the rock which continually falls back to its starting point. This comparison is not scientifically correct: a man of science rises ever, in seeking truth; and if he never finds it in its wholeness, he discovers nevertheless very significant fragments; and these fragments of universal truth are precisely what constitute science.

Bernard (1927 [1865], p. 222).

In contrast to Bernard and Darwin, the third founder of modern-day clinical and translational science labored in relative obscurity in a monastery in Brno, Hungary-Austria (now the Czech Republic). Gregor Mendel was an Augustinian monk, born in 1822 in Austria. He published his epic work *Experiments with Plant Hybrids* in the same year (1865) as Bernard wrote An Introduction to the Study of Experimental Medicine. However, it was not until a dozen years after he died that the significance of Mendel's work began to be appreciated, and well into the 20th century that he finally was accorded the honor of being the "Father of modern genetics." Mendel's work contrasted to the other two in several substantial ways. First, he did not have the same degree of formal education in his field as did Bernard and Darwin in theirs. Second, he performed his most important experiments on plants rather than animals. Third, there were few, if any, prior publications on genetics in contrast to the more extensive literature on evolution and experiments in medicine, albeit the latter were often poorly designed. These facts, and Mendel's relative isolation, since he was not at a university, likely explain why it took much longer for his genetic theories to be widely acknowledged and accepted.

In addition to the overall critical concepts of his studies, Mendel made several observations that have been underplayed in 21st-century genetic studies:

The value and utility of any experiment are determined by the fitness of the material to the purpose for which it is used, and thus in the case before us it cannot be immaterial what plants are subjected to what experiment and in what manner such experiment is conducted.

Mendel (1901 [1865], p. 2).

And:

Some characters [do not permit] sharp and certain separations since the differences of the more or less nature [are often difficult to define].

Mendel (1901 [1865], p. 4).

Indeed, he concluded his 1865 book with the following insightful fact: of the more than three dozen characteristics of peas, Mendel selected only seven characteristics that "stand out clearly and definitely in the plants." In searching for the needle (one or more genes) in the haystack (a population), the medical scientist needs to shrink the haystack as much as possible before attempting to establish cause-and-effect relationships. While there were advances in clinical and translational science in the early part of the 20th century, it was not until midcentury that the next major advances occurred, this time in the United States. It was recognized that specific training programs and clinical research laboratories were required to take the next step in advancing human science, resulting in the establishment of a few clinical research facilities, the most notable being the Rockefeller University. This movement was specifically accelerated in 1960 when the General Clinical Research Centers (GCRCs) and the Medical Scientist Training Program (MSTP), a program to train MD/PhD students, were established. Both were supported by the National Institutes of Health (NIH). Prior to this time, most clinical research training used the apprentice model, and the laboratories of clinical investigators were either their offices and/or the hospital wards and intensive care units. It was realized that these approaches were no longer adequate. Clinical investigators needed a specific, specially designed space to perform their studies like the bench scientists had for their experiments. Furthermore, it was hypothesized that by formally training physicians in the science of the experimental method, as Bernard advocated, physicians would be trained to conduct high-quality research studies when they returned to human research.

During the next 35 years, substantial progress was made in both areas. Seventy-six GCRCs were funded, and more than 1600 students had enrolled in an MSTP in medical schools across the country. However, the outcomes of these two programs had not matched the expectations of their founders. Only 5% of MSTP graduates remained engaged in clinical research, and only 8% of NIH investigator-initiated grants (R01s) were supporting clinical research projects. This led to the Clinical Research Enhancement Act of 2000 that created a more formal training program specifically for clinical and translational investigators (K-30), a separate early career salary support program (K-23), a new midcareer mentoring program (K-24), a specific, educational loan repayment program, and the mandate to NIH to create a level playing field to fund R01 clinical research in most cases by establishing Clinical Research Study Sections to review these applications. Since 2000, several programs have been established by foundations and governing bodies worldwide that have incorporated some of the features of the Clinical Research Enhancement Act of 2000.

The success of this experiment is still pending, and indeed the answers remain elusive as in less than 15 years dramatic changes have occurred. By the end of the first decade of the 21st century, the GCRC and K-30 programs were merged into a new expanded entity termed Clinical and Translational Science Awards (CTSAs) (see Chapter 34). These linked educational, early support of clinical scientists and infrastructure into a single entity at 60 sites across the United States. A little more than half a decade later NIH further modified the implementation of the Clinical Research Enhancement Act of 2000. Support for the human research laboratories (GCRCs) was terminated, and NIH's direct financial support for human research infrastructure was substantially reduced. It is unclear whether clinical research infrastructure support has been reduced in other countries. Thus, the next few years will be challenging as new models evolve to provide infrastructure support for the translational/clinical investigator.

ORGANIZATION OF THIS BOOK

With the explosion of knowledge concerning the tools, training, and infrastructure to support clinical and translational investigation, there is the need to capture and catalog these components. The purpose of this textbook is to provide in one volume the fundamental basis of clinical and translational research. Rather than being a traditional treatise on the subject, our aim is to capture the excitement and innovation in our discipline in an era when a critical mass of bright, young trainees are entering a landscape of almost unlimited scientific opportunity. However, we believe that appropriate didactic information has not kept pace with the rapid growth of clinical research as a reemerging discipline. The purpose of this book is to provide that information in a single volume in the way that Harrison's *Principles of Internal Medicine* does for the field of adult medicine. Whether they are located in universities, medical schools, institutes, pharmaceutical companies, biotech companies, or clinical research organizations, clinical researchers will find this compendium invaluable for filling that void.

The chapters are written to enlighten the novice and extend the knowledge base of the established investigator. None of the chapters is meant to be comprehensive of its subject matter as in many cases there are entire books written on the various topics. However, like standard textbooks of medicine, surgery, pediatrics, and obstetrics/gynecology, our intent is to cover the scope of information necessary for clinical/translational investigators. In each chapter the reader is referred to other sources that can provide a more thorough review.

The book is organized in an entirely new way, reflecting the broad challenges and opportunities in contemporary clinical research. Internet-accessible educational material supplements the material embodied in the text, creating resources that represent a seamless and continuously updated learning instrument. Where relevant, the power of infusing informatics into the study of genetic, molecular, and metabolic investigations at the bedside is emphasized.

The book begins in the tradition of Bernard by reviewing the fundamental principles of experimental design for both patient-oriented and population-oriented research. Then, it turns to approaches (tools) available to clinical/translational investigators ranging from statistics to questionnaires to imaging to information technology. The third section reviews the infrastructure available to support clinical/translational investigators—their laboratory. The following component elucidates educational and career opportunities for these scientists in a variety of venues. The fifth section details funding for clinical and translational investigations. The next section covers the rapidly expanding field of human genetics, building on the foundation created by Mendel and Darwin. The seventh section addresses the ever-expanding horizon of human pharmacology and is followed by one on the social and ethical issues involved in human research. The next section addresses the application of the aforementioned principles to special populations, such as children, the elderly, and patients with psychiatric, oncologic, or neurologic diseases or acute illnesses. The final section addresses the powerful tools developed by population scientists during the past third of a century. Thus, the first and last sections provide textural bookends for the two major approaches to clinical research. Many chapters in this book do not provide the in-depth information that has been provided by entire textbooks on these individual subjects. The reader is referred to these textbooks when appropriate for additional information.

The chapters reflect the combination of basic and clinical research (both patient- and population-oriented) principles. Importantly, we hope that they reflect the editors' view that these disciplines cannot be uniquely separated, but are part of a continuum that flows back and forth between them—often sharing similar tools and approaches to better understand the totality of human disease. As such, we believe the information contained herein will also be of value to our enlarging audience of Masters or PhD trainees in clinical science or public health, medical students, students in other biomedical professional disciplines, clinical scientists in industry, practicing clinical investigators and administrators in the broad field of clinical and translational research with a specific focus on the study of the individual subject. The disciplines of genetics and precision medicine will especially benefit from the new knowledge human inheritance.

In this second edition, to focus the mission and concept of this book, we have provided brief clarifying overviews of the key chapters and content to introduce and facilitate the major points that will facilitate learning. These are reflected in the abstract, key words, key concepts, and summary.

NOTE

Color versions of many of the illustrations reproduced in black and white are available on the *Clinical and Translational Science* companion website which can be accessed at www.elsevierdirect.com/companions/9780123736390.

REFERENCES

Bernard, C., 1927 [1865]. An Introduction to the Study of Experimental Medicine. (English translation), Greene, H.C., Macmillan & Co.

Darwin, C., 1889. On the Origin of Species by Means of Natural Selection, or the Preservation of Favoured Races in the Struggle for Life, seventh English ed. D. Appleton & Co., New York, p. 5.

Mendel, J.G., 1865. Versuche über Plflanzenhybriden. Verhandlungen des Naturforschenden Vereines in Brünn bd. IV, Abhandlungen, English translation, Druery, C.T., Bateson, W., 1901. Experiments in Plant Hybridization. J. Royal Hortic. Society 26.

Chapter 2

Patient-Oriented Research

Ellen W. Seely and Steven Grinspoon
Harvard Medical School, Boston, MA, United States

Chapter Outline

Key Points

- Patient-oriented research is a specific type of clinical translational research where the patient is the focus of the investigation.
- Patient-oriented research can include observational, mechanistic, and therapeutic studies as well as clinical trials.
- Special research tools are available for the patient-oriented researcher.

INTRODUCTION

Patient-oriented research (POR) is defined by the National Institutes of Health (NIH) as "research conducted with human subjects (or on material of human origin such as tissues, specimens and cognitive phenomena) for which an investigator (or colleague) directly interacts with human subjects." Excluded from this definition are in vitro studies that utilize human tissues that cannot be linked to a living individual. Subsets of POR according to NIH include (1) mechanisms of human disease, (2) therapeutic interventions, (3) clinical trials, or (4) development of new technologies (NIH Glossary). Therefore, POR represents a subset of clinical research. However, whereas in clinical research, studies can involve human cells, POR requires an intact alive human being as it focuses on the understanding of human physiology as a key to understanding the mechanism of disease processes in humans. In other words, as expressed by Brown and Goldstein, "…if the investigator shakes hands with the patient in the course of the research, that scientist is performing POR" (Goldstein and Brown, 1997). Through understanding the mechanism of human diseases, interventions can then be developed to both prevent and treat these diseases.

POR holds a central role in any research that is dedicated to improving human health and is a crucial step in research whether basic or clinical (Fig. 2.1). Importantly, the patient is the start of the research process by serving as the inspiration

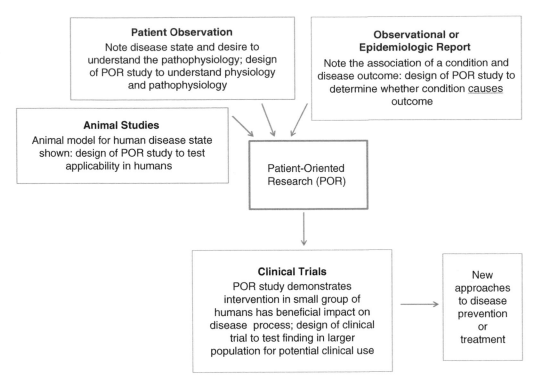

FIGURE 2.1 Position of patient-oriented research in biomedical research.

for a hypothesis (observation of a patient with an unusual phenotype) or as the motivation for the research (a patient for whom effective therapy does not exist for his/her condition), and the patient is the end of the process when the therapy developed through POR successfully benefits his/her health.

The results of human experimentation are almost always relevant, whereas data from basic research in in vitro and animal models, though often obtained in models with less confounding, may or may not be relevant to the human condition and thus require further testing in humans (see Role of the Patient in Patient-Oriented Research for specific examples). Studies of tissue and cells obtained from humans can be very informative and complement in vivo data, but cannot substitute for studies performed in humans. For example, the benefits of a therapeutic intervention demonstrated in an in vitro or animal study may not be seen in studies of humans. As a discipline, POR shares many of the basic principles of the methodology used in basic science, animal, or in vitro research. In this regard, formulation of a sound and reasonable hypothesis is the starting point, followed by good study design and rigorous hypothesis testing. However, unique to POR are a number of special considerations which form the basis of the subsequent chapters in this book, including unique statistical considerations, choice of relevant endpoints, feasibility, and ethical considerations in study design. Furthermore, there are unique safety and regulatory considerations in POR.

This chapter will discuss the types of POR, the role of POR in the translational research continuum, the unique considerations in study design, methodology for conducting POR, and specific funding opportunities for POR.

TYPES OF PATIENT-ORIENTED RESEARCH

A number of types of POR exist; these include (1) mechanistic studies of human disease also called physiological, involving the detailed investigation of a biological process, usually in a smaller number of subjects; (2) therapeutic studies, e.g., to improve a disease or condition; and (3) clinical trials, which are most often large, definitive studies. Much of POR is therefore interventional and can vary greatly in its scope, from small-scale physiological studies using a medicine or probe to delineate physiological mechanisms, to small treatment studies, to large-scale clinical trials and studies used to develop new technologies. POR can range from Phase I first in man studies to a definitive Phase III trial for a drug registration. In addition to these studies, observational studies, involving humans, for example, observational cohort studies, are not typically categorized as POR, but do involve human subjects and are often critical to the hypothesis generation.

Observational Studies

Observational studies are not always defined within the traditional framework of POR, but can be an important type of POR (see also Chapter 43). Observational studies often involve recruitment of patients and the interaction of the investigators with subjects to obtain natural history data. Observational studies are descriptive in nature and often lead to the determination of associations. Such studies are often important to generate the hypotheses, for subsequent interventional studies. These studies typically do not have well-defined mechanistic hypotheses, but rather have a stated goal to obtain data or determine an association. Because they are not definitive by their nature, observational studies must be large enough or collect data on adequately detailed clinically relevant endpoints to be viewed as rigorous. The Framingham Heart Study is an important example of a patient-oriented observational study. Data are collected on human subjects in this observational cohort at face-to-face study visits where measurements such as blood pressure and heart rate as well blood tests such as cholesterol levels are determined at regular intervals. Data obtained from study visits and observed outcomes led to the development of the Framingham Risk Score (Wilson et al., 1998), one of the most widely used cardiovascular risk prediction models.

However, observational studies demonstrate association not causality. Analysis of data from the Nurses' Health Study, another observational cohort, suggested a potential beneficial association between postmenopausal estrogen use and reduced cardiovascular disease (CVD) (Stampfer et al., 1991). Confounding can occur in observational studies and is the reason why such studies are not definitive and hypothesis generating rather than hypothesis testing. For example, the association of estrogen use with a lower risk of CVD in observational studies may be because younger, healthier women may chose to use estrogen. Analyses may be performed to control for confounders, e.g., in multivariate regression, but such analyses are not always able to account for all confounders.

Recently, with the determination of the human genome, genetic studies are becoming an increasingly important type of observational study. Such studies may show an association between a particular genotype and disease or may be useful to determine efficacy of a particular therapy in a pharmacogenomic analysis. For example, in non-small-cell lung cancer, studies suggested that mutations in the epidermal growth factor receptor (EGFR) predict response to the tyrosine kinase inhibitor, gefitinib (Lynch et al., 2004). As a result of these studies, gefitinib is FDA approved for use in the treatment of non-small-cell advanced lung cancers carrying EGFR mutations.

Limitations: Causality may be inferred but not proven if a covariate attenuates the relationship between other covariates. Formal PATH analysis may be undertaken, but causality can never be proven in observational studies (Asher, 1976). Indeed, a randomized, interventional study testing the use of estrogen and progestin in postmenopausal women suggested that estrogen and a progestin increased coronary artery disease rates, particularly during the first year after initiation (Manson et al., 2003). In addition, confounding may occur in observational studies.

Mechanism Studies

In contrast to observational studies, mechanistic studies are useful to determine causality. Such studies are hypothesis driven, and formulation of the hypothesis is the critical element for the success of the study. These studies are often smaller and are useful to determine physiology or disease mechanism, because they use sufficiently detailed endpoints which are assessed in response to a relevant perturbation. For example, in endocrine physiology, this is often achieved by blocking or stimulating a given pathway.

A good example of mechanistic studies can be seen in the elucidation of the role of leptin as a neuroendocrine modulator in undernutrition. Leptin was identified originally in mice, as an anorexigenic molecule produced in fat tissue (Zhang et al., 1994). Initial animal studies suggested that leptin could restore gonadal function in rats fed a very-low-calorie diet, suggesting that it may be an important signal for neuroendocrine function (Ahima et al., 1996). Two short-term, mechanistic studies determined that leptin restored luteinizing hormone (LH) pulsatility in both men and women undergoing a complete fast (Chan et al., 2003; Schurgin et al., 2004). Leptin replacement prevented the typical neuroendocrine response to starvation, thus proving leptin is an important adipogenic signal in this regard. In this case, the investigators hypothesized that short-term leptin replacement would restore normal gonadotropin function, and this hypothesis, based on animal studies, was proven correct in humans. Caveats to this type of study include the fact that sufficiently detailed endpoints must be ascertained to establish the hypothesis. For example, LH pulsatility was determined by frequent sampling, and this was more important than simply measuring estrogen or testosterone per se, as a surrogate. In these studies, it was critical to establish whether there were any changes in weight and body composition from leptin, to determine whether such changes may have confounded the results. Furthermore, trial design was an important consideration. In one study a crossover design was used, in which the patients served as their own controls, receiving leptin versus placebo treatment in a randomized order (Chan et al., 2003). This increased the power of the study and minimized the number of necessary patients. In the second study, a straightforward randomized, placebo-controlled design was used (Schurgin et al., 2004). The inclusion of a placebo comparator in each study was critical, as confounding might have prevented a true determination of causality in an open-label study.

In contrast, the initial hypothesized role of leptin to reduce weight in generalized obesity has not achieved success (Heymsfield et al., 1999). The story of leptin in this regard demonstrates an important lesson in study sequence. Leptin levels are low in animal models of leptin deficiency, and in these models, as well as in human models of leptin deficiency, leptin administration has been highly effective to reduce weight and appetite. However, in generalized human obesity, initial observations suggested that leptin levels were high, and thus leptin resistance occurs. Therefore, one might have hypothesized a priori that leptin may not result in weight loss or that very high, supraphysiological doses were needed. Indeed, initial studies suggest this is the case. Very high doses of leptin, compared to those used in the undernutrition studies, have been required in obesity studies and showed modest efficacy at best with respect to weight (Heymsfield et al., 1999). Thus two separate observations, one of restoration of gonadal function, and the other of leptin resistance, have informed the appropriate design of very different mechanistic studies to answer these important questions regarding a critical metabolic hormone.

Limitations: Mechanistic studies may not be definitive by their very nature because they are small, may use novel endpoints to assess detailed physiology, and may not be generalizable to large populations with complicated conditions. These studies may have less clinical relevance than large-scale trials, but are no less important as they may stimulate larger therapeutic studies or even definitive large-scale clinical trials.

Therapeutic Studies

Therapeutic studies are studies that determine the efficacy of a medicine or treatment approach to improve a condition in patients. Such studies involve many of the same issues of trial design outlined below for Clinical Trials including power, safety, and confounding but are often smaller and potentially more preliminary in nature. As such the distinction between therapeutic studies and clinical trials can be blurry, but may relate to differences in size, generalizability, and definitiveness of the study. For example, when testing a new drug in humans, Phase I studies are for safety, Phase II studies are the first therapeutic trials, whereas Phase III studies may be larger clinical trials. Even large Phase III studies for a new drug may not be as large as large-scale clinical trials, which often test a strategy for which there is some evidence already established for efficacy, but for which definitive proof is lacking. Clinical trials are thus one particular form of therapeutic studies. Stated differently, a clinical trial is not usually undertaken without some proof of potential efficacy, whereas a therapeutic study may be initiated to gather early efficacy data. As an example, patients with human immunodeficiency virus (HIV) have been shown to accumulate excess visceral fat in association with dyslipidemia as a result of new antiretroviral medications, which may increase cardiovascular risk in this population. Physiological studies in such patients have shown reduced growth hormone—releasing hormone—mediated growth hormone release that may contribute to overall reductions in growth hormone secretion (Koutkia et al., 2004). As such, a therapeutic study of growth hormone—releasing hormone was completed demonstrating a significant reduction in visceral adipose tissue with improvement in lipids (Falutz et al., 2007). The study was generalizable to the population with HIV and fat accumulation and was safe, particularly with respect to glucose, but was not powered to show a benefit in overall mortality.

Limitations: Such studies are generalizable only to the conditions of the study and patients being investigated. They may be less likely to assess hard endpoints and be less definitive than large-scale clinical trials, but are a critical component of POR. Therapeutic studies need to determine the appropriate risk—benefit ratio of any drug, which can be further explored in large-scale trials.

Clinical Trials

(Also see Chapter 3 for additional details.)

Clinical trials usually arise from either observational studies, in which case they are designed to determine when an association is causative, or from physiologic studies where they are designed to determine whether findings derived from a smaller number of individuals will hold for a larger population. Large-scale clinical trials are often more important in determining the clinical efficacy of a drug than elucidating physiology or disease mechanisms. In part, this relates to the limited endpoints one can use in a very large study. Clinical trials are always hypothesis driven and often stem from data derived from observational studies or mechanism studies. Such studies need to be adequately powered, e.g., large enough to know that a negative result is not the result of insufficient patients, but rather a true biological result. Such studies can only be generalized to the specific population and intervention studied, and may be misinterpreted or overgeneralized. For example, observational studies had demonstrated an association of estrogen in the form of hormone treatment with decreased incidence of CVD in postmenopausal women (Stampfer et al., 1991). Based on this association, a prospective randomized clinical trial was initiated to test the hypothesis that estrogen used for hormone treatment of postmenopausal women decreased CVD. The Women's Health Initiative (WHI) studied conjugated equine estrogen (CEE) and medroxyprogesterone in postmenopausal women, as well as CEE alone in hysterectomized women (Anderson et al., 2004). In contrast to the expected results, the study showed an increase in CVD with use of CEE and medroxyprogesterone. The study has been widely cited as demonstrating a negative cardiovascular effect of estrogen. However, a specific type of estrogen was used (CEE), and its effects could not be separated from the type of progesterone (medroxyprogesterone) in this study. Furthermore, secondary analyses suggest that the negative effects may be more pronounced in and possibly driven by older women (Rossouw et al., 2007). CEE alone reduced coronary calcium score in younger, hysterectomized women (Manson et al., 2007). This secondary analysis inspired a prospective randomized placebo-controlled study to determine if combined hormone treatment was beneficial as defined by less progression of carotid artery intima-media thickness in women closer to menopause. The study showed no benefit (Harman et al., 2014).

These studies raise a number of issues critical to successful trial design and interpretation (Grodstein et al., 2003). First, they demonstrate that observational studies are only hypothesis generating and such hypotheses need to be proven in interventional studies. Although these studies are large and involve an estrogen preparation commonly used at the time of study initiation, it remains unclear whether the results are due to the type of estrogen used (oral) or the combination with a specific type and dose of progesterone (medroxyprogesterone acetate). In addition, it has been questioned whether the investigation of coronary artery calcium (CAC) score in substudies, not originally planned in the primary study, is more hypothesis generating than definitive. Finally use of a surrogate CAC score cannot be equated per se with effects on more definitive hard endpoints (including events, such as myocardial infarctions).

Numerous other issues arise in the design of interventional studies. Is the study large enough, long enough, and are the endpoints and interventions adequate to answer the question at hand? Furthermore, the study results can only be generalized to the population and intervention studied. Most large-scale clinical trials are randomized and placebo-controlled to minimize confounding. For example, in the Diabetes Prevention Program (DPP), subjects were randomized to placebo, metformin, or lifestyle change (Knowler et al., 2002). Use of a nontreatment control group was critical to prove that the benefits of lifestyle change and metformin were above and beyond that expected from entry into a clinical study, e.g., healthier choices, placebo effects. Stratification may help to insure that potential confounding variables are spread equally across randomization groups.

Safety is another major issue in study design. Safety may be a primary endpoint, as in Phase I studies, or may be an important secondary endpoint. For example, in large-scale efficacy trials, safety may be an important secondary endpoint. Safety should be assessed in a uniform fashion, according to prespecified guidelines. Often, in large-scale clinical trials, this involves use of a data safety monitoring board (DSMB) to monitor events and may involve specific stopping rules either for safety or efficacy. Effects on safety are always important in clinical trials and may determine the fate of a drug or compound even if efficacy is shown. For example, the studies of torcetrapib, a cholesterol esterase transfer protein inhibitor, demonstrated highly significant effects to raise HDL, but increased blood pressure and did not result in improvement in atheroma volume or carotid intimal medial thickness (Kastelein et al., 2007; Nissen et al., 2007). Significant improvements in HDL must then be considered in the context of worsened blood pressure and lack of efficacy to improve atherosclerosis. To determine if raising HDL levels led to decrease in CVD events, a randomized placebo control trial was performed and demonstrated that the addition of niacin (to raise HDL) to a regimen of laropiprant (to lower LDL) was associated with no further benefit to decreasing major vascular events and increased serious adverse events such as exacerbation of diabetes as compared with laropiprant alone (Landray et al., 2014). However, the issue of generalizabilty also comes into play. The negative effects of torcetrapib may not be class-specific, e.g., generalizable to the class of CETP inhibitors, but drug specific, for example, through an effect on blood pressure. Niacin-induced increase in HDL may not be representative of other classes of agents to increase HDL. Further mechanistic studies are needed to clarify these important issues.

Another issue relates to adequacy of safety assessment in a single trial versus a combined pooling of multiple trials. In general, it is optimal to prespecify a rigorous safety endpoint in the study design. However, if an unexpected safety concern is raised, an individual study may be inadequately powered, and pooling may be necessary. This was seen in the recent meta-analyses of CVD events in relationship to rosiglitazone use (Nissen and Wolski, 2007). The advantage of such an analysis is that it can achieve the necessary power to discern an effect. The disadvantage is that the safety endpoint may not have been collected in a uniform fashion, making conclusions difficult to establish. If the results with respect to a prespecified safety analysis in an individual study are at odds with results from a meta-analysis of combined studies, the results may need further clarification.

Limitations: Large-scale clinical trials are often definitive, but only generalizable to the conditions, patients, and specific interventions studied. There may be specific subgroups of patients who are responders to the intervention, but the signal of their response may be lost in the heterogeneity of the study population. These studies take a long time to complete, and drugs or treatments that were state of the art at the time of study initiation may become dated by the time the results are reported. A negative result in such a clinical trial is problematic if the trial is not adequately powered. Because such studies are often multicentered, endpoints must be used that can be performed by all sites, e.g., the least common denominator approach. Clinical trials often do not define the precise mechanism(s) mediating the trial outcome. Nonetheless, placebo-controlled clinical trials are critical to definitively prove new treatment strategies and approaches, as well as to prove existing approaches in clinical medicine.

THE ROLE OF PATIENT-ORIENTED RESEARCH IN TRANSLATIONAL RESEARCH

Interaction of Basic Science and Patient-Oriented Research

Translational research has traditionally referred to the translation of basic research findings to the clinical level commonly termed "bench to bedside." POR plays a central role in this translation in demonstrating whether basic findings in cells or in animals apply to humans. For example, human epidermal growth factor receptor (HER2) was shown to be overexpressed in tumors of ~20% of women with breast cancer. Women with tumors overexpressing this receptor had poorer prognoses with greater tumor invasion and metastases. A human monoclonal antibody to HER2 (trastuzumab) was developed in the laboratory and then after clinical trials demonstrated trastuzumab added onto traditional chemotherapy led to improved survival in women with HER2-positive metastatic breast cancer; the US Food and Drug Administration approved this therapy (Sledge Jr., 2004).

Translational research is not unidirectional. Clinical observations from patients may inform both patient-oriented and basic investigation. Furthermore, another translational step has received increasing attention: the translation of research findings from POR such as clinical trials into daily clinical practice. The failure for translation into community care has been referred to as an additional translational block (Sung et al., 2003). The development of the Clinical and Translational Science Awards (CTSAs) (see Chapters 33, 34, and 39 for details) reflects the importance that the NIH is placing on translational research. Thus, POR plays a central role in the translation of basic research findings to eventual improvements in clinical practice and patient care.

Basic science studies often form the foundation, upon which POR is conducted. Basic science research can be molecular, often carried out in cells or other in vitro systems, or physiological, often performed in animals, and investigate genetic or mechanistic underpinnings of disease. Basic science studies are often very focused and can be constructed in a way to minimize confounding, e.g., knocking out a specific pathway, receptor or gene, in a specific cell line or animal system. Often informative, these studies may not recapitulate the complexity of the human system in which multiple overlapping physiological pathways contribute to homeostatic regulation. For example, in the regulation of hunger or weight, multiple overlapping pathways may not be adequately assessed in a single knockout model. On the other hand, such studies can critically shed light on pathways that do require further investigation and confirmation in human studies. Basic science studies are often very useful for the high-throughput testing of drugs to identify initial targets and therapeutic approaches. Recent advances, for example, in Alzheimer's research have resulted in manipulation and growth of stem cells to express plaques and tangles in vitro in a neural cell culture model (Choi et al., 2014). Such a system is critical to allow the rapid testing of compounds for potential treatment of Alzheimer's, but may be limited in terms of lacking the physiological regulation in human systems. POR interacts with basic science in the inverse as well. POR can elucidate physiologic mechanisms in response to a therapy, but it cannot provide the cellular mechanisms for the physiologic response. For example, patient-oriented investigators noted that the administration of estrogen to hypertensive postmenopausal women was associated with a fall in renal blood flow, a deleterious vascular effect, and the fall in renal blood flow appeared to be mediated by an activated renin–angiotensin–aldosterone system (Seely et al., 2004). As estrogen, in other studies, had been shown to have beneficial vascular effects, the investigators moved to an animal model to better understand the human findings. Instead of studying young healthy animals, they created a model of vascular aging by lowering vascular nitric oxide by administrating N-nitro-L-arginine methyl ester (L-NAME). In this model, animals that had undergone ovariectomy (done to simulate menopause) as compared to animals that had undergone ovariectomy but received estrogen treatment were administered angiotensin II. These groups had similar blood pressures, but those who received estrogen had evidence of vascular damage and proteinuria. Molecular studies demonstrated increased expression of the angiotensin II receptor type 1 receptor (AT1R) in the kidney of the animals that had received estrogen, and there was a significant positive relationship between the AT1R and the amount of proteinuria (Oestreicher et al., 2006). In this example, basic studies elucidated a cellular mechanism for a physiologic finding in POR.

Interaction of Patient-Oriented Research and the Community

There has been an increased recognition that many evidence-based therapies, demonstrated to be effective through POR, are not routinely implemented and used in patient care. One explanation for this is that POR, although including the patient as the research subject, has not included the patient or other end users such as the physician or other health-care stakeholders in the design of research studies. The term "patient-centered outcomes research" has arisen which posits that the inclusion of the end users in the research will increase the relevance of the research to the actual decisions patients make about their health (Frank et al., 2014). The end users can be involved in research from the step of making decisions about study funding to hypothesis generation through study design including recruitment and retention and data interpretation and importantly to dissemination of research results and increased update into clinical practice. The Patient Centered Outcomes Research Institute (PCORI) was created by the United States Patient Protection and Affordable Care Act of 2010 to fund patient-centered outcomes research (see Chapter 5 on PCOR).

THE ROLE OF THE PATIENT IN PATIENT-ORIENTED RESEARCH

That the patient is central to all research directed at advancing knowledge of human disease is the premise of POR. The patient plays a role in advancing research in several important ways. First, the patient is often the inspiration for a new line of investigation. Brown and Goldstein recall that their work on receptor control of cholesterol metabolism was inspired by their care of a 6-year-old girl with advanced atherosclerotic coronary artery disease, cutaneous xanthomas, and a very high LDL level (Goldstein and Brown, 1997). The phenotype of the disorder had been characterized through the POR of Donald Fredrickson at the NIH Clinical Center. Based on the detailed characterization of this very unusual phenotype through POR, Brown and Goldstein were inspired to determine the underlying molecular defect that would explain her phenotype. This research inspired by a patient with an unusual phenotype led to the discovery of the LDL receptor which had major implications for patients with unusual phenotypes but also for the general population.

Second, the experiments are carried out in patients/human subjects. Although numerous animal models may often suggest the potential utility of a drug strategy, human studies are needed to determine if the benefit observed in animal models applies to humans as well. Unfortunately, many strategies developed in animal models do not pan out in human studies. For example, in the cancer field, successful translation from animal studies to human trials is seen in less than 8% of cases (Mak et al., 2014). Indeed, it has been estimated that animal studies may overestimate the effects of cancer drugs by 30% because negative studies may not be published. In fact, animal models may even be misleading with respect to safety and toxicity. For example, anti-CD28 monoclonal antibodies, proposed for the use in autoimmune disorders, proved safe in animal studies at very high doses. However, this class of drug resulted in multisystem organ failure during initial human studies. In other examples, drugs that proved useful in animal studies, such as hedgehog inhibitors for chondrosarcoma proved futile in human studies. The complexity of the human system and obvious differences between human and animals in terms of genetics, system biology, and dosimetry often limit the conclusions drawn from animal studies and highlight the importance of testing in humans. In contrast, there are a number of examples in which animal models have proven to be highly valuable in aiding our progress toward understanding the physiology of a particular disease. One important example is fetal alcohol syndrome, in which key studies could not be conducted in humans out of safety concerns (Wilson and Cudd, 2011).

To determine whether therapies that have proven beneficial in animal models are beneficial in humans, patients are needed to participate in traditional trials, from Phase I to IV, with relative differences in the risk–benefit ratio. Patients participating in Phase I first in human studies to determine tolerability and toxicity are often healthy volunteers, who are not expecting direct benefit from participating and are exposing themselves to risk. These healthy volunteers play an important role in the development of therapies for patients with the conditions being investigated. Human subjects in more exploratory Phase II dose finding or proof of principle studies most commonly have the disease for which the therapy is directed and may see a benefit or experience harm, at rates that are unknown. Subjects in Phase III studies have more knowledge of anticipated risk and benefits, whereas subjects in Phase IV post marketing studies have very-well-outlined risks. In each case, informed consent is critical with proper explanation of risks and benefits. Properly designed studies carefully consider the role of the patient and insure that the correct population, most likely to benefit and least likely to experience harm, is recruited. In successful human studies, attention must be paid to calibrating the disease level in subjects, such that the study population can reasonably be expected to respond with minimal risk. For example, in a recent major study assessing the utility of a CT screening strategy for CVD, event rates were lower than anticipated and the role of CT as a screening strategy could not be easily discerned (Muhlestein et al., 2014). In contrast, too much disease may translate into excess risk, and thus a balance must be struck. Moreover, POR can only be generalized to the population studied and to the treatment used; thus proper choice of the correct patient population is not only a safety consideration, but also one of potential generalizability to the targeted population for drug approval and for postapproval use. As patient participants commonly do not directly benefit from participation in a given study but instead contribute to the better understanding of a condition to improve the care of others, the important role of a study participant in POR needs to be recognized and appreciated.

SEQUENCE OF INVESTIGATION

The sequence of POR, much like that of any scientific research, involves observation, generation of a testable hypothesis, the design of a study that will test this hypothesis, the performance of the study (hypothesis testing), the collection of data from the study, data analysis, interpretation of the data and formulation of conclusions (Fig. 2.2). The unique aspects of POR, with respect to this sequence, result from the special considerations of hypothesis generation and testing in humans. Observation, the first step in POR, results usually from a patient observation or from an observational or epidemiological study. Such observations generally relate to a public health concern or biological principle.

Hypothesis Generation

Formulation of the hypothesis is the most critical step in POR and is the cornerstone of the clinical research proposal (Fig. 2.2) (Inouye and Fiellin, 2005). Hypotheses can be generated from associational data in epidemiological or observational studies. Alternatively, hypotheses may be generated from animal or in vitro data, for example, relating to a new peptide that is shown to promote insulin signaling. Less often, but occasionally, a hypothesis may be generated from direct observations in a patient or class of patients, or from an already established, but as yet unproven, use of a strategy, e.g., lifestyle modification in diabetes. Three essential questions must be asked in the formulation of a hypothesis. First, is the question important? Second, will the answer to the question be relevant to public health, to elucidate a biological principle or mechanism, or to establish drug efficacy or safety? If the question is not important, then the study is much less likely to be successful and may not be worth the time, expense, and risk. Third, after importance and relevance, it is critical to establish if the hypothesis can be proven, given the constraints involved in POR. For example, is it safe to perform the study? Are endpoints well established to determine efficacy? Can an adequate number of patients be enrolled to determine a definitive result? If the hypothesis cannot be proven, the research should not be undertaken. Finally, it is important to determine whether the question already has been answered. Being familiar with the literature about the study question being asked is crucial. This familiarity will prevent the investigator from asking a question that has already been answered as well as allowing the incorporation of recent findings about the topic to better structure the study hypothesis.

There are many literature search engines now available and several should be used when designing a hypothesis (see Chapter 13). Literature searches, for example, on PubMed, are critical to help establish a hypothesis and determine whether there are sufficient data to justify a study; to define the study parameters, safety, and inclusion criteria; and study endpoints. As there is now a requirement to register all clinical trials, searching existing databases such as Clintrials.gov may also be useful to determine if other related studies exist. Direct discussion with other experts in the field is also useful for hypothesis generation.

Designing the Study

The most important step in POR after the development of a hypothesis is the design of the study to test the hypothesis (Fig. 2.2). Many interesting and promising hypotheses do not make it through the study design phase.

Study Population

Choosing the appropriate study population is a key step in study design. One approach is to start out by recruiting subjects with the well-established medical condition the investigator is interested in studying. However, subjects with long-standing medical conditions may already have complications of the disease that may influence the findings of the study. In addition, such subjects may be on medications that they cannot stop prior to study. Studying such subjects may give results that

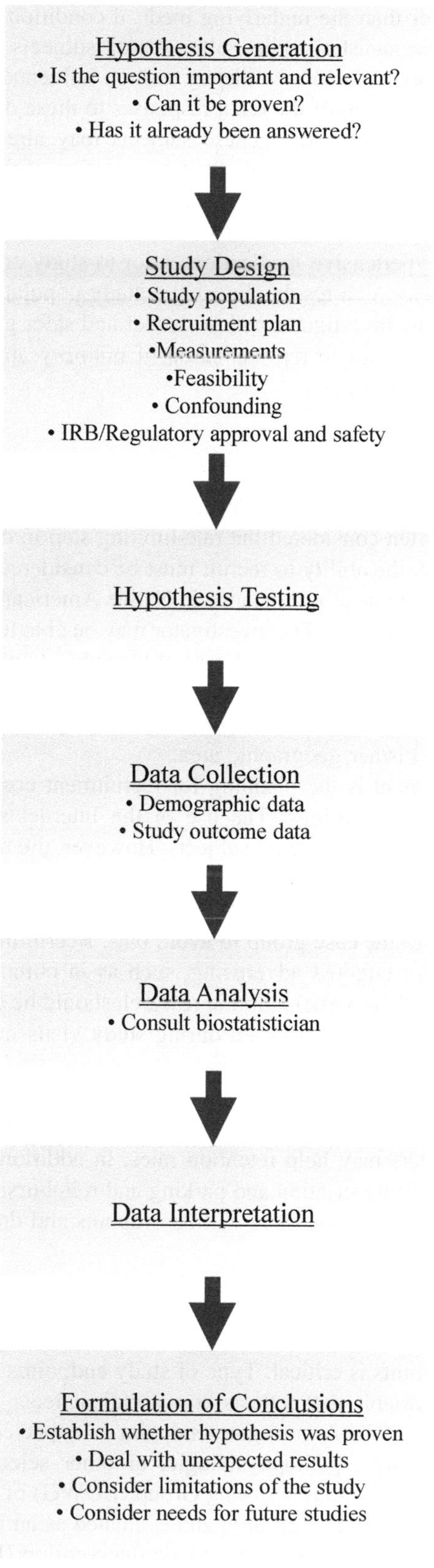

Hypothesis Generation
• Is the question important and relevant?
• Can it be proven?
• Has it already been answered?

Study Design
• Study population
• Recruitment plan
•Measurements
•Feasibility
• Confounding
• IRB/Regulatory approval and safety

Hypothesis Testing

Data Collection
• Demographic data
• Study outcome data

Data Analysis
• Consult biostatistician

Data Interpretation

Formulation of Conclusions
• Establish whether hypothesis was proven
• Deal with unexpected results
• Consider limitations of the study
• Consider needs for future studies

FIGURE 2.2   Steps in patient-oriented research.

reflect the medications they are on, rather than the underlying medical condition. As a result, patient-oriented studies of many medical conditions, particularly mechanistic studies, often recruit subjects with mild manifestations of the disease of interest. For example, if the goal is to understand how dietary sodium influences blood pressure through administration of high- and low-sodium diets and determination of the renal responses to these diets, a study population of subjects with severe hypertension may not be the best to consider. These subjects may already have end-organ disease (comorbid conditions) such as renal dysfunction that would complicate the interpretation of the renal handling of sodium. In addition, they might not be able to handle the high-salt diet due to hypertensive cardiomyopathy and so may be at risk for congestive heart failure through participation. Finally, since medications would influence the interpretation of the data, the subjects would need to be tapered off their antihypertensive medications prior to study and this may not be safe in subjects with severe hypertension and end-organ disease. As a result, choosing a study population with mild-to-moderate hypertension without end-organ disease may provide the investigator with a cleaner and safer group of subjects to study. It is critical to choose a representative population, with adequate representation of minority and female subjects, to allow appropriate generalization of the data to these important groups.

Recruitment and Retention of the Study Population

The success in studying a specific group of subjects is dependent on the investigator's ability to recruit that group of subjects into the study. Recruitment is often considered the rate-limiting step in clinical research (Association of Clinical Research Professionals, 1998). Therefore, the ability to recruit must be considered up front in the design of the study. An example would be a study designed to study insulin resistance in Native Americans in a city where this population did not exist. Solutions to this type of limitation may exist. The investigator may be able to collaborate with someone in a different geographic area or with a caretaker involved in the clinical care of the subjects of interest. For example, in the study of a condition that affects only a small number of individuals, working with a caregiver who specializes in the care of those patients may make recruitment successful. At times, the investigator may need to redefine the study question to be applicable to a population recruitable in his/her geographic area.

Another important issue with recruitment is the planning for recruitment costs. Recruitment costs may be relatively inexpensive such as fliers and postage for mailings. The use of the Internet sites such as Craigslist has become an increasingly more popular and effective method to recruit subjects. However, the use of the Internet as a recruitment tool is limited to the recruitment of subjects who have the Internet access. On the other hand, recruitment costs may be a substantial budget item when done by newspaper or radio ads. It is always important to remember to attempt to recruit a control group from a similar population to the case group to avoid bias. Recruitment of female and minority subjects can pose a special challenge and may require targeted advertising, such as in community newspapers and women's health magazines. In addition, special barriers to their participation in research should be considered (Gilliss et al., 2001; Swanson and Ward, 1995). Making special provisions for childcare during study visits may be useful to increase recruitment of female subjects.

Retention of subjects in clinical studies is another important focus. High dropout rates will decrease the power of the study to show meaningful results. Plans for retention of subjects are also crucial in designing clinical research studies. Maintaining frequent contact with subjects may help retention rates. In addition, budgeting for costs that subjects may accrue in the course of the study such as transportation and parking and reimbursing subjects for these costs may improve retention. However, even in the best planned study, there will be dropouts and dropout rates need to be considered in the initial power calculation for the study.

Deciding on Measurements

In all research, the choice of study endpoints is critical. Type of study endpoints include (1) clinical and (2) surrogate. A clinical endpoint is a characteristic or variable that reflects how a patient feels, functions, or survives and may include physiologic markers and events (Peck, 2007). A surrogate endpoint is expected to predict clinical benefit (or lack of benefit) based on epidemiologic, therapeutic, pathophysiologic, or other scientific evidence (Biomarkers Definitions Working Group, 2001). The Biomarkers Definitions Working Group (BDWG) of the National Institutes of Health defined a biomarker as "a characteristic that is objectively measured and evaluated as an indicator of normal biological processes, pathogenic processes, or pharmacologic responses to a therapeutic intervention (Biomarkers Definitions Working Group, 2001)." In POR, there are certain specific issues which affect the choice of study endpoints. For example, an investigator who has the goal of reducing the sequelae of hypertension might approach this goal in many different ways, which would involve the choice of different measurements.

If the goal is to determine whether a specific therapy reduces end-organ sequelae of hypertension (i.e., stroke and heart disease), the achievement of these clinical endpoints would require a long follow-up period and a large study population. There are several alternative questions which could be asked and which would then determine the choice of other endpoints. For example, the question could be rephrased as to whether a specific intervention lowers blood pressure. In this situation, the outcome could be the blood pressure response to the intervention. The decision would then need to be made as to the technique, which most accurately determines blood pressure, the major risk factor for stroke (Anonymous, 1982). One option would include manual determination with a blood pressure cuff and sphygmomanometer. The advantage of this technique would be its applicability to large numbers of subjects and its low cost. Another option would be the use of a 24-h blood pressure monitor, which has been shown to be a better predictor of stroke than office blood pressure (Verdecchia, 2000). The choice of this latter technique makes the blood pressure determination a better predictor of stroke but requires more cooperation of the subjects as subjects must wear a monitor for 24 h and increases the cost of the study as the 24 h blood pressure devices would be much more expensive than the use of a sphygmomanometer. Another approach would be to find a measurement, which has been shown to be associated with the endpoint. For example, abnormal brachial artery dilation as determined by brachial artery ultrasound response to ischemia reflects more diffuse coronary artery disease (Amir et al., 2006). Use of a technique such as this may also provide information as to the mechanism by which the intervention lowers blood pressure as this technique can be used to distinguish endothelial dependent versus independent vasodilation. Circulating blood markers are often used as surrogate endpoints. These can be determined on large numbers of subjects and involve only the intervention of phlebotomy. In this project, the response of C-reactive protein to the intervention could be chosen as a surrogate for heart disease since CRP has been shown to predict heart disease (Folsom et al., 2001; Ridker et al., 1998).

Feasibility

The successful completion of a clinical research study is dependent on its feasibility. The feasibility is based on several factors, which are common to research projects in general, and these factors must be considered in the design phase of the study. These key factors include availability of research subjects, ability to recruit research subjects into the study, availability of the techniques and personnel required for the measurement of study endpoints, expertise in the use of these techniques, and adequacy of funds to cover research costs.

Confounding

As with the design of any research study, it is important to determine whether the study that is designed will actually answer the question raised in the hypothesis. Thus, controlling for confounders is a central issue. In laboratory-based research, the control of confounders may be more easily achieved as the investigator can match for many baseline characteristics and have control over the environment during the course of the study. For example, an investigator can choose genetically identical animals, and the study can take place entirely within that one group of animals. In clinical research the ability to control for confounders is often more of a challenge. When confounders have not been controlled for, they are often dealt with through statistical adjustment, i.e., the use of regression analyses (Normand et al., 2005). However, assessing the impact of confounders by these techniques on study outcomes may be inadequate. Although achieving the level of control possible in the laboratory setting is not fully possible with human subjects, the lack of an attempt to control for as many confounders as possible commonly yields study results that are confusing to interpret. Consideration of subject age, gender, race, and co-morbidities can help the investigator match subjects in a control and an intervention group. Environmental conditions may be controlled by clinical investigators to varying degrees; for example; the control for circadian rhythm by performing the study measures at the same time of day in all subjects, the control for stress, by performing testing in a quiet room with the subject having rested before the intervention, the control for sodium balance by providing meals for subjects prior to a study measurement.

Subject Safety and the Institutional Review Board

(See Chapter 43 for additional details.)

The foremost concern for all clinical investigators must be ensuring and protecting the safety of each research subject. The design of the study should be safe, and exclusion criteria should be established that prevent subjects with a higher risk than average from participating in the study. All studies should be reviewed by an institutional review board (IRB). Many POR studies will also have independent data safety monitors who are investigators not involved in the study but who review all adverse events whether mild or serious as well as protocol deviations on a prespecified schedule, for example, every 6 months. Higher-risk POR studies may benefit from and may be required to have, via suggestion by NIH at the time of review or by insistence of the institutional IRB, a DSMB, which reviews data for preset safety cut offs at predetermined intervals. In addition a DSMB may be charged with reviewing efficacy analyses to determine if study endpoints are met prior to the study planned end date and may stop a study early if such endpoints are met or if risk is greater than expected (Slutsky and Lavery, 2004) (see Chapter 24 Ethical Issues for more details).

Database Development

In all studies there is a plethora of data produced. Therefore, it is critical for the investigator to develop a database system prior to the start of the study. Common information captured in databases for clinical research studies would include demographic data and study outcome data. Data in a database are only as good as the data entry. Ideal systems are those in which data can be electronically transferred from one database to another without manual reentry. However, manual data entry usually occurs at some point in the creation of a database. Some ways to decrease the inclusion of erroneous data include double manual entry, adequate training of personnel who perform data entry, and the regular review of the entered data for accuracy using an appropriate program to identify outliers. Because information in clinical research databases may contain medically sensitive information, it is critical that investigators set up the database in a secure system that complies with IRB and Health Insurance Portability and Accountability Act (HIPAA) regulations. HIPAA regulations have provided challenges to clinical research in maintaining deidentified databases (Kulynych and Korn, 2003).

Data Analysis Plan

(See Chapter 14 for additional details.)

Key in the design of a clinical research study is the data analysis plan. The study design will determine the data analysis plan, and the data analysis plan will in turn determine the required number of subjects. Consultation with a biostatistician is important. Statistical techniques for data analysis are discussed in detail in Chapter 9. Critical issues to assess include power, study design, endpoints, and method of analysis. Prior studies for endpoint variability must be considered. Minimization of confounding is best achieved by a randomized study, but this design, while generally useful, may not be appropriate for all studies, especially small physiology studies.

TOOLS OF THE PATIENT-ORIENTED RESEARCHER

At most academic medical centers, numerous tools exist for the patient-oriented researcher, including a clinical research center (CRC) or a similar clinical trials center, to perform detailed studies on patients (Fig. 2.3). The CTSAs also incorporate these resources (see Chapter 34). In addition, such centers usually make available statisticians to consult on study design and a core research laboratory to perform analyses on study endpoints using shared resources. Core lab facilities for genotyping and biomedical imaging may also be available, in addition to core nursing services, data safety monitoring services, alarmed storage facilities, and informatics. In addition to these services, coursework on statistics, regulatory issues, grant writing, and trial development may be offered by CRCs. Similar but more advanced coursework may also be available in master's level programs for the clinical researcher that offer in-depth resources for biostatistics, trial design, and other resources for POR. Given the complexity of human studies issues, coursework and certification

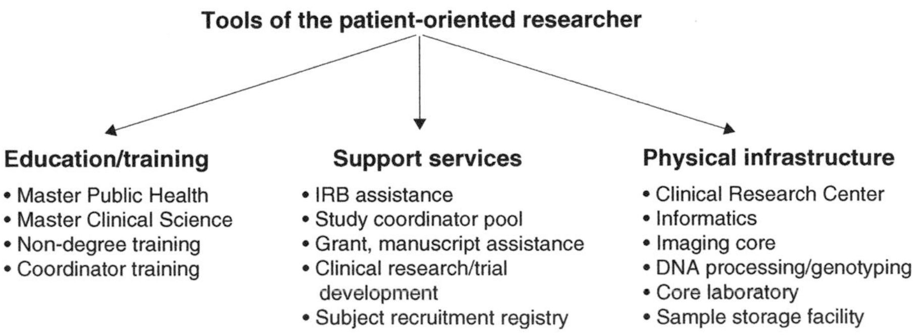

FIGURE 2.3 Tools of patient-oriented research.

Tools of the patient-oriented researcher

Education/training
- Master Public Health
- Master Clinical Science
- Non-degree training
- Coordinator training

Support services
- IRB assistance
- Study coordinator pool
- Grant, manuscript assistance
- Clinical research/trial development
- Subject recruitment registry

Physical infrastructure
- Clinical Research Center
- Informatics
- Imaging core
- DNA processing/genotyping
- Core laboratory
- Sample storage facility

programs are likely to be offered and required by the local IRB. CRCs may even offer a pool of people available to serve as study coordinators for individual studies. (See Chapters 15, 34, and 43 for additional details on CRC and translational CRC resources and additional resources and tools for the patient-oriented researcher.)

Increasingly, patient participation in human research takes the form of participation in a large database. Indeed, newer approaches to big data and databanking suggest the utility of large databanks of coded data and samples. For example, Partners Healthcare is collecting a large database of genetically coded samples from healthy and diseased patients. Researchers may gain access to this resource, for example, to compare rare genetic modifications in patients with and without multiple sclerosis, greatly accelerating the time it would take to perform a trial to enroll such patients (McCluskey, 2014).

Researchers can then compare information about disease course contained in the medical record with patient genotype in an anonymized fashion. Many of these patients consent to be contacted about future research. Patient-oriented researchers can then recruit patients with specific genotypes who have agreed to be contacted to recruit into studies of a response to an intervention or therapy according to genotype. In other research, patients may contribute anonymized data to a large clinical cohort, in which investigators may compare disease rates or characteristics in subjects with different medical conditions. For example, in the Partners System, the Research Patient Data Registry is a large registry of all ICD-9-coded visits. The patient-oriented researcher can then use the database to identify individuals who have agreed to be contacted and identify a patient population for recruitment into a study improving the efficiency of recruitment. Another use of the database would be to determine association of a disease with an outcome. For example, this registry was used to create a virtual cohort to establish that myocardial infarction rates are higher in HIV than non-HIV patients (Triant et al., 2007). This finding then led to a large NIH-funded patient-oriented multicenter study to determine if statins would reduce myocardial infarction and major adverse cardiac event rates in patients with HIV.

FUNDING FOR PATIENT-ORIENTED RESEARCH

Funding for POR can come from any source that funds biomedical research, NIH, foundation, industry, philanthropy, etc. Recognizing the importance of POR and its challenges (Vaitukaitis, 2000), NIH developed specific award mechanisms to support POR. The first, the K23 Mentored Patient-Oriented Research Career Development Award, is an award for early investigators to support up to 5 years of mentored POR. The second, the K24 Midcareer Investigator Award in Patient-Oriented Research, is to support midcareer investigators to mentor early career investigators toward independence in POR. Information about these awards can be found at the K Kiosk (NIH K Kiosk). In some other countries, similar programs are available.

CONCLUSIONS

POR focuses on the central role of the patient in the discovery of interventions or therapies that benefit human health. Patients commonly serve as the inspiration for the investigation, and the goal of the investigation is to improve the health of patients. The investigations are performed in patients/human subjects allowing determination of whether findings from basic cellular or animal studies are relevant for humans and whether associations seen in observational or epidemiologic studies can be proven to be causative as well as to demonstrate risk and benefits of intervention or therapies. The steps of patient-oriented investigation are similar to those of other types of investigation, but special considerations are needed when studying humans. Some of these considerations include choice of the study population, determination of study sample size, recruitment and retention of subjects, and subject safety. Because the stakes are very high in terms of patient safety and expense, designing studies that involve the fewest subjects possible to achieve the endpoint and formulating accurate, tailored conclusions that do not go beyond the data is a critical aspect of POR. Whether the involvement of patients and other stakeholders in the process of POR will improve the uptake of research findings into clinical care is being tested by grants from PCORI. Given today's emphasis on personalized patient-oriented medicine, POR is more important than ever.

REFERENCES

Asher, H.B., 1976. Causal Modeling. Sage Publications, Inc., Beverly Hills, CA.

Ahima, R.S., Prabakaran, D., Mantzoros, C., Qu, D., Lowell, B., Maratos-Flier, E., Flier, J.S., 1996. Role of leptin in the neuroendocrine response to fasting. Nature 382 (6588), 250−252.

Anderson, G.L., Limacher, M., Assaf, A.R., Bassford, T., Beresford, S.A., Black, H., Bonds, D., Brunner, R., Brzyski, R., Caan, B., Chlebowski, R., Curb, D., Gass, M., Hays, J., Heiss, G., Hendrix, S., Howard, B.V., Hsia, J., Hubbell, A., Jackson, R., Johnson, K.C., Judd, H., Kotchen, J.M.,

Kuller, L., LaCroix, A.Z., Lane, D., Langer, R.D., Lasser, N., Lewis, C.E., Manson, J., Margolis, K., Ockene, J., O'Sullivan, M.J., Phillips, L., Prentice, R.L., Ritenbaugh, C., Robbins, J., Rossouw, J.E., Sarto, G., Stefanick, M.L., Van Horn, L., Wactawski-Wende, J., Wallace, R., Wassertheil-Smoller, S., Women's Health Initiative Steering Committee, 2004. Effects of conjugated equine estrogen in postmenopausal women with hysterectomy: the Women's Health Initiative randomized controlled trial. JAMA 291 (14), 1701−1712.

Association of Clinical Research Professionals, 1998. Report on future trends. The Monitor 13−26.

Anonymous, 1982. Five-year findings of the hypertension detection and follow-up program. III. Reduction in stroke incidence among persons with high blood pressure. Hypertension Detection and Follow-up Program Cooperative Group. JAMA 247 (5), 633−638.

Amir, O., Jaffe, R., Shiran, A., Flugelman, M.Y., Halon, D.A., Lewis, B.S., 2006. Brachial reactivity and extent of coronary artery disease in patients with first ST-elevation acute myocardial infarction. Am. J. Cardiol. 98 (6), 754−757.

Biomarkers Definitions Working Group, 2001. Biomarkers and surrogate endpoints: preferred definitions and conceptual framework. Clin. Pharmacol. Ther. 69 (3), 89−95.

Chan, J.L., Heist, K., DePaoli, A.M., Veldhuis, J.D., Mantzoros, C.S., 2003. The role of falling leptin levels in the neuroendocrine and metabolic adaptation to short-term starvation in healthy men. J. Clin. Invest. 111 (9), 1409−1421.

Choi, S.H., Kim, Y.H., Hebisch, M., Sliwinski, C., Lee, S., D'Avanzo, C., Chen, H., Hooli, B., Asselin, C., Muffat, J., Klee, J.B., Zhang, C., Wainger, B.J., Peitz, M., Kovacs, D.M., Woolf, C.J., Wagner, S.L., Tanzi, R.E., Kim, D.Y., 2014. A three-dimensional human neural cell culture model of Alzheimer's disease. Nature 515 (7526), 274−278.

Falutz, J., Allas, S., Blot, K., Potvin, D., Kotler, D., Somero, M., Berger, D., Brown, S., Richmond, G., Fessel, J., Turner, R., Grinspoon, S., 2007. Metabolic effects of a growth hormone-releasing factor in patients with HIV. N. Engl. J. Med. 357 (23), 2359−2370.

Frank, L., Basch, E., Selby, J.V., 2014. The PCORI perspective on patient-centered outcomes research. JAMA 312 (15), 1513−1514.

Folsom, A.R., Aleksic, N., Park, E., Salomaa, V., Juneja, H., Wu, K.K., 2001. Prospective study of fibrinolytic factors and incident coronary heart disease: the Atherosclerosis Risk in Communities (ARIC) Study. Arterioscler. Thromb. Vasc. Biol. 21 (4), 611−617.

Goldstein, J.L., Brown, M.S., 1997. The clinical investigator: bewitched, bothered, and bewildered − but still beloved. J. Clin. Invest. 99 (12), 2803−2812.

Grodstein, F., Clarkson, T.B., Mason, J.E., 2003. Understanding the divergent data on postmenopausal hormone therapy. N. Engl. J. Med. 348 (7), 645−650.

Gilliss, C.L., Lee, K.A., Gutierrez, Y., Taylor, D., Beyene, Y., Neuhaus, J., Murrell, N., 2001. Recruitment and retention of healthy minority women into community-based longitudinal research. J. Women's Health Gend.-Based Med. 10 (1), 77−85.

Heymsfield, S.B., Greenberg, A.S., Fujioka, K., Dixon, R.M., Kushner, R., Hunt, T., Lubina, J.A., Patane, J., Self, B., Hunt, P., McCamish, M., 1999. Recombinant leptin for weight loss in obese and lean adults: a randomized, controlled, dose-escalation trial. JAMA 282 (16), 1568−1575.

Harman, S.M., Black, D.M., Naftolin, F., Brinton, E.A., Budoff, M.J., Cedars, M.I., Hopkins, P.N., Lobo, R.A., Manson, J.E., Merriam, G.R., Miller, V.M., Neal-Perry, G., Santoro, N., Taylor, H.S., Vittinghoff, E., Yan, M., Hodis, H.N., 2014. Arterial imaging outcomes and cardiovascular risk factors in recently menopausal women: a randomized trial. Ann. Intern. Med. 161 (4), 249−260.

Inouye, S.K., Fiellin, D.A., 2005. An evidence-based guide to writing grant proposals for clinical research. Ann. Intern. Med. 142 (4), 274−282.

Koutkia, P., Meininger, G., Canavan, B., Breu, J., Grinspoon, S., 2004. Metabolic regulation of growth hormone by free fatty acids, somatostatin, and ghrelin in HIV-lipodystrophy. Am. J. Physiol. Endocrinol. Metab. 286 (2), E296−E303.

Knowler, W.C., Barrett-Connor, E., Fowler, S.E., Hamman, R.F., Lachin, J.M., Walker, E.A., Nathan, D.M., 2002. Reduction in the incidence of type 2 diabetes with lifestyle intervention or metformin. N. Engl. J. Med. 346 (6), 393−403.

Kastelein, J.J.P., van Leuven, S.I., Burgess, L., Evans, G.W., Kuivenhoven, J.A., Barter, P.J., Revkin, J.H., Grobbee, D.E., Riley, W.A., Shear, C.L., Duggan, W.T., Bots, M.L., 2007. Effect of torcetrapib on carotid atherosclerosis in familial hypercholesterolemia. N. Engl. J. Med. 356 (16), 1620−1630.

Kulynych, J., Korn, D., 2003. The new HIPAA (Health Insurance Portability and Accountability Act of 1996) Medical Privacy Rule: help or hindrance for clinical research? Circulation 108 (8), 912−914.

Lynch, T.J., Bell, D.W., Sordella, R., Gurubhagavatula, S., Okimoto, R.A., Brannigan, B.W., Harris, P.L., Haserlat, S.M., Supko, J.G., Haluska, F.G., Louis, D.N., Christiani, D.C., Settleman, J., Haber, D.A., 2004. Activating mutations in the epidermal growth factor receptor underlying responsiveness of non-small-cell lung cancer to gefitinib. N. Engl. J. Med. 350 (21), 2129−2139.

HPS2-THRIVE Collaborative Group, Landray, M.J., Haynes, R., Hopewell, J.C., Parish, S., Aung, T., Tomson, J., Wallendszus, K., Craig, M., Jiang, L., Collins, R., Armitage, J., 2014. Effects of extended-release niacin with laropiprant in high-risk patients. N. Engl. J. Med. 371 (3), 302−312.

Manson, J.E., Hsia, J., Johnson, K.C., Rossouw, J.E., Assaf, A.R., Lasser, N.L., Trevisan, M., Black, H.R., Heckbert, S.R., Detrano, R., Strickland, O.L., Wong, N.D., Crouse, J.R., Stein, E., Cushman, M., 2003. Estrogen plus progestin and the risk of coronary heart disease. N. Engl. J. Med. 349 (6), 523−534.

Manson, J.E., Allison Matthew, A., Rossouw, J.E., Carr, J., Langer, R.D., Hsia, J., Kuller, L.H., Cochrane, B.B., Hunt, J.R., Ludlam, S.E., Pettinger, M.B., Gass, M., Margolis, K.L., Nathan, L., Ockene, J.K., Prentice, R.L., Robbins, J., Stefanick, M.L., for the WHI and WHI-CACS Investigators, 2007. Estrogen therapy and coronary-artery calcification. N. Engl. J. Med. 356 (25), 2591−2602.

Mak, I.W.Y., Evaniew, N., Ghert, M., 2014. Lost in translation: animal models and clinical trials in cancer treatment. Am. J. Transl. Res. 6 (2), 114−118.

Muhlestein, J.B., Lappé, D.L., Lima, J.A.C., Rosen, B.D., May, H.T., Knight, S., Bluemke, D.A., Towner, S.R., Le, V., Bair, T.L., Vavere, A.L., Anderson, J.L., 2014. Effect of screening for coronary artery disease using CT angiography on mortality and cardiac events in high-risk patients with diabetes: the FACTOR-64 randomized clinical trial. JAMA 312 (21), 2234−2243.

McCluskey, P.D., December 18, 2014. Partners to Collect Blood Samples for Researchers' Medical Studies. The Boston Globe.

NIH Glossary. http://grants.nih.gov/grants/guide/notice-files/not98-024.html.

Nissen, S.E., Tardif, J., Nicholls, S.J., Revkin, J.H., Shear, C.L., Duggan, W.T., Ruzyllo, W., Bachinsky, W.B., Lasala, G.P., Tuzcu, E.M., 2007. Effect of torcetrapib on the progression of coronary atherosclerosis. N. Engl. J. Med. 356 (13), 1304–1316.

Nissen, S.E., Wolski, K., 2007. Effect of rosiglitazone on the risk of myocardial infarction and death from cardiovascular causes. N. Engl. J. Med. 356 (24), 2457–2471.

Normand, S.L., Sykora, K., Li, P., Mamdani, M., Rochon, P.A., Anderson, G.M., 2005. Readers guide to critical appraisal of cohort studies: 3. Analytical strategies to reduce confounding. BMJ 330 (7498), 1021–1023.

NIH K Kiosk. http://grants.nih.gov/training/careerdevelopmentawards.htm.

Oestreicher, E.M., Guo, C., Seely, E.W., Kikuchi, T., Martinez-Vasquez, D., Jonasson, L., Yao, T., Burr, D., Mayoral, S., Roubsanthisuk, W., Ricchiuti, V., Adler, G.K., 2006. Estradiol increases proteinuria and angiotensin II type 1 receptor in kidneys of rats receiving L-NAME and angiotensin II. Kidney Int. 70 (10), 1759–1768.

Peck, C., 2007. Biomarkers for assessment of responses to therapies. In: Bridging the Gap between Preclinical and Clinical Evaluation of Therapeutic Candidates. National Institutes of Health, Washington, DC.

Rossouw, J.E., Prentice, R.L., Manson, J.E., Wu, L., Barad, D., Barnabei, V.M., Ko, M., LaCroix, A.Z., Margolis, K.L., Stefanick, M.L., 2007. Postmenopausal hormone therapy and risk of cardiovascular disease by age and years since menopause. JAMA 297 (13), 1465–1477.

Ridker, P.M., Buring, J.E., Shih, J., Matias, M., Hennekens, C.H., 1998. Prospective study of C-reactive protein and the risk of future cardiovascular events among apparently healthy women. Circulation 98 (8), 731–733.

Stampfer, M.J., Colditz, G.A., Willett, W.C., Manson, J.E., Rosner, B., Speizer, F.E., Hennekens, C.H., 1991. Postmenopausal estrogen therapy and cardiovascular disease. Ten-year follow-up from the nurses' health study. N. Engl. J. Med. 325 (11), 756–762.

Schurgin, S., Canavan, B., Koutkia, P., Depaoli, A.M., Grinspoon, S., 2004. Endocrine and metabolic effects of physiologic r-metHuLeptin administration during acute caloric deprivation in normal-weight women. J. Clin. Endocrinol. Metab. 89 (11), 5402–5409.

Sledge Jr., G.W., 2004. HERe-2 stay: the continuing importance of translational research in breast cancer. J. Natl. Cancer Inst. 96 (10), 725–727.

Sung, N.S., Crowley Jr., W.F., Genel, M., Salber, P., Sandy, L., Sherwood, L.M., Johnson, S.B., Catanese, V., Tilson, H., Getz, K., Larson, E.L., Scheinberg, D., Reece, E.A., Slavkin, H., Dobs, A., Grebb, J., Martinez, R.A., Korn, A., Rimoin, D., 2003. Central challenges facing the national clinical research enterprise. JAMA 289 (10), 1278–1287.

Seely, E.W., Brosnihan, K.B., Jeunemaitre, X., Okamura, K., Williams, G.H., Hollenberg, N.K., Herrington, D.M., 2004. Effects of conjugated oestrogen and droloxifene on the renin-angiotensin system, blood pressure, and renal blood flow in postmenopausal women. Clin. Endocrinol. 30 (3), 315–321.

Swanson, G.M., Ward, A.J., 1995. Recruiting minorities into clinical trials: toward a participant-friendly system. J. Natl. Cancer Inst. 87 (23), 1747–1759.

Slutsky, A.S., Lavery, J.V., 2004. Data safety and monitoring boards. N. Engl. J. Med. 350 (11), 1143–1147.

Triant, V.A., Lee, H., Hadigan, C., Grinspoon, S.K., 2007. Increased acute myocardial infarction rates and cardiovascular risk factors among patients with human immunodeficiency virus disease. J. Clin. Endocrinol. Metab. 92 (7), 2506–2512.

Verdecchia, P., 2000. Prognostic value of ambulatory blood pressure: current evidence and clinical implications. Hypertension 35 (3), 844–851.

Vaitukaitis, J.L., 2000. Reviving patient-oriented research. Acad. Med. 75 (7), 683–685.

Wilson, P.W.F., D'Agostino, R.B., Levy, D., Belanger, A.M., Silbershatz, H., Kannel, W.B., 1998. Prediction of coronary heart disease using risk factor categories. Circulation 97 (18), 1837–1847.

Wilson, S.E., Cudd, T.A., 2011. Focus on: the use of animal models for the study of fetal alcohol spectrum disorders. Alcohol Res. Health J. Natl. Inst. Alcohol Abuse Alcohol. 34 (11), 92–98.

Zhang, Y., Proenca, R., Maffei, M., Barone, M., Leopold, L., Friedman, J.M., 1994. Positional cloning of the mouse obese gene and its human homologue. Nature 372 (6505), 425–432.

Chapter 3

Clinical Trials

Robert M. Califf

U.S. Food and Drug Administration, Silver Spring, MD, United States

Chapter Outline

Key Points

- The primary tool used to generate definitive medical evidence is the randomized clinical trial (RCT), the essential basis of which is the allocation of a research subject to one intervention or another through a deliberate scheme that uses a table of random numbers to determine the assignment of the intervention.
- There has lately been a growing level of interest in assessing the quality of clinical trials, not only with regard to internal validity, but also in terms of a host of parameters that might guide the application of trial results to informing decisions about the development of medical products, medical practice, or health policy.
- When designing or interpreting the results of a clinical study, the purpose of the investigation is critical to placing the outcome in the appropriate context. Researchers and clinicians who design the investigation are responsible for constructing the project and presenting its results in a manner that reflects the intent of the study.

INTRODUCTION

Medical practice has entered an era of "evidence-based medicine" characterized by an increasingly widespread societal belief that clinical practice should be based on scientific information in addition to intuition, mechanistic reasoning, and opinion. The primary tool used to generate definitive medical evidence is the randomized clinical trial (RCT), the essential basis of which is the allocation of a research subject to one intervention or another through a deliberate scheme that uses a table of random numbers to determine the assignment of the intervention. This process ensures that underlying risks are randomly distributed between or among the arms of a study, thus enabling an internally valid comparison of the outcome of interest.

There has lately been a growing level of interest in assessing the quality of clinical trials, not only with regard to internal validity, but also in terms of a host of parameters that might guide the application of trial results to informing decisions about development of medical products, medical practice, or health policy. While the critical importance of clinical trials is widely acknowledged, as a scientific tool, they remain a work in progress.

HISTORY

The first randomization recorded in the published literature was performed by Fisher in 1926 in an agricultural experiment (Fisher and Mackenzie, 1923). In developing the statistical methods for analysis of variance, he recognized that experimental observations must be independent and not confounded to allow full acceptance of the statistical methodology. He therefore randomly assigned different agricultural plots to different applications of fertilizer. The first randomization of human subjects is credited to Amberson in a 1931 trial of tuberculosis therapy in 24 patients, in which a coin toss was used to make treatment assignments (Lilienfield, 1982). The British Medical Research Council trial of streptomycin in the treatment of tuberculosis in 1948 marks the beginning of the modern era of clinical trials (Medical Research Council, 1948). This study, which established principles for the use of random assignment in large numbers of patients, also set guidelines for the administration of the experimental therapy and objective evaluation of outcomes.

In recent years, computers have enabled rapid accumulation of data from thousands of patients in studies conducted throughout the world. Peto, Yusuf, Sleight, and Collins developed the concept of the *large simple trial* in the First International Study of Infarct Survival (ISIS-1) (ISIS-1 Collaborative Group, 1986), which stipulated that only by randomly assigning 10,000 patients could the balance of risks and benefits of beta blocker therapy be fully understood.

The development of client–server architecture in computer technology provides a mechanism for aggregating large amounts of data and distributing the data quickly to multiple users (see Chapter 13). Advances in the development of the internet provide opportunities for sharing information instantaneously throughout the world. Further, Web applications now allow transmission of massive clinical, biological, and imaging data sets for central analysis. Most recently, the creation and maintenance of comprehensive registries for clinical studies have become a firm expectation of society in general, and the posting of trial results in a forum freely available to the public is now a legal requirement for most types of clinical trials in the United States (Food and Drug Administration Amendments Act, 2007).

PHASES OF EVALUATION OF THERAPIES

Evaluating the results of a clinical trial requires an understanding of the investigation's goals. One important aspect of placing a trial in context is described in the common terminology of the *phase* of the clinical trial (Table 3.1). The first two phases focus on initial evaluation for evidence of frank toxicity, obvious clinical complications, and physiological

TABLE 3.1 Phases of Evaluation of New Therapies

Phase	Features	Purpose
I	First administration of a new therapy to patients	Exploratory clinical research to determine if further investigation is appropriate
II	Early trials of new therapy in patients	To acquire information on dose–response relationship, estimate incidence of adverse reactions, and provide additional insight into pathophysiology of disease and potential impact of new therapy
III	Large-scale comparative trial of new therapy versus standard of practice	Definitive evaluation of new therapy to determine if it should replace current standard of practice; randomized controlled trials required by regulatory agencies for registration of new therapeutic modalities
IV	Monitoring of use of therapy in clinical practice	Postmarketing surveillance to gather additional information on impact of new therapy on treatment of disease, rate of use of new therapy, and more robust estimate of incidence of adverse reactions established from registries

Adapted from Antman, E.M., Califf, R.M., 1996. Clinical trials and metaanalysis. In: Smith, T.W. (Ed.), Cardiovascular Therapeutics. Philadelphia, Saunders, p. 679.

measurements that would support or weaken belief in the therapy's hypothetical mechanism of action. In these phases, attention to detail is critical and should take priority over simplicity (although gathering detail for no specific purpose is a waste of resources, regardless of the phase of the trial).

The third phase, commonly referred to as the "pivotal" phase, evaluates the therapy in the relevant clinical context with the goal of determining whether the treatment should be used in clinical practice. For phase III studies, relevant endpoints include measures that can be recognized by patients as important: survival, major clinical events, quality of life, and cost. A well-designed clinical trial that informs the decisions that must be made by patients and health-care providers justifies serious consideration for changing clinical practice and certainly provides grounds for regulatory approval for sales and marketing.

After a therapy or diagnostic test is approved by regulatory authorities and is in use, phase IV begins. Traditionally, phase IV has been viewed as including a variety of studies that monitor a therapy in clinical practice with the accompanying responsibility of developing more effective protocols for its use, based on observational inference and reported adverse events. Phase IV is also used to develop new indications for drugs and devices already approved for a different use (see Chapter 7). The importance of this phase has grown with the recognition that many circumstances that arise in clinical practice will not have been encountered in randomized trials completed at the time the therapy receives regulatory approval. Phase IV studies may now include evaluation of new dosing regimens (Rogers et al., 1994; Forrow et al., 1992; Society of Thoracic Surgeons Database, 2005) and comparisons of one effective marketed therapy against another, giving birth to a discipline of comparative effectiveness (Tunis et al., 2003). In some cases, this need arises because of changing doses or expanding indications for a therapy; in other cases, a phase III study might not have provided the relevant comparisons for a particular therapeutic context; information that is only obtainable in the period after the therapy is approved for marketing.

CRITICAL GENERAL CONCEPTS

Purposes of Clinical Trials

Clinical trials may be divided into two broad categories: explanatory/scientific or probabilistic/pragmatic. The simplest but most essential concepts for understanding the relevance of a clinical study to clinical practice are validity and generalizability. Table 3.2 illustrates an approach to these issues, developed by the McMaster group, to be used when reading the literature.

Validity

The most fundamental question to ask of a clinical trial is whether the result is valid. Are the results of the trial internally consistent? Would the same result be obtained if the trial was repeated in an identical population? Was the trial design adequate; that is, did it include blinding, endpoint assessment, and statistical analyses? Of course, the most

TABLE 3.2 Questions to Ask When Reading and Interpreting the Results of a Clinical Trial

Are the Results of the Study Valid?

Primary Guides

Was the assignment of patients to treatment randomized?
Were all patients who entered the study properly accounted for at its conclusion?
Was follow-up complete?
Were patients analyzed in the groups to which they were randomized?

Secondary Guides

Were patients, their clinicians, and study personnel blinded to treatment?
Were the groups similar at the start of the trial?
Aside from the experimental intervention, were the groups treated equally?

What Were the Results?

How large was the treatment effect?
How precise was the treatment effect (confidence intervals)?

Will the results help me in caring for my patients?

Does my patient fulfill the enrollment criteria for the trial? If not, how close is the patient to the enrollment criteria?

Does my patient fit the features of a subgroup in the trial report? If so, are the results of the subgroup analysis in the trial valid?

Were all the clinically important outcomes considered?

Are the likely treatment benefits worth the potential harm and costs?

compelling evidence of validity in science is replication. If the results of a trial or study remain the same when the study is repeated, they are likely to be valid.

Generalizability

Given valid results from a clinical trial, it is equally important to determine whether the findings are generalizable. Unless the findings can be replicated and applied in multiple practice settings, little has been gained by the trial in terms of informing the choices being made by providers and patients. Since it is impossible to replicate every clinical study, it is especially important to understand the inclusion and exclusion criteria for the subjects participating in the study and to have an explicit awareness of additional therapies that the patients may have received. For example, studies done on "ideal" patients who lack comorbid conditions or on young patients without severe illness can be misleading when the results are applied to general clinical practice, since the rate of poor outcomes, complications, and potential drug interactions could be much higher in an older or more ill population. Of increasing concern in this regard are the age extremes (children and the very elderly) and patients with renal dysfunction or dementia (Alexander and Peterson, 2003; Roberts et al., 2003). In all of these categories, the findings of clinical trials that exclude these patients are unlikely to be easily extrapolated to effective clinical practice, especially with regard to dosing and expected adherence and harms.

Trade-off of Validity and Generalizability

A simple but useful way to conceptualize trial designs is in terms of a grid comparing the two constructs for a given trial (Fig. 3.1). To provide a clear answer to a conceptual question about disease mechanisms, it is often useful to limit the trial to a very narrow group of subjects in a highly controlled environment, yielding a trial that has high validity, but low generalizability. On the other hand, to test major public health interventions, it may be necessary to open up entry criteria to most patients with a general diagnosis and to place no restrictions on ancillary therapy, yielding a trial that is generalizable, but with open questions about the validity of the results according to issues such as the possibility of interactions between treatments. Of course, a trial that scores low in either characteristic would be practically useless, and the ideal would be to develop increasingly efficient tools that would allow trials to have high scores in both domains.

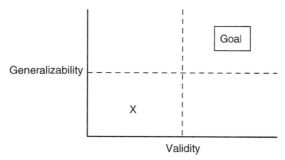

FIGURE 3.1   Grid for comparing validity and generalizability in clinical trial design.

EXPRESSING CLINICAL TRIAL RESULTS

The manner in which the results of clinical research are reported can have a profound influence on the perception of practitioners, who must weigh such information when deciding which therapies to use. The degree of enthusiasm with which a therapy is greeted by practitioners may be greatly affected by whether the results are presented in the most favorable light. To guard against this problem, investigators should report clinical outcome trials in terms of both relative and absolute effects on the risk of adverse outcomes and should include confidence intervals for point estimates. Even when exact results (in addition to the effect on risk of events) are provided so that the practitioner can reconstruct the results in different ways, the primary method of presentation has a major effect on perception (Forrow et al., 1992). Multiple studies have demonstrated that physicians are much more likely to recommend a therapy when the results are presented as a relative risk reduction rather than as an absolute difference in outcomes (Bobbio et al., 1994; Naylor et al., 1992). This appears to happen because relative risk reductions result in larger apparent differences, even though they are reporting exactly the same clinical phenomenon. This sobering problem points out a key issue of pragmatic trials: Because they are intended to answer questions that will directly affect patient care, the audience for the results will typically far exceed the local community of experts and often will include health-care providers with varying levels of expertise, lay people, and the press. Planning is critical to handle these issues appropriately.

One important metric for reporting the results of pragmatic clinical trials is the number of poor outcomes prevented by the more effective treatment, per 100 or 1000 patients treated. This measure, the *number needed to treat* (NNT), represents the absolute benefit of therapy and translates results for the specific populations studied into public health terms by quantifying how many patients would need to be treated to create a specific health benefit. The absolute difference can be used to assess quantitative interactions—that is, significant differences in the number of patients needed to treat to achieve a degree of benefit as a function of the type of patient treated. The use of thrombolytic therapy provides an example: the Fibrinolytic Therapy Trialists' (FTT) collaboration demonstrated that 37 lives are saved per 1000 patients treated when thrombolytics are used in patients with anterior ST segment elevation, whereas only 8 lives are saved per 1000 patients with inferior ST segment elevation (Fig. 3.2) (FTT Collaborative Group, 1994). The direction of the treatment effect is the same, but the magnitude of the effect is different.

Two other important aspects of the NNT calculation that deserve consideration are the duration of treatment needed to achieve the benefit and the *number needed to harm* (NNH). Depending on the circumstances, saving one life per 100 patients treated over 5 years versus saving one life per 100 patients treated in 1 week could be more or less important. The NNH can be simply calculated, just as the NNT is calculated.

This approach, however, becomes more complex with nondiscrete endpoints, such as exercise time, pain, or quality of life. One way to express trial results when the endpoint is a continuous measurement is to define the minimal clinically important difference (the smallest difference that would lead practitioners to change their practices) and to express the results in terms of the NNT to achieve that difference. Another problem with NNT and NNH occurs when the trial on which the calculation is based is not a generalizable trial that enrolled subjects likely to be treated in practice. Indeed, when clinically relevant subjects (e.g., elderly patients or those with renal dysfunction) are excluded, these simple calculations can become misleading, although the issue is usually *magnitude of effect* rather than *direction of effect*.

The *relative benefit of therapy*, on the other hand, is the best measure of the treatment effect in biological terms. This concept is defined as the proportional reduction in risk resulting from the more effective treatment, and it is generally expressed in terms of an odds ratio or relative risk reduction. The relative treatment effect can be used to assess qualitative interactions, which represent statistically significant differences in the direction of the treatment effect as a function of the

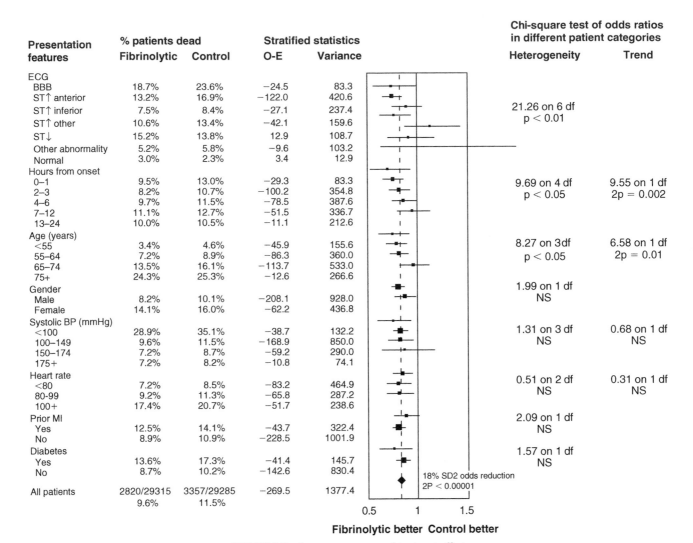

Presentation features	% patients dead		Stratified statistics			Chi-square test of odds ratios in different patient categories	
	Fibrinolytic	Control	O-E	Variance		Heterogeneity	Trend
ECG							
BBB	18.7%	23.6%	−24.5	83.3		21.26 on 6 df	
ST↑ anterior	13.2%	16.9%	−122.0	420.6		p < 0.01	
ST↑ inferior	7.5%	8.4%	−27.1	237.4			
ST↑ other	10.6%	13.4%	−42.1	159.6			
ST↓	15.2%	13.8%	12.9	108.7			
Other abnormality	5.2%	5.8%	−9.6	103.2			
Normal	3.0%	2.3%	3.4	12.9			
Hours from onset							
0–1	9.5%	13.0%	−29.3	83.3		9.69 on 4 df	9.55 on 1 df
2–3	8.2%	10.7%	−100.2	354.8		p < 0.05	2p = 0.002
4–6	9.7%	11.5%	−78.5	387.6			
7–12	11.1%	12.7%	−51.5	336.7			
13–24	10.0%	10.5%	−11.1	212.6			
Age (years)							
<55	3.4%	4.6%	−45.9	155.6		8.27 on 3df	6.58 on 1 df
55–64	7.2%	8.9%	−86.3	360.0		p < 0.05	2p = 0.01
65–74	13.5%	16.1%	−113.7	533.0			
75+	24.3%	25.3%	−12.6	266.6			
Gender						1.99 on 1 df	
Male	8.2%	10.1%	−208.1	928.0		NS	
Female	14.1%	16.0%	−62.2	436.8			
Systolic BP (mmHg)						1.31 on 3 df	0.68 on 1 df
<100	28.9%	35.1%	−38.7	132.2		NS	NS
100–149	9.6%	11.5%	−168.9	850.0			
150–174	7.2%	8.7%	−59.2	290.0			
175+	7.2%	8.2%	−10.8	74.1			
Heart rate						0.51 on 2 df	0.31 on 1 df
<80	7.2%	8.5%	−83.2	464.9		NS	NS
80-99	9.2%	11.3%	−65.8	287.2			
100+	17.4%	20.7%	−51.7	238.6			
Prior MI						2.09 on 1 df	
Yes	12.5%	14.1%	−43.7	322.4		NS	
No	8.9%	10.9%	−228.5	1001.9			
Diabetes						1.57 on 1 df	
Yes	13.6%	17.3%	−41.4	145.7		NS	
No	8.7%	10.2%	−142.6	830.4			
All patients	2820/29315	3357/29285	−269.5	1377.4			
	9.6%	11.5%					

18% SD2 odds reduction 2P < 0.00001

0.5   1   1.5

Fibrinolytic better Control better

FIGURE 3.2 Summary measures of treatment effect.

type of patient treated. In the FTT analysis, the treatment effect in patients without ST segment elevation is heterogeneous compared with that of patients with ST segment elevation (FTT Collaborative Group, 1994). Fig. 3.3 displays the calculations for commonly used measures of treatment effect.

A common way to display clinical trial results is the odds ratio plot (Fig. 3.4). Both absolute and relative differences in outcome can be expressed in terms of point estimates and confidence intervals. This type of display gives the reader a balanced perspective, since both the relative and the absolute differences are important, as well as the level of confidence in the estimate. Without confidence intervals, the reader will have difficulty ascertaining the precision of the estimate of the treatment effect. The goals of a pragmatic trial include (1) the enrollment of a broad array of patients so that the effect of treatment in different types of patients can be assessed and (2) the enrollment of enough patients with enough events to make the confidence intervals narrow and definitive. Using an odds ratio or risk ratio plot, the investigator can quickly create a visual image that defines the evidence for homogeneity or heterogeneity of the treatment effect as a function of baseline characteristics.

CONCEPTS UNDERLYING TRIAL DESIGN

As experience with multiple clinical trials accumulates, some general concepts deserve emphasis. These generalities may not always apply, but they serve as useful guides to the design or interpretation of trials. Failure to consider these general principles often leads to a faulty design and failure of the project.

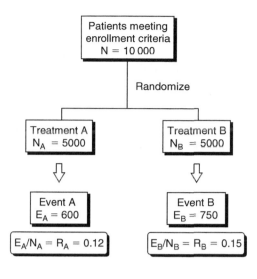

	Event	No event	
A	$E_A = 600$	4400	5000
B	$E_B = 750$	4250	5000
	1350	8650	10 000

Statistical tests of Rx effect

1. $\chi^2 = 19.268 \oslash p < 0.001$

2. Fischer Exact Test: $p < 0.001$

3. Comparison of proportions: $z = 4.360 \oslash p < 0.001$

Statements describing Rx effect

1. Relative Risk $R_A/R_B = 0.80$

2. Relative Risk Reduction = $(1 - \text{Relative Risk}) = 0.20$

3. Odds Ratio $= \dfrac{R_A/(1 - R_A)}{R_B/(1 - R_B)} = 0.77$

4. Absolute Risk Difference = $(R_B - R_A) = RD = 0.03$

5. Numbers Needed to Treat = (1/Abs. Risk Diff.) = 33

FIGURE 3.3   Measures of treatment effect in randomized controlled trials.

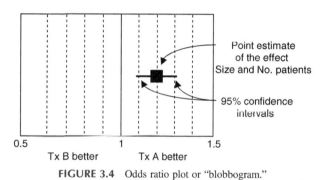

FIGURE 3.4   Odds ratio plot or "blobbogram."

Treatment Effects Are Modest

The most common mistake in designing pragmatic clinical trials is overestimating the expected treatment effect. Many researchers heavily involved in therapeutic development cannot resist assuming that the targeted biological or behavioral pathway is the most important contributor to a given patient outcome. Unfortunately, relative reductions in adverse clinical outcomes exceeding 25% are extremely uncommon.

When treatments affecting outcome are assessed, small trials typically overestimate the effect observed, a pattern that is frequently revealed in subsequent larger trials. The reasons for this observation are not entirely clear. One important factor is the existence of a publication bias against studies reporting negative findings (Olson et al., 2002). Of the many small studies performed, the positive ones tend to be published, although this problem may soon be abated because of the new legal requirement for posting trials in trial registries. A second factor could be analogous to regression to the mean in observational studies: When a variety of small trials are done, only those with a substantial treatment effect are likely to be continued into larger trials. Of course, in most cases there is so much uncertainty surrounding the estimate of the treatment effect in small trials that the true effect of many promising therapies is overestimated, whereas the effect of some therapies showing little promise based on point estimates from small studies is underestimated. Thus, when larger studies are completed, thereby giving a more reliable estimate of treatment effect, the estimate of benefit tends to regress toward average.

The **G**lobal **U**tilization of **S**treptokinase and **T**PA for **O**ccluded Coronary Arteries (GUSTO-I) trial used an extensive process to devise the expected sample size (GUSTO Investigators, 1993). An expected effect was calculated, using all previously published data on the relationship between coronary perfusion as shown on an angiogram and mortality in patients with ST-segment-elevation myocardial infarction. A panel of experts was then assembled both in Europe and the United States to determine the mathematical calculations and the differences that would be needed to create a clinically meaningful benefit. In the end, both approaches yielded a value of a 14% relative difference (1 life saved per 100 patients treated) or a 14% reduction in relative risk of death, whichever was smaller. The trial was then sized to detect these differences, and a difference of 15% on a relative basis and 1% on an absolute basis was observed when the trial was completed.

The major implication of this principle is that typical sample sizes will need to increase by a significant (perhaps logarithmic) amount. Indeed, there seems to be no logical substitute for the general view that we will need to enter many more patients into RCTs if we truly want to know the risks and benefits of a given treatment. Even if examining population subsets and restricting entry to only those subjects with biomarkers linked with treatment mechanisms leads to larger effects, patient populations will be subdivided into many small clusters that will require multiple trials.

Qualitative Interactions Are Uncommon

A reversal of treatment effect as a function of baseline characteristics historically has been unusual. Many training programs have taught clinicians that many therapies are effective only in very select subsets of the population, yet there are few examples demonstrating such targeted effects. Nonetheless, the literature is replete with examples of false-positive subgroup findings (Yusuf et al., 1991). There is a gathering consensus, however, that defining subgroups based on genomic characterization or functional imaging will identify patients who indeed respond differently to therapy. Interesting examples using gene expression analysis (Dressman et al., 2006) or genotyping (Schwarz et al., 2008) have demonstrated major differences in response to chemotherapy and anticoagulation therapy.

This principle has important implications for the amount of data collection in well-designed clinical trials. There is a tendency to collect voluminous amounts of information on the chance that the treatment may be effective only in a small group of patients; this rarely happens, however, and even if it did, the chances of detecting such an interaction are quite low. The main study is typically powered to detect a clinically meaningful effect, thereby leaving little power to detect the same effect in a smaller sample. Of course, when there is a compelling reason to look for a difference (e.g., response to therapy as a function of a known biological modifier of the disease response), it should be done. Before a trial starts, the researcher should direct considerable effort toward ensuring that the data collected will be of appropriate quality and utility, given the time, resources, and energy of the investigators and their staff.

An oft-overlooked component of this issue is the extraordinary expense of additional data items in large clinical trials when the costs of data collection, quality management, and analysis are aggregated. In a large trial, adding a single data point to be collected can add hundreds of thousands of dollars to the study budget.

Quantitative Interactions Are Common

When therapies are beneficial for specific patients with a given diagnosis, they are generally beneficial to most patients with that diagnosis. However, therapies commonly provide a differential absolute benefit as a function of the severity of the patient's illness. Given the same relative treatment effect, the number of lives saved or events prevented will be greater when the therapy is applied to patients with a greater underlying risk. Examples of this concept include the greater benefit of angiotensin-converting enzyme inhibitors in patients with markedly diminished left ventricular function, the larger benefit of thrombolytic therapy in patients with anterior infarction, and the greater benefit of bypass surgery in older patients compared to younger patients. Most often, these sorts of measures are the same ones that would be gleaned to characterize the population in clinical terms, so the extra cost of data ascertainment and recording is small. This principle has major implications for cost-effectiveness, as the greatest benefit typically occurs in sicker patients, yet the majority of patients with a disease at any given time have lower risk.

This same principle also seems to hold for harm. Elderly patients, patients with multiple comorbidities, and patients with renal dysfunction often have the highest risk of experiencing adverse drug effects. If they are excluded from clinical trials, the true risks will not be known when the treatment enters practice, and accurate assessment of risk through current methods of postmarketing assessment will be difficult, if not impossible (Gross and Strom, 2003).

Unintended Biological Targets Are Common

Therapies are appropriately developed by finding a pathway or "target" representing a biological process that could cause illness. Preliminary data are generated using a model that does not involve an intact human subject; if initial experiments appear promising, research in support of the pathway moves into various animal models and eventually into proof-of-concept testing in human subjects. Despite all good intentions, proposed therapies frequently either work via a different mechanism than the one for which they were devised, or affect a variety of different systems simultaneously. Examples of this abound: thrombolytic therapy for myocardial infarction was developed using coronary thrombosis models; unfortunately, this therapy also affects the intracranial vessels. Inotropic therapies for heart failure were developed using measures of cardiac function, but many of these agents, which clearly improve cardiac function acutely, also cause an increase in mortality, perhaps due to a detrimental effect on the neurohormonal system or apoptotic pathways. Several new agents for treatment of diabetes mellitus were developed to alter pathways of glucose uptake, but unanticipated effects on liver cells, aquaporin channels, and inflammatory pathways have been encountered (Califf and Kramer, 2008). The effect of the phentermine–fenfluramine combination on cardiac valves was unexpected. Major problems with myonecrosis led to the withdrawal of cerivastatin from the market (Davidson, 2004), and an extensive public debate resulted from the withdrawal of several COX-2 inhibitors after billions of dollars in sales (Topol, 2005). These examples point to the pressing need to evaluate therapies in broad populations of patients before making them available to the public, rather than relying on surrogate endpoints in small numbers of patients.

Interactions Among Therapies Are Not Predictable

Many common diseases can be treated with multiple therapies with some degree of benefit, yet clinical trials seldom evaluate more than one treatment at a time; evidence indicates that this may be an error. When abciximab was developed, its pharmacodynamic and pharmacokinetic interactions with heparin were easily characterized. However, the interaction of the two drugs with regard to clinical effect was unknown. A series of sequential clinical trials demonstrated that when full-dose abciximab was combined with a lower-than-normal dose of heparin, the bleeding rate in the setting of percutaneous intervention dropped to the same level as full-dose heparin alone, and the efficacy unexpectedly improved compared with full-dose abciximab and standard-dose heparin (Lincoff et al., 1997). This result was simply not predictable from the known biology and pharmacology of these agents.

Such testing of multiple therapies can also help avoid potentially dangerous effects, as in the case of mibefradil (Califf and Kramer, 1998), a calcium-channel blocker known to have significant drug interactions that could lead to toxicity. However, the extent of these interactions proved much more severe in practice than in clinical trials, leading to the drug being withdrawn from the market.

Long-Term Effects May Be Unpredictable

The concept that the short- and longer-term effects of therapy may differ is easiest to grasp when evaluating surgical therapy. Patients routinely assume a risk of operative mortality and morbidity to achieve longer-term gain. This principle also holds for some acute medical treatments. Fibrinolytic therapy actually increases the risk of death in the first 24 h following administration, but exerts a mortality benefit from that point forward. In both examples an "early hazard" must be overcome by later benefit for the treatment to be worthwhile. In the case of the COX-2 inhibitor rofecoxib, debate continues about whether harms related to cardiovascular events are an issue of concern only with longer periods of treatment.

It should be emphasized that unless trials include adequate follow-up times, important effects, either beneficial or detrimental, could be missed. An interesting recent case was the surprise finding of an overall mortality reduction with zoledronic acid, a bisphosphonate used to prevent second fractures (Lyles et al., 2007). Importantly, the survival curves separated only after 18 months of follow-up, so that the finding was entirely missed by previous shorter-term trials.

GENERAL DESIGN CONSIDERATIONS

When designing or interpreting the results of a clinical study, the purpose of the investigation is critical to placing the outcome in the appropriate context. Researchers and clinicians who design the investigation are responsible for constructing the project and presenting its results in a manner that reflects the intent of the study. In a small phase II study an improvement in a biomarker linked to a pathophysiological outcome is exciting, but enthusiasm can easily lead the investigator to overstate the clinical import of the finding. Similarly, "megatrials" with little data collection seldom provide useful information about disease mechanisms unless carefully planned substudies are performed. The structural characteristics of trials can be characterized as a function of the attributes discussed in the following sections.

Pragmatic Versus Explanatory

Most clinical trials are designed to demonstrate a physiological principle as part of the chain of causality of a particular disease. Such studies, termed *explanatory trials*, need only be large enough to prove or disprove the hypothesis being tested. Another perspective is that explanatory trials are focused on optimizing validity to "prove a point." Major problems have arisen because of the tendency of researchers performing explanatory trials to generalize the findings into recommendations about clinical therapeutics.

Studies designed to answer questions about which therapies should be used are called *pragmatic trials*. These trials should have a clinical outcome as the primary endpoint, so that when the trial is complete, the result will inform the practitioner and the public about whether using the treatment in the manner tested will result in better clinical outcomes than the alternative approaches. These trials generally require much larger sample sizes to arrive at a valid result, as well as a more heterogeneous population to be generalizable to populations treated in practice. This obligation to seriously consider generalizability in the design is a key feature of pragmatic trials.

The decision about whether to perform an explanatory or a pragmatic trial will have a significant effect on the design of the study. When the study is published, the reader must also take into account the intent of the investigators, since the implications for practice or knowledge will vary considerably depending on the type of study. The organization, goals, and structure of the pragmatic trial may be understood best by comparing the approach that might be used in an explanatory trial with the approach used in a pragmatic trial (Tunis et al., 2003). These same principles are important in designing disease registries.

Entry Criteria

In an explanatory trial, the entry criteria should be carefully controlled so that the particular measurement of interest will not be confounded. For example, a trial designed to determine whether a treatment for heart failure improves cardiac output should study patients who are stable enough for elective hemodynamic monitoring. Similarly, in a trial of depression, patients who are likely to return and who can provide the data needed for depression inventories are sought. In contrast, in a pragmatic trial, the general goal is to include patients who represent the population seen in clinical practice and whom the study organizers believe can make a plausible case for a benefit in outcome(s). From this perspective, the number of entry and exclusion criteria should be minimized, as the rate of enrollment will be inversely proportional to the number of criteria. In this broadening of entry criteria, particular effort is made to include patients with severe disease and comorbidities, since they are likely to be encountered in practice. An extreme version of open entry criteria is the uncertainty principle introduced by the Oxford group (Peto et al., 1995). In this scheme, all patients with a given diagnosis would be enrolled in a trial if the treating physician was uncertain as to whether the proposed intervention had a positive, neutral, or negative effect on clinical outcomes.

Thus, an explanatory trial focuses on very specific criteria to elucidate a biological principle, whereas a large pragmatic trial should employ entry criteria that mimic the conditions that would obtain if the treatment were employed in practice.

Data Collection Instrument

The data collection instrument provides the information on which the results of the trial are built; if an item is not included on the instrument, obviously it will not be available at the end of the trial. On the other hand, the likelihood of collecting accurate information is inversely proportional to the amount of data collected. In an explanatory trial, patient enrollment is generally not the most difficult issue, since small sample sizes are indicated. In a pragmatic trial, however, there is almost always an impetus to enroll patients as quickly as possible. Thus, a fundamental precept of pragmatic trials is that the data collection instrument should be as brief and simple as possible.

The ISIS-1 trial provides an excellent example of this principle: in this study, the data collection instrument consisted of a single-page FAX form (ISIS-1 Collaborative Group, 1986). This method made possible the accrual of tens of thousands of patients in mortality trials with no reimbursement to the enrolling health-care providers. Some of the most important findings for the broad use of therapies (beta blockers reduce mortality in acute myocardial infarction; aspirin reduces mortality in acute myocardial infarction; and fibrinolytic therapy is broadly beneficial in acute myocardial infarction) have resulted from this approach. Regardless of the length of the data collection form, it is critical to include only information that will be useful in analyzing the trial outcome, or for which there is an explicitly identified opportunity to acquire new knowledge.

At the other end of the spectrum, there is growing interest in patient-reported outcomes to assess quality of life and response to therapy (Weinfurt, 2003) (see also Chapter 12). While many more data items may be required, increasingly sophisticated electronic means of collecting patient–subject surveys are allowing more detailed data collection while reducing burdens on research sites.

Ancillary Therapy and Practice

Decisions about the use of nonstudy therapies in a clinical trial are critical to the study's validity and generalizability. Including therapies that will interact in deleterious fashion with the experimental agent could ruin an opportunity to detect a clinically important treatment advance.

Alternatively, the goal of a pragmatic trial is to evaluate a therapy in clinical, "real-world" context. Since clinical practice is not managed according to prespecified algorithms and many confounding situations can arise, evaluation of the experimental therapy in the setting of such an approach is likely to yield an unrealistic approximation of the likely impact of the therapy in clinical practice. For this reason, unless a specific detrimental interaction is known, pragmatic trials avoid prescribing particular ancillary therapeutic regimens. One exception is the encouragement (but not the requirement) to follow clinical practice guidelines if they exist for the disease under investigation.

Multiple Randomization

Until recently, enrolling a patient in multiple simultaneous clinical trials was considered ethically questionable. The origin of this concern is unclear, but seems to have arisen from a general impression that clinical research exposes patients to risks they would not experience in clinical practice, implying greater detriment from more clinical research and thus a violation of the principles of beneficence and justice, if a few subjects assumed such a risk for the benefit of the broader population. In specific instances, there are indeed legitimate concerns that the amount of information required to parse the balance of benefit and risk may be too great for subjects enrolled in multiple studies. Further, in some cases, there may be a known detrimental interaction of some of the required interventions in the trials. Finally, there are pragmatic considerations: in highly experimental situations, multiple enrollment may entail an unacceptable regulatory or administrative burden on the research site.

More recently, however, it has been proposed that when the uncertainty principle described previously is present, patients should be randomly assigned, perhaps even to two therapies. Stimulated by the evident need to develop multiple therapies simultaneously in HIV-AIDS treatment, the concept of multiple randomization has been reconsidered. Further, access to clinical trials is increasingly recognized as a benefit rather than a burden, in part, because the level of clinical care in research studies tends to be superior to that provided in general practice (Davis et al., 1985; Schmidt et al., 1999; Goss et al., 2006; Vist et al., 2007).

Factorial trial designs provide a specific approach to multiple randomizations, one that possesses advantages from both statistical and clinical perspectives. Because most patients are now treated with multiple therapies, the factorial design represents a clear means of determining whether therapies add to each other, work synergistically, or nullify the effects of one or both therapies being tested. As long as a significant interaction does not exist between the two therapies, both can be tested in a factorial design with a sample size similar to that needed for a single therapy. An increasingly common approach is to add a simple component to a trial to test a commonly used treatment, such as vitamin supplements. These trials have been critical in demonstrating the futility of vitamin supplements in many applications.

Adaptive Trial Designs

There is no effective way to develop therapies other than measuring intermediate physiologic endpoints in early phases and then making ongoing estimates of the value of continuing with the expensive effort of human-subjects research. However, other ways of winnowing the possible doses or intensities of therapy must be developed after initial physiological evaluation, since these physiological endpoints are unreliable predictors of ultimate clinical effects. One such method is the "pick-the-winner" approach. In this design (Fig. 3.5), several doses or intensities of treatment are devised, and at regular intervals during the trial an independent data and safety monitoring committee (DSMC) evaluates clinical outcomes with the goal of dropping arms of the study according to prespecified criteria.

Another form of adaptive design is the use of adaptive randomization, in which treatment allocation varies as a function of accruing information in the trial (Berry and Eick, 1995). In this design, if a particular arm of a trial is coming out ahead or behind, the odds of a patient being randomized to that arm can be altered.

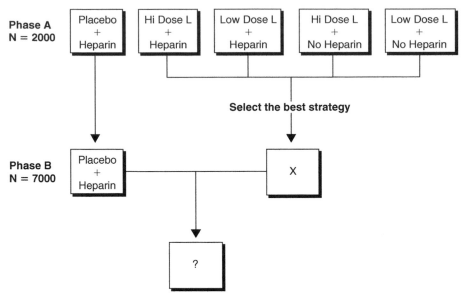

FIGURE 3.5 Pick-the-winner strategy.

An additional, albeit complex, form of adaptation is adjustment of the trial endpoint as new external information accrues. Particularly as we conduct longer-term trials to compare the capability of therapeutic strategies or diagnostic tests to lower rates of serious events, the ability to measure outcomes with greater sensitivity and precision will change the playing field of endpoint measurement in clinical trials. A recent example is the new definition of myocardial infarction adopted by professional societies in North America and Europe in a joint statement, which is based on the recognition that troponin measurements with improved operating characteristics are now routine on a global basis (Thygesen et al., 2007).

LEGAL AND ETHICAL ISSUES
Medical Justification

All proposed treatments in a clinical trial must be within the realm of currently acceptable medical practice for the patient's specific medical condition. Difficulties with such medical justification typically arise in two areas: (1) studies are generally performed because there is reason to believe that one therapeutic approach is better than another and (2) many currently accepted therapies have never been subjected to the type of scrutiny that is now being applied to new treatments. These factors create a dilemma for the practitioner, who may be uncomfortable with protocols that require a change in standard practice. The subject, of course, is given the opportunity to review the situation and make a decision, but for most patients, the physician's recommendation will be a critical factor in deciding whether to participate in a study. There is no escaping the basic fact that it remains a matter of judgment as to whether a potential new therapy can be compared with a placebo or an active comparator, or whether the case is not adequate for such a trial, based on previous data.

Groups of Patients Versus Individuals

The ethical balance typically depends on the good of larger numbers of patients versus the good of individuals involved in the trial. Examples are accumulating in which a therapy appeared to be better than its comparator based on preliminary results or small studies, but was subsequently shown to be inferior based on adequately sized studies (Lo et al., 1988). These experiences have led some authorities to argue that clinical practice should not change until a highly statistically significant difference in outcome is demonstrated (Peto et al., 1995). Indeed, the standard for acceptance of a drug for labeling by the Cardiorenal Group at the US Food and Drug Administration (FDA) is two adequate and well-controlled trials, each independently reaching statistical significance. If the alpha for each trial is set at 0.05, an alpha of 0.0025 (0.05 × 0.05) would be needed for both to be positive.

The counterargument is that the physician advising the individual patient should let that patient know which treatment is most likely to lead to the best outcome. In fact, Bayesian calculations could be used to provide running estimates of the likelihood that one treatment is better. In the typical general construct of large pragmatic trials, however, this approach is not taken: applying the ethical principles enumerated previously, an effort is made to accrue enough negative outcomes in a trial that a definitive result is achieved with a high degree of statistical significance and narrow confidence intervals.

An area of increasing confusion lies in the distinction between clinical investigation and measures taken to improve the quality of care as an administrative matter. The argument has been made that the former requires individual patient informed consent, while the latter falls under the purview of the process of medical care and does not require individual consent. This issue led to another major confrontation between the US Office of Human Research Protection (OHRP) and major Academic Health Centers (AHCs) when consent was waived by an IRB, but OHRP retrospectively ruled that waiving consent was not the correct decision (Pronovost et al., 2006; Miller and Emanuel, 2008).

Several special situations must be considered in studies conducted in the setting of emergency medical treatment, which often does not permit sufficient time for explaining the research project in exacting detail and for obtaining informed consent. In treating acute stroke or myocardial infarction, the time to administration of therapy is a critical determinant of outcome, and time spent considering participation in a protocol could increase the risk of death. Accordingly, the use of an abbreviated consent to participate, followed by a more detailed explanation later during the hospitalization, has been sanctioned. Collins et al. (1992) have made a compelling case that the slow, cumbersome informed consent form used in the United States in ISIS-2 actually resulted in the unnecessary deaths of a large number of patients with acute myocardial infarction.

An even more complex situation occurs in research concerning treatment of cardiac or respiratory arrest. Clinical investigation in this field almost came to a halt because of the impossibility of obtaining informed consent. After considerable national debate, such research is now being done only after careful consideration by the community of providers and citizens about the potential merits of the proposed research. A situation at least as complex exists for patients with psychiatric disturbances, and considerable discussion continues about the appropriate circumstances in which to obtain consent and to continue the patient in the trial as his or her clinical state changes.

Blinding

Blinding (or *masking*) is essential in most explanatory trials, since the opportunity for bias is substantial. In most pragmatic trials, blinding is also greatly preferred to reduce bias in the assessment of outcome. *Single blinding* refers to blinding of the patient, but not the investigator, to the therapy being given. *Double blinding* refers to blinding of both the patient and the investigator, and *triple blinding* refers to a double-blinded study in which the committee monitoring the trial is also blinded to which group is receiving which treatment. Despite the relative rarity of deceit in clinical research, examples of incorrect results due to bias in trials without blinding (Karlowski et al., 1975) and with single-blind studies reinforce the value of blinding (Henkin et al., 1976).

However, when blinding would prevent a true test of a treatment strategy, such as in surgical or behavioral interventions, other methods must be used to ensure objectivity. The clearest example is a trial of surgical versus medical therapy; in this situation, the patient and the primary physician cannot remain blinded. (Interestingly, in some circumstances, sham surgical incisions have been used successfully to ensure that high-cost, high-risk surgical procedures were being evaluated with maximum objectivity.) A similar situation exists when the administration of one therapy is markedly different than the other. In some cases, a "double-dummy" technique (in which the comparative therapies each have a placebo) can be used, but often this approach leads to excessive complexity and renders the proposed trial infeasible.

Given the large number of effective therapies, an increasing problem will be the lack of availability of placebo. Manufacturing a placebo that cannot be distinguished from the active therapy and that cannot affect the outcome of interest is a complex and expensive effort. Often, when a new therapy is compared with an old therapy, or two available therapies are compared, one of the commercial parties refuses to cooperate, since the manufacturer of the established therapy has nothing to gain by participating in a comparative trial with a new therapy. Since a placebo needs to mimic the active therapy sufficiently well that the blind cannot be broken, the successful performance of a placebo-controlled trial depends on the participation of the manufacturers of both therapies.

Endpoint Adjudication

The accurate and unbiased measurement of study endpoints is the foundation of a successful trials design, although many difficult issues may arise. Methods of endpoint ascertainment include blinded observers at the research sites and clinical events adjudication committees that can review objective data in a blinded manner independently of the site's judgment.

Since most important endpoints (other than death) require a judgment, unbiased assessment of endpoints is essential, especially when blinding is not feasible. This point has been made vividly in trials of cardiovascular devices. In the initial Coronary Angioplasty versus Excisional Atherectomy Trial (CAVEAT), comparing directional coronary atherectomy with balloon angioplasty, the majority of myocardial infarctions were not noted on the case report form, despite electrocardiographic and enzymatic evidence of these events (Harrington et al., 1995). Even in a blinded trial, recording of endpoints such as myocardial infarction, recurrent ischemia and new or recurrent heart failure is subjective enough that independent judgment is thought to be helpful in most cases (Mahaffey et al., 1997). Increasingly, central imaging laboratories are adding an element of objectivity to the assessment of images as clinical trial endpoints (Arias-Mendoza et al., 2004; Cranney et al., 1997).

Intensity of Intervention

When a therapeutic intervention is tested, one must always consider whether its intensity is appropriate. This issue is especially obvious in the dosing of drugs. In recent trials of erythropoietin-stimulating agents, aiming for higher hemoglobin targets has resulted in excessive cardiovascular morbidity (Phrommintikul et al., 2007). This same issue also exists in behavioral or policy interventions. A trial that used prognostic and normative information to assist in end-of-life decision making, the Study to Understand Prognoses and Preferences for Outcomes and Risks of Treatments (SUPPORT), failed to change behavior, perhaps because the strength of the intervention was not adequate to truly affect the practitioners (Covinsky et al., 2000). The major strategic question, then, is how to design appropriate explanatory studies to define the most likely effective strength of the intervention before embarking on a large pragmatic trial.

Surrogate Endpoints

The quest to circumvent the need for large sample sizes in clinical trials continues to fuel interest in surrogate markers. Researchers have hoped that small studies could be used to develop pathophysiological constructs capable of determining the strength of the intervention for definitive evaluation, or capable of eliminating the need for a definitive interventional trial altogether. Unfortunately, this approach has led to a number of therapeutic misadventures (Table 3.3). Antiarrhythmic drugs were developed based on their ability to reduce ventricular arrhythmias on ambulatory monitoring. When the Cardiac Arrhythmia Suppression Trial (CAST) was terminated prematurely because of a higher mortality with therapies that had been shown to reduce ventricular arrhythmias on monitoring, it became clear that this putative surrogate marker was inappropriate (Pratt and Moye, 1990). Similarly, studies aimed at developing dosing for heart failure therapies have used improvement in cardiac output as a surrogate marker. A succession of inotropic (milrinone and ibopamine) and vasodilator (flosequinan and prostacyclin) compounds have been shown to improve hemodynamics in the short term, but the results of long-term therapeutic trials have been disastrous (Packer, 1990). More recently, concern has emerged regarding blood-pressure-lowering drugs: two compounds that are equally effective in lowering blood pressure may have very different effects on mortality and other major clinical outcomes (ALLHAT Collaborative Group, 2000).

There is little debate about the value of lowering ambient glucose values in people with diabetes mellitus, yet not all drugs that lower glucose are beneficial (a point exemplified by the rosiglitazone controversy). An increasing number of cases are demonstrating that the endpoint of progression-free survival in cancer is not useful in all situations (Mietlowski and Wang, 2007). These lessons about surrogate endpoints have important implications for clinicians and for clinical trial design. Thus, when physiologic surrogates are proposed as therapeutic targets in individual patients, they should be validated in populations before they are accepted as standard practice.

Conflict of Interest

The investigator completely free of bias is a theoretical ideal that is likely not achievable in the real world. The (inevitable) degree of bias or conflict of interest, however, can be considered in a graded fashion. Investigators should not have a direct financial interest in an industry sponsor of a clinical trial. Paid consultancies are also considered to be beyond the scope of acceptable relationship with industry. Compensation for efforts on a clinical research project should be reasonable given the work performed and should be handled by means of an explicit contract. Perhaps, the most universal and common conflict in clinical investigation is the bias of the investigator that arises from a belief in a particular concept. Double blinding greatly reduces this risk, but the vast majority of clinical studies cannot be double blinded. Failure to maintain an open mind about the basic results of the investigation can cause the researcher to miss critical discoveries.

TABLE 3.3 Speculation on Reasons for Failure of Surrogate Endpoints

Disease and Intervention	Endpoints		Reason for Failure[a]			
	Surrogate	Clinical	A	B	C	D
Cardiologic Disorder						
Arrhythmia						
Encainide; flecainide	Ventricular arrhythmias	Survival		1		11
Quinidine; lidocaine	Atrial fibrillation	Survival		1		11
Congestive Heart Failure						
Milrinone; flosequinan	Cardiac output; ejection fraction	Survival		1		11
Elevated Lipid Levels						
Fibrates; hormones; diet; lovastatin	Cholesterol levels	Survival		1		11
Elevated Blood Pressure						
Calcium-channel blockers	Blood pressure	Myocardial infarction; survival		1		11
Cancer						
Prevention						
Finasteride	Prostate biopsy	Symptoms; survival	11[b]			
Advanced Disease						
Fluorouracil plus leucovorin	Tumor shrinkage	Survival		1		11
Other diseases						
HIV Infection or AIDS						
Antiretroviral agents	CD4 levels; viral load	AIDS events; survival		1	1	1
Osteoporosis						
Sodium fluoride	Bone mineral density	Bone fractures	1			1
Chronic Granulomatous Disease						
Interferon-γ	Bacterial killing; superoxide production	Serious infection			11	

AIDS, acquired immunodeficiency syndrome; HIV, human immunodeficiency virus; 1, likely or plausible; 11, very likely.
[a]A, surrogate endpoint not in causal pathway of the disease process; B, of several causal pathways of the disease, the intervention only affects the pathway mediated through the surrogate; C, the surrogate is not in the pathway of the intervention's effect or is insensitive to its effect; D, the intervention has mechanisms of action that are independent of the disease process.
[b]In settings in which only latent disease is prevented.
Adapted from Fleming, T.R., DeMets, D.L., 1996. Surrogate endpoints in clinical trials: Are we being misled? Ann. Intern. Med. 1996, 125.

Several documents have explicitly laid out the guidelines for governing conflict of interest, including the recent guidance issued jointly by the Association of American Medical Colleges and the Association of American Universities (Table 3.4). In addition, attention has focused on the responsibility of those who write editorials to be free of any conflict of interest (Angell and Kassirer, 1996).

Special Issues With Device Trials

Trials of medical devices raise special issues that deserve careful consideration. In comparing devices with other devices or with medical therapy, the orientation of the clinician implanting the devices is often complicated by the fact that the technical skill of the clinician is an integral component of the success or failure of the therapy. Therefore, failure of a therapy can be interpreted as a failure of the physician as well as the device. Obviously, in many device trials, blinding is also impossible. For these reasons, a particular focus on methodology is required in the assessment of clinical outcomes in device trials. Ideally, clinical outcomes should be assessed by a blinded review mechanism, and studies should be designed

TABLE 3.4 Definitions of Financial Interest in Research

Financial Interests in Research (Applicable to the Covered Individual, Spouse, and/or Dependents, or Any Foundation or Entity Controlled or Directed by the Individual or Spouse)

Consulting fees, honoraria (including honoraria from a third party, if the third party is a company with a financial interest) gifts, other emoluments, or in-kind compensation from a financial interested company for consulting, lecturing, travel, service on an advisory board, or any other purpose not directly related to the reasonable costs of conducting the research, as specified in the research agreement.

Equity interests, including stock options, in a non–publicly traded financially interested company.

Equity interests in a publicly traded financially interested company (see exceptions below).

Royalty income or right to future royalty income under a patent license or copyright, where the research is directly related to the licensed technology or work.

Any nonroyalty payments or entitlements to payments in connection with the research that are not directly related to the reasonable costs of the research, as specified in the research agreement, including bonus or milestone payments in excess of reasonable costs, whether received from a financially interested company or from the institution.

Service as an officer, director, or in any other fiduciary role for a financially interested company, whether or not remuneration is received.

Exceptions

Interests of any amount in publicly traded, diversified mutual funds.

Payments to the institution (or via the institution to the individual) that are directly related to reasonable costs incurred in the conduct of research as specified in the research agreement.

Salary and other payments for services from the institution.

Adapted from Appendix C: February 2008. Definition of financial interests in research. In: Protecting patients, preserving integrity, advancing health: Accelerating the implementation of COI policies in human subjects research. A report of the Association of American Medical Colleges–Association of American Universities Advisory Committee on Financial Conflicts of Interest in Human Subjects Research. Available at: https://services.aamc.org/Publications/ (accessed March 24, 2008).

by groups that include investigators who do not have a particular interest in the device-related outcomes but who have expertise in the disease-specific outcomes or clinical research methodology.

HYPOTHESIS FORMULATION

Primary Hypothesis

Every clinical study should have a primary hypothesis. The goal of the study design is to develop a hypothesis that allows the most important question from the viewpoint of the investigators to be answered without ambiguity. This issue is obvious in clinical trials, but in observational studies the appropriate approach to the problem is much less clear. Often, the investigator is tempted to "dredge" the data; there exists no method for tracking multiple analyses to develop considerations related to multiple hypothesis testing.

Recently, the concept of coprimary endpoints has gained acceptance in clinical research, the idea being that maintaining a constant study-wide alpha allows investigators to test several ideas within the same trial without sacrificing the standard level of susceptibility to random error. Another issue that is becoming increasingly important in the era of comparative effectiveness is the concept of outcome scorecards. In this scheme, treatments are compared with regard to a variety of outcomes that include different measures of effectiveness and the balancing of risk of toxicity or harm. These are commonly included in composite outcomes, which provide a basis for statistical evaluation but may lead to difficulty in interpretation, if different elements of the composite come out differently according to treatment assignment.

Secondary and Tertiary Hypotheses

A number of secondary hypotheses will be of interest to investigators, including analyses of the relationship between patient characteristics and treatment effect. In addition to answering questions about the therapy being evaluated, the study can address questions concerning other aspects of the diagnosis, treatment, or outcomes of the disease. Constructing pathophysiological substudies embedded in larger clinical outcome studies has been especially rewarding. The GUSTO-I trial convincingly demonstrated the relationship between coronary perfusion, left ventricular function, and mortality in a systematic substudy (GUSTO Angiographic Investigators, 1993).

Intention to Treat

One of the most important concepts in the interpretation of clinical trial results is that of *intention to treat*. Excluding patients who were randomized into a trial leads to bias that cannot be quantified; therefore, the results of the trial cannot be interpreted with confidence.

The purpose of randomization is to ensure the random distribution of any factors, known or unknown, that might affect the outcomes of the subjects allocated to one treatment or another. Any postrandomization deletion of patients weakens the assurance that the randomized groups are at equal risk before beginning treatment. Nevertheless, there are several common situations in which it may be reasonable to drop patients from an analysis.

In blinded trials, when patients are randomized but do not receive the treatment, it is reasonable to create a study plan that would exclude these patients from the primary analysis. The plan can call for substitution of additional patients to fulfill the planned sample size. When this happens, extensive analyses must be done to ensure that there was no bias in determining which subjects were not treated. In unblinded trials, dropping patients who do not receive the treatment is particularly treacherous and should not be allowed. Similarly, withdrawing patients from analysis after treatment has started cannot be permitted in trials designed to determine whether a therapy should be used in practice, since the opportunity to "drop out without being counted" does not exist when a therapy is given in practice.

PUBLICATION BIAS

Clinical trials with negative findings are much less likely to be published than those with positive results. Approximately 85% of studies published in medical journals report positive results (Dickersin and Min, 1993). In a sobering analysis, Simes (1987) found that a review of published literature showed combination chemotherapy for advanced ovarian cancer to be beneficial, whereas a review of published and unpublished trials together showed that the therapy had no significant effect. Dickersin et al. (1987) found substantial evidence of negative reporting bias in a review of clinical trial protocols submitted to Oxford University and Johns Hopkins University. In particular, industry-sponsored research with negative results was unlikely to be published.

Twenty years after these studies, we now require studies of human subjects in the United States to be posted in clinical trials registries. A registry of all clinical trials, publicly or privately funded, is needed so that all evidence generated from human clinical trials will be available to the public. This issue of a comprehensive clinical trials registry has been a topic of great public interest (DeAngelis et al., 2004). The National Library of Medicine (Zarin et al., 2005) is a critical repository for this registry (www.clinicaltrials.gov), and registration with this repository will in time presumably be required for all clinical trials, regardless of funding sources.

STATISTICAL CONSIDERATIONS

Type I Error and Multiple Comparisons

Hypothesis testing in a clinical study may be thought of as setting up a "straw man" that the effects of the two treatments being compared are identical. The goal of statistical testing is to determine whether this "straw man hypothesis" should be accepted or rejected based on probabilities. The *type I error* (alpha) is the probability of rejecting the null hypothesis when it is correct. Since clinicians have been trained in a simple, dichotomous mode of thinking (as if the p value was the only measure of probability), the type I error is generally designated at an alpha level of 0.05. However, if the same question is asked repeatedly, or if multiple subgroups within a trial are evaluated, the likelihood of finding a "nominal" p value of less than .05 increases substantially (Lee et al., 1980). When evaluating the meaning of a p value, clinicians should be aware of the number of tests of significance performed and the importance placed on the p value by the investigator as a function of multiple comparisons (see also this chapter and Chapter 14).

Type II Error and Sample Size

The *type II error* (beta) is the probability of inappropriately accepting the null hypothesis (no difference in treatment effect) when a true difference in outcome exists. The power of a study (1-beta) is the probability of rejecting the null hypothesis appropriately. This probability is critically dependent on (1) the difference in outcomes observed between treatments and (2) the number of endpoint observations. A common error in thinking about statistical power is to assume that the number of patients determines the power; rather, it is the number of outcomes.

The precision with which the primary endpoint can be measured also affects the power of the study; endpoints that can be measured precisely require fewer patients. An example is the use of sestamibi-estimated myocardial infarct size. Measuring the area at risk before reperfusion and then measuring final infarct size can dramatically reduce the variance of the endpoint measure by providing an estimate of salvage rather than simply of infarct size (Gibbons et al., 1994). As is often the case, however, the more precise measure is more difficult to obtain, leading to great difficulty in finding sites that can perform the study; in many cases, the time required to complete the study is as important as the number of patients needed. This same argument is one of the primary motivators in the detailed quality control measures typically employed when instruments are developed and administered in trials of behavioral therapy or psychiatry. For studies using physiological endpoints, using a continuous measure generally will increase the power to detect a difference.

A review of the *New England Journal of Medicine* in 1978 determined that 67 of the 71 negative studies had made a significant (more than 10% chance of missing a 25% treatment effect) type II error and that 50 of the 71 trials had more than 10% chance of missing a 50% treatment effect (Frieman et al., 1978). Unfortunately, the situation has not improved sufficiently since that time. The most common reasons for failing to complete studies with adequate power include inadequate funding for the project and loss of enthusiasm by the investigators.

A statistical power of at least 80% is highly desirable when conducting a clinical trial; 90% power is preferable. Discarding a good idea or a promising therapy because the study designed to test it had little chance of detecting a true difference is obviously an unfortunate circumstance. One of the most difficult concepts to grasp is that a study with little power to detect a true difference not only has little chance of demonstrating a significant difference in favor of the better treatment, but also that the direction of the observed treatment effect is highly unpredictable because of random variation inherent in small samples. There is an overwhelming tendency to assume that if the observed effect is in the wrong direction in a small study, the therapy is not promising, whereas if the observed effect is in the expected direction but the *p* value is insignificant, the reason for the insignificant *p* value is an inadequate sample size. We can avoid these problems by designing and conducting clinical trials of adequate size.

Noninferiority

The concept of noninferiority has become increasingly important in the present cost-conscious environment, in which many effective therapies are already available. Where an effective therapy exists, the substitution of a less expensive (but clinically noninferior) one is obviously attractive. In these positive control studies, substantial effort is needed to define noninferiority. Sample size estimates require the designation of a difference below which the outcome with the new therapy is noninferior to the standard comparator and above which one therapy would be considered superior to the other. Sample sizes are often larger than required to demonstrate one therapy to be clearly superior to the other.

Clinicians must be wary of studies that are designed with a substantial type II error resulting from an inadequate number of endpoints, with the result that the *p* value is greater than .05 because not enough events accrued, as opposed to a valid conclusion that one treatment is not inferior to the other. This error could lead to a gradual loss of therapeutic effectiveness for the target condition. For example, if we were willing to accept that a therapy for acute myocardial infarction with 1% higher mortality in an absolute sense was "equivalent," and we examined four new, less-expensive therapies that met those criteria, we could cause a significant erosion of the progress in reducing mortality stemming from acute myocardial infarction.

Another interesting feature of noninferiority trials is that poor study conduct can bias the result toward no difference. For example, if no subjects in either treatment group took the assigned treatment, within the boundaries of the fluctuations of random chance, the outcomes in the randomized cohorts should be identical.

Sample Size Calculations

The critical step in a sample size calculation, whether for a trial to determine a difference or to test for equivalence, is the estimate of the *minimally important clinical difference* (MID). By reviewing the proposed therapy in comparison with the currently available therapy, the investigators should endeavor to determine the smallest difference in the primary endpoint that would change clinical practice. Practical considerations may not allow a sample size large enough to evaluate the MID, but the number should be known. In some cases, the disease may be too rare to enroll sufficient patients, whereas in other cases the treatment may be too expensive or the sponsor may not have enough money. Once the MID and the financial status of the trial are established, the sample size can be determined easily from a variety of published computer algorithms or tables. It is useful for investigators to produce plots or tables to enable them to see the effects of small variations in event rates or treatment effects on the needed sample size. In the GUSTO-I trial (GUSTO Investigators, 1993), the sample size was set after a series of international meetings determined that saving an additional 1 life per 100 patients treated with a new thrombolytic regimen would be a clinically meaningful advance. With this knowledge, and a range of possible underlying mortality rates in the control group, a table was produced demonstrating that a 1% absolute reduction (a difference of 1 life per 100 treated) or a 15% relative reduction could be detected with 90% certainty by including 10000 patients per study arm.

METAANALYSIS AND SYSTEMATIC OVERVIEWS

Clinicians are often faced with therapeutic dilemmas, in which there is insufficient evidence to be certain of the best treatment. The basic principle of combining medical data from multiple sources seems intuitively appealing, since this approach results in greater statistical power. However, the trade-off is the assumption that the studies being combined are similar enough that the combined result will be valid. Inevitably, this assumption rests on expert opinion.

Table 3.5 provides an approach to reading metaanalyses. The most common problems associated with metaanalyses are combining studies with different designs or outcomes and failing to find unpublished negative studies. There is no question regarding the critical importance of a full literature search, as well as involvement of experts in the field of interest to ensure that all relevant information is included. Statistical methods have been developed to help in the assessment of systematic publication bias (Begg and Berlin, 1988). Another complex issue involves the assessment of the quality of individual studies included in a systematic overview. Statistical methods have been proposed for differential weighting as a function of quality (Detsky et al., 1992), but these have not been broadly adopted.

The methodology of the statistical evaluation of pooled information has recently been a topic of tremendous interest. The *fixed effects model* assumes that the trials being evaluated are homogeneous with regard to estimate of the outcome; given the uncertainties expressed previously, the assumption of homogeneity seems unlikely. Accordingly, a *random effects model* has been developed that considers not only the variation *within* trials but also the random error *between* trials (Berkey et al., 1995).

Another interesting approach to metaanalyses, termed *cumulative metaanalysis*, has been developed (Lau et al., 1992). As data become available from new trials, they are combined with findings of previous trials with the calculation of a cumulative test of significance. In theory, this should allow the medical community to determine the point at which the new therapy should be adopted into practice. Another variation on the theme of metaanalysis is *metaregression*, which allows evaluation of covariate effects within multiple trials to explain heterogeneity in observed results.

The degree to which metaanalysis should be used as a tool for informing the development of policies concerning the safety of therapeutics has been a particularly heated issue of late. The antidiabetic thiazoledinedione rosiglitazone was reported to cause an excess of cardiovascular events in a metaanalysis, when no definitive large trials were available to

---

TABLE 3.5 How to Read and Interpret a Metaanalysis

Are the Results of the Study Valid?

Primary Guides

Does the overview address a focused clinical question?
Are the criteria used to select articles for inclusion appropriate?

Secondary Guides

Is it unlikely that important, relevant studies were missed?
Is the validity of the included studies appraised?
Are the assessments of studies reproducible?
Are the results similar from study to study?

What Are the Results?

What are the overall results of the review?
How precise are the results?

Will the Results Help Me in Caring for My Patients?

Can the results be applied to my patient?
Are all clinically important outcomes considered?
Are the benefits worth the risks and costs?

confirm the finding (Nissen and Wolski, 2007). As clinical trials data increasingly become publicly available, the methods and standards for reporting compilations of data from different trials will need to be clarified.

The apparent lack of congruence between the results of metaanalyses of small trials and subsequent results of large trials has led to substantial confusion. Metaanalyses of small trials found that both magnesium therapy and nitrates provided a substantial (0.25%) reduction in the mortality of patients with myocardial infarction (Antman, 1995). The large ISIS-4 trial found no significant effect on mortality of either treatment (ISIS-4 Collaborative Group, 1995). Although many causes have been posited for these discrepancies, a definitive explanation does not exist, and the chief implication seems to be that large numbers of patients are needed to be certain of a given therapeutic effect.

UNDERSTANDING COVARIATES AND SUBGROUPS

Because of the insatiable curiosity of clinicians and patients regarding whether different responses to a treatment may be seen in different types of patients, an analysis of trial results as a function of baseline characteristics is inevitable. Traditionally, this analysis has been performed using a subgroup analysis, in which the treatment effect is estimated as a function of baseline characteristics examined one at a time (e.g., age, sex, or weight). This approach typically generates vast quantities of both false-positive and false-negative findings. By chance alone, a significant difference will be apparent in at least 1 in 20 subgroups, even if there is absolutely no treatment effect. In 1980, Lee et al. randomly split a population of 1073 into two hypothetical treatment groups (the treatments were actually identical) and found a difference in survival in a subgroup of patients with a p value of, .05 and then, using simulations, showed how frequently such random variation can misguide a naïve investigator.

At the same time, given the large number of patients needed to demonstrate an important treatment effect, dividing the population into subgroups markedly reduces the power to detect real differences. Consider a treatment that reduces mortality by 15% in a population equally divided between men and women, with a p value for the treatment effect of .03. If the treatment effect is identical for men and women, the approximate p value will be .06 within each subgroup, since each group will have about half as many outcomes. It would obviously be foolish to conclude that the treatment was effective in the overall population but not in men or women.

A more appropriate and conservative method would be to develop a statistical model that predicted outcome with regard to the primary endpoint for the trial and then evaluated the effect of the treatment as an effect of each covariate after adjusting for the effects of the general prognostic model. This type of analysis, known as a *treatment by covariate interaction analysis*, assumes that the treatment effect is homogeneous in the subgroups examined, unless a definitive difference is observed.

An example of this approach is provided by the **P**rospective **R**andomized **A**mlodipine **S**urvival **E**valuation (PRAISE) trial (Packer et al., 1996), which observed a reduction in mortality with amlodipine in patients with idiopathic dilated cardiomyopathy but not in patients with ischemic cardiomyopathy. This case was particularly interesting, because this subgroup was prespecified to the extent that the randomization was stratified. However, the reason for the stratification was that the trial designers expected that amlodipine would be ineffective in patients without cardiovascular disease; the opposite finding was in fact observed. The trial organization, acting in responsible fashion, mounted a confirmatory second trial. In the completed follow-up trial (PRAISE-2), the special benefit in the idiopathic dilated cardiomyopathy group was not replicated (Cabell et al., 2004).

THERAPEUTIC TRUISMS

A review of recent clinical trials demonstrates that many commonly held beliefs about clinical practice need to be challenged based on quantitative findings. If these assumptions are to be shown to be less solid than previously believed, a substantial change in the pace of clinical investigation will be needed.

Frequently, medical trainees have been taught that variations in practice patterns are inconsequential. The common observation that different practitioners treat the same problem in different ways has been tolerated because of the general belief that these differences do not matter. Clinical trials have demonstrated, however, that small changes in practice patterns in epidemic diseases can have a sizable impact. Indeed, the distillation of trial results into clinical practice guidelines has enabled direct research into the effects of variations in practice on clinical outcomes. The fundamental message is that reliable delivery of effective therapies leads to better outcomes.

Another ingrained belief of medical training is that observation of the patient will provide evidence for instances when a treatment needs to be changed. Although no one would dispute the importance of following symptoms, many acute therapies have effects that cannot be judged in a short time, and many therapies for chronic illness prevent adverse outcomes in patients with very few symptoms. For example, in treating acute congestive heart failure, inotropic agents improve cardiac output early after initiation of therapy but lead to a higher risk of death. Beta blockers cause symptomatic deterioration acutely but appear to improve long-term outcome. Mibefradil was effective in reducing angina and improving exercise tolerance, but it also caused sudden death in an alarming proportion of patients, leading to its withdrawal from the market. Most recently, erythropoietin at higher doses seems to provide a transient improvement in quality of life, but a subsequent increase in mortal cardiac events compared with lower doses.

Similarly, the standard method of determining the dose of a drug has been to measure physiological endpoints. In a sense, this technique resembles the use of a surrogate endpoint. No field has more impressively demonstrated the futility of this approach than the arena of treatment for heart failure. Several vasodilator and inotropic therapies have been shown to improve hemodynamics in the acute phase but subsequently were shown to increase mortality. The experience with heparin and warfarin has taught us that large numbers of subjects are essential to understanding the relationship between the dose of a drug and clinical outcome.

Finally, the imperative of "do no harm" has long been a fundamental tenet of medical practice. However, most biologically potent therapies cause harm in some patients while helping others. The recent emphasis on the neurological complications of bypass surgery provides ample demonstration that a therapy that saves lives can also lead to complications (Roach et al., 1996). Intracranial hemorrhage resulting from thrombolytic treatment exemplifies a therapy that is beneficial for populations but has devastating effects on some individuals. Similarly, beta blockade causes early deterioration in many patients with heart failure, but the longer-term survival benefits are documented in multiple clinical trials. The patients who are harmed can be detected easily, but those patients whose lives are saved cannot be detected.

STUDY ORGANIZATION

Regardless of the size of the trial being contemplated by the investigator, the general principles of organization of the study should be similar (Fig. 3.6). A balance of interest and power must be created to ensure that after the trial is designed, the experiment can be performed without bias and the interpretation will be generalizable.

Executive Functions

The Steering Committee

In a large trial, the steering committee is a critical component of the study organization and is responsible for designing, executing, and disseminating the study. A diverse steering committee, comprising multiple perspectives that include biology, biostatistics, and clinical medicine, is more likely to organize a trial that will withstand external scrutiny. This

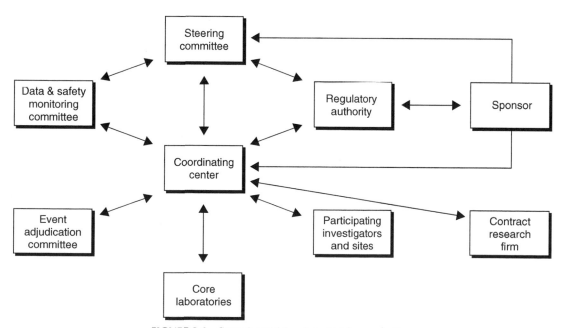

FIGURE 3.6   General model for clinical trials organization.

same principle holds for small trials; an individual investigator, by organizing a committee of peers, can avoid the pitfalls of egocentric thinking about a clinical trial.

The principal investigator plays a key role in the function of the trial as a whole, and a healthy interaction with the steering committee can provide a stimulating exchange of ideas on how best to conduct a study. The principal trial statistician is also crucial in making final recommendations about study design and data analysis. An executive committee can be useful, as it provides a small group to make real-time critical decisions for the trial organization. This committee typically includes the sponsor, principal investigator, statistician, and key representatives from the steering committee and the data coordinating center (DCC).

The Data and Safety Monitoring Committee

The DSMC is charged with overseeing the safety of the trial from the point of view of the participating subjects. The DSMC should include clinical experts, biostatisticians, and, sometimes, medical ethicists; these individuals should have no financial interest, emotional attachment, or other investment in the therapies being studied. Committee members have access to otherwise confidential data during the course of the trial, allowing decisions to be made on the basis of information that, if made available to investigators, could compromise their objectivity. The DSMC also shoulders an increasingly scrutinized ethical obligation to review the management of the trial in the broadest sense, in conjunction with each Institutional Review Board (IRB), to ensure that patients are treated according to ethical principles.

The role of the DSMC has become a topic of significant global interest. Little has been published about the function of these groups, yet they hold considerable power over the functioning of clinical trials. The first textbook on issues surrounding DSMCs was published only relatively recently (Ellenberg et al., 2003).

The Institutional Review Board

The IRB continues to play a central role in the conduct of all types of clinical research. Approval by an IRB is generally required for any type of human subjects research, even if the research is not funded by an external source. The IRB should consist of physicians with expertise in clinical trials and nonphysicians expert in clinical research, as well as representatives with expertise in medical ethics and representatives of the community in which the research is being conducted. As with the DSMC, the IRB function has come under scrutiny, especially from government agencies charged with ensuring the protection of human subjects.

Several types of studies are typically exempted from the IRB process, including studies of public behavior, research on educational practices, and studies of existing data in which research data cannot be linked to individual subjects. Surveys and interviews may also be exempted when the subjects are not identified and the data are unlikely to result in a lawsuit, financial loss, or reduced employability of the subject.

Regulatory Authorities

Government regulatory authorities have played a major role in the conduct of clinical research. The FDA and other national health authorities provide the rules by which industry-sponsored clinical trials are conducted. In general, regulatory requirements include interpretation of fundamental guidelines to ensure adherence to human rights and ethical standards. The FDA and equivalent international authorities are charged with ensuring that the drugs and devices marketed to the public are safe and effective (a charge with broad leeway for interpretation). Importantly, in the United States, there is no mandate to assess comparative effectiveness or cost-effectiveness, although the advent of organizations [such as the National Institute for Clinical Excellence (NICE) in the United Kingdom and the Developing Evidence to Inform Decisions about Effectiveness (DECIDE) Network in the United States] charged with government-sponsored technology evaluations has led to a resurgence of cost evaluation in the clinical trial portfolio.

A controversial organization that has recently become extremely powerful in the United States is the OHRP, which reports directly to the Secretary of Health and Human Services and has tremendous power to control studies and their conduct through its ability to summarily halt studies or forbid entire institutions from conducting trials. Several recent cases have caused considerable debate about whether the powers of this organization are appropriate.

Industry or Government Sponsors

Having provided funding for the study, the sponsor of a clinical trial understandably prefers to be heavily involved in the conduct of the study. Worldwide, the majority of clinical investigation is now done either directly by the pharmaceutical or medical device industry or indirectly by for-profit clinical research organizations. This approach may seem reasonable or even desirable for explanatory trials, but pragmatic trials, if not overseen by an independent steering committee, are at greater risk of bias, because the sponsor of a study has a significant financial stake in the success of the therapy being tested. Even in the case of government sponsorship, trials are frequently performed as a result of political agendas, with much to be gained or lost for individuals within the scientific community depending on the result. All of these issues underscore the desirability of a diverse steering committee to manage the general functioning of a large pragmatic clinical trial.

Coordinating Functions

The fundamental functions are intellectual and scientific leadership, site management, and data management. These core functions are supported by administrative functions that include information technology (IT), finance, human resources, contracts management, pharmacy and supplies distribution, and randomization services. Given the magnitude of large trials, each project is dependent on the administrative and leadership skills of a project manager and a principal investigator. A major weakness in any one of these functions can lead to failure of the entire effort, whereas excellence in all components creates a fulfilling and exciting experience.

Intellectual Leadership

The roles of the principal investigator and chief statistician are critical to the success of the trial organization. Not only must these leaders provide conceptual expertise, but their knowledge of successful, real-world operations is also vital to a trial's success. It is important to remember that in large studies, a small change in protocol or addition of just one more patient visit or testing procedure can add huge amounts to the study cost. The larger the trial, however, the greater the economies of scale in materials, supplies, and organization that can be achieved. For example, a simple protocol amendment can take months and cost hundreds of thousands of dollars (and even more in terms of the delay) to successfully pass through review by multiple national regulatory authorities and hundreds of local IRBs. If the intellectual leaders of a trial are not in touch with the practical implications of their decisions for study logistics, the trial's potential for providing useful information can be compromised.

Data Coordinating Center

The DCC is responsible for coordinating the collection and cleaning of data for the clinical trial. In this role, the DCC must comply with an increasing number of regulations governing both the quality of the data and maintenance of data confidentiality. Further, the DCC produces routine reports that allow the trial organization and the DSMC to oversee the conduct of the trial and ensure that the question the human subjects volunteered to answer is being addressed properly. The DCC must be capable of harnessing data from multiple formats, including standard paper data records, remote electronic data entry, and information from third-party computer sources.

Site Management Organization

Managing the conduct of the participating sites is a major undertaking, requiring training and communications programs, as well as knowledge of regulatory affairs, to ensure compliance with federal, state, local, and, in some circumstances, international guidelines. International patient enrollment is often a feature of large clinical trials, and the organization must be able to provide in-service education and study monitoring in multiple languages, while also complying with regulations of various national authorities.

Given the imperative to initiate and complete trials efficiently, site management groups are increasingly concerned with maintaining good relations with clinical sites that perform well in clinical trials. These relationships are often fostered by ongoing educational programs aimed at increasing the quality of participation at the sites and rewarding the personnel by supplementing their knowledge of conducting and interpreting clinical trials. In addition, metrics are being introduced that will assess functions such as recruitment rates, protocol deviations, data quality, and personnel turnover. Sites that perform well are selected for future trials to increase efficiency.

Supporting Functions

Information Technology

Large trials are increasingly dependent on a successful information platform. A competitive coordinating center is dependent on first-rate IT expertise to maintain communication, often on a global basis. Trials involving regulatory submission have significant requirements for validation so that any changes made to the study database(s) can be reviewed and audited.

Finance

Even in relatively simple, low-paying trials, an effective financial system is critical to success. Study budgets are typically divided, with approximately half of the funds going to the sites performing the study and half going to coordinating efforts, with this money frequently split again among multiple contractors and subcontractors. Since payments to the sites often depend on documented activities at the sites, the flow of cash needs to be carefully regulated to avoid either overpayment or underpayment. Furthermore, the coordinating staff should be carefully monitored to ensure that study funds are appropriately allocated so that the necessary tasks are performed without overspending.

Human Resources

The staff required to conduct large pragmatic trials comprise a diverse group of employees with a wide array of abilities and needs. IT expertise, in particular, is difficult to acquire and maintain in this very competitive environment. The second most difficult group of employees to find and retain is qualified project leaders; the knowledge base and skills needed for such a role are extraordinary.

Contracts Management

Clinical trials in every country increasingly operate within the strictures of legal contracts. In a typical multicenter clinical trial, a contract must exist with each research site; sometimes multiple parties may be involved if different practice groups participating in the trial share the same hospital facility. The sponsor will usually contract with entities to coordinate portions of the trial. The number of coordination contracts depends on whether the primary coordination is performed internally within an industry or government sponsor, or contracted to one or more contract research organizations. Each participating site then has a contract with the sponsor, the coordinating organization, or both. In addition to specifying a scope of work and a payment scheme, indemnification for liability is a major issue that must be resolved in the contract. A new area of contracting and liability is the DSMC (DeMets et al., 2004).

Pharmacy and Supplies

The production and distribution of study materials, including those required for in-service education and training, and actual supplies, such as investigational drugs and devices, requires considerable expertise. The knowledge needed ranges from practical skills such as knowing how to package materials to achieve maximum understanding by site personnel, to expertise in "just-in-time" distribution across international boundaries and working knowledge of the mountains of regulations regarding good clinical practice and good manufacturing practice for clinical trials.

Randomization Services

A fundamental principle of pragmatic trials is that proper randomization will balance for baseline risk, including both known and unknown factors, to allow for an unbiased comparison of treatments. In large multicenter trials, this issue takes on tremendous complexity. Because sealed envelopes are notoriously prone to tampering in large, geographically distributed trials, central randomization has come to be regarded as superior. This can be accomplished by either telephone randomization by a human staffer or, increasingly, an automated interactive voice randomization service (IVRS). IVRS has the advantage of providing instantaneous, around-the-clock access to global networks of investigators and automatic recording of patient characteristics at the time of randomization.

Project Management

Within the context of the sponsoring organization with its ongoing priorities, the coordinating entities with their ongoing priorities, and the sites with their ongoing priorities, someone must ensure that the individual project is completed in a timely fashion without unanticipated budget overruns. This responsibility is typically shared by the principal investigator, the project manager, and a sponsor representative (a project officer for government grants and contracts and a scientific or business manager for industry trials). This task should ideally fall to people with skills in organizational management, finance, regulatory affairs, medical affairs, leadership, and personnel management. People with such an array of skills are difficult to find. Interestingly, few formal educational programs are in place to train project managers, despite the critical shortage of qualified individuals.

INTEGRATION INTO PRACTICE

Because the ultimate goal of clinical investigation is to prevent, diagnose, or treat disease, integrating the findings of a clinical investigation into practice must be undertaken with care. The old method of each practitioner trying to extrapolate from reading journal articles to making individual decisions is clearly inadequate. Recognition of this deficit has led to a variety of efforts to synthesize empirical information into practice guidelines. These guidelines may be considered as a mixture of opinion-based advice and proven approaches to treatment that are not to be considered optional for patients who meet criteria. In addition, efforts such as the Cochrane collaboration (Cochrane Handbook for Systematic Reviews of Interventions, 2006) are attempting to make available systematic overviews of clinical trials in most major therapeutic areas, an effort further enhanced by new clinical trials registry requirements.

This effort has been integrated into a conceptual framework of a "cycle of quality," in which disease registries capture continuous information about the quality of care for populations (Califf et al., 2002). Within these populations, clinical trials, of appropriate size and performed in relevant study cohorts, can lead to definitive clinical practice guidelines. These guidelines can then form the basis for performance measures that are used to capture the quality of care delivered. Ultimately, gaps in clinical outcomes in this system can help define the need for new technologies and behavioral approaches. Increasingly, the linkage of interoperable electronic health records, professional-society-driven quality efforts, and patient/payer-driven interest in improving outcomes is leading to a system in which clinical trials are embedded within disease registries, so that the total population can be understood and the implementation of findings into practice can be measured (Welke et al., 2004).

THE FUTURE

As revolutionary developments in human biology and IT continue to unfold, and as the costs of medical therapies continue to climb, the importance of innovative, reliable, and cost-effective clinical trials will only grow in importance. During the next several years, practitioners will make increasing use of electronic health records that generate computerized databases to capture information at the point of care. Early efforts in this area, focusing on procedure reports to meet mandates from payers and quality reviewers, will be replaced by systems aimed at capturing data about the entire course of the patient's encounter with the health-care system. Multimedia tools will allow clinicians to view medical records and imaging studies simultaneously in the clinic or in the hospital.

To expedite the efficient exchange of information, the nomenclature of diagnosis, treatments, and outcomes is becoming increasingly standardized. In a parallel development, a variety of "disease registries" are evolving, in which coded information about patients with particular problems is collected over time to ensure that they receive appropriate care in a systematic fashion. This combination of electronic health records and disease registries will have dramatic implications for the conduct of clinical trials.

In many countries, the computerized management of information is an element of a coalescence of practitioners into integrated health systems. To efficiently care for populations of patients at a reasonable cost, practitioners are working in large, geographically linked, economically interdependent groups. This integration of health systems will enable rapid deployment of trials into the community. This includes not only trials of diagnostic strategies and therapies, but also evaluations of strategies of care using cluster randomization (randomization of practices instead of individual patients) to produce a refined and continuous learning process for health-care providers, an approach dubbed the "learning health system" by the Institute of Medicine.

Although integrated health-care systems will provide the structure for medical practice, global communications will provide mechanisms to quickly answer questions about diagnosis, prevention, prognosis, and treatment of common and uncommon diseases. The ability to aggregate information about thousands of patients in multiple health systems will change the critical issues facing clinical researchers. Increasingly, attention will be diverted from attempts to obtain data, and much effort will be required to develop efficient means of analyzing and interpreting the types of information that will be available.

Ultimately, leading practitioners will band together in global networks oriented toward treating illnesses of common interest. When a specific question requiring randomization is identified, studies will be much more straightforward, because the randomization can simply be added to the computerized database, and information that currently requires construction of a clinical trials infrastructure will be immediately accessible without additional work. Information systems will be designed to provide continuous feedback of information to clinicians, supporting rational therapeutic decisions. In essence, a continuous series of observational studies will be in progress, assessing outcomes as a function of diagnostic processes and therapeutic strategies.

All of the principles elucidated above will continue to be relevant; indeed, they will evolve toward an increasingly sophisticated state, given better access to aggregate results of multiple trials. As in many other aspects of modern life, the pace at which knowledge is generated and refined will occur at rates that a decade ago would not have even been imaginable.

ACKNOWLEDGMENT

Substantial portions of this chapter originally appeared in: Califf, R.M. (2007) Large clinical trials and registries: clinical research institutes, in *Principles and Practice of Clinical Research*, second edn (J.I. Gallin and F. Ognibene, eds). Burlington, MA, Academic Press (Elsevier), 2007, which was adapted and updated for the present publication with permission from the authors and the publisher.

REFERENCES

Alexander, K.P., Peterson, E.D., 2003. Evidence-based care for all patients. Am. J. Med. 114, 333—335.

ALLHAT Collaborative Research Group, 2000. Major cardiovascular events in hypertensive patients randomized to doxazosin vs chlorthalidone — the antihypertensive and lipid-lowering treatment to prevent heart attack trial (ALLHAT). JAMA 283, 1967—1975.

Angell, M., Kassirer, J.P., 1996. Editorials and conflicts of interest [Editorial]. N. Engl. J. Med. 335, 1055—1056.

Antman, E., 1995. Randomized trial of magnesium for acute myocardial infarction: big numbers do not tell the whole story. Am. J. Cardiol. 75, 391—393.

Arias-Mendoza, F., Zakian, K., Schwartz, A., Howe, F.A., Koutcher, J.A., Leach, M.O., Griffiths, J.R., Heerschap, A., Glickson, J.D., Nelson, S.J., et al., 2004. Methodological standardization for a multi-institutional in vivo trial of localized 31P MR spectroscopy in human cancer research. In vitro and normal volunteer studies. NMR Biomed. 17, 382—391.

Begg, C., Berlin, J., 1988. Publication bias: a problem in interpreting medical data. J. R. Stat. Soc. A 151, 419—445.

Berkey, C.S., Hoaglin, D.C., Mosteller, F., Colditz, G.A., 1995. A random-effects regression model for meta-analysis. Stat. Med. 14, 395—411.

Berry, D.A., Eick, S.G., 1995. Adaptive assignment versus balanced randomization in clinical trials: a decision analysis. Stat. Med. 14, 231—246.

Bobbio, M., Demichelis, B., Giustetto, G., 1994. Completeness of reporting trial results: effect on physicians' willingness to prescribe. Lancet 343, 1209—1211.

Cabell, C.H., Trichon, B.H., Velazquez, E.J., Dumesnil, J.G., Anstrom, K.J., Ryan, T., Miller, A.B., Belkin, R.N., Cropp, A.B., O'Connor, C.M., Jollis, J.G., 2004. Importance of echocardiography in patients with severe nonischemic heart failure: the second prospective randomized amlodipine survival evaluation (PRAISE-2) echocardiographic study. Am. Heart J. 147, 151—157.

Califf, R.M., Kramer, J.M., 1998. What have we learned from the calcium channel blocker controversy? Circulation 97, 1529—1531.

Califf, R.M., Kramer, J.M., 2008. The balance of benefit and safety of rosiglitazone: important lessons for our system of drug development. Pharma-coepidemiol. Drug Saf. 17, 782—786.

Califf, R.M., Peterson, E.D., Gibbons, R.J., Garson Jr., A., Brindis, R.G., Beller, G.A., Smith Jr., S.C., American College of Cardiology, American Heart Association, 2002. Integrating quality into the cycle of therapeutic development. J. Am. Coll. Cardiol. 40, 1895—1901.

(updated September 2006). In: Higgins, J.P.T., Green, S. (Eds.), 2006. Cochrane Handbook for Systematic Reviews of Interventions, 4.2.6. John Wiley & Sons, Chichester. Available at: http://training.cochrane.org/handbook.

Collins, R., Doll, R., Peto, R., 1992. Ethics of clinical trials. In: Williams, C. (Ed.), Introducing New Treatments for Cancer: Practical, Ethical and Legal Problems. John Wiley & Sons, Chichester, pp. 49—65.

Covinsky, K.E., Fuller, J.D., Yaffe, K., Johnston, C.B., Hamel, M.B., Lynn, J., Teno, J.M., Phillips, R.S., 2000. Communication and decision making in seriously ill patients: findings of the support project. The Study to Understand Prognoses and Preferences for Outcomes and Risks of Treatments. J. Am. Geriatr. Soc. 48 (Suppl. 5), S187—S193.

Cranney, A., Tugwell, P., Cummings, S., Sambrook, P., Adachi, J., Silman, A.J., Gillespie, W.J., Felson, D.T., Shea, B., Wells, G., 1997. Osteoporosis clinical trials endpoints: candidate variables and clinimetric properties. J. Rheumatol. 24, 1222—1229.

Davidson, M.H., 2004. Rosuvastatin safety: lessons from the FDA review and post-approval surveillance. Expert. Opin. Drug Saf. 3, 547—557.

Davis, S., Wright, P.W., Schulman, S.F., Hill, L.D., Pinkham, R.D., Johnson, L.P., Jones, T.W., Kellogg Jr., H.B., Radke, H.M., Sikkema, W.W., et al., 1985. Participants in prospective, randomized clinical trials for resected non-small cell lung cancer have improved survival compared with non-participants in such trials. Cancer 56, 1710—1718.

DeAngelis, C.D., Drazen, J.M., Frizelle, F.A., Haug, C., Hoey, J., Horton, R., Kotzin, S., Laine, C., Marusic, A., Overbeke, A.J., et al., 2004. Clinical trial registration: a statement from the international committee of medical journal editors. JAMA 351, 1250—1251.

DeMets, D.L., Fleming, T.R., Rockhold, F., Massie, B., Merchant, T., Meisel, A., Mishkin, B., Wittes, J., Stump, D., Califf, R.M., 2004. Liability issues for data monitoring committee members. Clin. Trials 1, 525—531.

Detsky, A.S., Naylor, C.D., O'Rourke, K., McGeer, A.J., L'Abbé, K.A., 1992. Incorporating variations in the quality of individual randomized trials into meta-analysis. J. Clin. Epidemiol. 45, 255—265.

Dickersin, K., Chan, S., Chalmers, T.C., Sacks, H.S., Smith Jr., H., 1987. Publication bias and clinical trials. Control. Clin. Trials 8, 343—353.

Dickersin, K., Min, Y.I., 1993. Publication bias: the problem that won't go away. Ann. N. Y. Acad. Sci. 703, 135—146.

Dressman, H.K., Hans, C., Bild, A., Olson, J.A., Rosen, E., Marcom, P.K., Liotcheva, V.B., Jones, E.L., Vujaskovic, Z., Marks, J., et al., 2006. Gene expression profiles of multiple breast cancer phenotypes and response to neoadjuvant chemotherapy. Clin. Cancer Res. 12 (3 Pt. 1), 819—826.

Ellenberg, S.S., Fleming, T.R., DeMets, D.L., 2003. Data Monitoring Committees in Clinical Trials: A Practical Perspective. John Wiley & Sons, Chichester.

Fibrinolytic Therapy Trialists' (FTT) Collaborative Group, 1994. Indications for fibrinolytic therapy in suspected acute myocardial infarction: collaborative overview of early mortality and major morbidity results from all randomised trials of more than 1000 patients. Lancet 343, 311—322.

Fisher, R.A., Mackenzie, W.A., 1923. Studies of crop variation: II. The manurial response of different potato varieties. J. Agric. Sci. 13, 315.

Food and Drug Administration Amendments Act of 2007, 2007. Title VIII, Section 801 (Public Law No. 110—85, 121 Stat 825).

Forrow, L., Taylor, W.C., Arnold, R.M., 1992. Absolutely relative: how research results are summarized can affect treatment decisions. Am. J. Med. 92, 121—124.

Frieman, J.A., Chalmers, T.C., Smith Jr., H., Kuebler, R.R., 1978. The importance of beta, the type II error and sample size in the design and interpretation of the randomized control trial: survey of 71 negative trials. N. Engl. J. Med. 299, 690—694.

Gibbons, R.J., Christian, T.F., Hopfenspirger, M., Hodge, D.O., Bailey, K.R., 1994. Myocardium at risk and infarct size after thrombolytic therapy for acute myocardial infarction: implications for the design of randomized trials of acute intervention. J. Am. Coll. Cardiol. 24, 616—623.

Goss, C.H., Rubenfeld, G.D., Ramsey, B.W., Aitken, M.L., 2006. Clinical trial participants compared with nonparticipants in cystic fibrosis. Am. J. Respir. Crit. Care Med. 173, 98—104.

Gross, R., Strom, B.L., 2003. Toward improved adverse event/suspected adverse drug reaction reporting. Pharmacoepidemiol. Drug Saf. 12, 89—91.

GUSTO Angiographic Investigators, 1993. The effects of tissue plasminogen activator, streptokinase, or both on coronary artery patency, ventricular function and survival after acute myocardial infarction. N. Engl. J. Med. 329, 1615—1622.

GUSTO Investigators, 1993. An international randomized trial comparing four thrombolytic strategies for acute myocardial infarction. N. Engl. J. Med. 329, 673—682.

Harrington, R.A., Lincoff, A.M., Califf, R.M., Holmes Jr., D.R., Berdan, L.G., O'Hanesian, M.A., Keeler, G.P., Garratt, K.N., Ohman, E.M., Mark, D.B., et al., 1995. Characteristics and consequences of myocardial infarction after percutaneous coronary intervention: Insights from the Coronary Angioplasty Versus Excisional Atherectomy Trial (CAVEAT). J. Am. Coll. Cardiol. 25, 1693—1699.

Henkin, R.I., Schecter, P.J., Friedewald, W.T., Demets, D.L., Raff, M., 1976. A double-blind study of the effects of zinc sulfate on taste and smell dysfunction. Am. J. Med. Sci. 272, 285—299.

ISIS-1 (First International Study of Infarct Survival) Collaborative Group, 1986. Randomised trial of intravenous atenolol among 16,027 cases of suspected acute myocardial infarction: ISIS-1. Lancet 2, 57—66.

ISIS-4 (Fourth International Study of Infarct Survival) Collaborative Group, 1995. ISIS-4: a randomised factorial trial assessing early oral captopril, oral mononitrate and intravenous magnesium sulphate in 48,050 patients with suspected acute myocardial infarction. Lancet 345, 669—685.

Karlowski, T.R., Chalmers, T.C., Frenkel, L.D., Kapikian, A.Z., Lewis, T.L., Lynch, J.M., 1975. Ascorbic acid for the common cold: a prophylactic and therapeutic trial. JAMA 231, 1038—1042.

Lau, J., Antman, E.M., Jimenez-Silva, J., Kupelnick, B., Mosteller, F., Chalmers, T.C., 1992. Cumulative meta-analysis of therapeutic trials for myocardial infarction. N. Engl. J. Med. 327, 248—254.

Lee, K.L., McNeer, J.F., Starmer, C.F., Harris, P.J., Rosati, R.A., 1980. Clinical judgment and statistics: lessons from a simulated randomized trial in coronary artery disease. Circulation 61, 508—515.

Lilienfield, A.M., 1982. Ceteris paribus: the evolution of the clinical trial. Bull. Hist. Med. 56, 1—18.

Lincoff, A.M., Tcheng, J.E., Califf, R.M., Bass, T., Popma, J.J., Teirstein, P.S., Kleiman, N.S., Hattel, L.J., Anderson, H.V., Ferguson, J.J., et al., 1997. Standard versus low-dose weight-adjusted heparin in patients treated with the platelet glycoprotein IIb/IIIa receptor antibody fragment abciximab (c7E3 Fab) during percutaneous coronary revascularization. PROLOG Investigators. Am. J. Cardiol. 79, 286—291.

Lo, B., Fiegal, D., Cummins, S., Hulley, S.B., 1988. Addressing ethical issues. In: Hulley, S.B., Cummings, S.R. (Eds.), Designing Clinical Research. Williams & Wilkins, Baltimore, MD, pp. 151—157.

Lyles, K.W., Colón-Emeric, C.S., Magaziner, J.S., Adachi, J.D., Pieper, C.F., Mautalen, C., Hyldstrup, L., Recknor, C., Nordsletten, L., Moore, K.A., et al., 2007. Zoledronic acid and clinical fractures and mortality after hip fracture. N. Engl. J. Med. 357, 1799—1809.

Mahaffey, K.W., Granger, C.B., Tardiff, B.E., For the GUSTO-IIb Investigators, et al., 1997. Endpoint adjudication by a clinical events committee can impact the statistical outcome of a clinical trial: results from GUSTO-IIb [Abstract]. J. Am. Coll. Cardiol. 29 (Suppl. A), 410A.

Medical Research Council, 1948. Streptomycin treatment of pulmonary tuberculosis. Br. Med. J. 2, 769—782.

Mietlowski, W., Wang, J., 2007. Letter to the re Yu and Holmgren. Traditional endpoint of progression-free survival (PFS) may not be appropriate for evaluating cytostatic agents combined with chemotherapy in cancer clinical trials. Contemp. Clin. Trials 28, 674.

Miller, F.G., Emanuel, E.J., 2008. Quality-improvement research and informed consent. N. Engl. J. Med. 358, 765—767.

Naylor, C.D., Chen, E., Strauss, B., 1992. Measured enthusiasm: does the method of reporting trial results alter perceptions of therapeutic effectiveness? Ann. Intern. Med. 117, 916—921.

Nissen, S.E., Wolski, K., 2007. Effect of rosiglitazone on the risk of myocardial infarction and death from cardiovascular causes. N. Engl. J. Med. 356, 2457—2471.

Olson, C.M., Rennie, D., Cook, D., Dickersin, K., Flanagin, A., Hogan, J.W., Zhu, Q., Reiling, J., Pace, B., 2002. Publication bias in decision making. JAMA 287, 2825—2828.

Packer, M., 1990. Calcium channel blockers in chronic heart failure: the risks of "physiologically rational" therapy [Editorial]. Circulation 82, 2254−2257.

Packer, M., O'Connor, C.M., Ghali, J.K., Pressler, M.L., Carson, P.E., Belkin, R.N., Miller, A.B., Neuberg, G.W., Frid, D., Wertheimer, J.H., et al., 1996. Effect of amlodipine on morbidity and mortality in severe chronic heart failure. N. Engl. J. Med. 335, 1107−1114.

Peto, R., Collins, R., Gray, R., 1995. Large-scale randomized evidence: large, simple trials and overviews of trials. J. Clin. Epidemiol. 48, 23−40.

Phrommintikul, A., Haas, S.J., Elsik, M., Krum, H., 2007. Mortality and target haemoglobin concentrations in anaemic patients with chronic kidney disease treated with erythropoietin: a meta-analysis. Lancet 369, 381−388.

Pratt, C.M., Moye, L., 1990. The cardiac arrhythmia suppression trial: implications for anti-arrhythmic drug development. J. Clin. Pharmacol. 30, 967−974.

Pronovost, P., Needham, D., Berenholtz, S., Sinopoli, D., Chu, H., Cosgrove, S., Sexton, B., Hyzy, R., Welsh, R., Roth, G., Bander, J., et al., 2006. An intervention to decrease catheter-related bloodstream infections in the ICU. N. Engl. J. Med. 355, 2725−2732.

Roach, G.W., Kanchuger, M., Mangano, C.M., Newman, M., Nussmeier, N., Wolman, R., Aggarwal, A., Marschall, K., Graham, S.H., Ley, C., 1996. Adverse cerebral outcomes after coronary bypass surgery. N. Engl. J. Med. 335, 1857−1863.

Roberts, R., Rodriguez, W., Murphy, D., Crescenzi, T., 2003. Pediatric drug labeling: improving the safety and efficacy of pediatric therapies. JAMA 290, 905−911.

Rogers, W.J., Bowlby, L.J., Chandra, N.C., French, W.J., Gore, J.M., Lambrew, C.T., Rubison, R.M., Tiefenbrunn, A.J., Weaver, W.D., 1994. Treatment of myocardial infarction in the United States (1990 to 1993) observations from the national registry of myocardial infarction. Circulation 90, 2103−2114.

Schmidt, B., Gillie, P., Caco, C., Roberts, J., Roberts, R., 1999. Do sick newborn infants benefit from participation in a randomized clinical trial? J. Pediatr. 134, 151−155.

Schwarz, U.I., Ritchie, M.D., Bradford, Y., Li, C., Dudek, S.M., Frye-Anderson, A., Kim, R.B., Roden, D.M., Stein, C.M., 2008. Genetic determinants of response to Warfarin during initial anticoagulation. N. Engl. J. Med. 358, 999−1008.

Simes, R.J., 1987. Publication bias: the case for an international registry of clinical trials. J. Clin. Oncol. 4, 1529−1541.

Society of Thoracic Surgeons Database, 2005. Available at: www.sts.org/sections/stsnationaldatabase/.

Thygesen, K., Alpert, J.S., White, H.D., For the Joint ESC/ACCF/AHA/WHF Task Force for the Redefinition of Myocardial Infarction, 2007. Universal definition of myocardial infarction. Circulation 116, 2634−2653.

Topol, E.J., 2005. Arthritis medicines and cardiovascular events − "House of coxibs". JAMA 293, 366−368.

Tunis, S.R., Stryer, D.B., Clancy, C.M., 2003. Practical clinical trials: increasing the value of clinical research for decision making in clinical and health policy. JAMA 290, 1624−1632.

Vist, G.E., Hagen, K.B., Devereaux, P.J., Bryant, D., Kristoffersen, D.T., Oxman, A.D., 2007. Outcomes of patients who participate in randomised controlled trials compared to similar patients receiving similar interventions who do not participate. Cochrane Database Syst. Rev. MR000009.

Weinfurt, K.P., 2003. Outcomes research related to patient decision making in oncology. Clin. Ther. 25, 671−683.

Welke, K.F., Ferguson Jr., T.B., Coombs, L.P., Dokholyan, R.S., Murray, C.J., Schrader, M.A., Peterson, E.D., 2004. Validity of the society of thoracic surgeons national adult cardiac surgery database. Ann. Thorac. Surg. 77, 1137−1139.

Yusuf, S., Wittes, J., Probstfield, J., Tyroler, H.A., 1991. Analysis and interpretation of treatment effects in subgroups of patients in randomized clinical trials. JAMA 266, 93−98.

Zarin, D.A., Tse, T., Ide, N.C., 2005. Trial registration at ClinicalTrials.gov between May and October 2005. N. Engl. J. Med. 353, 2779−2787.

Chapter 4

Introduction to Epidemiology

Donna K. Arnett[1] and Steven A. Claas[2]

[1]University of Kentucky, Lexington, KY, United States; [2]University of Alabama at Birmingham, Birmingham, AL, United States

Chapter Outline

Key Points

- Epidemiology concerns itself with the frequency, distribution, cause, control, and prevention of disease. In contrast to other branches of medical science that focus on the health status of *individuals*, epidemiology is the study of the health status of *populations*.
- *Descriptive* epidemiology seeks to answer fundamental questions related to the "where," "when," and "who" dimensions of a disease of interest. *Analytic* epidemiology explores specific questions or hypotheses and searches for factors that correlate with, modify the risk of, and cause disease in populations.
- Two fundamental measurements in the epidemiologist's toolkit are *incidence* and *prevalence*. Incidence refers to the number of new disease cases reported in a population over a certain period of time among those without the disease at first measurement. Prevalence is the proportion of people in a population who have a disease of interest at a particular time.
- *Risk* is another fundamental epidemiological concept. Risk can be interpreted as the probability of developing a disease and is measured as the number of new disease episodes during a period of time divided by the number of people followed during that time.
- Epidemiologists use *associations* to explore whether particular factors or exposures increase or decrease the risk of disease in a population. *Relative risk* (RR) compares the *probability* of being affected by disease between exposed and unexposed groups; *odds ratio* compares the relative *odds* of being affected by disease between exposed and unexposed groups.
- Epidemiologists use both observational and experimental studies to explore hypotheses. Study design is dictated by factors such as the disease in question; the hypothesis; and available time, fiscal, and population resources. The weight of evidence provided varies greatly by study design, from minimal (case study) to considerable (randomized control trial).
- Epidemiologists must remain aware of, guard against, and compensate for factors that might make their findings invalid and unreliable such as selection bias, information bias, and confounding.

Clinical and Translational Science. http://dx.doi.org/10.1016/B978-0-12-802101-9.00004-1

• Epidemiologists often collect data on study participants' sex, race, and ethnicity because these may be associated with (or be confounders of) outcomes of interest. Because sex, race, and ethnicity can be used as instruments of discrimination, great care must be taken to define these categorical distinctions, and their use in any study should be well justified.

INTRODUCTION: DEFINITION AND ROLE OF EPIDEMIOLOGY

Conceived somewhat narrowly, epidemiology concerns itself with the incidence, distribution, cause, control, and prevention of disease. Conceived more liberally, epidemiology's gaze extends beyond disease to include any marker of health status, such as disability, injury, mortality, hospitalization, or quality-of-life measures (see Chapter 12). In contrast to other branches of medical science that focus on the health status of *individuals*, epidemiology is the study of the health status of *populations*. Because epidemiology serves as the scientific basis of public health, it is—at its very core—an example of translational science (see Chapters 6 and 7). Epidemiology employs the tools of basic science (e.g., observation, hypothesis testing, statistics, correlation and causal reasoning) to protect, promote, and restore health and well-being in populations.

The practice of epidemiology can be conceptually organized into two approaches—descriptive and analytic. *Descriptive epidemiology* seeks to answer fundamental questions related to the "where," "when," and "who" dimensions of the disease of interest. (Throughout the remainder of this chapter, we will often use the shorthand term "disease" to represent the more verbose, albeit more accurate, term "health-related state or event.") The spatial distribution of a disease provides one descriptive measure of the extent of the problem. Characterizing the spatial distribution of a disease can also be an important first step in determining (or at least hypothesizing) how or if the disease might spread through a population. Depending on the disease under study, the scale of distribution can vary greatly, ranging from a hospital wing to the entire globe. The temporal distribution of a disease can also be important. As with spatial scale, the temporal scale relevant in an epidemiological inquiry varies widely depending on the disease and other circumstances of interest. For example, a recent study that followed patients admitted on weekends and those admitted on weekdays for a first myocardial infarction found that admission on weekends was associated with higher mortality (Kostis et al., 2007). At the other end of the temporal spectrum, epidemiologists often track disease occurrence over the course of many years to determine long-term (also known as "secular") trends of occurrence in populations. Not only can characterizing the incidence of disease relative to time suggest possible causal mechanisms, understanding the timing of disease can also help public health officials predict and possibly prevent future occurrences. Finally, describing who is affected by a disease is fundamentally important. Like spatial and temporal dimensions, the salience of any personal variable will depend on the disease of interest. Personal characteristics may be inherent (e.g., age, genotype) or acquired (e.g., immune status, educational level). Personal characteristics also include behavioral patterns (e.g., diet, exercise) and social circumstance (e.g., occupation, income). As discussed in the following, personal characteristics are often markers of risk, and collecting such data is the first step in determining which characteristics may be operative in the context of a particular disease.

It should be clear from the previous discussion that gathering descriptive data on the epidemiologic variables of space, time, and person can have intrinsic value for public health practitioners. However, the observations of descriptive epidemiology also frequently prompt questions or hypotheses that become the starting point of *analytic epidemiology*. For example, descriptive data may reveal that cases of a rare respiratory illness clustered around a particular industrial site. Although the descriptive data may raise flags for public health officials, these data say nothing about the causal role of the industrial site or any other possible relationship between the disease and the environment (other than the obvious spatial relationship). To investigate these questions, epidemiologists must draw upon the analytical tools available to them in the form of various study designs and statistical tests. These are described in greater detail in the Types of Epidemiological Studies section.

Analytic epidemiology (with sound descriptive underpinnings) can be put to many uses. For instance, epidemiology can be used to search for factors that correlate with, modify the risk of, and cause disease in populations. (For the moment, we suspend knotty epistemological issues associated with determining causation.) For example, in 1978 public health officials in a number of states reported to the US Centers for Disease Control (CDC) cases of a sometimes fatal condition dubbed toxic-shock syndrome (TSS). Although *Staphylococcus aureus* was (correctly) implicated as the causal pathogen early in the description of the syndrome (Todd et al., 1978), the factors underlying an apparent increase in the occurrence of the syndrome remained a mystery. As expected, descriptive data emerged first: in the first 55 cases reported to the CDC, 95% were women; the average age of cases was about 25 years; for the 40 patients with menstrual history data, 95% reported symptoms within 5 days of the onset of menses; *S. aureus* was isolated from vaginal cultures in two-thirds of patients (Centers for Disease Control, 1980). These observations prompted a flurry of TSS research conducted by epidemiologists, clinicians, and laboratory scientists. Two CDC-sponsored case–control studies found statistically significant associations between TSS and the use of tampons (Shands et al., 1982). Although these early TSS studies were fraught with methodological complications (Todd, 1982), after 10 years of epidemiologic surveillance and analysis, researchers reported that the incidence of menstrual TSS had decreased and the drop could be explained by the withdrawal of a particular brand of tampon from the market as well as changes in the composition of tampons on the market (Schuchat and Broome, 1991). Menstrual TSS and its association with tampon use remains incompletely understood, and it would be reckless to make simple claims of causation. However, it is clear that the work of epidemiologists (in collaboration with others) played a significant role in identifying factors that contributed to the increased occurrence of TSS in the US population in the late 1970s.

Although it is true that epidemiologists often rely on clinicians to provide reliable and consistent diagnoses, it is also true that epidemiologists generate knowledge useful for clinicians. Obviously, as members of multidisciplinary teams searching for the cause of newly identified diseases, epidemiologists make clinically valuable contributions. Beyond the search for causes, however, epidemiologists also help create an accurate and useful clinical picture of a disease. For example, during the late 1940s investigators began following a population of over 5000 individuals in Framingham, MA, in an effort to identify factors that contribute to cardiovascular disease. This long-term epidemiological work identified age, hypertension, smoking, diabetes, and hyperlipidemia as coronary risk factors. These risk factors have subsequently been incorporated into algorithms that return global risk scores used by clinicians to assess and characterize patients' risk of cardiovascular disease (Wilson et al., 1998; Hemann et al., 2007).

Epidemiology is also used to test hypotheses about the natural history and routes of exposure to disease. The tools of analytical epidemiology can also be used to plan and assess interventions, ameliorations, and other public health initiatives. Ultimately, the fruits of epidemiology are put to use by individuals—whenever someone chooses to stop smoking, start exercising, lower their dietary fat, or wash their hands during cold and flu season, they are using knowledge gained through epidemiology to make decisions that protect, promote, and restore health.

The remainder of this chapter will discuss the fundamental principles of descriptive and analytic epidemiology. Many of the topics discussed in the following sections (such as experimental design) are covered in more detail elsewhere in this volume. Therefore, our goal is not to provide definitive descriptions of all elements of population-based medical research, but rather to explain the rudiments of the science of epidemiology and create a context in which subsequent chapters can be understood.

MEASURING OCCURRENCE OF DISEASE

Defining Diseases, Outcomes, and Other Health-Related States or Events

Before any epidemiological data collection can begin, meaningful health-related states or events must be defined, characterized, and accepted by all those involved in the research. This condition may seem so obvious as to be trivial, but it is assuredly not. A number of factors can complicate this fundamental prerequisite of epidemiological research. In some instances the health-related state or event is poorly understood, and the clinical picture is incomplete, making definitions difficult to derive. For example, the metabolic syndrome is a group of conditions (including obesity, high blood sugar, high blood pressure, and high triglycerides) associated with cardiovascular disease and diabetes. Although there is consensus that these conditions often occur together and associate with disease (and, therefore, may be a valuable health-related state from a public health perspective), there is no consensus on the exact definition of the syndrome. Not only do major public health organizations such as the National Cholesterol Education Program, the World Health Organization, the International Diabetes Federation have slightly different definitions of metabolic syndrome, but—significantly—using these different definitions can lead to different estimates of the occurrence of the syndrome in populations (Cheung et al., 2006; Olijhoek et al., 2007).

In some instances epidemiologists must decide between using a continuous variable or a categorical variable based on the continuous measure. For example, blood pressure is frequently measured as systolic and diastolic pressure in nominal units of millimeters of mercury. For clinical purposes, however, these two continuous measures, in combination with treatment status, are often translated into a single categorical diagnosis such as "prehypertension" or "stage 2 hypertension" (Chobanian et al., 2003). Which type of variable to use in an epidemiologic study hinges on such considerations as which is likely to offer the greatest statistical power in the context of the hypothesis being tested and study design being used, which is most clinically relevant, and whether the thresholds that delimit the categories are empirically justified and widely accepted.

Epidemiologists must also sometimes choose between *primary* or *surrogate* outcomes. Primary outcomes, as the name suggests, are direct measures of the actual health-related event of interest. Surrogate outcomes are used in place of primary outcomes. For example, a decrease in blood pressure over time is a surrogate outcome often used as a predictor of decreased occurrence of cardiovascular disease, a primary outcome. The choice to use a surrogate measure often results from logistical constraints. For example, an epidemiologist may hypothesize that a particular diet maintained throughout life will be associated with fewer occurrences of fatal cardiovascular disease in a population. Ideally, a study would be designed to follow a population from birth to death, with detailed surveillance of diet, disease, and cause of death throughout. Such a study would be both complicated and costly. A more feasible alternative might be a shorter-term study using surrogate measures, such as blood lipid concentrations, which have well-documented associations with both diet and cardiovascular disease.

In summary, the following questions should be considered when defining health-related states or events and selecting outcome measures for epidemiologic studies:

- Is the definition/outcome suitable for the hypothesis being tested?
- How relevant is the definition/outcome to clinical and/or public health practice?
- Is the definition/outcome compatible with earlier studies, and is it acceptable to others who are conducting or planning to conduct similar studies?
- How feasible (both logistically and financially) is it to collect the data prescribed by the definition/outcome?

Calculating Incidence and Prevalence

Up to this point we have used the term *occurrence* to loosely refer to the burden of disease in a population. Epidemiologists have adopted a number of common measures to quantify how often a disease occurs in a population; which measure is used depends on the question being asked or the hypothesis being tested.

The term *incidence rate* refers to the number of new disease cases reported in a population over a certain period of time among those without the disease at first measurement. Thus, the incidence rate is calculated as shown in Eq. (4.1):

$$\text{Incidence rate} = \frac{\text{Number of new disease episodes}}{\text{Total time at risk for subjects followed}} \quad (4.1)$$

It is important to note that the denominator is neither an average time at risk for a population nor does it correspond to the entire follow-up period (i.e., time of surveillance) of the study—it is the sum of all time units at risk for all members of the population. For example, Chenoweth et al. (2007) sought to determine the incidence of pneumonia in patients receiving mechanical ventilation at home. They followed 57 patients from June 1995 through December 2001. During that period, the 57 patients accrued a combined total of 50,762 days on their ventilators, and there were 79 episodes of pneumonia [see Eq. (4.2)].

$$\text{Pneumonia incidence rate} = \frac{79 \text{ cases of pneumonia}}{50\,762 \text{ ventilator-days}} \quad (4.2)$$
$$= 0.00155 \text{ cases/ventilator-day}$$

Incidence rates are often converted to more easily interpreted units. For example, Chenoweth et al. reported 1.55 episodes per 1000 ventilator-days.

Incidence rate is a useful measure because it does not require a uniform follow-up period for all participants in a study; because time-at-risk is part of the equation, meaningful data can be produced even if participants begin the study late or drop out early.

Unlike incidence rate, which concerns itself with new cases of disease, *prevalence* deals with existing cases of disease. Simply put, prevalence is the proportion of people in a population who have the disease of interest at a particular time. Thus, prevalence is calculated as shown in Eq. (4.3).

$$\text{Prevalence} = \frac{\text{Number of subjects with disease at a particular time}}{\text{Total number of subjects in the population}} \quad (4.3)$$

For example, Bisgaard and Szefler (2007) sought to determine the prevalence of asthma like symptoms in preschool children in the United States and Europe. These researchers acquired data from 7251 households and found that 3077 of 9490 children met their criteria [see Eq. (4.4)].

$$\text{Asthma symptom prevalence} = \frac{3077 \text{ children with symptoms}}{9490 \text{ children in population}} = 32\% \tag{4.4}$$

Although prevalence is typically calculated using the number of subjects with the disease at a single time point (sometimes referred to as "point prevalence"), "period prevalence" is calculated using the number of subjects with the disease at a starting time point plus the number of new cases that develop over the study period. Period prevalence is a useful measure for intermittent conditions (such as lower back pain) because point prevalence tends to underestimate the number of individuals in a population with episodic but chronic disease. Because prevalence represents the abstract "burden" of disease in a population at a particular time, the measure can be used to make quality-of-life assessments or calculate the financial costs of disease in a population.

Deciding which measure of disease occurrence to use depends on both the natural history of the disease and the type of epidemiological question being asked. Prevalence is influenced by both the incidence and duration of the disease in individuals. This fact makes prevalence an especially meaningful measure of occurrence for chronic diseases. For example, herpes simplex virus type 2 infections are typically nonfatal but also incurable. The incidence rate may be low, but because the infections are lifelong, seroprevalence of the virus in a population can be considerably higher. In contrast, easily transmitted diseases of short duration, such as influenza, can have a low point prevalence but a high incidence and, therefore, are often quantified using the latter. An epidemiologist tracking the secular trends of atherosclerosis in a population would probably be best served by making a series of prevalence measures, whereas an epidemiologist tracking an outbreak of foodborne *Escherichia coli* O157:H7 would find it most useful to measure incidence.

MEASURING RISK AND ASSOCIATION

Incidence and prevalence are core epidemiological parameters used to answer the fundamental question, "How much disease exists in a population?" These measures of disease occurrence cannot, however, by themselves answer another fundamental epidemiological question, "What factors increase the chances of being affected by a disease?" To answer this question, epidemiologists turn to measures of *risk* and *association* to identify and quantify risk factors (see this chapter). Identifying risk factors is a crucial epidemiological task because these factors help create a clinical picture of disease and can be used in public health practice. Some risk factors may even play an etiological role in the development of a disease. The term *association* is used to describe a statistically established relationship between a risk factor and a disease. It is important to remember that *association does not imply causation*. We describe the general concept of risk and two important measures of association, RR and odds ratio, in the following sections.

Calculating Risk

Like incidence rate described previously, *risk* is a measure of new disease episodes. Unlike incidence rate, however, the new episodes are normalized with respect to the number of subjects followed rather than the time the subjects are at risk. Thus, risk is calculated as shown in Eq. (4.5).

$$\text{Risk} = \frac{\text{Number of new disease episodes during time period}}{\text{Number of subjects followed for the time period}} \tag{4.5}$$

It is important to note two differences between risk and incidence rate. First, the concept of risk assumes all subjects are followed and "at risk" for the same time period. Second, when reporting risk, it is always necessary to specify a time period to which the risk applies (although it need not be the actual time of follow-up, i.e., adjustments can be made to ease interpretation). The work of Semrad et al. (2007) offers an example of the risk of venous thromboembolism (VTE) in patients with malignant gliomas. During a 2-year period, 9489 subjects were monitored for VTE and 715 cases were reported [see Eq. (4.6)].

$$\begin{aligned}\text{VTE risk} &= \frac{715 \text{ cases of VTE}}{9489 \text{ subjects followed for 2 years}} \\ &= 7.5\% \ 2-\text{year risk}\end{aligned} \tag{4.6}$$

Risk is a useful measure because it can be interpreted as the probability of developing a disease. Simple risk estimates can be difficult to make, however, because subject death or other losses to follow-up effectively eliminate them (and their data) from the sample population. For these and other reasons, risk is often considered less useful as a measure of disease occurrence but more useful as a predictive tool. As we will see in the following, in situations where study designs allow it, risk is also used to compare groups with different exposures to risk factors.

Quantifying Associations

Given the risk parameter described previously, it follows that a simple means to quantify how strongly a suspected risk factor is associated with a disease is to compare the risk of a group of individuals exposed to the risk factor of interest with the risk of an unexposed group. Thus, RR (also known as the *risk ratio*) is defined as shown in Eq. (4.7).

$$\text{Relative risk} = \frac{\text{Risk of group exposed to factor}}{\text{Risk of group unexposed to factor}} \qquad (4.7)$$

At the most fundamental level, RR is simple to interpret:

- If RR = 1, then the risk factor shows no association with the disease.
- If RR > 1, then the risk factor is positively associated with the disease.
- If RR < 1, then the risk factor is negatively associated with the disease.

There are a number of disadvantages to using RR as a measure of association. For example, it is analytically tricky to adjust RR estimates for potentially confounding covariates (see the Threats to Validity and Reliability section). Certain types of study designs (such as case—control studies) preclude the calculation of RR. Finally, because they are expressed as ratios and because analysts decide how the outcome of interest is defined and how the ratio is set up, the same study data can be used to generate very different-looking (but perfectly legitimate) RR estimates. For example, imagine a sample population is monitored for 1 year for getting at least one new tooth cavity (the outcome of interest). During that time, 791 of 850 individuals exposed to a new toothpaste additive (the factor of interest) had no new cavities (probability or "risk" of no new cavities = 0.93 per year), and 860 of 1000 individuals not using the additive had no new cavities (risk = 0.86 per year). Using these figures, the RR of no new cavities for users of the new toothpaste additive is 1.08. Consider, however, if the event of interest was expressed as getting new cavities (as opposed to not getting them). In this case, the figures would be 59/850 for additive users and 140/1000 for nonusers. This results in a RR of new cavities for additive users of 0.49. If we express it as the inverse (the RR of new cavities for people who do not use the additive), the RR is 2.02. All of these RR estimates are correctly calculated and all point toward the same conclusion—that using a toothpaste with the new additive is associated with a lower risk of getting at least one new cavity during a year's time. However, the magnitudes of the RR estimates obviously appear quite different. This intrinsically tricky property of RR demands that, although the *direction* of an association is intuitively obvious from the measure, the *magnitude* of RR must always be interpreted with care.

Whereas RR compares the *probability* of being affected by disease between exposed and unexposed groups, as its name implies, an odds ratio compares the relative *odds* of being affected by disease between exposed and unexposed groups. Recall that probability expresses the ratio of the number of events of interest to the total number of events; odds expresses the ratio of the occurrences of one event to the occurrences of its corresponding nonevent. Although people's intuitive reasoning about the likelihood of events tends toward probability rather than odds, at a fundamental level, odds ratios are interpreted like RR—the direction and magnitude of an association is represented by direction (i.e., either greater than or less than 1) and degree that an odds ratio varies from 1 [see Eq. (4.8)].

$$\text{Odds ratio} = \frac{(\text{Probability of event in group 1})/(1 - \text{Probability of event in group 1})}{(\text{Probability of event in group 2})/(1 - \text{Probability of event in group 2})} \qquad (4.8)$$

We can use the previous toothpaste additive—dental cavity example to calculate an odds ratio. For toothpaste additive users, the odds were (new cavity) 59:791 (no new cavity), or 0.075:1. For additive nonusers, the odds were (new cavity) 140:860 (no new cavity), or 0.163:1. Thus, there is a 2.17-fold greater odds (0.163/0.075) of getting at least one new cavity for people who do not use the toothpaste additive compared to those who do.

Using multivariable logistic regression models, odds ratios can be adjusted for covariates (see the Threats to Validity and Reliability section). Unlike RR, odds ratios can be calculated from case—control studies. For relatively rare outcomes, odds ratios approximate RR because the denominator in the individual odds estimates approximates the total number of events. Readers who are interested in learning more about the calculation of RR and odds ratio and their interpretation should consult Davies et al. (1998) and Naylor et al. (1992) and the commentaries associated with these articles.

TYPES OF EPIDEMIOLOGICAL STUDIES

The careful design and implementation of public health surveillance programs underlie all descriptive epidemiological studies. Because this volume concerns itself with clinical and translational science, however, we will forbear further treatment of this topic and instead direct interested readers to the books edited by Teutsch and Churchill and Brookmeyer and Stroup (see Recommended Resources section).

Analytical epidemiological studies are often classified as either *experimental* or *nonexperimental*, where nonexperimental studies are those in which researchers make observations without manipulating independent variables. In practice, the range of complex and compound study designs available to researchers makes this distinction somewhat theoretical. For purposes of discussion, however, we will consider randomized and nonrandomized clinical trials to be experimental and refer readers to the Clinical Trials section of Chapter 3 and the detailed discussion of experimental design in Chapter 8. After eliminating the design of surveillance programs and clinical trials, we are left to describe three classic epidemiological study designs: cross-sectional, cohort, and case—control.

Cross-sectional Studies

A cross-sectional study observes a sample population at a nominal single point in time. Although the cross-sectional design is often used for descriptive prevalence studies, this design is also used to investigate associations between risk factors and diseases. Cross-sectional studies allow investigators to collect data on multiple diseases and risk factors simultaneously, and they permit considerable analytical freedom to define and compare subgroups within the sample population. Cross-sectional studies often offer a relatively inexpensive way to collect a lot of epidemiological data in a short period of time.

The flexibility and efficiency of cross-sectional studies do, however, come at a cost. Of the various designs available to researchers, cross-sectional studies provide the least robust evidence that a risk factor plays a causal role in disease etiology—hence the use of the word "association" to cautiously describe the relationship between a risk factor and a disease. Because cross-sectional studies provide a single snapshot in time of disease and risk factor status, it is impossible to determine if exposure to a risk factor occurred before, during, or after disease emergence. Cross-sectional studies are also susceptible to prevalence—incidence bias (see the Threats to Validity and Reliability section), which can cause the association between potentially significant risk factors and a disease to be underestimated. Despite these epistemological and analytical shortcomings, cross-sectional studies provide an expedient means to generate hypotheses which can be subsequently tested in studies of other design.

The study by Ha et al. (2007) provides an excellent example of a cross-sectional investigation. Using data from the US National Health and Nutrition Examination Survey, these investigators found statistically significant associations between serum concentrations of persistent organic pollutants (POPs) and self-reported history of cardiovascular disease in women. This research illustrates characteristics of an appropriate, well-designed, and carefully reported cross-sectional study: the study investigated a recently reported risk factor—disease association about which little is known (Mastin, 2005); the study used a robust ($n = 889$), population-based sample; 21 different risk factors (POPs) were simultaneously investigated; self-reported affectation data (although suboptimal) were probably relatively quick, easy, and inexpensive to collect; and the authors organized and analyzed their risk data to allow comparisons between POPs and between participant demographic (i.e., sex, age) strata. Finally, these researchers concluded their report by calling for cautious interpretation of their cross-sectional data and urged more research on the reported association.

Cohort Studies

Cohort studies are so named because they identify and follow a group of individuals through time (i.e., *longitudinally*). Individuals in the cohort are bound by a common characteristic, such as having been born in the same year, living in the same town, or having been exposed to the same risk factor. A defining characteristic of cohort studies is that the health-related event (or events) of interest is not present in the cohort at the start of follow-up. It would be somewhat misleading to say that a cohort is without disease at the beginning of a study, because sometimes the presence of a disease is, in fact, the characteristic that defines the cohort. However, in such cases, the disease itself is not the health-related state or event of interest. For example, a cohort of HIV-infected individuals may be followed to learn about specific events in the natural history of AIDS. Cohort studies can be open (i.e., the pool of participants changes during the study) or closed (i.e., the cohort remains static during the study), prospective (i.e., the cohort is assembled before outcome measures are made) or retrospective (i.e., the cohort is defined after outcome data have been collected).

Although cohort studies are observational, they are also analytical because statistical comparisons can be made between the cohort and the general population from which the cohort was drawn or among subgroups with different risk exposures or other characteristics of interest within the cohort. Cohort studies are useful for describing the incidence and natural history of a disease and for finding associations between purported risk factors and health outcomes. However, it is important to remember that, like cross-sectional studies, cohort studies cannot demonstrate that a risk factor positively associated with a disease is a factor that causes the disease. It is arguable, however, that because the longitudinal nature of cohort studies allows investigators to determine whether exposure to a risk factor occurred before, during, or after disease emergence, the cohort design provides slightly stronger evidence for (or against) causation than the cross-sectional design. Like cross-sectional studies, cohort studies can simultaneously track multiple health-related events in the same sample. Finally, cohort studies are useful when experimental studies (such as randomized clinical trials) are practically or ethically impossible. For example, it would be unethical to intentionally expose a clinical trial group to a suspected carcinogen (e.g., tobacco smoke, asbestos), but it would not be unethical to compare cancer incidence between cohorts of smokers or asbestos minors and the general population.

Cohort studies are useful only if the group can be followed for a length of time sufficient to observe a statistically adequate number of outcomes. For this reason, using a cohort design to study rare events (e.g., certain kinds of cancer) is often not practical. Of course, the problem associated with studying a rare event can be offset by following a larger cohort and/or following the cohort for a longer period of time. These strategies, however, can make cohort studies prohibitively expensive. Cohort studies are susceptible to a number of biases, including differential confounding with risk, differential loss to follow-up, and the potential for zero-time bias. (These are discussed in the Threats to Validity and Reliability section.)

RR is often used to quantify hypothesized associations in cohort studies. The data used to calculate RR can be conveniently laid out in a 2×2 matrix. Fig. 4.1 contains the data from the previously mentioned toothpaste example. Tests such as chi-square or Fisher's exact can be used to determine whether an association reaches the level of statistical significance.

The Framingham Study mentioned previously is an example of a cohort study that has contributed significantly to our understanding of cardiovascular disease (Dawber et al., 1951). The Nurses' Health Study was begun in 1976 with a cohort of about 120,000 women and has since provided data for more than 600 published articles concerning oral contraceptive and hormone use, smoking, menopausal status, and other health-related events (Belanger et al., 1978). Examples of other notable cohort studies include the British Doctors Study (which followed 34,439 participants and, among other findings, provided convincing data which linked smoking to lung cancer) (Doll and Hill, 1956); the Bogalusa Heart Study (which followed 11,796 participants to investigate cardiovascular disease in children and young adults) (Berenson et al., 2005); and the National Children's Study (which plans to follow 100,000 US children from birth to age 21, tracking a wide range of environmental factors and health markers) (Branum et al., 2003).

Case–Control Studies

In contrast to cohort studies—which select participants on the basis of *not having experienced* the state or event of interest—case–control studies begin by selecting a group of participants *who have experienced* the state or event (*cases*) and a group of participants who have not (*controls*). By retrospectively comparing the proportion of cases with a potential

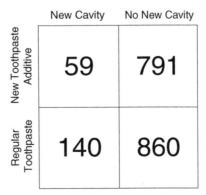

FIGURE 4.1 2×2 matrix showing results for a cohort study sorted by exposure and outcome.

risk factor to the proportion of controls with the same risk factor, these studies seek to identify variables that predict the outcome of interest. The predictive value of these factors is often quantified using odds ratios.

Because the case—control study begins with a sample of affected individuals, the design is particularly well suited for the study of rare diseases; it is not susceptible to the same problems related to statistical power that often afflict cohort studies (i.e., too few observed cases to make meaningful conclusions). In fact, for some diseases, a case—control study may offer the only feasible study design. Although each case—control study can concern itself with only one disease, many potential risk factors can be assessed with respect to that disease. The case—control design is often used for relatively quick, inexpensive studies intended to generate hypotheses about the significance of risk factors. Subsequent studies of different designs must be used to validate a risk factor or seek evidence to suggest the factor plays a causal role.

Ideally, cases should be randomly selected from the pool of all individuals in a population who have experienced the outcome. For many reasons, however, this is often impossible—including the frequently inescapable fact that not all affected individuals in a population may have been diagnosed at the time of case selection. Locating and selecting an appropriate control group can be even more difficult. For example, recruiting a "convenience" sample may result in a control group that systematically differs—in ways unknown and unmeasured—from the case group or that is otherwise not representative of the general population to which the cases should be compared. In an effort to minimize unanticipated differences between groups, controls can be individually matched to cases on the basis of basic demographic variables and potential confounders such as sex and age. This, however, can result in "overmatching" when cases and controls are matched on the basis of nonconfounding variables (i.e., those associated with an exposure but not the disease), potentially resulting in an underestimation of the predictive value of an important risk factor. Other case—control design measures, such as randomly drawing controls from a population-based sample or using multiple control groups, can be used to minimize some of these potential biases.

Reports of case—control studies in the medical literature are ubiquitous. (A search of NCBI PubMed database returned over 100,000 instances.) Despite the fact that many of these case—control studies undoubtedly provided valuable (albeit less than definitive) data associating risk factors with disease, it is often the controversial—and confuted—studies that are remembered. For example, from the late 1970s to the middle 1990s, many studies, including case—control studies, reported cardioprotective effects of hormone replacement therapy in postmenopausal women (Stampfer and Colditz, 1991). However, by the late 1990s and early 2000s, evidence—especially from randomized control trials—began to emerge suggesting hormone replacement had no cardioprotective benefits and, in fact, may have deleterious cardiovascular associations (Hulley et al., 1998; Herrington et al., 2000; Waters et al., 2002). How could this have happened? It is now generally accepted that a primary confounder of this association was socioeconomic status (see the Threats to Validity and Reliability section). That is, in the groups investigated, women undergoing hormone-replacement therapy were generally of higher socioeconomic status—and a higher socioeconomic status correlates with lower cardiovascular risk. Although some early studies statistically adjusted their analyses to account for socioeconomic status, these adjustments were inadequate, and the confounding by socioeconomic status long went undetected and unmitigated (Lawlor et al., 2004). Rather than bemoan the fact that epidemiology has suffered such discomfitures, some have suggested these incidents be remembered as "classics" in the history of the science (Smith, 2004). Certainly such controverted studies should serve to illustrate the limits of the case—control design and forewarn against overly ambitious interpretation.

Hybrid Study Designs

There are a number of study designs that incorporate elements of both cohort and case—control studies. These hybrid designs are used because they often offer both logistical and analytical advantages over any of the more basic designs.

Nested case—control studies are case—control studies built retrospectively upon cohort studies (Ernster, 1994). Nested case—control studies are typically used as a cost-efficient design to test a novel biomarker or other data collection instrument that is prohibitively expensive to administer in the cohort, but is feasible for testing in a smaller subset. After data are collected prospectively from the cohort, investigators retrospectively select samples of cases and controls and use the previously collected data and any newly acquired data collected from cases and controls as they would in the standard case—control design. As in a standard case—control study, cases and controls are usually matched on the basis of basic demographic parameters and potential confounders. The nested case—control design ameliorates the problem of recall bias which can be an issue in standard case—control studies. Recall bias is caused by systematic differences between cases and controls in the accuracy of their memory of affectation status, exposures, or other data collected by survey. Because cohort studies effectively detect cases as they occur, participant recall errors are eliminated. Including cases and controls from the same cohort can also reduce selection bias. However, the pool of cases selected retrospectively may not be fully representative of the original cohort due to death or losses to follow-up.

The *case–cohort* design is similar to a nested case–control study in that subsamples of a cohort are analyzed (Prentice, 1986). In a case–cohort study, the cases include all individuals in the cohort who meet the case criteria. This group is compared to a subcohort that is randomly drawn from the full cohort but does not include any cases. Unlike case–control studies, case–cohort studies allow direct estimation of prevalence and risk ratios, and different subcohorts can be drawn from the same cohort to be used as a control group for different disease outcomes. The case–cohort design can be more cost-effective than cohort studies because only a subsample of the noncases (i.e., the subcohort) needs to be assessed for risk exposure.

Choosing between the nested case–control and the case–cohort design depends on both logistical and analytical considerations. For example, a nested case–control study might be a better choice simply because sensitive biological specimens stored in batches need to be handled only once for analysis after all cases and controls are selected. In contrast, if case specimens are analyzed as they emerge in a case–cohort study, specimen batches may need to be handled repeatedly (e.g., thawed and refrozen) which could cause deterioration. Conversely, a case–cohort study might be the better choice because it is not restricted to the same analytical constraints as case–control-type studies. Interested readers are referred to Wacholder (1991) and Langholz and Thomas (1990) for more detailed comparisons of these study designs.

Study Design: Summary

Even a cursory review of the literature will reveal that the study designs described previously do not fully capture the complexity and sophistication of contemporary epidemiological research. For example, there are prospective case–control studies, retrospective case–cohort studies, *ecological studies* where group (rather than individual) data are used as the basis for analysis and comparison, and serial cross-sectional studies which comprise *pseudolongitudinal* studies. Descriptions of all of these are beyond the scope of this chapter. Keeping in mind the fundamental characteristics of basic study designs, however, can provide a context in which designs of greater complexity can be understood. To this end, we provide the following summary:

- Cross-sectional studies compare—at a single point in time—disease prevalence rates between individuals with different exposures to risk factors. A higher prevalence of disease among individuals with greater risk exposure provides evidence of association between risk factor and disease.
- Cohort studies compare—over a period of time—disease incidence rates between individuals with different exposures to risk factors. A higher incidence of disease among individuals with greater risk exposure provides evidence of association between risk factor and disease.
- Case–control studies compare exposures to risk factors between diseased (cases) and nondiseased (controls) individuals. A higher level of risk exposure in cases provides evidence of association between risk factor and disease.

THREATS TO VALIDITY AND RELIABILITY

The many occasions within this chapter when we have pointed to this section serve to emphasize an important fact: epidemiologists must remain aware of, guard against, and compensate for forces and factors that might make their findings invalid and unreliable. Although an exhaustive treatment of these forces and factors is outside the goals of this chapter (Delgado-Rodriguez and Llorca (2004) cataloged over 70 types of bias alone), a brief discussion of the major threats to validity and reliability is in order.

Defining and Measuring Threats to Validity and Reliability

A measurement (or study) is *valid if it measures what it sets out to measure*. A measurement (study) is *reliable if it produces the same result upon replication*. The relationship between these two concepts is best illustrated with the bull's-eye analogy. If we imagine that the center of a bull's eye represents the correct answer to a research question and the shots taken at the target represent repeated attempts to learn the correct answer, Fig. 4.2 illustrates the four possible combinations of valid–invalid and reliable–unreliable findings. It is important to remember that a reliable measurement does not guarantee a valid measurement, nor does an unreliable measurement imply an invalid measurement.

A study is said to have *internal validity* if it accurately measures what it intends to measure *within the study's sample*. A study is said to have *external validity* if its findings are generalizable to a wider population. Validity is compromised by bias—any factor that systematically distorts our apprehension of the truth and, therefore, results in incorrect measurements

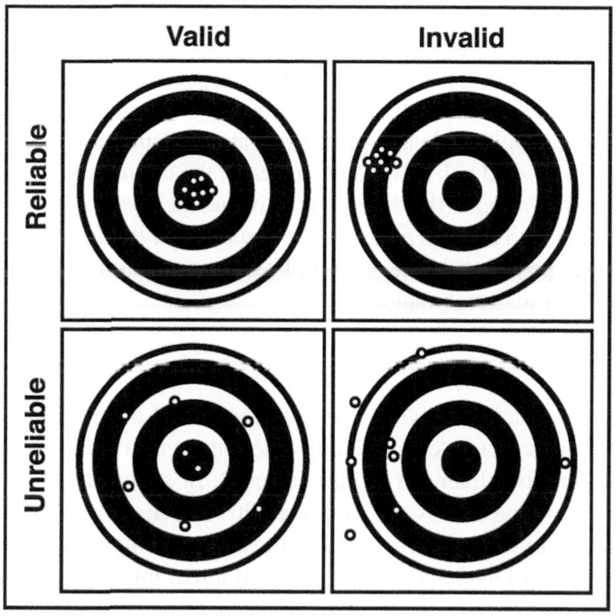

FIGURE 4.2   Relationship between validity and reliability.

or conclusions. As noted above, there are many types of bias, and these have been classified by a number of different taxonomies. We outline and offer specific examples of a few of the most important types of bias in the following section.

Selection Bias

Selection bias occurs when groups being compared in an analysis differ systematically in ways unknown or unintended. For example, in a cohort study, the exposed and unexposed groups may differ in ways other than their exposure to the risk factor under study (e.g., smokers might drink more sugared soft drinks per day than nonsmokers). A common form of selection bias is *ascertainment bias*, in which, for example, a cohort sample is chosen nonrandomly from a population (thus compromising external validity). A form of ascertainment bias known as *prevalence—incidence bias* (or Neyman bias) can occur when there is a gap between exposure to the risk factor of interest and recruitment of participants. In a cross-sectional study, for example, prevalent cases include only those who did not immediately die from the disease and, therefore, may possess better-than-average survival characteristics. Incident cases, on the other hand, would represent the full spectrum of diseased individuals. Similarly, *spectrum bias* results when only clear or obvious instances of a diseased individuals are recruited as cases and, therefore, do not represent the full spectrum of actually diseased individuals in the population. In prospective cohort studies, *zero-time bias* results when participants are recruited in a manner resulting in unintended systematic differences between groups at the beginning of the study. *Differential loss to follow-up* bias occurs when participants are lost or removed at different rates from exposure or outcome groups.

Information Bias

Information bias results from systematically incorrect observation, classification, or measurement of risk exposure and/or disease outcome. *Recall bias*, described previously, can be a problem in case—control studies because cases tend to spend more time and effort searching their memory about possible causes of their disease (i.e., exposures), but controls have no such incentive. *Observer bias* occurs when data collectors have knowledge of a study's hypothesis or participants' exposure or affectation status. For example, a researcher making diagnoses for a study may inadvertently perform more thorough clinical exams on individuals he/she knows have had the exposure of interest. In studies where participants supply information via surveys or interviews, *reporting bias* may happen if participants alter their responses because they know (or believe they know) what the researchers want to hear.

Confounding

In short, confounding is a type of bias caused by a misattribution of cause. That is, confounding occurs when an association between a presumed risk factor and the outcome of interest is, in fact, accounted for by another factor. Confounding results not from erroneous measurement, but from incorrect interpretation of what may be an accurate measurement. A variable can be a confounder only if it is distributed differently between compared groups. For example, early studies reported associations between oral contraceptive use and increased risk of myocardial infarction. Subsequent studies, however, determined that a large proportion of oral contraceptive users were also cigarette smokers. Cigarette smoking was, in fact, the actual factor responsible for increasing the likelihood of heart attack (Burchard et al., 2003). Note that for a factor to confound an association between a purported risk factor and a disease, the confounding factor must be associated both with the risk factor and the disease, and its relation to the disease must be independent of its association with the risk factor (i.e., it should not interact with the risk factor to have its effect on the disease; see the following section). Confounders may bias interpretation of an association either positively or negatively. Positive confounders result in an overestimated association while negative confounders lead to an underestimation.

Interaction

Unlike confounding, in which an association between a risk factor A and an outcome is merely a reflection of the association between risk factor B with risk factor A and the outcome, interaction results when the strength of the association between risk factor A and an outcome is influenced by risk factor B. In statistical terms, an interaction results when the effect of two (or more) variables on an outcome is not simply additive. Generally, when we speak of variables having an interactive effect on an outcome, we are not concerned with a threat to validity—interaction is "real" within our data set, not the result of mistaken interpretation of associations. However, if a real interaction between variables is not detected or is handled improperly during analysis, the true magnitude of certain associations may be obscured. Also, just as causal variables can interact, so can two or more threats to validity. For example, selection—maturation interaction occurs when comparison groups that entered a study with similar disease indicators are different in their rates of progressive worsening of the disease. This interaction of time and personal characteristic (i.e., disease maturation) may lead to a specious treatment effect.

Estimating and Avoiding Threats to Validity

Some forms of bias can be averted with proper study design while other forms must be addressed during data analysis.

Selection bias can be avoided by careful recruitment of study participants. For example, in choosing a control group for a case—control study, the goal is to capture controls from the same population that gave rise to the cases and, indeed, who would have been enrolled in the study as cases if they had developed the disease in question. For instance, a case—control study might be designed to test the association between coffee consumption and breast cancer. If cases are identified during routine mammography from a particular women's health center, an appropriate control group would be women seen for routine mammography from the same health center who do not have breast cancer. If a control group were recruited from the general population, selection bias might be introduced since women who get routine mammograms may be different in many ways related to the exposure and disease from those women who do not.

Careful study design is also necessary to reduce information bias. For example, if a case—control study seeks to determine whether exposure to a particular environmental factor is associated with lung cancer, a carefully chosen control group may be necessary to avoid recall bias. A control group that suffers from benign lung disease may be deemed appropriate since they would be expected to recall past exposures more likely than a healthy control group. A potential problem arises when choosing a control group with a disease other than the one being studied, however, if the exposure in question is associated with both diseases. To reduce observer bias, data collectors are often blinded to the disease/exposure status of the study participants. Thoughtful data collection instruments are necessary to avoid reporting bias. For example, if the information to be collected is deemed sensitive, such as information on sexual practices or illegal drug use, participants might be more honest if they are allowed to write their responses on paper rather than provide them in a face-to-face interview. Making the data collection process anonymous might also be necessary to get unbiased answers from participants.

Confounding is a major threat to validity, and potential confounders may be known and measured (either accurately or with error), known and not measured, or unknown. It is possible to minimize confounding during both the design phase and the statistical analysis phase of a study. For example, by design, randomized studies attempt to reduce confounding by distributing confounding factors (both known and unknown) equally between the groups to be compared. Matching is another way to reduce confounding during the design phase, and it also allows efficient management of confounding during the analysis phase. For example, if a researcher suspects that gender will confound the association of interest, comparison groups may be constructed so that they are matched on gender, with an equal ratio of women to men in each group. The ratio need not be 50% to achieve a matched design, but the ratio must be the same in each comparison group. It is also possible to eliminate confounding by a particular factor by restricting the study population. For example, if smoking is known to be a confounder in the association of interest, the study population can be limited to nonsmokers. During the analysis phase of a study, researchers can control for confounding factors by either computing standardized rates and risks of outcomes, conducting matched analysis in studies with a matched design, using stratified analyses (i.e., computing unbiased stratum-specific measures of association), or adjusting for confounding in a multivariable statistical model.

The nature of interactions must be uncovered to estimate the "true" association for each level of the interacting variable. For example, if the true effect of a particular antihypertensive drug is beneficial to men but has no effect on women and the analysis is conducted without regard to gender, the analysis will return the "average" effect, which is accurate for neither men nor women. If we statistically adjust for gender, we will get the "weighted average" effect, which, again, is accurate for neither men nor women. To assess an interaction, one may conduct a stratified analysis to uncover the true effect for each group individually, and to determine whether the measure of association (drug effect on blood pressure, in our example) differs across the levels of the third variable (gender, in our example). Using a multivariable statistical model, the interaction effect may be tested for statistical significance.

Estimating and Avoiding Threats to Reliability

Reliability is threatened whenever a system of measure is unable to produce consistent results when it should. For example, a large epidemiological study of blood pressure may enlist many practitioners in many clinics as data collectors. Even with standardized protocols, differences in equipment and personnel may lead to unreliable measurements.

Several methods are used in epidemiologic research in an attempt to achieve reliable results. A carefully prepared manual of study procedures is necessary for consistency. The manual should provide a detailed description of items such as participant eligibility requirements, variable definitions, clinic visit forms, and precise laboratory methods. Training sessions for data collection staff are necessary to provide instruction on standardized protocols, making sure all staff are aware of and able to perform the techniques uniformly. There are ways to measure inter- and intrarater reliability, which is the degree to which different personnel agree in their measurement of a particular variable and the degree to which an individual will get the same value upon measuring a variable twice (test–retest reliability), respectively. If several different personnel are collecting data, verification that consistent procedures are used is necessary. Laboratories can be checked for reliability by submitting blind replicates of samples and evaluating the concordance of the lab findings.

MOVING FROM ASSOCIATION TO CAUSATION

Throughout this chapter we have avoided "knotty epistemological issues" related to causation, warned against "overly ambitious" interpretation of epidemiological studies, and have assiduously applied the word "association" to describe the relationship between risk factor and disease. But what about *cause*? Although it should be clear from this chapter that the findings of epidemiologists are put to many uses other than discovering the causes of disease, discovering the causes of disease is a vitally important element of public health practice, clinical medicine, and epidemiology. How, then, do we move from association to causation?

This is a difficult question to answer for philosophical, semantic, and practical reasons. For our purposes, we can set aside the weightier philosophical issues (e.g., what, if any, evidentiary standards are sufficient to prove something is true and knowable?)—not because these are uninteresting questions, but because our goals are ultimately pragmatic, and we can at least come to some consensus on what constitutes a sufficient process to establish cause. The semantic issue is best illustrated by example: Do changes in the machinery that drives cell growth and death cause lung cancer? Do carcinogens in cigarette smoke cause lung cancer? Do cigarettes cause lung cancer? Does peer pressure and tobacco product advertising cause lung cancer? This simple web of component causes illustrates that the definition of *cause* in any particular case may not be easy to nail down. Ultimately, this issue can be addressed by not discussing causal relationships in simplistic or abbreviated terms.

We are left to discuss the practical demands of providing a sufficient process to establish a cause and effect relationship. Epidemiologists have typically approached this issue in two related ways. First, as we suggested in the previous study design section, some designs provide stronger evidence for causation than others. The following list ranks various designs from strongest to weakest with respect to their ability to provide evidence of causation:

1. Experimental studies:
   a. Randomized controlled clinical trial

   b. Randomized cross-over clinical trial
   c. Randomized controlled laboratory study
2. Observational studies:
   a. Cohort study
   b. Case—control study
   c. Cross-sectional study
   d. Case series
   e. Case report

This list ranges from little more than anecdotal observation (a case report) to what is considered the "gold standard" of research designs—the randomized controlled clinical trial. The differences between the end members of this continuum should be obvious. A case report might offer an observation of a single risk factor—disease association with potentially no information on the timing or extent of exposure or sources of confounding. Randomized controlled clinical trials generate data from a statistically sufficient sample population in which the degree of exposure to a risk factor is controlled by researchers, and confounding is avoided by randomly assigning participants to exposure groups. Differences in strength between other designs are less profound but are a function of similar considerations. This spectrum of study designs does not provide a definitive procedure for moving from association to causation, but it approximates the real-world progression from hypothesis-generating studies to studies that offer increasingly stronger evidence to support (or reject) a hypothesis of causation.

The other manner in which cause and effect is discussed in epidemiology is epitomized by the work of Sir Austin Bradford Hill. In his seminal 1965 article, Hill outlined nine different dimensions along which a purported causal relationship might be assessed (Hill, 1965). Some of Hill's dimensions will be familiar as they are the same criteria used to evaluate study designs and generate the spectrum of designs listed previously.

- **Strength**: Is the magnitude of association sufficiently large that other risk factors can be ruled out?
- **Consistency**: Has the association been repeatedly observed by different researchers using different experimental designs and analytical methods?
- **Specificity**: Is the association valid only for a specific population group, specific risk factor(s), and specific disease?
- **Temporality**: Does exposure to the risk factor precede the disease?
- **Biological gradient**: Does the association between risk factor and disease exhibit a dose—response relationship?
- **Plausibility**: Is there a credible biological mechanism that might explain the association?
- **Coherence**: Is the association consistent with the known natural history biology of the disease?
- **Experimental evidence**: Does experimentally manipulating the risk factor change the nature of the association?
- **Analogy**: Is there an established cause—effect relationship similar to the association?

Hill is quick to point out that none of these nine "viewpoints" (his own carefully chosen term) can provide indisputable evidence of a causal relationship. In fact, finding counterexamples in the literature to "debunk" Hill's criteria is a trivial exercise—and, given philosophical precedents, ultimately a needless one. Although it is true that a "checklist" approach to establishing causality in epidemiology is problematic (Lanes and Poole, 1984), it is also true that Hill's work [and similar efforts, e.g., Henle-Koch postulates (Evans, 1976)] ought not be considered a method but a heuristic.

Clearly, neither the study design spectrum nor Hill's viewpoints offer a peremptory path from association to causation. Given the complex, multicausal nature of disease and the apparent theoretical limits of science to "prove" causation, moving from association to causation becomes as much a social as a scientific process. In addition to conducting more research and gathering more and better evidence, epidemiologists and their colleagues must discuss and debate purported causal relationships and their translational significance; Hill's viewpoints provide useful sites for such interrogation and deliberation. Scientists who feel uncomfortable with this seemingly unscientific manner of reaching consensus would do well to remember Hill's words that "all scientific work is incomplete" and "all scientific work is liable to be upset or modified by advancing knowledge." Those who seek to protect, promote and restore health and well-being, Hill wrote, cannot "postpone the action[s]" dictated by current knowledge. Ultimately, any discussion of causation points to a quandary faced in one form or another by all those who seek to translate scientific knowledge into practice, including epidemiologists: the overly cautious epidemiologist who eschews public health action because causation has not been definitively proven commits as grave an error as the overly ambitious epidemiologist who initiates public health action predicated on a specious understanding of causation.

   Readers interested in learning more about the theoretical and practical aspects of determining cause in epidemiology are referred to the work of Karhausen (2000) and Parascandola and Weed (2001).

CLINICAL EPIDEMIOLOGY

Clinical epidemiology, as its name implies, represents the marriage of clinical medicine and traditional epidemiology. Clinical epidemiology is "clinical" in two senses—the research is usually conducted in a clinical setting using patients as study subjects, and the findings are typically used by clinicians to make better decisions about patient care. Clinical epidemiology is a major constituent of what has become known as "evidence-based medicine," that is, using the scientific method and its fruits to inform clinical practice. To younger clinicians, this may seem manifest. However, in addition to being a discipline driven by strong traditions, clinical medicine has historically been built upon extrapolations from the basic sciences of physics, chemistry, and biology as well as anatomy and physiology. Clinical epidemiology is often used to test whether the extrapolations from relatively simple model systems to the spectacularly complex systems of humans in their environments are valid.

All of the fundamental epidemiological principles introduced in this chapter apply to clinical epidemiology. Differences between clinical and more traditional epidemiology are largely driven by purpose and context and are best illustrated by example. A traditional epidemiological study of heart disease may follow a cohort of individuals to collect outcome and potential risk factor data. A successful study will find associations between risk factors (e.g., smoking, diet, exercise) and the occurrence of a disease or event (e.g., coronary heart disease, myocardial infarction). Clinical epidemiology, in a sense, picks up where traditional, public health epidemiology leaves off by assessing the consequences of disease and the efficacy of possible interventions. A clinical study may follow a group of patients with coronary heart disease and assess whether a novel treatment provides more relief from the disease than a traditional treatment. It should be obvious from this description that the work of clinical epidemiologists usually falls into the realm of experimental epidemiology. As a result, clinical epidemiologists have at their disposal randomized controlled and cross-over study designs, tools exquisitely suited to advancing both clinical knowledge and practice. Much of the remainder of this volume is devoted to topics central to or strongly allied with clinical epidemiology (see Chapters 3, 6, 7, 17 and 23).

SEX, GENDER, RACE, AND ETHNICITY IN EPIDEMIOLOGY

Epidemiologists often collect data on study participants' sex, race, and ethnicity. The motives for doing this are pragmatic: the constructs of sex, race, and ethnicity may be associated with (or be confounders of) outcomes of interest. Unfortunately, sex, race, and ethnicity have also been (and still are) used as instruments of subordination, subjugation, and discrimination. Many people are justifiably concerned that the very use of such categories in medical research potentially perpetuates harmful discrimination. We define these often misused terms, discuss the grounds for their contentiousness, and offer a reasoned perspective on their role in epidemiological research in the following.

Within the medical professions, "sex" is defined by fairly straightforward biological criteria: females have two X chromosomes; males have one X and one Y chromosome. In humans, there are forms of intersexuality with well understood biological bases; however, these are relatively rare (Sax, 2002). "Gender" refers to notions of what is feminine and what is masculine. Whereas the definition of sex is generally uniform across contemporary cultures, the traits and behaviors associated with gender vary from culture to culture. Although the purpose of some studies may warrant the collection of gender data, epidemiologists typically collect data on sex (albeit self-reported); care should be taken to use the correct term. A "race" is defined as a category of people who share characteristics that are presumed to be biologically inherited. "Ethnicity" describes the state of belonging to a social group defined by a shared cultural heritage. The US National Institutes for Health currently endorses the use of five racial categories (American Indian or Alaska Native, Asian, Black or African American, Native Hawaiian or Other Pacific Islander, White) and two ethnic categories (Hispanic or Latino, not Hispanic or Latino). It is important to note that both race and ethnicity are social constructs—people, after all, must decide which biological traits factor into definitions of race and which cultural criteria delimit ethnicity.

The use of these categories is contentious for a number of reasons. For example, some fear that assigning individuals to these categories places study participants at risk for future discrimination (either inadvertent or intentional). Some believe that maps of the human genome and the advent of genetic epidemiology make categorical distinctions such as race obsolete. We now understand genetic variation to be a continuous phenomenon that cannot demarcate discrete population groups, and genetic variation between traditional racial groups is smaller than within groups (Burchard et al., 2003). Ultimately the genetic characterization of differences or similarities between populations may provide significantly more useful epidemiologic data than crude phenotypic indicators such as skin color. However, others believe traditional racial categories ought to continue to play a role in medical research. Self-identified race is often associated with attitudes toward health and use of health services, and significant disparities in healthcare often occur among self-identified racial categories (Flores et al., 2001). To ignore traditional racial categories is to ignore real-world determinants of public health.

Given the concerns associated with collecting sex, race, and ethnicity data and the dubious veracity of some of these categorical distinctions in light of existing scientific evidence, what role do these concepts have in contemporary epidemiology? Ultimately, these categories have proven to be useful markers of risk in epidemiological studies, and their utility—even in the age of genomics—is unlikely to be surpassed in many circumstances (see Osborne and Feit, 1992 for the important distinction between "risk marker" and "risk factor" in this context). Ignoring these socially constructed categories may hamper discoveries that could reduce health disparities between groups. When making these categorical distinctions, however, epidemiologists must recognize they are simplifying complex and dynamic interactions among biological, social, geographic, and cultural forces. We offer the following recommendations [based on the race/ethnicity work of Kaplan and Bennett (2003), who offer a more thorough discussion] for those using these socially charged categories in epidemiological research:

- When race or ethnicity data are presented, category definitions should be described and these definitions should be justified.
- When sex, race, or ethnicity data are presented, their use should be justified.
- When sex, race, or ethnicity data are presented, all potentially relevant confounding variables (such as socioeconomic status) should be considered in the analyses.

CONCLUSION

Epidemiology is a critically important translational science that serves both public health and clinical medicine. By employing the tools of basic science, epidemiology—whether deployed at the population level or in the clinic—seeks to answer fundamental questions related to the "where," "when," and "who" dimensions of health-related states or events. Armed with such data, community health specialists can design and implement interventions to protect, promote, and restore health and well-being in populations, and clinicians can make better diagnoses and treatment decisions for their patients.

ACKNOWLEDGMENT

The authors thank Amy I. Lynch, PhD, for her valuable contributions to this chapter.

REFERENCES

Belanger, C.F., Hennekens, C.H., Rosner, B., Speizer, F.E., 1978. The nurses' health study. Am. J. Nurs. 78, 1039—1040.

Berenson, G.S., Srnivasan, S.R., Bogalusa Heart Study Group, 2005. Cardiovascular risk factors in youth with implications for aging: the Bogalusa Heart Study. Neurobiol. Aging 26, 303—307.

Bisgaard, H., Szefler, S., 2007. Prevalence of asthma-like symptoms in young children. Pediatr. Pulmonol. 42, 723—728.

Branum, A.M., Collman, G.W., Correa, A., Keim, S.A., Kessel, W., Kimmel, C.A., Klebanoff, M.A., Longnecker, M.P., Mendola, P., Rigas, M., et al., 2003. The National Children's Study of environmental effects on child health and development. Environ. Health Perspect. 111, 642—646.

Burchard, E.G., Ziv, E., Coyle, N., Gomez, S.L., Tang, H., Karter, A.J., Mountain, J.L., Perez-Stable, E.J., Sheppard, D., Risch, N., 2003. The importance of race and ethnic background in biomedical research and clinical practice. N. Engl. J. Med. 348, 1170—1175.

Centers for Disease Control, 1980. Toxic-shock syndrome — United States. MMWR 29, 229—230.

Chenoweth, C.E., Washer, L.L., Obeyesekera, K., Friedman, C., Brewer, K., Fugitt, G.E., Lark, R., 2007. Ventilator-associated pneumonia in the home care setting. Infect. Control Hosp. Epidemiol. 28, 910—915.

Cheung, B.M., Ong, K.L., Man, Y.B., Wong, L.Y., Lau, C.P., Lam, K.S., 2006. Prevalence of the metabolic syndrome in the United States National Health and Nutrition Examination Survey 1999—2002 according to different defining criteria. J. Clin. Hypertens. (Greenwich) 8, 562—570.

Chobanian, A.V., Bakris, G.L., Black, H.R., Cushman, W.C., Green, L.A., Izzo Jr., J.L., Jones, D.W., Materson, B.J., Oparil, S., Wright Jr., J.T., et al., 2003. Seventh report of the Joint National Committee on prevention, detection, evaluation, and treatment of high blood pressure. Hypertension 42, 1206—1252.

Davies, H.T., Crombie, I.K., Tavakoli, M., 1998. When can odds ratios mislead? BMJ 316, 989—991.

Dawber, T.R., Meadors, G.F., Moore Jr., F.E., 1951. Epidemiological approaches to heart disease: the Framingham Study. Am. J. Public Health Nations Health 41, 279—281.

Delgado-Rodriguez, M., Llorca, J., 2004. Bias. J. Epidemiol. Community Health 58, 635—641.

Doll, R., Hill, A.B., 1956. Lung cancer and other causes of death in relation to smoking; a second report on the mortality of British doctors. BMJ 2, 1071—1081.

Ernster, V.L., 1994. Nested case-control studies. Prev. Med. 23, 587—590.

Evans, A.S., 1976. Causation and disease: the Henle—Koch postulates revisited. Yale J. Biol. Med. 49, 175—195.

Flores, G., Fuentes-Afflick, E., Carter-Pokras, O., Claudio, L., Lamberty, G., Lara, M., Pachter, L., Ramos Gomez, F., Mendoza, F., Valdez, R.B., et al., 2001. Why ethnicity and race are so important in child health services research today. Arch. Pediatr. Adolesc. Med. 155, 1178—1179.

Ha, M.H., Lee, D.H., Jacobs, D.R., 2007. Association between serum concentrations of persistent organic pollutants and self-reported cardiovascular disease prevalence: results from the National health and Nutrition Examination survey, 1999–2002. Environ. Health Perspect. 115, 1204–1209.

Hemann, B.A., Bimson, W.F., Taylor, A.J., 2007. The Framingham Risk Score: an appraisal of its benefits and limitations. Am. Heart Hosp. J. 5, 91–96.

Herrington, D.M., Reboussin, D.M., Brosnihan, K.B., Sharp, P.C., Shumaker, S.A., Snyder, T.E., Furberg, C.D., Kowalchuk, G.J., Stuckey, T.D., Rogers, W.J., et al., 2000. Effects of estrogen replacement on the progression of coronary-artery atherosclerosis. N. Engl. J. Med. 343, 522–529.

Hill, A.B., 1965. The environment and disease: association or causation? Proc. R. Soc. Med. 58, 295–300.

Hulley, S., Grady, D., Bush, T., Furberg, C., Herrington, D., Riggs, B., Vittinghoff, E., 1998. Randomized trial of estrogen plus progestin for secondary prevention of coronary heart disease in postmenopausal women. Heart and Estrogen/progestin Replacement Study (HERS) Research Group. JAMA 280, 605–613.

Kaplan, J.B., Bennett, T., 2003. Use of race and ethnicity in biomedical publication. JAMA 289, 2709–2716.

Karhausen, L.R., 2000. Causation: the elusive grail of epidemiology. Med. Health Care Philos. 3, 59–67.

Kostis, W.J., Demissie, K., Marcella, S.W., Shao, Y.H., Wilson, A.C., Moreyra, A.E., Myocardial Infarction Data Acquisition System (MIDAS 10) Study Group, 2007. Weekend versus weekday admission and mortality from myocardial infarction. N. Engl. J. Med. 356, 1099–1109.

Lanes, S.F., Poole, C., 1984. Truth in packaging? The unwrapping of epidemiologic research. J. Occup. Med. 26, 571–574.

Langholz, B., Thomas, D.C., 1990. Nested case-control and case-cohort methods of sampling from a cohort: a critical comparison. Am. J. Epidemiol. 131, 169–176.

Lawlor, D.A., Smith, G.D., Ebrahim, S., 2004. Socioeconomic position and hormone replacement therapy use: explaining the discrepancy in evidence from observational and randomized controlled trials. Am. J. Public Health 94, 2149–2154.

Mastin, J.P., 2005. Environmental cardiovascular disease. Cardiovasc. Toxicol. 5, 91–94.

Naylor, C.D., Chen, E., Strauss, B., 1992. Measured enthusiasm: does the method of reporting trial results alter perceptions of therapeutic effectiveness? Ann. Intern. Med. 117, 916–921.

Olijhoek, J.K., van der Graaf, Y., Haffner, S.M., Visseren, F.L., For the SMART Study Group, 2007. Defining the metabolic syndrome: resolving unresolved issues? Eur. J. Intern. Med. 18, 309–313.

Osborne, N.G., Feit, M.D., 1992. The use of race in medical research. JAMA 267, 275–279.

Parascandola, M., Weed, D.L., 2001. Causation in epidemiology. J. Epidemiol. Community Health 55, 905–912.

Prentice, R.L., 1986. A case-cohort design for epidemiologic cohort studies and disease prevention trials. Biometrika 73, 1–11.

Sax, L., 2002. How common is intersex? A response to Anne Fausto-Sterling. J. Sex. Res. 39, 174–178.

Schuchat, A., Broome, C.V., 1991. Toxic shock syndrome and tampons. Epidemiol. Rev. 13, 99–112.

Semrad, T.J., O'Donnell, R., Wun, T., Chew, H., Harvey, D., Zhou, H., White, R.H., 2007. Epidemiology of venous thromboembolism in 9489 patients with malignant glioma. J. Neurosurg. 106, 601–608.

Shands, K.N., Schlech IIIrd, W.F., Hargrett, N.T., Dan, B.B., Schmid, G.P., Bennett, J.V., 1982. Toxic shock syndrome: case-control studies at the Centers for Disease Control. Ann. Intern. Med. 96, 895–898.

Smith, G.D., 2004. Classics in epidemiology: should they get it right? Int. J. Epidemiol. 33, 441–442.

Stampfer, M.J., Colditz, G.A., 1991. Estrogen replacement therapy and coronary heart disease: a quantitative assessment of the epidemiologic evidence. Prev. Med. 20, 47–63.

Todd, J., Fishaut, M., Kapral, F., Welch, T., 1978. Toxic-shock syndrome associated with phage-group-I Staphylococci. Lancet 2, 1116–1118.

Todd, J.K., 1982. Toxic shock syndrome: a perspective through the looking glass. Ann. Intern. Med. 96, 839–842.

Wacholder, S., 1991. Practical considerations in choosing between the case-cohort and nested case-control designs. Epidemiology 2, 155–158.

Waters, D.D., Alderman, E.L., Hsia, J., Howard, B.V., Cobb, F.R., Rogers, W.J., Ouyang, P., Thompson, P., Tardif, J.C., Higginson, L., et al., 2002. Effects of hormone replacement therapy and antioxidant vitamin supplements on coronary atherosclerosis in postmenopausal women: a randomized controlled trial. JAMA 288, 2432–2440.

Wilson, P.W., D'Agostino, R.B., Levy, D., Belanger, A.M., Silbershatz, H., Kannel, W.B., 1998. Prediction of coronary heart disease using risk factor categories. Circulation 97, 1837–1847.

RECOMMENDED RESOURCES

Armitage, P., Berry, G., Matthews, J.N.S., 2002. Statistical Methods in Medical Research. Blackwell Science, Malden, MA.

Aschengrau, A., Seage IIIrd, G.R., 2008. Essentials of Epidemiology in Public Health. Jones and Barlett, Sudbury, MA.

Brookmeyer, R., Stroup, D.F. (Eds.), 2004. Monitoring the Health of Populations: Statistical Principles and Methods for Public Health Surveillance. Oxford University Press, New York.

Teutsch, S.M., Churchill, R.E. (Eds.), 2000. Principles and Practice of Public Health Surveillance. Oxford University Press, New York.

Chapter 5

The Patient-Centered Outcomes Research Institute: Current Approach to Funding Clinical Research and Future Directions

Joe V. Selby and Danielle M. Whicher

Patient-Centered Outcomes Research Institute, Washington, DC, United States

Chapter Outline

Key Points

- Patient-centered research addresses questions and examines outcomes that are important to patients, their caregivers, and clinicians.
- Comparative effectiveness research (CER) compares the benefits and harms of two or more options available to patients and clinicians.
- Pragmatic research aims to produce practical information and outcomes that are important to patients and other end users without imposing few, if any, additional burdens on study participants and clinicians.
- Pragmatic patient-centered comparative effectiveness research compares two or more active well-defined interventions and generates results that are easier to interpret, apply, and reproduce, and in turn, the evidence generated is more likely to be implemented in clinical care.

- Innovative research designs such as cluster randomization, adaptive designs, and platform trials are recommended to increase study efficiency, reduce study costs, and make studies more feasible.
- Cluster randomization trials study the effect of interventions on an entire group of people instead of individuals.
- Adaptive trials allow changes to be made to the research design while it is underway in response to accumulating data collected from trial participants.
- Platform trials are adaptive research projects that study different interventions for the same clinical condition.
- Heterogeneity of treatment effect is the idea that specific interventions will have differential effectiveness, safety, or comparative effectiveness on different subgroups of patients.
- Patient partners are individuals with a lived experience with a condition of interest as well as their family members, caregivers, and advocacy organizations.
- Stakeholder partners are members of constituencies with knowledge of the health condition and interest in the research question based on professional, rather than personal, experience.
- The national patient-centered outcomes research network (PCORnet) supports collaborative multinetwork pragmatic research that is both observational and interventional, and patient-powered research networks.
- Multinetwork pragmatic research may utilize a *common data model* which transforms the data at each network site into a database with common architecture, data fields, and tables, and *distributed queries* yield summary reports of the data pooled at different sites which can then be analyzed.
- PCORnet is funded by PCORI (Patient-Centered Outcomes Research Institute) and involves 13 clinical data research networks, 20 patient-powered research networks, more than 90 million patients, and nearly 100 health-care delivery systems.

INTRODUCTION: THE PATIENT-CENTERED OUTCOMES RESEARCH INSTITUTE

The legislative language that created the Patient-Centered Outcomes Research Institute (PCORI) in 2010 signaled both political and scientific concern that the current clinical research enterprise was not adequately serving the needs of patients, clinicians, or other decision-makers. Rising health-care costs, rapid adoption of new diagnostic and treatment options without solid evidence of added benefit, evidence of wide geographic variations in clinical practice, and a growing recognition that clinical decisions are often made without good clinical evidence (Orszag and Ellis, 2007; Tricoci et al., 2009) despite huge annual US investments in clinical research set the stage for bipartisan political interest in the more systematic application of comparative clinical effectiveness research to inform practice and policy-making. Additional scientific concerns included the high costs of conducting traditional clinical research, especially randomized trials; the growing recognition that evidence from relatively small randomized studies in select populations was inadequate for guiding treatment decisions in diverse patient populations; and an appreciation that patients and clinicians worry about a broader range of outcomes than those typically measured in clinical studies.

The legislation specified that PCORI was to focus on funding patient-centered comparative effectiveness research (CER) (Compilation of Patient Protection and Affordable Care Act). It recognized that randomized trials, observational studies, and evidence syntheses would be part of the PCORI portfolio. Several additional specifications ensure that PCORI and PCORI-funded research will differ from that of other major funders, particularly the National Institutes of Health (NIH) and the life sciences industry. PCORI's purpose and mission were defined precisely as serving the evidence needs of patients, caregivers, clinicians, payers, and policy-makers. This differs greatly from the important missions of the NIH and industry-sponsored research. PCORI was instructed to regularly involve patients and all key stakeholders in the course of planning, conducting, and disseminating research findings (see Methodology Standards for Patient-Centered Comparative Effectiveness Research section). It was told to issue contracts for research rather than grants, reflecting the importance of getting the research questions right and ensuring that findings are delivered and shared in a timely manner. It was also instructed to consider the possibility that the relative effectiveness of alternative approaches to screening, diagnosis, treatment, or management might differ among patient subgroups—defined by age, gender, level of comorbid illness, education, or genetic or molecular markers. Lastly, PCORI was told that it would be evaluated in 2018 to see whether the research it funded was beginning to change practice, reduce variation and disparities, and improve patient outcomes.

PCORI's Board of Governors, Methodology Committee (Defining Features of Pragmatic Research section), and staff leaders have taken the authorizing legislation extremely seriously. Understanding and meeting these legislative requirements has required discussion and debate, innovation, and iterative improvements in how PCORI solicits, funds, and manages its research portfolio. This chapter defines patient-centered CER and PCORI's approach to funding high-quality research that is useful to patients and other stakeholders. It describes the results of the decisions made and implemented by PCORI as of the end of 2015, 5 years after PCORI's Board of Governors was first convened. These decisions have important implications for the conduct of clinical research and for the training of the next generation of clinical researchers.

PATIENT-CENTERED COMPARATIVE EFFECTIVENESS RESEARCH

Defining Patient-Centered Comparative Effectiveness Research

PCORI's authorizing legislation defines CER, but does not define "patient-centered outcomes research" (Compilation of Patient Protection and Affordable Care Act). The term "patient-centered" was coined in 1988 by the Commonwealth Picker Fund (Gerteis et al., 1993) to call attention to the need for clinicians, staff, and health-care systems to focus more on the needs of patients and family members in the setting of clinical care. The Institute of Medicine later defined patient-centeredness as "care that is respectful of and responsive to individual patient preferences, needs, and values, and ensuring that patient values guide all clinical decisions" (Committee on Quality of Health Care in America Institute of Medicine, 2001). The term "patient-centered research" was first seen almost as long ago (Noble, 1989; Feussner, 1999), but a definition was not developed. However, the definition of patient-centeredness as being responsive to individual patient needs can also be applied to the activity of clinical research. Building on this notion, PCORI has defined patient-centered research as research that addresses questions and examines outcomes that are important to patients, their caregivers, and clinicians (Table 5.1). The goal of patient-centered research is to generate evidence that will help patients, caregivers, and clinicians make more informed and personalized health-care decisions by addressing the following types of questions:

- Given my personal characteristics, conditions, and preferences, what should I expect will happen to me?
- What are my options, and what are the potential benefits and harms of those options?
- What can I do to improve the outcomes that are most important to me?
- How can clinicians and the care delivery systems they work in help me make the best decisions about my health and health care (Patient-Centered Outcomes Research Institute, 2013a)?

This definition conveys the importance of measuring a range of outcomes, including patient-reported outcomes, so that patients will have the information of most importance to them. It emphasizes the critical roles of the patient and of health-care systems in achieving the outcomes of most importance to patients. Finally, it recognizes that the comparative effectiveness of various options may differ for individual patients, depending on personal, social, clinical, and genetic characteristics, as well as personal preferences.

PCORI operationalizes this definition in its solicitations and reviews of research applications by focusing on two simple questions: is the comparative research question one that matters to patients; and do the outcomes being assessed matter to patients? Although not all patient-centered outcomes research is comparative, PCORI is limited to funding CER: research that compares the benefits and harms of two or more options available to patients and clinicians. PCORI funds comparative studies of interventions or their components that have been previously studied in at least one adequately powered efficacy study or alternatively that are currently used in clinical practice without solid evidence of efficacy. The options being compared must additionally represent realistic and well-defined clinical choices faced by patients, providers, or systems. In the case of multicomponent system-level interventions, exact constellations of components may not have been previously evaluated, but the components are options that have been used previously by systems and could reasonably be combined.

CER can be contrasted with efficacy research, which aims to establish whether a new intervention can have an effect on health outcomes under ideal conditions, i.e. in carefully controlled environments and, often times, in patients without other illnesses than that being treated in the study. Unlike efficacy research, CER aims to determine which of the interventions being compared is best for a general population of patients and for particular, previously hypothesized, subgroups of patients (Federal Coordinating Council for Comparative Effectiveness Research, 2009). To accomplish this, CER studies should enroll patients representative of all individuals with the clinical condition of interest who may be exposed to the interventions being studied. It should ideally be conducted in settings where those patients normally receive their medical care (Committee on Comparative Effectiveness Research Prioritization, 2009; Sox and Greenfield, 2009). To see if there

TABLE 5.1 Key Features of Patient-Centered Comparative Effectiveness Research

- Fills evidence gaps that are important to patients and other relevant stakeholders
- Compares two or more efficacious and commonly used options for prevention, diagnosis, or treatment or for organizing health-care services
- Considers the range of clinical outcomes relevant to patients
- Enrolls representative populations from settings where those patients typically receive their medical care
- Attends to differences in effectiveness and preferences across patient subgroups

are differences in the comparative effectiveness of interventions in subgroups of patients with common characteristics—also referred to as "heterogeneity of treatment effect"—CER studies should prespecify subgroups of interest and provide an analytic plan for examining possible differences. This usually requires enrolling larger study samples than in efficacy studies.

Establishing National Priority Areas for Patient-Centered Comparative Effectiveness Research

PCORI's authorizing legislation also required that the institute develop national priorities for research to guide its funding initiatives. Using the abovementioned definition of patient-centered CER, PCORI worked with its Board of Governors and stakeholder communities to develop a set of five research priority areas. Prior to finalizing its National Priorities for Research, the institute invited public comment on the set of draft priorities for a 53-day period from January through March, 2012 (Patient-Centered Outcomes Research Institute). PCORI used that feedback to further refine those priorities.

Rather than focusing on particular clinical conditions, these priorities recognize that there are cross-cutting areas of inquiry in which further research could lead to significant improvements in health-care delivery and health outcomes (Table 5.2). PCORI's National Priorities for Research call not only for research comparing different options for preventing, diagnosing, and treating clinical conditions in individual patients, but also for research that compares different approaches for organizing and delivering care at the health-system level; for reducing or eliminating health-care disparities; and for rapidly and effectively disseminating the findings from CER studies to patients, caregivers, and clinicians to help in making more informed health-care decisions. Finally, to efficiently implement methodologically robust CER, there is a need to strengthen both the research methods and the national infrastructure for conducting efficient CER. PCORI's research priorities recognize the need to build research capacity and to improve methods for engaging all stakeholders in research, for prioritizing research questions and for conducting and analyzing the results of CER (Table 5.1) (Fleurence et al., 2015).

PCORI uses these National Priorities for Research to shape its research agenda. Within each priority areas, PCORI works with a multistakeholder Advisory Panel to further identify and prioritize specific evidence gaps (Patient-Centered Outcomes Research Institute, 2015a).

PATIENT AND STAKEHOLDER ENGAGEMENT IN RESEARCH

PCORI's Efforts to Engage Patients and Stakeholders in Setting Research Priorities and Selecting Research Applications for Funding

PCORI became convinced early on that if its research was to meet the needs of patients, their caregivers, clinicians, and other important stakeholders, including hospitals and health-care systems, insurers and employers, policy-makers, and researchers, it was important to involve these stakeholders in every aspect of the research process. Put another way, if

TABLE 5.2 PCORI's National Priorities for Patient-Centered Comparative Clinical Effectiveness Research

Research Priority	Description
Assessment of options for prevention, diagnosis, and treatment	Research comparing interventions delivered to individual patients to prevent, diagnosis, or treat a specified medical condition
Improving health-care systems	Research comparing approaches to organizing and delivering health-care services to groups of patients
Addressing disparities	Research comparing interventions that aim to reduce or eliminate health disparities that exist between groups of patients based on characteristics such as race, geographic location, or socioeconomic status
Dissemination and communication research	Research comparing approaches to improving the communication and effective use of findings generated by patient-centered CER
Accelerating patient-centered outcomes research and methodology	Activities designed to improve the infrastructure and research methods available to support patient-centered CER

PCORI hoped that its research findings would be disseminated and implemented to change practice, the stakeholders who would do the implementing should be convinced that the study addressed the right questions and was conducted and analyzed appropriately. This can be accomplished most reliably by involving these stakeholders from the beginning. While other funders, notably the Agency for Healthcare Research and Quality, had pioneered the concept of involving stakeholders in identifying research needs, PCORI has extended the concept of engagement rather dramatically.

Engagement begins with PCORI's Board of Governors, a 21-person body recruited and appointed by the Comptroller General of the United States and comprised of representatives of stakeholder groups specified in PCORI's legislation. These include patients and health-care consumers; clinicians and health-care systems; insurers; pharmaceutical, device, and diagnostic manufacturers; researchers; and federal or state government agencies (Patient-Centered Outcomes Research Institute, 2013b). Thus, engagement and collaboration among stakeholders is modeled first at the level of the Board. PCORI actively solicits CER questions from patients and stakeholder groups through a variety of mechanisms (Selby et al., 2015). Patients, clinicians, and other nonscientist stakeholders then join with PCORI on its Advisory Panels, which work to prioritize the submitted research questions.

Consistent with its authorizing legislation, PCORI has formed Advisory Panels corresponding to its National Priorities for Research areas, including Advisory Panels on the Assessment of Prevention, Diagnosis, and Treatment Options, on Improving Healthcare Systems, on Addressing Disparities, and on Communication and Dissemination Research. When these advisory panels review research topics, a critical criterion considered is whether the proposed research topic, if addressed, would be likely to change current practice. Panelists consider whether there is a current gap in evidence that is leading to unwarranted practice variation; whether clinician specialists, payers, or others have asked for better evidence in this area; whether the question and the evidence generated would endure, or whether a new technology is expected to soon render the question moot; and whether other barriers to practice change would prevent the uptake of evidence, even if strong. PCORI has established other Advisory Panels to assist in achieving its mission. These include Advisory Panels on Clinical Trials, Rare Diseases, and Patient Engagement. These panels provide critical guidance to PCORI in these three areas but do not prioritize research topics.

The same stakeholder groups also serve on each peer review, or "merit review," panel convened by PCORI to review submitted research applications, comprising ~50% of reviewers. The merit review criteria include a criterion for patient-centeredness: whether the application addresses a question that is relevant to patients and their caregivers and whether it proposes to measure the outcomes that matter to patients; and a criterion on patient and stakeholder engagement: whether the applicant has involved the appropriate stakeholders, either as individuals or as organizations or both, in the design of the study and whether these stakeholders will continue to be involved as the study is conducted and results are analyzed and disseminated.

Engaging Patients and Stakeholders in Conducting Research

With the single exception of proposals for projects that address analytic methods for CER, PCORI requires that investigators submitting applications to PCORI engage with patients and other stakeholders when designing, conducting, and disseminating the results of their proposed research study. PCORI defines patient partners as individuals with a lived experience with the condition of interest as well as their family members and caregivers. Professional and advocacy organizations which represent those individuals also fall within this definition. There are advantages to involving individual patients in that their experiences can be brought into deliberations on formulating the study question, designing the study, selecting appropriate outcomes, and developing effective strategies for recruitment and retention. There are also important advantages to involving advocacy organizations (both patient-related and professional organizations) whose expertise includes an understanding of the entire disease spectrum, the policy implications of research questions, and the potential barriers to dissemination and implementation. These organizations can play critical roles in packaging study results in formats useful to those they represent. For PCORI's larger awards, the Pragmatic Clinical Studies, and the targeted funding announcements, engagement of both individuals and relevant organizations is required. PCORI works with awardees to help identify and involve the critical organizations.

In addition to patients and patient advocacy organizations, PCORI requires investigators to engage other stakeholder partners, defined as members of constituencies with knowledge of the health condition and interest in the research question based on professional, rather than personal experience (Patient-Centered Outcomes Research Institute, 2015b). These stakeholder partners represent other end users of the information generated by the study. Depending on the specific objective of the research, relevant stakeholders often include clinicians, delivery systems, insurers and, in some cases, employers, drug or device manufacturers, and other policy-makers. These representatives bring important perspectives on the significance of different evidence gaps, the types of evidence needed to address those gaps, and potentially, data that are available that could be leveraged in conducting the needed research.

PCORI's Rubric for Patient and Stakeholder Engagement

For clinical researchers, these requirements for engagement may at first appear daunting. For experienced researchers accustomed to thinking of patients principally as subjects of research or possibly as beneficiaries of research, the prospect of working with them as research partners is unfamiliar. Establishing and sustaining working relationships with existing, well-established advocacy organizations can be challenging. Additionally, engagement is time and resource-intensive. PCORI has taken multiple steps to support engagement in the research that it funds. An engagement rubric is available on PCORI's Website (Patient-Centered Outcomes Research Institute, 2015a,b,c,d,e,f) and within every funding announcement and online application form. The rubric outlines the rationale, types, and examples of engagement and can be seen as a "best practices" document. It identifies different decision points during the planning and conduct of a study, as well as during the dissemination of a study's results, where soliciting the input of patients and stakeholders can be influential. Additionally, the engagement rubric outlines a set of principles for effective engagement, which include reciprocal relationships, colearning, partnerships, transparency, honesty, and trust (Table 5.3). These principles are meant to guide researcher's decisions about how to structure meaningful engagement activities over the life cycle of a study.

In addition to the engagement rubric, PCORI provides training videos on the topic of engagement as well as a framework for structuring financial compensation of patients engaged in PCORI-funded studies (Patient-Centered Outcomes Research Institute). To facilitate successful partnerships between researchers, patients, and other stakeholders that can lead to the development of patient-centered research proposals, PCORI also offers specific funding under the Eugene Washington Engagement Awards program (Patient-Centered Outcomes Research Institute, 2015c) and also under its Pipeline to Proposals program (Patient-Centered Outcomes Research Institute).

While PCORI provides general guidance for effective engagement of patients and other stakeholders, it recognizes that the best engagement strategies will vary from project to project and that there is no one correct way to engage patients and other stakeholders in every study. The structure, level, and frequency of engagement activities depend on a number of factors including the phase of the research project, features of the research design, the health condition and interventions being examined, and the availability of the patient and stakeholder partners. Some examples of engagement strategies include fielding surveys or collecting qualitative data to inform the research question and study design; including patient

TABLE 5.3 PCORI Principles for Effective Engagement of Patient and Stakeholder Partners in Research

PCORI Principles	Description
Reciprocal relationships	This principle is demonstrated when the roles and decision-making authority of all research partners, including the patient and other stakeholder partners, are defined collaboratively and clearly stated.
Colearning	This principle is demonstrated when the goal is not to turn patients or other stakeholder partners into researchers, but to help them understand the research process; likewise, the research team will learn about patient-centeredness and patient/other stakeholder engagement and will incorporate patient and other stakeholder partners into the research process.
Partnerships	This principle is demonstrated when time and contributions of patient and other stakeholder partners are valued and demonstrated in fair financial compensation, as well as in reasonable and thoughtful requests for time commitment by patient and other stakeholder partners. When projects include priority populations, the research team is committed to diversity across all project activities and demonstrates cultural competency, including disability accommodations, when appropriate.
Trust, transparency, and honesty	These principles are demonstrated when major decisions are made inclusively and information is shared readily with all research partners. Patients, other stakeholders, and researchers are committed to open and honest communication with one another.

and stakeholder partners as coinvestigators and active members of the study team, sometimes with decision-making authority; and forming patient and stakeholder advisory panels that convene at critical decision points over the course of a study to provide guidance to the study team.

PCORI continues to learn from its own experiences with patient and stakeholder engagement in prioritizing research topics and selecting research proposals for funding as well as from the experience of PCORI-funded investigators. As knowledge and experience with engagement increases, PCORI works internally to improve its own processes for patient and stakeholder engagement and to further refine its guidance to the research, patient, and other stakeholder communities.

METHODOLOGY STANDARDS FOR PATIENT-CENTERED COMPARATIVE EFFECTIVENESS RESEARCH

PCORI's Methodology Committee

PCORI's authorizing legislation also created a Methodology Committee to promote the implementation and conduct of robust patient-centered CER. The Methodology Committee is comprised of 17 members who were appointed by the Comptroller General (Patient-Centered Outcomes Research Institute, 2015d). The role of this committee is to make recommendations to PCORI's Board of Governors and PCORI staff on the appropriate use of various methods in patient-centered CER, on methodological standards for patient-centered CER, and on priorities for improving research methods or their application (Patient-Centered Outcomes Research Institute, 2015e). Once the committee was established, one of its first tasks was to develop a Methodology Report that presented a set of Methodology Standards for the conduct of patient-centered CER. The first version of the Methodology Report was released in 2013. It is regularly revised and updated by the members of the Methodology Committee, with the support of PCORI staff.

PCORI's Methodology Standards

The Methodology Standards and accompanying report reflect current accepted best practice for identifying important research questions, designing high-quality research, implementing research, and analyzing its results. As a group, they are meant to serve as minimal standards for the conduct of patient-centered CER. The initial set of Methodology Standards was comprised of 47 unique standards, falling within 11 distinct categories (Table 5.4). Of these, five categories describe cross-cutting standards, which apply to any CER study. The remaining categories describe standards that apply to specific types of research designs and study methods. Each of these categories contains two to six individual standards.

The individual standards described in the initial version of PCORI's Methodology Report reflect areas where there are often inconsistencies or deficiencies in the ways in which research methods are applied and where there is evidence supporting the appropriate use of those methods (Hickam et al., 2013).

To promote the adoption of these standards, PCORI requires that investigators describe their plans for adhering to all relevant PCORI Methodology Standards in their research applications. The standards are used by the PCORI staff and by merit reviewers to assess the quality of research applications submitted to PCORI. Once studies are funded, PCORI staff overseeing each funded research project periodically assess whether those projects are adhering to the Methodology Standards. PCORI staff work with the Methodology Committee to identify areas where additional standards may be needed to support the conduct of methodologically robust CER.

TABLE 5.4 Categories of PCORI's Initial Methodology Standards

Cross-cutting Standards	Standards for Specific Study Designs and Methods
Formulating research questions	Data networks
Patient-centeredness	Data registries
Data integrity and rigorous analyses	Adaptive and Bayesian trial designs
Preventing and handling missing data	Causal inference
Heterogeneity of treatment effects	Studies of diagnostic tests
	Systematic reviews

PRAGMATIC RESEARCH

Defining Features of Pragmatic Research

The concept of pragmatic research is closely aligned with that of CER. Both CER and pragmatic research address practical questions faced by patients and clinicians (Luce et al., 2009; Tunis et al., 2003). The goal of both is to generate evidence that can be immediately implemented to improve health care. Research questions for each relate not to whether or how an intervention works in an ideal situation—often characterized as explanatory research (Schwartz and Lellouch, 1967)—but rather to whether interventions work in the routine clinical settings and in the broad range of patients found in real-world clinical practice (Mullins et al., 2010; Zwarenstein et al., 2008). For instance, patients with comorbidities and those less likely to adhere to the assigned or prescribed intervention are often excluded from explanatory studies, but would be included in pragmatic studies. Pragmatic research tends toward fewer exclusion criteria, a wider range of clinical settings, simpler data collection plans, and inclusion of a broader range of clinicians and delivery systems. It imposes few, if any, additional burdens on study participants and clinicians (Luce et al., 2009; Thorpe et al., 2009). As with CER, pragmatic studies are ideally conducted within and integrated into clinical practice. As with CER, pragmatic research includes the range of outcomes important to patients, clinicians, and other end users (Thorpe et al., 2009; Califf and Sugarman, 2015).

In an effort to characterize the various design features and trade-offs that researchers ought to consider when designing pragmatic research studies, Thorpe et al. (2009) developed a framework referred to as the pragmatic-explanatory continuum indicator summary (PRECIS) and more recently published a revised version of this framework referred to as PRECIS-2 (Loudon et al., 2015). The PRECIS-2 tool identifies nine domains that researchers should consider when designing pragmatic research and illustrates how design features can vary from one end to the other of the explanatory—pragmatic continuum. The tool demonstrates that most research will incorporate some more pragmatic and some more explanatory features depending on the research question and the needs of end users of the evidence generated (Loudon et al., 2015). While patient-centered CER can exist anywhere along this spectrum, PCORI encourages researchers to consider integrating pragmatic features when appropriate and feasible.

Because pragmatic research is generally designed to compare alternative interventions in diverse patient groups and clinical settings, these studies generally need large sample sizes so that they are adequately powered to detect small yet clinically meaningful differences in patient-centered outcomes and possible differences in treatment effects for various subgroups of the population. The involvement of diverse populations and diverse clinical settings, including community hospitals and practices and safety-net clinics, can create challenges for implementing pragmatic research. For example, investigators will need to determine how to harmonize clinical data collected across the different study sites, will have to train research staff at each of the sites in the study protocol, and may have to have research protocols and materials approved by multiple institutional review boards (IRBs) (Chalkidou et al., 2012) although requirements for using a single IRB in multisite studies are being strengthened at PCORI and elsewhere (National Institutes of Health (2015)). Additionally, community clinics and other nonacademic settings that do not typically participate in clinical research often lack the infrastructure to support it. Networks of delivery sites that adopt a common data platform and make arrangements for streamlined contracting and IRB oversight offer the potential for repeated use and marked reductions in the costs of conducting clinical research (Chalkidou et al., 2012).

PCORI's Efforts to Support Pragmatic Research

Despite an extensive literature on the concept of pragmatic research, relatively few pragmatic studies have actually been conducted in the United States to date. Given the central importance of this concept to the field of CER, PCORI has established a funding initiative that solicits and supports large-scale, pragmatic, patient-centered CER designed to address critical evidence gaps identified and prioritized by PCORI's Advisory Panels as well as by other key stakeholders including the IOM and AHRQ (Selby, 2013). Studies typically compare two or more active, well-defined interventions. Occasionally, and especially in studies of system-level interventions, the most meaningful comparator is "usual care." However, "usual care" must also be carefully defined and applicants must take care to ensure and demonstrate that patients in that study arm received care as defined. Research that compares well-defined interventions generates results that are easier to interpret, apply, and reproduce; as a result, the evidence generated is more likely to be implemented in clinical care (Patient-Centered Outcomes Research Institute, 2015a,b,c,d,e,f). To date, PCORI has invested over $175 million to support large pragmatic research studies. Funded studies (Table 5.5) address important evidence gaps in a wide variety of clinical areas from screening for breast cancer to treatment for acute low back pain.

In addition to PCORI's investment in the area of pragmatic research, the National Institutes of Health (NIH) currently funds an initiative, referred to as the NIH Health Care Systems Research Collaboratory, that supports a number of systems-based pragmatic trial projects (NIH Health Care Systems Research Collaboratory). These projects address important issues in public health and health care, such as how to prevent suicide in adults who experience suicidal thoughts (NIH Health Care Systems Research Collaboratory), and are also meant to facilitate the development of best practices for conducting pragmatic research moving forward. Further, to allow for colearning between the NIH and PCORI, the NIH has established a close partnership with the PCORI-funded national patient-centered clinical research network, commonly referred to as PCORnet (see The National Patient-Centered Outcomes Research Network section).

TABLE 5.5 First 14 Pragmatic Clinical Studies Funded by PCORI

Study Title	Institution	Study Design
Targeted Interventions to Prevent Chronic Low Back Pain in High Risk Patients: A Multi-site Pragmatic RCTs	University of Pittsburgh	Cluster randomized controlled trial
Early Supported Discharge for Improving Functional Outcomes after Stroke	Wake Forest University Health Sciences	Cluster randomized controlled trial
Enabling a Paradigm Shift: A Preference-Tolerant RCT of Personalized versus Annual Screening for Breast Cancer	University of California, San Francisco	Randomized controlled trial
Pragmatic Trial of More versus Less Intensive Strategies for Active Surveillance of Patients with Small Pulmonary Nodules	Kaiser Foundation Research Institute, Division of Kaiser Foundation Hospitals	Cluster randomized controlled trial
A Pragmatic Trial to Improve Colony Stimulating Factor Use in Cancer	Fred Hutchinson Cancer Research Center	Cluster randomized controlled trial
Pragmatic Randomized Trial of Proton versus Photon Therapy for Patients with Stage II or III Breast Cancer	University of Pennsylvania	Randomized controlled trial
Integrating Patient-Centered Exercise Coaching into Primary Care to Reduce Fragility Fracture	Pennsylvania State University Hershey Medical Care	Randomized controlled trial
Anti-TNF Monotherapy versus Combination Therapy with Low Dose Methotrexate in Pediatric Crohn's Disease	University of North Carolina at Chapel Hill	Randomized controlled trial
A Practical Intervention to Improve Patient-Centered Outcomes after Hip Fractures among Older Adults (the Regain Trial)	University of Pennsylvania	Randomized controlled trial
Mobility: Improving Patient-Centered Outcomes among Overweight and Obese Youth with Bipolar Spectrum Disorders Treated with Second-Generation Antipsychotics	University of Cincinnati	Randomized controlled trial
Comparing Outcomes of Drugs and Appendectomy (CODA)	University of Washington	Randomized controlled trial
Integrated versus Referral Care for Complex Psychiatric Disorders in Rural Federally Qualified Health Centers	University of Washington	Randomized controlled trial
Integrating Behavioral Health and Primary Care	University of Vermont and State Agricultural College	Cluster randomized controlled trial
Comparative Effectiveness of Pulmonary Embolism Prevention after Hip and Knee Replacement (PEP-PER): Balancing Safety and Effectiveness	Medical University of South Carolina	Randomized controlled trial

Study Designs for Pragmatic Research and Other Comparative Effectiveness Research

Depending on the research topic and question, pragmatic research may use randomized or observational study designs. PCORI encourages investigators to use randomized designs when feasible and ethically appropriate, and particularly when observational analyses would not provide evidence of sufficient validity to justify practice changes. Approximately two-thirds of PCORI's entire CER portfolio involves random assignment of interventions, and essentially all of the larger pragmatic clinical studies involve randomization.

However, there are many situations in which previously collected or prospectively collected observational data can meaningfully address certain evidence gaps, and there are situations in which randomization is simply not feasible. In these cases, only observational studies are available to guide care. For instance, patient or clinician preferences for one of two interventions may be so strong that recruiting sufficient numbers of patients willing to be randomized may not be feasible. When observational studies are proposed, PCORI requires that investigators adhere to the Methodology Standards related to causal inference and describe the methods they will use to control for confounding in their study sample and data analysis.

A relatively strong observational approach is one that capitalizes on variation in exposure to competing interventions that result from differences in policies or practice at a level higher than the individual patient, such as the clinic site, delivery system, health plan, or geographic area. That is, the choice of intervention is not determined primarily by patient characteristics. It must further be the case that the actual determinant of the variation in exposure is unrelated to patients' prognosis or to other characteristics predictive of outcomes. These research opportunities, sometimes called "natural experiments," can occur as a result of differences in insurance coverage, in clinical practice guidelines, or in the availability of a particular screening, diagnostic, or treatment technology between two or more otherwise similar patient populations. Natural experiments can potentially reduce or eliminate confounding or selection bias that would be of concern in an observational cohort study conducted in a single study population. PCORI expects applicants proposing such studies to demonstrate both that sizable differences in exposures between populations actually exist and that there are no other differences between these populations that would predict differences in study outcomes. PCORI currently funds a number of natural experiments based, for example, on variation in care programs at the level of states, variations in insurance benefits design, the gradual roll-out of a new chronic conditions program across regions within a state, and the introduction of a new program within a safety-net provider system.

PCORI also encourages investigators to consider the use of innovative randomized designs that can increase study efficiency, reduce study costs, and make studies more feasible and acceptable within care delivery systems. Examples of innovative designs that would be of interest include trials involving cluster randomization (Platt et al., 2010), trials relying on adaptive designs (Connor et al., 2013), and related platform trials. Cluster randomized trials randomize entire groups of people, instead of individual participants, to study interventions. Depending on the intervention, it may be possible to allow individual patients within the cluster to request and receive the alternative approach. In other cases, such as in a comparison of two system-level interventions, it may be impossible to afford patients in one cluster access to the other intervention. This type of design is particularly well-suited to comparisons of system-level and other low-risk interventions because it is often essential to waive or greatly simplify the individual informed consent process (Chalkidou et al., 2012). Cluster randomization may also be preferable in situations where the potential for contamination across intervention arms is high because of contact between subjects at the same office, clinic, or community. Power calculations and analytic techniques employed in cluster RCTs must account for the fact that individuals within clusters will likely be more closely related to one another than to individuals in other clusters (Tunis et al., 2010; Campbell et al., 2012). PCORI is currently funding at least 25 cluster randomized trials.

Adaptive trials allow for prespecified changes to be made to the trial design while it is under way in response to accumulating data collected from trial participants (Hickam et al., 2013; Kairalla et al., 2013; US Food and Drug Administration, 2010). Adaptive trials have great potential for reducing sample sizes or the duration of follow-up needed to obtain conclusive trial results, advantages that are particularly attractive in pragmatic research. Planned adaptations could include terminating a study arm early for futility or benefit based on planned interim looks at the data; changing the proportion of individuals randomized to each study arm so that the statistical significance of an apparent benefit in one study arm based on early findings is achieved sooner; or identifying effects (benefit, futility, or harm) in certain subgroups of patients sooner (Hickam et al., 2013). These adaptations can increase efficiency by reducing the number of participants required to generate precise estimates of the comparative effectiveness of the interventions being studied in relevant populations (Conner et al., 2013).

Recognizing the potential utility of adaptive designs for pragmatic research as well as the fact that these types of studies are methodologically complex, PCORI's Methodology Committee developed a number of standards on adaptive trial designs to encourage the appropriate use of these approaches (Hickam et al., 2013). These standards describe the importance of prespecifying all planned adaptations, including the timing of those adaptations as well as the interim analyses, statistical models, and decision rules that will support those adaptations. The standards also highlight the importance of ensuring that the study infrastructure is capable of supporting those interim analyses and planned adaptations (Hickam et al., 2013).

Building on the thinking behind adaptive trials, researchers have recently described the concept of platform trials (Berry et al., 2015). Platform trials are those that establish a trial infrastructure that can be leveraged to sequentially study a number of different interventions for the same clinical condition. By allowing sequential comparisons of interventions or combinations of interventions to leverage the same trial infrastructure, platform trials are another mechanism for encouraging efficiency in the generation of research evidence. As with other adaptive trials, platform trials rely on interim data analyses to inform future, prespecified adjustments to the trial comparisons (Berry et al., 2015). While PCORI's portfolio of funded research does not currently included any platform studies, the organization is working internally to understand clinical areas that can benefit from this type of study design. Greater use of cluster randomized trials as well as adaptive and platform trials will lead to improved understanding of their utility for addressing practical questions and of how these designs can be integrated in dynamic learning health-care systems (see Integrating Research Into the Learning Health-Care System section).

INTEGRATING RESEARCH INTO THE LEARNING HEALTH-CARE SYSTEM

The Case for Locating Clinical Research Within Health-Care Delivery

A characteristic of pragmatic clinical research is integration of research activities into clinical practice to the fullest extent compatible with maintaining the quality of the research. Conducting research within clinical practice increases the generalizability of findings, making it more likely that relevant lessons from research will be trusted and acted upon to implement practice changes. Many aspects of routine care delivery can now be harnessed to support the more efficient recruitment of study participants, deliver interventions, and monitor outcomes without interfering with or slowing care delivery. A stream of data sources, from electronic health records (EHRs), laboratory databases, claims, and billing data to interactive electronic exchanges with patients can contribute to identifying those patients who meet eligibility criteria for clinical trials or observational studies; to collecting medical history and other baseline information; to capturing outcomes prospectively; and to measuring adherence to randomized treatments over time. Replacing primary data collection with routinely collected clinical data holds great promise for reducing the costs of research. Less costly, more efficient research allows for the larger sample sizes needed for truly pragmatic studies and for the large numbers of such studies that will be needed to transform health care into evidence-based care and health-care delivery systems into learning health-care systems.

The support and involvement of health-care delivery systems and their clinicians in this research is essential (Whicher et al., 2015; Schmittdiel et al., 2010). Allowing access to clinical data streams, physician and staff participation in research, and randomization of patients each require the active assent and support of host systems. In turn, these systems should expect and require that the research they host is truly valuable to their system as well as to society. Such involvement is consistent with PCORI's concept of engagement, helping to ensure that the research done produces practical information that is useful to clinicians and patients, not only to funders and researchers. The CER to be done includes much work on identifying best practices in health-care systems. In the 2009 IOM Report on high priority CER questions, nearly half the 100 questions involved the development or improvement of systems approaches to improving the treatment and management of common illnesses (Institute of Medicine, 2009).

Electronic Health Records and Insurance Claims Data

The EHR is now in use in a majority of US hospitals and ambulatory care settings, thanks to major investments of the federal government over the past decade and the work of the Office of the National Coordinator for Health Information Technology in the Department of Health and Human Services (Blumenthal, 2010). This remarkable achievement lays the foundation for capturing much of the critical clinical data elements needed for conducting high-quality CER. These include physiological measures (blood pressures, temperatures, heart rates, body mass index), laboratory test results, routinely collected self-reported measures such as smoking status and physical activity measures, and, in some settings, condition-specific self-report measures such as depression scores in patients with recognized clinical depression or physical function measures following orthopedic surgeries. Other measures, for example, information from test reports for pulmonary and cardiac function, can be obtained with relative ease, and more complex information can be extracted from pathology, surgical, and hospital discharge reports using natural language processing. EHR systems also offer the capacity to reach out directly to patients through patient portals to collect a range of patient-reported information including outcomes that make the research more patient-centered (Wu et al., 2010).

Although EHR data is often criticized for its lack of accuracy, these criticisms relate more to its accuracy for clinical purposes in individual patients than to its research uses. Research is typically done with a certain amount of mismeasurement. Much highly useful research has already been performed using data from clinical records, including studies of colorectal cancer screening (Doubeni et al., 2013) and studies that use EHR data to identify undiagnosed hypertension or diabetes, to measure physician responsiveness to poor control in these conditions, and to measure patient adherence to treatment (Kharbanda et al., 2014). Such data have supported both observational and experimental studies. Rates of errors and the possible biases that various errors could introduce can be assessed and adjusted for if necessary in observational studies or identified and corrected when studies involve direct patient contact such as in clinical trials. Validation studies for endpoint data and the development of standardized "computable phenotypes" based on claims and EHR data that identify various conditions and events (Bayley et al., 2013; Xu et al., 2015; Richesson and Smerek, 2014) are rapidly enhancing the utility of data drawn from these sources. Nevertheless, much remains to be done to standardize the collection of EHR data and its storage for research use. It is easy to speculate that as systems and their clinicians begin to use the EHR and its data for research purposes, their resolve to invest time and resources in improving, expanding, and standardizing the data entered will increase.

Insurance claims data, from both commercial and public insurers, have been used for research purposes for many years. Claims data capture the diagnoses and procedures associated with hospitalizations and emergency or routine outpatient visits. Receipt of laboratory tests and the details of prescription dispensing are all recorded using standard nomenclatures. Unlike EHR, this information is captured regardless of where the encounter occurs, even for care received in disparate geographic locations. Claims data on insured populations also creates the "defined population" so necessary to conducting longitudinal outcomes research. To conduct a longitudinal study electronically, it is essential to know who remains under observation at every point in time. The EHR does not provide this information. It is typically not updated if a study subject dies, moves to another state, or changes insurance coverage (and therefore provider). Claims data have been available and used for much longer than EHR data.

Neither EHR data nor claims data are nearly so useful separately as they are together. The greater clinical detail of EHR data, when combined with the potential completeness of follow-up over time of claims-based data, creates a foundation for a much broader range of more sophisticated outcomes studies focused on both safety and effectiveness research. A current challenge is that insurance claims and EHR data are currently held by separate entities that typically are engaged in business transactions with each other. Sharing of detailed data has been done only rarely, in part, because of concerns that "proprietary" information could be inadvertently shared, harming one entity's business position in the relationship. The opportunities for delivery systems and health plans and public providers to share data and join forces in addressing clinical questions of importance to both partners are increasing, thanks to the evolution of health-care payment. With the spread of shared risk arrangements such as accountable care organizations and value-based purchasing, payers and providers of care now find reasons to share information for performance improvement more frequently. As trust and the appreciation of the utility of research increase, the barriers to such sharing for research purposes should decrease.

A second challenge is presented by the need to share individual-level data and to use individual identifiers to accomplish data linkage. This raises questions of data security, privacy, and the voice of the patient in decisions about whether data may be shared. In randomized clinical trials or smaller observational cohort studies where collecting individual informed consent is either required or at least feasible, consent for data sharing and linkage can be obtained at the same time. For larger observational outcomes studies, informed consent is usually infeasible. Yet the exclusion of those from whom consent cannot be obtained would undermine the value of these nonrandomized studies. Current US law, as contained in the Health Insurance Portability and Accountability Act (HIPAA), allows for waiving of informed consent under the oversight of an Institutional Review Board when its collection is deemed to be infeasible. However, these protections are not universally considered to be adequate, and the greater engagement of patient communities with the organizations that hold and wish to share clinical data will be necessary to protect the interests of all parties and enhance the use of these data.

The Use of Clinical Registries in Research

Data from electronic data sources can be augmented with more detailed information than usually found in the EHR or claims using standardized data collection procedures from clinical records or direct outreach to obtain patient-reported data.

Clinical registries (Gliklich and Dreyer, 2014) often focus on patients who have a specific procedure (e.g., coronary artery bypass graft surgery or a knee replacement implant) or who have a specific illness, such as cancer or diabetes mellitus. Registries serve many purposes, including performance improvement, safety monitoring, longitudinal disease surveillance, and clinical outcomes research. Participation in registries by physicians and systems is sometimes encouraged by linking payment to participation, and there are now hundreds of registries in the United States as well as many international registries.

Registries typically collect information for all eligible patients. Thus, they have the potential to support pragmatic research, either as longitudinal observational studies or as sampling frames for pragmatic clinical trials, adding data details not found in either EHRs or claims data. However, their ability to host such studies would usually require the continued collection of information from multiple sources over time. For many registries, this additional data collection can be prohibitively expensive. This situation again raises the possibility of linkage with existing data collected though EHRs and insurance claims. Interestingly, the data that would be sought are very similar from registry to registry, suggesting a broad solution in which collection and linkage of EHR and claims data is done once with subsequent linking to a variety of registries to support ongoing follow-up within the registries and CER studies.

The National Patient-Centered Outcomes Research Network

In 2013, PCORI's Board of Governors determined to establish a national clinical research network that would facilitate the efficient conduct of large pragmatic CER studies based on real-world health-care delivery settings (Selby et al., 2013; Fleurence et al., 2014). PCORnet is a "network of networks" that features access to EHR data on millions of persons from multiple care delivery settings and the engagement of patients, clinicians, delivery systems, and health plans in all aspects of network governance. To emphasize the critical importance of patient engagement, PCORI funded 18 Patient-powered research networks (PPRNs; Table 5.6) as a central part of PCORnet. Each PPRN includes one or more organizations of patients with a single condition. Across the 18 PPRNs, nearly 100 distinct illnesses are represented. In each organization, patients are heavily involved in governance activities and actively interested in research participation. PCORI also funded 11 clinical data research networks (CDRNs; Table 5.7) as part of PCORnet. Each network includes multiple health-care delivery systems and brings EHR data for at least 1 million persons and the capacity or willingness to collaborate with the other networks in standardizing their held data and in participating in multinetwork pragmatic research, both observational and interventional. As of October, 2015, the numbers of PPRNs and CDRNs were increased to 20 and 13, respectively.

The concepts of a common data model and of distributed data queries and analyses address several concerns frequently encountered in building research networks (Maro et al., 2009; Brown et al., 2010). In 2016, EHRs in the United States remain highly diverse in the ways data are captured and stored. Even when the same EHR vendor is used, there are widespread differences in how systems deploy and use the EHR and many distinctive local conventions for how data are entered into various fields. Thus, direct linkage or pooling of EHR data is impossible. A common data model (Brown et al., 2010) transforms the data at each network site into a database (model) with a common architecture, data fields, and tables. Each field contains data stored in standardized fashion. Once this is accomplished, the data stored in various sites can be evaluated using centrally written programs or queries that are distributed to each site to evaluate content and ranges of data values there. These "distributed queries" yield summary reports that can then be pooled across sites. Even complex analyses can be carried out without transferring any individual-level data (Brown et al., 2010). This approach greatly reduces concerns about data security and the proprietary interests of participating health systems. Currently, PCORnet has placed clinical data on ~90 million persons into local instances of the common data model across the CDRNs.

PCORnet adopted the same common data model architecture as that used by the US Food and Drug Administration's (FDA) Sentinel Initiative (Behrman et al., 2011), which is a parallel national research network established to conduct drug safety surveillance based on commercial insurance claims data for well over 100 million covered lives. Thus, PCORnet and Sentinel are well-positioned to create linked data on many millions of individuals included in both networks (Curtis et al., 2014). The NIH-sponsored Clinical and Translational Sciences Awards (CTSA) program is another natural partner for both activities. PCORI, the FDA, and the NIH work closely together to coordinate activities toward the goal of a larger national clinical research infrastructure. The capacities to link with both Medicare data sources and with national clinical registries for specific research projects will also be essential in this endeavor.

An important contribution of PCORnet has been the introduction of capacity for conducting randomized studies. Unlike safety research or surveillance, CER often requires randomized evaluations because of the high likelihood of selection biases in observational analyses of the outcomes of treatment choices. A demonstration trial is currently underway within PCORnet that compares two doses of aspirin daily (325 mg vs 81mg) for the secondary prevention of cardiovascular disease (Hernandez et al., 2015). Although aspirin use is almost universally recommended for prevention

TABLE 5.6 PCORnet's Patient-Powered Research Networks (PPRNs)

Network Name	Sponsor Institution	Condition(s)
iCounquerMS	Accelerated Cure Project for MS	Multiple sclerosis
ABOUT network	University of South Florida	Hereditary breast, ovarian, pancreatic, prostate, melanoma, and related cancers (HBOC)
AD-PCPRN	Mayo Clinic	Alzheimer's disease and dementia
Ar-PoWER	Global Healthy Living Foundation	Rheumatoid arthritis and spondyloarthritis
CCFA Partners	The University of North Carolina at Chapel Hill	Inflammatory bowel disease (including Crohn's disease and ulcerative colitis)
CENA	Genetic Alliance, Inc.	Multiple conditions
COPD	COPD Foundation, Inc.	Chronic obstructive pulmonary disease
CPPRN	Regents of the University of California, Los Angeles	Behavioral health in underresourced communities
DuchenneConnect	Parent Project Muscular Dystrophy	Duchenne and Becker muscular dystrophy
IAN	Hugo W. Moser Research Institute Kennedy Krieger	Autism spectrum disorder
ImproveCareNow	Cincinnati Children's Hospital Medical Center	Inflammatory bowel disease
Health eHeart	University of California, San Francisco	Cardiovascular health
MoodNetwork	Massachusetts General Hospital	Individuals with mood disorders
NephCure	Arbor Research Collaborative for Health	Primary nephrotic syndrome
PARTNERS	Duke University	Pediatric rheumatology
PI Connect	Immune Deficiency Foundation	Primary immunodeficiency
PMS_DN	Phelan-McDermid Syndrome Foundation	Phelan-McDermid Syndrome
PRIDEnet	University of California, San Francisco	Sexual and gender minorities
REN	Epilepsy Foundation	Rare epilepsies
Vasculitis-PPRN	University of Pennsylvania	Vasculitis

in persons who have already had an episode of cardiovascular disease, the recommended dose varies and there is little evidence on the comparative effectiveness or safety of the two doses. The trial involves seven CDRNs and one PPRN and will recruit 20,000 persons with a history of prior cardiovascular disease. Outcomes over 3 years include both recurrent cardiovascular events and bleeding events. The study provides PCORnet's first opportunity to evaluate the feasibility of using a central IRB for oversight of multiinstitutional research (Kaufmann and O'Rourke, 2015), of electronic recruitment through patient portals, and of linkage of EHR data to that held by both commercial and public (Medicare) payers.

PCORI's long-term vision for PCORnet is that it will contribute to development of an even larger and more comprehensive national infrastructure for conducting clinical research, from CER to safety studies, to surveillance and performance assessment work. Major funders, in addition to PCORI, would be the National Institutes of Health and the life sciences industry. These major stakeholders have been involved in PCORnet since its inception, serving on PCORnet's Steering Committee during the first 18 months, and since then on PCORnet's Advisory Committee. PCORnet's common data model is open-source so that other institutions may readily join or collaborate with PCORnet in the future by transforming their data into the common data model format. PCORnet is envisioned as a network that generates research ideas in consultation with its patient, clinician, and systems members; responds to solicitations from other funders both public and private; and serves as a resource for qualified researchers both inside and outside of the PCORnet. The existence and extension of a national infrastructure will provide new, more efficient ways for researchers to collaborate on important studies. Identifying data sources will occupy less of researchers' time, whereas engagement with stakeholders to get the questions right will likely take more time. An infrastructure of this size is not meant to replace research done at single institutions or in smaller networks, although the efforts toward data standardization may facilitate those collaborations.

TABLE 5.7 PCORnet's Clinical Data Research Networks (CDRNs)

Network Name	Lead Organization
ADVANCE	Oregon Community Health Information Network
CAPriCORN	The Chicago Community Trust
Greater Plains Collaborative	University of Kansas Medical Center
REACHnet	Louisiana Public Health Institute
LHSnet	Mayo Clinic
Mid-South CDRN	Vanderbilt University
NYC-CDRN	Weill Medical College of Cornell University
OneFlorida	University of Florida
PEDSNet	The Children's Hospital of Philadelphia
PORTAL	Kaiser Foundation Research Institute
pSCANNER	University of California, San Diego
PaTH	University of Pittsburgh

PCORnet is meant to allow the "right-sizing" of projects, even projects that are necessarily quite large, at an affordable cost, but most studies will likely not involve or require the entirety of the network. An exception may be studies of very rare diseases or surveillance studies where variation in practices or outcomes across sites is of primary interest.

CONCLUSION: VISION OF CLINICAL RESEARCH IN THE 21ST CENTURY

As the complexity, costs, and risks of health care increase, the need for high-quality clinical research has never been greater. The lack of credible evidence for much that is done in the name of clinical care is a liability that leads to poorer clinical outcomes and wasted resources. However, research itself is expensive and must prove its value if it is to be supported and funded. This chapter has pointed out a number of ways in which PCORI is trying to make research more effective, in terms of getting the clinical questions right so that they truly inform decisions that stakeholders must make.

The views we promote call for a changed and expanded role for clinical researchers and for greater involvement of patients, clinicians, and health system leaders. The researcher in this patient-centered scenario must develop and apply skills in communication, across research disciplines and especially with nonresearchers. More than ever, the researcher is expected to place his/her skills and training in the service of those who need information and to become a member of larger teams that include nonresearchers. The notion that clinical data are a competitive advantage or the property of a researcher or an institution is supplanted by the view that the data belong to the patients, clinicians, and systems who generated them. Qualified researchers pursuing important questions deserve access to these data, particularly when the resources of funding agencies such as PCORI or the National Institutes of Health (NIH) have contributed to the data collection or aggregation.

There is a particular need for researchers and care delivery systems to join forces in pursuit of improving health-care decision-making and delivery. Researchers possess the skills needed by systems to learn and profit from their growing data resources. Health systems control access to those data but are often unable to take full advantage for lack of analytic expertise. Health systems strive to build trusted relationships with their patients and enrollees. Participation in meaningful research together with patients can serve to enhance the trust.

Whether these ideas pertain only to PCORI-funded research or whether they have relevance to research funded by others, including the NIH, the FDA, and the life sciences industry, is a legitimate question. Certainly, the primary research agendas of these major funders differ from the stakeholder-driven agenda of PCORI. Yet the idea that engagement of patients and other key stakeholders can improve the utility of funded research applies to each of these domains of research activity. Happily, both industry funders, the FDA and the NIH, have moved rapidly in recent years toward greater involvement of patients in research. For example, the importance of patient involvement in drug development has been recognized by the FDA and industry for several years (Nyman, 2015; Sanofi Appoints, 2014; US Food and Drug Administration, 2015). The NIH's 2015 solicitations for the President's Precision-Medicine Initiative create a major role for patient participants (National Institutes of Health).

The need for training patients, researchers, clinicians, and systems leaders to take on these expanded responsibilities in clinical research cannot be overlooked. PCORI provides a range of training opportunities for patients and nonresearchers detailed in "Methodology Standards for Patient-Centered Comparative Effectiveness Research section." The Agency for Healthcare Quality and Research (AHRQ) offers research training in Patient-Centered Outcomes Research (Agency for Healthcare Research and Quality, 2014) using funds allocated to AHRQ from the PCOR Trust Fund. Many academic medical centers and schools of public health have instituted courses and traineeships focused on conducting patient-centered outcomes research. The Robert Wood Johnson Foundation, in 2015, brought its long-standing Clinical Scholars research training program to a close, but replaced it with an innovative National Leadership Program, which integrates researchers with clinicians, health systems, and community leaders with goals of improving the health of communities and the country.

All of these efforts reflect a growing appreciation of the need for change in how we plan and conduct research in the US PCORI hopes to continue its contribution toward aligning national research efforts and funding with the critical needs of stakeholders in the health-care enterprise, beginning with patients. These needs stretch from fundamental discovery in basic science, through translation, to comparative effectiveness and safety research and to the evaluation of health systems interventions. In each area, the chance to make research more relevant, more reliable, more informative, and more widely used is worthy of pursuit, even when pursuit requires transformative change.

GLOSSARY

Adaptive trials Trials that allow prespecified changes to be made to the trial design while it is underway in response to accumulating data collected from trial participants.

Cluster randomization Randomization of entire groups of people, instead of individual participants, to study interventions.

Comparative effectiveness research Research that compares the benefits and harms of two or more options available to patients and clinicians.

Heterogeneity of treatment effect The idea that specific interventions will have differential effectiveness, safety, or comparative effectiveness on different subgroups of patients.

Patient-centered research Research that addresses questions and examines outcomes that are important to patients, their caregivers, and clinicians.

Patient partners Individuals with a lived experience with the condition of interest as well as their family members and caregivers, and advocacy organizations who represent those individuals.

Platform trials Adaptive projects designed to establish a trial infrastructure that can be leveraged to sequentially study a number of different interventions for the same clinical condition.

PCORnet The PCORI-funded national patient-centered clinical research network, involving 13 clinical data research networks, 20 patient-powered research networks, more than 90 million patients, and nearly 100 health-care delivery systems. It is based primarily on electronic health record data and is designed to support both experimental and observational multicenter studies.

Pragmatic research Research that aims to produce practical information that can be generalized broadly by reducing the number of exclusion criteria, including a wider range of clinical settings and clinicians, that imposes few, if any, additional burdens on study participants and clinicians and that includes outcomes that are important to patients and other end users.

Stakeholder partners Members of constituencies with knowledge of the health condition and interest in the research question based on professional, rather than personal experience.

LIST OF ACRONYMS AND ABBREVIATIONS

AHRQ Agency for Healthcare Research and Quality
CDRN Clinical data research network
CER Comparative effectiveness research
CTSA Clinical and Translational Sciences Awards
EHR Electronic health record
FDA US Food and Drug Administration
IOM Institute of Medicine
IRB Institutional Review Board
NIH National Institutes of Health
PCORI Patient-Centered Outcomes Research Institute
PCORnet The national patient-centered clinical research network
PPRN Patient-powered research network

REFERENCES

Agency for Healthcare Research and Quality, 2014. AHRQ Projects Funded by the Patient-Centered Outcomes Research Trust Fund. Available at: http://www.ahrq.gov/funding/training-grants/pcor/pcortf-tcd.html.

Berry, S.M., Connor, J.T., Lewis, R.J., 2015. Platform Trials: an efficient strategy for evaluating multiple treatment. JAMA 313 (16), 1619—1621.

Blumenthal, D., 2010. Launching HITECH. N. Engl. J. Med. 362 (5), 382—385.

Bayley, K.B., Belnap, T., Savitz, L., Masica, A.L., Shah, N., Fleming, N.S., 2013. Challenges in using electronic health record data for CER: experience of 4 learning organizations and solutions applied. Med. Care 51 (8 Suppl. 3), S80—S86.

Brown, J.S., Holmes, J.H., Shah, K., Hall, K., Lazarus, R., Platt, R., 2010. Distributed health data networks: a practical and preferred approach to multi-institutional evaluations of comparative effectiveness, safety, and quality of care. Med. Care 48 (6 Suppl.), S45—S51.

Behrman, R.E., Benner, J., Brown, J.S., McClellan, M., Woodcock, J., Platt, R., 2011. Developing the Sentinel system — a national resource for evidence development. New Engl. J. Med. 364, 498—499.

Compilation of Patient Protection and Affordable Care Act: Extracted Sections Concerning Patient-Centered Outcomes Research and the Authorization of the Patient-Centered Outcomes Research Institute (PCORI). Available at: http://www.pcori.org/sites/default/files/PCORI_Authorizing_Legislation.pdf.

Committee on Quality of Health Care in America, Institute of Medicine, 2001. Crossing the Quality Chasm: A New Health System for the 21st Century. National Academies Press, Washington, DC, p. 6.

Committee on Comparative Effectiveness Research Prioritization, 2009. Initial National Priorities for Comparative Effectiveness Research. Institute of Medicines, Washington, DC.

Califf, R.M., Sugarman, J., 2015. Exploring the ethical and regulatory issues in pragmatic clinical trials. Clin. Trials 12 (5), 436—441.

Chalkidou, K., Tunis, S., Whicher, D., Fowler, R., Zwarenstein, M., 2012. The role for pragmatic randomized controlled trials (pRCTs) in comparative effectiveness research. Clin. Trials 9 (4), 436—446.

Connor, J.T., Elm, J.J., Broglio, K.R., ESETT, ADAPT-IT Investigators, 2013. Bayesian adaptive trials offer advantages in comparative effectiveness trials: an example in status epilepticus. J. Clin. Epidemiol. 66, S130—S137.

Campbell, M.K., Piaggio, G., Elbourne, D.R., Altman, D.G., CONSORT Group, 2012. Consort 2010 statement: extension to cluster randomised trials. BMJ 345, e5661. http://dx.doi.org/10.1136/bmj.e5661.

Conner, J.T., Luce, B.R., Broglio, K.R., Ishak, K.J., Mullins, C.D., Vanness, D.J., Fleurence, R., Saunders, E., Davis, B.R., 2013. Do Bayesian adaptive trials offer advantages for comparative effectiveness research? Protocol for the RE-ADAPT study. Clin. Trials 10 (5), 807—827.

Curtis, L.H., Brown, J., Platt, R., 2014. Four health data networks illustrate the potential for a shared national multipurpose big-data network. Health Aff. 33 (7), 1178—1186.

Doubeni, C.A., Weinmann, S., Adams, K., Kamineni, A., Buist, D.S., Ash, A.S., Rutter, C.M., Doria-Rose, V.P., Corley, D.A., Greenlee, R.T., Chubak, J., Williams, A., Kroll-Desrosiers, A.R., Johnson, E., Webster, J., Richert-Boe, K., Levin, T.R., Fletcher, R.H., Weiss, N.S., 2013. Screening colonoscopy and risk for incident late-stage colorectal cancer diagnosis in average-risk adults: a nested case-control study. Ann. Intern Med. 158 (5 Pt 1), 312—320.

Feussner, J.R., 1999. Priorities for patient-centered research. Med. Care 37 (9), 843—845.

Federal Coordinating Council for Comparative Effectiveness Research, 2009. Report to the President and to Congress. Department of Health and Human Services, Washington, DC, pp. 3—4.

Fleurence, R., Whicher, D., Dunham, K., Gerson, J., Newhouse, R., Luce, B., 2015. The Patient-Centered Outcomes Research Institute's role in advancing methods for patient-centered outcomes research. Med. Care 53 (1), 2—8.

Fleurence, R.L., Curtis, L.H., Califf, R.M., Platt, R., Selby, J.V., Brown, J.S., 2014. Launching PCORnet, a national patient-centered clinical research network. J. Am. Med. Inf. Assoc. 21, 578—582.

Gerteis, M., Edgman-Levitan, S., Daley, J., Delbanco, T. (Eds.), 1993. Through the Patient's Eyes: Understanding and Promoting Patient-centered Care. Jossey-Bass, San Francisco.

Gliklich, R.E., Dreyer, N.A. (Eds.), 2014. Registries for Evaluating Patient Outcomes: A User's Guide, third ed. AHRQ Publication. No. 13(14)-EHC111. Available at: http://effectivehealthcare.ahrq.gov/search-for-guides-reviews-and-reports/?pageaction=displayproduct&productid=972.

Hickam, D., Totten, A., Berg, A., Radar, K., Goodman, S., Newhouse, R. (Eds.), 2013. The PCORI Methodology Report. PCORI, Washington, DC. Available at: http://www.pcori.org/assets/2013/11/PCORI-Methodology-Report.pdf.

Hernandez, A.F., Fleurence, R.F., Rothman, R.L., 2015. The ADAPTABLE trial and PCORnet: shining light on a new research paradigm. Ann. Intern Med. 163, 635—636.

Institute of Medicine, 2009. 100 Initial Priority Topics for Comparative Effectiveness Research. Available at: http://iom.nationalacademies.org/~/media/Files/Report%20Files/2009/ComparativeEffectivenessResearchPriorities/Stand%20Alone%20List%20of%20100%20CER%20Priorities%20-%20for%20web.ashx.

Kairalla, J.A., Coffey, C.S., Thomann, M.A., Muller, K.E., 2013. Adaptive trial designs: a review of barriers and opportunities. Trials 13. http://dx.doi.org/10.1186/1745-6215-13-145.

Kharbanda, E.O., Parker, E.D., Sinaiko, A.R., Daley, M.F., Margolis, K.L., Becker, M., Sherwood, N.E., Magid, D.J., O'Connor, P.J., 2014. Initiation of oral contraceptives and changes in blood pressure and body mass index in healthy adolescents. J. Pediatr. 165, 1029—1033.

Kaufmann, P., O'Rourke, P.P., 2015. Central institutional review board review for an academic trial network. Acad. Med. 90 (3), 321—333.

Luce, B.R., Kramer, J.M., Goodman, S.N., Connor, J.T., Tunis, S., Whicher, D., Schwartz, J.S., 2009. Rethinking randomized clinical trials for comparative effectiveness research: the need for transformational change. Ann. Intern. Med. 151 (3), 206—209.

Loudon, K., Treweek, S., Sullivan, F., Donnan, P., Thorpe, K.E., Zwarenstein, M., 2015. The PRECIS-2 tool: designing trials that are fit for purpose. BMJ 350, h2147. http://dx.doi.org/10.1136/bmj.h2147.

Mullins, C.D., Whicher, D., Reese, E., Tunis, S., 2010. Generating evidence for CER using more pragmatic randomized clinical trials. Pharmacoeconomics 28 (10), 969–976.

Maro, J.C., Platt, R., Holmes, J.H., Strom, B.L., Hennessey, S., Lazarus, R., Brown, J.S., 2009. Design of a national distributed health data network. Ann. Intern. Med. 151 (5), 341–344.

Noble, J., 1989. Patient-centered research: through the looking glass in search of a paradigm. J. General Intern. Med. 4 (6), 555–557.

National Institutes of Health, 2015. Announcement of a Draft NIH Policy on the Use of a Single Institutional Review Board for Multi-site Research. Federal Register. Available at: https://www.federalregister.gov/articles/2015/01/06/2014-30964/announcement-of-a-draft-nih-policy-on-the-use-of-a-single-institutional-review-board-for-multi-site.

NIH Health Care Systems Research Collaboratory. About Us. Available at: https://www.nihcollaboratory.org/about-us/Pages/default.aspx.

NIH Health Care Systems Research Collaboratory. UH3 Project: Suicide Prevention Outreach Trial (SPOT). Available at: https://www.nihcollaboratory.org/demonstration-projects/Pages/SPOT.aspx.

Nyman, M., 2015. 2 Reasons Patient Engagement Is the Backbone of Scientific Discovery. PhRMA. Available at: http://catalyst.phrma.org/3-reasons-patient-engagement-is-the-backbone-of-scientific-discovery.

National Institutes of Health. About the Precision Medicine Initiative Cohort Program. Available at: https://www.nih.gov/precision-medicine-initiative-cohort-program.

Orszag, P.R., Ellis, P., 2007. Addressing rising health care costs – a view from the Congressional Budget office. N. Engl. J. Med. 357 (19), 1885–1887.

Patient-Centered Outcomes Research Institute, 2013a. Patient-Centered Outcomes Research. Available at: http://www.pcori.org/research-results/patient-centered-outcomes-research.

Patient-Centered Outcomes Research Institute. Public Comments for PCORI's National Priorities and Research Agenda. Available at: http://www.pcori.org/research-results/research-we-support/national-priorities-and-research-agenda/how-we-developed-our-0.

Patient-Centered Outcomes Research Institute, 2015a. Join an Advisory Panel. Available at: http://www.pcori.org/get-involved/join-advisory-panel.

Patient-Centered Outcomes Research Institute, 2013b. Appointment of the Board. Available at: http://www.pcori.org/assets/2013/07/PCORI-Appointment-of-the-Board-Policy-05062013.pdf.

Patient-Centered Outcomes Research Institute, 2015b. Engagement Rubric for Applicants. Available at: http://www.pcori.org/sites/default/files/Engagement-Rubric.pdf.

Patient-Centered Outcomes Research Institute. Financial Compensation of Patients, Caregivers, and Patient-Caregiver Organizations Engaged in PCORI-Funded Research as Engaged Research Partners. Available at: http://www.pcori.org/sites/default/files/PCORI-Compensation-Framework-for-Engaged-Research-Partners.pdf.

Patient-Centered Outcomes Research Institute, 2015c. Eugene Washington PCORI Engagement Awards. Available at: http://www.pcori.org/funding-opportunities/programmatic-funding/eugene-washington-pcori-engagement-awards.

Patient-Centered Outcomes Research Institute. Pipeline to Proposal Awards. Available at: http://www.pcori.org/funding-opportunities/programmatic-funding/pipeline-proposal-awards.

Patient-Centered Outcomes Research Institute, 2015d. Methodology Committee. Available at: http://www.pcori.org/about-us/governance/methodology-committee.

Patient-Centered Outcomes Research Institute, 2015e. Methodology Committee Charter. Available at: http://www.pcori.org/sites/default/files/PCORI-Methodology-Committee-Charter.pdf.

Patient-Centered Outcomes Research Institute, 2015f. Winter 2015 Funding Cycle – PCORI Funding Announcement: Large Pragmatic Studies to Evaluate Patient-Centered Outcomes. Available at: http://www.pcori.org/sites/default/files/PCORI-PFA-2015-Winter-Pragmatic-Studies.pdf.

Platt, R., Takvorian, S.U., Septimus, E., Hickok, J., Meedy, J., Perlin, J., Jernigan, J.A., Kleinman, K., Huang, S.S., 2010. Cluster randomized trials in comparative effectiveness research: randomizing hospitals to test methods for prevention of healthcare-associated infections. Med. Care 48 (6 Suppl.), S52–S57.

Richesson, R., Smerek, M., 2014. Electronic Health Records-Based Phenotyping. Rethinking Clinical Trials: A Living Textbook of Pragmatic Clinical Trials. Available at: http://sites.duke.edu/rethinkingclinicaltrials/ehr-phenotyping/.

Robert Wood Johnson Foundation, 2015. National Leadership Program Centers: 2015 Call for Proposals. Available at: http://www.rwjf.org/en/library/funding-opportunities/2015/national-leadership-program-centers.html.

Sox, H.C., Greenfield, S., 2009. Comparative effectiveness research: a report from the institute of medicine. Ann. Intern. Med. 151, 203–205.

Selby, J.V., Forsythe, L., Sox, H.C., 2015. Stakeholder-driven comparative effectiveness research: an update from pcori. JAMA 314 (21), 2235–2236.

Schwartz, D., Lellouch, J., 1967. Explanatory and pragmatic attitudes in therapeutical trials. J. Chronic Dis. 20 (8), 637–648.

Selby, J.V., 2013. Introducing a New PCORI Research Funding Initiative – Large Pragmatic Clinical Trials. PCORI, Washington, DC. Available at: http://www.pcori.org/blog/introducing-new-pcori-research-funding-initiative-large-pragmatic-clinical-trials.

Schmittdiel, J.A., Grumbach, K., Selby, J.V., 2010. System-based participatory research in health care: an approach for sustainable translational research and quality improvement. Ann. Fam. Med. 8, 256–259.

Selby, J.V., Krumholz, H.M., Kuntz, R.E., Collins, F.S., 2013. Network News: powering clinical research. Sci. Transl. Med. 5 (182), 182fs13.

Sanofi Appoints Dr. Anne C. Beal to the Newly Created Position of Chief Patient Officer, 2014. Global Newswire. Available at: http://globenewswire.com/news-release/2014/03/31/622536/10074582/en/Sanofi-Appoints-Dr-Anne-C-Beal-to-the-Newly-Created-Position-of-Chief-Patient-Officer.html.

Tricoci, P., Allen, J.M., Kramer, J.M., Califf, R.M., Smith Jr., S.C., 2009. Scientific evidence underlying the ACC/AHA clinical practice guidelines. JAMA 301 (8), 831−841.

Tunis, S.R., Stryer, D.B., Clancy, C.M., 2003. Practical clinical trials: increasing the value of clinical research for decision making in clinical and health policy. JAMA 290 (12), 1624−1632.

Thorpe, K.E., Zwarentstein, M., Oxman, A.D., Treweek, S., Furberg, C.D., Altman, D.G., Tunis, S., Bergel, E., Harvey, I., Magid, D.J., Chalkidou, K., 2009. A pragmatic-explanatory continuum indicator summary (PRECIS): a tool to help trial designers. J. Clin. Epidemiol. 62 (5), 464−475.

Tunis, S.R., Benner, J., McClellan, M., 2010. Comparative effectiveness research: policy context, methods development and research infrastructure. Stat. Med. 29, 1963−1976.

US Food and Drug Administration, 2010. Guidance for Industry: Adaptive Design Clinical Trials for Drugs and Biologics. Draft Guidance. Available at: http://www.fda.gov/downloads/Drugs/Guidances/ucm201790.pdf.

US Food and Drug Administration, 2015. Patient-focused Drug Development: Disease Area Meetings Planned for Fiscal Years 2013−2017. Available at: http://www.fda.gov/ForIndustry/UserFees/PrescriptionDrugUserFee/ucm347317.htm.

Whicher, D.M., Miller, J.E., Dunham, K.M., Joffe, S., 2015. Gatekeepers for pragmatic clinical trials. Clin. Trials 12 (5), 442−448.

Wu, A.W., Snyder, C., Clancy, C.M., Steinwachs, D.M., 2010. Adding the patient perspective to comparative effectiveness research. Health Aff. 29 (10), 1863−1871.

Xu, J., Rasmussen, L.V., Shaw, P.L., Jiang, G., Kiefer, R.C., Mo, H., Pacheco, J.A., Speltz, P., Zhu, Q., Denny, J.C., Pathak, J., Thompson, W.K., Montague, E., 2015. Review and evaluation of electronic health records-driven phenotype algorithm authoring tools for clinical and translational research. J. Am. Med. Inf. Assoc. 22, 1251−1260.

Zwarenstein, M., Treweek, S., Gagnier, J.J., Altman, D.G., Tunis, S., Haynes, B., Oxman, A.D., Moher, D., CONSORT group, Pragmatic Trials in Healthcare (Practihc) group, 2008. Improving the reporting of pragmatic trials: an extension of the CONSORT statement. BMJ 337, a2390. http://dx.doi.org/10.1136/bmj.a2390.

Chapter 6

Health-Care Technology Assessment (HTA)

Uwe E. Reinhardt

Princeton University, Princeton, NJ, United States

Chapter Outline

Key Points

- Just because individual patients—typically sick and anxious persons traditionally called "patients" and now called "health-care consumers"—are not clinically trained to perform proper benefit–cost analyses of proposed medical intervention does not mean that such analysis cannot be conducted from society's perspective.
- There are several alternative approaches to performing such benefit–cost analyses, each with its own strength and limitations. Together, they fall under the generic rubric of health-care technology assessment (HTA).
- It has been found possible and useful to collapse multidimensional clinical outcomes into one-dimensional metrics [quality-adjusted life years (QLAYs) or disability-adjusted life years lost (DALYs)] that can be used in such cost–benefit analyses.
- These one-dimensional metrics, however, have been a source of controversy, because they implicitly may discriminate against certain patients—e.g., the aged or the disabled. These controversies linger.

- Ultimately, the main barrier to full-fledged benefit analysis of alternative medical interventions—including new medical technology—is a pervasive hesitance in most societies to assign a monetary value to a QALY or a DALY, especially in the political forum of public health insurance programs.
- These methodological problems notwithstanding, health-care technology assessment (HTA) in one form or another is likely to become a routine feature of future health system, relying on ever more sophisticated approaches.

SUMMARY

This chapter surveys the terrain generally known as health-care technology assessment (HTA) in health care, that is, the economic evaluation of existing or new treatment alternatives. HTA had traditionally been eschewed by health-care systems around the world, on the theory that it is well-nigh impossible to estimate the benefits expected from medical interventions, especially of life-saving interventions. Pressed by the ever-rising cost of modern health care—especially of new health-care products—and constrained budgets, however, it is a safe bet that in the future more and more of modern medical practice will be subjected to HTA, mainly to arrive at decisions whether or not to include particular interventions in the benefit package covered by public or private health insurance.

The chapter begins by a brief description of how the producers of medical interventions and those who pay for them approach HTA. Thereafter the chapter surveys the main alternative approaches used for HTA by those who pay for health care, using a wider social perspective, rather than only that of patients or third-party payers. These methods are (1) full-fledged cost–benefit analysis (CBA), (2) cost-effectiveness analysis (CEA), and (3) cost–utility analysis (CUA).

To perform any of these analyses, it is necessary to collapse multidimensional clinical outcomes into one-dimensional metrics [quality-adjusted life years (QALYs) or disability-adjusted life years lost (DALYs)] that can be used to estimate unit costs in HTA, with the unit represented outcome rather than treatment (the latter being just an input measure). The chapter describes how QALYs and DALYs are defined and estimated.

Finally, this chapter focuses on some methodological and philosophical barriers faced by HTA. The metrics QALYs and DALYs, for example, can implicitly bias HTA against the aged and the disabled, unless carefully used. Furthermore, all of HTA stumbles over a major philosophical issue, namely, how to put monetary values on clinical outcomes, especially for life-saving medical interventions.

INTRODUCTION

One can think of an economy as a group of consenting adults exchanging favors with one another. The favor can be a good or a service.

In primitive societies, the actual favors were exchanged on a barter basis. It did not take long, however, for humans to invent "money," which is defined as anything that people accept in exchange for a favor—be it cattle, coins of precious metals or, as in prisoner of war camps in World War II, cigarettes.

Although the exchanges of favors for money typically are mutually beneficial exchanges, the offeror of a favor and its buyer do bring conflicting interests to the deal. It is considered acceptable that the offeror seeks to extract from the recipient of the favor the maximum amount of money that can possibly be extracted. Similarly, it is reasonable that the recipient of the favor seeks to surrender to the offeror as little money as is acceptable to the offeror.

What makes this tug-of-war possible and productive is a benefit–cost calculus both parties to the deal will instinctively perform before consummating it. To do so, however, several conditions must be met. First, both parties to the deal must fully understand the nature and quality of the favor being exchanged. Second, both parties to the deal must have full knowledge of a binding money price for the favor. Third, both parties to the deal must have roughly equal market power, which is not always the case.

Imagine, for example, two persons in a desert, one equipped with an ample supply of water, the other near death from thirst. In that case the offeror could most likely extract all of the thirsty person's wealth for a life-saving drink of water. We would not consider that to be a market in which a fair deal can be cut, even though the offeror describes his or her pricing policy by the mellow term "value pricing," where "value" refers to the value the near-death recipient puts on his or her own life.

Life-saving health care can lean in the direction of an inherently unfair market setting of this sort. A manufacturer of an on-patent drug that can extend a patient's life for months or years enjoys market power that reminds one of the offeror of water in the desert setting, where once again the "value pricing" the manufacturer adopts refers to the value the afflicted patients puts on his or her own life.

The question then is how compelling society views the benefit—cost calculus underlying this exchange of a favor—a life-prolonging pill for money. It is a question that increasingly troubles physicians and their patients today, along with health insurers and government officials who often pay for such drugs.

The purpose of this chapter will be to explore approaches to benefit—cost calculations in the examination. The endeavor goes under various generic labels, such as "technology assessment" or "cost—benefit analysis" or, when applied to pharmacological or biological products, "pharmacoeconomics" (a term still so novel among the laity that Microsoft's spell check sees red when encountering it). As a young field, its practitioners still debate among themselves a whole host of challenging methodological issues. Some of these are purely conceptual, others are rooted in the techniques of procuring relevant data—e.g., how individuals subjectively evaluate different health states—and some of them concern issues of social equity, e.g., whether the value of medical outcomes varies with age.

Although used only gingerly by public policy-makers so far, it is a safe bet that technology assessment will become the core of health policy in the coming decade, as no country can afford much longer the vast sums of money traditionally spent on dubious medical practices and products.

In societies that look to both government and investor-owned for-profit enterprises for the development of new medical technology, be it pharmacological or biological products or medical devices, the economic evaluation of such products can be made from several quite distinct perspectives:

- that of the product's developers and manufacturers;
- that of the product's end users; and
- in countries with comprehensive health insurance, that of the private or public third-party payer who effectively purchases the product on behalf of the end user and pays for it out of some collective financial pool, be it a public treasury of a private insurance pool.

Although these perspectives are not totally independent of one another, they nevertheless may come to different conclusions about the economic merits of a new medical technology. These differences are poorly understood in the debate over health policy—even among people who should know better—and utterly confuse the general public.

This chapter will touch only briefly on the first two perspectives, but emphasize the third perspective, which may be loosely labeled by the vague term "society's perspective." It is the vantage point commonly adopted in the literature on technology assessment in health care—for example, in the classic text *Methods for the Evaluation of Health Care Programs* by Michael Drummond et al. (2005). Although the title of the chapter is pharmacoeconomics, the discussion often will take the broader focus of economic evaluation of health care in general

It will become clear from the discussion that the economic evaluation of new medical technology is only partially based on scientific methodology. Either explicitly or implicitly, it involves strong doses of preferred social ethics, which in turn derive from widely shared theories of justice. That aspect of CBA is not always fully appreciated, in part because those who conduct economic valuations in the context of health care prefer to style their work as pure science, which it rarely is.

THE EVALUATION OF NEW MEDICAL TECHNOLOGY: THE PRODUCER'S PERSPECTIVE

Usually the development of new medical technology requires considerable up-front investment of human and material resources during a research and development (R&D) phase that can stretch over many years—sometimes in excess of a decade. The recovery of these investments through the future benefits yielded by the R&D effort can stretch over even more years. That circumstance makes spending on medical R&D a classic investment proposition.

Private, Investor-Owned Producers of New Medical Technology

When the R&D process for new medical technology is managed and financed by a private, investor-owned, for-profit entity, such as a research-oriented pharmaceutical manufacturer, the firm subjects the process to what is known in business circles as "capital budgeting." Specifically, the firm considers solely the opportunity costs borne by the firm's owners—for a public corporation, the shareholders—and the future revenues reaped by these owners. The firm will not take into account in any way the legally tolerated spillover costs that the project may visit on other members of society or the spillover benefits yielded by the process to individuals other than the firm's owners.

Usually the development of many new medical technologies requires up-front investments of hundreds of millions or even several billion of dollars. In investor-owned entities, the owners must advance for this purpose their own funds, or funds borrowed in their name from creditors, before a chemical compound or a biological or a new medical device becomes a product that is approved by government for sale to end users. In the United States, the agency charged with making that approval—or rejecting a new product—is the US Food and Drug Administration (FDA). Most other modern nations have similar government agencies. Many critics of the medical technology industry do not understand how many tests of clinical safety and effectiveness a new medical technology must pass during the R&D phase before the government will approve it for sale to the public.

To illustrate this process with a highly stylized set of assumptions, made solely for simplicity, suppose the development of a new prescription drug requires an R&D-oriented drug manufacturer to spend $40 million in cash per year for 10 years. As noted, these huge cash advances made by the corporation can come from but two sources: creditors and owners. Raising those funds is primarily the task of the firm's Chief Financial officer (CFO), who can be viewed as the firm's in-house banker. The CFO "lends" the required funds—in this case, $40 million a year for 10 years—to project teams composed of production and marketing people who have formally applied for these funds with detailed and lengthy investment proposals. The CFO, the firm's banker, first of all examines the proposals for realism and methodological soundness and, upon approval, lends the project team the requested funds, at an interest rate that is called the firm's "weighted average cost of capital (WACC)." (The weights in the WACC are the fraction of funds raised from creditors and the fraction raised from owners/shareholders.) A handy way for the CFO to track this loan would be to set up an amortization table, calculated at the WACC, of the sort bankers sometimes establish for regular mortgage loans.

The interest charged by the CFO to the project team is a weighted average of what it costs the firm for which the CFO acts, after taxes, to procure a dollar of funds from creditors (usually buyers of the corporation's bonds) and from owners (in the form of retained profits or in return for newly issued stock certificates.) The after-tax cost of debt financing is laid down by lawyers in detailed, legally enforceable contracts called "indentures." It is market-determined. The CFO's cost of raising funds from owners, on the other hand, is the owners' opportunity cost of those funds, that is, what the firm's owners (shareholders in a corporation) could have earned on their money, had they invested it in other corporations *with a similar risk profile*. This is called the *cost of equity capital*. Estimating what that opportunity cost is, is an art in its own right. As a general rule, the more risky that investments in the company's stock are in the eyes of shareholders, the higher will be the firm's cost of equity capital, that is, the rate of return to expect from owning the firm's stock.[1] In practice, the WACC for business firms currently ranges from 9% to 15% per year, depending upon the risk that the firm's creditors or owners shoulder when they entrust money to the firm.

Assuming now a WACC of 12% per year for the R&D project assumed above, what would be a corporation's total cumulative investment in bringing a drug to market by the time the drug is ready for launch into the market, if the firm had spent $40 million a year for 10 years on developing the drug? The figure turns out to be $702 million.[2] In the jargon of finance, it is called the *future value* of the cash flow. Of that total, $400 would be the sum of the actual cash outlays of $40 per year for 10 years—the number the firm's accounting would book as R&D expenses. The remaining $302 would be accumulated interest costs calculated at the WACC, that is, the firm's opportunity cost of those cash outlays over the entire R&D phase.

For the firm to break even on the R&D project in question, the future net after-tax cash profit stream yielded by it, converted to *present-value equivalents* at the WACC of 12%, would have to sum to $702 million. If that present-value equivalent summed to more than $702 million, adopting the proposed R&D program would enhance the shareholders' wealth. Finally, if, before launching the associated R&D effort, the present value of future net cash after-tax profits from the drugs at the point in time the product is launched were less than $702 million, the corporation's Board of Directors should never have approved going ahead with the project, because it would diminish the shareholders' wealth. Of course, one could also calculate the present values of the entire cost and cash profits stream as of the time the R&D project itself began and would come to the same conclusions.

Some thoughtless critics of the pharmaceutical industry have used not the accumulated cost at time of launch in this context, but the *present value* of the 10-year stream of $40 million a year at the time when the R&D project was started, which turns out to be $226 million.[3] It is an amount large enough such that, if it were in the bank at the time the R&D project is begun and if money in that bank account earned after-tax interest of 12% per year in any year on the amount still in that account at the beginning of the year, then that fund would be enough to support a stream of annual withdrawals of $40 million for 10 years. In other words, it is an entirely different concept than the $702 million calculated earlier off the same cash flow.

---

1. For a formal presentation on the WACC, see Brigham and Earhardt (2008, Chapter 10).
2. Calculated as $40 m $(1.12^{10} - 1)/0.12$.
3. Calculated as $40 m $(1 - 1.12^{-10})/0.12$.

The future and present values we had calculated earlier for a 10-year cash flow of $40 million a year are highly sensitive to two variables: the WACC and the length in years of the R&D process. For example, if the WACC were 12% but the R&D phase was only 6 years, then the future value of the project would be only $325 million and the present value only $164 million. On the other hand, if the R&D phase was 12 rather than 6 or 10 years, the future value would be $965 million and the present value $264 million. If with that R&D phase the WACC were only 8%, the future value would be $759 million and the present value $301 million, and so on.

In short, the total cost to the R&D-based manufacturer of bringing a new pharmaceutical product to market is a fluid construct not easily understood by the uninitiated. There is the added problem, beyond the compass of this paper, of what to do with the cash outlays on the R&D of projects that faltered along the way—the analog of dry holes in the oil business. These costs must be allocated somehow to and recovered by projects that lead to marketable products.

Whether or not a contemplated new product will in fact be developed by an investor-owned producer hinges solely on the expected profitability of that product, which in turn is driven by the price the producer can charge for the product. Here it must be noted that that price is set under the rules of Anglo-Saxon capitalism, which makes maximizing the wealth the firm's owners derive from their investment in the firm without breaking the laws of the land, the sole social responsibility of the firm's board of directors and the managers that board appoints.

Martin Shkreli, CEO of Turing Pharmaceuticals, which had purchased sole distribution rights of the 60-year-old drug *Daraprim* for the United States from a European pharmaceutical company, put it more bluntly at a Forbes Healthcare Summit that most like-minded CEO's would dare: "It's a business; we're supposed to make as much money as possible" (Ramsey, 2015). Shkreli had raised the price of *Daraprim* by 5000% from $13.50 to $750 per tablet. Whatever one may think of Shkreli, his gleeful bluntness does serve the purpose of drawing attention to the issue of price setting for products whose market turf is protected through a government-granted monopoly (Reinhardt, 2015)—patents and market exclusivity—that seems to know no upper bound yet (Rockoff, 2015).

Public Producers of New Technology

When a government undertakes the R&D for new medical technology—for example, the National Institutes of Health (NIH) in the United States—then it, too, should add up the cumulative opportunity costs to society of completing the R&D phase until a usable product emerges. To go ahead with the project, ideally these cumulative opportunity costs, calculated as the *future value* of these costs as of the point in time the product yielded by the R&D projects is ready for use, should be covered by the *present-value* sum of the future social benefits that use of the product will yield. In principle, if that condition is not met, the project should not have been undertaken.

In the public sector, this juxtaposition of costs and benefits is called "cost–benefit analysis." It resembles in many ways what is called "capital budgeting" in the private sector, although for public-sector investment projects both the social opportunity costs invested in the project and the social benefits expected to be yielded by the projects are calculated from a broad, societal perspective, which typically abstracts from which particular individuals in society bear the costs of the project and which individuals reap its benefits. It is a purely *collectivist* perspective.

The public-sector analog of the WACC used by private corporations in their capital-budgeting exercises is the so-called "social discount rate" whose proper magnitude has remained a controversy among economists for all public-sector investment projects, and in particular for projects yielding future health benefits. As will be explained further on, especially controversial is the idea to convert *physical* future health benefits—e.g., added life years—into present-value equivalents by means of discounting with a social discount rate. A full discussion of this issue, however, goes much beyond the compass of this chapter.

THE EVALUATION OF NEW MEDICAL TECHNOLOGY: THE END USER'S PERSPECTIVE

A *nouvelle vague* in health policy in many countries—certainly in the United States—is the idea that *commercial market forces*, rather than government regulation, should govern the allocation of health care in society. Embedded in that idea is the notion that "consumers" (formerly "patients"), the end users of new medical technology, are best situated to perform the requisite economic evaluation of new medical technologies, including pharmaceutical and biological products.

At first blush, that may appear as an appealing idea, especially to economists indoctrinated with a belief in the beneficence of markets. As has been pointed out by the current author, however, the approach has powerful ethical and practical implications either not well understood by the laity or delicately swept under the rug, so to speak (Reinhardt, 2001).

The Ethical Precepts Driving Markets

Commercial markets march to the ancient Roman dictum *res tantum valet quantum vendi potest*—in English "a thing is worth what you can sell it for." Practically, in a genuine market system, it means that a new pharmaceutical product would be worth what one can sell it for to end users, who would have to pay for it out of pocket. That is, in fact, the general idea behind the new movement of Consumer Directed Health Care (CDHC), which is a code word for insurance policies with very high deductibles—up to $10,500 a year per family, coupled with personally owned and tax-favored health savings accounts from which deductible payments can be made.

It should be immediately clear to anyone that on the market approach the "value" of a new medical product would vary not only with the end user's perceived medical need for the product, but also with her or his ability to pay for it out of pocket. Concretely, a novel product that controls hypertension or asthma, or a drug-eluting stent, or a new implantable defibrillator, would be deemed to have a higher value if used by, say, a lawyer, a professor, or a corporate executive rich enough to bid high prices for these products than it would if used by a lower-income taxi driver or a waitress who can afford to bid only lower prices for the same products. That proposition is not usually made clear by advocates of the market approach. If one does not accept the ethical underpinnings of this valuation, then one implicitly questions the ability of markets to allocate health-care resources in accordance with society's wishes, and someone other than the end user must perform the economic evaluation of medical products and treatment regimens. An additional implication of the market approach to health care and to the economic evaluation is that the products in question should be rationed among potential users on the basis of their income. The advocates of consumer-directed health care clearly have that in mind, although they tend to be hesitant to articulate their goal quite this bluntly.

It is not clear that the distributive social ethic implied in a genuine market approach to health care is as yet acceptable among the general public, even in the United States, let alone in countries that like their health systems to march to the principle of social solidarity. Yet, remarkably, one sees the market approach to economic valuations of new medical technology advocated with increasing frequency, possibly without the advocates' realization of the distinct social ethic they package into their prescription. In any event, the ethical implication of the approach should always be debated quite openly in discussion on health policy, such as discussions on Consumer-Directed Health Care, for example.

It is not argued here that market forces cannot play a constructive role in a health system whose distributive ethic and other rules of conducts are strongly regulated by government. Judiciously regulated, market forces certainly can play a productive role in health care. The point here merely is that trusting the economic evaluation of new medical technology to the market is problematic if it implies that the individual end user's ability to pay for the technology should drive this evaluation.

The Implications of the Market Approach for the Producers of New Medical Technology

The producers of new medical technology typically and understandably are enthusiastic about the contribution their innovations can make to humankind. They tend to speak glowingly about the great "value" their products represent—especially when their products contribute to saving lives. At the same time, few of them have any clear idea about *who* should determine that "value" and *how* it should be determined.

One often hears spokespersons of the producers of new medical technology decry government's role in health care and wax eloquent on the virtue of the marketplace. The question is whether they really mean it. Would they, for example, openly state that the value of their products rises with the income of the products' end users, or would they openly advocate that their products be rationed among human beings by market price and the recipients' ability to pay? If not, do they actually believe that the end users of their products are best suited to perform the economic evaluations of their products?

These blunt questions are provoked by the pharmaceutical industry's steadfast opposition to "reference pricing" for its products. Under reference pricing, private and public insurers group pharmaceutical products aiming at the same therapeutic target into so-called "therapeutic groupings" and then pay fully out of the insurer's collective funds only for one of the lower-cost products in the therapeutic grouping. The price of that product is the so-called "reference price." If a patient and his or her physician would prefer a higher-priced product that also is in the grouping—perhaps a more recent brand-name drug with fewer untoward adverse effects—that patient must pay out of pocket the full difference between the reference price for the grouping and the price charged for the higher-cost brand-name drug. In effect, reference pricing can be said to be a marriage of social solidarity, practiced up to the level of the reference-priced drug, with a raw market approach that relies on individual end users to evaluate the qualities of the higher-priced drugs in the therapeutic grouping. It is like a business firm that reimburses its employees for coach class airfare, but allows employees to upgrade with their own funds to business or first class, leaving it up to the employee to determine whether the added benefits from upgrading are worth the added costs borne by the employee.[4]

Reference pricing is used in a number of European nations, notably in Germany, which was the first to adopt it formally for its *statutory* health insurance system covering 90% of Germany's population. Reference pricing for prescription drugs is also used in Canada, Australia, and New Zealand. So far it has not been adopted in the United States, either by the private or public insurance sector, because it is vehemently opposed by America's pharmaceutical industry, as it is by the pharmaceutical industries in other countries that use reference pricing. Private insurers in the United States, however, now do lean heavily on the direction of full-fledged reference pricing through sundry three-tiered reimbursement systems for drugs.

Those who oppose reference pricing, including some economists who are supportive of the pharmaceutical industry, typically argue that the end users of drugs are not technically equipped to perform the required CBA for rival drugs in a therapeutic grouping and that physicians are too busy to undertake it for them.[5] They further argue that such a system is inequitable, because low-income patients often cannot afford to pay the additional out-of-pocket costs for higher-priced brand-name drugs that may be more effective than drugs priced at the reference price (Danzon, 2000, p. 25). While that argument may well be right, those making it in effect question the entire propositions that a market approach to health care relying on individual patients as evaluators of their own health care cannot work and, even if it could, would be inequitable. To the extent that the producers of new medical technology are among the opponents of reference pricing for their products, they implicitly reject the entire market approach as well.

This circumstance, then, leaves one with approaches in which the leaders of some collectivity—be it a government or a private insurance company—perform the required economic evaluation of new medical technology on behalf of patients. In the United Kingdom, the government-run *National Center for Clinical Excellence (NICE)* is such a body, performing economic evaluations for the country's government-run National Health Service (NHS). In Germany, it is the recently established government-funded *Institut fur Qualität und Wirtschaftlichkeit im Gesundheitswesen (IQWiG)*, also widely known in English as the *Institute for Quality and Efficiency in Health* Care.[6] It performs economic evaluations of drugs for Germany's statutory health insurance system. In Australia, it is the *Pharmaceutical Benefits Advisory Committee (PBAC)*.

THE EVALUATION OF NEW MEDICAL TECHNOLOGY: SOCIETY'S PERSPECTIVE

The thrust of the previous section is that, although much lip service is being paid these days to the virtue of the private market in health care, when the rubber hits the road, so to speak, neither patients nor the producers of medical technology seem willing to accept the often harsh verdicts of the marketplace. In the end, either explicitly or implicitly, they call for some larger collective—a government agency or private insurers—to regulate the health-care sector on behalf of patients and to perform the requisite economic evaluation of health care for them. Most textbooks on technology assessment in health care adopt this collectivist approach as well, as does the remainder of this chapter.

The General Framework for Technology Assessment in Health Care

The economic evaluation of new medical products, new treatment options, or new health policies in general always involves a comparison of the *negative* and *positive* consequences associated with two different courses of action, one of which may be called the "baseline" (B) and the other the proposed "alternative" course of action (A) that is to be evaluated. In other words, one evaluates one course of action (A) relative to the baseline (B), which requires one to evaluate the *change* in future costs and benefits when option A is adopted rather than the baseline B, assuming one of the two will be adopted for certain. Technically, we speak of the *incremental* costs and benefits of adopting option A rather than option B. Fig. 6.1 illustrates this process.

The first step in the economic evaluation of medical technology is to specify precisely what the baseline and alternative courses of actions are. Often the baseline is simply the status quo, that is, the currently practiced course of action that would continue if the new product, treatment option, or health policy were not adopted. The alternative course of action is the adoption of a new product, treatment option, or health policy. But sometimes the "baseline" may not be the status quo, but one of two *new* courses of actions of which one will in fact be pursued, because it has already been decided, for whatever reason, to depart from the status quo.

---

4. For a fuller description of reference pricing, see Kavanos and Reinhardt (2003).
5. See, for example, essays in Lopez-Casasnovas and Jönsson (2001).
6. See http://www.iqwig.de/iqwig-presents-a-concept-for-cost-benefit.738.en.html.

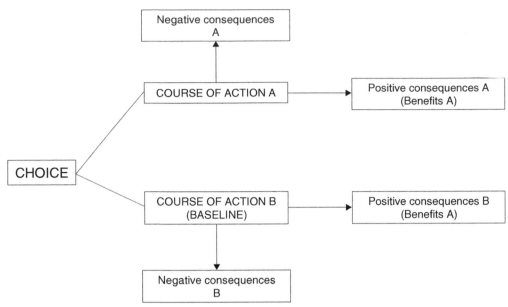

FIGURE 6.1 The general framework for economic evaluation. *Adapted from Drummond, M.F., Sculpher, M.J., Torrance, G.W., O'Brien, B.J., Stoddard, G.L., 2005. Methods for the Evaluation of Health Care Programmes. Oxford: Oxford University Press, Box 2.1.*

The next and most challenging step in the evaluation is identifying and measuring quantitatively—usually in terms of monetary values—*all* of the positive and negative, quantifiable consequences of each of the two courses of action. Quantifying these consequences is part of the art of CBA. Usually these positive and negative consequences are time-phased streams of costs and benefits that flow over long periods of time, which raises the issue of discounting, to be explored further on.

To the extent that some consequences simply cannot be quantified, they nevertheless should be considered by policy-makers or other decision-makers as well, after the formal, quantified CBA has been completed. These adjustments are called "qualitative" considerations. They may modify the recommendation for action emerging from the quantifiable CBA.

Defining Net Incremental Benefits

Given the analytic framework sketched out in Fig. 6.1, one may define the net incremental benefits (NIBs) yielded by the decision to follow of course of action A rather than course of action B as:

$$NIB = (\text{Benefits A} - \text{Benefits B}) - (\text{Costs A} - \text{Costs B}) \text{ or}$$

$$= (\text{Benefits A} - \text{Costs A}) - (\text{Benefits B} - \text{Costs B})$$

(6.1)

Either of these definitions of NIB are the best way to describe the economic merit of that decision, because the NIB is impervious to how one arrays the separate benefits and costs associated with each course of action independently.

Some decision-makers, however, prefer to think in terms of benefit—cost ratios (BCRs), and those ratios are sensitive to the way one arranges the costs and benefits of each course of action. Economists commonly use the concept of "opportunity costs" to define the cost of a course of action. In terms of Fig. 6.1, the total cost of the decision to adopt course of action A rather than B therefore would include costs directly incurred on course of action A plus the benefits given up by not following course of action B, because these foregone benefits of course of action B are an opportunity cost to be charged to course of action A. Similarly, one could define the benefits yielded by course of action A as the benefits yielded directly by that course of action plus the costs avoided by not following course of action B. One would then write the BCR of course of action A as:

$$BCR = (\text{Benefits A} + \text{Costs B})/(\text{Costs A} + \text{Benefits B})$$

(6.2)

Others, however, might prefer to write the ratio in terms of incremental benefits and costs as:

$$BCR = (\text{Benefits A} - \text{Benefits B})/(\text{Costs A} - \text{Costs B})$$

(6.3)

In sum, one has to be careful to describe in the analysis about how the separate benefit and cost figures are arrayed.

Whose Costs and Benefits?

Many times the economic evaluation of new medical products is made from the narrower perspective of a subunit of society—e.g., a business firm providing health insurance for its employees, an insurance company seeking to control its own outlays on health care for insured customers, or a family. Naturally, the decision-makers in this subunit will exclude from their benefit–cost calculus any costs they do not bear and any benefits that do not accrue to them.

From society's perspective, such narrowly based CBA in health care can be highly misleading. An insurance company, for example, may prefer to cover only a product that must be infused at a clinic or outpatient department, rather than a much more expensive new, rival product that can be infused continuously on the job, thus reducing substantially the cost of absenteeism from the job. Similarly, a family may consider only its own cost of treating migraine without taking into account that a more expensive product will reduce the employer's cost through reduced absenteeism or better productivity on the job.

Research entities, such as NICE in the United Kingdom or the IQWiG in Germany, that perform technology assessment in health care usually adopt an all-inclusive societal perspective, which means that they include in their benefit–cost analyses, for every relevant future year, all costs, on whomever they may fall, and all benefits, to whomever they accrue. Many health-care products or treatments, for example, are valued by their recipients more highly than the money price the recipient has to pay the providers of these products or treatments. This is certainly so for life-saving products and treatments. A proper societal CBA would use as a benefit measure the representative recipient's valuation. Economists have been able to estimate the average value people attach to added life years as revealed by their behavior in labor markets or *vis à vis* product safety (Viscusi, 1993). The providers of health care, on the other hand, would count in their benefit–cost calculus only the revenues they receive for that care.

The Issue of Discounting Costs and Benefits

As already noted, when the benefits and costs from a course of action occur over many years, as usually they do, the question arises whether, say, a benefit occurring many years hence in the future should be valued the same as the same benefit accruing in the very near future (and ditto for costs). It matters not here whether the benefits and costs are stated in monetary terms or physical units (e.g., added life years, fewer disabilities).

Suppose a quantitative monetary or physical measure of benefits in some future year t is $B(t)$. It could be an extra life year, for example. Then the idea of "discounting the future" is to treat that future benefit as the present-value equivalent (i.e., now, at time $t = 0$) as:

$$B_0(t) = B(t)/(1 + r)^t \tag{6.4}$$

where r is an annual compound discount rate, and similarly for all time-phased costs and benefits triggered by the course of action in question. Two questions arise in connection with this construct. First, should future costs and benefits be discounted at all and, second, if they should be discounted, at what discount rate r should they be discounted?

There are three distinct views on the first question.

One school would set $r = 0$, arguing that one should not count future benefits less than equally sized benefits accruing now or near term.

A second school argues that one should discount the future and use for it a discount rate r used for ordinary investments elsewhere in the economy. Those rates tend to be high—certainly above 5% per personal investments and, for business firms, usually in excess of 10%.

A third school argues that the future should be discounted, but not at observable rates used in business or in the financial markets, but instead at time preference rates people exhibit strictly in the context of health care. Those time preference rates might be gotten through contingent valuation techniques—loosely speaking, experiments in which respondents are asked to evaluate hypothetical scenarios.

These differences among these schools of thought remain controversial and have spawned a huge literature. In the meantime, textbooks on technology assessments generally do suggest that future cost and benefit streams should be discounted.[7]

---

7. In this connection, see Drummond et al. (2005, Section 4.2) and Sherman Rolland et al. (2007, Chapter 4).

Should Benefits Be Age Adjusted?

Another area of lingering controversy is the question whether society, or policy analysts working on behalf of society, should assign the same social value to a health benefit—e.g., to an added year of life—regardless of the age of the person to whom that benefit accrues or, for that matter, regardless of that person's socioeconomic characteristics. Is adding another life year to a person of working age worth more than adding that year to a nonworking 80-year-old?

The World Bank in conjunction with the World Health Organization (WHO) and the Harvard School of Public Health developed during the 1990s the concept of the "DALY" to calculate the global disease burden (Murray and Zachary, 1997). In its application to measure the global disease burden, DALYs are multiplied by the age-weighting equation:

$$W = (C)(x)e^{-\beta x} \tag{6.5}$$

where x = the age of the person benefiting from a change in DALYs and $e = 2.71$, the natural constant (whose natural logarithm is 1). Fig. 6.2 illustrates this weighing scheme. It is seen that the equation accords greater weight to life years of persons in their productive years and less weight to very young and older people. This adjustment may appeal in low-income societies in which human productivity is crucial for sheer survival. This approach has triggered numerous critical reviews over its inherent subjectivity and especially over its age-weighting.[8]

The late British health economist Alan Williams (1997) proposed in this connection the ethical doctrine of "fair innings." According to that doctrine a person is entitled to only so many "fair innings" over the course of life, beyond which the person can be thought to "live on borrowed time." From that premise it follows that if scarce resources must be rationed for life-saving medical interventions among people of different ages, greater weight should be given to adding another life year to a younger person of, say, age 40 than to a person aged 70 or 80. The argument is that the latter have already been blessed with a high number of "fair innings" and should take second place behind the younger person when added life years wrested from nature through medical treatments must be rationed.

At this time, the idea of age-weighting the benefits from health care probably could not even be discussed openly in the United States, let alone be implemented. Indeed, the entire topic of rationing health care is a taboo in the United States.

Like the issue of discounting, the issue of age-weighting the benefits from alternative courses of actions—e.g., treatments—in health care remains controversial and has spawned a large literature.

COST–BENEFIT, COST-EFFECTIVENESS, AND COST–UTILITY ANALYSIS

Given the researcher's decisions regarding discounting and age-weighting of benefits, economic evaluations of alternative courses of action in health care generally take one of the three distinct forms:

- Cost–benefit analysis (CBA)
- Cost-effectiveness analysis (CEA)
- Cost–utility analysis (CUA)

The second approach, CEA, usually is based on a one-dimensional measure of health outcome (e.g., reductions in blood pressure or added life years) from medical treatments while the third uses subjective, individual preference valuations of multidimensional health outcomes from clinical treatments (e.g., added life years with various attributes of the quality of life).

In the end, however, all of these approaches require policy-makers to come to terms with one of the most challenging problems in the economic evaluation of health care, namely, putting monetary values on the positive and negative consequences of medical interventions, including the use of new medical technology.

Cost–Benefit Analysis

In a full-fledged CBA, monetary values must be *explicitly* put on all of the negative and positive consequences associated with the two courses of action being evaluated, where the emphasis is on "explicitly." These consequences may be multidimensional, which complicates the analysis. The monetary value of these consequences sometimes can be objectively observed directly in the marketplace—e.g., as the prices or production costs of products. At other times, they must be specified.

One approach widely used for that purpose is rooted solidly in formal economic welfare analysis, that is, it seeks to obtain from samples of individuals their "willingness to pay" (WTP) to avoid particular negative consequences

---

8. See, for example, Anand and Hanson (1997).

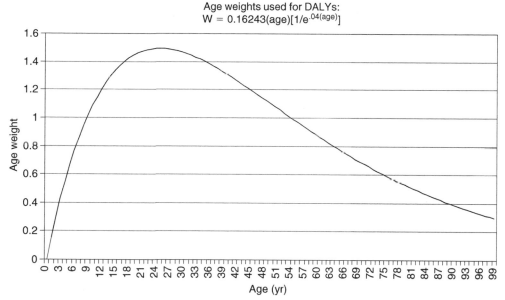

FIGURE 6.2 The age-weighting scheme for DALYs used by the WHO.

(e.g., adverse effects of a drug) or to pay for positive consequences (e.g., the cessation of pain). Sometimes this can be done through inferences from observable market behavior, for example, how much people are willing to pay to avoid bodily harm can be inferred from the extra pay they require to do dangerous work. At other times WTP is inferred by what is called "contingent valuation." Under that approach, researchers seek to infer WTP from experiments in which respondents are presented with hypothetical situations with negative and positive consequences and are then asked how much they would be willing to pay to avoid negative consequences or how much they would be willing to pay to enjoy positive consequences. Naturally, there always arises the question how accurate such information can be, because people will not always do in fact what they claim they would do under hypothetical scenarios.

The problem with the WTP method of valuing benefits, however, is that it is bound to vary with the income of the potential bidder, which means that the measured benefits would be smaller when poor people do the bidding than when rich people bid. In research, one would therefore have to find potential bidders who can be deemed to be validly representative of the society for which the societal CBA is being conducted. Smith and Richardson (2005) discuss these issues in greater detail.

As already noted earlier, two distinct criteria can be used with CBA. They are (1) the NIB from following the alternative course of action A rather than the baseline course of action B and (2) the associated BCR.

The NIB of course of action A over course of action B is defined as:

$$\text{NIB} = (\text{Benefits A} - \text{Benefits B}) - (\text{Costs A} - \text{Costs B})$$
$$= (\text{Benefits A} - \text{Costs A}) - (\text{Benefits B} - \text{Costs B})$$

(6.6)

where "Benefits A" and "Benefits B" in the equation may be the present (discounted) value of a time-phased stream of benefits, possibly age-weighted, and "Costs A" and "Costs B" would be the associated present (discounted) value of time-phased cost streams. If NIB is used as an evaluation criterion for action, one would prefer course of action A if NIB > 0 and baseline the baseline course of action B if NIB < 0.

The associated BCR is defined as:

$$\text{BCR} = [(\text{Benefits A} - \text{Benefits B})]/[(\text{Costs A} - \text{Costs B})]$$

(6.7)

or, as was explained earlier, as:

$$\text{BCR} = (\text{Benefits A} + \text{Costs B})/(\text{Costs A} + \text{Benefits B})$$

(6.8)

depending on how one defines the benefits and costs of course of action A. If the BCR > 0, one would go ahead with course of action A and if BCR < 1 one would prefer the baseline course of action B.

As a general rule, as already noted earlier, the NIB criterion is preferable to the BCR criterion, because the NIB is *not* sensitive to whether a negative consequence is treated as an increase in costs or a decrease in benefits, while the BCR ratio is very sensitive to such classifications.

Cost-Effectiveness Analysis[9]

Because it is so challenging to put *explicit* monetary values on all of the negative and positive consequences associated with alternative courses of action in health care, researchers often prefer to leave at least the benefit side in physical terms, monetarizing only the cost side. For example, one may try to estimate the average cost per added life year saved with different preventive-care strategies, or with different acute-care treatment strategies—e.g., surgical treatment of lower-back pain versus medical treatment. The idea then would be to favor the treatment method with the lowest total cost (per year, or in present-value terms) of achieving *given goals* in a change in physical health status.

The CEA works well when the physical health status measure or other treatment goal in question is one-dimensional—e.g., a reduction in systolic blood pressure by 10 points. It is problematic when the change in health status is multidimensional or, in the case of "added life years" or "change in life expectancy" as a physical outcome measure, when life years can be associated with different degrees of the quality of life (i.e., degrees of disability).

For such cases, researchers have developed "CUA," which seeks to collapse multidimensional health outcome changes into a one-dimensional metric, for example, health-adjusted life-year indices that collapse changes in mortality and morbidity triggered by a new medical technology into one unidimensional metric.

Cost−Utility Analysis: Quality-Adjusted Life Years and Disability-Adjusted Life Years Lost

Under CUA, subjective valuations of the various dimensions of a multidimensional health outcome from a medical treatment are collapsed into a unidimensional metric. Two such metrics are now in wide use among health services researchers worldwide: the QALY and the DALYs (Gold et al., 2002).

To explore these concepts, it will be useful to posit a very simple, stylized life-cycle model for the representative individual, which may be the average experience for a cohort of individuals. Thus, we shall imagine a 20-year-old person who is expected to live to age 45 in perfect health (without any degree of disability), and then spend the remaining years until death at age 60 with:

- a particular health status, which may have a number of dimensions, such as pain and mobility, such that 1 calendar year in that health status is scored subjectively by respondents in a survey as, on average, the equivalent of 0.6 calendar years in perfect health, so that each such year is set equivalent to 0.6 of a quality-adjusted life year or 0.6 QALYs, or equivalently,
- a degree of disability such that losing or gaining that year with that disability status is scored subjectively by a set of respondents as the equivalent of 0.4 of a healthy year. For example, if a treatment yields an additional life year in that disability status, then that benefits is scored as 0.4 healthy years, or 0.4 disability-adjusted life years, or 0.4 DALYs.

It is assumed in this stylized example that the maximum life expectancy of persons in this society is expected to be 80 years. Finally, for the sake of simplicity it is assumed initially that future life years enjoyed, or lost, are neither age-weighted nor discounted.

With these preliminaries, let us define what is meant by QALYs and DALYs.

Quality-Adjusted Life Years

Under this approach, life years lived in less than perfect health are converted into what the representative individual would consider the equivalent number of years in perfect health. For example, if a person said he or she would be indifferent between living 20 more years in a particular lower health status described to him or her and only 16 more years in perfect health, then each of the 20 years in less than perfect health would be considered by that person the equivalent of $16/20 = 4/5 = 0.8$ of a health year, or 0.8 QALYs. The fraction 0.8 would be the quality-of-life weight assigned to each year with the specified disability.

---

9. For a rigorous review of this method, see Garber (2000).

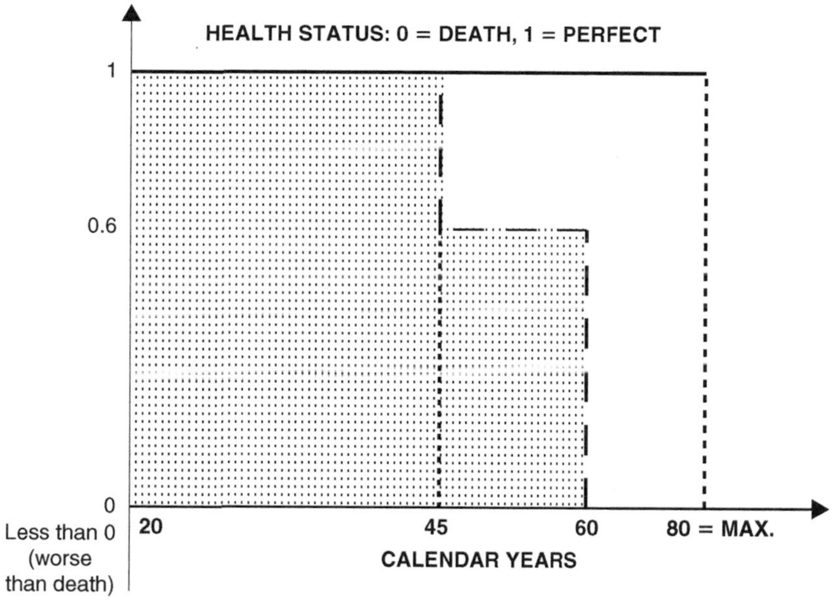

FIGURE 6.3 The definition and measurement of quality-adjusted life years (QALYs).

The beneficial outcome from a medical intervention or health policy is then the change in QALYs attributable to that course of action, which is then used as the denominator in the "incremental cost per incremental QALY" or simply "cost per QALY" attributed to that course of action.

Fig. 6.3 illustrates this concept. The vertical axis is defined as health status such that death is indexed as 0, perfect health as 1, and a health status considered "worse than death" (e.g., severe chronic pain) as less than zero.

At age 20 the person is expected to have another 40 calendar life years, of which, however, 15 years are expected to be in less than perfect health. To adjust for that lower quality of life, we calculate the person's expected QALYs as:

$$\text{QALYs} = (45 - 20) + 0.6(60 - 45) = 25 + 0.6(15) = 34 \tag{6.9}$$

Note that at age 20 the person is expected to live another 40 calendar years, but we would count that as only 34 QALYs.

It can be asked how the quality-of-life-year weights used to calculate QALYs are determined. Three methods have been widely used for that purpose, namely:

- Visual analog scales (VAS);
- Standard gambles (SG); and
- Time trade-off adjustors (TTT)

Under the *VAS* approach, a group of relevant interviewees are given a scale from 0 to 1 or from 0% to 100%, where "0" means "worst imaginable health status" and "1" means "perfect health." A health status is then described and the interviewee picks a number from the scale that reflects his or her assessment of it. That number then becomes the weight assigned to the added life year in the described health status to obtain the implied number of QALYs. For example, if on average the interviewees score the year in that health status as 0.7, then that added life year is scored as 0.7 of a QALY or 0.7 QALYs.

Under the *SG* approach the interviewee is given a choice between (1) a described health outcome, e.g., an added life year, with a described health status that will occur with *certainty*, or (2) a *risky* choice between one better health outcome (usually full health) and one worse outcome (usually "death"). The person then picks a probability of having the better outcome such that he/she would be indifferent between the *certain* outcome with the described less-than-perfect health status and taking the gamble of having either the better or the worse outcome. The chosen probability then becomes the weight assigned to the health outcome with the described health status. If, for example, the interviewees on average stated that to take the gamble rather than the certain outcome, the probability of the better outcome in the gamble would have to be 80%, then the certain outcome (added life year) with the described health status must be fairly tolerable and that extra year in the described less-than-perfect health status would be scored as 0.8 QALYs.

Under the *TTT* approach, the interviewees are asked to imagine themselves living in the described health status for T years and then to tell how many years in that health status they would be willing to trade off in exchange for life years in full health. For example, if T = 10 and the respondent says that that is equivalent to 8 years in full health, then 8/10 = 0.80 becomes the quality weight attached to the 10 life years to yield 0.8 QALYs.

One can imagine still other ways to tease quality weights out of people.

Quality-Adjusted Life Years as Quality-Adjusted Life Expectancy

For pedagogic reasons, the previous example of QALYs is very stylized and simple. In actual applications, QALYs usually are calculated as quality-adjusted life expectancy.

To understand this concept, consider a cohort of, say, 1 million 50-year-old persons at some point in time, which we shall call $t = 0$. If of this cohort 975,380 survive to time $t = 1$, we define the ratio 975,380/1,000,000 = 0.98538 as the 1-year survival probability of these 50-year-olds. We denote that survival probability as $P_{50,1}$. In similar fashion, if 781,453 of that cohort are alive at time $t = 24$, then $P_{50,24} = 0.781453$, and so on.

The algebraic sum of these survival probabilities up to and including $t = N$ is called the *life expectancy* of this cohort of 50-year-olds alive at $t = 0$, where N is the last year to which anyone of this cohort survives.

Now let $0 \leq Q_{50,t} \leq 1$ be the quality-of-life coefficient expected for someone who now, at $t = 0$, is 50 years old now t years hence, with $Q_{50,t}$ meaning death and $Q_{50,t} = 1$ "perfect health." For example, $Q_{50,24} = 0.71$ would be the quality-of-life coefficient of that 50-year old in the 24th year from now, i.e., in year 24. Then the quality-of-life-adjusted life expectancy for this now 50-year-old is defined as the sum

$$\mathbf{QALY} = \mathbf{P_{50,1}} \cdot \mathbf{Q_{50,1}} + \mathbf{P_{50,2}} \cdot \mathbf{Q_{50,2}} + \mathbf{P_{50,3}} \cdot \mathbf{Q_{50,3}} + \ldots + \mathbf{P_{50,N}} \cdot \mathbf{Q_{50,N,}}$$

where, once again, N is the last year in which anyone from that cohort of now, at $t = 0$, 50-year-olds is expected to be alive.

Application of a new medical technology in the treatment of a disease—e.g., the Gilead Science, Inc.'s drug *Harvoni* for Hepatitis C—can change either the survival probabilities P or the quality-of-life coefficients Q, or both. A change in QALYs experienced in treatments without Harvoni to QALYs experienced with treatments including *Harvoni* would be the added QALYs attributed to the new drug. If the change in total treatment cost between the two treatment approaches was then divided by the associated change in QALYs, we would obtain the *incremental cost-effectiveness ratio (ICER)* associated with the new drug. In principle, that cost-effectiveness ratio, a dollar amount, can then be compared to the monetary value of an added QALY in a decision whether or not the drug is "worth" applying.

The practical problem in our latitudes, of course, is that there is no consensus on what the monetary value of a QALY is. Attempts have been made to infer that value from previous estimates of the value of a statistical life (Hirth et al., 2000), but these values per QALY exhibited a wide range, from about $21,000 to about $900,000 (see Table 1 of Hirth et al., 2000). In many evaluations of new medical technology researchers in the United States therefore have used the rather arbitrary numbers of $100,000 to $200,000 (Cutler et al., 2006; Van Nuys et al., 2015).

Disability-Adjusted Life Years Lost

Under this method, one posits as the *ideal* a maximum number of calendar years lived in perfect health, e.g., 80 years, and then counts as DALYs the appropriate fraction of calendar years lived in less than perfect health, as well as the number of calendar years short of 80 not lived at all.

The beneficial outcome from medical interventions or particular health policies is then measured by the reduction in DALYs—that is, the reduction in DALYs—attributable to that course of action. Those DALYs are then used as the denominator in the "incremental cost per incremental DALY" or simply "cost per DALY" attributed to that course of action.

Fig. 6.4 illustrates this concept for the simple, stylized example we had posited at the outset.

At the person's age 20, he or she is expected to lose:

$$\text{DALYs} = (80 - 60) + 0.4(60 - 45) = 20 + 0.4(15) = 26 \tag{6.10}$$

from a maximum life of 80 years in perfect health (without any disabilities). The person is expected to die 20 years before the maximum expected life of 80, but we count those 20 premature calendar years lost as 26 DALYs because 15 calendar years are lived with disability degree 0.4.

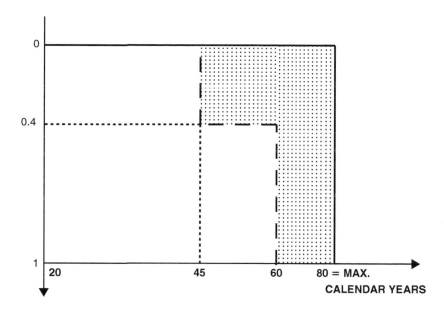

DEGREE OF DISABILITY: 0 = NONE, 1 = DEATH

FIGURE 6.4 The definition and measurement of disability-adjusted life years lost (DALYs).

In this simple illustration, we have counted a DALY as the same for any person at any age, and we did not discount future DALYs. In fact, as already noted the World Bank and the WHO give more weight to a DALY lost by a person of working age than to a DALY lost by a person too young to work or too old to work.

Change in Quality-Adjusted Life Years and Disability-Adjusted Life Years Lost Achieved Through Health Care

Suppose the projection depicted in the graphs previously was the baseline course of action B and then a new medical technology (course of action A) were applied to someone before age 45 with the following two consequences:

- The new treatment will extend the number of calendar years of life for this person by 10 years.
- At the same time, the new technology will enhance the person's health status in a way so that, from age 45 to age 70, 1 calendar year of the person's life will be scored as the equivalent of 0.70 of a calendar year in perfect health, rather than only 0.60 under the baseline treatment, or, equivalently, that the degree of disability suffered by this person is such that 1 calendar year of life with these disabilities is scored as a loss of only 0.30 calendar years without disabilities (in perfect health), rather than the 0.40 under the baseline.

The associated change in health status can then be calculated in three ways:

- *Unadjusted life years*: the intervention is expected to yield 10 added calendar life years.
- *QALYs gained*: The intervention is expected to yield 8.5 QALYs, calculated as $(60 - 45)$ $(0.70-0.60) + (10)$ $(0.70) = (15)$ $(0.10) + 7 = 8.5$.
- *Reduction in DALYs (added to life)*: After this intervention, the DALYs lost will be DALYs $= (80 - 70) + (70 - 45)$ $(0.3) = 10 + 25$ $(0.3) = 10 + 7.5 = 17.5$. We had previously calculated that, before the intervention, 26 DALYs would be lost. Thus there is a reduction in DALYs lost of 8.5.

Fig. 6.5 depicts the situation we have described previously. In this illustration, the shaded area depicts QALYs gained by the intervention, and also the reduction in DALYs achieved thereby. Is that always the case? It is not. In this stylized example, the number of DALYs and the number of QALYs yielded by the hypothesized medical intervention are equal only because age-weighting and discounting were not applied. Had the DALYs been age-weighted and had future QALYs and DALYs been converted to present values through discounting, the change in QALYs and DALYs would not have been identical.

HEALTH STATUS: 0 = DEATH, 1 = PERFECT   **DEGREE OF DISABILITY: 0 = NONE, 1 = DEATH**

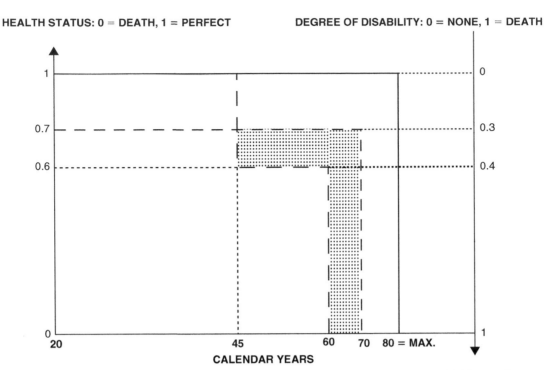

FIGURE 6.5   Added quality-adjusted life years and disability-adjusted life years lost from a medical intervention.

CAN ONE EVER AVOID PUTTING MONETARY VALUES ON HEALTH BENEFITS?

Because CEA and CUA merely rely on costs per some physical metric of outcome from alternative medical treatments, it may be thought that use of these methods avoids having to assign monetary values explicitly to the benefits produced by alternative treatments. In many applications, it may appear that it is so, but in fact it is not.

Comparing Alternative Treatments

A CEA could be based strictly on a comparison of the cost per unit of health outcome achieved with one treatment (A) with that of another (B). Dong et al. (2004), for example, compared thus two different approaches to treating type 1 diabetes mellitus with ACE inhibitors early or later, after the condition has been diagnosed. They calculated the incremental cost per added QALY yielded by earlier treatment.

Similarly, one might calculate the cost per unit reduction in the systolic blood pressure (bp) achieved with two alternative prescription drugs, A and B, on the assumption that the decision to lower pb by X units actually has been judged worth what it now costs, that is, that the average pb for a cohort of insured individuals is to be lowered by X one way or the other. If the cost per unit bp with product A is found to be lower than that with product B, the most *cost-effective treatment* would be the new pharmaceutical product. The rational, best clinical practice then would be to switch from product B to product A. In practice, of course, ignorance of the CBA and sheer inertia, coupled with the marketing tactics of different drug manufacturers (in some countries, including under-the-table payments to practitioners) might stand in the way of a quick move to the evidently best practice.

If, in the previous example, product A was associated with lower costs per unit reduction in bp, but had one or two somewhat riskier adverse not present in drug B, then one would resort to CUA to modify the physical outcome measure, as described previously. Here an implicit evaluation of benefits does creep into the analysis, but it is not a *monetized* one.

But even if one discovered through CEA or CUA the lowest-cost strategy among alternative treatment methods aimed at lowering the representative patient's blood pressure by a given number of points—the most "cost-effective" treatment—the decision-maker acting on behalf of society in societal benefit–cost analyses still had to confront at some prior time the fundamental question: "Is lowering average systolic blood pressure of patients by X points worth it at all, even with the most cost-effective drug treatment?" After all, many things are not worth doing even under the most cost-effective method of doing them. In our example, the decision-maker using the analysis still must *implicitly* put a monetary value on the benefits from lowering blood pressure by X points and compare that value with the least cost method of achieving that reduction. The only difference between CBA and CEA (or CUA) analysis, then, is that under CBA *exact* monetary values must be put on benefits, while under CEA (or CUA) one merely needs to think about the *minimum* monetary values a set of health benefits must have to justify the cost of achieving these benefits.

Cost-Effectiveness Analysis and Cost—Utility Analysis and National Health Policy

CEA or CUA can also be used to sharpen the public debate over national health policy—especially over cost-control strategies and the issue of rationing health care.

One can illustrate this proposition by thinking of a nation's health system as a giant business firm—e.g. Health USA, Inc. Its managers (doctors, hospital executives, and so on) may tell "society":

Taking as a state of nature how prudently (e.g., residents of Utah in the US) or sloppily (residents elsewhere in the US) you wish to live in regards to health maintenance, we in the health sector can wrest from nature additional, QALYs for you, at prices that range from very low (e.g., with immunizations, or with good prenatal care) to very, very high (e.g., with expensive biologicals that purchase at best a few added life-months for terminally ill patients.). Shown below is the current shape of our QALY-supply curve. Please tell us the maximum price at which you, the rest of society, are willing to purchase added QALYs from us, the health system, and we shall deliver them to you, up to that point.

The supply curve that might accompany this proclamation might be like Fig. 6.6. In the graph, point A may represent buying added QALYs purchased through low-cost preventive care (including furthering health literacy and education). Point B may represent a routine, life-saving surgery. Point C may represent more expensive chronic care, such as renal dialysis. Finally, point D may represent treatment of patients with biologicals such as Avastin and Erbitux. These products carry such a high price tag that they imply a price per added QALY of about $250,000.

The supply curve in the graph should be thought of as the locus of the most cost-effective clinical practices needed to achieve the level of QALYs on the horizontal axis. As noted in Fig. 6.6, health services researchers—really operations research—can help managers of the health system get from an inefficient treatment strategy—one that costs more than is necessary—onto the efficient QALY-supply curve. That task done, however, it is up to politicians or health insurance executives, as representatives of individual patients, to pick the maximum price at which they will purchase added QALYs for individuals, but with collective funds (taxes or insurance funds). Finally, it is up to the people to decide what they would like these representatives to do in this regard.

Fig. 6.6 is a pictorial rendering of one of the most vexing and intractable issues in modern health policy. The graph clearly poses the questions:

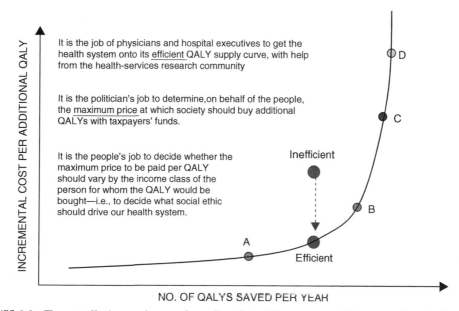

FIGURE 6.6 The cost-effective supply curve for quality-adjusted life years wrested from nature by a health system.

- Is there a maximum price per QALY above which those purchasing health care on behalf of insured patients (private or public third-party payers) should not go, to purchase additional QALYs out of the collective funds (taxes or premiums) entrusted to them?
- If there is such a price, should it be the *same* for all members of society—e.g., for a poor American insured by the government-run Medicaid program as for a corporate executive, or a young American and an old American?

Some nations have boldly tried to explore these issues as part of the debate on health policy. Former Governor Kitzhaber of the State of Oregon had tried to force his citizenry to come to terms with this issue as well, openly and honestly, like mature adults. But many other nations—and certainly most Americans—still steadfastly refuse even to engage on the issue. These nations do ration health care of course, but it is done implicitly, either by leaving some members of society uninsured, or by strict health-care utilization controls. In that context, economic evaluations of alternative health-care strategies, including new medical technology, can turn out to be just like howling into the wind.

UNRESOLVED CONTROVERSIES ON ECONOMIC VALUATIONS IN HEALTH CARE

In conclusion, it may be useful to examine briefly some of the lingering controversies on the economic evaluations in the context of health care.

Attempts by health services researchers during the past two decades to subject medical treatments—especially those based on new medical technology such as new drugs or medical devices—have remained controversial for on a variety of disparate arguments, each driven by a different a motivation. Some of the criticism of the technique rests of purely scientific methodology—e.g., how best to elicit valuations of health status from random samples of individuals. Others have rested on the ethical implications embedded in the application of technique which is, in effect, laden with subjective ethical values and not purely scientific. Finally, some objections are motivated strictly by narrow commercial interests—for example, the apprehension of drug and device manufacturers to see their products and revenues potentially threatened by transparency on cost-effectiveness.

Methodological Issues[10]

Like every endeavor to quantify the consequences of human activity, quantifying the positive and negative consequences of alternative medical treatments and using such metrics in CBA, which requires one to convert all benefits and costs into comparable units, is fraught with methodological difficulties. Pharmacoeconomics and all economic evaluations in health care are no exception. These methodological difficulties are described and debated continuously, at conferences around the world, by sincere health services researchers without any ideological or pecuniary axe to grind.

As this author has pointed out (Reinhardt, 1997) while there is great scientific merit in the open and sincere dueling over methodology among health services researchers at these conferences—and in the literature—it does carry with it the danger that policy-makers may distrust the entire approach and proceed, as usual, to make decisions on the basis of folklore, hunches, or their own ideological or pecuniary interests. Here it must be kept in mind that in their wrestling matches over measurement methodology, health services researchers have very good company.

For example, financial accountants, who face similar methodological difficulties of definition and measurement, have solved this problem by a dual posture. Intraprofessionally they engage in fierce arguments over issues such as the valuation of assets, depreciation, estimated bad debt expense, or financial derivative contracts. *Vis à vis* the users of their product, however, accountants have written down and sworn allegiance to a code of *generally accepted accounting principles* (the GAAP) that are accepted without question by the users of accounting information in government, finance, and business, even though every user educated in financial accounting knows that financial accounting data are systematically biased, highly imperfect, and often misleading metrics. The entire so-called real world thrives on such imperfection.

Objections on Ethical Grounds

More serious than problems purely of measurement science are the ethical implications embedded in the economic evaluation of human activity of any kind, especially so in the context of health care. Many of these concerns have been ably summarized in papers by Erik Nord et al. (1999) and Jeff Richardson (2002).

---

10. For a thorough review of these methodological issues, see Dolan (2000).

One genre of criticism revolves around the question just who should be asked to collapse life years spent in particular states of health into utility-equivalent QALYs or DALYs, as is done in CUA. A common argument is that persons in reasonably good health asked to imagine themselves to be in a much worse state of health would convert life years spent in that worse state of health into much lower utility-equivalent QALYs than would someone already in that worse state of health, because people tend to adapt to it. Implied in this line of argument is the proposition that economic evaluations of new medical technology based on QALYs or DALYs may be biased against individuals already in the low health status.

A similar argument is that the use of QALYs and DALYs implicitly biases economic evaluations against older people, because a successful medical intervention generally produces fewer added life years for older people than for younger people. This argument applies even if QALYs and DALYs are not explicitly age-weighted. If they are, as was illustrated earlier in this chapter in connection with DALYs, then the criticism applies with even greater force, especially in a country such as the United States which finds age discrimination abhorrent.

Finally, as Jeff Richardson (2002) points out, survey data have consistently revealed that the general public typically does not wish to see scarce health-care resources allocated to individual members of society so as to maximize the number of added QALYs or DALYs that can be wrestled from nature with a given national health-care budget—the seemingly efficient thing to do—but instead wishes to see those budgets directed mainly to the sickest members of society, even at the cost of a much higher number of QALYs that could be more cheaply purchased for other members of society. On the basis of that research, Richardson advocates so-called "empirical ethics," that is, a search for the socially most desired allocation of health-care budgets through iterative allocations and surveys of the public on the merits of the latest allocation.

Objection on Commercial Grounds

Although commonly paying lip service to the merits of economic evaluations in health care—and using such evaluations in their marketing when they turn out supportive of a new medical technology—the producers of medical treatments and the products going into them tend to look askance at these valuations when they threaten to detract from the top and bottom lines of their income statements. It can explain, for example, why any prior attempt by the US Congress to establish technology assessment centers in the United States—such as NICE in the United Kingdom or IQWiG in Germany—has been met with stiff opposition from the producers of medical technology and, so far, has faltered. In this effort the industry has been successful even to enlist entire political parties.

One can understand why the suppliers of health care and health-care products would be nervous about rigorous economic evaluations of their offerings. They are right in arguing that poorly performed evaluations can hurt not only these providers' economic position, but patients as well. And they are right to insist that any such evaluations be completely transparent and accessible to them for external audit, not only as to methodology, but also as to the raw data fed into the analysis. They are unlikely to be able to stop for much longer, however, the worldwide quest on the demand side for cost-effectiveness and benefit–cost analyses.

REFERENCES

Anand, S., Hanson, K., 1997. Disability-adjusted life years: a critical review. J. Health Econ. 16 (6), 685–702.

Brigham, E.F., Earhardt, M.C., 2008. Financial Management: Theory and Practice, twelfth ed. Thomson/South-Western.

Cutler, D., Rosen, A.B., Vijan, S., 2006. The value of medical spending in the United States 1960–2000. N. Engl. J. Med. 355 (9), 920–927.

Drummond, M.F., Sculpher, M.J., Torrance, G.W., O'Brien, B.J., Stoddard, G.L., 2005. Methods for the Evaluation of Health Care Programmes. Oxford University Press, Oxford.

Danzon, P.M., March/April 2000. Pharmaceutical benefit management: an alternative approach. Health Aff. 24–25.

Dolan, P., 2000. The measurement of health related quality of life for use in resource allocation decisions in health care. In: Culyer, A.J., Newhouse, J.P. (Eds.), Handbook of Health Economics, vol. 1B. North-Holland, Amsterdam (Chapter 32).

Dong, F.B., et al., 2004. Cost effectiveness of ACE inhibitors treatment for patients with type 1 diabetes mellitus. Pharmacoeconomics 22 (15), 1015–1027.

Garber, A., 2000. Advances in cost-effectiveness analysis. In: Culyer, A.J., Newhouse, J.P. (Eds.), Handbook of Health Economics, vol. 1A. North-Holland, Amsterdam (Chapter 4).

Gold, M.R., Stevenson, D., Fryback, D., 2002. HALYs and QALYs and DALYs, Oh My: similarities and differences in summary. Ann. Rev. Publ. Health 23, 115–134.

Hirth, R.A., Chernew, M.E., Miller, E., Fendrick, M., Weissert, W.G., July–September, 2000. Willingness to pay for a quality-adjusted life year. Med. Decis. Mak. 20 (3), 332–342.

Institute for Quality and Efficiency in Health Care (IQWiG), https://www.iqwig.de/en/home.2724.html.

Kavanos, P., Reinhardt, U., May/June 2003. Reference Pricing for Drugs: Is it Compatible with US Health Care. Health Affairs, pp. 16–30.

Lopez-Casasnovas, G., Jönsson, B. (Eds.), 2001. Reference Pricing and Pharmaceutical Policy. Springer Verlag Ibéria, Barcelona, pp. 82–83.

Murray, C.J.L., Zachary, A.Z., 1997. Understanding DALYs. J. Health Econ. 16 (6), 703–730.

Nord, E., Pinto, J.L., Richardson, J., Menzel, P., Ubel, P., 1999. Incorporating societal concerns for fairness in numerical valuations of health programmes. Health Econ. 8 (1), 25–39.

Ramsey, L., December 9, 2015. Why Hated Pharma CEO Martin Shkreli Is the Villain We Need. Business Insider. Available at: http://www.businessinsider.com/why-we-need-martin-shkreli-2015-12.

Reinhardt, U.E., 1997. Making economic valuations more respectable. Soc. Sci. Med. 45 (4), 555–562.

Reinhardt, U., 2001. Can efficiency in health care be left to the market? J. Health Policy Polit. Law 26 (5), 967–992.

Reinhardt, U., August 11, 2015. Probing our moral values in health care: the pricing of specialty drugs. J. Am. Med. Assoc. Available at: http://newsatjama.jama.com/2015/08/11/jama-forum-probing-our-moral-values-in-health-care-the-pricing-of-specialty-drugs/.

Richardson, J., April 2002. How Should We Decide Which Ethical Preferences to Include in the Economic Analyses of the Health Sector. Working Paper 129. Centre for Health Program Evaluation, Monash University, Australia. Available at: http://www.buseco.monash.edu.au/centres/che/pubs/wp129.pdf.

Rockoff, J.D., December 9, 2015. How Pfizer Set the Cost of its New Drug at $9,850 a Month. The Wall Street Journal. Available at: http://www.wsj.com/articles/the-art-of-setting-a-drug-price-1449628081.

Rolland, S., Goodman, A.C., Stano, M., 2007. The Economics of Health and Health Care. Pearson/Prentice Hall, Upper Saddle River, NJ (Chapter 4).

Smith, R.D., Richardson, J., 2005. Can we estimate the 'social' value of a QALY? Four core issues to resolve. Health Policy 74 (1), 77–84.

Van Nuys, K., Brookmeyer, R., Chou, J.W., Dreyfus, D., Dieterich, D., Goldman, D.P., October 2015. Broad hepatitis C treatment scenarios return substantial health gains, but capacity is a concern. Health Aff. 34 (10), 1666–1674.

Viscusi, W.K., 1993. The value of risk to life and health. J. Econ. Lit. 31, 1912–1946.

Williams, A., 1997. Intergenerational equity: an exploration of the 'fair innings' argument. Health Econ. 6 (2), 117–132.

Chapter 7

Health Services Research: Translating Discovery and Research Into Practice and Policy

J. Sanford Schwartz

University of Pennsylvania, Philadelphia, PA, United States

Chapter Outline

Key Points

- The goal of medical care is to improve patient and population health.
- Prevention, diagnosis, and treatment of disease require integrating knowledge and understanding across the lines of clinical, social, and management sciences and to the population.
- The delivery of medical care and medical decision-making is complicated by three fundamental problems: *variation* in disease and individuals' preferences and values; *uncertainty* of medical information in predicting disease natural history and response to therapy; and *scarcity* of resources.
- After the identification of a health problem, *translational research* fundamentals include the discovery of biological, physiological, and pathophysiological information and the subsequent development and rigorous evaluation of potential interventions.
- *Health services research* is a multidisciplinary scientific endeavor that studies and generates knowledge to facilitate improved translation of medical discoveries into practice to improve the health of patients and the public.
- Traditionally, assessment of medical interventions has focused on *safety* and *efficacy* (net benefit under optimal conditions). To guide clinical practice and policy, after safety and efficacy have been demonstrated, one needs to assess an intervention's *effectiveness* (net benefit under average or typical conditions) and *efficiency* (net benefit relative to cost).

Clinical and Translational Science. http://dx.doi.org/10.1016/B978-0-12-802101-9.00007-7

- *Randomized clinical trials* (RCTs) provide the strongest evidence of *safety*, and *efficacy* with *high internal validity* (likelihood of similar results with a similar group of patients), *data reliability*, and *reproducibility*.
- *Validity* and *strength of effect* are assessed in terms of *relative risk reduction* of the primary outcomes of interest with statistical significance commonly measured in terms of odds ratios and confidence intervals.
- *Effectiveness studies* assess net benefits, risks, and costs of using interventions in actual practice.
- To effectively provide the information required to inform and guide appropriate use of medical interventions, studies utilize a broad range of *experimental, quasiexperimental*, and *nonexperimental methods* (including randomized controlled trials), metaanalysis, observational data, functional status measures, economic assessment, and decision modeling.
- Medical information is characterized by a high level of *uncertainty* and *variation*; therefore, medical hypotheses and management require the application and integration of *probabilistic reasoning* (the estimation of statistical associations among clinical variables) as well as *causal reasoning* (the physiological cause and effect among clinical variables) and *deterministic categorical reasoning* (strategies compiled into well-defined rules).
- The first step in *decision modeling* is to structure the problem (from the decision being examined to the ultimate clinical outcome of interest), and thereafter, assign probabilities to each probabilistic event, attribute values to the outcomes of interest, and consider alternative strategies with sensitivity analyses.
- Most research is focused on *prescriptive decision-making* (how to improve medical decisions).
- Formal decision support models (such as decision analysis and simulation) inform clinical guidelines, practice, and policy decision-making.
- *Quality medical care* should (1) be accessible in a timely fashion; (2) consistently provide appropriate and effective care; (3) eliminate avoidable mistakes, errors, and adverse events causing harm; (4) reduce unjustified geographic variation in care and health outcomes; (5) be responsive to patient's needs and preferences; and (6) eliminate racial/ethnic, gender socioeconomic, and other disparities to ensure equal access and treatment.
- Quality can be assessed in terms of structure (characteristics of providers, institutions, health-care systems, etc.), process (content/timing of care received by patients), and outcome (result experienced by patient).

INTRODUCTION

In theory, there is no difference between theory and practice. But, in practice, there is.

Jan L.A. van de Snepscheut

The goal of medical care is to improve patient and population health. Prevention, diagnosis, and treatment of disease require integrating knowledge and understanding across the basic, clinical, population, social, and management sciences. While this process often is presented as orderly and structured (e.g., discovery of new knowledge, mechanisms, and techniques generated by advances in basic science research is subjected to rigorous clinical evaluation, with beneficial interventions adopted). In fact, the translational research process is complex and messy, characterized by bidirectional exchange of observations and findings. The delivery of medical care and medical decision-making is complicated by three fundamental problems: *variation* in disease among biological systems and variations in preferences and values across individuals; *uncertainty* of medical information and in predicting disease natural history and response to therapy; and *scarcity* of resources, with resulting need to allocate resources and address trade-offs among benefits, risks, and costs.

This chapter addresses the fundamental aspects of translational research that occur *after* identification of health problems; discovery of fundamental biological, physiological, and pathophysiological information; and subsequent development and rigorous evaluation of potential interventions. Too often, the scientific community has taken the attitude that, "If you build it, they will come"—that is, demonstration of evidence about medical innovations will diffuse, be adopted, and be effectively implemented widely, rapidly, and appropriately by providers, patients, and policy-makers. The evidence, however, clearly demonstrates that scientific discovery and evaluation represent but the first steps in the translation process and that without better understanding of and development of interventions for effectively and appropriately applying innovations, the fruits of scientific and clinical progress will be incompletely and inefficiently adopted and implemented and their potential benefits not fully realized.

WHAT IS HEALTH SERVICES RESEARCH AND WHAT ARE ITS GOALS?

Health services research is a multidisciplinary scientific endeavor that studies and generates knowledge to facilitate improved translation of medical discoveries into practice to improve the health of patients and the public. Health services research seeks to identify the most effective ways to organize, manage, finance, and deliver high-quality care (Agency for Health Research and Quality, 2002; AcademyHealth, n.d.). This, in turn, requires scientific understanding of behavioral, psychological, structural, organizational, social, and political factors that affect access to and effective and efficient delivery of quality, cost-effective health care to improve quantity and quality of life (Donabedian, 1980). Health services research examines individual, family, organizational, institutional, community, and population level factors that affect health behaviors and outcomes. Grounded in theory, the goal of health services research is to generate knowledge to inform and guide decision-makers and those responsible for delivering health care. Thus, health services research examines the trade-offs among health-care benefits, risks, and costs; factors influencing access to and adoption of medical and health-care interventions; and the impact of alternative care strategies on a broad range of health outcomes, including morbidity, mortality, and quality of life. The goal is to inform clinicians, institutions, and systems to improve the quality of the medical care they provide and, thereby, improve patient care and enhance individual health and the health of the public.

This chapter presents a broad overview of some of the central components of health services and policy research. Inherently multidisciplinary, health services research incorporates perspectives, theories, knowledge, and research methods of a broad range of clinical, social, behavioral, and management sciences, including clinical medicine, epidemiology, biostatistics, economics, political science, communication, psychology, public health, operations research, informatics, finance, and management. Thus, this chapter examines health services research through an overview of broad problem areas.

ASSESSING MEDICAL INTERVENTIONS: OUTCOMES, EFFECTIVENESS, AND COST-EFFECTIVENESS

Before new discoveries that have been translated into new technologies (e.g., drugs, devices, assays) and techniques and procedures can be adopted, their potential clinical benefit must be assessed to determine their clinical impact to guide their appropriate use. Rigorous, highly structured evaluations (discussed in greater detail in several other chapters in this book) are required to obtain regulatory approval for new drugs; regulatory requirements for new devices (e.g., diagnostic technologies; medical equipment; implants) are more variable. New medical practices and procedures generally do not require formal regulatory approval.

Traditionally, assessment of medical interventions has focused on safety (adverse events and adverse effects) and efficacy (net benefit under optimal conditions). However, while necessary, assessment of safety and efficacy is not sufficient to guide clinical practice and policy (e.g., reimbursement). Rather, once safety and efficacy have been demonstrated, one needs to assess an intervention's effectiveness (net benefit under average or typical conditions) and efficiency (net benefit relative to cost) (Box 7.1).

Efficacy is assessed through randomized clinical trials (RCTs) (see Chapter 3) designed to rigorously assess etiological and causal relationships. When carefully designed and conducted, efficacy RCTs provide the strongest evidence of safety (i.e., adverse events and effects) and clinical benefits because of high internal validity (likelihood of similar results if the same intervention is performed on a similar group of patients), data reliability, and reproducibility. As such, studies assessing safety and efficacy use a carefully prescribed protocol, conducted among well-defined populations, with well-specified inclusion and exclusion criteria, careful monitoring, and outcomes assessed by investigators blinded to the patient's randomly assigned treatment group using intention-to-treat analyses. Validity and strength of effect are assessed in terms of relative risk reduction of the primary (or, occasionally, surrogate) outcomes of interest, with statistical significance commonly measured in terms of odds ratios and confidence intervals, often compared with placebo. The goal of the study sponsor (who has invested substantial resources in product development) is to obtain regulatory approval as quickly as possible. Thus, regulatory approval-oriented RCTs often are targeted toward outcomes for which and enroll patients in whom the greatest net benefit can be demonstrated in the shortest time.

BOX 7.1 Assessment of Medical Interventions: Evaluation of Medical Care

Safety	Adverse effects acceptable?
Efficacy	Can it work?
Effectiveness	Does it work?
Efficiency	Is there sufficient value?
Efficacy	Net benefit *optimal* conditions
Effectiveness	Net benefit *average* conditions

In contrast, effectiveness studies assess net benefits, risks, and costs in actual practice. As such, patients frequently exhibit a broader range of inclusion criteria than in efficacy RCTs, with less intensive monitoring and more flexibility in management due to less extensive protocol-induced services and costs. Moreover, effectiveness studies focus on the *incremental* impact of the new intervention compared with standard therapy (as opposed to placebo, which is commonly used in efficacy trials). In addition to "real-world" or "practical" RCTs in which access to an intervention is randomized, following which patient care proceeds without a strictly prescribed protocol, effectiveness studies frequently use a range of quasi- and nonexperimental, observational designs. As such, effectiveness research provides estimates of the expected net benefits, safety, and costs in actual clinical practice. As with efficacy studies, the statistical significance of an intervention's incremental impact is assessed in terms of observed relative risk compared with standard therapy. However, in addition, absolute risk reduction is of fundamental importance to effectiveness and cost-effectiveness research. In practice, absolute risk reduction often affects effectiveness and cost-effectiveness to a greater extent than relative risk reduction. Thus, an intervention that provides small relative risk reduction for a medical problem that is common or has high morbidity and/or mortality may provide greater benefit than an intervention with greater relative risk reduction for a less common or less serious medical condition.

Assessing Medical Interventions

Where is the wisdom we lost in knowledge?

Where is the knowledge we lost in information?

T.S. Eliot, 'The Rock', 1934.

All medical interventions that obtain regulatory approval have some benefit. However, once adequate safety and efficacy have been established and regulatory approval has been obtained, clinicians, patients, payers, and other decision-makers require additional information about effectiveness and cost-effectiveness to determine:

- How much value?
- In which patients?
- Under what conditions?

Moreover, since some management options exist for virtually every medical problem, all assessments of medical interventions inherently are comparative. Thus, evaluation of new medical interventions requires assessment of their *incremental* value—i.e., what are the additional benefits, risks, and costs of the innovation compared with the next best alternatives, measured in terms of clinically relevant outcomes (survival, disease sequelae and complications, functional status, quality of life) and resource costs.

The reference standard for assessing medical interventions is the RCT (see Chapter 3). While well-conducted RCTs exhibit the highest level of validity, reliability, and reproducibility for the limited number of causal associations examined, they often have limited generalizability and transportability and frequently lack important clinically relevant information (Box 7.2).

In practice, patients often do not precisely meet the carefully specified inclusion and exclusion criteria that underlie RCTs. Thus, results must be generalized to a broader range of patients than typically studied. From a clinical perspective, evaluation of effectiveness requires assessment of *incremental* benefit and risk among a broad spectrum of patients in real-world settings, with adequate representation of patients from various clinically relevant subsets. Moreover, clinical management decisions often are driven by time horizons that exceed those of RCTs, or than can be practically addressed through RCTs, which are constrained in sample size and length (and which must, according to ethical standards, be terminated once definitive net benefits or harms are demonstrated). Medical decision-making also often is concerned with a

---

BOX 7.2 Limitations of Randomized Clinical Trials

- Patient selection
- Time horizon
- Outcomes assessed
- Clinically relevant subgroups
- Comparators (often placebo)
- Practice patterns/style of care
- Patient preferences RE: events/outcomes
- Resource use and cost

> **BOX 7.3 Common Methods to Assess Medical Interventions**
> - Randomized controlled trials
> - Metaanalysis
> - Administrative claims data
> - Functional status
> - Economic assessment
> - Decision modeling

broader set of clinical outcomes (e.g., functional status, quality of life, cost) and comparators than are usually accommodated within an RCT. Thus, important information required to determine an innovation's appropriateness in a given patient or population prior to adoption, use, and reimbursement often is not available from regulatory oriented RCTs or, often, even from postregulatory RCTs and other common epidemiological methods.

To effectively provide the information required to inform and guide appropriate use of medical interventions in general and medical innovations in particular requires studies using a broad range of experimental, quasiexperimental, and nonexperimental methods, including randomized controlled trials, metaanalysis, observational data (e.g., administrative claims, increasingly integrated with electronic clinical records and data sets); functional status and quality of life (e.g., cognition; emotional and psychological function; energy, fatigue, and vitality; physical function; role activities; social function; sexual function; sleep pattern); economic assessment (e.g., cost-effectiveness); and decision modeling. This, in turn, requires a broad range of methodological expertise and technical analytic capacity, drawn from such disciplines as clinical epidemiology, biostatistics, information technology, operations research, and economics (Box 7.3).

Metaanalysis and Data Synthesis

The rapid growth of medical scientific information has increased demand for more rigorous and objective methods of data analysis and synthesis. Metaanalysis uses a prospectively specified, structured approach to combine data from multiple studies and is a useful method for aggregation, integration, and reanalysis of existing data across studies to assess consistency and variability across study findings and increase information content, statistical power, and precision of estimates of uncertainty and, to some degree, generalizability. When used in conjunction with formal methods of combining expert clinical judgments, metaanalysis can increase rigor and transparency of clinical guidelines. However, the conclusions and strength of metaanalysis is limited to the range and quality of the studies upon which it is based and is no substitute for rigorous, valid and reliable research comparing outcomes of alternative clinical management strategies and interventions. Thus, expanding knowledge of efficacy and effectiveness and facilitating translation of medical advances require increased investment in rigorously designed clinical research.

Often addressing questions of clinical effectiveness and efficiency requires integrating prospective and retrospective observational data (administrative claims; electronic medical records; registries) from various practice settings (hospital and ambulatory; community based), assessing a broad range of clinical outcomes (including service utilization and cost; patient-reported functional status; quality of life), using active comparators, across a broad range of clinically relevant subgroups, over long time horizons, examining and accounting for patient adherence. Various study designs and statistical methods (e.g., case–control designs; case-mix adjustment for severity, comorbidity, functional status, sociodemographic factors, including various forms of regression, propensity scoring, and instrumental variables) are used to minimize (but not fully eliminate) the inherent residual confounding and bias that plague nonexperimental designs (see Chapter 4 on Epidemiology).

Administrative Claims Data

Administrative data on medical encounters to explore patterns of care are readily accessible and provide comprehensive measurement of events (at least uniformly measured events) and longitudinal follow-up at relatively low cost for clinically relevant episodes of illness. However, administrative data were developed for administrative reasons and thus generally do not provide important clinically relevant information, level of clinical detail, and data on some important outcomes of interest. Administrative data also frequently suffer from limited data accuracy, consistency, and completeness. Thus, causality and significance of observed findings are unclear. Administrative data are most useful when linked with complementary clinical databases, such as registries and electronic medical records. Creation of comprehensive, representative, integrated clinical, and administrative data that are widely accessible to clinical investigators and the development and validation of new analytical methods are important priorities for informing our understanding and facilitating translation of effective interventions to patients and populations.

BOX 7.4 Common Major Domains of Functional Status

- Cognition
- Emotional/psychological function
- Energy, fatigue, and vitality
- Physical activities
- Role activities/social activities
- Sexual function/sleep pattern

Functional Status and Quality of Life

Functional status has long been recognized as an important outcome, especially for disabling and chronic diseases that comprise an increasing portion of the disease burden (Box 7.4).

Functional status measures can be disease specific (e.g., New York Heart Cardiac Classification; Canadian Heart Failure Questionnaire; Karnofsky Index in patients with cancer) or assess general health and functional status [e.g., Sickness Impact Profile (SIP); General Well-being Index (GWB); Nottingham Health Profile (NHP); Quality of Well-Being (QWB); Medical Outcome Study (MOS)]. The importance of patient perspectives is discounted by some researchers and policy-makers in the belief that patients have very limited knowledge of what constitutes technical quality and because of the difficulty in measuring patients' views accurately and reliably. Therefore they are wary of using patient-reported outcomes to assess medical interventions, considering them "subjective." However, these measures, when well constructed and carefully administered, have demonstrated psychometric properties (i.e., validity, reliability, reproducibility) comparable to many physiological measures commonly used to assess medical interventions. Moreover, excluding such important outcomes from formal assessment of medical interventions is tantamount to not valuing such outcomes at all, despite the importance of the domains that they assess. Important active areas of research include enhancing the translation of functional status measures into practical tools that can be used by and are of value to clinicians and patients, demonstrating associations between functional status and response to therapy and linking functional status measures to patient preferences and utilities (see section Preference and Utility Assessment).

Cost-Effectiveness

In a cost-constrained environment, where resources are limited, wants and opportunities exceed resources and thus choices must be made, information on *incremental* clinical, economic value and cost-effectiveness (see Chapter 6) increasingly is required to guide reimbursement policy by insurers and other purchasers, as well as patients and providers.

Medical interventions that have received regulatory approval have some demonstrated potential benefit and value. The challenge facing providers, patients, payers, administrators, policy-makers, and other decision-makers is to determine how much value, in which patients, under which conditions. Cost-effectiveness analysis research (discussed in Chapter 6) provides a framework for addressing and informing questions of efficiency and value. The objective of economic analysis is to assess the most efficient use of available resources, defined in terms of patient outcome and cost. Cost-effectiveness analysis is useful when one intervention is more effective but more expensive than an alternative to determine whether the improved outcomes of a more effective intervention warrant its increased costs. Economic analysis therefore assesses opportunity costs—the value obtained or foregone from alternative deployment of available resources. Thus, cost-effectiveness analyses must take into account resource use, patient preferences and utilities, and the perspectives of the range of relevant decision-makers (e.g., patients, providers, payers, public).

Cost-effectiveness analyses should be performed from a societal perspective, calculating the costs and benefits to patients, providers, payers and the community, considering all relevant costs and benefits, regardless of which party incurs the various costs or accrues the various benefits. Subanalyses may examine the differential impact of the intervention across these various perspectives.

Economic analyses should consider direct medical costs (e.g., acquisition and administration of provider, hospital, and home health services, including drugs, devices, tests and procedures, including those attributable to adverse events, and disease sequelae affected by the interventions). Economic analyses should measure actual resource use and assign estimated actual costs of these resources, rather than charges that often have little relationship to true resource costs. In addition, the value of medical interventions also is a function of gains and losses in productivity ("indirect" costs) and important but nonfinancial "intangible" costs, such as pain, quality of life, and functional status.

Cost-effectiveness research is inherently comparative, assessing the *incremental costs* of the interventions being compared relative to the *incremental benefits* in outcomes. The cost of providing a service or intervention generally is relatively similar across patients, but incremental benefits may vary substantially as a function of disease stage, disease severity, and coexisting morbidity. Therefore, absolute risk reduction and cost-effectiveness often differ across clinically relevant subgroups. Thus, cost-effectiveness research is of particular value in informing how best to target innovative interventions to those patients and populations where the incremental benefit and value are greatest. Since adverse disease outcomes and sequelae increase in incidence as disease severity increases, effective medical interventions often provide greater incremental absolute medical benefit in higher risk populations. (The exception is conditions where there is a threshold beyond which the effectiveness of therapy declines.) Given that for most diseases there are many more patients in early stages of disease than more advanced stages, for many diseases and health problems, targeting interventions to patients and populations often can maximize health outcomes and save considerable money relative to more widespread, less selective adoption.

Decision Modeling

Even this broad range of experimental, quasi-, and nonexperimental methods is not adequate to address a subset of important clinically relevant questions that relate to unobserved variables, factors, and outcomes. In such situations, modeling (e.g., decision analysis; simulation) must be employed to extrapolate directly observed outcomes to time frames that extend beyond those of RCTs or observational data; incorporate relevant competing risks and integrate the broad range of data on risk, benefit, and outcomes from individual studies; incorporate a broad range of perspectives; and assess the sensitivity of findings and conclusions to the range of observed variation in key variables. When carefully performed using rigorous methods (i.e., when model assumptions and input estimated variability are driven by careful analysis of empirical data using rigorous epidemiological and statistical principles, informed by and consistent with biological knowledge and models), decision modeling has demonstrated high levels of validity. But expert clinical judgment always will be required to extrapolate from limited rigorous data on safety, efficacy, and effectiveness to the far broader spectrum of patients and clinical problems cared for by clinicians.

Medical Decision-Making

Most people would rather die than think. In fact, they do.

Bertrand Russell

Translating discovery and innovation into practice requires optimal decision-making and effectively translating information into intentions and behavior.

The Medical Decision-Making Task

Medical decision-making is complex (Kassirer, 1990). Medical information is characterized by high levels of *uncertainty* (imperfect information) and *variation* (biological; measurement) and thus requires application of probabilistic reasoning. Moreover, there is substantial variation across patients, providers, and other decision-makers with regard to risk and outcomes preferences and perspectives. Medical decision-making is characterized by large volumes of information often conflicting or of uncertain validity and reliability that evolves and changes rapidly and is distributed among disparate sources. Moreover, important information is often not available. Medical decisions often have high stakes; incorporate a variety of perspectives, preferences, values, and needs; and must be made within a fragmented, complex organization, financing, delivery system characterized by inadequate clinical information system and misaligned incentives in a complex social, political, and economic environment.

In theory, physician decision-making is hypothesis driven. Preliminary hypotheses are generated from initial observations and information selectively gathered and interpreted within the relevant clinical context, with specification of expected findings (or their absence). As additional data are acquired, hypotheses are revised and refined, with elimination of selected hypotheses and recruitment of new hypotheses. Hypotheses are then verified if they are coherent, parsimonious, and consistent with observed data. The resultant working hypothesis or hypotheses guide development of a patient management plan driven by the estimated risks and benefits of alternative intervention strategies.

Medical hypotheses and management require application and integration of probabilistic reasoning (i.e., estimation of statistical associations among clinical variables), causal reasoning (i.e., physiological cause and effect among clinical variables), and deterministic categorical reasoning, in which strategies are compiled into well-defined rules.

Types of Medical Decision-Making Reasoning

Probabilistic reasoning requires estimation using mathematical principles of a problem's prevalence (prior probability) and of a series of conditional probabilities related to the frequency of features associated with defined diseases (Pauker and Kassirer, 1980). However, because of shortcomings of data and inherent cognitive limitations, people (including physicians and other experts) frequently engage in widely recognized biases and heuristics when making probabilistic decisions—shortcuts that while often of value also lead to predictable incorrect estimates and errors in judgment (Teversky and Kahneman, 1974; Kahneman et al., 1982).

Causal reasoning unites findings in a common framework and provides consistency checks among related findings. Causal reasoning is particularly useful when abnormal findings or events violate normal physiologic expectations and for interpreting findings and relationships that do not fit an idealized pattern or are not obvious from probabilistic associations or previously compiled concepts. Causal reasoning provides important support for development of professional expert guidelines and expert systems, including structuring complex decision support models.

Deterministic reasoning compiles knowledge from any source into unambiguous rules, identifying rules that describe routine practice—"if (certain conditions met), then (certain action appropriate)." Deterministic reasoning is useful for common, routine, simple, well-defined, straightforward diagnostic problems when conditions are recognizable, meaningful information is available, and action is specified that has known, predetermined consequences. However, deterministic reasoning is not effective for problems characterized by significant uncertainty and thus is of limited applicability for multiple complaints or interacting diseases or complex clinical problems. Furthermore, care must be taken when using deterministic reasoning to assure that rules are based on data and evidence derived from rigorous observation and experimentation, as opposed to subjective (often faulty) opinion.

Descriptive Decision-Making Research

Research in descriptive medical decision-making (how people make decisions) is one of the major areas of focus of health services research. Cognitive psychologists, operations researchers, and clinicians have been informing and improving our understanding of cognitive decision-making, focusing on errors in probabilistic reasoning. This body of research involves identifying common cognitive errors, the circumstances under which they occur, and developing strategies to avoid or minimize their occurrence and impact (Tversky and Kahneman, 1974). Most of the research in this area to date has focused on physician decision-making, with much less attention to nonphysician providers, patients, and their families.

Prescriptive Decision-Making

Most research in medical decision-making is focused on prescriptive decision-making—i.e., how to improve medical decisions. Increasingly, formal decision support models (decision analysis; simulation) inform clinical guidelines and practice and policy decision-making.

The first step in decision modeling is to structure the problem, from the decision being examined to the ultimate clinical outcome of interest. Probabilities then are assigned to each probabilistic event, using the best information available. Values are assigned to the outcomes of interest. Once data values and their distributions are estimated, the expected likelihood and value of alternative strategies are calculated, with sensitivity analyses performed to estimate the robustness of the model results and to determine the impact of various levels of uncertainty for key variables on the findings.

Decision modeling has several advantages over more traditional forms of evidence synthesis and decision-making. Decision models force a systematic examination of the problem and assignment of explicit values. Therefore, even before a model is run, it makes explicit the thought processes, evidence, and valuation and weighting used by decision-makers. Thus, disagreements can be focused on questions that can be addressed and informed by evidence, as opposed to opinion. Finally, decision modeling controls complexity. All models are representations and thus, to some degree, simplifications of reality. However, as long as such representations and simplifications adequately represent reality, they are able to suffice for decision-making. This simplification of inherently complex problems, along with computer calculations of cognitively challenging tasks, avoids information processing errors that are common in complex decisions.

Decision modeling is difficult, time-consuming, and requires substantial clinical and technical expertise. A good model must represent the clinical scenario, including the biological model and causal reasoning underlying relevant alternative diagnostic and management options. This requires a good understanding of the clinical problem, its key variables and their interrelationships. Also, difficult decisions are required with regard to assessment and valuation of the evidence that underlies the point estimates and probabilistic distribution estimation of the uncertainty surrounding these estimates. Thus, modeling requires substantial expertise in evaluating and weighting clinical research evidence, as well as mathematical and statistical knowledge, skill, and expertise.

While increasingly complex models often represent closer approximation of reality, complexity has its problems. The results of complex models often are not intuitive and as model complexity increases, so does the difficulty of explaining the results to clinicians and policy decision-makers, with resulting declines in face validity.

Because decision models focus decisions on outcomes of interest, decision modeling is useful for organizing and making explicit the basis on which decisions are made; identifies the critical elements that drive decisions and for which more research is warranted to the extent that more precise estimates will clarify the relative value of alternative management options; and focuses on the task and values clinical management strategies in terms of their impact on improved clinical patient outcomes.

Preference and Utility Assessment

An important component of decision models and an increasingly active area of active clinical investigation is the assessment of patient preferences and utilities. For many clinical situations, the choice between alternative clinical management strategies is sensitive to differences in the patient preferences for various clinical outcomes. Yet assessment of utilities of patients, providers, and other decision-makers is complex, with trade-offs between respondent task simplicity and clarity versus estimate validity. Analog scales are easy to use and often give high test–retest values, but have lower levels of validity and mathematical cohesion. In contrast, probabilistic scales (standard gamble; time trade-off) increase respondent burden but display improved psychometric properties.

Another research challenge concerns mapping measures of functional status onto utility scales. As discussed above, functional status is increasingly being recognized as an important outcome of care that formally needs to be incorporated into medical decision-making. This remains an area of active and ongoing investigation, with most work focused on the standardized, more widely used utility assessment instruments (e.g., "Quality of Well-Being"; Euroqual scales).

Evidence-Based Medicine

Medical care should be based on strong scientific evidence to maximize translation of patient benefit from the most effective and cost-effective interventions and to protect patients from unnecessary harm and costs. As noted above, clinical medicine is characterized by inherent variability across patients, and time and data often are incomplete, uncertain, and conflicting. As with all research, validity, reliability, and confidence in results are greatest when there is high consistency and congruency across studies and methods. Given cognitive limitations, subjective integration of results from multiple studies often results in errors of judgment and attribution. Thus, increasingly, formal, standardized, quantitative methods are required to reduce and synthesize disparate information, an approach commonly referred to as "evidence-based medicine."

Facilitating Appropriate Behaviors

It is a vexing problem ... many innovations, technological advances, and proven new treatments are too slowly adopted ... At the same time, other innovations and new treatments get diffused too quickly, despite insufficient scientific evidence of their clinical utility.

Accelerating Quality Improvement in Health Care: Strategies to Speed the Diffusion of Evidence Based Innovations. NIHCM/NCQHC Conference Proceedings, 2003.

Ultimately, the value of information is to facilitate appropriate, desired behaviors. In a sense, the challenge of changing behavior in desired ways is the central goal of translational research initiatives to facilitate and stimulate translational research. Many effective medical practices are too slowly and incompletely adopted, while unproven interventions of little if any demonstrated value diffuse widely. Similarly, patients demonstrate poor adherence with prescribed medications and effective health promotion, disease prevention behaviors while widely adopting products and behaviors of undetermined benefit.

Ideally, diffusion of innovation is driven by science and evidence: innovation is rigorously assessed prior to adoption; scientific evidence narrowly constrains decisions; the physician or other decision-maker is aware of, knows, and correctly interprets the relevant scientific evidence; the clinician serves as the patient's agent, acting on the patient's behalf in the same fashion as the patient would if the patient had the requisite scientific and medical knowledge and judgment (this, of course, assumes the clinician knows the patient's preferences and utilities). In addition, increasingly physicians and other decision-makers must allocate scarce societal resources. In this ideal scenario, beneficial technologies are rapidly and widely adopted; discredited technologies are not adopted or discarded.

Thus, classic diffusion theory frequently represents the adoption and behavior change process as a rational, logically sequenced series of steps (Fig. 7.1). However, these criteria are rarely, if ever, fulfilled. While informed and guided by information, adoption and diffusion of medical innovation and practice also is influenced by social, cultural, organizational, and other environmental factors (Rogers, 2003). How one interprets and acts upon information also is a function of the decision-maker, the information source (the messenger), the nature of communication (the message), and characteristics of the innovation itself (e.g., cost, risk, reversibility, ability to try) (Box 7.5).

A decision-maker's attitude overall toward risk influences that person's propensity to adopt a medical innovation, with early adopters being less risk averse. On average, younger age, specialist status, group and multispecialty group membership, and urban or suburban location are associated with earlier trial and adoption of medical innovation. Other things being equal, the more extensive and diverse one's professional, social, and intellectual networks and degree of social integration, the earlier and more likely one is to be aware of and adopt a beneficial innovation.

An innovation's characteristics also influence its adoption. Advantages over available technology; ease of communication of perceived advantages; ease of trial, implementation, and observability of impact all facilitate earlier adoption. Early stage of development and risk of early obsolescence, patient demand, and increased cost or other economic disincentives inhibit early adoption.

Awareness of an innovation is a function not only of one's professional and social networks and exposures, but also is influenced by characteristics of the message, messenger, medium, audience, and the setting in which the message is delivered. While physicians often attribute their decisions to peer-reviewed published medical scientific literature, collegial interaction, especially with respected local and national opinion leaders, is more influential. In general, informal, personal interaction is more influential than more formal communication vehicles, such as continuing medical information, public media, advertisements, and direct marketing.

Environmental factors also affect technology diffusion and adoption. Large organizational size can variably facilitate adoption (e.g., when capital and acquisition costs are high) or inhibit innovation (e.g., when administrative processes slow

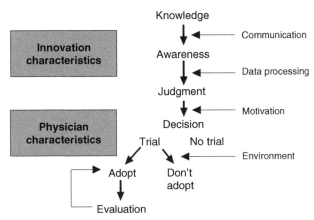

FIGURE 7.1   Simplified schematic of the classic paradigm of the adoption and diffusion of medical innovation.

BOX 7.5 Factors Influencing Diffusion and Adoption of Medical Innovation

- Physician characteristics
- Professional values and standards
- Organizational characteristics
- Market structure and competition
- Financial incentives

- Patient characteristics and values
- Channels of communication, awareness
- Manufacturer strategies
- Social and environmental factors

decision-making). Until recently, the earliest adoption of medical innovations commonly occurred within Academic Health Centers (AHCs). This occurs in part because of the perceived mission of academic medical centers and in part because of their large size and ability to finance expensive, capital-intensive technologies. Diffusion and adoption of medical innovation also is influenced by market competition, driven by patient demand or by provider financial incentives as a result of higher reimbursement or greater market share from patient perception of increased prestige and expertise. In recent years, competitive market forces and manufacturer strategies to reduce upfront capital expenditures have facilitated early adoption by community medical centers and physician groups.

Trial of innovations is, in part, a function of ease of experimentation with the intervention and reversibility of the trial. Thus, ease of discontinuation (e.g., pharmaceuticals; noninvasive diagnostic tests) and lower financial barriers to the physician and patient of initial use (e.g., capital investment required; unit cost) correspondingly lower the barrier to trial of the innovation. Risk of early obsolescence and legal, regulatory and administrative restrictions, or reporting requirements also inhibit early innovation, while the prestige from being perceived as an innovator or on the cutting edge of practice is associated with earlier adoption.

Judgment of innovation similarly is heavily driven by peers, especially opinion leaders, in addition to rigorous, peer-reviewed, medical scientific evidence. Respected regulatory (e.g., FDA), public health agencies (e.g., CDC, NIH, AHRQ), and professional societies also influence physician assessment of an innovation's benefit. Physician judgments also are influenced by patient feedback and peer experience with respect to patient outcomes and acceptance.

Changing Physician and Patient Behavior

A variety of interventions have been proposed to change physician behavior to encourage closer adherence with scientific evidence and recommended practices to improve processes and outcomes of care (Box 7.6).

Education is most effective when provided by influential peers ("opinion leaders"). Peer pressure in conjunction with data feedback (e.g., providing risk-adjusted outcomes comparing process and outcomes of physicians and hospitals with those of their peers), financial and nonfinancial incentives, and modification of the environment to support and facilitate desired behaviors have demonstrated success in modifying physician behavior. Innovation diffusion and provider behavior is most effectively modified by a combination of interventions (Fig. 7.2).

Changes in administrative structure and process and regulation often appear effective, at least in the short run. However, unless these interventions are supported by underlying values and reinforced by financial and nonfinancial incentives, they are easily bypassed and rapidly extinguished. Moreover, unless administrative restrictions are limited, they quickly become cumbersome. Education is often (but not always) necessary and frequently not sufficient to change behavior. New information is adopted and translated into practice faster and to a greater degree with greater persistence combined with supporting feedback, peer pressure, incentives, and social and organizational support. Feedback, especially when combined with peer pressure and financial and nonfinancial incentives, can be especially effective forces supporting behavior change. As noted above, education, especially when reinforced by opinion leaders, can be an effective modifier of

---

BOX 7.6 Interventions to Change Physician Behavior

- Education
- Administrative structure/process
- Regulation

- Feedback
- Peer pressure
- Incentives (financial; social; psychological/behavioral)

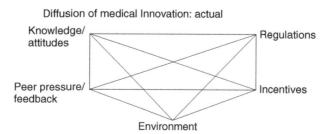

FIGURE 7.2 Factors affecting the adoption and diffusion of medical innovation.

behavior. Financial incentives also can be powerful motivators of behavioral change. Positive incentives (bonuses) may be more acceptable and effective than negative incentives (penalties). However, financial incentives must be carefully calibrated, balanced, and weighted and their effectiveness and impact modulated by coexisting structural elements. The environment strongly influences and modifies the context in which behavior and practice take place. Indeed, modifying the environment often is an important and effective way to facilitate and support desired behavioral change.

Studies of physician behavior have focused disproportionately on AHCs, particularly trainees, even though the generalizability of results from these studies to the vast majority of medical care which is provided in community-based ambulatory settings is unknown. To address this research shortcoming, the Agency for Health Research and Quality in recent years has funded the early development of research-based practice networks in a broad range of practice settings, but there is both need and opportunities for development of more and a broader range of community practice networks and to expand this research more broadly to community-based settings.

Studies of patient behavior are even more limited. What we know about influences on patient behavior is largely informed by public health, communications, and marketing theories. Perhaps the most influential have been the Health Belief Model and the Theory of Reasoned Action and variations thereof (Fishbein and Ajzen, 1975; Fishbein, 1980). These theories state that behavior is influenced by the perceived importance, perceived benefits, and perceived barriers to a given behavior. These attitudes and beliefs are influenced by family, friends, and social pressure, as well as by respected, trusted authorities such as physicians and other medical professionals. Central to the Theory of Reasoned Action is the concept of self-efficacy—that action is possible and will lead to the desired improvements in outcome—and its impact on intentions, which are closely tied to and must be modified to alter subsequent behavior. Informed, shared decision-making, which also incorporates elements of trust, is a relatively recent area of increasingly active investigation.

The content and design of patient messages should be targeted to influence knowledge, attitudes, and beliefs, which may vary at different stages of the patient's adoption and diffusion process. Thus, the nature of information and how it is communicated, and facilitators, action cues, and feedback support may need to be different for patients who are not even aware of the problem, for those who are aware but not committed to behavior change, for those who want to engage in the behavior and for those who are engaging in the behavior but who need to sustain it. Similarly, communication and various supports might need to be different for chronic as opposed to acute conditions, asymptomatic as opposed to symptomatic conditions, and therapeutic as opposed to preventive interventions.

Most studies of physician and patient behavior to date have been largely empirically based. Given the importance of understanding and affecting behavior change to accelerate the translation of discovery into practice, there is great need for more research in this area that is explicitly grounded in, integrates, and informs behavioral, social, and communication theory. Such research should be given much higher priority by agencies that fund translational research.

Given their impact on behavior, there is a great deal of interest in "pay for performance"—making a portion of the reimbursement and compensation conditional on attaining specified targets, related both to process of care and outcomes. Financial incentives are attractive because they encourage and support behavior change and are motivational as opposed to requiring direct intervention and intrusion into the patient—physician encounter. However, financial incentives also pose significant challenges. One needs to be alert to conflicts of interest. Moreover, they are difficult to calibrate, easily manipulated, and not transparent.

However, the enthusiasm for implementing such policies is not supported by empirical research. It is clear that financial incentives must be carefully designed, implemented, and calibrated to avoid undesired unintended consequences. Moreover, while financial incentives have been demonstrated to affect provider behavior, no rigorous RCT has yet demonstrated the effectiveness of "pay-for-performance" interventions in improving patient outcomes. Thus, rigorous research in this highly promising area of behavioral economics is required.

Communicating and Translating Information Into Decisions and Practice

Given the fundamental importance of communicating information that is effectively translated into provider and patient behavior, it is surprising how little attention has been devoted to this area and how low a priority this field has occupied in translational research. This area of investigation requires understanding individual, social, cultural, organizational, and financial facilitators of and barriers to initiating and sustaining desired intentions and behaviors and development of evidence-based, actionable messages; determining effective channels of communication and respected, credible, trusted messengers; and targeting the messages to relevant decision-makers and those who influence them. Much of what is known about this area derives from research and fields that have been largely outside the standard biomedical research enterprise—sociology, psychology, marketing, communication, organizational management, but too often this fundamental research has not benefited from adequate clinical input. Future research and funding in this area should integrate these currently largely independent fields and disciplines.

Quality Medical Care

As to diseases make a habit of two things — to help, or at least, to do no harm.

Hippocrates, Epidemics, book I, chapter 11.

It is, at first glance, somewhat surprising for concerns about quality of care in the United States to be increasing at a time when, from a technical and scientific standpoint, the nation's health-care system capabilities are extraordinary and unparalleled. The substantial public and private sector investments in fundamental biomedical science and applied research over the past 50 years have culminated in new forms of technology that have greatly improved medical care and health status. Despite this, the full potential of medical science has not been fulfilled. As our methods to measure and improve the quality of care have evolved and improved, it has become apparent that serious, extensive problems in health-care quality exist across all types of providers, delivery systems, and financing mechanisms.

The US medical care system is technologically advanced, with a unique focus on the consumer and relative ease of access for those with insurance coverage or ability to pay. However, access to this same system is limited for tens of millions of Americans without adequate insurance, and the system is plagued by glaring quality problems as evidenced by significant variations in service use and clinical outcomes (Wennberg and Gittelsohn, 1973; Fisher et al., 2008) and widespread disparities and inequities in health-care access and utilization and health outcomes. While US physicians, hospitals, technical capacity, and research programs are highly respected, poor quality medical care is widespread. Adoption of scientific knowledge into clinical practice is uneven and often delayed. The Institute of Medicine estimated that it takes an average of 17 years for new knowledge generated by randomized controlled trials to be incorporated into practice. Medical care is characterized by underuse of safe and effective services, overuse of interventions of marginal or no benefit, errors and mistakes leading to patient injury, and widespread inequities in access, outcomes, and health status (Institute of Medicine, 1998, 1999, 2001, 2002; President's Advisory Commission, 1998). Indeed, there is widespread consensus "that the quality of health care received by the people of the United States falls short of what it should be" and requires systematic change (Institute of Medicine, 2001).

Recent surveys demonstrate widespread erosion of confidence in and assessment of the quality of medical care. More than 50% of US physicians say that their ability to deliver quality care has worsened over the past 5 years; the majority of Americans believe that changes in insurance have worsened the care they receive; a slim majority rates the quality of health care as good or excellent, and a substantial percentage believes the health-care system needs to be redesigned.

While quality medical care is grounded in scientific discovery and evidence, it ultimately falls to health-care professionals and organizations operating within the context, constraints, and influences of our underlying culture and social system to translate these advances to patients and the public. Achieving the promise of scientific medical discovery requires understanding and effectively addressing complex social, organizational, and behavioral factors. Effective and timely translation of rigorous scientific evidence on risk and benefit is hindered by the absence of programs translating rigorous scientific evidence on risk and benefit into everyday clinical practice. While information-intensive, medical care is characterized by rudimentary clinical information systems. Health-care systems are highly fragmented, lacking continuity, and coordinated care. Thus, research for improving quality of care requires integration of elements of clinical research with the behavioral social sciences, organizational management, operations research, information sciences, and systems design and analysis to design, develop, and implement a safer health-care system.

The remarkable advances in medical science also have created significant challenges to the provision of high-quality medical care. The knowledge base and complexity of modern medicine is huge and rapidly growing, exceeding the ability of any individual to keep up-to-date in all relevant areas. Chronic diseases and major comorbidities have become more prevalent, thereby increasing the burden of illness, as medical advances have extended survival from these and more acute conditions. The financing and delivery of medical care has become much more fragmented and complex, a result of continued increasing specialization and a diverse and poorly coordinated insurance system. Investment in clinical information systems severely lags that of other information-based service industries, and information systems are not adequate for the needs and demands of an information-intensive endeavor. Communication both within the system and with patients and caregivers is poor, despite dramatically increased interest by patients in substantive involvement in decision-making surrounding their own care. Incentives to physicians, patients, hospitals, insurers, payers, and health-care managers are often misaligned, thereby undermining desired behaviors and practices.

Quality Measurement and Assessment

The definition of quality adopted is very important in that it lays out the framework and approaches for quality goals and measurement. The Institute of Medicine defines quality of health care as the "degree to which health services for individuals and populations increase the likelihood of desired health outcomes and are consistent with current professional knowledge." Other common definitions emphasize balancing expected gains and losses that result from care and improvement or maintenance of individual patient and population quality of life and life expectancy and efficiency of the care process.

Quality is a multiattribute construct. Quality medical care should (1) be accessible in a timely fashion, without significant access barriers; (2) consistently provide appropriate and effective care; (3) eliminate avoidable mistakes, errors, and adverse events causing harm; (4) reduce unjustified geographic variation in care and health outcomes; (5) be responsive to patient needs and preferences; and (6) eliminate racial/ethnic, gender, socioeconomic, and other nonbiological disparities and inequalities in access and treatment. In other words, quality medical care should be safe, effective, appropriate, patient centered, accessible, and delivered in a timely, efficient, fair, and equitable fashion (Institute of Medicine, 2000). Quality measurement must address each of the elements in Box 7.7 and their components in a valid, reliable way.

Even once the elements of quality are defined, definitions of quality and emphasis on the various components of quality will differ according to the perspective from which one examines a health-care encounter, individual subjective utilities and preferences, and how information generated about quality will be used. Physicians typically focus on technical excellence and skills; knowledge, judgment, and decision-making; and appropriateness and timeliness of service prescription and delivery, such as whether an accurate diagnosis is made, the correct surgical procedure is performed proficiently ("the right thing, the right way, at the right time"), as well as whether the patient's health status has improved. Patients judge the health-care system through a more personal lens by their access to health care, functional status, and quality of life, the nature of one's interpersonal interactions with physicians and other clinical providers (empathy, honesty, sensitivity, compassion, understanding, listening, responsive to patient preferences and values, communication) and office and institutional staff (respect, courtesy, efficiency, friendliness, convenience, access) and the degree to which care decisions are consistent with personal values and preferences. Health-care managers, payers, and purchasers (health plans, employers, and government programs) place emphasis on health of enrolled populations and attributes of care that reflect organizational performance, such as access, appropriateness, cost, and efficiency; to the degree that outcomes are assessed, the focus tends to be on process measures and avoidance of avoidable serious adverse events. Purchasers also value patient satisfaction; public health officials value impact on population health and equity. None of the common definitions specify the components of quality required to measure and improve quality.

Variations in Care

A man who falls sick at home or abroad is liable to get heroic treatment, or nominal treatment, random treatment, or no treatment at all, according to the hands into which he happens to fall.

Jacob Bigelow in speech to Massachusetts Medical Society, 1835.

There are substantial, widespread variations in physician practices and patient outcomes that are not explained by differences in disease severity, case mix, other clinical factors, outcomes, or patient preference across physicians, hospitals, communities, regions, and nations; in adherence with professional guidelines and standards; and in response to changes in financing and reimbursement. This variation reflects a failure to consistently practice in accordance with the best scientific evidence, as well as a lack of clear evidence in some situations on what approach works best. These unexplained variations suggest misutilization (overuse, underuse and misuse) of medical services and care, as both high and low areas of service

---

BOX 7.7 Components of Quality Medical Care

- Access to care
- Scientific evidence base
- Avoid errors and adverse events
- Congruent with patient values

- Competent health-care providers
- Appropriate technology
- Timely, accurate information
- Efficient, patient-centered health-care system

for similar patients are unlikely to be correct. The result is worsened patient outcomes and increased costs of care. However, one cannot determine from observed variations in care the appropriate level of care. Thus, detection of variations that are not explained by clinical or patient preference differences is an indicator of a potential problem that requires further thoughtful, careful, critical investigation, and assessment.

Quality measurement and improvement is a very active area of investigation, and methods to measure and improve quality continue to evolve. Quality can be assessed using implicit criteria (when there are no prior standards or agreements about what reflects good or poor quality) or explicit criteria (where a strong scientific evidence base allows development of guidelines and standards) to determine whether a process of care is acceptable, whether better care might have led to improved outcomes, and whether observed outcomes are consistent with those predicted by validated, evidence-based models.

Quality can be assessed in terms of *structure* (characteristics of providers, institutions, and health-care systems, such as certification, training, licensure credentials; ownership; procedural volume; administration; organization; technology and facilities; financial resources; and stability); *process* (content/timing of physician and system care received by patients); and *outcome* (result experienced by patient) (Donabedian, 1980). An initial focus on structure has been largely supplanted by a focus on process and most recently outcome, because while easier to define and measure with some face validity, there is poor correlation between most structural measures and outcomes. Process can be measured implicitly (by unstructured or semistructured expert review) or explicitly (using prespecified criteria, rigorously developed through expert review of the medical scientific evidence). Improved outcomes, the goal of medical care, are multifactorial and frequently are more expensive and require more time to measure, particularly for chronic diseases and when multiple medical problems coexist. Outcome measures are most appropriate when there is a strong evidence base linking quality of care and outcome, when differences in quality may not be revealed by process measures, outcomes are sufficiently frequent to detect differences in quality of care with acceptable sensitivity, and outcome events can be assessed in an acceptable time frame.

Process and outcomes are complementary. Process measures are easier, faster, and less expensive to measure and sometimes easier to improve than outcomes. Process measures are most useful in situations where outcomes evolve and become evident over long periods of time (e.g., treatment of hypertension), whereas outcome measures are most useful for assessing quality when the outcomes of interest evolve rapidly (e.g., coronary artery bypass surgery). Process and outcome measures also are inextricably linked—process measures are valid only to the extent they reflect changes in improved health outcomes (i.e., variations in the attribute measured are strongly associated with subsequent outcome); similarly, outcome measures are valid measures of quality only to the extent that they are affected by alternative processes of care over which the provider has control.

Currently, most quality measures use administrative claims-based process measures. While adequate for assessing preventive care, assessments of other types of quality require more detailed clinical data including, in many cases, patient-reported outcomes. In addition to assessing quality, process and outcome measures can be used to evaluate the impact of clinical, organizational, or policy initiatives (e.g., the impact of "pay for performance" on quality of care).

Outcomes measures should be clinically relevant, scientifically sound, predictable and feasible to obtain at low cost, exhibit strong psychometric properties, and have comparison benchmarks. Potential quality outcome measures are:

- Deaths
- Complications/events
- Failure-to-rescue
- Readmissions
- Length of stay
- Functional status

Individual outcome measures often vary significantly in terms of statistical, epidemiological, and psychometric attributes (Box 7.8). For example, while mortality generally is considered a valid outcome measure, it often is characterized by limited sensitivity because death rates are so low for many conditions, with resulting low statistical power to detect

---

BOX 7.8 Potential Quality Outcome Measures

- Deaths
- Complications/events
- Failure-to-rescue
- Readmissions
- Length of stay
- Functional status

clinically important differences (e.g., with an in-hospital death rate of 2.5%, more than 3000 deaths must occur per hospital to be 80% certain of being able to detect a 50% difference in mortality rates between hospitals at $p = .05$, a level rarely observed for a medical procedure or intervention).

Conversely, complications are more frequent and thus potentially more sensitive measures of quality. However, complication rates often are not well correlated with common quality measures with high face validity and appear to be more sensitive to patient than provider factors (and thus less valid measures of provider quality). Thus, complication rates, while frequent and convenient to identify, at times may not be as valid a measure of provider or institutional quality as measures such as death or failure-to-rescue (death given a complication), which better reflect provider characteristics. Further, complication and death rates may at times measure different elements of quality and thus cannot be used interchangeably.

To be valid and reliable, measures of quality need to meet a variety of criteria and require an integrated system perspective. Quality measures (Box 7.9) should be patient focused, reflect episodes of illnesses and disease natural history, include information on relevant process measures and clinical outcomes (including patient-reported outcomes, such as functional status measures and quality of life), resource utilization, and costs, provide longitudinal follow-up, be appropriately adjusted for case mix (disease severity, coexistent disease/comorbidity, baseline functional status, socio-demographics such as age, sex, race, income), and include relevant benchmarks. Method of standardization (direct or indirect) also affects measurement and interpretation of quality performance data. Valid outcome measures should not be susceptible to manipulation or variations in coding definitions and data elements. Careful attention must be paid to data element definition and criteria and consistency of coding. Thus, data should be carefully monitored and audited for accuracy and reproducibility to assure validity and consistency across patients, providers, system, and time.

Given the significant limitations of currently available measures, development of more robust and clinically meaningful outcome quality instruments is an important and undersupported research priority.

Data Sources

Diverse sets of data are used to measure and assess provider quality of care and the impact of programmatic and policy initiatives on quality, including administrative claims data (medical care insurance and pharmacy benefit management companies; hospitals, pharmacies, and health-care systems; clinical records maintained by health-care professionals; and survey data collected by payers and providers). Each data source has its own set of advantages and limitations. While the most appropriate data source depends on the question being addressed and the purpose for which the information will be used, generally assessments that integrate patient-level clinical, administrative, and patient-reported outcomes data provide the most rigorous, valid, and reliable estimates of quality. There is great need for development and maintenance of problem- and population-based integrated databases accessible to clinical investigators, health services, and policy researchers at low cost.

Assessments of quality require careful attention to adjustments for potential confounders, one of the most important of which is differences in case mix. Such adjustments require a conceptual model that links variables such as age, coexisting conditions, clinical signs, and symptoms to the outcomes being assessed. These variables must be measured and appropriate statistical tests used to adjust for differences among them across subjects when comparing performance with an explicit standard. Appropriate use of the explicit outcome method virtually always requires detailed clinical data.

While guidelines have been developed in an effort to facilitate more rigorous methods to assess quality of care, this remains an important and active area of investigation, and methods continue to evolve as research in this area proceeds and data sources and analytical capacity expand. As with all research, validity, reliability, and confidence in results are greatest when there is high consistency and congruency across data sources, study designs, and methods.

Increasing public awareness of, interest in, and concern about quality have led to the growing availability of information on physician, hospital, and managed care plan performance derived from a variety of prospective and retrospective data examining a variety of outcomes using primarily quasi- and nonexperimental designs. These include careful analyses of state level and Medicare of physician and hospital performance for selected surgical and medical procedures and interventions, as well as less rigorous assessments by managed care organizations; and lay press ratings of variable quality. The validity and reliability of these analyses are themselves variable and generally have not been validated. The impact of this information on patients, providers, and payers has not been well studied and requires more rigorous investigation.

Given the dependence on observational data, a major limitation, and therefore active area of and priority for quality measurement and improvement research, is the development of improved methods for case-mix adjustment and statistical methods to adjust for the selection biases and other sources of potential confounding inherent in nonrandomized observation data, with an emphasis on the increasing common registries and integrated clinical/administrative databases.

Medical Errors and Safety

To err is human, …

Alexander Pope, An Essay on Criticism, 1709.

Some degree of error is inherent in all human activity … In highly technical, complicated systems, even minor errors may have disastrous consequences.

Lucian Leape et al. (1991)

Preventable medical errors and injuries are common, accounting for tens of thousands of US deaths annually (more than from motor vehicle accidents, breast cancer, or workplace injuries) and an estimated 1.3 million cases of serious disability or prolonged hospitalization, incurring incremental costs in excess of $20 billion (Institute of Medicine, 2000; 2001). Thus, medical errors rank among the most widespread and serious public health problems. Therefore, reducing medical errors and improving patient safety (freedom from accidental injury) are an important priority for translating the potential benefits of medical progress into practice to benefit individual and public health. Fortunately, most medical errors are avoidable, and much of the knowledge base required to reduce or eliminate medical errors can be determined and effective interventions can be developed, evaluated, and implemented. But significant research and practical implementation challenges persist.

As outlined by the Institute of Medicine (2001), threats to patient safety (freedom from accidental or preventable injury) result from adverse events (injury resulting from a medical intervention) and errors (failure of a planned action to be completed as intended or use of a wrong plan to achieve an aim). Errors may take the form of either an error of execution (an otherwise correct action that does not proceed as intended) or an error of planning (the intended action is not correct). Errors increase opportunities for patient harm; the accumulation of errors increases accidents and harm.

Medication errors, most of which are preventable, are common, occurring in 2–12% of medications ordered and administered in hospitals, emergency departments, and in ambulatory practice. While relatively few result in adverse drug events (ADEs), in the aggregate more than 7000 US deaths each year are attributed to medication error, affecting between 2.4% and 6.5% of hospitalized patients, prolonging hospital length of stay on average by two days, and increasing costs by $2000–2600 per patient. At the other extreme are less common but extremely harmful wrong site surgery (e.g., amputation of the wrong limb) errors.

Medication errors occur at each step of the drug delivery process (diagnosis, prescribing, dispensing, administration, ingestion, monitoring, systems, and management control). Factors associated with hospital drug errors include altered drug metabolism (as may occur with renal or hepatic compromise), known patient allergy to drug of same medication class, and use of the wrong drug name, abbreviation, or incorrect dose or frequency of administration. Adverse drug reactions may be caused by exaggerated physiological effects, which typically are dose related (Type A) and idiosyncratic reactions (Type B). Type A reactions generally are less serious but because of their frequency in the aggregate account for significant patient harm and thus high attributable risk. Moreover, because they are potentially predictable and avoidable, Type A reactions are a good indicator of quality problems. Type B reactions are less common, idiosyncratic, and thus difficult to predict, but often more serious, with the result that relatively more attention is devoted to these drug reactions. (FDA requires reporting of postmarketing unexpected drug reactions; there are no comparable reporting mechanisms for Type A reactions.) The advent of improved pharmacogenetic databases offers the potential for identifying patients at increased risk for many Type B reactions currently considered "idiosyncratic." The elderly are at increased risk of ADEs because of the increased number of medications prescribed, physiological changes associated with aging, and underlying multiple chronic disease that may make the elderly more prone to ADEs or less able to experience or recognize early warning signs or make them more sensitive to drug effects.

Physician errors may be caused by inadequate education and training, but more important are system factors—inadequate access on a timely basis to important information on drugs; inadequate information systems and support; incomplete fragmented, inaccessible patient information; multiple and changing formularies; and time constraints and interruptions.

Most medical errors result from the convergence of multiple contributing elements, rather than resulting from a single, identifiable root cause. Most errors result not from substandard or negligent care but rather from failure of interactions among human, systems, and environmental factors. Thus, while potentially avoidable, prevention of these errors requires an integrated systems approach and a reorientation of medical culture, which often attributes errors to an individual oversight or mistake. Interventions to reduce medical errors range from simple process changes (e.g., computer physician order entry with automated detection of dose and drug interaction data; unit dosing, standardized doses, and administration times; automated dispensing, patient, and drug bar-coding; pharmacokinetic monitoring of high-risk drugs; standardizing common processes of care) to more complex systems to detect ADEs. In general, combinations of multiple interventions that address human cognitive and behavioral, technical, systems, and environmental factors are most effective (Institute of Medicine, 2006). Rigorous studies evaluating the incremental benefit of intervention components and cost-effectiveness are required to better inform theory and practice.

Patient Adherence

Adherence has been defined as the extent to which a person's behavior—taking medication, following a diet, and/or executing lifestyle changes—corresponds with agreed recommendations and guidelines. Patient adherence with prescribed medications and services and recommended behaviors is poor. It is estimated that, on average, only about 50% of medications for chronic diseases are taken as prescribed, even for serious conditions for which effective, once-daily, well-tolerated oral medications are prescribed (e.g., statin therapy among patients with coronary artery disease). Moreover, patients are most likely to discontinue therapy early. Thus, many patients do not receive effective therapy. Most treated patients are not adequately monitored and thus do not achieve treatment goals.

Patients must be adherent to receive the full benefit of effective prescribed medical care. For any intervention, health outcome events, quality of life, resource use, and financial expenditures are a function of disease severity and associated morbidity and mortality as modified by treatment efficacy. Efficacy is mediated by adherence, with imperfect adherence (whether by patient or provider) reducing potential benefit (Fig. 7.3).

Poor adherence with prescribed therapy is associated with worse physiological control, negative clinical outcomes (increased acute events, hospitalizations), and increased cost of care. For example, highly adherent patients are more likely to achieve better blood pressure, LDL cholesterol, and hemoglobin A1c control than patients with medium or low compliance and are less likely to be hospitalized; patients adherent with statin, aspirin, beta blocker, or heart failure therapy are less likely to develop recurrent acute myocardial infarction, be hospitalized for an acute cardiovascular event, or suffer more severe adverse outcomes once a cardiovascular event occurs. In the aggregate, the potential benefits of improving

FIGURE 7.3 Health outcomes cascade and adherence.

patient adherence are substantial. It has been estimated that increasing adherence with prescribed or recommended interventions may have a greater impact on population health than improvements in specific medical treatments.

Less well appreciated is that poor patient adherence with prescribed medications (overuse, underuse, erratic use) is associated with ADEs and drug failure. It is estimated that approximately 5% of hospital admissions are due to patient nonadherence with prescribed medications; 11% among the elderly. In addition, poor adherence commonly is accompanied by unnecessary and inappropriate medication titration by physicians (who often are unaware of patient nonadherence), thereby increasing drug risks and costs.

Nonadherence with prescribed medications occurs at all stages of the treatment cascade (Fig. 7.4).

Factors that have been found to be associated with decreased adherence include patient attitudes and beliefs; disease severity and symptoms; depression; patient age, gender, race, and ethnicity, insurance status; out of pocket health-care and medication expenditures; benefit design (deductible vs. copayment); number of comorbid conditions (especially depression, dementia); adverse events; dosing regimen and complexity (number of prescribed medications, pill burden, dosing frequency, duration of therapy, timing initiation, i.e., concurrent vs. sequential); fragmented care; and health-care system utilization and prior adherence with prescribed and recommended care.

An important dimension of patients' experience with their care is the degree to which patients and their health-care providers establish a partnership "to ensure that decisions respect patients' wants, needs, and preferences and that patients have the education and support they need to make decisions and participate in their own care" (Institute of Medicine, 2001). Patient experiences and their impressions of that experience, as much as the technical quality of care, affect how people use the health-care system and benefit from it. Only recently has valid, reliable data begun to be collected to measure patient experience by health plan members and hospitalized patients. Little of this has been subjected to careful analysis, including the relationship between these experiences and patient adherence with prescribed care and patient health outcomes.

A number of interventions have been proposed to improve patient adherence, including improved patient education (oral, written, and computer-based patient explanations and instructions re benefits, adverse effects, and adverse events); increased, improved communication and counseling (e.g., reminders, computer-assisted and telephone monitoring and follow-up); increased convenience (e.g., simplified dosing, tailored regimens, dose dispensers); timing of initiation of therapy; treatment of comorbid conditions, especially depression; increasing patient involvement in their care (e.g., self-care, shared decision-making); increased social support; and rewards (e.g., financial incentives).

There is only limited evidence of success of interventions to improve medication adherence. Interventions to improve medication adherence have produced only modest results, with even effective interventions associated with small improvements. Persistence of effects requires continued reinforcement. Most assessments of interventions have focused on intermediate outcomes (e.g., blood pressure, LDL cholesterol, hemoglobin A1c) as opposed to clinical endpoints (e.g., clinical events, disease sequelae, mortality). Almost all interventions with demonstrated effectiveness for chronic conditions are complex, involving combinations of interventions.

Adherence is strongly influenced by cultural and contextual influences, which are not well understood. Thus, advancing understanding of the factors associated with adherence and improving patient adherence requires advances in theoretical and applied research about behavior change (briefly discussed earlier in this chapter) and greater involvement of and collaboration with social and behavioral scientists.

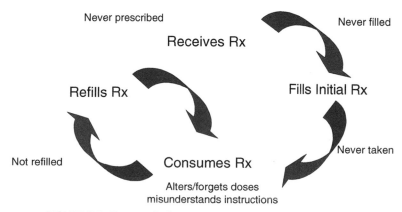

FIGURE 7.4 Process of adherence and opportunities for nonadherence.

Disparities and Inequities of Care

Healing is a matter of time, but it is sometimes also a matter of opportunity.

Hippocrates, Precepts, chapter 1.

Medical care should be provided based on an individual's needs and not on personal characteristics unrelated to the patient's condition, such as gender, race, age, ethnicity, income, education, disability, sexual orientation, or geographic residence (Institute of Medicine, 2001). Yet the benefits of health care are not distributed equally to all Americans. Even after adjustment for clinical factors, such as severity and stage of disease and comorbidity, significant disparities in access, utilization, treatment, and quality by income, age, education, race, and gender exist across a wide range of disease areas, clinical services, and clinical settings. Most importantly, these disparities are associated with worsened health outcomes. Those with low incomes and underserved racial and ethnic minorities fare worse than other Americans on a wide variety of indicators of health-care access, quality, and health outcomes.

Multiple factors contribute to disparities and inequities in medical care and health outcomes and patient and population health, which take place within the broader social, political and economic context. These include socioeconomic factors (e.g., poverty, lack of or less comprehensive medical insurance); provider, patient, and institutional factors (e.g., biases, prejudices, stereotypes, attitudes, personal and cultural beliefs, linguistic barriers, adherence); and structural health system factors (limited access, fragmented financing and delivery, financial incentives, episodic care) (Fig. 7.5).

As noted by the American College of Physicians, "the benefits of American medicine are available only to those with access to the health care system." Yet over the past three decades, there has been a steady increase in the number of Americans who lack health insurance (currently in excess of 40 million). Lack of insurance is a major reason for not obtaining access to care when and where needed. Those without insurance coverage are less likely to obtain needed medical care and preventive tests, present with more advanced disease, and have worse health status and disease outcomes. Uninsured nonelderly adults (especially those uninsured more than a year), those with chronic conditions or multiple diseases, and those in poor health are more likely to report not seeing a physician when needed and not receiving recommended preventive services than those with insurance coverage. But even with medical insurance, people may not be able to obtain care because of lack of an established relationship with a physician (having a regular place to go for health care is an even stronger predictor of preventive care services than health insurance) or because of language, cultural, transportation, geographical, or financial barriers.

Health-care disparities can be reduced and even eliminated with concerted effort if effective interventions are developed, implemented, rigorously evaluated, and targeted to identifiable medical, health-care system, and social root causes. However, evidence regarding the relationships among these various factors and their relative contributions to specific disparities is indirect and not well understood. Disparities and inequities in medical care delivery, quality, and individual and population health require explication of these relationships through multidisciplinary behavioral, social, and organizational science research.

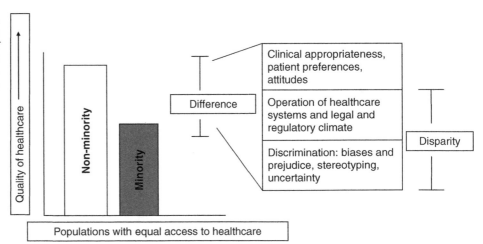

FIGURE 7.5 Differences, disparities, and discrimination: populations with equal access to health care. *Institute of Medicine, 2002. Unequal Treatment: Confronting Racial and Ethnic Disparities in Health Care. National Academy Press, Washington, DC.*

Quality Management and Improvement

One of the troubles of our age is that habits of thought cannot change as quickly as techniques, with the result that as skill increases, wisdom fades.

Bertrand Russell

The important, tangible advances resulting from scientific discovery and its translation to practice and patients have been accompanied by increased complexity of medical care, with medical care science and technologies advancing more rapidly than our ability to deliver them safely, effectively, and efficiently (Institute of Medicine, 2001). It is widely perceived that our health-care systems increasingly are unable to meet important patient and payer expectations and needs, and lag in the ability to translate knowledge into practice and apply new technology safely and appropriately in a timely fashion.

Improvement in medical care and health quality requires effective management which, in turn, requires a restructuring of how we deliver care—adoption of a patient-focused, clinically relevant, system-wide perspective; development and implementation of effective and clinically meaningful management information systems; improved alignment of financial and nonfinancial incentives to reinforce system objectives. Effective patient management requires providing clinicians with clinically relevant, case mix—adjusted, patient-focused, illness-oriented information, including information on resources provided across providers and sites of care. Without substantial changes in the ways, health care is organized and delivered, the problems resulting from the growing complexity of health-care science and technology likely will increase (Institute of Medicine, 2001).

One of the first steps in the quality improvement process is to assess practices, processes, and outcomes, absolutely and relative to peers, practice guidelines, and "best practices." This commonly requires using systems analysis techniques to better understand frequency and source of errors and to guide development of effective interventions (Berwick, 1989). However, rigorous evaluations of interventions have been limited, with most analyses using weak observational designs with limited statistical analysis with little or poor adjustment for case-mix and potential confounders. Even less research has been performed to better understand and inform mechanisms of action, both of causes of errors and effectiveness of interventions to mitigate these events.

An important area of investigation includes greater understanding of organizational and systems behavior. Particularly important are opportunities to identify and develop interventions to eliminate misaligned incentives that distort practice, impede adherence, and undermine quality of care and quality improvement. Clinician compensation should not discourage care of sicker or more complicated patients nor encourage provision of procedures. Compensation also should provide incentives for improved quality of care and patient outcomes. However, proposed interventions must be subjected to rigorous evaluation. For example, payers and policy-makers have exhibited widespread interest in and enthusiasm for pay-for-performance to better align physician and patient or system incentives. However, to date, few studies have demonstrated the ability to modify some physician behaviors, and no rigorous research or randomized trials have as yet demonstrated improvement in patient outcomes.

Information technology plays a central role in understanding and improving quality of care. Despite the information-intensive nature of medicine, health-care information technology investment lags that of many other industries. Since quality is a multifactorial construct, its measurement requires integrating a broad range of data (of varying completeness, accuracy, and consistency) from multiple sources across populations, a task for which computerized databases and analysis are essential. Similarly, quality measurement, and therefore improvement, is limited by the fragmented and widely dispersed information, especially for patients with multiple, chronic diseases. For example, the inability to determine patient prescription refill data (as opposed to what is prescribed by the physician) for chronic diseases limits assessment and improvement of medication adherence and related outcomes. In contrast, computer ordering systems with built-in data checks have demonstrated significant effectiveness for reducing drug prescribing and dosing errors and facilitating adherence with clinical protocols and guidelines.

In addition to investment and migration to information systems, individual, organizational, and cultural changes are required to optimize quality, including adapting to increased transparency and accountability. A particular research need is to facilitate development of and clinician and investigator access to high-quality integrated analytical, clinical, and administrative databases, including disease registries that permit comparison with peers and best practices. However, such analytical databases require great care with regard to assuring data validity, updating, and maintenance, while maintaining patient confidentiality.

Medical care increasingly is delivered within groups and organizations. Medical care also reflects underlying culture and society. Thus, delivery of more effective medical care and facilitating translation of medical advances to patients requires better understanding of organizational management and behavior and increased community-based research, including research directed toward patient beliefs, attitudes, intentions, and behaviors. This, in turn, requires conduct of collaborative research with investigators with expertise and experience in organizational management, sociology, and communications. In addition, leadership is required from clinicians to help define criteria, methods, processes, and standards for assessing outcomes and quality of care in conjunction with administrative leadership in providing required data and developing, rigorously evaluating and implementing effective management systems.

CONCLUSIONS

Translation of medical advances requires generating new information on the comparative effectiveness of alternative medical interventions and their application to patients and populations. This, in turn, requires generating new information from greatly expanded comparative studies of medical interventions using a broad spectrum of experimental, quasiexperimental, and observational data, in conjunction with improved analysis and synthesis of medical evidence and delineation of evidence-based practice guidelines (which will require investment in methodological development, as well as expanded empiric analysis). Concurrently, quality improvement requires development and validation of improved measures for assessing quality of care, development, and rigorous evaluation of innovative improved care processes, development, implementation, and rigorous evaluation of tools and systems to support providers and patients, and system redesign to more effectively implement effective interventions.

Translation of medical innovation requires methodological expertise, skill, time, and resources. Although difficult, effectively translating advances in scientific knowledge to practice and policy to improve the health of patients and the public requires research of the type discussed in this chapter. The first step is to recognize the importance of the problem and the complexity of the challenge. At present the system's failings and shortcomings are not widely or fully appreciated. Furthermore, there is little external pressure to engender the fundamental systems reform and redesign required. The knowledge base for assessment of medical interventions and quality improvement is rapidly evolving, and the importance of fundamental research in this area is not widely appreciated or assigned high priority. Resources for the conduct of basic and applied research in this area are extremely limited in relation to the scale and demands of the task. Moreover, improved translation of scientific evidence will, in many cases, require changes in attitudes and behavior by and realignment of incentives among physicians, hospitals, payers, patients, and other constituencies, and thus will encounter both passive (i.e., inertia) and active (e.g., litigation; lobbying) resistance. But without effectively addressing these issues, translation of medical advances to practice will be severely impaired, and patients will not receive the full potential benefits of medical progress.

In the midst of challenge lies opportunity.

Albert Einstein

REFERENCES

AcademyHealth, n.d. http://www.academyhealth.org/about/whatishsr.htm.

Agency for Health Research and Quality, 2002. http://www.ahrq.gov/about/whatis.htm.

Berwick, D.M., 1989. Continuous quality improvement as an ideal in health care. N. Engl. J. Med. 320, 53–56.

Donabedian, A., 1980. Explorations in Quality Assessment and Monitoring. In: The Definition of Quality and Approaches to Its Assessment, vol. 1. Health Administration Press, Ann Arbor, MI.

Fishbein, M., 1980. A theory of reasoned action: some applications and implications. Neb. Symp. Motiv. 27, 65–116.

Fishbein, M., Ajzen, I., 1975. Belief, Attitude, Intention, and Behavior: An Introduction to Theory and Research. Addison-Wesley, Reading, MA.

Fisher, E.S., Goodman, D.C., Skinner, J.S., Wennberg, J.E., 2008. Tracking the Care of Patients with Severe Chronic Illness: The Dartmouth Atlas of Health Care. The Dartmouth Institute of Health Policy & Practice, Hanover, NH.

Institute of Medicine, 1998. The Urgent Need to Improve Health Care Quality: A Statement by the Roundtable on Health Care Quality, Board on Health Care Services. National Academy Press, Washington, DC.

Institute of Medicine, 1999. To Err Is Human: Building a Safer Health System. National Academy Press, Washington, DC.

Institute of Medicine, 2001. Crossing the Quality Chasm: A New Health System for the Twenty-first Century. National Academy Press, Washington, DC.

Institute of Medicine, 2002. Unequal Treatment: Confronting Racial and Ethnic Disparities in Health Care. National Academy Press, Washington, DC.

Institute of Medicine, 2006. Preventing medication errors: quality chasm series. In: Aspden, P., Wolcott, J., Bootman, J.L., Cronenwett, L.R. (Eds.), Board on Health Care Services. National Academy Press 1999, Washington, DC.

Kahneman, D., Slovic, P., Tversky, A., 1982. Judgment Under Uncertainty: Heuristics and Biases. Cambridge University Press, Cambridge.

Kassirer, J.P., 1990. Diagnostic reasoning. In: Sox, H.C. (Ed.), Common Diagnostic Tests: Use and Interpretation. American College of Physicians, Philadelphia, pp. 1—15.

Leape, L.L., Brennan, T.A., Laid, N., Lawthers, A.G., Localio, A.R., Barnes, B.A., Hebert, L., Newhouse, J.P., Weiler, P.C., Hiatt, H., 1991. The nature of adverse events in hospitalized patients. N. Engl. J. Med. 324, 377—384.

Pauker, S.G., Kassirer, J.P., 1980. The threshold approach to clinical decision making. N. Engl. J. Med. 302, 1109.

President's Advisory Commission, 1998. Quality First: Better Health Care for All Americans. Final Report of the President's Advisory Commission on Consumer Protection and Quality in the Health Care Industry. US Government Printing Office, Washington, DC.

Rogers, E.M., 2003. Diffusion of Innovations, fifth ed. Free Press, New York.

Tversky, A., Kahneman, D., 1974. Judgment under uncertainty: heuristics and biases. Science 185, 1124—1131.

Wennberg, J., Gittelsohn, A., 1973. Small area variations in health care delivery. Science 182, 1102—1108.

Section II

Approaches

Chapter 8

Measurement of Biological Materials

Mark D. Kellogg

Boston Children's Hospital, Boston, MA, United States

Chapter Outline

Key Points

- Immunoassays represent the most frequently used and perhaps simplest approach to the analysis of biological materials for the translational researcher.
- Enzyme immunoassay (EIA) and enzyme linked immunoassay formats are the easiest to utilize and are widely commercially available. More sensitive methods of immunoassay (Electrochemiluminescence, RIA, fluorescence polarization immunoassay) require specialized instrumentation for use.
- The concept of buyer aware is very appropriate when selecting from the available commercial immunoassays. The "Research use only"–labeled kit requires the user to verify the performance of the assay before use with research samples. Numerous examples of commercially available immunoassays not measuring the stated molecule have been reported in the past few years.

- Mass spectrometry and the associated separation techniques of gas and liquid chromatography have become the de facto approach to the profiling and quantification of protein, lipids, carbohydrates, and even nucleic acids. Instrumentation and associated processes have resulted in the capability to generate significant amounts of data. This data deluge has created the need for informatics processes to change the data into useful information.
- Massively parallel sequencing of nucleic acids has been termed "next-generation sequencing" and has largely supplanted the use of microarray approaches to genomics. Similar to the mass spectrometric analysis of biological materials, this has resulted in huge amounts of data and the need for powerful informatics approaches to make these data useful.
- Multiomics is the combination of processes and data from the genomics, proteomics, lipidomics, transcriptomics, metabolomics, and other -omics areas. The concept is based on the recognition that the structure and function of the biological systems of interest to the translational research are interrelated and impacted by all of the -omics.

INTRODUCTION

This chapter summarizes various methods employed to characterize and quantify biological materials from human sources used in translational and clinical research. The measurement of biological compounds in body fluids and tissues is a critical component of clinical research and represents an objective endpoint for many trials, especially those involving therapeutic interventions. The purpose of this chapter is to provide an overview of select methods that are available to the clinical researcher to assess biological compounds from human material. The technologies that will be discussed are (1) immunochemical, including immunoassays and immunochemistry; (2) mass spectrometry (MS) as applied to small molecules, proteins, and metabolites; and (3) nucleic acid sequencing as applied to genomics. It should be noted that some of these methods, such as immunoassays, and some types of MS are significantly more mature technologies than are techniques used in genomics, proteomics, and metabolomics. Since the topics discussed herein provide an overview, the reader is referred to the references at the end of the chapter for a more detailed discussion of specific methodologies.

IMMUNOASSAYS AND IMMUNOCHEMISTRY

Background

Immunoassay methodologies represent, perhaps, the most frequently used approach to measure biological compounds in translational and clinical research. Assays exist, from either commercial or research sources, for both the qualitative and quantitative measurement of a plethora of naturally occurring small molecules such as lipid mediators and hormones as well as larger peptides and proteins that are present in human body fluids and tissues (Davies, 1994). In addition, a number of therapeutic agents can be measured by immunoassays. It is important to note that immunoassays can not only measure antigens but antibodies as well. Many immunoassays are extremely sensitive and can detect as little as 0.1 pg of compound per milliliter of body fluid (Ashihara et al., 2001). This section will focus primarily on the most commonly used immunoassay approach—enzyme immunoassay (EIA). Mention will also be made about other immunoassay methods including radioimmunoassay (RIA), fluorescent and chemiluminescent immunoassays, precipitation immunoassays, and particle immunoassays.

Basic Principles

The first immunoassays resulted from the pioneering work of Yalow and Berson in the late 1950s and utilized antibodies labeled with a radioisotope like ^{125}I. Regardless of the method used, all immunoassays rely upon the interaction of an antigen with an antibody (Davies, 1994). The extent to which this interaction occurs (the amount of antigen that is bound to antibody versus free) allows one to measure, either qualitatively or quantitatively, the amount of that particular antigen that is present in a biological fluid or tissue. Detection methods for particular assays vary and depend on the approach used to detect the antigen—antibody complex.

Antigens are defined as any substance that possesses antigenic sites (epitopes) which produce corresponding antibodies (Davies, 1994). Antigens can be small molecules such as haptens, hormones, etc. or they can be very large compounds such as glycolipids and proteins. Antibodies that are generated in response to antigens can be one of the five types and include IgG, IgM, IgA, IgE, and IgD. Antibodies consist of heavy chains and either κ or λ light chains and possess constant and variable regions. The hypervariable region can be assembled to recognize a wide variety of epitopes (Ashihara et al., 2001).

Although antibodies can serve as antigens, for purposes of immunoassays, they are reactants used to detect antigens. Different types of antibodies can be obtained from several sources. Polyclonal antibodies are generated by immunizing an animal with an antigen. In this case, multiple antibodies are generated, which recognize different epitopes. As a consequence, the affinity of polyclonal antibodies for a complex antigen is usually stronger than is that of a monoclonal antibody.

Monoclonal antibodies are generated using somatic cell fusion and hybridoma selection (Koehler and Milstein, 1975). The resulting established cell line generates a homogeneous antibody population that represents a single epitope (Ashihara et al., 2001). While specific for a certain epitope, it should be kept in mind that monoclonal antibodies may cross-react with different antigens that possess the same epitope. Nonetheless, the development of monoclonal antibodies has revolutionized immunoassay methodologies because monoclonal antibodies are well defined, and specific reagents and their production can yield a nearly limitless supply of an antibody (Zola, 1987). Further, they can be prepared through immunization of a nonpurified antigen.

A more recent approach to the development of antibodies for use in immunoassays is phage display, in which antibody fragments of predetermined binding specificity are encoded in a phage and expressed in bacteria (Winter et al., 1994). Table 8.1 shows the classification of the various immunoassays available and some of their key characteristics.

Enzyme Immunoassays

Overview

Immunoassays employing enzyme-mediated reactions as labels were developed as alternatives to RIAs. Commonly used techniques include the enzyme-linked immunosorbent assay (ELISA), the EIA, and the enzyme-multiplied immunoassay (EMIT). Heterogenous EIAs are essentially the same as RIAs except that enzymes are used as labels rather than radio-isotopes. Unlike RIAs, however, homogenous assays can be developed that eliminate washing steps for separation of bound and free molecules. EIAs have a number of advantages compared to other immunoassay methodologies in that highly sensitive assays can be developed because of the ability of enzymatic reactions to amplify antigen—antibody interactions. In addition, reagents are cheap and have a longer shelf life than RIAs. Further, no radiation hazards exist. Finally, a number of different types of assays have been developed. Disadvantages of EIAs include the fact that assays can be more complex, and enzyme activity may be affected by various substances in biological fluids (Ashihara et al., 2001; Nakamura and Kasahara, 1992).

Assay Methodology

In terms of methodologies, as noted, heterogeneous EIAs are similar to RIAs although detection of antigen—antibody interactions is afforded by cleavage of substrates by enzymes linked to antibodies. Heterogenous EIAs are at least as sensitive as RIAs and in some cases are more sensitive. Various enzymes can be utilized in EIAs. The most common are

TABLE 8.1 Classification of Various Immunoassays and Their Characteristics

	Labels (Reporter Groups)	B/F Separation	Signal Detection	Sensitivity
Precipitation immunoassays	Not required	Not required	Naked eye	~10 μg/mL
			Turbidity Nephelometry	
Particle immunoassays	Artificial particles (gelatin, particles, latex, etc.)	Not required	Naked eye Pattern analyzer Spectrophotometry Particle counting	~5 ng/mL
Radioimmunoassays	Radioisotopes (^{125}I, ^{3}H)	Required	Photon counting	~5 pg/mL
Enzyme immunoassays	Enzymes	Required	Spectrophotometry	~0.1 pg/mL
			Fluorometry Photon counting	
Fluorescent immunoassays	Fluorophores	Required	Photon counting	~5 pg/mL
Chemiluminescent immunoassays	Chemiluminescent compounds	Required	Photon counting	~5 pg/mL

alkaline phosphatase, β-galactosidase, glucose oxidase, urease, and catalase. The development of substrates cleaved by enzymes initially employed colorimetric and fluorometric detection and later chemiluminescent methods. EIAs are readily amenable to adaptation to fully automated techniques.

An important advantage of EIAs over RIAs is that the former can be developed as homogenous assays in which the tedious washing step to remove free antigen is eliminated, although homogenous EIAs are frequently less sensitive than RIAs or heterogeneous EIAs. The first homogeneous EIA developed was EMIT (Reubenstein et al., 1972).

In summary, EIAs are very frequently used immunoassay techniques that offer accurate quantification of a number of biological compounds in human biological fluids and tissues.

Other Types of Immunoassays

Fluorescent and Chemiluminescent Immunoassays

A number of other immunoassay methodologies exist to quantify biological substances that differ primarily on the detection method that is employed. These include fluorescent immunoassays that utilize fluorescent compounds as immunolabels. Fluorescent compounds emit photons that can be detected after irradiation with light at particular wavelengths. The assays can be either heterogeneous or homogeneous. A number of assays have been developed to detect very low concentrations (10^{-15} M) of compounds in biological fluids such as drugs and hormones (Nakamura, 1992).

A second important immunoassay approach is the use of chemiluminescent molecules as detectors (Ashihara et al., 2001). Commonly used reagents include luminol derivatives, acridinium esters, or nitrophenyl oxalate derivates and ruthenium tris-bipyridyl with tripropylamine. Labels of chemiluminescent compounds produce light electrochemically on the surface of electrodes that is subsequently detected. These assays are often highly sensitive and may surpass the sensitivity of many RIAs.

Precipitation and Particle Immunoassays

Two other types of immunoassays that are of potential use to translational and clinical researchers are precipitation immunoassays and particle immunoassays (Ashihara et al., 2001). These methods, unlike other immunoassays, are often qualitative in nature. They have also been in use longer than other assay types. Precipitation assays are used to measure immunoprecipitation reactions, which form when large complexes of antigens and antibodies combine to generate insoluble complexes. Detection of complexes can be afforded using light scattering instrumentation and is termed nephelometry. The lower limit of sensitivity using these methods is about 10 μg/mL. Common assays that use these techniques include the measurement of many major serum proteins.

A related method of immunoassay is particle agglutination in which either antibodies or antigens are detected in biological fluids using corresponding antigens or antibodies, respectively, bound to various particles. Commonly used particles are latex beads and gelatin particles. These assays have wide applicability and can measure biological molecules as diverse as human chorionic gonadotropin or antibodies to HIV.

Radioimmunoassay

Various competitive RIA methods have been developed to measure a plethora of different biological compounds. Fig. 8.1 shows one of the general methods that is routinely employed (Ashihara et al., 2001). Initially, a known amount of labeled antigen and an antigen from a biological specimen are combined and reacted with a known amount of antibody that is usually coated on a solid phase such as sepharose beads or on the inner wall of plastic tubes. After the mixture equilibrates, it is washed to remove unreacted antigens, and the immune complex containing both labeled and unlabeled antigen is trapped in the solid phase. The washing step is referred to as B/F (bound versus free) separation (Ashihara et al., 2001). Radioactivity can be detected by scintillation counting and is expressed as counts per minute (CPM). Applying the concept of competition between labeled and unlabeled antigen, the antigen-bound percentage of total radioactivity against logarithmic concentration of the antigen can be compared to a standard curve as shown in Fig. 8.1. The CPM plot on the standard curve gives the concentration of antigen. To prepare a standard curve, known amounts of both labeled and unlabeled antigen are reacted as detailed earlier.

Various other competitive RIA methods exist in which a second antibody is utilized to capture antigen—antibody complexes in the solid phase. In addition, noncompetitive assays are available that employ conditions of antibody excess. These include techniques that are termed immunoradiometric (or sandwich)-type assays. These latter approaches can

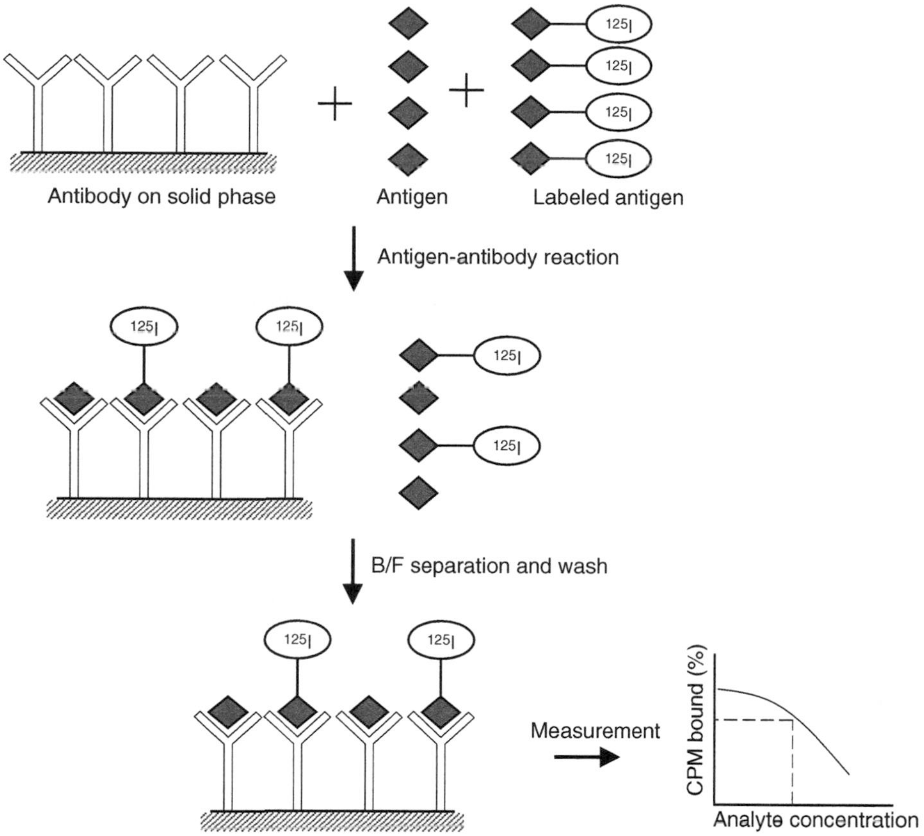

FIGURE 8.1 Assay principle of competitive radioimmunoassay (RIA) using an antibody on a solid phase. *Modified with permission from Ashihara, Y., Kasahara, Y., Nakamura, R.M., 2001. Immunoassays and immunochemistry. In: Henry, J.B. (Ed.), Clinical Diagnosis and Management by Laboratory Methods. W.B. Saunders, Philadelphia, pp. 821—849.*

increase greatly the level of sensitivity of detection of compounds that are present in biological samples at very low concentrations (Espinosa et al., 1987).

In summary, RIAs can offer a number of advantages over other immunoassays in that they are highly sensitive and precise, although the latest chemiluminescent-based methods typically surpass RIA limits of detection. Disadvantages include the fact that radioisotopes are highly regulated, may have a short half-life, and that RIAs are heterogeneous assays.

Multiplex Immunoassays

Technological advances in the analysis of materials have created a situation where hundreds of candidate biomarkers can be generated in the discovery phases of biomarker research. Each of the candidate markers needs to be confirmed in the verification/validation phase. Typically, single molecule—based immunoassay approaches were utilized. Several manufacturers have developed immunoassay formats that allow for the multiplex analysis of several biomarkers simultaneously. This improves throughput, usually decreases the amount of sample needed, and often improves the detection limits for the molecule(s).

There are two basic formats for the multiplex immunoassay. The first is a bead-based format utilizing a flow cytometric approach where beads are color-coded (fluorescent) and with unique antibody or antigen coatings. The second format utilizes an array of antibody (or antigen) in a microtiter plate well. Slide-based and reverse protein array platforms are also available (Tighe et al., 2013).

Bead-based approaches claim the capability of plexing dozens of molecules; microtiter plate—based arrays are space constrained and usually will plex 2—10 analytes. Some reports of 20 analytes in one well have been published (Tighe et al., 2013).

Several reviews are available describing the advantages and disadvantages of the technology (Tighe et al., 2013; Christiansson et al., 2014). While early versions of the technology appeared in the 1980s, most of the advances to improve detection limits and reproducibility took place in the late 2000's and early 2010's. While comparisons to ELISA and EIA are varied, users should be aware of significant differences between multiplex analysis and single molecule ELISA- or

EIA-based assays (Tighe, 2013; de Koning et al., 2012; Fu et al., 2010). The ability to combine multiple antibody—antigen interaction into one solution and achieve optimal conditions is a difficult task and still a fairly immature technology. Users should be sure to conduct verification studies (described later in the chapter) to assure the method is performing to expectations. de Koning noted that using the same manufacturer, a 4-plex cytokine assay that included IL-6 produced different results and study conclusions than the single plex IL-6 kit (de Koning et al., 2012).

Characterization of Immunoassay Performance

Most researchers will not have the expertise or time to develop their own immunoassays for use in translational or clinical research. As such they will rely upon commercially available kits. Given that the conduct of the typical commercial immunoassay is fairly simple and includes calibrators and often quality control material, many researchers are lulled into a false sense of security regarding the data produced from the commercial kits. Prassas et al. (2014) reported in 2014 that the commercial ELISA for CUZD1 was actually measuring Cancer Antigen CA125, and use of the CUZD1 kit costs their lab approximately US$500,000. Not only did the use of the assay cost money, but also significant delays in their project. Rifai had also reported similar questionable quality in several different ELISA kits (Rifai et al., 2012). Prassas proposed a series of experiments that should be conducted to assure antigen identity when using a new assay: (1) mass spectrometric identification of the target antigen in the assay calibrators; (2) verification that immune reactivity matches the expected molecular weight of the target antigen; (3) assessment of the assay's immune reactivity against (a) pure recombinant protein, (b) native form of the protein in biological samples, and (c) comparison with an assay from another vendor if available; (4) assessment of the assay's sensitivity to enzymatic protein deglycosylation or protease treatment; and (5) immuno-mass spectrometric verification of both capture and detection antibody specificities (Prassas et al., 2014). In addition to the above experiments to confirm antigen identity, all assays (immunoassay, spectrophotometric, mass spectrometric, etc.) should be validated to confirm the manufacturer's claims prior to use in studies (Linnet and Boyd, 2012; Rozet et al., 2011).

MASS SPECTROMETRY AND CHROMATOGRAPHY
Background

In most cases the analysis of biological materials involves samples taken from animal or human sources, and the samples are complex mixtures of water and biomolecules. These mixtures require separation of components prior to analysis to improve method sensitivity and precision. Simple techniques such as centrifugation, evaporation, filtration, or precipitation can be used to separate cellular, particulate, and liquid phases and specific components such as proteins, lipids, or carbohydrates. Additional steps to concentrate the molecule(s) of interest are also common, particularly when molecules are in picomolar or lower concentrations in the original sample. However, most MS techniques still require a chromatographic separation to improve specificity of detection.

The term chromatography encompasses a diverse group of methods that permit the separation of closely related components of complex mixtures. Chromatography, as a technique, was invented by the Russian botanist Mikhail Tswett over 100 years ago when he used a glass column packed with chalk to separate a solution of plant pigments (Scott, 2003a,b,c; Skoog and Leary, 1992a,b,c,d). By the middle of the 20th century, the basic concepts of chromatographic separation techniques were developed; however, it has been only in the past three decades that technology has advanced enough to make chromatographic techniques relevant and readily available to the biological researcher.

Two types of chromatography that are of particular interest in translational and clinical research include gas chromatography (GC) and liquid chromatography (LC). These techniques will be described in this section. GC and LC methodologies can be used to quantify small molecules including fatty acids, nucleic acids, drugs, and drug metabolites. Further, these techniques can be used to purify complex biomolecules such as proteins and peptides.

Basic Principles

GC and LC are either preparative or analytical in nature (Skoog and Leary, 1992a,b,c,d). Preparative chromatography is used to purify components of a mixture for further use. Analytical chromatography uses much smaller amounts of material than preparative chromatography and is generally used to quantify specific molecules.

In all chromatographic separations the sample is dissolved in a mobile phase, whether it is a gas or a liquid, and is then passed through a stationary phase, which is fixed in place on a solid surface, usually a column. After passing through the stationary phase, the components of the sample are analyzed by a detector. Complex mixtures are separated based upon differences in distribution between the mobile and stationary phases. Molecules that interact weakly with the stationary phase elute rapidly while compounds that are strongly retained by the stationary phase move along slowly with the mobile phase (Scott, 2003a,b,c; Skoog and Leary, 1992a,b,c,d; Niessen, 2006a,b,c,d). The retention is a measure of the speed at which an analyte moves through the chromatographic system and is usually measured by retention time, the time between injection and detection. Retention times can vary due to slight changes in the mobile phase, the stationary phase, temperature of the environment, and sample matrix (Skoog and Leary, 1992a,b,c,d). Therefore, it is important to test chemically similar standards under identical chromatographic conditions to the compound of interest.

The degree to which components of a mixture are separated using a particular chromatographic technique is dependent upon the mobile phase composition and flow rate, the shape and particle size of the stationary phase, the length of the column in which the stationary phase is packed, and the efficiency of the column (Scott, 2003a,b,c; Skoog and Leary, 1992a,b,c,d; Niessen, 2006a,b,c,d). The efficiency of the column is measured by the number of theoretical plates (N) (Scott, 2003a,b,c; Skoog and Leary, 1992a,b,c,d). The number of theoretical plates is the ratio of the length of the column (in centimeters) to the diameter of the stationary phase particle size (also termed "plate height"). Thus, the efficiency of the column increases as the stationary phase particle size decreases.

Gas Chromatography

Overview

In GC, the sample is first vaporized and then injected into the chromatographic column (Skoog and Leary, 1992a,b,c,d; Scott, 2003a,b,c). A chemically inert gas, such as helium, argon, nitrogen, carbon dioxide, or hydrogen, is used as the mobile phase (termed carrier gas) to transport the analyte(s) through the stationary phase, typically a microscopic layer of liquid or polymer on an inert solid support. The column is housed in a temperature-controlled oven. The optimum column temperature depends upon the boiling point of the sample and the degree of separation required. Typically, a temperature program in which the column temperature is increased either continuously or in steps is used to elute a complex mixture of analytes with a broad boiling range. A detector is used to monitor the outlet stream from the column.

Types of Stationary Phases

GC stationary phases are packed into capillary columns, composed of either fused silica or stainless steel, with inner diameters ranging between 100 and 300 μm. GC column efficiency is dependent upon length, ranging anywhere from 2 to 50 m long. Numerous stationary phases exist; most are high molecular weight, thermally stable polymers that are liquids or gums. The most common stationary phases are polysiloxanes as they are stable, robust, and versatile (Skoog and Leary, 1992a,b,c,d; Scott, 2003a,b,c). Polysiloxanes are characterized by the repeating siloxane backbone shown in Fig. 8.2A. Each silicon atom contains two functional groups; the type and number of functional groups make each type of column unique. In the most basic siloxane column, the silicon atoms are all substituted with methyl groups. These columns are nonpolar in nature and are commonly used to analyze hydrocarbons, drugs, and steroids. In other polysiloxanes, some of the methyl groups are replaced with phenyl, cyanopropyl, or trifluoropropyl moieties to increase the polarity. These polymers are less stable, less robust, and have a more limited temperature range, but are more polar in nature than the polysiloxanes. Common applications for these columns include the analysis of free acids, alcohols, and ethers.

Chiral stationary phases have also been developed for GC analysis (Skoog and Leary, 1992a,b,c,d; Scott, 2003a,b,c). These stationary phases are used to separate individual enantiomers, stereoisomers which differ only in the spatial arrangement of their atoms and in their ability to rotate the plane of polarized light. This type of chromatography has become increasingly important as many pharmaceutical compounds exist as enantiomers. Significant differences have been

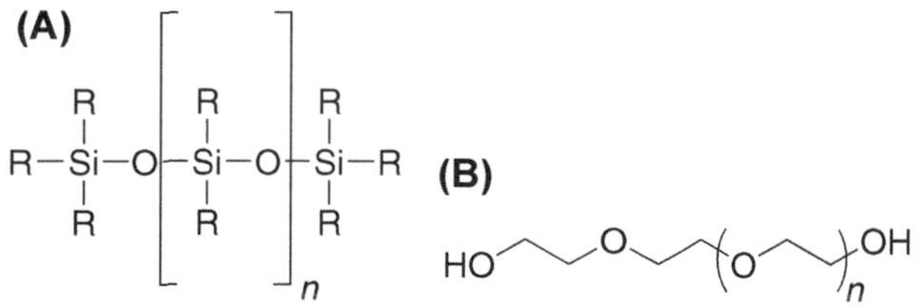

FIGURE 8.2 Stationary phases in gas chromatography (GC). (A) Polysiloxanes. The most common R group is a methyl. (B) Polyethylene glycols.

found in the pharmacokinetic and pharmacodynamic properties of the enantiomers of many drugs, and thus there is often a need to separate enantiomers of various drugs using chiral chromatography to better study their pharmacological properties as well as their metabolic disposition. Chiral stationary phases for GC are often amino acid—substituted polysiloxanes.

Types of Detectors

Typical detectors used in GC include flame ionization detectors (FIDs), thermal conductivity detectors (TCDs), thermionic detectors (TIDs), electron capture detectors (ECDs), and atomic emission detectors (AEDs) (Skoog and Leary, 1992a,b,c,d; Scott, 2003a,b,c). The most widely used and generally applicable detector is the FID. In this detector, effluent from the column is mixed with hydrogen and air and then ignited. Ions and electrons that can conduct electricity through the flame are produced. The FID is most useful for the detection of organic compounds; it is generally insensitive to carbonyl, alcohol, and amine functionalities as well as halogens and noncombustible gases such as water and carbon dioxide. The TCD or katharometer detector is also widely used. This detector responds to all analytes that have a different thermal conductivity and heat capacity from the carrier gas. Thus, it can be used for detection of a broad range of molecules, including both organics and inorganics. Additionally, the sample is not destroyed and can thus be collected for further use. The other types of detectors mentioned are more specific in nature. TIDs are selective for organic compounds containing phosphorus and nitrogen while ECDs measure X-radiation.

A mass spectrometer can also be coupled with a GC and used as a method of detection.

Liquid Chromatography

Overview

In LC, also referred to as high-performance liquid chromatography (HPLC), the sample is injected into a stream of solvent that is being delivered by a high-pressure pump and transported through the column where separation takes place (Skoog and Leary, 1992a,b,c,d). Solvent can be delivered either isocratically or using a gradient (McMaster, 2007). In an isocratic elution, the solvent remains at a constant concentration throughout the course of the separation. In a gradient method, the composition of the solvent is changed during the separation to elute compounds with differing chemical properties from a complex mixture. Sample output is monitored by a flow-through detector.

Typical analytical HPLC columns range in length from 20 to 300 mm and are packed with particles with a 3—5 μm internal diameter. In addition to column length and particle size, the internal diameter of the column affects the relative column efficiency (Table 8.2; Niessen, 2006a,b,c,d). Microbore, microcapillary, and nano-LC columns are typically used for characterization of biological molecules, often coupled with MS, where sample is limited. Columns with larger particles, lengths, and internal diameters exist for the preparative purification of large quantities of compounds, including synthesized organic molecules or larger biomolecules including proteins.

Types of Stationary Phases

Compounds can be separated by LC using a variety of mechanisms. Separation of molecules can be based upon their selective adsorption on a solid phase, differences in their ion exchange properties, or by differences in their molecular size (Niessen, 2006a,b,c,d; Skoog and Leary, 1992a,b,c,d). The most common mechanism of separation is adsorption and is

TABLE 8.2 Characterization of Quantitative Liquid Chromatography Columns With Various Internal Diameters

Column Types	Internal Diameter (mm)	Flow Rate (μL/min)	Injection Volume (μL)	Relative Maximal Detectable Concentration
Analytical (conventional)	4.6	1000	100	1
Narrowbore	2.0	200	19	5.3
Microbore	1.0	47	4.7	21.2
Microcapillary	0.32	4.9	0.49	207
Nano liquid chromatography	0.05	0.120	0.012	8464

based upon the relative polarities of the mobile and stationary phases. There are two types of adsorption stationary phases: normal phase and reversed phase. In normal-phase chromatography, the stationary phase is highly polar, typically water or triethylene glycol supported on silica, while relatively nonpolar solvents, such as hexane and isopropyl alcohol mixtures, are used as the mobile phase. In this system, the least polar compound in a mixture elutes first, because it is the most soluble in the mobile phase. Increasing the polarity of the mobile phase decreases the retention time (Skoog and Leary, 1992a,b,c,d). In reversed-phase chromatography, the stationary phase is nonpolar, typically a C-8 or C-18 hydrocarbon, while the mobile phase is relatively polar (mixtures of water and methanol or acetonitrile). In this system, the most polar compound in a mixture elutes first, as it is most compatible with the stationary phase. Increasing the polarity of the mobile phase increases the retention time (Skoog and Leary, 1992a,b,c,d).

Types of Detectors

Commonly used LC detectors include absorbance, evaporative light scattering (ELSD), fluorescence/chemiluminescence, electrochemical, and refractive index detectors (Skoog and Leary, 1992a,b,c,d; McMaster, 2007; Scott, 2003a,b,c). Absorbance detectors are the most widely used detectors in LC. Detection is based upon the ultraviolet absorbance of a chromophore in the analyte of interest. Instruments are available that can detect either one or two programmed wavelengths. In addition, photodiode array detectors can monitor a range of wavelengths between 200 and 400 nm. ELSD detectors are also widely used as they are a universal detector that responds equally to nonvolatile analytes and is not dependent on the presence of a chromophore in the molecule. Detection is based upon the ability of the sample to cause photon scattering when it traverses the path of a laser beam. Fluorescence and chemiluminescence detectors are also widely used, sensitive, and specific, but detection is dependent upon the fluorescent properties of the molecule or requires pre-column or postcolumn derivatization.

As with GC, mass spectrometers can be used as detectors for LC. MS will be discussed in another section.

Ultrahigh-Performance Liquid Chromatography

The primary limitation of HPLC is the lack of high column efficiency, especially when compared to GC. However, technological advances have made available LC columns packed with 1.7 μm particles (Churchwell et al., 2005; de Villiers et al., 2006). These stationary phases, along with the development of instrumentation which can operate at pressures greater than 10,000 psi, allow for separations with significantly greater resolution, speed, and sensitivity than traditional HPLC. Initial versions of this technique were called ultrahigh-performance liquid chromatography (UPLC), but this term was trademarked by the Waters Corporation. UHPLC and UPLC are essentially the same technology using sub-2μM particles in the chromatography column. Because the technology produces data in much shorter timeframes, it must be coupled with detectors that can acquire data quickly. As such, most applications of UHPLC are paired with mass spectrometers for rapid data acquisition.

Mass Spectrometry

Background

A mass spectrometer is an instrument that produces ions and separates them in the gas phase according to their mass-to-charge (m/z) ratio (Skoog and Leary, 1992a,b,c,d). MS is a widely used and diverse technique. It can be employed to quantify the components of both inorganic and organic complex mixtures, provide structural information about a specific molecular species, determine the isotopic ratio of atoms in a sample, or define the structure and composition of solid surfaces including human and animal tissues.

The steps of an MS analysis can be divided into four parts: (1) sample introduction, (2) ionization, (3) mass analysis, and (4) data analysis (Caprioli and Sutter, 1995, 2007). The discussion herein will focus primarily on the most commonly used methods of ionization and mass analysis. Briefly, samples can be introduced into the mass spectrometer as a gas, liquid, or solid, but liquids and solids must first be volatilized before, or concomitant with, ionization. Gases are commonly introduced into the mass spectrometer using GC. Liquids can be introduced either using direct infusion, where the sample is infused into tubing connected to the instrument with a syringe pump, or via HPLC or UPLC. Thin layers of solids are first plated on a sample slide, typically using a matrix which assists ionization, and are then inserted into the mass spectrometer. The sample is then either ablated with a laser beam or heated to produce ions. It is important to consider the matrix in which the solid is plated; it needs to have a high coefficient of absorption at the wavelength of the laser, be chemically inert in respect to reactivity with the analyte, and have a low sublimation rate.

Methods of Ionization

Electron Impact Ionization

Electron impact ionization (EI) is an ionization technique that is coupled with GC (Skoog and Leary, 1992a,b,c,d; Caprioli and Sutter, 2007). In this method, the gaseous analyte is bombarded by energetic electrons generated from a heated filament in the ion source. This interaction generates a radical cation ($M^{+\bullet}$) and two electrons:

$$M + e^- \rightarrow M^{+\bullet} + 2e^-$$

In this case $M^{+\bullet}$ is the molecular ion, and its m/z corresponds to the molecular mass of the analyte. The primary drawback to using this technique is that it can cause extensive molecule fragmentation, which reduces the sensitivity of detection. Fragmentation can be reduced by choosing an electron energy close to the ionization potential of the analyte.

Chemical Ionization

Chemical ionization (CI) relies on gas-phase chemical reactions that take place between the analyte of interest and ions generated from a reagent gas (Skoog and Leary, 1992a,b,c,d). Molecules can be ionized by transfer of an electron, a proton, or other charged species, and either positive or negative ions can result. When coupled with GC, CI is essentially EI with an excess of reagent gas. The primary difference is that the ionization chamber is kept at a higher pressure to increase the concentration of the reagent gas. Methane is the most commonly used reagent gas, and ions are generated as shown in the following reaction. The generated ions shown in bold react with the analyte:

$$CH_4^{+}\bullet + CH_4 \rightarrow \mathbf{CH_5^+} + CH_3\bullet$$

$$CH_4^{+}\bullet + CH_4 \rightarrow \mathbf{C_2H_5^+} + H_2 + H\bullet$$

When coupled with LC, the composition of the reagent gas is derived from the mobile-phase solvents. The initial electrons are produced by a discharge from a corona needle. These electrons then react with the mobile phase to generate ions, which react with the analyte(s) of interest. Unlike when coupled with GC, solvent-mediated CI takes place at atmospheric pressure (AP) and is commonly referred to as APCI (Niessen, 2006a,b,c,d). Both normal-phase and reversed-phase LC can be coupled with APCI, and this technique is particularly useful for less polar analytes.

Electrospray Ionization

Electrospray ionization (ESI) is the most common ionization technique used with LC. Like CI, ESI is a soft ionization technique, meaning that it produces very little ion excitation and thus little or no fragmentation of the analyte occurs (Skoog and Leary, 1992a,b,c,d; Niessen, 2006a,b,c,d). How ions are generated by ESI is not completely understood, but the technique is relatively simple. The dissolved sample, whether it is effluent from an LC column or from a syringe pump, is sprayed across a high-voltage field into a cone-shaped orifice on the mass analyzer while heat and gas flows are used to desolvate analyte ions. This ionization method is particularly useful because multiply charged species can be formed, thus allowing the analysis of molecules with molecular weights in excess of 50,000 Da. This technique has greatly enhanced the ability to study large biomolecules such as proteins because most mass spectrometers can only typically detect m/z ratios less than 2000 or 3000 Da and it can be coupled with LC.

Matrix-Assisted Laser Desorption Ionization

In matrix-assisted laser desorption ionization (MALDI), the analyte is plated on a sample probe in a solution containing an excess of a matrix compound with a chromophore, which absorbs at the wavelength of the laser (Caprioli and Sutter, 1995, 2007). The probe is then placed into the mass spectrometer and ablated with the laser. The matrix absorbs the energy from the laser producing a spray of plasma, which results in the vaporization and ionization of the analyte. Like ESI, this technique allows the analysis of large biomolecules with high molecular weights. MALDI also allows for protein analysis in intact tissue samples thus permitting the imaging of tissues using MS. One downside to this method of ionization is that the mass analyzer must be compatible with the pulsing of the laser.

Mass Analyzers and Modes of Analysis

Common mass analyzers include quadrupole, ion trap, fourier transform ion cyclotron resonance (FT-ICR), and time-of-flight (TOF) instruments (Skoog and Leary, 1992a,b,c,d; Niessen, 2006a,b,c,d). Quadrupoles are the most widely used mass analyzers because they can tolerate a wide *m/z* range and are of relatively low cost. These instruments contain four cylindrical rods placed in parallel in a radial array. A radio frequency (RF) and a positive or negative direct current are applied to the rods, and ions accelerate through the rods to a detector. Only ions of an exact *m/z* can pass through the quadrupole at a particular RF, thus experiments can be performed analyzing for a specific *m/z*. When a range of RFs are scanned, a broad range of *m/z* can be monitored. Ion trap mass analyzers are very similar to quadrupoles. However, instead of ions passing through the quadrupoles, ions are trapped in a circular electrode in an RF field. In this system, specific ions can be excited by energy from the electrode and ejected from the RF field to a detector. Being able to eject ions from the electrode is particularly useful because all ions except the *m/z* of interest can be ejected and further fragmentation experiments on this *m/z* can be performed. An FT-ICR mass spectrometer is much like an ion trap instrument in which the ions are trapped in a magnetic rather than an RF field. These instruments are key for the investigation of very large biomolecules where high mass accuracy is important (Chen et al., 1995). Additionally, these instruments have been used to detect attamoles of biomolecules in crude extracts of human blood and even in single cells when coupled with capillary electrophoresis (Hofstadler et al., 1996; Valaskovic et al., 1996). FT-ICR mass spectrometers, however, are incredibly expensive and historically not user-friendly.

Unlike other mass analyzers, in a TOF instrument, ions are accelerated into a field-free linear tube (Skoog and Leary, 1992a,b,c,d; Caprioli and Sutter, 2007; Niessen, 2006a,b,c,d). The time it takes for an ion to reach the detector at the far end of the tube is dependent upon the *m/z*. Pulsed ion introduction into the mass analyzer is necessary to prevent the simultaneous arrival of ions with different *m/z*, thus TOF is an ideal mass analyzer for MALDI. Because of the pulsed ion introduction, mass spectra are obtained from the accumulation of each ion introduction rather than by scanning. The result is a spectrum that has improved signal-to-noise because the random noise is averaged. In addition, most TOF instruments have an essentially unlimited mass range of *m/z* 10,000−20,000.

Experiments using MS can be performed using only a single mass analyzer or with multiple mass analyzers connected in tandem. A list showing the different modes of analysis used in MS and tandem MS and their common applications can be found in Table 8.3 (Niessen, 2006a,b,c,d). The most common tandem mass spectrometer is the triple quadrupole instrument (Fig. 8.3) (Niessen, 2006a,b,c,d). In this system, mass analysis is performed in the first and third quadrupoles. The second quadrupole is used as a collision cell to achieve ion fragmentation. Fragmenting ions in the collision cell is termed collision-induced dissociation and is used to obtain structural information about molecules with an *m/z* of interest. Other tandem mass spectrometers include the quadrupole ion trap, the quadrupole TOF, the ion trap TOF, and the TOF−TOF (Niessen, 2006a,b,c,d). All tandem mass spectrometers can be used to obtain structural information about molecules. In addition, these instruments can be used for selected reaction monitoring (SRM) experiments in which the transition of a specific precursor ion to a specific product ion is monitored. SRM is highly specific and sensitive and is often used for quantitation.

TABLE 8.3 Analysis Modes in Mass Spectrometry (MS)

Mode	MS-1	MS-2	Application
Full scan	Scanning a range of ions	Off	To obtain information about all of the mass-to-charge (*m/z*) ratio in a particular range in a sample
Selected ion monitoring (SIM)	Selecting one ion	Off	To monitor a specific *m/z* of interest
Product (or daughter) ion scan	Selecting one precursor ion	Scanning product ions	To obtain structural information about a molecule with a specific *m/z*
Precursor (or parent) ion scan	Scanning precursor ions	Selecting one product ion	To monitor ions which give identical product ions in collision-induced dissociation (CID)
Neutral-loss scan	Scanning precursor ions	Scanning product ions	MS-1 and MS-2 are scanned at a fixed *m/z* difference to monitor for compounds that lose a common neutral species
Selected reaction monitoring (SRM)	Selecting one precursor ion	Selecting one product ion	To monitor the transition of a specific precursor ion to a specific product ion generated from CID

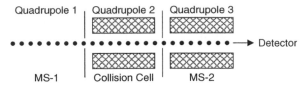

FIGURE 8.3 A triple quadrupole instrument used for tandem mass spectrometry (MS) analysis.

Mass Spectrometry and Clinical Research

Initially it was believed that MS was not useful for routine clinical research because it was too difficult and too expensive. These misconceptions have rapidly been proven wrong. Annesley et al. (2016) noted that the number of manuscripts published that include MS as part of the study grew three times that of the total number of articles published. These techniques are commonly used for many of the -omics approaches to clinical research. The basic molecules of DNA, proteins, lipids, and carbohydrates can all be analyzed using the variations of chromatography−MS platforms. Combined with informatics the MS-based interrogation of biological samples helped create the fields of proteomics, lipidomics, metabolomics, and when combined multiomics. This chapter is meant to be more of an overview of analytical techniques; thus, the reader is directed to several current reviews, special issues of journals, and book chapters that have been written on the use of MS in clinical and translation research (Annesley et al., 2016; Niessen, 2006a,b,c,d; Vogeser, 2003; Dooley, 2003; Adaway et al., 2015).

GENOMICS

Background

The field of genomics was initially focused on the comprehensive analysis of the expression of large numbers of genes simultaneously by measuring the relative amounts of RNA in biological samples (Chung et al., 2007). The technique was first described in the mid-1990s. In the late 1990s, great technological advances were made, and sophisticated robotic systems were developed that allowed for the automated spotting of DNA fragments onto fixed substrates at densities that could not be achieved manually. Subsequently, DNA microarrays were used extensively in translational and clinical research to identify the importance of certain genes in a variety of human diseases including cancer, asthma, cardiovascular disease, and neuropsychiatric disorders, among others (Quackenbush, 2006). However, arrays and early genome analyzers were relatively costly and slow. The short read, massively parallel sequencing techniques introduced in the late 2000s provided the "next generation" of technology and moved sequencing to genome and exome levels that could be accomplished in hours and for a few thousands of dollars (Corp, 2015). Indeed, over the past decade, the field of genomics has revolutionized our understanding of the genetics of human disease, and routine use of exome and whole genome data in the personalized care of patients (medical genomics) has become commonplace. While the analysis of the biological material is fairly straightforward, the analysis of the data is the challenge for the next decade. How to determine which variants are clinically relevant, how to present that data as useful information, and the ethical constraints associated with genetic information will be hurdles for translational and clinical researchers (Clayton, 2003; Pandey et al., 2012).

Basic Principles and Methodological Considerations

While microarray platforms are still available, the advent and rapid adoption of next-generation sequencing (NGS) technology has moved microarray to the background. Readers are referred to the first edition of this text for a review of the DNA microarray platforms.

Traditional Sanger sequencing is limited to 1000−1200 base pair (bp) reads and cannot surpass 2000 bp. To circumvent this limitation, the human genome project developed the shotgun approach to sequencing where the genomic DNA is fragmented and cloned into sequencing vectors that are individually analyzed. This concept formed the basis for the current massively parallel sequencing approach that is commonly called "next-generation sequencing" (Venter, 2003; Margulies et al., 2005).

Several variations of NGS have been developed and commercialized in the past 5 years including the Roche/545 FLX, Illumina/Solexa genome analyzer, Ion Torrent: Proton/PGM, and the Applied Biosystems SOLiD analyzer. While minor differences exist in the specifics, all rely on basic steps that include (1) library preparation, (2) cluster generation,

(3) sequencing, and (4) data analysis. The European Molecular Biology Laboratory (EMBL) offers online course in NGS, which can be found at the following website: https://www.ebi.ac.uk/training/online/course/ebi-next-generation-sequencing-practical-course/what-you-will-learn/what-next-generation-dna-.

Library preparation involves fragmenting the DNA or RNA to a desired length (usually 100−400 bp) and ligation of those fragments to sequencing adapters. In the Illumina technology the DNA-adapter complex is then captured onto a slide or beads in the 454 and Ion Torrent technology. The captured DNA is then amplified. Like any step in a multistep process, the quality of the library preparation is key to successful experiments. The library preparation phase is where the user chooses which type of sequencing is desired. Choices available here include between exome, whole genome (DNA-seq), RNA (mRNA or micro-RNA-seq), 16S amplicon, Chip-Seq, or single-cell DNA (or RNA). Preparation processes are specific for the sequencing platform that will be utilized (i.e., Illumina, 454, Ion Torrent: Proton/PGM, PacBio (SMRT), etc.), and commercial kits are readily available. Readers are referred to the manufacturer websites and the following reviews for details (Head et al., 2014; Roberts et al., 2013; van Dijk et al., 2014).

Following construction of the library the DNA−adapter complex is then amplified. Again variations in the amplification processes exist between manufacturers. The Roche/454 and Ion Torrent systems utilize an emulsion-based amplification system, while Illumina is a solid-phase amplification that generates clonal clusters through bridge amplification processes (Pareek et al., 2011).

Sequencing the amplified product involves flooding the reaction vessel (wells or slide) with nucleotides and polymerase. The addition of a nucleotide will result in an optical signal (454, Illumina) or the release of a hydrogen ion (Ion Torrent). This is repeated until all the fragments are read. The EBML website listed previously describes the sequencing technology for these platforms.

The sequencing machines will "call" the base pairs and compute some quality score (Phred quality score) generating a list of base data and quality score. This raw sequencing data is typically stored in a FASTQ file (FASTA is another available format). This FASTQ file is then fed into the secondary analysis "pipeline" where data is filtered using the quality scores and then aligned to create the sequence. If a datum is to be assembled (aligned) with no reference to align against, this is termed de novo assembly. This requires a greater number of reads (referred to as depth of coverage), which allows greater overlap of regions. If a reference genome is available, alignment is simpler. In either case the greater the number of reads that occur, the certainty of alignment is improved. The result of this processing typically generates a BAM (binary alignment map) file. These aligned data are then fed into the next "pipeline" where the variants present (single nucleotide polymorphisms, indels) are identified and VCF (variant call format) files containing the information to describe the variant(s) are generated. These variants can then be compared against publicly available databases, such as ClinVar, 1000 Genomes, and others, to further filter and prioritize the variants. The filtered list is then compared against the biology to describe the variants. Typically, the data analysis process is referred to as bioinformatics and the different steps and software used create "pipelines." Bao et al. (2014) describe the current methods for bioinformatics analysis of NGS data and Fig. 8.4 from that manuscript shows the overall process.

Summary and Applications to Clinical and Translational Research

Genomics continues to be one of the most powerful tools available to the clinical and translational researcher today and can be combined with other molecular tools to elucidate mechanisms of human disease (Liu and Karuturi, 2004; Collins et al., 2006). The data generated can be used to generate molecular signatures that define human physiology and pathophysiology, provide clinical correlates, and assess the impact of therapeutic interventions on human health. The field is rapidly progressing, and new technologies will likely continue to be brought forth that will expand our understanding of the role of differential gene expression in human disease.

PROTEOMICS, LIPIDOMICS, METABOLOMICS, AND MULTIOMICS

Background

The use of -ome and -omics as a suffix to describe an area of biological research has exploded in the past decade. While genomics and proteomics were early areas of research, we now have lipidomics, metabolomics, transcriptomics, and even subdisciplines with names such as functional genomics, proteogenomics, and immunoproteomics. The combination of data and processes from multiple areas is now referred to as multiomics (Ritchie et al., 2015). Descriptions of the similarities and differences in these areas is beyond the scope of this chapter, and readers are encouraged to read recent reviews on the areas (Gowda and Djukovic, 2014; Griffiths and Wang, 2009).

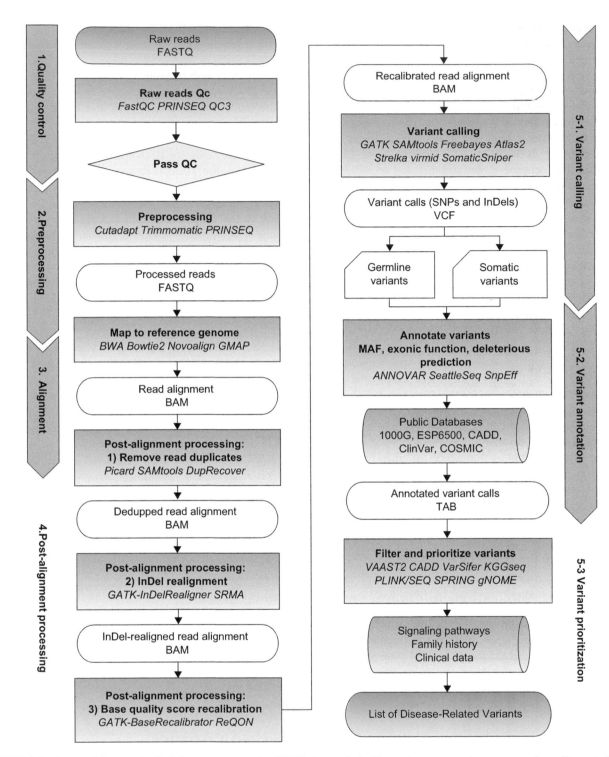

FIGURE 8.4 A general framework of whole exome sequencing (WES) data analysis. Five major steps are shown: raw reads quality control (QC), preprocessing, alignment, postprocessing, and variant analysis (variant calling, annotation, and prioritization). Notes: FASTQ, BAM (binary alignment map), VCF (variant call format), and TAB (tab-delimited) refer to the standard file format of raw data, alignment, variant calls, and annotated variants, respectively. A selection of tools supporting each analysis step is shown in italics.

While each area works with a different biological material (nucleotide, protein, lipid, carbohydrate), the MS-based approach to analysis of the biological molecules has emerged as the preferred analytical platform for proteomics, lipidomics, and metabolomics. The remainder of this chapter will highlight the differences in how MS is utilized in the different -omic areas.

Each of these -omics approaches is fraught with a number of limitations regarding characterization of the human -omes including technical issues and the fact that many molecules are present at very low concentrations and thus are not detectable with some of the approaches mentioned earlier. MS emerged as the method of choice to analyze and characterize the various human -omes because of its high sensitivity and has largely replaced other methods of proteomic analysis (Liebler, 2002). In addition, this approach has been facilitated by the availability of sequencing information from the human genome project that has led to the assembly of protein databases containing all putative human proteins and has thus become a very powerful bioinformatics tool for data mining. For the translational researcher, the field of -omics offers the potential to define the human -ome in association with human physiology and pathophysiology as well as to determine the effects of therapeutic interventions on it. The study and use of the various -omics processes attempts to bridge the gap between genomics and proteomics and biological endpoints (Fig. 8.5; Nicholson et al., 2002). While the analytical processes involved in -omics have advanced with technologies to improve the specific capture of biological molecules, detection at lower limits, better quantification, and improved throughput, the major advances in this area are largely related to the informatics. Like genomics, proteomic methods generate huge amounts of data, and the largest hurdle in proteomic experiments lies with the data processing and analysis (Kumar and Mann, 2009). The purpose of this section is to provide the reader with an overview of analytical aspects of -omics, particularly the MS-based approaches.

Basic Principles and Methodological Considerations

MS has revolutionized our ability to analyze proteins, lipids, and even carbohydrates. The two most common techniques employed are matrix-assisted laser desorption mass spectrometry (MALDI-MS) and electrospray ionization mass spectrometry (ESI-MS; Chung et al., 2007; Aebersold and Mann, 2003). These methods were discussed in detail in a previous section. The further discussion is the application of these methods to -omics fields. Coupled with bioinformatics and statistical software programs, MALDI- and ESI-MS allow the highly accurate determination of molecular weight, primary and higher-order structure, posttranslational modifications, quantification, and localization.

Matrix-Assisted Laser Desorption Ionization Mass Spectroscopy

MALDI-MS is useful for protein profiling of biological samples and for identification of proteins from purified samples containing no more than a few different species (Chung et al., 2007). The advantages of this approach for protein analysis are that sample preparation is simple, and it is amenable to high-throughput analysis. As previously discussed, MALDI is most often coupled to TOF analyzers that measure the mass of intact proteins and peptides. More recently, MALDI has been coupled to quadrupole ion trap mass spectrometers or TOF—TOF instruments to allow for fragmentation of MALDI-generated protein precursor ions (Aebersold and Mann, 2003).

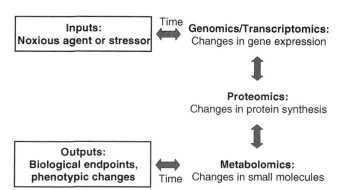

FIGURE 8.5 Relationships between real-world biological input and outputs to "-omic" responses. *Modified with permission from Chung, C.H., et al., 2007. Genomics and proteomics: emerging technologies in clinical cancer research. Critical Reviews in Oncology/Hematology 61 (1), 1—25.*

Electrospray Ionization Mass Spectroscopy

In ESI-MS, multiply charged peptide ions are directly analyzed in solution (Kicman et al., 2007). Peptides are generated from proteins of interest by pretreatment with proteases, most commonly trypsin, due to the fact that ESI-MS analysis can only detect compounds with an m/z of up to ~4000 and is centered around m/z 800–1000 (Chung et al., 2007). A potential limitation of ESI-MS is that because each peptide in a sample generates a distribution of ions, it becomes increasingly difficult to analyze and separate overlapping signal distributions from complex mixtures. This can be overcome by coupling ESI-MS to LC and by performing multidimensional separations in which, for example, proteins are initially separated on gels by molecular weight, and proteins with molecular weights in similar ranges are analyzed by LC-ESI-MS as a group after trypsinization. ESI-MS can also be used for protein profiling and to detect large numbers of proteins in complex biological fluids (Ahn et al., 2007).

Applications of -Omic Analysis

Profiling

The goal of profiling studies is to determine which markers potentially are indicative of disease and response to therapeutic interventions (Chung et al., 2007). In this regard, profiling is often undertaken by comparing tissue or body fluids from control and disease samples. Both human body fluids and tissues can be examined utilizing either MALDI-MS or ESI-MS approaches. Profiling may be useful for discovery research purposes, but information is generally limited to molecular weight information only. This approach is largely replaced by more sophisticated technologies, primarily involving ESI-MS that identify multiple molecules from complex mixtures.

Identification

Two strategies used to identify molecules are "top-down" and "shotgun" -omics (Chung et al., 2007; Ahn et al., 2007; Qian et al., 2006). It should be kept in mind that no method or MS instrument exists that can identify the components of a complex protein sample in a single step. As a consequence, individual components for separating and identifying the polypeptides as well as tools for integrating and analyzing the data are used together. The general methods employed in these two strategies are summarized in Fig. 8.6.

For "top-down" -omics, molecule of interest must first be isolated in relatively pure form. Techniques vary based on the molecule of interest, but chromatography or combinations of immunochromatography are most commonly utilized (Fredolini et al., 2016).

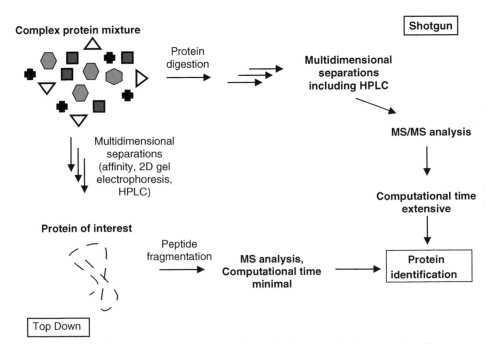

FIGURE 8.6 Top-down versus shotgun protein identification strategies from complex mixtures.

"Shotgun" -omics is the second method available for identification using MS and has replaced "top-down" methods in many cases (Chung et al., 2007; Aebersold and Mann, 2003; Drabik et al., 2007). The major advantage is that large numbers of molecules in a biological fluid or tissue extract can be analyzed simultaneously. A potential limitation of this method is that the amount of data collected is huge in some cases and the analysis is daunting. On the other hand, substantial progress has been made over the past several years in bioinformatics methods of data analysis. In addition, partial separation of molecules prior to MS analysis using techniques such as one-dimensional gel electrophoresis, immunoseparation, and chromatography has proven useful to reduce the complexity of data analysis.

Quantification

Methods to obtain quantitative information regarding molecules in biological fluids or tissues have improved significantly in the past few years (Pratsch et al., 2014; Schwudke et al., 2011; Wellhausen and Seitz, 2012). Relative quantification can be achieved using two-dimensional gel or capillary electrophoresis. Several MS-based approaches exist for the relative or absolute quantification of biological molecules. Relative quantification can be achieved in several ways including introducing stable heavy isotope tags onto proteins obtained from biological sources. These tags interact with various amino acid residues including sulfhydryl and amine groups. In addition, tags can also be introduced via the transfer of ^{18}O from water to peptides. Other methods have been described in which proteins are metabolically labeled in cells using, for example, ^{13}C-arginine. The proteins of interest are subsequently isolated and quantified. Relative quantification is achieved by comparing the ratio of peptides possessing the heavy isotope tags to those that do not. Absolute quantification of proteins can be achieved using stable isotope dilution methods in which stable isotope-labeled peptides are chemically synthesized and introduced into peptide mixtures from biological fluids as internal standards (Aebersold and Mann, 2003).

Summary and Applications to Clinical and Translational Research

While -omics as a field has evolved significantly over the past decade, the full potential of the discipline is yet to be recognized, particularly in clinical and translational research (Sinha et al., 2007). There have been multiple reports of the use of MS-based -omics for the possible diagnosis of disease and treatment responses, yet few have advanced to full clinical utility (Di Meo et al., 2014). For the translational researcher today, the applications of -omics to human disease are still largely discovery-based, and significantly more research is necessary to define the role of this technology in our discipline (Di Meo et al., 2014; Diamandis and Li, 2015).

CONCLUSION

Immunoassays represent the most frequently used and perhaps simplest approach to the analysis of biological materials for the translational researcher. Several different formats are available, each with differing performance characteristics. The concept of buyer aware is very appropriate when selecting from the available commercial immunoassays though. The "Research use only"—labeled kit requires the user to verify the performance of the assay before use with research samples. Numerous examples of commercially available immunoassays not measuring the stated molecule have been reported in the past few years. MS and the associated separation techniques of GC and LC have become the de facto approach to the profiling and quantification of protein, lipids, carbohydrates, and even nucleic acids. The massively parallel sequencing of nucleic acids has been termed "next-generation sequencing" and has largely supplanted the use of microarray approaches to genomics.

Multiomics is the combination of processes and data from the genomics, proteomics, lipidomics, transcriptomics, metabolomics, and other -omics areas. The concept is based on the recognition that the structure and function of the biological systems of interest to the translational research are interrelated and impacted by all of the -omics.

REFERENCES

Ashihara, Y., Kasahara, Y., Nakamura, R.M., 2001. Immunoassays and immunochemistry. In: Henry, J.B. (Ed.), Clinical Diagnosis and Management by Laboratory Methods. W.B. Saunders, Philadelphia, pp. 821—849.

Annesley, T.M., et al., 2016. Clinical mass spectrometry-achieving prominence in laboratory medicine. Clinical Chemistry 62 (1), 1—3.

Adaway, J.E., Keevil, B.G., Owen, L.J., 2015. Liquid chromatography tandem mass spectrometry in the clinical laboratory. Annals of Clinical Biochemistry 52 (Pt 1), 18—38.

Aebersold, R., Mann, M., 2003. Mass spectrometry-based proteomics. Nature 422 (6928), 198—207.

Ahn, N.G., et al., 2007. Achieving in-depth proteomics profiling by mass spectrometry. ACS Chemical Biology 2 (1), 39—52.

Bao, R., et al., 2014. Review of current methods, applications, and data management for the bioinformatics analysis of whole exome sequencing. Cancer Informatics 13 (Suppl. 2), 67–82.

Christiansson, L., et al., 2014. The use of multiplex platforms for absolute and relative protein quantification of clinical material. EuPA Open Proteomics 3, 37–47.

Churchwell, M.I., et al., 2005. Improving LC-MS sensitivity through increases in chromatographic performance: comparisons of UPLC-ES/MS/MS to HPLC-ES/MS/MS. Journal of Chromatography B 825 (2), 134–143.

Caprioli, R., Sutter, M., 2007. Mass Spectrometry Tutorial. http://www.mc.vanderbilt.edu/msrc/tutorials/ms/ms.htm.

Caprioli, R., Sutter, M., 1995. In: Glasel, J.A., Deutscher, M.P. (Eds.), Introduction to Biophysical Methods for Protein and Nucleic Acid Research. Academic Press.

Chen, R., et al., 1995. Trapping, detection, and mass determination of Coliphage T4 DNA ions by electrospray ionization fourier transform ion cyclotron resonance mass spectrometry. Analytical Chemistry 67 (7), 1159–1163.

Chung, C.H., et al., 2007. Genomics and proteomics: emerging technologies in clinical cancer research. Critical Reviews in Oncology/Hematology 61 (1), 1–25.

Corp, I., 2015. An Introduction to Next-Generation Sequencing Technology.

Clayton, E.W., 2003. Ethical, legal, and social implications of genomic medicine. New England Journal of Medicine 349 (6), 562–569.

Collins, C.D., et al., 2006. The application of genomic and proteomic technologies in predictive, preventive and personalized medicine. Vascular Pharmacology 45 (5), 258–267.

Davies, C., 1994. Principles. In: Wild, D. (Ed.), The Immunoassay Handbook. Stockton Press, New York, pp. 3–47.

Dooley, K.C., 2003. Tandem mass spectrometry in the clinical chemistry laboratory. Clinical Biochemistry 36 (6), 471–481.

van Dijk, E.L., et al., 2014. Ten years of next-generation sequencing technology. Trends in Genetics 30 (9), 418–426.

Drabik, A., et al., 2007. Proteomics in neurosciences. Mass Spectrometry Reviews 26 (3), 432–450.

Diamandis, E., Li, M., 2015. The side effects of translational omics: overtesting, overdiagnosis, overtreatment. Clinical Chemistry and Laboratory Medicine 54 (3), 389–396.

Espinosa, R.J., Brugues, M.J., Llanos, O.J., 1987. Technical and clinical performances of six sensitive immunoradiometric assays of thyrotropin in serum. Clinical Chemistry 33, 1439–1445.

Fu, Q., Zhu, J., Van Eyk, J.E., 2010. Comparison of multiplex immunoassay platforms. Clinical Chemistry 56 (2), 314–318.

Fredolini, C., et al., 2016. Immunocapture strategies in translational proteomics. Expert Review of Proteomics 13 (1), 83–98.

Gowda, G.A., Djukovic, D., 2014. Overview of mass spectrometry-based metabolomics: opportunities and challenges. Methods in Molecular Biology 1198, 3–12.

Griffiths, W.J., Wang, Y., 2009. Mass spectrometry: from proteomics to metabolomics and lipidomics. Chemical Society Reviews 38 (7), 1882–1896.

Hofstadler, S.A., et al., 1996. Analysis of single cells with capillary electrophoresis electrospray ionization Fourier transform ion cyclotron resonance mass spectrometry. Rapid Communications in Mass Spectrometry 10 (8), 919–922.

Head, S.R., et al., 2014. Library construction for next-generation sequencing: overviews and challenges. Biotechniques 56 (2), 61–64, 66, 68, passim.

Koehler, G., Milstein, C., 1975. Continuous cultures of fused cells secreting antibody of predefined specificity. Nature 256, 495–497.

de Koning, L., et al., 2012. A multiplex immunoassay gives different results than singleplex immunoassays which may bias epidemiologic associations. Clinical Biochemistry 45 (10–11), 848–851.

Kumar, C., Mann, M., 2009. Bioinformatics analysis of mass spectrometry-based proteomics data sets. FEBS Letters 583 (11), 1703–1712.

Kicman, A.T., Parkin, M.C., Iles, R.K., 2007. An introduction to mass spectrometry based proteomics-detection and characterization of gonadotropins and related molecules. Molecular and Cellular Endocrinology 260–262, 212–227.

Linnet, K., Boyd, J.C., 2012. Selection and analytical validation of methods – with statistical techniques. In: Burtis, C.A., Ashwood, E.F., Bruns, D.E. (Eds.), Tietz Textbook of Clinical Chemistry and Molecular Methods. Elsevier Saunders, Philadelphia, pp. 7–47.

Liu, E.T., Karuturi, K.R., 2004. Microarrays and clinical investigations. New England Journal of Medicine 350 (16), 1595–1597.

Liebler, D., 2002. Introduction to Proteomics. Humana Press, Totowa.

McMaster, M.C., 2007. HPLC: A Practical User's Guide, second ed. Wiley Interscience, Hoboken, NJ.

Margulies, M., et al., 2005. Genome sequencing in microfabricated high-density picolitre reactors. Nature 437 (7057), 376–380.

Di Meo, A., et al., 2014. What is wrong with clinical proteomics? Clinical Chemistry 60 (10), 1258–1266.

Nakamura, R.M., Kasahara, Y., 1992. Heterogeneous enzyme immunoassays. In: Nakamura, R.M., Kasahara, Y., Rechnitz, G.A. (Eds.), Immunochemical Assays and Biosensor Technology for the 1990's. American Society for Microbiology, Washington, pp. 149–167.

Nakamura, R.M., 1992. Fluorescence immunoassays. In: Nakamura, R.M., Kasahara, Y., Rechnitz, G.A. (Eds.), Immunochemical Assays and Biosensor Technology for the 1990's. American Society for Microbiology, Washington, pp. 205–227.

Niessen, W.M.A., 2006a. Liquid chromatography and sample pretreatment. In: Liquid Chromatography-Mass Spectrometry. CRC Press, Boca Raton, pp. 3–22.

Niessen, W.M.A., 2006b. Atmospheric-pressure ionization. In: Liquid Chromatography-Mass Spectrometry. CRC Press, Boca Raton, pp. 141–178.

Niessen, W.M.A., 2006c. Mass spectrometry. In: Liquid Chromatography-Mass Spectrometry. CRC Press, Boca Raton, pp. 23–52.

Niessen, W.M.A., 2006d. Clinical applications of LC-MS. In: Liquid Chromatography-Mass Spectrometry. CRC Press, Boca Raton, pp. 331–358.

Nicholson, J.K., et al., 2002. Metabonomics: a platform for studying drug toxicity and gene function. Nature Reviews Drug Discovery 1 (2), 153–161.

Prassas, I., et al., 2014. False biomarker discovery due to reactivity of a commercial ELISA for CUZD1 with cancer antigen CA125. Clinical Chemistry 60 (2), 381–388.

Pandey, K.R., et al., 2012. The curation of genetic variants: difficulties and possible solutions. Genomics Proteomics Bioinformatics 10 (6), 317–325.

Pareek, C.S., Smoczynski, R., Tretyn, A., 2011. Sequencing technologies and genome sequencing. Journal of Applied Genetics 52 (4), 413–435.

Pratsch, K., Wellhausen, R., Seitz, H., 2014. Advances in the quantification of protein microarrays. Current Opinion in Chemical Biology 18, 16–20.

Quackenbush, J., 2006. Microarray analysis and tumor classification. New England Journal of Medicine 354 (23), 2463–2472.

Qian, W.J., et al., 2006. Advances and challenges in liquid chromatography-mass spectrometry-based proteomics profiling for clinical applications. Molecular & Cellular Proteomics 5 (10), 1727–1744.

Reubenstein, K.E., Schneider, R.S., Ullman, E.F., 1972. Homogeneous enzyme immunoassay. A new immunological technique. Biochemical and Biophysical Research Communications 47, 846–851.

Rifai, N., Watson, I.D., Miller, W.G., 2012. Commercial immunoassays in biomarkers studies: researchers beware! Clinical Chemistry 58 (10), 1387–1388.

Rozet, E., et al., 2011. Advances in validation, risk and uncertainty assessment of bioanalytical methods. Journal of Pharmaceutical and Biomedical Analysis 55 (4), 848–858.

Roberts, R.J., Carneiro, M.O., Schatz, M.C., 2013. The advantages of SMRT sequencing. Genome Biology 14 (7), 405.

Ritchie, M.D., et al., 2015. Methods of integrating data to uncover genotype-phenotype interactions. Nature Reviews Genetics 16 (2), 85–97.

Scott, R.P.W., 2003a. Principles and Practices of Chromatography. http://www.library4science.com/.

Skoog, D.A., Leary, J.J., 1992a. An introduction to chromatographic separations. In: Principles of Instrumental Analysis. Saunders College Publishers, Fort Worth, TX, pp. 579–604.

Skoog, D.A., Leary, J.J., 1992b. Gas chromatography. In: Principles of Instrumental Analysis. Saunders College Publishers, Fort Worth, TX, pp. 605–627.

Scott, R.P.W., 2003b. Gas Chromatography. http://www.library4science.com/.

Skoog, D.A., Leary, J.J., 1992c. High-performance liquid chromatography. In: Principles of Instrumental Analysis. Saunders College Publishers, Fort Worth, TX, pp. 628–669.

Scott, R.P.W., 2003c. Liquid Chromatography. http://www.library4science.com/.

Skoog, D.A., Leary, J.J., 1992d. Mass spectrometry. In: Principles of Instrumental Analysis. Saunders College Publishers, Fort Worth, TX, pp. 420–461.

Schwudke, D., et al., 2011. Shotgun lipidomics on high resolution mass spectrometers. Cold Spring Harbor Perspectives in Medicine 3 (9), a004614.

Sinha, A., et al., 2007. Proteomics in clinical interventions: achievements and limitations in biomarker development. Life Sciences 80 (15), 1345–1354.

Tighe, P., et al., 2013. Utility, reliability and reproducibility of immunoassay multiplex kits. Methods 61 (1), 23–29.

de Villiers, A., et al., 2006. Evaluation of ultra performance liquid chromatography. Part I. Possibilities and limitations. Journal of Chromatography A 1127 (1–2), 60–69.

Valaskovic, G.A., Kelleher, N.L., McLafferty, F.W., 1996. Attomole protein characterization by capillary electrophoresis-mass spectrometry. Science 273 (5279), 1199–1202.

Vogeser, M., 2003. Liquid chromatography-tandem mass spectrometry—application in the clinical laboratory. Clinical Chemistry and Laboratory Medicine 41 (2), 117–126.

Venter, J.C., 2003. A part of the human genome sequence. Science 299 (5610), 1183–1184.

Winter, G., et al., 1994. Making antibodies by phage display technology. Annual Review of Immunology 12, 433–455.

Wellhausen, R., Seitz, H., 2012. Facing current quantification challenges in protein microarrays. Journal of Biomedicine and Biotechnology 2012, 831347.

Zola, H., 1987. In: Raton, B. (Ed.), Monoclonal Antibodies: A Manual of Techniques. CRC Press.

Chapter 9

Imaging Tools in Clinical Research: Focus on Imaging Technologies

Ruth M. Dunne, Ailbhe C. O'Neill and Clare M. Tempany

Harvard Medical School, Boston, MA, United States

Chapter Outline

Key Points

- Imaging is playing a much larger role in the detection, diagnosis, staging and restaging of disease, and monitoring response to therapy.
- Imaging has a unique ability to participate in individualized or "P4 medicine" (predictive, personalized, preemptive, and participatory medicine).
- Technical advances in computed tomography (CT) include multidetector CT generating "isotropic voxels" or data cells that are of the same dimension in axial, coronal, and sagittal planes creating three-dimensional (3D) images. Dual energy imaging is another advance in CT where dual energy sources and detectors are used, increasing temporal resolution and allowing for differentiation of tissues and discrimination of tissue compositions.
- Higher field strengths in magnetic resonance imaging (MRI) provide higher-resolution images, and introduction of parallel coil transmitter systems using multichannel radiofrequency (RF) coils can produce a more uniform image at higher field strengths. Ongoing research is being performed into molecular imaging by MRI using different contrast agents including iron oxide—based contrast media [ultrasmall particles of iron oxide (USPIOs) or superparamagnetic particles of iron oxide image contrast].
- Elastography is a new and promising ultrasound (US) technology relying on the premise that palpable abnormalities, such as malignancy, tend to be "harder" (less elastic) than nonmalignant tissue. The introduction of US contrast agents is another exciting development where microbubbles (MBs) produce strong US echoes and serve as contrast agents. US tomography, a recently defined concept, requires hundreds of transducer elements covering a large-angle area of the body and may diminish the operator-dependency of the procedure and improve reproducibility.
- Hybrid systems where a CT is coupled to a PET or SPECT system has been a major advance in nuclear medicine imaging providing both anatomical and functional information. Hybrid PET/MRI combines the physiologic information obtained by PET with the morphological and functional information from MRI and is a promising new technique. Multiple PET tracers are available for

Clinical and Translational Science. http://dx.doi.org/10.1016/B978-0-12-802101-9.00009-0

imaging of cancer, but molecular imaging has also made advances in fields outside oncology with the growing use of PET/CT in dementia work and for amyloid imaging.
- Optical imaging uses fluorescent and bioluminescent probes that emit radiation at visible or near-infrared wavelengths, which can be detected by optical camera. There have been significant technologic advances including new probes that respond to cellular activity, an array of specific optical imaging agents and optical coherence tomography.

INTRODUCTION

Imaging is playing a much larger role in the detection, diagnosis, staging and restaging of disease, and monitor response to therapy. Recent technical advances in imaging modalities have led to further improving and refining diagnostic capabilities. Imaging has also evolved from the traditional "anatomic-centric" approach focused on structure and morphology to visualization of biologic processes within the body through functional and molecular imaging. Molecular imaging has been described as having "potential to define itself as a core interdisciplinary science for extracting spatially and temporally resolved biological information at all physical scales from Angstroms to microns to centimeters in intact biological systems" (Zerhouni, 2007). Furthermore, imaging is becoming increasingly important as an approach to synthesize, extract, and translate useful information from large multidimensional databases accumulated in research frontiers such as functional genomics, proteomics, and functional imaging. This offers the field of imaging the unique ability to truly participate in individualized medicine also termed "P4 medicine," i.e., predictive, personalized, preemptive, and participatory medicine, which has been described as "where patients will be given clinical information from molecular imaging prior to symptoms being expressed and prior to loss of function" (Zerhouni, 2007).

IMAGING TECHNOLOGIES
Overview

Medical images are produced through different processes that necessitate the use of physical probes that are created or affected by the human body, and they are detected in different ways (Wolbarst and Hendee, 2006). Since the 1960s, the introduction of new biomedical imaging techniques has been accepted as one of the most important advances in medicine. These imaging technologies have contributed to both basic and clinical medical research by providing investigators with powerful new approaches to explore physiologic and pathologic processes (Ehman et al., 2007). The current developments in molecular medicine are taking the fields of morphologic and functional imaging further into the more relevant arenas of the imaging of molecular events within physiologic and pathologic processes. This has led to the expansion and further development of all medical imaging modalities.

Computed Tomography
Technological Advances

Technological advances in computed tomography (CT) over the last 25 years have led to increased utilization and applications of this imaging modality in all aspects of patient care and research. The earliest form of CT was axial or "step and shoot" where the patient was moved into position and then the X-ray tube rotated around to produce the X-ray beam, with both steps repeated and the table gradually moving distally, until the scan was completed (Alexander and Gunderman, 2010; Hounsfield, 1973). In contrast in the subsequent helical or spiral CT, both patient movement and rotation of the X-ray tube are continuous, producing shorter scanning time with images obtained in a single breath (Kalender et al., 1990). With the advent of multidetector CT, acquisition time decreased as multiple interleaving helical scans are simultaneously acquired (Hu et al., 2000). Multidetector CT acquires data as isotropic voxels, where each data cell or voxel is of the same dimension in axial, coronal, and sagittal planes. Isotropic voxels due to high resolutions allows visualization of the images in multiple planes creating three-dimensional (3D) images (Kalender, 1995). The number of detectors has continued to increase from the first 4-row detector CT to the current 320-row detector system that enables acquisition in the time of a single heartbeat on a stationary table (Einstein et al., 2010). Another technical advance was the creation of dual energy imaging that utilizes dual energy sources and dual detectors increasing temporal resolution (Flohr et al., 2006). Dual energy can acquire data using two different X-ray energy spectra that allow for differentiation of tissues and discrimination of tissue compositions (Marin et al., 2014).

For example, dual energy contrast-enhanced CT allows isolation of the iodine contribution that if subtracted can provide a virtual noncontrast CT or alternatively can be added on a color map to demonstrate areas of contrast enhancement (Marin et al., 2014; Graser et al., 2010). It has uses in decreasing metallic artifact reduction using virtual monochromatic images to decrease beam hardening and improve image quality (Bamberg et al., 2011).

However, increased CT utilization has been paralleled by an increase in radiation exposure, and CT accounts for the majority of radiation exposure in medical imaging (Lee et al., 2008). To counterbalance this, there has been a concerted effort to decrease the radiation dose while maintaining adequate diagnostic image quality. In CT coronary angiogram, it is possible to image when the heart is at rest in diastole (prospective) and not throughout the cardiac cycle (retrospective), which decreases the effective radiation dose to the patient (Earls et al., 2008). An advantage of retrospective gating is that imaging during the entire cardiac cycle allows for the performance of quantitative analysis such as left ventricular function and calculation of the ejection fraction.

In imaging with CT, there is a trade-off between image noise and radiation dose. Noise is present in all electronic systems, and in CT is dependent on the number of X-ray photons that reach the detector versus random quantum mottle. Signal-to-noise ratio (SNR) describes the measure of true signal reflected in the image versus the random quantum mottle of noise (Polacin et al., 1992). Automatic tube current modulation adjusts the radiation dose according to the patient's attenuation of X-ray beam, maintaining image quality and ultimately reducing dose. This is performed in one of the three ways: patient body habitus, z-axis, and rotational. In relation to *patient body habitus* the tube current is adjusted based on the overall size of the patient to maintain similar image quality but varying the dose between small and large patients. Patients are assessed prior to the CT examination and their weight and height are documented. For a given patient size, the appropriate tube current is chosen and used for the entire examination. Another method is using the "reference milliamperage-based automatic exposure control," where patients with larger body habitus are scanned with a higher tube voltage, but the tube current is reduced to decrease overall dose (Lee et al., 2008). The *z-axis* modulation uses information derived from the initial topogram or scout image. The current is modulated so that it is lower, for example, in the thorax and higher in the abdomen as there is increased attenuation of X-rays from solid organs in the abdomen. Lastly, *rotational* dose modulation occurs in asymmetric regions, for example, the neck versus the abdomen, to decrease fluctuations in image noise throughout the examination (Lee et al., 2008).

CT image reconstruction was most commonly performed using filtered back projection, as it is a fast process although it does produce relatively noisy images with less contrast detail. Currently, iterative reconstruction is favored as it has been shown to decrease image noise (random quantum mottle) allowing for lower tube currents and hence radiation dose reduction without the loss of diagnostic image quality (Willemink et al., 2013; Kubo et al., 2014).

Clinical Care and Research Applications

While primarily developed for clinical care, these applications have also seen increasing use in clinical research either primarily or secondarily.

Detection

CT Colonography Colorectal cancer screening aims to reduce mortality by the detection of colonic adenomas and early colorectal cancers. CT colonography is a noninvasive screening tool for visual assessment of the colon. The patient is scanned in both supine and prone positions to enable movement of colonic fluid and feces to the dependent surface and therefore increasing endoluminal surface visualization. Two-dimensional (2D) axial images and 3D endoluminal images of the colon are acquired for interpretation (Royster et al., 1997; Dachman et al., 1998). Multidetector scanning has lead to shorter acquisition times, and isotropic imaging decreases volume averaging. Volume averaging is a when a structure imaged on one slice becomes averaged on another slice producing an artifact. Isotropic imaging also improves image quality in the multiplanar reformats (images generated from the original axial images reformatted into coronal, sagittal, or oblique planes) and in the 3D endoluminal renderings, which is a type of virtual colonoscopy (Fig. 9.1). Fecal tagging of residual stool with an oral iodinated contrast agent either alone or in combination with oral barium allows electronic subtraction of labeled stool to help in the detection of polyps (Zalis et al., 2005, 2006).

Lung Cancer Screening Lung cancer is associated with the highest cancer-related mortality, accounting for a quarter of all cancer deaths (Siegel et al., 2014). Recently, low-dose helical CT has been introduced as a screening modality for high-risk patients [aged between 55 and 74 years with significant smoking history (≥30 pack years)]. When compared to chest radiography, annual low-dose CT in high-risk patients has been shown to reduce mortality by 20% (National Lung Screening Trial Research T et al., 2013; Black et al., 2014).

Coronary Artery Calcium Scoring Coronary artery disease is a leading cause of death worldwide. Noncontrast-enhanced CT is a noninvasive screening tool that can measure the amount of calcium in coronary arteries as a predictor of future cardiovascular events. The lack of coronary artery calcifications has a high negative predictive

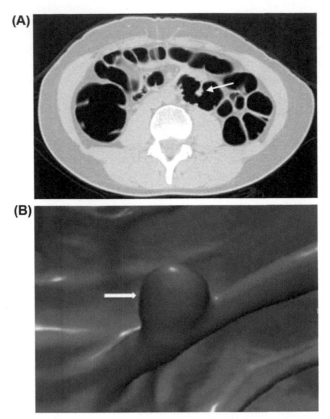

FIGURE 9.1   Computed tomography (CT) Colonography. (A) Axial noncontrast CT image demonstrating focal polypoid lesion arising from colonic haustral fold. Subsequent colonoscopic biopsy was consistent with an adenoma; (B) three-dimensional (3D) volume-rendered image or "virtual colonoscopy" image of the same lesion.

value (100%) for clinically significant coronary artery disease (Breen et al., 1992; Shaw et al., 2003). This is a useful study in low- to intermediate-risk patients presenting with acute chest pain and may obviate the need for an invasive cardiac catheterization. CT examinations gated to the cardiac cycle can also evaluate valve calcification prior to percutaneous replacement surgery and assess the amount of calcification in the aorta preoperatively (Hoey and Ganeshan, 2014a,b).

A commonly used scoring system is the Agatston scoring system which quantifies both the number and density of coronary artery calcifications present to provide an overall score related to severity of coronary artery calcification. This score has been shown to be an imaging biomarker for coronary artery disease (Agatston et al., 1990).

Diagnosis

CT Coronary Artery Angiography CT coronary artery angiogram is a noninvasive tool that has a range of applications (de Roos and Higgins, 2014; Vanhoenacker et al., 2007). More detailed evaluation of cardiac structures and decreased heart rate variability is possible with the high spatial resolution of multidetector 320-row CT that enables single heart beat acquisition, and the high temporal resolution of dual source CT provides high image quality irrespective of a patient's heart rate (Pelliccia et al., 2013; Sun et al., 2014).

High spatial resolution allows evaluation of the location and degree of coronary artery stenoses as well as assessing the origin and course of the coronary arteries to evaluate for coronary artery anomalies (Schoepf et al., 2004). Functional assessment of myocardial perfusion using CT can be performed not only to evaluate the presence of coronary artery disease but also to assess if there is a significant hemodynamic component causing perfusion defects (Ko et al., 2014; Rocha-Filho et al., 2010).

The use of retrospective gating allows for the functional evaluation of perfusion gradients and left ventricular function. Transcatheter aortic valve replacement is of increasing importance in patients with critical aortic stenosis who are not suitable for surgery. CT examinations gated to the cardiac cycle can carefully and precisely measure the aortic valve orifice (Hoey and Ganeshan, 2014a,b).

Brain Perfusion Imaging in acute stroke initially involves a noncontrast CT scan to exclude a hemorrhagic stroke or another process mimicking a stroke (for example, a space-occupying lesion or subdural hematoma). If a hemorrhagic stroke is excluded and if the patient is within the time cutoff, they may be eligible for thrombolysis. In these cases using CT angiography to determine the etiology and CT perfusion to quantify the infarction may help select patients who would have benefit from thrombolysis leading to improved outcomes (Provenzale and Wintermark, 2008).

CT angiography in acute stroke is used to assess the presence and location of a potentially treatable thrombus or an embolus causing arterial occlusion. Perfusion CT imaging provides quantification of stroke with a pictorial depiction of the nonviable infarcted tissue compared to the potentially salvageable ischemic but not yet infarcted tissue or penumbra (Hopyan et al., 2010). Consecutive multiphase images are obtained over a certain time period (50–60 s) during dynamic contrast injection to create cine sequences. This allows postprocessing software to calculate cerebral blood flow (CBF), cerebral blood volume (CBV), and time to peak producing both numerical values in different areas and color maps that allow differentiation between an ischemic core, ischemic penumbra, and area of infarction. Infarcted tissue compared to ischemic tissue is more at risk for hemorrhage secondary to reperfusion injury if blood flow is restored (Hopyan et al., 2010; Pan et al., 2007).

Lesion Characterization

CT has a major role in characterization of incidental and/or suspicious lesions. CT quantifies size and provides information of lesion density using Hounsfield units and postcontrast enhancement obtained at different phases to depict varying imaging features in arterial, portal, or excretory phases. Hounsfield units are calculated based on differential attenuation of the X-ray beam where fluid has a value of zero, fat has a negative value, soft tissue is positive, and calcium and metallic densities are also positive but with higher numbers in the hundreds to thousands.

The high spatial resolution of CT is useful in assessing both lesion morphology and its characteristics. An example of this is the evaluation of an incidental renal lesion. The lesion on CT can first be subdivided into a cystic or solid mass based on lesion density. Cystic lesions are predominantly composed of simple fluid but can have suspicious features on CT that can be worrisome for malignancy such as thick septations, mural nodules, higher attenuation, and postcontrast enhancement. Cystic renal lesions are graded on CT using a Bosniak classification with category I having benign features and category IV highly suspicious for renal cell carcinoma (Bosniak, 1997). Solid renal lesions are evaluated based on their size, the presence of fat on the unenhanced examination, and enhancement characteristics on nephrographic and delayed phase sequences where lack of fat and contrast washout are features of malignancy (Silverman et al., 2008).

In the assessment of hematuria, CT urography is performed for the detection of urothelial malignancy but this imaging technique due to the high spatial resolution is also useful in demonstrating the morphology of the collecting systems and variants including duplicated systems (Fig. 9.2).

Tumor Staging and Treatment Response

CT plays an integral role in the initial treatment paradigm of the oncology patient as well as their subsequent follow-up. At time of diagnosis, CT is used to evaluate the primary tumor size, characterize the local extent, and assess for metastatic disease. In addition, in patients with metastatic disease of unknown primary, CT may facilitate the planning of image-guided or surgical biopsy to confirm the diagnosis and to stage the patient.

CT is also the primary modality used in the follow-up of oncology patients to screen for recurrent or metastatic disease, and in patients with metastatic disease it is used to assess response to cancer treatments. The Response Evaluation Criteria in Solid Tumors (RECIST) is a commonly used system to assess for tumor treatment response based on tumor size (Therasse et al., 2000; Eisenhauer et al., 2009). Up to five lesions on the baseline CT scan are measured, and changes in size on follow-up imaging are calculated and divided into complete response, partial response, stable disease, or progression of disease based on the percentage change (Eisenhauer et al., 2009).

While conventionally imaging biomarker of tumor response has relied on interval tumor size as a marker of treatment response, studies have shown evaluation of tumor size alone may underestimate the tumor response as it evaluates only the size and not the changes in the tumor density. The Choi criterion was introduced initially to assess response of patients on molecular-targeted therapy where changes in both size and tumor density are assessed for treatment response (Choi et al., 2007). In addition to assessing for tumor response to treatment, CT can also evaluate treatment-related complications in the oncology patients relating to immunosuppression, drug toxicity, and coagulopathy (Chikarmane et al., 2012; Shinagare et al., 2011; Adelberg and Bishop, 2009).

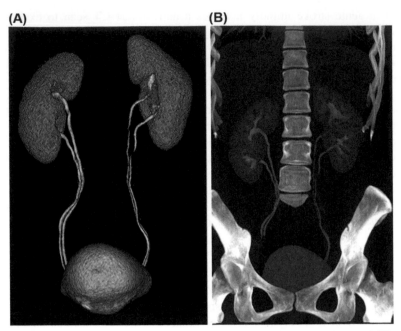

FIGURE 9.2   Computed tomography (CT) Urography. (A) Three-dimensional (3D) volume-rendered image of CT urogram 8 min following intravenous contrast injection demonstrating bilateral duplicated collecting systems and ureters; (B) maximum projection image of CT urogram 8 min following intravenous contrast injection demonstrating bilateral duplicated collecting systems and ureters.

Magnetic Resonance Imaging

Technological Advances

Magnetic resonance imaging (MRI) is a sophisticated imaging technique that has evolved since its introduction in the early 1980s with proven outcomes in both clinical medicine and biomedical research (Ai et al., 2012). The major strength of MRI is that it can provide excellent contrast definition of normal anatomy and pathologic processes. Initially MRI was time-consuming and expensive and, while its use in brain imaging had been proven, its utility in other parts of the body where significant motion occurred remained questionable (Tempany and McNeil, 2001). Therefore, one of the major objectives of MR researchers and manufacturers has classically been to reduce image acquisition time without loss of image quality.

The intrinsic contrast of the MR image is produced by differences in proton density (PD) and MR relaxation times (T1 and T2). However, by selection of appropriate magnetization preparation schemes, signal intensity can be modulated by other important processes such as flow effects (e.g., bulk flow for angiography or microflow for tissue perfusion), Brownian water motion (e.g., diffusion imaging to assess tissue microstructure), and tissue oxygenation (for functional information; Blamire, 2008). Furthermore, the local chemical environment of the nucleus of interest modulates the MR frequency (the chemical shift), allowing the identification of molecular groups using MR spectroscopy (MRS) and thus providing metabolic measurements (Blamire, 2008). The depth and complexity of its physics allow researchers to constantly develop new techniques that rapidly enter the clinical arena as prerequisite diagnostic approach and the biomedical research area for detailed functional and structural assessment in normals and clinical subjects.

For many years, clinical MRI practice has benefited from the additional specificity for certain disease processes that is afforded using injectable contrast media. These contrast agents highlight areas of abnormality via passive accumulation of agent within the tissue; usually leading to T1 enhancement. The future is likely to see a new generation of high- sensitivity agents that either are targeted to specific disease processes or provide specific information on regional metabolism. One such technique is the use of hyperpolarized compounds to overcome the low sensitivity of the nuclear resonance phenomenon which arises from the small difference in occupation levels of the nuclear spin states resulting in low bulk magnetization. Using prepolarization techniques, the population difference can be enhanced dynamically leading to an increase in effective nuclear magnetization (up to factor of 10^4; Blamire, 2008). Clinical and research applications of this technique include use of gaseous helium (^{3}He) and xenon (^{129}Xe) to evaluate lung structure and function. In the future, use of dynamic nuclear polarization using carbon compounds (^{13}C) will allow great polarization enhancement; in animal models angiographic images of the pig heart have been acquired using ^{13}C urea as a "contrast agent" (Blamire, 2008), and this technique is used to measure metabolic reactions in vivo, for clinical and research purposes, and to monitor the effects of therapy through their earliest impact on metabolism, e.g., response to targeted molecular oncology treatments. Another technique undergoing research is the use of molecular imaging by MRI to allow high-resolution, noninvasive mapping at the cellular or molecular level. Although MRI offers excellent resolution, it is inherently insensitive to anything other than the properties of tissue water. However, in animal studies, use of iron oxide—based contrast media [ultrasmall particles of iron oxide (USPIOs) or superparamagnetic particles of iron oxide] has allowed labeling of cultured cells which are visible on MRI due to susceptibility-based distortion of the magnetic field within the tissue (Blamire, 2008). An associated approach uses contrast agents with affinity for a specific receptor or cell surface marker. These agents use either gadolinium compounds or UPSIOs to provide image contrast. Both in the clinical and research arenas, promise has been shown in use of such agents in demonstration of neuroinflammation (using E-selectin), fibrous areas of intravascular plaque (using agents that bind to fibrin), and active inflammation in carotid atherosclerotic plaques (using UPSIOs that label macrophages via phagocytosis; Blamire, 2008).

For all of these MR measurements, the scanning hardware remains essentially constant. The MRI system consists of four essential components: the magnet, the gradient system, the radiofrequency (RF) system, and the digital electronics, each of which has undergone major advances in the last number of years. In terms of field strength, clinical MRI systems in use today range in magnetic field strengths from 0.2 to 7 T with use of 3 T MRI systems increasing; however, magnets with field strength of up to 9.4 and 11.7 T are being used in animal imaging systems in the research arena (Cherry, 2004). Typically, high—field strength magnets are introduced in small-bore animal imaging systems much earlier when compared with larger-bore human applications (Cherry, 2004) and so have been used as models for developing high-resolution, high-speed, and high-sensitivity pulse sequences for human imaging (Cherry, 2004). In conventional MRI, the static magnetic field strength determines the polarization of the magnetization, which, in turn, directly determines the achievable SNR of the image (i.e., image quality) that can be traded for speed, image resolution, or a combination of the two. In general, higher field strength provides higher-resolution images with new image contrast. However, use of higher—field strength magnets is not without its challenges including main field inhomogeneity, fat suppression, and sequences that achieve contrast based on steady-state behavior of the magnetization and that rely on relatively uniform precessional frequency (e.g., fast imaging with steady-state precession), differences in relaxation (T1 and T2), and increased RF power deposition in the patient (specific absorption rate). Such challenges provide research opportunities as higher—field strength magnets continue to enter the clinical arena.

The RF system refers to a sensor whose role is to pick up the weak nuclear magnetic resonance signals, generated by the scanned subject's body, for further processing and image generation and has undergone major developments since MRI system inception. First, use of phased array or RF receiver resonator containing multiple elements that are electrically isolated and feed into separate, autonomous data processing channels allows for increased SNR, improved sensitivity, and extended field of view that come from multiple receiver coil design. The extension of such modern systems uses at least 8 channel devices and is often equipped with the capability for at least 16 channels. Research systems commonly use 32 channels and, in some cases, approach 100 channels, particularly for brain imaging investigations where the small field of view (and hence low depth penetration) of the individual coil elements is acceptable (Blamire, 2008). Moreover, careful RF coil design and introduction of parallel coil transmitter systems (Vaughan et al., 2006) using multichannel RF coils to both excite and receive the imaging signal overcome inhomogeneous RF excitation at higher field strengths owing to the presence of dielectric effects within the tissue (Tofts, 1994) leading to a more uniform image. It is anticipated that further developments in coil technology and range of coil options for specific clinical investigations available will continue.

Regardless of the field strength of the magnet, consistency in image acquisition is paramount with imaging parameters chosen mostly depending on the clinical question or the diagnosis being addressed. Choices in magnet type, coil type, exogenous contrast administration, and combination of sequences are the core issues; the remainder of this section will provide an overview of this process through examples of utility of MRI in disease detection, lesion characterization, tumor staging, and treatment response evaluation.

Clinical and Research Applications

Detection

Breast MRI Breast MRI may be used to screen patients at high risk of breast cancer to aid early detection (ACR, 2014). Breast MR is also applied in other clinical situations such as evaluations of the extent of disease and staging in patients with a new breast malignancy and evaluation of patients with breast augmentation, prior lumpectomy with positive

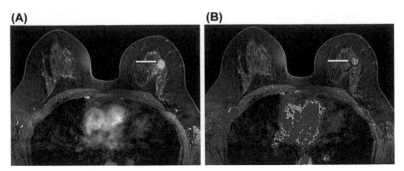

FIGURE 9.3 Breast magnetic resonance imaging (MRI). (A) Three-dimensional (3D) dynamic contrast-enhanced MRI sequence demonstrating enhancing mass (*white arrow*) in the left breast; (B) color map of dynamic contrast-enhanced MRI demonstrating same mass with contrast kinetics suggestive of breast cancer (rapid uptake and washout) depicted by red color (*white arrow*), confirmed by subsequent biopsy.

margins, and before and after neo-adjuvant chemotherapy to assess response prior to surgery (ACR, 2014). MRI-targeted biopsy can also be performed as discussed in Chapter 10.

The standard breast MRI protocol includes T2-weighted sequences (anatomy and signal analysis), T1-weighted gradient-echo sequences, and injected dynamic 3D sequences for performing volume and multiplanar reconstructions (Thomassin-Naggara et al., 2012). Nonfat-suppressed T2-weighted sequences are useful for detecting cysts or microcysts due to their fluid content. Nonfat-suppressed T1-weighted gradient-echo sequences are useful for detecting the presence of fat within a lesion, which usually predicts its benign nature. They also allow susceptibility artifact of metal biopsy markers to be detected. Dynamic sequences allow imaging of tumor angiogenesis, based on studying dynamic uptake of a contrast agent resulting in T1 shortening, which varies with the microvascularization characteristics of tumors and as with perfusion imaging requires good temporal resolution (<2 min) and good spatial resolution (1 mm isotropic pixel; Thomassin-Naggara et al., 2012). 3D T1-weighted fat-suppressed gradient-echo sequences with a moderate flip angle are generally used as they allow thinner slices with a better SNR than 2D sequences (Thomassin-Naggara et al., 2012). Detected lesions are categorized based on imaging features such as signal intensity, enhancement pattern, and morphology using a standardized reporting system, e.g., Breast Imaging Reporting and Data System (BI-RADS; ACR, 2013) which assigns a likelihood of malignancy to each lesion (Fig. 9.3). New functional imaging sequences including MR spectroscopy are now appearing in an attempt to increase the specificity of MRI. Detection of an abnormal choline peak provides early information (24 h after beginning neoadjuvant treatment) on the chemosensitivity of a breast tumor (Thomassin-Naggara et al., 2012).

Lesion Characterization

Prostate MRI Multiparametric (mp)-MRI has emerged as a robust method for localizing prostate tumors and determining their size, aggressiveness, and invasiveness, thereby improving disease detection, staging, and risk stratification (Hoeks et al., 2011). Mp-MR can be performed on either 1.5- or 3-T scanners and using pelvic phased-array coil with or without the addition of an endorectal coil to maximize the SNR, thus improving spatial resolution and/or acquisition time of MR images (Barentsz et al., 2012; Dickinson et al., 2011; Hegde et al., 2013). Mp-MRI combines conventional anatomical T2-weighted sequences and functional MRI (fMRI) sequences such as diffusion-weighted (DW) MRI, dynamic contrast-enhanced MRI, with or without MR spectroscopy to best characterize prostate cancer (Hoeks et al., 2011; Hricak et al., 2007; Umbehr et al., 2009). T2-weighted fast spin-echo imaging is typically performed in three orthogonal planes (axial, coronal, and sagittal) using small field of view (12—16 cm) and slice thickness of 3 mm (Hegde et al., 2013). Cancer foci appear as areas of low signal intensity on T2-weighted sequences. Moreover, T2-weighted images can be used to assess for extraglandular spread of tumor. DW imaging is acquired using a single-shot spin echo planar imaging (EPI) sequence. Diffusion imaging provides functional information about tissue microstructure by quantifying the Brownian motion of free water protons within a tissue through the application of a series of magnetic gradients or b-values (Turner et al., 1991). Water molecule diffusion is inversely proportion to cell membrane integrity and tissue cellularity. Normal prostate tissue allows free diffusion of water molecules and, therefore, appears hypointense on high-b-value DW images. Cancerous tissue has more restricted diffusion because the high cellular density inhibits movement of water molecules resulting in markedly hyperintense foci on high-b-value DW images (Magnetic Resonance, 2015). Native DW imaging has some amount of T2 contrast that may be reduced through calculation of apparent diffusion coefficient (ADC) maps, which provide a quantitative assessment of water diffusion. Cancerous foci appear as focal markedly hypointense areas on ADC maps (Magnetic Resonance, 2015). Dynamic contrast-enhanced imaging is discussed in the previous section (Fig. 9.4).

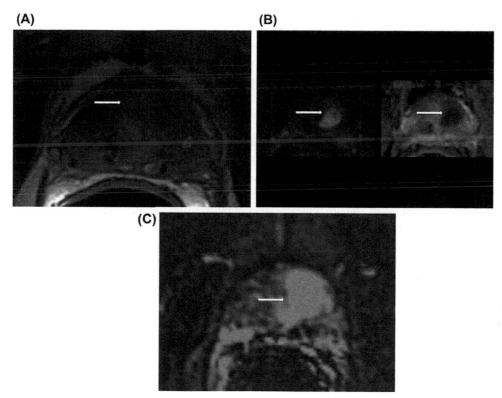

FIGURE 9.4   Prostate magnetic resonance imaging (MRI). (A) Axial T2-weighted image of prostate demonstrating focal poorly circumscribed lesion with homogenous low signal intensity in the left transition zone (*white arrow*); (B) axial diffusion-weighted image demonstrating low signal intensity in the same lesion on B1400 image on the left and corresponding low signal intensity on apparent diffusion coefficient (ADC) map on the right (*white arrows*) in keeping with diffusion restriction seen in tumor; (C) axial dynamic contrast-enhanced MRI image demonstrating rapid enhancement of the same lesion. Subsequent MRI guided biopsy demonstrated high-grade prostate cancer.

Chest MRI MRI of chest may be used to characterize mediastinal, chest wall, and pleural masses using a variety of different sequences. Axial in-phase and out-of-phase fast spoiled gradient-echo T1-weighted sequences using small field of view and contiguous thin slice (3—4 mm) are useful in evaluation of thymic masses distinguishing normal thymus and thymic hyperplasia from thymic neoplasms, lymphoma, or nodal metastases. Such sequences may also be used as non-contrast T1-weighted images. Axial double inversion recovery fast spin-echo T2- or T1-weighted sequences or black blood sequences are helpful in evaluating mediastinal and hilar masses and their relationship to vascular structures. Electrocardiographic-gating may be used to minimize cardiac motion. Alternatively, axial balanced steady-state free precession gradient-echo sequences may be used. This short-time-to-repetition (TR) sequence refocuses the transverse magnetization at the end of each TR interval, allowing single-breath-hold image acquisition that is relatively motion insensitive. The inherent T2/T1 contrast of the sequence makes it highly effective for imaging the thoracic vasculature in the absence of gadolinium contrast medium (bright blood sequence). DW imaging sequences may be used to help characterize tumor subtype differentiation prior to surgery. In mesothelioma, ADC values of epithelioid subtype have been shown to be higher than those of biphasic or sarcomatoid subtypes (Gill et al., 2010). DW MRI is also a promising tool for differentiating between benign and malignant pleural disease.

Brain MRI Conventional spin-echo MR imaging is more sensitive and more specific than CT for the detection of acute cerebral ischemia within the first few hours after the onset of stroke. It has the additional benefit of depicting the pathologic entity (stroke and its mimics) in multiple planes. The MR sequences typically used in the evaluation of acute stroke include T1-weighted spin-echo, T2-weighted fast spin-echo, fluid-attenuated inversion recovery, T2*-weighted gradient-echo, and gadolinium-enhanced T1-weighted spin-echo sequences. Hyperacute cerebral ischemia demonstrates hyperintense signal in white matter on T2-weighted images and fluid-attenuated inversion recovery images, with a resultant loss of gray matter—white matter differentiation analogous to the loss at CT. T2*-weighted gradient-echo images depict an acute intracranial hemorrhage as an area of abnormal blooming. Susceptibility-weighted imaging is a recently developed technique that uses both magnitude and phase images from a high-resolution, 3D, fully velocity-compensated gradient-echo sequence. Compared with CT and other MR imaging methods, this technique may be a powerful new approach for detecting a cerebral hemorrhage in a patient with acute stroke (Wycliffe et al., 2004).

Conventional MR imaging is less sensitive than DW MR imaging in the first few hours after a stroke (hyperacute phase) and may result in false-negative findings. DW imaging is an essential component of acute stroke MR imaging (Schaefer et al., 2000). In acute stroke, there is an alteration of homeostasis causing excess intracellular water accumulation, or cytotoxic edema, with an overall decreased rate of water molecular diffusion within the affected tissue. Therefore, areas of cytotoxic edema, in which the motion of water molecules is restricted, appear brighter on DW images because of lesser signal losses.

Statistically significant correlations have been demonstrated repeatedly between the acute infarct volume on DW images and various neurologic scales for the assessment of acute and chronic stroke (Schaefer et al., 2000; Fig. 9.5).

While DW MR imaging is most useful for detecting irreversibly infarcted tissue, perfusion-weighted imaging may be used to identify areas of reversible ischemia as well. Perfusion-weighted MR imaging techniques most commonly relies on administration of an MR contrast agent and are typically susceptibility based and depend on T2* effects, but they may be T1 weighted instead. Dynamic susceptibility-weighted (T2*-weighted) sequences involve tracking of the tissue signal changes caused by susceptibility (T2*) effects to create a hemodynamic time—signal intensity curve. As in dynamic CT perfusion imaging, perfusion maps of CBV and mean transit time can be calculated from this curve by using a deconvolution technique (Edelman et al., 2000). It also has been shown that patients who have lesions with a larger volume on perfusion-weighted MR images than on DW MR images have worse outcomes and larger final infarct volumes (Barber et al., 1998). Thus, the evaluation of images for a diffusion—perfusion mismatch at a very early stage of stroke may help predict the clinical outcome (Fig. 9.5).

Like CT angiography, MR angiography is useful for detecting intravascular occlusion due to a thrombus and for evaluating the carotid bifurcation in patients with acute stroke. Time-of-flight MR angiography and contrast-enhanced MR angiography are commonly used to evaluate the intracranial and extracranial circulation.

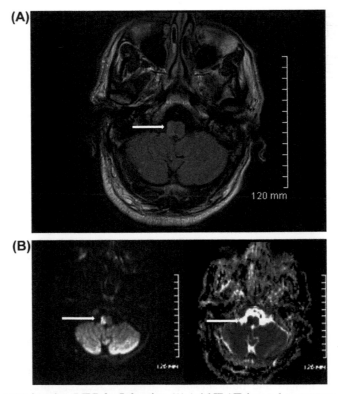

FIGURE 9.5 Brain magnetic resonance imaging (MRI) for Infarction. (A) Axial FLAIR image demonstrates subtle increased signal intensity in the right medulla (*white arrow*); (B) axial diffusion-weighted image demonstrating low signal intensity in the same lesion on high B-value image on the left and corresponding low signal intensity on apparent diffusion coefficient (ADC) map on the right (*white arrows*) in keeping with diffusion restriction and infarction that is much more readily evident on conventional FLAIR sequence.

Staging and Surgical Planning

Brain MRI fMRI using the blood oxygenation–dependent (BOLD) contrast mechanism is a potentially important technique in presurgical planning in oncology and epilepsy (Blamire, 2008). Recently, improvements in scanner hardware (such as availability of high-quality EPI sequences), computer processing speed, and standardization of processing techniques have allowed routine, real-time analysis of fMRI data for clinical use (Blamire, 2008). However, localization of regional brain activity via BOLD is based on contributions from CBF, CBV, oxygen extraction, and local metabolic rate, and so is only related indirectly to local brain activity. Although some researchers have shown a benefit of using fMRI as part of the wider armamentarium for neurosurgical planning (Blamire, 2008), further research to improve our understanding of the processes controlling the neurovascular mediation of the BOLD response will allow fMRI to achieve its true clinical potential.

An alternative technique for localizing and quantifying brain function by MRI is the measurement of CBF using arterial spin labeling (ASL). ASL uses "spin-tagging" of the inflowing blood by direct preparation or inversion of water magnetization outside the slices of interest and subsequently delivered by flow processes into the imaged volume over a period of several seconds (Blamire, 2008). Subtracting paired image data with and without spin labeling, which are acquired separately, reveals the flow signal. The major advantage of ASL is that it is completely noninvasive and does not require injection of contrast material when compared to dynamic contrast sequences. The main disadvantage of ASL is the reduced sensitivity due to the inherently small volume of flowing blood within the brain and the T1 relaxation recovery of the spin label. Although somewhat overcome by high–field strength systems with higher SNR and longer T1 values, the transition to clinical use requires further work to improve the sensitivity and define the optimum clinical protocol.

Diffusion tensor imaging is an extension of DW imaging that allows data profiling based upon white matter tract orientation. DW imaging is based on the measurement of Brownian motion of water molecules. This motion is restricted by membranous boundaries. In white matter, diffusion follows the "pathway of least resistance" along the white matter tract also known as anisotropic diffusion; this direction of maximum diffusivity along the white-matter fibers is projected into the final image. This technique is used to study white matter diseases, neurodegenerative disorders, and in presurgical planning to delineate relationship of white matter tracts to tumor being resected.

Ultrasound

Technological Advances

Ultrasound (US) has been described as a "womb-to-tomb" clinical imaging modality (Wolbarst and Hendee, 2006) with its ability to be used on everyone from fetuses to the elderly, excellent image quality, and lack of exposure to ionizing radiation. The basic principle of US imaging is the visualization of different organs and tissues within the body as reflecting surfaces for the sound waves transmitted from an oscillating transducer pressed up against the body surface (McVeigh, 2006).

Today, with frequencies in the range of 20–60 MHz, the imaging resolution is improved sufficiently to reconstruct dynamic signals from cylindrical volumes about 43 µm diameter and 66 µm depth, over a field of view of 10 mm (width) and 5 mm (depth). Numerous advances in US technology over the past few decades, such as color and power Doppler, speckle reduction, compound imaging, harmonic imaging, and three/four-dimensional imaging have become the standard of care (Whitsett, 2009). More recently, innovations in US technology, such as elastography, US contrast media, and US tomography, have been introduced into clinical practice and research; other new technologies such as capacitive micromachined ultrasonic transducers (CMUTs) are in development and are nearly ready for clinical use.

Pulsed Doppler imaging is used to image vascular blood flow. A sample volume is placed interactively by the operator on the target structure. The pulsed Doppler system can then visualize blood flow using Doppler effect, or the shift in frequency and wavelength of waves which results from a source moving with respect to the medium, a receiver moving with respect to the medium, or even a moving medium (McVeigh, 2006).

Speckle artifact reduction is a US processing tool that uses an algorithm to reduce image degradation due to noise or scatter, improving spatial and contrast resolution and allowing operators to detect differences in tissue planes by defining the differences between tissue types.

Compound imaging uses electronic beam steering to produce multiple overlapping scan lines that originate from different angles from the transducer crystal, improving visualization of tissue boundaries and reducing noise thus improving contrast resolution. Compound imaging is most useful in imaging rounded structures and calcifications and may also be used in vascular imaging for plaque characterization.

Harmonic imaging has evolved to become an integral part of any US examination. Harmonic waves are generated from nonlinear distortion of an acoustic signal as an US wave passes through tissues of the body. The harmonic frequency is higher on its return to the transducer, which aids in improved axial and lateral resolution and reduces the effect of side lobe artifacts. The returning echoes or harmonic frequency is double the original frequency sent into the body. The harmonic imaging improves resolution, but because the harmonic frequency is higher, the trade-off to imaging is decreased penetration. New advances in engineering techniques have allowed harmonic frequencies to be applied in most applications with little effect in penetration (Whitsett, 2009). Harmonic imaging is useful in the detection of cystic or fluid-filled structures or solid masses within an area of interest.

3D images are created by using real-time 2D imaging to acquire a volume of images through an area of interest using mechanical or freehand scanning methods. US system software reproduces the images in three orthogonal planes: sagittal, transverse, and coronal. The coronal plane is the one dimension in US that has not previously been able to be imaged prior to 3D technology. Key prerequisites of volume image acquisition are appropriate transducer movement and speed and frame rates to acquire enough image frames for quality reconstruction (Whitsett, 2009). Similar to CT, 3D postprocessing methods may be used on the volumetric data set including multiplanar formatting (to allow viewing of a single plane in the 3D image volume), surface rendering (to depict the surface of objects imaged), and volume rendering. 3D imaging is most useful in imaging of vascular structures, gynecology, and obstetrics. Four-dimensional (4D) imaging has become one of the newest diagnostic and research tools used in US imaging especially fetal imaging (Whitsett, 2009). 4D imaging requires a transducer that can scan simultaneously in three planes with the fourth dimension of time combined with volumetric sampling to produce the 4D and surface image. A mechanical multicrystal array transducer is used which allows the electronic steering of the beam to create an arc of scan lines, which are then converted to a 4D image. New developments incorporating solid-state technology in 4D imaging transducers will eliminate the need for mechanical moving parts in acquisition of 4D images and result in smaller, high-definition imaging transducers for 2D imaging (Whitsett, 2009).

Elastography is a new and promising US technology. Elastography relies on the premise that palpable abnormalities, such as malignancy or other pathology, tend to be "harder" (less elastic) than nonmalignant tissue. Ultrasonic elasticity imaging is currently implemented using one of two methods: strain imaging (or how readily tissue deforms in response to an applied compression) and shear-wave elastography (or differences in speed of sound through tissues of varying stiffness). The fundamental principle of elasticity imaging is that stiffer tissues deform less (low strain) and propagate sound faster than softer tissues. Strain imaging involves application of external compression of a region of interest either manually or automatically using transducer typically used for B-mode imaging; the region's dimensions are then measured before and after compression. Alternatively, a compression-release cycle can be generated from an internal source, such as vessel pulsation or cardiac motion. Changes in the echo pattern before and after compression are analyzed using the US system software algorithm and used to generate an "elastogram"—a qualitative map of relative tissue stiffness (Barr, 2012). The elastogram is typically displayed adjacent to the original B-mode image with an overlay in either gray scale or color depicting whether the lesion is hard or soft relative to the other tissues in the insonation field of view. To overcome operator dependence of compression techniques for strain imaging, newer techniques, such as acoustic radiation force imaging (ARFI), have been developed. In this technique, a short high-intensity compression pulse is applied to a region of interest along the axis of the US beam, and the strain displacement of the tissue is measured (Barr, 2012). Shear-wave elastography also incorporates ARFI technology for quantitative assessment of tissue elasticity by measuring the speed of a shear wave generated in the tissue perpendicular to the axis of the "push pulse" by the tissue displacement. The shear-wave propagation velocity is then used to calculate the tissue stiffness in region of interest (Garra, 2007). Advantages of shear-wave elastography over strain imaging include being quantitative, reproducible, and less operator-dependent. However, user identification of regions of interest can result in varying quantitative values depending on the region imaged (Garra, 2007). Newer US systems provide color-coded shear-wave velocities overlaid upon a gray-scale view to aid with this selection. These newer techniques have been increasingly used to address research questions particularly related to the cardiovascular system.

The introduction of US contrast agents is another exciting development. Although US methods, such as Doppler imaging, are very effective at visualizing and measuring blood flow in large vessels with rapidly moving blood, they are unable to accurately detect flow in smaller vessels and capillaries. The development of microbubbles (MBs) as contrast agents for US has subsequently allowed for the imaging of small vessels with slowly moving blood. In their simplest form, MBs consist of a lipid or albumin shell that is filled with a gas such as perfluoropropane (typically $1-10$ μm in diameter). MBs produce very strong US echoes due to their ability to scatter acoustic waves as well as the unique harmonics that they create when exposed to US. In fact, a standard clinical US system can detect the echo from an individual MB; thus, in principle, the sensitivity for detection of such contrast agents is extremely high (Cherry, 2004).

US tomography, a recently defined concept, requires hundreds of transducer elements covering a large-angle area of the body and perhaps firing and gathering echoes separately. This can diminish the operator-dependency of the procedure and improve reproducibility. Transmission US tomography has been shown to have diagnostic potential in relatively homogeneous tissues such as the female breast, where variations in the speed and/or absorption of sound are measured and processed to create images (Wolbarst and Hendee, 2006). Reflection US tomography, with its multiplicity of omnidirectional echoes, continues to pose major technical problems and remains an area of continued research (Wolbarst and Hendee, 2006).

Since its inception, US imaging has been implemented with transducers made of piezoelectric material in which application of a mechanical stress induces a voltage and, conversely, an applied voltage induces a mechanical stress. Ceramics made with lead zirconate titanate (PZT) are the most common type of piezoelectric materials used in US applications. Piezoelectric transducers require several labor-intensive steps to manufacture, and their fabrication is particularly demanding for small-scale applications such as intravascular and intracardiac imaging. CMUTs are an alternative US transducer. Each CMUT array element is composed of several unit cells (or capacitor composed of a fixed plate electrode and a flexible plate electrode separated by a vacuum cavity, which can be of any shape). To transmit a US wave, an alternating current electric field is applied across the two CMUT electrodes, causing the flexible plate to vibrate and transmit a wave of ultrasonic frequency into the surrounding medium. Conversely, a US wave impacting the CMUT induces vibration of the flexible plate, which can then be detected. Using electronic microfabrication techniques, CMUTs can be manufactured with greater flexibility in transducer design in a wide range of frequencies and geometries and do not require backing materials of piezoelectric transducers to achieve bandwidth required for high axial resolution needed for clinical imaging.

Clinical Applications

Detection

Obstetric Ultrasound Due to its noninvasive nature and lack of ionizing radiation, prenatal screening US is part of routine care in pregnancy to assess fetal growth and development. Determination of gestational age and assessment of fetal size can be performed using measurements including crown-rump length, biparietal diameter, femur length, and abdominal circumference. In early pregnancy, fetal heart rate can be assessed using M-mode Doppler and acts as an important biomarker for fetal well-being. Later in pregnancy, color Doppler is used in the evaluation of heart and vascular structures including umbilical vasculature. More recently, 4D imaging has become an important technique for the detection and diagnosis of life-altering fetal abnormalities, proving useful information in detecting facial and cranial anomalies and limb abnormalities, imaging documentation of multiple pregnancies, and fetal heart assessment through special cardiac software.

Diagnosis

Gynecological Ultrasound Pelvic US is the first choice imaging modality in evaluation of uterus and ovaries in females. Typically, scanning is performed using both a transabdominal and transvaginal approach. Transvaginal images are acquired using a specially designed endocavity probe placed in the vagina of the patient and provides high-resolution images as a result of the scan head's closer proximity to the uterus and the higher frequency used in the transducer array resulting in higher resolving power. Transvaginal US may also be used in assessment of early pregnancy as fetal cardiac pulsation can be clearly observed as early as 6 weeks of gestation. Moreover, 3D imaging can provide additional information. It is a useful tool for reconstructing the coronal plane for diagnosing uterine anomalies, locating gestational sacs in a uterus with a uterine anomaly, and for imaging of the endometrial lining in sonohysterography.

Lesion Characterization

The ultrasonic arsenal for lesion detection and characterization includes B-mode imaging (imaging of acoustic impedance differences), Doppler imaging (imaging of flow and movement), and more recently elasticity imaging (imaging of differential tissue stiffness).

In addition, use of US contrast agents can further aid lesion characterization through assessment of tissue vascularity and perfusion.

Breast Ultrasound Breast US is frequently used to evaluate breast abnormalities that are found with screening or diagnostic mammography or palpable breast abnormalities and is the first-line imaging modality in evaluation of dense

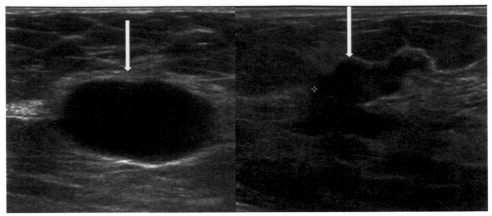

FIGURE 9.6 Breast Ultrasound. B-mode image demonstrates characteristic appearance of simple cyst on the left (*white arrow*), which is well circumscribed with an imperceptible margin, anechoic, and demonstrates through transmission. This is in contrast to the breast cancer on the right (*white arrow*), which is poorly marginated, irregular in shape, hypoechoic, and demonstrates no through transmission.

breast tissue in younger patients. In general, B-mode imaging using a high-frequency linear array transducer is performed. Detected lesions are categorized based on imaging features such as echogenicity, compressibility, and morphology using a standardized reporting system, e.g., BI-RADS (ACR, 2013) which assigns a likelihood of malignancy to each lesion (Fig. 9.6). More recently, elastography has been used to add incremental value to B-mode imaging in evaluation of breast masses as malignant masses have lower strain than normal breast tissue, fibrocystic areas, or benign masses such as fibroadenomas. Breast cancers also appear larger on elastography than gray-scale imaging, which is thought to be due to the presence of an adjacent desmoplastic reaction.

Liver Ultrasound Liver biopsy is considered the gold standard in determining degree of liver fibrosis in patients with chronic liver disease, helping to predict patient prognosis and guide treatment decisions. However, as biopsies are invasive with inherent risk of complications, elastography appears to be a promising and noninvasive alternative. Transient elastography uses a piston-mounted US transducer to apply a shear wave that propagates through the liver. The shear-wave velocity measurement is then used to calculate liver stiffness. Alternative elastography methods have also been studied in liver fibrosis evaluation including ARFI and real-time elastography (Barr, 2012; Garra, 2007).

US contrast agents have also been used in the evaluation of focal liver lesions by obtaining useful information about vascularity, perfusion rates, and potential tumor detection. Owing to the vascular structure of focal liver lesions, differences in filling time, and washout of the MB, contrast agent can facilitate not only the identification of focal lesions from within the surrounding tissue, but its patterns of enhancement can also aid characterization of focal liver masses allowing differentiation of benign entities from tumor.

Treatment Response

Another promising potential application of US contrast agents comes from the molecular targeting of the MB contrast agents, enabling not only identification of anatomical and vascular structures, but also imaging of biological processes including angiogenesis and inflammation (Willmann et al., 2008). Once coated with targeting molecules, these MBs can be targeted toward specific cell-surface receptors, allowing them to bind and accumulate in areas expressing the complementary protein (Cherry, 2004). Among the targets being explored are proangiogenic factors, including VEGFR2, endoglin, and integrin $\alpha v\beta 3$, that are used to target, visualize, and monitor angiogenesis, and P-selectin or mucosal addressin cellular adhesion molecule to target and monitor inflammation (Deshpande et al., 2010). Use of MBs for targeted drug delivery will be discussed in Chapter 10.

Nuclear Medicine

Nuclear medicine is a unique modality that differs from other imaging techniques such as CT and MRI. While routine CT and MRI mainly focus on the anatomic structures and are dependent on structural changes in morphology or size, nuclear medicine has the ability to yield data on the metabolic function of the investigated organ, tissue, or system, at the molecular level. Nuclear medicine imaging involves tracking of injected or ingested radiopharmaceuticals that are designed to map physiologic or pathophysiologic processes. In this context, radiopharmaceuticals are described as radiotracers.

The unique potential of nuclear medicine in investigating any particular organ or disease is based on the great diversity and biochemical properties of the available radiotracers. The majority of radiotracers consist of two distinct moieties—a carrier, which is responsible for the biodistribution of the tracer [e.g., fluoro-2-deoxy-D-glucose (FDG), a glucose analog], and a radioactive marker, which enables the external detection of it (e.g., fluorine-18). Some radiotracers do not require a carrier (e.g., gallium-67, an iron analog, or thallium-201, a potassium analog) because of their biochemical properties. Nevertheless, regardless of their structure, the common characteristic of all radiotracers is that they can be metabolized through certain physiologic or pathophysiologic processes that are correlated with physiologic or pathologic conditions of target organs or tissues (Maisey, 1998; Munley et al., 2001; Valdes Olmos et al., 2001).

In fact, a very large number of radiolabeled tracers have been developed to probe specific biological targets and functions, and a growing number are moving into clinical use. An important strength is the availability of radiotracers of biologically relevant elements, particularly carbon, allowing contrast agents to be labeled by direct isotopic substitution (Cherry, 2004). This allows small biomolecules (many drugs, receptor ligands, etc.) to be labeled without changing their biochemical properties. Another powerful approach is to create analog radiotracers, in which deliberate chemical changes are made to a biologically active molecule to isolate specific pathways or cause specific trapping of a radiotracer in cells expressing the target of choice (Cherry, 2004).

The nuclear medicine imaging methods are divided into single photon imaging that utilize radionuclides with single or multiple uncorrelated gamma ray emissions and positron emission tomography (PET) in which radionuclides decay by positron emission resulting in two simultaneous annihilation photons emitted back-to-back (Cherry, 2004).

Single Photon Emission Computed Tomography

The simplest nuclear medicine imaging technique is planar imaging which provides functional images in a 2D format that is similar to that obtained with X-ray imaging. The planar images are obtained on gamma camera systems that typically contain scintillation detection material, such as sodium iodide crystals. In general, single photon imaging is accomplished using a gamma camera in which the radioactivity distribution within the imaged object is projected through a collimator to form an image on the detector (Cherry, 2004). Resolution and signal-to-noise are generally dominated by the collimator design. Multiple gamma camera heads can be mounted on a gantry to improve sensitivity, and by rotating the heads around the object, cross-sectional functional images (single photon emission computed tomography or SPECT) can be obtained (McVeigh, 2006).

Positron Emission Tomography

PET with FDG (a glucose analog that is labeled with the cyclotron-produced positron-emitting radioisotope fluorine-18), or FDG-PET, is routinely used for the staging and restaging of malignancy, metabolic characterization of malignancy, and monitoring of response to therapy (Hoffman and Gambhir, 2007; Margolis et al., 2007). The metabolic and molecular information provided from PET imaging with FDG can be considered one of the first validated and clinically useful tomographic "molecular imaging" techniques (Hoffman and Gambhir, 2007). Some of the first observations of the response to molecularly targeted drugs were made using this powerful technique in the context of imatinib and sunitinib therapy of gastrointestinal stromal tumors. Tumor metabolic responses to these drugs were observed as early as 24 h or just a few days following initiation of therapy and preceded significant morphologic changes on CT (Van den Abbeele and Badawi, 2002; Demetri et al., 2002; Holdsworth et al., 2007; Joensuu et al., 2001).

FDG-PET takes advantage of the fact that malignant tumors have an increased rate of aerobic glycolysis compared to normal tissues. Most tumors demonstrate an increase in glucose transporters such as GLUT 1 as well as increased hexokinase and decreased glucose-6-phosphatase activity resulting in the retention of the glucose analog FDG (Brown et al., 2002; Brown and Wahl, 1993). FDG is taken up by metabolically active tumor cells using facilitated transport similar to that used by glucose (Kapoor et al., 2004). The rate of uptake of FDG by tumor cells is proportional to their metabolic activity. Since FDG is a radiopharmaceutical analog of glucose, it also undergoes phosphorylation to form FDG-6-phosphate like glucose; however, unlike glucose, it does not undergo further metabolism, thereby becoming "trapped" in metabolically active cells (Kapoor et al., 2004). In short, FDG-PET provides a functional metabolic map of glucose uptake in the whole body (Zealley et al., 2001).

Positrons emitted by F18-FDG are positively charged subnuclear particles that, in the tissue, collide with an electron very close to the point of positron emission (Tutt et al., 2004). During this collision, two high-energy photons of 511 keV

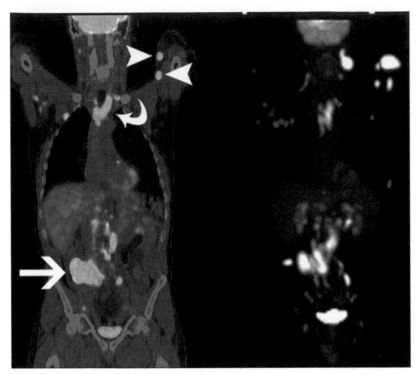

FIGURE 9.7   Fluoro-2-deoxy-D-glucose (FDG)–positron emission tomography (PET) computed tomography (CT). Coronal fused PET–CT image on the left and coronal maximum intensity projection image on the right demonstrating foci of increased tracer uptake in the cecum, neck, and left upper extremity in this patient with lymphoma. Increased metabolic activity in malignant tissue is accompanied by increased glucose uptake relative to that of surrounding normal tissue. This focal increase in glucose uptake can be detected with FDG–PET, which allows identification of malignant tumor foci.

energy are emitted at 180 degrees to each other. These may be detected by a PET scanner and used to determine their point of origin; this forms the basis of PET imaging (Kapoor et al., 2004; Tutt et al., 2004). Increased metabolic activity in malignant tissue is accompanied by increased glucose uptake relative to that of surrounding normal tissue. This focal increase in glucose uptake can be detected with FDG-PET, which allows identification of malignant tumor foci (Erturk et al., 2006; Fig. 9.7).

Unlike SPECT, PET does not require physical collimation due to the back-to-back nature of the annihilation photons that define the direction of emission (Cherry, 2004). PET scanners typically consist of rings of small detector elements in a cylindrical geometry, simultaneously providing projection views of the radioactivity distribution from many different directions that are then reconstructed into tomographic images. Because no physical collimation is required, the sensitivity is much higher (typically 10^2 to 10^3 times higher) than a SPECT system. However, single-photon emitting radionuclides are generally more readily available than positron-emitting radionuclides, as many of the latter have short half-lives. Because of FDG-PET's ability to assess normal and abnormal function at the molecular level, it is being increasingly used in a variety of clinical research areas.

Technical Advances

Hybrid Systems Hybrid systems that couple a CT scanner to a PET or SPECT system are now in routine clinical and research use, and the integrated scanners provide both anatomical and functional information (Garcia, 2012). PET/CT has been particularly important in oncology imaging for evaluating the staging and restaging of cancer and evaluation of treatment response. Although the most commonly used radiotracer in clinical practice and in the study of cancers is currently F18-FDG, other F18-labeled tracers could be available soon. These include F18-sodium fluoride for imaging of skeletal metastases, F18-choline to image prostate cancer, F18-fluorothymidine to measure tumor cell proliferation, F18-DOPA for imaging of primary and metastatic neuroendocrine tumors as well as low-grade brain tumors, F18-fluoromisonidazole to assess tumor hypoxia, and others (Groves et al., 2007). Other positron-emitting radionuclides such as oxygen-15, nitrogen-13, and carbon-11 can also be used for PET imaging. However, their short half-life (20 min or less) is a major drawback since they require an on-site cyclotron facility. With its half-life of ~110 min, F-18 can be produced in off-site cyclotron facilities and easily distributed to imaging centers located within a few hours of a cyclotron facility.

Molecular imaging has also made advances in fields outside oncology. For example, there have been considerable advances in Alzheimer disease (AD) with PET radiotracers that image β-amyloid burden in patients aiding in the diagnosis and clinical management of patients at risk of developing or with AD (Vlassenko et al., 2012; Bateman et al., 2012).

Hybrid PET/MR combines the physiologic information obtained by PET with the functional and morphological information from MRI. MRI has better soft tissue contrast than CT and is preferred in imaging the brain, breast, and liver, among other structures. MRI can also provide functional information with the use of diffusion, perfusion, fMRI, and spectroscopy (Gaertner et al., 2013). MRI has the added advantage over CT in that it does not use ionizing radiation, but multiple MRI sequences do contribute to longer examination times compared to CT. PET/MR scanners, depending on the vendor, may be either fully integrated in one single system or divided into two stand-alone systems that enable the MRI to be used for other studies separate to PET/MR but has the disadvantage of patient motion artifact between the two separate scans (Torigian et al., 2013). PET/MR due to the high soft tissue contrast provided by MR could prove to be superior to PET/CT in the oncologic evaluation of the brain, head and neck, soft tissue sarcomas, breast, and parenchymal abdominal organs (Antoch and Bockisch, 2009). PET/MR may also have higher accuracy in detection metastases in the liver, brain, and bone. MRI is important in the evaluation of prostate cancer, and PET/MR using 18F-fluorocholine may have a role in the evaluation of patients with biochemical failure and in the assessment of treatment response (Wetter et al., 2014).

Other potential clinical and research applications of PET/MR are in neurology and cardiology, among others. There is a growing use of PET/CT in dementia work-up as PET agents are used for amyloid imaging (Yang et al., 2012). The combined use of PET/MR would allow pathophysiologic information from the PET to be complemented by the anatomical information from MRI as well as the use of fMRI. PET has been used to assess myocardial viability and perfusion in patients with coronary artery disease while MRI is performed for evaluation of cardiac anatomy, structure, and ventricular function (Nekolla et al., 2009). Therefore the two modalities combined could potentially have many applications in cardiac imaging.

Molecular Imaging Molecular imaging describes in vivo assessment of biological processes at a cellular or molecular level and can provide greater insight into disease processes (Britz-Cunningham and Adelstein, 2003). Nanoparticles in particular may be used in nuclear imaging, and many varieties have been formulated including proteins, lipids, carbon-based, and metal based (Abou et al., 2015). Nanoparticles must be radiolabeled to allow visualization with nuclear medicine imaging. For imaging with SPECT this can be performed by direct labeling through addition of radioisotope during synthesis, a two-step process using a chelator, or via covalent radioiodination (Abou et al., 2015). Nanoparticles developed for clinical use with SPECT include ^{99m}Tc-colloid and ^{99m}Tc-sulfur nanocolloid, which are used for intraoperative guidance of lymph node resection in melanoma, prostate, and breast cancer (Ganswindt et al., 2011; Alazraki et al., 1997).

Radiolabeling nanoparticles for PET can be performed by covalent chemistry with radioisotopes such as F18, C11, Br76, and I-124, with radiometals bound via chelation or incorporation of radiometals directly onto the nanoparticle (Abou et al., 2015). Radiolabeled nanoparticles for PET are still in the investigation stage and not currently in clinical use. In Chapter 10 we discuss the local targeted delivery of nanotechnology using US to guide and facilitate the local delivery of drugs in a noninvasive manner inside the body. Future developments using nanoparticles with PET are in image-guided drug delivery. This involves a drug formulation that has both diagnostic and therapeutic properties allowing in vivo assessment of drug biodistribution, target site visualization, and target site accumulation with monitoring of outcome (Chakravarty et al., 2014).

Optical Imaging

Optical imaging uses fluorescent and bioluminescent probes that emit radiation at visible or near-infrared (NIR) wavelengths, which can be detected by optical cameras (Ehman et al., 2007). This can be described as the medical counterpart to shining a bright flashlight to display the bone. The major drawback of visible fluorescence imaging is that it can penetrate tissue only to a depth of 1−2 mm. However, fortunately, fluorescence in the NIR region can penetrate up to several centimeters of the tissue, thus enabling imaging at greater depths (Fragioni, 2003). The development of new probes and new imaging modalities, as well as advances in genomic sciences, that make it possible to produce genetically targeted probes, has made it possible to develop optical imaging into a powerful imaging tool. Today, probes that respond to cellular activity and probes that emit in the NIR region are available to the researcher (Ehman et al., 2007; Wilkinson et al.,

2004). New optical imaging techniques such as optical coherence tomography have emerged as powerful research and even clinical tools (Ehman et al., 2007).

Optical Imaging Approaches

A number of optical imaging approaches have recently been described; these techniques rely on fluorescence and bioluminescence, or absorption and reflectance as the source of contrast (Weissleder and Mahmood, 2001).

Fluorescence

Fluorescence results from a process that occurs when certain molecules, such as polyaromatic hydrocarbons or heterocycles, called fluorophores, fluorochromes, or fluorescent dyes absorb light. This results in a rise in the molecule's energy to an excited state. As these molecules decay from this excited state, they emit fluorescent light. NIR fluorescence imaging, in particular, relies on light with a defined bandwidth as the source of energy that raises the energy of a fluorescent molecule, which eventually emits a signal with different spectral characteristics that can be resolved with an emission filter and captured with a high-sensitivity charge-coupled device camera. Weissleder et al. recently developed autoquenched, NIR fluorescent probes that become detectable after enzyme (e.g., protease) activation (Weissleder and Mahmood, 2001; Weissleder, 2001). The same group has also shown the power and specificity of this imaging approach for imaging certain transgenes in mice (Tung et al., 2000).

Fluorescent imaging is advantageous for two main reasons. First, as many of the fluorophores used have high quantum yields, this leads to robust signals when using appropriate illumination and acquisition times (Cherry, 2004). Second, the fluorescence emission can be activated by specific biologic molecules or events which allow very low concentrations of enzymes to be detected (Cherry, 2004). However, autofluorescence from the tissue may limit detection sensitivity. Moreover, as light has to get into the imaged tissue (to excite the fluorescent molecules) and back out again (to reach the detector), this can lead to high degree of scattering of both the excitation and emission light (Cherry, 2004).

Bioluminescence

Bioluminescence involves introducing reporter genes that encode for enzymes (known as luciferases) that can catalyze a light-producing reaction. Luciferases are found in organisms such as fireflies, glowworms, and jellyfish (Contag and Bachmann, 2002; Contag et al., 1998; Greer and Szalay, 2002). Luciferase genes have been cloned from a large number of organisms, including bacteria, fireflies (*Photinus pyralis*), coral (*Renilla*), jellyfish (*Aequorea*), and dinoflagellates (*Gonyaulax*) (Contag and Bachmann, 2002; Contag et al., 1998; Greer and Szalay, 2002). The most commonly used reporter gene for imaging is the one that encodes for firefly luciferase (Contag et al., 1998). In the firefly, luciferase utilizes energy from adenosine triphosphate (ATP) to convert its substrate, luciferin, to oxyluciferin, with the emission of a detectable photon (Contag et al., 1998). Immediately prior to imaging, animals are injected with the substrate for the enzyme. In the case of the firefly luciferase the substrate is luciferin, a small molecule that rapidly distributes throughout the whole body of the subject after intravenous injection. In cells that are expressing the luciferase reporter gene, the luciferin, in the presence of oxygen and ATP, is converted to oxyluciferin with the emission of light (peak at 560 nm). The reaction is catalyzed by the luciferase enzyme and does not occur to any significant extent when luciferase is absent. The generation of an optical signal is therefore specific to cells that contain the reporter gene (Contag and Bachmann, 2002).

The main advantage of this approach is that there is no need for external light stimulation (Cherry, 2004). As there is no autofluorescence, there is no problem in depth of penetration of the excitation light. However, this approach is limited to studying genetically manipulated cells, as the reporter gene has to be introduced into the organism that is to be studied. In addition, the bioluminescence signals are typically very weak. Thus, to date, this approach has only been used in animal research.

Optical Imaging Techniques

Recently, optical imaging has experienced significant technologic advances. Simultaneously, an array of specific optical imaging agents has brought new capabilities to biomedical research and is edging toward clinical use. This section will review the progress in the translation of macroscopic optical imaging including fluorescence-guided surgery and endoscopy, intravascular fluorescence imaging, diffuse fluorescence and optical tomography, and multispectral optoacoustics (photoacoustics), that are used for applications ranging from tumor resection and assessment of atherosclerotic plaques to dermatologic, eye, and breast examinations.

Fluorescence-Guided Surgery

Use of fluorescent agents has gained attention for enhancing surgical vision and providing guidance toward accurate tumor delineation and identification of sentinel lymph nodes. Specialized camera systems with appropriate illumination to excite the fluorochromes and suitable emission filters are required. Such systems allow the fluorescence distribution to be displayed in real time using a color or black-and-white monitor (Taruttis and Ntziachristos, 2012). Numerous animal studies have demonstrated the value of the use of fluorescence-guided surgery using, for example, models of breast cancer (Themelis et al., 2011) and ischemic bowel (Matsui et al., 2011). In humans, indocyanine green (ICG), an approved NIR dye, has been used for lymphatic mapping (Kitai et al., 2005), liver cancer resection (Ishizawa et al., 2009), and endoscopic examination of superficial gastric tumors which passively accumulate ICG (Kimura et al., 2007). More recently, clinical trials using aminolevulinic acid (ALA) to enhance generation of porphyrin in tumors, which provided visible fluorescence for cystoscopic resection of bladder cancer (Hermann et al., 2011) and brain glioma resection (Stummer et al., 2006), have been reported. Further research opportunities include use of active tumor-targeting agents to allow fluorescence-guided surgery to have a broader clinical impact.

Intravascular Fluorescence Imaging

NIR fluorescence imaging has the potential to depict inflammatory and other atherosclerotic molecular biomarkers in vivo. As diseased arteries such as the coronary arteries are not accessible to noninvasive optical techniques, optical imaging of atherosclerosis involved optical fiber—based catheter systems similar to intravascular US. Initial animal studies in rabbits have achieved 2D imaging of the arterial wall to show inflammation in plaques and stent-induced injury and have used optical frequency domain imaging to allow microstructure characterization (Yoo et al., 2011). A limiting factor for intravascular fluorescence imaging remains the lack of clinically approved agents for molecular imaging. To this end, progress in clinical translation of agents for fluorescence-guided surgery could also benefit intravascular imaging and diagnostics for atherosclerosis in general because some biologic processes are of interest in both cases (Taruttis and Ntziachristos, 2012).

Diffuse Optical and Fluorescence Tomography

To allow deeper tissue penetration, tomographic approaches using fully diffusive photons have been investigated. In tomography, the tissue is illuminated at a multitude of surface points sequentially, and the diffusive photons that emanate are measured in parallel at various positions or projections. Photon propagation through the tissue of interest is modeled, typically using a solution to the diffusion equation, and appropriate inversion schemes are applied to obtain quantitative images. This general scheme can equally apply to resolving optical attenuation (absorption, scattering) or fluorescence biodistribution. Early research has been performed using this technique in breast cancer imaging to aid diagnosis and evaluate treatment response (Taruttis and Ntziachristos, 2012). By combining concurrent MRI and diffuse optical tomography with ICG injection, lesions were found to be detected by the absorption enhancement provided by ICG. Using spectroscopic capabilities by multiwavelength illumination, maps of oxyhemoglobin can be used to assess chemotherapy response (Taruttis and Ntziachristos, 2012). Further applications of tomographic technique include imaging of finger joints to detect disease activity in rheumatoid arthritis and neuroimaging using diffuse optical tomography to study functional connectivity for diagnostic purposes in preterm infants (Taruttis and Ntziachristos, 2012).

Multispectral Optoacoustic Imaging

Optoacoustics uses US waves to resolve the absorption of light; US waves are emitted as a result of local thermal expansion at the site of each absorber. In other words, optoacoustic images display optical contrast at US resolutions. Optoacoustic methods have been applied to biomedicine to overcome photon scattering and the resultant decrease in accuracy in image formation and degradation of spatial resolution in deep tissues. Because US scatters orders of magnitude less than light in the tissue, when clinically relevant ultrasonic frequencies are used, high spatial resolutions can be preserved through millimeters to centimeters of penetration (Taruttis and Ntziachristos, 2012). Currently, the use of optoacoustics lies in the combination of diffuse illumination with US detection at multiple positions around the patient using mechanical scanning or detector arrays in US probes. This allows for dynamic imaging such as organ and tumor perfusion, or heartbeat (Taruttis and Ntziachristos, 2012).

Multispectral optoacoustic tomography (MSOT) uses multiple illumination wavelengths to identify detected absorbers in the tissue based on their unique spectral profile. Hemoglobin is a dominant absorber of light in the tissue; therefore, MSOT can be used to image blood vessels without the use of exogenous contrast agent. Moreover, as oxyhemoglobin and deoxyhemoglobin have different optical absorption spectra, MSOT can be used to characterize the oxygenation saturation in these vessels (Taruttis and Ntziachristos, 2012) and may have potential clinical applications in evaluation of peripheral artery disease or tumor perfusion. Other potential endogenous targets or sources of optical absorption include lipids, as found in atherosclerotic plaques, and melanin, as found in melanoma tumors whereby MSOT can be used as biomarker for the presence of disease. Additionally, exogenous optical absorbers can be used such as simple organic dyes such as ICG or gold and carbon nanoparticles.

CONCLUSION

Recent advances in biomedical imaging have not just focused on imaging of structure and morphology but emphasized on visualization of the individual biologic processes underlying disease. As our molecular knowledge increases and individualized treatments are being offered, imaging must continue to evolve at a molecular level to provide a tailored approach to each clinical and research scenario thus contributing to more accurate diagnostics and improved treatment efficacy.

REFERENCES

Abou, D.S., Pickett, J.E., Thorek, D.L., 2015. Nuclear molecular imaging with nanoparticles: radiochemistry, applications, and translation. Br. J. Radiol. 20150185.

ACR, 2013. American College of Radiology BI-RADS Atlas. http://wwwacrorg//Quality-Safety/Resources/BIRADS.

ACR, 2014. ACR Practice Parameter for the Performance of Contrast-Enhanced Magnetic Resonance Imaging of the Breast. http://wwwacrorg/w/media/2a0eb28eb59041e2825179afb72ef624pdf.

Adelberg, D.E., Bishop, M.R., 2009. Emergencies related to cancer chemotherapy and hematopoietic stem cell transplantation. Emerg. Med. Clin. North Am. 27 (2), 311−331.

Agatston, A.S., Janowitz, W.R., Hildner, F.J., Zusmer, N.R., Viamonte Jr., M., Detrano, R., 1990. Quantification of coronary artery calcium using ultrafast computed tomography. J. Am. Coll. Cardiol. 15 (4), 827−832.

Ai, T., Morelli, J.N., Hu, X., et al., 2012. A historical overview of magnetic resonance imaging, focusing on technological innovations. Invest Radiol. 47 (12), 725−741.

Alazraki, N.P., Eshima, D., Eshima, L.A., et al., 1997. Lymphoscintigraphy, the sentinel node concept, and the intraoperative gamma probe in melanoma, breast cancer, and other potential cancers. Semin. Nucl. Med. 27 (1), 55−67.

Alexander, R.E., Gunderman, R.B., 2010. EMI and the first CT scanner. J. Am. Coll. Radiol. 7 (10), 778−781.

Antoch, G., Bockisch, A., 2009. Combined PET/MRI: a new dimension in whole-body oncology imaging? Eur. J. Nucl. Med. Mol. Imaging 36 (Suppl. 1), S113−S120.

Bamberg, F., Dierks, A., Nikolaou, K., Reiser, M.F., Becker, C.R., Johnson, T.R., 2011. Metal artifact reduction by dual energy computed tomography using monoenergetic extrapolation. Eur. Radiol. 21 (7), 1424−1429.

Barber, P.A., Darby, D.G., Desmond, P.M., et al., 1998. Prediction of stroke outcome with echoplanar perfusion- and diffusion-weighted MRI. Neurology 51 (2), 418−426.

Barentsz, J.O., Richenberg, J., Clements, R., et al., 2012. ESUR prostate MR guidelines 2012. Eur. Radiol. 22 (4), 746−757.

Barr, R.G., 2012. Sonographic breast elastography: a primer. J. Ultrasound Med. 31 (5), 773−783.

Bateman, R.J., Xiong, C., Benzinger, T.L., et al., 2012. Clinical and biomarker changes in dominantly inherited Alzheimer's disease. N. Engl. J. Med. 367 (9), 795−804.

Black, W.C., Gareen, I.F., Soneji, S.S., et al., 2014. Cost-effectiveness of CT screening in the National Lung Screening Trial. N. Engl. J. Med. 371 (19), 1793−1802.

Blamire, A.M., 2008. The technology of MRI − the next ten years? Br. J. Radiol. 81, 601−617.

Bosniak, M.A., 1997. Diagnosis and management of patients with complicated cystic lesions of the kidney. AJR Am. J. Roentgenol. 169 (3), 819−821.

Breen, J.F., Sheedy 2nd, P.F., Schwartz, R.S., et al., 1992. Coronary artery calcification detected with ultrafast CT as an indication of coronary artery disease. Radiology 185 (2), 435−439.

Britz-Cunningham, S.H., Adelstein, S.J., 2003. Molecular targeting with radionuclides: state of the science. J. Nucl. Med. 44 (12), 1945−1961.

Brown, R.S., Goodman, T.M., Zasadny, K.R., Greenson, J.K., Wahl, R.L., 2002. Expression of hexokinase II and Glut-1 in untreated human breast cancer. Nucl. Med. Biol. 29 (4), 443−453.

Brown, R.S., Wahl, R.L., 1993. Overexpression of Glut-1 glucose transporter in human breast cancer. An immunohistochemical study. Cancer 72 (10), 2979−2985.

Chakravarty, R., Hong, H., Cai, W., 2014. Positron emission tomography image-guided drug delivery: current status and future perspectives. Mol. Pharm. 11 (11), 3777−3797.

Cherry, S.R., 2004. In vivo molecular and genomic imaging: new challenges for imaging physics. Phys. Med. Biol. 49 (3), R13−R48.

Chikarmane, S.A., Khurana, B., Krajewski, K.M., et al., 2012. What the emergency radiologist needs to know about treatment-related complications from conventional chemotherapy and newer molecular targeted agents. Emerg. Radiol. 19 (6), 535−546.

Choi, H., Charnsangavej, C., Faria, S.C., et al., 2007. Correlation of computed tomography and positron emission tomography in patients with metastatic gastrointestinal stromal tumor treated at a single institution with imatinib mesylate: proposal of new computed tomography response criteria. J. Clin. Oncol. 25 (13), 1753−1759.

Contag, C.H., Bachmann, M.H., 2002. Advances in in vivo bioluminescence imaging of gene expression. Ann. Rev. Biomed. Eng. 4, 235−260.

Contag, P.R., Olomu, I.N., Stevenson, D.K., Contag, C.H., 1998. Bioluminescent indicators in living mammals. Nature Med. 4 (2), 245−247.

Dachman, A.H., Kuniyoshi, J.K., Boyle, C.M., et al., 1998. CT colonography with three-dimensional problem solving for detection of colonic polyps. Am. J. Roentgenol. 171 (4), 989−995.

Demetri, G.D., von Mehren, M., Blanke, C.D., et al., 2002. Efficacy and safety of imatinib mesylate in advanced gastrointestinal stromal tumors. N. Engl. J. Med. 347 (7), 472−480.

Deshpande, N., Needles, A., Willmann, J.K., 2010. Molecular ultrasound imaging: current status and future directions. Clin. Radiol. 65 (7), 567−581.

Dickinson, L., Ahmed, H.U., Allen, C., et al., 2011. Magnetic resonance imaging for the detection, localisation, and characterisation of prostate cancer: recommendations from a European consensus meeting. Eur. Urol. 59 (4), 477−494.

Earls, J.P., Berman, E.L., Urban, B.A., et al., 2008. Prospectively gated transverse coronary CT angiography versus retrospectively gated helical technique: improved image quality and reduced radiation dose. Radiology 246 (3), 742−753.

Edelman, R.R., Mattle, H.P., Atkinson, D.J., et al., 2000. Cerebral blood flow: assessment with dynamic contrast-enhanced T2*-weighted MR imaging at 1.5 T. Radiology 176, 211−220.

Ehman, R.L., Hendee, W.R., Welch, M.J., et al., 2007. Blueprint for imaging in biomedical research. Radiology 244 (1), 12−27.

Einstein, A.J., Elliston, C.D., Arai, A.E., et al., 2010. Radiation dose from single-heartbeat coronary CT angiography performed with a 320-detector row volume scanner. Radiology 254 (3), 698−706.

Eisenhauer, E.A., Therasse, P., Bogaerts, J., et al., 2009. New response evaluation criteria in solid tumours: revised RECIST guideline (version 1.1). Eur. J. Cancer 45 (2), 228−247.

Erturk, S.M., Ichikawa, T., Fujii, H., Yasuda, S., Ros, P.R., 2006. PET imaging for evaluation of metastatic colorectal cancer of the liver. Eur. J. Radiol. 58 (2), 229−235.

Flohr, T.G., McCollough, C.H., Bruder, H., et al., 2006. First performance evaluation of a dual-source CT (DSCT) system. Eur. Radiol. 16 (2), 256−268.

Fragioni, J., 2003. In vivo near infra-red fluorescence imaging. Curr. Opin. Chem. Biol. 7, 626−634.

Gaertner, F.C., Furst, S., Schwaiger, M., 2013. PET/MR: a paradigm shift. Cancer Imaging 13, 36−52.

Ganswindt, U., Schilling, D., Muller, A.C., Bares, R., Bartenstein, P., Belka, C., 2011. Distribution of prostate sentinel nodes: a SPECT-derived anatomic atlas. Int. J. Radiat. Oncol. Biol. Phys. 79 (5), 1364−1372.

Garcia, E.V., 2012. Physical attributes, limitations, and future potential for PET and SPECT. J. Nucl. Cardiol. 19 (Suppl. 1), S19−S29.

Garra, B.S., 2007. Imaging and estimation of tissue elasticity by ultrasound. Ultrasound Q 23 (4), 255−268.

Gill, R.R., Umeoka, S., Mamata, H., et al., 2010. Diffusion-weighted MRI of malignant pleural mesothelioma: preliminary assessment of apparent diffusion coefficient in histologic subtypes. AJR Am. J. Roentgenol. 195 (2), W125−W130.

Greer 3rd, L.F., Szalay, A.A., 2002. Imaging of light emission from the expression of luciferases in living cells and organisms: a review. Luminescence 17 (1), 43−74.

Graser, A., Becker, C.R., Staehler, M., et al., 2010. Single-phase dual-energy CT allows for characterization of renal masses as benign or malignant. Invest. Radiol. 45 (7), 399−405.

Groves, A.M., Win, T., Haim, S.B., Ell, P.J., 2007. Non-[18F]FDG PET in clinical oncology. Lancet Oncol. 8 (9), 822−830.

Hegde, J.V., Mulkern, R.V., Panych, L.P., et al., 2013. Multiparametric MRI of prostate cancer: an update on state-of-the-art techniques and their performance in detecting and localizing prostate cancer. J. Magn. Reson. Imaging 37 (5), 1035−1054.

Hermann, G.G., Mogensen, K., Carlsson, S., Marcussen, N., Duun, S., 2011. Fluorescence-guided transurethral resection of bladder tumours reduces bladder tumour recurrence due to less residual tumour tissue in Ta/T1 patients: a randomized two-centre study. BJU Int. 108, E297−E303.

Hoeks, C.M., Barentsz, J.O., Hambrock, T., et al., 2011. Prostate cancer: multiparametric MR imaging for detection, localization, and staging. Radiology 261 (1), 46−66.

Hoey, E.T., Ganeshan, A., 2014a. Multi-detector CT angiography of the aortic valve-Part 2: disease specific findings. Quant. Imaging Med. Surg. 4 (4), 273−281.

Hoey, E.T., Ganeshan, A., 2014b. Multi-detector CT angiography of the aortic valve-Part 1: anatomy, technique and systematic approach to interpretation. Quant. Imaging Med. Surg. 4 (4), 265−272.

Hoffman, J.M., Gambhir, S.S., 2007. Molecular imaging: the vision and opportunity for radiology in the future. Radiology 244 (1), 39−47.

Holdsworth, C.H., Badawi, R.D., Manola, J.B., et al., 2007. CT and PET: early prognostic indicators of response to imatinib mesylate in patients with gastrointestinal stromal tumor. AJR Am. J. Roentgenol. 189 (6), W324−W330.

Hopyan, J., Ciarallo, A., Dowlatshahi, D., et al., 2010. Certainty of stroke diagnosis: incremental benefit with CT perfusion over noncontrast CT and CT angiography. Radiology 255 (1), 142−153.

Hounsfield, G.N., 1973. Computerized transverse axial scanning (tomography). 1. Description of system. Br. J. Radiol. 46 (552), 1016−1022.

Hricak, H., Choyke, P.L., Eberhardt, S.C., Leibel, S.A., Scardino, P.T., 2007. Imaging prostate cancer: a multidisciplinary perspective. Radiology 243 (1), 28−53.

Hu, H., He, H.D., Foley, W.D., Fox, S.H., 2000. Four multidetector-row helical CT: image quality and volume coverage speed. Radiology 215 (1), 55−62.

Ishizawa, T., Fukushima, N., Shibahara, J., et al., 2009. Real-time identification of liver cancers by using indocyanine green fluorescent imaging. Cancer 115, 2491−2504.

Joensuu, H., Roberts, P.J., Sarlomo-Rikala, M., et al., 2001. Effect of the tyrosine kinase inhibitor STI571 in a patient with a metastatic gastrointestinal stromal tumor. N. Engl. J. Med. 344 (14), 1052−1056.

Kalender, W.A., 1995. Thin-section three-dimensional spiral CT: is isotropic imaging possible? Radiology 197 (3), 578−580.

Kalender, W.A., Seissler, W., Klotz, E., Vock, P., 1990. Spiral volumetric CT with single-breath-hold technique, continuous transport, and continuous scanner rotation. Radiology 176 (1), 181−183.

Kapoor, V., McCook, B.M., Torok, F.S., 2004. An introduction to PET-CT imaging. Radiographics 24 (2), 523−543.

Kimura, T., Muguruma, N., Ito, S., et al., 2007. Infrared fluorescence endoscopy for the diagnosis of superficial gastric tumors. Gastrointest. Endosc. 66 (1), 37–43.

Kitai, T., Inomoto, T., Miwa, M., Shikayama, T., 2005. Fluorescence navigation with indocyanine green for detecting sentinel lymph nodes in breast cancer. Breast Cancer 12 (3), 211–215.

Ko, S.M., Song, M.G., Chee, H.K., Hwang, H.K., Feuchtner, G.M., Min, J.K., 2014. Diagnostic performance of dual-energy CT stress myocardial perfusion imaging: direct comparison with cardiovascular MRI. AJR Am. J. Roentgenol. 203 (6), W605–W613.

Kubo, T., Ohno, Y., Kauczor, H.U., Hatabu, H., 2014. Radiation dose reduction in chest CT–review of available options. Eur. J. Radiol. 83 (10), 1953–1961.

Lee, C.H., Goo, J.M., Ye, H.J., et al., 2008. Radiation dose modulation techniques in the multidetector CT era: from basics to practice. Radiographics 28 (5), 1451–1459.

Magnetic Resonance Prostate Imaging Reporting and Data System (MR PI-RADS). http://wwwacrorg/w/media/ACR/Documents/PDF/QualitySafety/Resources/PIRADS/PIRADS.

Maisey, M., 1998. Radionuclide imaging in cancer management. J. R. Coll. Phys. London 32 (6), 525–529.

Marin, D., Boll, D.T., Mileto, A., Nelson, R.C., 2014. State of the art: dual-energy CT of the abdomen. Radiology 271 (2), 327–342.

Margolis, D.J., Hoffman, J.M., Herfkens, R.J., Jeffrey, R.B., Quon, A., Gambhir, S.S., 2007. Molecular imaging techniques in body imaging. Radiology 245 (2), 333–356.

Matsui, A., Winer, J.H., Laurence, R.G., Frangioni, J.V., 2011. Predicting the survival of experimental ischaemic small bowel using intraoperative near infrared fluorescence angiography. Br. J. Surg. 98, 1725–1734.

McVeigh, E.R., 2006. Emerging imaging techniques. Circ. Res. 98 (7), 879–886.

Munley, M.T., Marks, L.B., Hardenbergh, P.H., Bentel, G.C., 2001. Functional imaging of normal tissues with nuclear medicine: applications in radiotherapy. Semin. Radiat. Oncol. 11 (1), 28–36.

National Lung Screening Trial Research T, Church, T.R., Black, W.C., et al., 2013. Results of initial low-dose computed tomographic screening for lung cancer. N. Engl. J. Med. 368 (21), 1980–1991.

Nekolla, S.G., Martinez-Moeller, A., Saraste, A., 2009. PET and MRI in cardiac imaging: from validation studies to integrated applications. Eur. J. Nucl. Med. Mol. Imaging 36 (Suppl. 1), S121–S130.

Pan, J., Konstas, A.A., Bateman, B., Ortolano, G.A., Pile-Spellman, J., 2007. Reperfusion injury following cerebral ischemia: pathophysiology, MR imaging, and potential therapies. Neuroradiology 49 (2), 93–102.

Pelliccia, F., Pasceri, V., Evangelista, A., et al., 2013. Diagnostic accuracy of 320-row computed tomography as compared with invasive coronary angiography in unselected, consecutive patients with suspected coronary artery disease. Int. J. Cardiovasc. Imaging 29 (2), 443–452.

Polacin, A., Kalender, W.A., Marchal, G., 1992. Evaluation of section sensitivity profiles and image noise in spiral CT. Radiology 185 (1), 29–35.

Provenzale, J.M., Wintermark, M., 2008. Optimization of perfusion imaging for acute cerebral ischemia: review of recent clinical trials and recommendations for future studies. AJR Am. J. Roentgenol. 191 (4), 1263–1270.

Rocha-Filho, J.A., Blankstein, R., Shturman, L.D., et al., 2010. Incremental value of adenosine-induced stress myocardial perfusion imaging with dual-source CT at cardiac CT angiography. Radiology 254 (2), 410–419.

de Roos, A., Higgins, C.B., 2014. Cardiac radiology: centenary review. Radiology 273 (Suppl. 2), S142–S159.

Royster, A.P., Fenlon, H.M., Clarke, P.D., Nunes, D.P., Ferrucci, J.T., 1997. CT colonoscopy of colorectal neoplasms: two-dimensional and three-dimensional virtual-reality techniques with colonoscopic correlation. AJR Am. J. Roentgenol. 169 (5), 1237–1242.

Schaefer, P.W., Grant, P.E., Gonzalez, R.G., 2000. Diffusion-weighted MR imaging of the brain. Radiology 217 (2), 331–345.

Schoepf, U.J., Becker, C.R., Ohnesorge, B.M., Yucel, E.K., 2004. CT of coronary artery disease. Radiology 232 (1), 18–37.

Shaw, L.J., Raggi, P., Schisterman, E., Berman, D.S., Callister, T.Q., 2003. Prognostic value of cardiac risk factors and coronary artery calcium screening for all-cause mortality. Radiology 228 (3), 826–833.

Shinagare, A.B., Guo, M., Hatabu, H., et al., 2011. Incidence of pulmonary embolism in oncologic outpatients at a tertiary cancer center. Cancer 117 (16), 3860–3866.

Siegel, R., Ma, J., Zou, Z., Jemal, A., 2014. Cancer statistics, 2014. CA: A Cancer J. Clin. 64 (1), 9–29.

Silverman, S.G., Israel, G.M., Herts, B.R., Richie, J.P., 2008. Management of the incidental renal mass. Radiology 249 (1), 16–31.

Stummer, W., Pichlmeier, U., Meinel, T., et al., 2006. Fluorescence-guided surgery with 5-aminolevulinic acid for resection of malignant glioma: a randomised controlled multicentre phase III trial. Lancet Oncol. 7 (5), 392–401.

Sun, Z., Al Moudi, M., Cao, Y., 2014. CT angiography in the diagnosis of cardiovascular disease: a transformation in cardiovascular CT practice. Quant. Imaging Med. Surg. 4 (5), 376–396.

Taruttis, A., Ntziachristos, V., 2012. Translational optical imaging. AJR Am. J. Roentgenol. 199 (2), 263–271.

Tempany, C.M., McNeil, B.J., 2001. Advances in biomedical imaging. JAMA 285 (5), 562–567.

Themelis, G., Harlaar, N.J., Kelder, W., et al., 2011. Enhancing surgical vision by using real-time imaging of $\alpha v \beta 3$-integrin targeted near-infrared fluorescent agent. Ann. Surg. Oncol. 18, 3506–3513.

Therasse, P., Arbuck, S.G., Eisenhauer, E.A., et al., 2000. New guidelines to evaluate the response to treatment in solid tumors. European Organization for Research and Treatment of Cancer, National Cancer Institute of the United States, National Cancer Institute of Canada. J. Natl. Cancer Inst. 92 (3), 205–216.

Thomassin-Naggara, I.T., Lalonde, L., David, J., Péloquin, L., Chopier, J., 2012. Astuces et nouveautés techniques en IRM mammaire. J. Radiol. Diagn. Interventionnelle 93 (11), 876–889.

Tofts, P.S., 1994. Standing waves in uniform water phantoms. J. Magn. Reson. 104, 143–147.

Torigian, D.A., Zaidi, H., Kwee, T.C., et al., 2013. PET/MR imaging: technical aspects and potential clinical applications. Radiology 267 (1), 26–44.

Tung, C.H., Mahmood, U., Bredow, S., Weissleder, R., 2000. In vivo imaging of proteolytic enzyme activity using a novel molecular reporter. Cancer Res. 60, 4953–4958.

Turner, R., Le Bihan, D., Chesnick, A.S., 1991. Echo-planar imaging of diffusion and perfusion. Magn. Reson. Med. 19 (2), 247–253.

Tutt, A.N., Plunkett, T.A., Barrington, S.F., Leslie, M.D., 2004. The role of positron emission tomography in the management of colorectal cancer. Colorectal Dis. 6 (1), 2–9.

Umbehr, M., Bachmann, L.M., Held, U., et al., 2009. Combined magnetic resonance imaging and magnetic resonance spectroscopy imaging in the diagnosis of prostate cancer: a systematic review and meta-analysis. Eur. Urol. 55 (3), 575–590.

Valdes Olmos, R.A., Tanis, P.J., Hoefnagel, C.A., et al., 2001. Improved sentinel node visualization in breast cancer by optimizing the colloid particle concentration and tracer dosage. Nucl. Med. Commun. 22 (5), 579–586.

Van den Abbeele, A.D., Badawi, R.D., 2002. Use of positron emission tomography in oncology and its potential role to assess response to imatinib mesylate therapy in gastrointestinal stromal tumors (GISTs). Eur. J. Cancer 38 (Suppl. 5), S60–S65.

Vanhoenacker, P.K., Heijenbrok-Kal, M.H., Van Heste, R., et al., 2007. Diagnostic performance of multidetector CT angiography for assessment of coronary artery disease: meta-analysis. Radiology 244 (2), 419–428.

Vaughan, T.D.L., Snyder, C., Tian, J.F., Akgun, C., Shrivastava, D., et al., 2006. 9.4T human MRI: preliminary results. Magn. Reson. Med. 56, 1274–1282.

Vlassenko, A.G., Benzinger, T.L., Morris, J.C., 2012. PET amyloid-beta imaging in preclinical Alzheimer's disease. Biochim. Biophys. Acta 1822 (3), 370–379.

Weissleder, R., 2001. A clearer vision for in vivo imaging. Nat. Biotechnol. 19 (4), 316–317.

Weissleder, R., Mahmood, U., 2001. Molecular imaging. Radiology 219 (2), 316–333.

Wetter, A., Lipponer, C., Nensa, F., et al., 2014. Evaluation of the PET component of simultaneous [(18)F]choline PET/MRI in prostate cancer: comparison with [(18)F]choline PET/CT. Eur. J. Nucl. Med. Mol. Imaging 41 (1), 79–88.

Willemink, M.J., de Jong, P.A., Leiner, T., et al., 2013. Iterative reconstruction techniques for computed tomography Part 1: technical principles. Eur. Radiol. 23 (6), 1623–1631.

Willmann, J.K., van Bruggen, N., Dinkelborg, L.M., et al., 2008. Molecular imaging in drug development. Nat. Rev. Drug Discovery 7 (7), 591–607.

Wilkinson, J.M., Kuok, M.H., Adamson, G., 2004. Biomedical applications of optical imaging. Med. Dev. Technol. 15 (10), 22–24.

Whitsett, M.C., 2009. Ultrasound imaging and advances in system features. Ultrasound Clin. 4 (3), 391–401.

Wolbarst, A., Hendee, W.R., 2006. Evolving and experimental technologies in medical imaging. Radiology 238 (1), 16–39.

Wycliffe, N.D., Choe, J., Holshouser, B., Oyoyo, U.E., Haacke, E.M., Kido, D.K., 2004. Reliability in detection of hemorrhage in acute stroke by a new three-dimensional gradient recalled echo susceptibility-weighted imaging technique compared to computed tomography: a retrospective study. J. Magn. Reson. Imaging: JMRI 20 (3), 372–377.

Yang, L., Rieves, D., Ganley, C., 2012. Brain amyloid imaging—FDA approval of florbetapir F18 injection. N. Engl. J. Med. 367 (10), 885–887.

Yoo, H., Kim, J.W., Shishkov, M., et al., 2011. Intra-arterial catheter for simultaneous microstructural and molecular imaging in vivo. Nat. Med. 17 (12), 1680–1684.

Zalis, M.E., Barish, M.A., Choi, J.R., et al., 2005. CT colonography reporting and data system: a consensus proposal. Radiology 236 (1), 3–9.

Zalis, M.E., Perumpillichira, J.J., Magee, C., Kohlberg, G., Hahn, P.F., 2006. Tagging-based, electronically cleansed CT colonography: evaluation of patient comfort and image readability. Radiology 239 (1), 149–159.

Zealley, I.A., Skehan, S.J., Rawlinson, J., Coates, G., Nahmias, C., Somers, S., 2001. Selection of patients for resection of hepatic metastases: improved detection of extrahepatic disease with FDG pet. Radiographics 21 (Spec No), S55–S69.

Zerhouni, E., 2007. Major Trends in the Imaging Sciences. http://wwwrsnaorg/Media/rsna/rsna_newsrelease_targetcfm??&id5342.

Chapter 10

Imaging Tools in Human Research: Focus on Image-Guided Intervention

Ruth M. Dunne, Ailbhe C. O'Neill and Clare M. Tempany

Harvard Medical School, Boston, MA, United States

Chapter Outline

Key Points

- Image-guided procedures may be performed using anatomic, molecular, or functional imaging and many are suited to outpatient settings as they are minimally invasive.
- Image-guided biopsies provide tissue diagnosis but their role is evolving. Lung cancer biopsies now play a role in personalized treatments. The use of multiparametric MRI in guiding prostate biopsy allows targeting of suspicious area in lieu of random sampling.
- Image-guided ablation is a method of directly eradicating a tumor using energy-based methods including thermal methods (radiofrequency ablation, microwave ablation, and cryoablation) or nonthermal/chemical methods (ethanol ablation).
- Focused ultrasound uses high-frequency ultrasound to cause local tissue changes, which can lead to heating and thermocoagulation and cell death. MR-guided focused ultrasound for the treatment of uterine fibroids is an example of its clinical application.
- Targeted drug delivery has been developed for local delivery of small molecules in the field of nanomedicine. Imaging has multiple roles from evaluating biodistribution to target site accumulation.
- Hybrid operating rooms combine full operating rooms with intraoperative imaging. In neurosurgery they can optimize resection margins while limiting neurological morbidity. There is also evolving use with minimally invasive thoracotomies for small lung tumors with wire placement performed for localization.
- Image-guided fiducial marker placements and image-guided brachytherapy are preformed to aid in targeted radiation therapy treatment.

Clinical and Translational Science. http://dx.doi.org/10.1016/B978-0-12-802101-9.00010-7

INTRODUCTION

Extraordinary advances in imaging have taken place with novel new techniques providing unique information no longer limited to anatomic imaging but with new molecular and functional data. These data sets are proving very valuable in many diseases and now also in treatment and particularly in clinical research. In keeping with an overall desire to be less invasive and perform more therapies as outpatient procedures, image guidance really enables this approach. This chapter will review and describe several image-guided procedures and illustrate the value-added of the addition of imaging in all stages of diagnosis and therapy.

IMAGE-GUIDED BIOPSY

Breast

Imaging and imaging-guided biopsies are the cornerstone of diagnosis, treatment, and staging of patients with breast cancer. Percutaneous sampling may be performed in a patient with a palpable mass, or an abnormality on imaging such a calcification or architectural distortion using ultrasound (US), mammography, magnetic resonance imaging (MRI), or tomosynthesis guidance (Schrading et al., 2015). Core needle biopsies may be performed with either an automated throw device or a vacuum-assisted needle with sizes ranging from 7 to 14 gauge. An automated throw needle takes a single core of tissue on biopsy compared to the vacuum-assisted device that takes multiple cores of tissue in a single biopsy allowing for a larger sample size (Park and Hong, 2014). Axillary node staging may be done by fine needle aspiration or core needle biopsy of an enlarged axillary lymph node.

US is the initial imaging test in a patient with a palpable mass under 30 years or suspected lesions on mammography or MRI. Advantages include real-time imaging that can guide biopsies and also the axilla may be staged at the time of initial breast US. However, limitations of ultrasonic guidance include lack of visualization of calcifications; therefore biopsy sampling of calcifications is performed using stereotactic guidance. This involves a biopsy performed under mammography guidance. Disadvantages of stereotactic biopsies are patient discomfort as the breast remains in the mammographic machine under compression and any patient movement may lead to incorrect site sampling (Huang et al., 2014). MRI-guided biopsies require intravenous gadolinium contrast to identify the lesion for biopsy and are performed with the patient in the prone position on the MRI table.

Small radiopaque markers are often placed following biopsy and are visible on all breast imaging techniques (Li et al., 2005). If the lesion is for excision, the marker is used to localize the site for wire-guided excision. Preoperatively, wires are placed with either US or stereotactic guidance in lesions that are not palpable to guide surgical excision in breast conserving surgery (Rasmussen and Seerup, 1984).

If a lesion is shown to be benign and has had a radiopaque marker placed, this is useful for follow-up imaging, demonstrating a site of prior biopsy.

Lung

In addition to its traditional diagnostic role, CT-guided lung biopsy also plays a key role in personalized treatment. With recent advances in target therapy, obtaining tumor tissue to analyze for the molecular fingerprint of somatic mutation for personalized medicine has become an increasing indication for lung biopsy (Paez et al., 2004).

Lung cancer is a heterogeneous, complex, and challenging disease to treat. In the past, lung cancer was divided into non-small-cell and small-cell subtypes. More recently and with the advent of personalized medicine, multiple advances have been made in understanding the underlying biology and molecular mechanisms of lung cancer (Hensing et al., 2014). Lung cancer is no longer considered a single disease entity and is now being subdivided into molecular subtypes based on genotyping and genomic profiling with dedicated, targeted, and chemotherapeutic strategies. The concept of using information from a patient's tumor to make therapeutic and treatment decisions has revolutionized the landscape for cancer care and research in general. Management of non-small-cell lung cancer, in particular, has seen several of these advances, with the understanding of activating mutations in EGFR, fusion genes involving ALK, rearrangements in ROS-1, and ongoing research in targeted therapies for K-RAS and MET mutations (Hensing et al., 2014). Future directions will involve incorporation of molecular characteristics and next-generation sequencing into screening strategies to improve early detection, while also having applications for joint treatment clinical decision-making with patients and physicians.

Image-guided biopsy of pleural and lung parenchymal abnormalities detected on diagnostic imaging is usually performed percutaneously using CT and/or CT-fluoroscopic guidance. Image guidance using CT is preferred to US as it allows better visualization of lung parenchyma. US requires a medium through which an acoustic beam can be transmitted, reflected, and recorded. As aerated lung or bones are not such media, lesions surrounded by aerated lung or protected by overlying osseous structures cannot be visualized and thus sampled. US guidance may, however, be considered for larger pleural-based lesions. For central lung lesions, planning CT with IV contrast is performed to ensure avoidance of hilar vessels. Alternatively, bronchoscopic biopsy can be performed.

CT guided lung biopsy is performed using a low-dose axial scan with 120 kVp, 30 mA per slice, 0.5- to 1-s rotation time, and collimation of 2.5—5 mm (Tsai et al., 2009). CT fluoroscopy is a technical advancement, which enables real-time visualization of a lesion during needle manipulation (Paulson et al., 2001). This technique is especially useful for targeting small lung lesions, juxtaphrenic lesions, and patients with poor breath-holding capacity (Fig. 10.1). CT fluoroscopy is more accurate than conventional CT in diagnostic biopsy of pulmonary lesions with a significant reduction in complication rates (Paulson et al., 2001). Additional advantages are the simplification of the biopsy process and decreased procedure time. However, this technique is associated with a small radiation dose to the operator with the mean estimated effective doctor dose ranging from 0.025 to 0.054 mSv per procedure (Paulson et al., 2001). Percutaneous biopsy may be performed using fine needle aspiration biopsy (22—25 gauge) for cytological and microbiological analysis or using core/cutting needle biopsy (18—20 gauge needle) for histological evaluation (Fig. 10.1). The diagnostic accuracy of aspiration biopsy is almost as good as core biopsy for the diagnosis of malignant lesions, especially if an onsite cytopathologist is present (Lal et al., 2012). However, for the diagnosis of benign lesions and lymphoma and for molecular analysis, core biopsy is preferred (Lal et al., 2012).

Prostate

Transrectal US (TRUS)-guided biopsy is considered the standard of care for diagnosis of prostate cancer and is indicated by suspicious findings at screening with serum prostate-specific antigen testing or digital rectal examination. Traditional TRUS-guided prostate biopsy schemes rely on an approach that is largely centered on the peripheral zone of the prostate and consist of random sampling (10—12 core biopsies) from different areas of the peripheral zone. However, although approximately 70—80% of cancers originate in the peripheral zone, this approach leads to substantial undersampling of the prostate. Moreover, US has only limited ability to help detect prostate cancer and is mainly used for anatomic reference during needle placement; therefore, acquisition of individual biopsy samples is not directed toward areas that are suspicious for cancer but simply toward different anatomic locations in the prostate (Halpern, 2000). Anterior tumors are frequently missed at routine TRUS-guided biopsy leading to an elevated false negative rate of up to 47% (Taira et al., 2010). In addition, preoperative biopsies may fail to reveal the most aggressive component of tumor in 33—50% of patients or lead to detection of incidental cancers of little or no clinical relevance (Pinthus et al., 2006).

With increasing recognition of the capability of multiparametric MRI to detect and characterize prostate cancer foci, the use of multiparametric MRI to guide prostate biopsy is the focus of many ongoing clinical trials (van de Ven et al., 2013; Tokuda et al., 2012; Hambrock et al., 2010; Overduin et al., 2013; Penzkofer et al., 2015). MR-guided biopsy of prostate can be performed using a number of different approaches: direct "in-bore" MR-guided biopsy (either transperineal or transrectal), an exclusively MR-based approach, in which preprocedure multiparametric MRI is used to define targets and real-time MR used to provide image guidance for needle placement during biopsy (Tokuda et al., 2012; Overduin et al., 2013; Penzkofer et al., 2015) and "out-of-bore" MR-guided biopsy whereby preprocedure multiparametric MRI is used to define targets for subsequent TRUS-guided biopsy either with use of software based image coregistration, also known as "MRI-TRUS fusion" prostate biopsy, or without automated image coregistration, also known as "cognitive fusion" prostate biopsy.

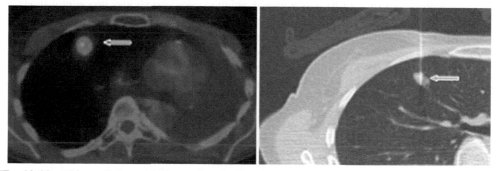

FIGURE 10.1   CT-guided lung biopsy. Patient with history of renal cell carcinoma and left pneumonectomy for primary lung adenocarcinoma. Fused PET-CT image demonstrates a new right middle lobe nodule with increased FDG uptake (*blue arrow*). 22-G biopsy needle sample of the nodule was successfully performed (*orange arrow*).

Clinical uses of MR-guided biopsy include patients with suspected cancer and previous negative TRUS-guided biopsy, patients with known cancer for whom active surveillance is an option, patients with known cancer to determine disease status during active surveillance, and candidates for focal prostate cancer therapy including ablation and high-intensity focused US discussed later in this chapter.

IMAGE-GUIDED THERAPY

Ablation

Image-guided ablation is a technique that utilizes chemical or energy-based methods (thermal and nonthermal) to directly destroy or eradicate a tumor. The decision to perform ablation in lieu of surgery is based upon multiple factors including patient age, life expectancy, comorbidities, and the presence of metachronous lesions.

Image guidance may be CT, MRI, PET-CT, or US depending on the organ targeted and type of ablation. The use of imaging is beneficial for visualizing the tumor for planning and targeting, intraprocedural monitoring, and modification (Ahmed et al., 2011, 2014). Imaging is also used to assess response immediately during the procedure and also on follow-up examinations. It has been proven to be a safe and effective method offering targeted treatment that spares uninvolved tissue (Georgiades and Rodriguez, 2013; Rodriguez et al., 2011; Dodd et al., 2000). A common phenomenon related to all types of ablation is the "post ablation syndrome" consisting of low-grade fever, malaise, myalgia, delayed pain, nausea, and vomiting. This is self-limiting and resolves after 7−10 days (Carrafiello et al., 2007).

Thermal ablation techniques include radiofrequency ablation (RFA), microwave ablation (MWA), and cryoablation and aim to destroy the tumor through heat or cold techniques (Dodd et al., 2000; Vogl et al., 2014).

RFA aims to produce coagulation necrosis using an alternating electric current between the 375−500 KHz range. MWA utilizes high-frequency waves between 900 MHZ and 2.4 GHz that cause oscillation of water molecules creating friction that results in heat production and tissue destruction via coagulative necrosis. The thermal ablation techniques that use heat aim to raise the temperature of the tissue to a level that produces coagulative necrosis (typically between 60°C and 100°C) while avoiding charring or vaporization of tissue (Dodd et al., 2000; Vogl et al., 2014). Cryoablation destroys tissue by applying freezing temperatures or alternating freezing and thawing that produces cytotoxic effects by disrupting the cell membrane and inducing cell death (Rubinsky et al., 1990). Chemical ablation was one of the first ablative techniques used and produces coagulative necrosis by direct instillation of an agent such as ethanol or acetic acid.

Radiofrequency Ablation

RFA delivers high-frequency alternating electric currents via an applicator placed into the tumor under direct image guidance. Power is adjusted to keep the target temperatures constant, and thermometers are incorporated into the tips of the electrodes to allow continuous monitoring of temperature during the ablation. Multiple ablations zones can be overlapped to ensure adequate tumor coverage. RFA is used most often in the treatment of hepatocellular carcinoma (HCC) and can be used to ablate several tumors in one session (Willatt et al., 2012). RFA is also the preferred ablation technique for osteoid osteomas due to the small ablation zone and no associated collateral damage (Hinshaw et al., 2014). A limitation of RFA and other types of thermal ablation methods is the "heat sink" effect of nearby blood vessels or bronchi in the lung that dissipate the thermal energy limiting the size of the ablation zone (Goldberg et al., 1998).

Microwave Ablation

The main advantage of MWA compared to RFA is a greater ablation zone and shorter procedural time. In comparison to RFA, higher temperatures are seen in MWA decreasing the "heat sink" effect from nearby vessels or airways (Fig. 10.2). MWA is the primary ablation technique for both primary and metastatic disease in the lung due to the speed and ability to obtain large ablation zone. It may also be used in the treatment of HCC where heat-based ablation techniques are preferred due to superior hemostasis in coagulopathic cirrhotic patients (Hinshaw et al., 2014). MWA in the kidney is preferred in the ablation of angiomyolipoma due to the high bleeding risk and can also be safely performed in malignant exophytic renal lesions (Hinshaw et al., 2014).

Cryoablation

Cryoablation consists of cycles of freezing and thawing using a compressed cryogen (argon gas or liquid nitrogen) and helium, respectively. The standard protocol is a 10-min freeze, an 8-min thaw, and a second 10-min freeze. A major advantage of cryoablation is that the ablation zone is visible as an "ice ball" during the procedure. Approximately one

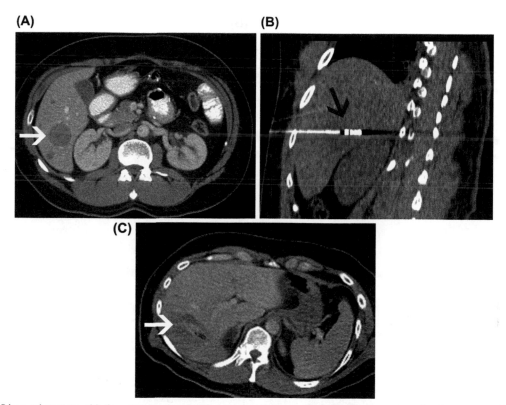

FIGURE 10.2   Liver microwave ablation. (A) Contrast-enhanced axial CT image demonstrates hypoattenuating liver metastasis in patient with gastrointestinal stromal tumor (*white arrow*). (B) Intraprocedural noncontrast sagittal CT reconstruction demonstrates microwave applicator deployed within the tumor (*black arrow*). (C) Intraprocedural contrast-enhanced axial CT image following ablation illustrates microwave applicator within the tumor with surrounding larger area of hypoattenuation in keeping with ablation zone consisting of tumor and surrounding rim of normal liver parenchyma (*white arrow*).

applicator is used per centimeter of tumor and applicators come in different shapes allowing the operator the ability to modify the size and contour of the ice ball and therefore the ablation zone during the procedure. The "ice ball" is best demonstrated on MRI but is also visible on CT and US (Fig. 10.3).

Cryoablation is preferred for the ablation of small renal tumors compared to heat-based thermal ablation especially in the ablation of endophytic tumors or tumors in close proximity to vulnerable structures such as the ureter. Cryoablation can also be performed in the lung if the tumor is close to central structures or the chest wall. There is also evidence that cryoablation of painful skeletal metastases is associated with fast and durable pain relief (Thacker et al., 2011). Cryoablation is not recommended in the ablation of HCC in cirrhotic patients due to the bleeding risk and a form of disseminated intravascular coagulation-like reaction termed cryoshock (Jansen et al., 2010).

Ethanol Ablation

This is one of the earliest ablative techniques that was used primarily in the treatment of HCC (Livraghi et al., 1995). Fine needle injection of 95% ethyl alcohol is injected into the tumor under US or CT guidance. The hard cirrhotic liver around the soft tumor limits the flow of alcohol into the adjacent parenchyma. Advantages include low cost and low risk of complications, but the disadvantage is that multiple procedures are required to be effective. It has now been superseded by other methods of ablation but still has use in coagulopathic cirrhotic patients and in exophytic tumors where other methods could damage the hepatic capsule causing bleeding or bile leaks (Lin et al., 2004).

Focused Ultrasound

High-frequency ultrasound (HIFU) can cause localized tissue changes, which include thermocoagulation (due to local delivery of HIFU with high energy causing molecular vibrations resulting in heat delivery), HIFU combined with intravenous microbubbles can cause local drug delivery, and similar combination delivered to the cranium can cause

(A)

(B)

(C)

(D)

(E)

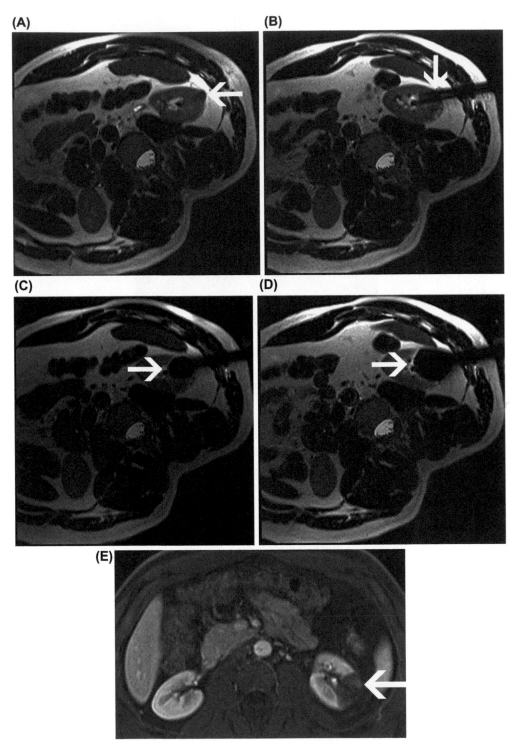

FIGURE 10.3 Cryoablation of renal tumor. (A) Preprocedural MR axial T2-weighted image of 1.9 × 1.6 cm biopsy proven papillary RCC with patient in right posterior oblique prior to ablation (*white arrow*). (B) Intraprocedural MR axial T2-weighted image demonstrates 2 × 17 G cryoablation needles anterior and posterior to mass (*white arrow*). (C) Intraprocedural MR axial T2-weighted image with ice ball at 3 min. A major advantage of cryoablation is the ablation zone is visible as an "ice ball" during the procedure. (D) Intraprocedural MR axial T2-weighted image with ice ball at 15 min. (E) Contrast-enhanced MRI 1 day after ablation—axial T1-weighted image at 60 s after contrast administration demonstrates ablation zone, which is hypoenhancing relative to remainder of kidney (*white arrow*).

blood—brain barrier disruption. Thus this technology can have far-reaching effects, many of which are still under active investigation today. The application best studied, applied, and tested is the delivery of local heat to focal tissues, to cause thermocoagulation and when the temperature is high enough (>55°C) can cause cell death. This technique has been used for many years using diagnostic US as the image guidance modality. It is used clinically in many places worldwide to treat liver, kidney, prostate, and gynecological diseases. It has not, however, seen widespread acceptance in the United States. It is generally though that this is due to the limitations of US as a guidance modality. The pioneering work of Jolesz, Hynynen, and others in the 1990s resulted in the development of MR-guided focused US systems. The advantage of MR is the ability of MR to provide temperature data—using differences in proton resonance frequency induced by temperature changes the MR phase contrast sequences can display the change from baseline to normal body temperature. Thus for the first time this combination of MRI guidance and focused US allows a complete closed-loop feedback technique with MR defining the target, guiding the sonication delivery, and imaging the immediate temperature change. At the end of the procedure, intravenous contrast-enhanced MRI will show the areas of tissue necrosis as those with no perfusion.

MR-guided focused US (MRgFUS) received FDA clearance for treatment of uterine fibroids in 2004, later for palliation of bone metastases, with multiple other applications in clinical trials (Tempany, 2011).

Targeted Drug Delivery

Targeted drug delivery techniques are of great interest for safer and more effective pharmaceutical therapies. The local targeted delivery of small molecules has been developed in the field of nanomedicine. These approaches have been enhanced and augmented by imaging. US can also guide and facilitate the local delivery of drugs in a noninvasive manner inside the body. The bioeffects of US can lead to local tissue heating, cavitation, and radiation force, which can be used for (1) local drug release from nanocarriers circulating in the blood, (2) increased extravasation of drugs and/or carriers, and (3) enhanced diffusivity of drugs. When using nanocarriers sensitive to mechanical forces (the oscillating US pressure waves) and/or sensitive to temperature, the content of the nanocarriers can be released locally (Deckers and Moonen, 2010). The role of imaging encompasses multiple aspects including visualization of biodistribution in real time, analysis of drug distribution at the target site, can predict drug response, evaluate drug efficacy longitudinally, facilitate triggered drug release, and noninvasively assess target site accumulation. In image-guided theranostics the disease diagnosis and therapy are combined together (Lammers et al., 2010).

Hybrid Operating Room

Hybrid operating rooms combine a full operating room with a mobile ceiling-mounted MR scanner that can be moved into the operating room (OR) for intraoperative imaging (Tempany et al., 2015). All the anesthetic and surgical devices in the OR are MRI compatible.

Neurosurgery

Intraoperative MRI in neurosurgery is used to aid obtaining optimal resection margins and limiting neurologic morbidity. The delineation of tumor margins and localization of critical brain structures are two main issues in neurosurgery, and MRI can provide both the structural and functional information. Functional MRI and diffusion tensor imaging are the two primary ways of obtaining functional information. Functional MRI assesses blood oxygen levels in cortical regions during specific tasks and if performed preoperatively allows for increased use of awake surgery and highlights key cortical structures to minimize morbidity (Belliveau et al., 1991). Diffusion tensor imaging maps subcortical anatomy and white matter tracks. These white matter tracts can change in location during surgery, so this technique requires an update during surgery (Wu et al., 2007; Elhawary et al., 2011).

Newer minimally invasive methods for neurosurgical tissue ablation include laser interstitial therapy and focused US that create thermal ablation zones to target treatment of brain tumors and epileptic foci (Tempany et al., 2015).

Minimally Invasive Thoracotomy for Lung Nodules

The advent of lung cancer screening has shown a possible reduction in lung cancer mortality, and the detection of smaller tumors increases the numbers that can be managed with localized surgery, for example segmentectomy in lieu of lobectomy (Swanson, 2010; Black et al., 2014). As such disease may be nonpalpable or not visualized at the time of surgery, image-guided preoperative or intraoperative localization of smaller tumors is a new and advancing technique for tumor localization. Percutaneous placement of wires into the tumor can be performed with high rates of accuracy (Gossot et al., 1994).

Treatment Planning in Radiation Oncology—The Role of the Radiologist

Image guidance in radiation therapy is vital to provide precise anatomical detail as radiation therapy is evolving to a targeted focused treatment. There is increased radiation dose per fraction to the specific site to maximize efficacy and the targeted nature reduces irradiation of surrounding structures and secondary complications (Kothary et al., 2009a,b).

Real-time tumor tracking radiation therapy systems are used to focus the radiation dose on the primary tumor (Schweikard et al., 2004). Fiducial markers are often placed to facilitate this process (Kitamura et al., 2002). Fiducial markers are commonly subcentimeter, cylindrical gold markers. These are usually placed using CT guidance with at least one fiducial marker within the tumor and a minimal of two more fiducial markers along the perimeter of the tumor at least 1—2 cm apart to avoid obscuration of a marker secondary to artifact (Kothary et al., 2009a,b). These are especially useful in tracking tumors that are within structures that move with respiration, for example, tumors in the liver or lung, whereas pelvic structures are less affected by respiratory motion (Kitamura et al., 2002; Shirato et al., 2003; Imura et al., 2005; Kothary et al., 2009a,b).

Image-Guided Brachytherapy in Prostate and Cervical Cancer

Brachytherapy involves placing a radioactive source in close proximity to the tumor to be treated, allowing for a high dose to the tumor with relative sparing of the surrounding normal structures.

Prostate brachytherapy involves insertion of radioactive pellets or seeds (^{125}Iodine or ^{103}Palladium) into the prostate and aims to ensure treatment of the cancer while sparing unaffected local structures. The seeds are deployed via transperineally inserted needles guided by a template guide under transrectal US, CT, or MRI, with MRI providing the best image guidance (Langley and Laing, 2002; Yu and Hricak, 2000).

Image-guided insertion of brachytherapy allows for higher precision and accurate seed placement, so the tumor is adequately treated reducing relapse rates (Song et al., 2010; Patriciu et al., 2007; Potters et al., 2005).

Brachytherapy in cervical cancer may be intracavitatory, interstitial, or both. Intracavitatory brachytherapy involves placing a radioactive source via an applicator through the vagina for treatment of the upper vagina, cervix, and uterus (Viswanathan et al., 2012). In contrast interstitial brachytherapy is the placement of catheters via a transperineal or vaginal approach. This is used when there is involvement of the lower vagina, lateral extension of tumor, or very large tumors (Viswanathan and Thomadsen, 2012). Image-guided brachytherapy with 2D and 3D planning is shown to improve local control and decrease the toxicity to adjacent normal tissue structures compared to external beam radiation (Potter et al., 2011; Rijkmans et al., 2014).

CONCLUSION

Imaging modalities are now being used and applied in many new and exciting methods of image-guided intervention. This chapter has described some of these to illustrate the range of applications. The use of imaging for precise biopsies, tissue characterization, tumor ablation, and drug delivery provides heretofore unavailable information and guidance.

REFERENCES

Ahmed, M., Brace, C.L., Lee Jr., F.T., Goldberg, S.N., 2011. Principles of and advances in percutaneous ablation. Radiology 258 (2), 351—369.

Ahmed, M., Solbiati, L., Brace, C.L., et al., 2014. Image-guided tumor ablation: standardization of terminology and reporting criteria-a 10-year update. J. Vasc. Interv. Radiol. 25 (11), 1691—1705. e4.

Belliveau, J.W., Kennedy Jr., D.N., McKinstry, R.C., et al., 1991. Functional mapping of the human visual cortex by magnetic resonance imaging. Science 254 (5032), 716—719.

Black, W.C., Gareen, I.F., Soneji, S.S., et al., 2014. Cost-effectiveness of CT screening in the National Lung Screening Trial. N. Engl. J. Med. 371 (19), 1793—1802.

Carrafiello, G., Lagana, D., Ianniello, A., et al., 2007. Post-radiofrequency ablation syndrome after percutaneous radiofrequency of abdominal tumours: one centre experience and review of published works. Australas. Radiol. 51 (6), 550—554.

Deckers, R., Moonen, C.T., 2010. Ultrasound triggered, image guided, local drug delivery. J. Control. Release 148 (1), 25—33.

Dodd 3rd, G.D., Soulen, M.C., Kane, R.A., et al., 2000. Minimally invasive treatment of malignant hepatic tumors: at the threshold of a major breakthrough. Radiographics 20 (1), 9—27.

Elhawary, H., Liu, H., Patel, P., et al., 2011. Intraoperative real-time querying of white matter tracts during frameless stereotactic neuronavigation. Neurosurgery 68 (2), 506−516 discussion 16.

Georgiades, C., Rodriguez, R., 2013. Renal tumor ablation. Tech. Vasc. Interv. Radiol. 16 (4), 230−238.

Goldberg, S.N., Hahn, P.F., Tanabe, K.K., et al., 1998. Percutaneous radiofrequency tissue ablation: does perfusion-mediated tissue cooling limit coagulation necrosis? J. Vasc. Interv. Radiol. 9 (1 Pt. 1), 101−111.

Gossot, D., Miaux, Y., Guermazi, A., Celerier, M., Friga, J., 1994. The hook-wire technique for localization of pulmonary nodules during thoracoscopic resection. Chest 105 (5), 1467−1469.

Halpern, E.S., Strup, S.E., 2000. Using gray-scale and color and power Doppler sonography to detect prostatic cancer. AJR Am. J. Roentgenol. 174 (3), 623−627.

Hambrock, T., Somford, D.M., Hoeks, C., et al., 2010. Magnetic resonance imaging guided prostate biopsy in men with repeat negative biopsies and increased prostate specific antigen. J. Urol. 183 (2), 520−527.

Hensing, T., Chawla, A., Batra, R., Salgia, R., 2014. A personalized treatment for lung cancer: molecular pathways, targeted therapies, and genomic characterization. Adv. Exp. Med. Biol. 799, 85−117.

Hinshaw, J.L., Lubner, M.G., Ziemlewicz, T.J., Lee Jr., F.T., Brace, C.L., 2014. Percutaneous tumor ablation tools: microwave, radiofrequency, or cryoablation−what should you use and why? Radiographics 34 (5), 1344−1362.

Huang, M.L., Adrada, B.E., Candelaria, R., Thames, D., Dawson, D., Yang, W.T., 2014. Stereotactic breast biopsy: pitfalls and pearls. Tech. Vasc. Interv. Radiol. 17 (1), 32−39.

Imura, M., Yamazaki, K., Shirato, H., et al., 2005. Insertion and fixation of fiducial markers for setup and tracking of lung tumors in radiotherapy. Int. J. Radiat. Oncol. Biol. Phys. 63 (5), 1442−1447.

Jansen, M.C., van Hillegersberg, R., Schoots, I.G., et al., 2010. Cryoablation induces greater inflammatory and coagulative responses than radiofrequency ablation or laser induced thermotherapy in a rat liver model. Surgery 147 (5), 686−695.

Kitamura, K., Shirato, H., Shimizu, S., et al., 2002. Registration accuracy and possible migration of internal fiducial gold marker implanted in prostate and liver treated with real-time tumor-tracking radiation therapy (RTRT). Radiother. Oncol. 62 (3), 275−281.

Kothary, N., Dieterich, S., Louie, J.D., Koong, A.C., Hofmann, L.V., Sze, D.Y., 2009a. A primer on image-guided radiation therapy for the interventional radiologist. J. Vasc. Interv. Radiol. 20 (7), 859−862.

Kothary, N., Heit, J.J., Louie, J.D., et al., 2009b. Safety and efficacy of percutaneous fiducial marker implantation for image-guided radiation therapy. J. Vasc. Interv. Radiol. 20 (2), 235−239.

Lal, H., Neyaz, Z., Nath, A., Borah, S., 2012. CT-guided percutaneous biopsy of intrathoracic lesions. Korean J. Radiol. 13 (2), 210−226.

Lammers, T., Kiessling, F., Hennink, W.E., Storm, G., 2010. Nanotheranostics and image-guided drug delivery: current concepts and future directions. Mol. Pharm. 7 (6), 1899−1912.

Langley, S.E., Laing, R., 2002. Prostate brachytherapy has come of age: a review of the technique and results. BJU Int. 89 (3), 241−249.

Li, Y., Wang, J., Holloway, C., Plewes, D.B., 2005. Development of an MRI/x-ray/ultrasound compatible marker for pre-operative breast tumour localization. Phys. Med. Biol. 50 (14), 3349−3360.

Lin, S.M., Lin, C.J., Lin, C.C., Hsu, C.W., Chen, Y.C., 2004. Radiofrequency ablation improves prognosis compared with ethanol injection for hepatocellular carcinoma < or =4 cm. Gastroenterology 127 (6), 1714−1723.

Livraghi, T., Giorgio, A., Marin, G., et al., 1995. Hepatocellular carcinoma and cirrhosis in 746 patients: long-term results of percutaneous ethanol injection. Radiology 197 (1), 101−108.

Overduin, C.G., Futterer, J.J., Barentsz, J.O., 2013. MRI-guided biopsy for prostate cancer detection: a systematic review of current clinical results. Curr. Urol. Rep. 14 (3), 209−213.

Paez, J.G., Janne, P.A., Lee, J.C., et al., 2004. EGFR mutations in lung cancer: correlation with clinical response to gefitinib therapy. Science 304 (5676), 1497−1500.

Park, H.L., Hong, J., 2014. Vacuum-assisted breast biopsy for breast cancer. Gland. Surg. 3 (2), 120−127.

Patriciu, A., Petrisor, D., Muntener, M., Mazilu, D., Schar, M., Stoianovici, D., 2007. Automatic brachytherapy seed placement under MRI guidance. IEEE Trans. Biomed. Eng. 54 (8), 1499−1506.

Paulson, E.K., Sheafor, D.H., Enterline, D.S., McAdams, H.P., Yoshizumi, T.T., 2001. CT fluoroscopy−guided interventional procedures: techniques and radiation dose to radiologists. Radiology 220 (1), 161−167.

Penzkofer, T., Tuncali, K., Fedorov, A., et al., 2015. Transperineal in-bore 3-T MR imaging-guided prostate biopsy: a prospective clinical observational study. Radiology 274 (1), 170−180.

Pinthus, J.W., Witkos, M., Fleshner, N.E., et al., 2006. Prostate cancers scored as Gleason 6 on prostate biopsy are frequently Gleason 7 tumors at radical prostatectomy: implication on outcome. J. Urol. 176 (3), 979−984.

Potter, R., Georg, P., Dimopoulos, J.C., et al., 2011. Clinical outcome of protocol based image (MRI) guided adaptive brachytherapy combined with 3D conformal radiotherapy with or without chemotherapy in patients with locally advanced cervical cancer. Radiother. Oncol. 100 (1), 116−123.

Potters, L., Morgenstern, C., Calugaru, E., et al., 2005. 12-year outcomes following permanent prostate brachytherapy in patients with clinically localized prostate cancer. J. Urol. 173 (5), 1562−1566.

Rasmussen, O.S., Seerup, A., 1984. Preoperative radiographically guided wire marking of nonpalpable breast lesions. Acta Radiol. Diagn. 25 (1), 13−16.

Rijkmans, E.C., Nout, R.A., Rutten, I.H., et al., 2014. Improved survival of patients with cervical cancer treated with image-guided brachytherapy compared with conventional brachytherapy. Gynecol. Oncol. 135 (2), 231−238.

Rodriguez, R., Cizman, Z., Hong, K., Koliatsos, A., Georgiades, C., 2011. Prospective analysis of the safety and efficacy of percutaneous cryoablation for pT1NxMx biopsy-proven renal cell carcinoma. Cardiovasc. Interv. Radiol. 34 (3), 573−578.

Rubinsky, B., Lee, C.Y., Bastacky, J., Onik, G., 1990. The process of freezing and the mechanism of damage during hepatic cryosurgery. Cryobiology 27 (1), 85−97.

Schrading, S., Distelmaier, M., Dirrichs, T., Detering, S., Brolund, L., Strobel, K., Kuhl, C.K., 2015. Digital breast tomosynthesis-guided vacuum-assisted breast biopsy: initial experiences and comparison with prone stereotactic vacuum-assisted biopsy. Radiology 274 (3), 654–662. http://dx.doi.org/10.1148/radiol.14141397. [Epub 2014], PMID: 25386875.

Schweikard, A., Shiomi, H., Adler, J., 2004. Respiration tracking in radiosurgery. Med. Phys. 31 (10), 2738–2741.

Shirato, H., Harada, T., Harabayashi, T., et al., 2003. Feasibility of insertion/implantation of 2.0-mm-diameter gold internal fiducial markers for precise setup and real-time tumor tracking in radiotherapy. Int. J. Radiat. Oncol. Biol. Phys. 56 (1), 240–247.

Song, S.E., Cho, N.B., Fischer, G., et al., 2010. Development of a pneumatic robot for MRI-guided transperineal prostate biopsy and brachytherapy: new approaches. In: IEEE International Conference on Robotics and Automation: ICRA: [Proceedings], 2010, pp. 2580–2585.

Swanson, S.J., 2010. Segmentectomy for lung cancer. Semin. Thorac. Cardiovasc. Surg. 22 (3), 244–249.

Taira, A.V., Merrick, G.S., Galbreath, R.W., et al., 2010. Performance of transperineal template-guided mapping biopsy in detecting prostate cancer in the initial and repeat biopsy setting. Prostate Cancer Prostatic Dis. 13 (1), 71–77.

Tempany, C.M., Jayender, J., Kapur, T., Bueno, R., Golby, A., Agar, N., Jolesz, F.A., 2015. Multimodal imaging for improved diagnosis and treatment of cancers. Cancer 121 (6), 817–827. http://dx.doi.org/10.1002/cncr.29012. [Epub 2014], PMID: 25204551. [Free PMC Article].

Tempany, C.M., 2011. Focused ultrasound surgery in oncology: overview and principles. Radiology 259 (1), 39–56.

Thacker, P.G., Callstrom, M.R., Curry, T.B., et al., 2011. Palliation of painful metastatic disease involving bone with imaging-guided treatment: comparison of patients' immediate response to radiofrequency ablation and cryoablation. AJR Am. J. Roentgenol. 197 (2), 510–515.

Tokuda, J., Tuncali, K., Iordachita, I., et al., 2012. In-bore setup and software for 3T MRI-guided transperineal prostate biopsy. Phys. Med. Biol. 57 (18), 5823–5840.

Tsai, I.C., Tsai, W.L., Chen, M.C., et al., 2009. CT-guided core biopsy of lung lesions: a primer. AJR Am. J. Roentgenol. 193 (5), 1228–1235.

van de Ven, W.J., Hulsbergen-van de Kaa, C.A., Hambrock, T., Barentsz, J.O., Huisman, H.J., 2013. Simulated required accuracy of image registration tools for targeting high-grade cancer components with prostate biopsies. Eur. Radiol. 23 (5), 1401–1407.

Viswanathan, A.N., Thomadsen, B., American Brachytherapy Society Cervical Cancer Recommendations Committee, American Brachytherapy Society, 2012. American Brachytherapy Society consensus guidelines for locally advanced carcinoma of the cervix. Part I: general principles. Brachytherapy 11 (1), 33–46.

Viswanathan, A.N., Creutzberg, C.L., Craighead, P., et al., 2012. International brachytherapy practice patterns: a survey of the Gynecologic Cancer Intergroup (GCIG). Int. J. Radiat. Oncol. Biol. Phys. 82 (1), 250–255.

Vogl, T.J., Farshid, P., Naguib, N.N., et al., 2014. Thermal ablation of liver metastases from colorectal cancer: radiofrequency, microwave and laser ablation therapies. Radiol. Med. 119 (7), 451–461.

Willatt, J.M., Francis, I.R., Novelli, P.M., Vellody, R., Pandya, A., Krishnamurthy, V.N., 2012. Interventional therapies for hepatocellular carcinoma. Cancer Imaging 12, 79–88.

Wu, J.S., Zhou, L.F., Tang, W.J., et al., 2007. Clinical evaluation and follow-up outcome of diffusion tensor imaging-based functional neuronavigation: a prospective, controlled study in patients with gliomas involving pyramidal tracts. Neurosurgery 61 (5), 935–948; discussion 48–49.

Yu, K.K., Hricak, H., 2000. Imaging prostate cancer. Radiol. Clin. North Am. 38 (1), 59–85, viii.

Chapter 11

Nanotechnology in Clinical and Translational Research

James R. Baker, Jr., Brent B. Ward and Thommey P. Thomas

University of Michigan, Ann Arbor, MI, United States

Chapter Outline

Key Points

- Nanomaterials are similar in size to proteins and can uniquely interact with cellular systems and structures.
- Nanomaterials have unique capabilities because of their small size.
- Nanomaterials will improve clinical development of therapeutics because they can combine diagnostic imaging and quantitation of drugs in development.

Nanomaterials have the potential to revolutionize therapeutic development and improve clinical research by allowing simultaneous drug delivery, imaging, and quantifying therapeutic effect.

INTRODUCTION AND HISTORICAL PERSPECTIVE

History of Nanotechnology in Medicine

Professor Noro Taniguchi of the Tokyo University of Science coined the term nanotechnology in 1974. Nanotechnology refers to molecular devices smaller than one micron (1 μm) on the "nano" scale. One nanometer (nm) is one-billionth or 10^{29} of a meter. The field was originally inspired by a talk "There's plenty of room at the bottom," by Richard Feynman in 1959 at the American Physical Society. Feynman suggested a number of concepts, including print font size that would permit the *Encyclopedia Britannica* to fit on the head of a pin, a feat since accomplished. The broader concept was that because of their small size, nanomaterials have unique qualities that are not found in the same materials at larger sizes. Principles developed from nanotechnology research are being used to develop everything from the next generation of computer chips to fluid-handling devices that will markedly miniaturize current devices. Importantly, the field of nano-electromechanical systems (NEMS) will be important in implantable devices for a range of biological systems from stress sensors in aneurysms to neural implants.

Soon after the development of mechanical and electrical approaches in nanotechnology, biologists began to explore direct applications using this technology. "Biological Approaches and Novel Applications for Molecular Nanotechnology" held in 1996 was the first scientific conference on the topic. The initial focus was small robots that "create billions of tiny factories small enough to work within a single cell," but this proved to be more a dream than a scientific endeavor. However, it became clear that biological systems are organized at nanoscale dimensions and synthetic nanomaterials correlated in size with biological structures such as proteins, glycolipids, and DNA. Unique interactions between synthetic nanomaterials and more complex biological systems were also observed, most likely due to their size. These ranged from good (delivery of materials across the gut) to potentially dangerous (ability of nanoparticles to enter the brain). It was also discovered that the detrimental activities of some types of environmental materials, such as diesel exhaust, were due to their nanoscale dimensions. Building on these discoveries, scientists are now using nanostructures for biological applications based on their unique capabilities to traverse and interact with similarly sized biological materials. Nanotechnology now remains at the forefront of medicine and biological technologies from a research perspective, and "nanomedicine" was identified as one of the "Roadmap Initiatives" of NIH (NIH Roadmap for Medical Research http://nihroadmap.nih.gov/nanomedicine/).

Rationale for Nanotechnology in Medicine and Research

Nanomedicine is nanotechnology focused on medical intervention at the molecular scale for repair of tissues and treatment of disease. This field combines the expertise of medicine, mathematics, biology, engineering, chemistry, and computer science for the creation of devices for human application. Much of the research in nanomedicine revolves in the oncology realm, with a desire for specifically targeting disease as well as enhancing prevention through earlier diagnosis via superior imaging techniques. According to the National Cancer Institute (2004), "nanotechnology will serve as multifunctional tools that will not only be used with any number of diagnostic and therapeutic agents, but will change the very foundations of cancer diagnosis, treatment, and prevention."

The focus of this chapter is how nanotechnology might impact clinical and translational research. While it is not outlandish to believe that this field might affect all areas of the research endeavor, such a statement will not help the reader understand the potential for nanotechnology. Therefore, we will focus on two examples of how nanotechnology might specifically impact research. One example is a combination of therapeutic and imaging material based on the use of nanoparticles for drug delivery. It will combine imaging with cell-specific therapeutic delivery to allow better evaluation of how medication is reaching its intended target. The other is the impact of nanoparticles in diagnostic assays that allow precise measurements of concentrations that are orders of magnitude lower than current techniques. Both approaches give an insight into the potential breakthroughs in clinical research that might be accomplished with nanotechnology.

NANOTECHNOLOGY IN BASIC RESEARCH APPLICATIONS SUPPORTING CLINICAL TRANSLATION

Our ability to create, analyze, and interact with biological molecules on the nanoscale has merged with our understanding of molecular processes in biology. Nanotechnology has given us excellent tools that allow the delineation of processes in complex biological systems to a degree that was previously impossible. Here are three examples of these tools and how their use will aid translational research.

Knockout of Specific Biomarkers and Genes

One of the best ways to determine the function of a particular biomolecule or gene is to inactivate the molecule and determine the effect of the deletion. This has been accomplished mainly through genetic manipulations, but these manipulations can alter other functions of cells and are very problematic in whole animals because they can alter development. Nanotechnology provides methods to inactivate specific proteins or genes in cells without genetic changes, to better isolate their function. A metal nanoparticle can be targeted to a specific protein or gene within a cell, and then the cell can be exposed to frequencies of light that induce inductive heating of the particle (Liao et al., 2006; Csaki et al., 2007). This inactivates the specific protein/gene target without altering other cell functions. This type of technique is easier and can be accomplished with less effort than genetic manipulation, and it allows for rapidly functional analysis.

Structural Analysis of Proteins

Characterizing the physical structure of proteins has been an arduous but necessary effort in understanding structure—function relationships. It normally involves the production and purification of large amounts of protein followed by crystal formation and X-ray diffraction analysis. This also suffers from the limitation that crystal formation may not accurately represent the native structure of the protein in fluid phase. Single proteins can now be visualized directly by atomic force microscopy. This device works like a record player, with a needle-like probe scanning across the surface of a mica chip where protein has been immobilized (Woolley et al., 2000). When a smaller tip extension, using a carbon nanotube, has been attached, one can obtain resolution of only a few nanometers, which allows the structure of proteins to be visualized. Antibodies have been imaged in this way and demonstrate their familiar "Y" structure (Wong et al., 1998). It is likely this approach will also allow analysis of complexes of proteins that bind together for functional activities, such as activating transcription.

Artificial Substrates to Examine Cellular Functions

One would like to rapidly test therapeutics or efficacy and toxicity in a system that mimics the function of an organ or animal. Unfortunately, tissue culture systems are often not adequate for this task. Nanostructured tissues populated with cells may be one answer to this problem. By laying down exact nanoarrays of extracellular matrix proteins, cells can be made to form organelles in vitro (Smith and Ma, 2004; Goldberg et al., 2007). This saves animals and time and may better represent native structures.

CLINICAL APPLICATIONS OF NANOTECHNOLOGY FOR RESEARCH

The ability to better control and monitor clinical trials is one of the major goals in improving translational research. Nanotechnology provides a number of ways to improve the monitoring of therapeutic trials, both by better analysis of activity and by pharmacokinetics in vivo, and analyzing patient samples ex vivo. Nanoparticles provide opportunities for accomplishing these goals, and we will present work on drug delivery, in vitro assays, and in vivo monitoring that shows the potential of nanomedicine to aid human clinical research.

Nanoparticle-Targeted Drug Delivery

A literature search of the term "targeted therapy" or "targeted drug delivery" reveals an explosion of research in this field, with nearly 1500 articles from 2000 to 2006. While several concepts of "targeting" are presented in these articles, most deal with potential solutions for cancer treatment. Cancer chemotherapeutics are small drug molecules that can easily diffuse through vascular pores. Therapeutics that involve delivery mechanisms, or imaging and sensing applications, are much more complex and have tended to be much larger macromolecules. This raises issues as to what sized molecule can escape the vasculature, find the tumor cells, and enter these cells to deliver a therapeutic or imaging agent. While the exact size of molecules that easily pass through vascular pores and go into tissues where tumors are located is unclear, it is probably the same size as most proteins (20 nm). Studies have documented that molecules 100 nm in diameter do not effectively diffuse across the endothelium (Kong et al., 2001) and even molecules 40 nm in diameter are problematic unless the endothelium is traumatized by radiation or heating (Kong et al., 2000). The vasculature in early neoplastic lesions may be even more restrictive (Bilbao et al., 2000). Thus, creating a multifunctional therapeutic that is still small enough to exit the vasculature to interact with and specifically eliminate abnormal cells may be one of the most important achievements of nanotechnology.

Specific Delivery of Agents to Cells That Abnormally Express or Overexpress a Cell-Surface Molecule

Advances in cancer research have resulted in significant understanding of the cellular and molecular changes leading to malignancy. The concepts of tumor progression were elaborated by Foulds in the 1950s (Foulds, 1957) which was shortly followed by evidence from cytogenetics in the 1960s. Molecular techniques evolved and revealed that tumorogenesis results from a single altered cell (Nowell, 1976). Ultimately a multistep model with acquisition of various cellular abnormalities was proposed (Vogelstein and Kinzler, 1993). While all of the steps involved at present remain unknown, our understanding of the cancer cell and of the role of the surrounding network of tissues is increasing (Hanahan and Weinberg, 2000). Markers of premalignancy and malignancy have been identified in some cases and targeted approaches are being made to utilize these markers in treatment strategies. In addition, other nontransforming markers that are overexpressed for a variety of reasons may be targeted on cancer cells to offer some selective advantages in treatment.

Nanotherapeutics are complex molecules that can identify specific tumor markers on cancer cells in vivo. These are engineered molecules that can also recognize cellular targets of specific pathophysiologic changes in particular cells. The analysis of cancer signature(s) must be coupled to one or more therapeutic agents that can be efficiently delivered to specifically kill the abnormal cells without collateral damage. This is difficult to accomplish since most chemotherapeutic agents will stop cell growth or induce apoptosis if inadvertently delivered or absorbed into normal cells (Rosenberg, 1992; Culver, 1994). Tumors often require higher doses of cytotoxic drugs since they have developed mechanisms of evading anticancer drugs, such as molecular pumps. Therapeutic agents will also require several different mechanisms of action, working in parallel to prevent the development of resistant neoplasms. It is also important if the therapeutic agent could monitor the response to therapy by identifying residual disease immediately after treatment. This is a crucial need, since even a few remaining cells may result in regrowth, or worse, lead to a tumor that is resistant to therapy. Identifying residual disease at the end of therapy (rather than after tumor regrowth) will facilitate eradication of the few remaining tumor cells. Thus, an ideal therapeutic must have the ability to target cancer cells and specifically deliver a therapeutic, image the extent of the tumor, and monitor for a therapeutic response.

Nanotechnology provides technical advances that underlie the potential for successful targeted therapeutics. We will focus our example of a nanoparticle therapeutic on one type of scaffold. This is monodispersed dendritic polymers, or dendrimers, as a backbone for multifunctional nanodevices (Kukowska-Latallo et al., 2005). A second aspect of this type of therapeutic is the development of chemical linkers that allow targeting, therapeutic, and imaging molecules to be linked to the surface of the dendrimer (Choi et al., 2004; Choi and Baker, 2005). These applications also rest on the ability to design and manufacture these molecules in a consistent manner. In this section, we will discuss these advances and how they could aid clinical research.

Dendrimers as Nanoparticle Scaffolds

The achievement of nanotechnology for targeted therapeutics involves the development of particles small enough to escape vascular pores, such as dendrimers. These polymers are synthesized as well-defined spherical structures ranging from 1 to 10 nm in diameter. Molecular weight and the number of terminal groups increase exponentially as a function of generation (the number of layers) of the polymer (Fig. 11.1). Different types of dendrimers can be synthesized based on the core

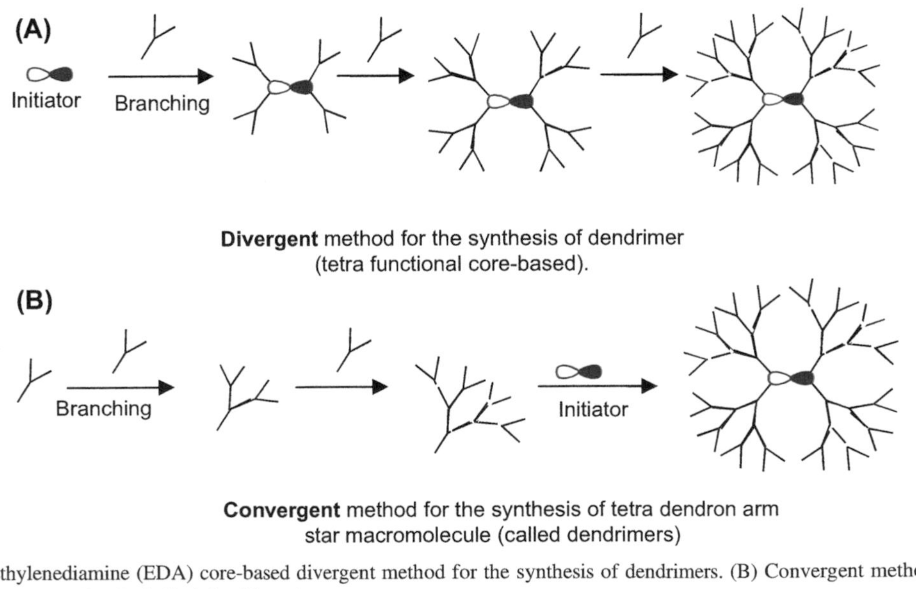

FIGURE 11.1 (A) Ethylenediamine (EDA) core-based divergent method for the synthesis of dendrimers. (B) Convergent method for the synthesis of tetra dendron arm star macromolecule (called dendrimers).

structure that initiates the polymerization process (Tomalia et al., 1990). Poly(amidoamine) spherical dendrimers (PAMAM) with ethylenediamine (EDA) as a tetravalent initiator core are used in our studies (Fig. 11.2). These dendritic macromolecules are available commercially in kilogram quantities. We have now produced 100-g lots of this material under current good manufacturing processes for biotechnology applications. The size range, aspect ratio, and solubility of dendrimers mirror those of proteins (Fig. 11.3).

Dendrimers are characterized by a number of analytical techniques, including electrospray ionization mass spectroscopy (ES-MS), ^{13}C nuclear magnetic resonance spectroscopy (NMR), high-performance liquid chromatography (HPLC), size exclusion chromatography (SEC) with multiangle laser light scattering, capillary electrophoresis (CE), and gel permeation chromatography (GPC), and a variety of gel electrophoresis techniques. These tests assure the uniformity of the polymer population and are essential to monitor quality control of dendrimer manufacture for human therapeutic applications. Importantly, extensive work has been completed with PAMAM dendrimers that have shown no evidence of toxicity when administered intravenously (Bourne et al., 1996; Roberts et al., 1996).

Work from many investigators using dendrimers has documented essentially every desired component required for a viable anticancer nanostructure. PAMAM dendrimers have been used as a scaffold for the attachment of several types of biologic materials. This work has focused on the preparation of dendrimer—antibody conjugates for use in in vitro diagnostic applications (Singh et al., 1994) for the production of dendrimer—chelant—antibody constructs and for the development of boronated dendrimer—antibody conjugates (for neutron capture therapy); each of these latter compounds is envisioned as a cancer therapeutic (Barth et al., 1994; Wiener et al., 1994; Wu et al., 1994). Some of these conjugates have also been employed in the magnetic resonance imaging of tumors (Wiener et al., 1994; Wu et al., 1994). Results from these

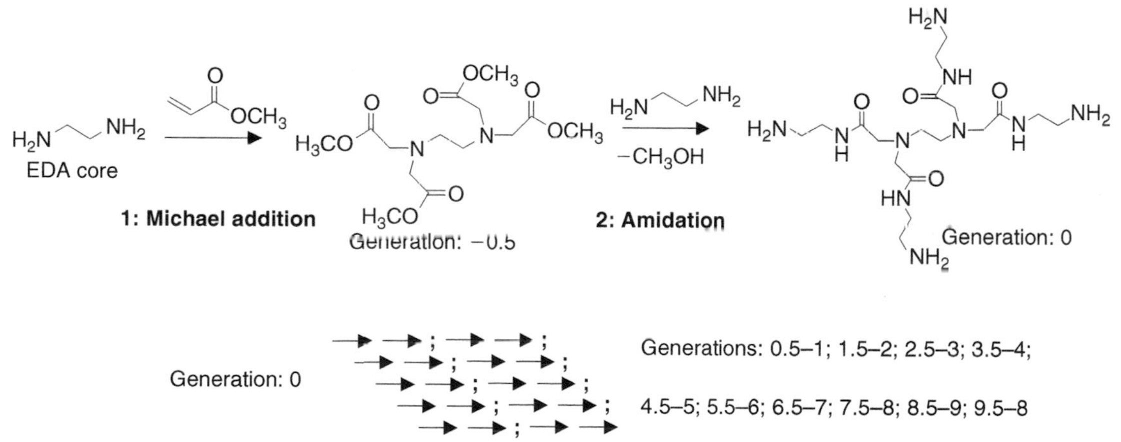

FIGURE 11.2 Poly(amidoamine) (PAMAM) dendrimer synthesis by repeating of Michael addition and amidation.

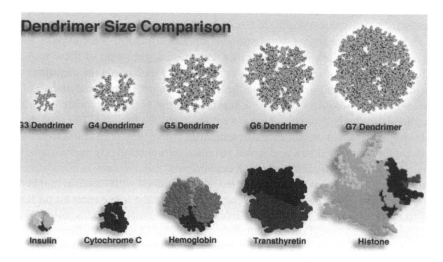

FIGURE 11.3 Dendrimer models compared to similar-sized proteins to give an idea of their size and molecular weight.

studies have documented that, when administered in vivo, antibodies can direct dendrimer-associated therapeutic agents to antigen-bearing tumors. Dendrimers also have been reported to enter tumors and carry either chemotherapeutic agents or genetic therapeutics. In particular, current studies show that cisplatin solubilized by dendrimer polymers has increased efficacy and is less toxic (Duncan et al., 2006). Dendrimers have also been conjugated to fluorochromes and shown to enter specific cells. They can then be detected within the cell in a manner compatible with sensing apparatus for evaluation of physiologic changes within cells (Barker et al., 1997). Finally, dendrimers have been constructed as differentiated block copolymers where the outer portions of the molecule may be digested with either enzyme- or light-induced catalysis (Urdea and Horn, 1993). This would allow the controlled degradation of the polymer to release therapeutics at the disease site and could provide a mechanism for an external trigger to release the therapeutic agents.

Multifunctional Single Dendrimer Nanodevices for In Vitro Testing

Over the past several years, we have made great progress in developing nanomolecular, multifunctional therapeutics based on dendrimers. We have produced a dendrimer that has molecules attached to provide the functions necessary for targeting, active sensing, imaging, and a therapeutic drug. This involves coupling of functional groups such as sensing units, MRI contrast agents, triggering devices, and targeting molecules to the surface of a generation 5 dendritic polymer (MW 25,000 Da, diameter 5 nm). This has proven to be an arduous synthetic endeavor given the multiple conjugation steps that are required. Several challenging protect–deprotect steps are needed to produce such a multifunctional agent. This dendrimer-based agent has multiple functions and can be used as a testing platform to evaluate a nanostructure and its function.

Because of prior reports where folic acid (FA) was coupled to proteins or drugs as a means of targeting cells (through the high-affinity folate receptor), we adopted folate as an initial targeting ligand to attach to the dendrimer. In addition, because we had experience with coupling FITC to proteins, we used FITC as a signaling device to follow the fate of the dendrimer complex within cells. The techniques for this are described extensively in the publication by Quintana et al. (2002). While this was not an optimized system as FITC bleaches and the material could only be analyzed once, it provided a direct means to evaluate structural aspects of the polymer scaffold, cell delivery, and cell internalization.

Research had suggested that when coupling FA to proteins or polymers, a net overall negative charge (predominantly carboxyl moieties) on the molecule surface was important for cellular targeting. However, PAMAM dendrimers are synthesized with a surface of primary amines, so we attempted to simply conjugate folate and FITC to amino surface dendrimers as a means to initiate our studies. When we placed this material in vitro with c cells, which have elevated levels of the high-affinity folate receptor, we saw very poor uptake that required several hours. More disconcerting, there was nonspecific uptake in cell lines that lacked the high-affinity folate receptor.

To improve this performance we could have attempted many manipulations of the polymer surface or structure, all of which would have been synthetically complex and require considerable time to accomplish. Instead, we used computer modeling to evaluate several proposed polymer modifications that would improve delivery (Fig. 11.4). Because differences in polymer surface charge were thought to be of paramount importance in the biological function of this scaffold, we first examined changes in the terminal groups of the polymer. Three different surface modifications were evaluated: carboxyl, hydroxyl, and acetamide substitutions (Fig. 11.4). The carboxyl modification previously reported to achieve efficient delivery to cells did appear to improve accessibility of FA molecules conjugated to the surface of the polymer. However, modeling studies also suggested that at higher concentrations of nanostructure, the carboxyl surface molecules would interact with secondary and tertiary amines in the dendrimer and cause aggregation of these molecules. In contrast, both acetamide and hydroxyl surface modifications resulted in more compact structures in the dendrimer scaffold due to the loss of repulsion from adjacent surface-charged molecules (Fig. 11.4). In addition, it appeared that all of the FAs on the surface of the molecule were accessible for binding with cellular receptors in the acetamide molecule, while approximately two-thirds of the targeting moieties in the hydroxyl molecule were available for binding. The modeling of the amine surface molecule suggested that none of the FA molecules were externalized where they could bind to cellular receptors (Quintana et al., 2002). This provided confidence that an FA-coupled polymer with neutralizing surface modifications might be able to efficiently target cells through the high-affinity folate receptor.

Polymers with the three modified surfaces were synthesized and substituted with identical numbers of FA and fluorescein isothiocyanate (FITC) moieties. The activity of these molecules in targeting the folate receptor on KB cells in vitro was then evaluated and compared to that predicted in the molecular modeling. The results were dramatic (Quintana et al., 2002). The cellular uptake of the acetamide and hydroxyl surface polymers was very rapid and very efficient. It occurred within minutes, peaking at approximately 20 or 30 min and appeared to resume after another 30 min, which corresponds to folate receptor recycling. Uptake was faster for the acetamide surface rather than the hydroxyl surface, confirming the modeling prediction that there were more surface folate groups available for binding. Also in accordance with the modeling, the carboxyl surface

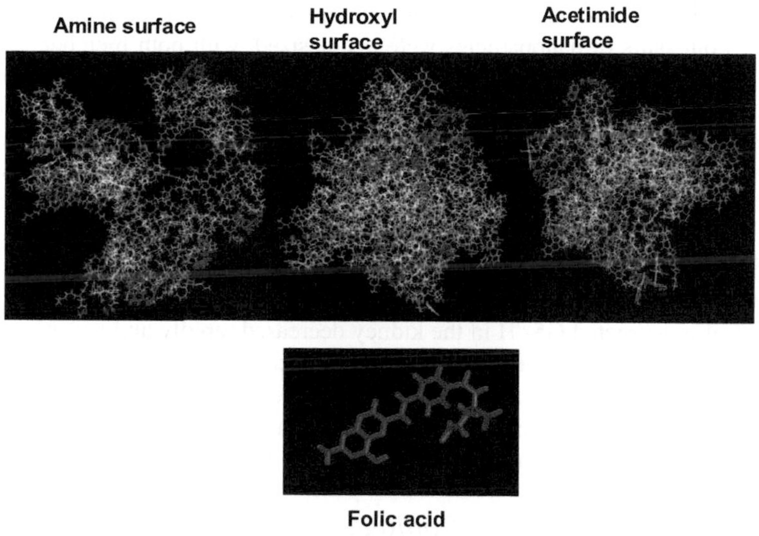

FIGURE 11.4 Modeling of folate-conjugated dendrimer nanostructure. Folate is exposed on the surface of the amine-surfaced polymer (*left panel*), but nonspecific interactions between surface amines and cell membranes caused nontargeted uptake. Both acetamide-surfaced (*right panel*) and hydroxyl-surfaced dendrimers (*middle panel*) lacked nonspecific interactions; however, the acetamide was predicted to have the folate in a surface position that was likely to interact with receptors on cells. This was proven to be correct in studies with intact cells expressing the receptor.

molecules initially had rapid uptake, but as the concentration of the nanostructure was increased the uptake appeared to stop. This suggested that the molecules were aggregating and no longer available for binding to the cell surface receptor. The acetamide surface polymer continued to accumulate within the cells over time and appeared by confocal microscopy to internalize after approximately 4–6 h (Quintana et al., 2002).

The next step was to develop a more complex device where fluorescein, folate, and drug were conjugated on a single polymer. This was accomplished by first acetylating ~80% of the primary amines on the surface of the polymer followed by the sequential attachment of folate, fluorescein, and one of two different drugs—either paclitaxel (Taxol) or methotrexate. The drugs were attached through two linker mechanisms: an ester linkage and an amide linkage. The ester linkage should hydrolyze once the device internalized within the KB cell, while the amide linkage should retain drug and serve as a control. These conjugates were produced and tested in the KB cells. Polymers with the drug were internalized as efficiently as polymers that had only fluorescein on their surface. In addition, this targeted material was tested for the ability to induce cytotoxicity in the KB cells, by both an assay of mitochondrial activity (in an MTT assay) and by clonogenic assay. The ester-linked drug delivered with the folate was active as a cytotoxic agent, whereas the amide-linked drug was not. This suggested that the ester-linked drug was active, because the drug was released approximately 4–6 h after internalization within the cell. The concentration of dendrimer-delivered methotrexate that induced cytotoxicity was compared to free methotrexate. In this culture system the folate-targeted material was 5- to 10-fold more active than free drug. We believe that this might be due to polyvalent receptor interactions and that the data suggest that targeting this drug to resistant cancer cells might be more effective than using free drug.

Additionally, 50% of bound nanostructure remained after acidic wash at 0°C, while free folate is easily removed by this treatment. This suggested that the folate–dendrimer conjugates have increased binding stability on the cell surface, possibly through multiple receptor–folate interactions. This observation has been subsequently proven in extensive studies, where the avidity of these molecules for cells is greater than 100,000 times that of free FA (Hong et al., 2007). This suggests that low-affinity but highly specific ligands can be used to develop high-affinity targeting scaffolds using dendrimers and offers promise for devices based on single chain Fab or other small molecules.

Multifunctional Single Dendrimer Nanodevices

Further studies have now demonstrated tumor therapeutic efficacy of drug–dendrimer conjugates in vivo. We have examined the biodistribution and elimination of tritiated G5-^{3}H-FA to test its ability to target the (folate receptor) FR-positive human KB tumor xenografts established in immunodeficient nude mice. The mice were maintained on a folate-deficient diet for the duration of the experiment to minimize the circulating levels of FA (Mathias et al., 1998). The free FA level achieved in the serum of the mice prior to the experiment approximated human serum levels (Belz and Nau, 1998; Nelson et al., 2004). Four pairs of nanoparticles were synthesized, with both members of each pair containing the same tracer. One member of each pair contained FA—one pair with and one pair without the antineoplastic drug MTX—while the other conjugate lacked FA and served as a nontargeted control with or without the drug.

Mice were evaluated at various time points (5 min to 7 days) following intravenous administration of the conjugates. Two groups of mice received either control nontargeted tritiated G5-^{3}H dendrimer or targeted tritiated G5-^{3}H-FA conjugate (Kukowska-Latallo et al., 2005). To summarize, the cumulative clearance of the targeted G5-^{3}H-FA over the first 4 days was lower than that of G5-^{3}H, which may reflect retention of G5-^{3}H-FA within tissues expressing folate receptors. The FA-conjugated dendrimer accumulated in the tumor and reached a maximum in 24—28 h, whereas the control dendrimer failed to do so. While the kidney is the major clearance organ for these dendrimers, it is also known to express high levels of the FR on its tubules. The level of nontargeted G5-^{3}H in the kidney decreased rapidly and was maintained at a moderate level over the next several days. In contrast, the level of G5-^{3}H-FA increased slightly over the first 24 h, most likely due to FR present on the kidney tubules. This was followed by a decrease over the next several days as the compound was cleared by the kidney (Kukowska-Latallo et al., 2005). This showed that the nontargeted dendrimer was eliminated rapidly from the blood through the kidney, while the targeted material accumulated in the tumor.

Biodistribution of Fluorescent Dendrimer Conjugate

To further confirm and localize the dendrimer nanoparticles within tumor tissue, dendrimers conjugated with the red fluorescent dye 6-TAMRA (6T) were employed. Confocal microscopy images of tumor samples were obtained at 15 h following intravenous injection of the targeted G5-6T-FA and the nontargeted G5-6T conjugates (Fig. 11.5). The tumor tissue demonstrated a significant number of fluorescent cells with FA-targeted dye-conjugated dendrimer (right panel, Fig. 11.5) compared to those with nontargeted dendrimer (center panel, Fig. 11.5). Flow cytometry analysis of a single-cell suspension isolated from the same tumors showed higher mean channel fluorescence for tumor cells from mice receiving G5-6T-FA.

Targeted Drug Delivery to Tumor Cells Through the Folate Receptor

The efficacy of different doses of conjugates was tested on SCID C.B-17 mice bearing subcutaneous human KB xenografts and was compared to equivalent and higher doses of free MTX. Six groups of immunodeficient SCID mice with five mice in each group were injected subcutaneously on one flank with 5 3 10^6 KB cells in 200 μL PBS suspension. The therapeutic dose was compared to three different (cumulative) doses of free MTX equivalent to 33.3, 21.7, and 5.0 mg/kg accumulated in 10—15 injections, based on the survival of the mouse. Saline and the conjugate without MTX (G5-Fl-FA) were used as controls. The body weights of the mice were monitored throughout the experiment as an indication of adverse effects of the drug, and mice demonstrated acute and chronic toxicity in the highest and in the second highest cumulative doses of free MTX equal to 33.3 and 21.7 mg/kg, respectively. The survival of mice from groups receiving G5-Fl-FA-MTX or G5-FA-MTX conjugate indicates that tumor growth based on the endpoint volume of 4 cm^3 can be delayed by at least 30 days. We have achieved a complete cure in one mouse treated with G5-FA-MTX conjugate at day 39 of the trial. The tumor in this

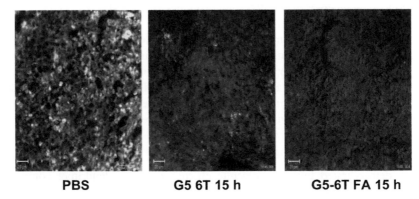

PBS   **G5 6T 15 h**   **G5-6T FA 15 h**

FIGURE 11.5 Fluorescence images of mouse KB cell tumors, harvested 15 h after injection of a dendrimer conjugated to a red fluorescent dye. No localization is observed in tumors where the dendrimer has no folic acid (*center panel*) while there is greatly enhanced uptake with folic acid–targeted material (*right panel*).

mouse was not palpable for the next 20 days up to the 60th day of the trial. At the termination of the trial, there were three (out of eight) survivors receiving G5-FA-MTX and two (out of eight) survivors receiving G5-FI-Fa-MTX. There were no mice surviving in the group receiving free MTX or in any other control group. Fig. 11.6 shows the mice dosed with MTX and those with MTX conjugated to the targeted dendrimer. The lack of toxicity of the targeted drug is apparent, as compared to the free drug.

Future of Nanoparticle Targeting Technology

Using this and other types of nanoparticles, one can envision a multitude of therapeutics whereby both the targeting ligand and the chemotherapeutic agent utilized could be specifically modified to meet the needs of an individual tumor. In support of this concept, additional dendrimers targeting epidermal growth factor receptor (EGFR), prostate specific membrane antigen (PSMA), and RGD peptides have been fabricated and shown to target specific cells, as well as demonstrating the feasibility of conjugates with other drugs such as Taxol. In this scenario, patient tumors could be screened for their cell surface receptor expression to make decisions regarding the targeting scheme that would have greatest therapeutic potential. Targeting other cellular receptors, such as EGFR, as is done with monoclonal antibodies, may have direct therapeutic effects in colorectal and head and neck cancer (HNSCC) as well as with a number of other tumors. EGFR is overexpressed in ~92% of HNSCC of specimens studied (Grandis and Tweardy, 1993). EGFR overexpression in HNSCC is the result of both decreased receptor downregulation and increased mRNA synthesis. EGFR inhibition with antibody binding (cetuximab) can suppress the growth of overexpressing tumors as an adjunct to radiation therapy (Bonner et al., 2006). This could also be achieved with ligands for EGFR coupled to dendrimer, providing a nonbiological alternative to antibodies to block EGFR function.

In summary, these results demonstrate the ability of dendrimers to target tumor cells and specifically to monitor uptake. The use of therapeutics coupled to dendrimers or other nanoparticles could provide better outcomes in clinical trials and could also allow better estimates of pharmacokinetics and tumor uptake of experimental drugs. This will aid the performance of translational research.

Nanoparticles for Analytical Techniques

Rapid and low-cost quantification of specific proteins and genes has become important in the early diagnosis and treatment of many human diseases and has become crucial to the monitoring of clinical research. Studies over the past two decades have led to the development of many assays for the quantification of nucleic acids and proteins. These methods include extremely sensitive assays for molecular identification such as the polymerase chain reaction (PCR) for nucleic acids and enzyme-linked immunosorbent assays (ELISA) for proteins. These techniques have become routine in many laboratories due to the commercial availability of a vast array of nucleic acid probes and antibodies offered as convenient "kits" which include all the needed reagents. Despite their high sensitivity and efficiency, these methods still suffer several limitations such as the need for sample purification and the requirement of experienced laboratory personnel and setup. These methods are complex, laborious, impracticable, and cost-ineffective for a quick "on-the-site" diagnosis in places such as an outpatient clinic or in a clinical research organization setting. Moreover, most of these methods cannot be used for real-time in vivo diagnostic purposes.

FIGURE 11.6 Mice on left dosed with free methotrexate lose weight, lose hair, and appear sick. Mice on right dosed with dendrimer-transported folate-targeted methotrexate exhibit no adverse effects from the chemotherapy. Drug-induced necrosis of the tumor on the flank of the mouse in the upper right corner.

Nanotechnology has recently impacted several fields of diagnostic assays. Nanoparticles offer a unique structural platform for diagnostic purposes that can overcome some of the limitations of conventional diagnosis. Owing to their structural stability, large surface area, and achievable surface functionalities, multiple biological molecules such as peptides and oligonucleotides can be linked on to a single nanoparticle. Unlike larger-sized rigid microparticles and other macroscopic substrates used in conventional assays, the nano-sized particle would allow the retention of the native molecular interaction of the conjugated ligands, owing to the size, shape, and flexibility of many nanoparticle platforms. It is possible to control the physical parameters and the surface functionalities of nanoparticles allowing specific biologic recognition of other biomolecules. Moreover, the development of an engineered nanodevice that has a long half-life, but is slowly excreted from the body, would allow the in vivo identification of overexpressed disease-specific proteins and nucleic acids.

Recent studies have utilized metal-based nanomaterials (1–100 nm in size) as platforms for the in vitro quantification of proteins and genes, showing very high sensitivity and specificity versus conventional diagnostic techniques (Hirsch et al., 2006). The metal nanoparticles consist of a dielectric core such as silica, surrounded by a thin metallic shell such as reduced gold or silver. They show specific optical absorption/scattering properties and can be surface functionalized by biomolecules such as DNA and protein through sulfhydryl moieties. The pioneering work of Mirkin et al. has led to the development of several novel metal nanoparticle-based in vitro biodiagnostic assay systems for screening nucleic acids, proteins, and other biomolecules (Rosi and Mirkin, 2005; Thaxton et al., 2006). Our studies have shown the applicability of dendrimer-based nanoparticles (3–7 nm in size) for the in vivo quantification of proteins such as a cell surface receptor, or a cellular event such as apoptosis, using an optical fiber–based detection system (Thomas et al., 2007, 2008). Here we briefly review the recent development in the rapidly evolving nanomaterial-based diagnostic systems.

Nanoparticle-Based In Vitro Quantification of DNA

Mirkin et al. have modified gold nanoparticle (13 nm) surface with multiple molecules on an oligonucleotide, which in the presence of the corresponding complementary oligonucleotide forms higher-order nanoparticle assembly units consequent to hybridization and cross-linking. The polymeric networks thus formed elicit a red shift in the optical resonance that allows colorimetric quantification of the added complementary nucleotide, demonstrating for the first time a nanoparticle-based detection method for nucleic acids (Mirkin et al., 1996; Elghanian et al., 1997). Consequent to the cooperativity of the cluster formation, the melting temperature range of the DNA in the cluster is significantly narrower than that of the native DNA (Jin et al., 2003). This property allows increased selectivity versus conventional fluorescence-based assay and the easy and rapid identification of base-pair mismatches in DNA. Nonetheless, this assay method has the limitation of low detection sensitivity (nanomolar range) versus conventional fluorescence-based detection (picomolar range).

Recently, Mirkin et al. have developed a more sensitive chip-based "scanometric" assay system using gold nanoparticles that also allows "multiplexing" for the simultaneous detection of multiple DNA species. In this assay, DNA molecules complementary to one-half of a target DNA are bound to a glass slide, which is then allowed to anneal with the target DNA in the presence of gold nanoparticles containing DNA complementary to the second half of the target DNA (Fig. 11.7) (Taton et al., 2000). The signal of the sandwich formed is amplified by reducing the gold surface using silver, the nanometer-sized gold particles grow into micrometer size, and is visualized using a flatbed scanner. Because of the

FIGURE 11.7 Scanometric chip-based sandwich assay for DNA employing gold nanoparticles.

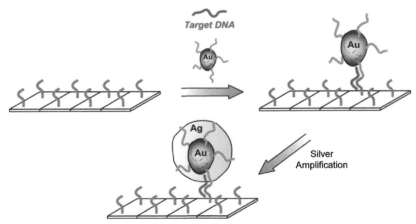

stringent annealing temperature of the cluster formed, the background can be greatly reduced, enabling the detection of femtomole levels of DNA, without the need for prior PCR amplification (Storhoff et al., 2004b). By using "spectro-scopically coded" Raman dye–labeled DNAs attached to gold nanoparticles, this method also allows multiplexing for the detection of multiple DNA species in a single assay format (Cao et al., 2002).

A further improvement in the gold nanoparticle-based assay is the development of "bio-bar-code" assay in which gold nanoparticles are loaded with thousands of "bar-code" DNAs and a few complementary chains of the target DNAs (Nam et al., 2004) (Fig. 11.8). The target DNA is initially allowed to form a bridge between the gold nanoparticle and another magnetic particle onto which a complementary nucleotide to a different region of the target DNA is attached. This complex is separated magnetically and washed to remove the bar-code DNA for detection by scanometric method. By using different bar codes for different targets, this method allows the simultaneous detection of different DNAs at an incredibly low level of 5×10^{219} M (500 zM; 10 DNA strands in solution).

The development of an "anisotropic" synthesis of gold nanoparticles in which a complementary oligonucleotide is preferentially attached to only one side of the nanoparticle allows the controlled synthesis of more defined polymeric assembly units (Xu et al., 2006). A nanoparticle-based assay by the measurement of electrical current has also been reported (Park et al., 2002). This assay, with a sensitivity of 500 fM and very high selectivity, is based on detection of electrical signals resulting from the flow of current across electrodes in the presence of target DNA linked onto a metal nanoparticle.

In fluorescence-based detection, silica nanoparticles encapsulated with a large payload of fluorophores offer a sensitive method for chip-based DNA assay that allows detection of even a single base-pair DNA mismatch (Zhao et al., 2003; Wang et al., 2006), or a single bacterium (Tan et al., 2004). Despite their high sensitivity, the fluorophore-based methods have the limitation of fluorescence quenching, and the simultaneous combinatorial assays for multiple DNAs are difficult because of fluorescence overlap of the multiple fluorophores encapsulated into the nanoparticle. Light scattering–based measurement of selenium nanoparticles has been shown to be suitable for the quantification of DNAs using cDNA

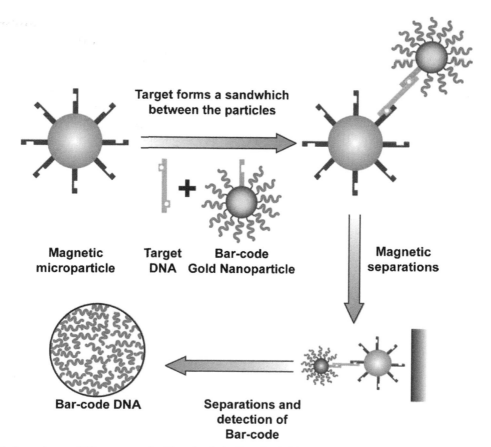

FIGURE 11.8 "Bio-bar-code assay." The target nucleotide molecule is allowed to hybridize at its two halves with complementary nucleotides that had been linked to either magnetic microparticles or gold nanoparticles. The gold nanoparticle also carries several hundred copies of a bar-code DNA. The hybridized complex is magnetically separated and the bar code corresponding to the desired target DNA is identified by PCR.

microarrays (Bao et al., 2002; Stimpson et al., 1995; Storhoff et al., 2004a). A microarray-based technology using gold nanoparticles has recently been reported for the screening of duplex and triplex DNA-binding drugs such as doxorubicin and daunorubicin (Lytton-Jean et al., 2007). In addition to the gold nanoparticles, silver-based nanoparticles are also highly suitable for DNA-conjugation and detection (Lee et al., 2007).

Nanoparticle-Based In Vitro Quantification of Proteins

Antibody conjugated onto a nanoparticle can serve as a suitable platform for the detection and quantification of proteins. West and coworkers have utilized antibody-conjugated gold nanoparticles that aggregate upon binding to the corresponding antigen. The aggregate formed is quantified due to its specific light absorption profile, with a sensitivity similar to conventional ELISA method (Hirsch et al., 2005, 2006). This method allows detection of a protein in intact biological fluids without the need for sample purification and preparation. In another method, complementary nucleotides are utilized to allow hybridization and aggregation to gold nanoparticles, allowing the sensitive detection of multiple proteins simultaneously, using Mirkin's "bar-code" format previously described for DNA detection (Nam et al., 2003; Georgakopoulos et al., 2005; Goluch et al., 2006; Hill and Mirkin, 2006; Stoeva et al., 2006). This is done by using two different antibodies that recognize two different epitopes of the target protein. The two antibodies linked to the gold nanoparticle, or to a magnetic particle, are allowed to form a sandwich in the presence of the target protein, followed by bar-code DNA-based detection using protocols similar to that depicted in Fig. 11.8. The sensitivity of these gold nanoparticle-based antibody—antigen aggregation methods can be enhanced by the "Raman dye color coding," in a multiplexing setup (Cao et al., 2003; Grubisha et al., 2003).

Nanoparticle-Based In Vivo Quantification of Proteins

Although the metal nanoparticle-based assays described earlier are highly suitable in in vitro diagnostics, a more biocompatible and aqueous soluble system is preferred for in vivo diagnostics. Studies from our laboratory have shown the applicability of polyamidoamine dendrimer nanoparticles which have the potential for in vivo diagnostic applications. This is due to the biomacromolecule-mimicking and biocompatible properties of the dendrimer (Thomas et al., 2007). We have demonstrated the applicability of fluorophore-tagged generation 5-polyamidoamine dendrimers (5 nm size) for the quantification of cancer-specific proteins such as FA receptor and human epidermal growth factor receptor 2 (HER 2), overexpressed in certain tumors (Thomas et al., 2004). This is achieved by the specific dendrimer-based targeting of ligands such as FA and Herceptin, intravenously injected into mice followed by quantification by measuring the fluorescence in the tumor using a two-photon optical fiber—based detection system. Our recent studies have utilized a "double clad optical fiber" that enabled the detection of nanomalor quantities of FA-targeted dendrimer nanoparticles in FA receptor—expressing cells (Fig. 11.9) (Thomas et al., 2008). Our recent studies have also shown the synthesis of biocompatible dendrimer—gold hybrid nanoparticles that may have potential future application for in vivo imaging (Shi et al., 2007).

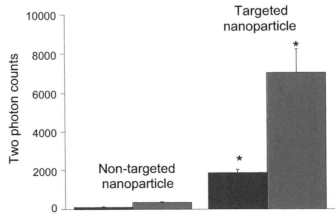

FIGURE 11.9 In vivo targeting of folic acid (FA)-conjugated ("targeted nanoparticle") and control ("non-targeted nanoparticle") dendrimers into FA receptor—expressing mice tumors. Tumor-bearing mice were intravenously injected with the nanoparticles. Fifteen hours later the tumors were isolated and the two-photon optical counts were taken by inserting a single mode (*red*) or double clad (*green*) optical fiber probe. The data shown are corrected for the background fluorescence of tumors from saline-injected mice. *Significantly different from the respective nontargeted tumor.

The studies described previously demonstrate the potential applicability of engineered nanoparticles for in vitro and in vivo bio-diagnostics. These nanotechnology-based methods are emerging as a significant step forward for the easy and rapid analysis of the genetic basis and treatment of diseases.

CONCLUSION AND FUTURE DIRECTIONS IN NANOMEDICINE

There is clear promise for the role of nanomedicine in clinical and translational research. The most near-term applications will involve molecular targeting. Current applications are focused on cancer; however, as our understanding of the complex processes involved in diseased cells increases, new opportunities for targeting will emerge. When normal cellular processes are involved, selective advantage can be offered by an upregulation of specific normal processes with associated molecules. Abnormal receptors and processes also offer an opportunity for targeting in that they do not have counterparts within normal tissue. Delineation of altered proteins from new sources, such as alternative splicing, will add to the targets for many diseases. Given the potential, most therapies may become targeted.

The real opportunities may lie in techniques that are just emerging. The "molecular surgical" technique using inductive heating of targeted nanoparticles holds particular promise. This approach could address illnesses where abnormal proteins are expressed in high quantities (Alzheimer's disease) or where inflammatory molecules are overexpressed (rheumatoid arthritis). In addition, specifically inactivating an oncogene could treat cancer without cytotoxic drugs. The limited number of strategies reviewed in this chapter only provide a glimpse of where this field could evolve.

REFERENCES

Bao, P., Frutos, A.G., Greef, C., Lahiri, J., Muller, U., Peterson, T.C., Warden, L., Xie, X., 2002. High-sensitivity detection of DNA hybridization on microarrays using resonance light scattering. Anal. Chem. 74, 1792−1797.

Barker, S.L., Shortreed, M.R., Kopelman, R., 1997. Utilization of lipophilic ionic additives in liquid polymer film optodes for selective anion activity measurements. Anal. Chem. 69, 990−995.

Barth, R.F., Adams, D.M., Soloway, A.H., Alam, F., Darby, M.V., 1994. Boronated starburst dendrimer-monoclonal antibody immunoconjugates: evaluation as a potential delivery system for neutron capture therapy. Bioconjug. Chem. 5, 58−66.

Belz, S., Nau, H., 1998. Determination of folate patterns in mouse plasma, erythrocytes and embryos by HPLC coupled with a microbiological assay. Anal. Biochem. 265, 157−166.

Bilbao, R., Bustos, M., Alzuguren, P., Pajares, M.J., Drozdzik, M., Qian, C., Prieto, J., 2000. A blood−tumor barrier limits gene transfer to experimental liver cancer: the effect of vasoactive compounds. Gene Ther. 7, 1824.

Bonner, J.A., Harari, P.M., Giralt, J., Azarnia, N., Shin, D.M., Cohen, R.B., et al., 2006. Radiotherapy plus cetuximab for squamous-cell carcinoma of the head and neck. N. Engl. J. Med. 354, 567−578.

Bourne, M.W., Margerun, L., Hylton, N., Campion, B., Lai, J.J., Derugin, N., Higgins, C.B., 1996. Evaluation of the effects of intravascular MR contrast media (gadolinium dendrimer) on 3D time of flight magnetic resonance angiography of the body. J. Magn. Reson. Imaging 6, 305−310.

Cao, Y.C., Jin, R., Mirkin, C.A., 2002. Nanoparticles with Raman spectroscopic fingerprints for DNA and RNA detection. Science 297, 1536−1540.

Cao, Y.C., Jin, R., Nam, J.-M., Thaxton, C.S., Mirkin, C.A., 2003. Raman dye-labeled nanoparticle probes for proteins. J. Am. Chem. Soc. 125, 14676−14677.

Choi, Y.S., Baker, J.R., 2005. Targeting cancer cells with DNA-assembled dendrimers: a mix and match strategy for cancer. Cell Cycle 4, 669−671.

Choi, Y.S., Mecke, A., Orr, B.G., Banaszak Holl, M.M., Baker, J.R., 2004. DNA-directed synthesis of generation 7 and 5 PAMAM dendrimer nanoclusters. Nano Lett. 4, 391−397.

Csaki, A., Garwe, F., Steinbruck, A., Maubach, G., Festag, G., Weise, A., Riemann, I., Konig, K., Fritzsche, W., 2007. A parallel approach for subwavelength molecular surgery using gene-specific positioned metal nanoparticles as laser light antennas. Nano Lett. 7, 247−253.

Culver, K.W., 1994. Clinical applications of gene therapy for cancer. Clin. Chem. 40, 510.

Duncan, R., Ringsdorf, H., Satchi-Fainaro, R., 2006. Polymer therapeutics − polymers as drugs, drug and protein conjugates and gene delivery systems: past, present and future opportunities. J. Drug Target 6, 337−341.

Elghanian, R., Storhoff, J.J., Mucic, R.C., Letsinger, R.L., Mirkin, C.A., 1997. Selective colorimetric detection of polynucleotides based on the distance-dependent optical properties of gold nanoparticles (see comment). Science 277, 1078−1081.

Foulds, L., 1957. Tumor progression. Cancer Res. 17, 355−356.

Georgakopoulos, D.G., Chang, L., Nam, J.-M., Thaxton, C.S., Mufson, E.J., Klein, W.L., Mirkin, C.A., 2005. Nanoparticle-based detection in cerebral spinal fluid of a soluble pathogenic biomarker for Alzheimer's disease (see comment). Proc. Natl. Acad. Sci. U. S. A. 102, 2273−2276.

Goldberg, M., Langer, R., Jia, X., 2007. Nanostructured materials for applications in drug delivery and tissue engineering. J. Biomater. Sci. Polym. Ed. 18, 241−268.

Goluch, E.D., Nam, J.-M., Georgakopoulos, D.G., Chiesl, T.N., Shaikh, K.A., Ryu, K.S., Barron, A.E., Mirkin, C.A., Liu, C., 2006. A bio-barcode assay for on-chip attomolar-sensitivity protein detection. Lab Chip 6, 1293−1299.

Grandis, J.R., Tweardy, D.J., 1993. Elevated level of transforming growth factor alpha and epidermal growth factor receptor messenger RNA are early markers of carcinogenesis in head and neck cancer. Cancer Res. 53, 3579–3584.

Grubisha, D.S., Lipert, R.J., Park, H.-Y., Driskell, J., Porter, M.D., 2003. Femtomolar detection of prostate-specific antigen: an immunoassay based on surface-enhanced Raman scattering and immunogold labels. Anal. Chem. 75, 5936–5943.

Hanahan, D., Weinberg, R.A., 2000. The hallmarks of cancer. Cell 100, 57–70.

Hill, H.D., Mirkin, C.A., 2006. The bio-barcode assay for the detection of protein and nucleic acid targets using DTT-induced ligand exchange. Nat. Protoc. 1, 324–336.

Hirsch, L.R., Gobin, A.M., Lowery, A.R., Tam, F., Drezek, R.A., Halas, N.J., West, J.L., 2006. Metal nanoshells. Ann. Biomed. Eng. 34, 15–22.

Hirsch, L.R., Halas, N.J., West, J.L., 2005. Whole-blood immunoassay facilitated by gold nanoshell-conjugate antibodies. Methods Mol. Biol. 303, 101–111.

Hong, S., Leroueil, P., Majoros, I., Orr, B., Baker Jr., J.R., 2007. The binding avidity of a nanoparticle-based multivalent targeted drug delivery platform. Chem. Biol. 14, 107–115.

Jin, R., Wu, G., Li, Z., Mirkin, C.A., Schatz, G.C., 2003. What controls the melting properties of DNA-linked gold nanoparticle assemblies. J. Am. Chem. Soc. 125, 1643–1654.

Kong, G., Braun, R.D., Dewhirst, M.W., 2000. Hyperthermia enables tumor-specific nanoparticle delivery: effect of particle size. Cancer Res. 60, 4440–4445.

Kong, G., Braun, R.D., Dewhirst, M.W., 2001. Characterization of the effect of hyperthermia on nanoparticle extravasation from tumor vasculature. Cancer Res. 61, 3027–3032.

Kukowska-Latallo, J.F., Candido, K.A., Cao, Z., Nigavekar, S.S., Majoros, I.J., Thomas, T.P., Balogh, L.P., Khan, M.K., Baker Jr., J.R., 2005. Nanoparticle targeting of anticancer drug improves therapeutic response in animal model of human epithelial cancer. Cancer Res. 65, 5317–5324.

Lee, J.-S., Lytton-Jean, A.K.R., Hurst, S.J., Mirkin, C.A., 2007. Silver nanoparticle-oligonucleotide conjugates based on DNA with triple cyclic disulfide moieties. Nano Lett. 7, 2112–2115.

Liao, H., Nehl, C.L., Hafner, J.H., 2006. Biomedical applications of plasmon resonant metal nanoparticles. Nanomedicine 1, 201–208.

Lytton-Jean, A.K.R., Han, M.S., Mirkin, C.A., 2007. Microarray detection of duplex and triplex DNA binders with DNA-modified gold nanoparticles. Anal. Chem. 79, 6037–6041.

Mathias, C.J., Wang, S., Waters, D.J., Turek, J.J., Low, P.S., Green, M.A., 1998. Indium-111-DTPA-folate as a potential folate-receptor-targeted radiopharmaceutical. J. Nucl. Med. 39, 1579–1585.

Mirkin, C.A., Letsinger, R.L., Mucic, R.C., Storhoff, J.J., 1996. A DNA-based method for rationally assembling nanoparticles into macroscopic materials. Nature 382, 607–609.

Nam, J.-M., Stoeva, S.I., Mirkin, C.A., 2004. Bio-bar-code-based DNA detection with PCR-like sensitivity. J. Am. Chem. Soc. 126, 5932–5933.

Nam, J.-M., Thaxton, C.S., Mirkin, C.A., 2003. Nanoparticle-based bio-bar codes for the ultrasensitive detection of proteins (see comment). Science 301, 1884–1886.

National Cancer Institute, 2004. Cancer Nanotechnology: Going Small for Big Advances. US Department of Health and Human Services, NIH/NCI.

Nelson, B.C., Pfeiffer, C.M., Margolis, S.A., Nelson, C.P., 2004. Solid-phase extraction-electrospray ionization mass spectrometry for the quantification of folate in human plasma or serum. Anal. Biochem. 325, 41–51.

Nowell, P.C., 1976. The clonal evolution of tumor cell populations. Science 194, 23–28.

Park, J.M., Greten, F.R., Li, Z.-W., Karin, M., 2002. Macrophage apoptosis by anthrax lethal factor through p38 MAP kinase inhibition. Science 297, 2048–2051.

Quintana, A., Raczka, E., Piehler, L., Lee, I., Myc, A., Majoros, I., Patri, A.K., Thomas, T., Mule, J., Baker Jr., J.R., 2002. Design and function of a dendrimer-based therapeutic nanodevice targeted to tumor cells through the folate receptor. Pharm. Res. 19, 1310–1316.

Roberts, J.C., Bhalgat, M.K., Zera, R.T., 1996. Preliminary biological evaluation of polyamidoamine (PAMAM) Starburst dendrimers. J. Biomed. Mater. Res. 30, 53–65.

Rosenberg, S.A., 1992. Karnofsky Memorial Lecture. The immunotherapy and gene therapy of cancer. J. Clin. Oncol. 10, 180.

Rosi, N.L., Mirkin, C.A., 2005. Nanostructures in biodiagnostics. Chem. Rev. 105, 1547–1562.

Shi, X., Wang, S., Meshinchi, S., Van Antwerp, M., Bi, X., Lee, I., Baker Jr., J.R., 2007. Dendrimer-entrapped gold nanoparticles as a platform for cancer-cell targeting and imaging. Small 3, 1245–1252.

Singh, P., Moll III, F., Lin, S.H., Ferzli, C., Yu, K.S., Koski, R.K., Saul, R.G., Cronin, P., 1994. Starburst dendrimers: enhanced performance and flexibility for immunoassays. Clin. Chem. 40, 1845–1849.

Smith, L.A., Ma, P.X., 2004. Nano-fibrous scaffolds for tissue engineering. Colloids Surf. B Biointerfaces 39, 125–131.

Stimpson, D.I., Hoijer, J.V., Hsieh, W.T., Jou, C., Gordon, J., Theriault, T., Gamble, R., Baldeschwieler, J.D., 1995. Real-time detection of DNA hybridization and melting on oligonucleotide arrays by using optical wave guides. Proc. Natl Acad. Sci. USA 92, 6379–6383.

Stoeva, S.I., Lee, J.-S., Smith, J.E., Rosen, S.T., Mirkin, C.A., 2006. Multiplexed detection of protein cancer markers with biobarcoded nanoparticle probes (see comment). J. Am. Chem. Soc. 128, 8378–8379.

Storhoff, J.J., Lucas, A.D., Garimella, V., Bao, Y.P., Muller, U.R., 2004a. Homogeneous detection of unamplified genomic DNA sequences based on colorimetric scatter of gold nanoparticle probes. Nat. Biotechnol. 22, 883–887.

Storhoff, J.J., Marla, S.S., Bao, P., Hagenow, S., Mehta, H., Lucas, A., Garimella, V., Patno, T., Buckingham, W., Cork, W., Muller, U.R., 2004b. Gold nanoparticle-based detection of genomic DNA targets on microarrays using a novel optical detection system. Biosens. Bioelectron. 19, 875–883.

Tan, W., Wang, K., He, X., Zhao, X.J., Drake, T., Wang, L., Bagwe, R.P., 2004. Bionanotechnology based on silica nanoparticles. Med. Res. Rev. 24, 621−638.

Taton, T.A., Mirkin, C.A., Letsinger, R.L., 2000. Scanometric DNA array detection with nanoparticle probes. Science 289, 1757−1760.

Thaxton, C.S., Georganopoulou, D.G., Mirkin, C.A., 2006. Gold nanoparticle probes for the detection of nucleic acid targets. Clin. Chim. Acta 363, 120−126.

Thomas, T.P., Myaing, M., Ye, J.Y., Candido, K.A., Kotlyar, A., Beals, J.L., Cao, P., Keszler, B., Patri, A.K., Norris, T.B., Baker, J.R., 2004. Detection and analysis of tumor fluorescence using a two-photon optical fiber probe. Biophys. J. 86, 3959−3965.

Thomas, T.P., Shukla, R., Majoros, I.J., Myc, A., Baker Jr., J.R., 2007. Polyamidoamine dendrimer-based multifunctional nanoparticles. In: Mirkin, C.A., Niemeyer, C.M. (Eds.), Polyamidoamine Dendrimer-based Multifunctional Nanoparticles. Wiley-VCH Press, New York, pp. 305−319.

Thomas, T.P., Ye, J.Y., Chang, Y.-C., Kotlyar, A., Cao, Z., Majoros, I.J., Norris, T.B., Baker Jr., J.R., 2008. Investigation of tumor cell targeting of a dendrimer nanoparticle using a double-clad optical fiber probe. J. Biomed. Opt. 13 (1).

Tomalia, D.A., Naylor, A.M., Goddard III, W.A., 1990. Starburst dendrimers: molecular-level control of size, shape, surface chemistry, topology, and flexibility from atoms to macroscopic matter. Angew. Chem. Int. Ed. Engl. 29, 138−175.

Urdea, M.S., Horn, T., 1993. Dendrimer development. Science 256, 534.

Vogelstein, B., Kinzler, K.W., 1993. The multistep nature of cancer. Trends Genet. 9, 138−141.

Wang, L., O'donoghue, M.B., Tan, W., 2006. Nanoparticles for multiplex diagnostics and imaging. Nanomedicine 1, 413−426.

Wiener, E.C., Magnin, R.L., Gansow, O.A., Brechbeil, M.W., Brothers II, H.M., Tomalia, D.A., Lauterbur, P.C., 1994. Dendrimer-based metal ion chelates and their applications to magnetic resonance imaging. Magn. Reson. Med. 31, 1.

Wong, S.S., Joselevich, E., Woolley, A.T., Cheung, C.L., Lieber, C.M., 1998. Covalently functionalized nanotubes as nanometre-sized probes in chemistry and biology. Nature 394, 52−55.

Woolley, A.T., Cheung, C.L., Hafner, J.H., Lieber, C.M., 2000. Structural biology with carbon nanotube AFM probes. Chem. Biol. 7, R193−R204.

Wu, C., Brechbiel, M.W., Kozak, R.W., Gansow, O.A., 1994. Metal-chelate-dendrimer-antibody constructs for use in radioimmunotherapy and imaging. Bioorg. Med. Chem. Lett. 4, 449−454.

Xu, X., Rosi, N.L., Wang, Y., Huo, F., Mirkin, C.A., 2006. Asymmetric functionalization of gold nanoparticles with oligonucleotides. J. Am. Chem. Soc. 128, 9286−9287.

Zhao, X., Tapec-Dytioco, R., Tan, W., 2003. Ultrasensitive DNA detection using highly fluorescent bioconjugated nanoparticles. J. Am. Chem. Soc. 125, 11474−11475.

Chapter 12

The Use of Questionnaires and Surveys

Marcia A. Testa[1] and Donald C. Simonson[2]

[1]Harvard T. H. Chan School of Public Health, Boston, MA, United States; [2]Brigham and Women's Hospital, Harvard Medical School, Boston, MA, United States

Chapter Outline

INTRODUCTION

Certain types of information required for testing clinical research hypotheses can only be obtained by observing or asking patients how they feel and function. For many areas of clinical investigation, the interpretation of research results derived solely from clinical and laboratory data can be enhanced by information that reflects the patient's perspective of a disease condition or state of health. The application of surveys and questionnaires in clinical research is most commonly used when conducting health outcomes research (HOR) focusing on the measurement of outcomes such as health status, health-related quality of life (HRQOL), disease-specific symptoms, treatment-related side effects, and patient satisfaction. Furthermore, patient-centered outcomes increasingly are being included as the primary and secondary endpoints in comparative effectiveness research as funding for this area has grown substantially since the formation of the Patient-Centered Outcomes Research Institute (PCORI) in 2010 (Selby et al., 2012). Patient-centered outcomes research (PCOR) addresses the questions and concerns most relevant to patients and also involves caregivers, clinicians, and other health-care stakeholders throughout the process. PCORI funding is projected to be $650 million per year through 2019. Given this directed research funding, a large number of PCORI-funded research will require methods involving the use of questionnaires and surveys.

Studies evaluating the patient-centered risks and benefits of new therapies should always include a description of risk and benefit from the patient's perspective. Outcomes such as pain, discomfort, psychological distress, mobility, vitality, and general well-being cannot be measured by laboratory or clinical evaluation. Subjective descriptions of these outcomes as reported directly by the patient or trained observer provide a unique dimension of health assessment for use in clinical and comparative effectiveness research. Regardless of whether such data are obtained through structured self-assessment questionnaires, open-ended surveys conducted by a trained interviewer, or daily diaries recorded on a mobile device, methods for measuring and analyzing patient functioning, feelings, behavior, knowledge, and attitudes should be carried out with the same high degree of rigor and precision required for all other areas of scientific investigation. The purpose of this chapter is to describe the scientific framework, as well as the more commonly used methodological issues and techniques, for collecting data from clinical research subjects using questionnaires and surveys.

There are numerous comprehensive textbooks devoted to the qualitative, quantitative, and psychometric methods required for developing questionnaires and surveys to measure health outcomes such as symptoms, health status, health behaviors and attitudes, HRQOL, and patient satisfaction. These textbooks focus on statistical issues in measurement (Lord et al., 2008), psychometric theory (Nunnally and Bernstein, 1994), clinimetrics (Feinstein, 1987), health survey and questionnaire design and development (Fowler, 1995; Streiner and Norman, 2008; DeVellis, 2012; Johnson, 2015), and measuring disease and health (Bowling, 2001, 2005; McDowell, 2006). The topics covered in this chapter will not attempt to summarize or provide a bibliography of this textbook material, but rather will describe the high priority issues and methods most relevant to the conduct of *clinical trials and comparative effectiveness research*. Most importantly, we emphasize how critical it is that investigators understand that while constructing and asking questions in an ad hoc, spontaneous and unstructured manner is often appropriate for clinical practice, such methods typically do not perform well in research practice. A primary goal of this chapter is to promote an awareness of the scope and breadth of the scientific disciplines underlying questionnaire and survey development that should guide the design and analysis phases of a clinical research study. The following sections will present measurement, design, and analysis issues with which all clinical investigators should be familiar prior to employing patient surveys and questionnaires in their research studies.

THE PRACTICE OF QUESTIONNAIRE AND SURVEY MEASUREMENT

The Emergence of Questionnaire and Survey Measurement in Clinical Investigation

An historical perspective of the evolution of the use of questionnaires and surveys in clinical research is very useful for understanding the current issues, methods, and techniques used for employing patient questionnaires in research studies. At the *individual patient level* the most comprehensive and informative method for obtaining subjective aspects of a patient's functioning, health state, or disease condition is simply to ask the patient to describe what they are able to do and how they feel both physically and emotionally, and then to transcribe the patient's verbal responses into a clinical case report for further evaluation (Beckman and Frankel, 1984; Smith and Hoppe, 1991; Coulehan and Block, 2006). Analyzing qualitative descriptions of a patient's condition has been a common practice in clinical diagnosis since the origins of the physician-patient medical interview. While such descriptions are unstructured and might include numerous pages of text, they certainly provide a greater wealth of information as compared to a structured survey question with a numeric response, such as "How do you feel on a scale of 1 to 10, with 10 being the best you have ever felt, and 1 being the worst you have ever felt?" Structured questions such as this often appear awkward and foreign to respondents, especially those who are not familiar with translating feelings into numbers. However, what is gained in comprehensiveness and scope when using literal transcription restricts the ability to pool information across individuals and analyze testable research hypotheses. Given that qualitative descriptions provide a richer source of data, let us first briefly review why *quantifiable* and *numeric* assessments are preferred for research purposes, and how questionnaires try to capture the essence of qualitative descriptions in their numeric translations.

Clinical medicine has established that the individual case presentation, using a relatively free-format approach to presenting evidence, is the most desirable and cognitively sophisticated method for diagnostic assessment and evaluation. Feinstein commented "History taking, the most clinically sophisticated procedure of medicine, is an extraordinary investigative technique: in few forms of scientific research does the observed object talk" (Feinstein, 1967). Even though this prose-based, qualitative assessment technique clearly provides the most comprehensive and sophisticated level of information, using this technique it is difficult to assemble and summarize numerous individualized textual reports to quantitatively test a research hypothesis. This is the challenge that clinical investigators face when designing clinical research studies that require assessing outcomes that are not directly observable, often referred to as "latent constructs." Paradoxically, during the 1970s, as clinical inquiry moved away from individual case presentations to more experimental clinical trial methods requiring comparisons of groups of patients, these more cognitively sophisticated, qualitative assessments became viewed as the more inferior *soft* measures. On the other hand, patient characteristics that were capable of representation by a numeric value, such as a lab assay, a reading from a medical device, or a simple dichotomous diagnostic observation, were elevated to the more superior status of *hard* measures. The preference for using these hard measures and avoiding soft measures is rooted in early medical training when students are exposed to a medical curriculum that values *objective data* involving numbers, while *subjective data* obtained from the patient's own story lack value because they lack straightforward, numeric quantification. Furthermore, as parametric statistical tests became the gold standard for interpreting the results of research, quantitative variables that were interval scaled and normally distributed became sought after as the coin of the realm for primary endpoint variable selection, while the qualitative, clinical reports faced near extinction in the emerging field of clinical trials.

In striking contrast, during the last years of the 19th century, methods for experimental study in psychology embraced the measurement of subjective assessments and focused on developing methods to measure constructs that were not directly observable. Constructs such as intelligence, stress, coping, and personality were at the center of intellectual debate and scientific investigation and launched an entire discipline focused on subjective assessment, measurement, and statistics. At the beginning of the 20th century, new research techniques in these disciplines were thriving. During this same period, physicians were still disputing the association between maternal mortality and hand washing, and the experimental method was virtually absent from the rudimentary clinical investigation techniques employed at the time. It was not until the 1970s that the principles of randomized clinical trials and statistical hypothesis testing became commonly accepted requirements for conducting clinical research. These quantitative methods requirements were largely responsible for launching interval-scaled and dichotomous variables to dominant endpoint status in clinical trials research.

As the therapeutic clinical trials movement expanded rapidly during the 1970s and 1980s, the data obtained from medical interviews and clinical case reports were almost universally excluded as an important and useful source of information. However, decades earlier, in the 1940s, an evolutionary branch of clinical investigation rooted in the earlier psychometrics movement made an appearance in oncology that would ensure the survival of measures of patient functioning into the 21st century. At that time, a health status performance index was developed by Karnofsky and Burchenal (1949) to measure the more subjective side of the outcomes of cancer treatment. The basic concepts underlying this early translation of health states into numerical scales are still quite relevant today.

As shown in Table 12.1, this index classified health functioning into four broad categories:

(1) normal functioning with no limitations; (2) unable to work; (3) unable to care for one's self; and (4) terminal states.

These four broader categories were further divided into 11 hierarchical levels with a corresponding numeric indicator ranging from 0 to 100, heuristically representing the percent of human functioning between death (0%) and full health (100%). Each of these 11 levels was associated with a qualitative description of the corresponding health state. To score an individual according to this index, the investigator decided in which of the four categories the individual belonged, selected the most appropriate description within that category, and recorded the corresponding percentage associated with that description. The index was both simple and elegant, adapting the basic psychometric principle of moving from a qualitative-based description of health to a quantitative measure of health status. Still commonly used today, the Karnofsky index was reported as an outcome variable in thousands of studies since its inception (Schag et al.,

TABLE 12.1 The Karnofsky Performance Index

General Category	Percentage (%) of Full Functioning	Specific Criteria
Able to carry on normal activity	100	Normal general status; no complaint; no evidence of disease
No special care required	90	Able to carry on normal activity; minor signs or symptoms of disease
Normal activity with effort	80	Some signs or symptoms of disease
Unable to work	70	Able to care for self, unable to carry on normal activity or do work
Able to live at home and care for most personal needs	60	Requires occasional assistance from others, but able to care for most needs
Various amount of assistance needed	50	Requires considerable assistance from others; frequent medical care
Unable to care for self	40	Disabled, requires special care and assistance
Requires institutional or hospital care or equivalent	30	Severely disabled, hospitalization indicated, death not imminent
Disease may be rapidly progressing	20	Very sick, hospitalization necessary, active supportive treatment necessary
Terminal states	10	Moribund
	0	Dead

1984) and has stood the test of time, appearing in more than 2500 published clinical oncology research studies between 2010 and 2016.

Karnofsky's relatively simple performance index laid the early groundwork for measuring health functioning by categorizing human functioning qualitatively, and then transforming this qualitative assessment into a quantitative measure ranging from 0 (death) to 100 (full health). This concept of quantification is similar to the "quality-adjusted life year" (QALY), which would make its formal appearance some 30 years later, and which is commonly used today in cost-effectiveness and cost-utility studies. The Karnofsky index preserved the allegiance of the research community to quantitative measures, and by using a percentage-based scale also strived to approximate an interval-scaled, linear measure. Although the assumption of linearity might not be followed exactly, many psychometric analyses have shown that well-developed scales and indices are summarized quite accurately by means and standard deviations, and also can be analyzed using parametric inferential statistical techniques. Such indices and scales make it possible not only to quantify health functioning, but also to test differences between treatment groups with regard to the mean level of health functioning for groups of individuals. Although the scale and distributional assumptions must still be verified individually, by the beginning of the 1970s the stage was set to accept the quantification of unobservable health constructs as part of clinical research.

Beyond Karnofsky's original contribution, relatively little progress was made in the development of indices and scales for measuring health in clinical medicine until the 1970s when extraordinary medical interventions could be used to prolong life for individuals who were *functionally* dead. Finding the answers to questions regarding the *quality* versus the *quantity* of life required new measures. In the years that followed, symptom indices, health status measures, and quality-of-life scales gained prominence in clinical and health-care research. The need to quantify and evaluate health outcomes using these new tools also laid the foundation for the field of *health outcomes research* which was further expanded by the creation of the US Agency for Health Care Policy and Research in 1989 and renamed the Agency for Healthcare Research and Quality (AHRQ) in 1999.

Nearly 35 years after the Karnofsky index was developed, Feinstein coined the term "clinimetrics" to describe a new clinical research discipline "concerned with indexes, rating scales and other expressions that are used to describe or measure symptoms, physical signs, and other distinctly clinical phenomena in clinical medicine" (Feinstein, 1987). Feinstein's interest in clinimetrics was built upon his prior work in clinical judgment, decision-making, and biostatistics that created the foundation for critiquing the role of measurement and quantification in clinical research. What was unique about Feinstein's conceptualization of the field is that he remained true to those applications useful for clinical medicine and investigation. While other authors have contributed excellent and voluminous works concerned with measuring disease, health, and quality of life (QOL) more broadly for the social sciences, Feinstein approached the topic from the narrower perspective of developing indices that would enhance clinical investigation, an approach that was consistent with Karnofsky's original efforts. It is worthwhile for the clinical investigator to keep Feinstein's clinimetric focus in mind when sorting through research methods for developing and using questionnaires and surveys in clinical research. There are many common areas of measurement methodology shared by the social and clinical sciences; however, for the clinical investigator it is advantageous to restrict the scope, keeping in mind the specific clinical research question of interest.

Items, Scales, Questionnaires, and Instruments

As with any diagnostic-, laboratory-, or physiologic-based health monitoring instrument or device, the questionnaire, survey, or interview schedule is the instrument that allows the researcher to translate the patient's condition into quantifiable, reliable, valid, and accurate measures. In fact, in survey research, a collection of questions used to measure a construct that cannot be measured directly is typically referred to as the *survey instrument*. For structured surveys, a formula that substantively combines the individual's responses to the questions or "items" is used to estimate the construct's true value. The scoring is typically programmed electronically, and the resulting computation is referred to as the *scale score*. Therefore, for any individual or respondent who answers the items in the instrument, it is possible to obtain an estimated value of the unobservable construct. Since the process often requires entering the questionnaire data into a computer to obtain the scale scores, structured surveys are more aligned with medical research's affinity for hard data as compared to more qualitative surveys. However, as will be discussed hereafter, the scale scores from structured surveys are still intrinsically subjective, representing an intangible attribute, and often are still regarded as soft measures.

The fact that one is measuring an unobservable construct based on patient responses formed by a cognitive human process (answering the questions) often makes physical scientists wary of the resulting values, thinking of them as somehow inferior to hard measures such as laboratory values. This distinction is intriguing. Certainly no one has ever *seen* a blood glucose value (Davidoff, 1996), but clinicians consider this a hard measure primarily because the information comes from a sample of the patient's blood, rather than a sample of the patient's cognitive processes. Yet, similar to deriving a scale score from a questionnaire, a glucose reading is derived from an assay formula. Admittedly, there is a difference, but it has little to do with the sample content or rigor of the computations required to arrive at an estimate. Ensuring the precision, reliability, validity, standardization, and interpretability of the scale score estimate with respect to the value of the true construct is a major part of instrument development and the science of measuring disease, health, and other unobservable aspects of the human condition. The primary distinction is that patient-reported, questionnaire-derived values are often obtained from processes that lack adequate guidelines for development, standardization, quality control, and general acceptance. In contrast, scientists universally agree on how to measure blood glucose. Laboratory engineers have standardized the process, manufacturers have received regulatory and government approval for marketing blood glucose meters, and finally, through widespread use, normative data are available to interpret the readings. Although regulatory agencies such as the US Food and Drug Administration (FDA) have recently created mechanisms for approving clinical outcome assessments (COAs) that "measure a patient's symptoms, overall mental state, or the effects of a disease or condition on how the patient functions" as part of qualification for drug development tools (United States Food and Drug Administration, 2014), in general, all questionnaires and indices used in health survey research must be individually reviewed by the user to guarantee that the appropriate quality standards regarding the appropriateness, accuracy, and validity have been met.

It is the broad nature of the human condition and the nonexhaustive scope of what researchers want to measure that seem to have impeded the progress in standardization and scientific acceptance of the measures obtained from health surveys and questionnaires. However, in 2004, the National Institutes of Health (NIH) roadmap to reengineer clinical research recognizing the lack of standardized instruments available for assessing patient-reported outcomes initiated a multicenter cooperative group to develop the Patient Reported Outcomes Measurement Information System (PROMIS) (http://www.healthmeasures.net/explore-measurement-systems/promis). The PROMIS group's primary mission was to "build and validate common, accessible item banks to measure key symptoms and health concepts applicable to a range of chronic conditions, enabling efficient and interpretable clinical trial and clinical practice applications of patient-reported outcomes (PROs)" (Cella et al., 2007). The recognition by NIH to include patient assessment methods as part of the tools required to conduct clinical research was as important as the actual questionnaires and tools produced by the PROMIS collaborative group. To use the PROMIS instruments for clinical research studies, one may obtain a user account at www.assessmentcenter.net. In many respects the arrival of PCORI to the medical research scene has further enhanced the value of the type of research and accelerated the need to develop standardization and quality control of the instruments and tools used in this type of research.

The Role of Psychometrics

One of the primary scientific disciplines of survey and questionnaire measurement methodology is *psychometrics*. Psychometrics is the field of study concerned with the theories and methods of psychological measurement. Just as the design and manufacturing of medical devices require an engineering level of training and knowledge, psychometrics can be considered the engineering-level discipline associated with questionnaire and survey development. As mentioned earlier, there are several excellent textbooks on psychometrics to which the reader can refer at the introductory (Fowler, 1995; DeVellis, 2012), intermediate (Streiner and Norman, 2008), and advanced levels (Lord et al., 2008; Nunnally and Bernstein, 1994). While the psychometrician must be familiar with the engineering techniques for questionnaire development, such as conducting focus groups, developing item pools, performing factor analyses, and estimating the reliability and validity of scales, the clinical investigator does not need to be expert in these areas to effectively utilize questionnaires in research practice. However, the clinical investigator does need to know how to select the most appropriate instrument and how to interpret the resulting scores. If it is ultimately determined that a new questionnaire, instrument, scale, or index must be developed, the clinical investigator should recognize that the development process will take a substantial amount of time and considerable expertise. An advanced psychometric understanding is not necessary to successfully employ survey instruments and questionnaires as part of a clinical research protocol; however, an appreciation and understanding of the precision, quality, and standards for evaluation and interpretation are required for effectively drawing conclusions from the data such instruments provide.

Questionnaires Used to Assess Patient-Reported Outcomes

The most commonly asked question from clinical researchers undertaking patient-reported assessment is "Which questionnaire should I use for my study?" The answer, of course, depends upon the purpose of the research. The most frequent application of questionnaires and surveys in clinical studies is to obtain information on patient-reported health outcomes which is the patient's own description or report of an outcome relating to a health state, medical condition, treatment impact, or satisfaction rating. The data derived from such reports reflect the individual's intrinsic interpretation of experiences that are not observable by others. The methodological issues relating to questionnaires used for this purpose can be found in the discipline of "health outcomes research", and more recently within the narrower field of "patient-centered outcomes research". The field of HOR was popularized in the 1980s and spans a broad spectrum of topics ranging from clinical trials evaluating the effectiveness of therapeutic interventions to assessing the impact of reimbursement policies on the outcomes of care. HOR specifically refers to the types of studies that evaluate the *end results* of health-care practices. This field evolved within the more general discipline of health services research. The disciplines that provide the quantitative and design methods by which HOR is conducted can be found primarily in the fields of epidemiology, biostatistics, psychology, and sociology.

Within this broader field of HOR, the focus on patient-reported outcomes methods revolves around collecting data measuring the effects that people experience and care about, such as change in the ability to function or QOL, especially for chronic conditions where cure is not possible. To be precise, the term "quality of life" is a broad multidimensional concept that usually includes subjective evaluations of both positive and negative aspects of life. The term "health-related quality of life" includes those aspects of overall QOL that can be clearly shown to affect health—either physical or mental. Within clinical and health research, it is accepted that both terms represent HRQOL. Traditionally, clinical researchers have relied primarily on mortality, morbidity, and biomedical measures to determine whether a health intervention is necessary and whether it is successful. However, since the early 1980s the number of publications that incorporate the concepts of "QOL," "health status," and "patient satisfaction" has increased exponentially.

While questionnaires of health status had been used in health-care services research in the late 1970s and early 1980s, the use of questionnaires to measure health outcomes in clinical research gained prominence in the mid-1980s, when the first large-scale, multicenter clinical trials incorporated their use in evaluating the impact of drug therapies on QOL and health status. One of the first such trials where QOL was a primary endpoint differentiated three antihypertensive medications—captopril, propranolol, and methyldopa—on the basis of their impact on patient-reported QOL while showing similar effects on blood pressure control (Croog et al., 1986). The more favorable QOL effects of captopril as compared to the other two treatment arms were considered important for explaining why patients often failed to comply with their antihypertensive therapy, thereby potentially increasing their risk of cardiovascular morbidity and mortality. While the differences were explained on the basis of the different mechanisms of action, a follow-up study found QOL differences between two antihypertensive drugs within the same class, supporting the hypothesis that the pharmacodynamics of the drug might also play an important role in determining effects on patient QOL (Testa et al., 1993).

Beyond the obvious importance of knowing that patients experience improved QOL and reduced distress with medications having fewer side effects, it was the description of the potential pathway between feelings, behavior, morbidity, and mortality that captured the attention of clinical investigators. If a certain pattern of patient behavior required for adherence to therapies known to decrease morbidity and mortality could be explored using questionnaires, then the field was no longer simply measuring how patients feel as an end in and of itself, but rather measuring actual predictors of morbidity and mortality. Furthermore, these predictors were potentially modifiable.

The questionnaires used in both of the hypertension clinical trials mentioned previously were extensive, consisting of hundreds of questions to cover many domains of QOL and to provide the sensitivity required for estimating within-individual longitudinal changes that were salient to patients. These more sensitive QOL measures were able to assess the clinically unobservable consequences of drug therapy. The realization that questionnaires could play an important role in therapeutic drug research fostered the growth of their application in clinical trials research for future decades (Testa and Simonson, 1996). From the 1990s onward, clinical trial researchers, investigating treatments for many therapeutic areas, wanted measures that could differentiate on the basis of patient-reported QOL. Unfortunately, the motivation and desire to include these measures were not met with the same level of enthusiasm to include the exhaustive and comprehensive questionnaires and scales required to detect within-individual changes. The mismatch between the instrument required for detecting therapeutic differences among varying active control arms and the perception that *burdensome* questionnaires could be effectively replaced by much shorter instruments led to thousands of clinical studies concluding null QOL treatment effects when, in actuality, the instruments were too crude, not sensitive enough, and often not appropriate or specific enough to detect treatment-related, within-individual changes over the course of treatment.

Evidence of the requirements for sensitivity and specificity was demonstrated in a clinical study by Testa and Simonson (1998) that showed the relatively strong relationship between improvements in QOL and improvements in HbA_{1c} during treatment in persons with type 2 diabetes. Prior studies of this relationship failed to demonstrate this association, often concluding that the lack of adherence to diabetes therapies was due in part to the fact that patients did not feel any short-term benefits of glucose lowering. These previous findings seemed counterintuitive to what clinicians were seeing in their clinical practices. Upon further analysis (Testa et al., 1998) comparing the instrumentation used by Testa and Simonson to those used in the previous studies, it was revealed that the earlier instruments were not developed with adequate sensitivity to detect within-individual differences that were important to patients. Instruments that can detect changes on the *operative scale* that are sensitive enough to detect many states of health between the "lowest level at which it would be possible for an individual to live and work as they currently do" and "full health" are required. This is in contrast to *absolute scales* that measure fewer states of health between death and full health. Evaluative instruments with operative scales are usually required for detecting QOL differences due to therapeutic intervention, and such evaluative instruments are neither short nor simple.

The demand for short, simple measures of health status and QOL for clinical research is evident from the large number of publications citing the use of the Short Form 36, commonly referred to as the SF-36 (Ware and Sherbourne, 1992). The SF-36 was developed from the full-scale assessment used in the Medical Outcomes Study (MOS) (Tarlov et al., 1989) and contains 36 questions. The various items in the SF-36 had been used by the Rand Corporation prior to its development and were produced after experience with earlier 18- and 20-item MOS short forms. For example, the mental health component of the SF-36 consists of only five questions that were selected from a much longer instrument, the Mental Health Inventory (MHI-38), which consists of 38 questions and was originally used in the RAND Health Insurance Experiment (Brook et al., 1983). The full length 38-item MHI consists of six lower level constructs—depression, anxiety, loss of behavioral and emotional control, positive affect, emotional ties, and life satisfaction. The five questions from the Mental Health Inventory (MHI-5) used in the mental health component of the SF-36 utilizes only three lower level constructs, namely, positive affect (2 items), anxiety (1 item), and depression (2 items). It eliminates three constructs altogether in addition to removing 33 items. In this case, reducing the number of items also reduced the number of lower level constructs covered under the higher level mental health construct.

The SF-36 was originally designed primarily for use in general population surveys. It assesses the following eight health concepts:

1. limitations in physical activities because of health problems
2. limitations in social activities because of physical or emotional problem
3. limitations in usual role activities because of physical health problems
4. bodily pain
5. general mental health (psychological distress and well-being)
6. limitations in usual role activities because of emotional problems
7. vitality (energy and fatigue)
8. general health perceptions

The survey was constructed for self-administration by persons 14 years of age and older or for administration by a trained interviewer in person, or by telephone. The desire to have even briefer methods of assessment led to the SF-12, consisting of only 12 of the original 36 items, and capable of measuring the physical and mental health components of the SF-36, but not the individual eight domains. The extensive use of the SF-36 and SF-12 is documented by over 22,000 publications between 1992 and 2015 citing its use. While there is a desire to reduce patient burden using fewer questions, reducing the number of items comes at a cost that varies depending upon the construct's distance from a more observable biophysiologic mechanism.

Fig. 12.1 depicts the theoretical relationship between the measurement complexity (represented here simply by the number of items required to obtain a valid measure) of specific questionnaires and the distance of the construct from the biophysiologic mechanism. A scale measuring a broad construct such as "emotional well-being" is both distant from an observable biophysiologic mechanism and includes many different facets such as positive affect, emotional control, anxiety, depression, and life satisfaction, requiring many distinct items. A Symptom Distress Index is closer to the biophysiologic mechanism, but needs to include many items to cover all possible symptoms. A disease-specific index specifically for gastroesophageal reflux disease would require fewer items. Pain can typically be measured by one item and is usually associated with mechanisms such as swollen joints, bone fracture or headache, near the biophysiologic mechanism.

Both the SF-36 and the SF-12 are multidimensional in terms of the constructs they measure. However, the desire for even shorter summary measures that could still assess multilevel health states is evidenced by the development of the SF-6D, which cross-tabulates the responses to 11 items from the original SF-36 into a six-dimensional health state classification that includes physical functioning, role limitations, social functioning, pain, mental health, and vitality (Brazier et al., 1998). An SF-6D *health state* is obtained by selecting one level from each dimension, which may contain between two and six levels each. This produces 9000 possible unique health state profiles from the original 11 responses. From these profiles, a single value "utility" measure scored on a 0 to 1 scale is obtained with 0 indicating "death" and

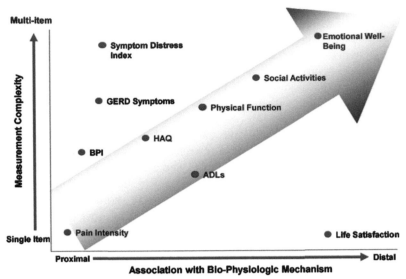

FIGURE 12.1 The theoretical relationship between the measurement complexity (represented here by the number of items required to obtain a valid measure on the Y-axis) of specific questionnaires (represented by the *bulleted* (X, Y) coordinates) and the distance of the construct from the biophysiologic mechanism (X-axis).

1 indicating "full health." The result was appealing to those desiring to reduce the investigator's burden of longer assessments; it also resulted in the investigator having a much simpler task for data collection and analysis. By asking only 11 questions, one could get a single number for health status that in theory represented 9000 different health conditions. However, as attractive as these *concentrated* survey instruments appear to be for clinical research, the problem is that the SF-36, SF-12, and the SF-6D were all developed as measures for cross-population comparisons, and not for detecting longitudinal differences in treatment impact within disease-specific groups—the more common focus of clinical research. For the most part, these utility indices are not sensitive to therapeutic-specific, within-individual changes that are meaningful to patients. As the number of clinical studies employing inappropriate instruments producing falsely negative or null results for health status and QOL outcomes grew, researchers questioned the validity of the instruments and attempted to increase the sensitivity of shorter instruments by opting for *disease-specific*, rather than *generic*, questionnaires. A summary showing the typology of measurement for these two classes of instruments is given in Table 12.2.

TABLE 12.2 Typology of Questionnaire Assessment and Evaluation

Instrument Coverage	Typology		
	Discriminative	**Predictive**	**Evaluative**
	Distinguish between individuals or groups when no external "gold standard" is available	Classify individuals into predefined categories when an external "gold standard" is available	Estimate the magnitude of longitudinal change in an individual or group
Generic: Broad coverage for a general population	Used primarily in large-scale surveys to describe populations	Used to screen populations for purposes of surveillance or predicting risk of diseases or conditions	Used to evaluate the comparative effectiveness of health care and public health programs at a population level
Disease-specific: Narrow coverage for a target population with a particular disease, condition, or individual characteristics	Used to distinguish between individuals according to disease or condition severity	Used to predict outcomes or screen individuals at high risk	Used to measure comparative changes in an outcome as a result of an intervention or treatment

Disease-specific instruments are tailored to the characteristics and dimensions that individuals with the same disease have in common, while generic instruments tap the dimensions of health that are common to all individuals regardless of their levels of health, disease, or condition. The primary trade-off between using disease-specific versus generic instruments is that disease-specific instruments are shorter and more targeted, but they are not applicable for comparisons across populations of individuals with different diseases. However, for therapeutic clinical investigations, studies typically are conducted within a particular disease condition and, therefore targeting the instrument to the population is an advantage, rather than a disadvantage. With disease-specific instruments, the universe of potential questions for the general population is reduced to a smaller subset focusing on a narrower disease-specific target population allowing one to spend *questionnaire capital* on questions that are relevant and sensitive to the disease condition and therapeutic intervention, while not wasting resources on questions that are not.

A web-based database resource for selecting patient questionnaires is the Patient-Reported Outcome and Quality of Life Instruments Database (PROQOLID; https://eprovide.mapi-trust.org/), which provides information on 885 instruments as of October, 2014. When selecting published instruments, there are several types that should be considered, including (1) generic instruments that can be used in general populations to assess a wide range of domains applicable to a variety of health states and conditions, (2) disease-specific instruments that are grouped by pathology or disease, and (3) population-specific instruments based upon demographic and patient characteristics. Questionnaires are often categorized by patient-specific factors such as age [pediatric patients (neonate, infant, and child), adolescent, adult, geriatric patients] and gender or type of respondent, such as patient or caregiver. Other questionnaires are classified on the major outcome domain such as coping, disability/physical functioning, health status, psychosocial, psychological, quality of end of life, QOL, satisfaction, social functioning, symptom/functioning, utility, and work performance. The various modes of administration should also be considered, including caregiver-administered, computer and mobile device delivery formats, interactive voice response, interviewer-administered, nurse-rated, physician-rated, proxy-administered, proxy-rated, self-administered, and telephone-administered. Searching through the literature, one finds that there are thousands of instruments available. Two questions are addressed in the next sections, namely, "How does one choose among the vast array of patient survey instruments?" and "Are there newer more sophisticated ways of addressing the *quantity of items* versus the *quality of items* dilemma?"

Choosing the Appropriate Questionnaire

As described briefly earlier, the range of applications using surveys and questionnaires in medical and health-care research is extremely broad, and the selection of instruments is continually expanding. The quest to find short, yet sensitive, instruments led the authors of such instruments to develop a large number of targeted instruments, drilling down to such a narrow level that there was literally an explosion of diverse instruments even within relatively narrow target populations, diseases, and conditions. In an attempt to regulate the proliferation and standards for instrument use and development in the area of therapeutic drug research, standards and guidelines were proposed by the US FDA in the form of an industry guidance for using patient-reported outcomes in clinical trials of new drugs for the purpose of making therapeutic claims (United States Food and Drug Administration, 2009). While this guidance adopts a framework specific to drug development and therapeutic claims, the topics outlined in the guidance provide a good framework covering the most important aspects that a clinical investigator should consider when developing or choosing a specific questionnaire.

For instrument selection, it helps to organize the selection process in terms of the overall goal of the research, the design employed, and the areas or domains to be assessed. It is also necessary to describe, as part of research planning, these components to gain a better understanding of the general measurement landscape required for questionnaire and survey assessment. Almost any type of study design may employ questionnaires and surveys in clinical investigation, including cross-sectional surveys, case-control studies, observational and prospective studies, clinical trials, quasiexperimental designs, and program evaluation. The specific types of questionnaires used in the research study should be selected based primarily on the research hypothesis, and secondarily, by issues relating to practicality, efficiency, subject burden, and budget. While the researcher may make a choice based upon disease condition, mode of administration, and target population, one also needs to consider the underlying conceptual model of the research hypothesis. When using questionnaires and surveys in clinical research, one should first keep in mind that the *questionnaire selection* should be one of the last components that is addressed.

In all cases, the researcher should always begin with a conceptual health model that outlines the relationships among the variables being studied and the causal or associative pathways among the independent and dependent variables. The dependent variables are those outcome measures that the researcher is interested in changing or modifying, while the

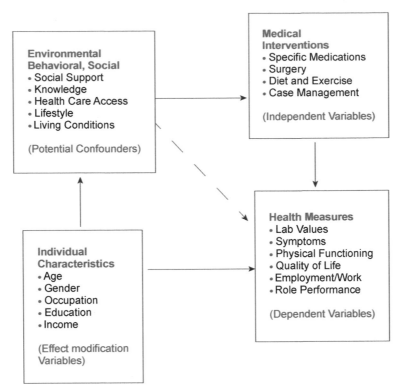

FIGURE 12.2 A conceptual model for health outcomes research showing one of many potential causal pathways between the impact of medical interventions on health outcomes as modified or further explained by patient individual characteristics and environmental, behavioral, and social factors.

independent variables are typically those interventions that cause the change in the dependent variable. Commonly, the measures or variables collected will fall into one of three general categories, namely, outcomes (dependent variables), patient characteristics, or independent variables which the investigator wishes to control (independent variables used as covariates) and treatments or interventions (independent variables of interest). A diagram of a generic conceptual health model is shown in Fig. 12.2.

The questionnaire used to measure a construct for one type of study design might not be suitable or appropriate for a different type of application or study design. For example, the questionnaire appropriate for measuring the *outcome* or *dependent variable* "physical health status" in a clinical trial might be different from one measuring the same construct, but used primarily as an *independent variable* classifying individuals for purposes of prediction or risk adjustment. The reason for the distinction is that certain scales are developed specifically to be *evaluative*, while others are developed to be *discriminative* or *predictive* as described earlier in Table 12.2.

When measuring physical health status as an outcome or dependent variable, it is necessary to use an instrument that is capable of detecting changes in physical health *within individuals* due to the intervention or exposure. This type of within-individual measurement scale is referred to as an *evaluative scale*. If the goal of the measurement of physical health is to explain or control for differences in another outcome, then the instrument must be able to distinguish differences *between individuals*, and a *discriminative* scale might be more appropriate. The scale's sensitivity required for detecting within-individual changes is considerably higher than that required for detecting between-individual differences. This difference has implications for not only the content and coverage of the items and questions, but also for the number of *ticks* on the scale's *ruler*, which determines the precision of a measure, and ultimately, the sensitivity of the instrument for detecting changes in the construct of interest.

It should be mentioned that a more sensitive scaling or metric can always be rescaled into broader categories when detailed precision is not necessary, but the opposite is not true. For example, many researchers use a dichotomous categorization of the Karnofsky performance index of less than 80% versus 80% or higher for determining eligibility criteria or as an independent variable to stratify analyses. In contrast, all 11 levels are used when it is desired to measure health functioning as an outcome of therapy, such as improvement in performance due to an efficacious therapeutic intervention.

The Different Types of Questionnaires

Questionnaires and survey instruments that simply collect objective information such as age, gender, height, weight, and medications must be formatted for purposes of clarity for recording responses; however, the issues of scale measurement in these circumstances are not of concern since all information is contained by answering one question. Researchers should be aware that a data collection form such as a case report form used in a clinical trial is not a questionnaire—it is simply a data recording document. To the novice the difference between the two may appear subtle, but it is extremely important. Often the naïve clinical researcher delegates the task of developing and formatting a patient-reported outcome questionnaire inappropriately to a data manager simply because they do not appreciate this difference.

The answer to a single item response or scale is the analyzable unit or measure. On the other hand, if the variable of interest cannot be measured at the single item level, it can only be measured indirectly. It is this case that requires the science of scale measurement and psychometrics. Although the intricacies for establishing a balance between content, coverage, and sensitivity are rather complex, adopting a standard framework for choosing an instrument for a particular purpose is often useful. Several key components and taxonomy of this framework are given in Table 12.3.

TABLE 12.3 Taxonomy of the Measurement of Patient-Reported Outcomes Used in Clinical Trials

Attribute	Types
Research objectives	• To define entry criteria for study populations • To describe the level of functioning in a study population and to serve as covariates (predictors or independent variables) in the analyses • To serve as outcome measures (endpoints or dependent variables) to assess and evaluate effectiveness of therapies and treatments • To serve as outcome measures (endpoints or dependent variables) to assess and evaluate the side effects and adverse events associated with treatments
Domains and constructs	• Overall health status • Symptoms/signs, individually or as a syndrome associated with a medical condition • Functional status (physical, psychological, or social) • Health perceptions (e.g., self rating of health, or worry about condition) • Satisfaction with treatment or preference for treatment • Adherence to medical treatment
Target population or target condition	• Generic—used for the general population • Condition-specific—used for specific diseases and health conditions • Population-specific—defined by age, disability, cultural, and socioeconomic levels
Modes of data collection	• Interviewer-administered • Self-administered, with or without supervision • Computer-administered or computer-assisted • Interactively administered (e.g., interactive voice response systems or web-based systems) • Electronic diaries, smartphones and mobile devices
Timing and frequency of administration	• Dynamically as events occur [e.g., electronic, handheld (e.g., smartphone) mobile recording device]—usually for numerous measures throughout the study interval • At regular intervals throughout a study (e.g., at regularly scheduled clinic visits)—usually to determine the overall impact during the entire study • Baseline and end of treatment, when it is assumed that steady state in the observed outcome is only available at the end of the treatment
Types of scales or scores	• Single rating on a single concept (e.g., pain severity) • Index—single score combining multiple ratings of related domains or independent concepts • Profile—multiple separate scales of related domains • Battery—multiple separate scales reflecting independent concepts • Composite—a combination of multiple related scales reflecting a unified higher level construct
Weighting of items or concepts	• All items and domains are equally weighted (summative scales) • Items are assigned variable weights—based upon a factor or principal components analysis • Domains are assigned variable weights—based upon a factor or principal components analysis

In addition, instruments are often divided based upon the domains and type of individuals being measured. Generic measures focus on broad areas of coverage useful in general populations, while disease-specific measures focus on a more restricted set of domains and individuals with specific characteristics, diseases, or conditions.

Another issue that distinguishes among questionnaires is the general format of the questions and whether or not the questionnaires are used with interviewer assistance or are self-administered. Interview-based assessment can be either *open ended*, *semistructured*, or *structured*. Open-ended, interviewer-assisted questionnaires allow the respondent a great deal of latitude with regard to how a state of health or feeling is described. The amount of information is extremely rich, but the free text requires qualitative analysis and extensive coding if quantitative methods of analysis are to be used. The semistructured interview allows for some free text, but does not require as much qualitative analysis and coding. The structured interview has a set of fixed questions and fixed response selections that may not vary, although skip-logic is often employed. If an interviewer is not available or required, then the questionnaire can be administered as a self-assessment. Different terminology is often used to describe the format of the tool that is used, including survey, questionnaire, instrument, and checklist. This varying terminology is a matter of style rather than substance. However, there are some distinctions worth mentioning. A checklist is typically used when the questions can be answered by a simple yes or no. The "yes's" are usually assigned a value of 1, the "no's" a value of 0, and then the total score is referred to as an "index." The Karnofsky performance index is an example of a checklist. The term "survey" is used primarily in the social sciences when large-scale population assessments are undertaken. The terms "patient questionnaire" and "assessment instrument" are more often used when conducting clinical or health-care research.

Types of Health Outcomes Assessed Using Questionnaires

There are certain types of health outcomes that are definitive or relatively nondisputable in their assessment. Death, severe complications, myocardial infarction, stroke, abnormal blood tests, and the diagnoses of certain diseases or conditions all fall into this category. The measurement of mortality and morbidity are relatively straightforward; however, the accuracy of the data obtained from reports about the cause of death of parents and relatives, and self-reports of diseases, conditions, and medications can be affected by the way questions are asked and how the data are recorded. The general rule is to structure the question to be clearly understood by the target population, to make sure that only one phenomenon is being assessed with each question, to use skip-logic to ease the burden of administration, and to make the multiple choices clearly nonoverlapping.

While these brief rules cover most cases of data collection, they do not cover the issues that need to be addressed for measuring nonobservable or intangible constructs. There is a hierarchy of health outcomes assessment ranging from the most objective to the most subjective, which includes mortality and morbidity at the highest level of objectivity. Moving from the more objective to the least objective includes physiologic assessment, symptom evaluation, functional assessments, health status, QOL, and finally satisfaction with health care and treatments. There are also those measures that do not focus on health outcomes; rather, they focus on health-related behaviors and patient characteristics, such as smoking, alcohol use, diet, lifestyle, exercise, education, socioeconomic status, and occupation. In addition, the distinction needs to be made at the outset as to whether the questionnaire scales will be used as dependent or independent variables since this will determine the requirements for the scale's sensitivity and responsiveness.

Evaluating Questionnaires and Survey Instruments

The researcher must be able to select an appropriate instrument for the purposes outlined previously and also be able to evaluate its measurement properties as well as interpret the meaning of the resulting scores. This section will summarize the components of questionnaires and surveys including items, scales, composite measures, and summary scales. It also will give a general overview of measurement performance properties, including reliability, validity, responsiveness, and sensitivity.

The manner in which a questionnaire is developed will ultimately determine how well it performs in the field. Questionnaire development typically begins with an initial statement of the principal measurement goal. This is often part of the work of an *expert committee* assembled to decide what type of information should be collected and for what purpose, although in the context of PCOR the "committee" may be community representatives of the disease. When unobservable constructs are being evaluated, the committee (whether expert or community based) must often come to a consensus using guided, formal brainstorming and group consensus methods such as the Delphi (Weaver, 1971) or nominal group techniques (Dunnette et al., 1963; Delbecq and VandeVen, 1971). Once agreement is reached, the constructs and areas of desired measurement need to be dissected and thoroughly discussed so that an item pool can be developed. An item pool is a collection of questions that are thought to represent the particular construct that is being surveyed. A critical step in the development process is to talk to a group of individuals who are representatives of the target population who will be answering the questions contained in the instrument. This is particularly critical for PCOR. At this stage, the general methodology is to assemble groups of such individuals to obtain information using formal focus group methodology (Stewart and Shamdasani, 2015). Several focus groups should be undertaken so that, within each focus group, members are relatively homogeneous to foster greater communication within the group without dominance by one particular faction or individual. The group members are asked questions by a trained facilitator about the construct to be measured by the questionnaire in relation to the disease or condition understudy.

At the level of the focus group, the measurement process is all qualitative. A focus group facilitator is responsible for guiding discussion along the lines of a predetermined script, but does not contribute to the content of the discussion. The focus group may be filmed but, at a minimum, there should always be an audio recording. The transcripts of the focus group discussion are then subjected to a formal qualitative analysis where common themes are identified and further parsed into sub themes that ultimately form the building blocks of the item pool. The text from the transcripts is also used to build the content of the items themselves. Once a good set of potential questions is created from the focus group qualitative analysis, a decision must be made as to the format and structure of the questions, as well as the standardized response formats that will be used. Table 12.4 shows a set of potential responses and formats as described by the 2009 FDA patient-reported outcomes guidance document cited previously.

The questions commonly referred to as "item stems" can be arranged in a variety of different formats. One such format is a matrix design where a set of ordered responses is displayed across the top row, the item stems are displayed along the left-hand side, and the response values are marked by the cell values within the item response by an item stem two-way matrix. While this format conserves space and time for the respondent in completing the items in contrast to writing out each individual question with a separate response category, it may be subject to a potential bias whereby the respondent fills columns in the matrix without paying close attention to the individual item stems. Even at this stage, decisions as to format and style can involve a trade-off between sensitivity, efficiency, and burden.

TABLE 12.4 Types of Item Response Formats Used In Questionnaires and Surveys

Response Format	Description
Visual analog scale (VAS)	A line of fixed length (usually 100 mm) with words that anchor the scale at the extreme ends and with no words describing intermediate positions. Patients are instructed to place a mark on the line corresponding to their perceived health state.
Anchored or categorized VAS	A VAS that has the addition of one or more intermediate marks positioned along the line with reference terms assigned to each mark to help patients identify the locations (e.g., halfway) between the ends of the scale.
Likert	A multiple choice bipolar response format, measuring either positive or negative response to statements using an *ordered* set of discrete terms or statements from which patients are asked to choose the response that best describes their state or experience. Responses often range from "strongly agree" to "strongly disagree" and are usually arranged horizontally and anchored with consecutive integers.
Rating	A multiple choice set of numerical categories from which patients are asked to choose the category that best describes their state or experience. The ends of rating scales are anchored with words, but the categories do not have labels.
Adjectival	A multiple choice set of ordered adjectival categories from which patients are asked to choose the category that best describes their state or experience. Each category contains a label, and when it also includes a corresponding numeric value, it is called a Juster scale.
Event log	Specific events are recorded as they occur using a patient diary or other reporting system (e.g., electronic diary using a mobile device such as a smartphone or tablet).
Pictorial	A set of pictures applied to any of the types of response options mentioned earlier. This response format is suitable for children, individuals with cognitive impairments or who are unable to read.
Checklist	Provides a simple choice between a limited set of options, such as Yes, No, or Don't know; True or False; Agree or Disagree; Choice A versus Choice B.

Once a suitable number of items have been generated as the pool for a single construct, it is necessary to perform a quantitative analysis to eliminate unclear, redundant, and noninformative items, as well as those that measure something distinctly different. This is done by administering all the items in the pool to a *development sample* of potential respondents. The size of the sample will depend upon the available resources, but typically includes between 100 and 300 individuals. While this number might appear large, one should be reminded that these are the data that will justify not only which items will remain, but also will supply the input for the initial psychometric analysis used for determining the reliability and validity of the scales. One uses a variety of analytical methods to determine which items are retained and which are deleted. As a first pass, in addition to answering the actual items, development pool subjects are often asked to comment on the quality of the questions. That is, they rate the *quality* (i.e., ease of answering, clarity, relevance) of each question in addition to answering it. Initially, there will be items that respondents will mark as "poor" because they do not understand what is being asked, or the language used is not clear or specific. These items are immediately rejected. Further item pool reduction requires statistical analyses. Although a description of the types of statistical analyses required is beyond the scope of this chapter, basically they involve eliminating items that (1) have a small variance, (2) show a strong ceiling or floor effect with the majority responses at the upper or lower ends in a population that spans the ranges of intended answers, (3) do not *hang together* with other items in the scale as determined by a test of internal consistency with and without the item included, and (4) reduce the strength of the correlation between the computed scale and other measures or scales known to be related to the underlying construct. It is only after all these stages of item selection have been completed that the instrument is field tested to gather information about the psychometric properties of the final scale, which includes information about the scale's reliability and validity. The resources required to develop a well-performing scale are directly proportional to both the extent to which the underlying construct is unobservable and the broadness of the construct, as depicted previously in Fig. 12.1. The more deeply hidden the attributes of the construct, the greater the effort required to ascertain and measure those attributes.

Reliability, Validity, Sensitivity, and Responsiveness

The standards by which questionnaire scales are judged include performance measures of accuracy, precision, and validity—similar to any medical device or instrument. The only difference is that in most instances there is no gold standard for this type of measurement. The absence of the gold standard is not merely a technical problem that could be overcome with greater technical resources; rather, it is due to the fact that the construct being measured is, in fact, the gold standard itself. For constructs such as satisfaction, symptom distress, health perceptions, and QOL, each individual respondent has his or her own intrinsic gold standard. There is no external way that an investigator can evaluate the scale scores against some external standard. As such, the type of methods by which scale performance is judged must rely heavily on criteria demonstrating precision, reliability, internal consistency, and associations with other measures known to be related to the same underlying construct. Table 12.5 summarizes three primary performance criteria—reliability, validity, and sensitivity.

Reliability and validity are the two performance measures that are well established in psychometrics. To these performance measures, we can add *sensitivity*. Sensitivity and specificity are diagnostic and screening testing attributes with which most clinical investigators are quite familiar. Sensitivity is defined as the probability of a diagnostic or screening test detecting disease when disease is present, reflecting the test's ability to detect a true positive. When used to judge a health outcome scale, sensitivity can be particularly important for evaluating a scale's ability to detect treatment and medical intervention effects. As shown in Table 12.5, scales are first judged based upon their reliability, defined as the ability to remain stable over repeated assessments during a time when external influences are negligible. Furthermore, it is not sufficient that they simply remain constant; they should remain constant in the presence of large intersubject variability. For example, a scale score that always remains constant simply because everyone has the same score regardless of what is happening to the true construct is quite stable, but essentially provides no information to distinguish among individuals.

Much has been written about methods for establishing the validity of a scale score. Once again, the familiar problem of not having a gold standard makes the assessment of validity more problematic as compared to other types of measures for which a gold standard exists. Without a gold standard, one can only judge validity by observing the scale's relationship with other measures. In the field of questionnaire and survey research, one often hears the question "Is this a valid scale?" To the knowledgeable student of measurement, there is only one correct answer to this question. The correct answer is "For which particular measure X and in which population would you like me to evaluate its validity?" Validity requires two measures. No scale possesses validity without reference to what is being measured. To be estimable, a scale Y can only be valid with respect to some other measure X. Furthermore, validity might vary depending on the characteristics of the population under study. If one has undertaken the steps as described earlier in the scale development process, and has demonstrated the reliability and content, criterion-related, and construct validity, then the scale can be said to be valid with

TABLE 12.5 Scale Performance Properties, Tests, and Criteria for Evaluation

Scale Performance Property	Test of Performance	Performance Criteria
Reliability	**Test-retest**: Intraclass correlation coefficient should be high in the presence of significant between-individual variance, and the mean levels should not differ between assessments taken during steady state	Assesses the ability of the scale to remain stable during a period when external influencing factors are negligible (steady state)
	Internal consistency: Within-item correlation should be relatively high as measured by an internal consistency statistic such as Cronbach's alpha	Assesses the degree to which items in the scale are measuring the same construct, or constructs related to the same phenomena.
Validity	**Content**: Items and response options are relevant and are comprehensive measures of the domain or concept; items should be from a randomly chosen subset of the universe of appropriate items	Easiest to determine when the domain is well defined, and much more difficult when measuring attributes such as beliefs, attitudes, or feelings due to the complexity of determining the range of potential items and whether or not the sample of items is representative
	Criterion-related: Items or scale is required to have only an empirical association with some criterion or "gold standard" (also called predictive validity)	Establishes the strength of the *empirical* relationship between two events which should be associated
	Construct: Concerned with the theoretical relationship of the scale score to other variables	Assesses the extent to which a measure behaves the way that the construct it purports to measure should behave with regard to established measures of other constructs
Sensitivity	**Metric or scale**: Has enough precision to accurately distinguish cross-sectionally between two levels on the scale known to be important to patients, often referred to as the minimal clinically important difference (MCID)	Determines whether there are a sufficient number and accurate *ticks* on the scale's ruler not to miss a difference which is considered important
	Responsiveness: Has enough precision to accurately distinguish between two measures at different times longitudinally to estimate changes known to be important to patients—the MCID	Determines after taking everything together in terms of reliability, validity, and precision that, when a change occurs in the underlying construct, there is a corresponding change in the measurement scale

respect to the construct of interest. The fact that there is no gold standard, directly observable value of X makes testing the validity more complex, but not impossible. Validity is also variable with biologic measurements. Often when describing a biologic response to a low salt intake, the question is raised: "What is the level on a normal salt intake?" This question assumes there is a "normal" level (the gold standard), while the validity of the response can only be determined by also knowing the level of salt intake.

The best test of the validity of a scale for clinical research is whether or not the scale is able to measure changes when an intervention reasonably known to affect the underlying construct is tested. This performance property comes under the heading of sensitivity and, when referring to longitudinal changes, is often referred to as the *responsiveness* of a scale score. Responsiveness represents the ability of a scale to change when the underlying construct changes and, as such, is also related to the scale's validity. One should keep in mind that except by chance, a poorly performing scale will not demonstrate treatment effects when treatment effects do not exist, nor will they detect effects when they do exist. The greatest risk associated with poorly developed scales is that the research study will have been done in vain. For example, use of an unreliable scale would be similar to using a blood pressure measurement instrument that had a 20 mmHg margin of error. Efficacy studies of a new antihypertensive medication using this instrument would simply yield null results. A schematic representation of the interrelationships among these scale performance measurement properties is shown in Fig. 12.3. Further discussion of these performance measures can be found in a review by Aaronson et al. (2002).

The quantitative discipline of measurement is related to, but also somewhat distinct from, statistical inference. Measurement error is often treated as a nuisance parameter in a statistical model. As long as the measurement errors are unbiased, normally distributed, and independent, the statistician believes that the analysis may proceed without question.

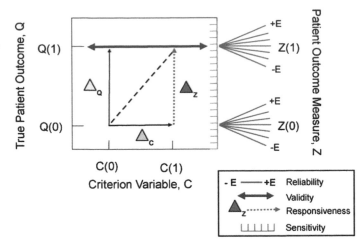

FIGURE 12.3 A schematic representation of the definitions and interrelationships among the scale performance measurement properties of reliability, validity, sensitivity, and responsiveness. Adapted from Testa, M.A., Simonson, D.C., 1996. Assessing quality-of-life outcomes. N. Engl. J. Med. 334 (13), 835–840.

However, knowing in advance that the questionnaire you are using may not be reliable or valid is extremely important. Increasing sample size cannot increase the power of the study to detect important effects if the measurement instrument associated with the dependent variable does not produce an accurate, reliable, and valid measure of the construct being evaluated. In fact, clinical investigators are often surprised that they could not find a treatment effect for constructs such as health status or QOL in large-scale multicenter clinical trials with hundreds and sometimes thousands of subjects. They are often schooled only in the belief that by increasing sample size, one can increase the power of a study to detect a treatment effect when one exists. Since most medical devices have a relatively small margin of measurement error, this is often the case for clinical and laboratory endpoints. Unfortunately, the same cannot be said for scale scores that measure constructs such as health status and QOL. Being aware of measurement performance issues can save the researcher both time and resources by taking the appropriate steps to be selective during the design phases of research planning. Considering the importance of measurement, the next section briefly describes some more recent applications of measurement methods that might be useful to clinical researchers.

Item Response Theory, Dynamic Questionnaires, and Computer Adaptive Testing

As noted previously, effective measurement of patient-reported outcomes poses a significant challenge for clinical researchers, especially when deciding among the hundreds of available questionnaires. Very few questionnaires have used item banks consistently over a long enough period of time to establish the performance measurement of items in clinical research. The PROMIS initiative was initially established to construct item banks that would be useful for measuring key health status concepts such as emotional distress, fatigue, pain, physical functioning, and social role participation across a wide variety of diseases and chronic conditions. In addition, the goal was to offer both *static* and *dynamic* questionnaires. While static questionnaires have a fixed number and content of items, dynamic questionnaires are constructed from statistical models based on item response theory (IRT) producing calibrations associated with answers to questions. These calibrations are used in algorithms to select the most informative *next question* in a sequence of questions delivered by a computer. This process, known as computer adaptive testing (CAT), *adapts* to the individual patient based on his or her responses to the previous question. When appropriate, CAT has the ability to yield more highly sensitive metrics as compared to those obtained from static questionnaires.

Dynamic questionnaires bridge the gap between the burden of long static forms and the lack of sensitivity of short static forms by using CAT. The basic idea behind dynamic assessment is that given a large bank of items that covers the entire range of potential responses of the general population, each individual respondent will be administered only those questions that are pertinent to his or her level of functioning. Based upon the same type of measurement theory that developed educational computerized assessment, after the subject answers the first question, the computer analyzes the response to select the next best question for estimation purposes. Rather than having to answer all questions that might provide coverage for the general population, or selecting only 10 questions for a disease-specific targeted population, all individuals can answer from the same large *pool* of questions using a dynamic computerized assessment approach. The item bank and IRT permit tailoring the set of questions to the individual, thereby reducing the burden of static forms and providing the sensitivity of longer forms. Whereas many disease-specific instruments are used once or twice and then relegated to the shelf, the proposed item pool approach has the potential for both sustainability and standardization in the field of clinical research. Today the use of IRT and CAT is growing. Rose et al. (2014) used PROMIS to demonstrate that its item bank provided a common metric that improved the measurement of the SF-36 physical function (PF) scale by facilitating the standard of patient-reported outcome measures and implementation of CATs for more efficient PF assessments. Of the over 2200 medical journal publications using IRT or CAT since 1980 through 2015, approximately 85% had been published in the past 10 years and 60% in the past 5 years. The medical literature continues to demonstrate that the use of IRT and CAT in clinical research is becoming more prevalent as more item banks are developed using IRT and more CAT applications launched using electronic and mobile devices.

STATISTICAL AND ANALYSIS CONSIDERATIONS

Analysis and Interpretation of Multiple, Hierarchical, and Correlated Scales

The statistical methods associated with analyzing surveys and questionnaires often generate multiple scales that may be independent or correlated, and either linearly or hierarchically related. For example, if one defines the construct of mental and emotional health as was done in the MHI-38, there are two subdomains of the construct—psychological well-being, and psychological distress. Furthermore, psychological distress includes the scales of depression, anxiety, and loss of emotional and behavioral control, while psychological well-being includes the scales of positive affect, life satisfaction, and emotional ties. The overall construct, scales, and subscales are structured hierarchically and are highly correlated. If one simply has a measure that represents a unidimensional construct with one numerical score, the statistical considerations would be similar to any other dependent variable. However, if the scale were developed such that there were several layers to the global construct and multiple dimensions that the researcher wants to explore, then the analysis should take this correlated and often hierarchical multiplicity structure into account. For example, most QOL and patient satisfaction measures contain many different domains that are highly correlated. One should avoid adjusting for multiple comparisons in this instance unless a modification is made to the adjustment that takes into account the correlated, hierarchical structure of the scales.

There are several ways that one can deal with the issue of multiplicity. For example, one can create a summary measure by performing a principal components analysis to identify the best way to weight each of the subscales. This method conserves the type I alpha error simply by combining measures into one overall scale, thereby avoiding the multiplicity issue altogether. If one adopts the overall summary measure as the primary endpoint, then one can still use the subscales to explain the reasons why the overall scale might be changing and why a particular treatment effect was observed. When using this approach, one may also use a *modified Bonferroni alpha level* which accounts for the intercorrelations among scales as a way to provide a confidence level for the treatment effects associated with the correlated subscales. An example of this modified Bonferroni method is given in a comparative effectiveness clinical trial comparing the impact of basal-bolus versus premix analog insulin on glycemic variability and patient-centered outcomes during insulin intensification in type 1 and type 2 diabetes (Testa et al., 2012).

There are other approaches that can be used to address the issue of multiplicity, including performing global statistical tests. A global test combines the information, not at the scale level, but at the inference level. One simple approach is to first perform a factor analysis using an orthogonal rotation to identify scale scores that not only represent the different subareas of a summary scale, but also have the property of independence. Given this independence, it is possible to conduct a global test using an analysis of variance by combining the mean squares in such a way as to yield one overall significance test (Salsburg, 1992).

Interpreting a Minimal Clinically Important Difference

It is important to consider whether the observed scale changes are clinically meaningful. Usually detecting statistically small, but clinically nonmeaningful, changes with patient-reported health outcomes is rare since the relatively larger measurement error associated with scale estimates as compared to clinical measures actually makes it much harder to detect mean differences. What is important is to acquire an appreciation of the variability of the scale scores in association with the within- and between-patient differences. As with any measure, it is important to distinguish between mean population shifts and within-individual changes (Testa, 2000a,b). It should be recognized that there is a critical distinction between a stochastic shift in the outcome of interest from a population health perspective versus the change that would be considered important for an individual patient. Criticism of scale score changes as being *too small* to be important usually arises because of a lack of understanding of the distinction between mean population shifts and the degree of change required to categorize an individual as a "responder."

This issue is relevant to all types of clinical research. For example, for many widely used measures (e.g., blood pressure, pain, exercise tolerance), the ability to show a difference between treatment groups on the basis of being classified as a "responder" versus "nonresponder" has been considered evidence of a relevant treatment effect. Establishing responder criteria for a scale score improvement based upon a minimal clinically important difference for the individual can serve as a benchmark for interpreting the population treatment effects observed. There are several ways that have been proposed to map changes in patient scale scores to enhance the interpretation of the importance of observed changes. However, it is important to recognize that the issue is not mapping the scores to gain a better understanding of defining a mean shift; rather, it is to gain an appreciation of how a mean improvement can translate into a large percentage of individuals moving into a greater functional operational range. Sometimes, the issue of what constitutes a meaningful important difference for measures such as health perceptions and satisfaction seems counterintuitive to the measurement of subjective phenomena. By definition, the subjectivity makes the assessment of what is important the decision of the respondent, an important aspect of measurement to keep in mind when engaging in debates concerning this topic.

SUMMARY AND CONCLUSIONS

The use of questionnaires and survey instruments can greatly expand the types of data that can be obtained from clinical research. The primary purpose of a questionnaire is to better understand how patients feel and behave as a group, usually after some form of intervention or therapy. If the desire is simply to determine how an individual patient is feeling or functioning, questionnaires may not be necessary. The empirical requirements of research have led investigators to quantify that which seems unquantifiable. Concepts such as perceptions, feelings, and attitudes would appear at first glance to be immeasurable. Furthermore, one might consider such concepts as so soft that even if one accepted that they could be measured accurately and reliably, interpretation would be too difficult. In this chapter, we have attempted to dispel this belief. While it is important to understand the various technical aspects and requirements for using questionnaires and surveys in clinical research, probably the most important message is that the methods are highly scientific and should be treated as such. If proper care is taken to select, evaluate, and utilize the scientific tools of questionnaires and surveys in clinical research, the results gained from their use should be regarded with the same degree of confidence that clinical and laboratory measure possess. In the end, the key to understanding the complexities and potential problems associated with the use of questionnaires and surveys lies more with obtaining a good understanding and appreciation for the scientific disciplines of measurement and human perception than with the technical aspects of questionnaire selection and administration.

GLOSSARY

Clinimetrics The practice of assessing or describing symptoms, signs, and laboratory findings by means of scales, indices, and other quantitative instruments.

Computer adaptive testing A form of computer-based testing that adapts to the examinee's ability level.

Delphi method A structured communication technique, originally developed as a systematic, interactive forecasting method which relies on a facilitator and panel of experts who answer questionnaires in two or more rounds to reach consensus.

Discriminative scale A scale designed to distinguish between individuals or groups when no external "gold standard" is available.

Evaluative scale A scale designed to be sensitive to detecting minimal clinically important longitudinal changes in outcomes.

Item response theory A psychometric model for scoring questionnaires that does not assume that each item has comparable difficulty, but rather models the response of each examinee of a given ability to each item in the test.

Latent variable A variable that is not directly observed, but is inferred (estimated) through a mathematical model from other variables that are directly measured.

Minimal clinically important difference (MCID) The smallest change in a treatment outcome that a patient or physician (depending on whose perspective is being assessed) would identify as important.

Nominal group method A group process involving problem identification, solution generation, and decision-making that is used to make a decision quickly, as by a vote, allowing everyone's opinion to be taken into account.

Operative scale A scale that has the ability and sensitivity to the MCID in states of health between the "lowest level at which it would be possible for an individual to live and work as they currently do" and "full health" in contrast to an *absolute scale* capable of detecting broader (not MCID) states of health between "death" and "full health."

Patient-centered outcomes Outcomes of medical care that are inclusive of the patient's preferences, autonomy, and needs, focusing on outcomes that people notice and care about such as survival, function, symptoms, and health-related quality of life.

Patient-reported outcomes Health outcomes that are measured through direct reports from patients using interview, questionnaire, survey, and diary methods and instruments.

Predictive scale A scale designed to classify individuals into predefined categories when an external "gold standard" is available.

Psychometrics Scientific field of study concerned with the theory and technique of psychological measurement.

Quality of life and health-related quality of life Quality of life (QOL) is a broad multidimensional concept that includes subjective evaluations of both positive and negative aspects of life, and in medical and health research; the term "health-related quality of life" includes those aspects of overall quality of life that can be clearly shown to affect physical, mental, and social health.

Reliability The ability of a scale to remain stable during a period when external influencing factors are negligible as measured by reliability estimates including interrater or interobserver reliability (degree to which different raters/observers give consistent estimates of the same phenomenon); test—retest reliability (consistency of a measure from one time to another); parallel-forms reliability (consistency of the results of two tests constructed in the same way from the same content domain), and internal consistency reliability (consistency of results across items within a test).

Responsiveness The ability of a scale to change when the underlying construct changes as measured by several statistics reflecting longitudinal change relative to the standard deviation effect size, standard error of measurement, and the MCID.

Sensitivity The degree to which a scale is able to detect states of health that correspond to MCIDs and is conceptually analogous to having sufficient *ticks* on a ruler to capture an important difference or change.

Validity The degree to which a scale measures what it claims to measure as determined by estimates of validity, including construct validity (convergent and discriminant); content validity (representation and face); and criterion validity (concurrent and predictive).

LIST OF ACRONYMS AND ABBREVIATIONS

AHRQ Agency for Health Research and Quality
CAT Computer adaptive testing
COA Clinical outcomes assessment
FDA Food and Drug Administration (United States)
HOR Health outcomes research
IRT Item response theory
MCID Minimal clinically important difference
MHI-38 Mental Health Inventory (38 items)
MOS Medical Outcomes Study
PCOR Patient-centered outcomes research
PCORI Patient-Centered Outcomes Research Institute
PRO Patient-reported outcome
PROMIS Patient-Reported Outcomes Measurement Information System
SF-36 Short Form 36 (36 Items)

REFERENCES

Aaronson, N., Alonso, J., Burnam, A., Lohr, K.N., Patrick, D.L., Perrin, E., Stein, R.E., 2002. Assessing health status and quality-of-life instruments: attributes and review criteria. Qual. Life. Res. 11 (3), 193—205.

Beckman, H.B., Frankel, R.M., 1984. The effect of physician behavior on the collection of data. Ann. Intern. Med. 101 (6), 692—696.

Bowling, A., 2001. Measuring Disease, second ed. Open University Press, Printed by Saint Edmundsbury Press Limited, Buckingham.

Bowling, A., 2005. Measuring Health, third ed. Open University Press, McGraw-Hill House, Berkshire, England.

Brazier, J., Usherwood, T., Harper, R., Thomas, K., 1998. Deriving a preference-based single index from the UK SF-36 Health Survey. J. Clin. Epidemiol. 51 (11), 1115—1128.

Brook, R.H., Ware Jr., J.E., Rogers, W.H., Keeler, E.B., Davies, A.R., Donald, C.A., et al., 1983. Does free care improve adults' health? Results from a randomized controlled trial. N. Engl. J. Med. 309, 1426—1434.

Cella, D., Yount, S., Rothrock, N., Gershon, R., Cook, K., Reeve, B., Ader, D., Fries, J.F., Bruce, B., Rose, M., PROMIS Cooperative Group, 2007. The patient-reported outcomes measurement information system (PROMIS): progress of an NIH roadmap cooperative group during its first two years. Med. Care. 45 (5 Suppl. 1), S3—S11.

Coulehan, J.L., Block, M.L. (Eds.), 2006. The Medical Interview: Mastering Skills for Clinical Practice, fifth ed. F.A. Davis Company, Philadelphia.

Croog, S.H., Levine, S., Testa, M.A., Brown, B., Bulpitt, C.J., Jenkins, C.D., Klerman, G.L., Williams, G.H., 1986. The effects of antihypertensive therapy on the quality of life. N. Engl. J. Med. 314 (26), 1657—1664.

Davidoff, F., 1996. Who has seen a blood sugar? The shaping of the invisible world. In: Davidoff, F., Deutsch, S., Egan, K., Ende, J. (Eds.), Who Has Seen a Blood Sugar: Reflections on Medical Education. American College of Physicians, Philadelphia, pp. 96—100.

Delbecq, A.L., VandeVen, A.H., 1971. A group process model for problem identification and program planning. J. Appl. Behav. Sci. 7 (4), 466—491.

DeVellis, R.F., 2012. Scale Development: Theory and Applications. In: Applied Social Research Methods Series, third ed., vol. 26. Sage, Thousand Oaks, CA.

Dunnette, M.D., Campbell, J.D., Jaastad, K., 1963. The effect of group participation on brainstorming effectiveness for two industrial samples. J. Appl. Psychol. 47 (1), 30—37.

Feinstein, A., 1967. Clinical Judgment. Williams & Wilkins, Baltimore, MD.

Feinstein, A.R., 1987. Clinimetrics. Yale University Press, New Haven, CT.

Fowler, F.J., 1995. Improving Survey Questions. Design and Evaluation. Sage, Thousand Oaks, CA.

Johnson, T.P., 2015. Handbook of Health Survey Methods. John Wiley and Sons, Inc., Hoboken, NJ.

Karnofsky, D.A., Burchenal, J.H., 1949. The clinical evaluation of chemotherapeutic agents in cancer. In: MacLeod, C.M. (Ed.), Evaluation of Chemotherapeutic Agents. Columbia University Press, New York, pp. 196–200.

Lord, F.M., Novick, M.R., Birnbaum, A., 2008. Statistical Theories of Mental Test Scores, second ed. In: Addison-Wesley Series in Behavioral Science: Quantitative Methods. Information Age Pub, Charlotte, NC.

McDowell, I., 2006. Measuring Health. A Guide to Rating Scales and Questionnaires, third ed. Oxford University Press, New York.

Nunnally, J.C., Bernstein, I.H., 1994. Psychometric Theory, third ed. McGraw-Hill, New York.

Rose, M., Bjorner, J.B., Gandeka, B., Bruce, B., Fries, J.F., Ware, J.E., 2014. The PROMIS physical function item bank was calibrated to a standardized metric and shown to improve measurement efficiency. J. Clin. Epidemiol. 67 (5), 516–526.

Salsburg, D.S., 1992. The Use of Restricted Significant Tests in Clinical Trials. Springer-Verlag, New York.

Schag, C.C., Heinrich, R.L., Ganz, P.A., 1984. Karnofsky performance status revisited: reliability, validity, and guidelines. J. Clin. Oncol. 2 (3), 187–193.

Selby, J.V., Beal, A.C., Frank, L., 2012. The Patient-Centered Outcomes Research Institute (PCORI) national priorities for research and initial research agenda. JAMA 307 (15), 1583–1584.

Smith, R.C., Hoppe, R.B., 1991. The patient's story: integrating the patient- and physician-centered approaches to interviewing. Ann. Intern. Med. 115 (6), 470–477.

Stewart, D.W., Shamdasani, P.N., 2015. Focus Groups: Theory and Practice (Applied Social Research Methods), third ed. SAGE Publications, Inc., Thousand Oaks, CA.

Streiner, D.L., Norman, G.R., 2008. Health Measurement Scales: A Practical Guide to Their Development and Use, fourth ed. Oxford University Press, Oxford.

Tarlov, A.R., Ware Jr., J.E., Greenfield, S., Nelson, E.C., Perrin, E., Zubkoff, M., The Medical Outcomes Study, 1989. An application of methods for monitoring the results of medical care. JAMA 262 (7), 925–930.

Testa, M.A., 2000. Quality of-life assessment in diabetes research: interpreting the magnitude and meaning of treatment effects. Diabetes Spectr. 13 (1), 36–41.

Testa, M.A., 2000. Interpretation of quality of life outcomes: issues that affect magnitude and meaning. Med. Care 38 (Suppl. II), II-166–II-174.

Testa, M.A., Anderson, R.B., Nackley, J.F., Hollenberg, N.K., 1993. Quality of life and antihypertensive therapy in men: a comparison of captopril with enalapril. N. Engl. J. Med. 328 (13), 907–913.

Testa, M.A., Gill, J., Su, M., Turner, R.R., Blonde, L., Simonson, D.C., 2012. Comparative effectiveness of basal-bolus versus premix analog insulin on glycemic variability and patient-centered outcomes during insulin intensification in type 1 and type 2 diabetes: a randomized, controlled, crossover trial. J. Clin. Endocrinol. Metab. 97 (10), 3504–3514.

Testa, M.A., Simonson, D.C., 1996. Assessing quality-of-life outcomes. N. Engl. J. Med. 334 (13), 835–840.

Testa, M.A., Simonson, D.C., 1998. Health economic benefits and quality of life during improved glycemic control in patients with type 2 diabetes mellitus: a randomized controlled, double-blind trial. JAMA 280 (17), 1490–1496.

Testa, M.A., Simonson, D.C., Turner, R.R., 1998. Valuing quality of life and improvements in glycemic control in people with type 2 diabetes. Diabetes Care 21 (Suppl. 3), C44–C52.

United States Food and Drug Administration, December, 2009. Guidance for Industry: Patient-Reported Outcome Measures: Use in Medical Product Development to Support Labeling Claims. Retrieved from: http://www.fda.gov/downloads/drugs/guidancecomplianceregulatoryinformation/guidances/ucm071975.pdf.

United Stated Food and Drug Administration, January, 2014. Guidance for Industry and FDA Staff Qualification Process for Drug Development Tools. U.S. Department of Health and Human Services, Food and Drug Administration, Center for Drug Evaluation and Research (CDER). Obtained from: http://www.fda.gov/cder/guidance/index.htm.

Ware Jr., J.E., Sherbourne, C.D., 1992. The MOS 36-item short-form health survey (SF-36). I. Conceptual framework and item selection. Med. Care 30 (6), 473–483.

Weaver, W.T., 1971. The Delphi forecasting method. Phi Delta Kappan 52 (5), 267–271.

Chapter 13

Information Technology

Shawn N. Murphy, Henry C. Chueh and Christopher D. Herrick

Massachusetts General Hospital, Boston, MA, United States

Chapter Outline

Key Points

- The clinical database used by the electronic health record is not interchangeable with a research data warehouse. The clinical database generally cannot support queries across patients.
- A clinical data registry is a form of data repository, which is usually oriented around a specific disease entity or a narrow disease domain and is usually maintained by hand.
- When finding medical concepts in the data warehouse, the role of metadata becomes crucial. Metadata represent data about the data and how medical knowledge is represented in the data warehouse.
- Medical concepts are organized hierarchically from least granular to most granular and need to be grouped to be utilized in a practical way in queries in a data warehouse.
- Different approaches to organizing the health-care data warehouse allow for faster query formulation and performance or easier maintenance.
- Finding eligible patients for clinical trials is a key function of the research data warehouse.
- Capturing data from patients and providers at the point of care is an appealing option for putting data into a data warehouse but can be difficult to implement because it is very person intensive.
- Data standards exist for capturing health-related research data.
- Cloud computing is having a significant impact on the ability of clinical researchers to work with large data sets and run complex analyses.

Clinical and Translational Science. **http://dx.doi.org/10.1016/B978-0-12-802101-9.00013-2**

INTRODUCTION

Clinical research information technology spans diverse areas. It would be difficult to cover all of them in one source. This chapter will focus on the role of clinical data repositories (CDRs), information technology support of participant recruitment, data warehousing, principles of data collection, and data standards. The discussion will also touch briefly on clinical trial management systems software, publicly available biomedical literature databases, and new approaches to data integration using emerging Web technologies. The evolving science of clinical research informatics covers a combination of all of these areas. There is an overlap with clinical informatics in many fundamental ways. For example, design approaches for large-scale CDRs share many similarities regardless of their intended use. But clinical research poses its own set of data management challenges that are unique. Precision and rigor in data definition is a primary concern in research data, but usually only a secondary issue for clinical systems for data outside laboratory and diagnostic reporting. We often prioritize the primary subjective voice in clinical documentation, but repeated validation by objective observers is a hallmark of good clinical research data systems.

CLINICAL DATA REPOSITORIES

Harnessing the Clinical Enterprise as a Data Source

All clinical institutions have a CDR, whether it is in paper or in electronic format. However, the term "clinical data repository" has come to represent an electronic database of clinical information aggregated at some level. Typically a CDR might host a combination of laboratory results, diagnostic reports, and clinical documentation of various forms. The CDR is often one of the first places clinical investigators tap into for data to support their research. There are many reasons for this: the data are already collected as part of routine clinical care; in electronic form, the data are often searchable or at least exportable; and CDR systems are supported comprehensively by information systems personnel.

Clinical Data Repositories: A Problematic Data Source

CDRs suffer from a variety of problems when used as a research data source, in part because of the confusion surrounding the nature of CDRs themselves. For example, a common misconception about CDRs is that they are interchangeable as a clinical data warehouse (CDW), but this is generally not the case. Data warehouses, as discussed in detail in an upcoming section of this chapter, have distinguishing characteristics such as support for data query or analysis "in place," and the straightforward transformation of data into smaller "data marts." The creation of a true CDW requires that the clinical data be reorganized at a minimum, and often be deconstructed into more granular forms. A true clinical repository, on the other hand, collects information as it exists in its primary form, albeit usually with additional indexing. In theory, a CDR can be an excellent data source for a CDW. In reality, CDR designs exist along the spectrum from a true repository to a true data warehouse.

Consequently, the use of CDR data poses unique challenges for the clinical researcher. The context in which the data were collected is often variable and may not even be suitable to answer clinical questions, much less research ones (Stein et al., 2000). Clinical notes from general care clinics might sit alongside notes from highly specialized clinical research units. There may be hidden assumptions about how the data have been manipulated prior to entry into the CDR. The degree of manipulation prior to their entry into any CDR may even vary between data types. For example, to help normalize reporting across different sites, nearly similar—but not exactly the same—clinical concepts may have been merged. Most importantly, terminology that may appear identical across the CDR is typically not curated with a rigor that research requires. As evidence of this complexity in extracting reliable information, specialized techniques for mining the variable data structures in a CDR continue to be explored (Prather et al., 1997; Nigrin and Kohane, 1998; Altiparmak et al., 2006; Mullins et al., 2006; Bellazzi and Zupan, 2008).

The Expanding Use of Preresearch Clinical Registries

CDRs remain a common source of electronic clinical research data, but because of the challenges in using the data, as discussed earlier, there is a trend toward the use of preresearch clinical registries. These registries are databases of patient populations that include detailed clinical information. Registry populations are usually oriented around specific disease entities or narrow disease domains. This allows a well-defined data set to be established for the registry. A key aspect of these registries is that the information collected in them is justifiably captured as part of routine clinical care. Although

Institutional Review Board (IRB) approval of such registries may still be desired or required, acquiring consent from patients individually is not necessary. Unlike most routine clinical data in a CDR, the data in preresearch clinical registries have several important characteristics:

Data terminology is defined in advance—by establishing explicit data fields and clear definitions for these fields, consistency of meaning is established.

Many data fields have explicitly enumerated choices—by limiting the possible answers to each data field, comparability is enhanced.

Data policies are often established—typically the registry data are *comprehensive*, and a *complete* data set is captured for every patient.

Use of clinical registry data for research, of course, always requires IRB review of a specific research protocol. Preresearch clinical registries also are often used for quality assurance and other clinical practice management activities. Care must be taken to establish clear policies of use to avoid inadvertent research use and subsequent breach of privacy.

A question that arises commonly with these registries is whether they should be designed as a transactional clinical system or as a form of data warehouse. Since registry databases are updated continuously and incrementally by users (whether by clinicians or by data collection staff), a transactional database design is usually the most appropriate. Some authors have suggested hybrid models that incorporate elements of both designs (Foster et al., 2000). The details of transactional database design are well beyond the scope of this chapter, and many references are available for this purpose (Teorey, 1999; Silberschatz et al., 2005). Research data warehouse design is described in detail in the following section.

Design for Research Data Warehouses

The art of data warehousing has taken the industry by storm, and many of the same principles can be applied to the health-care enterprise. The data warehouse allows for rapid querying and reporting across patients, which unexpectedly is not available in most transactional systems such as CDRs. Rather, transactional systems are optimized for single patient lookups. The design of a CDW differs in major ways from a transactional health-care system. Up to this point, the discussion of health-care databases has implicitly referred to transactional systems. Logically enough, such a system is concerned primarily with processing transactions. A transaction is a read, insert, update, or delete to the data in a database. Inserts and updates usually occur in bursts during the day, such as when a single patient's laboratory test results are sent to the database. Although transactional databases are usually updated by small amounts of data at a time, these data may arrive with extremely high volume, often tens or hundreds of transactions per second. Therefore the database must be optimized to handle these transactions, and the resultant design typically involves numerous smaller data tables (Kimball, 2002).

A data warehouse is typically not optimized to handle transactions. Without this requirement to handle high volumes of transactions, a data warehouse can be optimized for rapid, aggregate searches. Optimal searching of a database requires very large database tables. Consequently, the design of data warehouses nearly always adopts a model of only a few tables that can hold nearly all the available data. Many forms of health-care data can be managed in a single table through the use of the classical entity—attribute—value schema, or EAV (Murphy et al., 1999; Nadkarni and Brandt, 1998).

The EAV schema forces one to define the fundamental fact of health care (Kimball, 2002). The fundamental fact of health care will be the most detailed rendition possible of any health-care event as reported from the data warehouse. In the authors' experience, we have defined this as an **observation** on a patient, made at a specific time, by a specific observer, during a specific event. The fact may be accompanied by any number of values or modifiers. Each observation is tagged with a specific concept code, and each observation is entered as a row in a **Fact table**. This Fact table can grow to billions of rows, each representing an observation on a patient. The Fact table is complemented by at least an Event table, a Patient table, a Concept table, and an Observer table (Murphy et al., 1999) (see Fig. 13.1).

The **Patient table** is straightforward. Each record in the table represents a patient in the database. The table includes common fields such as gender, age, race, etc. Most attributes of the patient dimension table are discrete (i.e., Male/Female, Zip code, etc.) or dates.

The **Event table** represents a clinical "session" where observations were made. This session can involve a patient directly (such as a visit to a doctor's office), or it can involve the patient indirectly (such as performing diagnostic tests on a patient) or collecting a specimen (such as a tube of blood). One or more observations can be made during a visit. Visits have a start and end date—time. The visit record also contains specifics about the location of the session, such as in which hospital or clinic the session occurred, and whether the patient was an inpatient or outpatient at the time of the visit.

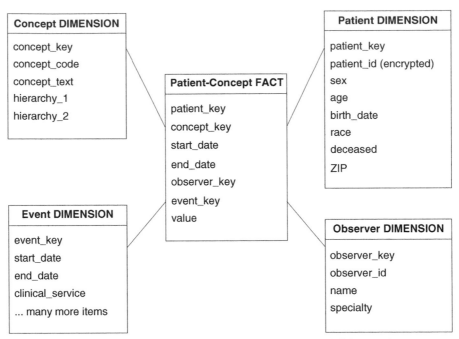

FIGURE 13.1 A basic entity—attribute—value schema for a clinical data warehouse.

The **Observer table** is a list of observers. Generally, each row in this dimension represents a clinical provider at an institution but, more abstractly, it may be an observing machine, such as a continuous blood pressure monitor in an intensive care unit.

The **Concept table** is the key to understanding how to search the fact table. A concept specifies exactly what observation was made on the patient and is represented in a particular row of the fact table. A code is used to represent the concept in the fact table, and the concept table links it to a human-readable description of the code.

Metadata

When looking at concepts in the table, the role of metadata becomes crucial. Metadata represent data about the data and is where medical knowledge is represented in the data warehouse. The primary form of representation is in the grouping of health-care terminologies. Terms are grouped into hierarchies with each level up usually expressing a more general concept. Many diverse concepts about a patient can exist in the fact table. In a CDW, 100,000—500,000 different concepts may exist. Diverse concepts including ICD-9 and ICD-10 codes (International Classification of Diseases, 9th and 10th Editions, the most common terminologies used in hospitals to classify diagnoses), CPT codes (Current Procedural Terminology, the most common terminology used in hospitals to classify procedures), NDCs (National Drug Codes, the most common terminology used in hospitals to classify medications), and LOINC (Logical Observation Identifiers Names and Codes, the most common terminology used in hospitals to classify laboratory tests), as well as numerous local coding systems are used to describe the patient. The challenge is in maintaining and updating the classification of these concepts. A sustainable classification needs to absorb new codes seamlessly and be compatible with previous coding and classification systems.

Concepts need to be organized hierarchically to be utilized practically. For example, there are about 120 diagnosis codes in the terminologies described previously that represent every type of diabetes. In many instances, one would like to query them as a group rather than having to select every code individually. This becomes increasingly important as systems evolve and hundreds of thousands of different codes are added to the warehouse.

The hierarchical classification system used in the concept table closely parallels the model used in modern computer file systems, such as Linux or Windows. The concept is placed on a path with the parent term positioned in the "folder" position of the path, and the child term in the "file" position. For a former child to become a parent, the child is added next to the parent on a new row. For example, in the following concept dimension table, the parent "antiinfectives" can have the three children "penicillin," "ampicillin," and "Bactrim" added to it. The new rows not only map the new codes (NDCs in this case) to those in the fact table, but also show that they are types of antiinfectives, and most importantly, allow group queries to be performed easily in the warehouse. A table will begin to emerge, which looks like as follows:

C_fullpath	C_basecode
Med-V2\	L_027
Med-V2\antiinfectives\	L_028
Med-V2\antiinfectives\penicillin\	00002032902
Med-V2\antiinfectives\ampicillin\	60429002340
Med-V2\antiinfectives\Bactrim\	00003013850
Med-V2\antihypertensives\	L_029
Med-V2\antihypertensives\propranolol\	00005310223
Med-V2\antihypertensives\nifedipine\	00026881118

This human-maintained hierarchy is the starting point for adding additional codes automatically from larger lookup tables. For simplicity, hierarchical relationships are all expressed in a limited number of rows. If a change to the hierarchical organization of these rows must be made, a new version name is added to the top position (in the previous example the next version would be named "Med-V3"). The old hierarchy is retained in the table for backward compatibility, but a complete new hierarchy will be included with all items now starting with the name "Med-V3."

Straight mappings of other coding systems into the previous Base table are achieved by adding the coding system as an additional layer of names onto the path. For example, if one wanted to add mappings of local charge codes to the NDCs in the Base table, then the following piece of the Charge Code table would be joined to the Base table:

C_basecode	Name	C_localcode
00002032902	V-Cillin K V potassium	B00908765
00002032902	Pen-Vee K V potassium	B00908766
60429002340	Ampicillin	B00908767
00005310223	Propranolol hydrochloride	B00908903
00,005310223	Inderal	B00908904

and the following rows would then appear in the Lookup table. They would be added to both versions of the hierarchy, but only one is shown for simplicity:

C_fullpath	C_basecode
Med-V2\	L_027
Med-V2\antiinfectives\	L_028
Med-V2\antiinfectives\penicillin\	00002032902
Med-V2\antiinfectives\penicillin\V-Cillin K V potassium\	B00908765
Med-V2\antiinfectives\penicillin\Pen-Vee K V potassium\	B00908766
Med-V2\antiinfectives\ampicillin\	60429002340
Med-V2\antiinfectives\ampicillin\ampicillin\	B00908767
Med-V2\antiinfectives\Bactrim\	00003013850
Med-V2\antihypertensives\	L_029
Med-V2\antihypertensives\propranolol\	00005310223
Med-V2\antihypertensives\propranolol\propranolol hydrochloride\	B00908903
Med-V2\antihypertensives\propranolol\inderal\	B00908904
Med-V2\antihypertensives\nifedipine\	00026881118

The organization of concepts described previously allows the user to navigate the hierarchies and use the concepts in a query. Like a file path in Windows or Linux, the path of the hierarchy indicates in which groups the concept belongs, with the most general group listed on the far left and each group to the right increasingly specific. A user interface to present this

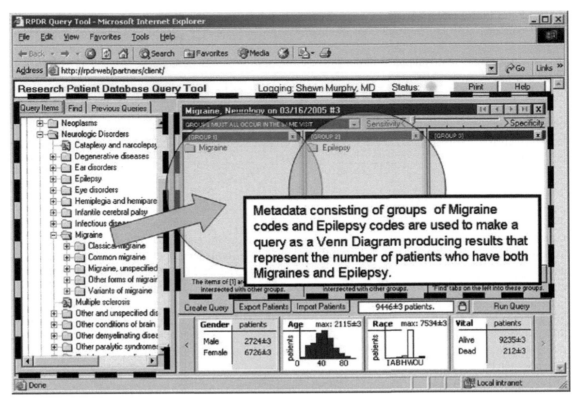

FIGURE 13.2   A visual query interface to a clinical data warehouse for research.

concept representation is shown in Fig. 13.2. The use of this interface has been described in detail (Murphy et al., 2003), and it is essentially a way of building queries visually using concepts represented in the Concept and Provider dimension tables.

Aggregating Data

In health care, the term "data warehouse" is generally reserved for those databases optimized for analysis and integrated queries across patient populations, as opposed to transactional databases that are optimized for rapid updates and highly specific kinds of retrieval (like those based upon a specific patient identifier).

There appear to be three fundamentally different approaches to organizing the health-care data warehouse. The first is to extract tables from the transaction systems of the health-care organization and load them into the database platform of the data warehouse with minimal transformation of the data model. The codes present in the columns are usually transformed to make them compatible with codes from other systems. For example, an ICD-9 diagnosis code stored as "27.60" in one system may be transformed to a common format of 02760. However, the tables are essentially left in the same schema as the transaction system (Inmon, 2005).

The second approach is more ambitious, where not just the codes from different systems are transformed to look the same, but the data are transformed to look the same as well. The diverse data coming from different systems must be made to fit into new tables. This involves a considerable amount of data transformation, but queries against the warehouse are then much less complex and can be performed much faster (Kimball, 2002).

The third approach is to keep the data local at its source in a "distributed" data warehouse. The local systems may choose to rehost the data in some special tables, or may support queries directly against the transaction data. Queries are distributed to the local databases across a network. In this environment, patient data are distributed across the local databases so that queries cannot be fully resolved in any single system. Thus, massive amounts of data will need to be aggregated across the network to be compared centrally. If hundreds of comprehensive queries need to be performed daily, the data traffic could stress or exceed the network capacity; they also provide poor query performance. So this architecture is usually used only if a centralized approach is not considered feasible, as when there is a need to provide a virtual data warehouse across many institutions (Drake et al., 2007; McMurry et al., 2007).

Regardless of the approach taken previously, the data warehouse will usually distribute data back to clinical investigators as a "data mart." These are subsets of the data in the data warehouse. The advantage of this approach is that the data can be prepared as per request, and do not have to exist in an investigator-friendly state in the warehouse itself. The EAV scheme is notoriously unfriendly to the casual user of data (Nadkarni and Brandt, 1998). Furthermore, the metadata exist in tables that are not obviously connected to the patient data. Thus, tables in the data warehouse often contain no humanly readable content. Finally, the data in the warehouse are usually updated once a day, which means that direct analysis of data in the warehouse would occur against constantly shifting data. The result of all of these factors is that the data are often exported into a data mart for further analysis. This also limits the set of clinical data that an investigator can view in detail, which is an important method for managing privacy concerns.

The health-care data warehouse should be built with patient privacy in mind. The most common strategy is to separate the data warehouse into two databases. The clinical data goes into one database, and the identifiers of the patients go into a second database. Access to the second, identified, database is strictly controlled and only accessed during data loading and the building of the data marts. The patients are given codes in the clinical database, and these codes can only be looked up in the identified database. Additional measures such as transparent database encryption (TDE) and field encryption may be implemented in both databases as an added measure of security. In this way, customers can use the clinical database and not have access to the patient identifiers.

Work Processes

Data generally flows into the data warehouse by loading it from transactional systems, or by duplicating the feeds of data that are going to the transaction systems. Clinical transactional systems usually provide the data as large "data dumps" or downloads. It is quite common for the download to consist of everything of interest in the transactional system available in the system across all time, as opposed to only new or incremental data since the last download. The reason for this approach is twofold. First, the transactional system may contain millions of records, but with current technology this can usually be written in just hours. Reloading all these data into the data warehouse similarly takes only a few hours. Thus, the simplicity of this model, as opposed to the complexity of the update models, makes this a much more desirable process. The risk of an incremental update process is that the errors in update flags will cause the data warehouse to become unsynchronized with the transactional system. Second, many transaction systems simply do not have a way to provide incremental updates.

When new data are obtained from the transactional systems, the data are analyzed for any new codes or coding systems. In principle, the transaction systems should have no difficulties providing the last "data dictionaries" of their codes to the data warehousing team. In practice, actual up-to-date data dictionaries can be surprisingly hard to come by. The codes that come from the transaction systems must be compared to those codes recognized by the data warehouse after every load. Any unrecognized codes must be reconciled with the transaction system to understand their meaning, and then published with the rest of the data warehouse metadata. This usually causes a delay in the data warehouse being able to recognize the latest codes from the transaction systems. The constant updating of the metadata is one of the most intensive tasks of the data warehousing team. The number of hours that need to be invested in this operational activity should not be underestimated.

When the data are loaded from the transactional systems, they are usually first loaded to a "staging area." As previously discussed, the data structure usually differs considerably between the transactional system and the data warehouse. Loading the transactional data into a staging area allows the data to be reviewed for quality assurance before proceeding with the complexity of transforming the data into the format of the data warehouse. Because the teams from the transactional systems are usually very familiar with the data in this form, it is desirable to have those teams responsible for transferring and loading such data into their corresponding staging area.

Considerations for Warehousing Genomic Data [Also See Chapters 16–19 (Human Genetics Section)]

Genomic data usually populate the CDW from laboratory systems. Increasing numbers of institutions have laboratories for molecular medicine as part of their infrastructure since genetic results can make up an important part of today's clinical decision-making. The genomic data that is sent to the clinical record often consist of variants expressed as single nucleotide polymorphisms (SNPs), patterns of RNA expression, and interpretative reports on these data by expert geneticists. Often the SNPs are obtained using microarray technology, resulting in thousands of reported values. Each SNP (of the thousands reported) can be entered into the CDW as a "fact" about the patient, such that each microarray adds thousands of new pieces of information on a patient. Current data warehouse design is able to handle this massive amount of information, which is accumulating at enormous rates. In fact, in January 2014, Illumina launched the HiSeq X Ten Sequencer which provides a complete sequencing of the genome for $1000. The raw storage to keep these data will require an average laboratory, processing 10000 patients a year, to add over 500 terabytes of storage per year by 2010 (Chan, 2005; Gullapalli et al., 2012).

Although the capability of determining a person's genotype has advanced by many orders of magnitude over the past 20 years, the capability to determine a person's phenotype has not advanced in a similar manner. One area of phenotyping that has advanced is the use of radiology. Images of the internal organs are now available in great detail from modern magnetic resonance imaging (MRI) and computed tomography (CT). Many of these images are stored in DICOM (Digital Imaging and Communications in Medicine), an imaging storage and communication standard which contains not only the pixel data of the image, but also a rich set of metadata tags that give additional information about the patient, study, scanner, and imaging protocol. However, the use of the data in these images presents its own challenges. DICOM does not necessarily adhere to an easily structured format and usually requires a programmer or third-party application to extract the DICOM metadata. Interpretation of the images often requires a clinician to read the image and enter the findings into a report. Other forms of phenotyping require a similar clinical interpretation to make use of the data. A patient's phenotype in even sophisticated clinical systems may consist of only a handful of useful concepts. Paradoxically, the biggest challenge in the context of this age of expanding genetic data is to acquire an equally detailed phenotype. "Sequencing the phenome" will be critical to taking advantage of genetic data (Freimer and Sabatti, 2003). The phenome is likely to be infinitely more complex and costly to sequence than the genome.

Research Data Warehouse Versus Population Management Databases

The function of the research data warehouse is often compared to large databases maintained for missions pertaining to hospital operations improvement or population management. The question is often whether a single database could satisfy both functions. Because of the obvious economic advantages of only needing to maintain one data store, we will consider this in detail.

The necessity to perform large queries that cut through many patients is common to both types of databases. Both types of databases require a high degree of structure and organization in the data, and both require extensive metadata to aggregate groups of concepts (such as all the codes pertaining to type 2 diabetes) when formulating queries. Neither works well as a transactional system (with very frequent updates) because they both need large indexes to perform large queries rapidly.

However, a common architecture is usually not chosen for both databases, thus often requiring them to live separately. There are several reasons for this observation. First, the goal of the research data warehouse is to integrate data which encourages organization of data into a large fact table as described previously. The hospital operations database is often not focused on putting together all the data on a patient, but rather on very specific aspects of the patient, for example, who is the primary care provider of the patient? Second, research data warehouses are often built around deidentified information, while hospital operations databases are usually only accessed by individuals who have full access to patient-protected health information. Third, research data often need to be integrated in a specific manner into a specific data model so that it can support data exchange between hospitals. Fourth, most researchers request large downloads of data from the warehouse, which they will then use for developing computed phenotypes and epidemiological research. Therefore, the ability to extract data easily to perform free-form analytics is the fulcrum of most research databases, while running specific, targeted reports is the primary output of most operational systems. Fifth, the research database usually needs to contain unstructured data, such as pathology reports and electrocardiograms, while operations databases are focused almost exclusively on structured data. Sixth, research databases often want "Big Data" included, such as genomics and imaging data, which is unusual in operations databases. Finally, the refresh cycles often need to be much higher with operations databases than research databases.

Data validation is important for both types of databases, but the emphasis is often different on the kind of validation that is required. The research database is focused mostly on patients' disease states which often demand complex computation algorithms to determine the disease status of the patient. The operations databases also will often determine data with complex algorithms, but the emphasis is usually on an aspect of billing or patient management, such as associating patients with the correct primary care provider. One of the primary activities of research databases is to identify patients with diseases, so they can be placed in a registry of patients to be recruited for clinical trials. This will be discussed next.

INFORMATION TECHNOLOGY SUPPORT OF PARTICIPANT RECRUITMENT

Strategies for Participant Recruitment

Participant recruitment remains one of the most difficult and time-consuming phases of performing a clinical trial. Advertising of new trials in traditional media such as newspapers and public billboards is still the mainstay of many trial recruitment efforts. But the use of information technology is beginning to show its impact in several ways. Conceptually, participant recruitment can be divided into three main categories based on two key prerequisites for successful enrollment: participant eligibility and participant interest. The first determines whether a participant is suitable for a study, and the second determines whether a participant would be willing to participate in a study. The ordering of these prerequisites determines the approach to recruitment. One can argue that there is no need to screen participants who have not displayed interest, and hence the strategy behind mass advertising as well as anticipatory participant contact registries. Conversely, one can argue that there is no need to reach out to participants who will not meet the eligibility criteria, and hence the strategy behind targeted recruitment. Both strategies, as well as hybrid strategies, are suitable for information technology enhancements. Several strategies that take advantage of information technology are highlighted in the following sections.

Reaching Only Interested Participants

If mass advertising does ensure participant interest at the outset (since presumably only those who respond are interested), the cost of such advertising can be steep, and the return on investment can be quite low. Approaches to improving the rate of response by narrowing the denominator are therefore desirable. One such approach is through the use of online advertising. Although some potential exists for targeted advertising in traditional media (selection of a particular journal or newspaper, use of public transport billboards, or even the decision to locate an advertisement in a specific newspaper section such as sports), the instruments are fairly blunt and nonspecific. The advent of online advertising opens new possibilities. Advertising can be based on keywords within Web pages or Web search terms in quite specific ways. For example, advertising can be placed for a diabetes trial whenever someone searches with terms such as "diabetes," "weight loss," "insulin," "sugarless," or other terms that might suggest a diabetic patient.

Another approach gaining increased use is the use of participant contact registries. These are registries prepopulated through various methods with individuals who have voluntarily self-identified themselves as interested in clinical research, usually in particular disease areas. This approach mirrors the trend in patient involvement in their health care, in that it empowers individuals to self-organize in their interests for study participation. Several academic health centers have such registries available for interested volunteers, including Massachusetts General Hospital and Brigham and Women's Hospital (RSVP for Health: http://www.rsvpforhealth.org/), and Washington University School of Medicine in St Louis (Research Participant Registry: https://vfh.wustl.edu/), among others.

Identifying Only Eligible Participants

An alternative approach to recruitment is to first identify eligible participants and then contact those individuals for potential interest in a clinical study. While this strategy can be used with basic paper chart review, this process is quite tedious. With the advent of electronic systems, the process can be automated in very efficient ways. One common approach now is to simply mine accessible electronic repositories to identify potentially eligible participants. This "static" method is particularly suitable for those clinical enterprises that have access to some form of clinical databases. These databases include basic demographics at a minimum. At the most, they contain rich clinical data in the form of laboratory and other test results, diagnoses, medications, and other patient attributes. Cancer or tumor registries are also commonly used as sources for this approach. Patient lists generated from these digital queries are then provided to investigators who subsequently contact these potential subjects individually. These methods do raise legitimate concerns about patient privacy, and, consequently, contact with patients is usually mediated through patients' personal physicians (Beskow et al., 2006).

An emerging and novel approach is the dynamic identification of potential participants from electronic health records (EHRs). The increasing prevalence of EHRs makes it practical to highlight patients who are potential candidates for clinical studies. Use of clinical data in the form of electronically documented diagnoses, medications, laboratory results, and routine demographics provides the basis for automated searches. Notification of potential candidates can then be done in a variety of ways. Email or pager notification has been used (Afrin et al., 2003), as has notification at the point of care through an EHR. The latter approach in some studies resulted in a 10-fold increase of referrals and a doubling of referral rate (Embi et al., 2005).

PRINCIPLES OF DATA COLLECTION

Automation and the Human Element

Guidelines for the design of case report forms are applicable to their online counterparts. It is critical to collect all data necessary (and nothing more) in a practical and standard format that allows for the appropriate aggregation of results and a statistically valid analysis (Madison and Plaunt, 2003). While online forms can suffer from the same inconsistencies and complexities that paper forms have, technology introduces the potential for additional confusion because of the human—computer interface. Strategies have been put forth to help alleviate this issue (Shneiderman, 1998). Allowing unwanted interactions between electronic systems and humans can introduce bias during data collection activities. Several examples of common design flaws specific to online data collection applications include the following:

- Difficult user interfaces that cause users to misread questions or skip over fields.
- Default selections that do not require mandatory review.
- Mandatory fields that force a user to select an inappropriate option because of no alternatives.
- Batch selection ("Check all") becomes a path of least resistance.

Data Validity: Capturing Data With Consistent Metadata

Designers of data collection applications may falsely assume that a clear and usable user interface will ensure valid data capture. In striving to reduce technology barriers, the need to accurately assess user knowledge and incorporate well-understood terminology may be overlooked. If nothing else, valid data collection is dependent on maintaining and using consistent metadata, or data about the data being collected. Such metadata may include explicit definitions, allowable ranges or choices, and information about the contexts in which the data will be used. Without accurate metadata, data that appears comparable superficially may not be valid at all. Use of appropriate metadata can also be a useful asset for data transformation for analytic use (Brandt et al., 2002).

Continuous Quality Assurance

Data collected without quality assurance is not only likely to suffer from inaccuracies at any point in time, but also will likely foster a culture where data entry errors increase over time due to the lack of feedback to users. Therefore, it is imperative that quality assurance is a pervasive and continuous feature of all data collection activities. A classic approach to ensure quality and validity in many electronic data capture (EDC) applications is the use of double data entry for case report forms. Despite the obvious appeal of this simple approach, studies over time have shown that the benefit is slight and the extra time incurred is significant (Gibson et al., 1994; Buchele et al., 2005). This is because errors caught with double entry are limited to keying errors, while many other categories of errors will simply be keyed in wrong—twice. An increased focus on making metadata as visible as possible to users as well as active through range and value checks, and techniques for a full cycle of quality assurance, such as postentry sampling and verification, is likely to be of more value (King and Lashley, 2000).

"Stand-alone" Electronic Data Capture

The most common approach today for computer-based clinical research data collection is through the use of "stand-alone" EDC systems. "Stand-alone" in this context refers to the fact that these systems stand independent from clinical systems. In fact, while many small clinical studies do use true stand-alone, "homegrown" systems based on spreadsheets, word processing documents, or other similar methods to collect data, larger studies invariably use EDC modules that are part of a comprehensive clinical trial management software system.

Integration With Clinical Documentation

It is well understood to those who do clinical investigation that routine clinical documentation is not immediately suitable for rigorous clinical research. Variability among clinicians, both in their approach to and detail of documentation, as well as the lack of consistent definitions, means that routine clinical data is often not comparable without additional human review.

On the other hand, the dedicated collection of clinical research data apart from clinical care is an expensive proposition. The majority of studies, however, must take this approach out of necessity. The holy grail of clinical research data collection is a true integration of clinical research data collection with routine clinical documentation. Although this goal is rarely achieved, and solutions have not been generalized, there are some basic models for attacking this problem that will be described in the following paragraph.

Integrated electronic data collection using a one-pass model means that a clinician will take a single pass at documentation. This single pass will collect all relevant clinical information as well as any detailed clinical research data. An alternative approach that is often used is to have a clinician perform clinical documentation first, and then supplement it with an additional set of case report forms for research—a two-pass model. In this model the clinical documentation will typically follow the form of the local standard for the enterprise—anything from the EHR. The additional case report forms are then implemented using a similar format. Workflow considerations will dictate which approach may work better in any particular situation.

Problems With Integrated Models

Integrated models for data collection have fundamental challenges that must be addressed even if basic workflow issues have been resolved. One of the most taxing issues is the balance between expressiveness and precision. Many outstanding clinicians will testify to the importance of expressiveness in their clinical documentation, both to truly characterize a patient's clinical condition in detail and also to support their case with full human context. Conversely, expressiveness is anathema to clinical research data collection where precision is needed. Templates to limit choices and categorize patients more precisely are desirable for clinical research. If these templates are used in place of clinical narrative, they will often result in dry-sounding prose where one patient is nearly indistinguishable from another. Clearly, there are trade-offs in this tension, as well as strategies such as juxtaposing templates for precision with free-form fields for expressivity, but in practice any approach taken inevitably requires compromises.

One issue is the conflict between time and costs and research valid data. Health-care providers are under increasing pressure to reduce costs, which usually means spending less time with a patient or obtaining patient data. If one has to complete a patient visit and write up every 20–30 min, inaccurate and incomplete data for research are likely to be common. With clinical research increasingly needing accurate, individual environmental data, e.g., salt intake, to precisely interpret research finding, this conflict will be accelerated. Developing effective compromises likely will result in a substantial increase in cost. Thus, the major advantage, lower cost, of an integrated over a stand-alone model of EDC research system may be substantially diminished.

Another issue often overlooked in the development of integrated clinical research data collection is the need for validation as a postclinical process. Assuming that the primary data can be collected reasonably by clinicians, it remains critical that those data be validated from a research perspective before being entered as "research ready" into a final research repository. This step also allows for quality assurance such as checking for missing fields, unexpected variability, and confirmation of appropriate data collection through comparison with other clinical data sources. Tools to help make this postclinical process efficient are important, since rapid review and sampling techniques will be necessary.

Finally, it is critical to understand the basic difference between the nature of clinical documentation and clinical research data collection. Clinical documentation must be preserved in a stable way. All versions of a clinical document must be attributed clearly to authors, and guarantees need to be made about final documents being unalterable. Research data, on the other hand, will often require correction or updates when new information is available and may be entered by numerous investigative staff. For example, a clinician may identify a patient as having "renal failure" as a preoperative risk factor. Later, if review of the data reveals that by strict research criteria the patient does not qualify as having renal failure (e.g., creatinine does not exceed a specified threshold), the research data field will need correction. It would be inappropriate for such corrections to modify the historical clinical documentation in any way since that documentation may have been viewed by other clinicians and formed the basis of other clinical decisions. This implies that while integrated data collection models will usually have detailed databases with specific data fields to hold data, the clinical documentation generated from these fields will need to be preserved in its entirety as a "snapshot" with a life cycle independent from the way the data fields are used in subsequent research data flow. Such clinical snapshots can be implemented in a variety of ways—paper output, digital storage of a complete document, or careful versioning of individual data fields such that alterations for research purposes do not impact clinical documentation. Data flow associated with collection of clinical research data in the context of clinical care is represented in Fig. 13.3.

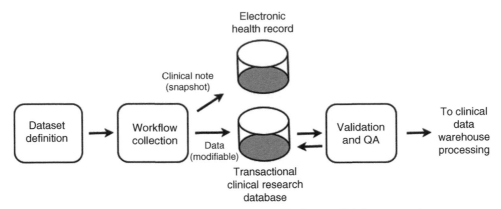

FIGURE 13.3   Data flow diagram for integrated data collection for clinical and research purposes.

DATA STANDARDS

Standards unique to data management in clinical research are relatively new compared with standards for clinical data management. The major organization promoting standards in this arena is the Clinical Data Interchange Standards Consortium (CDISC), "a global, open, multidisciplinary, non-profit organization that has established standards to support the acquisition, exchange, submission and archiving of clinical research data and metadata." Health Level Seven (HL7) is a well-established standards organization that develops specifications broadly for health care and is known for its clinical document architecture (CDA) (http://www.hl7.org). These two organizations have collaborated to promote standards for "single source" data capture for both the clinical record and clinical research (Kush et al., 2007).

Existing clinical research data standards focus primarily on interchange standards for the automation of data exchange and data aggregation. For example, the electronic common technical document (eCTD) is a specification that allows for the electronic transmission of the common technical document from an applicant (e.g., pharmaceutical company) to a regulator (e.g., the FDA). The eCTD was developed through a working group of the International Conference on Harmonization. For those using CDISC standards, the FDA accepts CDISC's case report tabulation data definition specification (also known as "define.xml") for the submission of study data. This trend toward interoperability between standards, rather than the need to grow one standard to meet everyone's needs, is particularly important (Lueders, 2005). It means that different standards organizations can focus on different areas in need of standards, knowing that ultimately these multiple standards can be used together. In some domains, organizations have emerged to address the interoperability of standards specifically (e.g., the Web Services Interoperability Organization). The clinical research enterprise appears to be embracing an approach that acknowledges that "best of breed" solutions can lead to more effective use of standards.

Use of Web Standards

One can argue that much of the recent innovation within clinical research information technology is driven by the continuing development of new data standards and exchange protocols built on top of the Internet and the World Wide Web ("Web"). The Web at its core has always been a collection of standards driven technologies. Over the last 10 years, many standards built for data exchange revolved around the use of the Web's eXtensible Markup Language, or XML (http://www.w3.org/XML/). XML allows for flexibility in the specification of document formats while retaining the ability to have such documents automatically processed and validated field by field (see Fig. 13.4). Unfortunately, XML also tends to be excessively verbose which can be a deterrent for the exchange of large data sets. Newer data exchange standards such as JSON (JavaScript Object Notation) look to address this issue by providing a streamlined language syntax while maintaining the general readability of the structure and the data contained within. Continued innovation has been focused on finding better and faster ways to compress and serialize JSON data to minimize the amount of data that is transferred at any one time. Web protocols such as HyperText Transfer Protocol (HTTP) and Simple Object Access Protocol (SOAP) continue to be the standard by which data is usually exchanged. These protocols have largely replaced the use of more traditional modes of electronic data exchange such as dedicated point-to-point connections between systems.

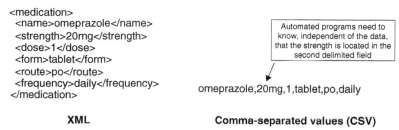

```
<medication>
  <name>omeprazole</name>
  <strength>20mg</strength>
  <dose>1</dose>
  <form>tablet</form>
  <route>po</route>
  <frequency>daily</frequency>
</medication>
```

Automated programs need to know, independent of the data, that the strength is located in the second delimited field

omeprazole,20mg,1,tablet,po,daily

XML   **Comma-separated values (CSV)**

FIGURE 13.4   A comparison of data in XML format versus a standard delimited format.

The key aspect of these Web protocols is that they can be specified in terms, so they can be implemented without lengthy negotiations between exchanging parties.

Content Standards

Another major area is the standards that specify data sets to cover clinical specialties for clinical research. While some of the major standards organizations are starting down this path (e.g., CDISC has some work in development for cardiovascular data), most of this work has been done within the context of professional societies. In part, this is because many professional societies have a vested interest in the advancement of their clinical field through clinical research. Practically speaking, these societies are also where clinical experts can be called on to define the data sets that are most relevant for a particular clinical area or disease.

An emerging problem with these activities is that varying content standards are being established by different organizations for the same clinical area. For example, for carotid arterial disease and its associated interventions, the American College of Cardiology (ACC) and the Society of Vascular Surgery (SVS) are just two of the many that have established prospective registries. While the ACC and SVS (and others) are collaborating to make data elements across their registries as interoperable as possible, the coordination of multiple registries, their related data sets, and their multiple methods of data submission will remain a significant challenge to clinical research data management for the foreseeable future.

CLINICAL TRIAL MANAGEMENT SYSTEMS

A full discussion of clinical trial management systems (CTMSs)—software to help investigators manage the process of running a clinical trial—is beyond the scope of this chapter. Suffice it to say that numerous solutions now exist to help manage clinical trials end to end. These software solutions span a variety of approaches, including the following:

- Traditional turnkey CTMS from large commercial vendors (e.g., Oracle Clinical, Phase Forward Clintrials).
- Newer Web-based hosted solutions.
- Add-ons to general productivity software such as Microsoft Office, Cognos.
- The emerging breed of open source software (e.g., TrialDB, OpenClinica).

The scope of what a CTMS can do is broad and varies depending on the actual software implementation, but all CTMS have two key overarching characteristics that are in balanced tension: consistency and compliance, and customizability. The former is critical to establish invariate processes that drive regulatory requirements as well as good scientific practice. The latter is mandatory because all trials differ in their design and the type of data that need to be captured and managed. CTMS are sometimes inappropriately equated with clinical research information technology generally. Whereas CTMS's primary role is managing all aspects of clinical trials, clinical research information technology must support a wide variety of clinical studies and research.

PUBLICLY AVAILABLE DATABASES

With the advent of the World Wide Web (WWW) and commonality of Web browser software, databases to support clinical research have proliferated. Many of these databases, such as Online Mendelian Inheritance in Man (OMIM), are related to diverse aspects of genomics and are discussed elsewhere in this book (Chapter 17). Several prominent, freely available databases of primary biomedical information will be discussed here briefly. Many other free databases exist, as do many fee-based databases, and there will be no attempt to list them comprehensively, since such a list would become hopelessly outdated almost immediately.

Biomedical Literature

One of the most prominent public sources is PubMed (http://www.ncbi.nlm.nih.gov/pubmed/), a Website that provides access to the MEDLINE database of citations and abstracts of biomedical research publications. PubMed and its associated databases are provided by the National Library of Medicine (NLM) and the National Institutes of Health (NIH), and it has become a mainstay with investigators and the general public alike. In addition, the NLM has also established PubMed Central, a freely available collection of full text articles. Recently, the US government has established a policy requiring all NIH-funded research publications to be submitted to PubMed Central in full text form. Unlike PubMed that has enjoyed broad support almost uniformly, the concept of the government as a publisher has raised a lively debate about copyright, economics, and other related issues (Greenberg, 1999; Markovitz, 2000).

Clinical Trials and Research

Another source with growing importance is the NLM's ClinicalTrials.gov resource (McCray and Ide, 2000). ClinicalTrials. gov "offers up-to-date information for locating federally and privately supported clinical trials for a wide range of diseases and conditions" (http://clinicaltrials.gov). Specific information about clinical trials is submitted directly by study sponsors or their representatives. Information available for each trial includes the disease or treatment studied, a study design and description, eligibility requirements, study sites, contact information, and relevant links to other Web-based resources such as PubMed.

CenterWatch (http://wwwcenterwatch.com) is another publicly available online clinical trials listings database that should be mentioned because it preceded ClinicalTrials.gov. CenterWatch provides some additional services not available through ClinicalTrials.gov (partly because of some of the requirements placed on government agencies if any user information is to be collected). For example, patients can register their email with CenterWatch to receive email notification about new trials in specific therapeutic areas.

The Computer Retrieval of Information on Scientific Projects (CRISP) (http://crisp.cit.nih.gov) is a searchable database of federally funded biomedical research projects. Maintained by the Office of Extramural Research at the NIH, it includes projects funded by the NIH and several other government agencies including, but not limited to, the Food and Drug Administration (FDA), Centers for Disease Control and Prevention (CDCP), and Agency for Healthcare Research and Quality (AHRQ).

Evidence-Based Medicine

The Cochrane Collaboration has made available free abstracts of its Cochrane Reviews of evidence-based health care (http://www.cochrane.org). It also has fee-based access to its complete collections. The National Cancer Institute supports a comprehensive cancer database called PDQ (Physician Data Query). It is a combination of three components that contain "peer-reviewed summaries on cancer treatment, screening, prevention, genetics, and supportive care, and complementary and alternative medicine; a registry of more than 6000 open and 17,000 closed cancer clinical trials from around the world; and a directory of professionals who provide genetics services" (http://www.cancer.gov/cancertopics/pdq/cancerdatabase).

THE GROWING IMPACT OF BIG DATA AND THE CLOUD

As has been touched upon throughout this chapter, information technology has an ever-growing impact on clinical research. In the early 2000s, the WWW created the foundation for many innovations, from enabling data exchange, to hosting complex software and large publicly available databases. A subsequent explosion in Web development for clinical research ensued with the advent of the so-called "Web 2.0" technologies. These technologies allowed software developers to publish software not only with specific capabilities, but also with open protocols that allow other developers to incorporate or leverage such software with relatively little effort. A well-known example of this type of Web-based software is Google Maps. Other developers can leverage Google Maps through its published Application Programming Interface, or API. The difference here is that, traditionally, the use of APIs in programming required integration of software at the source-code level—a relatively complex and tightly coupled approach limited to expert software engineers. Web-based APIs allow developers to integrate software in a loosely coupled fashion over the Web, using basic Web protocols. This means that those who have access to a lot of domain expertise in clinical research, but may only have access to relatively amateur programmers, can build rapidly upon software written by others. For example, the history of Basic Local Alignment Search Tool (BLAST), the well-known sequence similarity search program, has followed the trajectory from a stand-alone set of BLAST algorithms and programs, to a proliferation of Web-based user interfaces to BLAST engines, and finally to an integration of BLAST methods across publicly available data sets (Bhadra et al., 2006).

The growth of cloud computing or "the cloud" has also had a significant impact on the ability of clinical researchers to work with large data sets and run complex analyses. Instead of allocating a physical server to a single use, the cloud enables the attributes of this server (storage, CPU, memory) to be pooled with other servers that then get shared among many applications or users. In many instances, these resources also get virtualized whereby the representation of the server presented to a user is completely abstracted from the physical hardware that supports it; a user may think they are running an analysis on a desktop machine when they are actually running as part of a smaller application hosted across servers. The benefit is that virtualized environments can be customized to specific tasks, easily scaled up or down based on needs, and created or destroyed on demand. Although usually offered for a fee, the clinical researcher pays only for the resources he/she need and has instant access to large quantities of storage and computing resources that previously had to be allocated through an information technology department, or paid for and set up independently. Of course, clinical researchers need to exercise caution when moving protected health information into a shared environment as this opens up the possibility of a data breach.

Beyond cloud computing and virtualized environments, clinical researchers also have access to a new generation of data storage technologies aimed at storing enormous amounts of data for quick analysis. In 2004, Google paved the way for "Big Data" by publishing a seminal paper introducing the world to MapReduce, a new way of analyzing terabytes of data on a distributed network of commodity machines (Dean and Ghemawat, 2004). By splitting the data into small chunks and then distributing these chunks over a large network of computers, Google found a way to create a massively parallel environment that could churn through Weblogs at incredible speeds. Google's framework was later adopted by Doug Cutting who created Hadoop, a scalable open source platform for the storage and analysis of big data. Hadoop and HDFS, the file system at the core of the Hadoop platform, as well as similar technologies created by companies such as Facebook and Yahoo, have challenged the limits of traditional relational database management systems and pushed the upper bounds of data that can be analyzed at any one time. In response, large relational database vendors such as Microsoft and Oracle have created their own distributed database systems. Other data warehousing companies have focused on integrating traditional database structures with new HDFS storage capabilities. As the database landscape continues to evolve, the promise of storing and analyzing petabytes of data has become a reality and offers untold potential for clinical research discovery, especially in the age of inexpensive genetic sequencing.

With these newfound abilities to integrate software capabilities across the WWW network through cloud and big data technologies, the advantages for accelerating data exchange as well as algorithm and computing exchange are clear. These advantages will also come with new challenges, primarily in the area of privacy concerns as patient data become virtualized across wide geographic areas in digital form. As patients become increasingly involved in their personal health care through the use of personal health information technologies, they will also become empowered to become full-fledged members of the clinical research enterprise. No longer only subjects for recruitment, patients will be personally providing data and consent online, contributing personally in active ways to clinical research data sets by sharing treatment experiences (see http://www.patientslikeme.com) and, quite possibly, self-organizing to perform their own clinical research as investigators in their own right.

CONCLUSION

Although related to clinical information technology, clinical research information technology poses its own data management challenges that are unique. The problems of using CDRs as a data source for research are examined, as are the advantages of using preresearch clinical registries. Clinical data warehousing principles are described, including detail on database design, the organization of metadata, and work processes. Techniques for applying information technology to the vexing problem of study participant recruitment are discussed. A review of the principles of data collection for clinical research covers issues of automation, data validity, and the integration of research data collection with clinical care. The discussion also touches briefly on data standards in clinical research, clinical trial management systems, publicly available biomedical literature databases, and emerging approaches to data integration using new cloud and big data technologies.

REFERENCES

Altiparmak, F., Ferhatosmanoglu, H., Erdal, S., Trost, D.C., 2006. Information mining over heterogeneous and high-dimensional time-series data in clinical trials databases. IEEE Trans. Inf. Technol. Biomed. 10 (2), 254–263.

Afrin, L.B., Oates, J.C., Boyd, C.K., Daniels, M.S., 2003. Leveraging of open EMR architecture for clinical trial accrual. AMIA ... Annual Symposium Proceedings/AMIA Symposium 16–20.

Bellazzi, R., Zupan, B., 2008. Predictive data mining in clinical medicine: current issues and guidelines. Int. J. Med. Inf. 77 (2), 81–97 (Review, 108 refs).

Beskow, L.M., Sandler, R.S., Weinberger, M., 2006. Research recruitment through US central cancer registries: balancing privacy and scientific issues. Am. J. Publ. Health 96 (11), 1920–1926.

Brandt, C.A., Morse, R., Matthews, K., Sun, K., Deshpande, A.M., Gadagkar, R., Cohen, D.B., Miller, P.L., Nadkarni, P.M., 2002. Metadata-driven creation of data marts from an EAV-modeled clinical research database. Int. J. Med. Inf. 65 (3), 225–241.

Buchele, G., Och, B., Bolte, G., Weiland, S.K., 2005. Single vs. double data entry. Epidemiology 16 (1), 130–131.

Bhadra, R., Sandhya, S., Abhinandan, K.R., Chakrabarti, S., Sowdhamini, R., Srinivasan, N., July 1, 2006. Cascade PSI-BLAST web server: a remote homology search tool for relating protein domains. Nucleic Acids Res. 34 (Web Server issue), W143–W146.

Chan, E.Y., 2005. Advances in sequencing technology. Mutat. Res. 573 (1–2), 13–40.

Drake, T.A., Braun, J., Marchevsky, A., Kohane, I.S., Fletcher, C., Chueh, H., Beckwith, B., Berkowicz, D., Kuo, F., Zeng, Q.T., Balis, U., Holzbach, A., McMurry, A., Gee, C.E., McDonald, C.J., Schadow, G., Davis, M., Hattab, E.M., Blevins, L., Hook, J., Becich, M., Crowley, R.S., Taube, S.E., Berman, J., The Shared Pathology Informatics Network, 2007. A system for sharing routine surgical pathology specimens across institutions: the Shared Pathology Informatics Network. Hum. Pathol. 38 (8), 1212–1225.

Dean, J., Ghemawat, S., 2004. MapReduce: simplified data processing on large clusters. In: OSDI'04, 6th Symposium on Operating Systems Design and Implementation, Sponsored by USENIX, in Cooperation with ACM SIGOPS, pp. 137–150.

Embi, P.J., Jain, A., Clark, J., Bizjack, S., Hornung, R., Harris, C.M., 2005. Effect of a clinical trial alert system on physician participation in trial recruitment. Arch. Intern. Med. 165 (19), 2272–2277 (see comment).

Foster, N.L., Gombosi, E., Teboe, C., Little, R.J., 2000. Balanced centralized and distributed database design in a clinical research environment. Stat. Med. 19 (11–12), 1531–1544.

Freimer, N., Sabatti, C., 2003. The human phenome project. Nat. Genet. 34 (1), 15–21.

Gullapalli, R.R., et al., 2012. Next generation sequencing in clinical medicine: challenges and lessons for pathology and biomedical informatics. J. Pathol. Inf. 3, 40.

Gibson, D., Harvey, A.J., Everett, V., Parmar, M.K., 1994. Is double data entry necessary? The CHART trials. CHART Steering Committee. Continuous, hyperfractionated, accelerated radiotherapy. Control. Clin. Trials 15 (6), 482–488 (see comment).

Greenberg, D.S., 1999. National Institutes of Health moves ahead with 'PubMed central'. Lancet 354 (9183), 1009.

Inmon, W.H., 2005. Building the Data Warehouse, fifth ed. John Wiley & Sons, New York.

Kimball, R., 2002. The Data Warehouse Toolkit: The Complete Guide to Dimensional Modeling, second ed. John Wiley & Sons, New York.

King, D.W., Lashley, R., 2000. A quantifiable alternative to double data entry. Control. Clin. Trials 21 (2), 94–102.

Kush, R., Alschuler, L., Ruggeri, R., Cassells, S., Gupta, N., Bain, L., Claise, K., Shah, M., Nahm, M., 2007. Implementing single source: the STARBRITE proof-of-concept study. J. Am. Med. Inf. Assoc. 14 (5), 662–673.

Lueders, H., 2005. Interoperability and Open Standards for eGovernment Services. White Paper, Computing Technology Industry Association.

Mullins, I.M., Siadaty, M.S., Lyman, J., Scully, K., Garrett, C.T., Miller, W.G., Muller, R., Robson, B., Apte, C., Weiss, S., Rigoutsos, I., Platt, D., Cohen, S., Knaus, W.A., 2006. Data mining and clinical data repositories: insights from a 667,000 patient data set. Comput. Biol. Med. 36 (12), 1351–1377.

Murphy, S.N., Morgan, M.M., Barnett, G.O., Chueh, H.C., 1999. Optimizing healthcare research data warehouse design through past COSTAR query analysis. AMIA ... Annual Symposium Proceedings/AMIA Symposium 892–896.

Murphy, S.N., Gainer, V., Chueh, H.C., 2003. A visual interface designed for novice users to find research patient cohorts in a large biomedical database. AMIA ... Annual Symposium Proceedings/AMIA Symposium 489–493.

McMurry, A.J., Gilbert, C.A., Reis, B.Y., Chueh, H.C., Kohane, I.S., Mandl, K.D., 2007. A self-scaling, distributed information architecture for public health, research, and clinical care. J. Am. Med. Inf. Assoc. 14 (4), 527–533.

Madison, T., Plaunt, M., 2003. Clinical data management. In: Chow, S.C. (Ed.), Encyclopedia of Biopharmaceutical Statistics, second ed. Informa Healthcare.

Markovitz, B.P., 2000. Biomedicine's electronic publishing paradigm shift: copyright policy and PubMed central. J. Am. Med. Inf. Assoc. 7 (3), 222–229.

McCray, A.T., Ide, N.C., 2000. Design and implementation of a national clinical trials registry. J. Am. Med. Inf. Assoc. 7 (3), 313–323.

Nigrin, D.J., Kohane, I.S., 1998. Data mining by clinicians. AMIA ... Annual Symposium Proceedings/AMIA Symposium 957–961.

Nadkarni, P.M., Brandt, C., 1998. Data extraction and ad hoc query of an entity-attribute-value database. J. Am. Med. Inf. Assoc. 5 (6), 511–527.

Prather, J.C., Lobach, D.F., Goodwin, L.K., Hales, J.W., Hage, M.L., Hammond, W.E., 1997. Medical data mining: knowledge discovery in a clinical data warehouse. Proceedings/AMIA Annual Fall Symposium 101–105.

Stein, H.D., Nadkarni, P., Erdos, J., Miller, P.L., 2000. Exploring the degree of concordance of coded and textual data in answering clinical queries from a clinical data repository. J. Am. Med. Inf. Assoc. 7 (1), 42–54.

Silberschatz, A., Korth, H.F., Sudarshan, S., 2005. Database System Concepts, fifth ed. McGraw-Hill, New York.

Shneiderman, B., 1998. Designing the User Interface: Strategies for Effective Human-Computer Interaction, third ed. Addison–Wesley, Reading, MA.

Teorey, T., 1999. Database Modeling and Design: The Fundamental Principles, third ed. Morgan Kaufmann Press, San Francisco.

Chapter 14

Principles of Biostatistics

Kush Kapur

Harvard Medical School, Boston, MA, United States

Chapter Outline

Key Points

- Descriptive statistics and univariate analysis provide an initial summary for all the variables.
- Central limit theorem is the key to understanding the validity of any statistical procedure.
- Most of the model building in the medical field relies on generalized linear modeling framework which encompasses a very broad class of models such as linear, logistic, Cox proportional hazard, multinomial, and Poisson regression models.
- False discovery rate (FDR) is a less conservative approach for multiple comparisons than Bonferroni's adjustment.
- In statistical literature, the missing mechanisms are classified into three categories: missing completely at random (MCAR), missing at random (MAR), and missing not at random (MNAR).
- Generalized linear mixed-effects models are an extension of generalized linear model which incorporates random effects in the linear predictors to account for correlations within the observations parsimoniously.

INTRODUCTION

Over the past three decades, randomized controlled trials (RCTs) have evolved tremendously making them one of the standard and effective approaches for determining the efficacy and safety of a medical treatment or procedure. RCTs that entail sound study design principles such as identification of homogenous patient population, a well-defined hypothesis, appropriate estimates of sample sizes, etc., require only simple inferential statistical procedures for establishing the positive treatment effects. However, there are instances where a carefully designed RCT can be challenging to implement due to inherent nature of the disease. Moreover, investigators are often faced with issues of slow recruitment, subjects' dropouts, and noncompliance. To avoid future pitfalls to answer primary clinical questions, it should be required that clinical trial investigators have a thorough understanding of basic statistical principles. This understanding helps them effectively establish a communication bridge with the study biostatistician during the design stage of study.

In this chapter, we will first cover the basic statistical principles before we introduce readers to the linear modeling framework. We also will provide details on the linear mixed-effects modeling framework to handle nested/hierarchical data. We have included basic statistical assumptions and definitions of missing data which are commonly employed nowadays during the sensitivity analysis stage. In the summary section, we have provided a few excellent references for the readers who are interested in developing a deeper understanding of the methods discussed in this chapter.

For illustration purposes throughout this chapter, we will use a clinical trial in the area of epilepsy, specifically conducted in patients with status epilepticus (SE). SE is defined as an epileptic seizure of greater than 5 min or more than one seizure within a 5 min period without the subject returning to consciousness during this period. It is a life-threatening condition particularly if treatment is delayed.

Example: Alldredge et al. (2001) published their findings in the New England Journal of Medicine (NEJM) about a randomized, double-blind trial for the treatment of out-of-hospital SE using intravenous benzodiazepines. In this trial, subjects above the age of 18 were randomized to receive intravenous diazepam (5 mg), lorazepam (2 mg), or placebo in case they had prolonged or repetitive generalized convulsive seizures. If needed, subjects received an identical injection up to a maximum of 10 mg of diazepam or 4 mg of lorazepam. The study defined termination of SE by the time of arrival at the emergency department as the primary endpoint. The time period of SE before arrival at the hospital was defined as the interval between administration of the study drug and termination of SE. The duration was considered as censored in case the seizures were ongoing on arrival to the emergency department or in case open-label drug treatment had to be administered before arrival to the emergency department. We will discuss the details of censoring in the sections focusing on the types of outcome and survival analysis. In addition, there were five secondary outcomes defined in the study: (1) out-of-hospital complications related to the occurrence of respiratory or cardiovascular complication after the study drug administration, (2) presence of cardiorespiratory complications at the time of transfer to the emergency department, (3) duration of SE before arrival at the hospital defined as the interval between administration of the study drug and the termination of SE, (4) neurologic outcome at discharge categorized as unchanged new neurologic deficits or death, and (5) location to which the patient was transferred from the emergency department. Some of the important analysis techniques discussed in this chapter will be explained thoroughly with help of the statistical methods used in this published trial.

Prior to focusing on the statistical analysis techniques, we will first introduce readers to different kinds of variables. The choice of a specific kind of outcome variable in any study can have a huge impact on the sample size requirements. Furthermore, the statistical procedures needed for the analysis depend on this particular choice.

TYPES OF DATA

We can broadly classify the measured variables as continuous, discrete, and categorical. The most common one that is encountered in RCTs is the continuous data type. Continuous variables can take any value on a real line between negative infinity and positive infinity. However, in any real-world setting, continuous data are restricted to take only finite values within some specified range. In our illustration, age is a continuous data type, and it is restricted to 18 years and above.

The second type of data is known as discrete, and it can take only integer values. Few examples of such data in our example are number of study drugs, number of seizures, or number of hospital visits, etc. These variables take integer values such as 0, 1, 2,…6, etc. The methods used to analyze such data rely on the assumptions of the counting nature of the distribution and appropriate techniques to handle this data type by incorporating Poisson or negative binomial distributions. Discrete variables can be treated as continuous for the analysis whenever the values taken by them are very large. Examples of such discrete outcome variables which can be treated as continuous are quality of life scores, number of cases exposed to a certain disease, population in a certain region, etc.

The third data type is known as categorical. It does not take any numerical value and is defined in terms of categories. For the analysis, a numerical pseudocode is applied to each category, and a data dictionary is created to store information about these assignments. These variables can be further classified into two kinds: ordinal and nominal. Ordinal variables have sequential ordering imbedded in them. For example, the ratings of patient health, quality of care, severity of seizures, etc., are all ordinal variables. These variables are ordered (either in ascending or descending order) in terms of categories; however, the distance between each category is unknown, and it is the main focus of statistical inference. The assignment of numerical value such as 0, 1, 2, etc., for each category is usually set to retain the ordering. However, the actual assignments of the numerical scheme are only for the ease of collecting and organizing the variables before the analysis.

On the other hand, nominal variables do not contain any specific ordering for each category. Binary variables are a subclass of nominal variables with only two categories. A very common binary variable encountered in most studies is gender. Other common nominal variables are race, hospital sites, centers, etc.

Censoring is defined as the mechanism which causes continuous data type to remain unknown beyond (greater than and/or less than) a specific value due to some practical limitations of the measurement process. There are also a few instances where censoring occurs in the intermittent intervals. In most cases, the censoring mechanism is assumed to be completely independent of the main continuous variable of interest.

In our example, Alldredge et al. (2001) have presented the characteristics of the patient population by each treatment arm in Table 1 of their manuscript. The primary outcome variable which is the duration of SE before arrival to the hospital in minutes is both continuous and censored at the time of arrival to the emergency department. The remaining variables such as gender, race, prior history of seizure, and cause of SE in Table 1 are all nominal variables.

Descriptive statistics are used to provide a summary of the data sets collected in an RCT. In the next section, we will provide an overview of descriptive statistics and discuss central limit theorem (CLT) which plays a fundamental role in the inferential statistics.

DESCRIPTIVE STATISTICS

Descriptive statistics allows an initial assessment of all the variables collected in a study. The statistics used to describe the central most position of a sampled data are the mean, median, and mode of a frequency distribution. Similarly, the statistics used to describe the spread of a distribution are standard deviation, variance, absolute deviation, interquartile range (IQR), and difference between maximum and minimum observation. The mean is computed by summing up all the observations in a sample and dividing the sum by the total number of observations. The median represents the middle observation if all the observations in the sample are arranged in an ascending or descending order. The mode is the peak of the population, and it is estimated from a sample as the observations which appear a maximum number of times. The central tendency of the continuous or discrete data is generally provided using all the three statistics (mean, median, and mode). They all coincide to a single value in case of a symmetric distribution with only one mode. The central tendency of a nominal variable is defined by the mode, whereas ordinal variable is summarized using the median. The range of a distribution is denoted by the minimum and the maximum observations in a sample. The standard deviation provides an estimate of the spread of the population. It is computed as the square root of the average squared deviations of all observations from the mean. The standard deviation is in the same scale of the measurement as the original variable. The variance is the square of the standard deviation. The IQR is defined by the 25th and 75th percentile cut points.

The plots and graphs such as histograms, bar plots, scatter plots, box plots, and mean curves allow a visual inspection of variable distributions. Fig. 14.1A displays various forms of parametric density functions for the continuous data that are encountered in a clinical trial. The four most common parametric forms of distributions shown in this plot are normal, gamma, log-normal, and uniform. The box plots of three continuous variables with normal, gamma, and log-normal distributions are shown in Fig. 14.1B. The center line in the box denotes the median, and the box itself represents IQR. The whisker shows the range which either goes from minimum to maximum or captures the cut points defined as the first quartile subtracted by a factor of 1.5 times IQR and the third quartile added by the same factor. The observations falling outside the latter defined whisker ranges can represent most likely outliers for symmetric distributions. However, if the sample is generated from a log-normal distribution, then there is no specific reason to deem such observations as the probable outliers. Fig. 14.1C shows histograms of a positively skewed distribution. The histograms in the top and bottom figures are shown using sample sizes of 50 and 500, respectively. A kernel density smoother based on these sample sizes is overlaid along with the histogram. From the plots, it is apparent that the true population density is not accurately captured with 50 samples. The appropriate selection of bin width of the bars makes the histogram visually appealing and allows revelation of the true features of underlying population density. Fig. 14.1D shows the bar plots for nominal data with three categories for two groups A and B. The bar plots can be presented in terms of the observed counts or percentages of total samples in each group. It can be readily seen from the bar plots that both the groups A and B differ significantly in their responses.

The normal distribution is a bell-shaped distribution, and it plays a fundamental role in the inferential statistics by virtue of CLT. It is described by only two parameters: location which defines the mean or central tendency and scale which captures the spread or standard deviation of the distribution.

In Fig. 14.2, the "standard" normal distribution is plotted. The standard normal distribution has mean of 0 and standard deviation equal to 1. The Student's t-distributions are shown in the same plot along with varying degrees of freedom (df). It is apparent from the plot that Student's t-distributions approach the standard normal distribution as their df gets larger.

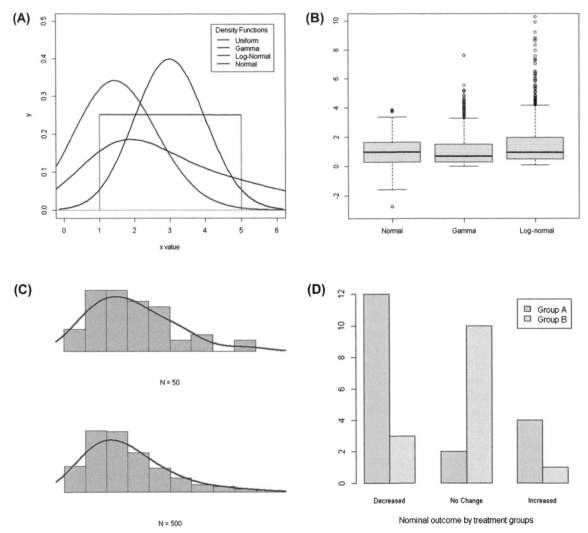

FIGURE 14.1 Plots for the visualization of distribution of the variables: (A) continuous distributions, (B) box plots of continuous distributions, (C) positively skewed distribution shown for two different sample sizes, and (D) bar plot of a nominal variable for two groups.

Central Limit Theorem

Most of the statistical testing procedures which are used to make inference about a population mean (and parameters) are based on the CLT. CLT states that the "estimate" of mean of a population follows a normal distribution (bell-shape curve) with large sample sizes irrespective of the shape of underlying population distribution. The estimate of a population mean is also known as the "sample mean." CLT also states that the mean and standard deviation of the sample mean, which define its normal distribution, can be written in terms of the true population mean and standard deviation. Specifically, it states that the mean is equal to the true population mean, and the standard deviation is equal to the standard deviation of population divided by square root of the sample size. To avoid confusion, standard deviation of the sample mean is also known as the standard error. CLT principle holds true for large sample sizes. The sample mean (denoted by $\bar{x}$) and its standard error (denoted by $SE_{\bar{x}}$) are estimated from the samples $x_1, x_2,...,x_N$ of size N as $\bar{x} = \frac{1}{N} \sum_{i=1}^{N} x_i$ and $SE_{\bar{x}} = \frac{s}{\sqrt{N}}$ where s is the estimate of standard deviation of the population given by $s = \sqrt{\frac{\sum_{i=1}^{N}(x_i - \bar{x})}{(N-1)}}$.

The principle of CLT in general is applicable to any parameter that is estimated from the data. CLT states that an estimate of a parameter will always follow a normal distribution with large sample sizes. The mean of the distribution of parameter estimate will be the true population value which we are trying to identify based on our experiment, and the

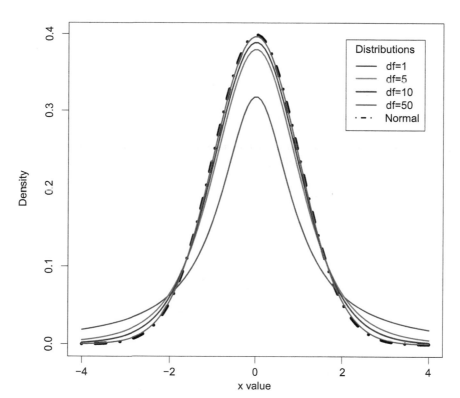

FIGURE 14.2 Normal distribution and Student's t-distributions with varying degrees of freedoms.

standard error is some function of the sample based on our specified model and always inversely proportional to the square root of the sample size. To make the idea clear, let us first consider a few examples of "parameters" in any given model:

1. An estimate of the population mean given by sum of all the observations divided by the number of observations is a parameter. Similarly, an estimate of standard deviation calculated as an average of square root of sum of squared deviations is another parameter.

2. The correlations coefficients (denoted generally by rho "ρ") such as Pearson's or Spearman's correlation describe the strength of relationship between two continuous variables.

3. In multiple regression models, the beta parameter "β" captures the relationship between the response and explanatory variables. These parameters generalize the concept of correlation to multiple regression problems.

4. The odds ratio (OR) in logistic regression models for the binary variable is used to provide estimates of the odds of switching from the reference category to the main category of interest for one unit change in the explanatory variable.

CLT guarantees that we will be able to determine the true relationship between the variables given that we have a large sample of observations and a correctly specified model of inference.

A Cautionary Note on the Application of Central Limit Theorem

Let us consider two extreme examples to better understand the application of CLT:

1. Suppose that in our experiment, we only obtain a single sample from our population of interest. The sample mean in this scenario is nothing but the single observation itself as we have no further information about this population from the data. The variability of this estimate is as large as the variability of the population itself. Therefore in this setting, we have no idea about the location of the true mean.

2. Let us pretend that we have infinite resources for our second scenario, and we are able to collect all the possible samples in the population. As the sample size goes to infinity in this latter scenario, the standard error almost goes to zero. Hence, we are completely certain about the true population mean. In fact we do not need to conduct any formal

hypothesis testing for the second scenario as we have all the samples, and we are completely certain about the distribution of the underlying population.

However, often, we are faced with a situation that is somewhere in between these two extremes. The testing procedures developed using CLT which rely on the normality of the parameter estimates perform very well for moderate-to-large sample sizes irrespective of the underlying population distribution. A key to understanding the CLT is to focus on the meaning of moderate-to-large sample sizes. Sample sizes depend on the skewness of the underlying distribution. If the population distribution is extremely skewed, then the requirements of sample size in comparison to the distribution, which is symmetric and has thin tails, can be very large. These requirements can worsen if the data are rather uniformly distributed or have very long and thick tails as all the observations are very spread out.

Throughout this discussion, we have assumed that the underlying population is represented by only one distribution. Any application of CLT becomes questionable if the population has a mixture of distributions, and a proper care must be taken while trying to analyze such data using the approaches discussed in this chapter. Fig. 14.3 shows a hypothetical example of a mixture of two normal distributions with x-axis representing the observation range. Note how the distribution is bimodal with modes at -2 and $+2$.

TESTING AND SUMMARIZING RELATIONSHIP BETWEEN TWO VARIABLES

To understand the fundamental notion behind statistical testing procedures and model building, we need to also develop a thorough understanding of p-value, confidence intervals (CIs), and Type I and II error rates. In what follows, we will cover these concepts in detail before providing methods to summarize relationship between two variables.

p-Value and Confidence Intervals

p-value which is an output of a formal testing procedure summarizes the significance of the parameters in a model. It is defined as the probability of obtaining the observed data or relationship if the null hypothesis holds true. The larger the p-value, the more "unsure" we are that the sample came from some other alternate distribution rather than our hypothesized null distribution. All the regulatory bodies now widely accept a cutoff point of 0.05 for testing the null hypothesis. This choice of cutoff simply implies that if we observe a p-value less than 0.05, then it is extremely likely that the sample originated from the alternate rather than the null distribution. In most situations, the null hypothesis is stated as the parameter which captures the relationship between two variables or groups is equal to zero. In such cases, the p-value of less than 0.05 denotes that the relationship is strong and that the parameter is highly significant. The cutoff point is also known as the Type I error rate (or alpha "α" value). We will cover this in detail later, but we first provide a comparison between p-value and CI. If relationship between the outcome and independent variable is significant (i.e., we have ascertained that the parameter is not zero given the observations using a formal testing procedure), then the parameter estimate itself denotes the strength of this association. Using the principle of

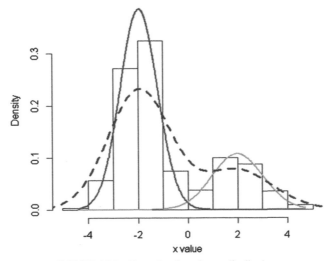

FIGURE 14.3 Example of a mixture distribution.

CLT, an interval can be constructed around this parameter estimate to denote the uncertainty associated with this parameter given our sample size. This interval is known as the CI of the parameter. Most often, an interval is constructed to cover 95% of the probability symmetrically about the point estimate. Hence, 95% CIs are the intervals which capture the uncertainty in the point estimate and provide the range of values for the most probable location of the true parameter. In fact, both p-value and CI are complementary to each other except for few complex problems which are beyond the scope of this chapter. Hence, a significant p-value implies that the 95% CI does not contain the null value stated in the null hypothesis. An informal way of thinking about CIs is that if in case we had resources to repeat our experiment at least 100 times with a fixed sample size of N for each experiment, then 95 out of 100 CIs constructed for the parameter will contain the true value of the parameter somewhere within those intervals and only 5 CIs will provide erroneous findings. The p-value has a very similar interpretation; it is also closely tied to the concept of Type I error rate.

Type I and II Error Rates

The performance of any hypothesis testing is defined in terms of the Type I and II error rates. Type I error is the probability of rejecting the null hypothesis in case it holds true, i.e., if the samples are obtained from the null distribution, then Type I error rate is the probability of concluding that samples were obtained from the alternate distribution instead. Similarly, Type II error is the probability of accepting the null hypothesis in case the data are generated from the alternate hypothesis distribution. Therefore, if the samples are obtained from the alternate distribution, then Type II error rate is the probability of concluding that samples were obtained from the null distribution.

Type I error rate is usually set to 0.05 as discussed in the previous subsection. Consider a hypothetical situation where we repeat our experiment 100 times with a fixed sample size of N. In each experiment, we make certain that our samples are generated from the null hypothesis distribution. By setting Type I error rate to 0.05 for all the experiments, we are simply making sure that we allow incorrect conclusions only 5 times out of 100 ($\alpha \times 100$) that the samples originated from the alternate distribution instead of the null distribution. Vis-à-vis we are guaranteeing that we are 95 times correct about the true null distribution.

Type II error rate is denoted by beta "β," and it is generally set to a less stringent criterion of 0.20. A complementary notion of Type II error rate is known as power. It is the probability of rejecting the null if the data were generated from the alternate hypothesis. It is simply equal to $1 - \beta$. Please note that the Type II error rate and parameters in a linear model are both denoted by the same Greek symbol "β," but they denote totally two very different statistical concepts.

The hypothesis testing depends on the concept of Type I error rate; however, the sample size requirement for any experiment is also based on the notion of Type II error rate in addition to the Type I error rate. A little more simplistically, we test the null to retain the Type I error rate, but we collect a sample of size N to ensure that we have enough power to reject the null in case the samples originate from the alternate hypothesis.

BASELINE COMPARISONS AND PRIMARY OUTCOME ANALYSIS

Prior to carrying out the main analysis in an RCT, it is expected that the baseline variables with respect to each treatment group will be compared. This first step provides a summary of the population and allows for checking the imbalances in the randomization procedure. If the randomization is successfully implemented with respect to the baseline variables and possible confounders, then the analysis of primary outcome variable is completely straightforward. In what follows, we will provide details about these common testing procedures.

If the response variable is continuous and there are only two treatment groups, then inference about the efficacy is generally given in terms of the mean difference and significance is determined by constructing the t-test statistic. This t-test statistic is computed by taking the ratio of mean difference and its standard error. This t-test statistic is known to follow a Student's t-distribution with df equal to $(N - 1)$ where N is the total sample size.

If there are more than two treatment groups and the response variable is continuous, then one-way analysis of variance (ANOVA) method gives a way to express the total variation in the outcome in terms of between and within treatment group variations. The ratio of these two variations (between divided by within sum of squares) adjusted by corresponding numerator and denominator dfs provides a method to test the significance of the treatment main effect. This test statistic is known as F-test, and since it is the ratio of two variances it has two dfs. In case of one-way ANOVA, numerator df is 1 and denominator df is $(N - 1)$. The hypothesis for the main effect in one-way ANOVA is written as

$$H_0: \mu_1 = \mu_2 = \cdots = \mu_k \quad \text{versus} \quad H_\alpha: \text{NULL NOT TRUE} \ (\mu_1 = \mu_2 = \cdots = \mu_k)$$

where H_0 and H_α denote the null and alternate hypotheses, respectively. In the notation aforementioned, the alternate hypothesis specifies that the mean of any one treatment group out of k groups is different from the rest of the $(k-1)$ groups. The posthoc testing procedures mostly focus on pairwise, one-to-many, or preidentified set of comparisons of various treatment groups. These posthoc comparisons are usually based on the t-tests after adjusting them for multiple comparisons. There are various methods for performing multiple comparison adjustments. We will discuss a few later in the chapter.

If the response variable is highly skewed and sample sizes are very small, then CLT-based large sample parametric testing procedures perform very poorly in terms of retaining Type I error rate at its nominal value (for example, $\alpha = 0.05$). Instead, their nonparametric counterparts guarantee control of Type I error rate even for small sample sizes. Wilcoxon rank-sum test which replaces the t-test for comparing continuous outcome for two groups is defined based on the ranks of the observations and sum of the ranks instead of the actual observed values. The hypothesis is generally written in terms of the comparison of medians of the two groups for historical reasons; however, the test is constructed in terms of comparison of all the possible percentile points using the ranks of the two distributions. Kruskal–Wallis test statistic generalizes the Wilcoxon rank-sum test for the k treatment groups.

For matched data, parametric testing procedure for two group comparisons is the two-sample paired t-test, and its nonparametric counterpart is the Wilcoxon signed-rank test. For more than two groups, repeated measures ANOVA is employed.

Chi-square tests or Fisher's exact tests are used to find associations between two or more categorical outcome variables. The results are summarized in terms of the contingency tables containing actual counts or proportions in each cell. The strength of association for the contingency tables such as relative risk, risk difference, and OR is reported based on the design of the study.

The relationship between two continuous measures is summarized using the Pearson's correlation or simple linear regression models. Pearson's correlation estimate is sensitive to outliers, and it is completely invalid if the relationship between the two variables is highly nonlinear. Spearman's correlation provides an estimate of the association nonparametrically

TABLE 14.1 Summary of Testing Procedures for Bivariate Analysis

Dependent Variable	Independent Variable	Parametric Test	Nonparametric Test	Summary Statistics
Continuous (independent or unpaired data)	Binary (two groups or time points)	Two-sample t-test	Wilcoxon rank-sum test (similar to Mann Whitney U-Statistics)	Mean difference, standard error, CI, median, and IQR.
Continuous (matched or paired data)	Binary (two groups or time points)	Paired two-sample t-test	Wilcoxon signed-rank test	Mean difference, standard error, CI, median, and IQR.
Continuous	Categorical (many groups)	ANOVA (F-test for the main effect)	Kruskal–Wallis test for the main effect	Test for main effect, posthoc comparison adjusting for multiplicity, mean differences, standard errors, adjusted CIs, medians, and IQRs
Continuous	Continuous	One-sample t-test using Fisher's z-transform for the Pearson's correlation	One-sample t-test based on Fisher's z-transform for the Spearman's correlation	Pearson's or Spearman's correlation, standard error, CI
Categorical (independent or unpaired in all the groups)	Categorical	z or t-test	Chi-square test for testing association or independence. Fisher's exact test in case sample size in any cell ≤5	Cohort study: risk difference or risk ratio Case–control study: odds ratio For both the designs: standard error, CI
Binary (matched or paired data)	Binary	z or t-test	McNemar's test	Odds ratio, standard error, CI

ANOVA, analysis of variance; *CI*, confidence interval; *IQR*, interquartile range.

using the ranks of the observations. It is generally used to provide an estimate of correlation when the variables follow a highly skewed distribution or the relationship between them is highly nonlinear. Table 14.1 provides a summary of testing procedures used to determine relationships between the two variables depending on different kinds of data.

Alldredge et al. (2001) presented an initial assessment of the patient characteristic among three treatment groups in Table 1 of their manuscript. They determined that the recruitment of subjects was well balanced in terms of the continuous variable age as it did not differ significantly among the three groups using ANOVA. Similarly, they found that two nominal variables, gender and history of seizures, were not statistically different among three treatment groups using chi-square tests. However, they did conclude that there was an imbalance in the racial group as this variable was found to be significantly different across three treatment groups using a chi-square test. The two continuous outcome variables defined as interval from the onset of SE to administration of the study treatment and duration of SE before the study treatment were also found to have significant treatment group effects using ANOVA.

As a part of primary analysis, they also analyzed termination of SE by the time of arrival at the emergency department by treating it as a binary variable. In their sample of 66 subjects who were administered lorazepam, 59.1% of patients had the termination of SE by the time of arrival at the emergency department. Similarly, 42.6 (29/68)% of subjects had the termination in the diazepam group, and 21.1 (15/71)% of patients responded in the placebo group. The chi-square test was highly significant with a p-value = 0.001. The results of their findings were summarized in terms of OR. The OR estimate for lorazepam in comparison to placebo group was 5.4 (95% CI 2.3—13.2), and for diazepam in comparison to placebo group was 2.8 (95% CI 1.2—6.7). The OR comparing lorazepam versus diazepam was not statistically significant as its p-value was greater than 0.05 and its CI contained 1 in the interval.

GENERALIZED LINEAR MODELS

The main aim of "inferential" statistics is to understand the phenomenon of variation in the dependent variable in terms of the variation in the independent variable. If the relationship is strong, then the variation in the response variable will follow a pattern which is very similar to the explanatory variable. The generalized linear model framework allows for explaining variation in the response variable in terms of several explanatory variables. If the relationship is strong between the response and any one explanatory variable conditional on all the other covariates, then the variation in the response variable will also be explained by the variation in the explanatory variable holding all the other covariates to a constant value. The ideas of mediation, confounding, and interactions also can be tested by building appropriate models using this framework. As we noted previously, the primary response variable in a well-designed and balanced RCT requires very simple statistical procedures; however, in several scenarios, the biological mechanism necessitates covariate-adjusted analysis. These analyses are based on the assumption of additive effects of covariates on some transformed scale of the response variable.

In the following subsections, we will cover models for continuous, categorical, and censored variables.

Linear Regression Model for the Continuous Data

In a linear model framework, the relationship between the average value of continuous outcome and the average values of explanatory variables is written using the additive principle:

$$E(y) = \beta_0 + \beta_1 E(x_1) + \beta_2 E(x_2) + \ldots + \beta_p E(x_p)$$

In this equation, the variable y denotes the response variable and the variables $(x_1, x_2,\ldots,x_n)$ denote p explanatory variables. $E(.)$ notation denotes the average of each variable. The strength of relationship between the response variable with a specific explanatory variable x_k in this model is estimated by the parameter β_k holding all the other variables to a constant value. The linear model approximately captures the variation in outcome in terms of each specific explanatory variable x_k using the slope parameter β_k. Ideally, we would like to explain all the variability in the response; however, in real-world applications especially in medical field, we also need to take into account the measurement error.

Essentially, the linear modeling approach is based on the following assumptions: (1) linearity of the response variable with respect to the specific explanatory variable holding all the other variables constant. This implies that the average of the outcome changes by a factor of β_k units if the explanatory variable x_k changes by 1 unit after setting all the other explanatory variables to some fixed value; (2) normality of the response variable which implies that the error in the linear model is also normally distributed with the variance parameter equal to the response variable's variance parameter (often denoted by σ_e^2); (3) the response of subject i is independent of any other subject j; and (4) the number of explanatory variables in the model p is much smaller than the sample size N.

Based on these assumptions, the linear model for outcome of a particular subject can be expressed in terms of the weighted summation of the independent variables belonging to that subject and the error term as follows

$$y_i = \beta_0 + \beta_1 x_{1i} + \beta_2 x_{2i} + \dots + \beta_p x_{pi} + \varepsilon_i$$

As the response variable is assumed to follow a normal distribution, hence the mean and the variance of its distribution can be derived from the aforementioned model specification. The equation for the mean response can be derived from this linear equation by taking the averages on both sides of the equation. As the coefficients are assumed to be constant in this equation, hence the average of the right side simply turns into averages of each explanatory variable.

The error is assumed to have a mean of 0; hence it vanishes from the mean response equation.

The parameters in this model are estimated using either ordinary least squares or maximum likelihood approach. The parameters in the linear model are the coefficients β_k of each explanatory variable and the variance of the error σ_ε^2. Both of these approaches in the linear modeling framework provide "unbiased" estimates with "minimum variance" if the assumptions stated earlier hold true. The ordinary least squares approach is based on minimization of the squared residual, whereas the maximum likelihood method is based on maximization of the likelihood of parameters given the observed data. The unbiased property means that if the samples are collected from a population several times, then the average value of estimated parameters will approach the true value as the sample size grows large. The uncertainty in the parameter is represented by its variance. The square root of this variance is known as the standard error of the parameter. The minimum variance property implies that the parameter estimates have the smallest standard errors (square root of the variance).

To test the null hypothesis whether any parameter is zero denoted as H_0: $\beta_k = 0$ versus H_α: $\beta_k \neq 0$, a Wald test statistic is generally constructed as the ratio of the square of estimate of the parameter and its variance. The Wald test statistic follows a chi-square distribution. If the p-value is less than prespecified Type I error rate (for example, 0.05), then the null hypothesis is rejected. The strength of relationship is defined by the estimate of the parameter and the range of plausible values representing this strength is provided using CIs. The Wald test statistic can also be used to test several parameters at the same time. The other test statistics which are also commonly employed to test the significance of parameters are known as likelihood ratio test and score test.

In case we only have two continuous variables with one representing the outcome and other as the explanatory variable, the beta parameter representing the relationship between the two is equivalent to the Pearson's correlation coefficient. The only difference is that Pearson's correlation coefficient is scale invariant; however, the parameter in the linear model is dependent on underlying scale of both the measures. To be precise, the correlation coefficient can be obtained from beta parameter estimate by multiplying it by the standard deviation of the explanatory variable and dividing it by the standard deviation of the outcome.

Logistic Regression Model for the Binary Data

For binary outcome variables, the relationship between the average value of binary outcome and explanatory variables cannot be simply written in terms of the additive principle as in the case of continuous data. The reason is that the average value in binary outcome provides an estimate for the probability of success (or failure depending on the choice of reference category) and the probability measure by definition is restricted to the range of 0−1. Some of the brilliant statisticians and economists in the early 19th century figured out a way to tackle this problem by taking the log of ratio of the probability of success with the probability of failure. This transformed variable can vary from − infinity to + infinity. Hence, the relationship of multiple explanatory variables with this transformed variable is defined using the additive principle as follows:

$$\log\left(\frac{P(z_i = 1)}{P(z_i = 0)}\right) = \beta_0 + \beta_1 x_{1i} + \beta_2 x_{2i} + \dots + \beta_p x_{pi}$$

In this model the new outcome variable y_i is defined as $\log(P(z_i = 1)/P(z_i = 0))$. z_i is the observed binary outcome for the ith subject, and the probabilities of success and failure are denoted by $P(z_i = 1)$ and $P(z_i = 0)$, respectively. The interpretation of β_k in this model is in terms of change in odds (on the log scale) of the binary outcome turning from 0 to 1 for every one unit change in the continuous explanatory variable x_k holding all the remaining explanatory variables to a constant value. For categorical variables, β_k is simply change in log odds for a change in the category of x_k from the reference category to the one defined by this indicator. The exponentiation of each parameter, i.e., $\exp(\beta_k)$, turns into OR in the original scale instead of the log scale. If the parameter $\beta_k < 0$, then the OR of x_k given by $\exp(\beta_k)$ is less than 1. On the other hand, if the parameter $\beta_k > 0$, then the OR is greater than 1. The former case implies that one unit increase in the explanatory variable reduces the odds of having an event by a multiplicative factor $\exp(\beta_k)$, whereas the latter one implies that it increases the odds by a multiplicative factor $\exp(\beta_k)$. Since this new variable y_i is derived from the binary outcome z_i, there is no error term in the model. The transformation in the logistic model fixes the overall scale of y_i to the variability of a "standard" logistic distribution which is equal to $\pi^2/3$. The hypothesis in this model is stated in terms of the parameters as $H_0: \beta_k = 0$ versus $H_a: \beta_k \neq 0$, and the Wald test statistic can be used to test its significance.

Alldredge et al. (2001) used a logistic regression model to analyze the outcome variable z_i which is the presence or absence of SE on arrival at the emergency department after adjusting for the imbalances in their design due to race and the interval from the onset of SE to administration of the study treatment as follows:

$$\log\left(\frac{P(z_i = 1)}{P(z_i = 0)}\right) = \beta_0 + \beta_1 G_{1i} + \beta_2 G_{2i} + \beta_3 \text{Time}_i + \beta_4 R_{1i} + \beta_5 R_{2i} + \ldots + \beta_m R_{mi}$$

In this model, G_{1i} and G_{2i} are the indicator variables for lorazepam and diazepam groups, respectively. The placebo group in this model is treated as the reference. The indicator variables simply take a value 1 if the subject was assigned to the respective treatment group and 0, otherwise. The terms $\exp(\beta_1)$ and $\exp(\beta_2)$ are the corresponding "adjusted" estimates of ORs for lorazepam group versus the placebo and diazepam group versus the placebo, respectively. The continuous covariate Time_i is the interval from the onset of SE to administration of the study treatment for each subject. The indicator variables $(R_{1i}, R_{2i}, \ldots, R_{mi})$ are the indicator terms for nominal variable race with m categories. In Table 1 of Alldredge et al. (2001), there are seven categories of race variable, and hence m in this model could be equal to 6. In certain situations, it is necessary to combine a few categories given the small sample size in those respective categories. Using this model, they estimated that the "adjusted" odds of termination of SE on arrival at the emergency department were 4.8 (95% CI 1.9–13.0) times as high for the lorazepam group in comparison to the placebo. Similarly, the adjusted odds were 2.3 (95% CI 1.0–5.9) times as high for the diazepam group as for the placebo group, and 1.9 (95% CI 0.8–4.4) times as high for the lorazepam group as for the diazepam group.

Ordinal Outcomes

To model ordinal outcomes, the logistic regression transformation is generalized by taking the log of ratio of the probability of event being greater than pth category divided by probability of event being less than pth category. The model now contains $(m - 1)$ equations defining the relationship between this derived variable from the ordinal variable with m categories. All the explanatory variables are included in these $(m - 1)$ equations with the same parameters, i.e., a parameter for any particular explanatory variable is fixed to take the same value in all the equations. This approach leads to simplifying assumption that the odds of change from any category p to the immediately next one (i.e., $p + 1$) on a log scale is set to a constant value with every unit change in explanatory variable x_k conditionally on all the other variables. Hence, this model is known as proportional odds regression model.

Nominal Outcomes

Further generalization of proportional odds model is known as multinomial logistic regression model which can accommodate nominal outcomes. The number of parameters in these models is generally much higher than that of proportional odds model. Hence, the sample size requirements in these models are much higher to have improved precision (i.e., low variability of the parameter estimates). The multinomial model is built by setting one of the categories as the reference category and then calculating the transformation similar to the binary logistic model for the remaining categories. This model is also defined by $(m - 1)$ equations where each equation is constructed for the log of the ratio of probability of the outcome being in the pth category to the probability of the outcome in the reference category. The relationship between the outcome variable and any particular explanatory variable is now captured using $(m - 1)$ parameters which define the log odds of response jumping from the reference category to the pth category.

Survival Analysis for the Censored Data

In survival analysis, the interest not only lies in characterizing whether any event has occurred but also the time of its occurrence. Subjects generally enter RCTs at a particular time point and are followed up only to a prespecified time due to the design or budget constraints. The primary outcome in the survival analysis is defined by two variables: (1) the binary event of interest and (2) the time of occurrence of this binary event. Some of the subjects in these analyses contribute toward censored time periods, i.e., the binary event of interest does not occur from the point of randomization for these subjects to the prespecified follow-up period or to the dropout time. The status whether they experienced the binary event after the end of follow-up period is unknown. This kind of censoring is also known as left censoring as we have no idea about the status of the subject beyond a certain time period of follow-up. Most of the examples found in the medical literature assume that the left censoring mechanism is independent of the probability of occurrence of the binary event.

Although the methods developed for the survival analysis fall within a linear modeling framework, the model building and results interpretation require careful consideration. We have ignored the presentation of right censoring to abridge the discussion. The first step in the survival analysis is to calculate risk set and estimate survival curve for each treatment arm. The risk set $R(t)$ at any time t contains total number of subjects who have "not" experienced the binary event of interest until time t. They are still considered at risk of experiencing the event until the follow-up period. The survival curve as a function of time t is denoted by $S(t)$. It represents the probability of survival (or having no event) up to time t. The survival function is generally estimated using the nonparametric approach known as Kaplan–Meier (KM) curve. It is based on the computation of the product of the conditional survival probabilities. The conditional probability at time point t denotes the probability of surviving in consecutive next short interval (between time t and $t + dt$) given that the individual has survived up to this specified time t. The estimate of conditional probability is the complement of proportion of individuals who had the event at time t. This proportion of individuals can be estimated as the ratio of the number of individuals who had the event at time t and the individuals in the risk set $R(t)$.

The median of the survival time can be easily obtained from the KM curve after taking into account censoring. It is the point where KM curve reaches a value of 0.50, i.e., the product of conditional probabilities is equal to 0.50. In case there is no censoring, then median survival time is the sample median that is obtained by sorting the observed failures in any order and picking up the middle observation. The comparison of survival curves between two or more groups is based on either the generalized Wilcoxon test or the log-rank test. The null hypothesis in these comparisons is denoted by

$$H_0: S_1(t) = S_2(t) = \ldots = S_k(t) \quad \text{versus} \quad H_\alpha: \text{NULL NOT TRUE } \{S_1(t) = S_2(t) = \ldots = S_k(t)\}$$

where $S_1(t)$, $S_2(t)$,…,$S_k(t)$ are k survival curves. The alternate hypothesis simply means any one of the survival curves for the k groups is different from the rest. Although it is a better practice to report the results of both generalized Wilcoxon and the log-rank tests, there are few notable differences in these testing procedures. The generalized Wilcoxon test places more weight on differences in shorter survival times of the curve, whereas log-rank test favors differences toward the longer survival times. The log-rank test is considered the most powerful under the assumption of proportional hazards. This assumption implies that the two survival curves when plotted in the same graph should not cross each other at any time point. The log-rank test provides spurious p-values in case the proportional hazards assumption fails. Finally, if there is no censoring in any survival curve, then the results can be obtained using Wilcoxon rank-sum for two groups or Kruskal–Wallis tests for more than two groups.

To perform multivariate analysis for time to event data, we need to understand the meaning of hazard function $h(t)$. The hazard function is rate of change of the conditional probability of survival, and it is obtained as the first-order derivative of the probability of having an event in a very short interval just following time t conditional on the fact that the event has not occurred until time t. It can be shown that hazard function is the negative rate of change over time of the survival function $S(t)$ on a natural log scale. Note that the hazard can be interpreted as the short-term probability of having an event. Using the assumption of proportionality, the model for the survival data is written in terms of the log transformation of hazard ratio (HR) and the explanatory variables as follows:

$$\log\left(\frac{h_x(t)}{h_0(t)}\right) = \beta_1 x_{1i} + \beta_2 x_{2i} + \ldots + \beta_p x_{pi}$$

In this model the right-hand side contains the sum of linear explanatory variables $(x_1, x_2,\ldots,x_n)$. The parameter β_k captures the relationship between the explanatory variable and changes in hazard function for a unit change in x_k. Specifically, the parameter β_k is change in hazard function (on the log scale) from the baseline hazard function for a unit change in the explanatory variable x_k holding all the other variables to a constant value. The exponentiation of each parameter, i.e., $\exp(\beta_k)$, translates the HR in the original scale instead of log scale. There are several important aspects in this model which are different from the models presented in earlier subsections. (1) There is no term to describe intercept (β_0) in the linear part of the model. The baseline hazard function $h_0(t)$ replaces the intercept term, and it is estimated nonparametrically using KM estimation approach. (2) The baseline hazard $h_0(t)$ is also a function of time and represents the hazard of the control or reference population. (3) The survival model can also be written in terms of the multiplicative form as $h_x(t) = h_0(t) \times \exp(\beta_1 x_{1i}) \times \exp(\beta_2 x_{2i}) \times \ldots \times \exp(\beta_p x_{pi})$. Based on this multiplicative form, it is

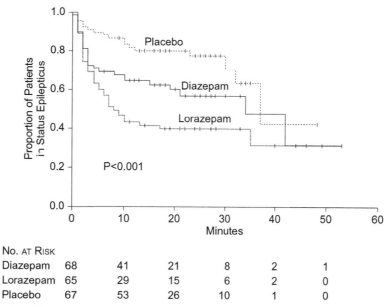

FIGURE 14.4   Kaplan—Meier plots of durations of out-of-hospital status epilepticus in three treatment groups. *Alldredge, B.K., Gelb, A.M., Isaacs, S.M., Corry, M.D., Allen, F., Ulrich, S., et al., 2001. A comparison of lorazepam, diazepam, and placebo for the treatment of out-of-hospital status epilepticus. N. Engl. J. Med. 345, 631—637.*

apparent that the impact of each covariate on the baseline hazard function is multiplicative. (4) The ratio of log hazard for a unit change of explanatory variable x_k is a constant conditional on all the other independent variables which implies the HR is not a function of time. This model was introduced in the medical field by Professor D.R. Cox (Cox, 1972) in his seminal work and has since been referred to as Cox proportional hazards regression model. In terms of the interpretation of parameters, the HR $\exp(\beta_k)$ of less than 1 (i.e., parameter $\beta_k < 0$) implies that one unit increase in the explanatory variable reduces the hazard of having an event by the multiplicative factor $\exp(\beta_k)$. Similarly, the HR $\exp(\beta_k)$ (i.e., $\beta_k > 0$) greater than 1 implies that hazard increases by the multiplicative factor $\exp(\beta_k)$ for one unit increase in the explanatory variable x_k.

Fig. 14.4 (obtained with permission from Alldredge et al., 2001) shows the plots of KM survival curves for the duration of SE before arrival at the hospital in the three treatment groups. The event of interest in this analysis is termination of SE which is a positive outcome unlike failure or death. Therefore, the curves that are steeper represent a faster improvement (faster termination of SE) in comparison to curves that are relatively flat. From this plot, we can see that the lorazepam group had the steepest survival curve whereas the placebo group was comparably flat. Using the log-rank test, these curves were found to be significantly different from each other with $p < .001$. Using the assumption of proportional hazards, Alldredge et al. (2001) also provided the results of a Cox proportional hazard regression model after adjusting for the imbalances in the study,

$$\log\left(\frac{h_x(t)}{h_0(t)}\right) = \beta_0 + \beta_1 G_{1i} + \beta_2 G_{2i} + \beta_3 \text{Time}_i + \beta_4 R_{1i} + \beta_5 R_{2i} + \ldots + \beta_m R_{mi}$$

Similar to the logistic regression model presented in the previous subsection, G_{1i} and G_{2i} were the indicator variables for lorazepam and diazepam groups, respectively. The terms $\exp(\beta_1)$ and $\exp(\beta_2)$ are the corresponding adjusted estimates of HRs for lorazepam versus placebo group and diazepam versus placebo group, respectively. They reported adjusted HR of ongoing SE to be 0.34 (95% CI 0.17—0.71) for the lorazepam group in comparison to the placebo group. The HR of the lorazepam versus the diazepam group was 0.65 (95% CI 0.36—1.17), but it was not significant.

MODEL BUILDING

In the previous section, we have covered various models for analyzing different kinds of data using a linear model framework. Before discussing some advanced concepts, let us review some important ideas behind successful model building.

Distribution of the Explanatory Variables

In a linear modeling framework, no specific assumption is made about the distribution of explanatory variables. They can be of any type: continuous, discrete, or categorical. The nominal variable in the model with m categories is included by recoding it in terms of the $(m - 1)$ indicator variables with one of them assigned as a reference category.

Confounding

Confounding occurs when the relationship between the outcome and primary explanatory changes (usually attenuates) in the presence of an additional explanatory variable known as a confounder. Confounding sometimes occurs due to poor design and imbalances in the sampling. It is most common in early phase trials due to lack of prior knowledge in the field. The treatment effect in these cases is estimated after adjusting for the identified confounding variables. The assumption, in such cases, is that the variability in the observed outcome that was present due to the confounding is removed by adding a linear term of the confounding variable and the remaining variability that is explained solely by the treatment is correctly captured in this adjusted model.

Interaction Effects

Effect modification occurs when the treatment effects are different across various strata of the effect modifying variable. The analysis in such scenarios is carried out by stratifying with respect to the effect modifier. Another way of handling it is by inclusion of the interaction term of the effect modifier with the primary explanatory variable in a linear model. The final output contains contrast statements for the treatment effects across different levels of the effect modifier.

Nonlinearity

The nonlinear relationships of response variable with explanatory variables are dealt within a linear modeling framework either by searching for some form of transformation of variables or by the inclusion of higher-order terms of the explanatory variables. Some of the advanced approaches are beyond the scope of this chapter but they are worth mentioning. These approaches incorporate nonparametric estimation procedures of the true underlying relationships using local neighborhood smoothing, splines, kernels, wavelet decompositions, or functional principal components (see, for example, Hastie et al., 2009). Besides the complexity, the interpretation of parameters using these approaches presents another challenge for researchers in the medical field.

Model Selection

The final model selection relies on two separate but very closely knit concepts: inference and prediction. If the investigator is interested in building a prediction model, then the final model selection is based around the ideas of reducing the prediction error, classification error rate, or the deviance of the partial likelihood. Since the main aim of RCTs is determining treatment effects, the final models in these cases are built based on identifying the true nature of the relationship that is biologically plausible and medically meaningful. From the inference standpoint, the final models contain relevant confounders, effect modifiers, or any other mediating variables which lie in the causal pathway defining the mechanism of treatment on the outcome even if these models do not have minimum mean squared error.

Collinearity

Collinearity is a severe problem in the modeling world. It can occur if the researcher is not aware of the mechanism of how each explanatory variable has been derived or collected. It occurs when two or more explanatory variables are highly correlated and have very similar relationship with the outcome. Special care is needed to justify collinear explanatory variables in a single model based on substantive research as collinearity significantly blurs the relationship and increases the variance of the estimate. It can also be easily confused with the idea of confounding if prior derivations about the explanatory variables are not understood clearly.

In the next section, we revisit the issue of multiple comparisons and provide a few additional technical details.

MULTIPLE COMPARISONS

In any RCT, a strict requirement of maintaining Type I error rate at its nominal level (e.g., 0.05) necessitates methods to control for multiple comparisons. The reason is that if we have more than one hypothesis, then the probability of observing all of these together is the product of their individual probabilities assuming that they are all independent. As an example, let us assume that we are conducting five tests and we set Type I error rate to 0.05 for each hypothesis. The probability of all the five tests holding true (i.e., null is not rejected in case the samples originate from null distributions in all the five hypotheses) is given by the product as $(0.95 \times 0.95 \times 0.95 \times 0.95 \times 0.95) = 0.77$. This implies that probability that the null will be rejected for any one of the five hypotheses will now be $(1 - 0.77) = 0.23$. This particular calculation indicates that the overall Type I error rate of our five hypotheses is much greater than 0.05 for this example. If instead we set the Type I error rate for any one single hypothesis after dividing 0.05 by the number of hypotheses 5, i.e., $= 0.01$, then the probability that all the five tests hold true under null will be $(0.99 \times 0.99 \times 0.99 \times 0.99 \times 0.99) = 0.95$. Hence, the overall Type I error rate is strictly maintained at $(1 - 0.95) = 0.05$ for all the five hypotheses. The procedure of dividing Type I error rate of a single hypothesis by the number of the hypotheses in a single experiment to retain the stringent criteria acceptable to regulatory bodies is known as Bonferroni's correction. A more precise method for retaining this family-wide error rate (FWER) for all the hypotheses is known as Dunn-Šidák's procedure (Šidák, 1967). If the number of hypotheses is greater than 3, then there is hardly any difference between these two procedures.

Recent advancements in the microarray experiments have led to development of new techniques which are not as stringent as FWER discussed in the previous paragraph. One of the procedures known as false discovery rate (FDR), replaces the strong criteria of protecting the probability of making incorrect decision if null holds (Type I error rate) to the average number of incorrect decisions when the null is true. The procedure begins by first setting the criterion for average allowable number of false discoveries (denoted by q) in an experiment. For example, $q = 0.05$ would imply that, on an average, 5 out of 100 hypotheses are allowed to be incorrectly rejected if the data are generated from null for all the 100 hypotheses. Note that the main caveat in using this approach is that the concept of probability is replaced by the average, which makes this criterion less stringent. It could happen that in any particular experiment or RCT, either only 1 out of 100 or even 50 out of 100 hypotheses are incorrectly rejected using this criterion when the data are generated from null for all the 100 hypotheses. The simple method to implement FDR proposed by Benjamini and Hochberg (1995) is to sort all the p-values in an ascending order and compare each with the factor given by $q \times \frac{i}{n}$, where i is the index for the p-values sorted in an ascending order and n is the total number of hypotheses. The hypotheses that have p values less than this factor are all rejected in favor of alternate hypothesis.

MISSING DATA

In several RCTs, the subjects are mostly followed up repeatedly even though the study is designed to evaluate efficacy at one single endpoint. In such studies, some of the subjects miss their regular visits at intermediate time points, and at other times, they completely drop out of the study. The protocol of any well-designed RCT clearly identifies the possible mechanisms that can lead to missing observations and include sections on the sensitivity analysis to address various scenarios. Historically, the medical field community was forced to rely heavily on two approaches: the complete case analysis and the last observation carry forward (LOCF). In complete case analysis approach, all the subjects with missing observations are deleted from the analysis. On the other hand, all the missing observations are replaced by the last available observation for each subject in LOCF approach. The analysis in the latter approach is carried out using this imputed data set as if no subject had any missing observation. Both these approaches provide biased and inefficient estimates of the treatment effect. In addition, complete case analysis method leads to wastefulness of resources. Lastly, LOCF approach is extremely strong in its assumption, and the replacement of missing observation with the previously observed values leads to arbitrary reduction in the variability of outcome variable.

Rubin (1976) has defined three kinds of missing mechanisms: missing completely at random (MCAR), missing at random (MAR), and missing not at random (MNAR). As the name implies in MCAR, the probability of missing observation is not dependent on any observed and/or any missing value. Simplistically, it implies that "missingness" is completely independent of the ongoing treatment, i.e., it is exactly the same for all the subjects. MAR is defined as the probability of missingness as being dependent only on the observed outcome and completely independent of any unobserved values. MNAR is defined as the probability of missingness being dependent also on the missing values instead of just the observed outcomes. This assumption is hard to validate or justify in an RCT, and very few trials or studies address it appropriately. The methods developed in the literature for handling missing data are based on multiple imputations, expectation—maximization algorithm, propensity score matching, or inverse probability weighting of observed outcomes. The details of these methods are complicated, and hence relevant references are provided in the summary section.

LINEAR MIXED-EFFECTS MODELS (CLUSTERED OR LONGITUDINAL STUDIES)

Mixed-effects modeling framework provides a powerful machinery of handling nested data by inclusion of one or two random effects at each level to explain between subject variability as well as capture correlations within nested observations. These models are based on the principle of conditional independence which imposes independence on any two clustered or repeated observations conditional on their random effects. The statistical methods that ignore the repeated or clustered nature of the observations are inefficient in providing inference about the efficacy.

The random effects in the model enforce a specific correlation structure of the outcome based on the random-effect design matrix chosen by the statistician. To explain the heterogeneity of the response profile, the design matrix of the random effects is usually taken to be a subset of the design matrix for the average response profile. The average response profile design matrix usually consists of overall intercept, indicator variables for the treatment groups, time effect in case the observations are of longitudinal type, covariates, and their interactions.

Laird and Ware (1982) have shown that mixed-effects model can be used for longitudinal data based on ignorable nonresponse which falls under MAR assumption. Hence, the mixed-effects modeling approach, unlike complete case analysis procedures, allows unbalanced and missing observations in the data. The final estimate of the efficacy using this approach is based on actual observations contributed by each subject. It has the small variance when compared to other statistical approaches. Hence, this approach offers several advantages over regular linear regression models or other approaches such as repeated measures and multivariate ANOVA (MANOVA).

Mixed-effects models separate out the variance components within each cluster and between various clusters of observation. A mixed-effect model is based on the simple inclusion of one or two additional terms which vary randomly at different levels of nesting and are independent of the error in the model. Let us consider the random-intercept models that are commonly used for clustered continuous data. The relationship between subject's outcome and the independent variables can be written as follows

$$y_{ij} = \beta_0 + \beta_1 x_{1ij} + \beta_2 x_{2ij} + \ldots + \beta_p x_{pij} + \vartheta_{0i} + \varepsilon_{ij}$$

where y_{ij} denotes the outcome for subject i, and subscript j denotes the repeated observation or time point of measurement. The p-independent variables in this model are denoted by $(x_{1j}, x_{2j},\ldots,x_{pj})$. In this model, the β_0 represents overall mean of all the subjects, and the parameters corresponding to the independent variables represent the overall average relationship or the average slope of the covariate x_k. These are also known as the fixed effects as they represent the population-averaged relationship. The error in this model is generally assumed to follow a normal distribution with mean 0 and variance σ_ε^2. The random effect ϑ_{0i} in this model denotes the subject-specific deviation from the overall population mean, and it is assumed to be independent of the error. This random-effect deviation captures the variability of clusters around the overall population mean. The most common assumption is that ϑ_{0i} is normally distributed with mean 0 and variance σ_ϑ^2. The inclusion of one random effect in the model imposes a compound symmetry structure for the correlations within each subject's measurements. The compound symmetry structure implies that the correlation between any two observations within a subject or cluster is constant irrespective of the separation between these two observations. In other words, it indicates that the observations within each subject can be permuted in any order prior to the analysis without changing the final results. Finally, the conditionally independent assumption in the mixed-effects modeling framework makes the two observations of the subject or cluster independently conditional on the random effect.

The compound symmetry structure is an effective approach for defining the correlations within nested observation of the clustered data as we do not have any further information in the changes of the correlations within these nested observations. However, this assumption is highly restrictive for the longitudinal data since the correlations do not tend to stay constant within different time points. The correlations can actually be dependent on the separation of these nested observations over the time points. Therefore, a linear mixed-effects model based on the extension of previous model can be written as

$$y_{ij} = \beta_0 + \beta_1 x_{1ij} + \beta_2 x_{2ij} + \ldots + \beta_p x_{pij} + \vartheta_{0i} + \vartheta_{1i} z_{1ij} + \ldots + \vartheta_{ri} z_{rij} + \varepsilon_{ij}$$

The model presented here allows for more generic correlation structure within observations by the inclusion of r random effects $(\vartheta_{0i}, \vartheta_{1i},\ldots,\vartheta_{ri})$. These r random effects are modeled as random variation using multivariate normal distribution with mean 0 and variance–covariance matrix Σ_ϑ. The response profile in this model for each subject is free to vary around the overall population-averaged trajectory.

In case of longitudinal studies, few covariates are generally used to model the growth trajectory of the response. Examples include $x_{ij} = t_{ij}$ to capture the linear trend; inclusion of $x_{ij} = t_{ij}^2$ along with linear trend t_{ij} to model quadratic changing trajectory; and sometimes a transformed variant of t_{ij} such as $\log(t_{ij})$, $\exp(t_{ij})$, $\text{sqrt}(t_{ij})$, etc., to approximately allow for nonlinearity in the response trajectory parsimoniously.

In these models, however, the interpretation of the parameters defining the relationship between the outcome and the time-varying covariates other than one defining the growth trajectory can be suspected as it is assumed to be constant conditional on the specified value of the covariate. Usually, the time-varying covariate is included in terms of the overall average covariate for the subject along with the time-varying deviations from the averages to separate out the constant and time-dependent relationships. More sophisticated approaches rely on the inclusion of nonparametric expansion of time-varying covariates in terms of the orthogonal basis functions such as splines, functional principal components, or Fourier series expansions (see, for example, Wood, 2006).

Let us consider three very simple examples of mixed-effects models for longitudinal data analysis to gain deeper insight of this modeling framework:

$$\text{Model 1:} \; y_{ij} = \beta_0 + \vartheta_{0i} + \varepsilon_{ij}$$

$$\text{Model 3:} \; y_{ij} = \beta_0 + \beta_1 t_{ij} + \vartheta_{0i} + \varepsilon_{ij}$$

$$\text{Model 3:} \; y_{ij} = \beta_0 + \beta_1 t_{ij} + \vartheta_{0i} + \vartheta_{1i} t_{ij} + + \varepsilon_{ij}$$

Model 1 defines the overall mean of the population using the intercept parameter β_0 which is a constant. The deviation of each cluster from the overall mean is given by the random effect of each cluster ϑ_{0i}. Hence, the mean of each cluster is estimated as $\beta_0 + \vartheta_{0i}$ in the model. The variances of the error and random effect in the model are defined by σ_ε^2 and σ_ϑ^2, respectively. The covariance within the clusters in this model is given by the variance of random effect (generally denoted by σ_ϑ^2) which also captures the between-clusters variability. Finally, in this model the within-cluster variability is captured by error variance σ_ε^2. The total variability of the outcome in this model is estimated by the sum of both the variance components as equal to $\sigma_\varepsilon^2 + \sigma_\vartheta^2$.

Model 2 is a slight extension of the previous model. It only differs from the previous model in terms of allowing the overall mean values to change linearly by a factor of β_1 across different time points. The variance covariance matrix for the outcomes in this model is exactly the same as in the Model 1.

Model 3 allows for the variations at the baseline and across different time points by letting the slope of time effect to have random effect in the model. The variances of random intercept and slope are given by $\sigma_{\vartheta 0}^2$ and $\sigma_{\vartheta 1}^2$, respectively. The model also allows an inclusion of covariance parameter between the random intercept and slope term (denoted by $\sigma_{\vartheta 01}$). Based on this model specification, the variance–covariance matrix of the outcomes is now defined by the variance terms as $\sigma_\varepsilon^2 + \sigma_{\vartheta 0}^2 + \sigma_{\vartheta 1}^2 t_{ij}^2 + \sigma_{\vartheta 01} t_{ij}$ and the covariance terms as $\sigma_{\vartheta 0}^2 + \sigma_{\vartheta 1}^2 t_{ij}^2 + \sigma_{\vartheta 01} t_{ij}$. Hence, the variance and covariance parameters in this model are functions of time point of measurements. Again, the overall population-averaged response profiles are allowed to change linearly by a factor of β_1. In addition, the subject-specific trajectories can now vary randomly around the population-averaged response profiles in terms of the baseline location and rate of change with respect to t_{ij}.

Estimation of the random effects is generally based on methods relying on either empirical or full Bayesian approaches. The mixed-effects model framework has a shrinkage effect built into it due to the nature of estimation procedures. The shrinkage effect causes the estimates of response profiles to fit very close to the population-averaged response for those individuals whose response variability is very high in comparison to the overall average variability. On the other hand, the trajectories estimates follow close to the actual observed individual response profiles for the individuals whose responses have smaller variability in comparison to the overall variability (see Fitzmaurice et al., 2004; Hedeker and Gibbons, 2006 for details on estimation and shrinkage effect).

It is worthy to note that the methods such as ANOVA, analysis of covariance (ANCOVA), MANOVA, and repeated measures ANOVA can all be handled within a linear mixed-effects modeling framework.

CONCLUSION

In this introductory chapter, we have covered the basic principles behind the statistical analysis techniques. However, there is a vast literature of methods that have been developed over the last few decades specifically for RCTs. We hope that this chapter inspires the readers to dig deeper into the statistical texts that cover these ideas in detail without getting uneasy with the mathematical notations. Additionally, there are now widely available commercial statistical packages such as SPSS,

SAS, and Stata to carry out the analysis using the methods presented in this chapter. Some of the references for the material included in this chapter are from books by Rosner (2011), Kutner et al. (2004), Kleinbaum and Klein (2010, 2012), Fitzmaurice et al. (2004), Hedeker and Gibbons (2006), Schafer (1997). Rosner (2011) presents a lot of fundamental ideas of biostatistics discussed in the beginning of this chapter. Kutner et al. (2004) have covered extensive details of linear modeling and experimental designs in their book containing almost 1400 pages. The two introductory textbooks for logistic and survival regression models are authored by Kleinbaum and Klein (2010, 2012). Fitzmaurice et al. (2004), and Hedeker and Gibbons (2006) are the two excellent sources for the applied researchers hoping to gain further insight on longitudinal data analyses. Missing data concepts require some technical notations, and the books by Schafer (1997) and Little and Rubin (1987) thoroughly cover these concepts.

REFERENCES

Alldredge, B.K., Gelb, A.M., Isaacs, S.M., Corry, M.D., Allen, F., Ulrich, S., et al., 2001. A comparison of lorazepam, diazepam, and placebo for the treatment of out-of-hospital status epilepticus. N. Engl. J. Med. 345, 631–637.

Benjamini, Y., Hochberg, Y., 1995. Controlling the false discovery rate: a practical and powerful approach to multiple testing. J. R. Stat. Soc. Ser. B 57 (1), 289–300.

Cox, D.R., 1972. Regression models and life-tables. J. R. Stat. Soc. Ser. B 34 (2), 187–220.

Fitzmaurice, G.M., Laird, N.M., Ware, J.H., 2004. Applied Longitudinal Analysis. Wiley, New York.

Hastie, T., Tibshirani, R., Friedman, J., 2009. The Elements of Statistical Learning: Data Mining, Inference, and Prediction, second ed. Springer-Verlag, New York.

Hedeker, D., Gibbons, R.D., 2006. Longitudinal Data Analysis. Wiley, New York.

Kutner, M., Nachtsheim, C., Neter, J., Li, W., 2004. Applied Linear Regression Models, fifth ed. McGraw-Hill/Irwin, New York.

Kleinbaum, D.G., Klein, M., 2010. Logistic Regression-A Self-learning Text, third ed. Springer, New York.

Kleinbaum, D.G., Klein, M., 2012. Survival Analysis-A Self-learning Text, third ed. Springer, New York.

Laird, N.M., Ware, J.H., 1982. Random effects models for longitudinal data. Biometrics 38, 963–974.

Little, R.J.A., Rubin, D.B., 1987. Statistical Analysis with Missing Data. John Wiley & Sons, New York.

Rubin, D.B., 1976. Inference and missing data. Biometrika 63, 581–592.

Rosner, B., 2011. Fundamentals of Biostatistics. Brooks/Cole, Boston.

Šidák, Z.K., 1967. Rectangular confidence regions for the means of multivariate normal distributions. J. Am. Stat. Assoc. 62 (318), 626–633.

Schafer, J.L., 1997. Analysis of Incomplete Multivariate Data. Chapman & Hall, New York.

Wood, S., 2006. Generalized Additive Models: An Introduction with R. Chapman & Hall/CRC, Boca Raton, FL.

Chapter 15

Good Clinical Practice and Good Laboratory Practice

Nathalie K. Zgheib[1], Stephanie L. Tomasic[2] and Robert A. Branch[3]

[1]American University of Beirut, Beirut, Lebanon; [2]The Pitt-Bridge: Gateway to Success, Pittsburgh, PA, United States; [3]University of Pittsburgh, Pittsburgh, PA, United States

Chapter Outline

Key Points

- International Conference on Harmonization—Good Clinical Practice (ICH-GCP) is a guidance document meant to provide assistance to industry and health-care professionals in conducting ethical and credible clinical research.
- Title 21 of the Code of Federal Regulations (CFR) (most relevant: 21 CFR 50, 21 CFR 56, and 21 CFR 312) is generally consistent with ICH-GCP, but also embraces additional regulations such as rules for financial disclosure.
- Title 21 CFR part 58 sets the guidelines for Good Laboratory Practice (GLP).
- Key participants in clinical research are the sponsor, the investigator, and the institutional review board (IRB).
- Documentation and record keeping are crucial for the evaluation and monitoring of study compliance with rules and regulations.
- Standard operating procedures ensure that research is conducted in compliance with rules and regulations.
- Data management and presentation can be facilitated by information technology.
- Elements of quality assurance for GLP include instrument validation, certification of analysts, reagents and laboratory facility, and adequate sample tracking.

Clinical and Translational Science. http://dx.doi.org/10.1016/B978-0-12-802101-9.00015-6

OVERVIEW

Clinical research has become a highly sophisticated endeavor that merits and has a high degree of regulatory oversight provided at institutional, state, and federal levels within the United States. It is also an international endeavor, and there has been a major attempt to integrate different national requirements through the International Conference on Harmonization (ICH). The primary driving force to acquire this uniformity has been new drug or device development, and this industry's need for codification of all aspects of their industry. These quality standards have been extended and applied, with minor modifications, to all clinical research irrespective of funding source using the premise that the voluntary contribution of research time, effort, and risk call for the highest ethical and procedural quality that can be applied.

Clinical and laboratory data used in national regulatory decisions are legally mandated to meet the written specification detailed under the rubric of "Good Clinical Practice" (GCP) and "Good Laboratory Practice" (GLP). Information that is collected for research purposes other than regulatory review are expected to, but not legally required to, meet the letter of the law. Thus, these practices should provide the premise for all clinical research and for all laboratory research that contributes to clinical research.

To provide clear distinction between the two disciplines supporting clinical investigation, the first section, Good Clinical Practice, is written to advise the clinician investigator, while the second, Good Laboratory Practice, is written for the laboratory director with emphasis on specialized laboratory services that have not yet been incorporated into routine clinical practice.

GOOD CLINICAL PRACTICE

Introduction

Let us say for the sake of discussion that you are a physician recently hired in an academic institution, and you have either been asked to "do research," or you are a curious person with a lot of unanswered scientific questions in mind and would like to "do research." You also think that there are wonderful opportunities out there and that this would increase your funding.

Although this looks to be a really interesting and entertaining thing to do, you have absolutely no idea about where to start! So far in your career you have been a very successful clinician and patients simply *love you*. You think that doing clinical research should probably be similar to doing clinical work. Furthermore, you have heard of "GCP," the investigator's responsibilities, and Food and Drug Administration (FDA) Form 1572, but you think that there should not be a problem since you are a *good person*.

Unfortunately, it is actually not that simple, and therefore, this Good Clinical Practice section of this chapter is written to provide insight into conducting high-quality clinical research and directs you to additional important resources.

Definition

GCP is "an international ethical and scientific quality standard for designing, conducting, recording and reporting trials that involve the participation of human subjects. Compliance with this standard provides public assurance that the rights, safety and well-being of trial subjects are protected" (FDA, 1996). It is important to stress that GCP relates to research only, and hence a more appropriate title would be **Good Clinical Research Practice** (Grimes et al., 2005).

The guidelines were labeled ICH-GCP as they were developed in 1996 at the ICH to "provide clinical trials with a unified standard across the European Union, Japan and the United States." They have been prepared for clinical trials which are planned to be submitted to regulatory authorities; however, they may also be applied to any kind of clinical research.

ICH-GCP has 13 principles, summarized in Table 15.1. Its main objectives are to ensure the ethics, quality, and integrity of research trials. It is also meant to avoid problems that may arise while obtaining marketing authorizations for drugs across different countries where different versions of GCP are being adhered to.

Rules and Regulations

ICH-GCP is a **guidance** document meant to provide assistance to industry and health-care professionals in conducting ethical and credible clinical research. Therefore, there are several sets of **rules and regulations** in place to govern clinical research most of which have adopted the ICH-GCP guidelines. Although the various regulations take different forms and

TABLE 15.1 Principles of International Conference on Harmonization—Good Clinical Practice (ICH-GCP) (FDA, 1996)

Ethics	Declaration of Helsinki, GCP, and other regulatory authorities are followed
Risks and benefits	Study risks are justified by anticipated benefits
Trial subjects	Rights, safety, and well-being of research subjects are secured
Investigational product	Adequate supporting information is provided
Science	Protocol is scientifically sound and well written
Compliance	Trial is conducted in compliance with approved protocol
Qualified physician	Medical care and decisions are made by a qualified physician
Trial staff	Trial is conducted by qualified staff
Informed consent	Informed consent is given before recruitment
Data	Data are securely recorded, managed, and stored
Confidentiality	Research subject's privacy and confidentiality are protected
Good manufacturing practice	Investigational products are appropriately manufactured, handled, and stored
Quality assurance	Systems and procedures for trial quality assurance are in place

sometimes get to be overlapping and confusing, they all have the same goal which is "protection of rights, integrity, and confidentiality of human subjects" (FDA, 1996).

In the United States there have been a series of events that have led to the development of several rules and regulations for conducting research on human subjects embodied in the Belmont Report (National Commission for the Protection of Human Subjects of Biomedical and Behavioral Research, 1979; U-S History.com, 2007; FDA, 2009a) (Table 15.2). Currently the US Department of Health and Human Services (DHHS; http://www.hhs.gov/) is the main regulatory body that has the role of protecting the health of all Americans. It has published the Code of Federal Regulation (CFR) Title 45 part 46 (DHSS, 2009), which focuses on protection of human subjects. It is also known as the Common Rule because, in 1991, 17 federal departments that sponsor research adopted it. DHHS combines several agencies, including the National Institutes of Health (NIH; http://www.nih.gov) and the FDA (http://www.fda.gov).

NIH is the US focal point for supporting medical research. It defines **research** as "a systematic investigation, including research development, testing and evaluation designed to develop or contribute to generalizable knowledge." Therefore, it does not exclusively focus on "clinical investigation" which, as defined by the FDA, "involves a test article which could be biologic products, drugs, electronic products, food and color additives, and medical devices." The Office of Human Research Protections (OHRP; http://www.hhs.gov/ohrp) is responsible for supervision of 45 CFR Part 46 Federal Policy Regulations. It provides guidance for compliance and negotiates assurance agreements with institutional review boards (IRBs).

FDA is the body where pharmaceutical marketing and registration are submitted. It maintains Title 21 of the CFRs (most relevant: 21 CFR 50, 21 CFR 56, and 21 CFR 312) (FDA, 2014a, 2013a,b), which are generally consistent with ICH-GCP, but also embraces additional regulations such as rules for financial disclosure developed in 1998 (21 CFR 54) (FDA, 2014b). These regulations apply to **clinical investigations**, defined as "any experimentation that involves a test article and one or more human subjects" (Table 15.3).

On an international level, clinical research regulations differ, and the OHRP has compiled a list of regulations for many countries. This compilation can be accessed online, and most of the parties refer to ICH-GCP (OHRP, 2014).

Clinical Practice and Research

One single most important issue is the distinction between research and clinical practice, especially as many clinical trials are done on subjects who are the investigator's own patients. The ethical issue involved is the inherent tension between optimal medical care for the individual patient versus acquiring new data and information with a minimum variation in therapy to permit generalizable knowledge. This has the potential to lead to research subject misunderstanding of the intent of a medical decision in addition to the investigator unintentionally conducting unethical research.

TABLE 15.2 Milestones in US Food and Drug Law History (U-S History.com, 2007; FDA, 2009a)

Triggering Event	Outcome Title	Outcome Description	Date
Alcohol main ingredient of "miracle cures"	The Pure Food and Drug Act	Ingredients disclosure Standards for purity and strength	1906
Meatpacking plant in Chicago	Gould Amendment to the Food and Drug Law of 1906	Numerical count of ingredients on package	1913
Antifreeze with sulfonamide elixir	The Federal Food, Drug and Cosmetic (FDC) Act	Drug safety studies	1938
Nazi experimentation on concentration camps inmates	The Nuremberg Code	Informed consent Animal studies Qualified personnel	1947
Birth defects due to thalidomide	Kefauver–Harris Drug Amendments	Drug efficacy and safety studies Adverse drug events reporting	1962
The Tuskegee Syphilis Study	The National Research Act	Human subjects protection	1974
National Commission for the Protection of Human Subjects of Biomedical and Behavioral Research	The Belmont Report	Three ethical principles Respect for persons Beneficence Justice	1979

TABLE 15.3 Current US Regulations and the International Conference on Harmonization (ICH) guidelines (FDA, 1996, 2014a,b, 2013a,b; DHSS, 2009)

Body	Definitions	Regulation	Topics
National Institute of Health—Office of Human Research Protections (NIH-OHRP)	**Research:** "A systematic investigation, including research development, testing, and evaluation, designed to develop or contribute to generalizable knowledge"	Title 45 Part 46	*Protection of human subjects* Basics Pregnant women, fetuses, and neonates Prisoners Children
Food and Drug Administration (FDA)	**Clinical investigation:** "Any experiment that involves a test article and one or more human subjects"	Title 21	*Part 50 Protection of Human Subjects* General provisions Informed consent Reserved Children *Part 56 IRBs* General provisions Organization and personnel Functions and operations Records and reports Noncompliance *Part 312 Investigational New Drug Application* General provisions Investigational New Drug application Administrative actions Responsibilities of sponsors and investigators Life-threatening and severely debilitating diseases Miscellaneous Lab research, animals, or in vitro tests
ICH	**Clinical trial/study:** "Any investigation in human subjects intended to discover or verify the clinical, pharmacological, and/or other pharmacodynamic effects of an investigational product with the object of ascertaining its safety and/or efficacy"	ICH Guidance E6	Institutional review board Investigator Sponsor Protocol and amendment(s) Investigator's brochure Essential documents

Part A of the Belmont Report differentiates between clinical research and practice; it states: "Research and practice may be carried on together when research is designed to evaluate the safety and efficacy of a therapy. This need not cause any confusion regarding whether or not the activity requires review; the general rule is that if there is any element of research in an activity, that activity should undergo review for the protection of human subjects." Therefore, it becomes necessary to develop ethical research protocols that are continuously supervised by the regulatory authorities. This leads to a common clinical perception that research appears to be made complicated by a lot of **paperwork** that is necessary for documentation and reporting. Fig. 15.1 shows a schematic of differences in oversight mechanisms between clinical practice and research. In clinical practice (Fig. 15.1A), the physician prescribes treatment to patients based on clinical judgment, the information already available within the constraint of the local standard on medical practice, and the knowledge that the FDA has approved the drug. These are enforced by the host clinical facility, but the FDA has no jurisdiction. In this scenario, the FDA jurisdiction is confined to oversight of the pharmaceutical industry in the provision of the approved drug or device. In clinical research overseen by the FDA, there is a multilevel supervision in each step from drug manufacture, drug distribution from manufacturer to institution, to the patient, to the IRB responsible for approval of the consent form used, and independent of the institution through Form 1572 to the investigator responsible for the study (Fig. 15.1B). For drugs already approved by the FDA, but being used in NIH-sponsored clinical trials, the oversight is somewhat simpler, with a line of responsibility from the OHRP to the IRB to the principal investigator (PI) (Fig. 15.1C). It is clear, therefore, that the real conflict of interest that each PI has between ideal clinical care and ideal clinical research merits and receives rigorous oversight.

Despite appearing complicated, research is actually "fun," challenging, and rewarding. It is, however, essential to follow the guidelines of GCP to answer interesting scientific questions, generate credible results that can be extrapolated to a generalizable conclusion, and provide ethical clinical care.

Key Participants in Clinical Research

Three main parties are involved in the conduct of research, and both the ICH-GCP guidelines and the federal regulations highlight the responsibilities of each.

Sponsor

The sponsor is "an individual, company, institution, or organization which takes responsibility for the initiation, management, and/or financing of a clinical trial" (FDA, 1996). A sponsor is usually a pharmaceutical or a biotechnological company, but can be a local investigator who has obtained an investigator-initiated Investigational New Drug (IND) from the FDA.

A sponsor with a conventional IND has the following main responsibilities:

1. Develop and update the "investigator's brochure," which is "the compilation of the clinical and nonclinical data on the investigational product relevant to the study" (FDA, 1996).
2. Maintain IND application and records.
3. Develop the study protocol.
4. Develop case report forms (CRF), which are printed or electronic documents designed to record all of the protocol-required information to be reported to the sponsor on each trial subject.
5. Select qualified investigators.
6. Provide investigational articles to investigators.
7. Ensure adherence to protocol by doing regular monitoring and audits.
8. Inform the FDA and investigators of safety information and adverse events. Note that it is the investigators' responsibility to report safety information to the IRB and to the sponsor; however, it is the sponsor's responsibility to inform the FDA.
9. Submit reports to the FDA and other regulatory agencies.

The sponsor also allocates a data safety and monitoring board (DSMB), which is a group of independent people whose job is to review adverse event (AE) information, do interim analysis, and make decisions on the risk/benefit ratio and whether the study may continue, require modifications, or be terminated. When the study is of low risk and not too complicated, an internal data safety and monitoring plan (DSMP) is sufficient (NIH, 1998). A sponsor often transfers some or all of its responsibilities to an academic research organization or a contract research organization.

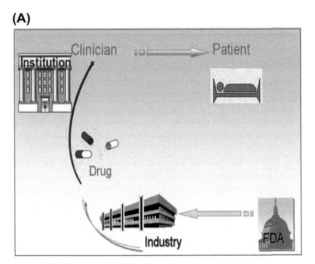

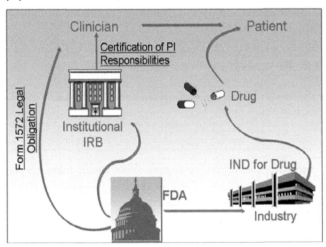

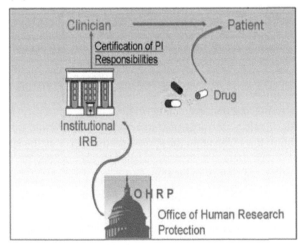

FIGURE 15.1 Parties involved in clinical practice and research. (A) Clinical practice; (B) Clinical research and FDA; and (C) Clinical research and NIH.

Institutional Review Board

The IRB, also known as independent ethics committee (IEC), is "an independent body constituted of medical, scientific, and non-scientific members, whose responsibility is to ensure the protection of the rights, safety and well-being of human subjects involved in a trial. Among other things, its functions include reviewing, approving, and providing continuous review of trial protocol, amendments and methods and material to be used in obtaining and documenting informed consent of trial subjects" (FDA, 1996). Federally funded human research is required to have IRB review and approval. Most research institutions, professional organizations, and academic journals apply the same requirements.

A sponsor can elect to use the local IRB of each of the study sites or an outside (regional, national, or international) IRB depending on the study complexity and the local institution's willingness to accept an outside IRB peer view over research conducted within the institution.

Investigator

An investigator is "a person responsible for the conduct of the clinical trial at a trial site. If a trial is conducted by a team of individuals at a trial site, the investigator is the responsible leader of the team and may be called the principal investigator (PI)" (FDA, 1996). Sometimes an investigator is the sponsor; in this case the individual is called an "investigator–sponsor" and is responsible for both roles. Every investigator involved in drug or biologic research under IND regulations is obligated to sign an FDA Form 1572. Basically, the PI signs a contract with the FDA that documents his/her agreement to follow the protocol and study obligations and alleges the investigator's knowledge in conducting the trial (DHHS, 2013).

FDA Form 1572 contains several sections of information, but the most important is Section 9 which contains the investigator's commitments summarized in Table 15.4. Table 15.4 also contains some additional responsibilities highlighted by ICH-GCP.

One important investigator's responsibility explicitly described with extensive details in ICH-GCP and in the federal regulations is the management and reporting of **AEs**. Table 15.5 introduces some important ICH-GCP definitions (FDA, 1996).

Fig. 15.2 gives some guidance on the thinking process and the actions that need to be taken should an AE occur. The investigator should evaluate the clinical significance of the event, assess causality, and then report. Generally, he/she should inform the sponsor about all AEs that occur (both anticipated and unanticipated) according to the research protocol guidelines. The sponsor is then responsible for notifying the FDA. IND safety reporting can be provided to the FDA yearly as a narrative description of the most frequent and serious AEs by body system and listing of causes of death and withdrawals due to adverse experiences, or on the FDA MedWatch Form 3500. This form is mandatory for FDA-approved drugs adverse drug event reporting (FDA MedWatch, 2014). FDA regulations require that sponsors report AEs within 7 days for life-threatening or fatal events, and within 15 days for other serious events. It is the sponsor's responsibility to report back to the investigator any unexpected serious AEs that are probably or definitely associated with the study that occur at the other investigator sites and ensure that it gets reported to the local IRB within 30 days, and the research subjects are informed of the risk as "new information." As for events that occur on site, it is the investigator's responsibility to immediately (usually within 24 h) report unexpected life-threatening or fatal events that may be related to the intervention, and unexpected other serious study-related ones within 10 days. The rest of unanticipated (i.e., unexpected) AEs should be listed in the IRB yearly progress report. The AEs are also reviewed by the DSMPs or DSMBs that issue the appropriate recommendations. Note, however, that DSMBs meet periodically and are designed to review aggregate data rather than respond to individual reports in real time; therefore, it sometimes becomes the responsibility of the investigative team to react to safety reports and determine if the study is safe to continue as is, should be modified, or terminated. Finally, in most clinical trials, the research subject is also a patient, and hence the investigator has a state obligation to report to institutional risk committees, usually within 24 h of the event occurrence (FDA, 1996, 2009b; NIH, 1999; ICH, 1994).

Conflict of interest is a relevant issue that has recently attracted negative attention. Today's research environment is highly competitive and places traditional ethical principles at increased risk of being ignored, to the detriment of protection of human research subjects. Conflict of interest in research potentially leads to lack of objectivity in research subjects' recruitment and clinical management and will lower scientific credibility of the research results. It is therefore necessary and mandatory to disclose any potential (real or perceived; personal or direct kin relative) conflict of interest, both financial

TABLE 15.4 FDA and ICH-GCP (International Conference on Harmonization—Good Clinical Practice) Investigator's Responsibilities

Items	Description	The Investigator Should
Food and Drug Administration (FDA) Form 1572 (DHHS, 2013)		
Knowledge base	Scientific background	Be qualified by education, training, and experience
	Investigator brochure	Know and understand the pharmacokinetics and dynamics of the trial product
	Personnel	Ensure that all personnel are well trained in clinical research and know GCP
	Financial conflict	Disclose his/her and the personnel's potential financial conflict
Protocol review and approval	Consent form and process	Ensure that research participants are informed about the study details and risks
	Institutional review board (IRB)	Get review and approval from IRB
Compliance	Protocol adherence	Do exactly what is written in the protocol and report and explain unanticipated problems or deviations
	Record keeping	Check accuracy, completeness, legibility, and timeliness of records and reports
	Audit	Permit monitoring and audits by the FDA
Adverse events reporting		Inform the IRB and sponsor about adverse drug events
Medical care and decisions		Provide medical care to trial participants in case of adverse events
ICH-GCP Additional Items (FDA, 1996)		
Resources	Time	Have sufficient time to conduct the study
	Recruitment	Demonstrate a potential to be able to recruit adequate numbers of research subjects
Investigational product	Responsibility	Be responsible of the product at the trial site and keep a log of use
	Storage	Store the product according to requirements
New information	Adverse events	Inform research participants about new risks
	Premature suspension or termination of a trial	Inform research participants about termination decisions
Reports	Progress report	Submit written summaries of the study progress
	Final report	Provide a final report at the end of the study

TABLE 15.5 Adverse Events (FDA, 1996; ICH, 1994)

Terms	Definition
Adverse event (AE)	"Any untoward medical occurrence … which does not necessarily have to have a causal relationship with this treatment"
Adverse drug reaction (ADR)	"All noxious and unintended responses to a medicinal product related to any dose"
Unexpected ADR	"An adverse reaction, the nature or severity of which is not consistent with the applicable product information"
Serious adverse drug event or ADR	"Any untoward medical occurrence that at any dose results in death [or] is life-threatening"

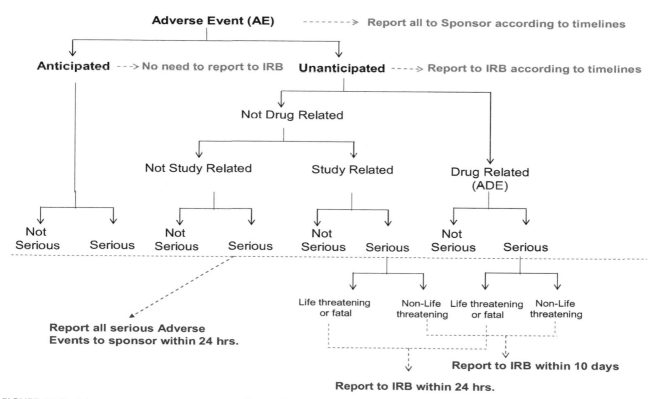

FIGURE 15.2 Adverse events: assessment and reporting requirements by the principle investigator to the institutional review board (IRB) (FDA, 2009b).

and nonfinancial. Currently, both the NIH's Office of Extramural Research (DHHS Office of Extramural Research, 2008) and the FDA (21 CFR 54) (FDA, 2014a,b) require that personnel involved in research projects submit a declaration of "significant financial interests" likely to affect the proposed research. Table 15.6 lists important terms defined by 21 CFR 54. The position of academic institutions and regulatory authorities is not to penalize investigators' financial and other interests; instead, they only require investigators to complete disclosure, evaluate their potential conflict, and appropriately manage the conflict.

Although the investigator's responsibilities seem to be "all things to all people," the good news is that investigators can and should **delegate**. A key person or personnel on whom the PI relies is the clinical research coordinator (CRC). His/her role is to ensure the quality and timely completion of study under direct supervision of the PI. The PI may also rely on institutional mechanisms such as the Investigational Drug Service, within the pharmacy, whose role is "to be responsible for storing and dispensing the study drug, in addition to maintaining adequate records and documentation" (FDA, 1996). It is important that delegation be accompanied with adequate supervision and training, because liability ultimately always resides with the PI. Two of the FDA Form 1572 investigator's commitments state: "I agree to personally conduct or supervise the described investigation(s) … I agree to ensure that all associates, colleagues and employees assisting in the conduct of the study(ies) are informed about their obligations in meeting the above commitments" (DHHS, 2013). Since May 2000, the DHSS has made it obligatory for all investigators submitting new grants and their "key personnel" to show proof of adequate training and knowledge in human subjects' protection, a regulation that led to the initiation of elaborate research training modules and certificates available in most institutions throughout the United States (NIH, 2000).

Finally, one important aspect, which is not highlighted in the guidelines, is for the investigator to be aware of the study budget and make sure that all expenditures are consistent with the approved budget for the project. It is particularly important to include ethical contingency planning in case there is an AE that results in additional medical expenses. This requires coordination between the sponsor and health-care institutional fiscal review offices where clinical care is provided.

TABLE 15.6 Financial Disclosure: 21 Code of Federal Regulation Part 54 Definitions (FDA, 2014b)

Terms	Definition
Compensation affected by the outcome of clinical studies	"Compensation that could be higher for a favorable outcome than for an unfavorable outcome, such as compensation that is explicitly greater for a favorable result or compensation to the investigator in the form of an equity interest in the sponsor of a covered study or in the form of compensation tied to sales of the product, such as a royalty interest."
Significant equity interest	"Any ownership interest, stock options, or other financial interest whose value cannot be readily determined through reference to public prices (generally, interests in a nonpublicly traded corporation), or any equity interest in a publicly traded corporation that exceeds $50,000 during the time the clinical investigator is carrying out the study and for 1 year following completion of the study."
Proprietary interest in the tested product	"Property or other financial interest in the product including, but not limited to, a patent, trademark, copyright, or licensing agreement."
Significant payments of other sorts	"Payments made by the sponsor of a covered study to the investigator or the institution to support activities of the investigator that have a monetary value of more than $25,000, exclusive of the costs of conducting the clinical study or other clinical studies, (e.g., a grant to fund ongoing research, compensation in the form of equipment, or retainers for ongoing consultation or honoraria) during the time the clinical investigator is carrying out the study and for 1 year following the completion of the study."

Documentation and Record Keeping

Adequate documentation is crucial in clinical trials. Research records permit the PI, sponsor, and monitors to evaluate if the study conduct is in compliance with GCP guidelines. They also assist the investigator in the oversight of all aspects of study progress, in addition to management and assessment of the quality and integrity of generated data. Records should be maintained in such a manner that an individual not associated with the research project can easily track the study progress. Hence, the ICH-GCP guidelines describe essential documents for the conduct of a clinical trial. They group them into three sections according to the stage of the trial during which they would normally be generated: (1) before, (2) during (individual CRFs), and (3) after the trial (FDA, 1996).

Table 15.7 lists essential documents that should be kept in the regulatory binder. Original source documents, such as flow sheets, clinical and lab reports, physician notes, etc., which are then transcribed into the CRFs, should also be kept up to date. Documents can be organized in a convenient way to suit either the investigator or CRC preferences as long as documents are complete, easy to retrieve, and correct. As is the case for clinical progress notes, when an error is discovered, it must be corrected without obscuring the original data, without using "whiteout" and being countersigned as to by whom and when the change was made. Documents are expected to remain available long after the study is complete; a general rule is a minimum of 7 years after the study.

Research paper records can be transferred into an electronic system, or data can be directly entered into an electronic database. The latter is accepted by the FDA, and the electronic documents would serve as source documents. The FDA has issued a guidance document about "computerized systems that are used to create, modify, maintain, archive, retrieve, or transmit clinical data required to be maintained and/or submitted to the FDA" (FDA, 2007). Nine principles are recommended, most importantly: electronic records should be as compliant with the regulatory guidelines as paper systems; an audit tracking system must be in place to identify who made what changes and when; and the system should be validated, secure, regularly backed up, and readily available for FDA review. FDA also outlines the process of electronic signatures certification.

Whether research records are stored in paper form, electronic form, or both, the two most important points to remember are as follows:

1. Things that have been said or done and have not been recorded are considered not done (e.g., the informed consent process).
2. Research subjects' confidentiality should always be a concern; therefore all identifiable records should be kept secure, and access should be limited to specified study personnel only.

TABLE 15.7 Essential Documents for the Regulatory Binder (FDA, 1996)

Protocol

Informed consent document

Curriculum vitae

Financial disclosure

Signature log

Sponsor correspondence

Institutional review board correspondence

Investigator's brochure

Drug accountability documentation

Food and Drug Administration Form 1572

Certification of relevant training

Laboratory certification

Range of normal laboratory values

Subjects' data
    Subject identification code list
    Screening and enrollment randomization log
    Visit log
    Communication: letters, telephone contacts, informed consent process
    Source documents and case report forms
    Adverse events

Fiscal reports

Reports from data safety and monitoring plan or data safety and monitoring board

Auditing reports

Progress and final reports

Data Management and Presentation

Data management is essential for the success of any research project and is usually best handled by professional information technology (IT) support. Data management includes all the components of data acquisition, analysis, storage, security, and sharing. It also includes moving CRFs and other regulatory documents to electronic format in accordance with the FDA guidance document (FDA, 2007). Good management is important to ensure confidentiality (within the American Health Insurance Portability and Accountability Act—HIPAA—rules) (DHSS Office for Civil Rights, 2014), research integrity, and intellectual property protection. The latter is especially relevant when the research project is industry-sponsored, and hence the sponsor is perceived as *owner* of the data which may lead to delay in data publication despite academic pressure on the investigator to do so. Currently, many national and international data banks are available for researchers to deposit their data.

Good data management is a prerequisite for **data presentation** into a manuscript publication. Currently, many scholarly journals request supplementary data with the published article, including the raw (nonanalyzed) data, and routinely request a note stating that the study has been approved by IRB, a statement that provides the journal editor with a relatively fair assurance that the study results have been generated by adherence to GCP. Journals are also concerned about the ethics of **authorship**, and hence they are following the uniform requirements developed by the International Committee of Medical Journal Editors (ICMJE, 2008). Guidelines mainly define authorship and authors' contributions; in addition, they help determine who the first author should be.

Monitoring and Compliance

ICH-GCP defines monitoring as "the act of overseeing progress of a clinical trial, and of ensuring that it is conducted, recorded and reported in accordance with the protocol and applicable regulatory requirement(s)." It is important to remember that just being a *good person* is not enough; it is mandated to follow regulations, and any attempt to deviate from requirements and to falsify data or records can be expected to lead to serious litigation. FDA 21 CFR 56 subpart E deals with issues of IRB noncompliance, and the FDA website (http://www.fda.gov) regularly posts updated "warning letters" and "debarment lists" of individual investigators. OHRP (http://www.hhs.gov/ohrp/) also posts a guidance document for compliance in addition to "determination letters" to institutions.

Standard Operating Procedures (SOPs) are the linchpin mechanism to ensure that research is conducted in compliance with rules and regulations. ICH-GCP defines SOPs as "detailed, written instructions to achieve uniformity of performance for a specific function" (FDA, 1996). It states that all parties involved in the conduct of research should have SOPs in place. SOPs should be as detailed as possible and be constantly revised and updated. They should describe everyday practice standards, including administrative, clinical, and laboratory issues. They should state what is actually done, and not what is supposed to be or is intended to be done. By using SOPs, it is established that the investigator is knowledgeable of his/her practice standards and aware of regulations. Compliance with availability and use of SOPs helps sponsors feel more comfortable in funding an individual PI because research performance is likely to be reliable. SOPs are important guides for how to structure an audit if there is cause for concern, and many resources are available for help in developing them in concert with GCP guidelines.

Conclusion

In conclusion, even if the process used to maintain compliance appears tedious and time-consuming, it remains straightforward to follow established guidelines of GCP. The major points to keep in mind are as follows:

1. Research is a dynamic process, and hence the research team has to be adaptable to change.
2. Applying GCP guidelines provides a consistent framework to make research easier and decreases the risk for liability should an unanticipated event happen.
3. Audits should be anticipated and welcomed as they are essential and helpful not only to the sponsor and the research subject but also to the investigator.
4. Delegation of tasks coupled with constant supervision and adequate training are essential.

And always remember that research, like clinical practice, is a lot of fun, personally rewarding, and contributes to future generalizable knowledge that can be applied in clinical practice.

GOOD LABORATORY PRACTICE

Introduction

GLP is a formal legally defined term that encapsulates the practice of using a standardized set of guidelines to generate reproducible results that can be validated by replicating experimental conditions. GLP is an essential ingredient for any professional scientist.

Scientific measurements, whether they pertain to monitoring clinical determinations of blood sugar, contaminants in pharmaceutical products, or testing materials for integrity, purity, or sterility affect important decisions. Scientists have traditionally adopted sound laboratory practices directed at assuring the quality of their data as a personal acknowledgment of their responsibility. However, until recently these practices were not consistently adopted, enforced, or audited, and laboratories used to be in a situation where they have had to interpret regulations and develop procedures on an ad hoc basis.

In the mid-1970s, FDA reached the alarming conclusion that the proof of the safety of many regulated products was based on invalid studies. After convening a joint task force that included the FDA, the US Congress, the public, and industry as formal stake holders, GLP regulatory requirements for assuring a study's validity were finally proposed on November 19, 1976. The new regulations were designated as a new part, 3.e., of Chapter 21 of the CFRs. These final regulations, entitled Good Laboratory Practice for Nonclinical Laboratory Studies, were codified as Part 58 (21CFR) (FDA, 2013c).

Definition and Scope

GLP is a set of guidelines that govern the process, organization, and conditions under which laboratory studies are conducted. GLP is defined by principles that provide a framework within which laboratory studies are planned, performed,

TABLE 15.8 When Is Good Laboratory Practice (GLP) Needed? (FDA, 2013c)

GLP Is Needed For	GLP Is Not Needed For
Nonclinical safety studies of drug development	Basic research
Agricultural pesticide	Studies to develop new analytical methods
Development of toxic chemicals including new chemical entities	Chemical tests used to derive the specifications of a marketed food product
Food additives	
Testing of substances for explosive hazards	

monitored, recorded, reported, and archived. GLP provides an assurance to regulatory authorities that the data submitted are a true reflection of the results and can be relied upon while making risk/safety and efficacy assessments.

Although original GLP regulations were intended for toxicity testing, their applicability to any analytical instrument and method enables implementation to all scientific disciplines regulated by the FDA.

GLP regulates all nonclinical safety studies that support applications for research or marketing new and approved products regulated by the FDA or similar national legislation. This includes medicinal and veterinary drugs, aroma and color additives in food, nutrition supplements for livestock, and biological products. The definition of a nonclinical laboratory study "means in vivo or in vitro experiments in which test articles are studied prospectively in test systems under laboratory conditions to determine their safety." The term does not include studies utilizing human subjects, clinical studies, or field trials in animals. The term also does not include basic exploratory studies carried out to determine whether a test article has any potential utility or to determine physical or chemical characteristics of a test article. Table 15.8 broadly outlines the types of studies where GLP is required.

The definition of a testing facility denotes that the facility shall permit an authorized employee of the FDA to inspect the facility and to inspect all records and specimens required to be maintained regarding the study (FDA, 2013c). The FDA will not consider a nonclinical laboratory study valid if the testing facility refuses to permit inspection. Certification of laboratory facilities is normally done by an external agency. For example, an analytical laboratory might be audited by representatives of a federal agency with which they have a contract. An independent laboratory might file documentation with a responsible state or federal agency.

Organization and Personnel

Personnel are defined as each individual engaged in the conduct of or responsible for the supervision of a nonclinical laboratory study.

1. Each member will have education, training, and experience to conduct the assigned functions.
2. Each testing facility has to maintain a current summary of training and experience for each member engaged in or supervising the conduct of a nonclinical laboratory study.
3. There must always be a sufficient number of personnel for the timely and proper conduct of the study according to the protocol.
4. All personnel are required to take appropriate health and safety precautions and be free of medical conditions that would have an adverse effect on a nonclinical laboratory study.

Management of the Testing Facility

Each nonclinical laboratory is required to have a study director and a quality assurance unit (QAU).

Study Director

The study director is responsible for study control. The study cannot have an assistant director but can have an alternate study director who serves only in the director's absence. The study director has overall responsibility for all work conducted in that laboratory. This includes the technical conduct of the safety studies, as well as for the interpretation, analysis, documentation, and reporting of the results. He or she is designated by and receives support from management.

The responsibilities of the study director are as follows:

1. Approving of protocols and any subsequent changes.
2. Ensuring that the current revision of the protocol is followed.
3. Collating records and verifying all experimental data, including observations and AEs.
4. Ensuring that all GLP regulations are followed.
5. Creating a final statement on GLP compliance.
6. Ensuring timely archiving.

Quality Assurance Unit

The QAU serves an internal control function. It is responsible for monitoring each study to assure management that facilities, equipment, personnel, methods, practices, records, controls, SOPs, final reports (for data integrity), and archives are in conformance with the GLP regulations. For any given study, the QAU is entirely separate from and independent of the personnel engaged in the direction and conduct of that study.

The QAU is responsible for immediately reporting any problems to the director. The GLP regulations also require the QAU to maintain and periodically submit to laboratory management: comprehensive written records listing findings and problems, actions recommended and taken, and scheduled dates for inspection. A designated representative from the FDA or Environmental Protection Agency (EPA) may ask to see the written procedures established for the QAU's inspection and may request the laboratory's management to certify that inspections are being implemented and followed in accordance with the regulations governing the QAU (FDA, 2001).

The FDA mandates that responsibilities, procedures applicable to the QAU, records maintained by the QAU, and the method of indexing such records be maintained in writing. The primary responsibilities of a QAU are as follows:

1. Maintain a copy for all studies. This includes a description of the study to be conducted, the objectives and design of the study, the date when the study is initiated, the current status of each study, identity of the sponsor, and name of the study director.
2. Maintain copies of all protocols pertaining to the studies for which QAU is responsible.
3. Inspect the documentation for each study periodically to ensure the integrity of the study with respect to internal laboratory activities. Properly written and signed records of each periodic inspection must be maintained. The records must include the date of the inspection, the study inspected, the phase or segment of the study inspected, the person performing the inspection, findings and problems, action recommended and taken to resolve existing problems, and any scheduled date for reinspection.
4. Determine whether deviations from protocols and SOPs were made with proper authorization and documentation.
5. Review the final study report to assure that it accurately describes the methods and SOPs and that the reported results accurately reflect the raw data of the study.
6. Prepare and sign a statement to be included with the final study report that specifies the dates of audits and dates of reports to management and to the study director.
7. Audit the correctness of statements, made by the study director, on GLP compliance of the study.
8. Audit laboratory equipment.

Quality Assurance

The various elements of quality assurance are described as follows:

1. **SOPs** are procedures that have been tested and approved for conducting a particular study. These procedures must be evaluated and/or be published by the regulatory agency involved (e.g., EPA or FDA); these agencies may not accept analytical data obtained by other procedures (FDA, 2001). Within any commercial laboratory, SOPs should either be available or developed to acceptable standards, so that any analytical data collected and reported can be tied to a documented procedure.
2. **Statistical procedures** for data evaluation are specific to the particular field the study is being conducted in. Each field has its own standards that are deemed acceptable within that field or they may adopt specific statistical analysis procedures for defining detection limits, confidence intervals, analyze measurement units, etc. Regulatory agencies often describe acceptable statistical procedures.

3. **Instrumentation validation** is a process inherently necessary for any analytical laboratory. Data produced by "faulty" instruments may give the appearance of valid data. These events are particularly difficult to detect with modern computer-controlled systems which tend to remove the analyst from the data collection/instrument control functions. Thus, it is essential that some objective procedures be implemented for continuously assessing the validity of instrumental data. These procedures, when executed on a regular basis, will establish the continuing acceptable operation of laboratory instruments within prescribed specifications. After "control limits" are assigned to threshold values of upper and lower ranges around the expected instrumental output use of that instrument, quality assurance procedures will require that whenever an instrument's performance is outside these limits, analytical reports are discontinued and the cause of the problem is determined. Continued reporting only occurs when the instrument is certified to be operating again within control limits. Equipment must be adequately inspected, cleaned, and maintained. Equipment used for the generation, measurement, or assessment of data must be adequately tested, calibrated, and/or standardized. Written records of inspection, maintenance, testing, calibrating, and/or standardizing operations must be maintained.

4. **Reagents and materials certification** must follow accepted procedures and must be adequately documented. Each and every container for laboratory reagents/materials must be labeled with information related to its certification value, date, and expiration time. This policy is meant to ensure that reagents are used as outlined in the SOPs.

5. **Analysts' certification** is a required part of QA. Acceptable proof of satisfactory training and/or competence with specific laboratory procedures must be established for each analyst.

6. **Laboratory facilities certification** is normally done by an external agency. The evaluation is based on space (amount, quality, and relevance), ventilation, equipment, storage, hygiene, etc. The FDA has implemented a program of regular inspections and data audits to monitor laboratory compliance with the GLP requirements.

7. **Sample tracking** is an aspect of quality assurance which has received a great deal of attention with the advent of computer-based laboratory information management systems. Sample tracking is a crucial part of quality assurance. Procedures for sample tracking must maintain the unmistakable connection between a set of analytical data and the specimen and/or samples from which they were obtained.

Test, Reference, and Control Articles

Control articles are very important because they are commonly used to calibrate instruments. The accuracy of the reference substances determines the accuracy of the analytical method. The control substance has to be well defined with respect to identity, strength, purity, composition, and/or other characteristics which will appropriately define the test or control article. Methods of synthesis, fabrication, or derivation of test and control articles must be documented and must be available for inspection. The stability of each test and control article has to be determined through periodic reanalysis of each batch.

Protocol for and Conduct of a Nonclinical Laboratory Study

Each study requires an approved written protocol that clearly indicates the objectives and all methods for the conduct of the study. The protocol needs to contain the following elements:

1. A descriptive title and statement of the purpose of the study.
2. Details regarding the sponsor, the identification of the test, and control articles.
3. The number, body weight range, sex, source of supply, species, strain, and age of the test system.
4. The procedure for identification of the test system.
5. A description of the experimental design and reagents used in the protocol.
6. Details regarding doses, type, and frequency of tests, analyses, and measurements.
7. Records of the forementioned details and the date of approval of the protocol by the sponsor and the dated signature of the study director.
8. A statement of the proposed statistical methods to be used.

The nonclinical laboratory study has to be conducted in strict accordance with the specified protocol.

All data that are generated during the conduct of a nonclinical laboratory study, except those that are generated by automated data collection systems, have to be recorded, dated, and signed by the person entering the data. Changes made to a record cannot obscure the original entry. In automated data collection systems, the individual responsible for direct data input has to be identified at the time of data input.

Reporting of Nonclinical Laboratory Study Results

The final report, signed by both the study director and the QAU, should include the following details:

1. Name of the study director and facility.
2. Objectives, procedures, statistical methods, and changes to the protocol if any.
3. Identification of the test articles, with details on purity, composition, stability, etc.
4. A description of the method, dosage, duration, and route of administration of the test substance.
5. A description of the calculations and a summary of the data analysis with a conclusion.
6. The signatures of all the personnel involved in the study, including the study director.
7. The location of the final report and corrections or additions to the final report.

Record Keeping

Documentation and Maintenance of Records

A central feature of GLP guidelines is the maintenance of records. Maintenance of instrument and reagent certification records provides the primary resource for postevaluation of results, even after the passage of several years. Maintenance of all records specified provides documentation that may be required in the event of legal challenges due to repercussions of decisions based on the original analytical results.

Storage and Retrieval of Records and Data

All raw data, documentation, protocols, final reports, and specimens generated as a result of a nonclinical laboratory study have to be retained in an archive that is secure and accessible only to authorized personnel.

Retention of Records

Legal requirements for record storage can vary depending on use of the information. Records need to be retained for a period of at least 2 years following the date on which an application for a research or marketing permit, in support of which the results of the nonclinical laboratory study were submitted, receives FDA approval. Records have to be maintained for a period of at least 5 years after the date of FDA approval. Biological specimens can be stored as long as they can be maintained in good quality. Protocols, quality assurance documents, personnel records, etc. are kept for a minimum of 2 years following the date of FDA approval. (When in doubt, store it.)

Disqualification of Laboratory Facility Resources

The FDA and the EPA both conduct audits and provide certification to GLP laboratories. Audits involve the inspection of the facility, equipment records, and specimens and may also include the investigation of an experiment in depth from raw data to final reports.

The FDA usually conducts two types of inspections:

1. The **routine inspection** consists of a periodic evaluation of the compliance of a laboratory. A data audit is done.
2. **For-cause inspections** are conducted less frequently. These inspections are sometimes initiated by routine inspections when serious noncompliance with GLP regulations is observed, or because of unexpected data.

Purposes and Goals of Disqualification

The purposes of disqualification are to exclude studies that were conducted in a laboratory not compliant with GLP, *unless* it is demonstrated that the noncompliance did not occur during the study period and therefore did not affect the validity of the results.

Grounds for Disqualification

The grounds for disqualification are as follows:

1. The testing facility failed to comply with one or more of the regulations set forth in Title 21 CFR.
2. Noncompliance adversely affected the validity of the nonclinical laboratory study.

Reinstatement of a Disqualified Testing Facility

A facility can be reinstated after providing adequate proof during an inspection that it is in compliance with GLP regulations.

Conclusion

GLP is a set of guidelines that govern the process, organization, and conditions under which laboratory studies are conducted. GLP is defined by principles that provide a framework within which laboratory studies are planned, performed, monitored, recorded, reported, and archived. GLP provides an assurance to regulatory authorities that the data submitted are a true reflection of the results and can be relied upon while making risk/safety and efficacy assessments.

GLOSSARY

Clinical investigation Any experiment that involves a test article and one or more human subjects.
Good Clinical Practice (GCP) An international ethical and scientific quality standard for designing, conducting, recording, and reporting trials that involve the participation of human subjects.
Good Laboratory Practice (GLP) A formal legally defined term that encapsulates the practice of using a standardized set of guidelines to generate reproducible results that can be validated by replicating experimental conditions.
Institutional review board (IRB) An independent body constituted of medical, scientific, and nonscientific members, whose responsibility is to ensure the protection of the rights, safety, and well-being of human subjects involved in a trial.
Investigator A person responsible for the conduct of the clinical trial at a trial site.
Research A systematic investigation, including research development, testing and evaluation, designed to develop or contribute to generalizable knowledge.
Sponsor An individual, company, institution, or organization which takes responsibility for the initiation, management, and/or financing of a clinical trial.

LIST OF ACRONYMS AND ABBREVIATIONS

ADE Adverse drug event
ARO Academic Research Organization
CFR Code of Federal Regulation
CRC Clinical research coordinator
CRF Case report forms
CRO Contract Research Organization
DHSS (US) Department of Health and Human Services
DSMB Data safety and monitoring board
DSMP Data safety and monitoring plan
EPA Environmental Protection Agency
FDA Food and Drug Administration
FDC The Federal Food, Drug and Cosmetic Act
GCP Good Clinical Practice
GLP Good Laboratory Practice
HIPAA American Health Insurance Portability and Accountability Act
ICH International Conference on Harmonization
IDS Investigational Drug Service
IND Investigational New Drug
IRB Institutional review board
IT Information technology
NIH National Institutes of Health
OHRP Office of Human Research Protections
PI Principal investigator
QAU Quality assurance unit
SOP Standard Operating Procedures

REFERENCES

DHSS, 2009. CFR Title 45. Public Welfare. Part 46. Protection of Human Subjects. http://www.hhs.gov/ohrp/humansubjects/guidance/45cfr46.htm.
DHHS, 2013. Statement of Investigator. http://www.fda.gov/downloads/AboutFDA/ReportsManualsForms/Forms/UCM074728.pdf.

DHHS Office of Extramural Research, 2008. Conflict of Interest. http://oig.hhs.gov/oei/reports/oei-03-06-00460.pdf.

DHSS Office for Civil Rights, 2014. Medical Privacy — National Standards to Protect the Privacy of Personal Health Information. http://www.hhs.gov/ocr/hipaa/.

FDA, 1996. Guidance for Industry. E6 Good Clinical Practice: Consolidated Guidance. http://www.fda.gov/downloads/Drugs//guidance/ucm073122.pdf.

FDA, 2009a. FDA Backgrounder: Milestones in US Food and Drug Law History. http://www.fda.gov/AboutFDA/WhatWeDo/History/Milestones/.

FDA, 2014a. CFR Title 21: Food and Drugs. Part 50: Protection of Human Subjects. http://www.accessdata.fda.gov/scripts/cdrh/cfdocs/cfcfr/CFRSearch.cfm?CFRPart=50.

FDA, 2013a. CFR Title 21: Food and Drugs. Part 56: Institutional Review Boards. http://www.accessdata.fda.gov/scripts/cdrh/cfdocs/cfcfr/CFRSearch.cfm?CFRPart=56.

FDA, 2013b. CFR Title 21: Food and Drugs. Part 312: Investigational New Drug Application. http://www.accessdata.fda.gov/scripts/cdrh/cfdocs/cfcfr/CFRSearch.cfm?CFRPart5312.

FDA, 2014b. CFR Title 21: Food and Drugs. Part 54: Financial Disclosure by Clinical Investigators. http://www.accessdata.fda.gov/scripts/cdrh/cfdocs/cfcfr/CFRsearch.cfm?CFRPart554.

FDA, 2009b. Guidance for Clinical Investigators, Sponsors, and IRBs. Adverse Event Reporting to IRBs —Improving Human Subject Protection. http://www.fda.gov/downloads/RegulatoryInformation/Guidances/ucm126572.pdf.

FDA, 2007. Guidance for Industry: Computerized Systems Used in Clinical Trials. http://www.fda.gov/ohrms/dockets/98fr/04d-0440-gdl0002.pdf.

FDA, 2013c. CFR Title 21: Food and Drugs. Part 58: Good Laboratory Practice for Nonclinical Laboratory Studies. http://www.accessdata.fda.gov/scripts/cdrh/cfdocs/cfcfr/CFRsearch.cfm?CFRPart=58.

FDA, 2001. Bioresearch Monitoring: Good Laboratory Practice. http://www.fda.gov/downloads/ICECI/EnforcementActions/Bioresearch/UCM133765.pdf.

Grimes, D.A., Hubacher, D., Nanda, K., Schulz, K.F., Moher, D., Altman, D.G., 2005. The Good Clinical Practice guideline: a bronze standard for clinical research. Lancet 366, 172—174.

ICH, 1994. Clinical Safety Data Management: Definitions and Standards for Expedited Reporting E2A. http://www.ich.org/fileadmin/Public_Web_Site/ICH_Products/Guidelines/Efficacy/E2A/Step4/E2A_Guideline.pdf.

ICMJE, 2008. Uniform Requirements for Manuscripts Submitted to Biomedical Journals: Writing and Editing for Biomedical Publication. http://www.icmje.org/recommendations/archives/2004_urm.pdf.

FDA MedWatch, 2014. The FDA Safety Information and Adverse Event Reporting Program. http://www.fda.gov/medwatch/.

National Commission for the Protection of Human Subjects of Biomedical and Behavioral Research, 1979. The Belmont Report: Ethical Principles and Guidelines for the Protection of Human Subjects of Research. Office of Human Subjects Research. http://www.hhs.gov/ohrp/humansubjects/guidance/belmont.html.

NIH, 1998. NIH Policy for Data and Safety Monitoring. 10 June 1998. http://grants.nih.gov/grants/guide/notice-files/not98-084.html.

NIH, 1999. Guidance on Reporting Adverse Events to Institutional Review Boards for NIH-supported Multicenter Clinical Trials. http://grants.nih.gov/grants/guide/notice-files/not99-107.html.

NIH, 2000. Required Education in the Protection of Human Research Participants. 25 August 2000. grants.nih.gov/grants/guide/notice-files/NOT-OD-00-039.html.

OHRP, 2014. International Compilation of Human Subject Research Protections 2007.

U-S History.com, 2007. The Pure Food and Drug Act 1906. http://www.u-s-history.com/pages/h917.

Human Genetics

Chapter 16

Introduction to Human Genetics*

Bruce R. Korf

University of Alabama at Birmingham, Birmingham, AL, United States

Chapter Outline

Key Points

- Single gene traits are transmitted in accordance with Mendel's laws as dominant or recessive.
- Common, complex traits may cluster in families according to the principles of multifactorial inheritance.
- Genetic information is stored in the form of the sequence of DNA and encodes proteins as well as some noncoding RNAs.
- The human genome sequence has been determined, and the mechanisms of gene expression and regulation are now being studied at the genomic level.
- The tools of genetics and genomics can be applied to the study of both rare single gene and common multifactorial phenotypes.

---

* Based on chapter from first edition written by Achara Sathienkijkanchai, MD, and Bruce R. Korf, MD, PhD.

INTRODUCTION

Genetics is the science that deals with the storage of information within the cell, its transmission from generation to generation, and variation among individuals within a population. Human genetics research has a long history, dating from the study of quantitative traits in the 19th century and the study of human Mendelian traits in the first decade of the 20th century. Medical applications have included such landmarks as a newborn screening for inborn errors of metabolism, cytogenetic analysis, molecular diagnosis, and therapeutic interventions, such as enzyme replacement. For the most part, however, medical applications had been limited to diagnosis and management of rare disorders caused by mutations in individual genes or structural abnormalities of chromosomes. Recent advances have opened the possibility of understanding genetic contributions to more common disorders, such as diabetes and hypertension. Genetic approaches are now being applied to conditions in virtually all areas of medicine. Moreover, the power of diagnostic technology has advanced significantly and conditions not previously amenable to therapy now are potentially treatable. This chapter will review the basic principles of human genetics to serve as a basis for other chapters that will deal with specific genetic approaches in clinical research.

BASIC MOLECULAR GENETICS

DNA Structure

Genetic information is stored in the cell as molecules of **deoxyribonucleic acid (DNA)**. Each DNA molecule consists of a pair of helical deoxyribose-phosphate backbones connected by hydrogen bonding between nucleotide bases. There are two types of nucleotide bases, purines [adenine (A) and guanine (G)] and pyrimidines [cytosine (C) and thymine (T)] (Fig. 16.1). Purines pair with pyrimidines in complementary A-T and G-C base pairs (Fig. 16.2).

Each DNA strand has polarity that results from the way the sugars are attached to each other. The phosphate group at position C5 (the 5′ carbon) of one sugar joins to the hydroxyl group at position C3 (the 3′ carbon) of the next sugar by a phosphodiester bridge. The sequence of nucleotide bases is written in the 5′ to 3′ direction based on a free 5′ phosphate group at one end of the chain and a free 3′ hydroxyl group at the other end. The sequence of the nucleotide bases on one strand of DNA (in the 5′ to 3′ direction) is complementary to the nucleotide base sequence of the other strand in the 3′ to 5′ direction. Thus if we know the sequence of nucleotide bases on one strand, we can infer the sequence of bases on the other strand (Fig. 16.3).

The DNA double helix serves two major functions: (1) serving as a template for replication; (2) serving as a template for the production of RNA and proteins.

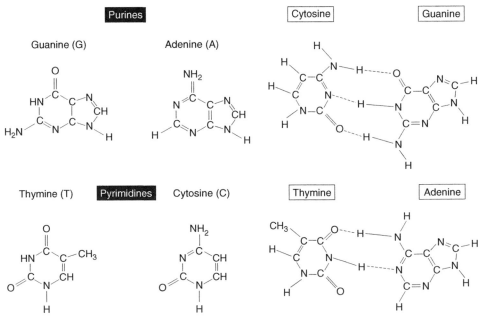

FIGURE 16.1 DNA nucleotide bases and base pairing.

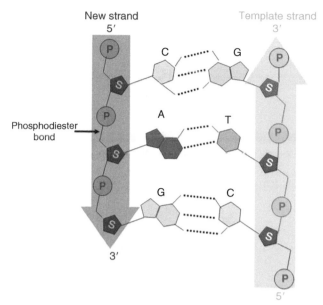

FIGURE 16.2   Double strand of DNA and hydrogen bonds between bases (P = phosphate, S = sugar).

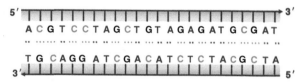

FIGURE 16.3   The sequence of nucleotide bases on one strand of DNA is complementary to the nucleotide base sequence of the other strand.

DNA Replication

Cell division requires the replication of each strand of DNA. During DNA replication, the two strands separate and unwind, with each strand serving as a template to generate a new strand (Fig. 16.4). This process results in two identical daughter molecules, each composed of one parental strand and one newly synthesized strand. DNA replication occurs at every cell division, and then genetic information is transmitted to both daughter cells. There is a complex proofreading mechanism for the identification and repair of copy errors to improve the accuracy of DNA replication.

DNA replication occurs simultaneously at multiple sites along a DNA molecule. Each site consists of a replication fork, in which bases are added in a 5′ to 3′ direction, catalyzed by a DNA polymerase. This occurs continuously from one strand where the 5′ bases are exposed at the replication fork, but in discontinuous fragments ("Okazaki fragments") for the other strand (Fig. 16.5). These Okazaki fragments are synthesized from short RNA primers copied from the DNA by an RNA polymerase. The RNA primer is then removed, and the fragments are ligated together by DNA ligase to form a continuous strand.

Transcription

A **gene** consists of a segment of DNA that codes for the synthesis of an RNA molecule, which in turn may serve as a template for the synthesis of a polypeptide. **RNA** (ribonucleic acid) is a single-stranded nucleotide polymer that is similar to a strand of DNA except that the sugar is ribose instead of deoxyribose and uracil substitutes for thymine. Classically, each gene has been conceptualized as encoding a specific protein, although now it is known that some genes encode RNA that is not translated into protein and that some genes encode multiple proteins due to alternative sites of initiation of transcription or mRNA processing. Genes vary in size from hundreds of base pairs to more than 2 million base pairs. The process of copying information from DNA to RNA is referred to as **transcription**.

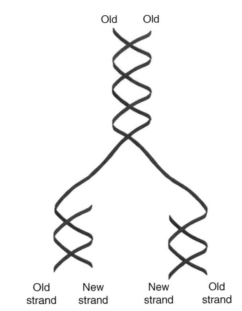

FIGURE 16.4 DNA replication: two strands unwind and replicate.

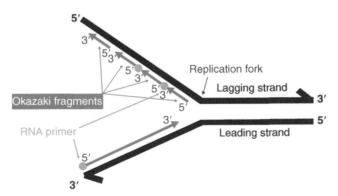

FIGURE 16.5 DNA replication fork.

Transcription is catalyzed by the enzyme **RNA polymerase**, which binds to a regulatory region at the 5′ end of the coding strand called the **promoter**. Specific regulatory molecules bind to the promoter and are responsible for controlling the activation or repression of transcription, assuring that gene expression is under tight control. Most protein-encoding genes consist of coding regions called **exons** interrupted by noncoding regions called **introns**. The first and the last exons usually contain sequences that are not translated, called the 5′ untranslated region (5′UTR) of exon 1 and 3′ UTR at the 3′ end of the last exon (Fig. 16.6). The DNA strand in the 3′ to 5′ direction serves as the template for the transcription of RNA in a 5′ to 3′ direction. Therefore, RNA is an exact copy of the 5′ to 3′ untranscribed strand of DNA (**sense strand**), except for the substitution of U for T. The 3′ to 5′ transcribed strand of DNA is called the noncoding or **antisense strand** (Fig. 16.7).

RNA is processed by addition of a 7-methylguanosine **cap** to the 5′ end soon after transcription begins. Transcription continues through the entire coding sequence of the gene. At the 3′ end just downstream from the end of the coding sequence, the RNA is cleaved, and a **poly-A tail** consisting of 200–300 bases is enzymatically added. The 5′ cap and poly-A tail appear to increase the stability of the mRNA molecule and promote its transport to the cytoplasm. The RNA transcript (primary RNA) is processed into mature mRNA by the removal of introns and splicing together of exons. Introns usually start with the nucleotides GT (GU in RNA) at the 5′ end (called the splice donor site) and end with the nucleotides AG (called splice acceptor site). These serve as signals for a complex system that recognizes the beginning and the end of each intron and splices together the two adjacent exons (Fig. 16.7). The splicing process occurs in the nucleus, and then the mature mRNA is exported to the cytoplasm where translation takes place.

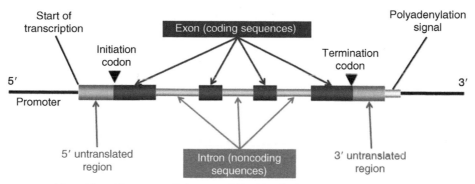

FIGURE 16.6   Gene structure. *Adapted from Thompson and Thompson, 2001. Genetics in Medicine, sixth ed. W.B. Saunders, Pennsylvania, PA, p. 20, Figs. 16.3–16.6, with permission of Saunders Elsevier Inc.*

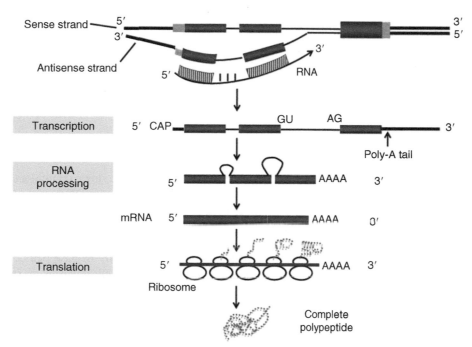

FIGURE 16.7   DNA transcription and translation. *Adapted from Thompson and Thompson, 2001. Genetics in Medicine, sixth ed. W.B. Saunders, Pennsylvania, PA, p. 23, Figs. 16.3–16.8, with permission of Saunders Elsevier Inc.*

Translation

The sequence of bases in DNA provides the information that determines the structure of proteins, which consist of chains of amino acids. The basic unit of genetic information is a triplet of bases called a **codon**. Each codon corresponds with a specific amino acid. There are 64 codons, which encode 20 amino acids and 3 "stop codons" that signal the end of a protein synthesis (UAA, UAG, and UGA). Therefore, some amino acids are encoded by more than one codon, e.g., two codons, CAA and CAG, define glutamine (Table 16.1). This genetic code is essentially universal; all organisms (bacteria and viruses, animals, and plants) use the same code, with few exceptions.

The initial step of translation is the formation of a complex consisting of mRNA, the ribosome, which contains rRNA (ribosomal RNA) and several proteins, and tRNA (transfer RNA). The sequence of mRNA is read in triplets beginning at the 5′ end, which always is AUG (**start codon**), and translation ends when a **stop codon** is reached (UAA, UAG, or UGA). The nucleotide sequence of the mRNA is encoded into the amino acid sequence by the action of tRNA molecules. tRNA molecules bind specific amino acids defined by their anticodon sequence, which is complementary to each codon of mRNA. The translation process includes codon recognition by tRNA, peptide binding of the next amino acid, and movement of the ribosome three nucleotides further in the 3′ direction of the mRNA. Finally, the completed polypeptide is

TABLE 16.1 Genetic Code for all Amino Acids in mRNA

		Second nucleotide				
		Uracil (U)	Cytosine (C)	Adenine (A)	Guanine (G)	Third nucleotide
First nucleotide	U	UUU ⎤ Phe UUC ⎦ UUA ⎤ Leu UUG ⎦	UCU ⎤ UCC ⎥ Ser UCA ⎥ UCG ⎦	UAU ⎤ Tyr UAC ⎦ UAA Stop UAG Stop	UGU ⎤ Cys UGC ⎦ UGA Stop UGG Trp	U C A G
	C	CUU ⎤ CUC ⎥ Leu CUA ⎥ CUG ⎦	CCU ⎤ CCC ⎥ Pro CCA ⎥ CCG ⎦	CAU ⎤ His CAC ⎦ CAA ⎤ Gln CAG ⎦	CGU ⎤ CGC ⎥ Arg CGA ⎥ CGG ⎦	U C A G
	A	AUU ⎤ AUC ⎥ Ile AUA ⎦ AUG Met	ACU ⎤ ACC ⎥ Thr ACA ⎥ ACG ⎦	AAU ⎤ Asn AAC ⎦ AAA ⎤ Lys AAG ⎦	AGU ⎤ Ser AGC ⎦ AGA ⎤ Arg AGG ⎦	U C A G
	G	GUU ⎤ GUC ⎥ Val GUA ⎥ GUG ⎦	GCU ⎤ GCC ⎥ Ala GCA ⎥ GCG ⎦	GAU ⎤ Asp GAC ⎦ GAA ⎤ Glu GAG ⎦	GGU ⎤ GGC ⎥ Gly GGA ⎥ GGG ⎦	U C A G

released from the ribosome (Fig. 16.7). The specific ordering of the amino acid sequence of the polypeptide or protein determines the unique properties of each protein.

Chromosome Structure and Function

DNA in the nucleus is organized into chromosomes, each chromosome representing a single continuous DNA strand. Each cell contains two haploid chromosome sets of about 3.15 billion base pairs of genomic DNA, measuring around 1.8 m. Packaging this length of DNA into the nucleus requires substantial compaction. DNA does not exist in the cell in a naked form, but rather as a chromatin complex with a family of basic chromosomal proteins called histones. An octomer of histones (H2A, H2B, H3, and H4) forms a disc-shaped core structure around which about 150 base pairs of DNA are wrapped to form a nucleosome. Nucleosomes are separated from one another by 50−70 base pairs of linker DNA, like beads on a string. The long strings of nucleosomes are compacted into a secondary helical chromatin structure. Chromatin is further compacted into the highly condensed structures comprising each chromosome (Fig. 16.8).

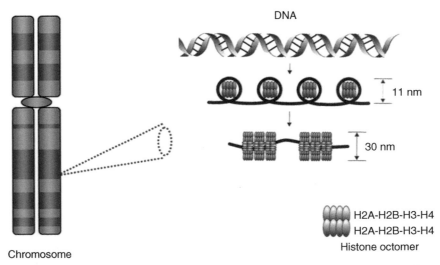

FIGURE 16.8 Structural organization of the DNA. *Adapted from Korf, B.R., 2007. Human Genetics, third ed. Blackwells, Malden, MA, p. 8, Fig. 1.7, with permission of Blackwell Publishing Inc.*

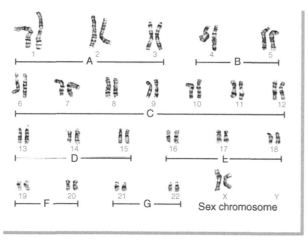

FIGURE 16.9 Normal female karyotype. *Courtesy of Cytogenetics Laboratory, University of Alabama at Birmingham.*

Humans have 46 chromosomes that are arranged in 23 pairs. The first 22 pairs are called **autosomes**; the remaining pair, called **sex chromosomes**, includes the X and Y chromosomes. Females have two X chromosomes, males an X and a Y. Since chromosomes contain the individual genes and we inherit two copies of every chromosome, one from each parent, we also inherit two copies of every gene (except for some on the sex chromosomes).

Chromosomes consist of a **short (p)** and **long (q) arm** separated by an area of constriction known as the **centromere**. Chromosomes can be visualized under a light microscope during mitosis. With special staining techniques, chromosomes can be seen as containing alternating transverse light and dark bands, which assist in the identification of individual chromosomes, since the banding pattern is unique to each chromosome. The bands also serve as geographic landmarks for localization of genetic markers and genes. A **karyotype** is a picture of chromosomes taken with a light microscope during metaphase, in which chromosomes are arranged in pairs by size and designed by number from 1 to 22 with the last pair as X/Y (Fig. 16.9).

Mitosis and Meiosis

There are two types of cell division: **mitosis** and **meiosis**. Mitosis is a somatic cell division, whereas meiosis involves division of germ cells. Mitotic division results in two genetically identical daughter cells (46 chromosomes or diploid), whereas meiosis results in the formation of reproductive cells (gametes) in which each cell contains only 23 chromosomes, one chromosome of each pair (haploid).

Mitosis

Mitosis is divided into four phases: prophase, metaphase, anaphase, and telophase. **Interphase** is the interval from the end of mitosis until the beginning of the next. Each cell division begins with a phase of DNA replication, referred to as S phase. DNA replication results in two sister chromatids for each chromosome. **Prophase** is marked by gradual condensation of the chromosomes, disappearance of the nucleolus and nuclear membrane, and the beginning of the formation of the mitotic spindle. At **metaphase**, the chromosomes become arranged on the equatorial plane, but homologous chromosomes do not pair. In this stage, chromosomes also reach maximum condensation. In **anaphase**, the chromosomes divide at the centromeric regions, and the two chromatids separate and migrate to opposite poles. **Telophase** begins with the formation of the nuclear membranes and division of the cytoplasm (Fig. 16.10).

Meiosis

Meiosis consists of one round of DNA replication and two rounds of chromosome segregation. In meiosis, there are two steps: meiosis I and meiosis II. The differences between meiosis and mitosis are (1) homologous chromosomes pair at prophase of meiosis I; (2) genetic recombination, called meiotic crossing over, occurs regularly at prophase of meiosis I;

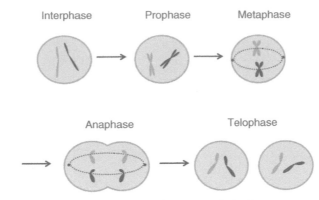

FIGURE 16.10 The process of mitosis.

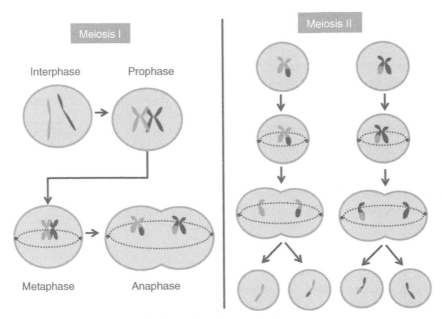

FIGURE 16.11 The process of meiosis.

and (3) the chromosome number is reduced to half after meiosis I, so that the daughter cells resulting from meiosis I are haploid (23 chromosomes) (Fig. 16.11).

Gametogenesis

Gametes (germ cells) are produced in the gonads. In females, this is called oogenesis and, in males, spermatogenesis. Most new mutations occur during gametogenesis, but there is a higher mutation rate in males, especially with increased paternal age. In females, meiosis I begins about 4 weeks before birth and then arrests in prophase, resulting in a primary oocyte. The primary oocyte persists in this stage until ovulation (after puberty). The arrest of meiosis I may contribute to the increased rate of chromosomal nondisjunction with advanced maternal age.

PATTERNS OF GENETIC TRANSMISSION

The laws of inheritance were discovered by Gregor Mendel, a 19th-century monk, who conducted breeding experiments on garden peas. He recognized that heredity is based on individual factors that are transmitted from generation to generation in a predictable pattern, and each factor is responsible for an observable trait (phenotype). Similar laws of inheritance govern genetic transmission in humans, and the study of these patterns has made major contributions to the understanding of health and disease.

Single Gene Inheritance (Mendelian Inheritance)

There are three patterns of single gene inheritance: **autosomal dominant (AD), autosomal recessive (AR),** and **sex linked**. The term "autosomal" refers to genes on chromosomes 1−22 (autosomes). Genes on chromosome X or Y are referred to as sex linked (X-linked or Y-linked).

Alleles are the different versions of the same gene at a given gene locus. **Homozygosity** occurs when two alleles at a particular gene locus are identical; the term **heterozygosity** is used if the alleles are different. **Genotype** refers to the genetic information in the alleles at a locus, whereas **phenotype** refers to the physical manifestations that result from the action of the genotype.

Autosomal Dominant Inheritance

Autosomal dominant inheritance involves phenotypes that occur when an allele is either homozygous or heterozygous. AD disorders comprise more than 50% of genetic disorders due to single gene mutations. An individual with an AD disorder needs to only have one variant allele to be affected. AD disorders are often more clinically severe or lethal in the homozygous state than in the heterozygous state. The characteristics of an AD disorder include:

1. both males and females are affected;
2. affected individuals usually have one affected parent;
3. a child of a heterozygous affected individual has a 50% chance of being affected.

Pedigrees of families with AD traits usually show a vertical transmission pattern (Fig. 16.12). Since there is a possibility of a new mutation in AD disorders, some affected individuals may not have an affected parent. Some AD disorders have a high rate of new mutation, such as achondroplasia and neurofibromatosis type 1 (NF1); approximately 80% of individuals with achondroplasia and 50% of those with NF1 result from a new mutation and have unaffected parents.

Penetrance and Expressivity

Penetrance is the probability of phenotypic expression of a specific allele. A highly penetrant allele will express itself almost regardless of the effects of environment. In some AD disorders, individuals do not clinically express the phenotype, even if they have a mutant allele. This is referred to as nonpenetrance (incomplete penetrance) and can lead to apparent skipping of generations in pedigrees. Consequently, we may see that an affected child has an affected grandparent, but the parent is not affected.

The manifestation of some traits is age related. An example is multiple endocrine neoplasia 1 (MEN 1), a disorder characterized by parathyroid hyperplasia, pancreatic islet cell, and pituitary adenomas, which usually present in adulthood. In one study, the age-related penetrance of MEN 1 was 7% by age 10 years and nearly 100% by age 60 years. Therefore, a 10-year-old child with a MEN 1 gene mutation will most likely have a normal clinical phenotype, although the child will be at risk of developing tumors over time.

Expressivity is the degree of phenotypic expression of an allele. Expressivity can differ in individuals who have the same mutation, referred to as variable expressivity. As a result, the severity of a dominantly inherited disorder may vary from mild to severe within a single family or among unrelated families. An example is NF1. NF1 is characterized by

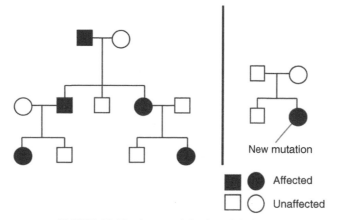

FIGURE 16.12 Autosomal dominant inheritance.

multiple café-au-lait spots, axillary and inguinal freckling, dermal neurofibromas, and iris Lisch nodules. NF1 presents extreme clinical variability but 100% penetrance; therefore, some NF1 patients may have only café-au-lait spots and freckles, whereas others may have life-threatening tumors or clinically severe plexiform neurofibromas.

Autosomal Recessive Inheritance

Autosomal recessive inheritance occurs when a phenotype is expressed only in the homozygous state of particular alleles. Consequently, individuals with AR disorders need to have two mutant alleles to be affected. An individual who has only one allele (heterozygote) is a **carrier**, and that person is not clinically recognizable because one wild-type allele can compensate for the variant allele. It is estimated that everyone carries multiple genetic variants that have no clinical effect in the heterozygous state, but which would be deleterious if homozygous.

The characteristics of an AR trait include:

1. both sexes can be affected;
2. an affected individual usually has unaffected parents, but parents are heterozygous carriers for the mutant gene;
3. the recurrence risk for each sibling of the affected individual is 25%.

Pedigrees of families with AR disorders usually present as with one or more affected siblings and, for rare disorders, no other family history of the condition (Fig. 16.13). There may be an increased frequency of consanguinity, where a variant allele is inherited from a common ancestor. Examples of AR disorders include cystic fibrosis, sickle cell anemia, and most inborn errors of metabolism.

X-Linked Inheritance

X-linked inheritance occurs when a phenotype is expressed in either the homozygous or heterozygous state of the particular alleles on X chromosome. Since males receive one X chromosome from their mothers and one Y chromosome from their fathers, males are **hemizygous** for genes on the X chromosome. Thus, males have only one copy of each X-linked gene, whereas females have two.

To achieve equivalence of the expression of X-linked genes in males and females (dosage compensation), only one X chromosome in females is transcriptionally active while the other one is inactive. This mechanism is called **X-inactivation**. X-inactivation in female somatic cells occurs randomly for the paternally or maternally inherited X chromosome. There are some genes on X and Y chromosomes that escape the X-inactivation, including genes at both ends of each chromosome, referred to as the **pseudoautosomal regions**.

The characteristics of an X-linked trait include:

1. males are more likely to be affected;
2. there is no male-to-male transmission (no mutant gene transmission from father to son), which helps to differentiate X-linked disorders from AD disorders;
3. heterozygous females (carriers) are unaffected, except if the trait is dominant, or if a female carrier has only one X chromosome (as in Turner syndrome patients), or in cases of nonrandom X-inactivation;
4. all daughters of affected males are heterozygous carriers.

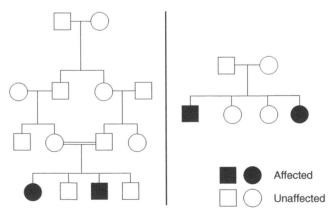

FIGURE 16.13 Autosomal recessive inheritance.

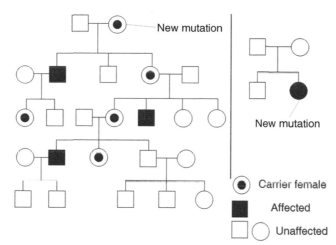

FIGURE 16.14   X-linked inheritance.

The rate of new mutation of lethal X-linked disorders is generally about one-third of cases, since one-third of the alleles would be lost in affected males. The recurrence risk of an X-linked disorder depends on whether the mother is a carrier of the X-linked variant gene or not. If the mother is a carrier, every son will have 50% chance of being affected and every daughter will have 50% chance of being a carrier. In some instances, an affected individual may represent a new mutation. Pedigrees of families with X-linked disorders usually present a transmission pattern of carrier females having affected sons (Fig. 16.14). Examples of X-linked disorders include Duchenne muscular dystrophy and hemophilia A.

Non-Mendelian Inheritance

Genomic Imprinting

In general, the two copies of any gene are either expressed or silenced equally in the genome. Genomic imprinting is a biological phenomenon observed in some genes in which there is different expression of two copies of a gene depending on the parental origin. Whether the maternal or paternal copy is expressed is a property of a specific imprinted gene. Several hundred imprinted genes have been found in the human genome.

Imprinting is associated with methylation of cytosine bases to 5-methylcytosine in promoter regions where cytosine is followed by guanine (CpG islands), which results in transcriptional inactivation (gene silencing). During male and female gametogenesis, the methylation imprint is erased by demethylation, after which a new sex-specific methylation imprint is reset on specific genes. Consequently, the specific gene copy is either expressed or silenced, regardless of whether that particular gene copy was active or inactive in the previous generation.

In Fig. 16.15, a chromosome carries two genes, A and B. A is imprinted in the female germ line, and B is imprinted in the male germ line, as indicated by asterisks. In somatic cells, A is imprinted when present on a maternally inherited chromosome, and B is imprinted when present on a paternally inherited chromosome. In the male germ line, sperm contains either a chromosome from the mother (A*/B) or the father (A/B*), and the old imprint is erased. Then the B locus of both paternal- and maternal-derived chromosomes is newly imprinted. In contrast, during oogenesis, oocytes contain either chromosome from the mother (A*/B) or the father (A/B*) in which the old imprint is erased. Then the A locus of both paternal- and maternal-derived chromosomes is newly imprinted.

There are several human genetic disorders that result from abnormally imprinted genes, such as Prader—Willi syndrome (PWS), Angelman syndrome (AS), and Beckwith—Wiedemann syndrome (Robertson, 2005). PWS is characterized by obesity, small hands and feet, hypogonadism, and mild-to-moderate intellectual disability. AS is characterized by characteristic facial features, severe intellectual disability, and seizures. Seventy percent of patients with PWS have a microdeletion of paternally derived chromosome 15q11-q13 (Fig. 16.16). Approximately 20—25% of patients with PWS do not have the microdeletion, but have two copies of the maternal chromosome 15 (maternal **uniparental disomy**) and no copy of the paternal chromosome. In contrast, 70% of patients with AS have the same microdeletion on chromosome 15q11-13 as PWS patients, but the deletion occurs on the maternally derived chromosome. Hence, patients with AS have genetic information in 15q11-q13 that derives only from their fathers. Approximately 3—5% of patients with AS have two

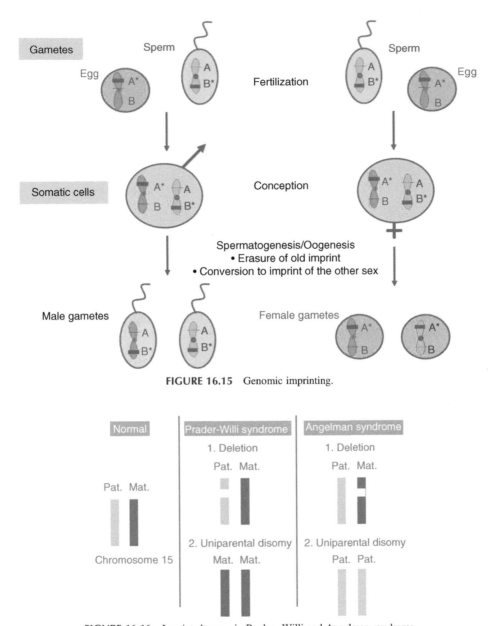

FIGURE 16.15 Genomic imprinting.

FIGURE 16.16 Imprinted genes in Prader–Willi and Angelman syndrome.

copies of the paternal chromosome 15 (paternal uniparental disomy). A small minority have mutation of the imprinted gene *UBE3A* in this region of chromosome 15.

Mitochondrial Inheritance

DNA is found not only in the nucleus, but also within mitochondria. A mitochondrion is a membrane-enclosed organelle in the cytoplasm of most eukaryotic cells and is the site of cellular energy production (ATP). The number of mitochondria in a cell varies widely by organism and tissue type. An average eukaryotic cell contains 1000−10,000 mitochondria.

Mitochondrial DNA (mtDNA) in humans is a double-stranded circular DNA molecule, composed of 16,569 base pairs. Human mtDNA contains 37 genes with introns, encoding two types of rRNA, 22 tRNAs, and 13 polypeptides involved in oxidative phosphorylative enzymes that are important for cellular energy metabolism (ATP synthesis; Fig. 16.17).

In the zygote, virtually all mitochondria are derived from the oocyte. Therefore, mutations in mtDNA are inherited from the mother (**maternal inheritance**). All of the children of a female with a mutation in mtDNA inherit the mutation, whereas a male carrying the same mutation cannot pass it to his offspring (Fig. 16.18). Unlike nuclear DNA, which

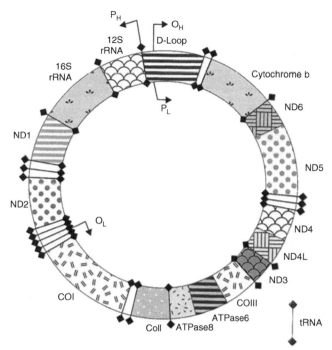

FIGURE 16.17   Mitochondrial genes in man. *Reproduced from Korf, B.R., 2007. Human Genetics, third ed. Blackwells, Malden, MA, p. 56, Fig. 3.29, with permission of Blackwell Publishing Inc.*

undergoes genetic recombination during meiosis and biparental inheritance, mtDNA follows uniparental inheritance, without an opportunity for genetic recombination. This near absence of genetic recombination makes mtDNA useful for studies of population genetics and evolutionary biology.

As noted earlier, each cell contains thousands of mtDNA molecules which reside in hundreds of individual mitochondria, whereas there are only two complete copies of nuclear DNA in each cell. If there is a mutation in an mtDNA molecule, when cell division occurs, both normal mtDNA molecules and mutant mtDNA molecules will replicate and then randomly distribute between two daughter cells. The daughter cells may receive only normal mtDNA molecules or only mutant mtDNA molecules (**homoplasmy**), but also may receive a mixture of normal and mutant mtDNA (**heteroplasmy**). If this occurs in oogenesis, it will result in different oocytes containing different proportions of mutant and normal mtDNA molecules. Thus, an individual offspring of a mother carrying an mtDNA mutation will inherit different numbers of mutant mtDNA molecules, resulting in variable expression and incomplete penetrance of mitochondrial disorders (Fig. 16.19).

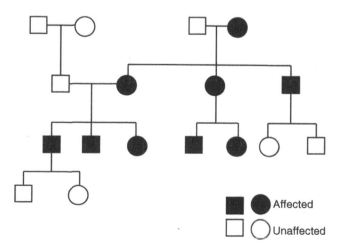

FIGURE 16.18   Mitochondrial inheritance.

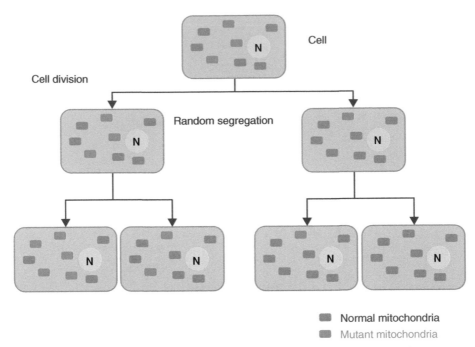

FIGURE 16.19   Replicative segregation of a heteroplasmic mitochondrial mutation.

Digenic Inheritance

Digenic inheritance occurs when two or more variant alleles in two different unlinked genes are required for the expression of a particular phenotype (Lupski, 2012). Known examples of digenic inheritance disorders include retinitis pigmentosa (RP), holoprosencephaly, and hereditary hearing impairment. In RP, individuals will develop symptoms when they have heterozygous mutations for both *ROM1* and peripherin genes, or homozygous mutations of each gene. Parents who carry only one mutation of either gene do not have RP.

Anticipation and Triplet Repeat Disorders

Anticipation is a phenomenon characterized by the progressively earlier onset and increased severity of certain phenotypes in successive generations within a family. This phenomenon is associated with expansion of the number of simple sequence repeats (SSRs or microsatellites), usually triplet repeats, within specific genes. The phenotype results from expansion beyond a threshold level. As the repeat size increases, the severity increases and age of onset decreases. Furthermore, the likelihood of further expansion increases as the repeat size increases, thereby accounting for anticipation.

Nearly 20 triplet repeat disorders have been described so far, and all of them affect the central or peripheral nervous system or muscle. The threshold of repeat number size causing a pathological phenotype is different for each disorder, and the expanded repeats can occur in an exon, in an intron, or in the 5′/3′ UTR of a gene. Very long expansions (up to 1000 or more repeats) usually occur outside the coding region of a gene, whereas more modest expansions occur within coding regions.

There are two types of triplet repeat disorders. Type I trinucleotide diseases are characterized by CAG trinucleotide expansions within the coding regions of different genes (Huntington disease, spinocerebellar atrophies, Kennedy disease, and dentatorubral-pallidoluysian atrophy). This CAG codes for glutamine. Thus, these are referred to as polyglutamine disorders. Type II trinucleotide diseases are characterized by expansions of CTG, GAA, GCC, or CGG trinucleotides within a noncoding region of the gene involved, either at the 5′ UTR (CGG in fragile X syndrome type A, FRAXA; CGG in FRAXE), at the 3′ UTR (CTG in myotonic dystrophy, DM), or in an intron (GAA in Friedreich ataxia, FRDA; Fig. 16.20).

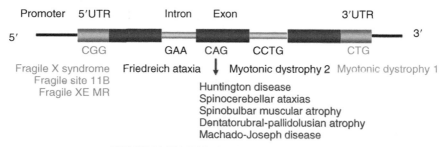

FIGURE 16.20   Triplet/quadruplet repeat disorders.

Multifactorial Inheritance (Polygenic Inheritance)

Multifactorial inheritance describes the action of multiple genes that interact with one another and with environmental factors to lead to a phenotype. Multifactorial disorders are believed to constitute the majority of common traits affecting humans, such as diabetes mellitus, essential hypertension, coronary heart disease, and cancer, as well as some of the common isolated birth defects, including cleft lip/palate (CL/P), neural tube defects, congenital heart disease, and club foot. Multifactorial inheritance does not follow a simple Mendelian transmission pattern. This apparent non-Mendelian pattern may be secondary to the condition being a syndrome rather than a disease, however. For example, there is evidence that hypertension and type 2 diabetes are syndromes with distinctly different pathophysiologic pathways leading the distant phenotype, i.e., elevated blood pressure or glucose. It is likely that the genetic variations underlying the altered physiology have strong interactions with the environment. Thus, this environment—gene interaction linked to a heterogeneous distant phenotype may have the appearance of a polygenic inheritance, but Mendelian transmission may actually be the case for the intermediate/proximate phenotype. Nevertheless, the recurrence risk of multifactorial disorders is higher among relatives of an affected individual than in the general population. The empirical risks of a multifactorial condition are based on a collection of large population studies. For example, in an isolated CL/P, the recurrence risk in siblings of affected individuals is about 3—5%, whereas the incidence of CL/P in the general population is only 1:700. Multifactorial disorders also reveal greater concordance for a disease in monozygotic twins than dizygotic twins or full siblings.

The characteristics of the recurrence risk of a multifactorial condition include:

1. risk is higher for first-degree relatives of affected family members than for more distant relatives;
2. risk increases with the presence of more than one affected relative;
3. risk increases with a more severe form or early onset of the disorder;
4. an affected individual of the sex less likely to be affected has a higher risk of having affected offspring and siblings. For example, congenital pyloric stenosis is a disorder that occurs five times more commonly in boys than in girls. If an affected proband is female, her relatives will have higher recurrence risks than if the affected proband is male.

CYTOGENETICS AND CHROMOSOMAL DISORDERS

Abnormalities of chromosome number or structure may result in congenital malformations, intellectual disability, miscarriage, stillbirth, infertility, and risk of malignancy. Human chromosomal abnormalities were first identified in the 1950s, when simple methods for analysis of metaphase chromosomes in dividing cells were introduced. Over the years there has been a gradual improvement in the resolution of chromosomal analysis, especially in the past decade with the advent of genomic approaches (Fig. 16.21).

Methods of Chromosomal Analysis

The most basic approach to chromosomal analysis involves culture of cells, most often peripheral blood lymphocytes, harvesting of mitotic cells at metaphase, when the chromosomes are maximally condensed, spreading the chromosomes by treatment of cells with hypotonic saline, fixation, spreading onto a microscope slide, and then staining. This is the approach that led to the discovery that Down syndrome is due to having an extra copy of chromosome 21 in 1959. Chromosome banding methods, introduced in the late 1960s, permitted more subtle changes of chromosome structure to be discerned, allowing for diagnosis of individuals with chromosomal abnormalities that would previously have escaped detection. The

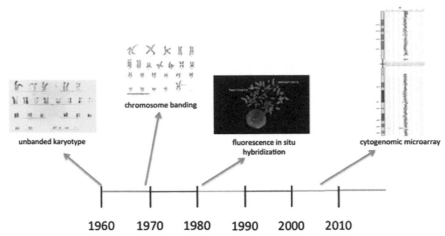

FIGURE 16.21   Progression of cytogenetic technologies from unbanded karyotype, to chromosome banding, to fluorescence in situ hybridization, to cytogenomic microarray.

subsequent development of **fluorescence in situ hybridization (FISH)**, where a fluorescent-tagged DNA probe is hybridized to fixed chromosomes on a microscope slide, allowed deletions or duplications in the range of a few million base pairs to be visualized. This revealed a whole new class of genetic disorders due to copy number changes, such as chromosome 22q11 deletion syndrome, characterized by congenital heart defects, palatal abnormalities, immune deficiency, parathyroid hypoplasia, and neurocognitive abnormalities.

Since the completion of the sequencing of the human genome, genomic technologies have been introduced that greatly increase the precision of detection of copy number variations (CNVs). Collectively, these approaches are referred to as **cytogenomics** (Miller et al., 2010). One approach, **comparative genomic hybridization (CGH)**, involves competitive hybridization of fragmented DNA from a patient sample and a reference sample, each labeled with a different color fluorophore, to oligonucleotides from across the genome in a microarray. The other major approach uses quantitative analysis of hybridization of patient DNA to oligonucleotides known to include polymorphic DNA sequences (**SNP array**). Either method is able to detect copy number changes down to several thousand base pairs, well below the limit of resolution of the light microscope. SNP arrays also reveal regions of homozygosity indicative of inbreeding or phenomena such as uniparental disomy.

Chromosomal Abnormalities

Chromosomal abnormalities consist of (1) numerical abnormalities (aneuploidy and polyploidy); (2) structural abnormalities (structural rearrangement).

Abnormalities of Chromosome Number

Aneuploidy is a deviation of the normal chromosome number leading to the loss or gain of one or several individual chromosomes from the diploid set, such as monosomy and trisomy. Aneuploidy commonly results from **nondisjunction** during meiosis. The first human chromosomal disorder to be recognized was Down syndrome, also known as trisomy 21 (Fig. 16.22). Down syndrome is the most common autosomal trisomy compatible with live birth; it is present in about 1:700 newborns. Turner syndrome is an example of a monosomy, where the individual is born with only one X chromosome. In humans, there are only three full autosomal trisomies occurring in live-born infants, trisomy 13, 18, and 21, and they all are associated with advanced maternal age.

Polyploidy is a condition in which the cell has more than two copies of the haploid genome, such as triploidy (69 chromosomes) and tetraploidy (92 chromosomes). Polyploidy is generally not compatible with life, except in patients with mosaic polyploidy.

Abnormalities of Chromosome Structure

Structural chromosome abnormalities result from a break or breaks that disrupt the continuity of a chromosome, followed by reconstitution in an abnormal combination. These are present in about 0.7—2.4 per 1000 intellectually disabled

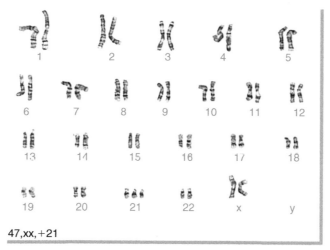

47,xx,+21

FIGURE 16.22   Female karyotype with trisomy 21. *Courtesy of Cytogenetics Laboratory, University of Alabama at Birmingham.*

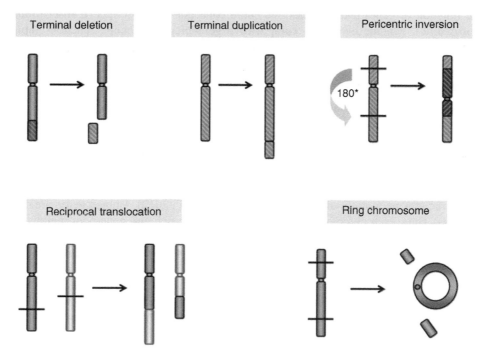

FIGURE 16.23   Abnormalities of chromosome structure.

individuals. There are several forms of structural abnormalities, including deletion, duplication, isochromosome, inversion, ring chromosome, and translocation (Fig. 16.23).

A **deletion** involves loss of a chromosome segment resulting from one single break, with loss of the distal fragment, or from two breaks and loss of the interstitial segment. A **duplication** involves addition of a chromosomal segment. Deletion results in monosomy for a group of genes, and duplication in trisomy for the genes. To be identified by chromosome study, a deletion or duplication segment must be large enough, about at least 2−5 Mb; however, use of FISH or cytogenomic microarrays permits detection of copy number changes down to a resolution of thousands of base pairs.

An **isochromosome** is an inverted duplication of one arm of a chromosome with loss of the other arm. An **inversion** is a chromosomal segment that is flipped 180 degrees, caused by a break at two different sites on one chromosome followed by reunion of the inverted segment. **Translocation** refers to an exchange of material between chromosomes. If no material is lost or gained, the translocation is referred to as **balanced**. Balanced translocations and inversions usually cause no phenotype, but they can lead to unbalanced chromosomes after crossing over in meiosis, which results in abnormal offspring or miscarriage.

Microdeletion Syndromes

Microdeletion syndromes involve very small deletions of adjacent gene loci (contiguous genes) in a defined chromosome region, resulting in clinically recognizable syndromes. This microdeletion segment is usually less than a few million base pairs, too small to be identified by routine chromosomal study. Use of FISH, or, more recently, cytogenomic microarray, is necessary to diagnose most microdeletions. There have been more than 20 different microdeletion syndromes identified, such as velocardiofacial/DiGeorge syndrome (22q11 deletion) and Williams syndrome (7q11 deletion).

THE HUMAN GENOME

The human genome consists of approximately 3.15 billion base pairs per haploid set of chromosomes. The human genome was sequenced by an international collaborative group in an effort referred to as the **Human Genome Project (HGP)**. The main goal of the HGP was to determine the entire sequence of human genome and to find all the genes within it. A preliminary draft of the genome sequence was published in 2001 and the complete version in 2003. The HGP is also concerned with the function of genes and genomes, the entire collection of transcripts (**transcriptome**), analysis of all human proteins (**proteome**), genome evolution (**comparative genomics**), epigenetic functions (**epigenome**), and the development of new techniques for management of data (**bioinformatics**).

Structure and Organization of the Genome

Before the HGP, it was estimated that the human genome contained about 100,000 genes; however, current estimates place the number around 22,000 protein-coding genes. Coding DNA in genes (exons) accounts for only 1.5% of the total sequence, referred to collectively as the **exome**. Therefore, much of the human genome does not encode protein. The human genome also contains thousands of genes that encode several types of RNAs that are not translated into protein, such as rRNA, tRNA, and RNA involved in gene regulation and the mRNA splicing process (Table 16.2). The ENCODE project has cataloged coding sequences in the human genome and revealed that a high proportion of sequences are transcribed in some tissue at some point in development (Consortium EP, 2012). Human genes are distributed unequally across the chromosomes. Thus, each chromosome contains various gene-rich and gene-poor regions, which are, to some extent, correlated with the banding of chromosomes and GC content.

Much of the human genome comprises **pseudogenes**, **repeated sequences**, and **transposons**. Pseudogenes are genomic DNA sequences that are similar to normal genes but have lost their protein-coding ability. Pseudogenes contain fascinating biological and evolutionary histories within their sequences because a pseudogene shares ancestry with a functional gene. Pseudogenes can complicate molecular genetic studies. A researcher who wishes to amplify a gene by polymerase chain reaction (PCR) may simultaneously amplify a pseudogene that shares similar sequences with its parental gene.

TABLE 16.2 Major Forms of RNA Transcribed From Genome

Designation	Type of RNA	Function
tRNA	Transfer RNA	Protein synthesis
rRNA	Ribosomal RNA	Protein synthesis
snRNA	Small nuclear RNA	Splicing
snoRNA	Small nucleolar RNA	RNA modification
miRNA	Micro RNA	Gene regulation
siRNA	Small interfering RNA	Viral defense
lncRNA	Long noncoding RNA	Gene regulation/unknown

At least 50% of the human genome consists of repeated sequences. There are several groups of repeated sequences. SSRs are polymorphic loci present in nuclear DNA that consist of repeating units of 1—10 base pairs (di-, tri-, or tetra-nucleotide repeats, etc.). They are typically neutral and codominant, and are used as molecular markers with wide-ranging applications in genetics, including kinship and population studies (Although, as noted previously, some triplet repeats may be involved in neurological disorders characterized by genetic anticipation.). **Segmental duplications** (SDs, or **low copy repeats**) are blocks of 1—300 kb that are copied from DNA at one site and move to other sites through evolution. SDs play a major role in chromosomal rearrangement because unequal crossing-over may occur between partially homologous duplicated units during meiosis or in somatic cells (**nonallelic homologous recombination**). This predisposes to duplication and deletion, leading to genomic disorders.

The final group of repeated sequences is referred to as interspersed repetitive DNA or transposon-derived repeats. This group includes long interspersed nuclear elements (LINEs), short interspersed nuclear elements (SINEs), LTR retroposons, and DNA transposons (Figs. 16.24 and 16.25). A transposon is a DNA sequence with the ability to move and be inserted in a new location of the genome. LINEs are DNA sequences ranging from 6—8 kb that represent reverse-transcribed RNA molecules that integrate into the genome. A typical LINE contains a 5' UTR, two open reading frames (ORFs), and a 3' UTR. One ORF encodes a reverse transcriptase. This enzyme copies both LINE transcripts and others into DNA. Approximately 900,000 LINEs, which account for 21% of the genome, are dispersed throughout the human genome. Importantly, transposition results in genetic disease if a sequence is inserted within a gene. SINEs are 100—400 base pairs long with tandem duplication of CG-rich segments separated by AT-rich segments. They are mostly derived from tRNA genes. SINEs do not encode protein and rely on other elements (LINEs) for transposition. The most common type of SINE in humans is the Alu family. Alu elements do not contain coding sequences and are specific for primate genomes. While previously believed to be "junk DNA," recent research suggests that both LINEs and SINEs have a significant role in gene evolution, structure, and transcriptional regulation. The distribution of these elements has been implicated in some genetic diseases and cancers.

GENETIC VARIATION

Mutation and Polymorphism

Types of Mutations

A mutation is a change in the sequence of DNA. Mutations can be subdivided into **germ line mutations**, which is any detectable, heritable variation in the lineage of germ cells that can be passed on to offspring, and **somatic mutations**, in which mutations occur only in a subset of cells from only certain tissues and cannot be transmitted to offspring.

Mutations can occur at the level of entire chromosomes or parts of chromosomes, or at the level of the individual gene. Gene mutation can originate either from an error in DNA replication (spontaneous mutation) or a failure to repair DNA damage (induced mutation). During DNA replication, errors occur at a rate of 1 in every 10^5 base pairs. Proofreading

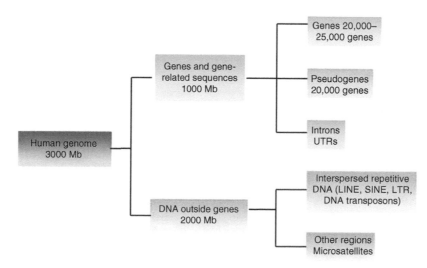

FIGURE 16.24 Types of sequence in the nuclear human genome.

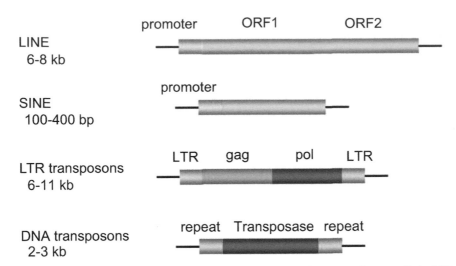

promoter   ORF1   ORF2

LINE
6-8 kb

promoter

SINE
100-400 bp

LTR   gag   pol   LTR

LTR transposons
6-11 kb

repeat Transposase repeat

DNA transposons
2-3 kb

FIGURE 16.25   Four classes of transposable genetic elements. *Adapted from Korf, B.R., 2007. Human Genetics, third ed. Blackwells, Malden, MA, p. 80, Fig. 4.23, with permission of Blackwell Publishing Inc.*

mechanisms reduce this to about 1 in 10^7 to 10^9 base pairs. Several genetic disorders caused by defects of the DNA repair system (**DNA repair disorders**) have been described, such as ataxia-telangiectasia, Bloom syndrome, Fanconi anemia, and xeroderma pigmentosum. Since abnormalities of DNA repair contribute to the pathogenesis of cancer due to accumulation of mutations in cells, individuals with DNA repair disorders have an increased risk of malignancy. Furthermore, there are chemical, physical, and biological circumstances (mutagens) that increase the likelihood of DNA damage leading to gene mutation. Common known mutagens include ultraviolet light and ionizing radiation.

There are three types of gene mutations: **point mutations** (single nucleotide substitutions), **insertions**, and **deletions** (Fig. 16.26).

Point mutations occur by a single nucleotide substitution in a DNA sequence (Fig. 16.27). A transition is an alteration that exchanges a purine for a purine (A to G or G to A) or a pyrimidine for a pyrimidine (C to T or T to C). Less common is transversion, which exchanges a purine for a pyrimidine or a pyrimidine for a purine (e.g., C to A or T to G). If a new codon encodes the same amino acid, there will be no change in the coding protein, referred to a silent mutation. A missense mutation is an alteration of a codon to one that encodes a different amino acid. The vast majority of detected mutations are missense mutations, which account for about 50% of disease-causing mutations. A nonsense mutation is an alteration that produces a stop codon, resulting in premature termination of translation of a protein. In most cases, the mRNA carrying a stop mutation is unstable, leading it to be rapidly degraded within the cell (**non−sense-mediated mRNA decay**). A splice site mutation is a gene mutation that alters nucleotides at splice control sequences, changing the patterns of RNA splicing.

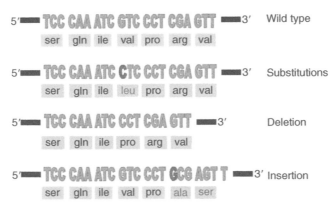

5′ TCC CAA ATC GTC CCT CGA GTT 3′   Wild type
   ser gln ile val pro arg val

5′ TCC CAA ATC CTC CCT CGA GTT 3′   Substitutions
   ser gln ile leu pro arg val

5′ TCC CAA ATC CCT CGA GTT 3′   Deletion
   ser gln ile pro arg val

5′ TCC CAA ATC GTC CCT GCG AGT T 3′   Insertion
   ser gln ile val pro ala ser

FIGURE 16.26   Gene mutations. *Adapted from Korf, B.R., 2007. Human Genetics, third ed. Blackwells, Malden, MA, p. 22, Fig. 2.6, with permission of Blackwell Publishing Inc.*

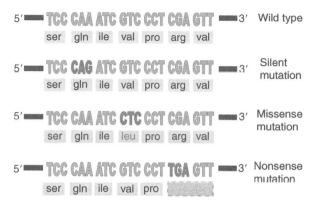

FIGURE 16.27 Point mutations. *Adapted from Korf, B.R., 2007. Human Genetics, third ed. Blackwells, Malden, MA, p. 22, Fig. 2.5, with permission of Blackwell Publishing Inc.*

Small deletions and insertions result from addition or deletion of a small number of bases. These alterations account for one-quarter of all disease-causing mutations in humans. When the number of bases involved is not a multiple of three, the reading frame is changed, referred to as a frameshift mutation. Insertions or deletions in the coding region of a gene may alter the splicing of mRNA, resulting in an abnormal gene product. If the number of bases involved is a multiple of three, it will cause loss or gain of codons and lead to an insertion or deletion of amino acid in the gene product.

Large deletions and insertions are alterations of gene structure that involve large segments of genes, and sometimes groups of genes. For example, 60% of individuals with Duchenne muscular dystrophy have a large deletion of part of the dystrophin gene on the X chromosome.

Single Nucleotide Polymorphisms and Copy Number Variation

Base sequence variation is common, occurring once in every several hundred bases between any two individuals. A **polymorphism** is defined as the occurrence of more than one allele at a gene locus where the most common allele has a frequency ≤ 0.99. **Single nucleotide polymorphisms** (SNPs) are differences in a single nucleotide at particular sites along chromosomes. It is estimated that SNPs occur every 1 in 1000 base pairs in human genome. SNPs can occur both in coding DNA (cSNPs) and noncoding DNA (noncoding SNPs).

Other types of DNA polymorphisms include **SSRs** or microsatellites, **variable number of tandem repeats** (VNTRs or minisatellites), and **CNVs**. Microsatellites can be used as highly informative markers since they are scattered throughout the genome with a high frequency. The commonly used microsatellite markers are dinucleotide repeats (e.g., CA_n). VNTRs or minisatellites consist of repeat units of 20–500 base pairs. Both micro- and minisatellites usually occur in noncoding DNA (Fig. 16.28).

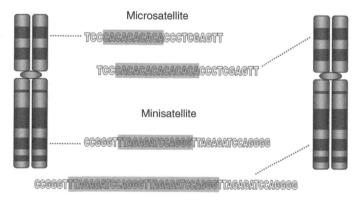

FIGURE 16.28 DNA polymorphism; microsatellite and minisatellite.

CNV is defined as a change in copy number involving a DNA segment that is 1 kb or larger. The frequencies of most CNVs have not yet been well defined in human populations. It is believed that CNVs influence gene expression, phenotypic variation, and adaptation by disrupting genes and altering gene dosage, or confer risk of a complex disease trait such as HIV-1 infection or glomerulonephritis. As previously noted, CNVs have been identified within the regions commonly deleted in many syndromes, such as DiGeorge, Williams, Prader—Willi, and Smith—Magenis. Previously, SNPs were thought to be the predominant form of genomic variation and to account for much normal phenotypic variation. It is now clear that there are also CNVs in normal individuals, underscoring the potential importance of CNV in genetic diversity and evolution.

Genotype—Phenotype Correlations

There are about 3000 currently defined single gene diseases. There are four possible effects of disease-causing mutations on protein function (Fig. 16.29).

1. **Loss-of-function mutations** are the result of a gene product being reduced in quantity or having no function. When the allele produces a reduced amount or activity of product, it is called **hypomorph**. When the allele has a complete loss of function (null allele), it is called **amorph**. Phenotypes associated with such mutations are most often inherited in a recessive manner. Examples are most inborn errors of metabolism and α-thalassemia. For some gene products, 50% of the normal level is not enough for normal function, and such **haploinsufficiency** produces an abnormal phenotype, which is therefore inherited in a dominant manner.
2. **Gain-of-function** mutations change the gene products by enhancing the function of proteins. When the allele produces increased amount or activity of product, it is called **hypermorph**. If the allele produces a novel activity or product, it is called a **neomorph**. Examples of diseases caused by gain-of-function mutations include achondroplasia and Huntington disease. Achondroplasia is caused by a mutation in *FGFR3* gene and is characterized by short stature resulting from short limbs. Fibroblast growth factor receptor 3 negatively regulates bone growth. Ninety-eight percent of patients with achondroplasia have a G380R mutation which results in constitutive activation of the FGF receptor, leading to negative control of bone growth.
3. **Dominant negative mutations** result in an altered gene product that acts antagonistically to the product of the wild-type allele. This mutant allele is called an **antimorph**. These mutations are usually characterized by a dominant or semidominant phenotype. In humans, Marfan syndrome and osteogenesis imperfecta (OI) are examples of dominant negative mutations occurring in an AD pattern. In Marfan syndrome, the defective product of the fibrillin gene (*FBN1*) antagonizes the product of the normal allele. In OI, the defects occur in type I collagen structure.
4. **Epigenetic changes** are caused by alterations in DNA that do not change base sequence, usually consisting of a change in the DNA methylation pattern. Inappropriate methylation can cause a pathogenic loss of function. For example, in some tumors, function of the p16 (*CDKN2A*) tumor suppressor gene is negated by methylation of the promoter region rather than by mutation of its DNA sequence.

Sometimes loss-of-function and gain-of-function mutations in the same gene can cause different phenotypes. For example, loss-of-function mutations in the *RET* gene cause Hirschsprung disease, whereas some gain-of-function

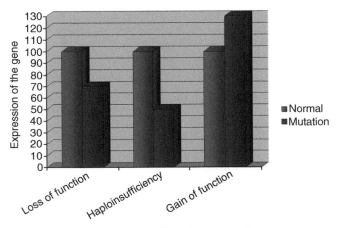

FIGURE 16.29 Effects of gene mutations.

mutations of this gene can lead to multiple endocrine neoplasia type 2 (MEN 2). There have been some patients with missense mutations affecting cysteine 618 or 620 who suffer from both thyroid cancer and Hirschsprung disease.

Variation in clinical phenotype may be due to one of two types of genetic variation: allelic heterogeneity and locus heterogeneity. **Allelic heterogeneity** is referred to as the presence of multiple different alleles at a single locus. These different alleles can result in either similar clinical phenotypes or very distinctive clinical presentations. In some conditions, there is a clear genotype—phenotype correlation between a specific allele and a specific phenotype. For example, cystic fibrosis is caused by two disease-causing mutations of *CFTR* gene and is inherited in an AR manner. More than 1000 alleles have been identified in the *CFTR* gene. The best correlation between genotype and phenotype is seen in terms of pancreatic function. A455E and R117H mutations are associated with pancreatic sufficiency. In contrast to CFTR mutations, different mutant alleles in *FGFR3* gene cause distinctive genetic conditions, such as achondroplasia, hypochondroplasia, thanatophoric dysplasia, Muenke syndrome, and SADDAN syndrome.

Locus heterogeneity is the production of similar phenotypes by mutations at two or more different loci/genes. For example, RP is caused by different mutations in more than 10 genes and is inherited in an AD, AR, or X-linked manner. Two distinct genes, *TSC1* and *TSC2*, account for tuberous sclerosis complex in different individuals. These genes encode two proteins (hamartin and tuberin) that form a complex; deficiency of either protein inactivates the complex, resulting in the tuberous sclerosis phenotype.

MEDICAL APPLICATIONS

Molecular Diagnostic Testing

Genetic testing is the analysis of human DNA, RNA, chromosomes, proteins, and certain metabolites to detect heritable disease-related genotypes, mutations, phenotypes, or karyotypes for clinical purposes (Korf and Rehm, 2013). Several thousand genetic tests are currently in use, and more are being developed. Genetic tests can be used to establish a diagnosis in a symptomatic individual, predict disease in a fetus by prenatal testing or after birth by presymptomatic testing, determine carrier status for an AR disorder, or assess risk of multifactorial disease. It can also be used to assess how an individual will metabolize certain drugs, permitting drug dosage or choice to be customized.

Approaches to Genetic Testing

The use of cytogenetic testing has already been described, including the recent addition of cytogenomic microarrays. Cytogenomic testing is now viewed as the first-line approach to the assessment of an individual with developmental disability, autism spectrum disorder, or congenital anomalies, due to its higher sensitivity than standard cytogenetic analysis. It must be remembered, though, that cytogenomic testing does not detect balanced rearrangements, so it would not, for example, be the test of choice for a couple, who have experienced multiple miscarriages, at risk for carrying a balanced translocation.

Molecular diagnostic testing involves examination of DNA or RNA taken from cells in a sample of blood or, occasionally, from other body fluids or tissues. The results of a diagnostic test can inform medical decision-making about health care and the management of an individual with a genetic disorder. The approach to molecular diagnosis depends on the specific gene being tested, although the recent application of next-generation sequencing may lead toward a more generic test that will address changes in almost any gene.

Targeted testing involves analysis of a specific gene or genes suspected to be the cause of a patient's problem, or for which a person is at risk based on family history. Various technologies can be used to detect specific mutations, including sequencing of the entire gene, or at least the entire coding sequence of a gene (perhaps excluding introns). In some cases, such as sickle cell anemia, a single specific nucleotide substitution accounts for all cases, so mutation detection can be highly targeted. The region of the gene near the mutation can be amplified using the PCR (Fig. 16.30) and analyzed for the mutation by Sanger sequencing (Fig. 16.31). In others, such as cystic fibrosis, a limited number of mutations account for a majority of cases, making it possible to design approaches that target multiple mutations. In still other cases, for example, hereditary breast and ovarian cancer due to *BRCA1* or *BRCA2* mutation, there is a large diversity of mutations that may be seen, in which case, sequencing the entire gene is necessary.

Next-generation sequencing is a general term for DNA sequencing technologies that permit very rapid determination of DNA sequence for the entire genome (Bamshad et al., 2011). Various different approaches are in use, with more in development, but generally these have lowered the cost of sequencing an entire genome from hundreds of millions of dollars to thousands. The approach can be targeted to the entire collection of protein-encoding exons, referred to as the

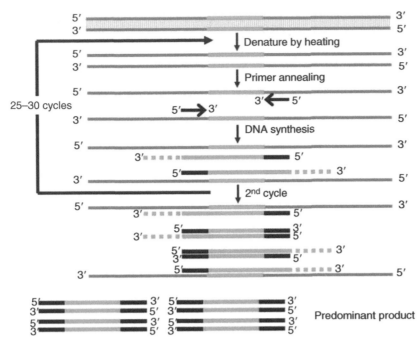

FIGURE 16.30   Polymerase chain reaction.

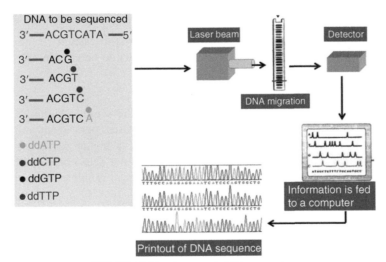

FIGURE 16.31   Automated DNA sequencing.

exome, which comprises about 1.5% of the genome, or can include the entire genome (Fig. 16.32). In either case, some genes are not fully sequenced, so some gene mutations might be missed. Also, the approach is not able to reliably detect certain types of mutations, including structural changes (e.g., translocation), copy number changes, or triplet repeat expansions.

Next-generation sequencing is rapidly being deployed as a diagnostic test. It offers several advantages over targeted testing in instances where the specific gene of interest is difficult to predict, such as in a child with intellectual disability. It can identify mutations in genes that might not have been predicted to be involved, and therefore would not have been tested for individually. It can also be an efficient approach to testing large sets of genes in conditions such as cardiomyopathy or hereditary deafness, where dozens, if not hundreds, of genes might explain a phenotype, well more than can be tested one at a time. In some cases, targeted panels of exons may be tested; in others, the entire exome or genome sequenced, but analysis can be focused on a list of genes of interest.

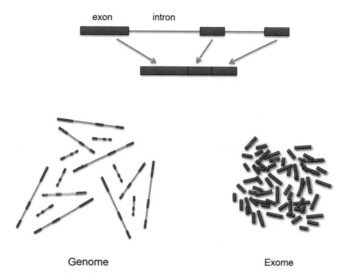

FIGURE 16.32 Exome versus genome sequencing. The exome consists of the set of protein-encoding sequences, which represents ~1.5% of the genome.

The major challenge in exome or genome sequencing is sorting through the tens of thousands (exome) or millions (genome) of genetic variants to determine which, if any, correlates with the phenotype. This requires bioinformatics tools to filter variants and identify those that may be pathogenic. Common variants that are annotated in various genome databases can be excluded if the disorder is rare. Sometimes a variant will be found that has already been demonstrated to be pathogenic. In other cases it is necessary to make a judgment of the likelihood of pathogenicity based on various factors (Richards et al., 2008). The likely effect of the variant on protein function can sometimes be inferred from the mutation (e.g., a frameshift or stop mutation will lead to lack of protein) or estimated from computer algorithms or analysis of evolutionary conservation of the specific nucleotide change. Mode of inheritance can be inferred—a dominant may be present in one parent, or in neither if the mutation is de novo. A recessive will be homozygous in the child but heterozygous in both parents. Hence it is helpful to sequence a child in both parents ("trio" testing). In many cases, it is not possible to definitively conclude whether a variant is pathogenic, defining a **variant of unknown significance**.

Exome or genome sequencing has the potential to identify pathogenic mutations in genes that are unrelated to the indications for testing, yet are medically significant. These are referred to as **incidental findings**. An example would be finding a pathogenic mutation in the *BRCA1* gene associated with hereditary breast and ovarian cancer in a child with autism spectrum disorder. The American College of Medical Genetics and Genomics has published recommendations regarding the return of results of incidental findings (Green et al., 2013). They have identified 56 genes that should be scrutinized for pathogenic variants that would be returned to the referring physician unless the patient (or parents for a child) opts out of learning of such findings. The list includes genes associated with clinically significant phenotypes that are amenable to medical intervention, such as risk of cancer or cardiac arrhythmia. The list is likely to evolve as additional genes are discovered that fit the criteria for inclusion, or as additional interventions are devised for known genetic disorders.

Indications for Genetic Testing

Genetic testing offers the advantage that testing can be done from DNA obtained from any tissue, and therefore does not require biopsy of affected tissue. Hence, diagnostic testing can be done from a blood specimen, rather than, for example, a muscle biopsy if the clinical concern is muscular dystrophy. It can also be used for testing prior to the onset of clinical signs and symptoms. This permits prenatal diagnosis or presymptomatic diagnosis in an entity associated with age-dependent penetrance.

Prenatal testing is classically done on fetal cells obtained by amniocentesis between 15 and 18 weeks of gestation, or chorionic villus biopsy between 10 and 12 weeks. Preimplantation testing can now be done in some cases. This involves ovulation of multiple oocytes, in vitro fertilization, biopsy of single blastomeres from multiple embryos, genetic testing on these single cells, and implantation of embryos not found to carry the mutation for which the embryo is at risk. This can provide very early diagnosis, although this is expensive. Recently, noninvasive approaches to prenatal screening and even diagnosis have been devised based on the analysis of fetal cell–free DNA found in mother's blood during pregnancy. Next-generation sequencing can be used to assess gene dosage on chromosomes 13, 18, and 21 to screen for pregnancies with a trisomic fetus; in principle, it can be used to detect specific gene mutations and even sequence the entire fetal genome, although this is not yet routinely done.

Presymptomatic testing is offered to individuals at risk, usually based on family history; common examples are risk of cancer, such as breast and ovarian cancer or colon cancer, or AD neurodegenerative disorders, such as Huntington disease. Genetic testing can identify pathogenic mutations in some such individuals, indicating increased risk of disease, although typically the result is probabilistic rather than diagnosis, since nonpenetrance is possible. It is important that thorough genetic counseling be offered to individuals in these cases (see later discussion).

Common disorders such as hypertension or diabetes are the result of both genetic and nongenetic factors, so genetic testing usually is nondiagnostic. However, as the underlying intermediate/proximate phenotypes of these syndromes are identified, genetic testing may be useful to direct specific therapy. Genome-wide association studies (GWASs), however, have revealed genetic markers, usually SNPs, that are associated with increased odds of disease. In most cases, the changes in risk for those with a specific genotype are small, raising question about the clinical utility of such testing. On the other hand, it is argued that finding an increase in risk can lead to effective interventions, such as motivating weight loss in an individual at risk of type 2 diabetes. Genetic risk assessment is not commonly employed yet in medical practice, but some companies have offered direct-to-consumer testing; at present, the Food and Drug Administration has restricted medical information being provided on a direct-to-consumer basis, although this may change as evidence is provided for clinical validity of these tests.

Pharmacogenetic testing is based on the fact that the absorption, activation, inactivation, and excretion of drugs may be influenced by specific genes, and testing can be used to customize the choice or dosage of a drug. For example, the antiplatelet drug clopidogrel is a prodrug that is activated by the cytochrome enzyme CPY2D19. Individuals with a genetic polymorphism that inactivates this enzyme are unresponsive to clopidogrel, and therefore require use of a different antiplatelet therapy. Pharmacogenetic testing is currently not widely used, mainly due to concerns about cost and lack of systems to insure rapid turnaround and provide decision support to clinicians. This is likely to be an area of major evolution in medical practice in the coming years.

One place where genetic testing is now used to guide therapy is cancer. Cancer cells tend to have multiple genetic mutations, including both chromosomal rearrangements and gene mutations, which are the source of phenotype changes that characterize cancer cells. Genetic testing, including the use of cytogenomic microarrays and next-generation sequencing, is increasingly being used to define cancer genotypes, since some tumors with specific genomic changes are known to respond to specific therapies. Such testing offers the possibility of customizing treatment to the genomic signature of the tumor, increasing the likelihood of successful treatment, and avoiding the use of toxic and expensive treatments in tumors that can be predicted not to respond.

Therapy of Genetic Disease

In the coming decades, molecular biology, protein engineering, and the HGP will have an enormous impact on the treatment of genetic diseases. Treatment of genetic disease may take many forms, including surgery and pharmacological therapy, and can be performed at many levels, including:

1. clinical phenotype (surgical and medical intervention)
2. metabolic or biochemical dysfunction (dietary and pharmacologic)
3. replacement of defective protein (protein or enzyme infusion)
4. organ transplantation
5. modulation of disrupted cellular function (small molecule)
6. mutant gene (restoration or correction of function)

Clinical Phenotype

Historically, most of the interventions for genetic disorder have been focused on amelioration of the phenotype through anticipatory guidance, surveillance for treatable complications, symptomatic therapy, and surgery. Although most genetic disorders cannot be "cured," the patient and family may derive major benefit from a specific diagnosis in terms of knowledge of the natural history, targeted surveillance for complications, and recognition of familial risks of recurrence.

Metabolic Therapies

Many inborn errors of metabolism are amenable to treatment that can significantly alter the natural history of the condition. Children with phenylketonuria (PKU) can be treated with a phenylalanine-restricted diet, and in some cases, coenyzme can be administered to bolster enzyme activity. Accumulation of ammonia in individuals with urea cycle disorders can be reduced by administration of sodium benzoate and sodium phenylacetate, which complex with ammonia to produce water-soluble compounds that can be excreted in the urine. The possibility of treatment of inborn errors of metabolism through dietary and/or medical therapies is the rationale for newborn screening, which permits diagnosis and institution of treatment in infancy prior to irreversible damage.

Replacement of Defective Protein

For some disorders it is possible to replace the defective protein or enzyme, usually by intravenous infusion. This has long been practiced for hemophilia A, where factor VIII is administered intravenously to augment blood clotting following injury. Many lysosomal enzyme deficiency disorders can be treated by intravenous infusion of purified enzyme that is chemically targeted to be taken up by cells through receptor-mediated endocytosis and then transported to the lysosome. This is highly effective for disorders such as Gaucher disease and Fabry disease, although this is less effective for conditions with substantial central nervous system involvement, since the enzyme does not penetrate the blood−brain barrier. Alpha-1-antitrypsin can be administered to augment the missing protein, ameliorating the chronic lung disease associated with this condition.

Organ Transplantation

Transplantation of bone marrow, liver, kidney, heart, etc., can be therapeutic in conditions that target particular organs. Urea cycle disorders, for example, can be treated by liver transplantation, since the urea cycle is mainly a hepatic function. Hemoglobinopathies can be treated with bone marrow transplantation. Although organ transplantation can be curative, the problems of rejection and the need for lifelong immune suppression complicate their routine use. Recently, the ability to create induced pluripotent stem cell lines from a person's own cultured cells offers the possibility of autologous transplantation either to replace cells destroyed by autoimmunity or degeneration; this approach may be coupled with in vitro correction of a genetic defect in some instances (see later discussion).

Small Molecule Therapy

As the cellular mechanisms of pathogenicity for genetic disorders come to light, it becomes possible to institute treatment by use of small molecules that target the defective cellular processes. Imatinib mesylate occupies the ATP binding site on the bcr-abl fusion protein in chronic myelogenous leukemia, with substantial clinical effectiveness. The mTOR inhibitor everolimus has been shown to be effective in treatment of some complications of tuberous sclerosis complex, since the tuberin−hamartin protein complex normally functions as an mTOR regulator. The risk of aortic dissection may be reduced in Marfan syndrome by treatment with losartan, which block TGF-β receptors, given that fibrillin deficiency in this condition leads to excessive TGF-β signaling. It is likely that many additional small molecule approaches will be developed in the years to come as the pathogenesis of genetic disorders is better understood.

Gene-Based Therapies

Gene-based therapies either target-specific mutated genes or proteins to restore function, or actually correct underlying mutations. In some cases, expression of a mutated gene can be augmented. For example, stop mutations (i.e., mutations that substitute a stop codon for an amino acid) can be overridden by specific drugs, including aminoglycoside antibiotics. In other cases, oligonucleotides can be introduced into a cell to alter patterns of RNA splicing, for example, skipping an exon with a mutation. Small molecules have also been identified that interact with a mutated protein, restoring partial function. This has been accomplished, for example, in cystic fibrosis, where ivacaftor can restore chloride channel function to protein with the pathogenic G551D mutation.

Replacement of a defective gene with a normal copy has long been a therapeutic goal. Most approaches involve introduction of a normal copy of the coding sequence using a vector, such as a modified virus. In some instances, the inserted gene functions extrachromosomally and is not transmitted when a cell divides; in other cases, the inserted gene integrates into the chromosome, and so is transmitted to future cell generations. There have been many challenges with this approach, including immune reactions to the new gene or the vector or insertion of DNA sequences near protooncogenes, leading to cancer. Achieving physiologically relevant levels of expression in the correct cell types is also a challenge.

As noted earlier, the use of induced pluripotent stem cell lines has generated interest as a possible therapeutic approach. A patient's own cells can be corrected in vitro using either homologous recombination-based approaches, or, more recently, CRISPR technology that allows gene editing.

GENETIC COUNSELING

Genetic counseling is defined by the NSGC (National Society of Genetic Counselors) as the process of helping people understand and adapt to the medical, psychological, and familial implications of the genetic contributions to disease (National Society of Genetic Counselors' Definition Task F et al., 2006). This process integrates:

1. interpretation of family and medical histories to assess the chance of disease occurrence or recurrence;
2. education about inheritance, testing, management, prevention, resources, and research; and
3. counseling to promote informed choices and adaptation to the risk or condition.

Genetic counseling can occur before conception to identify couples at risk of having a child with a genetic disorder (e.g., an AR trait if both parents are carriers), prenatally, or at any time in life for diagnostic or presymptomatic testing.

As part of the genetic counseling process, it is important to examine the relationship of individuals in a family based on their kinship by drawing a pedigree. Standard symbols are used, as shown in Fig. 16.33, to depict the inheritance of traits within a family tree (Bennett et al., 2008). The individual who brings the family to medical attention is called the **index patient** or **proband** (**proposita** if female and **propositus** if male). The person who seeks genetic counseling is called the **consultand**. Index patient and consultand are often different persons.

A common reason for affected individuals seeking genetic counseling is to ascertain the risk for heritable disease in offspring. Many disease-causing mutations can now be detected directly in carriers and affected individuals; however, there are many genetic disorders for which prenatal diagnosis is not feasible or the gene responsible is unknown. Genetic

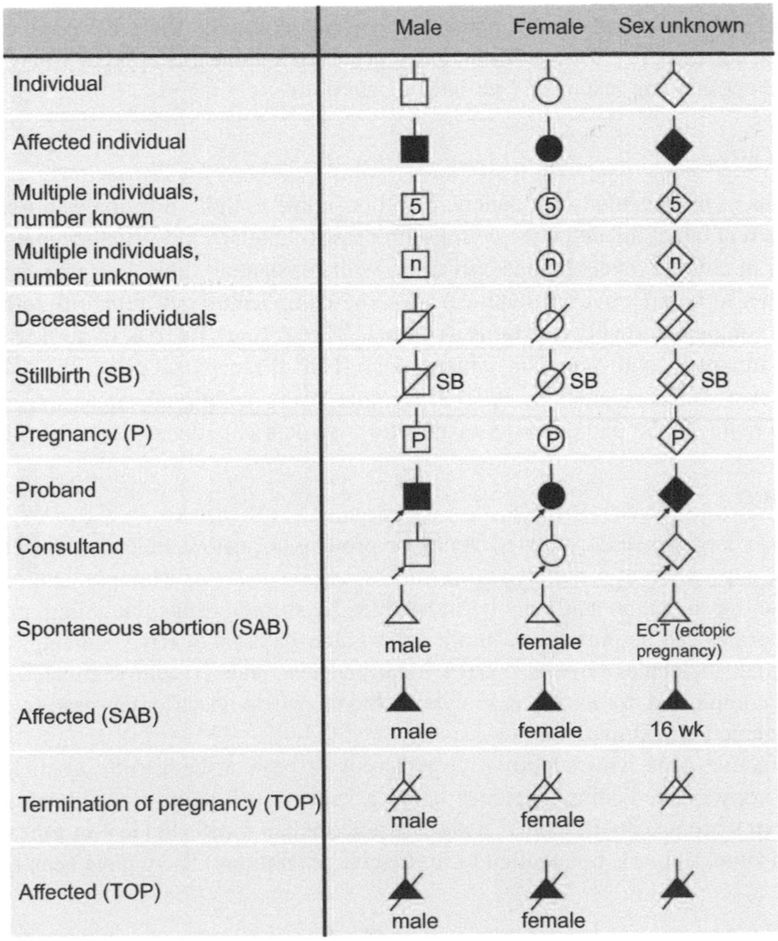

FIGURE 16.33 Common pedigree symbols. *Adapted from Bennett et al., 1995. Recommendations for standardized human pedigree nomenclature. Am. J. Hum. Genet., p. 746, Fig. 16.1.*

counselors must be aware of available options and provide information to affected individuals or couples regarding choices such as adoption, artificial insemination, and use of a donated egg. If the genotype of an affected individual is identified by DNA analysis and the inheritance pattern of the disorder is known, it is possible to offer prenatal diagnosis as well as to predict the recurrence risk for that person precisely.

Estimation of recurrence risks is based on knowledge of the inheritance pattern of a particular genetic condition as well as thorough analysis of the family history. If a disorder is known to obey a pattern of Mendelian inheritance, the recurrence risk for specific family members can be determined based on Mendelian principles. For example, a child of a heterozygous affected individual with an AD disorder with complete penetrance has a 50% chance of being affected and the recurrence risk for each sibling of the affected individual with an AR disorder is 25%. In contrast to Mendelian traits, the underlying mechanisms of inheritance for most chromosomal and multifactorial disorders are unknown, so estimates of recurrence risk are based on empirical data. For example, the recurrence risk for a subsequent sibling of a child with isolated CL/P is approximately 3—5%. If a child has Down syndrome, the empiric risk of recurrence would be approximately 1% if the karyotype were trisomy 21. Regarding genetic counseling for consanguineous couple, it is estimated that the risk of abnormal offspring in first cousins increases to approximately 4.5—5%, whereas it is up to 3% for any child of any unrelated parents.

In certain circumstances, premarital genetic testing is already a fact of life, such as in West Africans from countries with a high occurrence of sickle cell disease, in Asian people from countries with a high incidence of thalassemias, and in Jewish people of Eastern European Ashkenazi background with a higher carrier rate of Tay—Sachs disease and many other recessive conditions. The list of conditions for which preconceptual carrier screening is gradually increasing is, to some extent, dependent on the ancestry of the parents.

The advent of next-generation sequencing of the exome or genome raises new issues in genetic counseling. Challenges of finding variants of unknown significance or incidental findings have been noted previously. These issues, and the sheer complexity of analysis of genomic data, make it highly advisable for genetic counseling to be offered prior to initiation of testing, and then again when results are available.

PHENOTYPING AND CLINICAL RESEARCH

The genetic approach has been most successful in the study of disorders with well-defined patterns of inheritance, typically "single-gene" Mendelian disorders. Although phenotypic variability is common, if a disorder has high penetrance it is usually possible to identify affected and unaffected individuals in a family. Armed with this information, genetic analysis is possible, often leading to the identification of the gene responsible for the disorder. The approach is much more difficult when dealing with disorders in which penetrance is incomplete, where inheritance patterns are more complex, and where phenotype depends on gene—environment interactions. In this section, we will consider two major issues: genotype—environment interactions and phenotypic complexity.

Genotype—Environment Interaction

Although there are many examples of highly penetrant phenotypes, in reality, no genotype acts in total isolation from the effects of the environment. For example, PKU is an AR disorder characterized by an inability to metabolize the amino acid phenylalanine. The disorder is fully penetrant in terms of enzyme deficiency, but the phenotype of abnormal brain development is only expressed after exposure to phenylalanine. Although this is inevitable in undiagnosed children, the phenotype has been practically eliminated in developed regions by implementation of newborn screening. Children diagnosed early in life are placed on a phenylalanine-restricted diet and do not develop the full-blown phenotype. Other examples of "single-gene" disorders with prominent genotype—environment interactions are infections in children with sickle cell anemia and emphysema in carriers of mutations leading to alpha-1-antitrypsin deficiency when exposed to cigarette smoke.

The challenge is greater for traits where phenotypes are less dramatic or environmental exposures less likely. Individuals at risk of type I diabetes, for example, may never express the disorder if they are not exposed to a viral infection that may trigger an autoimmune process. Similarly, individuals at risk of certain types of cancer may require a concomitant environmental exposure, such as ultraviolet exposure for skin cancer or melanoma. Furthermore, some phenotypes are intrinsically defined in terms of genotype—environment interactions. Pharmacogenetic traits, for example, require drug exposure to be expressed. Similarly, genetic predisposition to infection is only expressed in individuals who are exposed to the infectious agent.

Accounting for environmental exposures raises many challenges. The number of variables is potentially infinite, and relevant variables may not be known in advance. Studies may be confounded by problems of recall bias and spurious associations due to multiple comparisons. Controlling for large numbers of environmental variables may create impossible demands on sample size to achieve adequate study power. There are no simple answers to these challenges, but recognition of the pitfalls calls attention to the importance of careful accounting of environmental variables in study design, and recognizing the potential for environmental factors to confound interpretation of phenotype.

Phenotypic Complexity

Aside from environmental interactions, determination of phenotypes is complicated by genetic heterogeneity and the occurrence of phenocopies, as well as ambiguity in the assignment of phenotypes from one study to another. These problems are particularly troublesome in large cohort studies, where differences in phenotypic assignment can impede the ability to replicate studies to verify findings.

Gene locus heterogeneity means that alleles in distinct loci can give rise to similar phenotypes. There are many well-known examples from highly penetrant disorders. For example, as noted previously, tuberous sclerosis complex can result from mutation in either of two genes, one on chromosome 9 (*TSC1*) and one on chromosome 16 (*TSC2*). Oculocutaneous albinism similarly results from mutation in any one of several genes that encode proteins required for the synthesis of melanin. Gene locus heterogeneity is a particular problem when data are pooled across multiple families, where the genetic etiology may be different in different families. In some cases, careful phenotyping may reveal subtle differences that provide clues to different genetic mechanisms, but not always.

Phenocopies are environmentally determined traits that mimic those that are genetically determined. Examples might include prenatal injuries that mimic genetically determined congenital anomalies or cancers due to environmental causes in individuals who are part of families with hereditary risk of cancer. The possibility of phenocopies must be considered in assignment of phenotypes, especially if the phenotype is known to be the one that can be environmentally induced.

The third issue may be described as **phenotypic complexity**. This is particularly problematic when dealing with common multifactorial traits. Such traits are the result of interaction of multiple genetic and environmental factors, and therefore invoke all of the cautions under consideration in this section. That is, phenotypic assignment may be confounded by environmental interaction, genetic heterogeneity, and the occurrence of phenocopies. Failure to account for these factors may underlie instances where genetic association findings cannot be replicated from one study to another.

The challenges can, in part, be ameliorated by careful documentation and control for environmental variables, as noted earlier. In addition, there is a need for definitions of the phenotype to be as precise as possible. In part, this is accomplished by careful attention to detail in assignment of phenotypes and documentation of criteria for phenotyping in reporting the study results. In addition, there may be value to using precise physiologically defined phenotypes. These are sometimes referred to as "endophenotypes" or intermediate phenotypes and represent subsets of patients with similar findings on a carefully defined physiological test. An example might be tests of renal sodium absorption as a measure in studies of hypertension, or EEG findings in studies of epilepsy. Use of such endophenotypes increases the likelihood that phenotypes can be compared from one study subject to another or from one study to another.

Approaches to Gene Discovery

The future understanding of the pathophysiology of genetic disorders is dependent on elucidation of the gene or genes that contribute to pathology. There has been an exponential increase in knowledge of genes associated with clinical phenotypes since the application of recombinant DNA technologies in the 1970s. Genes that encode proteins that were already known to be the cause of specific phenotypes, such as tyrosinase in oculocutaneous albinism or beta globin in sickle cell anemia, were identified quickly. Beginning in the 1980s, linkage analysis was combined with gene cloning to discover genes based on their location in the genome, referred to as **positional cloning**. Major successes were achieved with this approach in identifying genes where the underlying gene product was not known in advanced, e.g., in Duchenne muscular dystrophy, cystic fibrosis, and neurofibromatosis. With the advent of next-generation sequencing, gene discovery has reached a new level, permitting direct analysis for mutations in affected individuals, sometimes leading to discovery of genes not previously known to be associated with an abnormal phenotype.

Molecular genetic approaches have also been applied to understanding the genetic contributions to common disorders. As previously noted, case—control association studies can be used to identify association of a particular SNP with disease. Major improvements in genotypic technology, along with the discovery that haplotypes of 10,000 or so bases have been preserved in human evolution, have made it possible to do an unbiased search of SNPs across the genome for association, referred to as **genome-wide association study**. This approach has generated a large catalog of association results, in some cases identifying new genetic pathways not previously recognized as being involved in specific diseases. As previously noted, the clinical utility of genetic risk assessment may be limited, but the knowledge of genetic mechanisms contributing to common disorders may be very important for future understanding of pathogenesis and development of new treatments.

CONCLUSION

Genetic factors contribute to the pathogenesis of both rare and common disorders. The clinical researcher should be alert to recognizing familial disorders, as these can be of key importance in gene discovery. Chromosomal analysis is still revealing abnormalities of structure, especially microdeletion or duplication syndromes that can be identified using microarray technology. Since the sequencing of the human genome, new tools have been developed that add increasing power to the approaches to gene identification. Single gene variants can now be identified even in single individuals or small families by next-generation sequencing, and GWASs have revealed variants that are associated with risk of common disease. Elucidation of the genetic basis of disease can be a prelude to new approaches to diagnosis, counseling, and even treatment, making it increasingly important to incorporate genetic and genomic approaches into clinical research.

REFERENCES

Bamshad, M.J., Ng, S.B., Bigham, A.W., et al., 2011. Exome sequencing as a tool for Mendelian disease gene discovery. Nat. Rev. Genet. 12 (11), 745—755.

Bennett, R.L., French, K.S., Resta, R.G., Doyle, D.L., 2008. Standardized human pedigree nomenclature: update and assessment of the recommendations of the National Society of Genetic Counselors. J. Genet. Couns. 17 (5), 424—433.

Consortium EP, 2012. An integrated encyclopedia of DNA elements in the human genome. Nature 489 (7414), 57—74.

Green, R.C., Berg, J.S., Grody, W.W., et al., 2013. ACMG recommendations for reporting of incidental findings in clinical exome and genome sequencing. Genet. Med. 15 (7), 565—574.

Korf, B.R., Rehm, H.L., 2013. New approaches to molecular diagnosis. JAMA 309 (14), 1511—1521.

Lupski, J.R., 2012. Digenic inheritance and Mendelian disease. Nat. Genet. 44 (12), 1291—1292.

Miller, D.T., Adam, M.P., Aradhya, S., et al., 2010. Consensus statement: chromosomal microarray is a first-tier clinical diagnostic test for individuals with developmental disabilities or congenital anomalies. Am. J. Hum. Genet. 86 (5), 749—764.

National Society of Genetic Counselors' Definition Task F, Resta, R., Biesecker, B.B., et al., 2006. A new definition of genetic counseling: National Society of Genetic Counselors' Task Force report. J. Genet. Couns. 15 (2), 77—83.

Robertson, K.D., 2005. DNA methylation and human disease. Nat. Rev. Genet. 6 (8), 597—610.

Richards, C.S., Bale, S., Bellissimo, D.B., et al., 2008. ACMG recommendations for standards for interpretation and reporting of sequence variations: revisions 2007. Genet. Med. 10 (4), 294—300.

BIBLIOGRAPHY

Korf, B.R., Irons, M.B., 2013. Human Genetics and Genomics. Wiley-Blackwell, Sussex.

Nussbaum, R.L., et al., 2007. Thompson & Thompson Genetics in Medicine. Saunders/Elsevier, Philadelphia.

Strachan, T., et al., 2011. Human Molecular Genetics. Garland Science, New York.

Chapter 17

Epidemiologic and Population Genetic Studies

Angela J. Rogers[1] and Scott T. Weiss[2,3,4]

[1]Stanford University, Stanford, CA, United States; [2]Harvard Medical School, Boston, MA, United States; [3]Partners HealthCare Personalized Medicine, Boston, MA, United States; [4]Channing Division of Network Medicine, Boston, MA, United States

Chapter Outline

Key Points

- Intermediate phenotypes may be more objectively defined, statistically powerful, and homogeneous.
- The advantages of family-based studies are that they are immune to population stratification but they cost more and do not work for late-onset diseases.

Clinical and Translational Science. http://dx.doi.org/10.1016/B978-0-12-802101-9.00017-X

- Candidate gene studies are almost not done since genome-wide association studies are relatively inexpensive and have a much broader range of genes to be tested. Stringent *p* values must be observed with replication to avoid false positive associations.
- Violation of Hardy–Weinberg equilibrium is a symptom of potential genotyping error.
- There are multiple ways to correct for multiple comparisons: a stringent *p* value, Bonferroni correction, permutation testing.
- Population stratification is defined as differences in allele frequency due to evolutionary history but not disease.
- There are a variety of ways to control for population stratification, one is matching on ethnicity, and another is principle components or genomic control.
- Combining SNP data with other forms of genomic data, such as epigenetic data and gene expression data in integrative analysis, is the future of genetic association studies.
- There is no one ideal study design for a genetic association study. The choice of design depends on the disease, allele frequency, and the research question.
- In interpreting the results of a genetic association study there are three important questions. Is there genotyping error? Was population stratification controlled for? Was the problem of multiple comparisons addressed?

INTRODUCTION

Since the completion of the Human Genome Project, we have witnessed an explosion in the ease and cost-effectiveness of genotyping. With high-throughput analytic techniques, we are on the verge of a "$1000 personal genome" — with the potential that in the very near future, patients will have access to their DNA sequence (or at least the 1,000,000 or so single nucleotide polymorphisms (SNPs) of greatest interest).

Physicians and scientists will need to analyze and interpret these vast amounts of data, to determine which genetic mutations modulate disease pathogenesis. Great strides were made in the 20th century in identifying the etiology of Mendelian genetic disorders (generally caused by mutations in one or several genes). Because of the relatively simple inheritance of these disorders and the ease of identifying the discrete phenotype, methodologies like twin studies or sib-pair analysis frequently sufficed to identify loci of interest that could then be mapped by positional cloning, in a small number of pedigrees.

In the 21st century we have turned our attention increasingly toward complex disorders, such as asthma, hypertension, and diabetes—diseases that are relatively common across human populations and likely involve interaction of hundreds of genes and their mutations. Indeed, they are likely not even "diseases" but syndromes with a variety of pathways leading to the final common symptom, e.g., an elevated blood pressure or an elevated glucose level. Given the complex inheritance of these diseases, familial aggregation analysis that was invaluable for Mendelian diseases is no longer sufficient.

Instead, modern genetic analysis relies heavily on genetic association studies, in which particular alleles (or groups of alleles) occur more frequently in diseased individuals than in those who do not carry the alleles. Fig. 17.1 gives an overview of complex trait genetics where once a trait or phenotype is shown to be genetic in origin, the next step is a genetic association study to localize the susceptibility allele or gene. This is what this chapter is about. It is worth noting that once the association study is performed, the next step is to determine the functional variants in that gene and then use the results to explain a variation in the disease phenotype, predict clinical events, or design new treatments.

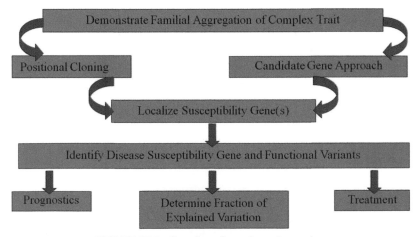

FIGURE 17.1 Overview of complex trait genetics.

The focus of this chapter is on the design and interpretation of these genetic association studies. We begin with a focus on epidemiologic study design, including population issues, such as defining phenotype and type of population, then move on to genetic study design with a focus on marker selection and bioinformatics tools. We then discuss issues of interpretation of genetic association studies, with a particular focus on dealing with genotyping error, correction for multiple comparisons, and population stratification and power. Finally, we conclude with a brief discussion of the future of genetic association studies, including newer biologic methods/targets (including sequencing, expression, copy number variation, and epigenetics) and analytic strategies (gene/environment interaction and pathway analysis).

DESIGN ISSUES IN GENETIC ASSOCIATION STUDIES

Population Issues: Defining a Phenotype and Selecting Epidemiologic Study Design

Definition of Phenotype

As with any epidemiologic study, an appropriate phenotypic definition is critical. Misclassification of cases and controls can lead to marked loss of power to appreciate true genetic associations. One reason for lack of replication between genetic studies may be different phenotypic definitions. For example, a study of subjects with diabetes and control comparisons who actually have negative hemoglobin a_1c may have much more power than another study of equal sample size that used a random population of controls who simply did not self-report diabetes.

In addition to dichotomous outcomes (such as "asthma"), many studies incorporate intermediate phenotypes; in the case of asthma, intermediate phenotypes would include spirometry (a measure of lung function), or level of methacholine challenge (a measure of airways hyperresponsiveness). The use of intermediate phenotypes can lead to a more powerful test statistic, as these continuous measures contain more variability than simple dichotomous outcomes. Additionally, intermediate phenotypes create a more homogeneous subgroup of the disease syndrome with the potential of applying classical Mendelian analytical techniques to the intermediate phenotype. Furthermore, as discussed later, this approach may also allow for a better definition of the pathway leading to the distant phenotype, i.e., disease syndrome.

Is the Disease Phenotype Actually a Syndrome?

Efforts of the past 15 years have been directed at improving our understanding and structure of the human genome. The assumption has been that this step is critical in understanding the genetic basis of human disease. However, little progress has been made in understanding the other major required component of identifying the genetic underpinnings of common human diseases. Most current approaches assumed that the phenotype selected is a disease rather than a syndrome. If these phenotypes are actually a collection of different diseases, then as stated earlier the use of classic Mendelian approaches will not work. Thus, heterogeneity of the phenotype could reflect a polygenic underpinning of the disease, i.e., the genetic underpinning of the disease in the individual is caused by small variations in a large number of genes. Alternatively, the heterogeneity of the phenotype may be a manifestation that the phenotype chosen is a collection of diseases, but in the individual only one or very few genes underlie that individual's "disease". If the latter is the case, then greater efforts need to be directed at developing and using intermediate phenotypes of the "disease" that are closer to the actual genetic alteration. For example, many studies have documented that using anemia as a phenotype has not been successful in defining the underlying genetic contributions to this "disease". Anemia would be a distant phenotype that has a substantial environmental interaction component to it. To understand the genetic underpinnings of anemia, intermediate phenotypes had to be identified. For example, not only was anemia required, but also the red blood cells had to sickle either spontaneously or under certain laboratory conditions. The sickled red blood cell in the patient with anemia was the intermediate phenotype that allowed for the identification of the genetic underpinnings of sickle cell anemia.

Several advantages of this approach are readily apparent:

1. Using intermediate rather than distant phenotypes as one's cohort allows for the control of the effects of the disease itself on the phenotype independent of genetic effects. Using this paradigm, the subjects with the distant phenotype that do not have the intermediate phenotype under investigation can serve as a control group rather than, or in addition to, normal subjects.
2. Since the cohort is more homogeneous, a smaller sample size will be needed to achieve the appropriate power.

3. Since the characteristics of the cohort are closer to the actual genetic perturbation, it is possible that classical Mendelian statistical approaches can be used.
4. This approach is more likely to be driven by a biologic pathway analysis rather than a nonhypothesis approach.

Thus the actual mechanisms causing the distant phenotype will be more apparent with ramifications for specific individualized preventive and treatment strategies based on genetic markers. Most common diseases with a genetic component—e.g., hypertension, diabetes, asthma, osteoporosis, and depression—fit a syndrome rather than a disease model. Thus the most will be amenable to an intermediate phenotype approach.

There are disadvantages of the intermediate phenotype approach:

1. It is likely to be much more expensive to identify intermediate phenotypes in the disease under the investigation. For example, in hypertension, environmental factors are likely to highly confound determining an intermediate phenotype. Salt intake has a profound effect on the level of blood pressure elevation in some, but not all, hypertensive subjects. Thus, control or measurement of salt intake would be critical to identify an intermediate phenotype. Additionally, manipulation of salt intake may be a very effective tool to accomplish this end since some, but not all, hypertensives have a defect in salt handling. Thus, because of the need for strict environmental control and potential manipulation, the identification of a hypertensive intermediate phenotype can cost thousands of dollars.
2. Since this approach is hypothesis-driven, some, and perhaps many, genetic causes of the disease (distant phenotype) may be missed.
3. Usual epidemiologic approaches, including traditional longitudinal cohort studies, are not likely to be successful since they are unlikely to be able to have the rigid environmental control that is needed. Thus, a special facility like a clinical research center, or in the United States, a Clinical and Translational Science Award (CTSA) program, will be required (see Chapter 34 for additional details).

EPIDEMIOLOGIC STUDY DESIGN

There are three basic types of study design used for genetic association studies. These include case—control studies, population (or cohort) studies, and family-based studies. Each design has its own strengths and weaknesses for use in a genetic association study, and is worth a brief review.

Case—Control Studies

These are among the most frequently performed genetic association studies, in which allele frequencies among affected individuals (cases) are compared to unaffected individuals (controls). Case and control individuals should be drawn from the same population—this is most obvious in terms of environmental exposure, for example, when cases and controls are drawn from the same zip code or school. It is equally critical for genetic exposure, however—for example, if case definition requires that all four grandparents are from the same ethnic background. Failure to draw from the same source population is a frequent cause of spurious associations. When populations from disparate genetic backgrounds are differentially represented among cases and controls, this can cause false positives, termed "population stratification"; this important concept is discussed in more detail later in this chapter.

Cohort or Population Studies

A "cohort" is defined epidemiologically as "any designated group of individuals who are followed or traced over a period of time" (Last et al., 1995). Cohort studies are a frequent source population for genetic association studies; for example, such large cohorts as the Nurses' Health Study or the Framingham Heart Study have followed thousands of subjects for decades, and collected blood samples on a substantial subset of those individuals. Because these individuals joined the study prior to development of disease, there is less risk of recall or reporting bias. Additionally, because both diseased and nondiseased individuals are members of the same group, there is less risk of false associations than with case—control studies. Disadvantages to this study design include the cost inherent in longitudinally following individuals, many of whom will not develop disease and the fact that the results from cohort studies may not be applicable to the general population. For example, results from the Nurses' Health Study may be biased by the fact that the individuals chose to be nurses. Thus, whether or not similar conclusions would apply to factory workers or farmers, for example, always is uncertain.

Family-Based Studies

Family-based genetic association tests (FBAT) are based on the transmission-disequilibrium test (TDT). Families are recruited based on a disease phenotype in the child (proband). If a particular disease locus is uninvolved in the disease, one would expect the parents' allele at that locus to be transmitted randomly (i.e., Mendelian, transmitted 50% of the time). On the other hand, if the allele is actually associated with disease, and the proband was selected based on the disease, there will be apparent overtransmission of the allele—it will differ significantly from the 50% expected.

Only trios in which at least one parent is heterozygous at the marker of interest can be used in the test statistic. For example, if one parent is AA, and the other at that locus, the child must be Aa; this is a noninformative trio. The method has been extended to analyze continuous phenotypes as well.

Comparison of Family-Based Versus Case–Control Studies

Family-based association testing has several key advantages to the case–control or cohort studies. First, because each child's expected genotype frequency is conditioned on its parents (who are obviously of the same genetic background); the family-based association study is immune to population stratification. Second, within each trio, only one person (the proband or subject) must be phenotyped. This is particularly useful when phenotyping is very expensive or invasive; for example, if liver biopsy is required for disease diagnosis, this would ideally be performed only on cases (rather than on control subjects as well). Family-based testing also offers an important method of assessing genotyping quality because data can be analyzed for Mendelian errors (this concept is covered in more detail in the quality control section later).

These advantages must be weighed against several key disadvantages versus case–control testing. First, family-based tests are ideal for diseases that arise early in life (such as autism or type I diabetes mellitus), as parents are more likely to be alive and thus available for genotyping. In late-onset diseases such as Alzheimer's disease or coronary artery disease, parents are often unavailable. While statistical methods have been developed to impute parental genotypes (based on sibling genotypes, for example), these methods are much less powerful. In family-based testing, three people must be genotyped (the subject and at least two family members) versus two for most case–control studies—these extra genotyping costs should be weighed against the extra phenotyping costs in case–control studies discussed earlier. Finally, many trios in family-based studies will be noninformative at a given allele if the parents are not heterozygous; this substantial loss of information must be accounted for in power calculations.

GENETIC STUDY DESIGN: GENOME-WIDE ASSOCIATION STUDY VERSUS HYPOTHESIS-DRIVEN (CANDIDATE GENE) APPROACHES

In addition to adequate population selection (including both meticulously defined phenotypes and selection of the appropriate epidemiologic study), attention must be focused on the genetic study design. We focus in this section on SNP-based studies, which are currently by far the most common targets of genetic association study. Other targets used more commonly in the past, included restriction fragment length polymorphisms (RFLPs) and short tandem repeats (STRs). Emerging fields of genetic study, including structural variation, particularly copy number variation (CNV) and expression data, are covered in more detail in the future directions section of this chapter. Fig. 17.2 summarizes some of the differences between a candidate gene study and a genome-wide association study (GWAS).

Extensive background on human genetics is beyond the scope of this chapter (for an introduction, see Chapter 16). Several concepts are so integral to comprehension of this chapter that we will briefly review them here. The human genome comprises approximately three billion base pairs. The vast majority of these loci are shared between all humans, with variability among humans explained by approximately 10 million SNPs (Frazer et al., 2007) and also structural variation. SNPs provide information not just about the particular locus being assessed, but also neighboring regions because of linkage disequilibrium, or LD, which is a measure of recombination in the genome. When two SNPs are on separate chromosomes, they will segregate randomly (i.e., carrying the minor allele at SNP A does not affect your chances of carrying the minor allele at SNP B). If, on the other hand, the SNPs are in "high LD", with little recombination between them, the genotype at SNP B can be quite informative about SNP A.

Thus, comprehensive testing of all SNPs is unnecessary; instead, a subset can be selected. SNP selection in genetic association studies involves two major approaches. The first is a hypothesis-free approach, used in GWAs. The second involves focusing on either a specific "candidate gene" or region of the genome based on a prior work in humans or animals. For the latter approach, marker selection is study-specific, and we will focus on use of bioinformatics to aid in this process.

Candidate Gene Scan	Genome
based on pathophysiology	no prior knowledge necessary
cheap	expensive
technology nonintensive	technology intensive
association study	allele sharing study
gene known	gene unknown

FIGURE 17.2 Candidate gene versus GWAS.

Hypothesis-Free Testing: Genome-Wide Association Study

In genome-wide association studies, SNPs are selected to assess virtually the entire genome. Instead of hypothesizing about a certain gene of interest, SNPs across all chromosomes and genes are studied for association with a phenotype of interest. There are two major platforms used for GWAS, Affymetrix (which selects SNPs in random distribution across the genome) and Illumina (which selects haplotype-tagging SNPs across the genome, based on the HAPMAP CEPH population). Early GWAS platforms included approximately 100,000 SNPs; the most recent release from Affymetrix contains one million. With increasing number of SNPs examined, the genome coverage becomes more and more complete (0.80% coverage of common variants with current platforms), and thus there is less chance of missing genetic associations.

A major strength of this study design is its hypothesis-free approach. Early GWASs have shown consistent associations with disease across multiple populations in areas with no known gene nearby (Libioulle et al., 2007)—clearly such "gene deserts" would never have been targeted had we relied on prior hypotheses about a gene function. Because of the genotyping method and the fact that SNPs on the platforms are prespecified, the per-SNP genotyping cost is very low. In 2015 a subject can be genotyped at 500,000 markers for around $200 using the current Illumina MEGA chip. The cost of whole exome sequencing is ~$600 per subject, and whole genome sequencing for research is ~$1200 per subject, and the cost continues to fall. This is less than one cent per SNP and much cheaper than smaller-scale genotyping methods discussed later.

Downsides to the GWAS approach include the per-subject cost of genotyping—even $200 per subject multiplied by 10,000 cases and 10,000 controls quickly gives a budget $5 million. Additionally, testing of so many SNPs causes severe issues with multiple testing—this is discussed further in the interpretation section of this chapter.

Candidate Gene or Region-Based Testing, With a Focus on SNP Selection

This is the methodology used almost exclusively in genetic association studies prior to 2005, and remains the mainstay of smaller genetic studies today and in pathway determined intermediate phenotype studies. Even when new genes or regions of interest are identified by GWAS, follow-up studies often assess not only the SNP or SNPs identified in the GWAS, but also do more fine-mapping of SNPs in the region of interest.

The number of SNPs studied depends greatly on budgetary constraints, and can vary from simple replication of one to two SNPs previously identified, to full sequencing of a region of interest to identify a new variation and potentially functional SNPs. SNP selection also depends on the genotyping platform available, with methods like Taqman ideal for studies of single or a few SNPs or Sequenom for several dozen. Choosing the best SNP is critical for a successful association study. Many bioinformatics tools are available in a web-based format to assist with this process. Given the importance of SNP selection, we will focus on several of the major bioinformatics tools available currently, with the caveat that the field continues to advance rapidly, and thus while these are a starting point, they should by no means be considered comprehensive.

Reviewing the Past Literature

One of the most critical steps in selecting SNP for an association study is reviewing what is already known about a particular gene or genes. If an SNP has previously been identified as being associated with your disease of interest (or a similar disease), it makes sense to include it in your panel as well.

NCBI (www.ncbi.nlm.nih.gov)

We find the National Center for Biotechnology Information (NCBI) to be a particularly useful place to start. One can quickly link from the NCBI home page to multiple useful databases that can be interrogated for information on a particular gene. These databases include:

- Online Mendelian Inheritance in Man (OMIM) (www.ncbi.nlm.nih.gov/omim). This is a catalog of genes and genetic disorders that is maintained by researchers at Johns Hopkins. The site includes an overview of the gene's structure and function. It also reviews molecular genetics, with a focus on previously identified genetic variants that cause disease in humans. The site contains links to high-profile publications about the gene and also links to particular genetic variants that have been associated with disease.
- PubMed (www.ncbi.nlm.nih.gov/pubmed). While OMIM is an excellent place to start one's literature review, it is an overview, not meant to be all-inclusive. Furthermore, the site is updated intermittently (thus may not reflect some of the latest literature). Pubmed is a useful site for a more comprehensive review of prior associations. With the move toward Open Access by many major medical journals, many articles are freely available soon after publication. This can give a more complete list of previously associated SNPs that should be included in the panel.

Locating SNP Within the Gene

Several sites can be critical for identifying known genetic variability in the gene of interest and its flanking region.

- UCSC genome bioinformatics site (http://genome.ucsc.edu/). The UCSC genome browser contains a very user-friendly interface between the reference sequence of a gene and extensive information on known variability within it. One can quickly access an information on location of SNP (exons, introns, and boundary), on functionality of SNP (silent or changing an amino acid), and whether the region is highly conserved between humans with other organisms.
- SNPper (http://snpper.chip.org/bio/snpper). This is another useful database that combines information from the UCSC genome browser and dbSNP. One can search a particular SNP of interest or retrieve a list of all SNPs with a gene that have been reported to dbSNP. The site contains links for primer design, links back to dbSNP (part of NCBI), and offers a quick tabular overview of the number of SNPs in exons, number of coding SNPs, etc.

Identifying Linkage Disequilibrium—Tagging SNP

As discussed earlier, because of LD in many regions of a gene, every SNP in a gene does not need to be genotyped and tested for association. Instead, LD-tagging SNPs can be selected that cover the major LD blocks in the gene within a given population.

- HapMap (www.hapmap.org). Identifying LD-tagging SNP has become much easier since the completion of the HapMap project (Frazer et al., 2007; International HapMap et al., 2010). Over one million SNPs were genotyped in more than 1000 individuals from 11 diverse populations of European, Asian, and African ancestry. For any given gene, the LD map is available, as is the minor allele frequency at a particular locus.

The genotypes of the gene of interest and their flanking sequence can be extracted. These data can then be uploaded into Haploview (a program freely available online at http://www.broad.mit.edu/mpg/haploview/). Based on HapMap frequencies, the optimal LD-tagging SNPs can be selected using the Tagger feature of Haploview.

Summary on Study Design

In summary, like all human research, a successful genetic association study requires extensive upfront planning. This includes a focus on the epidemiologic approach (including selection of study design and population [s]) that is most appropriate for the disease of interest. It also requires selecting the best approach from a genetic standpoint, including extensive bioinformatics review for SNP selection if using a hypothesis or candidate gene-based approach (starting, perhaps, with some of the resources outlined earlier), or selecting the best platform for GWAS.

INTERPRETING RESULTS OF GENETIC ASSOCIATION STUDIES

No matter how carefully a genetic association study is designed, it risks false positive or negative conclusions in the analysis phase of the study. While this is true for all epidemiologic studies, we focus here on four major concepts that must be addressed in genetic association studies. First, we highlight the importance of quality control prior to analysis of samples. Next, we discuss methods for dealing with multiple comparisons, which are especially critical in the context of GWASs that can involve hundreds of thousands of comparisons. We address potential confounding due to population stratification and methods for adjusting for this. Finally, we stress the importance of adequate sample size and power to avoid false negative results.

The early literature in human genetics is littered with findings that could not be replicated by other groups and are now felt to represent false positive associations. On the other hand, well-accepted associations fail to replicate in many published studies. The four issues we discuss here likely contributed significantly to the problem of both false positive and negative results. Well-designed studies in the future must account for all of them.

Genotype Quality Control

It is critical to perform thorough review of genotyping quality prior to any analysis of the data. Genotyping error has been associated with both false positive and false negative results. Genotype misclassification reduces power in case—control studies, thus increasing the risk of false negatives. In family-based studies, on the other hand, genotyping error has been shown to bias away from the null, with apparent overtransmission of the major allele (Mitchell et al., 2003). Multiple safeguards should be in place to recognize genotyping errors and remove faulty markers or subjects from analysis.

Genotype Completion Rate

The genotype completion rate refers to the percentage of samples that were "called" (as either homozygous major, minor, or heterozygous) at a given locus. Most genotyping platforms in use today should achieve results of at least 0.90%, with most 0.95%. Failure to achieve this level of completion does more than simply reduce power to find an association because nongenotyped subjects cannot be evaluated. Instead, it suggests a problem with genotyping at that locus (for example, failure of primers to anneal well, inadequate PCR amplification), and thus is likely associated with a high error rate in those subjects that were called. Those markers that do not achieve a sufficient completion rate should be thrown out and not analyzed.

Reproducible Genotyping

Several methods exist to repeat genotyping at a given locus. Many investigators include duplicate samples of 5—10% of the cohort in their initial genotyping. An acceptable rate of discordance is <1%. Higher rates again indicate errors in genotyping at that locus. Re-genotyping the cohort using a different platform is another method—for example, if an association was found using Illumina, to repeat genotyping at a few key SNPs using Taqman to ensure that this was not an error. This is one of the most thorough QC measures, but can be both costly and time-consuming.

Hardy—Weinberg Equilibrium

Hardy—Weinberg equilibrium (or HWE) simply states that at a given locus with allele frequencies p and q, a randomly mating population should have genotype frequencies of p^2, $2pq$, and q^2. One major cause of deviation from HWE is associated with disease, i.e., if a is the risk allele and genotype aa is associated with osteoporosis and subjects were recruited with osteoporosis, aa will be overrepresented in cases (in a case—control study) or be over-transmitted (in a family-based study). Thus, not all deviation from HWE is bad!

On the other hand, deviation from HWE in control subjects or in founders (parents) can also exist for multiple reasons other than association with disease, and lead to false positive results if not eliminated. A major cause is genotyping error—many genotyping platforms undercall heterozygotes, for example. Such loci should ideally be excluded from analysis. While many studies simply exclude all loci with deviation from HWE in controls or parents, this may indicate CNV (discussed later in this chapter) at that locus and may merit further evaluation. Inbreeding is another obvious cause of departure from HWE; it is important to account for inbreeding in analysis.

Analysis of HWE is not a particularly powerful method of recognizing genotyping error, even in large samples (Cox and Kraft, 2006). Departures from HWE also do not necessarily mean that a genotyping error has occurred. But large deviations from HWE should merit further analysis (examining genotyping call plots, looking at HWE in neighboring markers, etc.).

Mendelian Errors and Nonpaternity in Family-Based Studies

Several additional methods are specific to studies involving related individuals. All loci should be checked for Mendelian inconsistencies (i.e., if parental genotypes are aa and AA, the child's genotype must be aA—a genotype call of AA or aa indicates genotype error). Inconsistencies should be 1%. Analytic tools such as PedCheck will typically exclude results for the effected subject at that locus, so that they do not contribute to analysis. We would advocate a more stringent approach, similar to that taken with highly discordant repeat genotyping explained earlier—if a locus reveals 0.1% Mendelian errors, it suggests a high genotyping error rate at that site, and the entire locus should be removed from analysis.

In addition to evaluation of specific loci, each family trio should be evaluated for nonpaternity or now nonmaternity with advances in assisted reproduction. In such cases, the offspring would differ from the reported father or mother at multiple loci (i.e., many Mendelian inconsistencies). Such trios are generally removed from analysis.

Correction for Multiple Comparisons

In epidemiologic studies in general, we typically accept a type I error rate of $\alpha = 0.05$, i.e., we accept a 1/20 likelihood that the results that we consider statistically significant are in fact due to chance (false positive). In genetic association studies, we typically examine multiple loci; with GWAS, the number of SNPs tested can rise to as high as 500,000 to one million. With 500,000 tests, the expected number of false positive results is 25,000; clearly the 0.05 significance threshold is too liberal in this case. Even in smaller studies that examined only 10 loci, using a threshold of 0.05 as positively associated with disease without correction for 10 separate tests no doubt contributed to many false positive results in early genetic association studies. Many methods for multiple comparisons correction are currently utilized and merit further discussion. However, currently there is no agreement among genetic statisticians as to how to handle this issue, particularly with 500,000 to one million SNPs.

Traditional Method: Controlling Family-Wise Error Rate

The most famous of these methods is the Bonferroni correction, developed by Carlo Bonferroni; this is the most conservative correction for multiple comparisons. The 0.05 significance threshold noted earlier is maintained, but divided among the number of tests performed. Thus, if 10 independent tests are performed, a p value of 0.05/10 or 0.005 is required to declare or 1×10^{-7}.

The Bonferroni method is widely considered to be too conservative for a threshold. First, genetic markers are often highly correlated because of LD (this is true in both a GWAS or a test of multiple markers within a gene). Second, the significance threshold of $p = 1 \times 10^{-7}$ for GWASs requires huge sample sizes and often large effect sizes (which are rare in complex genetic diseases in which GWASs are generally utilized). Thus such a conservative approach will reject many promising loci that would have merited further study. Other multiple-testing strategies that maintain the family-wise error rate, but account for some of the correlation among markers include the Nyholt and Holm corrections.

False Discovery Rate

The false discovery rate (FDR) is a less conservative approach to multiple comparisons correction than the traditional methods described earlier. While the Bonferroni false positive rate of 0.05 means that 5% of all results will be truly negative, the FDR value of 0.05 means that 5% of "declared" positive results are truly negative. If many p values fall into the range where the null hypothesis of no association should be rejected, the FDR is much less conservative. It thus adjusts for the actual p value distribution of the data, and balances type II versus type I error (common methods used include those described by Benjamini and Hochberg (1995) and Storey and Tibshirani (2003)).

Permutation Testing/Empirical Distribution

Permutation testing is frequently used in genome-wide association studies to determine a range of p values that could be expected under the null hypothesis of no association. Phenotype and genotype data are unlinked from one another (though within-genotype correlation is preserved), then randomly paired in a new data set and analyzed to form a test statistic. This procedure is repeated thousands of times across all loci, resulting in an empirical distribution of p values that could be expected by chance alone (Doerge and Churchill, 1996).

These data can be used in two ways. First, an "empirical" p value can be determined for a given locus. Thus if a p value at SNP A is 0.0001 and values of 0.0001 or lower occurred 150 times in 10,000 permutations, then the "empirical" p value at SNP A is 150/10,000 or 0.015. Additionally, the empirical distribution can be used to set a significance threshold for the study. This empirical distribution is compared to the actual distribution of p values in the data; if the two distributions diverge below a certain threshold, p values in this range are less likely to occur by chance alone. SNPs with p values below this threshold are often moved forward for replication testing (see later).

Conditional Power in Family-Based Testing

This is a newer method of multiple-comparisons corrections, which decreases the number of tests performed. As mentioned in the introduction to family-based study design earlier, only families with at least one heterozygous parent are informative at a given locus. Thus even in large studies, many SNPs with low minor allele frequency and therefore few heterozygous parents will be severely underpowered to test for any genetic association. Laird and Lange developed a conditional power screening method to estimate power to find association to avoid testing these loci (Lange et al., 2003). In the screening step, the child's genotype at each locus can be used in noninformative families. In informative families, however, each locus is evaluated using the parental genotypes to impute the child's genotype. Because the imputation is based on expected genotype under the null hypothesis of random transmission rather than actual genotype, this does not bias the test statistic. The phenotypes of the noninformative children can be used to estimate the effect size of the genotypes.

SNPs are ranked by conditional power, and only the p values of the top candidate SNPs are evaluated. Thus, one must only correct for the number of the most powerful SNPs actually tested. The number of SNPs tested varies; in a GWAS of 500,000 SNPs, one may choose to evaluate the top 1000 most powerful SNPs and Bonferroni corrects those; other weighted measures of assigning power have also been developed (Ionita-Laza et al., 2007).

Replication

The techniques described earlier use statistical methods to reduce the false positives in genetic association studies that result because of multiple comparisons. Despite wide implementation of these measures, many initially exciting SNPs have been evaluated in independent cohorts by independent investigators, who were unable to replicate the association, suggesting that the initial findings were still due to chance. The most rigorous form of "multiple comparisons" adjustment is to show that the same SNP, in the same direction and genetic model, is associated with the same phenotype in several populations.

An early form of replication involved a "two-step" approach to GWAS. A given sample was split into a testing cohort and replication cohort. In the testing cohort, a large number of SNPs were tested (with inherent high genotyping costs per individual). The most promising of these SNPs were then carried forward for testing in the "replication" part of the sample. A stringent multiple corrections adjustment (such as Bonferroni) is often employed in the second stage. The two-stage method is particularly useful when genotyping costs for the first stage are prohibitive. As per-SNP genotyping costs have plummeted, the two-stage approach loses this advantage, and the loss of power inherent in this method (Skol et al., 2006) means that it is now much less frequently employed.

Thus an initial genetic association study is performed in a population, and a subset of promising SNP can be moved forward into other populations. These SNPs are often chosen based on a combination of the methods discussed earlier (permutation testing, conditional power screen, or a certain percentage of top p values, generally more lenient than the Bonferroni correction to avoid as many false negative results). Those SNPs are then assessed in one or several replication populations, and only those that replicate are reported as positive findings. Many journals that require new genetic associations have to be replicated in at least one independent cohort for publication. Additional replication by independent groups lends further support to the likelihood of true association.

Population Stratification

Population stratification is a major cause of false positive results in genetic association studies, particularly in case–control studies. It can be thought of as a specific type of confounding by race or ethnicity. If cases and controls are drawn from different ethnic backgrounds, SNP frequency will vary because of genotype frequency in the population rather than because of an association with disease. A classic example of this was described by Knowler et al. (1988), who described an apparent association between an SNP and type II diabetes in a case–control cohort of Pima Indians. When the cohort is stratified by percentage of Indian versus European ancestry, the association is found to be entirely spurious. Population stratification is particularly problematic in more recently admixed populations (such as African American or Hispanic). Multiple methods have been developed for adjusting for population stratification.

Ethnic Matching of Cases and Controls

At a minimum, case—control studies must carefully assess the ethnic background of their subjects. A requirement of "all four grandparents are of Puerto Rican ancestry" as a requirement for study entry in both cases and controls, for example, would help to decrease the likelihood of significant population differences between the groups. This method helps to limit major ethnic differences between groups, but subtle population structural differences may persist. Campbell et al. (2005), for example, showed an apparent association between lactase gene and height in a "white" or "Caucasian" cohort whose grandparents were born either in the United States or Europe. The genetic effect size was greatly diminished when cases and controls were matched based on country of grandparental origin. The association was completely eliminated, however, when only Polish or Scandinavian individuals were genotyped. This suggests that self-described ethnicity can diminish the effects of population stratification, but does not necessarily eliminate them.

Structured Association

A commonly used method for assessing samples for population admixture is to use genetically independent (unlinked) markers that are randomly distributed in the genome and not thought to be associated with disease. Assessment of as few as 30 markers can help to identify whether population stratification is present (Pritchard and Rosenberg, 1999). Using structured association, these SNPs can be used to group subjects into a certain number of subpopulations and tests for an association can be performed within them. Programs such as STRUCTURE implement this method and are widely used and freely downloadable on the web (http://pritchardlab.stanford.edu/structure.html). These methods are best for identifying major divergence between individuals in the sample subjects who are self-declared white but fit better with Asian, for example. Gross outliers such as these, can be identified by "spiking in" the data with subjects of known ethnicity (HapMap subjects, for example). These methods are less good at detecting subtle differences in ethnicity (such as intra-European as described earlier).

Genomic Control

Genomic control similarly uses data from unlinked genetic markers to assess for population stratification. A χ^2 distribution is developed for each locus tested, with only a very small minority expected to truly be associated with disease. If subtle population substructure exists, the χ^2 distribution will differ from the expected by what is called a "variance inflation factor (λ)" (Devlin and Roeder, 1999). This factor will vary with marker number and the number of subjects (with increasing λ as sample size increases). The actual test statistic can be adjusted by the variance inflation factor to give a more conservative χ^2 estimate and p value.

Family-Based Study Design

Family-based studies are often referred to as being "immune" to population stratification. This is because the test statistic is based on the transmission of an allele from parent to offspring. Because parents and offspring come from the same genetic background, population stratification cannot occur. If a particular allele is less frequent in subjects of Caucasian ancestry than African, for example, that will be reflected in the parental genotype and does not bias the test statistic.

The one exception to this rule is that the conditional power screening step in FBAT is conditioned only on parental genotypes; thus if population stratification occurs at a particular locus, the estimate of genetic effect size may be biased, distorting the power calculation. On the other hand, the test statistic in the testing step remains robust to population stratification. With this one caveat, family-based testing avoids the population stratification problem and is considered a major advantage of the method.

Power

In this section, we have focused on three major issues in interpretation of genetic association studies—genotyping error, multiple comparisons testing, and population stratification—that frequently cause false positive results when not adjusted appropriately. A major cause of false negative results is also worth addressing here: inadequate sample size leading to low power.

Small sample size is felt to be a major cause of failed replication of candidate genes. A key concept in genetic association studies is the "Winner's Curse" phenomenon, in which an original publication overestimates the genetic effect size of a particular polymorphism (similar to the concept that a winning auction bid will likely be higher than the actual value of an object; Ioannidis et al., 2001; Lohmueller et al., 2003). Thus if an original publication reports an OR of 2.0, replication studies that power based on effect size of 2.0 will be unlikely to replicate the finding, even if it is real.

Positionally cloned candidate genes will often have slightly larger effect sizes (which allowed their detection via linkage). For other candidate genes, however, more modest effect sizes will be the norm (with OR in the 1.2−1.5 range). To pick up such modest effects, very large sample sizes are needed. In addition to strength of genetic association, power depends very much on minor allele frequency, LD in the region (unless the causative SNP itself is being tested), and accuracy of genotyping and phenotyping.

Multiple well-designed websites are available to help with power calculation. Several of our favorite sites include QUANTO (http://biostats.usc.edu/Quanto.html; Quanto, 2006) which is available for free download; for family-based testing, we use the power calculations available as part of the PBAT package (C.G Lambert, HelixTree Genetics Analysis Software, Golden Helix, Inc., http://www.goldenhelix.com, Bozeman, MT).

FUTURE DIRECTIONS

We focused thus far on some of the major design and interpretation issues in genetic association studies. We chose to focus on a particular type of polymorphism (SNPs), the most common target of genetic association studies thus far. One of the most exciting aspects about working in genetic epidemiology is the rapid movement of the field, limiting to SNP association is likely only scratching the surface. In coming years, we will see increasing emphasis on novel biology. Many of the design and analyses sections discussed in this chapter will be true for emerging areas of genetic research as well (the importance of accurate phenotyping/disease classification and adequate statistical power, for example); others will require development of new methods for appropriate analysis.

Emerging Biologic Targets

Whole Genome Sequencing

SNPs are one of the major sources of DNA variation (in addition to structural variation, discussed later). The SNP association methods that we have discussed thus far make use of LD to avoid needing to assess every single polymorphism in the genome. One problem with this approach is that even the newer > million SNP arrays are mainly targeted toward common variations (usually SNPs with 0.5% minor allele frequency). This approach fails to capture most rare variants, which are actually more likely to be functional (as opposed to common variants, which are more frequently regulatory in function).

Given plummeting genotyping costs, it is widely assumed that we will enter the era of the "$1000 genome" soon, with individual whole genome sequencing available. This will enable rare variant detection and thus brings the prospect of discovering many true associations, but methods of adequately correcting for multiple comparisons while maintaining power to appreciate genetic association are even more critical for sequencing studies than that of the GWASs we have emphasized in this chapter. While a full discussion of analytic approaches for whole genome sequence studies are beyond the scope of this chapter, several are worth noting. First, given that rare variants are more likely functional, upweighting, or filtering for variants most likely to impact gene expression and translation is particularly powerful. Programs like Polyphen (http://genetics.bwh.harvard.edu/pph2/; Adzhubei et al., 2010), or incorporation of DNase hypersensitivity sites available on the UCSC genome browser can be used among others. In addition, multiple statistical programs have been developed that incorporate information from multiple nearby SNP into a gene- or region-based test to improve power (example approaches include burden-based tests (Li and Leal, 2008) and SKAT-O (Lee et al., 2012)).

Structural Variation (With Emphasis on Copy Number Variations)

Genetic structural variation includes larger-scale genetic polymorphism, involving 1 kb to millions of kb. The most frequent type is CNV, though it also includes chromosomal translocations, inversions, and balanced insertions. CNVs, by nature of their sheer size, are responsible for a substantial fraction of genetic variability, and are more likely to disrupt gene function and expression levels than SNPs.

CNVs have been increasingly found to be associated with human disease, but the methods for genetic association study of CNVs are still in the developmental phase. As novel statistical and genotyping methods are developed that are designed specifically for CNVs (rather than simply implementing methods that were designed for SNP-based association studies), this field will continue to provide novel biological associations.

Expression

mRNA or protein expression levels are another focus area of genetic association studies. Functional variation in DNA sequence (via SNP or larger-scale structural variation) alters downstream expression level. One defines an eQTL or expression quantitative trait locus where an SNP is explaining some proportion of transcript abundance. The expression level can now be interrogated on a particular gene level (i.e., expression levels of mRNA is higher in diseased than in nondiseased individuals), or genome-wide (hypothesis-free) level using micro-array technology.

Epigenetics

Another emerging focus of association studies is on so-called "epigenetics", genetic variation that is not related to DNA sequence, but rather other processes (such as differences in DNA methylation or histone modification) that modify gene function or expression level.

Integrative Statistical Approaches

In addition to the more recent biological targets for association study described earlier, statistical methods are being developed to evaluate the roles of multiple genes or pathways simultaneously in determining a genetic effect. In complex trait genetics, most disease phenotypes are likely modulated by dozens to even hundreds of genes simultaneously; thus studying one gene at a time is not a powerful approach. Instead, novel statistical methods often use a Bayesian approach to evaluate which genes together best explain variability. Additional methods to better incorporate gene—gene or gene—environment interaction are also in development.

From Association to Functional Variant

The focus of this chapter was on using genetic association studies to identify genes or genomic regions that cause human disease. A positive association study, though, is just the beginning; as we have discussed, there are numerous causes of false positive associations, and "proving" that a locus actually causes disease is an arduous process. Replicating the results in other human populations is a starting point, but ideally we expand from mere association to identifying function. This could include finding a functional variant in LD with an identified SNP that changes protein expression levels, or creating a knockout mouse for a target gene that mimics the disease originally identified in humans. Such work requires extensive collaboration with investigators in molecular biology, epidemiology, and statistics, and is a part of what makes studying human genetics so exciting.

CONCLUSION

In this section, we outlined some of the major causes of spurious results in genetic association studies—including genotyping error, failure to account for multiple comparisons testing, and population stratification—and described some of the most common methods used to adjust for those issues. We also discussed the issue of inadequate power in genetic association studies, which frequently causes false negatives.

Even a perfectly designed study can result in some false positive or negative associations, however. We have focused in this chapter on genetic association using SNPs, which are the genetic polymorphisms that are most frequently assessed in today's genetic association studies. In our final section, we move away from SNP/phenotype associations to discuss emerging areas of investigation. By combining the information on SNP and other types of genetic polymorphism with higher-level gene expression (called genetical genomics), we will develop a more complete understanding of the genetic underpinnings of human disease.

REFERENCES

Adzhubei, I.A., Schmidt, S., Peshkin, L., et al., April 2010. A method and server for predicting damaging missense mutations. Nat. Methods 7 (4), 248—249.

Benjamini, Y., Hochberg, Y., 1995. Controlling the false discovery rate: a practical and powerful approach to multiple testing. J. R. Stat. Soc. B 57, 289—300.

Cox, D.G., Kraft, P., 2006. Quantification of the power of Hardy-Weinberg equilibrium testing to detect genotyping error. Hum. Hered. 61 (1), 10—14.

Campbell, C.D., Ogburn, E.L., Lunetta, K.L., et al., August 2005. Demonstrating stratification in a European American population. Nat. Genet. 37 (8), 868—872.

Doerge, R.W., Churchill, G.A., January 1996. Permutation tests for multiple loci affecting a quantitative character. Genetics 142 (1), 285–294.

Devlin, B., Roeder, K., December 1999. Genomic control for association studies. Biometrics 55 (4), 997–1004.

Frazer, K.A., Ballinger, D.G., Cox, D.R., et al., October 18, 2007. A second generation human haplotype map of over 3.1 million SNPs. Nature 449 (7164), 851–861.

International HapMap, C., Altshuler, D.M., Gibbs, R.A., et al., September 2, 2010. Integrating common and rare genetic variation in diverse human populations. Nature 467 (7311), 52–58.

Ionita-Laza, I., McQueen, M.B., Laird, N.M., Lange, C., September 2007. Genomewide weighted hypothesis testing in family-based association studies, with an application to a 100K scan. Am. J. Hum. Genet. 81 (3), 607–614.

Ioannidis, J.P., Ntzani, E.E., Trikalinos, T.A., Contopoulos-Ioannidis, D.G., November 2001. Replication validity of genetic association studies. Nat. Genet. 29 (3), 306–309.

Knowler, W.C., Williams, R.C., Pettitt, D.J., Steinberg, A.G., October 1988. Gm3;5,13,14 and type 2 diabetes mellitus: an association in American Indians with genetic admixture. Am. J. Hum. Genet. 43 (4), 520–526.

Last, J.M., Abramson, J.H., Association, I.E., 1995. A Dictionary of Epidemiology. Oxford University Press.

Libioulle, C., Louis, E., Hansoul, S., et al., April 20, 2007. Novel Crohn disease locus identified by genome-wide association maps to a gene desert on 5p13.1 and modulates expression of PTGER4. PLoS Genet. 3 (4), e58.

Lange, C., Lyon, H., DeMeo, D., Raby, B., Silverman, E.K., Weiss, S.T., 2003. A new powerful non-parametric two-stage approach for testing multiple phenotypes in family-based association studies. Am. J. Hum. Genet. 73 (5), 613 (abstract).

Lohmueller, K.E., Pearce, C.L., Pike, M., Lander, E.S., Hirschhorn, J.N., February 2003. Meta-analysis of genetic association studies supports a contribution of common variants to susceptibility to common disease. Nat. Genet. 33 (2), 177–182.

Li, B., Leal, S.M., September 2008. Methods for detecting associations with rare variants for common diseases: application to analysis of sequence data. Am. J. Hum. Genet. 83 (3), 311–321.

Lee, S., Wu, M.C., Lin, X., September 2012. Optimal tests for rare variant effects in sequencing association studies. Biostatistics 13 (4), 762–775.

Mitchell, A.A., Cutler, D.J., Chakravarti, A., March 2003. Undetected genotyping errors cause apparent overtransmission of common alleles in the transmission/disequilibrium test. Am. J. Hum. Genet. 72 (3), 598–610.

Pritchard, J.K., Rosenberg, N.A., July 1999. Use of unlinked genetic markers to detect population stratification in association studies. Am. J. Hum. Genet. 65 (1), 220–228.

Quanto 1.1: A Computer Program for Power and Sample Size Calculations for Genetic-Epidemiology Studies (Computer Program), 2006.

Storey, J.D., Tibshirani, R., August 5, 2003. Statistical significance for genomewide studies. Proc. Natl. Acad. Sci. USA 100 (16), 9440–9445.

Skol, A.D., Scott, L.J., Abecasis, G.R., Boehnke, M., February 2006. Joint analysis is more efficient than replication-based analysis for two-stage genome-wide association studies. Nat. Genet. 38 (2), 209–213.

Chapter 18

Pharmacogenetics of Drug Metabolism

Zeruesenay Desta and David A. Flockhart

Indiana University, Indianapolis, IN, United States

Chapter Outline

Key Points

- The goal of effective and safe therapy of many drugs is made difficult by large interpatient variability in response and toxicity, and this problem is a substantial burden for patients, their caretakers, and the health-care system.
- Pharmacogenetic tests that are most valuable are those for specific drugs for which the prediction of activity or adverse effects is important and difficult to anticipate given our current clinical tools and technologic capability.
- Although data from a variety of platforms documenting a wide range of genetic variability are being accumulated rapidly and clinical genetic tests have been recommended or implemented for drugs such as mercaptopurine, azathioprine, warfarin, irinotecan, and tamoxifen, there is still a great need for well-designed, prospective clinical trials that test pharmacogenetic approaches versus standard practice.

INTRODUCTION

The goal of effective and safe therapy of many drugs is made difficult by large interpatient variability in response and toxicity, and this problem is a substantial burden for patients, their caretakers, and the health-care system. For all major classes of drugs, a substantial proportion of patients may not respond or respond only partially, or experience adverse drug reactions when given standard doses, and differences in pharmacokinetic (affecting drug concentrations) and pharmacodynamic (affecting drug targets) contribute to this variable response. While potential causes for variable drug efficacy and toxicity do include a host of intrinsic and extrinsic factors, inherited differences in drug-metabolizing enzymes, drug transporters, and drug targets in general account for a significant proportion of the interindividual differences in drug disposition and response. For certain drugs or classes of drugs, genetic factors can account for substantial and clinically meaningful interindividual variability in drug disposition and effects. The pharmacogenetics of drug transporters and drug targets are addressed in other chapters. In this chapter, common genetic polymorphisms affecting pharmacokinetics via effects on drug metabolism are outlined and their clinical relevance is discussed.

For many drugs, the response to chronic administration is determined by the area under the plasma concentration time curve (AUC) during a dosing interval at steady state, a measure of drug exposure. The AUC divided by the dosing interval is equivalent to the average concentration during chronic drug administration. By far the most important variants in drug metabolism are those that affect the AUC by causing a change in either the oral bioavailability (F_{PO}) or the apparent oral clearance (CL). For an orally administered drug, the AUC_{PO} is given by:

$$AUC_{PO} = F_{PO}*Dose_{PO}/CL$$

Drug concentrations can vary by up to 600-fold between two individuals of the same weight on the same drug dosage. While other aspects of pharmacokinetics (e.g., absorption and distribution) can potentially be controlled by genetic variation, those that influence drug clearance (elimination) have been shown to be most important for chronic dosing. Clearly, excretion by renal, hepatobiliary, and other routes is an important determinant of clearance for some drugs and their metabolites. A majority of the drugs in current use are eliminated primarily by metabolism. Therefore, clearance by metabolism via enzymes located in the liver and for some drugs in the gut wall represents the single most common and major cause for variable drug response. There are two major categories of metabolic reactions.

Phase I reactions (oxidation, hydrolysis, and reduction) result in relatively small chemical changes that often make compounds more hydrophilic. Although phase I metabolism can be mediated by a wide range of enzymes that include flavin-containing monooxygenases, epoxide hydrolases, oxidoreductases, dehydrogenases, esterases, oxidoreductases, and amidases, research on most of these enzymes is relatively young, and the potential for genetic influence on the activity of these enzymes is evolving. The best understood pathway of phase I drug metabolism is oxidation by members of the superfamily of hemeproteins called the cytochrome P450 (CYP) enzymes, located primarily in the endoplasmic reticulum of hepatocytes and enterocytes. More importantly, a significant proportion of drugs cited in studies of adverse drug reactions and drug interactions are metabolized by CYPs, and these adverse events represent important causes of morbidity and mortality that compromise public health (Kohn et al., 2000). In humans, there are 18 families of CYP genes, 42 subfamilies, and 57 P450 genes, but only 8 enzymes that belong to the CYP1, CYP2, and CYP3 subfamilies are responsible for xenobiotic and drug metabolism: CYP1A2, CYP2A6, CYP2B6, CYP2C8, CYP2C9, CYP2C19, CYP2D6, and CYP3A4/5. Approximately 80% of oxidative metabolism and almost 50% of the overall elimination of commonly used drugs can be attributed to one or more of these enzymes (Goodman et al., 1990). Almost all the genes that code for these enzymes are highly polymorphic and contribute to the large interindividual variability in the activity of these CYPs.

Phase II reactions involve acetylation, methylation, and conjugation with glucuronic acid, amino acids, or inorganic sulfate. The main enzymes involved in these reactions include UDP-glucuronyltransferases (UGTs), sulfotransferases, glutathione-S-transferases, N-acetyltransferases (NATs), and methyltransferases. Often, metabolism by phase II enzymes leads to more water-soluble molecules that are more easily eliminated by biliary or renal elimination. Evidence is accumulating that these reactions, particularly UGT-mediated conjugation, play important roles as the primary routes of clearance for a number of commonly used drugs.

PHARMACOGENETICS OF DRUG METABOLISM: HISTORICAL ASPECTS

A number of key observations have led to the idea that response to xenobiotics might be controlled by genetics and ethnic differences. The concept of variable response to chemicals (chemical individuality in man) was first suggested by Garrod in 1906 (Garrod, 1909). Experimental evidence in support of this concept was later provided in 1931 by Fox and subsequently by others who showed that some people are unable to taste phenylthiocarbamide while others taste it bitter (Fox, 1932). It was also then shown that the frequency of nontasters varies among ethnic groups. The idea that inheritance might contribute to variation in drug response in an ethnic-dependent fashion was first reported during World War II with the observation that hemolytic crises induced by primaquine or other chemically related antimalarial drugs are much more common in African Americans than in Caucasians. This observation was later explained by genetic defects discovered in glucose-6-phosphate dehydrogenase. This deficiency also appear to be the cause for the observation that was made around

510 BC by Pythagoras that ingestion of fava beans resulted in a potentially fatal reaction in some, but not in all, individuals (Meyer, 2004).

Landmark discoveries showing a link between deficiencies in drug metabolism and unexpected adverse drug effects were made in the early 1950s. Interindividual variability in isoniazid metabolism was observed in 1953. In 1954, a link between isoniazid neurotoxicity and its metabolism was suggested. Subsequent population studies by Evans et al. (1960) identified slow and rapid acetylators of isoniazid, although it took almost 40 years for the molecular basis for these phenotypes to be discovered (Meyer, 2004). Another example was the prolonged apnea (1956) observed in a small number of patients receiving succinylcholine, which was later (1957) (Kalow and Genest, 1957) shown to be due to plasma pseudocholinesterase deficiency. The discovery of these two clinically important genetic deficiencies spurred further research into variable drug response, and in 1959, Vogel coined the term "pharmacogenetics."

The field of drug metabolism pharmacogenetics was energized and given further prominence by the groundbreaking discoveries of the CYP2D6 and CYP2C19 polymorphisms in the 1960s and 1970s. Sjöqvist et al. observed large inter-individual variability in nortriptyline and desipramine plasma concentrations after a fixed dose and identified two groups of patients based on phenotypes (Hammer and Sjöqvist, 1967). A subsequent study in twins suggested that this difference is genetic in nature. Later on, Smith et al. in London (Mahgoub et al., 1977) and Eichelbaum et al. in Germany (Eichelbaum et al., 1979) independently reported unexpected adverse reactions to debrisoquine and sparteine, respectively, in volunteers that were participating in clinical trials. Upon analysis of the plasma concentrations of these drugs, it was shown that the exposure to these drugs was substantially higher in those subjects who developed serious adverse effects. The inheritance was monogenic as an autosomal recessive trait for both debrisoquine and sparteine. Subsequently it was shown that the same enzyme deficiency accounted for the differences in phenotypes of debrisoquine and sparteine, and the responsible enzyme is now known to be CYP2D6, which also metabolizes desipramine, nortriptyline, and a large number of other drugs.

At the end of the 1970s, the first poor metabolizer (PM) of mephenytoin was discovered when a patient who was enrolled in a clinical mephenytoin trial experienced extreme sedation after taking 300 mg/day mephenytoin. This effect was shown to be due to deficiency of mephenytoin hydroxylation, and the enzyme responsible is now known to be CYP2C19. In the ensuing years, other clinically important drug metabolisms in pharmacogenetics—for example, toxicity of 6-mercaptopurine, 5-fluorouracil, irinotecan, and tolbutamide caused by polymorphisms in TPMT (1980), DPD (1988), UGT1A1 (1997), and CYP2C9 (1980), respectively—have been discovered. The discovery of genetic polymorphisms in drug metabolism was driven by observations of unusual drug response by alert clinicians: a phenotype to genotype approach that remains valuable. Many years after the differences in phenotypes were discovered, the underlying molecular genetic causes of drug metabolism deficiencies were elucidated through cloning of the genes—for example, CYP2D6 in 1988, NAT-2 in 1991, TPMT in 1993, CYP2C19 in 1994. All these discoveries, together with completion of the human genome project and technological advances in molecular biology and increasingly comprehensive platforms, have led to the current wave of interest in inherited differences of drug metabolism and response.

GENETIC POLYMORPHISMS OF INDIVIDUAL DRUG-METABOLIZING GENES

CYP1A2

CYP1A2 accounts for ~15% of the hepatic P450 content and is the main clearance mechanism for clinically important drugs such as theophylline, caffeine, clozapine, olanzapine, tizanidine, duloxetine, and ramelteon. This enzyme also biotransforms procarcinogens such as food-derived heterocyclic and aromatic mutagens, N-heterocyclics found in tobacco smoke, and difuranocoumarins to reactive carcinogens, as well as endogenous substances (e.g., estradiol, melatonin, and uroporphyrinogen).

The in vivo activity of CYP1A2, as measured by caffeine metabolism, varies widely (up to 70-fold) among subjects. This variability is in part due to exposure to drugs (and other chemicals) that induce or inhibit its activity. There is also evidence that genetic polymorphisms influence CYP1A2 activity. In the mid-1980s, slow and fast metabolizer phenotypes were reported based on the metabolism of the chemical carcinogen acetylaminofluorene in a panel of human liver microsomal preparations. Using the metabolism of caffeine as a marker of activity in vivo, several investigators demonstrated high interindividual and interethnic variability in CYP1A2 activity. A study on monozygotic and dizygotic twins revealed the genetic basis of this variability (Rasmussen et al., 2002). The human *CYP1A2* gene is located on chromosome 15 and contains seven exons. The coding region starts at nucleotide 10 of exon 2. Exons 2—6 are highly conserved among the human, mouse, and rat. To date, 16 alleles (and many more variants) of the *CYP1A2* gene have been identified (see http://www.imm.ki.se/CYPalleles/cyp1a2.htm). Single nucleotide polymorphisms (SNPs) that change the amino acid have been reported, but their frequency in the population is very low (<1%). Several variants have been found in the *CYP1A2* upstream sequence and in intron 1; some of these variants have been suggested to alter expression and activity at baseline and/or after exposure to inducers. As a result, differences in inducibility by these variants, normal, diminished, or hyperinduction phenotypes may be observed. For example, the *CYP1A2*1C* allele (−3860G > A located in the enhancer region) is associated with reduced CYP1A2 expression, while *CYP1A2*1F* (−163C > A) is associated with higher inducibility.

Clozapine

Clozapine is a prototypical atypical antipsychotic whose metabolism covaries with CYP1A2 activity. Because clinical studies indicate a positive association between clozapine plasma concentration and antipsychotic response and since the extent of antipsychotic response varies markedly among patients, polymorphism of CYP1A2 and/or its regulators [e.g., aromatic hydrocarbon (AH) receptor] has been studied in an attempt to individualize clozapine therapy through identification of responders and nonresponders to the drug. There appears to be an association between the *CYP1A2*1F* allele and enhanced clozapine clearance, increased dose requirement and nonresponsiveness, particularly in smokers, and increased plasma concentrations and adverse effects after discontinuation of smoking. A few studies did not find this relationship. *CYP1A2*1C* and *CYP1A2*1D* (−2467delT) appear to be associated with increased clozapine exposure and adverse effects.

Caffeine

The major role that CYP1A2 plays in caffeine metabolism is well established. Caffeine metabolism is one of the best markers of CYP1A2 activity available in vitro and in vivo and has been used as an important tool to study the influence of genetic and nongenetic factors influencing CYP1A2 function. The *CYP1A2*1F* variant appears to increase the clearance of caffeine in smokers; other SNPs, alleles, or haplotypes might increase or decrease caffeine metabolism. Increased risk of recurrent pregnancy loss induced by caffeine (Sata et al., 2005) and more recently decreased risk of nonfatal myocardial infarction (MI) induced by caffeine (Cornelis et al., 2006) have been associated with *CYP1A2*1F* allele. If validated, these data suggest that CYP1A2 polymorphisms may produce important clinical consequences.

Summary

Although the precise molecular mechanism remains unknown, the change inherent in the *CYP1A2*1F* allele appears to have clinical consequences for CYP1A2 substrates such as caffeine and clozapine in smokers since this allele seems to influence the inducibility of the enzyme. However, no genetically polymorphic site in the *CYP1A2* gene can be used currently to predict the extensive interindividual variation in metabolic phenotype between individuals, despite the identification of many variant alleles and extensive resequencing efforts. Further studies will be required to define the contribution of the CYP1A2 genetic variations or regulatory elements to the variability of CYP1A2 and eventually to the clinical response to drugs metabolized primarily by this route.

CYP2B6

Historically, CYP2B6 was thought to play a minor role in human drug metabolism because earlier studies reported none or very low levels of CYP2B6 expression in human livers (Ekins et al., 1998). In addition, the lack of suitable in vitro and in vivo tools to characterize its function in the past also impeded research. Despite this history, it is now established that CYP2B6 is a significant component of the P450 enzyme system, with an average relative abundance of ≈6% of the total hepatic P450 content, in sharp contrast to earlier reports that it represents ≈0.2% (Ward et al., 2003). Notable progress has also been made toward the development of specific probes of activity. The development of these new biochemical tools, the ability to heterologously express CYP2B6, the discovery of extensive variations of the *CYP2B6* gene, improved understanding of the roles of nuclear receptors in CYP2B6 molecular regulation, the potential effects of ethnic differences and other factors in its expression, and more importantly the emerging evidence for clinical relevance have all contributed to an appreciation of the considerable importance of this enzyme. A growing number of clinically important drugs across several therapeutic classes (≈8% of drugs on the market) are fully or partially metabolized by CYP2B6. These drugs include efavirenz, nevirapine, cyclophosphamide, bupropion, ifosfamide, thioTEPA, methadone, meperidine, artemisinin, ketamine, and propofol. CYP2B6 can also metabolize several drugs of abuse, procarcinogens, and numerous environmental chemicals/toxins.

CYP2B6 expression and activity vary extensively among individual human livers in vitro, which likely results in large changes in activity in humans and may account for the large interpatient variability in the pharmacokinetics of and response often seen with drugs primarily metabolized by CYP2B6. Exposure to drugs and other chemicals can influence CYP2B6 activity: the enzyme is highly inducible by structurally diverse xenobiotics and endogenous compounds, and several compounds that inhibit or inactivate its catalytic activity have been identified. In addition, CYP2B6 genetic polymorphisms contribute to the interindividual variability in CYP2B6 activity. The *CYP2B6* gene is located within the CYP gene cluster on chromosome 19 (at 19q13.2) and contains nine exons that code for a 48-kDa microsomal protein consisting of 491 amino acids. The lack of selective substrate probes to characterize CYP2B6 activity hampered identification of functionally important variants of the *CYP2B6* gene until recently. Thus, a systematic discovery of SNPs preceded clinical pharmacogenetics (genotype to phenotype). The first such approach was described in 2001 by Ariyoshi et al., who reported a G516T common SNP in exon 4 (Gln to His at codon 172), with 19.9% allelic frequency in Japanese subjects. This variant was shown to increase the catalytic activity in expression systems in vitro (Ariyoshi et al., 2001). Lang et al. (2004) carried out a systematic and extensive search for sequence polymorphisms of the *CYP2B6* gene and reported five SNPs in the coding region that change the amino acid (in addition to multiple promoter and synonymous SNPs): C64T in exon 1 (Arg22Cys), G516T in exon 4 (Gln172His), C777A in exon 5 (Ser259Arg), A785G in exon 5 (Lys262Arg), and C1459T in exon 9 (Arg487Cys). These SNPs alone or in combination defined the first six distinct alleles [CYP2B6*2 (R22C), *3 (S259R), *4 (K262R), *5 (R487C), *6 (Q172H and K262R), and *7 (Q172H, K262R, and R487C)]. Subsequent studies have confirmed these initial findings but also discovered numerous additional DNA sequence variations across the entire *CYP2B6* gene, with the exception of key response elements of the gene. The *CYP2B6* gene, with over 28 alleles and over 100 SNPs whose haplotype structure is not determined, is one of the most highly polymorphic P450 genes (http://www.imm.ki.se/CYPalleles/cyp2b6.htm). Many of these variants are in linkage disequilibrium (LD), forming distinct haplotypes that have a wide range of functional consequences in human liver, in heterologous expression systems, and/or in in vivo human studies. The spectrum of functional changes of CYP2B6 genetic polymorphisms include phenotypically null alleles (rare nonsynonymous coding cSNPs), partially diminished function caused by other cSNPs (e.g., the *6 allele), noncoding SNPs such as the 15631G > T intronic variant that appears to trigger formation of a nonfunctional splice variant, deletion of exons 1−4 of *CYP2B6* (*29), deletion polymorphisms at 1132C > T (R378stop) (*28); gain in function variants that include a −82T > C change in the 5′-flanking region (*22) that relocates the transcriptional start site, and variants that cause increases in CYP2B6 expression, including the *4 allele (K262R) and the *6 allele for some substrates (e.g., cyclophosphamide; Zanger et al., 2007). More recent studies provide evidence that some of these polymorphisms alter drug exposure and in certain cases, drug response.

Efavirenz

This nonnucleoside reverse transcriptase inhibitor—based therapy is often the preferred initial therapy of HIV infection, but there is high interpatient variability in its pharmacokinetics and clinical response at the usual therapeutic dose (600 mg/day oral dose). Efavirenz plasma concentrations are good predictors of efavirenz response: high concentrations (>4 mg/L) are associated with central nervous system toxicity and concentrations <1 mg/L are associated with treatment failure. The role that CYP2B6 genetic polymorphisms might play became clear subsequent to the discovery that CYP2B6 is the principal enzyme catalyzing efavirenz metabolism. Two clinical studies conducted in HIV patients were published in 2004, which demonstrated that the plasma efavirenz concentrations of patients with CYP2B6 *6/*6 (or the T516T) genotype are significantly higher than those of patients with genotypes *1/*6 (or C516T) or *1/*1 (or C516C) genotypes (see review in Zanger et al., 2007). These studies also showed that carriers of the *6/*6 allele (or T516T) are at increased risk for efavirenz-induced central nervous system symptoms and hepatic injury. In addition to the *6 allele, CYP2B6*18, which is quite common in black population, significantly increases plasma exposure. Subsequently, clinical studies in HIV patients have consistently confirmed a relationship between CYP2B6 genetic polymorphisms (typically *6/*6 and *18/*18) and efavirenz exposure and/or response, extent of autoinduction and drug interactions, and further expanded the association to other rare genetic variations. Some studies recommend reduction of dose of efavirenz from the usual therapeutic dose (600 mg/day) to 400 mg/day when they carry these genotypes. Collectively, these studies suggest that individuals with *6/*6 and *18/*18 genotype may have over threefold higher efavirenz concentrations and may be predisposed to adverse effects. In addition, these data provide direct evidence that CYP2B6 is relevant in vivo and that efavirenz can be used as a model phenotyping drug to further understand the role CYP2B6 plays in human drug metabolism.

Other Substrates

In limited clinical studies, *CYP2B6* genetic polymorphisms were associated with pharmacokinetic or response variability of certain substrates (e.g., methadone, bupropion, and nevirapine). The *CYP2B6*4 (K262R) allele was associated with modest but significantly higher (1.4-fold) median bupropion clearance and a slightly higher AUC of hydroxybupropion. A recent study shows decreased 4-hydroxybupropion exposure in carriers of the *6/*6 genotype in healthy volunteers, which was then associated with reduced efficacy because this metabolite is pharmacologically active. However, the significance of CYP2B6 genetics to bupropion therapy needs further confirmation as the drug has alternative metabolic pathways, and because both the metabolite and parent drug are pharmacologically active. Moreover, bupropion and its metabolite have chiral centers that produce multiple enantiomers or diastereomers. Until the chiral metabolism and effects of bupropion are elucidated, it may be difficult to fully understand the role of CYP2B6 genetics in bupropion response. The *6 allele of *CYP2B6* appears to be associated with a significantly increased risk of methadone-induced QTc prolongation. This observation was attributed to decreased metabolism of (S)-methadone, an enantiomer that seems more arrhythmogenic in vitro. These data, if validated, implicate CYP2B6 in methadone metabolism in vivo and may offer a means of anticipating the risk of cardiac arrhythmias and sudden death. There are documented associations between *CYP2B6* genotypes, typically the G516T SNP, and the pharmacokinetics and response to nevirapine (a nonnucleoside reverse transcriptase inhibitor widely used in HIV prevention and treatment) in adults and children.

Summary

A range of translational studies have consistently demonstrated that the *CYP2B6*6 allele or the tag SNP defining this allele is by far the most frequent variant associated with functional changes in activity (40–60% in certain populations, with wide interethnic variability). Since the *6 allele identifying SNP G516T also serves as a tag SNP for other relatively rare haplotypes, it seems reasonable to suggest that this SNP defines most of the genetic contribution to variability in CYP2B6 activity that informs clinical significance. It is also clear that the distribution of some of functionally important variants in the *CYP2B6* gene (e.g., *18 allele) is highly dependent on the ethnic background of the population. The contribution of other *CYP2B6* variants is limited by their low frequency, lack of functional significance, or both. Although it is clear that important advances have been observed in the last few years with respect to *CYP2B6* genetic polymorphisms, our current knowledge is insufficient to provide us with efficient tools to predict the large interindividual variability in CYP2B6 activity with confidence.

CYP2C8

CYP2C8 accounts for ~7% of CYP content in the liver and oxidizes about 5% of drugs cleared by phase I metabolism. This enzyme is important in the elimination of drugs used in the treatment of diabetes (repaglinide, rosiglitazone, pioglitazone), cancer (paclitaxel, trans retinoic acid, enzalutamide), antianemic (daprodustat), malaria (amodiaquine and chloroquine), statins (cerivastatin), asthma (montelukast), viral HEPC (dasabuvir), and arrhythmias (amiodarone). The activity of CYP2C8 varies widely among individuals, but the molecular mechanisms involved are less apparent because, due to the lack of a robust standardized phenotyping probe, it has been difficult to estimate the in vivo variation. However, evidence exists that this variability is in part due to exposure to induction (the enzyme is PXR- and CAR-inducible) and to coprescribed inhibitors. The withdrawal of cerivastatin from the market due to severe or fatal rhabdomyolysis resulting from drug interactions (mostly with gemfibrozil, a strong inhibitor of CYP2C8) is supportive of the role environmental factors play in CYP2C8 activity.

Although the consequences are yet to be studied, there is also evidence that genetic variations of the *CYP2C8* gene might contribute to its variable activity. Currently, nearly 100 nonsynonymous single nucleotide variations and short deletions, as well as essential splice site variants, have been found in the *CYP2C8* gene. Fourteen CYP2C8 variant alleles, most of which are coding region SNPs, have been identified (http://www.cypalleles.ki.se/cyp2c8.htm), and their frequencies differ significantly both between and within continental populations. Of these, *CYP2C8*2, *CYP2C8*3, and *CYP2C8*4 are relatively common. The *CYP2C8*2 (Ile269Phe in exon 5) allele is present mainly in Africans with an allelic frequency of 18%. *CYP2C8*3 (Arg139Lys and Lys399Arg in exons 3 and 8; allelic frequency of 10–23%) and *CYP2C8*4 (Ile264Met in exon 5; allelic frequency of 7.5–11%) are mainly found in Caucasians. The frequency of both these alleles is relatively low in Asians (0–5%). Other variants leading to amino acid changes are extremely rare. In addition, two SNPs have been reported in the promoter region of *CYP2C8* (*1B and *1C).

In vitro studies using mammalian and bacterial expression systems have shown that *CYP2C8*3* had reduced enzyme activity as measured by substrate metabolic turnover (e.g., paclitaxol, amodiaquine, or arachidonic acid). However, the in vivo relevance of *CYP2C8*3* is difficult to interpret, particularly when substrates of both *CYP2C8* and *CYP2C9* are used as probes of its activity as there appears a strong LD between *CYP2C8*3* and *CYP2C9*2*. In contrast, *CYP2C8*3* was reported to be associated with increased metabolism of certain substrates (e.g., repaglinide). In human liver microsomes, a small decrease in paclitaxel 6α-hydroxylation was found for *CYP2C8*3*-positive samples. The in vivo relevance of *CYP2C8*3* and the molecular mechanisms underlying *CYP2C8* interindividual variability remain uncertain. The *CYP2C8* gene is relatively highly conserved, and, except for the rare *CYP2C8*5* that codes for a truncated protein, the contribution of the described *CYP2C8* alleles to the observed interindividual variability remains unclear. Many issues remain to be solved: in vitro and in vivo functional characterization of CYP2C8 genetic variants that have been identified so far is needed. To do this, a selective and sensitive probe of activity in vivo would be required.

CYP2C9

The CYP2C9 enzyme is mainly expressed in the liver, accounting for approximately 15–20% of the hepatic total P450 content, and to some extent in the gut wall. It metabolizes fully or partially over 100 clinically important drugs ($\approx$ 10–20% of the currently marketed drugs). Drugs primarily metabolized by CYP2C9 include oral anticoagulants (e.g., warfarin), oral hypoglycaemics (e.g., tolbutamide, glyburide, glimepiride, glipizide, and gliclazide), angiotensin II blockers (e.g., losartan, valsartan, candesartan, and irbesartan), loop diuretics (torsemide), nateglinide, phenytoin, fluvastatin, many nonsteroidal antiinflammatory drugs (e.g., diclofenac, meloxicam, oxicam), and COX2 inhibitors (e.g., celecoxib). Some of these drugs are of narrow therapeutic range (e.g., warfarin, phenytoin, and tolbutamide). In addition, CYP2C9 metabolizes the endogenous substrates, arachidonic acid and linolenic acid.

There is extensive intersubject variability in CYP2C9 activity. This variability is associated with difficulties in dose adjustment or with life-threatening adverse effects of drugs such as warfarin and phenytoin. While drug interactions that influence its activity are well known, genetic polymorphisms also importantly contribute.

A possible genetic polymorphism of *CYP2C9* was first suggested in 1979 when investigators observed wide differences in tolbutamide elimination parameters and suggested the existence of slow metabolizers of tolbutamide (approximately 1 in 500). This was confirmed in the 1980s and early 1990s when cDNAs were cloned (Romkes et al., 1991). A clinical study conducted in 1985 described an individual with a tolbutamide half-life of 37 h (five times the normal population value) and a reduction in plasma clearance to almost 20%, but population studies have shown that the incidence of this phenotype (slow metabolizer) may be rare. This individual was later (1996) shown to be homozygous for the variant causing Ile to Leu substitution at residue 359 (this variant is designated as *CYP2C9*3*). Impaired (S)-warfarin metabolism catalyzed by the R144C variant of CYP2C9 was reported in 1994, and this variant is now designated as the *2 allele. In 1996, the variant Ile359/Leu359 was reported and found to be associated with slow elimination of tolbutamide. Systematic investigation that ensued (1996) confirmed the existence of the *CYP2C9*2* and *CYP2C9*3* variants at significant frequencies (close to 10%) in a Northern European population.

Although the gene coding for the CYP2C9 protein is highly polymorphic, over 30 alleles and allelic variants have been identified to date within the regulatory and coding regions and most of them are associated with reduced CYP2C9 activity (http://www.imm.ki.se/CYPalleles/cyp2c9.htm). *CYP2C9*3* and *CYP2C9*2* have the most clinical relevance because they are relatively frequent, particularly in Caucasians (10–15% and 8–13% respectively), although they are relatively rare in other populations. In fact, *CYP2C9*2* has not been detected in Asians. Up to 40% of Caucasians possess one or both of the *CYP2C9*2* and *CYP2C9*3* alleles. This high frequency has prompted numerous studies aimed at determining the functional effects of these common CYP2C9 variants.

The *CYP2C9*2* allele (Arg144Cys) appears to reduce the metabolism of substrates by 30–50%, but some substrates (e.g., warfarin, acenocoumarol, and phenytoin) are significantly affected in vivo, while others are not. The *CYP2C9*3* allele includes a single base substitution resulting in an amino acid change at residue 359 (Ile to Leu). The CYP2C9.3 protein has substantially reduced catalytic activity toward CYP2C9 substrates (e.g., by $\approx$ 95% of warfarin intrinsic clearance) in vitro. In vivo investigations show that individuals heterozygous or homozygous for the *CYP2C9*3* allele have reduced intrinsic clearance of narrow therapeutic drugs such as warfarin, phenytoin, glipizide, acenocoumarin, and tolbutamide. *CYP2C9*2* and/or *3* have been shown to affect the oral clearance of over 16 different drugs. Consistent with the in vitro functional assays, *CYP2C9*3* results in a larger reduction change in substrate clearance than the *CYP2C9*2* allele. For most substrates, *CYP2C9*3* heterozygous individuals had ~50% of the wild-type total oral clearance and

*CYP2C9*3* homozygous individuals had a 5- to 10-fold reduction. A moderate but significant effect of *CYP2C9*2* was found only for the clearance of certain substrates (*S*-warfarin, acenocoumarol, tolbutamide, and celecoxib). This suggests differences in substrate specificity among the CYP2C9.1, CYP2C9.2, and CYP2C9.3 proteins. Patients who carry these two common variant alleles are at more risk for clinical toxicity from CYP2C9 substrates including phenytoin and warfarin (see later discussion).

Other variant alleles described in the literature that include *CYP2C9*4* (Ile359Thr), *CYP2C9*5* (Asp360Glu), *CYP2C9*7* (Leu19Ile), *CYP2C9*9* (His251Arg), *CYP2C9*10* (Glu272Gly), *CYP2C9*11* (Arg335Trp), *CYP2C9*12* (Pro489Ser), and *CYP2C9*13* (Leu90Pro) are also associated with reduced CYP2C9 activity, but they occur very rarely in the general population. Only one allele, *CYP2C9*6*, results in a frameshift due to deletion at del818A and thus leads to absent CYP2C9 activity. This allele, which was only identified in African Americans at an allelic frequency of 0.6% (but not in Caucasians and Asians), has been shown to drastically slow the elimination of phenytoin and warfarin and probably other substrates. On the basis of in vitro data, the *CYP2C9*8* allele (Arg150His) is believed to increase the activity of the enzyme, but this does not appear to be the case in vivo. In addition to variants in the coding region of the CYP2C9 gene, several other variants have been identified in the promoter, 3'-untranslated and intron of the gene, but they do not seem to significantly influence CYP2C9 function in vivo in the absence of induction.

Warfarin

Prospective clinical trials have unequivocally demonstrated the effectiveness of warfarin in the prevention and treatment of venous and arterial thrombosis, and it is not surprising that this drug is the most widely prescribed oral anticoagulant worldwide (30 million prescriptions in 2004). However, warfarin therapy causes a significant risk of bleeding. Approximately 0−4.8%, 1.2−7%, and >9.6% annual incidence of fatal (intracranial), major, and minor bleeds, respectively, have been reported, and these adverse effects account for 1 in 10 hospital admissions. Major hemorrhages occur most frequently during the first year, and although the risk persists with time, the risk is most prominent during the first month of warfarin therapy. Although warfarin prevents ~20 strokes for each bleed, the fear of intracranial hemorrhagic complications often causes physicians to avoid prescribing warfarin to patients who are likely to benefit from such therapy. In addition, patients vary widely in their response to warfarin, necessitating a wide range of doses (0.5−60 mg/day) to achieve optimal therapeutic anticoagulation responses [International Normalized Ratio (INR) of 2−3]. Therefore, optimizing beneficial effects, while ensuring adequate safety, has been a major challenge of warfarin therapy. A number of factors, including interaction with other drugs, age, diet, and disease, influence a patient's response to warfarin, but inherited differences in warfarin metabolism are also established as an important contributor to this variable response.

Warfarin is a chiral drug that is marketed as a racemate consisting of equal amounts of R- and S-enantiomers. The S-enantiomer is 7- to 10-fold more potent as an anticoagulant than the R-isomer. The major metabolic pathway of S-warfarin, 7-hydroxylation, is catalyzed primarily by hepatic CYP2C9. CYP2C9 variants that reduce S-warfarin metabolism in vitro were first described in 1994. In 1995, evidence was provided that patients who carry the Arg/Cys144 variant genotype (*1/*2) required lower doses of warfarin for anticoagulant control compared to Arg/Arg144 homozygotes (*1*1). In 1996, another variant of CYP2C9 (the Ile359/Leu359 variant, *1*3) was shown to substantially increase the Michaelis−Menten constant (Km), substrate concentration at which velocity is half of the maximum velocity, of warfarin in vitro. By 1997, a patient who exhibited unusual sensitivity to warfarin therapy and could tolerate no more than 0.5 mg of the racemic drug per day had been found to be homozygous for the CYP2C9*3 allele. Warfarin clearance in this patient was substantially decreased compared to control patients receiving 4−8 mg warfarin per day (plasma S:R ratios of 3.9: 1 versus 0.50 ± 0.25:1). That *3 significantly reduces warfarin metabolism in vitro and in vivo was confirmed in 1998.

A strong association of *CYP2C9* genotype with warfarin dosage, risk for minor and major bleeding, INR > 4, and difficulties during induction of therapy, was reported in 1999 ($n = 36$ low dose, $n = 52$ random dose, $n = 100$ control). Subsequently, numerous studies, mostly retrospective in nature, have been conducted and essentially further confirmed the initial findings that patients with *2 or *3 genotypes require lower maintenance and cumulative induction doses and are at an increased risk for supratherapeutic INR (i.e., INR > 3) and bleeding during warfarin induction. More recently, evidence from limited randomized prospective trials suggest that *CYP2C9* genotype−guided warfarin therapy is more efficient and safer than the "average-dose" protocol. The same effect of CYP2C9 genetic polymorphisms on acenocoumarol, a drug used as a substitute to warfarin in some countries, is documented. Collectively, these data have moved CYP2C9 testing toward clinical practice, and it now appears that a combination of multiple retrospective studies and a few prospective trials that demonstrate the superiority of a genotype-driven approach has been sufficient for the FDA to approve a warfarin label change. This may in turn result in more widespread adoption of clinical pharmacogenetic testing before the use of warfarin, and in general.

Phenytoin

Phenytoin remains in wide clinical use for the treatment and prevention of seizures. The drug is widely recognized as being difficult to dose because of its low therapeutic index, which hinders its optimal and safe use. The major and rate-limiting clearance mechanism of phenytoin (80–90%) is 4'-para-hydroxylation primarily by CYP2C9 to 5-(4p-hydroxyphenyl)-5-phenylhydantoin (p-HPPH), with preferential production of the S-enantiomer of p-HPPH, with a minor contribution from CYP2C19 at higher concentrations. Phenytoin has complex nonlinear pharmacokinetics. This property, together with its narrow therapeutic range and the fact that phenytoin therapeutic response and CNS toxicity (e.g., ataxia and nystagmus) are closely related to its plasma concentrations, suggests that small changes in CYP2C9 activity may be clinically important for phenytoin exposure and response, making personalized therapy with phenytoin using pharmacogenetic approaches attractive.

Insufficient para-hydroxylation as a cause for phenytoin toxicity was reported in 1964. That phenytoin metabolism exhibits bimodal distribution in a population study was first suggested in 1997. The first evidence that CYP2C9 genetic variants (typically *1/*3) are associated with decreased maximum elimination rate (by 33–42%) was provided in the mid-1990s. Plasma concentrations of phenytoin were substantially increased in patients carrying the *1/*3 genotype. Subsequent case reports reported substantial elevations (four- to five-fold) in phenytoin AUCs in those with the CYP2C9*3/*3 genotype. Generally, the presence of at least one CYP2C9*2 or *3 allele is associated with one-third lower mean dose requirements for phenytoin. A "gene–dose" effect appears to exist. For example, dose requirements of phenytoin are higher in *1/*1 > *1/*2 or *1/*3 > *2/*3 genotypes. In addition, several case reports indicate associations between CYP2C9 variants and profound changes in phenytoin pharmacokinetics and/or toxicity. These include a fivefold increase in exposure of phenytoin in an African American subject with a CYP2C9*6/*6 genotype, markedly reduced phenytoin elimination, and increased toxicity in carriers of CYP2C9*3 allele (*1/*3 or *3/*3 genotype); more individuals with the CYP2C9*1/*3 genotype had skin reactions to phenytoin compared with nonexposed controls.

That CYP2C9 genetic polymorphisms influence phenytoin pharmacokinetics, dose requirement, and central nervous system adverse effects is compelling. The drug is widely used and has nonlinear pharmacokinetics and a narrow therapeutic range. These properties of phenytoin together with the link between drug phenytoin exposure and response or adverse effects point toward the potential of implementing CYP2C9 genetic polymorphisms in personalized phenytoin therapy. However, further prospective studies are required to better characterize prediction of clinical outcomes of safety and efficacy by each specific genotype. It has to be also demonstrated that pharmacogenetic approaches are helpful compared to conventional approaches (e.g., clinical assessment and therapeutic drug monitoring). Clearly, CYP2C9 alone cannot explain phenytoin response variability, and its role should be studied in the context of other genetic variation (e.g., target or transport) and nongenetic factors (e.g., drug interaction).

CYP2C19

The CYP2C19 gene is mapped to chromosome 10 (10q24.1-q24.3) and contains nine exons that code for a protein consisting of 490 amino acids. This important enzyme is expressed primarily in human liver, and to a smaller extent in several extrahepatic tissues that include the gut wall. It is important in the metabolism of widely used drugs such as the proton pump inhibitors (PPIs) (omeprazole, esomeprazole, lansoprazole, and pantoprazole), several antidepressants, diazepam, phenytoin, proguanil, clopidogrel, voriconazole, nelfinavir, thalidomide, and cyclophosphamide (Desta et al., 2002). As a result of genetic polymorphisms in the CYP2C19 gene and nongenetic factors (e.g., drug interactions), wide interindividual variability is seen in the in vivo activity of CYP2C19. Therefore, from a translational perspective, identification of the mechanisms and causes for interindividual variability in CYP2C19 activity and developing means to prospectively predict them is important to optimize therapy with its substrates.

It is now firmly established that the high interindividual and interethnic variability in the pharmacokinetics of CYP2C19 substrates is due in significant part to polymorphisms in the CYP2C19 gene (Goldstein, 2001; Desta et al., 2002). The first individual to be identified as a PM of mephenytoin (later identified as being metabolized via CYP2C19) was in 1979. Subsequent studies confirmed that S-mephenytoin hydroxylase is polymorphically expressed and that other drugs such as omeprazole and proguanil have metabolism that cosegregates with that of mephenytoin. The molecular basis of CYP2C19 genetic variability was shown first in 1994 subsequent to the cloning of the gene. With over 20 alleles and allelic variants, the CYP2C19 gene is one of the highly polymorphic P450 genes (http://www.cypalleles.ki.se/cyp2C19.htm). These alleles include those that lead to complete absence of enzyme function (CYP2C19*2 to *8); those that are associated with decreased function (CYP2C19*9, *11, and *13) (Desta et al., 2002); and a novel polymorphism

(*CYP2C19**17 consisting of −806C > T and −3402C > T), with a frequency of 18% in Caucasians, that is associated with an approximately 40% increase in the metabolism of S-mephenytoin, omeprazole, and citalopram (Sim et al., 2006). Therefore, individuals can be categorized phenotypically as poor, intermediate, extensive, or "ultrarapid" metabolizers. Of the null alleles, *CYP2C19**2 and *3 are by far the most common and account for the majority of PMs. For example, the allelic frequency of the most common variant, the *CYP2C19**2, has been reported to be approximately 30%, 17%, and 15% in Chinese, African Americans, and Caucasians, respectively: the *CYP2C19**3 allele is more frequent in Chinese (≈5%) than in Caucasians and African Americans (<1%). The *CYP2C19**2 allele together with the *3 allele accounts for almost all PMs in Asians, while the *2 allele accounts for 75−85% PMs in other populations (Desta et al., 2002). The incidence of CYP2C19 PMs (determined by phenotype and genotype) carrying two defective *CYP2C19* genes is 1−8% in Caucasian and African populations, 13−23% in Asians, and 38−70% in Vanuatu Islands in eastern Melanesia (Desta et al., 2002).

Proton Pump Inhibitors

The clinical relevance of CYP2C19 genetic polymorphisms can now be assessed from data presented in a number of different populations. It is well established that in extensive metabolizers (EMs) approximately 80% of doses of the PPIs such as omeprazole, lansoprazole, and pantoprazole are cleared by CYP2C19 (Andersson, 1996); about sixfold higher exposure to these drugs is observed in PMs than in EMs of CYP2C19. As a result, PMs achieve greater acid suppression as well as higher eradication of *Helicobacter pylori* infection and healing rates for both duodenal and gastric ulcers. Therefore, genotyping for the common alleles of CYP2C19 before initiating PPIs for the treatment of reflux disease and *H. pylori* infection has been suggested to be a cost-effective tool to select appropriate duration of treatment and dosage regimens.

Clopidogrel

Clopidogrel is widely used in the prevention and treatment of thrombotic complications following stroke, unstable angina, MI, and coronary stent placement. It is a prodrug that requires conversion to its active metabolite by the CYP system before it exerts its antiplatelet effects. The lack or diminished antiplatelet response in a substantial number of patients receiving clopidogrel is a major problem during its clinical use. The in vitro oxidative metabolism of clopidogrel is complex, but researchers should not be daunted: in vivo multiple studies have now shown that *CYP2C19**2 decreases the antiplatelet response of clopidogrel (Brandt et al., 2007; Giusti et al., 2007). The influence of CYP2C19 genetic variation on clopidogrel response has major implications because clopidogrel nonresponsive platelets are at risk for thrombotic events, with devastating potential outcomes to patient. Genotype-guided studies that demonstrate an effect of CYP2C19 genotype on clinical outcomes beyond the surrogate of in vitro platelet aggregation, such as stroke and stent restenosis, would be required to implement CYP2C19 genotyping in clinical practice, but these have not yet been reported.

Cyclophosphamide

Cyclophosphamide is a prodrug that requires metabolic activation by CYP enzymes to 4-hydroxycyclophosphamide. Multiple CYPs have been implicated in this activation, including CYP2A6, 2B6, 2C19, 2C9, 3A4, and 3A5, but CYP2C19 appears to be a key enzyme, particularly at low cyclophosphamide concentrations. Thus, in a recent study we conducted in lupus nephritis patients taking cyclophosphamide, we found that *CYP2C19**2 is a predictor of premature ovarian failure and progression to end-stage renal disease (Takada et al., 2004), findings that have now been validated by other authors (Singh et al., 2007). While the precise mechanism of this effect is unclear, these observations suggest a line of in vitro investigation to determine the cause of ovarian toxicity, and thus to prevent it in women at risk, an example of "reverse" translational research.

CYP2D6

CYP2D6 is by far the most studied genetically polymorphic enzyme involved in human drug metabolism. Conceptually, the first observations of multiple genetic variants contributing to one phenotype, of the possibility of genetic "ultrarapid" metabolism, of variants that alter some but not all substrates, and of functionally significant copy number polymorphisms have all been made first with CYP2D6. This key hepatic enzyme is responsible for the metabolism of a large number of clinically important drugs, and its history carries a significant lesson for the study of other genetic variants that influence drug response. The history of the discovery of this genetic polymorphism is instructive to translational scientists, in that it derived from the observation by astute physicians of unusual "outlier" responses to medications prescribed for hypertension, arrhythmias, and depression. This continues to be the case for this enzyme, as new substrates such as tamoxifen (Stearns et al., 2003) continue to be identified in this way. In 1967, Sjöqvist and coworkers in Sweden identified two groups of patients based on the plasma concentrations of nortriptyline and desipramine after a fixed dose, and later showed that this difference was genetic in nature (Hammer and Sjöqvist, 1967). In the mid-1970s, Smith's group in the United Kingdom discovered that 4-hydroxylation of debrisoquine in humans is polymorphic and that the PM phenotype is inherited as a single autosomal recessive trait (Mahgoub et al., 1977). The terms "extensive metabolizer" and "poor metabolizer" described at that time continue to be used today. At the same time, the group led by Michel Eichelbaum identified the sparteine polymorphism after the clinical observation of two patients with defective metabolism in 1975. Ten to fifteen years after these differences in phenotypes were discovered, the genetic basis underlying the bimodal debrisoquine—sparteine phenotype was elucidated by collaborative work between Urs Meyer and Frank Gonzalez (Gonzalez et al., 1988) in which they cloned the human CYP2D6 cDNA, documented altered RFLP patterns in slow metabolizers of debrisoquine, and later identified what we now refer to as the CYP2D6*3 and CYP2D6*4 alleles.

The *CYP2D6* gene locus is notably complex. The coding exons are located on chromosome 22 and are flanked by a pseudogene (*CYP2D7*) that has complicated sequencing efforts. Genotyping efforts by a wide range of investigators continue to identify increasingly rare new alleles, but it is of note that no organized resequencing effort in the major human ethnic groups has been carried out and as a result the haplotype structure of the gene remains unclear in 2008. A total of 67 discrete alleles have been identified by the human CYP nomenclature committee (www.cypalleles.ki.se/cyp2d6.htm). The alleles designated as *3—*8 and *11—*16 result in no enzyme activity, whereas other alleles such as *2, *10, *17, and *41 (Cai et al., 2006), which are common in specific ethnic groups, result in a CYP2D6 enzyme whose activity is reduced, but not eliminated (http://medicine.iupui.edu/flockhart/2D6.jpg). The diversity of CYP2D6 alleles suggests that nucleotide arrays or similar methods are needed to efficiently test for the most important CYP2D6 alleles in humans, but in general it is clear that more than 95% of PMs can be identified with methods that test for seven or eight alleles, including specifically the *3—*9 common "knockout" alleles, the *10 allele in Asians, and the *17 and *41 alleles in Africans. A significant number of alleles have now been reported to have multiple copy number polymorphisms, including the *2, *4, and *17 (Cai et al., 2006).

Substrates

A total of 48 drugs have been reported to be metabolized at least in part by this enzyme as of 2008 (www.drug-interactions. com). It is surprising how few of these have been studied sufficiently to recommend CYP2D6 testing in clinical practice, but the reasons for this relate to the difficulty of obtaining germ line DNA in clinical trials and of conducting randomized, prospective trials that compare a genotype-driven approach to standard care. As with all pharmacogenetic tests, the most valuable are those that address a situation where the prediction of response is particularly difficult. These include tests used specifically in cancer and in psychiatry.

Tamoxifen

After three decades of study in which tamoxifen was believed to be metabolized primarily by CYP3A, the group led by Desta et al. (2004) performed a comprehensive evaluation of the primary and secondary metabolism of tamoxifen in humans and showed that metabolism to the active metabolite endoxifen is catalyzed almost exclusively by CYP2D6. These observations led to clinical trials which showed that the concentration of endoxifen was associated with CYP2D6 genotype (Jin et al., 2005) and to retrospective examination of randomized prospective trials in which CYP2D6 was associated with breast cancer recurrence (Goetz et al., 2007), such that PMs had approximately twofold greater risk of recurrence, a risk that was confirmed in biobank studies in Germany (Schroth et al., 2007) and in Japan (Kiyotani et al., 2008). Large prospective trials involving thousands of patients have still not been subjected to this analysis, however, and there remains no prospective trial that tests the hypothesis that a CYP2D6-guided approach to the endocrine treatment of breast cancer is superior to a standard of care in which aromatase inhibitors are generally considered superior to tamoxifen for treatment of hormone-receptor-positive breast cancer in postmenopausal women. In situations like this, where new and expensive randomized trials are unlikely to be conducted, modeling approaches can be a useful translational tool, and thus, a Markov model has been proposed, which predicts that tamoxifen would be superior to aromatase inhibitors in many CYP2D6 extensive metabolizers (Punglia et al., 2008).

Codeine

The metabolism of codeine to the active analgesic morphine is mediated by CYP2D6, and morphine cannot be detected in individuals who carry the PM phenotype (Caraco et al., 1996). A number of studies have demonstrated that the pharmacodynamic response to codeine is associated with CYP2D6 genotype (Stamer and Stuber, 2007; Caraco et al., 1999), and a number of case reports have made clear the difficulty of managing pain in patients who are poor (Foster et al., 2007) or ultrarapid metabolizers (Madadi et al., 2007). A study in patients treated with codeine and hydroxyurea for painful sickle cell crises indicated that failing codeine therapy for a pain crisis while taking hydroxyurea is associated with an increase in reduced-functioning CYP2D6 alleles. These investigators recommend genetic analysis or trial of a non-CYP2D6 analgesic for these children (Alander et al., 2002). While no organized prospective trials of CYP2D6 genotyping in patients with pain have been reported, CYP2D6 genotyping of patients who experience inadequate response to codeine is becoming more commonplace, the rationale being to help narrow the differential diagnosis of inadequate analgesic response and allow consistent treatment with effective analgesics when possible.

Antidepressants

Similarly, although it has been clear for 30 years that tricyclic antidepressants such as desipramine and nortriptyline are metabolized primarily by CYP2D6, it remains the case that no large randomized trial has tested the possibility that pharmacogenetic testing might reduce the toxicity of these drugs, or improve outcomes. That said, recent data suggest that CYP2D6 PMs may switch antidepressant more and that they may use reduced doses (Bijl et al., 2008), and it has been clear for many years that CYP2D6 PMs may experience more adverse reactions to antidepressants, including venlafaxine (McAlpine et al., 2007) and mirtazapine (Brockmoller et al., 2007). Since patients often can perceive the anticholinergic effects of high serum concentrations of antidepressants, the presence of the ultrarapid CYP2D6 phenotype may be more clinically important because of the possibility of subtherapeutic dosing (de Leon, 2007).

CYP3A5

The CYP3A subfamily of enzymes is the most abundant in the liver as well as in the gut wall and metabolizes 45—60% of currently used drugs, as well as many endogenous and structurally different environmental chemicals. The human CYP3A locus comprises four functional genes (CYP3A43-CYP3A4-CYP3A7-CYP3A5), with three pseudogenes (CYP3AP1—3), all located in one locus at chromosome 7q21-q22.1. CYP3A4, which was identified in 1986, is by far the most abundant and the clearance mechanism for the majority of CYP3A substrates. CYP3A5 shows catalytic activity toward almost all CYP3A4 substrates in vitro, but it represents the dominant clearance mechanism for very few drugs (Lamba et al., 2002). CYP3A7 is predominantly expressed in the fetus while the level of expression in adults is small. CYP3A43 was identified in 2000, but its role in human drug metabolism or whether it is fully translated to protein remains unclear. An interesting characteristic of all substrates of CYP3A is high intersubject variability in pharmacokinetics. This likely changes response to these drugs. Although several genetic variants have been identified (over 39 CYP3A4 alleles, comprising 65 SNPs), none of these variants explains the large intersubject variability of CYP3A4. A-392G transition (CYP3A4*1B) located in the 5′-regulatory region (0% in Asian, 5% in Caucasians, and 54% in Africans) has been studied most, but functional studies are inconclusive or with minimal impact in vivo. Other rare variant alleles with some functional consequences have been identified, but none of these variants is the major cause of interindividual differences in CYP3A-mediated drug clearance in the general population. In general, the cause for variable CYP3A4 activity appears due to environmental exposure (inducer and inhibitor drugs and chemicals).

Compared to CYP3A4, CYP3A5 plays a minor role in the clearance of most drugs (Lamba et al., 2002). Although many of the substrates of CYP3A4 are metabolized in vitro by CYP3A5, the role of CYP3A5 in vivo remains largely unknown, and this is in part because of the lack of appropriate phenotyping tools (inhibitors or substrates) that specifically distinguish this enzyme from CYP3A4. However, the availability of CYP3A5 "expressors" and "nonexpressors" in human liver tissues has been instrumental in dissecting the contribution of this enzyme to human drug metabolism. CYP3A5 expression is highly polymorphic. Heterogeneity in CYP3A5 expression was observed immediately after its discovery because protein or mRNA expression was found in only about 20% of human liver and extrahepatic tissues. The main reason for this variable expression is variants in the CYP3A5 gene. Upon sequencing of cDNA sequences complementary to CYP3A5 mRNA in human liver tissues using reverse transcriptase PCR, evidence of retention of introns was obtained, the most common of which was retention of intron 3. This is the result of splice variation due to an A > G change (intron 3) at position 6986 (*3 allele) of the CYP3A5 gene (Kuehl et al., 2001). To date, 10 CYP3A5 alleles, consisting of 22

SNPs, have been identified. Of the variants reported, CYP3A5*3 (g.6986G) is the only one found in all ethnic groups tested. The allelic frequency of *3 varies from 35% to 48% in African Americans and Africans, 85—98% in Caucasians, and 60—75% in Asians. The CYP3A5*6 and CYP3A5*7 variants that affect the CYP3A5 expression are relatively frequent in African subjects but are rare in Caucasian subjects. The remaining CYP3A5 genetic variants are rare.

In contrast to CY3A4, CYP3A5 expression in humans exhibits a bimodal distribution, with the proportion of CYP3A5 "high expressers" and "low expressers" varying depending on the ethnic background. In Caucasians the concordance between the presence of the *3 allele and increased CYP3A5 expression is high. In African Americans the relationship is less robust because of additional polymorphisms (e.g., *6 and *7 which occur at frequencies of 13% and 10%, respectively). Both these variants have been found in samples homozygous for the *3 allele.

Tacrolimus

Tacrolimus, a potent immunosuppressive macrolide lactone, is a relatively specific inhibitor of T lymphocyte proliferation and is widely used in the prophylaxis of organ rejection after allogenic solid organ transplantation. However, the optimal clinical use of this drug is hampered by a number of pharmacological challenges: narrow therapeutic range; dose-related adverse effects, including, for example, the risk of nephrotoxicity and increased susceptibility to infections; and transplant rejection and graft loss or damage at low plasma concentrations. This situation is further complicated by the large inter- and intraindividual variations in drug pharmacokinetic characteristics. These characteristics make it an ideal candidate for efforts to improve therapy using pharmacogenetic testing. To avoid over- or underimmunosuppression, therapeutic drug monitoring is performed. Dosage regimens are usually adjusted to reach predefined target blood concentrations that have been associated with the optimal balance between efficacy and toxicity, but pharmacogenetic predictors have received focus recently.

It is well known that a significant part of the interindividual pharmacokinetic variability for tracrolimus results from variability in the expression and/or function of the CYP3A enzymes and of the multidrug transporter P-glycoprotein. There appears a significant contribution of CYP3A5 to tacrolimus metabolism. Several studies in organ transplant patients (mostly retrospective in nature) have shown that the CYP3A5 genotype is a good predictor of dose requirements and plasma concentrations. Summarizing these results, it is now well established that carriers of CYP3A5*1 alleles (CYP3A5 "high expressers") are associated with relatively higher doses of tacrolimus (30—50% higher in patients expressing the CYP3A5 enzyme) to achieve target blood concentrations than homozygous carriers of the CYP3A5*3 allele (CYP3A5 "low expressers"). As with tacrolimus, CYP3A5 genotype appears to influence sirolimus dose requirement. In contrast to the findings for tacrolimus, most studies do not support a relationship between the CYP3A5 genotype and cyclosporin disposition, dose requirements, or clinical response, suggesting the effect of this gene on cyclosporine is minimal.

Despite the clear statistical relationship between CYP3A5 expression and tacrolimus (or sirolimus) pharmacokinetics, there remains considerable variability in the dose-adjusted concentrations of tacrolimus achieved within nonexpressors (*3/*3) and overlap frequently exists between the genotype groups. Data correlating clinical outcome measures are lacking, but lower trough concentrations of tacrolimus have been reported in carriers of the CYP3A5*1 allele (Dai et al., 2006). Available data suggest that genotyping may prove to be a useful adjunct to, but not a replacement for, therapeutic drug monitoring because CYP3A5 genetic variation cannot explain all variability to tacrolimus. Many gene variants implicated in absorption, metabolism, transport, distribution, drug receptors, and targets must be considered and the quantitative contribution of CYP3A5 genetic variation in this context must be evaluated. Moving retrospective studies to prospective randomized studies should formally test whether a genetic approach, in association with therapeutic drug monitoring, may limit the pharmacokinetic variations observed among individuals.

Vincristine

The metabolic pathways of vincristine and the enzymes catalyzing them have been elucidated recently after the drug has been on the market for more than 30 years. In contrast to many other drugs, there is compelling evidence that CYP3A5 is the main enzyme catalyzing vincristine metabolism to M1 at least using in vitro expressed enzymes and human liver microsomal preparations (Dennison et al., 2006). Clinical studies in African and Caucasian patients hint at an important effect of the CYP3A5 polymorphism, which has a greater incidence of the rapid metabolizer genotype in Africans. Four percent of total vincristine doses administered to Caucasian patients were reduced due to vincristine-related neurotoxicity compared to 0.1% given to African Americans ($p < .0001$; Renbarger et al., 2008). While much work remains, polymorphic expression of CYP3A5 may well be an important determinant of response to vincristine. Prospective trials to quantify the association between CYP3A5 genotype and vincristine pharmacokinetics are ongoing. Well-designed clinical trials to test the hypothesis that this genotype might associate with clinical outcomes of response and neurotoxicity will be required to demonstrate clinical utility in this setting.

N-acetyltransferase 2

Human arylamine NAT catalyzes the transfer of an acetyl group from acetyl coenzyme A to drugs and several other chemicals with aromatic amines, heterocyclic amines, or hydrazine in their structure. Although humans express both forms of NAT and both exhibit genetic heterogeneity, the clinical significance of the polymorphism is better understood with NAT-2 than for NAT-1. Acetylation is an important route of metabolism and elimination for a large number of clinically important drugs, including isoniazid, dapsone, procainamide, sulfonamides, hydralazine, and phenelzine. In addition, NAT-2 acetylates several aromatic and heterocyclic amines from the environment and diet into carcinogenic and mutagenic intermediates, thus implicating this enzyme function with the occurrence of disease.

The *NAT-2* gene is located in the short arm of chromosome 8 (region 8p22). *NAT-2* is an unusual gene because it consists of only two exons, the second of which is an open reading frame (i.e., protein-coding regions) with no introns. The molecular basis of the *NAT-2* polymorphism was elucidated almost 40 years after the initial isoniazid acetylation defect was observed in 1954. The gene was cloned in 1991, which allowed the identification of two common alleles (now known as *NAT-2*5* and *6*) that are associated with slow acetylator phenotype. As of February 29, 2008 (see http://louisville.edu/medschool/pharmacology/Human.NAT2.pdf), there are 12 alleles and over 50 allelic variant alleles of the *NAT-2* gene, with different functional consequences. Most variant *NAT-2* alleles involve two or three point mutations. Besides the wild-type reference haplotype *NAT-2*4*, other variant alleles (e.g., *NAT-2*12A-C, *13A, *18*) define rapid acetylator phenotype. However, other alleles that include *NAT-2*5A-J, *6A-E, *7, *12D, *14A-G, *17,* and *19* clusters are associated with slow acetylation phenotype.

Isoniazid

NAT-2 is the rate-limiting step in acetylating isoniazid to acetylisoniazid, which is further hydrolyzed to acetylhydrazine and then inactivated via acetylation by *NAT-2* to diacetylhydrazine. Acetylhydrazine exists in equilibrium with hydrazine, which appears to be implicated in isoniazid-induced hepatotoxicity. The discovery of genetic defects in acetylation capacity is intimately related with the introduction of isoniazid for the treatment of tuberculosis in 1952 and the subsequent demonstration of extensive variability between individuals and populations in the pharmacokinetics of the drug (Meyer, 2004) due to differences in the individual's ability to convert isoniazid to acetylisoniazid. By 1954, the slow acetylator phenotype, which was shown to be an autosomal recessive trait based on family studies, was linked with increased risk for isoniazid neurotoxicity (peripheral neuropathy). The biomodal distribution of isoniazid acetylation was first reported by Evans et al. (1960), in which the authors identified two distinct phenotype groups: rapid and slow acetylators. The incidence of slow acetylator phenotype differs with different ethnic group (e.g., 40–70% of Caucasians and African Americans, 10–20% of Japanese and Canadian Eskimos, over 80% of Egyptians, and certain Jewish populations). Clearly, these early data hold historical importance within the field of pharmacogenetics of drug metabolism. However, except for isoniazid-induced hepatotoxicity, the relevance of *NAT-2* genetic polymorphism to isoniazid response is not clear. Although slow acetylators have been shown to be at increased risk for isoniazid adverse effects (neurotoxicity), genetic tests are not required to predict this adverse effect because this clinical problem can be effectively prevented by coadministration of pyridoxine, a cheap and safe supplement. Neither are there sufficient data to support genetic tests to anticipate the efficacy of the drug in the treatment of tuberculosis. To achieve similar isoniazid exposure, current standard doses presumably appropriate for patients with one high-activity *NAT-2* allele may be decreased or increased by ~50% for patients with no or two such alleles, respectively (Kinzig-Schippers et al., 2005), but there are no trials available that test the clinical value of this approach. Several studies have now shown that isoniazid-induced hepatitis is higher in slow acetylators, probably due to excessive accumulation of acetylhydrazine and eventually hydrazine.

Thiopurine Methyltransferase

The genetic polymorphism in the thiopurine methyltransferase (TPMT) gene first reported by Weinshilboum et al. in 1980 is widely viewed as the first clinically important example of a pharmacogenetic variant. The basis for this lies in the important clinical consequences of this relatively rare polymorphism. Approximately 89% of Caucasian subjects are homozygous for the trait of high levels of TPMT activity, approximately 11% are heterozygous, with intermediate activity, and 1 out of every 300 subjects is homozygous for the inherited trait of extremely low or undetectable activity. The human gene was first cloned, and the first common polymorphism characterized 16 years after the discovery of the phenotypic polymorphism in 1996 (Szumlanski et al., 1996). Subsequent population-based studies have demonstrated that the *3A allele accounts for 55—75% of all variant *TPMT* alleles in Caucasians. At least eight separate polymorphisms associated with very low TPMT activity have now been reported. Seven of those alter encoded amino acids, and one involves a mutation at the acceptor splice site between *TPMT* intron 9 and exon 10 (Weinshilboum et al., 1999). The biochemical mechanisms underlying these changes in activity have been studied intensively and are clearly diverse, including a number of mechanisms that reduce protein stability and compromise enzyme function.

The clinical importance of this enzyme and the inherited variability in its activity derive in part from the narrow therapeutic ranges of the drugs metabolized—6-mercaptopurine and azathioprine—and in part from the severity of the illness they are used to treat, namely acute lymphoblastic leukemia in children. Low activity of TPMT results in a shunting of thiopurine metabolism down a route toward the toxic thioguanine nucleotides, and as a result patients can experience life-threatening granulocytopenia when routine doses of 6 mercaptopurine or azathioprine (Lennard and Lilleyman, 1989) are used. It is also possible that children with lower TPMT have a higher relapse rate (Lennard et al., 1990) and this adds particular urgency to the argument for pharmacogenetic testing in this context. Large clinical trials in which pharmacogenetic-guided therapy is compared with standard therapy have not been conducted, but the severity of the consequences of not testing has made a strong argument for clinical testing.

It is important to note, however, that the clinical use of TPMT pharmacogenetic testing has been more widely adopted by adult gastroenterologists treating inflammatory bowel disease than by pediatric oncologists. This may in part reflect the inverse relationship between the willingness of physicians to tolerate adverse drug reactions and the morbidity of the illness under treatment: adverse effects seem more tolerable in patients with more life-threatening diseases. Since pediatric oncologists are willing to use the simple measure of the number of white cells in the blood as a surrogate for this toxicity, this illustrates an important principle in any predictive laboratory testing in therapeutics: pharmacogenetic tests will be most valuable, and therefore most worth investing time and energy in researching, when our collective clinical ability to predict toxicity is currently limited, and when the toxicity that results is unacceptable.

UDP-Glucuronosyltransferase

UGTs are quantitatively the most important phase II enzymes in mammals. These enzymes are expressed in the liver but also in extrahepatic tissues. UGTs catalyze the transfer of the glucuronic acid (sugar) moiety from the cosubstrate uridine 5'-diphospho-α-D-glucuronic acid to the substrate (often lipophilic aglycone) bearing a suitable functional group (e.g., OR, -SR, -NR'R'' or CR) and forms ß-D-glucuronide. Such a glucuronide metabolite has often increased aqueous solubility and increased recognition by biliary and renal transporters and is inactive. Rarely, glucuronides maintain or increase activity (e.g., morphine and retinoic acid glucuronides).

The human UGT superfamily comprises two families (UGT1 and UGT2) and three subfamilies (UGT1A, UGT2A, and UGT2B). Classification is based on amino acid sequence homology where enzymes in each family and subfamily share at least 50% and 60% homology in their amino acid sequences. At least 18 such human enzymes have been identified. A single UGT1A gene locus located on chromosome 2 (2q37) encodes nine functional proteins UGT1A1, 1A3—1A10. The UGT1A gene consists of at least 13 variable exons (alternative splicing variants of exon 1) at 5'-end that encode unique N-terminus that confers specificity and four 3'-end exons common for every UGT1A RNA. UGT2A and 2B enzymes are coded by individual genes clustered in chromosome 4 (4q13). As described previously, these enzymes are mainly expressed in the liver (UGT1A1, 1A3, 1A4, 1A6, 1A9, 2B4, 2B7, 2B10, 2B11, 2B15, 2B17, and 2B28). Others are expressed in extrahepatic tissues (one-third of people express 1A1, 1A3, and 1A6 in the gastrointestinal tract; UGT1A8 and UGT1A10 exclusively in the gastrointestinal tract and colon; UGT1A7 in the esophagus, stomach, and lung; UGT2A1 mainly in the nasal epithelium). All UGTs investigated to date display marked interindividual variation in expression among tissues or in vivo.

UGTs metabolize clinically important drugs (or their metabolites). Approximately 35% of drugs eliminated by phase II are metabolized by UGTs. Often metabolism by UGTs renders the glucuronide inactive and more easily excreted from the body via the urinary and biliary tracts. They are also important in the detoxification of environmental chemicals as well as endogenous compounds such as bilirubin, biogenic amines, steroid and thyroid hormones, fatty acids, and bile acids. Therefore, UGTs are important in drug efficacy and toxicity, chemical toxicity, and homeostasis of endogenous molecules/disease.

UGT1A1 metabolizes several drugs and endogenous compounds (e.g., bilirubin and estrogen) and is the most extensively studied enzyme of the UGT superfamily. The expression of this enzyme is highly variable, and genetic polymorphisms in the gene contribute to this. Over 30 genetic variants, many of which influence its function, have been reported. UGT1A1 is the primary enzyme responsible for the glucuronidation of bilirubin. Functional deficiencies of the UGT1A1 enzyme result in accumulation of serum level of total bilirubin, producing hyperbilirubinemic syndromes. Three grades of UGT1A1 deficiency occur in humans:

1. Crigler—Naijar syndrome (type I, reported in 1952), which occurs due to complete absence of UGT1A1 activity, is characterized by excessive accumulation of bilirubin (serum 20—50 mg/dL), and is associated with fatal encephalopathy.
2. Crigler—Naijar syndrome (type 2, reported in 1962), which occurs due to genetic lesions in the exon sequences of UGT1A1, results in severe but incomplete lack of UGT1A1 activity. The impact on bilirubin plasma levels is intermediate (7—20 mg/dL).
3. Gilbert syndrome (the mildest form, bilirubin normal to 5 mg/dL), which is due to reduction in hepatic UGT1A1 activity by 30%, is by far the most common syndrome related to UGT1A1 deficiency (0.5—19% in different populations). The classical picture of Gilbert syndrome is usually associated with TA repeats (5, 6, 7, and 8) in the UGT1A1 promoter polymorphism, the binding site for the transcription factor IID. Presence of seven repeats (TA7) compared to the normal genotype of six (TA6) repeats results in the variant allele UGT1A1*28 (allelic frequency $\approx$ 38%), which is associated with reduced gene transcription efficiency and overall enzyme activity. Variability in the number of TA repeats affects the expression levels of UGT1A1. Thus, the UGT1A1*28 is associated with reduced enzyme expression. Another variant that is associated with Gilbert syndrome and is common in Asians (11—13%) is UGT1A1*6.

The clinical consequences of UGT1A1 genetic polymorphisms can be best illustrated using irinotecan as an example. Irinotecan (CPT-11), a semisynthetic analog of the natural alkaloid camptothecin, is a topoisomerase I inhibitor and a widely used drug in the treatment of solid tumors, such as metastatic colon cancer and lung cancer. Despite its proven efficacy (prolongs survival), irinotecan has a narrow therapeutic range, and about 20—35% of patients experience dose-limiting severe diarrhea and myelosuppression. Irinotecan is a prodrug, which requires hydrolysis by carboxylesterases in normal tissues and tumors to its active metabolite, SN-38, a potent topoisomerase I inhibitor. Since irinotecan toxicities and efficacy are associated with SN-38 concentrations and this metabolite in plasma is highly variable among patients, it was important to understand factors that control not only its formation but also elimination.

SN-38 is primarily cleared by metabolism UGT-mediated glucuronidation. Slowed elimination as a cause for irinotecan toxicity was reported in 1994. A link between UGT1A1 deficiency and irinotecan-related toxicity was reported in 1997 when two patients with Gilbert syndrome exhibited enhanced risk for irinotecan toxicity. Since it was known at that time that mild hyperbilirubinemia was the cause for Gilbert syndrome and that this was due to deficiency of hepatic UGT1A1 activity, it was thought that both irinotecan and bilirubin may be metabolized by the same enzyme. Indeed, subsequent in vitro studies showed that glucuronidation of irinotecan and bilirubin correlate significantly in a panel of human liver microsomes. In 2000, the association of UGT1A1 polymorphisms and the accumulation of high levels of SN-38 and the associated diarrhea and leukopenia have been reported. Typically, UGT1A1*28 was suggested to be a significant risk factor for severe toxicity by irinotecan. It is now known that UGT1A1*28 and/or other variants are associated with reduced UGT1A1 expression and inducibility, increased SN-38 exposure and adverse effects, and probably with reduced efficacy.

Butyrylcholinesterase

Deficiency in butyrylcholinesterase (pseudocholinesterase) (BCHE) is probably one of the first and most widely recognized clinical examples of pharmacogenetics of drug metabolism that affects drug response. This enzyme hydrolyzes succinylcholine (suxamethonium), a short-acting and widely used depolarizing neuromuscular blocking agent that is used as a muscle relaxant as an adjuvant to general anesthesia during surgery. Succinylcholine is rapidly hydrolyzed by the human butyrylcholinesterase located primarily in the plasma and liver. This drug is popular because of its short duration of action ($\sim$ 30 min) as a result of rapid metabolic degradation by this enzyme, but some patients treated with succinylcholine experience prolonged muscular relaxation and apnea, a serious and potentially lethal adverse response that requires oxygen and artificial ventilation, lasting as long as several hours (6—8 h) after discontinuation of the infusion. A similar adverse effect was also noted with mivacurium, another muscle relaxant and a substrate for the same enzyme. The exaggerated effect of these drugs is due to the inheritance of an "atypical" form of BCHE, due to polymorphisms in the *BCHE* gene. The *BCHE* gene is located at chromosome 3, 3q26.1-q26.2. The gene consists of four exons separated by three large introns and codes for a protein with 574 amino acids. So far more than 58 alleles and allelic variants have been identified at the cholinesterase gene locus, but not all of them have been fully studied. In general, these mutations produce enzymes with different levels of catalytic activity. Variants that encode the most common atypical form of the enzyme include a cSNP, G209 > A, of the BCHE gene. This variant causes an amino acid substitution at codon 70 from aspartic acid for a glycine residue (D70G), forming the "atypical" variant of BCHE with altered active site and insensitive to inhibition by dibucaine. The atypical variant has about 30% lower enzymatic activities. Patients homozygous for this variant experience prolonged apnea after administration of the muscle relaxants.

About 96% of Caucasians are homozygous for the most common (usual) *BCHE* allele (UU) that codes for normal (typical) enzyme and are sensitive to dibucaine inhibition, while ≈4% have at least one abnormal allele that causes production of an enzyme with either altered affinity or decreased quantity. Homozygote variants that result in BChE activity that is low enough to prolong apnea meaningfully occur with a frequency of 1 in 3500 patients, while the S variant (atypical) associated with no cholinesterase activity is even more rare (1 in 100,000). The heterozygous form of this variant has about 30% lower activity than the wild-type. Other genotypes have modestly prolonged responses that will significantly increase the duration of action of the muscle relaxants only if acquired deficiency occurs concurrently with other conditions. Recently, two novel mutations in *BCHE* were identified in three families, a member of which has experienced severely prolonged duration of action of succinylcholine.

Inherited deficiency can be evaluated by assay for plasma cholinesterase activity with different substrates and testing the degree of inhibition of this activity with a well-known inhibitor, such as dibucaine and fluoride. The principle of this approach is based on the fact that atypical cholinesterase hydrolyzes various substrates at considerably reduced rates, and the pattern of affinities of these enzymes to substrates and inhibitors differs from that of normal cholinesterase; there may be as much as a 100-fold decreased affinity of this abnormal enzyme for succinylcholinesterase when compared with the normal enzyme. Unlike other drug-metabolizing enzymes, which require administration of a probe drug, these tests involve a single blood test. For example, dibucaine, a local anesthetic agent, is stable in the presence of cholinesterases. The typical (normal) cholinesterase is 20-fold more sensitive to inhibition by dibucaine than the atypical enzyme, which is relatively resistant to inhibition. This has led to the use of the dibucaine number as a method of quantitating cholinesterase activity. This number represents the percentage inhibition by dibucaine under standardized conditions using benzoylcholine as the substrate and 10 μM dibucaine. Using this system, most individuals cluster in a normal distribution around a high dibucaine number (about 80), while a small group of individuals will have intermediate levels (40–70) and very few individuals will have very low levels (<20). Other variants of cholinesterase include one sensitive to dibucaine but resistant to fluoride inhibition, one that is qualitatively altered so that it is functionally inactive (the so-called silent allele), and one whose enzyme activity is two to three times higher than normal and associated with succinylcholine resistance. Although genotyping tests are available to identify variants in the *BCHE* gene and use for research purposes, they are not routinely used clinically.

CONCLUSIONS

An increasing number of drug-metabolizing enzymes have been shown to result in large pharmacokinetic changes through a variety of different mechanisms. Drugs that are most affected are those that have a dominant route of clearance by a genetically polymorphic enzyme. The effects of such pharmacokinetic changes are most important in settings where clinically important pharmacodynamic change ensues. Pharmacogenetic tests that are most valuable are those for specific drugs for which the prediction of activity or adverse effects is important and difficult to anticipate given our *current* clinical tools and technologic capability. Although data from a variety of platforms documenting a wide range of genetic variability are being accumulated rapidly and clinical genetic tests have been recommended or implemented for drugs such as mercaptopurine, azathioprine, warfarin, irinotecan, and tamoxifen, there is still a great need for well-designed, prospective clinical trials that test pharmacogenetic approaches versus standard practice. There remains great translational value in the simple, careful observation of clinical outlier phenotypic responses to drug therapy and in research that attempts to identify pharmacokinetic, pharmacodynamic, and genetic mechanisms underlying such variability.

REFERENCES

Alander, S.W., Gaedigk, A., Woods, G.M., Leeder, J.S., 2002. CYP2D6 genotype as predictor of failed outpatient pain therapy in sickle cell patients. Clin. Pharmacol. Ther. 71, TPH–TPH109.

Andersson, T., 1996. Pharmacokinetics, metabolism and interactions of acid pump inhibitors. Focus on omeprazole, lansoprazole and pantoprazole. Clin. Pharmacokinet. 31 (1), 9–28.

Ariyoshi, N., Miyazaki, M., et al., 2001. A single nucleotide polymorphism of CYP2b6 found in Japanese enhances catalytic activity by autoactivation. Biochem. Biophys. Res. Commun. 281 (5), 1256–1260.

Bijl, M.J., Visser, L.E., et al., 2008. Influence of the CYP2D6*4 polymorphism on dose, switching and discontinuation of antidepressants. Br. J. Clin. Pharmacol. 5 (4), 558–564.

Brandt, J.T., Close, S.L., et al., 2007. Common polymorphisms of CYP2C19 and CYP2C9 affect the pharmacokinetic and pharmacodynamic response to clopidogrel but not prasugrel. J. Thromb. Haemost. 5 (12), 2429–2436.

Brockmoller, J., Meineke, I., et al., 2007. Pharmacokinetics of mirtazapine: enantioselective effects of the CYP2D6 ultra rapid metabolizer genotype and correlation with adverse effects. Clin. Pharmacol. Ther. 81 (5), 699–707.

Cai, W.M., Nikoloff, D.M., et al., 2006. CYP2D6 genetic variation in healthy adults and psychiatric African-American subjects: implications for clinical practice and genetic testing. Pharmacogenomics J. 6 (5), 343–350.

Caraco, Y., Sheller, J., et al., 1996. Pharmacogenetic determination of the effects of codeine and prediction of drug interactions. J. Pharmacol. Exp. Ther. 278 (3), 1165–1174.

Caraco, Y., Sheller, J., et al., 1999. Impact of ethnic origin and quinidine coadministration on codeines disposition and pharmacodynamic effects. J. Pharmacol. Exp. Ther. 290 (1), 413–422.

Cornelis, M.C., El-Sohemy, A., et al., 2006. Coffee, CYP1A2 genotype, and risk of myocardial infarction. JAMA 295 (10), 1135–1141.

Dai, Y., Hebert, M.F., et al., 2006. Effect of CYP3A5 polymorphism on tacrolimus metabolic clearance in vitro. Drug Metab. Dispos. 34 (5), 836–847.

de Leon, J., 2007. The crucial role of the therapeutic window in understanding the clinical relevance of the poor versus the ultrarapid metabolizer phenotypes in subjects taking drugs metabolized by CYP2D6 or CYP2C19. J. Clin. Psychopharmacol. 27 (3), 241–245.

Dennison, J.B., Kulanthaivel, P., et al., 2006. Selective metabolism of vincristine in vitro by CYP3A5. Drug Metab. Dispos. 34 (8), 1317–1327.

Desta, Z., Zhao, X., et al., 2002. Clinical significance of the cytochrome P450 2C19 genetic polymorphism. Clin. Pharmacokinet. 41 (12), 913–958.

Desta, Z., Ward, B.A., Soukhova, N.V., Flockhart, D.A., 2004. Comprehensive evaluation of tamoxifen sequential biotransformation by the human cytochrome P450 system in vitro: prominent roles for CYP3A and CYP2D6. J. Pharmacol. Exp. Ther. 310 (3), 1062–1075.

Eichelbaum, M., Spannbrucker, N., et al., 1979. Defective N-oxidation of sparteine in man: a new pharmacogenetic defect. Eur. J. Clin. Pharmacol. 16 (3), 183–187.

Ekins, S., Vandenbranden, M., et al., 1998. Further characterization of the expression in liver and catalytic activity of CYP2B6. J. Pharmacol. Exp. Ther. 286 (3), 1253–1259.

Evans, D.A., Manley, K.A., et al., 1960. Genetic control of isoniazid metabolism in man. BMJ 2 (5197), 485–491.

Foster, A., Mobley, E., et al., 2007. Complicated pain management in a CYP450 2D6 poor metabolizer. Pain Pract. 7 (4), 352–356.

Fox, A.L., 1932. The relationship between chemical constitution and taste. Proc. Natl. Acad. Sci. USA 18 (1), 115–120.

Garrod, A.E., 1909. The Inborn Errors of Metabolism. Oxford University Press, Oxford.

Giusti, B., Gori, A.M., et al., 2007. Cytochrome P450 2C19 loss-of-function polymorphism, but not CYP3A4 IVS10 1 12G/A and P2Y12 T744C polymorphisms, is associated with response variability to dual antiplatelet treatment in high-risk vascular patients. Pharmacogenet. Genomics 17 (12), 1057–1064.

Goetz, M.P., Knox, S.K., et al., 2007. The impact of cytochrome P450 2D6 metabolism in women receiving adjuvant tamoxifen. Breast Cancer Res. Treat. 101 (1), 113–121.

Goldstein, J.A., 2001. Clinical relevance of genetic polymorphisms in the human CYP2C subfamily. Br. J. Clin. Pharmacol. 52 (4), 349–355.

Gonzalez, F.J., Skoda, R.C., et al., 1988. Characterization of the common genetic defect in humans deficient in debrisoquine metabolism. Nature 331 (6155), 442–446.

Goodman, L.S., Gilman, A., et al., 1990. Goodman and Gilman's the Pharmacological Basis of Therapeutics. Pergamon Press, Elmsford, NY.

Hammer, W., Sjöqvist, F., 1967. Plasma levels of monomethylated tricyclic antidepressants during treatment with imipramine-like compounds. Life Sci. 6 (17), 1895–1903.

Jin, Y., Desta, Z., et al., 2005. CYP2D6 genotype, antidepressant use, and tamoxifen metabolism during adjuvant breast cancer treatment. J. Natl. Cancer Inst. 97 (1), 30–39.

Kalow, W., Genest, K., 1957. A method for the detection of atypical forms of human serum cholinesterase: determination of dibucaine numbers. Can. J. Biochem. Physiol. 35 (6), 339–346.

Kinzig-Schippers, M., Tomalik-Scharte, D., et al., 2005. Should we use N-acetyltransferase type 2 genotyping to personalize isoniazid doses? Antimicrob. Agents Chemother. 49 (5), 1733–1738.

Kiyotani, K., Mushiroda, T., Sasa, M., Bando, Y., Sumitomo, I., Hosono, N., Kubo, M., Nakamura, Y., Zembutsu, H., 2008. Impact of CYP2D6*10 on recurrence-free survival in breast cancer patients receiving adjuvant tamoxifen therapy. Cancer Sci. 99 (5), 995–999.

Kohn, L.T., Corrigan, J., et al., 2000. To Err Is Human: Building a Safer Health System. National Academy Press, Washington, DC.

Kuehl, P., Zhang, J., et al., 2001. Sequence diversity in CYP3A promoters and characterization of the genetic basis of polymorphic CYP3A5 expression. Nat. Genet. 27 (4), 383–391.

Lamba, J.K., Lin, Y.S., et al., 2002. Genetic contribution to variable human CYP3A-mediated metabolism. Adv. Drug Deliv. Rev. 54 (10), 1271–1294.

Lang, T., Klein, K., et al., 2004. Multiple novel nonsynonymous CYP2B6 gene polymorphisms in Caucasians: demonstration of phenotypic null alleles. J. Pharmacol. Exp. Ther. 311 (1), 34–43.

Lennard, L., Lilleyman, J.S., 1989. Variable mercaptopurine metabolism and treatment outcome in childhood lymphoblastic leukemia. J. Clin. Oncol. 7 (12), 1816–1823.

Lennard, L., Lilleyman, J.S., Van Loon, J., Weinshilboum, R.M., 1990. Genetic variation in response to 6-mercaptopurine for childhood acute lymphoblastic leukaemia. Lancet 336, 225–229.

Madadi, P., Koren, G., et al., 2007. Safety of codeine during breastfeeding: fatal morphine poisoning in the breastfed neonate of a mother prescribed codeine. Can. Fam. Physician 53 (1), 33—35.

Mahgoub, A., Idle, J.R., et al., 1977. Polymorphic hydroxylation of Debrisoquine in man. Lancet 2 (8038), 584—586.

McAlpine, D.E., O'Kane, D.J., et al., 2007. Cytochrome P450 2D6 genotype variation and venlafaxine dosage. Mayo Clin. Proc. 82 (9), 1065—1068.

Meyer, U.A., 2004. Pharmacogenetics — five decades of therapeutic lessons from genetic diversity. Nat. Rev. Genet. 5 (9), 669—676.

Punglia, R.S., Burstein, H.J., Winer, E.P., Weeks, J.C., 2008. Pharmacogenomic variation of CYP2D6 and the choice of optimal adjuvant endocrine therapy for postmenopausal breast cancer: a modeling analysis. J. Natl. Cancer Inst. 100 (9), 642—648.

Rasmussen, B.B., Brix, T.H., et al., 2002. The interindividual differences in the 3-demethylation of caffeine alias CYP1A2 is determined by both genetic and environmental factors. Pharmacogenetics 12 (6), 473—478.

Renbarger, J.L., McCammack, K.C., et al., 2008. Effect of race on vincristine-associated neurotoxicity in pediatric acute lymphoblastic leukemia patients. Pediatr. Blood Cancer 50 (4), 769—771.

Romkes, M., Faletto, M.B., et al., 1991. Cloning and expression of complementary DNAs for multiple members of the human cytochrome P450IIC subfamily. Biochemistry 30 (13), 3247—3255.

Sata, F., Yamada, H., et al., 2005. Caffeine intake, CYP1A2 polymorphism and the risk of recurrent pregnancy loss. Mol. Hum. Reprod. 11 (5), 357—360.

Schroth, W., Antoniadou, L., et al., 2007. Breast cancer treatment outcome with adjuvant tamoxifen relative to patient CYP2D6 and CYP2C19 genotypes. J. Clin. Oncol. 25 (33), 5187—5193.

Sim, S.C., Risinger, C., et al., 2006. A common novel CYP2C19 gene variant causes ultrarapid drug metabolism relevant for the drug response to proton pump inhibitors and antidepressants. Clin. Pharmacol. Ther. 79 (1), 103—113.

Singh, G., Saxena, N., et al., 2007. Cytochrome P450 polymorphism as a predictor of ovarian toxicity to pulse cyclophosphamide in systemic lupus erythematosus. J. Rheumatol. 34 (4), 731—733.

Stamer, U.M., Stuber, F., 2007. Codeine and tramadol analgesic efficacy and respiratory effects are influenced by CYP2D6 genotype. Anaesthesia 62 (12), 1294—1295 author reply 1295—1296.

Stearns, V., Johnson, M.D., et al., 2003. Active tamoxifen metabolite plasma concentrations after coadministration of tamoxifen and the selective serotonin reuptake inhibitor paroxetine. J. Natl. Cancer Inst. 95 (23), 1758—1764.

Szumlanski, C., Otterness, D., Her, C., Lee, D., Brandriff, B., Kelsell, D., Spurr, N., Lennard, L., Wieben, E., Weinshilboum, R., 1996. Thiopurine methyltransferase pharmacogenetics: human gene cloning and characterization of a common polymorphism. DNA Cell Biol. 15 (1), 17—30.

Takada, K., Arefayene, M., et al., 2004. Cytochrome P450 pharmacogenetics as a predictor of toxicity and clinical response to pulse cyclophosphamide in lupus nephritis. Arthritis Rheum. 50 (7), 2202—2210.

Ward, B.A., Gorski, J.C., et al., 2003. The cytochrome P450 2B6 (CYP2B6) is the main catalyst of efavirenz primary and secondary metabolism: implication for HIV/AIDS therapy and utility of efavirenz as a substrate marker of CYP2B6 catalytic activity. J. Pharmacol. Exp. Ther. 306 (1), 287—300.

Weinshilboum, R.M., Sladek, S.L., 1980. Mercaptopurine pharmacogenetics: monogenic inheritance of erythrocyte thiopurine methyltransferase activity. Am. J. Hum. Genet. 32 (5), 651—662.

Weinshilboum, R.M., Otterness, D.M., Szumlanski, C.L., 1999. Methylation pharmacogenetics: catechol O-methyltransferase, thiopurine methyltransferase, and histamine N-methyltransferase. Annu. Rev. Pharmacol. Toxicol. 39, 19—52.

Zanger, U.M., Klein, K., et al., 2007. Polymorphic CYP2B6: molecular mechanisms and emerging clinical significance. Pharmacogenomics 8 (7), 743—759.

Chapter 19

Statistical Techniques for Genetic Analysis

Jessica Lasky-Su

Harvard Medical School and Brigham and Women's Hospital, Boston, MA, United States

Chapter Outline

Key Points

- The field of statistical genetics is changing rapidly due to the technological advancements that have resulted in increasing quantities of genetic data.
- Twin, family, and adoption studies are used to determine the extent to which a complex disease has genetic determinants.
- Both common and rare genetic variants likely contribute to complex disease etiology and require different statistical methodologies to analyze these data appropriately.
- Case–control, population-based, and family-based designs can all be used to identify genetic determinants for disease and have different strengths and weaknesses.
- Properly analyzing all forms of genomewide genetic data requires careful data cleaning, quality control assessments, and adjusting for both population stratification and multiple comparisons.
- Integrating multiple forms of omics data into one analysis is becoming more common, and statistical methods in this realm continue to emerge.
- Using network medicine methods to create biological disease networks has great potential to increase the overall understanding of the biological pathways and variants that are involved in disease pathogenesis.

Clinical and Translational Science. http://dx.doi.org/10.1016/B978-0-12-802101-9.00019-3

INTRODUCTION

In the last 20 years, the field of genetics has undergone a revolution that has resulted in an explosion of genetic data. The increase is a direct result of ever-changing technologies through which genetic data are extracted. As a direct result of these ever-changing technologies, the type of data generated, the quality of the data, and the time taken to generate the data have also changed drastically. With each change comes new statistical challenges that statisticians have been rapidly at work to address. In this chapter, we review some of the established statistical genetics methodologies that have been developed to bring the field to where it is today. We will then review the current data that are being generated, the statistical methods that have been developed to analyze these data, and the important epidemiological and statistical considerations that are necessary for proper data analysis. While this chapter will provide a conceptual view of the statistical and analytical approaches that have been developed and are currently being used to analyze human genetic data, the methodological details of each approach are only briefly reviewed and references are provided for more rigorous statistical descriptions of the summaries described here.

GENETIC DETERMINATION OF COMPLEX DISEASE

Identifying the origin of a disease is an essential part of disease prevention. Although some human diseases are monogenic, being caused by a single genetic variant, most human diseases are complex in nature and result from the complex interactions of environmental conditions and multiple genetic variants. In many illnesses that have been termed "diseases," e.g., diabetes mellitus, hypertension, asthma, bipolar disorders, these illnesses are actually syndromes that comprises several distinct pathophysiologic mechanisms (diseases). The first step in complex trait genetics is to determine the extent to which the "disease" is actually genetic. This step is done by calculating the **heritability** of a disease, *the overall proportion of the variability of a disease that can be attributed to genetic factors.* Heritability is estimated using several different epidemiological studies.

Twin Studies

Twin studies are particularly useful for determining the heritability of a disease for several reasons, most notably because we know the differences in genetic makeup for different types of twins. Identical twins are monozygotic because they are formed from one fertilized egg that divides into two during the process of meiosis. Therefore, identical twins share 100% of their genetic information. In contrast, fraternal twins are dizygotic meaning that two separate eggs are fertilized. As a result, fraternal twins share only 50% of their genetic information, which is no more than any siblings. The key to twin studies is that both identical and fraternal twins also share a similar environment, albeit slightly more similar for monozygotic twins (e.g., monozygotic twins are more often dressed identically, etc.). Therefore if the rate of a disease is higher in the identical twins when compared to the fraternal twins, this difference can be attributed to an underlying genetic component. The concordance rate for a disease trait is therefore calculated for both identical and fraternal twins and compared. If the concordance rate is significantly higher in the identical twins compared with the fraternal twins, then we conclude that the disease has a genetic component. There is extensive statistical methodology that has been developed with twin data and that examines these relationships in more detail, enabling the estimation of measures of heritability, shared environment, and unique environment (Neale and Cardon, 1992).

Family Studies

Family studies are also used to evaluate the genetic component of a disease. Relying on Mendel's laws, a child inherits half of its genes from each parent. Using this mathematical relationship, it can be deduced that siblings share 50% of their genes, grandparents and grandchildren share 25% of their genes, and first cousins share 12.5% of their genes in common. Therefore, if there is a genetic component to the disease, the trait should be more often observed in siblings, who share a larger percentage of their genetics, than grandparents or cousins, who are more distantly related. These relationships are therefore used to determine the extent to which a disease is genetic.

Adoption Studies

Adoption studies are also used to examine the environmental influence on a disease of interest. Adoptive children typically share none of their genetics with their adoptive parents. Therefore if a disease trait is prevalent among the adoptive family as well as the adoptive child who is not biologically related, this would indicate that there is a strong environmental rather than genetic influence on the disease.

GENETIC LINKAGE STUDIES

Once a trait/disease is identified as having a genetic component, epidemiological studies are used to identify the genetic determinants that influence the disease of interest. Historically the search for disease genes began with linkage analysis, an approach that was used to identify the rough location of a disease gene. The general principal of linkage analysis is that alleles located closely together on a chromosome are more likely to be inherited together through the meiotic process. That is, genes that are close together are more likely to be "linked," meaning that there is a lower chance of a chromosomal crossover to occur between the disease allele and the genetic marker. When this is the case, the marker and the disease allele are in linkage disequilibrium (LD) with each other. Linkage analysis therefore uses various forms of family data to determine the degree to which a disease phenotype cooccurs with genetic markers that are typed throughout the genome.

The typical approach in linkage studies is to generate genetic markers throughout the genome. Initially the genetic markers that were used were microsatellites. **Microsatellite** markers contain *a DNA sequence (a combination of ACGT) that varies in length and is repeated a large number of times. The number of times this sequence is repeated is counted and recorded as the microsatellite value.* These markers were then measured throughout the genome and assessed for linkage to disease phenotypes. A distinct advantage of microsatellites is that they are highly polymorphic markers, with high mutation rates that made these markers popular in early linkage studies. However, few microsatellites exist throughout the genome and even fewer microsatellites were measured for linkage analyses (e.g., 400−1000), which resulted in insufficient genome coverage and low statistical power. **Single nucleotide polymorphisms (SNPs)**, *the actual measurement of the nucleotide itself*, are the most common genetic markers that are currently collected on a genomewide scale. The advantage of using SNPs is that there is a consistent numeric count to these markers (0, 1, 2) that results in an easy and uniform way to summarize the genetic data.

Statistical methodology for linkage analysis exists using both parametric and nonparametric approaches. Parametric linkage analysis was the first method to be developed (Fujii et al., 1955; Morton, 1955). Using this approach the probability that a disease variant is linked to a genetic marker is calculated using a LOD (logarithm (base 10) of odds) score. The LOD score uses the underlying family structure and Mendel's laws to compare the observed cosegregation of a marker and a disease with what is expected by chance. The higher the LOD score, the more likely the disease variant is linked to the genetic marker. There are several limitations of parametric linkage analysis. Most notably, parametric linkage analysis assumes a specific mode of inheritance (i.e., additive, dominant, recessive) that is used to calculate the LOD score. The resulting LOD score is highly sensitive to these assumptions, and in most instances this information is not known with certainty. In contrast to parametric linkage analysis, nonparametric linkage analysis uses information on the identity by descent (IBD) of two individuals and does not rely on assumptions of the disease model. **IBD** measures *the probability that two related individuals share 0, 1, or 2 alleles that were transmitted from the same common ancestor, implying that this chromosomal segment is from a common ancestor.* In linkage analyses therefore, the observed IBD relationship between two family members, most often affected sibling pairs, is compared with what the expected IBD transmission would be by chance. Several algorithms and programs are available to evaluate the IBD relationship in sibling pairs and entire pedigrees (Lander and Green, 1987; Abecasis et al., 2002; Elston and Stewart, 1971). One of the fundamental properties of linkage analysis is to measure the likelihood that a genetic marker is in LD with the disease variant. This is in direct contrast with genetic association analyses that test whether the genetic marker is directly associated with the disease of interest. Although linkage methods have been widely used, historically these studies were found to be severely underpowered, both in the context of an insufficient number of genetic markers and in a limited sample size. Linkage analyses have been successful in identifying monogenic diseases (Myers et al., 1993) and complex diseases with a limited number of strong genetic variants (Wijsman et al., 2005; Bird, 2005); however, linkage analysis has failed to identify disease susceptibility loci for most complex diseases that only have disease variants with modest genetic effects. However, as noted previously, this "failure" may be the result of assuming an illness is a disease when it is really a syndrome, e.g., assuming fever is a disease rather than a syndrome.

COMMON GENETIC STUDY DESIGNS AND STATISTICAL TESTS

Genetic association studies aim to assess the relationship between a genetic variant and a phenotype (or a disease of interest) where the genetic variant under investigation could be several things, including an allele, genotype, or haplotype.

In contrast to linkage studies that rely on LD, association studies assess the direct relationship between a disease and a genetic variant. Genetic association studies are typically performed in the context of a specific study design, including case—control, population-based, and family-based studies.

Case—Control Studies

Attributes of Case—Control Data

Case—control studies are one of the fundamental designs utilized in epidemiologic studies (see also Chapters 4 and 17). In this study design, a group of individuals with the disease of interest are used in the study. A group of individuals without the disease of interest and from the same base population as the cases are also identified and used in the study. The frequency of the genetic variant is then compared between cases and control subjects, and if the difference is large enough to pass a statistical significance threshold, the genetic variant is determined to be associated with the disease. There are several advantages of the case—control design. Case—control designs are cost efficient, particularly if the disease is rare. Case—control studies can also have limitations, most notably, the selection of the controls is often subject to population stratification, and the only phenotype that can be examined is limited to the disease status.

When a case—control study is underway, there are several important characteristics that should be considered when selecting the cases and controls. First, the precision of the phenotype is important. That is, the case definition should be very clear with explicit inclusion and exclusion criteria. When selecting controls, there should be appropriate exclusion criteria to insure that cases are not included in this group. To avoid selection biases, the controls should be derived from the same source population as the cases, and the covariate distribution should be the same in the cases and controls.

Statistical Tests for Case—Control Data

All analyses for case—control data have a binary outcome to reflect case—control status and the genetic markers that are genotyped are most often SNPs. The most basic test for this approach is the allelic association test. This is a standard chi-square test that compares the allele frequency in cases and controls and calculates a test statistic based on 1 degree of freedom. A Cochran—Armitage trend test is the genotype test for association. The Cochran—Armitage trend test modifies the chi-square test to incorporate a suspected ordering in the effects of the three categories, reflecting the three possible genotypes. The Pearson X^2 (Fujii et al., 1955) test then compares observed number of 11, 12, 22 genotypes to expected numbers under the null hypothesis. The trend test will have higher power than the chi-square test when the additive trend is correct. A logistic regression is another option for analyzing case—control data. In this case, the genetic information is the predictor variable and case—control status was the response variable. This test has many advantages, including the ability to adjust for possible confounders and population stratification adjustments. If the study is a matched case—control design, then a conditional logistic regression can be employed while also adjusting for the necessary covariates. While these statistical tests can be used on individual SNPs, these tests can also be applied on a genomewide scale when SNPs or other genetic markers are genotyped on a genome level. *The complete set of association analyses on a genome scale comprises a* **genomewide association analysis**.

Population-Based Studies

Attributes of Population-Based Data

Population-based studies are also commonly used for genetic studies (see also Chapter 17). Population-based studies have the advantage of being prospective and are often longitudinal in nature. Population cohorts are typically not selected for a disease of interest, so incident cases of disease can be collected. Case—control studies can also be generated within a population-based cohort, where incident cases are identified and matched with the appropriate controls. Nested case—control studies constitute a strong study design because the source population from which the cases arise is easily identifiable, thereby minimizing several sources of bias. Other advantages of population-based cohorts include the large range of phenotypes that can be studied, most often large sample sizes, and similar environmental exposures that are typically dictated by the ascertainment approaches. As noted previously the major problem with these types of studies is defining a clear, clean control group. For example, in illnesses that take time to develop, e.g., diabetes and hypertension, what age is used to identify a subject as a control rather than an illness one.

Common Statistical Tests for Population-Based Data

A distinct advantage of population-based data is the diverse range of phenotypes that can be assessed for genetic association. When there is interest between evaluating the relationship between these phenotypes and genetic variants, generalized linear models are most often used to analyze the association. For binary phenotypes, logistic/conditional logistic regression analyses are commonly used where linear regression and mixed models are most often used to assess the association with genetic variants and quantitative phenotypes. Many extensions of these analytic approaches exist that can incorporate longitudinal data, fixed and random effects, and covariates.

Family-Based Association Studies

Attributes of Family-Based Association Studies

Family-based association studies are an alternative study design to case—control or population-based studies that are implemented in human genetics research. These study designs typically consist of identifying a **proband**, *an individual who typically is affected with the disease of interest and serves as the starting point for a family to enter the study.* The family members of the affected proband, most often the parents, are then collected for the genetic study. This results in genetic data on families, often including siblings, parents, grandparents, aunts, uncles, and cousins. The most common study design is the collection of "trio" data, where genetic data are collected on an affected offspring and his/her parents.

Common Statistical Tests for Family-Based Data

Family-based association tests (FBATs) were first popularized with the transmission disequilibrium test (TDT; Spielman et al., 1993), which compares the rates of the alleles that are transmitted and untransmitted to the affected offspring from the parents. The TDT is identical to the McNemar statistical test for matched studies. The success of the TDT motivated statisticians to develop methodological extensions of the test, as the TDT has many assumptions and limitations. The TDT was quickly extended into other FBATs that incorporate different modes of inheritance, multiallelic marker data, quantitative and qualitative phenotypes, longitudinal data, and larger pedigrees (Sham and Curtis, 1995; Curtis and Sham, 1995; Bickeboller and Clerget-Darpoux, 1995; Spielman and Ewens, 1996). In contrast to the TDT, where parent—offspring trios are used, FBATs can use any type of pedigree structure including missing parental data, multiple siblings, and extended pedigrees (Curtis and Sham, 1995; Rabinowitz and Laird, 2000; Spielman and Ewens, 1998; Schaid and Li, 1997; Fulker et al., 1999; Horvath and Laird, 1998; Lake et al., 2000). Many extensions have also been proposed to incorporate the use of quantitative traits. Extensions to quantitative traits are described elsewhere (Fulker et al., 1999; Abecasis et al., 2000; Rabinowitz, 1997; Horvath et al., 2001; Laird et al., 2000; Laird and Lange, 2006). A benefit of using family-based data is that it is robust to population stratification, and therefore the concerns of the control group not being representative of the case group is minimized.

GENOMEWIDE ASSOCIATION STUDIES

There are various types of data generation that influence what statistical analysis will be performed. Genetic markers refer to any known sequence of DNA that can be identified throughout the genome that can be localized to a specific chromosomal region. One of the most common genetic markers is the SNP that represents a single base-pair of DNA. Longer regions of DNA are often referred to as microsatellites. Initial genomewide studies identified a limited number of microsatellites, often less than 1000, sparsely measured throughout the genome. These genetic markers were initially used in linkage analyses. More commonly today comprehensive genomewide genotyping is performed for genetic studies using 500,000 or more SNPs. Those data comprise a genomewide association study (GWAS). GWAS data are most often analyzed using genetic association techniques that are described previously for case—control, population-based, and family-based designs. Once the association tests are performed and *P* values are generated, these associations can be presented on a genomewide level using a Manhattan plot, where the x-axis represents the regions on the genome and the y-axis represents the $-\log_{10}(P$ value). Fig. 19.1 shows a Manhattan plot for a metaanalysis of rheumatoid arthritis (RA) GWAS studies, including over 100,000 individuals and over 10 million SNPs in European, Asian, and transethnic populations. In this example, some of the strongest genetic variants are visually identifiable in the Manhattan plots for both ethnicities separately and combined together, most notably the MHC region and *PTPN22* locus (Okada et al., 2014).

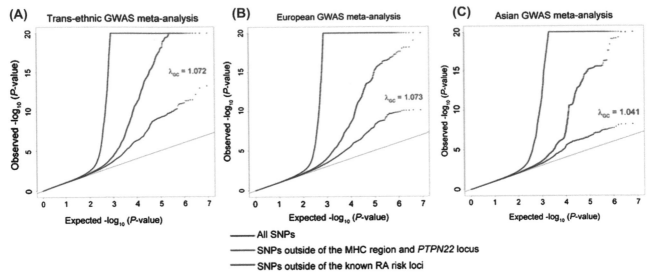

FIGURE 19.1 Illustration of the quantile–quantile plots (QQ plots) for the rheumatoid arthritis (RA) genomewide association study (GWAS) analyses for the transethnic, European, and Asian ancestry groupings. The x-axis indicates the expected −log (P values). The y-axis indicates the observed −log (P values). In (A), (B), and (C), there are three lines in the QQ plot. The *black line* reflects the P values when all single nucleotide polymorphisms (SNPs) are included, the *blue line* represents the P values when all SNPs outside of the MHC region and *PTPN22* locus are included, and the *red line* excludes all known risk loci for RA. As these corrections are made, the genomic inflation factor (GIF) comes closer to 1.

Population Stratification Adjustment

One of the major limitations to all genetic studies is the possible confounding that occurs when there are underlying subpopulations within the population being studied due to different ancestries, and those ancestries have systematic differences in their allelic distributions. When combined together, this combination can result in false-positive associations. Several methodologies have been developed to adjust for and assess population stratification. One of the first methods proposed was genomic control where unlinked markers throughout the genome are randomly selected in cases and controls and used to adjust the overall analysis (Devlin and Roeder, 1999). Population stratification and other departures from normality are also often assessed through the generation of a quantile–quantile plot (QQ plot) and the genomic inflation factor (GIF). GIFs and QQ plots are often used to compare the genomewide distribution of the test statistic to what is expected under the null hypothesis. Specifically, the QQ plots compare the observed −$\log_{10}(P$ value) to the expected −$\log_{10}(P$ value) under the null hypothesis and enable a good visualization of deviations in the results on a genomewide level. Significant deviations in the QQ plot and GIF may indicate several things, most notably population stratification, but also significant departures from what is expected including other errors such as problems with the samples themselves, technical biases, inappropriate analysis strategies, and the existence of unknown family relationships. Fig. 19.2 shows examples of the QQ plots and the GIF measurements that were generated for the European, Asian, and transethnic samples for the RA GWAS metaanalysis described previously. This demonstrates how the QQ plot and GIF change when the known causal regions are removed from the assessment. Another method used to adjust for this is principal component analysis (PCA), which identifies the largest sources of variability throughout the genetic data. These principal components can then be extracted and adjusted for in the analyses (Price et al., 2006). Although these methods have been readily used for common genetic variation, adjustment for stratification for rare genetic variants is still emerging.

Multiple Testing Adjustment

When performing genomewide-level analyses, association analyses are performed for each SNP that was genotyped, resulting in an inordinate number of statistical tests. Because we are simultaneously testing multiple SNPs at one time, the threshold for statistical significance must be adjusted. For example, if a significance threshold of 0.05 is used to declare statistical significance and there are 1,000,000 SNPs, 50,000 SNPs will be identified as statistically significant which identifies a large number of false-positive associations. Therefore a multiple testing correction that considers the simultaneous testing of multiple SNPs (often referred to as a family-wise error rate) and controls the overall significance level is necessary. A Bonferroni (1937) correction maintains the overall significance level, alpha, by using an individual

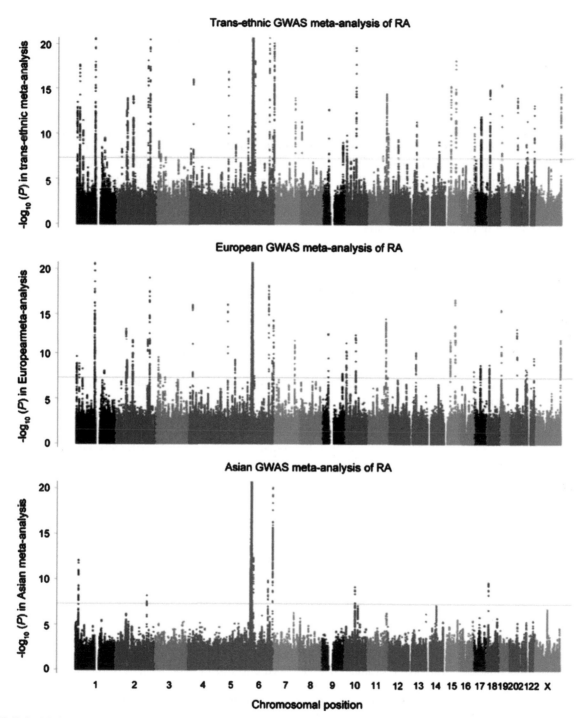

FIGURE 19.2   Manhattan plots of *P* values for the transethnic, European, and Asian ancestry rheumatoid arthritis (RA) genome-wide association study (GWAS) metaanalyses. The y-axis indicates the $-\log$ (*P* values) of genome-wide single nucleotide polymorphisms (SNPs) in each GWAS metaanalysis and the x-axis represents the chromosomes. The *horizontal gray line* represents the genomewide significance threshold of $P = 5.0 \times 10^{-8}$. These Manhattan plots illustrate how easily linked regions, such as the MHC region and the *PTPN22* locus, are associated with RA.

significance level of alpha/K for each of the K tests being performed. Although this correction is intuitive and simple to calculate, it is overly conservative and may miss true-positive associations. The false discovery rate (FDR) (Benjamini and Hochberg, 1995) is a popular method that is often implemented in genetic studies to control for multiple testing and was originally proposed to combat the overly conservative nature of other multiple testing adjustments. The FDR procedure corrects for the expected proportion of incorrectly rejected SNPs (i.e., false positives) under the null hypothesis. The

Q value was proposed as an analogous measure to the *P* value individual and refers the minimum FDR threshold by which a statistical test is declared significant.

NEXT-GENERATION SEQUENCING

Next-generation sequencing (NGS) refers to high-throughput parallel DNA sequencing technology that has enabled the generation of sequence data in a time-efficient manner, including whole-genome, whole-exome, and targeted genomic regions (Burgess, 2011; Metzker, 2010). Prior to NGS, SNP genotyping only covered a fraction of the genome, while whole genome sequencing covered the entire genome. Despite the rapid fall in price and the increase in use of NGS, analysis techniques to handle these massive quantities of data have been limited. Standardized methods for variant calling are needed and platform-specific problems such as sequencing and mapping errors need to be addressed. Although reference genome sequences exist, further ethnic refinement for the sequences is necessary. Further bioinformatic advancements and statistical methodology in areas related to NGS are necessary. Some of the methodological developments to date are summarized in the following section.

Rare Genetic Variant Analysis

Rare Variants Defined

GWAS are typically performed in genetic variants that have common minor allele frequencies (MAFs > 5%). These studies have been successful in identifying thousands of variants associated for several complex diseases (Hindorff et al., 2010). Despite these successes, much of the genetic component of disease remains to be identified, leaving several possible explanations for the "missing heritability" (Manolio et al., 2009). Because GWAS have focused on common genetic variants (MAF > 5%), rare genetic variants could explain a significant portion of this missing heritability. *Rare genetic variants are those with a low MAF, typically with a frequency of less than 5 or 1%.* The rapid advance of NGS has enabled comprehensive assessment of rare genetic variants, and statistical methodology has emerged for the analysis of these data. One of the most difficult challenges with rare variants is that, by definition rare variants are only found in a small fraction of the overall population and as a result extremely large sample sizes are necessary to study these variants. Large population-based cohorts are ideal for these studies. Nested case–control designs are also ideal because this design can be used to generate a large number of cases, and variants that are rare in the overall population but are directly associated with the disease will be overrepresented (i.e., associated) among these cases.

As described pereviously for GWAS, genetic associations are typically assessed using single-variant tests and because of the multiple testing, a significance threshold of 5×10^{-8} is typically used to identify significant genetic associations. This approach works well for genetic variants that are of low frequency as long as the sample size and the genetic effect size are large. However, these approaches can have limited success when the sample sizes or genetic effect sizes are small. In this case, rare-variant tests are necessary.

Rare-Variant Tests

Rather than testing each genetic variant individually, rare-variant methods have been developed that are based on collapsing/aggregate strategies designed to evaluate the combined effect of multiple genetic variants that are most often very close together in a chromosomal region or gene of interest. The genetic information from multiple variants is included in one test statistic and evaluated together, thereby increasing the overall statistical power when multiple genetic variants are associated with the phenotype/disease of interest. The statistical power is also increased by reducing the number of statistical tests. These statistical tests are generally regression-based and therefore can adjust for covariates and population stratification. The statistical power of each test depends on several conditions, including the genetic model, the number of actual disease variants, and the direction of their effect. Rare-variant tests can be divided into four general categories (Lee et al., 2014) that we now review. Burden tests (Asimit et al., 2012; Li and Leal, 2008; Madsen and Browning, 2009; Morgenthaler and Thilly, 2007; Morris and Zeggini, 2010) represent one of the most common analytical strategies for assessing rare genetic variants and are based on collapsing rare genetic variants together, counting the number of rare variants, and using this information in one statistical test. The method is powerful when there are a large portion of rare genetic variants that are associated with a disease in the same direction but are less powerful with the genetic variants that are associated with the disease in different directions. Adaptive burden tests (Hoffmann et al., 2010; Han and Pan, 2010;

Liu and Leal, 2010; Lin and Tang, 2011; Price et al., 2010) have expanded on the initial test to allow for null variants and variants of opposite effects by calculating individual marginal tests and then combining these tests together using individual weights to generate an overall test statistic. As a result, these tests tend to be more robust than the traditional burden tests that have fixed thresholds and weights; however, they are often more computationally intensive. Variance component tests that incorporate a random-effects model have also been proposed that evaluate the distribution of genetic effects for a group of rare variants. As with adaptive burden test, variance component tests are more powerful than burden tests when there are risk variants of opposite effects combined with null variants (Pan, 2009; Wu et al., 2011; Neale et al., 2011); however, burden tests are more powerful than variance component tests when there are a large number of causal rare genetic variants all with the same direction of effect. Therefore when both burden tests and variance component tests are combined, optimal power will be achieved. As such, combined burden and variance component tests were proposed (Derkach et al., 2013; Lee et al., 2012; Sun et al., 2013) that combine the two methods together. They build on burden tests and are more robust to genetic variants that influence the disease in opposite directions. The exponentially combined score test (Sun et al., 2013) is the most powerful approach when there are a small number of causal variants but is less powerful when a moderate or large number of variants are causal. Although the statistical methodology for exponentially combined score tests has been developed, there is no current software that implements this approach.

METAANALYSIS TECHNIQUES

For all genetic variants, common and rare, metaanalytic techniques are an efficient way to combine data from multiple studies together. Many metaanalytic techniques have been developed (Evangelou and Ioannidis, 2013). The two most common approaches to combining information across studies are through the use of P values and Z scores (Evangelou and Ioannidis, 2013). Although several P value approaches have been developed, one of the most common methods was proposed by Fisher (1925). Fisher's method combines P values into one statistical test to generate an overall combined P value by adding the $-2\ln(P$ value) for K tests. This sum has a chi-square distribution with 2k degrees of freedom. This approach is highly flexible because it only uses P values and therefore it does not depend on any other aspects of the individual studies; however, this approach can also be underpowered as it does not incorporate other important information such as the genetic effect size and the sample size that was used to generate the P value. Stouffer proposed something similar to the Fisher approach but instead of using P values, it uses Z scores. Several extensions of the Stouffer method can be incorporated by using weights in the calculation (Liptak, 1958). Metaanalytic techniques have been proposed for rare-variant data that combines score statistics and also requires information on the between-variant covariance matrices that described the linkage disequilibrium between the variants (Hu et al., 2013; Lee et al., 2013; Tang and Lin, 2013). Other metaanalytic approaches include fixed (e.g., Cochran–Mantel–Haenszel approach) and random (method proposed by DerSimonian and Laird, 1986) metaanalysis. Generating optimal weights when combining multiple data sets together is another optimal strategy to maximize statistical power. The inverse of the variance is often utilized as a weight when combining studies together (Cochran, 1954).

GENE-BY-ENVIRONMENT ANALYSIS

Environmental factors may act in conjunction with genetic variants to jointly influence the development of a complex disease through a **gene-by-environment interaction**. Gene-by-environment interactions describe how various genetic determinants can be modified by different environmental exposures (Hunter, 2005). Studies of gene–environment interactions require collecting information on both the environmental exposure and the genetic variant. Environmental exposures may be either quantitative or qualitative. The simplest gene by environmental model dichotomizes the genotype (genetic variant yes/no) and the environmental variable (environmental exposure yes/no), thereby forming a two by two table that can be assessed by a simple chi-square test. More complex scenarios occur when the environmental variable is measured using ordinal or continuous measures, resulting in more complex table structures. One of the primary approaches to analyzing gene by environmental interactions is through the use of regression models. In these analyses, the genetic variant, the environmental exposure, and the interaction of these two variables are used as predictors in a regression model, and the phenotype of interest is the response variable. The P value and beta coefficient of the interaction term in this model assess the strength and direction of the gene-by-environment interaction. The assessment of gene-by-environment interactions is often underpowered, as they are dependent on the MAF and the frequency of the environmental exposure of interest, both of which can be infrequent. Therefore, to adequately assess gene-by-environment interactions often requires large sample sizes or study designs that circumvent this problem.

MULTIVARIANT APPROACHES

Multiple Regression Analyses

Thus far the discussion of common genetic variants has focused on individual marker analysis followed by a multiple testing correction. While this is a standard and easily implemented statistical approach, it fails to consider the underlying genetic and biologic architecture for which genetic variants exist. For example, it is likely that multiple genetic variants within a gene or within the genome may act together to result in some change in biologic function. While we have discussed rare-variant methods, methods also exist to combine common genetic variants. Genetic variants that are in linkage disequilibrium with each other or another causal variant will also exhibit high correlation with one another. Moreover, when considering various established biological pathways, genes most often do not act in isolation, but are affected by other genes. Therefore, statistical tests that assess multiple genetic variants simultaneously are useful for determining the combined effect of multiple variants. To date many multivariant methods have been developed. Multivariate regression analysis is an initial approach to assess the joint effect of multiple genetic variants on a phenotypic outcome. In this case, the genetic variants, in addition to other relevant covariates, are used as predictor variables, and the clinical phenotype of interest is the response variable. Forward or backward stepwise regression is often utilized to reduce the number of genetic variants that are included in the statistical model, only including those variants that are significantly associated with the phenotype of interest.

Discrimination Methods

Statistical approaches that can use genetic data to classify cases and controls include PCA, discriminant analysis, Bayesian methods, random forests, and support vector machine learning methods. In addition to being a key methodology for representing population stratification, PCA is also a common data reduction technique that can be utilized with omics data to summarize a large number of genetic variants more directly. *PCA is a simple nonparametric method that can project the genetic variants into a lower-dimensional space, revealing inherent data structure, and providing a reduced dimensional representation of the original data.* Specifically, it uses orthogonal transformations to convert a set of correlated data to a set of uncorrelated "principal components." The orthogonal transformations are generated in such a way that the principal components are ordered such that the first component accounts for the largest amount of variation and the last component accounts for the least amount of variation. Once PCA is performed, the individual components can be correlated with clinical phenotypes to identify groups of genetic variants (i.e., those genetic variants that load onto the component being tested) that are associated with the phenotype of interest. An alternative, but similar approach to PCA is partial least square discriminant analysis (PLS-DA) that takes a group of genetic variants; it first transforms these variables to become uncorrelated and then includes these variables in a regression analysis with the case−control status as the predictor variable. PLS-DA then selects the set of genetic variants that best discriminates between case and controls. Random forest methods use a series of "decision trees" to classify individuals into groups of cases and controls. The "root node" is the initial node separating the data into two groups, followed by inner nodes that further separate the data. The final node, or leaf node, represents the final decision utilized to classify the individuals into cases and controls. Support vector machine learning is an approach that discriminates between two groups by defining an optimal hyperplane that has a maximal margin between points in either group. A training set is first used to generate the parameter estimates for the hyperplane and then testing set is used to evaluate the discriminatory accuracy of the parameters (Hastie et al., 2009). With all of these approaches, the ability to differentiate cases and controls can be assessed using receiver operating curves, sensitivity, and specificity.

NETWORK MEDICINE

Basic Network Theory

Complex diseases by definition arise from the interactions of multiple variables that include underlying genetics and environmental perturbations. Generating networks that can describe the relationship between multiple genetic variants and disease pathogenesis has the potential to explain the underlying biological relationships that exist. Network medicine is an emerging field that utilizes network theory to describe disease networks using omics data (Barabasi et al., 2011). By considering the joint effect of multiple variants, network medicine takes a holistic, rather than reductionist, approach. Network theory relies on the concept that networks can be described using a scale-free distribution where most variants within a network have few connections, while a small number of variants have a large number of connections (Barabasi et al., 2011; Barabasi and Bonabeau, 2003; Barabasi and Oltvai, 2004). Networks have nodes that represent each variant that is input into the model (gene, transcript, metabolite, protein, disease), and the edges represent the relationship between the variants under investigation, most often determined by statistical measures of correlation. While most nodes have few edges with other variants, few nodes have many edges and are the primary connectors that maintain the network and are called hubs. A characteristic of hubs is that they are vital to the network, and if the hubs are removed, the network is destroyed. One of the attractions of the network approach is that if a few disease components are identified, other disease-related components are likely to be found in their network-based vicinity that will elucidate the novel genetic variants and biological pathways.

Correlation Coexpression Networks

Correlation measurements are one of the most commonly used measured to describe the relationship between multiple nodes within a network and can be summarized in a matrix. Multiple correlation measures may be used, including the Pearson and the Spearman rank correlation (Zhang and Horvath, 2005). These correlation measures can be calculated for each pair of genetic variants that may subsequently be used to define an adjacency matrix, which described the extent to which any two genetic variants are connected (Horvath). Once the adjacency matrix is calculated, the coexpression network can be generated and hierarchical clustering and topological dissimilarity measures are used to generate clustering trees that then define different modules within the coexpression network (Yip and Horvath, 2007; Langfelder et al., 2008). Genetic variants that have a strong correlation with many of the variants can then be identified as the network hubs.

Other methodological approaches can also be used to generate networks. **Gaussian graphical models (GGMs)** have a distinct advantage over correlation approaches in that they are able to differentiate between direct and indirect effects through the use of conditional dependencies, which results in partial correlations. One genetic variant is regressed on another genetic determinant while adjusting for the other genetic variants to insure that a direct correlation is measured (Krumsiek et al., 2011). Because GGMs only consider direct associations, the networks are more sparse and often easier to interpret than correlation-based networks. Bayesian methodology can also be used to generate networks. In **Bayesian networks**, the genetic variants in the domain are modeled as random variables represented by nodes, and the edges between them represent the statistical dependence of one genetic variant on the other. Each node is annotated with the marginal distribution of the genetic variant (child node) conditioned on the values of the genetic variants (parent node) with edges pointing to it, and this information can be used to answer questions about the most probable values of genetic variables in the Bayesian networks given assignments to other genetic variables in the Bayesian network. Through these methodologies, graphical representations of the relationships between genetic variants can be illustrated to depict the nodes and hubs of the modules.

Network methods such as these have been utilized to describe the interrelationships between many complex diseases, and several visual representations of the "human diseasome" have been generated using various forms of genetics and other omics data. Fig. 19.3 illustrates one depiction of the human diseasome, demonstrating the underlying biological connections that exist between various complex diseases that were identified using genetic variants (Barabasi et al., 2011). This example demonstrates that while some diseases are hubs and connect many diseases together, most diseases are nodes with few connections. Fig. 19.3 also demonstrates that similar diseases are connected with similar underlying genetics, depicted by classifying a general set of diseases with the same color (e.g., cancers are blue, cardiovascular diseases are red, gastrointestinal disorders are yellow, psychiatric disorders are green). As network methods continue to evolve, enabling the understanding of how various complex diseases are connected, the depiction of the human diseasome will become more accurate and more detailed, informing disease development on many levels.

Pathway Analyses

Using the methods described previously, sets of genes or SNPs can be selected for further pathway analyses to identify biological pathways that are overrepresented in the phenotype of interest. There are several types of pathway analyses that have evolved over time (Khatri et al., 2012). Overrepresentation analysis (ORA) was first developed using a select set of variants and/or genes and comparing these genes to gene annotation databases, such as Gene Ontology (GO) and the Kyoto encyclopedia of genes and genomes (KEGG; Wixon and Kell, 2000), to identify biologic pathways that are overrepresented in the disease groups being studied. The fundamental principle underlying these methods is that disease phenotypes will result in abnormal biological processes that result in the change of groups of genes that can be identified through the use of function annotation databases. Functional class scoring (FCS) was developed to address one of the

Human disease network

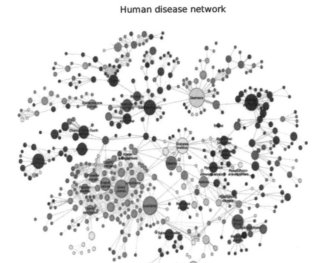

FIGURE 19.3 This figure depicts the "human diseasome" that was modeled using gene coexpression networks. This figure depicts that some diseases are highly connected to many others (e.g., colon cancer, leukemia, cardiomyopathy, deafness) and represents hubs of the network while other diseases have much fewer connections. As the field of network medicine continues to evolve, the understanding of the human diseasome will continue to improve.

primary limitations of ORA, namely that only a select group of variants can be used in the analysis, and the selection of these variants is arbitrary (Khatri et al., 2012). Furthermore, ORA only considers a list of genes but does not include strength of the association of these variants with the disease (e.g., *P* values, fold changes, etc.). FCS methods use information on all of the genes being studied as well as an association statistic to determine an enrichment score for particular groupings of SNPs. Gene set enrichment analysis (GSEA) is one of the most popular FCS methods that calculates an enrichment score to identify when a gene set is overrepresented in the top or bottom ranked genes. GSEA is a Web-based tool that can easily be accessed (http://www.broadinstitute.org/gsea/index.jsp). Topology-based methods are used in the newest pathway approaches and integrate information on how genes may interact with each other (Khatri et al., 2012).

INTEGRATIVE OMICS

The central dogma of molecular biology describes how DNA is transcribed into RNA and translated into proteins and then metabolized, forming the basis for disease development. Scientific advances have enabled the measurement of RNA, proteins, and metabolites on a genomewide scale, generating transcriptomic, proteomic, and metabolomic data. As a result of the generation of multiple data types, statistical methods that integrate various types of data have emerged. The premise behind the integration of multiple omics data is that it provides information about the disease process at multiple points along the disease pathway, something that no singular data type can do alone. Statistical methodology for integrating multiple omics data types is currently emerging. Several of the methodologies that we described earlier can be applied to other omics data types, both individually and integrated together. One of the most common integrative approaches is expression quantitative trait locus (*eQTL*) analysis. The transcriptome represents the measurement of all the RNA molecules in a cell type. The expression levels of mRNA are the direct result of a specific gene. eQTLs identify the genetic variants that are associated with expression levels in mRNAs. eQTLs that are associated with the gene proximal to the origin of the gene expression measurement are called *cis*-**eQTLs**, whereas the eQTLs that map far from the gene of origin are called *trans*-**eQTLs**. eQTLs are typically measured using a regression equation with the genetic variant as the predictor variable and the expression measurement as the response variable. Analogous approaches can be applied where the transcript is substituted for other downstream omic data measurements, including proteomic and metabolomic data. Network methods provide an ideal strategy for integrating multiple omics data together through either the inclusion of several omic data types into one network or the use of individual networks for each omic type followed by an integration across the networks, forming one large multilayered network.

PHENOTYPIC LIMITATIONS

Large omic data sets are often assembled using (1) large databases, such as electronic medical records (EMRs), (2) several cohorts that are assembled through large consortiums, or (3) large population-based cohorts. In many of these cases, maximizing the overall number of individuals is a primary focus. While large samples are typically successfully obtained, this is often at the sacrifice of detailed phenotypic information on the samples. For example, using approaches such as EMRs to obtain samples can provide access to a large number of individuals, but disease diagnosis is limited to the diagnoses that are recorded in the EMR. Similarly, when a sample is obtained using disease-specific consortiums, information collected on individuals and the ascertainment criteria will vary by study. Also, when longitudinal cohorts are collected, detailed clinical information is often only obtained on a few disease phenotypes. Therefore, obtaining large sample sizes often for genetic studies often comes at the sacrifice of detailed and consistent phenotypic information.

COMPUTER PROGRAMS

There are many statistical programs available for the analysis of human genetics data using the methods that are described in this chapter. We provide a brief review of some of the most widely used programs here. Plink (Purcell et al., 2007) is a free open source code that has extensive capabilities for the analysis of genetic data (http://pngu.mgh.harvard.edu/~purcell/plink/). This program can be utilized for many things including data management, data cleaning, population stratification assessment, metaanalysis, and genomewide-level data analysis. R is a flexible freeware statistical analysis program with many functions that can be utilized for various genetic data types. As analytic methods develop and genetic data evolve, additional functions are written and available for use with these new data types in R (Team, 2012). Therefore, R has great utility and evolves with the changing needs and ongoing developments with human genetics data. Haploview (Barrett et al., 2005) is another open source code that is designed to assess LD and haplotype analysis and interacts easily with other genetic analysis packages.

SUMMARY AND CONCLUSIONS

Due to advance in laboratory technologies, statistical genetics and genetic epidemiological approaches for complex diseases have been rapidly changing fields to address the enormity of genetic data. In this chapter, we reviewed established statistical genetic approaches that have brought the field to where it is today and we have described the emerging analytic fields that are being developed for current human genetics data. While twin, adoption, and family studies were first used to estimate the degree to which a disease is genetic, linkage studies were the first methods used to identify genetic regions that were associated with these complex diseases and genetic association tests quickly followed. As laboratory technology changed, the ability to generate more comprehensive genomewide genetic data developed, starting with GWAS data and now with whole genome sequence data. With this new challenges emerged, namely issues with multiple testing and the emergence of rare genetic variants that were underpowered using traditional statistical methods, resulting in the development of rare-variant analysis methods. Current research is not only focused on the analysis of individual genetic variants, but are also focused on analyzing multiple genetic variants together and/or integrating them with other relevant omics data types, particularly focusing on network methods. In addition, pathway and enrichment analyses are being used as a tool to identify novel biological pathways and variants associated with disease. These current areas of research have a unique set of challenges, in particular, identifying the most effective and powerful ways to integrate immense amounts of omics data that inform the disease etiology in unique ways.

GLOSSARY

Gene-by-environment interaction (G×E) Environmental exposures acting in conjunction with genetic variants to jointly influence the development of a complex disease.

Genomewide association analysis The complete set of association analyses on a genome scale.

Heritability The proportion of variance in a phenotype or disease that can be attributed to genetic factors (broad-sense heritability). Narrow-sense heritability can be expressed as the *ratio of the additive genetic variance to the total phenotypic variance* (V_A/V_P).

IBD It measures the probability that two related individuals share 0, 1, or 2 alleles that were transmitted from the same common ancestor, implying that this chromosomal segment is from a common ancestor.

Microsatellite Markers that contain a DNA sequence (a combination of ACGT) that varies in length and is repeated a large number of times. The number of times this sequence is repeated is counted and recorded as the microsatellite value.

Next-generation sequencing (NGS) High-throughput parallel DNA sequencing technology.

Principal component analysis (PCA) A nonparametric method that can project the genetic variants into lower dimensional space, revealing inherent data structure, and providing a reduced dimensional representation of the original data.

Proband An individual who typically is affected with the disease of interest and serves as the starting point for a family to enter the study.

Single nucleotide polymorphisms (SNPs) The actual measurement of the nucleotide itself (i.e., A, C, G, T).

LIST OF ACRONYMS AND ABBREVIATIONS

G×E Gene by environment
GSEA Gene set enrichment analysis
GWAS Genomewide association study
MAF Minor allele frequency
NGS Next-generation sequencing
ORA Overrepresentation analysis
PCA Principal component analysis
SNP Single nucleotide polymorphism

REFERENCES

Abecasis, G.R., Cherny, S.S., Cookson, W.O., Cardon, L.R., 2002. Merlin—rapid analysis of dense genetic maps using sparse gene flow trees. Nat. Genet. 30 (1), 97—101.

Abecasis, G.R., Cardon, L.R., Cookson, W.O., 2000. A general test of association for quantitative traits in nuclear families. Am. J. Hum. Genet. 66 (1), 279—292.

Asimit, J.L., Day-Williams, A.G., Morris, A.P., Zeggini, E., 2012. ARIEL and AMELIA: testing for an accumulation of rare variants using next-generation sequencing data. Hum. Hered. 73 (2), 84—94.

Bird, T.D., 2005. Genetic factors in Alzheimer's disease. N. Engl. J. Med. 352 (9), 862—864.

Bickeboller, H., Clerget-Darpoux, F., 1995. Statistical properties of the allelic and genotypic transmission/disequilibrium test for multiallelic markers. Genet. Epidemiol. 12 (6), 865—870.

Bonferroni, C., 1937. Teoria statistica delle classi e calcolo delle probability Volime in Onore di Ricardo dlla Volta. Universita di Firenze.

Benjamini, Y., Hochberg, Y., 1995. Controlling the false discovery rate: a practical and powerful approach to multiple testing. J. R. Stat. Soc. 57 (1), 289—300.

Burgess, D.J., 2011. Human disease: next-generation sequencing of the next generation. Nat. Rev. Genet. 12 (2), 78.

Barabasi, A.L., Gulbahce, N., Loscalzo, J., 2011. Network medicine: a network-based approach to human disease. Nat. Rev. Genet. 12 (1), 56—68.

Barabasi, A.L., Bonabeau, E., 2003. Scale-free networks. Sci. Am. 288 (5), 60—69.

Barabasi, A.L., Oltvai, Z.N., 2004. Network biology: understanding the cell's functional organization. Nat. Rev. Genet. 5 (2), 101—113.

Barrett, J.C., Fry, B., Maller, J., Daly, M.J., 2005. Haploview: analysis and visualization of LD and haplotype maps. Bioinformatics 21 (2), 263—265.

Curtis, D., Sham, P.C., 1995. A note on the application of the transmission disequilibrium test when a parent is missing. Am. J. Hum. Genet. 56 (3), 811—812.

Cochran, W.G., 1954. The combination of estimates from different experiments. Biometrics 10, 101—129.

Devlin, B., Roeder, K., 1999. Genomic control for association studies. Biometrics 55 (4), 997—1004.

Derkach, A., Lawless, J.F., Sun, L., 2013. Robust and powerful tests for rare variants using Fisher's method to combine evidence of association from two or more complementary tests. Genet. Epidemiol. 37 (1), 110—121.

DerSimonian, R., Laird, N., 1986. Meta-analysis in clinical trials. Control. Clin. Trials 7 (3), 177—188.

Elston, R.C., Stewart, J., 1971. A general model for the genetic analysis of pedigree data. Hum. Hered. 21 (6), 523—542.

Evangelou, E., Ioannidis, J.P., 2013. Meta-analysis methods for genome-wide association studies and beyond. Nat. Rev. Genet. 14 (6), 379—389.

Fujii, T., Moloney, W.C., Morton, N.E., 1955. Data on linkage of ovalocytosis and blood groups. Am. J. Hum. Genet. 7 (1), 72—75.

Fulker, D.W., Cherny, S.S., Sham, P.C., Hewitt, J.K., 1999. Combined linkage and association sib-pair analysis for quantitative traits. Am. J. Hum. Genet. 64 (1), 259—267.

Fisher, R.A., 1925. Statistical Methods for Research Workers. Edinburgh.

Horvath, S., Laird, N.M., 1998. A discordant-sibship test for disequilibrium and linkage: no need for parental data. Am. J. Hum. Genet. 63 (6), 1886—1897.

Horvath, S., Xu, X., Laird, N.M., 2001. The family based association test method: strategies for studying general genotype—phenotype associations. Eur. J. Hum. Genet. 9 (4), 301—306.

Hindorff, L.A., Junkins, H.A., Mehta, J., Manolio, T., 2010. A Catalog of Published Genome-Wide Association Studies. National Human Genome Research Institute. http://www.genome.gov/gwastudies.

Hoffmann, T.J., Marini, N.J., Witte, J.S., 2010. Comprehensive approach to analyzing rare genetic variants. PLoS One 5 (11), e13584.

Han, F., Pan, W., 2010. A data-adaptive sum test for disease association with multiple common or rare variants. Hum. Hered. 70 (1), 42—54.

Hu, Y.J., Berndt, S.I., Gustafsson, S., et al., 2013. Meta-analysis of gene-level associations for rare variants based on single-variant statistics. Am. J. Hum. Genet. 93 (2), 236—248.

Hunter, D.J., 2005. Gene-environment interactions in human diseases. Nat. Rev. Genet. 6 (4), 287–298.

Hastie, T., Tibshirani, R., Friedman, J.H., 2009. Separating Hyperplanes. The Elements of Statistical Learning: Data Mining, Inference, and Prediction, second ed. Springer.

Horvath, S. Extended Overview of Weighted Gene Co-expression Network Analysis (WGCNA).

Krumsiek, J., Suhre, K., Illig, T., Adamski, J., Theis, F.J., 2011. Gaussian graphical modeling reconstructs pathway reactions from high-throughput metabolomics data. BMC Syst. Biol. 5, 21.

Khatri, P., Sirota, M., Butte, A.J., 2012. Ten years of pathway analysis: current approaches and outstanding challenges. PLoS Comput. Biol. 8 (2), e1002375.

Lander, E.S., Green, P., 1987. Construction of multilocus genetic linkage maps in humans. Proc. Natl. Acad. Sci. USA 84 (8), 2363–2367.

Lake, S.L., Blacker, D., Laird, N.M., 2000. Family-based tests of association in the presence of linkage. Am. J. Hum. Genet 67 (6), 1515–1525.

Laird, N.M., Horvath, S., Xu, X., 2000. Implementing a unified approach to family-based tests of association. Genet. Epidemiol. 19 (Suppl. 1), S36–S42.

Laird, N.M., Lange, C., 2006. Family-based designs in the age of large-scale gene-association studies. Nat. Rev. Genet. 7 (5), 385–394.

Lee, S., Abecasis, G.R., Boehnke, M., Lin, X., 2014. Rare-variant association analysis: study designs and statistical tests. Am. J. Hum. Genet. 95 (1), 5–23.

Li, B., Leal, S.M., 2008. Methods for detecting associations with rare variants for common diseases: application to analysis of sequence data. Am. J. Hum. Genet. 83 (3), 311–321.

Liu, D.J., Leal, S.M., 2010. A novel adaptive method for the analysis of next-generation sequencing data to detect complex trait associations with rare variants due to gene main effects and interactions. PLoS Genet. 6 (10), e1001156.

Lin, D.Y., Tang, Z.Z., 2011. A general framework for detecting disease associations with rare variants in sequencing studies. Am. J. Hum. Genet. 89 (3), 354–367.

Lee, S., Wu, M.C., Lin, X., 2012. Optimal tests for rare variant effects in sequencing association studies. Biostatistics 13 (4), 762–775.

Liptak, T., 1958. On the combination of independent tests. Magy. Tud. Akad. Mat. Kutato Int. Kozl. 3, 171–197.

Lee, S., Teslovich, T.M., Boehnke, M., Lin, X., 2013. General framework for meta-analysis of rare variants in sequencing association studies. Am. J. Hum. Genet. 93 (1), 42–53.

Langfelder, P., Zhang, B., Horvath, S., 2008. Defining clusters from a hierarchical cluster tree: the dynamic tree cut package for R. Bioinformatics 24 (5), 719–720.

Morton, N.E., 1955. Sequential tests for the detection of linkage. Am. J. Hum. Genet. 7 (3), 277–318.

Myers, R.H., MacDonald, M.E., Koroshetz, W.J., et al., 1993. De novo expansion of a (CAG)n repeat in sporadic Huntington's disease. Nat. Genet. 5 (2), 168–173.

Metzker, M.L., 2010. Sequencing technologies the next generation Nat. Rev. Genet. 11 (1), 31–46.

Manolio, T.A., Collins, F.S., Cox, N.J., et al., 2009. Finding the missing heritability of complex diseases. Nature 461 (7265), 747–753.

Madsen, B.E., Browning, S.R., 2009. A groupwise association test for rare mutations using a weighted sum statistic. PLoS Genet. 5 (2), e1000384.

Morgenthaler, S., Thilly, W.G., 2007. A strategy to discover genes that carry multi-allelic or mono-allelic risk for common diseases: a cohort allelic sums test (CAST). Mutat. Res. 615 (1–2), 28–56.

Morris, A.P., Zeggini, E., 2010. An evaluation of statistical approaches to rare variant analysis in genetic association studies. Genet. Epidemiol. 34 (2), 188–193.

Neale, M.C., Cardon, L.R., 1992. North Atlantic Treaty Organization. Scientific Affairs Division. Methodology for Genetic Studies of Twins and Families. Kluwer Academic Publishers, Dordrecht, Boston.

Neale, B.M., Rivas, M.A., Voight, B.F., et al., 2011. Testing for an unusual distribution of rare variants. PLoS Genet. 7 (3), e1001322.

Okada, Y., Wu, D., Trynka, G., et al., 2014. Genetics of rheumatoid arthritis contributes to biology and drug discovery. Nature 506 (7488), 376–381.

Price, A.L., Patterson, N.J., Plenge, R.M., Weinblatt, M.E., Shadick, N.A., Reich, D., 2006. Principal components analysis corrects for stratification in genome-wide association studies. Nat. Genet. 38 (8), 904–909.

Price, A.L., Kryukov, G.V., de Bakker, P.I., et al., 2010. Pooled association tests for rare variants in exon-resequencing studies. Am. J. Hum. Genet. 86 (6), 832–838.

Pan, W., 2009. Asymptotic tests of association with multiple SNPs in linkage disequilibrium. Genet. Epidemiol. 33 (6), 497–507.

Purcell, S., Neale, B., Todd-Brown, K., et al., 2007. PLINK: a tool set for whole-genome association and population-based linkage analyses. Am. J. Hum. Genet. 81 (3), 559–575.

Rabinowitz, D., Laird, N., 2000. A unified approach to adjusting association tests for population admixture with arbitrary pedigree structure and arbitrary missing marker information. Hum. Hered. 50, 211–223.

Rabinowitz, D., 1997. A transmission disequilibrium test for quantitative trait loci. Hum. Hered. 47 (6), 342–350.

Spielman, R.S., McGinnis, R.E., Ewens, W.J., 1993. Transmission test for linkage disequilibrium: the insulin gene region and insulin-dependent diabetes mellitus (IDDM). Am. J. Hum. Genet. 52, 506–516.

Sham, P.C., Curtis, D., 1995. An extended transmission/disequilibrium test (TDT) for multi-allele marker loci. Ann. Hum. Genet. 59 (Pt 3), 323–336.

Spielman, R.S., Ewens, W.J., 1996. The TDT and other family-based tests for linkage disequilibrium and association. Am. J. Hum. Genet. 59 (5), 983–989.

Spielman, R.S., Ewens, W.J., 1998. A sibship test for linkage in the presence of association: the sib transmission/disequilibrium test. Am. J. Hum. Genet. 62 (2), 450–458.

Schaid, D.J., Li, H., 1997. Genotype relative-risks and association tests for nuclear families with missing parental data. Genet. Epidemiol. 14 (6), 1113−1118.

Sun, J., Zheng, Y., Hsu, L., 2013. A unified mixed-effects model for rare-variant association in sequencing studies. Genet. Epidemiol. 37 (4), 334−344.

Tang, Z.Z., Lin, D.Y., 2013. MASS: meta-analysis of score statistics for sequencing studies. Bioinformatics 29 (14), 1803−1805.

Team, R.C., 2012. R: a language and environment for statistical computing. In: Computing RFfS. Vienna, Austria.

Wijsman, E.M., Daw, E.W., Yu, X., et al., 2005. APOE and other loci affect age-at-onset in Alzheimer's disease families with PS2 mutation. Am. J. Med. Genet. B Neuropsychiatr. Genet. 132B (1), 14−20.

Wu, M.C., Lee, S., Cai, T., Li, Y., Boehnke, M., Lin, X., 2011. Rare-variant association testing for sequencing data with the sequence kernel association test. Am. J. Hum. Genet. 89 (1), 82−93.

Wixon, J., Kell, D., 2000. The Kyoto encyclopedia of genes and genomes−KEGG. Yeast 17 (1), 48−55.

Yip, A.M., Horvath, S., 2007. Gene network interconnectedness and the generalized topological overlap measure. BMC Bioinforma. 8, 22.

Zhang, B., Horvath, S., 2005. A general framework for weighted gene co-expression network analysis. Stat. Appl. Genet. Mol. Biol. 4, Article17.

Human Pharmacology

Chapter 20

Introduction to Clinical Pharmacology

Rommel G. Tirona and Richard B. Kim
The University of Western Ontario, London, ON, Canada

Chapter Outline

Key Points

- Drug-metabolizing enzymes and transporters continue to be key determinants of variation in drug responsiveness.
- A number of drug transporters, particularly efflux transporters, such as P-glycoprotein and the breast cancer resistance protein (BCRP), are proving to be clinically relevant to the disposition and organ-specific entry of many drugs in clinical use.
- Drug interactions involving drug transporters are proving to be more common and clinically important.
- A handful of cytochrome P450 (CYP) enzymes, particularly CYP3A4, CYP2D6, CYP2C9, and CYP2C19 account for the metabolism of most of the clinically prescribed drugs. Therefore understanding the expression and function of these enzymes have major utility in predicting drug interaction, toxicity, and lack of efficacy.
- Commonly occurring genetic variation in CYP enzymes such as CYP2D6, CYP2C19, and CYP2C9 have proven to be key predictors of variation in the response to a number of drugs that are metabolized by such enzymes.
- Coordinate regulation of drug-metabolizing enzymes and transporters by xenobiotic-sensing nuclear receptors, such as pregnane X receptor (PXR) and constitutive androstane receptor (CAR), are essential to the metabolism and clearance of a large number of endogenous as well as xenobiotic chemicals that humans are exposed to.
- Pharmacokinetics is defined as quantitative approaches to describe and predict the time course of drug concentrations in the body. Pharmacokinetic parameters such as clearance, volume of distribution, and half-life are key concepts that allow for the systematic and accurate prediction of drug elimination in humans.

INTRODUCTION: MECHANISMS OF DRUG DISPOSITION AND INTERACTIONS

Pathways governing drug disposition have been broadly defined by the terms absorption, distribution, metabolism, and excretion (ADME). For many drugs in clinical use, enzymatic biotransformation to either an inactive metabolite, or in some cases, bioactivation to the therapeutically relevant molecule has long been noted as the critical step in overall dispositions of most drugs (Murray, 1992). Indeed, there is now a wealth of clinical information that supports the important role of drug-metabolizing enzymes on drug disposition and the importance of variation in the expressed level or activity in such enzymes to the observed intersubject variation in drug responsiveness (Guengerich, 1995). However, there is now an increasing appreciation of the role of cell membrane-bound carrier proteins, referred to as transporters, in the absorption, distribution, and excretion of drugs in clinical use (Ho and Kim, 2005). Traditionally, simple physicochemical properties of the drug—such as lipophilicity, pKa, ionization, solubility, and molecular weight—had been considered to be major determinants governing the movement of drug across cellular compartments or organs (Lipinski, 2000). However, emerging evidence from molecular studies clearly demonstrates that targeted and often organ-specific expression of drug uptake and efflux transporters define the extent of drug entry, tissue distribution, and elimination by organs, such as the liver and kidney (Fig. 20.1). Therefore the coordinated expression and function of drug disposition genes in organs such as the intestine, kidney, and liver confer an individual's capacity for drug elimination while inhibition or induction of such pathways result in unexpected drug toxicity or loss of efficacy. Moreover, a number of drug transporters have now been recognized as clinically important for drugs in clinical development (Giacomini et al., 2010).

TRANSPORTERS AND DRUG ABSORPTION, DISTRIBUTION, AND EXCRETION

Until the relatively recent molecular cloning and identification of membrane-bound carrier proteins broadly referred to as drug transporters, there had been little appreciation of such processes as major determinants affecting the pharmacokinetic profile of a given drug. We know that for most drugs in clinical use today an array of transporters are importantly involved in the processes that determine their absorption, distribution, and excretion. Drug transporters can be broadly categorized into two major classes: uptake and efflux transporters (Fig. 20.1, Tables 20.1 and 20.2). Uptake transporters facilitate the translocation of drugs into cells. There are many types of uptake transporters, but key transporters of relevance to cellular drug uptake appear to be members of the organic anion transporting polypeptide (OATP; *SLCO*) (Hagenbuch and Meier, 2003), organic anion transporter (OAT; *SLC22A*) (Russel et al., 2002), and organic cation transporter (OCT; *SLC22A*) families (Table 20.1) (Jonker and Schinkel, 2004). By contrast, efflux transporters function to export drugs from the intracellular to the extracellular milieu, often against high-concentration gradients. Most efflux transporters are members of the ATP-binding cassette (ABC) superfamily of transmembrane proteins that utilize energy derived from ATP hydrolysis

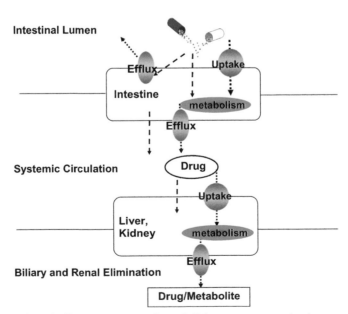

FIGURE 20.1 Schematic of drug uptake and efflux transporters, and metabolizing enzyme expression in organs such as the intestine and liver that governs drug absorption, distribution, metabolism, and excretion (ADME).

TABLE 20.1 Key Human Drug Uptake Transporters

Family	Member	Tissue Distribution	Cellular Localization	Examples of Typical Substrates
SLCO	OATP1A2	Brain, kidney, liver	Basolateral	Fexofenadine, rosuvastatin, bile salts, deltorphin, pitavastatin, methotrexate, hormone conjugates, eicosanoids
	OATP2B1	Liver, intestine, placenta	Basolateral	Bile salts, digoxin, fexofenadine, hormone conjugates
	OATP1B1	Liver	Basolateral	Pravastatin, rosuvastatin, atorvastatin, pitavastatin, cerivastatin, bile salts, methotrexate, bilirubin, enalapril, valsartan, rifampicin, hormone conjugates, eicosanoids, benzylpenicillin
	OATP1B3	Liver	Basolateral	Digoxin, methotrexate, rifampicin, bile salts, hormone conjugates, eicosanoids, statins, temocaprilat, telmisartan, repaglinide
SLC22	OAT1	Kidney, brain	Basolateral	Cidofovir, pulmonary arterial hypertension (PAH), acyclovir, tetracycline
	OAT3	Kidney, brain	Basolateral	Cimetidine, PAH, methotrexate, salicylate, valacyclovir, tetracycline
	OAT4	Kidney, placenta	Apical	PAH, tetracycline
	OCT1	Liver, brain, small intestine	Basolateral	Cimetidine, metformin, corticosteroids, quinidine, quinine
	OCT2	Kidney, brain, small intestine	Basolateral	Amantadine, metformin, choline, dopamine, histamine, norepinephrine, serotonin

TABLE 20.2 Key Human Drug Efflux Transporters

Family	Member	Tissue Distribution	Cellular Localization	Examples of Typical Substrates
ABCB	MDR1 (P-gp)	Kidney, liver, brain, small intestine	Apical	Digoxin, cyclosporine, taxol, vinca alkaloids, doxorubicin, loperamide, erythromycin, HMG-CoA reductase inhibitors, HIV-1 protease inhibitors
ABCC	MRP1	Ubiquitous	Basolateral	Vinca alkaloids, methotrexate, etoposide
	MRP2	Liver, kidney, small intestine	Apical	Vinca alkaloids, methotrexate, pravastatin, ampicillin, ceftriaxone, cisplatin, irinotecan, hormone conjugates
	MRP3	Liver, kidney, small intestine	Basolateral	Doxorubicin, vincristine, methotrexate, cisplatin
	MRP4	Kidney, brain, liver	Apical, basolateral	PAH, bile acids, AZT, methotrexate, 6-mercaptopurine
ABCG	BCRP	Placenta, liver, small intestine	Apical	Mitoxantrone, doxorubicin, topotecan, methotrexate, irinotecan (SN-38)

to actively extrude substrate drugs across biological membranes (Schneider and Hunke, 1998). Included within this class of transporters of particular relevance to drug elimination are P-glycoprotein (P-gp) (MDR1; *ABCB1*), certain members of the multidrug resistance-associated protein (MRP; *ABCC*) family, and the breast cancer resistance protein (BCRP; *ABCG*) (Table 20.2).

Important to our understanding of transporter-mediated drug disposition is the dynamic interplay between uptake and efflux transporters within any given epithelial cells, where the net uptake of drugs across such cells may be impeded or facilitated by the localization of transporters on apical or basolateral membranes (Fig. 20.1). For many drugs that undergo extensive metabolism, drug transporter expression may also impact the extent of intracellular concentration of the substrate drug available for metabolic biotransformation. Thus in many ways the coordinated function of transporter proteins with drug-metabolizing enzymes in organs such as the liver and intestine appears to be the way by which most drugs are ultimately eliminated from the body (Kim, 2002b).

Intestinal Transporters

In the small intestine, enterocytes possess a number of transporters critical for absorption of dietary constituents and drugs (Oude Elferink and de Waart, 2007). An important function of the intestinal epithelia appears to be in preventing the absorption of potential toxins and xenobiotics. Indeed, many drugs that have low or minimal oral bioavailability appear to interact with efflux transporters expressed in the intestine, and are thus unable to translocate across the intestinal epithelial cells. Two key efflux transporters that account for the bulk of the observed loss in oral drug absorption are noted later.

Multidrug Resistance Protein 1 (P-Glycoprotein)

P-gp, the gene product of MDR1, is a widely studied efflux transporter known for its ability to limit drug entry into various organ compartments. Human P-gp is a phosphorylated and glycosylated transmembrane protein that is 1280 amino acids long and is composed of two homologous and symmetrical sequences, each of which contains six transmembrane domains and an ATP-binding motif (Schneider and Hunke, 1998). P-gp functions as an efflux pump thereby facilitating the intracellular to extracellular movement of drugs. ATP hydrolysis provides the energy for active drug transport, enabling the transporter to function against steep concentration gradients. The key role of P-gp to drug absorption has been exemplified by studies in knockout mice with disruption of the homologous *mdr1a* gene (Schinkel, 1997). The oral bioavailability of paclitaxel (Sparreboom et al., 1997), digoxin (Schinkel et al., 1995), and HIV-1 protease inhibitors (Kim et al., 1998) is markedly increased in *mdr1a* knockout mice in comparison to wild-type mice indicating P-gp mediated drug efflux by enterocytes prevents adequate drug absorption. In humans, an extent of intestinal P-gp expression and activity has been shown to influence drug levels after administration of cyclosporin (Lown et al., 1997a,b) and digoxin (Drescher et al., 2003). Given the broad substrate specificity of P-gp for many structurally diverse drugs in clinical use (Kim, 2002a), it is likely that clinicians have underestimated the importance of this efflux transporter oral drug bioavailability and drug interactions.

Breast Cancer Resistance Protein (Also Known as ABCG2)

This efflux transporter was first cloned from mitoxantrone- and anthracycline-resistant breast and colon cancer cells (Miyake et al., 1999). Since BCRP has only one ATP-binding cassette and six putative transmembrane domains, it was suggested that BCRP is a half-transporter. However, it has recently been determined that BCRP may function as a homotetramer (Rocchi et al., 2000). In addition to the intestine, BCRP is expressed in the liver, placenta, heart, ovary, kidney, and brain (Eisenblatter et al., 2003). Interestingly, Bcrp1-deficient mice develop protoporphyria and diet-dependent phototoxicity (van Herwaarden et al., 2003). It appears that BCRP prevents heme or porphyrin accumulation in cells, enhancing hypoxic cell survival. There appears to be a remarkable substrate overlap with P-glycoprotein. Indeed, many anticancer drugs are substrates of this efflux transporter. In addition, nutraceuticals including the vitamin folic acid are known substrates of this transporter (Allen and Schinkel, 2002; Jonker et al., 2005). An unexpected function of this transporter has been in the mammary gland where the expression of this transporter has been implicated in the secretion of drugs and potential toxins and drugs into breast milk (Jonker et al., 2005; Merino et al., 2006).

Hepatic Transporters

Organic Anion Transporting Polypeptide

In the liver, an efficient extraction of drugs from the portal blood into hepatocytes is often mediated by uptake transporters expressed on the sinusoidal (basolateral) membrane (see Table 20.1; Fig. 20.2). Emerging evidence strongly supports the role of OATP transporters that are highly expressed in the liver, such as OATP1B1 (previously known as OATP-C or OATP2), OATP1B3 (previously known as OATP8), and OATP2B1 in the hepatic uptake of drug and hormone conjugates (Tirona and Kim, 2007). For example, the hepatic uptake of the HMG-CoA reductase inhibitor pravastatin is dependent on OATP1B1 and its activity is thought to be the rate-limiting step in pravastatin hepatic clearance (Nakai et al., 2001). Once a

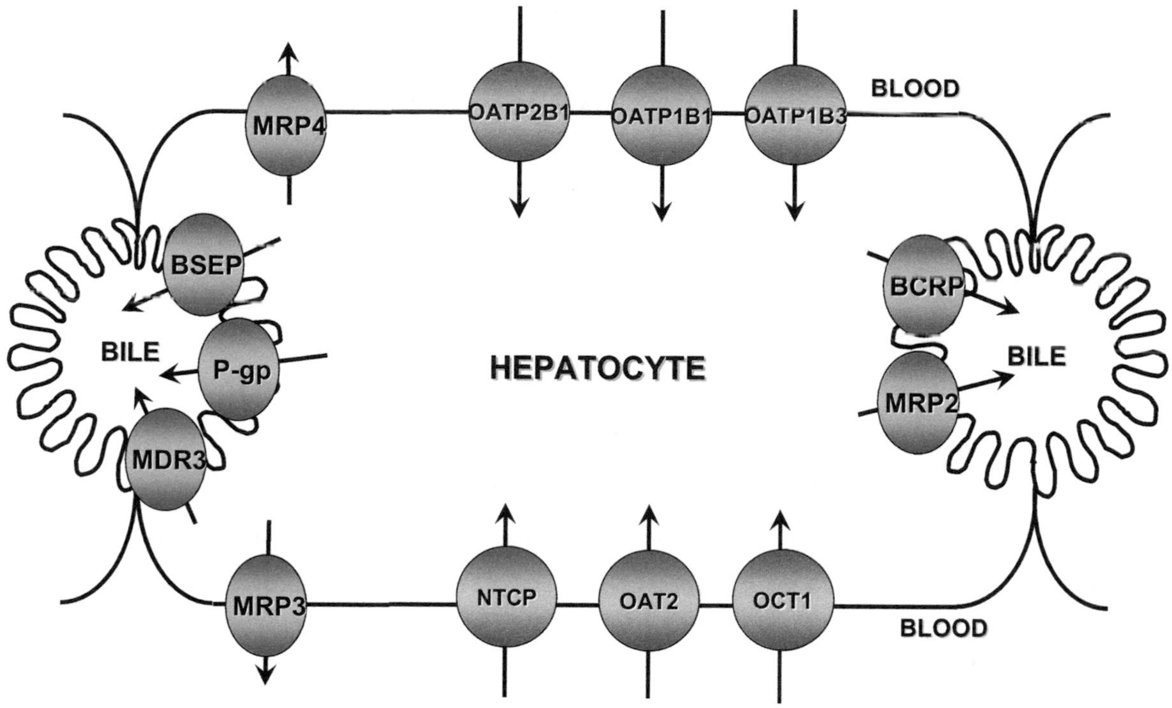

FIGURE 20.2 Expression of drug uptake and efflux transporters on the basolateral and canalicular membrane domain of hepatocytes. Coordinate expression and function of liver-enriched uptake and efflux transporters are critical to the extraction of drugs from the portal circulation and excretion of drug or drug metabolites into bile.

drug gains access into hepatocytes, it often undergoes metabolism mediated by phase I and II enzymes or may be secreted unchanged. Efflux transporters localized on the canalicular (apical) membrane of the hepatocyte, such as MDR1, MRP2, and BCRP, represent the final step in the vectorial transport of drugs from portal circulation into bile (Keppler and Konig, 1997; Silverman and Schrenk, 1997; Bohan and Boyer, 2002; Kato et al., 2002).

Among the various OATP transporters, OATP1B1, encoded by the *SLCO1B1* gene, has been found to be of particular relevance to statin-induced myopathy. Specifically, functional loss of single nucleotide polymorphisms (SNPs) in this transporter, c.521T>C (rs4149056), first reported in 2001 (Tirona et al., 2001), in a recent large clinical trial using genome-wide analysis, was shown to be the key predictor of myopathy among patients on high-dose simvastatin (Link et al., 2008). Genetic variation in this transporter has been shown to be of relevance to other statins as well as other drug substrates (Degorter et al., 2013; Gong and Kim, 2013).

Multidrug Resistance Protein 2 (ABCC2)

In addition to MDR1, a transporter previously referred to as the canalicular multispecific OAT (cMOAT), now referred to as MRP2, is responsible for the biliary excretion of numerous endogenous organic anions including bilirubin glucuronides as well as drugs such as methotrexate (Kruh et al., 2001), irinotecan (CPT-11) (Chu et al., 1997), and pravastatin (Sasaki et al., 2002). MRP2 was first cloned from rat and originally designated as a cMOAT due to its predominant localization in canalicular membranes of hepatocytes (Ito et al., 1997). MRP2 was also found in the apical membranes of enterocytes (Nakamura et al., 2002) and epithelial cells of proximal tubules in the kidney (Schaub et al., 1997). Rats lacking functional Mrp2 expression (i.e., Wistar TR⁻ and Eisai rat strains) are hyperbilirubinemic as a result of their inability to excrete bilirubin conjugates into bile (Paulusma et al., 1995; Ito et al., 1997), thus suggesting that bilirubin glucuronide conjugates are important substrates of MRP2 (Jager et al., 2003). Similarly, the absence of MRP2 in humans results in Dubin—Johnson syndrome (DJS), a relatively benign condition characterized by conjugated hyperbilirubinemia (Materna and Lage, 2003).

Renal Transporters

In the kidney, drug secretion represents the coordinate function of uptake and efflux transporters localized to the basolateral and apical membranes of proximal tubular cells. Members of the OAT family appear to be important renal transporters for

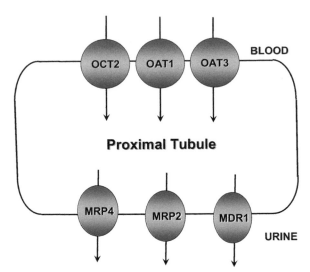

FIGURE 20.3 Expression of drug uptake and efflux transporters on the basolateral and apical membrane domain of renal tubular cells. For many renally eliminated drugs, net urinary secretion as well as extent of tubular reabsorption is often defined by expression and interplay of kidney-enriched drug uptake and efflux transporters. *BSEP*, bile salt efflux pump; *MDR3*, multidrug resistance 3.

uptake of organic anions (Table 20.2, Fig. 20.3). OAT substrates include a wide variety of clinically important anionic drugs, such as -lactam antibiotics, diuretics, nonsteroidal antiinflammatory drugs (NSAIDs), nucleoside antiviral drugs, and anticancer agents (Marzolini et al., 2004). For the most part, OATs are expressed on the basolateral side of renal tubular cells, thus facilitating the extraction of drugs from circulation (Lee and Kim, 2004). Drugs that are taken up by renal tubular cells are typically secreted into urine by efflux transporters. Like liver, a complement of efflux transporters are expressed on the apical (luminal) side of renal tubular cells and efflux substrate drugs to the tubular lumen. Of the various efflux transporters noted to be expressed in the kidney, major transporters of clinical relevance are MDR1 and members of the MRP (ABCC) family. Specifically, MRP2 and MRP4 are highly expressed and they participate in the urinary secretion of many drug—drug conjugates (Sampath et al., 2002; Sandusky et al., 2002; Ritter et al., 2005; Nies and Keppler, 2007). MRP4 appears to be involved in the efflux transport of monophosphate metabolites of nucleoside analogs, HIV reverse transcriptase inhibitors (Adachi et al., 2002), and anticancer agents such as 6-mercaptopurine (Ritter et al., 2005).

Blood—Brain Barrier

In organs such as the brain, targeted transporter expression is critical to the maintenance of barrier function. For example, the blood—brain barrier (BBB) serves a protective function by limiting access of drugs and toxic substances into the central nervous system (CNS). The BBB function is maintained by brain capillary endothelial cells, whose tight junctions effectively limit the entry of drugs via the paracellular route, and is further aided through the expression of key efflux transporters (Lee et al., 2001). Indeed, transporters such as MDR1, localized to the luminal side of the BBB endothelial cells, prevent the CNS entry of many substrate drugs (Cordon-Cardo et al., 1989). The importance of MDR1 expression at the level of the BBB has been shown in studies using *mdr1a* knockout mice (Schinkel et al., 1994). These mice are viable and fertile, with no apparent phenotypic abnormalities. However, *mdr1a* knockout mice have been shown to be 50- to 100-fold more sensitive to the neurotoxic pesticide ivermectin and the accumulation of this drug in the brain tissue of *mdr1a* (−/−) mice was noted to be 80- to 100-fold greater when compared to control mice (Schinkel et al., 1994). Additional studies have demonstrated that the CNS entry of a number of MDR1 drug substrates, such as digoxin, quinidine, tacrolimus, and HIV-1 protease inhibitors, is profoundly limited by MDR1 expression at the BBB (Kim et al., 1998; Fromm et al., 1999; Yokogawa et al., 1999).

DRUG-METABOLIZING ENZYMES

Drug-metabolizing enzymes are broadly categorized into two groups: phase I enzymes such as the cytochrome P450 (CYP) monooxygenases and phase II enzymes responsible for drug conjugation. For the most part, studies of drug disposition and interaction have focused primarily on the members of CYP enzymes (Guengerich, 1995). This is not surprising, given the extensive and often critical role CYP enzymes play in the metabolic biotransformation of drugs. Currently, hundreds of CYP enzymes from bacteria to man have been identified (Nelson et al., 1996). However, in humans, only a few key CYPs

TABLE 20.3 Drug Substrates and Inhibitors of Cytochrome P450 (CYP) Enzymes

Enzyme	Drug Substrates	Inhibitors
CYP1A2	Caffeine, clomipramine, theophylline	Cimetidine, fluvoxamine, ticlopidine, fluoroquinolones
CYP2C9	Diclofenac, ibuprofen, piroxicam, losartan, irbesartan, celecoxib, tolbutamide, tamoxifen, warfarin	Fluconazole, fluvastatin
CYP2C19	Omeprazole, lansoprazole, pantoprazole, (S)-mephenytoin, nelfinavir, diazepam	Cimetidine, fluvoxamine
CYP2D6	Amitriptyline, desipramine, imipramine, paroxetine, haloperidol, thioridazine mexiletine, propafenone, propranolol, metoprolol, timolol codeine, dextromethorphan, hydrocodone, tamoxifen	Cimetidine, fluoxetine, paroxetine, quinidine, amiodarone, ritonavir
CYP3A4	Diltiazem, felodipine, nimodipine, nifedipine, nisoldipine, nitrendipine, verapamil, cyclosporine A, tacrolimus, budesonide, cortisol, 17 β-estradiol, progesterone, testosterone, clarithromycin, erythromycin, cyclophosphamide, tamoxifen, vincristine, vinblastine, ifosfamide, alprazolam, midazolam, triazolam, alfentanyl, fentanyl, sufentanil, lovastatin, simvastatin, atorvastatin, indinavir, nelfinavir, ritonavir, saquinavir, amprenavir, atazanavir, quinidine, sildenafil	Ketoconazole, itraconazole, fluconazole, clarithromycin, erythromycin, nefazodone, ritonavir, grapefruit juice

appear to participate in the metabolism of drugs in clinical use today (Wrighton and Stevens, 1992). As shown in Table 20.3, CYP enzymes with the greatest potential for drug metabolism are CYP2C9, CYP2C19, CYP2D6, CYP3A4, and CYP3A5. Nevertheless, it has been noted that CYP3A4 alone may be involved in the metabolism of nearly 50% of all drugs currently prescribed (Thummel and Wilkinson, 1998), and its inhibition appears to be the basis of a number of clinically important drug interactions.

CYP2C9/CYP2C19

CYP2C9 and CYP2C19 are homologous enzymes important to human drug metabolism. In fact, CYP2C9 is among the most highly expressed CYP enzymes in human liver, second only to CYP3A4. While the CYP2C enzymes are all found in liver, they are also in the intestine (Paine et al., 2006). Despite sequence homology, the CYP2C enzymes differ in substrate specificities. CYP2C9 is well known for its ability to metabolize the anticoagulant, S-warfarin, and the NSAIDs. On the other hand, the anticonvulsant drug phenytoin, the gastric acid suppressor omeprazole, and the antiplatelet agent clopidogrel are significantly metabolized by CYP2C19. Genetic variation in *CYP2C9* and *CYP2C19* are the cause of clinically important variation in the drug response. Nonsynonymous polymorphisms associated with *CYP2C9*2* and *CYP2C9*3* alleles are approximately present in 5% of individuals globally and are associated with reduced enzymatic activity (Sullivan-Klose et al., 1996; Rettie et al., 1994). They influence the dose of substrate drugs including warfarin and phenytoin (Rettie et al., 1999; Furuya et al., 1995). *CYP2C19* genetic polymorphisms were originally described as S-mephenytoin polymorphism (Kupfer and Preisig, 1984). The major genetic variants causing null function are *CYP2C19*2* (aberrant splicing) and *CYP2C19*3* (premature stop codon), while a promoter variant named *CYP2C19*17* results in increased enzyme expression (de Morais et al., 1994; Wrighton et al., 1993; Sim et al., 2006). CYP2C19 poor metabolism affects approximately 25% of Asians and 5% of Caucasians. In contrast, an increased CYP2C19 activity afforded by the *CYP2C19*17* allele is common in Caucasians (25%) but rare in Asians. Genetic variation in *CYP2C19* has a strong influence in the efficacy of proton pump inhibitors for the treatment of gastric *Helicobacter pylori* infection and ulcers with extensive metabolizers (EMs) having dramatically poorer cure rates than poor metabolizers (PMs) (Furuta et al., 1998). Moreover, the effectiveness of the clopidogrel in the prevention of stent thrombosis after percutaneous coronary intervention is dependent on the formation of an active metabolite and a CYP2C19 genotype (Mega et al., 2009).

CYP2D6

CYP2D6 is expressed mainly in liver, and although this enzyme represents ~3% of the hepatic CYP content, it metabolizes ~20% of drugs. Antidepressants, antiarrhythmics, beta-blockers, and opioid analgesics are typical substrates of CYP2D6. There is a tremendous variability in liver CYP2D6 content where in some individuals no protein is present (Zanger et al., 2001). Among the CYP enzymes, CYP2D6 stands out for its relative inability to be induced by xenobiotic exposure. Clinically significant inhibition of CYP2D6 leading to decreased metabolism of substrate drugs occurs with cotreatment with quinidine, cinacalcet, and serotonin reuptake inhibitors. By far the major determinant of CYP2D6 activity and expression in liver is genetics. Indeed, the pharmacogenetics of *CYP2D6* has a long history dating from the 1960s with the description of genetic control of nortriptyline plasma levels (Alexanderson et al., 1969) and the discoveries of hereditary deficiencies in the metabolism of debrisoquine and sparteine and the eventual identification of the molecular mechanisms (Distlerath et al., 1985; Gonzalez et al., 1988). There are over 100 documented alleles of *CYP2D6* with a number of variants more common in different ethnicities (www.cypalleles.ki.se). *CYP2D6*6* (null function, splice aberration) occurs in one in four Caucasians and is the major cause of CYP2D6 poor metabolism. In Asians, *CYP2D6*10* (decreased activity, nonsynonymous polymorphism) and *CYP2D6*41* (decreased expression, intronic SNP causing aberrant splicing) are found in up to 50% and 10% of this population, respectively. Moreover, *CYP2D6*17* (decreased activity, nonsynonymous polymorphism) is present in 30% of Africans. A *CYP2D6* gene deletion (*CYP2D6*5*) is present in 5% of all populations. *CYP2D6* gene duplications (up to 13 copies) have been described for functional and nonfunctional alleles with ultrarapid metabolizer phenotype being particularly common in Africans. Such complex CYP2D6 genotype–phenotype relationships among various populations have rendered a personalized medicine approach to tailoring pharmacotherapy of CYP2D6 substrate drugs rather than a challenging one (Gaedigk et al., 2008). It has been reported that a drug dose, response, and toxicity risk of beta-blockers and antidepressants is highly dependent on CYP2D6 pharmacogenetics (Hicks et al., 2013). Moreover, the bioactivation of codeine to morphine requires CYP2D6 leading to reports of genetic influence of drug safety (Koren et al., 2006). Where there has been significant interest in CYP2D6 pharmacogenetics has been in the use of the estrogen receptor blocker, tamoxifen, in the adjuvant treatment of postmenopausal breast cancer (Brauch et al., 2013) as the active metabolite, endoxifen, is formed by two-step processes involving CYP2D6 (Dehal and Kupfer, 1997).

CYP3A4/CYP3A5

CYP3A4 is often considered the most important drug-metabolizing enzyme, given its relatively high expression in liver and intestine. Certainly, CYP3A4 is among the most abundant CYP enzymes in liver composing approximately 15−20% of hepatic CYP content (Kawakami et al.) and is clearly the key CYP enzyme present in small intestinal enterocytes (Kolars et al., 1992; Paine et al., 2006). Hence, CYP3A4 is a major component of the oral first-pass effect. There is a high interindividual variability in hepatic CYP3A4 expression ranging up to 100-fold (Westlind-Johnsson et al., 2003). Interestingly, CYP3A4 expression in liver and intestine do not appear coregulated (Thummel et al., 1996). It is estimated that up to 50% of all drugs are metabolized by CYP3A4 and that substrate drugs can be found in almost all therapeutic drug classes (Wilkinson, 2005). Variation in CYP3A4 activity is unimodal (Wilkinson, 2005) and despite a significant environmental component contributing to enzyme expression, it remains considered that genetics plays an important role in interindividual difference in CYP3A4-mediated drug metabolism (Ozdemir et al. 2000). An intron 6 polymorphism in the *CYP3A4* gene (*CYP3A4*22*) explains some of this heritability as this variation is associated with reduced hepatic CYP3A4 expression and altered plasma drug levels (Wang et al., 2011).

CYP3A5 is expressed in liver in only 10% of Caucasians but up to 60% of Africans. On an average, CYP3A5 is expressed at 20-fold lower levels than CYP3A4 in individuals who express this CYP isoform and harbor the *CYP3A5*1* allele. Nonexpression of CYP3A5 is commonly caused by genetic polymorphisms that produce aberrant splicing and truncated protein (*CYP3A5*3*, *CYP3A5*6*) (Kuehl et al., 2001). CYP3A5 genetic variation is an important contributor to tacrolimus dose requirements (Macphee et al., 2002) and a risk for renal toxicity (Tang et al., 2011).

DRUG–DRUG INTERACTIONS

Clinically Important Drug-Metabolism–Associated Interactions

For some drugs, multiple pathways of disposition or wide margins between therapeutic and toxic blood levels mean that clinically important drug toxicity or interactions rarely occur (see also Chapter 18). For drugs that rely on a single metabolic pathway, genetic or drug-related changes in function are increasingly recognized as causes for sporadic but occasionally life-threatening drug toxicity. Of particular importance, drug–drug interactions occurring at the level of CYP3A4 are a significant cause of adverse drug effects (Dresser et al., 2000). A number of commonly used medications are potent in vivo inhibitors of

CYP3A4 including azole antifungal agents, macrolide antibiotics, and calcium channel blockers. Mechanism-based inhibition of CYP3A4 has been purposeful therapeutic strategy in the treatment of HIV infection, whereby low-dose ritonavir is coadministered with other HIV protease inhibitors to improve bioavailability and pharmacokinetic profile (Kempf et al., 1997). The well-known grapefruit juice interaction with medication is the result of mechanism-based inactivation and loss of intestinal CYP3A4 by furanocoumarin grapefruit constituents (Bailey et al., 1991; Edwards et al., 1996; Lown et al., 1997a,b; Schmiedlinren et al., 1997; Bailey et al., 1998). This food–drug interaction can cause up to an 800% increase in blood levels of drugs metabolized by CYP3A4 (Lown et al., 1997a,b). Other drug interactions involving CYP3A4 relate to the enzyme's profound inducibility. Rifampin, antiepileptic drugs, and St. John's Wort are typical drugs that can strongly upregulate the intestinal and hepatic expression of CYP3A4. A dramatic example of CYP3A4 induction is the 96% reduction of midazolam blood level after oral administration following pretreatment with rifampin (Backman et al., 1996). Another example of metabolic drug interaction involves the nonsedating antihistamine terfenadine that undergoes rapid CYP3A4-mediated metabolism to its pharmacologically active metabolite. Normally, the blood level of the parent drug, terfenadine, is minimal. However, when a drug inhibitor of CYP3A4 such as azole antifungal agents (e.g., ketoconazole, itraconazole), or macrolide antibiotics such as erythromycin are coingested, the blood level of terfenadine rises substantially. Unfortunately, terfenadine at high blood levels exert deleterious effects on the cardiac conduction system and can lead to a frequently lethal form of ventricular arrhythmia known as torsades de pointes (Woosley et al., 1993). Not surprisingly, it has been replaced with the active noncardiotoxic metabolite fexofenadine. Similar toxic effects were also observed upon an inhibition of astemizole (Woosley, 1996) and the prokinetic agent cisapride metabolism (Sekkarie, 1997). Another class of drugs frequently cited for their predilection for interactions, especially with CYP3A4 inhibitor drugs, is the HMG-CoA reductase inhibitors such as lovastatin, atorvastatin, and simvastatin. Profound elevation in their blood levels can lead to rhabdomyolysis (Farmer and Gotto, 1994; Grunden and Fisher, 1997; Schmassmann-Suhijar et al., 1998).

Transporters and Drug Interactions

It is increasingly apparent that in addition to metabolism-associated drug–drug interactions, drug-induced alteration in transporter function can also manifest as unexpected drug–drug interactions. For example, digoxin, a drug that does not undergo significant metabolism, is handled by P-gp (Schinkel et al., 1995). Accordingly, the inhibition of P-gp transporter, by compounds such as quinidine and verapamil, may be the basis for the observed increase in digoxin levels when these agents are coadministered (Levêque and Jehl, 1995; Fromm et al., 1999).

Perhaps, the most widely appreciated drug interaction is that of penicillin and probenecid. It has been widely appreciated for over 50 years that coadministration of probenecid resulted in elevated penicillin serum levels (Burnell and Kirby, 1951). Given the overall therapeutic index of penicillin, this type of interaction was viewed as a beneficial and cost-effective drug–drug interaction. More recent studies have shown that the high renal clearance of penicillins due to avid active secretion can be decreased by inhibition of OAT-mediated transport (Fig. 20.3) on the basolateral membrane of proximal tubular cells with coadministration of probenecid (Jariyawat et al., 1999). Similar inhibitory effects of probenecid coadministration have now been extended to other anionic drugs to include certain angiotensin converting enzyme inhibitors and a number of HIV antiviral drugs (Wada et al., 2000). Another well-known kidney-associated drug interaction relates to methotrexate, a drug widely used in the treatment of various malignancies and rheumatoid arthritis. Methotrexate renal elimination occurs via glomerular filtration and active tubular secretion, and this drug is eliminated unchanged (Shen and Azarnoff, 1978). Interactions between methotrexate and drugs such as NSAIDs, probenecid, and penicillin have been reported and have resulted in severe complications including bone marrow suppression and acute renal failure (Basin et al., 1991; Ellison and Servi, 1985; Thyss et al., 1986). Accordingly, for many anionic compounds eliminated by the kidney, inhibition of OAT function by concomitantly administered drugs appears to be one key mechanism to account for the clinically observed interactions (Lee and Kim, 2004; Takeda et al., 2002). Inhibition of hepatic OATPs (OATP1B1 and OATP1B3) is increasingly recognized as a major contributor to drug interactions. Clinical relevance has been shown from studies of statin pharmacokinetics where there is a sevenfold increase in rosuvastatin plasma exposure by coadministration of cyclosporine A (Simonson et al., 2004) and an eightfold higher plasma level of atorvastatin when combined with telaprevir (Lee et al., 2011). Inhibition of other uptake transporters expressed in the kidney such as OCT2 (see Fig. 20.3, Table 20.1) is also likely an underestimated mechanism in the observed drug–drug interactions of substrate drugs.

INDUCTION AND REGULATION OF DRUG-METABOLIZING ENZYMES AND TRANSPORTERS

The extent of intersubject variation in the drug response or plasma levels results from states of enzyme or transporter inhibition or intrinsic lack of activity to increased activity or expression of such proteins. While the loss of transporter or enzyme activity often results in unexpected drug toxicities, the converse is associated with the loss of drug efficacy. The adaptive response to drug exposure that triggers an increase in enzymatic capacity for drug removal was first described in 1960 where pretreatment of rats with the barbiturate drug phenobarbital increased hepatic drug metabolic activity and shortened the duration of hypnotic effects (Conney et al., 1960). These findings led to the seminal observations in 1963 that, in humans, phenobarbital pretreatment lowers plasma levels of coumarin and phenytoin (Cucinells et al., 1963). However, it was not until molecular studies in the late 1990s that identified the ligand-activated transcription factors PXR (Kliewer et al., 1998; Lehmann et al., 1998) and constitutive androstane receptor (CAR) (Forman et al., 1998; Sueyoshi et al., 1999) that the molecular basis of induction-type drug—drug interactions become clarified. PXR and CAR are members of the nuclear receptor 1 (NR1) family of transcription factors that include hormone receptors such as the estrogen receptor (Giguere et al., 1988), progesterone receptor (Giguere et al., 1988), bile acid receptor, farnesoid X receptor (Wang et al., 1999), peroxisome proliferator-activated receptors α and γ, (Forman et al., 1996) and the vitamin D receptor (McDonnell et al., 1987; Baker et al., 1988). PXR and CAR are currently viewed as the major xenobiotic-activated modulators of drug disposition gene expression (Urquhart et al., 2007). Each of these nuclear receptors shares a common signaling mechanism involving ligand binding to the receptor, heterodimerization with the 9-cis-retinoic acid receptor (RXR), binding of the RXR heterodimer to response elements of target genes, release of corepressor proteins, and recruitment of coactivators and the general transcription machinery (Fig. 20.4) (Tirona and Kim, 2005).

Not surprisingly, a number of key transporters and CYP enzymes have been shown to possess conserved DNA sequence elements in their promoter regions that are recognized by PXR and CAR. Thus, there is both drug and target selectivity in the induction of drug transporters and metabolizing enzymes. The CYP enzymes particularly susceptible to induction by PXR and CAR include CYP2B6, CYP2C9, CYP2C19, and CYP3A4. In this context, the regulation of CYP3A4 by PXR and CAR is of special clinical relevance in that, as noted previously, CYP3A4 is responsible for the majority of human drug metabolism and also the most sensitive to the inducing effects of PXR and CAR agonist drugs (Quattrochi and Guzelian, 2001; Sueyoshi and Negishi, 2001; Tirona et al., 2003). It is therefore not surprising that inductive drug interactions are most commonly observed with CYP3A4 substrate drugs. The most widely used probe drug to assess CYP3A4 activity is the benzodiazepine midazolam. The extent of the inductive response in terms of plasma drug concentrations by different PXR activators is evidenced by the 16- and 2.7-fold decrease in oral midazolam AUC with rifampin (Floyd et al., 2003) and St. John's Wort (Dresser et al., 2003) pretreatment.

Given the frequent colocalization of transporters with CYP enzymes, it is not surprising to note that a number of drug transporters are also regulated by PXR and CAR. Molecular studies have shown that the MDR1 gene is regulated by PXR (Geick et al., 2001) and CAR (Burk et al., 2005). AUC of the MDR1 substrate drug digoxin decreased by 30% when subjects were pretreated with rifampin (Greiner et al., 1999), confirming the clinical relevance of PXR activation to MDR1 expression and drug effects. Note that although the magnitude of the change in digoxin plasma levels may appear modest, this drug has a narrow therapeutic index. Similar to when CYP3A4 is induced, induction of MDR1 would be predicted to

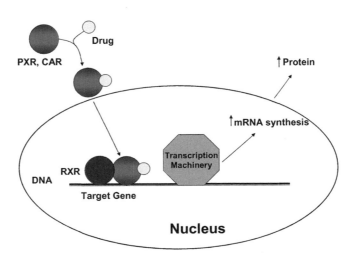

FIGURE 20.4 A schematic that outlines the mechanism that defines drug-induced induction of enzymes and transporters. Intracellular xenobiotic-sensing nuclear receptor proteins preganane X receptor (PXR) and constitutive androstane receptor (CAR) are involved in the transcriptional activation of most drug disposition genes. Drug bound PXR or CAR result in conformational changes in the nuclear receptor that result in the nuclear translocation and heterodimerization with its cognate partner, retinoid X receptor (RXR), recruitment of coactivators, and target gene-specific transcriptional activation.

result in lower substrate drug plasma levels and loss of drug effect. MRP2 expression has also been noted to be induced by treatment with rifampin (Fromm et al., 2000). Like MDR1, MRP2 also appears to be regulated by PXR and CAR (Kast et al., 2002). However, since most of the known substrates of MRP2 tend to be hormone or drug conjugates, the clinical relevance of MRP2 induction has not yet been clarified.

Clinically relevant examples and reports of drug therapy failures due to induction often involve rifampin coadministration. These include reports of oral contraceptive failure in women taking rifampin (Anonymous, 1974; Gupta and Ali, 1980; LeBel et al., 1998; Reimers and Jezek, 1971), where subsequent studies demonstrated that ethinylestradiol plasma levels were significantly reduced by rifampin therapy (Bolt et al., 1977; LeBel et al., 1998). Other reports include those of withdrawal symptoms in patients taking methadone (Kreek et al., 1976a,b; Bending and Skacel, 1977). In addition to methadone, the analgesic effect of morphine appears to dissipate more quickly among subjects treated with rifampin (Fromm et al., 1997). The loss of opioid effect may be compounded by the induction of MDR1 at the blood–brain barrier, thereby further limiting the CNS entry of morphine. In addition, we now know that widely ingested herbal remedies, such as St. John's Wort, can induce drug metabolism and transport. Case reports of St. John's Wort treatment causing transplant rejection in patients taking cyclosporin (Karliova et al., 2000; Ruschitzka et al., 2000) have appeared. Hyperforin, a constituent in St. John's Wort, has been identified as the high-affinity activator of PXR, thereby resulting in the marked induction of genes involved in the metabolism and elimination of cyclosporin (Moore et al., 2000). Like rifampin, St. John's Wort use has also been linked to methadone withdrawal (Eich-Hochli et al., 2003; Izzo, 2004; Zhou et al., 2004).

Although it is generally recognized that induction-associated drug interactions are far less common than inhibition-associated drug interactions, the recent elucidation of the molecular basis governing induction of drug-metabolizing enzymes and transporters has meant that a more predictive approach to drug synthesis and design can now be utilized during the drug discovery and development process that preemptively minimizes the risk for induction-related drug interactions.

PRINCIPLES OF PHARMACOKINETICS

Introduction to Pharmacokinetics

The fundamental principle in clinical pharmacology is that the drug response in individuals is related to the concentration of drug at the target site. For the most part, drug levels in target tissues, such as the brain, liver, or tumors, are not practically accessible for analysis and hence venous blood or plasma drug concentrations are typically obtained. Plasma drug concentrations are an appropriate surrogate measure of target site levels since, after all, the drug is delivered to organs via the bloodstream and for each drug a relationship exists in the tissue-to-plasma partitioning. Pharmacokinetics is defined as quantitative approaches to describe and predict the time course of drug concentrations in the body or, in most cases, plasma. The purpose of pharmacokinetics is to define the dose–response relationship, be it pharmacologic or toxic effects, for application to the therapeutic management of patients.

Overall drug disposition is determined by the net effects of the biochemical processes that govern cell membrane permeability, biotransformation, and protein binding of drugs together with the physiologic variables such as gastrointestinal motility, organ blood perfusion, and urine flow. The resulting magnitude and time course of drug levels can be quantitatively characterized by four basic pharmacokinetic parameters. **Clearance** is a parameter that describes the efficiency of drug removal in the eliminating organs such as the liver and kidney. **Volume of distribution** is a term that relates to the apparent extent of drug distribution away from plasma and into tissues. **Half-life** is simply the time required for drug concentrations in plasma to decline by half. Lastly, the fraction of the drug dose that reaches the systemic circulation when administered nonintravenously is parameterized by the term **bioavailability**.

Pharmacokinetic Concepts

Clearance

An important parameter that reflects how well the body eliminates drug from the circulation is clearance (Wilkinson, 1987). The concept of clearance in pharmacokinetics is similar to that first used in renal physiology. By definition, clearance is the volume of blood removed of drug per unit of time. This parameter has units of volume per time (e.g., L/h). Depending on whether one measures the drug concentrations in whole blood or plasma, the parameter that describes the efficiency of drug removal by the whole body is denoted as either blood or plasma clearance (CL_b or CL_p, respectively). With the understanding that the value of clearance is dependent on assessment of drug levels in blood or plasma, one can define the term systemic clearance (CL_S). For the majority of drugs, CL_S is the proportionality factor that relates the rate of

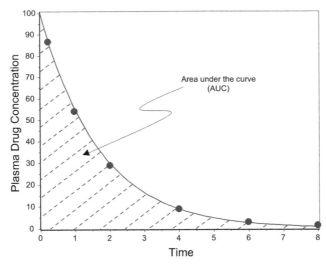

FIGURE 20.5 Area under the curve (AUC) is a quantitative measure of drug exposure that incorporates the magnitude of drug levels as well as the time course. In this case, a drug is administered as an intravenous bolus and drug levels are determined at various intervals (*red circle*). The AUC is calculated from the "area" under the concentration–time profile (*dashed red area*) typically using sums of trapezoids (trapezoid rule).

drug removal by the body via metabolism or excretion with the blood or plasma drug concentration (C_p or C_b, respectively) at any given time (Eq. (20.1)).

$$\text{Rate of drug removal in blood} = CL_S \times C_b \tag{20.1}$$

For most drugs, CL_S remains constant unless the elimination pathways in a patient are altered, for example, by renal or hepatic dysfunction. While blood levels of drug can be measured at any given time, it is often impractical or impossible to measure the overall rate of drug removal in the body by the liver and kidney, and so we must estimate the value of CL_S by other methods. Typically, CL_S is obtained after intravenous dosing of a quantity of drug and measurement of the resulting area under the blood/plasma concentration–time profile (AUC) (Fig. 20.5). AUC has units of amount/volume $\times$ time. The mathematical derivation is beyond the scope of this chapter; however, one can calculate CL_S from Eq. (20.2).

$$CL_S = \frac{\text{Dose}}{\text{AUC}} \tag{20.2}$$

From this relationship, it also becomes clear that when CL_S remains constant, the AUC, which is often termed "exposure," is directly proportional to dose.

Alternatively, after constant rate intravenous infusion of a drug, a steady state is achieved whereby the rate of drug delivery is equivalent to the rate of drug removal. Upon reaching the steady state, the plasma levels of the drug reach a constant value (C_{ss}). Based on Eq. (20.1), CL_S can be determined using Eq. (20.3).

$$CL_S = \frac{\text{Rate of intravenous drug infusion}}{C_{ss}} \tag{20.3}$$

Since most drugs are eliminated from the body by the organs and not through metabolism by the blood elements (red and white blood cells) themselves, there are limits to the values of CL_b. When the body is incapable of removing the drug, CL_b would have a value of 0 L/h. Since clearance is the volume of blood removed of drug per unit of time and since the organs can maximally clear drug from the volume of blood being perfused at any time, the greatest value of CL_b is that of cardiac output (0.5 L/min blood for an adult). Therefore, $0 < CL_b <$ cardiac output. However, when drug is measured in plasma, CL_p may have values greater than cardiac output due to distribution of drug into erythrocytes.

An important principle in the clearance concept is that of additivity of clearance. That is, CL_S is the sum of clearances by each eliminating organ. For example,

$$CL_S = CL_{liver} + CL_{kidney} + CL_{intestine} \tag{20.4}$$

If one were to determine each organ clearance for any drug (see next section), the systemic clearance (CL_S) would be their sum. When a drug is known to be eliminated from the body strictly by renal excretion, $CL_S = CL_{kidney}$. Whereas when a drug is removed solely by hepatic metabolic activity, $CL_S = CL_{liver}$. This principle is often used to determine organ drug clearances that are not practical to measure, such as CL_{liver}. In the case where a drug is eliminated by only liver and kidney, CL_{liver} can be estimated by taking the difference of CL_S and that of CL_{kidney} (i.e., $CL_{liver} = CL_S - CL_{kidney}$).

At the level of the eliminating organ, the efficiency of drug removal can be characterized by the extraction ratio (E), which is the arteriovenous difference in drug concentrations ($C_A - C_V$) divided by arterial drug concentration (C_A).

$$E = \frac{C_A - C_V}{C_A} \tag{20.5}$$

Inefficient drug removal by an organ would give E values approaching 0, while efficient removal would be reflected in values of E close to 1 ($0 \le E \le 1$). Clearance at any organ is the product of the blood flow rate to the organ (Q) and E:

$$Cl_{organ} = Q \times E \tag{20.6}$$

In practice, the extraction ratio of drugs by some eliminating organs such as the liver cannot be easily determined due to inability to access drug concentrations in blood entering via the portal and hepatic arterial circulation. However, for renally eliminated drugs, CL_{kidney} can be readily determined when no metabolism occurs by the kidney. On the basis of the relationship in Eq. (20.1), CL_{kidney} can be calculated from Eq. (20.7).

$$CL_{kidney} = \frac{\text{rate of urinary excretion}}{C_b} \tag{20.7}$$

The rate of urinary excretion can be determined by urine collection over a defined time interval. The most accurate measure of CL_{kidney} is obtained when the drug concentrations in blood used in the calculation are obtained at the midpoint time of the urine collection interval. By far the most common method of determining CL_{kidney} is through collection of urine over the entire time required for the drug to be eliminated completely by the body. In doing so, one can measure the total amount of drug excreted in urine after a single dose by any administration route from time 0 to infinity ($Ae_{urine, 0-\infty}$). By simultaneously characterizing the time course of drug in the blood to obtain the AUC, CL_{kidney} can be estimated (Eq. (20.8)):

$$CL_{kidney} = \frac{Ae_{urine,0-\infty}}{AUC_{0-\infty}} \tag{20.8}$$

For many drugs, the liver is a major site of drug elimination via metabolism. How the biochemical processes of drug metabolism interplay with hepatic perfusion rate has been the subject for the development of hepatic clearance models (Pang and Rowland, 1977). Such models have utility in understanding how, in a quantitative fashion, alterations in liver enzymatic activity and blood flow in diseases, such as cirrhosis, hepatitis, and sepsis, would alter the efficiency of hepatic drug elimination. In addition, hepatic clearance models have proven useful to predict the effects of metabolic enzyme inhibition and induction as encountered with drug–drug interactions. The so-called "well-stirred" model of hepatic clearance has found the most acceptance based on the experimental validity and mathematical simplicity (Wilkinson and Shand, 1975). In this model, E required to calculate CL_{liver} is a function of Q and the drug metabolic activity of the liver as defined by the term intrinsic clearance (CL_{int}). Hence,

$$CL_{liver} = \frac{Q \times CL_{int}}{Q + CL_{int}} \tag{20.9}$$

From such a relationship one can observe that CL_{liver} for drugs that are efficiently metabolized in liver and therefore have high E ($E \ge 0.7$) is most sensitive to changes in hepatic blood flow (Q), such as those observed in cirrhosis or heart failure rather than alterations in metabolic activity. In the case of low extraction ratio drugs ($E \le 0.3$), CL_{liver} is sensitive to changes in hepatic metabolic activity, such as that caused by enzyme inhibition or induction.

Volume of Distribution

Upon entry into the body, drug distributes from the blood into the tissues as a result of organ perfusion and movement across the cellular membranes. The rate and extent of drug distribution into tissues is largely determined by a combination of factors including drug physicochemical properties that either promote or hinder passive diffusion across the endothelial and parenchymal cell layer, the degree of protein binding to blood elements and tissues, and the contribution of facilitated membrane permeability provided by transporter proteins. The drug concentration achieved in the sampling fluid (plasma) after distribution is complete (equilibrium) and depends on the size of the administered dose and the extent of extravascular distribution of drug into tissues. This extent of distribution is parameterized as the volume of distribution and can be determined by relating the concentration measured with the known amount of drug in the body.

$$\text{Volume of distribution} = \frac{\text{Amount of drug in the body}}{C_p} \tag{20.10}$$

The so-called (apparent) volume of distribution is, in effect, a dilutional space. It simply is a proportionality constant relating drug mass and concentration. Volume of distribution is a particularly useful parameter in estimating the dose required to achieve a desired plasma concentration ($C_{desired}$). In the specific example where a drug rapidly distributes from the blood to tissues, the so-called one-compartment model, the product of volume of distribution and $C_{desired}$, gives the required initial dose. After a single intravenous bolus dose of a drug which exhibits one-compartment distribution, the volume of distribution (V) can be determined after an estimation of the plasma drug concentration immediately after the dose is administered (C_o).

$$V = \frac{\text{Dose}}{C_0} \qquad (20.11)$$

It is important to consider that the value of volume of distribution does not relate to any physiological volume. For comparison, the total water volume of a 70 kg human is 42 L, while the extracellular water volume is 12 L; the blood volume is 5 L and plasma volume is 2.5 L. The volume of distribution of drugs can be no less than that of the plasma volume but can be much greater than that of the total body water volume. For instance, the antiarrhythmic drug amiodarone has a volume of distribution of 5000 L.

In the special case of one-compartment drug distribution, the decline of drug levels in plasma after intravenous bolus injection falls monoexponentially because clearance for most drugs is a first-order process, that is, concentration-independent. However, for the majority of drugs, plasma drug concentrations fall in a multiexponential fashion, indicating that immediately after intravenous administration, the drug rapidly distributes throughout an initial space including the plasma space; then a slower, equilibrative distribution to extravascular spaces occurs. In this case, drug distribution is said to be multicompartmental.

For drugs with multicompartment distribution, there is the additional volume of distribution parameters to be considered. The first is the "central" volume (V_C), which is calculated again as the intravenous bolus dose divided by the drug concentration immediately after the first dose. V_C is the volume space whereby the drug achieves instantaneous distribution. But after the rapid equilibration phase, the drug begins to distribute to slowly equilibrating tissues causing multiphasic decline in drug concentrations in plasma. Drug distribution will eventually reach "whole body" equilibration and at that time the decline of drug concentrations in plasma remains monophasic. Hence, for drugs with multicompartment distribution, after the initial distribution volume of V_C, the volume of distribution changes with time, increasing to a constant value termed V_{AREA} when distribution equilibrium occurs. For completeness, V_{AREA} can be calculated from CL_S and the elimination rate constant obtained after analysis of the terminal log-linear phase of the drug concentration–time profile.

$$V_{AREA} = \frac{CL_S}{\text{Terminal elimination rate constant}} \qquad (20.12)$$

From a practical point, V_{AREA} is a useful parameter to determine the amount of drug in the body at any time when a drug distributes in a multicompartment manner. Following Eq. (20.10), the amount in the body is equal to the product of V_{AREA} and C_p. The drawback to V_{AREA} as a parameter is that it is dependent on terminal rate constant, a parameter that does not solely reflect the extent of extravascular drug distribution. For this reason, another volume term that does not suffer from this disadvantage, called volume of distribution at steady state (V_{SS}), is commonly calculated. V_{SS} is typically calculated using "noncompartmental" or moment analysis that requires estimation of the area under the first-moment curve (AUMC) (Benet and Galeazzi, 1979). The first-moment curve is a plot of the $C_p \times$ time versus time:

$$V_{SS} = \frac{\text{Dose} \times \text{AUMC}}{\text{AUC}^2} \qquad (20.13)$$

Not only is V_{SS} useful to relate the differences in extravascular distributions of different drugs, but it has practical utility in evaluating the changes in the extent of drug distribution that can occur within a patient due to pathological processes or protein-binding changes.

It must be noted that the volume of distribution parameters presented describes the extent of drug distribution in the body, but do not relate to the rate of drug distribution. The rate of drug distribution is dependent on the blood perfusion rates to the tissues, whereby the brain, liver, and kidneys are highly perfused, whereas the skin, bone, and fat are poorly perfused. Lipophilic drugs with good membrane permeability will rapidly distribute into the highly perfused tissues but slowly enter those poorly perfused. The concept of rate versus extent of distribution is an important consideration because the volume of distribution parameters is not particularly useful in explaining the rate of onset of drug effects.

Half-Life

The time required for drug levels to fall by 50% of an initial value is termed the half-life ($t_{1/2}$). For drugs administered intravenously as a bolus with one-compartment distribution, $t_{1/2}$ is constant throughout drug exposure. The $t_{1/2}$ can be directly calculated from the line of best fit "slope" or from two points along the log-linear phase of the drug concentration—time profile. In the example, when two plasma concentration determinations are used (C_1 and C_2 which are obtained at times t_1 and t_2, respectively), one can obtain a value for the "slope" that is equivalent to the negative elimination rate constant (k_e).

$$k_e = \frac{\ln C_2 - \ln C_1}{t_2 - t_1} \tag{20.14}$$

The $t_{1/2}$ can then be calculated by:

$$t_{1/2} = \frac{0.693}{k_e} \tag{20.15}$$

After intravenous bolus injection of a drug with one-compartment distribution, the time required after the dose for essentially all the dose to be eliminated by the body is three to five times the $t_{1/2}$. With drugs that distribute in a multicompartment fashion, $t_{1/2}$ is usually considered as that found at the terminal phase of the drug concentration—time profile. For the simpler one-compartment model, an important pharmacokinetic relationship exists that relates $t_{1/2}$ with volume of distribution and clearance:

$$t_{1/2} = \frac{0.693 \times V}{CL_S} \tag{20.16}$$

This relationship is particularly useful since it allows one to predict the effect of changes in volume of distribution or clearance on the $t_{1/2}$ of a drug in a patient. For example, if a drug's volume of distribution decreased because of dehydration, the $t_{1/2}$ would be expected to also decrease. Alternatively, if a patient was coadministered a drug that inhibits drug metabolism, the CL_S would decrease as it is a measure of elimination efficiency, and the $t_{1/2}$ as expected would increase. A more complicated scenario would be that where both volume of distribution and clearance is altered in a patient such as that which may occur with a drug—drug interaction that causes plasma protein binding displacement. Here, decreased plasma protein binding would increase volume of distribution because there would be less sequestration of drug in the vasculature. At the same time, clearance would be expected to increase because a large amount of drug that was freed from plasma proteins is now able to access the drug eliminating enzymes in liver or be filtered at the glomerulus. Therefore based on the relationship in Eq. (20.15), there may be no alterations in $t_{1/2}$ in the event of simultaneous increase in volume of distribution and clearance.

Bioavailability

Not all the drug dose that is administered extravascularly (e.g., oral, transdermal, inhalational) reaches the systemic circulation. For the most common situation of oral administration, there are several factors that play a role in impeding drug entry into the body. Some of these include drug solubility and formulation variables, degradation in the gastrointestinal tract by bacteria and acid, poor permeability through the enterocyte membrane, finite gastrointestinal transit time, and metabolism during first-pass transit through the gut and liver. Bioavailability (F) is the parameter that relates the actual drug dose with the so-called "effective dose" that was capable of entering the systemic circulation.

$$\text{Effective dose} = F \times \text{Dose} \tag{20.17}$$

The values of F can range from 0 to 1. When none of the administered dose is absorbed into systemic circulation, $F = 0$, whereas $F = 1$ when the entire dose reaches the posthepatic blood compartment. When a drug is taken orally, the bioavailability is the product of the "availabilities" at each individual step leading to drug reaching the systemic circulation. Hence,

$$F = F_{abs} \times F_{gut} \times F_{liver} \tag{20.18}$$

where F_{abs} is the fraction of the drug dose that is absorbed by enterocytes, F_{gut} is the fraction of the drug dose absorbed by enterocyte and escapes gut metabolism, and F_{liver} is the fraction of drug entering the liver from the portal circulation that leaves through the hepatic vein. F_{liver} can also be defined as $1-E_{liver}$. The process by which drug absorbed by enterocytes is eliminated by the gut and liver to prevent systemic drug exposure is called "first-pass effect."

Understanding bioavailability has several practical implications. For example, whether or not there is dose equivalency in brand-name and generic drug preparations, intravenous to oral drug dose conversions, and predicting the impact of drug–drug interactions that alter the efficiency of the first-pass effect.

The actual bioavailability of drug can only be determined after pharmacokinetic analysis of an intravenously and extravascularly administered drug, typically in the same subject. With the assumption that clearance is not changed between the time the drug is administered by different routes, oral bioavailability can be calculated by:

$$F = \frac{AUC_{oral} \times Dose_{intravenous}}{AUC_{intravenous} \times Dose_{oral}} \tag{20.19}$$

In the similar fashion that we have parameterized drug distribution, F relates only to the extent of drug dose entering the systemic circulation and not the rate by which drug enters the circulation. The rate of drug entry into the body is dependent on dosage form (e.g., tablet, suspension, enteric coating, sustained-release preparation), effect of food, and influence of facilitative absorption of drugs by enterocyte transport proteins. The rate of drug absorption is important in determining when and how high the plasma levels of drugs are achieved and hence the time course and magnitude of drug effects. Typically after oral drug dosing, plasma drug levels begin to rise after an initial lag time; because the dosage form has not disintegrated, the drug remains unabsorbed in the stomach or the drug has not yet dissolved. Eventually, plasma drug levels will peak to the maximal concentration (C_{max}), which occurs at time (t_{max}), then declines as drug absorption is complete and drug removal by the body becomes the significant drug disposition process. In the scenario where the bioavailability of a drug preparation is constant but the rate of absorption differs because of the effect of food, for example, one can expect the time course of the orally administered drug to follow a pattern. When the rate of absorption increases, the C_{max} increases and the t_{max} decreases. By contrast, as the rate of absorption decreases, C_{max} decreases and t_{max} increases (Fig. 20.6).

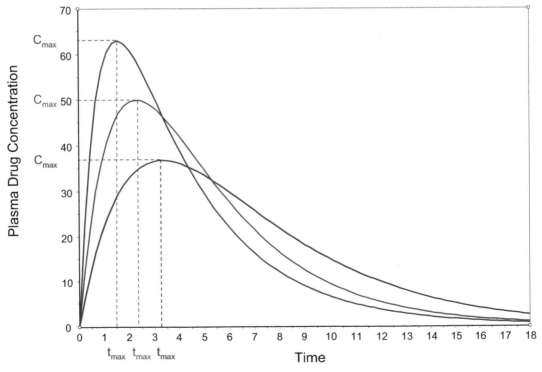

FIGURE 20.6 Effect of oral drug absorption rate on plasma concentration–time profiles. In the case of a drug that is formulated in three dosage forms containing similar drug content and identical bioavailabilty, the concentration–time profile differs depending on the rate of absorption. The relative absorption rates of the dosage forms is Purple > Red > Blue. C_{max} decreases and t_{max} increases with decreasing absorption rates while AUCs remain similar.

Dose Selection

The primary therapeutic goal is to achieve and maintain drug efficacy and minimize untoward adverse effects. Often, drug effects are observed when plasma drug levels reach a minimum effective concentration. Similarly, drug toxicity is concentration-dependent and for many drugs the toxic drug levels are significantly greater than those required for therapeutic benefit and hence are said to have a wide therapeutic window. When the therapeutic window is narrow, therapeutic concentrations are close to those that cause adverse effects. Therapeutic individualization remains largely a trial-and-error approach whereby changes in dose are dictated by whether one achieves the desired outcomes or provokes adverse effects. In certain cases, in particular drugs with a narrow therapeutic window, drug level monitoring can be a useful tool to titrate doses. Whether by monitoring pharmacological/toxic effects or guided by plasma drug levels, understanding pharmacokinetic principles is a requirement for rational dose selection. Each pharmacokinetic parameter impacts the two variables in dose administration; those being the dose and the dosing interval. For continuous infusions, one is concerned with a single variable, dose rate. For each drug, clearance will define the dose rate, half-life will determine the dosing interval, volume of distribution will play a significant role in dose, loading dose, and dosing interval, while bioavailability affects relative dose given by different routes of administration.

Continuous Intravenous Infusion

During drug administration by continuous (rate) intravenous infusion, the plasma drug concentrations rise in a hyperbolic manner until a constant drug level is maintained (Fig. 20.7). It is at this point steady state is achieved, whereby the rate of drug delivery equals the rate of drug elimination. When a constant rate infusion is given, the time required for steady state to be achieved is three to five times the $t_{1/2}$ of the drug. Importantly, this time to steady state is independent of the dose rate administered. When the dose rate is changed (increased or decreased), drug levels will rise or fall to a new steady-state level. Again the time required to reach this new steady state is three to five times the $t_{1/2}$. Based on Eq. (20.3), it can be noted that when CL_S remains constant in a patient, the steady-state plasma concentration of drug is directly proportional to the dose rate. For instance, doubling the dose rate would double the steady-state plasma drug levels.

Intravenous Loading Dose

In certain cases, the acuteness of the patient's condition necessitates that therapeutic drug levels are achieved rapidly. Loading doses are a single or multiple set of doses given to a patient to attain desired drug levels more rapidly than the

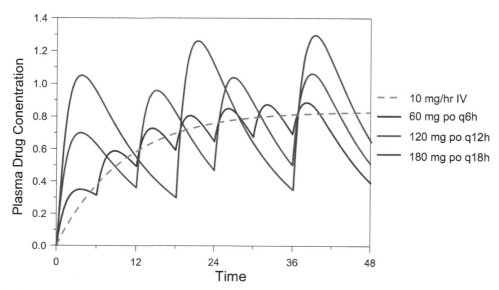

FIGURE 20.7 Continuous or intermittent drug administration on plasma concentration—time profiles. A drug with 100% oral bioavailability is administered frequently in small oral doses to less frequently in large oral doses or by continuous intravenous infusion. The dose rate is equivalent for all dosing regimens. The half-life of the drug is 7 h. Note that the time required to reach steady-state plasma levels (five half-lives) for all regimens is similar; so is the average plasma concentrations over the dosing intervals for orally administered drug. Differences are apparent in the magnitude of plasma concentration fluctuations with different dosing regimens.

three to five half-lives that occur in a continuous intravenous infusion. The loading dose for a one-compartment model drug can be determined with knowledge of the volume of distribution and target drug level (C_{target}):

$$Loading\ Dose = V \times C_{target} \qquad (20.20)$$

The intravenous loading dose can be administered as a bolus or as a short infusion. The short infusion is useful when trying to minimize the potential toxic effects of rapid, high-drug levels with a large loading dose. Particularly, the short infusion has utility for drugs that have multicompartment distribution where the central volume V_C is relatively small and an intravenous bolus loading dose would create very high initial drug levels that could have deleterious effects. An example of loading dose infusion for the antiarrhythmic drug lidocaine is shown in Fig. 20.8.

Intermittent Dose Administration

Maintenance dose regimens are commonly taken as continuous intermittent doses separated by hours to days. In contrast to continuous infusions, with intermittent drug dosing, levels rise and fall between drug doses. When the interval of drug dosing is shorter than that required for the drug to be completely eliminated by the body between doses, the plasma drug levels accumulate. In many respects, the accumulation of drugs administered in multiple doses is the same as that observed following a constant-rate intravenous infusion. Recall that the steady-state plasma drug level after a constant-rate intravenous infusion is dependent on the dose rate and CL_S (Eq. (20.3)). For intermittent dosing, a similar principle applies in that the average plasma concentration ($C_{ss,avg}$) that occurs with the rise and decline of drug levels is dependent on the dose and dosing interval (τ), which defines the dose rate of the intermittent dosing regimen.

$$C_{ss,avg} = \frac{F \times Dose}{CL_S \times \tau} \qquad (20.21)$$

Therefore, any combination of dose and dosing interval that gives the same dosing rate (Dose/τ) will attain the same average steady-state plasma concentration. The difference between such regimens is the degree of difference in the peak and trough plasma concentrations with each successive dose. Small doses given at short intervals gives rise to smaller plasma level fluctuations than large doses administered at longer intervals (see Fig. 20.7). Again, it takes three to five half-lives to reach a steady state during intermittent dosing.

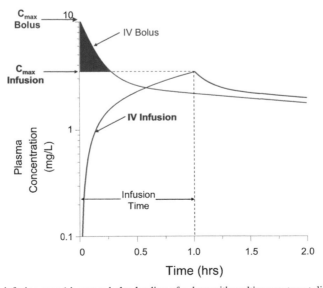

FIGURE 20.8 Impact of intravenous infusion over 1 h versus bolus loading of a drug with multicompartment distribution given at the same total dose. Bolus administration achieves rapid high plasma concentrations while short infusion provides slower rise of drug levels reaching maximum at the end of infusion. Short intravenous infusion loading doses may have advantages over bolus loading by minimizing high and potentially toxic drug plasma levels that may occur with drugs with multicompartment distribution.

CONCLUSION

Clinical pharmacology is a discipline that seeks to define the mechanistic basis of drug action through an in-depth understanding of human pharmacology and therapeutics. The key concepts relating to drug absorption, distribution, metabolism, and excretion (ADME) have been widely accepted and studied in relation to pharmacokinetics of drug disposition for many decades. However, with the relatively recent molecular identification and functional characterization of the key proteins that mediate drug uptake, efflux, and metabolism in organs of importance to drug disposition and response, such as the intestine, liver, kidney, and brain, there is now an emerging recognition of the need for multidisciplinary, mechanistic, and hypothesis-driven studies of genetic variation (pharmacogenomics), mRNA expression (transcriptomics), protein expression (proteomics), and metabolism (metabolomics) in an integrated fashion to predict intersubject variation in drug levels or response. Taken together, clinical pharmacology as a discipline is now poised to expand and become a major contributor to translational clinical research in general and personalized medicine in particular.

REFERENCES

Adachi, M., Reid, G., Schuetz, J.D., 2002. Therapeutic and biological importance of getting nucleotides out of cells: a case for the ABC transporters, MRP4 and 5. Adv. Drug Deliv. Rev. 54, 1333–1342.

Allen, J.D., Schinkel, A.H., 2002. Multidrug resistance and pharmacological protection mediated by the breast cancer resistance protein (BCRP/ABCG2). Mol. Cancer Ther. 1, 427–434.

Alexanderson, B., Evans, D.A., Sjoqvist, F., 1969. Steady-state plasma levels of nortriptyline in twins: influence of genetic factors and drug therapy. Br. Med. J. 4, 764–768.

Anonymous, 1974. Rifampicin, "pill" do not go well together. JAMA 227, 608.

Backman, J.T., Olkkola, K.T., Neuvonen, P.J., 1996. Rifampin drastically reduces plasma concentrations and effects of oral midazolam. Clin. Pharmacol. Ther. 59, 7–13.

Bailey, D.G., Spence, J.D., Munoz, C., Arnold, J.M., 1991. Interaction of citrus juices with felodipine and nifedipine. Lancet 337, 268–269.

Bailey, D.G., Malcolm, J., Arnold, O., Spence, J.D., 1998. Grapefruit juice–drug interactions. Br. J. Clin. Pharmacol. 46, 101–110.

Baker, A.R., McDonnell, D.P., Hughes, M., Crisp, T.M., Mangelsdorf, D.J., Haussler, M.R., et al., 1988. Cloning and expression of full-length cDNA encoding human vitamin D receptor. Proc. Natl. Acad. Sci. U.S.A. 85, 3294–3298.

Basin, K.S., Escalante, A., Beardmore, T.D., 1991. Severe pancytopenia in a patient taking low dose methotrexate and probenecid. J. Rheumatol. 18, 609–610.

Bending, M.R., Skacel, P.O., 1977. Rifampicin and methadone withdrawal. Lancet 1, 1211.

Benet, L.Z., Galeazzi, R.L., 1979. Noncompartmental determination of the steady-state volume of distribution. J. Pharm. Sci. 68, 1071–1074.

Bohan, A., Boyer, J.L., 2002. Mechanisms of hepatic transport of drugs, implications for cholestatic drug reactions. Semin. Liver. Dis. 22, 123–136.

Bolt, H.M., Bolt, M., Kappus, H., 1977. Interaction of rifampicin treatment with pharmacokinetics and metabolism of ethinyloestradiol in man. Acta Endocrinol. (Copenh.) 85, 189–197.

Brauch, H., Schroth, W., Goetz, M.P., Murdter, T.E., Winter, S., Ingle, J.N., Schwab, M., Eichelbaum, M., 2013. Tamoxifen use in postmenopausal breast cancer: CYP2D6 matters. J. Clin. Oncol. 31, 176–180.

Burk, O., Arnold, K.A., Geick, A., Tegude, H., Eichelbaum, M., 2005. A role for constitutive androstane receptor in the regulation of human intestinal MDR1 expression. Biol. Chem. 386, 503–513.

Burnell, J.M., Kirby, W.M., 1951. Effectiveness of a new compound, benemid, in elevating serum penicillin concentrations. J. Clin. Invest. 30, 697–700.

Chu, X.-Y., Kato, Y., Niinuma, K., Sudo, K.-I., Hakusui, H., Sugiyama, Y., 1997. Multispecific organic anion transporter is responsible for the biliary excretion of the camptothecin derivative irinotecan and metabolites in rats. J. Pharmacol. Exp. Ther. 281, 304–314.

Conney, A.H., Davison, C., Gastel, R., Burns, J.J., 1960. Adaptive increases in drug-metabolizing enzymes induced by phenobarbital and other drugs. J. Pharmacol. Exp. Ther. 130, 1–8.

Cordon-Cardo, C., O'Brien, J.P., Casals, D., Rittman-Grauer, L., Biedler, J.L., Melamed, M.R., et al., 1989. Multidrug-resistance gene (P-glycoprotein) is expressed by endothelial cells at blood–brain barrier sites. Proc. Natl. Acad. Sci. 86, 695–698.

Cucinells, S.A., Koster, R., Conney, A.H., Burns, J.J., 1963. Stimulatory effect of phenobarbital on the metabolism of diphenylhydantoin. J. Pharmacol. Exp. Ther. 141, 157–160.

Dehal, S.S., Kupfer, D., 1997. CYP2D6 catalyzes tamoxifen 4-hydroxylation in human liver. Cancer Res. 57, 3402–3406.

Degorter, M.K., Tirona, R.G., Schwarz, U.I., Choi, Y.H., Dresser, G.K., Suskin, N., et al., 2013. Clinical and pharmacogenetic predictors of circulating atorvastatin and rosuvastatin concentrations in routine clinical care. Circ. Cardiovasc. Genet. 6 (4), 400–408.

De Morais, S.M., Wilkinson, G.R., Blaisdell, J., Nakamura, K., Meyer, U.A., Goldstein, J.A., 1994. The major genetic defect responsible for the polymorphism of S-mephenytoin metabolism in humans. J. Biol. Chem. 269, 15419–15422.

Distlerath, L.M., Reilly, P.E., Martin, M.V., Davis, G.G., Wilkinson, G.R., Guengerich, F.P., 1985. Purification and characterization of the human liver cytochromes P450 involved in debrisoquine 4-hydroxylation and phenacetin O-deethylation, two prototypes for genetic polymorphism in oxidative drug metabolism. J. Biol. Chem. 260, 9057–9067.

Drescher, S., Glaeser, H., Murdter, T., Hitzl, M., Eichelbaum, M., Fromm, M.F., 2003. P-glycoprotein-mediated intestinal and biliary digoxin transport in humans. Clin. Pharmacol. Ther. 73, 223–231.

Dresser, G.K., Spence, J.D., Bailey, D.G., 2000. Pharmacokinetic-pharmacodynamic consequences and clinical relevance of cytochrome P450 3A4 inhibition. Clin. Pharmacokinet. 38, 41—57.

Dresser, G.K., Schwarz, U.I., Wilkinson, G.R., Kim, R.B., 2003. Coordinate induction of both cytochrome P4503A and MDR1 by St. John's wort in healthy subjects. Clin. Pharmacol. Ther. 73, 41—50.

Edwards, D.J., Bellevue 3rd, F.H., Woster, P.M., 1996. Identification of 6',7'-dihydroxybergamottin, a cytochrome P450 inhibitor, in grapefruit juice. Drug Metab. Dispos. 24, 1287—1290.

Eich-Hochli, D., Oppliger, R., Golay, K.P., Baumann, P., Eap, C.B., 2003. Methadone maintenance treatment and St. John's Wort—a case report. Pharmacopsychiatry 36, 35—37.

Eisenblatter, T., Huwel, S., Galla, H.J., 2003. Characterisation of the brain multidrug resistance protein (BMDP/ABCG2/BCRP) expressed at the blood—brain barrier. Brain Res. 971, 221—231.

Ellison, N.M., Servi, R.J., 1985. Acute renal failure and death following sequential intermediate-dose methotrexate and 5-FU: a possible adverse effect due to concomitant indomethacin administration. Cancer Treat. Rep. 69, 342—343.

Farmer, J.A., Gotto, A.M.J., 1994. Antihyperlipidaemic agents. Drug interactions of clinical significance. Drug Saf. 11, 301—309.

Floyd, M.D., Gervasini, G., Masica, A.L., Mayo, G., George Jr., A.L., Bhat, K., et al., 2003. Genotype-phenotype associations for common CYP3A4 and CYP3A5 variants in the basal and induced metabolism of midazolam in European- and African-American men and women. Pharmacogenetics 13, 595—606.

Forman, B.M., Chen, J., Evans, R.M., 1996. The peroxisome proliferator-activated receptors: ligands and activators. Ann. N.Y. Acad. Sci. 804, 266—275.

Forman, B.M., Tzameli, I., Choi, H.S., Chen, J., Simha, D., Seol, W., et al., 1998. Androstane metabolites bind to and deactivate the nuclear receptor CAR-beta. Nature 395, 612—615.

Fromm, M.F., Eckhardt, K., Li, S., Schanzle, G., Hofmann, U., Mikus, G., et al., 1997. Loss of analgesic effect of morphine due to coadministration of rifampin. Pain 72, 261—267.

Fromm, M.F., Kauffmann, H.M., Fritz, P., Burk, O., Kroemer, H.K., Warzok, R.W., et al., 2000. The effect of rifampin treatment on intestinal expression of human MRP transporters. Am. J. Pathol. 157, 1575—1580.

Fromm, M.F., Kim, R.B., Stein, C.M., Wilkinson, G.R., Roden, D.M., 1999. Inhibition of P-glycoprotein-mediated drug transport: a unifying mechanism to explain the interaction between digoxin and quinidine. Circulation 99, 552—557.

Furuta, T., Ohashi, K., Kamata, T., Takashima, M., Kosuge, K., Kawasaki, T., Hanai, H., Kubota, T., Ishizaki, T., Kaneko, E., 1998. Effect of genetic differences in omeprazole metabolism on cure rates for *Helicobacter pylori* infection and peptic ulcer. Ann. Intern. Med. 129, 1027—1030.

Furuya, H., Fernandez-Salguero, P., Gregory, W., Taber, H., Steward, A., Gonzalez, F.J., Idle, J.R., 1995. Genetic polymorphism of CYP2C9 and its effect on warfarin maintenance dose requirement in patients undergoing anticoagulation therapy. Pharmacogenetics 5, 389—392.

Gaedigk, A., Simon, S.D., Pearce, R.E., Bradford, L.D., Kennedy, M.J., Leeder, J.S., 2008. The CYP2D6 activity score: translating genotype information into a qualitative measure of phenotype. Clin. Pharmacol. Ther. 83, 234—242.

Geick, A., Eichelbaum, M., Burk, O., 2001. Nuclear receptor response elements mediate induction of intestinal MDR1 by rifampin. J. Biol. Chem. 276, 14581—14587.

Giacomini, K.M., Huang, S.M., Tweedie, D.J., Benet, L.Z., Brouwer, K.L., Chu, X., et al., March 2010. Membrane transporters in drug development. Nat. Rev. Drug Discov. 9 (3), 215—236.

Giguere, V., Yang, N., Segui, P., Evans, R.M., 1988. Identification of a new class of steroid hormone receptors. Nature 331, 91—94.

Gong, I.Y., Kim, R.B., 2013. Impact of genetic variation in OATP transporters to drug disposition and response. Drug Metab. Pharmacokinet. 28 (1), 4—18.

Gonzalez, F.J., Skoda, R.C., Kimura, S., Umeno, M., Zanger, U.M., Nebert, D.W., Gelboin, H.V., Hardwick, J.P., Meyer, U.A., 1988. Characterization of the common genetic defect in humans deficient in debrisoquine metabolism. Nature 331, 442—446.

Greiner, B., Eichelbaum, M., Fritz, P., Kreichgauer, H.P., von Richter, O., Zundler, J., et al., 1999. The role of intestinal P-glycoprotein in the interaction of digoxin and rifampin. J. Clin. Invest. 104, 147—153.

Grunden, J.W., Fisher, K.A., 1997. Lovastatin-induced rhabdomyolysis possibly associated with clarithromycin and azithromycin. Ann. Pharmacother. 31, 859—863.

Guengerich, F.P., 1995. Human cytochrome P450 enzymes. In: Ortiz de Montellano, P.R. (Ed.), Cychrome P450: Structure, Mechanism, and Biochemistry. Plenum Press, New York, pp. 473—535.

Gupta, K.C., Ali, M.Y., 1980. Failure of oral contraceptive with rifampicin. Med. J. Zamb. 15, 23.

Hagenbuch, B., Meier, P.J., 2003. The superfamily of organic anion transporting polypeptides. Biochim. Biophys. Acta 1609, 1—18.

Hicks, J.K., Swen, J.J., Thorn, C.F., Sangkuhl, K., Kharasch, E.D., Ellingrod, V.L., Skaar, T.C., Muller, D.J., Gaedigk, A., Stingl, J.C., 2013. Clinical pharmacogenetics Implementation Consortium Guideline for CYP2D6 and CYP2C19 genotypes and dosing of Tricyclic antidepressants. Clin. Pharmacol. Ther. 93, 402—408.

Ho, R.H., Kim, R.B., 2005. Transporters and drug therapy: implications for drug disposition and disease. Clin. Pharmacol. Ther. 78, 260—277.

Ito, K., Suzuki, H., Hirohashi, T., Kume, K., Shimizu, T., Sugiyama, Y., 1997. Molecular cloning of canalicular multispecific organic anion transporter defective in EHBR. Am. J. Physiol. 272, G16—G22.

Izzo, A.A., 2004. Drug interactions with St John's Wort (*Hypericum perforatum*): a review of the clinical evidence. Int. J. Clin. Pharmacol. Ther. 42, 139—148.

Jager, W., Gehring, E., Hagenauer, B., Aust, S., Senderowicz, A., Thalhammer, T., 2003. Biliary excretion of flavopiridol and its glucuronides in the isolated perfused rat liver: role of multidrug resistance protein 2 (MRP2). Life Sci. 73, 2841—2854.

Jariyawat, S., Sekine, T., Takeda, M., Apiwattanakul, N., Kanai, Y., Sophasan, S., et al., 1999. The interaction and transport of beta-lactam antibiotics with the cloned rat renal organic anion transporter 1. J. Pharmacol. Exp. Ther. 290, 672–677.

Jonker, J.W., Schinkel, A.H., 2004. Pharmacological and physiological functions of the polyspecific organic cation transporters: OCT1, 2, and 3 (SLC22A1-3). J. Pharmacol. Exp. Ther. 308, 2–9.

Jonker, J.W., Merino, G., Musters, S., van Herwaarden, A.E., Bolscher, E., Wagenaar, E., et al., 2005. The breast cancer resistance protein BCRP(ABCG2) concentrates drugs and carcinogenic xenotoxins into milk. Nat. Med. 11, 127–129.

Karliova, M., Treichel, U., Malago, M., Frilling, A., Gerken, G., Broelsch, C.E., 2000. Interaction of *Hypericum perforatum* (St John's wort) with cyclosporin: a metabolism in a patient after liver transplantation. J. Hepatol. 33, 853–855.

Kast, H.R., Goodwin, B., Tarr, P.T., Jones, S.A., Anisfeld, A.M., Stoltz, C.M., et al., 2002. Regulation of multidrug resistance-associated protein 2 (ABCC2) by the nuclear receptors pregnane X receptor, farnesoid X-activated receptor, and constitutive androstane receptor. J. Biol. Chem. 277, 2908–2915.

Kato, Y., Suzuki, H., Sugiyama, Y., 2002. Toxicological implications of hepatobiliary transporters. Toxicology 181/182, 287–290.

Kawakami, H., Ohtsuki, S., Kamiie, J., Suzuki, T., Abe, T., Terasaki, T., 2011. Simultaneous absolute quantification of 11 cytochrome P450 isoforms in human liver microsomes by liquid chromatography tandem mass spectrometry with in silico target peptide selection. J. Pharm. Sci. 100 (1), 341–352.

Kempf, D., Marsh, K., Kumar, G., Rodrigues, A., Denissen, J., McDonald, E., et al., 1997. Phamacokinetic enhancement of inhibitors of the human immunodeficiency virus protease by coadministration with ritonavir. Antimicrob. Agents Chemother 41, 654–660.

Keppler, D., Konig, J., 1997. Hepatic canalicular membrane 5: expression and localization of the conjugate export pump encoded by the MRP2 (cMRP/cMOAT) gene in liver [Review] [39 refs] FASEB J. 11, 509–516.

Kim, R.B., 2002a. Drugs as P-glycoprotein substrates, inhibitors, and inducers. Drug Metab. Rev. 34, 47–54.

Kim, R.B., 2002b. Transporters and xenobiotic disposition. Toxicology 181/182, 291–297.

Kim, R.B., Fromm, M.F., Wandel, C., Leake, B., Wood, A.J., Roden, D.M., et al., 1998. The drug transporter P-glycoprotein limits oral absorption and brain entry of HIV-1 protease inhibitors. J. Clin. Invest. 101, 289–294.

Kliewer, S.A., Moore, J.T., Wade, L., Staudinger, J.L., Watson, M.A., Jones, S.A., et al., 1998. An orphan nuclear receptor activated by pregnanes defines a novel steroid signaling pathway. Cell 92, 73–82.

Kolars, J.C., Schmiedlin-Ren, P., Schuetz, J.D., Fang, C., Watkins, P.B., 1992. Identification of rifampin-inducible P450IIIA4 (CYP3A4) in human small bowel enterocytes. J. Clin. Invest 90, 1871–1878.

Koren, G., Cairns, J., Chitayat, D., Gaedigk, A., Leeder, S.J., 2006. Pharmacogenetics of morphine poisoning in a breastfed neonate of a codeine-prescribed mother. Lancet 368, 704.

Kreek, M.J., Garfield, J.W., Gutjahr, C.L., Giusti, L.M., 1976a. Rifampin-induced methadone withdrawal. N. Engl. J. Med. 294, 1104–1106.

Kreek, M.J., Gutjahr, C.L., Garfield, J.W., Bowen, D.V., Field, F.H., 1976b. Drug interactions with methadone. Ann. N.Y. Acad. Sci. 281, 350–371.

Kruh, G.D., Zeng, H., Rea, P.A., Liu, G., Chen, Z.S., Lee, K., et al., 2001. MRP subfamily transporters and resistance to anticancer agents. J. Bioenerg. Biomembr. 33, 493–501.

Kuehl, P., Zhang, J., Lin, Y., Lamba, J., Assem, M., Schuetz, J., Watkins, P.B., Daly, A., Wrighton, S.A., Hall, S.D., Maurel, P., Relling, M., Brimer, C., Yasuda, K., Venkataramanan, R., Strom, S., Thummel, K., Boguski, M.S., Schuetz, E., 2001. Sequence diversity in CYP3A promoters and characterization of the genetic basis of polymorphic CYP3A5 expression. Nat. Genet. 27, 383–391.

Kupfer, A., Preisig, R., 1984. Pharmacogenetics of mephenytoin: a new drug hydroxylation polymorphism in man. Eur. J. Clin. Pharmacol. 26, 753–759.

LeBel, M., Masson, E., Guilbert, E., Colborn, D., Paquet, F., Allard, S., et al., 1998. Effects of rifabutin and rifampicin on the pharmacokinetics of ethinylestradiol and norethindrone. J. Clin. Pharmacol. 38, 1042–1050.

Lee, G., Dallas, S., Hong, M., Bendayan, R., 2001. Drug transporters in the central nervous system: brain barriers and brain parenchyma considerations. Pharmacol. Rev. 53, 569–596.

Lee, J.E., van Heeswijk, R., Alves, K., Smith, F., Garg, V., 2011. Effect of the hepatitis C virus protease inhibitor telaprevir on the pharmacokinetics of amlodipine and atorvastatin. Antimicrob. Agents Chemother. 55 (10), 4569–4574.

Lee, W., Kim, R.B., 2004. Transporters and renal drug elimination. Annu. Rev. Pharmacol. Toxicol. 44, 137–166.

Lehmann, J.M., McKee, D.D., Watson, M.A., Willson, T.M., Moore, J.T., Kliewer, S.A., 1998. The human orphan nuclear receptor PXR is activated by compounds that regulate CYP3A4 gene expression and cause drug interactions. J. Clin. Invest. 102, 1016–1023.

Levêque, D., Jehl, F., 1995. P-glycoprotein and pharmacokinetics. Anticancer Res. 15, 331–336.

Link, E., Parish, S., Armitage, J., Bowman, L., Heath, S., Matsuda, F., et al., 2008. SLCO1B1 variants and statin-induced myopathy—a genomewide study. N. Engl. J. Med 359, 789–799.

Lipinski, C.A., 2000. Drug-like properties and the causes of poor solubility and poor permeability. J. Pharmacol. Toxicol. Methods 44, 235–249.

Lown, K.S., Mayo, R.R., Leichtman, A.B., Hsiao, H.L., Turgeon, D.K., Schmiedlin, R., et al., 1997a. Role of intestinal P-glycoprotein (mdr1) in interpatient variation in the oral bioavailability of cyclosporine. Clin. Pharmacol. Ther. 62, 248–260.

Lown, K.S., Bailey, D.G., Fontana, R.J., Janardan, S.K., Adair, C.H., Fortlage, L.A., Brown, M.B., Guo, W., Watkins, P.B., 1997b. Grapefruit juice increases felodipine oral availability in humans by decreasing intestinal CYP3A protein expression. J. Clin. Invest. 99, 2545–2553.

Macphee, I.A., Fredericks, S., Tai, T., Syrris, P., Carter, N.D., Johnston, A., Goldberg, L., Holt, D.W., 2002. Tacrolimus pharmacogenetics: polymorphisms associated with expression of cytochrome p4503A5 and P-glycoprotein correlate with dose requirement. Transplantation 74, 1486–1489.

Marzolini, C., Tirona, R.G., Kim, R.B., 2004. Pharmacogenomics of the OATP and OAT families. Pharmacogenomics 5, 273–282.

Materna, V., Lage, H., 2003. Homozygous mutation Arg768Trp in the ABC-transporter encoding gene MRP2/cMOAT/ABCC2 causes Dubin-Johnson syndrome in a Caucasian patient. J. Hum. Genet. 48, 484–486.

McDonnell, D.P., Mangelsdorf, D.J., Pike, J.W., Haussler, M.R., O'Malley, B.W., 1987. Molecular cloning of complementary DNA encoding the avian receptor for vitamin D. Science 235, 1214–1217.

Mega, J.L., Close, S.L., Wiviott, S.D., Shen, L., Hockett, R.D., Brandt, J.T., Walker, J.R., Antman, E.M., Macias, W., Braunwald, E., Sabatine, M.S., 2009. Cytochrome p450 polymorphisms and response to clopidogrel. New Engl. J. Med. 360, 354–362.

Merino, G., Alvarez, A.I., Pulido, M.M., Molina, A.J., Schinkel, A.H., Prieto, J.G., 2006. Breast cancer resistance protein (BCRP/ABCG2) transports fluoroquinolone antibiotics and affects their oral availability, pharmacokinetics, and milk secretion. Drug Metab. Dispos. 34, 690–695.

Miyake, K., Mickley, L., Litman, T., Zhan, Z., Robey, R., Cristensen, B., et al., 1999. Molecular cloning of cDNAs which are highly overexpressed in mitoxantrone-resistant cells: demonstration of homology to ABC transport genes. Cancer Res. 59, 8–13.

Moore, L.B., Goodwin, B., Jones, S.A., Wisely, G.B., Serabjit-Singh, C.J., Willson, T.M., et al., 2000. St. John's wort induces hepatic drug metabolism through activation of the pregnane X receptor. Proc. Natl. Acad. Sci. U.S.A. 97, 7500–7502.

Murray, M., 1992. P450 enzymes. Inhibition mechanisms, genetic regulation and effects of liver disease. Clin. Pharmacokinet. 23, 132–146.

Nakai, D., Nakagomi, R., Furuta, Y., Tokui, T., Abe, T., Ikeda, T., et al., 2001. Human liver-specific organic anion transporter, LST-1, mediates uptake of pravastatin by human hepatocytes. J. Pharmacol. Exp. Ther. 297, 861–867.

Nakamura, T., Sakaeda, T., Ohmoto, N., Tamura, T., Aoyama, N., Shirakawa, T., et al., 2002. Real-time quantitative polymerase chain reaction for MDR1, MRP1, MRP2, and CYP3A-mRNA levels in Caco-2 cell lines, human duodenal enterocytes, normal colorectal tissues, and colorectal adenocarcinomas. Drug Metab. Dispos. 30, 4–6.

Nelson, D.R., Koymans, L., Kamataki, T., Stegeman, J.J., Feyereisen, R., Waxman, D.J., et al., 1996. P450 superfamily: update on new sequences, gene mapping, accession numbers, and nomenclature. Pharmacogenetics 6, 1–42.

Nies, A.T., Keppler, D., 2007. The apical conjugate efflux pump ABCC2 (MRP2). Pflugers Arch. 453, 643–659.

Oude Elferink, R.P., de Waart, R., 2007. Transporters in the intestine limiting drug and toxin absorption. J. Physiol. Biochem. 63, 75–81.

Ozdemir, V., Kalow, W., Tang, B.K., Paterson, A.D., Walker, S.E., Endrenyi, L., Kashuba, A.D., 2000. Evaluation of the genetic component of variability in CYP3A4 activity: a repeated drug administration method. Pharmacogenetics 10, 373–388.

Paine, M.F., Hart, H.L., Ludington, S.S., Haining, R.L., Rettie, A.E., Zeldin, D.C., 2006. The human intestinal cytochrome P450 "pie". Drug Metab. Dispos. 34, 880–886.

Pang, K.S., Rowland, M., 1977. Hepatic clearance of drugs. I. Theoretical considerations of a "well-stirred" model and a "parallel tube" model. Influence of hepatic blood flow, plasma and blood cell binding, and the hepatocellular enzymatic activity on hepatic drug clearance. J. Pharmacokinet. Biopharm. 5, 625–653.

Paulusma, C.C., Bosma, P.J., Bakker, C.T.M., Otter, M., Zaman, G.J.R., Oude Elferink, R.P.J., 1995. Cloning of a liver-specific MRP-homologue which is deficient in the TR-rat. Hepatology 22, 295.

Quattrochi, L.C., Guzelian, P.S., 2001. CYP3A regulation: from pharmacology to nuclear receptors. Drug Metab. Dispos. 29, 615–622.

Reimers, D., Jezek, A., 1971. [The simultaneous use of rifampicin and other antitubercular agents with oral contraceptives] (in German). Prax. Pneumol. 25, 255–262.

Rettie, A.E., Wienkers, L.C., Gonzalez, F.J., Trager, W.F., Korzekwa, K.R., 1994. Impaired (S)-warfarin metabolism catalysed by the R144C allelic variant of CYP2C9. Pharmacogenetics 4, 39–42.

Rettie, A.E., Haining, R.L., Bajpai, M., Levy, R.H., 1999. A common genetic basis for idiosyncratic toxicity of warfarin and phenytoin. Epilepsy Res. 35, 253–255.

Ritter, C.A., Jedlitschky, G., Meyer zu, S.H., Grube, M., Kock, K., Kroemer, H.K., 2005. Cellular export of drugs and signaling molecules by the ATP-binding cassette transporters MRP4 (ABCC4) and MRP5 (ABCC5). Drug Metab. Rev. 37, 253–278.

Rocchi, E., Khodjakov, A., Volk, E.L., Yang, C.H., Litman, T., Bates, S.E., et al., 2000. The product of the ABC half-transporter gene ABCG2 (BCRP/MXR/ABCP) is expressed in the plasma membrane. Biochem. Biophys. Res. Commun. 271, 42–46.

Ruschitzka, F., Meier, P.J., Turina, M., Luscher, T.F., Noll, G., 2000. Acute heart transplant rejection due to Saint John's wort. Lancet 355, 548–549.

Russel, F.G., Masereeuw, R., Van Aubel, R.A., 2002. Molecular aspects of renal anionic drug transport. Annu. Rev. Physiol. 64, 563–594.

Sampath, J., Adachi, M., Hatse, S., Naesens, L., Balzarini, J., Flatley, R.M., et al., 2002. Role of MRP4 and MRP5 in biology and chemotherapy. AAPS Pharm. Sci. 4, E14.

Sandusky, G.E., Mintze, K.S., Pratt, S.E., Dantzig, A.H., 2002. Expression of multidrug resistance-associated protein 2 (MRP2) in normal human tissues and carcinomas using tissue microarrays. Histopathology 41, 65–74.

Sasaki, M., Suzuki, H., Ito, K., Abe, T., Sugiyama, Y., 2002. Transcellular transport of organic anions across a double-transfected Madin–Darby canine kidney II cell monolayer expressing both human organic anion-transporting polypeptide (OATP2/SLC21A6) and Multidrug resistance-associated protein 2 (MRP2/ABCC2). J. Biol. Chem. 277, 6497–6503.

Schaub, T.P., Kartenbeck, J., Konig, J., Vogel, O., Witzgall, R., Kriz, W., et al., 1997. Expression of the conjugate export pump encoded by the MRP2 gene in the apical membrane of kidney proximal tubules. J. Am. Soc. Nephrol. 8, 1213–1221.

Schinkel, A.H., 1997. The physiological function of drug-transporting P-glycoproteins. Sem. Cancer Biol. 8, 161–170.

Schinkel, A.H., Smit, J.J.M., van Tellingen, O., Beijnen, J.H., Wagenaar, E., van Deemter, L., et al., 1994. Disruption of the mouse mdr1a P-glycoprotein gene leads to a deficiency in the blood–brain barrier and to increased sensitivity to drugs. Cell 77, 491–502.

Schinkel, A.H., Wagenaar, E., van Deemter, L., Mol, C.A.A.M., Borst, P., 1995. Absence of the mdr1a P-glycoprotein in mice affects tissue distribution and pharmacokinetics of dexamethasone, digoxin, and cyclosporin A. J. Clin. Invest. 96, 1698–1705.

Schmassmann-Suhijar, D., Bullingham, R., Gasser, R., Schmutz, J., Haefeli, W.E., 1998. Rhabdomyolysis due to interaction of simvastatin with mibefradil. Lancet 351, 1929–1930.

Schmiedlinren, P., Edwards, D.J., Fitzsimmons, M.E., He, K., Lown, K.S., Woster, P.M., et al., 1997. Mechanisms of enhanced oral availability of cyp3a4 substrates by grapefruit constituents—decreased enterocyte CYP3A4 concentration and mechanism-based inactivation by furanocoumarins. Drug Metab. Dispos. 25, 1228–1233.

Schneider, E., Hunke, S., 1998. ATP-binding-cassette (ABC) transport systems: functional and structural aspects of the ATP-hydrolyzing subunits/domains. FEMS Microbiol. Rev. 22, 1–20.

Sekkarie, M.A., 1997. Torsades de pointes in two chronic renal failure patients treated with cisapride and clarithromycin. Am. J. Kidney Dis. 30, 437–439.

Shen, D.D., Azarnoff, D.L., 1978. Clinical pharmacokinetics of methotrexate. Clin. Pharmacokinet. 3, 1–13.

Silverman, J.A., Schrenk, D., 1997. Hepatic canalicular membrane 4: expression of the multidrug resistance genes in the liver. FASEB J. 11, 308–313.

Sim, S.C., Risinger, C., Dahl, M.L., Aklillu, E., Christensen, M., Bertilsson, L., Ingelman-Sundberg, M., 2006. A common novel CYP2C19 gene variant causes ultrarapid drug metabolism relevant for the drug response to proton pump inhibitors and antidepressants. Clin. Pharmacol. Ther. 79, 103–113.

Simonson, S.G., Raza, A., Martin, P.D., Mitchell, P.D., Jarcho, J.A., Brown, C.D., et al., August 2004. Rosuvastatin pharmacokinetics in heart transplant recipients administered an antirejection regimen including cyclosporine. Clin. Pharmacol. Ther. 76 (2), 167–177.

Sparreboom, A., van Asperen, J., Mayer, U., Schinkel, A.H., Smit, J.W., Meijer, D.K., et al., 1997. Limited oral bioavailability and active epithelial excretion of paclitaxel (Taxol) caused by P-glycoprotein in the intestine. Proc. Natl. Acad. Sci. U.S.A. 94, 2031–2035.

Sueyoshi, T., Kawamoto, T., Zelko, I., Honkakoski, P., Negishi, M., 1999. The repressed nuclear receptor CAR responds to phenobarbital in activating the human CYP2B6 gene. J. Biol. Chem. 274, 6043–6046.

Sueyoshi, T., Negishi, M., 2001. Phenobarbital response elements of cytochrome P450 genes and nuclear receptors. Annu. Rev. Pharmacol. Toxicol. 41, 123–143.

Sullivan-Klose, T.H., Ghanayem, B.I., Bell, D.A., Zhang, Z.Y., Kaminsky, L.S., Shenfield, G.M., Miners, J.O., Birkett, D.J., Goldstein, J.A., 1996. The role of the CYP2C9-Leu359 allelic variant in the tolbutamide polymorphism. Pharmacogenetics 6, 341–349.

Takeda, M., Khamdang, S., Narikawa, S., Kimura, H., Hosoyamada, M., Cha, S.H., et al., 2002. Characterization of methotrexate transport and its drug interactions with human organic anion transporters. J. Pharmacol. Exp. Ther. 302, 666–671.

Tang, H.L., Xie, H.G., Yao, Y., Hu, Y.F., 2011. Lower tacrolimus daily dose requirements and acute rejection rates in the CYP3A5 nonexpressers than expressers. Pharmacogenet. Genomics 21, 713–720.

Thummel, K.E., Wilkinson, G.R., 1998. In vitro and in vivo drug interactions involving human CYP3A. Annu. Rev. Pharmacol. Toxicol. 38, 389–430.

Thummel, K.E., O'shea, D., Paine, M.F., Shen, D.D., Kunze, K.L., Perkins, J.D., Wilkinson, G.R., 1996. Oral first-pass elimination of midazolam involves both gastrointestinal and hepatic CYP3A-mediated metabolism. Clin. Pharmacol. Ther. 59, 491–502.

Thyss, A., Milano, G., Kubar, J., Namer, M., Schneider, M., 1986. Clinical and pharmacokinetic evidence of a life-threatening interaction between methotrexate and ketoprofen. Lancet 1, 256–258.

Tirona, R.G., Leake, B.F., Merino, G., Kim, R.B., 2001. Polymorphisms in OATP-C: identification of multiple allelic variants associated with altered transport activity among European- and African-Americans. J. Biol. Chem. 276 (38), 35669–35675.

Tirona, R.G., Kim, R.B., 2005. Nuclear receptors and drug disposition gene regulation. J. Pharm. Sci. 94, 1169–1186.

Tirona, R.G., Kim, R.B., 2007. Organic anion transporting polypeptides (OATPs). In: You, G., Morris, M.E. (Eds.), Drug Transporters. Wiley, New York, pp. 75–104.

Tirona, R.G., Lee, W., Leake, B.F., Lan, L.B., Cline, C.B., Lamba, V., et al., 2003. The orphan nuclear receptor HNF4alpha determines PXR- and CAR-mediated xenobiotic induction of CYP3A4. Nat. Med. 9, 220–224.

Urquhart, B.L., Tirona, R.G., Kim, R.B., 2007. Nuclear receptors and the regulation of drug-metabolizing enzymes and drug transporters: implications for interindividual variability in response to drugs. J. Clin. Pharmacol. 47, 566–578.

van Herwaarden, A.E., Jonker, J.W., Wagenaar, E., Brinkhuis, R.F., Schellens, J.H., Beijnen, J.H., et al., 2003. The breast cancer resistance protein (BCRP1/ABCG2) restricts exposure to the dietary carcinogen 2-amino-1-methyl-6-phenylimidazo[4,5-b]pyridine. Cancer Res. 63, 6447–6452.

Wada, S., Tsuda, M., Sekine, T., Cha, S.H., Kimura, M., Kanai, Y., et al., 2000. Rat multispecific organic anion transporter 1 (rOAT1) transports zidovudine, acyclovir, and other antiviral nucleoside analogs. J. Pharmacol. Exp. Ther. 294, 844–849.

Wang, D., Guo, Y., Wrighton, S.A., Cooke, G.E., Sadee, W., 2011. Intronic polymorphism in CYP3A4 affects hepatic expression and response to statin drugs. Pharmacogenomics J. 11, 274–286.

Wang, H., Chen, J., Hollister, K., Sowers, L.C., Forman, B.M., 1999. Endogenous bile acids are ligands for the nuclear receptor FXR/BAR. Mol. Cell 3, 543–553.

Westlind-Johnsson, A., Malmebo, S., Johansson, A., Otter, C., Andersson, T.B., Johansson, I., Edwards, R.J., Boobis, A.R., Ingelman-Sundberg, M., 2003. Comparative analysis of CYP3A expression in human liver suggests only a minor role for CYP3A5 in drug metabolism. Drug Metab. Dispos. 31, 755–761.

Wilkinson, G.R., 1987. Clearance approaches in pharmacology. Pharmacol. Rev. 39, 1–47.

Wilkinson, G.R., Shand, D.G., 1975. A physiological approach to hepatic drug clearance. Clin. Pharmacol. Ther. 18, 377–390.

Wilkinson, G.R., 2005. Drug metabolism and variability among patients in drug response. New Engl. J. Med. 352, 2211–2221.

Woosley, R.L., 1996. Cardiac actions of antihistamines. Annu. Rev. Pharmacol. Toxicol. 36, 233–252.

Woosley, R.L., Chen, Y., Freiman, J.P., Gillis, R.A., 1993. Mechanism of the cardiotoxic actions of terfenadine. JAMA 269, 1532–1536.

Wrighton, S.A., Stevens, J.C., 1992. The human hepatic cytochromes P450 involved in drug metabolism. Crit. Rev. Toxicol. 22, 1–21.

Wrighton, S.A., Stevens, J.C., Becker, G.W., Vandenbranden, M., 1993. Isolation and characterization of human liver cytochrome P450 2C19: correlation between 2C19 and S-mephenytoin 4'-hydroxylation. Arch. Biochem. Biophys. 306, 240–245.

Yokogawa, K., Takahashi, M., Tamai, I., Konishi, H., Nomura, M., Moritani, S., et al., 1999. P-glycoprotein-dependent disposition kinetics of tacrolimus: studies in mdr1a knockout mice. Pharm. Res. 16, 1213–1218.

Zanger, U.M., Fischer, J., Raimundo, S., Stuven, T., Evert, B.O., Schwab, M., Eichelbaum, M., 2001. Comprehensive analysis of the genetic factors determining expression and function of hepatic CYP2D6. Pharmacogenetics 11, 573–585.

Zhou, S., Chan, E., Pan, S.Q., Huang, M., Lee, E.J., 2004. Pharmacokinetic interactions of drugs with St. John's wort. J. Psychopharmacol. 18, 262–276.

Chapter 21

Adverse Drug Events

Dan M. Roden
Vanderbilt University Medical Center, Nashville, TN, United States

Chapter Outline

An 84-year-old woman develops atrial fibrillation (AF) 3 days after an uneventful coronary artery bypass grafting. Warfarin is started, and she is discharged on 5 mg daily. The international normalized ratio (INR) 3 days after discharge is 5.7, but over the next several weeks the warfarin dose is adjusted downward and ultimately a dose of 0.5 mg/day achieves the desired INR of 2–3. A month later, she is readmitted for heart failure, and the INR is 2.1. She is uncertain of her current warfarin dosage so the discharge dosage from her previous admission, 5 mg/day, is prescribed. Four days later, she has a massive GI bleed, and her INR is 12.

THE MULTIFACTORIAL NATURE OF ADVERSE DRUG EVENTS

All drugs produce adverse effects in some patients. However, some drugs and some patients are especially at high risk, and the problem is widespread (Lazarou et al., 1998; Pirmohamed et al., 2004). An Institute of Medicine report in 1999 estimated that adverse drug events (ADEs) were between the fourth and sixth most common cause of hospitalizations in the United States and accounted for 98,000 deaths annually (Kohn et al., 1999). The case described here illustrates the problem that serious adverse responses to drugs are often a combination of multiple contributing factors in an individual patient: altered pathophysiology (age, disease), genetics, and—especially in this case—a "system" error. Accordingly in this chapter, the term "ADE," rather than "reaction," is used to emphasize that these occurrences are usually multifactorial and often include both biologic and system issues.

No drug is free of ADE risk, and prescription of a drug therefore requires the physician to estimate that the risks of therapy are outweighed by the benefits. The list of ADEs for any given drug, let alone for all drugs commonly used by one practitioner, is enormous, and so the problem of ADEs presents challenges not only in guiding prescriptions to maximize benefit and minimize risk, but also in educating health practitioners in how to make such judgments in an increasingly complex molecular and genetic environment. Clearly, information tools must play an increasing role in future drug prescribing.

The consideration of risk versus benefit also applies to decisions by drug companies to develop drugs and by regulators to approve them. Indeed, ADEs are not only a major public health problem but also are a leading cause of drug relabeling or withdrawal after development and marketing (Giacomini et al., 2007b). Often ADEs are sufficiently rare so as to elude detection during the drug development process (that typically involves no more than 5000 patients), but sufficiently alarming—when they are detected—that the perception of risk versus benefit is altered. The nonsedating antihistamine terfenadine was withdrawn after it had been used by over 75,000,000 patients, because there was a small risk (probably well under 1/100,000) of the potentially fatal arrhythmia torsades de pointes, particularly with certain drug interactions. The judgment was that the benefit of the drug was exceeded by even this small risk, especially given availability of other less-dangerous alternate therapies (Roden, 2004). However, tools to accurately estimate true risk and true benefit are now only in their infancy.

Should the patient described earlier have been started on warfarin in the first place? AF is a widely recognized risk factor for thromboembolic disease, and in patients with chronic AF, warfarin clearly reduces stroke risk (Singer, 1996; Ezekowitz and Levine, 1999). While this patient was elderly and had heart failure, which increase the risk of AF, data that warfarin reduces stroke risk, when the only documented episode of the arrhythmia is a single brief self-terminating event after a cardiac procedure, are not available. In addition, while advanced age increases the risk of AF-related stroke, it also clearly increases the risk of warfarin-related bleeding, and recent data suggest that bleeding risk in the elderly has been generally underestimated (Wyse, 2007). This patient ended up taking an unusually low dose of warfarin, 0.5 mg daily, to achieve therapeutic anticoagulation. This could reflect age, concomitant drug therapy, or genetically determined impairment of warfarin metabolism. The error in drug dosage on the second admission clearly was the proximate cause of the severe adverse drug reaction. Thus, in this case, contributors to the ADE were the uncertain indication of the drug; underlying biologic factors including age, concomitant disease, and possibly genetic factors that rendered the patient at high risk; and a system error (i.e., one entirely unrelated to the biology).

TYPES OF ADVERSE DRUG EVENTS

System Errors

There are many well-recognized types of "system errors" that can generate severe ADEs. The wrong patient may receive the drug. The wrong drug may be dispensed. Notorious examples involving misinterpretation of physician handwriting abound (Varkey et al., 2007): Lasix for Prilosec or Compazine for Coumadin. Similarly, the biologic justification for prescribing a drug to an individual patient may be lacking: whether the patient presented in the case should have received warfarin is open to debate. The dosage administered may be incorrect, again for biologic reasons or because of "system" problems.

Linking Events to Drug Administration—the "Easy" Examples

Some events are easy to link to drug administration, whereas others are less straightforward. ADEs that involve extensions of the drug's desired pharmacologic effects are usually easy to recognize. Bleeding with anticoagulants or antiplatelet drugs, low blood pressure with antihypertensives, or severe cytopenia with antineoplastics are examples. Other ADEs are "off target," but are sufficiently unusual that an individual practitioner can readily relate them to drug administration: anaphylaxis with penicillin, rhabdomyolysis with an HMG-CoA reductase inhibitor, or QT prolongation and arrhythmias that occur rarely with a large number of drugs (antibiotics, antipsychotics, antihistamines, etc.) are examples.

Linking Events to Drug Administration—the "Hard" Examples

A more difficult situation is an increase in the incidence of a common event such as stroke, myocardial infarction, cancer, or sudden death. If the drugs are used in patient populations in whom such events are expected, identifying a specific role for a drug as a contributor may be exceedingly difficult, and often recognition requires a randomized clinical trial. The fact that antiarrhythmic drugs can increase sudden death when used in patients at risk for this event was first incontrovertibly demonstrated in the Cardiac Arrhythmia Suppression Trial in the late 1980s (Investigators, 1989) and confirmed in subsequent trials (Epstein et al., 1993; Waldo et al., 1996). More recently, an increased incidence of myocardial infarction has been described with the cyclooxygenase-2 inhibitor rofecoxib (Bombardier et al., 2000; Topol, 2004; Bresalier et al., 2005; Kerr et al., 2007) and the PPARγ agonist antidiabetic rosiglitazone (Nissen and Wolski, 2007). The rosiglitazone issue is particularly controversial, and not all evidence supports increased risk (Diamond et al., 2007), illustrating how difficult making this assessment can be.

A common thread in this situation appears to be that the suspect drugs have been approved on the basis of endpoints accrued during short clinical trials, and often such endpoints are "surrogates" for underlying presumed pathophysiology

(Temple, 1999). Thus, many antiarrhythmic drugs were approved because they could suppress isolated premature beats; the premature beats themselves were not a compelling indication for drug therapy, but rather were thought to represent "markers" (or surrogates) for increased risk of sudden death. Similarly, rosiglitazone was approved because it controlled blood sugar, and better antidiabetic control was presumed to produce a beneficial effect on cardiovascular endpoints to which diabetics are susceptible. The recognition that drugs may increase the incidence of common serious events such as myocardial infarction or death presents important and new challenges both to the drug development and regulatory sectors (Temple, 2002; Temple and Himmel, 2002) as well as to translational scientists to identify these events and to understand their underlying mechanisms.

Perhaps the most common "ADE" is the failure of a drug to achieve its desired therapeutic effect. Such failure can be analyzed in terms very similar to those used to analyze more manifest ADEs: Was the correct drug for the underlying pathophysiology prescribed? Was the correct dose prescribed? Did the patient take the drug? Were there interacting drugs or genetic factors that might affect the amount of drug delivered to the target sites? Could a higher dose of drug have been used?

Desired or adverse drug actions occur when a compound with pharmacologic activity (the administered drug or an active metabolite) interacts with specific molecular targets. These targets can be located in plasma, in extracellular spaces, at the cell surface, or within cells. Table 21.1 lists potential mechanisms underlying ADEs. One common mechanism underlying unexpected drug effects is variability in drug concentrations through aberrant drug disposition: failure to achieve effective concentrations of pharmacophores at target sites of action may account for failure of drug efficacy, and excessive concentrations often underlie unusual drug effects. This is termed variability in **pharmacokinetics**. A second mechanism is an unusual drug response in the face of usual drug concentrations termed variable **pharmacodynamics**. Variability in pharmacokinetics or in pharmacodynamics may reflect coexisting physiologies (e.g., age) or disease, concomitant drug therapy, or genetic factors.

Adverse Drug Events Due to Aberrant Drug Disposition (Variable Pharmacokinetics)

Drug disposition is a general term that encompasses the four processes that determine drug and metabolite concentrations in plasma, in tissue, and within cells: absorption, distribution, metabolism, and excretion (usually biliary or renal). Metabolism is generally accomplished by "phase I" enzymes that generate more polar (i.e., more easily water soluble) metabolites; the liver is the site of most drug metabolism, although some can occur in extrahepatic loci, such as the gut, kidney, or plasma. Importantly, biotransformation of parent drug may generate metabolites that have the same, or occasionally somewhat different, pharmacologic effect. Thus, metabolite-mediated beneficial and adverse effects can occur. Phase II drug metabolism conjugates parent drug or polar metabolites with other groups (methyl, acetyl, and glucuronide) to further enhance water solubility and promote excretion. Drug uptake into or efflux from intracellular sites of action may be passive or may be controlled by drug transport molecules. Variabilities in drug metabolism or transport are the well-described causes of variability in drug actions in general, including susceptibility to ADEs.

TABLE 21.1 Mechanisms Underlying Adverse Drug Events

Source	Example
System issues	Wrong patient, dose, or drug
Pharmacokinetics	Dysfunction of excretory organs
	Drug interactions
	Genetically determined altered drug disposition
Pharmacodynamics	Allergic/immunologic reactions
	Drug interactions
	Molecular target dysfunction
	Altered biologic context in which drug–target interaction occurs
Pharmacogenomics	Multiple loci modulate risk

High-Risk Pharmacokinetics

The greatest risk of ADEs due to unusually high concentrations occurs when the following two conditions are *both* met:

1. the drug is eliminated by a single pathway; and
2. there is a narrow margin between the concentrations required for efficacy and those that will produce toxicity.

There is high risk in this situation because inhibition of the single pathway for drug elimination (1) will lead to marked elevation of drug concentration and (2) will then result in high risk of ADEs. Common mechanisms for such inhibition include genetic factors (Meyer, 2000; Evans and Johnson, 2001; McLeod and Evans, 2001; Eichelbaum et al., 2006; Roden et al., 2006; Giacomini et al., 2007a,b), disease-related excretory organ dysfunction, and interacting drugs.

Disease

Disease-related dysfunction of eliminating organs is a well-recognized cause of aberrant drug disposition. QT-interval prolongation by the antiarrhythmics sotalol or dofetilide (both eliminated largely unchanged by urinary excretion) is much more exaggerated, and the risk of long QT-related arrhythmias correspondingly increased, if usual doses are prescribed to patients with renal failure.

Pharmacogenetics, Drug Disposition, and ADEs

A number of antidepressants are eliminated primarily by CYP2D6, an enzyme that is functionally absent (due to homozygosity or compound heterozygosity for loss-of-function alleles) in 5—10% of Caucasian and African populations (Meyer and Zanger, 1997; Meyer, 2000). In these "poor metabolizer" (PM) subjects, ADEs due to tricyclic antidepressants are common. Similarly, some beta blockers (timolol, metoprolol) and antiarrhythmics (propafenone) are also eliminated almost exclusively by CYP2D6-mediated metabolism, and excessive bradycardia and bronchospasm may occur in PMs. The active enantiomer of warfarin (which is prescribed as a racemate) is eliminated by a different enzyme, CYP2C9, and individuals with loss-of-function alleles for CYP2C9 require decreased dosages of warfarin to achieve stable anticoagulation and are at risk for bleeding with "usual" doses (Aithal et al., 1999; Taube et al., 2000; Higashi et al., 2002; Rieder et al., 2005); the case presentation may be an example.

A rare defect in the thiopurine methyltransferase (*TPMT*) gene results in total loss-of-function of enzyme activity in 1 in 300 individuals (Weinshilboum and Sladek, 1980; Evans et al., 1991; McLeod et al., 2000; Stanulla et al., 2005). Such patients are at high risk of bone marrow aplasia during treatment with azathioprine or 6-mercaptopurine. These drugs are bioinactivated by *TPMT*, and when not bioinactivated, are shunted toward 6-thioguanines, which are cytotoxic. Thus, individuals with *TPMT* deficiency shunt much greater drug concentrations toward 6-thioguanines and are susceptible to severe toxicity. In addition, subjects with wild-type TPMT activity may require higher than average doses to achieve maximum therapeutic efficacy. A common polymorphism in the *UGT1A1* gene, whose protein product underlies glucuronidation both of drugs and of bilirubin, results in decreased glucuronyl transferase activity in 2—3% of individuals (Ando et al., 2000; Danoff et al., 2004; Innocenti et al., 2004; Ratain, 2006). This may manifest in normal subjects as mild hyperbilirubinemia (Gilbert syndrome). These patients are also at increased risk of severe ADEs (primarily diarrhea) during therapy with the anticancer drug irinotecan and for hyperbilirubinemia with the antiretroviral agent atazanavir (Ribaudo et al., 2013). Irinotecan is a prodrug and is bioactivated to an active metabolite, SN-38, which is eliminated by glucuronidation. Patients with decreased *UGT1A1* activity therefore generate higher than normal concentrations of SN-38, which produces gastrointestinal and bone marrow toxicity.

Drug Interactions

Some drugs are especially likely to cause ADEs because they inhibit specific pathways for drug elimination. In this situation, ADEs are likely if a second "high-risk" drug using that pathway is also administered. The activity of the enzyme system CYP3A4, the most common pathway used to eliminate drugs, is highly variable among individuals for reasons that are not completely understood, although individuals with absent activity (such as CYP2D6 "PMs") have not been described. Many drugs are potent CYP3A4 inhibitors; thus, these inhibitors can generate a "high-risk" situation by markedly increased concentrations of CYP3A4 substrates to produce toxicity. The antirejection drug cyclosporine is a CYP3A4 substrate and its toxicity is enhanced by coadministration of potent CYP3A4 inhibitors, such as certain azole antifungals (ketoconazole, itraconazole) or macrolide antibiotics (erythromycin, clarithromycin). On the other hand, inhibition of CYP3A4 has been used deliberately to lower cyclosporine dose requirements, and thus cost (Valantine et al., 1992). These drugs also inhibit terfenadine metabolism to its active non−QT-prolonging metabolite fexofenadine (now marketed as a drug in its own right), and this interaction was responsible for most cases of terfenadine-related torsades de pointes (Woosley et al., 1993). Similarly, some tricyclic antidepressants, fluoxetine, paroxetine and quinidine, are potent CYP2D6 inhibitors, and can therefore confer the "PM" phenotype in individuals who are genetically extensive metabolizers.

Digoxin is eliminated by an active drug efflux pump, P-glycoprotein (encoded by the *MDR1* or *ABCB1* gene). Drugs that inhibit P-glycoprotein inhibit digoxin elimination, elevate serum digoxin concentrations, and produce toxicity; amiodarone, quinidine, verapamil, and itraconazole are examples (Fromm et al., 1999).

Adverse Drug Events Not Due to Elevated Drug or Metabolite Concentrations

Drugs can produce ADEs at low or therapeutic concentrations. Many of these reactions are thought to be immunologic: anaphylaxis, serum sickness, or immune-related thrombocytopenia or liver damage. In some cases the antigen and immunologic mechanism is well recognized, whereas in others the antigen is not well understood.

Other ADEs in this category may simply reflect an increased pharmacologic response due to concomitant physiologic factors. Perhaps the most common reactions that fall into this category are those occurring in the elderly, in whom ADEs may arise as a result of a combination of altered drug disposition and multiple other physiologic alterations, including inability to adjust blood pressure with changes in posture, decreased renal blood flow, and vascular "stiffening." In some cases gender is a risk factor; for reasons that remain incompletely understood, the incidence of drug-induced long QT-related arrhythmias is much higher in women than in men (Makkar et al., 1993). Similarly, drugs can potentiate each others' anticipated pharmacologic effects to cause ADEs: combined therapy with aspirin and warfarin to increase bleeding risk is an example.

Genetics of Variable Pharmacodynamics

Lack of response to the beta blocker bucindolol in patients with heart failure has been attributed to variations of the "target" molecule, encoded by the beta-1 adrenergic receptor gene *ADRB1* (Liggett et al., 2006). A beneficial response was seen in a large clinical trial only among subjects carrying an arginine residue at position 389, and no response was seen among patients with the glycine variant at this position. These data suggest that variations in the drug target molecule may be responsible for some ADEs, including failure of drug response.

A variation of this theme has been described in patients with drug-induced prolongation of the QT interval (Roden, 2004). The QT interval in normal individuals represents the aggregate effect of multiple ion currents (each reflecting the activity of multiple ion channel genes) in the cardiomyocyte membrane. Virtually all drugs that produce QT prolongation and long QT-related arrhythmias block one particular ion current, termed I_{Kr}, and encoded by the *KCNH2* gene (also termed *HERG*). Patients with *HERG* variants may display unusual responses to I_{Kr}-blocking drugs. In addition, patients with variations in other ion currents that also contribute to normal QT interval have been well described. The concept is that such genetic variations may remain subclinical (in part because I_{Kr} remains a major determinant of QT interval) until I_{Kr}-blocking drugs are administered, uncovering the previously subclinical lesion. Thus, as with many other ADEs, the development of drug-induced arrhythmias likely requires multiple "hits," including administration of a drug and a substrate that is in some way primed (by DNA variants or by acquired heart disease). Another way of thinking about this is that the interaction between a drug and its molecular target does not occur in isolation, but rather in a complex biologic milieu. Thus, for example, the risk of drug-induced torsades de pointes is exaggerated by hypokalemia, which appears to modulate the drug−HERG interaction (Yang and Roden, 1996; Yang et al., 1997).

While CYP2C9 variants have been invoked in variability in warfarin responses, a second gene, VKORC1 (encoding a component of the vitamin K-dependent target for the drug), also contributes; indeed, available data suggest that VKORC1 variants actually contribute more to variability ion warfarin dose than do CYP2C9 variants (Rieder et al., 2005). The warfarin story thus illustrates both pharmacokinetic and pharmacogenetic variability. Interestingly, more than half of the variability in warfarin dose remains unexplained, suggesting other (as yet unidentified) genes may play a role. Randomized clinical trials to compare early outcomes (assessed as time in therapeutic range) with genetically guided warfarin therapy versus clinical regimens have provided conflicting results (Kimmel et al., 2013; Pirmohamed et al., 2013). On the other hand, variants in *CYP2C9* and in *CYP4F2* (which encodes an enzyme involved in vitamin K metabolism) have been implicated as risk factors for bleeding during long-term warfarin therapy (Kawai et al., 2014; Roth et al., 2014).

GENETICS TO GENOMICS

The identification of the role of individual gene variants in mediating certain ADEs is described earlier: *TPMT*, *UGT1A1*, *CYP2D6, CYP2C9, KCNH2,* and *ADRB1* are examples, and there are many others. Such a focus on single genes in which variants may play a large role in determining ADEs is one underpinning of the field of **pharmacogenetics**.

The 21st century has seen the completion of the first human genome sequence, the identification of tens of millions of polymorphisms across individuals and populations, and the identification of the "architecture" of the genome, notably the definition of large areas of linkage disequilibrium (haplotype blocks). These tools are now enabling a revolution in modern genomics (see Chapters 16−18). For many years, physicians have been exposed to the "mantra" that common diseases such as cancer, Alzheimer disease, or atherosclerosis include genetic components but the specific causative genes have not been identified until very recently, when these tools have been used to identify multiple genetic loci conferring risk for these common diseases. In some cases, the increased risk can be substantial, up to twofold.

These genomic approaches are only now being implemented to study the genomic determinants of beneficial and adverse events to drugs, the field of **pharmacogenomics** (Roden et al., 2006; Giacomini et al., 2007a). As with gene hunting for diseases, it is likely that these technologies will not only identify new loci determining susceptibility to ADEs, but also that these loci will be in "unexpected" regions of the genome. A challenge to modern biology, in which translational scientists must play a major role, is the elucidation of the mechanisms whereby genetic variants translate to increased susceptibility to ADEs. More generally, ADEs provide a unique window to understanding normal and abnormal drug responses. Once the basic mechanisms underlying these effects are understood, systems to deliver the right drug to the right patient in the right dose at the right time can be developed. Understanding ADEs can not only help develop such systems but also point to ways in which new drugs can be developed to incorporate molecular mechanisms to avoid ADEs or to target entirely new biologic pathways for the safe and effective treatment of human disease.

REFERENCES

Aithal, G.P., Day, C.P., Kesteven, P.J., Daly, A.K., 1999. Association of polymorphisms in the cytochrome P450 CYP2C9 with warfarin dose requirement and risk of bleeding complications. Lancet 353, 717−719.

Ando, Y., Saka, H., Ando, M., Sawa, T., Muro, K., Ueoka, H., Yokoyama, A., Saitoh, S., Shimokata, K., Hasegawa, Y., 2000. Polymorphisms of UDP-glucuronosyltransferase gene and irinotecan toxicity: a pharmacogenetic analysis. Cancer Res. 60, 6921−6926.

Bombardier, C., Laine, L., Reicin, A., Shapiro, D., Burgos-Vargas, R., Davis, B., Day, R., Ferraz, M.B., Hawkey, C.J., Hochberg, M.C., Kvien, T.K., Schnitzer, T.J., 2000. Comparison of upper gastrointestinal toxicity of rofecoxib and naproxen in patients with rheumatoid arthritis. VIGOR Study Group N. Engl. J. Med. 343, 1520−1528.

Bresalier, R.S., Sandler, R.S., Quan, H., Bolognese, J.A., Oxenius, B., Horgan, K., Lines, C., Riddell, R., Morton, D., Lanas, A., Konstam, M.A., Baron, J.A., 2005. Cardiovascular events associated with rofecoxib in a colorectal adenoma chemoprevention trial. N. Engl. J. Med. 352, 1092−1102.

CAST Investigators, 1989. Preliminary report: effect of encainide and flecainide on mortality in a randomized trial of arrhythmia suppression after myocardial infarction. N. Engl. J. Med. 321, 406−412.

Danoff, T.M., Campbell, D.A., McCarthy, L.C., Lewis, K.F., Repasch, M.H., Saunders, A.M., Spurr, N.K., Purvis, I.J., Roses, A.D., Xu, C.F., 2004. A Gilbert's syndrome UGT1A1 variant confers susceptibility to tranilast-induced hyperbilirubinemia. Pharmacogenomics J. 4, 49−53.

Diamond, G.A., Bax, L., Kaul, S., 2007. Uncertain effects of rosiglitazone on the risk for myocardial infarction and cardiovascular death. Ann. Intern. Med. 147, 578−581.

Eichelbaum, M., Ingelman-Sundberg, M., Evans, W.E., 2006. Pharmacogenomics and individualized drug therapy. Annu. Rev. Med. 57, 119−137.

Epstein, A.E., Hallstrom, A.P., Rogers, W.J., Liebson, P.R., Seals, A.A., Anderson, J.L., Cohen, J.D., Capone, R.J., Wyse, D.G., 1993. Mortality following ventricular arrhythmia suppression by encainide, flecainide, and moricizine after myocardial infarction. The original design concept of the Cardiac Arrhythmia Suppression Trial (CAST). JAMA 270, 2451−2455.

Evans, W.E., Horner, M., Chu, Y.Q., Kalwinsky, D., Roberts, W.M., 1991. Altered mercaptopurine metabolism, toxic effects, and dosage requirement in a thiopurine methyltransferase-deficient child with acute lymphocytic leukemia. J. Pediatr. 119, 985−989.

Evans, W.E., Johnson, J.A., 2001. Pharmacogenomics: the inherited basis for interindividual differences in drug response. Annu. Rev. Genom. Hum. Genet. 2, 9−39.

Ezekowitz, M.D., Levine, J.A., 1999. Preventing stroke in patients with atrial fibrillation. JAMA 281, 1830−1835 (In process citation).

Fromm, M.F., Kim, R.B., Stein, C.M., Wilkinson, G.R., Roden, D.M., 1999. Inhibition of P-glycoprotein-mediated drug transport: a unifying mechanism to explain the interaction between digoxin and quinidine. Circulation 99, 552−557.

Giacomini, K.M., Brett, C.M., Altman, R.B., Benowitz, N.L., Dolan, M.E., Flockhart, D.A., Johnson, J.A., Hayes, D.F., Klein, T., Krauss, R.M., Kroetz, D.L., McLeod, H.L., Nguyen, A.T., Ratain, M.J., Relling, M.V., Reus, V., Roden, D.M., Schaefer, C.A., Shuldiner, A.R., Skaar, T., Tantisira, K., Tyndale, R.F., Wang, L., Weinshilboum, R.M., Weiss, S.T., Zineh, I., 2007a. The pharmacogenetics research network: from SNP discovery to clinical drug response. Clin. Pharmacol. Ther. 81, 328−345.

Giacomini, K.M., Krauss, R.M., Roden, D.M., Eichelbaum, M., Hayden, M.R., Nakamura, Y., 2007b. When good drugs go bad. Nature 446, 975−977.

Higashi, M.K., Veenstra, D.L., Kondo, L.M., Wittkowsky, A.K., Srinouanprachanh, S.L., Farin, F.M., Rettie, A.E., 2002. Association between CYP2C9 genetic variants and anticoagulation-related outcomes during warfarin therapy. JAMA 287, 1690−1698.

Innocenti, F., Undevia, S.D., Iyer, L., Chen, P.X., Das, S., Kocherginsky, M., Karrison, T., Janisch, L., Ramirez, J., Rudin, C.M., Vokes, E.E., Ratain, M.J., 2004. Genetic variants in the UDP-glucuronosyltransferase 1A1 gene predict the risk of severe neutropenia of irinotecan. J. Clin. Oncol. 22, 1382−1388.

Kawai, V.K., Cunningham, A., Vear, S.I., Van Driest, S.L., Oginni, A., Xu, H., Jiang, M., Li, C., Denny, J.C., Shaffer, C., Bowton, E., Gage, B.F., Ray, W.A., Roden, D.M., Stein, C.M., 2014. Genotype and risk of major bleeding during warfarin treatment. Pharmacogenomics 15, 1973−1983.

Kerr, D.J., Dunn, J.A., Langman, M.J., Smith, J.L., Midgley, R.S., Stanley, A., Stokes, J.C., Julier, P., Iveson, C., Duvvuri, R., McConkey, C.C., 2007. Rofecoxib and cardiovascular adverse events in adjuvant treatment of colorectal cancer. N. Engl. J. Med. 357, 360−369.

Kimmel, S.E., French, B., Kasner, S.E., Johnson, J.A., Anderson, J.L., Gage, B.F., Rosenberg, Y.D., Eby, C.S., Madigan, R.A., McBane, R.B., Abdel-Rahman, S.Z., Stevens, S.M., Yale, S., Mohler, E.R., Fang, M.C., Shah, V., Horenstein, R.B., Limdi, N.A., Muldowney, J.A.S., Gujral, J., Delafontaine, P., Desnick, R.J., Ortel, T.L., Billett, H.H., Pendleton, R.C., Geller, N.L., Halperin, J.L., Goldhaber, S.Z., Caldwell, M.D., Califf, R.M., Ellenberg, J.H., 2013. A pharmacogenetic versus a clinical algorithm for warfarin dosing. N. Engl. J. Med. 369, 2283−2293.

Kohn, K.T., Corrigan, J.M., Donaldson, M.S., 1999. To Err Is Human: Building a Safer Health System. National Academy Press, Washington, DC.

Lazarou, J., Pomeranz, B.H., Corey, P.N., 1998. Incidence of adverse drug reactions in hospitalized patients: a meta-analysis of prospective studies. JAMA 279, 1200−1205.

Liggett, S.B., Mialet-Perez, J., Thaneemit-Chen, S., Weber, S.A., Greene, S.M., Hodne, D., Nelson, B., Morrison, J., Domanski, M.J., Wagoner, L.E., Abraham, W.T., Anderson, J.L., Carlquist, J.F., Krause-Steinrauf, H.J., Lazzeroni, L.C., Port, J.D., Lavori, P.W., Bristow, M.R., 2006. A polymorphism within a conserved {beta}1-adrenergic receptor motif alters cardiac function and {beta}-blocker response in human heart failure. PNAS 103, 11288−11293.

Makkar, R.R., Fromm, B.S., Steinman, R.T., Meissner, M.D., Lehmann, M.H., 1993. Female gender as a risk factor for torsades de pointes associated with cardiovascular drugs. JAMA 270, 2590−2597.

McLeod, H.L., Evans, W.E., 2001. Pharmacogenomics: unlocking the human genome for better drug therapy. Annu. Rev. Pharmacol. Toxicol. 41, 101−121.

McLeod, H.L., Krynetski, E.Y., Relling, M.V., Evans, W.E., 2000. Genetic polymorphism of thiopurine methyltransferase and its clinical relevance for childhood acute lymphoblastic leukemia. Leukemia 14, 567−572.

Meyer, U.A., 2000. Pharmacogenetics and adverse drug reactions. Lancet 356, 1667−1671.

Meyer, U.A., Zanger, U.M., 1997. Molecular mechanisms of genetic polymorphisms of drug metabolism. Annu. Rev. Pharmacol. Toxicol. 37, 269−296.

Nissen, S.E., Wolski, K., 2007. Effect of rosiglitazone on the risk of myocardial infarction and death from cardiovascular causes. N. Engl. J. Med. 356, 2457−2471.

Pirmohamed, M., James, S., Meakin, S., Green, C., Scott, A.K., Walley, T.J., Farrar, K., Park, B.K., Breckenridge, A.M., 2004. Adverse drug reactions as cause of admission to hospital: prospective analysis of 18 820 patients. BMJ 329, 15−19.

Pirmohamed, M., Burnside, G., Eriksson, N., Jorgensen, A.L., Toh, C.H., Nicholson, T., Kesteven, P., Christersson, C., Wahlström, B., Stafberg, C., Zhang, J.E., Leathart, J.B., Kohnke, H., Maitland-van der Zee, A.H., Williamson, P.R., Daly, A.K., Avery, P., Kamali, F., Wadelius, M., 2013. A randomized trial of genotype-guided dosing of warfarin. N. Engl. J. Med. 369, 2294−2303.

Ratain, M.J., 2006. From bedside to bench to bedside to clinical practice: an odyssey with irinotecan. Clin. Cancer Res. 12, 1658−1660.

Ribaudo, H.J., Daar, E.S., Tierney, C., Morse, G.D., Mollan, K., Sax, P.E., Fischl, M.A., Collier, A.C., Haas, D.W., 2013. Impact of UGT1A1 Gilbert variant on discontinuation of ritonavir-boosted atazanavir in AIDS clinical trials group study A5202. J. Infect. Dis. 207, 420−425.

Rieder, M.J., Reiner, A.P., Gage, B.F., Nickerson, D.A., Eby, C.S., McLeod, H.L., Blough, D.K., Thummel, K.E., Veenstra, D.L., Rettie, A.E., 2005. Effect of VKORC1 haplotypes on transcriptional regulation and warfarin dose. N. Engl. J. Med. 352, 2285−2293.

Roden, D.M., 2004. Drug-induced prolongation of the QT interval. N. Engl. J. Med. 350, 1013−1022.

Roden, D.M., Altman, R.B., Benowitz, N.L., Flockhart, D.A., Giacomini, K.M., Johnson, J.A., Krauss, R.M., McLeod, H.L., Ratain, M.J., Relling, M.V., Ring, H., Shuldiner, A.R., Weinshilboum, R.M., Weiss, S.T., Pharmacogenetics Research Network, 2006. Pharmacogenomics: challenges and opportunities. Ann. Intern. Med. 145, 749−757.

Roth, J.A., Boudreau, D., Fujii, M.M., Farin, F.M., Rettie, A.E., Thummel, K.E., Veenstra, D.L., 2014. Genetic risk factors for major bleeding in warfarin patients in a community setting. Clin. Pharmacol. Ther. 95, 636−643.

Singer, D.E., 1996. Anticoagulation for atrial fibrillation: epidemiology informing a difficult clinical decision. Proc. Assoc. Am. Phys. 108, 29−36.

Stanulla, M., Schaeffeler, E., Flohr, T., Cario, G., Schrauder, A., Zimmermann, M., Welte, K., Ludwig, W.D., Bartram, C.R., Zanger, U.M., Eichelbaum, M., Schrappe, M., Schwab, M., 2005. Thiopurine methyltransferase (TPMT) genotype and early treatment response to mercaptopurine in childhood acute lymphoblastic leukemia. JAMA 293, 1485−1489.

Taube, J., Halsall, D., Baglin, T., 2000. Influence of cytochrome P-450 CYP2C9 polymorphisms on warfarin sensitivity and risk of over-anticoagulation in patients on long-term treatment. Blood 96, 1816−1819.

Temple, R., 2002. Policy developments in regulatory approval. Stat. Med. 21, 2939−2948.

Temple, R., 1999. Are surrogate markers adequate to assess cardiovascular disease drugs? JAMA 282, 790−795.

Temple, R.J., Himmel, M.H., 2002. Safety of newly approved drugs: implications for prescribing. JAMA 287, 2273−2275.

Topol, E.J., 2004. Failing the public health − rofecoxib, Merck, and the FDA. N. Engl. J. Med. 351, 1707−1709.

Valantine, H., Keogh, A., McIntosh, N., Hunt, S., Oyer, P., Schroeder, J., 1992. Cost containment: coadministration of diltiazem with cyclosporine after heart transplantation. J. Heart Lung Transpl. 11, 1−7.

Varkey, P., Aponte, P., Swanton, C., Fischer, D., Johnson, S.F., Brennan, M.D., 2007. The effect of computerized physician-order entry on outpatient prescription errors. Manag. Care Interface 20, 53–57.

Waldo, A.L., Camm, A.J., DeRuyter, H., Friedman, P.L., MacNeil, D.J., Pauls, J.F., Pitt, B., Pratt, C.M., Schwartz, P.J., Veltri, E.P., 1996. Effect of d-sotalol on mortality in patients with left ventricular dysfunction after recent and remote myocardial infarction. Lancet 348, 7–12.

Weinshilboum, R.M., Sladek, S.L., 1980. Mercaptopurine pharmacogenetics: monogenic inheritance of erythrocyte thiopurine methyltransferase activity. Am. J. Hum. Genet. 32, 651–662.

Woosley, R.L., Chen, Y., Freiman, J.P., Gillis, R.A., 1993. Mechanism of the cardiotoxic actions of terfenadine. JAMA 269, 1532–1536.

Wyse, D.G., 2007. Bleeding while starting anticoagulation for thromboembolism prophylaxis in elderly patients with atrial fibrillation: from bad to worse. Circulation 115, 2684–2686.

Yang, T., Roden, D.M., 1996. Extracellular potassium modulation of drug block of I_{Kr}: implications for Torsades de Pointes and reverse use-dependence. Circulation 93, 407–411.

Yang, T., Snyders, D.J., Roden, D.M., 1997. Rapid inactivation determines the rectification and $[K^1]_o$ dependence of the rapid component of the delayed rectifier K^1 current in cardiac cells. Circ. Res. 80, 782–789.

Societal Context of Human Research

Chapter 22

Translating Science to the Bedside: The Innovation Pipeline

Seema Basu

Partners HealthCare Innovation, Cambridge, MA, United States

Chapter Outline

Key Points

- Because technologies developed in academia are at the earliest stage and the investment of time and money is high to develop them further to bring them to market, the hurdles to licensing new technologies are high.
- Attracting resources to move opportunities forward remains the perpetual barrier. Funding is one key resource, but commercialization expertise is equally important. Other required resources may include marketing and sales channels, complementary technologies or intellectual property, and the human capital to focus exclusively on developing and bringing the product to market.
- To attract the resources necessary for commercialization, technologies need to address large revenue opportunities relative to the cost of development. A rule of thumb in venture capital is that at a *minimum*, a software company should have a $30 million annual market for its initial product, $100 million for a medical device or diagnostic, and several hundred millions of dollars for a novel therapeutic. Most successful projects have opportunities that are multiples of those minimums.
- In the context of translational and clinical research, intellectual property arises from discovery research and the practice of clinical medicine that forms the foundation of improved products and services for patient care.
- The broad vision for academic—industry relationships is to accelerate the convergence of activities in the research and clinical arenas and to facilitate the translation of ideas to real-world products and solutions to improve patient care.
- Philanthropy has become an important source of translational research funding, and this trend is increasing. In the past, foundations tended, like the government, to focus on basic research. Many foundations now are funding projects later in the development cycle, and even beginning to partner with companies in a "venture philanthropy" model.
- The most significant source of funding for translational research has been and remains industry.
- The pathway to commercialization depends on the size of the market opportunity, the risks of failure (e.g., due to safety, development time, or product costs), competitive pressures, strength of intellectual property, and the ability of the commercialization team of individuals to work together productively.
- Successful strategic collaborations are difficult to establish and even more challenging to sustain. When managed effectively, however, academic—industry relationships have the potential to yield great value to the respective parties.
- The key asset of any new venture is its intellectual property, comprising patent filings as well as the accumulated knowledge and know-how around the technology. A successful entrepreneur will have a strategy for securing all the intellectual property rights needed to practice the technology and protect it against imitators. That may involve in-licensing, cross-licensing, patent prosecution, and always a demonstrable understanding of the patent landscape with potential hazards identified.
- The crucial decision for clinical research: Do the benefits of an innovative product outweigh the potential risks to the human subject involved in the product's evaluation?
- The existence of divided loyalties due to personal and institutional financial interests in the commercial success of research results has become an increasingly serious concern in the decades following the passage of the Bayh—Dole Act of 1980.

REALITIES OF THE MARKETPLACE

With the 1980 enactment of the Patent and Trademark Law Amendments Act, which is more commonly known as the Bayh—Dole Act, universities and medical centers were allowed for the first time to elect to keep title to inventions made with federal grant funding. Prior to 1980, title to inventions remained with the federal government, and relatively few were licensed (Fig. 22.1).

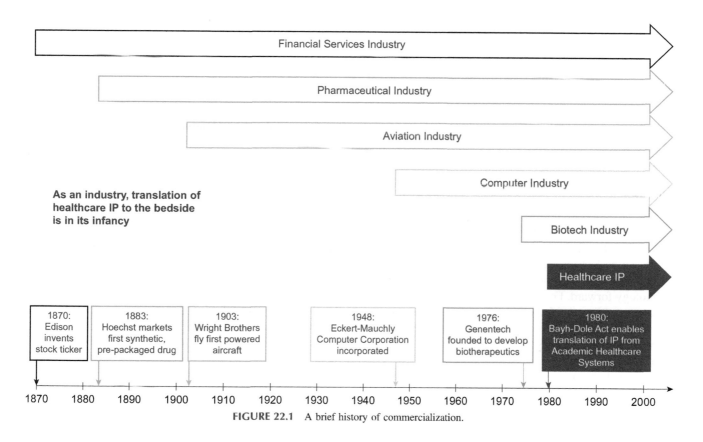

FIGURE 22.1   A brief history of commercialization.

Much has changed, all because of Bayh—Dole. A 2015 independent study, commissioned by the Biotechnology Industry Organization, documented the significant impact academia industry technology transfer makes to the US economy. The report (*The Economic Contribution of University/Nonprofit Inventions in the United States: 1996—2013*) estimates that academia—industry patent licensing bolstered US gross industry output during this 18-year period by up to $1.18 trillion, US gross domestic product by up to $518 billion, and supported up to 3,824,000 US jobs.

The annual licensing impact survey of 2015 conducted by the Association of University Technology Managers (AUTM) showed that:

- 818 start-up companies formed around academic patents (up by 16% from 2012)—or more than two new companies created every working day of the year
- 4200 start-ups in operation, mostly located in the same state as the parent research institution, creating regional economic development
- $22.8 billion in product sales from commercialized academic inventions
- 719 new products introduced into the market (up by 22% from 2012) or more than two new products introduced every day of the year

Commercialization is the pathway that transforms ideas and technologies into medical products that can be used in clinical practice. Academic institutions excel at research but require collaborations with industry to turn science into products. Because technologies developed in academia are at the earliest stage and the investment of time and money is high to develop them further to bring them to market, the hurdles to licensing new technologies are also high. Those ideas that attract initial investment (from companies or investors) have a strong business case—the reward is believed to be worth the investment, even when considering the risk.

While all technologies are different, they do share certain commercialization challenges. Attracting resources to move opportunities forward remains the perpetual barrier. Funding is one key resource, but commercialization expertise is equally important. Other required resources may include marketing and sales channels, complementary technologies or intellectual property (IP), and the human capital to focus exclusively on developing and bringing the product to market.

What Are the Phases Involved in Taking Ideas/Discoveries to the Patient's Bedside? What Are the Critical Issues to Consider at Each Phase?

Whatever the technology, the first step in the process is analyzing the opportunity. Great science is sometimes of limited commercial value (or else the commercial value will come far in the future). Some of the questions that must be asked are:

- Is there a sufficient commercial market to attract the resources needed?
- Are the benefits important enough to motivate clinicians (or others) to alter long-standing accepted practice and adopt the new technology?
- Is the technology developed enough?
- Is the technology simply an improvement, or is it truly a breakthrough?
- Can the idea be used or is there IP belonging to others that would block the freedom to operate?

The next step in the process (often concurrent with the first) is to protect the technology. While patents are the most common form of protection, different types of technologies require different strategies, and all require specific expertise in the chosen strategy. Engaging the technology transfer office early in the process is extremely important, as valuable IP rights can be lost through seemingly benign interactions with outside parties.

Once the idea is protected, it is time to further define the opportunity and find the necessary resources to bring the technology forward. For a *platform technology* (typically defined as one that differs significantly from the previous state of the art, providing the foundation for additional innovations), this would include selecting the applications that are to be developed. The key is to define the opportunities from the commercial rather than the academic perspective. Each technology is special and presents unique issues; sometimes problems can arise from thinking too big, other times the problem is thinking too small.

Resources to assist in this process can come from relationships with companies that already have an interest in developing and marketing the technology, or in the case of a start-up, from the entrepreneurs and investors. Increasingly, disease foundations are taking on the early role of defining the opportunity and providing translational funding to help expedite the process of getting products to market.

As projects move from the research phase into development and commercialization, true technology transfer needs to take place. Academic labs are optimized for research, while companies are optimized for product development, sales, and marketing. Although the inventor remains an important part of the process, at some point the development of the existing ideas becomes the responsibility of the company. In an ideal setting, the relationship can be structured so that the academic inventor researches next-generation technologies while the company develops products from the current discoveries.

How Does This Differ for Various Types of Technology? What Are the Market Pressures for Devices, Software, Diagnostics, Therapeutics, and Research Tools?

In many ways, software and therapeutics have opposite market dynamics. For example, therapeutics have very high technical risk, are slow to market, and patents are the method to protect IP. If brought to market, however, the chances for success are usually high.

Software has low technical risk, can be brought to market quickly, relies on copyright and upgrades (rarely patents) to protect the IP, but the market adoption risk is relatively high.

Devices, on the other hand, tend to look like a hybrid. Like software, the technical risk tends to be in the field of engineering rather than biology, and the greater risk is in market adoption (developing marketing and sales channels is often the single most difficult task for start-up device companies), and the US Food and Drug Administration (FDA) risk is low. Like therapeutics, patents are the main form of IP protection for devices. While the FDA barrier is much lower, considerable clinical data are usually key to adoption of novel devices in clinical practice.

Diagnostics present a difficult challenge for commercialization. The clinical evidence required for adoption can be nearly as expensive as the development of a therapeutic, yet markets tend to be much smaller and margins lower. As a result, many venture capital funds will not, in general, invest in any diagnostic technologies.

The keys to success are a clear path to productization (being able to use the test in real-world settings at real-world volumes), a path to clinical validation for a reasonable expenditure, and a clear change in clinical practice based on the outcome of the test.

To attract the resources necessary for commercialization, technologies need to address large revenue opportunities relative to the cost of development. A rule of thumb in venture capital is that at a *minimum*, a software company should have a $30 million annual market for its initial product, $100 million for medical device or diagnostic, and several hundred millions of dollars for a novel therapeutic. Most successful projects have opportunities that are multiples of those minimums.

What Funding Sources Support Development and Creation of Products and Services?

Government grants are clearly the most important source of funding for the earliest stages of the development of new technologies. With certain exceptions, however, government funding rarely extends into the realm of translational medicine. As projects become more advanced and attempt to cross the bridge from science to product, other sources of funding are required.

This translational funding typically comes from:

- Companies
- Venture capitalists
- Foundations
- Internal investments by the academic institution itself

What Are the Roles of Government, Philanthropy, and Industry?

While government funding is most important in the early phases of discovery (far too risky for companies and investors to even consider), the Small Business Innovation Research (SBIR) and Small Business Technology Transfer (STTR) grant mechanisms are important sources of funds for more developed projects. In these cases, the investigator must have a collaboration with a company.

For an SBIR, the principal investigator (PI) is part of the company, who performs the majority of the work. If the PI is at the academic health-care center, the STTR program is available and the research effort is more equal. While the STTR program is often a better fit for academics, the SBIR program is far larger. For some SBIR grantees, the National Institutes of Health (NIH) even provides commercialization assistance and market research consulting.

Philanthropy has become an important source of translational research funding, and this trend is increasing. In the past, foundations tended, like the government, to focus on basic research. Many foundations now are funding projects later in the development cycle, and even beginning to partner with companies in a "venture philanthropy" model. Disease foundations are not only a source of funding, however. They can also serve as a catalyst to validate specific projects and bring in additional resources from their partners.

Venture capital can be a source of funding for some opportunities, but these situations are relatively rare. Venture capital is a specific funding mechanism and is not appropriate in most cases. What might be an exciting project for a device company with an existing product line in the field may be completely inappropriate for a start-up. Venture funding is also rare—venture funds typically invest in less than 1% of the business plans they receive.

One commonly held misconception about venture capitalists is that they enjoy taking on risk. Venture capitalists typically are risk averse but understand that often high levels of risk are inherent in their world and they attempt to manage risk effectively. Any strategy that mitigates the risk of a project without reducing its value can dramatically improve the chances of a start-up receiving funding. As the venture industry has grown, particularly since the late 1990s, the amount invested per project has grown as well, leaving a funding gap for interesting but smaller opportunities.

For start-ups, that gap is increasingly being filled by angel investors (wealthy individuals who invest in private companies). As the venture industry grew and largely abandoned smaller financing opportunities, groups of angels banded together to fill the gap. The amount of funding for private companies from angels consistently exceeds the amount invested by the entire venture capital industry, and the variety of funded projects is much greater. (See **below** for more on angel investors.)

While the previously mentioned are important sources of funding, the most significant source of funding for translational research has been and remains industry. The majority of license agreements executed are with established companies, large and small, which bring not only funding but also the entire array of resources needed to bring the product to market.

Types of Relationship: Sponsored Research, Grants, Codevelopment, and Gifts

Once a company and an academic institution have agreed to work together to develop an invention, the relationship can take several forms. The early stages of the relationship may be based on a sponsored research agreement, where the company funds additional research in the investigator's lab to continue to develop and validate the new technology. This approach has the advantage of leveraging the expertise and equipment of those who know the technology best.

Generally, in exchange for such funding, the company receives an option to license any new IP developed during the project. In other cases, the company and the investigator may work more closely together in a codevelopment agreement, with the company and investigator each being responsible for specific deliverables. A few companies have developed novel mechanisms and infrastructure, including in-house incubators.

Another model is "open innovation," which is based loosely on the open source model in software, where larger companies try to work alongside academic centers and start-ups by openly sharing information and resources in exchange for potential access to technologies as they are developed.

What Resources Are Available at Academic Health-Care Systems to Help Translational Researchers Bring Products and Technologies to Market for Patient Benefit?

In almost every academic institution, resources to facilitate commercialization are clustered around the technology transfer office. Known by many names (licensing office, tech transfer office, IP office, or, in our case, Partners Innovations), these offices evaluate technologies for commercialization potential, develop and implement IP protection strategies, and guide investigators through conflict of interest rules and other institutional operating procedures.

Traditional technology transfer offices are often supplemented by additional commercialization resources. These resources can include business development groups to assist in marketing technologies to companies and build strategic relationships, internal funding sources, such as captive venture funds or institutional grant-making mechanisms, and development offices which can help to match donor interests with specific research projects.

A few institutions have developed incubators, which provide office and laboratory space and equipment to start-ups spinning out of the institution in an attempt to foster more new venture activity (see later discussions for a more detailed description). Still, others have invested in coordinating their clinical research administration into central offices to facilitate patient-oriented research and trials.

Translational research and the associated commercialization is a difficult process. Ideas and inventions are viewed through a completely different lens than traditional scientific discovery. Resources are scarce, and the failure rate is high, even for projects that are successfully licensed. The potential rewards, however, are great. The financial rewards and research resources that come with successful commercialization are attractive, but for most academic investigators the process of seeing the conception become reality is equally—if not more—exciting. These results, combined with the fundamental goal of improving patient care, are what drive the serial inventor.

IDEAS AND INNOVATIONS

Property is an intellectual production.

Ralph Waldo Emerson

What Is Intellectual Property?

Two dictionary definitions:

1. Any intangible asset that consists of human knowledge and ideas.
2. A product of the intellect that has commercial value.

In plain English, IP is a novel idea, or a new set of ideas. In the context of translational and clinical research, IP arises from discovery research and the practice of clinical medicine that forms the foundation of improved products and services for patient care.

From the legal perspective that gave rise to the term "intellectual property," it is worth remembering that intangible assets such as IP may freely be used by anyone, *unless* they are protected under federal law. These laws fall into three broad areas:

- Patents
- Copyrights
- Trademarks

For the purpose of this chapter, we will focus on patent laws and rights and, to a lesser extent, copyright laws and rights.

What Is a Patent?

A patent is granted by the government and gives its owner the right to exclude others from making, using, or selling the invention for a limited period of time. The concept of patent rights was developed in Europe in the mid-1600s to promote the commercial introduction and development of new ideas and to encourage the public dissemination of novel ideas. While certain variations exist in patent laws and practices from country to country, a patentable invention must meet the following criteria:

- New (not available to the public or having been offered for sale)
- Useful (sufficiently described to ascertain its specific commercial use)
- Nonobvious (subject matter is not obvious to those who are skilled in the art to which the subject matter pertains)

In addition, the invention must be adequately described in the patent application to allow people of ordinary skill in the art to recognize what is claimed and to provide enough information to practice the idea.

To obtain a patent right, the owner of a patent right must submit a patent application to the appropriate administrative office (in the United States, it is the US Patent and Trademark Office). The patent office evaluates whether the claimed invention is patentable and administers the granting of patent rights. If the patent owner is able to persuade the patent office that the invention meets the requirements of a patentable discovery, then the government will grant a patent for 20 years from the initial filing date of the application.

During the active life of the patent, the owner is provided the right to enforce the patent against others who practice the invention without permission (called infringement). Frequently, patent owners allow others to practice inventions with permission in exchange for compensation and other consideration. These types of agreements are called *licenses*.

How Do I Protect My Idea?

If an investigator has a new idea (e.g., in the form of scientific discovery), he or she will disclose it to their employer (typically an academic center or company), consistent with the obligation of an employment contract or with governing IP policies. The individual responsible for managing IP at the organization will evaluate the invention for patentability and discuss its potential for commercialization with the investigator. If an invention shows promise, the organization will go forward and file a patent application with the appropriate government IP office.

Who Owns My Invention?

Ownership of inventions is determined by employment law and the patent guidelines that determine inventorship. Those who contribute to the conception of the invention must be identified as inventors. The inventor determines what, if any, obligations he or she has to their employer through contract or policy. In the United States inventions are typically assigned to the organization that employs the inventor.

Inventions that result from research sponsored by a federal agency (e.g., an NIH grant) represent a special case. Historically, ownership of the work products stemming from federally sponsored research was unclear; there was no incentive to patent, and, in fact, very little was ever patented. In 1980, however, the Patent and Trademarks Act Amendments were cosponsored by Senators Birch Bayh and Robert Dole. The Bayh−Dole Act cleared the title for academic research institutions to own IP created by their investigators with federal research dollars, to commercialize the IP through licensing, and to share the fruits of the IP with its inventors. As a direct result of Bayh−Dole, protection and licensing of IP has become a significant function of academic research institutions.

What Is a Copyright?

As a patent protects an invention, a copyright protects certain rights of the author(s) of a specific written work, such as text (e.g., academic publications), drawings, plans, pictures, music, software, and Internet content. The first copyright laws were developed more than 300 years ago to protect the creators of maps and books. Copyright protection differs from patent protection in that it protects the creator's work immediately upon its creation. For copyright protection in the United States, the work must be:

- an original work of authorship (independently created, not copied from another work) and
- fixed in a tangible medium of expression (the work must be documented in a way that can be reproduced).[1]

The owner of a copyright may exclusively reproduce, adapt, distribute, perform, or display the work that is subject to the copyright for a period of time equal to the life of the author plus 50 or 70 years, depending on the exact circumstances. The owner may also transfer these rights to others through license agreements. The owner or licensee of a copyright cannot prevent the use of a work that does not fall under this protected bundle of rights.

Why Are Patents and Copyrights Important?

Governments support the protection of IP by patents and copyrights for several reasons. First, patents and copyrights provide economic incentives for inventors and creators. In effect, they give inventors a head start on potential competitors. By granting the patent owner or licensee the right to exclude others from making, using, or selling a product based on the invention for a period of time, patents and copyrights provide inventors and creators the incentive to invest time and effort to develop new ideas that might be commercially valuable. Even after the invention is known to the public, it cannot be adopted and used by others who have not made similar investments.

Second, for an idea to be commercialized, it must be developed into a product that can be sold to the public. Product development is carried out in partnership with industry, which invests in the development, testing, manufacturing, distribution, sale, and support for the product. Patents and copyrights assure the investor that these activities can be undertaken and that investments in the activities could be worthwhile. This is especially important for products that have long research and development timelines (e.g., pharmaceutical and biotechnology products) that may require hundreds of millions of dollars of investment prior to the commercial sale of a product. If IP protection did not exist, it would be more difficult to attract the investment capital required to research and develop these products.

Third, patents and copyrights prompt the public dissemination of ideas that help sustain a creative and vibrant economy. The publication of protected IP allows others to read and understand new ideas and encourages their improvement and expansion. Without government-supported IP protection, individuals and companies are less willing to share their ideas and societal progress suffers.

Lastly, the protection of IP helps to govern both an innate fairness and a reward system in society. That is, individuals and organizations who generate important new ideas should and do have the opportunity to benefit economically through recognition of their contributions.

What Are the Options for Getting an Idea to Market?

An owner of IP has several options for commercialization. An idea or IP and its associated rights may be developed into a product by the company or organization that made the invention. A company or an organization may choose not to develop the invention into a product, because product development is not aligned with its mission or corporate goals. The protected IP may then be transferred through license or sale (also called assignment) to a commercial entity that will develop it into a product. On occasion, an idea is so significant that a company will be formed for the sole purpose of commercializing the invention.

The pathway to commercialization depends on the size of the market opportunity, the risks of failure (e.g., due to safety, development time, or product costs), competitive pressures, strength of IP, and the ability of the commercialization team of individuals to work together productively.

How Far Must an Idea Progress Before It Is Ready for Commercialization?

An idea and its applications must be developed enough so that someone interested in commercializing the idea can evaluate the benefits of the product opportunity against its costs and can assess risks and the management of such. A proof of concept, or a demonstration of the feasibility of the product, is often useful in this regard. The expected benefits of the product should outweigh the expected costs.

What Is a Royalty?

When IP rights are transferred from the inventor to a licensee, the licensing agreement may include a provision that the licensee will provide a payment to the inventor every time a product is sold. This financial term is called a royalty. A royalty is usually expressed as a percentage of the sales of a product, but it can also be captured as a set dollar fee that is transferred to the royalty recipient every time a unit of product is sold by the provider. The size of the royalty is negotiated by the two parties to the license agreement and depends on the product benefits the licensor is transferring to the licensee and the anticipated profit margins the licensee will enjoy for selling the product.

For example, when a licensor grants a patent right that covers a product sold by a licensee, it would be common to have the licensee pay the licensor a royalty for the products sold by the licensee. If the negotiated royalty were 1% in this example, the licensor would receive $1 for every $100 of product sold by the licensee.

How Do Companies Decide Economic Feasibility?

To evaluate the commercial viability of a product opportunity, a company will want to understand the technology that makes the product valuable, the ability to protect and differentiate this technology from competitors (and their preexisting IP rights), the resources that will be required to develop and sell the product, and the associated risks, both technical and in the marketplace.

WORKING WITH INDUSTRY

The Academic Health-Care System as a "Living Laboratory"

How Can Academic–Industry Relationships Yield Mutual Value and Benefit Through Joint Intellectual Effort?

Academic–industry relationships have emerged as a key force in the development and commercialization of biomedical and health-care innovations. For the purpose of this discussion, academic–industry relationships can be defined as arrangements between academic health-care systems—comprising investigators, clinicians, and administrators—and profit corporations ("industry").

In these relationships, something of value, often in the form of academic-based IP or cutting-edge research, is exchanged for financial support, compensation, or access to proprietary technology and services from industry. This type of public–private collaboration is based on the promise of potential opportunities to develop and bring technologies to the marketplace for the benefit of the broader health-care community.

Fundamentally, academic–industry relationships are driven by the recognition that academia and industry are each unique and both come to the table with different but complementary perspectives and skill sets. It is one learning from the other that adds great value. Academic health-care systems can realize significant value from industry in that such collaborators can provide commercially validated products and solutions to advance the core mission of academic health-care systems—research, education, patient care, and charitable community service.

In collaborating, industry can provide financial and personnel resources in addition to proprietary technologies and services to accelerate the development of academic science. The market and industry insight offered by industry leadership is also instrumental in driving the progress of research and its ultimate translation into patient care.

The broad vision for academic–industry relationships is to accelerate the convergence of activities in the research and clinical arenas and to facilitate the translation of ideas to real-world products and solutions to improve patient care. The integrated health-care delivery component of an academic health-care system may be considered a "living laboratory" in which industry firms can leverage the clinical expertise of world-class physicians, researchers, and leadership; identify critical health-care challenges; develop hypotheses, and—in collaboration with academic health system investigators—design, test, and implement solutions that address those challenges.

Industry contribution can promote a more rapid development cycle and offer insight into the commercial development process. In a sustained partnership, the academic health system can gain solutions and services that meet the needs of its internal environment, and the industry partner can gain cutting-edge products to commercialize for widespread, enhanced patient benefit. Ultimately, these relationships contribute to the broader mission of supporting innovative research and the broad translation of academic science to the patient bedside.

What Makes Academic Health-Care Systems a Unique Resource for Industry?

An academic health system is a unique environment and resource for industry because it spans the continuum of health-care delivery and most every facet of a health-care provider network. An academic health-care system has direct access to patients and clinical care providers (physicians, nurses, and other clinicians), who are two vital constituents/customers within the overall health-care ecosystem.

An academic health system, through its research and clinical communities, provides domain expertise and clinical context for industry solutions where knowledge and understanding of clinical, research, and administrative workflows are critical to the product development process, which includes planning, design, testing, validation, and optimization of the end commercial product (Fig. 22.2).

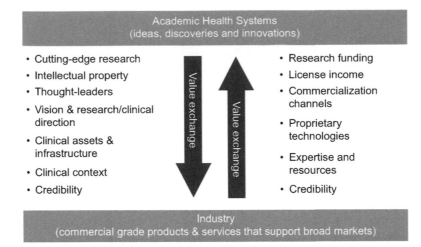

FIGURE 22.2 An illustrative model of the value proposition for academic—industry relationships.

In addition, academic health-care systems have the skills, ability, resources, and infrastructure to do the full spectrum of basic, clinical, and outcomes research. This not only generates new product ideas by identifying unmet medical need, but also demonstrates potential improvement of existing products.

What Are the Characteristics of Effective Strategic Alliances?

Academic—industry relationships can be characterized predominantly by "one-off" agreements in which the value exchange is narrowly defined. This, in turn, limits the scope of the transaction (e.g., sponsored research agreements, licenses, or consulting). While these arrangements can be of great value, they typically do not have natural extensions or value beyond the single focused deliverables, which makes them transactional by definition. In the case of strategic alliances, however, the relationship between the academic health-care system and industry is predicated on shared vision and broad collaborative efforts that require longer-term commitments or relationships and significant investments of resources, funds, and time from both parties.

Successful strategic collaborations are difficult to establish and even more challenging to sustain. When managed effectively, however, academic—industry relationships have the potential to yield great value to the respective parties.

As with any relationship, the success of the collaboration relies on clear alignment of goals and vision. In the case of *strategic alliances*, such alignment is even more critical. Strategic alliances typically consist of multiple projects that involve multiple constituencies and contributors from both academia and industry. They should be substantial and meaningful in scope, but not so onerous that they are unwieldy to manage. One tool to manage the alliance is the establishment of a formal joint governance structure, which includes representation from both the academic health-care system and the industry partner, to provide overall strategic direction and general oversight, to ensure alignment of goals, and to allocate funds and resources to further the goals of the relationship.

To compete more effectively, academic health-care systems and industry are beginning to recognize that strategic collaborations are essential to maintaining their competitive advantages. Industry looks to academic health-care systems not only for early-stage, high-risk research innovation but also for repurposing robust products to the health-care market for the improvement of patient care. On the other hand, as academic health-care systems strive to offer patient-oriented care to cost-conscious, knowledgeable consumers, strategic alliances with industry allow efficient adoption of mature products and technologies that can be readily adapted to health care.

It is imperative that academic health-care systems adopt and learn new competencies and skills, build on their internal infrastructure, and innovate in support of their core research and clinical missions. Strategic alliances can bring in competencies and skills from outside health care, build infrastructure through collaboration, and support system-wide innovation to keep academic health-care systems at the forefront of research, education, and clinical care.

How Are Strategic Academic—Industry Relationships Developed and Structured?

Academic—industry relationships are established at many different levels depending on the needs of the interested parties, which can range from transactional (e.g., sponsored research, licenses, and consulting agreements) to strategic alliances

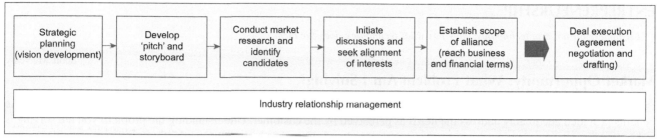

FIGURE 22.3 Process for developing a strategic alliance.

(e.g., broad, long-term, visionary codevelopment relationships generating next-generation solutions). The process often begins (Fig. 22.3) with something of value, such as a specific, tangible asset (e.g., IP) or high-level vision (e.g., thought leadership in an emerging area of shared interest).

The valuable asset or vision is then translated into a coherent story, otherwise known as the "pitch," which clearly articulates the value proposition of a potential relationship. Viable industry candidates are then identified through internal academic/clinical champions, market research, and due diligence, or through existing industry relationships. Once appropriate industry partners are identified, discussions are initiated to seek alignment of interests, with the goal of defining and determining the scope of the agreement as well as business and financial terms. Lastly, a legal contract is drafted, negotiated, and executed to memorialize the final agreement.

A continuum of academic—industry relationships, summarized as follows, articulates value ranging from short term to long term, with increasing intensity and investment at each growth stage of the relationship.

- *Consulting arrangement.* Industry compensates academic health-care system staff or faculty for expert guidance, counsel, and advice.
- *Sponsored research support.* Industry sponsors research to support academic research, often accompanied by an option to license new IP, and have access to data.
- *Licensing transaction.* Industry obtains rights to commercialize academic IP. Industry tracks academic discoveries by their publication, as well as existing relationships and marketing.
- *Strategic alliance.* Industry and academic institutions establish broad long-term relationships, often consisting of significant financial or resource contributions from industry or from both parties. Broader strategic alliances are typically more complex to establish, manage, and sustain.

How Can I Get Outside Companies Interested in My Idea?

Academic exercises and outreach can often generate industry interest, especially through publications in top scientific journals, and participation in various conferences, symposia, and colloquia. In general, any networking and relationship-building with industry allows investigators to keep industry abreast of important biomedical and technological development that may have commercial potential.

Aside from an investigator's individual effort and outreach, some academic health-care systems have marketing or business development professionals as part of the technology licensing organization who can serve as conduits to industry. These professionals can raise awareness of academic technology and initiatives, establish and cultivate relationships with industry, and execute the contractual agreements necessary to protect the investigator and the academic institution to ensure that academic science and technology is translated to the broader health-care community for widespread benefit.

Common Challenges

While relationships between academia and industry can produce many positive outcomes, there are also many challenges. The most fundamental difference between academia and industry is an environment of openness and knowledge-sharing in alignment with a charitable mission, versus one of confidentiality and protection of proprietary, competitive, and financial interests aimed at creating value for shareholders. Successful balancing of these interests can lead to exciting and fruitful relationships that ultimately drive great patient care advances.

ENTREPRENEURSHIP

Customers do not buy technology; they buy what technology does for them.

Scott McNealy

Market Opportunity: What Problem Am I Solving?

Products and services offered represent a value proposition, whereby the customer will pay the asking price only if the benefits provided adequately address the needs as perceived by the customer. Understanding the problem you are solving is at least as important as having the purported solution in hand. Usually, that means conducting extensive interviews with users to develop a deep understanding of how the proposed offering will address the problem in the actual constraints of their environment in a way that will be considered superior to their current practice. There must be a compelling case based on features, performance, or economics before a customer—especially a clinical provider—will change behavior or replace a satisfactory supplier to adopt a new offering.

To be commercially viable, any technological solution must squarely address a customer problem. Business problems may be difficult for the technologist to articulate, but answers to the following set of questions can define the unmet need of the customer indirectly:

- Is it a "nice to have" or a "need to have"?
- How is this problem currently addressed?
- What is unique about my solution?
- What is the size and growth rate of the target market?
- Who are my customers and how do I reach them?
- How does my product compare to competitors'?
- What is driving the demand for my solution?

Consider, for example, how new technologies such as microfluidics and disposable electronics have made feasible the development of point-of-care diagnostic tests that can be performed by less skilled personnel in a physician's office or even in the home setting. While this segment has been growing, market penetration is still quite low despite decades of promotion.

If one considers the customer value proposition in some detail, there are several factors that can be identified to account for this resistance. Rapid turnaround time is a key selling point, but in how many instances will the clinician be in a position to decide and act immediately on the outcome of a single analyte determination rather than wait for the full battery of lab results to be available?

The other often overlooked key consideration is reimbursement. Payers have established rates of reimbursement for outpatient testing that are based on competitive prices from high-volume commercial laboratory suppliers. It is most difficult to manufacture and distribute an individually packaged and self-contained kit at a price that will match the conventional laboratory version where the sample is sent out with many others to be batch processed in a highly optimized production facility.

Technology: What Is the Status of the Core Technology?

The term "platform technology" originated in the auto industry but is a useful concept to distinguish the potential scope of a new invention and therefore its most appropriate commercialization pathway.

Platforms are generally more fundamental discoveries that can support multiple applications and future improvements or derivatives. Platforms generally require more development resources to extract value, but the ultimate value attained may be much higher than for inventions that represent substantial improvement on an existing product. Platforms are more often suitable for start-up ventures with dedicated resources and focus, while improvements are usually better suited for out-licensing to a market incumbent who can use the technology for a next-generation version or extension of their existing product line.

To be fundable, there needs to be a clear product concept that embodies the technology with a clear development pathway, including discrete milestones, funding requirements to reach those points and timing. You should have a good sense of what are the principal technical risks and unknowns, along with a plan for how to address or mitigate them. The ratio of "R" to "D" will affect your valuation and the type of investor you can attract.

The key asset of any new venture is its IP, comprising patent filings as well as the accumulated knowledge and know-how around the technology. The successful entrepreneur will have a strategy for securing all the IP rights needed to practice the technology and protect it against imitators. That may involve in-licensing, cross-licensing, patent prosecution, and always a demonstrable understanding of the patent landscape with potential hazards identified. Keep in mind that freedom to operate is at least as important as patentability but much harder to ascertain.

These considerations and more can be summarized through answering the following set of questions:

- Is the technology a platform or an improvement?
- What are the scientific discovery and product development milestones?
- What is your IP status and strategy?

Financial: How Much Money Will It Take to Get This Idea to Market? Where Will It Come From?

A phased approach with financing requirements through each milestone is much easier for investors to accept than an overconfident "all-or-nothing" approach. Different options should be available for investors to enter and eventually exit in various circumstances. Prior to the collapse of the Internet bubble in 2000, an initial public offering was a realistic financing scenario, even without a clear path to profitability. That is no longer the case and it is now more common for ventures to follow the acquisition route, whereby large pharmaceutical and device companies will source innovation from new ventures that need the access and resources of a larger parent to reach the market.

This is especially true for therapeutics or Class-II devices that require substantial funding to support later-stage clinical trial, market introduction, and distribution. In addition to the nearly $1 billion expense estimated to bring the average therapeutic to market (not accounting for failures), there is too much pressure on investment firms to deliver returns in less than the time required (10 +years) to go all the way from bench to bedside.

Investors will look for exit scenarios before they put in the first dollar. Ventures must be able to identify target acquirers and point to analogs and precedents of start-ups that have sold at valuations that were attractive multiples on sales or funds invested. Financing of new ventures is guided by the cruel calculus of portfolio management wherein the promised return on any new investment must be sufficiently high to offset the risk (the expected high failure rate) on the prior investments—that target ROI is typically 400−1000%. This is a high hurdle, which requires a large market potential, large competitive market share, and generous gross margin to be delivered by an appropriate business model with excellent management.

The US capital markets have evolved into a somewhat complex ecology with a different species of investors occupying different niches. The following is a very brief thumbnail of different investor categories:

- *Friends and family.* Even at the earliest stages of company organization, funds are needed to cover basic expenses for activities such as patent filing/licensing, prototype development, conference presentations, legal incorporation, consultants, travel, and prototyping. This is necessary to support business plan development up to the first round of seed funding. This can be done as convertible debt financing, whereby repayment can be in the form of equity at the time of company formation.
- *Angels.* These are high-net-worth individuals who may have a personal connection with the venture participants or have a special interest in the venture's product aspirations. As a group, angels are idiosyncratic with respect to their interest areas, degree of formalism they will apply, and the level of funding they can provide.
- *Organized angel groups.* It can be difficult to find and gain access to individual angels, but some regions of the country have spawned angel organizations led by professional or volunteer staff that maintain some operations for soliciting business plans, evaluating them, syndicating investors, and executing the transactions. These groups usually maintain some public profile and meet on a regular basis to review submitted business plans. A listing of many of these groups can be found at the Angel Capital Association website (http://www.angelcapitalassociation.org).
- *Incubators.* As an economic development function, many state governments have granted funds and created programs to administrate them for the purpose of stimulating new company formation. Often these entities are closely affiliated with local university systems and their tech transfer offices. They usually focus on seed-level investment and may also provide some consulting assistance for organizing the business.
- *Venture capital firms.* These come in all shapes and sizes. Most specialize in certain industry sectors with the larger companies covering multiple sectors. Some will do seed capital and take a major role in the company organization, management, and strategy development, while others prefer to come in as more passive secondary investors. They offer more than money—their partners and staff represent an accumulated wisdom of much experience that can be applied to their portfolio company's advantage. On the other hand, their assistance comes at the price of control—founders must be prepared to give up majority ownership and authority in return for the investment.

- *Private equity firms.* These investors generally hold a more purely financial perspective and are less participatory in the planning and management of the company than venture capital firms. They are usually recruited in later stages of company development when companies are preparing for market introduction or some major strategic adjustment.
- *Corporate venture groups.* A number of the large technology-dependent health-care companies have established subsidiaries that are charged with finding emerging technology opportunities that relate to their core businesses. This strategy is the equivalent of buying an option on a developing technology that they may wish to acquire in the future, and in the interim, provides them with a close-up view to monitor the field as it progresses.
- *Philanthropic venture capital.* This is a relatively new phenomenon whereby some disease foundations (e.g., the Juvenile Diabetes Foundation) have decided to explore alternative ways to accelerate innovation. In addition to the traditional grant proposal route through academic institutions, some are now taking equity in start-ups that are developing promising technologies on a commercial pathway.

Management: How Do I Think About the Right People, Skills, and Change of Control?

Academic institutions generally encourage entrepreneurial start-ups arising from their research labs but also limit the degree to which their staff can participate in outside ventures. Hospitals, in particular, have rather strict rules of conduct to avoid any potential conflict of interest involving use of institutional resources or any activities that may affect patients either directly or indirectly. For these reasons and many others, inventors should partner up as early as possible with an experienced business person to act as the project manager. That person should be someone who has an adequate background to understand the technology as well as the communication skills to prepare and deliver the "pitch" to potential investors.

This is a first step in the gradual relinquishment of control that the inventors must be prepared to accept on the business development path. To attract additional management talent, the company will need to grant founders equity or reserve stock options for future equity distribution. Some of the senior roles to consider in the early recruitment stage are:

- research director
- clinical affairs
- regulatory affairs
- product development
- engineering design
- marketing/sales
- business development
- finance/administration

Depending on the circumstances, it may be possible to obtain some of these skills on a consulting or outsource basis and thereby reduce the capital requirement for company start-up. The model of a virtual company is increasingly popular, especially in regions such as Boston and San Francisco, where this type of expertise is concentrated.

Not surprisingly, for technology-driven ventures, quality technological leadership is critical to winning the confidence of investors. At the earliest stages, it can be very helpful to engage a Scientific Advisory Board comprising well-known experts in the field. Most of the time, these people are willing to lend their names, if they like the concept, in exchange for future stock options.

Professional investors will typically expect to be represented on the Board of Directors—the venture governing body—as a condition of their share purchase. Venture capital firms usually assign a partner for this duty who will bring valuable experience equivalent to a high-priced strategy consultant. VCs extract a large price in terms of equity ownership and control, but their expertise adds value in addition to the capital they contribute.

Legal and Regulatory Affairs: What Kind of Help Do Entrepreneurs Need at the Early Stages?

Starting at the point where significant assets (e.g., capital, IP, management time) are being committed, there needs to be some formal agreement in place to ensure that the involved parties have a common understanding and basis for fair treatment as they move forward. Attorneys with a specialization in business law have the expertise to advise on what kinds of agreements and arrangements are appropriate under different circumstances.

Many firms are willing to provide "scholarships," whereby fees are deferred until financing or equity distribution. The kinds of matters where counsel is called for include incorporation, equity allocation, patenting and licensing, and debt issuance.

Health care is a highly regulated industry, so nearly all contemplated products or services require a regulatory strategy, if not direct FDA approval. Drugs and devices are regulated by different bodies and under different rules, which often require careful interpretation and additional understanding of "case law" to plan an appropriate clinical trial design and market introduction. The FDA, through its website (http://www.fda.gov) does provide good access to their guidelines, transaction process, filings, and rulings, as well as educational material. There is also a well-developed regulatory consulting industry, mostly comprised of small firms and individuals that help companies formulate their strategy.

CLINICAL EVALUATION OF INNOVATIVE PRODUCTS

As to diseases, make a habit of two things—to help, or at least to do no harm.

Hippocrates

How Does Clinical Evaluation Differ From Earlier Stages of Innovation?

Human subjects are involved in clinical trials, and their safety is paramount. Taken to an extreme, the safest, lowest-risk course would be to change nothing and do no innovation. Yet innovation in the clinic is precisely how today's unmet medical needs will be treated in the future, and innovation necessarily incurs risk. The crucial decision for clinical research: Do the benefits of an innovative product outweigh the potential risks to the human subject involved in the product's evaluation?

How Is the Complex Relationship Between Safety and Innovation Managed?

There are multiple players on the team supporting the patient–doctor relationship, or in this case, the relationship between the subject and the investigator (see Fig. 22.4). The corporate sponsor of the clinical research, the academic health-care system, the Federal Food and Drug Administration, and the Institutional Review Board (IRB) each assure safety and promote innovation with a particular focus on their role and responsibilities. It is the unique role of the investigator to exercise scientific understanding, best clinical practices, and professional judgment while serving as the point of integration in the safe conduct of clinical trials for new treatments and diagnostics.

What Are the Role and Responsibilities of the Principal Investigator?

In the increasingly complex environment of clinical research, the PI must balance critical and often competing demands. There is no room for error. If the PI were to let one of these demands drop, the results could be grave: termination of important research, loss of professional credibility, or worse, an injury to a study participant.

Imagine an investigator juggling the requirements of the FDA, the IRB, finance, contracting, sponsor needs, and research staff while keeping his or her eyes ever vigilant on the reason they have undertaken this role: Elevating the practice of medicine. If it was not for the role of the PI, many innovative procedures, drugs, devices, and equipment would never be realized by patients and the scientific community.

PIs are responsible for all aspects of the clinical trial. They must ensure that the protocol, including all clinical aspects of the protocol, is followed by their research team, while making sure the conduct of the trial adheres to regulatory and good clinical practice requirements. While the PI may delegate study tasks to the research team, the responsibility for the overall conduct of the trial lies with the PI.

To carry out this role effectively, PIs must have the experience and education in the clinical area to conduct the clinical study. They must be free of any conflict of interests in the outcome. That is, at most, they should have *minimal* financial interest in the sponsoring company. Financial interest includes ownership, a consulting relationship, or royalties—all of which should be disclosed. As academic researchers, PIs must be free of any constraints on the analysis and publication of the data from the clinical trial. The integrity of the research data must be upheld. Any compromise might expose patients to increased risks or delay needed treatment and diagnostic products to the market, causing an unacceptable regression in clinical research.

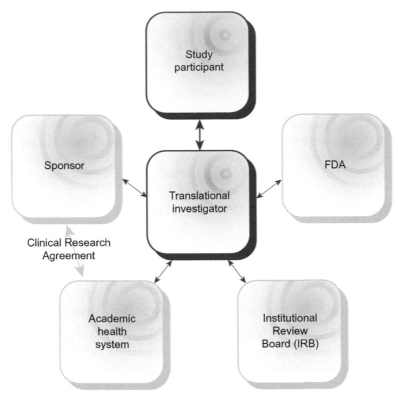

FIGURE 22.4 The relationship between the subject and the investigator.

What Are the Roles and Responsibilities of the Sponsor?

Clinical research may be funded ("sponsored") by different types of entities, including corporations, government (NIH), foundations, and health-care systems. Most nongovernment sponsorship comes from biotechnology, pharmaceutical, medical device, and equipment companies, but an increasing amount of sponsorship comes from other health/wellness vendors that desire scientific research on their product. Whatever the basis of sponsorship, if human subjects are involved, government regulatory authorities as well as individual IRBs and data safety monitoring boards will be responsible for the oversight of human subject protection.

The role(s) the sponsor takes are varied. The sponsor most often initiates the trial, finances it, develops the protocol, and submits the necessary application to the FDA—for example, an application for an Investigational New Drug (IND). The sponsor also monitors the study for the efficacy and safety, provides the investigational product, collects and analyzes data, and manages the clinical study. The sponsor is obligated to conduct its role in accordance with regulatory requirements. When an investigator initiates a clinical study, that investigator assumes all the responsibilities of a sponsor.

Clinical research organizations (CROs) have emerged to assist the corporate sponsor with all levels of their clinical study activity. CROs can provide very specific roles, such as collecting imaging for the sponsor in support of the study, or they may take on the majority of the responsibilities from contracting to monitoring the clinical study.

What Is the Role and Responsibility of the Food and Drug Administration?

Like an academic health-care system, the FDA, a federal agency of the US Department of Health and Human Services, has multiple distinct missions:

> *The FDA is responsible for protecting the public health by assuring the safety, efficacy, and security of human and veterinary drugs, biological products, medical devices, our nation's food supply, cosmetics, and products that emit radiation. The FDA is also responsible for advancing the public health by helping to speed innovations that make medicines and foods more effective, safer, and more affordable; and helping the public get the accurate, science-based information they need to use medicines and foods to improve their health.*

Like those of the clinical investigator, the actions of the FDA must balance and integrate the responsibility for assuring safety and efficacy of medical products in development, with the responsibility for helping speed innovations that make new products safer and more effective.

In the context of clinical and translational research, the FDA reviews and approves IND applications prior to first-in-human studies. They conduct audits and inspections of study sites and sponsors to ensure conformity with regulations; require safety reporting on serious adverse effects in clinical trials; and have been increasing pressure on sponsors to register and publish *all* data from clinical studies. The FDA disseminates safety information and reviews and approves New Drug Applications for sales and marketing in the United States.

To expedite the translation of innovation to the patient, the FDA has adopted mechanisms such as *Fast Track* interactions during clinical development, *Priority Review* of a completed application, and *Accelerated Approval* of drugs for life-threatening diseases. Certain drugs and devices that have been demonstrated to be safe and effective may also be made available for compassionate or humanitarian use prior to formal approval in patients who have no other therapeutic recourse.

The full range of the FDA's role and its responsibilities is far beyond this overview; please see Chapters 23 and 24 for a more in-depth treatment.

How Do Academic Health Centers Support and Guide These Roles?

Each academic health center has an IRB, sometimes called the independent ethics committee. IRB approval is required by federal regulation prior to research being performed on human subjects. The mandate of an IRB is to protect the safety and privacy rights of subjects. The IRB oversees these clinical trials by (1) reviewing the credentials of investigators, assuring experience and education appropriate to conduct the study; (2) assessing safety and efficacy of the clinical trial protocol; and (3) reviewing the informed consent form for disclosure of the risks in the trial and to ensure a complete disclosure to the subject.

Clinical studies can begin after IRB approval. The IRB continues to monitor the clinical study by providing continuing review and oversight of the study. Any change to the protocol, study staff, or increased risks must be presented to the IRB for ongoing approval. An IRB is normally academically based but; in recent years, private, independent IRBs have emerged.

An academic health center provides the extensive infrastructure that is required to support each aspect of the clinical trial. Benefits of working with an academic health center include:

- large cohorts of well-phenotyped patients and patient/medical record databases
- physicians with specialized clinical and research expertise
- protocol development
- peer support
- data analysis
- the capability to respond to immediate/urgent subject medical needs
- ancillary support services, including diagnostic services such as imaging and laboratory facilities, dispensing operations through the pharmacy, legal, budgetary, and account/billing assistance
- information technology (IT) hardware, software, and support infrastructure to help manage complex tracking and reporting

Clinical trial development, implementation and management are also becoming increasingly involved due to growing legal and regulatory requirements. The purpose of these specifications is primarily for the protection of human subjects and to ensure the integrity and communication of research data. To comply with these laws and regulations, academic health centers must have well-developed and dynamic clinical research programs focused on quality assurance and control, as well as continuing education programs for the staff. These programs must keep pace with the rapidly evolving environment.

Why Is a Clinical Trial Agreement Needed?

The Clinical Trial Agreement (CTA) is the key document that binds the sponsor of a clinical trial to the academic health center. The CTA spells out roles, responsibilities, and obligations of the academic health center sponsor and PI (including the CRO, when used).

One of the functions of the CTA is to ensure the protection of human subjects—from a financial perspective—in case of injury due to participation in the clinical study. Responsibilities are defined and allocated between the sponsor, the academic health center, and other parties. The indemnification and subject injury CTA language addresses these roles. Indemnification is relevant when a legal action is brought by the subject for some injury that occurs during the clinical trial, while subject injury language defines financial coverage for medical care to treat subject injuries before a claim being brought.

A key component of the CTA is attestations by the academic health center and sponsor that each will have enough insurance/money to cover their obligations under the contract. The CTA must also describe the commitment by the academic health center and sponsor to comply with the subject privacy rights such as the Health Insurance Portability and Accountability Act (HIPAA).

In clinical research HIPAA plays a very important role by defining the rights of the sponsor and academic health center to use, control, and disclose subject data, including protected health information (PHI). Subjects are made aware of this use and disclosure when signing their informed consent form or a separate form describing the use of their private information. All uses and disclosures must be defined and approved by the subject before the subject can be enrolled in the clinical research study.

Increasingly, in addition to the legal and regulatory compliance considerations addressed in the CTA, academic health centers are voluntarily joining professional, clinical research associations, such as the Association for the Accreditation of Human Research Protection Programs (AAHRPP), an accreditation association for IRBs. Membership in such organizations may require a heightened level of human subject protection which may involve stricter guidelines in the CTA. The CTA also defines the ownership and rights to IP and protection of confidential and proprietary information.

Publication rights are an essential part of the academic health center contract. The publication section defines when a researcher can publish and the amount of access to data from other sites participating in a multisite study. Publication is the mechanism for ensuring that the public is informed of the type and outcome of all research being performed. Publication should ensure that unsuccessful research is not duplicated, thereby eliminating increased risk to subjects and delaying public access to much-needed drugs, devices, and equipment.

This CTA discussion presents a broad overview of some of the considerations in contracting between the academic health center and the sponsor. While we can list some key issues that may be addressed in a CTA, each academic health center must make decisions about the appropriate inclusions and exclusions for themselves, based on federal, state, and/or local regulations and academic health center (hospital and medical school) policies and their interpretations of academic freedom and research integrity.

How Do I Get Started?

The steps and procedures for initiating a clinical trial are summarized in Fig. 22.5.

CONFLICTS OF INTEREST

What Are Conflicts of Interest?

The collaboration between academic health-care systems and industry in the long and complex process of bringing a discovery to the marketplace entails the interweaving of two very different cultures, with very different legal constraints:

- Industry must answer to its shareholders by realizing profit; to realize profit, it may choose to withhold information from its competitors and the public.
- The academic health center enjoys a different position in society, benefiting from its special status as a charitable, tax-exempt organization, and lives in a culture that rests on the open, unbiased pursuit of three missions: research, education, and patient care.

In the context of commercialization of technology, these two different cultures can, and do, place individuals and institutions in a position of having divided loyalties, which is the essence of conflicts of interest.

For example, an investigator at an academic medical center holds stock in a small company formed to develop the center's new cancer drug; he/she is also the investigator in a clinical trial of the drug. A question arises as to whether an individual patient's deteriorating condition is attributable to the drug or to other factors.

The investigator attributes the deterioration to other factors, thereby—consciously or not—preserving the potential success of the drug, of the company, and of his/her stake in it.

When this judgment proves incorrect—when the experimental drug proves to be the culprit—the investigator's erroneous conclusion will appear to many to be evidence that he/she gave priority, inappropriately, to his/her

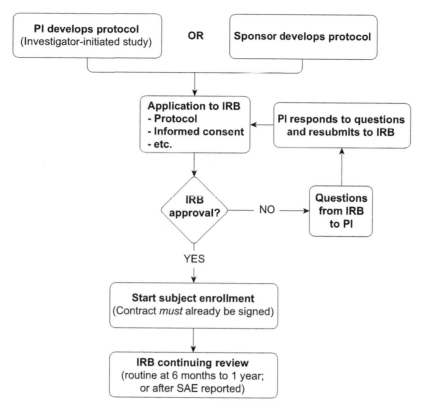

FIGURE 22.5   A flow chart of the process for initiating a clinical trial.

financial interests over those of her patient and the objectivity of research. And there will be no way to dispel that appearance.

Just as the individual investigator may have conflicts, so may the institution. Institutional conflicts would exist in the previously mentioned example if the institution or a senior administrator within the institution with authority over the research also held equity in the company. Failure of the institution—the IRB, the department chair, or other institutional official—to respond to what, in retrospect, may appear to be clear indications that the drug was culpable will be viewed in hindsight as having resulted from the institution having placed the prospects of financial gain over the interests of the patient.

The existence of divided loyalties due to personal and institutional financial interests in the commercial success of research results has become an increasingly serious concern in the decades following the passage of the Bayh—Dole Act of 1980. That landmark legislation permitted institutions to retain ownership of inventions made by their investigators with federal funding, and thereby gave the recipient institutions and the investigators a powerful incentive to collaborate with industry to commercialize those inventions.

Unlike activities such as teaching and publishing academic research, commercialization activities may give rise to financial conflicts. A few notable cases involving financial conflicts at academic health-care systems have generated much press and government interest in this issue (*Gelsinger v. Trustees of U. Penn.*, 2000 and *Wright v. Fred Hutchison Cancer Center*, 2002). This fuels concerns that conflicts of interest threaten the integrity of the academic research enterprise, the welfare of students, the safety of patients, and ultimately, the public's trust in academic medicine itself.

Required Components of Conflict of Interest Policies

Food and Drug Administration

The FDA requires that any company submitting an application for approval for new human drugs and biological products, and marketing applications and reclassification petitions for medical devices includes with its submission a disclosure of financial interests held by investigators participating in the clinical studies that support the submission. The FDA has the authority to disregard a study entirely if it determines that conflicts of interest raise serious questions about the integrity of the data (21 CFR, Part 54).

Public Health Service and National Science Foundation Regulations

The Public Health Service (PHS) regulations (42 CFR Part 50, Subpart F and 45 CFR Part 94) and National Science Foundation (NSF) policy (National Science Foundation, and 45 CFR 680) are intended to promote objectivity in funded research. They require disclosure of significant financial interests (SFI) from all individuals who are responsible for the design, conduct, or reporting of research to be funded by any PHS agency and the NSF. The regulations require that the applicant institution has a conflicts policy and that it "manages, reduces, or eliminates" any conflicts that are identified "when the designated official reasonably determines that SFI could directly and significantly affect the design, conduct, or reporting of the PHS-funded research."

The regulations contain no prohibitions, leaving that determination to the applicant institution. For PHS, all such conflicts must be reported to the awarding agency, with assurances that the conflicts have been appropriately handled.[2]

False Claims Act

Knowingly giving false statements in applications for federal funding, including false statements regarding financial interests, may subject both the investigator and the institution to legal liability with monetary penalties under the False Claims Act (31 U.S.C. Sec. 3730(b) and *U.S. ex rel. Cantekin v. University of Pittsburgh*, 1999 and 2000).

Fraud and Abuse Laws

The Stark Law (42 U.S.C. Sec. 1395 nn) and antikickback statutes (42 U.S.C. Sec. 1320a−7b) prohibit payments and receipt of payments given with an intent to influence the purchase of a product or services for which Medicare or Medicaid reimbursement is sought. If a payment is made that cannot be shown to have been fair market value for the service/product delivered, it is suspect under these criminal laws. Thus, for example, an academic health-care system investigator who prescribes or recommends to others the use of a company's product and who consults for that company on the design of a clinical trial would be in danger of violating these laws if he or she receives more than fair market value for consulting.

Tax-Exempt Status

Any arrangement with a company that restricts the ability of the academic health-care system to disseminate the results of research, or that otherwise leads to the activities of the academic health-care system being characterized as commercial product development rather than academic research, can create tax issues for the academic health-care system under its tax-exempt status, including the generation of "unrelated business income" and, in some cases, possibly jeopardizing the academic health-care system's tax-exempt status.

Guidance From Professional Organizations

A number of professional organizations have produced guidance documents on conflicts of interest; for example, the Association of American Medical Colleges (AAMC), the American Society of Clinical Oncologists (Journal of Clinical Oncology 2006), and FASEB (FASEB Office of Public Affairs, online), which issued a call for uniform national conflicts of interest standards. The AAMC's proposal that institutions adopt a "rebuttable presumption" against certain financial interests, both individual and institution, has provided a standard adopted by many academic medical institutions. The American Heart Association and the American Cancer Society have fully adopted the Public Health Service regulations as their own policy (American Heart Association, online; American Cancer Society, 2007, online).

Professional Journals

The *New England Journal of Medicine*, the *Journal of the American Medical Association*, and other peer-reviewed medical journals have policies on disclosure of financial interests, both individual and institutional, in connection with a proposed publication. Journals may refuse to accept a publication if they consider that the disclosure is not sufficient to address concerns about the impact of the disclosed financial interest on the objectivity of the publication.

Institutional Policies of Academic Health-Care Systems

Virtually all US academic medical centers have conflicts of interest policies, because receipt of federal research awards is contingent on the institution having such a policy. The content of those policies—including who must disclose, what must be disclosed, and what action is required in response to the disclosures—varies widely among institutions.

A commonly adopted standard is that of the AAMC (online): holding certain financial interests while conducting human subjects research will not be permitted unless circumstances warrant an exception. Some institutions, however, have adopted outright prohibitions on certain individual and institutional financial interests held concomitantly with the conduct of research, both human subjects research and research that does not involve human subjects (President and Fellows of Harvard College, amended 2004, online). In addition, IRBs at some academic health-care systems have adopted their own policies of disclosure and review to address risks to human subjects that may be posed by financial interests that would otherwise be permitted by the institution.

TRANSLATING SCIENCE TO THE BEDSIDE CORES IN ACADEMIC HEALTH CENTERS

With the increasing complexity of establishing effective academic/industry relationship, several academic health centers have established cores and/or offices specifically tasked with accomplishing this task. While they vary in scope, all are designed to assist the investigator to identify and develop an appropriate and fruitful relationship with industry and to assist him/her to negotiate the various requirements and pitfall inherit in those relationships. An example of such a core is "Partners HealthCare Innovation" a program at Partners HealthCare, Inc. in Boston, MA.

Partners HealthCare accounts for nearly $1.5 billion in laboratory and clinical research annually. The magnitude and breadth of its research and clinical capabilities makes Partners HealthCare a compelling source for new insights into disease mechanisms, technology breakthroughs, and patient care. Its hospitals have had enormous historic success transferring technology to companies that have led to substantial market outcomes.

How the Innovation Office Help Partners HealthCare Researchers

Innovation, Partners HealthCare's 85-person business development and investment arm, is charged with enhancing those outcomes and the pathways that enable them. It is designed to be the catalyst that strategically connects the enormous strength of the research community at Partners to external collaborators. In doing so, the Innovation team aims to deliver a broader and more consistent flow of new products and services to enhance patient benefits. Its new "Market Sector" model was established to significantly increase opportunities for collaborative innovation.

Nine market sectors act as clinical "verticals" in core specialties:

- Anesthesia
- Cancer and Pathology
- Cardiovascular and Pulmonary/Critical Care
- Dermatology and Plastic Surgery
- Medicine and Pediatrics
- Neuroscience
- Orthopedics/Regenerative Medicine
- Radiology
- Surgery

Highly experienced industry executives were recruited to lead the market sectors. Their expertise includes senior key roles at companies such as Baxter, Bristol-Myers Squibb, Monsanto, Boston Scientific, and PureTech.

Each sector is built around discrete strategies, goals, and metrics—all drawing on deep domain knowledge. The market sector leaders are charged with developing a "thesis for commercial investment" in Partners HealthCare research and clinical capabilities. To achieve successful engagements, market sectors focus on the attributes that industry seeks from academic medical centers, including disease insights, therapeutic options, technology, clinical expertise, compatibility with the company's portfolio and strategic direction, and a viable matching timeline for development.

The Important Link With Industry

Partners HealthCare offers a number of complementary strengths to industry collaborators, including broad, scientific and clinical expertise; patient access; rapid point-of-care progression; and longitudinal, phenotyped patient data to inform studies of intervention, cost, outcomes, patient behavior, and population health management. Partners also has substantial expertise in preclinical, clinical, postapproval, and observational research.

Its paradigm-shifting initiatives such as the Biobank, Big Data Platform, Connected Health, and Computational Pathology, as well as its other repositories, offer industry an opportunity to drive the delivery of value-based care. Finally, Partners hospitals offer unique insights as an accountable care organization and major purchaser and user of thousands of medical products daily.

The market sectors are designed to better align Partners HealthCare's ability to meet industry needs. market sector leadership engages at the decision-making level of key companies relevant to each sector. The sectors deliver cohesive, tailored offerings to industry and engage hospital leadership while efficiently conveying industry perspectives to Partners faculty and innovators to foster a lively, entrepreneurial culture.

Team members in each sector are mentored to enhance understanding of the related IP portfolio strengths and gaps, effectively engage in industry collaborations, and experience the notion of "derisking" product development through the industry lens. The market sectors analyze and share information on pattern of spend, state of pipeline, recent partnering decisions, IP portfolio, and, importantly, desire to be a true partner.

How the Partners Innovation Funds Assist Innovators

Partners Innovation also seeks to bring more investment capital to support the commercialization of faculty and employee discoveries. The impact of its highly successful "Partners Innovation Fund" is rapidly increasing. More than $200 million in new capital is being pursued through multiple new funds linked in novel ways to the technology strengths of the Partners family. If investment campaigns are successful, the funds will likely comprise the largest pool of capital directly tied to the technology of any academic center in the United States—no surprise, considering the size and quality of the Partners HealthCare research enterprise.

The Commercialization of Information Technology

Another substantial priority is further organizing the commercial application of a large portfolio of IT solutions that Partners has developed and implemented for its own use. Spearheaded by the Innovation Business Development/ Healthcare IT team, these technologies are being packaged into high-impact industry offerings.

In parallel, leading health IT centers throughout Partners and its hospitals play a core role in the offerings provided to industry. This helps leverage the unique environment and broad resources available through Partners' health delivery system. Domain expertise and clinical context for industry solutions are provided in situations where understanding clinical, research, and administrative workflows are critical to the product development process.

Material Transfers, Research Grants, and Patents

Innovation is also responsible for myriad functions that enable groundbreaking research and translation. This includes the transaction of more than 1500 material transfers and nearly 200 industry-sponsored research agreements each year. Innovation handles an enormous patent portfolio with upwards of 300 US filings per year and total patent management several times that amount. With more than 2400 agreements in place, Innovation also helps ensure full compliance with thousands of discrete financial requirements. Recent focused process improvements and tools have enhanced reliability, increased speed, and reduced inventory—in one case resulting in a greater than 85% enhancement.

The Operating Philosophy of Partners HealthCare Innovation

As a matter of operating philosophy, Innovation seeks to run its operations like a business, organizing for outcomes and increased capabilities. The overriding goal is to ensure that Partners is the optimal collaborator with context of real-world success always in the forefront. This translates to organization-wide performance measures and better management of workflows, expectations, processes, and approvals.

The Innovation Advisory Board provides invaluable real-time, independent, commercial guidance and contacts. The Board is complemented by the Commercialization Council, consisting of top investigators from across Partners—committed to ensuring that the capabilities, insights, and needs of the research community are priorities in all Innovation activities and plans.

Innovation's entire approach is based on executing a discrete and measurable strategy in each of its operating units with an overriding sense of urgency, mirroring the stakes for patients and providers during this time of rapid transformation.

SUMMARY

With the enactment of the Bayh–Dole Act in 1980, universities were allowed for the first time to elect to keep title to inventions made with federal grant funding. Prior to 1980, title to inventions remained with the federal government, and relatively few were licensed. Since 1980, many companies based on university research have been formed, and numerous products have been brought to market.

Commercialization is the pathway that transforms ideas and technologies into products that can be used in clinical practice. Academic institutions excel at research but require collaborations with industry to turn science into products.

In this chapter, we introduced translational and clinical investigators to principles of translating science to the patient bedside—the innovation pipeline—by responding to frequently asked questions. Given the realities of the marketplace and the protection of ideas and innovations, academic health centers can act as living laboratories by working with industry. These centers can also engage in entrepreneurship and the clinical evaluation of innovative products. A heightened level of activity between academic health centers and industry creates a strong need to manage conflict of interest issues.

Finally, a thumbnail sketch was provided for Partners HealthCare Innovation, which described key components of this highly effective technology commercialization office.

STATUTES AND FEDERAL REGULATIONS

CFR (Code of Federal Regulations) Title 21: Food and Drugs. Part 54: Financial Disclosure by Clinical Investigators.

CFR (Code of Federal Regulations) Title 42: Part 50 Subpart F: Responsibility of Applicants for Promoting Objectivity in Research for Which PHS Funding Is Sought.

CFR (Code of Federal Regulations) Title 45: Public Welfare. Part 94: Responsible Prospective Contractors.

CFR (Code of Federal Regulations) Title 45: Public Welfare. Part 680: National Science Foundation Rules of Practice and Statutory Conflict-of-Interest Exemptions.

U.S.C. 17, Copyright Law of the United States, Sec. 102 Subject Matter of Copyright.

U.S.C. 31, False Claims Act, Sec. 3730(b) Civil Actions for False Claims: Actions by Private Persons.

U.S.C. 42, The Public Health and Welfare, Sec. 1320a–7b Criminal Penalties for Acts Involving Federal Health-care Programs.

U.S.C. 42, The Public Health and Welfare, Sec. 1395 nn Limitation on Certain Physician Referrals.

CASES

Gelsinger v. Trustees of U. Penn., No. 001885 (Pa.C.P. settled, 11/2/00).

U.S. ex rel. Cantekin v. University of Pittsburgh 192F.3d 402 (third Cir. 1999), (Cert denied by U.S ex rel. Cantekin v. *University of Pittsburgh* 531 U.S. 880, 121S.Ct. 192 (October 2, 2000)).

Wright v. Fred Hutchison Cancer Center, Wash Super Ct 269F. Supp. 2d 1286 (W.D. Wash., 2002).

ENDNOTES

1. The Copyright Law of the United States (U.S.C. 17 Sec. 102) defines works of authorship as being *fixed in a tangible medium* that can be perceived, reproduced, or otherwise communicated. Works of authorship include literary, musical, dramatic, choreographic, pictorial, audiovisual, or architectural works. Notably, an *idea* is not a work of authorship.
2. Individual NIH agencies may have more restrictive requirements—see, e.g., National Heart Lung and Blood Institute's *Guidelines for Avoiding Conflicts of Interest in Multicenter Clinical Trials*. Updated September 6, 2000. Available at: http://www.nhlbi.nih.gov/funding/policies/coi-res.htm.

REFERENCES

American Cancer Society, January 2007. Research Scholar Grants Policies & Instructions Effective [online]. Available at: http://www.cancer.org/downloads/RES/RSG_Policies_Instructions_Jan_2007_Final.pdf.

American Heart Association, Policies Governing All Research Awards [online]. Available at: http://www.americanheart.org/presenter.jhtml?identifier 5 12150#Investigator.

Association of American Medical Colleges, Protecting Subjects, Preserving Trust, Promoting Progress II: Principles and Recommendations for Oversight of an Institution's Financial Interests in Human Subjects Research [online]. Available at: http://www.aamc.org/research/coi/2002coireport.pdf.

Association of American Medical Colleges, Protecting Subjects, Preserving Trust, Promoting Progress – Policy and Guidelines for the Oversight of Individual Financial Interests in Human Subjects Research [online]. Available at: http://www.aamc.org/research/coi/firstreport.pdf.

Association of University Technology Managers (AUTM) [online]. Available at: http://www.autm.net/about/BDTalkPts031407.pdf.

FASEB Office of Public Affairs, COI Toolkit [online]. Available at: http://opa.faseb.org/pages/Advocacy/coi/Toolkit.htm.

J. Clin. Oncol. 24 (3), January 28, 2006. http://dx.doi.org/10.1200/JCO.2005.04.8926.

National Science Foundation, Conflict of Interest Policy [online]. Available at: http://www.nsf.gov/od/ogc/coipolicy.jsp.

President and Fellows of Harvard College (amended 2004) Harvard University Policy on Conflicts of Interest and Commitment [online]. Available at: http://www.hms.harvard.edu/integrity/conf.html.

Chapter 23

Regulatory Environment

Christine Nguyen, Audrey Gassman and Hylton V. Joffe

U.S. Food and Drug Administration, Silver Spring, MD, United States

Chapter Outline

Key Points

- The FDA plays a key role in determining whether a medical product is safe for testing in humans in the United States.
- The FDA provides critical input and actively guides development of medical products intended for marketing in the United States.
- The FDA determines whether a medical product is safe and effective for initial and continued use as a marketed therapy in the United States.

INTRODUCTION

The Food and Drug Administration (FDA) plays a key role in determining whether a medical product is safe for testing in humans and whether the product is safe and effective for initial and continued use as a marketed therapy for patients in the United States. In addition, the FDA provides critical input and actively guides development of medical products. The FDA publishes general and drug-class-specific guidance documents, has meetings with drug developers at key time points during product development, provides feedback on the appropriateness of study protocols and other scientific and regulatory aspects related to investigational products under development, and monitors postmarket safety data during use in real-world settings. However, most clinical investigators and others involved in clinical research have a limited understanding of the role that the FDA plays in the development and surveillance of medical products throughout their life cycle. The aim of this chapter is to provide an overview of FDA's involvement in this process. This chapter will deal primarily with drug development. Biologics and medical devices are regulated somewhat differently, but practices and principles are similar.

This chapter will also briefly discuss key postapproval activities for which the FDA has enforcement authority. These activities are crucial for ensuring continued safety of medical products as clinical experience accrues.

THE US FOOD AND DRUG ADMINISTRATION

Overview

The FDA is one of several agencies within the Department of Health and Human Services (DHHS) in the executive branch of the United States government. Other agencies in DHHS include the National Institutes of Health (NIH), the Centers for Disease Control and Prevention (CDC), and the Centers for Medicare and Medicaid Services (CMS). The FDA has over 15,000 employees (some of whom are stationed across more than 200 offices, posts, and laboratories throughout the United States) and is headed by a commissioner who is appointed by the President and confirmed by the Senate. FDA-regulated products account for one-quarter of consumer spending (http://www.fda.gov/oc/history/default.htm). The FDA has regulatory authority over prescription drugs, biologics (products derived from living sources, such as vaccines, blood products, and gene therapy), medical devices (ranging from simple items such as tongue depressors to complex technology like heart–lung bypass machines), tobacco-containing products, most foods (except meat, poultry, and alcohol), and radiation-emitting electronic products (such as microwave ovens, cell phones, and magnetic resonance imaging equipment) (http://www.fda.gov/AboutFDA/CentersOffices/default.htm). The FDA is also responsible for ensuring the safety of cosmetics and truthful labeling of nonprescription drugs. The FDA has limited legal authority over dietary supplements (vitamins, minerals, herbs, and botanicals). Manufacturers of dietary supplements are responsible for ensuring safety of their products and truthfulness of the information in the product label, and, unlike drugs, dietary supplements do not require FDA approval prior to marketing unless the manufacturer of the dietary supplement wishes to make a specific disease claim (Dietary Supplement Health and Education Act, 1994) such as "prevents colon cancer" or "treats osteoporosis," whereupon the dietary supplement is viewed as a drug and regulated as such.

Organization

The FDA consists of the Office of the Commissioner and four directorates overseeing the core functions of the agency. These four directorates are the Office of Medical Products and Tobacco, the Office of Foods and Veterinary Medicine, the Office of Global Regulatory Operations and Policy, and the Office of Operations (http://www.fda.gov/aboutFDA/Centersoffices/default.htm). The Office of the Commissioner is responsible for effectively implementing FDA's mission. The Office of Regulatory Affairs, which is housed within the Global Regulatory Operations and Policy directorate, is the lead office for all field activities of the FDA (http://www.fda.gov/AboutFDA/CentersOffices/OfficeofGlobalRegulatoryOperationsandPolicy/ORA/default.htm). Personnel from this office inspect product manufacturing facilities and sites of clinical trials, analyze tens of thousands of product samples yearly to ensure adherence to FDA standards, and respond rapidly to public health emergencies caused by product problems and natural disasters.

The FDA has six centers within two of the directorates that provide specialized expertise for the various types of products falling under FDA's legal authority.

- The Center for Drug Evaluation and Research (CDER) assures that all prescription and nonprescription drugs are safe and effective (http://www.fda.gov/Drugs/default.htm). CDER is also responsible for regulating therapeutic biologics, such as monoclonal antibodies, proteins, and enzymes.
- The Center for Biologics Evaluation and Research (CBER) regulates other types of biologics, including gene therapy, cellular products such as pancreatic islet cells, blood, and vaccines (http://www.fda.gov/BiologicsBloodVaccines/default.htm).
- The Center for Devices and Radiological Health (CDRH) evaluates a diverse group of devices, such as pacemakers, contact lenses, glucose monitors, cell phones, televisions, and microwaves (http://www.fda.gov/MedicalDevices/default.htm).
- The Center for Tobacco Products (CTP) is responsible for implementing the Family Smoking Prevention and Tobacco Control Act passed in 2009. CTP regulates the manufacturing, distribution, and marketing of tobacco products (http://www.fda.gov/TobaccoProducts/default.htm).
- The Center for Veterinary Medicine (CVM) regulates food and additive products given to animals (http://www.fda.gov/AnimalVeterinary/default.htm).
- The Center for Food Safety and Applied Nutrition (CFSAN) is responsible for the safety of most food consumed in the United States (http://www.fda.gov/Food/default.htm).

Because most clinical researchers are predominantly involved with testing of drugs and therapeutic proteins (as opposed to other biologics or devices), the remainder of this chapter will focus on CDER and the drug regulatory environment.

The Office of New Drugs (OND) within CDER has primary regulatory responsibility for review and approval of all drugs and therapeutic biologics in the United States. This includes not only approval of new products, but also applications for expanded indications and new formulations of existing products. Within OND, there are six "Offices," each of which has responsibility over three to four review divisions. The divisions are organized by therapeutic area, much like those that exist in clinical medicine. This organization ensures that reviewers work within their areas of expertise. For example, neurologists in the Neurology Division review Alzheimer's drugs and antiseizure drugs, while endocrinologists in the Division of Metabolism and Endocrinology Products review diabetes, weight-loss, and lipid-altering drugs. Members of these review divisions, which include regulatory project managers, physicians and nonclinical pharmacologists/toxicologists, work closely with chemists (from the Office of Pharmaceutical Quality), biostatisticians and clinical pharmacologists (from the Office of Translational Sciences), and members of other divisions and offices (e.g., the Office of Surveillance and Epidemiology), as appropriate.

Legal Authority

The FDA celebrated its centennial in 2006. The 1906 Pure Food and Drug Act was FDA's founding statute that was created in response to scandals in the meat-packing industry that were widely exposed in Upton Sinclair's book, *The Jungle*. Sinclair described appalling situations such as dead rats shoved into sausage grinders and bribery of inspectors, who permitted diseased cows to be slaughtered for beef. The 1906 law provided the federal government with authority to remove dangerous or falsely labeled foods and drugs from the market but did not limit claims made by manufacturers and did not require approval or premarketing testing of these products.

FDA's modern legal authority began with the 1938 Federal Food, Drug, and Cosmetic Act, spurred by the elixir of sulfanilamide tragedy. Sulfanilamide, a bitter-tasting antimicrobial used for the treatment of streptococcal infections, was dissolved in a sweet-tasting liquid to appeal to pediatric patients. The manufacturer did not perform toxicity testing on this new solvent, which contained ethylene glycol (antifreeze) and caused more than 100 deaths, including many children. The 1938 Act required manufacturers to establish safety prior to marketing of new drugs (these requirements for safety are identical to those in current law) and required submission of a new drug application (NDA) to the FDA prior to marketing. The drugs would be approved if the FDA did not object within 180 days after receiving the marketing application.

FDA's drug authority was further strengthened in 1962 with the passage of the Kefauver–Harris Drug Amendments (http://www.fda.gov/oc/history/default.htm), which was spurred by the thalidomide tragedy. Thalidomide was marketed as a sleep-aid to pregnant women outside the United States and caused major birth defects (mainly phocomelia) in thousands of children. Thalidomide was never marketed in the United States, thanks to FDA reviewer Dr. Frances Kelsey, but some exposure did occur when it proved impossible to find all of the drug product that had been distributed for investigational use. The 1962 amendments required that manufacturers of new drugs establish efficacy in addition to safety, which was a response to hearings prior to Kefauver–Harris that established the poor quality of studies that had been performed to support the effectiveness of drugs and the absurd labeling that had resulted from these studies. The amendments also required that the FDA give positive approval before a new drug could be marketed, a major change in the review/approval dynamics. In addition, the amendments required that the Secretary of DHHS promulgate regulations (e.g., informed consent) to cover drugs for investigational uses. The amendments also established Good Manufacturing Practices (GMP), which require high standards during the manufacturing process of drugs (e.g., sanitation, qualified personnel, appropriate equipment, and validation and documentation of the manufacturing process).

To determine whether there is substantial evidence of effectiveness, the regulations require reports of "adequate and well-controlled investigations" (21 CFR 314.126). The use of the plural form of the word "investigation" is interpreted, in accordance with legislative history, as a requirement for at least two well-controlled trials to support efficacy (the 1997 Food and Drug Administration Modernization Act permits demonstration of efficacy based on a single pivotal study, where appropriate and when agreed to by the FDA). The process of independent substantiation with at least two well-controlled clinical trials is important, because a positive finding from one trial is more likely to represent a chance finding and the single trial may have undetected biases. The trials do not need to be identical in design (e.g., for the second study, the drug could be evaluated in a different phase of the disease or in a different patient population), but each trial must include a clear statement of the study objectives; a valid comparison with a control; an adequate number of research subjects with the condition of interest; a design that minimizes bias and assures comparability of the treatment groups; objective, well-defined, and prespecified methods for assessing response to treatment; and appropriate, predefined analyses (21 CFR 314.126; http://www.fda.gov/downloads/Drugs/GuidanceComplianceRegulatoryInformation/Guidances/UCM072008.pdf).

The law does not require a new drug to be better than, or even as good as, an already approved drug (although such comparisons may be informative), but the effect of the new drug must be clinically meaningful (Warner-Lambert v. Heckler, 1986).

The 1962 amendments also called upon the FDA to evaluate the effectiveness of all drugs approved solely on the grounds of safety between 1938 and 1962. To accomplish this mandate, the FDA contracted with the National Academy of Sciences/National Research Council to carry out the Drug Efficacy Study. Implementation of this study is called DESI (Drug Efficacy Study Implementation) (http://www.fda.gov/oc/history/default.htm). Over 3000 DESI drugs have been reviewed. Approximately one-third were found to be ineffective and have subsequently been withdrawn from the market.

In 1992, Congress passed the Prescription Drug User Fee Act (PDUFA), which has subsequently been reauthorized every five years. Under PDUFA, sponsors (individuals or organizations who are responsible for the research studies) pay user fees to the FDA when certain applications are submitted to the FDA for review. These fees, together with federally appropriated funds, support the drug review process and have been used to hire additional review and support staff, upgrade data systems, and improve procedures that make reviews more rigorous, consistent, and predictable. In exchange for the user fees, the FDA has agreed to meet performance goals (see later discussion for details) during drug development and review, ensuring timely review of drug applications, while preserving the integrity of the review process, with the goal of assuring that safe and effective drugs are made available to the public in a timely manner. PDUFA has established expectations of FDA review timelines for marketing applications, developed processes for interactions with sponsors during all phases of drug development, and implemented programs to ensure drug development in unmet medical conditions (e.g., pediatric drug development). Importantly, PDUFA performance goals do not apply to approval times (i.e., the FDA can require as many review cycles as it deems necessary to ensure an adequate demonstration of efficacy and safety prior to considering approval of the medical product).

In 1997, Congress passed the FDA Modernization Act (FDAMA), which established the Fast Track process, permitted evidence of efficacy based on a single clinical trial (where appropriate), and reauthorized PDUFA. The Fast Track process was designed to facilitate the development and expedited review of drugs that address unmet needs for the treatment of serious or life-threatening conditions. The Fast Track process includes an existing regulation, "Subpart H" (also known as accelerated approval), which allows certain products meeting the above criteria to be approved based on a surrogate endpoint that is not as well established as regularly used surrogates, such as blood pressure, but is reasonably likely to predict clinical benefit. An approval based on a positive response in the surrogate endpoint requires postapproval studies to confirm clinical benefit. The FDA can withdraw approval if the sponsor fails to satisfy the postapproval requirement, if safety issues arise, or if no benefit is confirmed.

The 2007 Food and Drug Administration Amendments Act (FDAAA) reauthorized PDUFA and also provided the FDA with important new regulatory authorities that significantly enhanced FDA's authority to oversee drug safety after approval. Some important FDAAA provisions included the authority to require Risk Evaluation and Mitigation Strategies (REMS) and to require certain studies or trials be conducted to address specific safety concerns. FDAAA also provided FDA with a new authority to require labeling changes for important safety information during the postmarketing period. Manufacturers not complying with these new requirements may face civil penalties and fines. These important authorities are discussed in more detail later in the chapter.

In 2012, Congress passed the Food and Drug Administration Safety and Innovation Act (FDASIA). FDASIA reauthorized PDUFA and also created a novel expedited drug development program called "breakthrough therapy" designation. This designation commits the FDA to help expedite the development and review of new drugs when there is preliminary clinical evidence that the drug may offer substantial improvement over available therapies for patients with serious or life-threatening illnesses. FDASIA also established the Generic Drug User Fee Amendments (GDUFA), which for the first time established user fees that drug companies must pay to support the review of generic drugs. Funding through GDUFA will provide the FDA with the necessary resources for more predictable and timely review of generic drug applications.

Nonclinical Testing

Nonclinical testing, which includes in vitro and animal studies, is the first stage in drug development. The purpose of nonclinical testing is to show that a drug is reasonably safe for testing and use in humans and to help identify the toxicities that should be assessed and monitored in clinical testing (21 CFR 312.23). Sponsors are not required to notify the FDA or obtain FDA's approval before initiating nonclinical studies. FDA reviewers perform independent analyses of the submitted data from completed studies and critically assess whether these data adequately support initial and continued testing in humans. If these data are insufficient, the FDA could place the clinical development program on "clinical hold" (see The Investigational New Drug Application Review Process section).

Nonclinical studies must comply with Good Laboratory Practices (GLP) (http://www.fda.gov/ICECI/ EnforcementActions/BioresearchMonitoring/ucm133789.htm) if the sponsor plans to submit the results of these studies to the FDA to support safety of the drug in humans. The FDA established GLP to ensure high standards for laboratories conducting nonclinical studies and reliability of the results. GLP covers aspects related to laboratory organization, physical structure, equipment, operating procedures, and handling of the test substance and animals. To confirm compliance with GLP, the FDA performs announced and unannounced inspections of these facilities on a routine basis or if FDA reviewers identify suspicious data.

The Investigational New Drug Application

Federal law requires that a drug be FDA approved for use in humans prior to transfer across state lines. How then is an unapproved new drug tested in clinical trials, which are often conducted in multiple states? In theory, it might be possible to confine the research subjects and the manufacture and sale of every component of the unapproved drug to one state, but this approach is impractical. In practice, the testing of unapproved drugs is carried out under an exemption to the above federal law, once called a "Notice of Claimed Investigational Exemption for a New Drug," now called an Investigational New Drug application (IND) (http://www.fda.gov/drugs/developmentapprovalprocess/ howdrugsaredevelopedandapproved/approvalapplications/investigationalnewdrugindapplication/default.htm; Feinsod and Chambers, 2004). The provisions governing INDs can be found in Section 505(i) of the Federal Food, Drug, and Cosmetic Act, and the relevant regulations are in the Code of Federal Regulations (CFR) Title 21 Food and Drugs: Part 312 Investigational New Drug Application (http://www.accessdata.fda.gov/scripts/cdrh/cfdocs/cfcfr/CFRsearch.cfm? CFRPart=312).

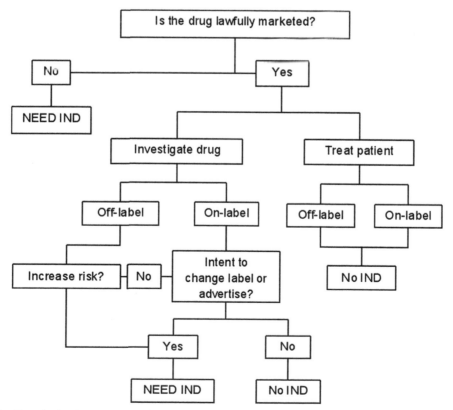

FIGURE 23.1 Should an Investigational New Drug Application (IND) be submitted to the FDA? *Adapted with permission from Feinsod, M., Chambers, W.A., 2004. Trials and tribulations: a primer on successfully navigating the waters of the Food and Drug Administration. Ophthalmology 111, 1801–1806.*

Is an Investigational New Drug Application Required?

In the United States, drugs that are not FDA approved require an IND prior to any use, including clinical testing (Fig. 23.1) (Feinsod and Chambers, 2004). A physician who wishes to use an FDA-approved drug to treat a patient in the course of medical practice does not require an IND, even if the drug will be used off-label (i.e., used to treat a condition for which the drug does not have an FDA-approved indication), because such use is not considered investigational. Use of an approved drug for clinical research may or may not require an IND. An IND is required if the proposed use significantly increases the risk to research subjects beyond the risk that occurs with on-label use, if the risk is less acceptable (e.g., use of an oncologic drug to treat a benign condition), or if there is intent to change the labeling of the drug based on study results (e.g., obtain a new indication) or modify its promotion.

The FDA has published a Guidance document to help clinical investigators, sponsors, and institutional review boards determine whether human research studies must be conducted under an IND (http://www.fda.gov/downloads/Drugs/ Guidances/UCM229175.pdf). In addition, the FDA website (http://www.fda.gov/drugs/developmentapprovalprocess/ howdrugsaredevelopedandapproved/approvalapplications/investigationalnewdrugindapplication/default.htm#FDA%20 Guidances%20for%20Investigational%20New%20Drugs) has extensive information on the IND process, including copies of the required IND forms, detailed instructions for filing an IND, and responses to frequently asked questions, as well as contact information for obtaining further guidance regarding IND procedures.

Types of Investigational New Drug Applications

Commercial INDs are submitted to the FDA by corporate entities (e.g., pharmaceutical companies), usually with the ultimate goal of obtaining marketing approval so that the drug product can be commercialized. Many INDs submitted by the National Institutes of Health (NIH) are also considered "commercial," because these INDs involve large clinical trials. Research (noncommercial) INDs are submitted to the FDA by clinical investigators and academicians who wish to initiate and conduct clinical studies, generally without the goal of obtaining marketing approval or changing the product label (http://www.fda.gov/Cder/regulatory/applications/ind_page_1.htm). Typically, a commercial IND initiates a long series of submissions of nonclinical data, clinical trial protocols, and clinical trial results, as well as a series of official meetings and communications concerning drug development. In contrast, the noncommercial IND generally involves a single clinical trial.

The INDs described earlier are intended for conducting clinical trials and obtaining efficacy or safety information about the tested drug. The FDA has another mechanism—known as the expanded access program—for making promising experimental drugs available to patients not participating in clinical trials. This program provides patients with serious or life-threatening diseases who have no alternative treatment options (e.g., patients with an advanced form of cancer who have failed other therapies), with access to experimental drugs, not for the purpose of clinical trials but rather to diagnose, monitor, or treat the patient's medical condition (http://www.fda.gov/downloads/Drugs/GuidanceComplianceRegulatoryInformation/ Guidances/UCM351261.pdf).

All INDs are reviewed by a team of professionals with expertise in the drug review process. The therapeutic area of the proposed clinical trial(s) is typically used to identify the division where the IND will be reviewed. This method of assignment assures that reviews are conducted by specialists who have expertise appropriate for the subject matter. The typical review team consists of regulatory project managers, chemists, nonclinical pharmacologists/toxicologists, and clinical reviewers known as medical officers. The team's primary focus during the IND stage is to ensure reasonable safety for research subjects participating in the clinical trials. The team also plays a major role in guiding drug development to ensure that the studies are adequate to fulfill their objectives.

The regulatory project manager serves as the administrative focal point for INDs and is the primary point of contact for sponsors of applications. Chemists focus on the manufacturing process of the drug to ensure that the product can be reliably synthesized at the proposed strength and with reasonable purity without toxic components. Nonclinical pharmacologists/toxicologists evaluate the in vitro and animal studies to ensure there is adequate evidence of safety to permit initiation and continuation of human studies. These specialists review acute and chronic animal toxicity studies, as well as animal pharmacology studies, and confirm that there are adequate safety margins between the proposed exposures in humans and the highest exposures not associated with toxicity in animals. The medical officer (physician) focuses on the proposed clinical study and determines whether the study population, study design, drug doses, and safety assessments are adequate based on potential signals identified in the nonclinical studies and the anticipated pharmacologic action of the drug.

The initial studies proposed under the IND are usually designed to assess safety and tolerability of single and sometimes multiple doses of the test product, often in healthy volunteers. Because early IND studies do not typically assess efficacy, rigorous hypothesis testing is not used, and the statistical review team is generally not involved at this stage. Clinical pharmacologists (specialists who review human pharmacokinetic and pharmacodynamic data) generally also have limited involvement during review of the initial studies submitted under an IND, because no human data have been generated (unless prior clinical studies have been conducted outside the United States). However, the clinical pharmacologists may offer input regarding the timing or other characteristics of the proposed pharmacokinetic assessments in these studies.

The Investigational New Drug Application Review Process

The FDA typically receives over 1000 INDs each year and completes a thorough safety review of each IND within 30 calendar days following receipt (http://www.fda.gov/Drugs/DevelopmentApprovalProcess/HowDrugsareDevelopedand Approved/ApprovalApplications/InvestigationalNewDrugINDApplication/default.htm). After the initial IND submission, the sponsor must wait for these 30 days to pass before initiating the trial. Usually, the FDA will contact the sponsor within that time period to advise of any problems. Sponsors are legally permitted to initiate the proposed study if there has been no response from the FDA within the 30 days. In the absence of communication from the FDA, it may be prudent to contact the review division to confirm that the FDA has received the IND and that the review team has concluded that the proposed study is sufficiently safe for initiation.

If the review team identifies a significant concern, the FDA will notify the sponsor and attempt to resolve the issue, if possible, during the 30-day review cycle. If this is not possible, the FDA will inform the sponsor that the study will be placed on clinical hold (21 CFR 312.42). A clinical hold means that the FDA will not allow the clinical investigation to proceed under the IND. Grounds for imposing a clinical hold are stipulated in the regulations and include unreasonable risk to subjects, unqualified clinical investigators, or insufficient information in the IND to assess risk to subjects. In the clinical hold letter, the FDA details the deficiencies identified during the review and the steps needed to resolve these deficiencies. When sponsors submit a detailed, complete response to the clinical hold, the FDA will review the data and either remove the clinical hold (if the concerns have been fully addressed) or maintain the clinical hold (explaining why the deficiencies have not been addressed). The FDA acts on these responses to clinical hold within 30 days of receipt. At any point in development, the FDA may put aspects of a proposed study on hold or may put certain clinical studies on hold while others proceed (e.g., the FDA may permit single dose studies but not studies with repeated dosing of study drug). This is called a "partial hold." Later studies may be put on hold if new unacceptable safety findings arise or the protocol for the investigation is clearly deficient in design for the study to meet its stated objectives.

Maintaining the Investigational New Drug Application

Table 23.1 summarizes the responsibilities of sponsors and investigators with regard to clinical trials of investigational products (21 CFR 312.50, 312.52, 312.53, 312.55–312.62, 312.64, 312.66, 312.68, 312.69).

Once an IND becomes active, the sponsor is responsible for submitting protocol amendments, information amendments, safety reports, and annual reports to the IND (21 CFR 312.30–312.33). Protocol amendments include new clinical trial protocols (the study may begin once the protocol has been submitted to the FDA and approved by an institutional review board), changes to existing clinical trial protocols, and notification that new investigators have been added to ongoing studies. Information amendments are used to submit essential information that is not within the scope of a protocol amendment (e.g., completed animal or clinical study reports). The sponsor is required to report to the FDA all suspected adverse reactions to study treatment that are serious and unexpected. This reporting must take place within 15 days of the sponsor receiving this information and within 7 days if the event is fatal or life-threatening (http://www.fda.gov/ downloads/Drugs/.../Guidances/UCM227351.pdf). Safety information that does not require expedited reporting is submitted in an annual report, which contains information that has been learned about the product since the prior annual report.

A sponsor may withdraw an IND at any time (21 CFR 312.38). The FDA may inactivate an IND at the sponsor's request, if no subjects are entered into the studies for over 2 years, or if all investigations have been on clinical hold for at least 1 year (21 CFR 312.45). The FDA can terminate an IND if the IND remains inactive for more than 5 years, if there are deficiencies in the IND (e.g., repeated failure to submit annual reports), or if there are issues with study conduct (21 CFR 312.44). Sponsors are given 30 days to respond to FDA's proposal to terminate the IND except if there is immediate and substantial danger to the study participants.

TABLE 23.1 Major Responsibilities of Sponsors and Investigators

Entity	Responsibilities
Sponsors[a]	• Select qualified investigators • Ensure proper monitoring of clinical trials • Submit required documents to the FDA to maintain the IND • Inform the FDA and investigators of new information (especially safety findings) • Monitor the progress of the clinical trials • Maintain adequate records, including drug disposition and financial conflict of interest statements from investigators • Permit FDA inspection of records • Dispose of unused supply of investigational drug
Investigators	• Conduct each study according to the study protocol • Obtain informed consent • Control of drug to ensure that only authorized people receive it • Maintain adequate records, including drug disposition and case histories • Submit progress, safety, financial disclosure, and final reports to sponsor • Assurance of institutional review board (IRB) review • Report anticipated problems to IRB • Permit FDA inspection of records

FDA, Food and Drug Administration; *IND*, Investigational new drug application.
[a]*A sponsor can transfer obligations to a contract research organization.*
Source: 21C.F.R. 312.50, 312.52, 312.53, 312.55–312.62, 312.64, 312.66, 312.68, and 312.69.

Guiding Drug Development

The FDA has several principal roles during clinical drug development. The FDA ensures that the rights and safety of the research subjects participating in the clinical trials are protected. The FDA also actively guides drug development by determining the necessary testing required to establish the efficacy and safety of the investigational product for the purposes of marketing. In addition, FDA's Critical Path initiative (http://www.fda.gov/ScienceResearch/SpecialTopics/CriticalPathInitiative/default.htm) has the goal of using new scientific discoveries to modernize drug development by improving the efficiency and accuracy of testing used to establish the safety and efficacy of investigational medical products. Despite major advances in basic and clinical sciences, drug development is facing several challenges including high failure rates (10% of products entering phase 1 testing reach the market) and increasing costs of bringing a new product to market (with some estimates of over $1 billion spent per drug approved). Because regulatory authorities such as the FDA see the full spectrum of drug development, including successes, failures, delays, and barriers, we are uniquely positioned to work with academia and drug companies to help identify and address the challenges contributing to this pipeline problem.

The FDA has published many documents (alone and in partnership with international regulatory agencies) designed to provide valuable assistance to sponsors throughout the clinical development life cycle. Guidances reflect our current thinking on a wide range of regulatory and scientific issues and are publicly available on the FDA website (http://www.fda.gov/Drugs/GuidanceComplianceRegulatoryInformation/Guidances/). Examples include "Content and Format of Investigational New Drug Applications (INDs) for Phase 1 Studies of Drugs," "Providing Clinical Evidence of Effectiveness for Human Drug and Biological Products," "Clinical Trial Endpoints for the Approval of Cancer Drugs and Biologics," and "Establishment and Operation of Clinical Trial Data Monitoring Committees." Sponsors are permitted to deviate from these nonbinding documents, provided that the alternative plan complies with relevant statutes and regulations and is scientifically valid.

International Conference on Harmonization (ICH) guidelines (http://www.ich.org) provide additional guidance for sponsors. The ICH was established by regulatory authorities and pharmaceutical experts from the United States, Europe, and Japan. The goal of the ICH is to harmonize the requirements needed to obtain drug approval in these three regions of the world. This process reduces the need for duplicate testing during clinical development, ensures more economical use of resources, and reduces the delay in availability of new drugs in light of global drug development and marketing strategies. Examples of ICH documents include "E1A—The extent of population exposure required to assess clinical safety of drugs intended for the long-term treatment of non-life-threatening conditions," "E3—Guideline on the Structure and Content of

Clinical Study Reports," "E4—Dose-Response Information to Support Drug Registration," "E10—Choice of Control Group and Related Issues in Clinical Trials," and "E14—Clinical Evaluation of QT/QTc Interval Prolongation and Proarrhythmic Potential for Non-Antiarrhythmic Drugs." The ICH also sets high standards for the conduct of clinical trials through its Good Clinical Practices (GCP) guidance (http://www.ich.org), which defines the clinical responsibilities of sponsors, investigators, study monitors, and institutional review boards. Table 23.2 summarizes the currently available major clinical ICH documents. The full set of ICH guidelines (including nonclinical guidelines) is publicly available on the ICH website (http://www.ich.org/products/guidelines.html).

In addition to guidance and ICH documents, the FDA actively guides clinical drug development through meetings (face-to-face or teleconferences) and written correspondence with sponsors at key time points during the life cycle of the drug development process (Fig. 23.2).

Prior to meetings, the sponsor submits a briefing package that contains background information for the meeting and a corresponding set of questions addressed to the FDA. These questions may cover the full spectrum of the regulatory and scientific aspects of drug development, ranging from specific questions about the proposed clinical trials to questions about whether the proposed development plan will fulfill specific regulatory requirements for marketing of the drug. The review team critically evaluates the briefing package, meets internally to discuss the sponsor's questions, and then sends

TABLE 23.2 International Conference on Harmonization (ICH): Major Clinical Guidelines

Code	Title	General Overview
E1	The Extent of Population Exposure to Assess Clinical Safety for Drugs Intended for Long-Term Treatment of Non-Life-Threatening Conditions	Discusses the number of patients and duration of exposure for the safety evaluation of drugs intended for the chronic treatment of non-life-threatening conditions
E2E	Pharmacovigilance Planning	Focuses on pharmacovigilance plans that may be submitted at the time of the licensing application for the early post-marketing period
E3	Structure and Content of Clinical Study Reports	Describes the format and content of clinical study reports that are accepted in all three ICH regions
E4	Dose-Response Information to Support Drug Registration	Discusses the design and conduct of studies to assess the relationship between doses, blood levels, and clinical response of a new drug
E5 (R1)	Ethnic Factors in the Acceptability of Foreign Clinical Data	Discusses intrinsic and extrinsic factors that may affect the results of clinical studies conducted in different regions of the world
E6 (R1)	Good Clinical Practice	Describes the responsibilities of all participants involved in the conduct of clinical trials
E7	Studies in Support of Special Populations: Geriatrics	Discusses the design and conduct of clinical trials for drugs that are likely to have significant use in the elderly
E9	Statistical Principles for Clinical Trials	Discusses statistical considerations, with a focus on clinical trials that form the basis for demonstrating effectiveness
E10	Choice of Control Group and Related Issues in Clinical Trials	Discusses the ethical and inferential properties and limitations of different kinds of control groups
E11	Clinical Investigation of Medicinal Products in the Pediatric Population	Addresses the conduct of clinical trials in children
E14	The Clinical Evaluation of QT/QTc Interval Prolongation and Proarrhythmic Potential for Non-Antiarrhythmic Drugs	Discusses the clinical assessment of the potential for a drug to delay cardiac repolarization
E15	Terminology in Pharmacogenomics	Defines key terms in pharmacogenomics and pharmacogenetics
E16	Qualification of Genomic Biomarkers	Describes context, structure, and format of regulatory submissions for qualification of genomic biomarkers

Source: http://www.ich.org.

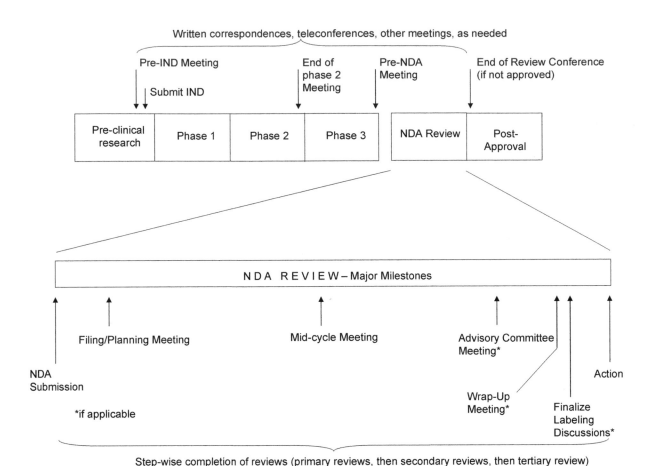

FIGURE 23.2   An overview of the drug development process and the Food and Drug Administration (FDA) review process. *IND,* Investigational new drug; *NDA,* New drug application. Also applies to biologics license applications (BLAs) for therapeutic biologics.

preliminary responses to the sponsor a few days prior to the scheduled meeting. The sponsor can cancel the meeting if there is full understanding and general agreement with the FDA responses. Typically, the meeting still occurs and FDA's initial written responses help focus the discussion on areas of disagreement or on issues needing further clarity. After the meeting, the FDA revises and finalizes the previously prepared written responses and sends official meeting minutes to the sponsor who then confirms that these comments appropriately capture the discussions and agreements reached at the meeting. These discussions and meeting minutes are very important—they represent the views of the FDA on critical issues of study design (e.g., study endpoints, duration, population, analyses, safety monitoring) so that the sponsor is clearly aware of expectations. They also represent views that the FDA will expect to maintain, barring a change in circumstances or recognition of a significant public health concern.

FDA—sponsor communications also occur at other times during the drug development life cycle. For example, the FDA may request additional information in response to a safety report submitted by the sponsor or based on an emerging safety signal in the ongoing studies or with related products. The Special Protocol Assessment (SPA) (http://www.fda.gov/downloads/Drugs/.../Guidances/ucm080571.pdf) is another critical FDA—sponsor communication for phase 3 protocols intended to confirm efficacy, animal carcinogenicity protocols, and chemistry stability protocols. Under an SPA, the sponsor of a drug development program submits a detailed study protocol with specific questions that usually relate to study design or statistical analysis. For example, the sponsor may ask the FDA to comment on the trial's proposed duration, patient population, comparator arms, doses of study medication, or the frequency and types of safety assessments. The review team sends written comments responding to the sponsor's questions within 45 days of receipt. Agreements reached on the SPA are binding for the FDA except in certain circumstances (e.g., if the FDA and sponsor agree in writing to modify the protocol or if the FDA identifies a scientific issue essential to determining the safety or efficacy of the drug after testing has begun).

The Marketing Application

Sponsors submit an NDA (http://www.fda.gov/Drugs/DevelopmentApprovalProcess/HowDrugsareDevelopedandApproved/ApprovalApplications/NewDrugApplicationNDA/) to the FDA to obtain permission to market and sell a drug in the United States for a specific indication. A biologics license application or BLA is submitted for biologics. Supplemental NDAs or BLAs (efficacy supplements) are applications submitted to the FDA to obtain authorization to modify the labeling of an already approved drug (e.g., new formulation, new indication, or new population).

Efficacy, Safety, and the Marketing Application

Nowadays, the NDA or BLA is almost always submitted in electronic form. The submission consists of as many as tens of thousands of pages of regulatory and scientific information to support approval of the new drug, including complete protocols and detailed results for all the nonclinical and clinical studies (21 CFR 314.50). In addition, the submission contains all case report forms for any patient who dies or who leaves the study because of an adverse event and, in many cases, for other patients with adverse events of interest, as well as narratives for patients with selected adverse events. The submission also contains case report tabulations (data sets with patient-level data used by the FDA to confirm the sponsor's analyses and perform additional analyses). The raw data and study reports with full details are crucial for allowing the review team to critically determine whether there is adequate evidence of efficacy and safety to support approvability of the drug.

The criteria for assessing efficacy are stringent. These criteria are based on the regulations describing the characteristics of adequate and well-controlled studies (21 CFR 314.126), the only legal evidentiary basis for deciding whether a drug is effective. The characteristics include a clear protocol, methods to reduce bias, such as randomization and blinding, and a clear, prespecified plan of analysis. In addition, experience has led to additional standards, many of which are described in ICH "E9—Statistical Principles for Clinical Trials" and "E10—Choice of Control Group and Related Issues in Clinical Trials" (http://www.ich.org/products/guidelines.html). The statistical plan for a study on which effectiveness is to be based must be prespecified in detail prior to database lock (i.e., prior to any unblinding), and it is useful for the sponsor and FDA biostatisticians to agree on the statistical tests and methodology that will be used to analyze the study. Critical decisions include defining the statistical population (e.g., intent-to-treat, modified intent-to-treat, per protocol), handling of dropouts and missing data, specifying the study endpoints and the appropriate statistical tests, and controlling the type 1 error rate by using appropriate corrections for multiple testing and interim analyses. A particular challenge is analysis of noninferiority studies, where choosing the appropriate margin and ensuring assay sensitivity can be difficult (see ICH E10). Although the FDA is fully aware of the multiplicity problems posed by subset analyses, the regulations specifically require demographic subgroup (gender, age, and race) analyses of effectiveness and safety data in the NDA, usually in the required integrated analyses of safety and effectiveness (21 CFR 314.50).

The safety review integrates information from nonclinical studies (e.g., drug manufacturing and components, cardiac conduction studies in animals, and animal toxicology studies), early clinical studies (e.g., QTc prolongation assessment, drug—drug interactions, and the effects of age, gender, ethnicity, renal impairment, and liver disease on the pharmacokinetics of the drug), and phase 2 and phase 3 clinical trials. The FDA performs broad explorations of the safety data, because the critical safety endpoints are often not known in advance, although potential safety concerns are identified based on the pharmacology of the drug and known safety concerns with other products in the drug class (if available). The safety review is an integrated analysis, typically pooling data from phase 2 and phase 3 studies to improve the precision of risk and enhance the power for detecting group differences. During the safety review, the clinical reviewer evaluates the extent of exposure to study drug (number of research subjects exposed and duration of exposure) and performs critical analyses of the deaths, serious adverse events (e.g., untoward effects that are life-threatening, lead to hospitalization, or cause congenital abnormalities), patients who drop out because of adverse events, other adverse events, laboratory data, vital signs, electrocardiograms, unintended pregnancies, overdose experience (e.g., phase 1 dose ranging studies that have tested higher doses than those typically used in the phase 3 studies), and any other investigations deemed necessary. In addition to an overall assessment, the FDA evaluates the impact of patient characteristics and risk factors (e.g., dose dependency, time dependency, drug—demographic interactions, drug—disease interactions, and drug—drug interactions) on the incidence of adverse events. The FDA focuses on treatment-emergent adverse events, which are defined as untoward effects that occur during a clinical trial in participants who have received at least one dose of study medication, regardless of whether the investigators attribute the events to study drug (investigator assessments of causality tend to correlate with known or anticipated side effects and are not likely to identify novel adverse events). The adverse events are converted from the investigators' verbatim terms to standardized terminology (preferred terms) using medical dictionaries that are used for sorting and grouping of like events, but for serious adverse events and patient dropouts due to adverse events it remains critical to review the investigator's description of individual cases. Because all drugs have risks, our determination that a drug is "safe" within the context of the law means that the benefits of the drug have been determined to outweigh the risks for the uses recommended in labeling.

The review team focuses not only on results generated from the submitted studies but also on whether the studies themselves were adequately designed to identify and characterize efficacy and safety issues (e.g., whether the correct safety evaluations were conducted at appropriate time intervals and whether the clinical trials included adequate numbers of patients at high risk for adverse effects who are likely to use the product, if approved). Finally, the review team evaluates whether any additional safety data should be collected during the postmarketing period.

The Marketing Application Review Process

Fig. 23.2 provides an overview of the NDA (or BLA) timeline and review process (http://www.fda.gov/downloads/aboutfda/centersoffices/cder/manualofpoliciesprocedures/ucm218757.pdf). If an application is deficient on its face, the FDA can refuse to file it (the sponsor can still insist on a full review and require the FDA to file the application over protest). An internal filing meeting is held ~45 days after NDA or BLA receipt to discuss whether there are any major deficiencies that would prevent formal review of the marketing application (e.g., omission of critical data needed to assess efficacy or safety). If the NDA or BLA is acceptable for review, the filing meeting is also used to formally assign a review priority classification (standard or priority review, the latter assigned if the drug treats a serious condition and appears to offer significant improvement over marketed therapies) and to set internal timelines to ensure the review is completed within the prespecified time period. The PDUFA time clock for reviewing and acting on the NDA or BLA is 10 months for standard reviews, 6 months for priority reviews, and 6 months for supplemental NDAs or BLAs containing qualifying pediatric study reports. For some products (e.g., NDAs for new molecular entities), the PDUFA time clock begins 60 days after the application receipt date (if the application is filed). For other products, the PDUFA time clock begins on the FDA receipt date (http://www.fda.gov/ForIndustry/UserFees/PrescriptionDrugUserFee/UCM2005475.htm).

During the review process, reviewers from each discipline critically evaluate the NDA or BLA, frequently discuss findings with each other and with supervisors (including team leaders, division directors, and office directors, depending on the type of application under review), and prepare detailed written reviews of their findings (often hundreds of pages in length). Additional information and analyses are requested from the sponsor, as needed. NDAs or BLAs that involve difficult or complex regulatory decisions may be presented at a regulatory briefing (internal, center-wide meeting with senior center officials) and/or at an advisory committee meeting (generally public forum with external experts) (http://www.fda.gov/AdvisoryCommittees/default.htm). The FDA review division and office carefully consider all recommendations obtained at regulatory briefings and advisory committee meetings, but are not bound by them.

Concurrent with the efficacy and safety review of the NDA or BLA, the Office of Scientific Investigations performs physical inspections of select study sites to confirm data integrity (e.g., whether adverse events have been appropriately captured and included in the safety database, whether case report forms match the laboratory printouts and electronic databases, and whether inclusion and exclusion criteria have been appropriately applied). If significant data integrity problems are identified, the FDA evaluates whether this is a widespread issue substantially affecting the integrity of the entire trial or whether it is an isolated issue that can be addressed by excluding the implicated study sites from the analyses.

If the drug is to be approved, the sponsor and the FDA discuss the contents of the label (also known as the prescribing information or package insert). The FDA ensures that the efficacy and safety findings are accurately communicated in the label (including limitations of our knowledge), that the label only contains scientific information substantiated by the data available, and that the label contains pertinent information for prescribers to use the drug appropriately. Sponsors of NDAs submitted on or after June 30, 2006 must provide labels in a new format known as the Physician Labeling Rule (PLR) format (http://www.fda.gov/Drugs/GuidanceComplianceRegulatoryInformation/LawsActsandRules/ucm084159.htm). The goal of this revised format is to make information more accessible to prescribers. The PLR format includes a new "Highlights" section (usually a half-page summary of the most crucial information for the safe and effective use of the drug), a table of contents, and a reordering of the full prescribing information so that prescribing information is more accessible to the health-care provider.

FDA Decision Regarding Approvability of the Marketing Application

The signatory authority makes the final decision regarding approvability of the NDA or BLA and communicates this official FDA decision to the sponsor in an action letter. For new molecular entities, the signatory authority is delegated to the director of the office who oversees the relevant review division. The signatory authority's decision is based on an integration of his/her findings and all the findings and conclusions from the review team (including primary reviewers, team leaders, and the division director or deputy division director). When deciding on the approvability of an NDA or BLA, the FDA chooses between one of two possible decisions: "Approval" or "Complete Response." An "Approval" permits marketing of the product in the United States for the agreed-upon indications as of the date of the Action Letter. A "Complete Response" letter is issued if the FDA determines that the application cannot be approved in its present form. This letter describes the specific deficiencies preventing approvability and, when possible, recommends actions that the sponsor could take to resolve these deficiencies. After receiving a "Complete Response" letter, the sponsor must resubmit the application, addressing all identified deficiencies, withdraw the application, or request an opportunity for a hearing on the question of whether there are grounds for denying approval of the application. The sponsor has the right to an end-of-review conference to discuss with the FDA the deficiencies of the application and how these deficiencies can be resolved.

Expanded Regulatory Authority on Postmarket Drug Safety

As mentioned previously, the FDA obtained new authorities and responsibilities with the passage of FDAAA in 2007 that strengthen its ability to ensure drug safety after drugs reach the marketplace. The new statute contained three important provisions to address specific and serious safety concerns or signals seen at the time of approval or during the post-marketing setting. These new authorities are described below.

Risk Evaluation and Mitigation Strategies

A REMS is a safety strategy, in addition to labeling, that helps manage or mitigate a known or potential serious risk of a drug (http://www.fda.gov/Drugs/DrugSafety/postmarketdrugsafetyinformationforpatientsandproviders/ucm111350. htm#information). The FDA will require a REMS if it determines that a REMS is needed to ensure that the benefits of the drug outweigh the risks. A REMS can be required at the time of drug approval (if a serious risk is recognized during the NDA or BLA review) or after approval (if the serious risk becomes evident in the postmarketing setting). A REMS may consist of a Medication Guide (an FDA-approved patient handout that pharmacists are required to distribute to the patient when a prescription is filled or refilled), a communication plan (e.g., a Dear Healthcare Provider letter that alerts prescribers of an important safety concern), Elements to Assure Safe Use (for example, medical interventions or other actions that health-care providers must take prior to prescribing or dispensing the drug to the patient), or some combination of these three tools. Sponsors must periodically assess whether the REMS is successful at meeting its goal(s) and must submit reports of these assessments to the FDA at certain time intervals. If the FDA determines that the REMS is not meeting its goals, the FDA may require the sponsor to revise the REMS and conduct assessments of the modifications.

Required Postmarketing Studies or Trials

Prior to 2007, the FDA could only require sponsors to conduct certain types of postmarketing studies (e.g., pediatric studies required under the Pediatric Research Equity Act and studies to confirm clinical benefit after accelerated approval under Subpart H). With new authorities under FDAAA, the FDA can now also require clinical studies or trials to assess a known serious risk or a signal of a serious risk (http://www.fda.gov/Drugs/GuidanceComplianceRegulatoryInformation/Post-marketingPhaseIVCommitments/). These studies are specifically designed to address outstanding safety question(s) that could have important public health implications. The FDA must agree with the study design and sets timelines for study completion and submission of the final study report(s). Sponsors must provide status updates on the study progress and ensure timely completion or potentially face civil penalties.

Postmarketing Safety Labeling Changes

New safety concerns may be uncovered after approval of a drug and can come from various sources, such as published literature, voluntary reports of adverse outcomes, new clinical trials, or reanalysis of previously analyzed safety data. When the FDA determines that new safety information about a serious risk should be included in drug labeling, it can require the drug sponsor(s) to update the labeling with the safety information. Both the FDA and the sponsor(s) must act promptly and work within short timelines to complete the evaluation of the new safety information and communicate the new safety risk in labeling to adequately inform prescribers and patients. Timelines may be further shortened if the FDA determines that the safety issue is a public health threat. Under certain circumstances, the FDA has authority to order a safety labeling change, for example, if the FDA determines that the sponsor's proposed labeling changes do not adequately address the new safety information (http://www.fda.gov/downloads/Drugs/GuidanceComplianceRegulatoryInformation/Guidances/UCM250783.pdf).

OTHER REGULATORY AGENCIES

Major counterparts of the FDA outside the United States include the European Medicines Agency (EMA) and Japan's Pharmaceuticals and Medical Devices Agency (PMDA) (http://www.pmda.go.jp/english; http://www.ema.europa.eu/ema/). The EMA provides a centralized process so that sponsors of eligible products are able to submit a single application to obtain marketing authorization that is valid throughout the European Union.

As discussed previously, regulatory authorities and pharmaceutical experts from Europe, Japan, and the United States established the ICH to harmonize requirements needed to obtain drug approval in these regions of the world and to reduce the need for duplicate testing during drug development. In addition, the FDA and other regulatory agencies have established memoranda of understanding and administrative processes for sharing findings, planned actions, and perspectives, when appropriate.

CONCLUSIONS

This is both an exciting and challenging time for medical product development and for management of medical products throughout their life cycle. On the one hand, there is a pipeline problem with high failure rates and increasing and substantial costs of development. The rapid advances in the basic sciences, however, have generated excitement over the possibility of new discoveries in the applied sciences, new targets for medical therapies, new drugs and biologics to prevent and treat human diseases, and widespread application of precision medicine to product development. These types of advances may improve efficacy and safety by allowing selection of individuals who are likely to respond to a drug or biologic and by identifying those who are likely to have an untoward effect from the medication. In addition, new regulations (such as breakthrough therapy designation) help expedite development and approval of important medical advances. Regardless of the times, public servants in agencies such as the FDA are constantly striving for ways to improve, streamline, and modernize the review process of medical products in all stages of development and continuously emphasize integrity, excellence, and the protection of the public health.

GLOSSARY

Advisory committee Advisory committees provide independent advice and recommendations to the FDA on scientific and technical matters related to the development and evaluation of products regulated by the Agency.

Biologics license application (BLA) A biologics license application is a submission to the FDA that contains comprehensive information on the manufacturing processes, chemistry, pharmacology, clinical pharmacology, and the medical effects of the biologic product. If the information meets FDA requirements, the application is approved and a license is issued allowing the company to market the product.

Complete Response letter A Complete Response letter is issued if the FDA determines that the marketing application cannot be approved in its present form. This letter describes the specific deficiencies preventing approvability and, when possible, the actions that the sponsor could take to resolve these deficiencies.

Expanded access program FDA's expanded access program makes promising experimental drugs available to patients not participating in clinical trials. This program provides patients with serious or life-threatening diseases who have no alternative treatment options (e.g., patients with an advanced form of cancer who have failed other therapies), with access to experimental drugs, not for the purpose of clinical trials but rather to diagnose, monitor, or treat the patient's medical condition.

Guidances Guidances are publically available documents written by the FDA that reflects the Agency's current thinking on a wide range of regulatory and scientific issues.

Investigational New Drug application (IND) Current Federal law requires that a drug be the subject of an approved marketing application before it is transported or distributed across state lines. Because a sponsor will probably want to ship the investigational drug to clinical investigators in many states, it must seek an exemption from this legal requirement. The IND is the means through which the sponsor technically obtains this exemption from the FDA.

New drug application (NDA) A new drug application is a submission to the FDA that contains comprehensive information on the manufacturing processes, chemistry, pharmacology, clinical pharmacology, and the medical effects of the drug product. If the information meets FDA requirements, the application is approved and the product may be marketed in the United States.

New molecular entity (NME) A new molecular entity is an active ingredient that has never before been marketed in the United States in any form.

Physicians' Labeling Rule (PLR) The Physician Labeling Rule (PLR) format is a revised format for the professional labeling of drugs and therapeutic biologics. This format includes a "Highlights" section (typically a half-page summary of the most crucial information for the safe and effective use of the drug), a table of contents, and a reordering of the full prescribing information so that prescribing information is more accessible to the health-care provider.

Postmarketing requirement (PMR) A postmarketing requirement is a study or clinical trial that the FDA requires the sponsor to conduct after the FDA has approved the drug for marketing or licensing.

Prescription Drug User Fee Act (PDUFA) Under PDUFA, sponsors pay user fees to the FDA when certain applications are submitted for review. These fees, together with federally appropriated funds, support the drug review process and improve procedures that make reviews more rigorous, consistent, and predictable. In exchange for the user fees, the FDA has agreed to meet performance goals during drug development and review, ensuring timely review of drug applications, while preserving the integrity of the review process, with the goal of assuring that safe and effective drugs are made available to the public in a timely manner.

Risk Evaluation and Mitigation Strategies (REMS) REMS are required risk management plans that use risk minimization strategies beyond the professional labeling to ensure that the benefits of certain prescription drugs outweigh their risks.

Special Protocol Assessment (SPA) A sponsor can request a special protocol assessment for animal carcinogenicity protocols, chemistry stability protocols, and phase 3 protocols intended to confirm efficacy. Under a special protocol assessment, the sponsor submits a detailed study protocol with specific questions that usually relate to study design or statistical analysis. The FDA sends written comments responding to the sponsor's questions within 45 days of receipt. Agreements reached on the SPA are binding for the FDA except in certain circumstances.

Sponsor A sponsor is an individual, pharmaceutical company, academic institution, or other organization that takes responsibility for and initiates a clinical investigation.

LIST OF ACRONYMS AND ABBREVIATIONS

BLA Biologics license application
CBER Center for Biologics Evaluation and Research
CDC Centers for Disease Control and Prevention
CDER Center for Drug Evaluation and Research
CDRH Center for Devices and Radiological Health
CFR Code of Federal Regulations
CFSAN Center for Food Safety and Applied Nutrition
CMS Centers for Medicare and Medicaid Services
CTP Center for Tobacco Products
CVM Center for Veterinary Medicine
DHHS Department of Health and Human Services
EMA European Medicines Agency
FDA Food and Drug Administration
FDAAA Food and Drug Administration Amendments Act
FDAMA Food and Drug Administration Modernization Act
FDASIA Food and Drug Administration Safety and Innovation Act
GDUFA Generic Drug User Fee Amendments
GLP Good Laboratory Practices
GMP Good Manufacturing Practices
ICH International Conference on Harmonization
IND Investigational new drug application
NDA New drug application
NIH National Institutes of Health
OND Office of New Drugs
PDUFA Prescription Drug User Fee Act
PLR Physician Labeling Rule
PMDA Japan's Pharmaceuticals and Medical Devices Agency
REMS Risk Evaluation and Mitigation Strategies
SPA Special Protocol Assessment

ACKNOWLEDGMENTS

We thank Julie Beitz, MD, for her critical review of this chapter. We also thank Mary Parks, MD, for her contributions to a prior version of this chapter used for an earlier edition of the book.

REFERENCES

Feinsod, M., Chambers, W.A., 2004. Trials and tribulations: a primer on successfully navigating the waters of the Food and Drug Administration. Ophthalmology 111, 1801—1806.

Cases

Warner-Lambert v. Heckler, 787 F2d 147 (3rd Cir. 1986).

Statutes and Regulations

CFR (Code of Federal Regulations) Title 21: Food and Drugs. Part 312 Investigational New Drug Application: 312.20, 312.23, 312.30—312.33, 312.38, 312.42, 312.44, 312.45, 312.50, 312.52, 312.53, 312.55—312.62, 312.64, 312.66, 312.68, 312.69. http://www.accessdata.fda.gov/scripts/cdrh/cfdocs/cfcfr/CFRSearch.cfm?CFRPart=312.

CFR (Code of Federal Regulations) Title 21: Food and Drugs. Part 314 Applications for FDA Approval to Market a New Drug: 314.50, 314.105, 314.110, 314.120, 314.126. http://www.accessdata.fda.gov/scripts/cdrh/cfdocs/cfcfr/CFRSearch.cfm?CFRPart=314.

Dietary Supplement Health and Education Act of 1994. http://www.fda.gov/regulatoryinformation/legislation/significantamendmentstothefdcact/ucm148003.htm.

Federal Food, Drug, and Cosmetic Act 1938. http://www.fda.gov/AboutFDA/WhatWeDo/History/ProductRegulation/ucm132818.htm.

Food and Drug Administration Modernization Act 1997. http://www.fda.gov/RegulatoryInformation/Legislation/SignificantAmendmentstotheFDCAct/FDAMA/default.htm.

Food and Drug Administration Amendments Act 2007. http://www.fda.gov/regulatoryinformation/legislation/significantamendmentstothefdcact/foodanddrugadministrationamendmentsactof2007/default.htm.

Food and Drug Administration Safety and Innovation Act 2012. http://www.fda.gov/regulatoryinformation/legislation/significantamendmentstothefdcact/fdasia/ucm20027187.htm.

Prescription Drug User Fee Act (PDUFA) 1992; PDUFA II 1997; PDUFA III 2002; PDUFA IV 2007; PDUFA V 2012. http://www.fda.gov/ForIndustry/UserFees/PrescriptionDrugUserFee/ucm144411.htm.

Pure Food and Drug Act 1906. http://www.fda.gov/AboutFDA/WhatWeDo/History/.

Online Sources

Center for Biologics Evaluation and Research. http://www.fda.gov/AboutFDA/CentersOffices/OfficeofMedicalProductsandTobacco/CBER/default.htm.

Center for Drug Evaluation and Research. http://www.fda.gov/AboutFDA/CentersOffices/OfficeofMedicalProductsandTobacco/CDER/default.htm.

Center for Drug Evaluation and Research (CDER) Drug and Biologic Approval and IND Activity Reports Center for Drug Evaluation and Research (CDER) Drug and Biologic Approval and IND Activity Reports. http://www.fda.gov/Drugs/DevelopmentApprovalProcess/HowDrugsareDevelopedandApproved/DrugandBiologicApprovalReports/default.htm.

Center for Drug Evaluation and Research (CDER) Guidance Page. http://www.fda.gov/Drugs/GuidanceComplianceRegulatoryInformation/Guidances/default.htm.

Center for Food Safety & Applied Nutrition. http://www.fda.gov/AboutFDA/CentersOffices/OfficeofFoods/CFSAN/default.htm.

Center for Tobacco Products. http://www.fda.gov/AboutFDA/CentersOffices/OfficeofMedicalProductsandTobacco/AbouttheCenterforTobaccoProducts/default.htm.

Center for Veterinary Medicine. http://www.fda.gov/AboutFDA/CentersOffices/OfficeofFoods/CVM/default.htm.

Critical Path Initiative. http://www.fda.gov/scienceresearch/specialtopics/criticalpathinitiative/default.htm.

Office of Scientific Investigations. http://www.fda.gov/AboutFDA/CentersOffices/OfficeofMedicalProductsandTobacco/CDER/ucm090085.htm.

European Medicines Agency. http://www.ema.europa.eu/ema.

Food and Drug Administration (FDA) Advisory Committees. http://www.fda.gov/advisorycommittees/default.htm.

Food and Drug Administration (FDA) History. http://www.fda.gov/aboutfda/whatwedo/history/default.htm.

Food and Drug Administration (FDA) Organization. http://www.fda.gov/AboutFDA/CentersOffices/OrganizationCharts/ucm393168.htm.

Food and Drug Administration (FDA) Organization Charts. http://www.fda.gov/AboutFDA/CentersOffices/default.htm.

Good Laboratory Practice Compliance Program. http://www.fda.gov/ICECI/Inspections/NonclinicalLaboratoriesInspectedunderGoodLaboratoryPractices/default.htm.

Guidance for Industry: Providing Clinical Evidence of Effectiveness for Human Drug and Biological Products. http://www.fda.gov/downloads/drugs/guidancecomplianceregulatoryinformation/guidances/ucm072008.pdf.

Guidance: The Special Protocol Assessment. http://www.fda.gov/downloads/drugs/guidancecomplianceregulatoryinformation/guidances/ucm080571.pdf.

International Conference on Harmonisation (ICH) Guidelines. http://www.ich.org.

Investigational New Drug (IND) Application. http://www.fda.gov/Drugs/DevelopmentApprovalProcess/HowDrugsareDevelopedandApproved/ApprovalApplications/InvestigationalNewDrugINDApplication/default.htm.

National Center for Toxicological Research. http://www.fda.gov/AboutFDA/CentersOffices/OC/OfficeofScientificandMedicalPrograms/NCTR/default.htm.

New Drug Application (NDA) Process. http://www.fda.gov/Drugs/DevelopmentApprovalProcess/HowDrugsareDevelopedandApproved/ApprovalApplications/NewDrugApplicationNDA/default.htm.

New Requirements for Prescribing Information. http://www.fda.gov/drugs/guidancecomplianceregulatoryinformation/lawsactsandrules/ucm084159.htm.

Office of Regulatory Affairs. http://www.fda.gov/AboutFDA/CentersOffices/OfficeofGlobalRegulatoryOperationsandPolicy/ORA/default.htm.

Pharmaceuticals and Medical Devices Agency (Japan). http://www.pmda.go.jp/english.

Postmarketing Study Requirements and Commitments. http://www.fda.gov/Drugs/GuidanceComplianceRegulatoryInformation/Post-marketingPhaseIV Commitments/ucm064633.htm.

Chapter 24

Ethical Issues in Translational Research and Clinical Investigation

Greg Koski[1,2]

[1]Harvard Medical School, Boston, MA, United States; [2]Massachusetts General Hospital, Boston, MA, United States

Chapter Outline

Key Points:

- The tools of science generate evidence as the basis for understanding.
- Science is not inherently ethical—the expectation that science be done in a socially responsible and ethical manner is imposed by society upon the scientific endeavor and those who engage in it.
- Over the past 25 years, concerns have grown that not all scientists are in fact as virtuous as once believed.
- The three principles of the Belmont Report are respect for persons, beneficence, and justice.
- Translational research, by its very nature, may explore questions that are fundamental to all living things and may lead to manipulations and modifications of those processes so critical to nature and biology that they create genuine concerns among scientists and society alike.
- Translational scientists should be sensitive to societal concerns and thoughtfully be guided by six guiding principles.
- The cardinal rule for all human research, whether stated in regulations or guidelines, is that all proposed studies must be reviewed and approved by a duly constitute ethical review committee prior to initiation of proposed studies, and that all changes to proposed research be reviewed and approved prior to implementation.
- It is expected that all investigators and other individuals engaged in human subjects research have a working knowledge of the principles underlying ethical conduct of human subjects research and the regulatory requirements adopted to ensure adequacy of protections for human subjects.

INTRODUCTION

There is a lot of talk about ethics these days—ethics in government, business, medicine, journalism, the workplace, society, and yes, science. Recognizing that many divergent views probably exist as to why ethical issues have garnered so much attention, one view is that just as many people rediscover religion when the prospect of having a life after death becomes germane, people turn to ethics when they sense that things are going wrong, that someone or some entity has done something it ought not to have done that may have harmed someone. One might argue well and hope that ethical considerations be prospectively addressed, but these are all too uncommon. Unfortunately, modern life seems so full of challenges and perceptions of wrongdoing, so an ethical revival is perhaps necessary and valuable.

Disclosure of behaviors that violate social norms of conduct and the values upon which those societal expectations are founded inevitably results in claims that the behaviors in question are *unethical*. What constitutes unethical behavior, like what it means to be pornographic, is often subjective—"I can't define it but I know it when I see it" is the retort one hears often, and it is not difficult to understand why.

Ethics is that branch of philosophy that deals with moral issues, including questions about what is right (or wrong) to do and other intangibles such as whether the *intentions* behind an action determine their goodness or whether the actual *outcome* is what that is important. Philosophers themselves debate whether there are any universal ethical rules, even when they do agree on a fundamental ethical *principle* such as justice. While ethics deals with moral issues, religion and government too do so, as all do in the course of each one's daily living.

In morality, as in ethics, right and wrong are not black and white; one person's sense of immoral or unethical conduct may be considered perfectly natural and acceptable in a different culture or set of circumstances. Sitting naked on a bench in Central Park is really not at all that different behaviorally from nude figure modeling for an art class, but one gets arrested for indecent exposure for the former, while the exquisiteness of human form is exalted in the latter. In our complex, multicultural world, ethical dilemmas are common and challenging in every field, biomedical research being no exception. Dr. Michael Grodin points out that "ethical dilemmas are difficult because they are difficult—*there are no easy ethical dilemmas*" (personal communication). The objective of this chapter is to provide a framework for ethical thinking and conduct in science, especially as applied to translational research and clinical investigation.

The Ethical Dimension of Scientific Inquiry

Science, as a discipline of inquiry, endeavors to understand or to better understand how the natural world works. The tools of science generate evidence as the basis for that understanding. These tools include observation of the natural world and/or behavior of living creatures and experimentation, a method of studying events under controlled conditions that can reduce variability and ambiguity of interpretation. Experimentation generally involves manipulation, an interaction that perturbs the natural order of things in a controlled way to see how the system responds. Analysis of data provides evidence that through interpretation and reasoning, scientists can draw conclusions about nature and the forces and processes that govern it. In essence, science is the pursuit and realization of knowledge about the natural world and the creatures that inhabit it. Science, while based in empiricism and founded on objectivity, is still very much a human endeavor that cannot be dissociated from its social and ethical context (Medewar, 1982).

It is entirely possible for science to be conducted rigorously without regard for the consequences that possession of knowledge or the processes of acquiring it may entail. Indeed, there are numerous examples of science being conducted without regard for its consequences on individuals or society, and there were no moral or ethical dimension to science, it is entirely likely that no one would be disturbed by such events in the least. But because science is a human endeavor that is carried out within a social context, society holds scientists accountable for their actions.

In the modern world, acquisition of knowledge is generally perceived as being good, unlike during the Dark Ages, when scientists were persecuted or even murdered for asking questions about things concerning which mortals had no reason or right to inquire. Over the centuries, science and religion have frequently been at odds; often, it is the juxtaposition of the two that brings the moral and ethical dimensions of science to the fore. Beyond that, secular society itself imposes certain expectations and limits on scientific inquiry, intentional or not. One might argue that society has a right to do so because the public is often called upon to enable the conduct of science, whether through the provision of resources or through direct participation as subjects of study.

While many believe that science should be conducted in the interests of society, no such prerequisite exists, nor a requirement that science be altruistic or even beneficial from a societal perspective. Simply put, *science is not inherently ethical*—the expectation that science be done in a socially responsible and ethical manner is imposed by society upon the scientific endeavor and those who engage in it. Given the special distinction between inanimate objects and the world of living things, special considerations arise in the life sciences, particularly when human beings are the subject of scientific inquiry.

The responsible scientist is always cognizant of the ethical dimension of science and its impact on society and is willing and able to internalize these societal expectations and normative behaviors.

RESPONSIBILITY IN SCIENCE AND SOCIETY

The pursuit of knowledge is generally considered to be a noble one, and scientists are considered to be, or at least expected to be, *virtuous* individuals. The concept of virtue, or more specifically civic virtue, was first popularized in ancient civilizations and took on renewed importance during the period of Enlightenment in the 18th century, which is the period of our development when the thirst for knowledge and discovery were probably unprecedented. Among scientists, certain attitudes, beliefs, and values have traditionally fostered virtuous behavior—hard work, perseverance, respect for others, collegiality, objectivity, and honesty— to name but a few. All of these contribute to the ethical dimensions of science, but in today's context, other factors seem to have acquired huge importance, namely, accountability and with it the concept of responsibility.

Responsibilities of Scientists

Today's scientist is generally a highly trained professional who acquires not only the necessary knowledge and skills to pursue the complex questions that arise, but also who understands and accepts the responsibility to ensure the validity and integrity of one's work, to ensure proper design and conduct of the research and to ensure that the data are properly recorded, analyzed, and reported.

Over the past 25 years, concerns have grown that not all scientists are in fact as virtuous as once believed, eliciting the promulgation of policies and regulations for oversight of the scientific process and investigation of alleged scientific misconduct. Some attribute this phenomenon at least in part to a growing entrepreneurial incentive in science, incentives that can be overpowering for some who find themselves facing competing and conflicting interests (Gross, 2001). These competing interests are not only financial in nature, but also may involve personal advancement, fame, politics, and can be of such scope and magnitude as to impact the future of our world as we know it—as the scientists of Los Alamos have already attested. Thus, the responsibilities of today's scientists involve not only how they may be impacted personally by the consequences of one's work, but also the broad impact and consequences of one's work for the world in which we live.

Societal Responsibilities

Although an ethics discussion is about what *should or ought to be done* in a given set of circumstances, more commonly it is a discussion of what was done and/or what should have been done. This distinction made by use of the past tense is nontrivial because scientific discovery can open a Pandora's box of both treasures and horrors. One can argue that an important role of public discourse on scientific questions is a careful consideration of the risks and benefits of acquiring knowledge. Some scientists counter that the public at large is not sufficiently well educated to comprehend the complexity of the techniques and questions scientists deal with today. There is probably truth on both sides, the take-home message being that a concerted effort be made by scientists and society, to remain informed and engaged, and that the potential knowledge and communication gaps between the lay public and the scientist be bridged appropriately. That the public be better informed is also nontrivial because it requires greater scientific literacy, something that does not develop passively. Our educational programs must include more effective instruction in science for the nonscientist if we are to avoid an intellectual dichotomy.

ETHICS AND TRANSLATIONAL RESEARCH

In its Belmont Report, the National Commission on Protection of Human Subjects in Biomedical and Behavioral Research (1979) reviewed and reaffirmed the ethical principles that should guide everyone engaged in research involving human subjects. These three principles, *respect for persons, beneficence, and justice* are the subjects of extensive writings regarding their origin, interpretation, and application, but none is succinctly or with greater wisdom and clarity than the original report.

Many ethicists say that the Belmont Report should be a required reading for every scientist involved in human research. Simply put, this recommendation is true but inadequate. Everyone involved in research with human subjects must do more than just reading the Belmont Report; they must have, at a minimum, an effective working knowledge of the principles identified and explained as a prerequisite for engaging in this endeavor. Even more importantly, these principles must be internalized. It is not only sufficient to know them—one must live by them. They provide the normative basis for the responsible scientist engaged in human subjects research, and any scientist unwilling or unable to be guided by them should not be permitted by society or his peers to participate in human research.

As mentioned earlier, one might as well add to these traditional principles another principle, that of *caring*. The ethics of care remind us that it is often necessary to subjugate one's own interests to those of another for whose interests and well-being one bears responsibility (Noddings, 1984).

Responsibility for the well-being of another individual is assumed in many types of care-giving relationships, including parenting, firefighting, nursing, medicine, and other professions. In these types of relationships, caring can be characterized as a social contract established by societal norms. Caring is a form of altruism, a personal character trait greatly prized when observed in others, but often difficult to achieve personally, particularly in situations where strong competing interests create ambivalence about the proper course of action.

Reconciling the entrepreneurial spirit so common in science today with a spirit of altruism is one of the great challenges facing scientists in both industry and academia, as evidenced by the vigorous discussions of conflicts of interest at every level of the scientific endeavor.

While the principles referenced above are certainly applicable to all clinical research, and while one might reasonably presume that they would also be appropriate for translational research, it is likely that they too are necessary but insufficient. Translational research, those critical studies in which advances made in the laboratory are first brought to bear in experiments performed on human beings, requires even more zealous attention to ethics than most clinical research, primarily because of uncertainty.

The death of the renowned explorer Sir Edmund Hillary almost a decade ago reminds us that while climbing Mt Everest will always be a monumental accomplishment accompanied by great risk, he who did it first faced far-greater risk because of the uncertainty about whether it could even be done. The translational scientist, whether exploring normal physiology, pathophysiology of disease, its diagnosis, prevention, or treatment, is akin to that first climber in some respects, but rarely he is the one who actually subjects to the associated risks—the risk is borne primarily by others: individuals, populations, or in the extreme, all of humankind.

Nuclear physicists Robert Oppenheimer and Hans Bethe, instrumental figures in development of the first atomic bomb, acknowledged the vexing uncertainty that accompanied the first detonation of a nuclear device in the atmosphere, including the prospect of actually igniting the atmosphere, starting combustion of nitrogen with oxygen, with potentially devastating immediate consequences, not to mention the long-term consequences for humanity (Broad, 2005). While not biomedical in nature, this was certainly an example of translational research, some would say of the very worst kind, because it translated scientific knowledge of the atom to the power of destruction. Although Oppenheimer and Bethe admitted to "no regrets" about having helped to achieve the technical success of creating the atomic bomb, they and some of their colleagues, as they watched the events of the Cold War unfold, expressed a sense of concern about the consequences of what they had done, collectively and individually, even if it was for what they believed at the time to be a good and necessary cause.

The translational biomedical scientist should heed and learn from this lesson. Fortunately, some have, as demonstrated by the Asilomar Conference on Recombinant DNA in 1975 (http://en.wikipedia.org/wiki/Asilomar_conference_on_recombinant_DNA), during which leading geneticists and molecular biologists voluntarily developed and adopted recommendations to forego certain types of genetic manipulation research until the potential risks, biohazards, and benefits were better understood (Berg et al., 1975). Today's ongoing debate within the scientific community and outright arguments among scientists, ethicists, religious leaders, governments, and others about human cloning illustrates the ongoing need for both dialog and restraint.

The recent scandal in South Korea, in which a renowned cellular biologist seemed so anxious to claim priority for the first successful cloning of a human that he would actually fabricate data for publication, is one of the more egregious examples of scientific misconduct, irresponsibility, and unethical behavior ever observed in the history of science (Hwang et al., 2005). That any scientist could so willingly disregard the norms of scientific and ethical conduct is most disturbing and gives everyone in science good cause to reevaluate the cultural and environmental factors that would drive a scientist to such lengths and permit him to succeed, even if that "success" was fraudulent and fleeting.

The extraordinarily powerful tools of cell biology, genomics, bioinformatics, nanotechnology, cybernetics, and functional brain imaging have opened some of the most important frontiers of biology to detailed inquiry and manipulation once believed to be the stuff of science fiction. Concurrently, society seems increasingly concerned that our readiness to deal with the consequences of exploration in these domains, be they environmental, social, or moral in nature, has not kept pace with our ability to ask questions. Albert Einstein once said that "Science without ethics is lame, and ethics without science is blind." To avoid being either blind or lame, science and ethics must walk hand-in-hand.

The rapidity of scientific and technological advancement since the Enlightenment has made it very difficult for ethics to keep pace, and the current public outcry to ban human cloning is just one modern-day example of the public anxiety and even fear that is bred of misunderstanding and uncertainty. The message here is that science must take care not to get too far out in front of public expectation and concern, even if that means slowing down in some areas of inquiry until a proper ethical framework, and where appropriate guidelines, regulations, and oversight mechanisms are in place to ensure safety and accountability.

Carol Levine's observation that our system for protection of human subjects of research was "born of abuse and reared in protectionism" underscores the reactive nature of a concerned public and the likely actions of policy-makers, a message that all translational scientists should listen to very carefully as the age of genomics and nanotechnology rolls on. One cannot doubt that failure of scientists to be sensitive to societal concerns about what they are doing will be met with not only resistance, but also with restrictions by law and regulation, neither of which is in the interests of either science or society.

GUIDING PRINCIPLES FOR THE RESPONSIBLE TRANSLATIONAL INVESTIGATOR

Translational research, by its very nature, may explore questions that are fundamental to all living things, and may lead to manipulations and modifications of those processes so critical to nature and biology that they create genuine concern among scientists and society alike.

As long as translational scientists behave in an ethical responsible manner, societal pressure, whether brought to bear through government regulation, funding restrictions, or radical activism (even violence, as by extremists of the animals rights movement), will be less likely to pose impediments to scientific inquiry and progress.

Accordingly, translational scientists should be sensitive to societal concerns and thoughtfully be guided by the following principles:

1. The questions asked and the approaches proposed to answer them should be soundly justifiable both ethically and scientifically to one's scientific peers as well as to a reasonable, well-informed public. The adage "you can please all of the people some of the time, and some of the people all of the time, but you can't please all of the people all of the time" is applicable to the scientific endeavor generally and most certainly to translational research. Translational scientists must be aware of and sensitive to the ethical, legal, and societal implications of their research—this means making the commitment and taking the time to acquire education and training as necessary to ensure full understanding and appreciation of the ethical, legal, and societal implications of one's scientific work.

2. A cardinal rule for guiding the behavior and activities of translational scientists is that no scientific studies that may seriously harm the environment, individuals, groups of individuals, or populations at large should be undertaken unless the risks are predictable and controllable. This requires that appropriate policies and procedures for reviewing, approving, overseeing, and monitoring research activities including mechanisms to detect and warn potential serious consequences, whether immediate, short-term or long-term, and safety procedures to contain any mishap that might occur. Rather than looking to ethicists, policy-makers, and legislators, translational scientists should take the lead in establishing the ethical framework and appropriate oversight infrastructure for responsible conduct of their research activities.

3. Particular caution must be exercised when consequences of the proposed research are not reversible. In such cases, no scientist should proceed without a broad-based consensus within the scientific community and the public at large that the risks are justified by the potential benefits of the research, fully considering the magnitude and likelihood of the associated risks. The pursuit of scientific knowledge cannot be justified solely by a desire to know without full consideration of the social and environmental context in which science is performed.

4. When proposed experiments will use living creatures as a means to advance science, the use of animal or human subjects should be justified and the subjects should be treated with dignity and respect. All reasonable efforts must be taken to prevent harm, injury, pain, and suffering in a manner consistent with all applicable ethical principles and good research and clinical practices. Abuse of living creatures in the name of science is unacceptable and intolerable.

5. Knowledge acquired through scientific studies should be shared and used in a manner that optimizes its utility and benefit for all. Fabrication or falsification of data, misrepresentation of results, knowing disregard for objectivity, withholding of information, and failure to conduct research in concert with all applicable ethical principles undermines the integrity of science and trust in the scientific endeavor. Openness, transparency, and honesty are as essential to good science as rigor of experimental design, responsible conduct, and full objective reporting of results.

6. Scientists must be willing to be personally accountable for the quality and integrity of their work and its consequences. By adopting a culture of conscience in science, each scientist affirms a personal commitment to responsible conduct research and accountability for its consequences, including appropriate compensation for harm done to others, directly or indirectly, or to the environment.

Beyond Ethics: Regulations, Compliance, and Professionalism in Translational Research

Far too often, discussions of ethics in research degenerate into discussions of regulations and compliance. This is unfortunate and misguided, because there is a very fundamental difference between ethics and law.

Laws are established by societies to place limits on unacceptable behaviors, telling members of the society what may or may not be done, and failure to comply with the law results in penalties and sanctions for the transgressor. Ethics, on the other hand, guides individuals toward what should or ought to be done in specific situations based upon the principles and values deemed important to society.

There may be times when an unethical behavior is legal and when an illegal behavior is ethical, but in general, the responsible translational scientist will be working in a domain that encompasses both ethical guidelines and legal requirements. While compliance with both is expected and appropriate, unethical conduct shreds the very fabric of integrity in science, and compliance with laws and regulations cannot and should not be the primary motivation for responsible scientific conduct. Science, as a profession, should be responsible for establishing and enforcing codes of conduct for its practitioners. That it should be necessary to regulate science, that is, to require scientists to behave responsibly and to set limits for a scientific community apparently unable or unwilling to restrain itself, is an affront to the dignity of the profession and all who take part in it.

Existing regulations pertain to management of grants and contracts, use of animals and humans as research subjects, and reporting and management of competing and conflicting interests in science, and for dealing with alleged scientific misconduct, narrowly defined as fabrication, falsification, and plagiarism. The intent of the regulations is largely to require operational procedures for identification and management of important issues in each of these areas, and the guidance issued by federal funding and oversight agencies are intended to establish the norms of conduct expected of institutions and their scientists. Thus, the regulations and guidance serve both an educational and normative function.

Commonly and regrettably, the requirements set forth in regulation are viewed by the scientific community as ethical guidelines, as if compliance with the regulations equates with ethical conduct. In reality, the requirements set forth in regulation are minimum requirements, not the goals to which responsible and professional scientists should aspire.

Until recently, education and training of scientists woefully neglected nontechnical issues, including ethics, the social context of research, and responsible conduct of science. In the 1990s programs for training in these areas were appended to training programs for scientists, generally requiring a few hours of training done on a weekend or evening as a condition of funding for the training program. Young scientists attended these sessions, often reluctantly, only to hear an apologetic introduction from some well-intended but equally reluctant research administrator promising to make these sessions relevant and interesting, as if they were not expected to be. Attendee evaluations of these sessions often expressed a notion of being "not as useless as I had expected," and sometimes indicating that "I actually learned something relevant," as if against all odds, ethical considerations were actually relevant to one's education and training.

This "add-on approach" to educating young scientists may reflect a sense of not needing to do so, based on a belief that the traditional system of education through mentoring would suffice, or it may reflect the reality of prioritization within the scientific community. Recent surveys of scientists at various stages of their career development demonstrate that sensitivity to these kinds of issues may actually wane as one progresses up the academic ladder, suggesting to some that the environment in which science is conducted today is toxic—that entrepreneurial drive and competitiveness, the need to get ahead and finish first, and to benefit personally from one's discoveries have overtaken the sense of collegiality, honesty, openness, integrity, and altruism once believed to prevail among scientists.

Needless to say, mentoring alone was never and will never be sufficient, and the need to fully incorporate training in ethical, legal, and societal issues into the core training of young scientists is fully recognized. Previously, established training programs for physician scientists (e.g., NIH's funded K30 and the Clinical and Translational Science Award programs) created at leading biomedical research centers and academic hospitals throughout the United States, and a new initiative for integrated training in responsible conduct of research at our nation's graduate schools demonstrate a resurgence of interests in rigorously training translational scientists in all dimensions of the scientific endeavor.

Increasingly, education and training of scientists focus on achieving *competency*—an expectation that every individual emerges from a training and educational program has mastered the skills and knowledge to be a competent scientist. This concept has evolved to included both cultural competency and ethical competency within the overall construct and rightfully so. Importantly, cultural and ethical competency is an intrinsic expectation that the scientist's education, training, and professionalism reflect today's growing recognition and acceptance of the social engagement and responsibility of science. In the end, the goodness of science and its benefits for mankind will only be as good as the people who enter the profession and the values and characteristics they bring to their work. Creating an environment that fosters a culture of conscience in science will do more to ensure integrity and professionalism among scientists than any training programs we offer, no matter how rigorous, it is not what one knows that determines how one behaves, it is what one believes.

JUSTICE, BENEFICENCE, AND RESPECT FOR PERSONS: FROM PRINCIPLES TO PRACTICE

The fundamental ethical principles upon which responsible conduct of human studies is based require that all members of the research team, and particularly investigators, give the highest possible priority to ensuring the well-being of their subjects of study. While this point derives naturally from the principles of respect, justice, and beneficence, the trend today is to couch the expectation within the mantel of "patient centricity." While the concept of patient centricity is obviously intended to put the interests of patient-subjects at the very center of the research agenda, the concept is too frequently contorted by asking what not research can do to be more respectful and sensitive to the needs and priorities of patient-subjects, but rather, how can we change our behaviors as scientists to enhance the probability of patient-subjects participating in research. To paraphrase President John F. Kennedy's famous line, science must "ask not what our patient-subjects do for us, but what can we do for our patient-subjects."

In medicine, the notion that health-care providers and organizations should be accountable to those for whom they have the responsibility for care has given rise to the concept of accountable care organizations. Recently, this paradigm has been extended to the clinical research realm as the concept of *accountable research*—that the entire research endeavor should be accountable to patient-subjects for responsible professional conduct of all research activities. Encompassed with the concept are expectations not only of ethical conduct and competency, but also of compliance, financial, scientific, and social accountability.

Translating from principles to actions, the investigator is obligated to incorporate into the design and conduct of his or her research specific provisions to minimize the likelihood and magnitude of harm, be it physical, emotional, psychological, or economic in nature. Protections for privacy are often paramount and very challenging, particularly when translational studies based on genetic information are considered. At present, protections against genetic discrimination are inadequate and simply having knowledge of one's genetic predisposition to disease can impair one's ability to acquire insurance, to be gainfully employed, and negatively impact one's relationships with family and friends. Passage of the Genetic Information Nondiscrimination Act in the United States (National Human Genome Research Institute, 2008) was a positive step, but much remains to be done. Accordingly, an investigator must take pains to ensure that all medical and health information, and especially genetic information, is properly used and protected. Investigators who value their ability to do science using genetic information ought to consider exercising their social responsibility by encouraging their elected officials to pass important legislation that protect individuals from genetic discrimination before abuses occur that could result in a public outcry for stringent restrictions on this type of scientific inquiry.

Research Risks, Harm, and Injury

At best, risk assessment in translational research is highly subjective and inexact, especially when humans are being exposed to compounds or biologic agents for the first time, and the full spectrum of potential adverse events is largely unknown, as is commonly the case for immunological agents and other modifiers of biological responses. A case in the United Kingdom, commonly referred to as the TeGenero case (Batty, 2006; Citizens for Responsible Care and Research, 2008), exemplifies better than any other the way *not* to conduct critical first-in-human trials of potent biological agents.

When planning a protocol, the responsible investigator considers several potential approaches and then weighs the potential benefits of each design against the potential risks. This evaluation should be heavily weighted toward safety until such time that enough is known about the compound and its biological effects. In the TeGenero example, subjects received injections of the active compound at fixed intervals less than an hour apart, even though the pharmacological half-life and the duration of the potential biological response was considerably longer. This ill-conceived and irresponsibly executed study evoked severe reactions that nearly killed some of the normal healthy volunteers and caused irreversible injury in others. The study elicited a formal UK government inquiry (Department of Health, 2006) that revealed failures at every step of the research process, including not only the design and execution of the study, but the oversight processes at the levels of the regulatory agencies and the ethical review committees.

In all translational research, safety must be the primary consideration, and as stated both in the Declaration of Helsinki (DoH) and the International Commission of Harmonization Good Clinical Practice guidelines, the interests of science and society must never be given priority over the interests and safety of the research volunteers.

Benefits

True, direct benefits in early translational research are rare and commonly overstated. The nature of research is one of the inquiries, not treatment, and care should be taken not to conflate the two. A treatment is a proven, safe, and effective approach to amelioration of a pathological condition. Research is the process of finding out whether or not a novel, potentially therapeutic approach is safe and effective. Everyone involved in clinical and translational research must understand and respect the difference. Participation in research may be beneficial to some subjects, but often those benefits are subjective or slow in coming, particularly in translational studies where understanding physiology or pathophysiology of a disease process is the pretext for the research, with only the potential for tangible benefits for individuals at some future, and often, remote stage of the research progression.

So-called "gene therapy" is a case in point. While the ultimate hope and goal for research that involves gene transfer or manipulation may be the emergence of a safe and effective therapeutic approach, to call the research a therapy is misleading, even dishonest and disrespectful to study participants. Sponsors and investigators must be careful not to overestimate benefits, or to minimize risks in relation to potential benefits, when presenting their proposed studies to either review boards or prospective subjects. Doing so undermines confidence in both the investigator and the proposed research, especially if something untoward happens. A direct, honest, and realistic representation of what is known, what is not, what good might come from the work, and what harm might occur is undoubtedly the best policy and practice for the research community and the public upon whom the research depends for support and participation.

Special Populations

Translational research is frequently directed toward understanding conditions and diseases that have not been amenable to effective diagnosis, prevention, and treatment. Patients and potential subjects afflicted by such conditions are vulnerable by virtue of their lack of options. The patient with recurrent neuroblastoma multiforme has few, if any, options for a dreaded and malignant condition, the treatment of which continues to elude modern medicine despite decades of intensive research. The prospect of raising false hopes in such a condition is high, and care must be taken by investigators and sponsors not to do so. Similarly, patients with dementias or other conditions that impair decision-making capacity ought to be afforded the highest measure of respect and protection in every aspect of their involvement in a research study.

Indeed, any study that seeks to enroll individuals under conditions that limit their ability to render a free, well-informed, and voluntary decision whether or not to participate must be conducted with sensitivity and caution to their special situations and need. Among these potential subjects are children of all ages, pregnant women and their fetuses, prisoners, students, soldiers, the elderly, and the critically ill. When encountering these situations in the course of one's research, the responsible investigator will find that seeking broad counsel from knowledgeable individuals outside of the research domain will inform and strengthen the approaches developed to deal with these special populations respectfully during recruitment, enrollment, and actual conduct of the study, while at the same time reducing the likelihood of harm and the perception of exploitation of vulnerable individuals for the sake of science and society. The translational investigator must never forget that the regulation and oversight of science as we know it today is predicated on abuses, atrocities, and irresponsible behaviors of fellow scientists—we need not be doomed to repeat history.

One can well argue that all patient-subjects are special, of course, by virtue of their willingness to undertake risks of personal harm for little if any direct benefit, as enablers in the pursuit of knowledge and potential new therapies. All in the scientific community will be well served by frequent reminders of this fact—patient-subjects are more than a resource—they are the essential means by which translational science can advance and must be respected as partners in the research endeavor.

Issues in Collection of DNA and Other Biospecimens for Research

In what has been called the "omics age," biological specimens collected from normal individuals and patients with pathological conditions are extremely valuable as sources of information and as sources of research materials, including proteins, lipids, and other complex molecules, most importantly, DNA.

These specimens are powerful research tools, and like all things powerful, the consequences of misuse are at least equally great, perhaps even greater. While biological specimens have been collected extensively and stored in vast repositories for decades, the more recently acquired ability to apply the profoundly powerful analytical tools of modern bioinformatics in a way that increasingly is able to identify individuals and families from whom specimens originated is at once promising and problematic, particularly in a society so driven by economic considerations.

Whether a scientist, institution, or company seeks to make money through discovery and innovation of new products or an insurance company seeks to improve its profit margin by controlling risk exposure, the potential for violations of privacy, psychosocial and economic harm, and discrimination are very real concerns when personal genetic information is accessible. Existing approaches to prevention of such harms are not really up to the task at hand. Laudable efforts by some institutions, such as the National Heart, Lung, and Blood Institute (NHLBI) of the National Institutes of Health (Nabel, 2008), to develop capability to widely share genetic information to promote science have emphasized development of policies and procedures to protect privacy, to respect the wishes and intent of the specimen donors, and to sanction scientists who behave irresponsibly. To date, the approach of NHLBI has garnered broad support among scientists and research participants, and public acceptance. The NHLBI approach is based on a shared belief that the intentions of those who will use the information made available will do so in an ethical and responsible manner. In the eyes of many, NHLBI is the goose that laid this golden egg. While the goose is not likely to be killed, just one irresponsible act could easily break the egg to the detriment of all.

Translational scientists bear a responsibility to science and society to engage in continuing discussions about the sharing and use of personal information in the omics age, and should take the lead in insuring that individuals are not harmed as a result of sharing their personal information. For its part, government should act now to adopt broad protections against genetic discrimination in all aspects of life, for the day when every living individual can be genetically identified is no longer that far-off. Already, individuals may be reluctant to allow analysis of their genetic information due to fears that they could lose insurance, employment, or be subject to familial or social stigmatization. These fears are real and justifiable, and unless they can be properly addressed in a timely manner, they will multiply much to the detriment of scientific inquiry.

REGULATION OF RESEARCH AND PROTECTION OF SUBJECTS

That it should be necessary to protect human subjects and society from scientists is a peculiar concept. Protection from criminals seems appropriate, as does the protection of the homeland. Regulation of research and a requirement for ethical review and protection of human subjects are relatively recent developments in the world of science and an unflattering commentary on the scientific endeavor as it has evolved. Not all nations have gone so far as to actually pass laws and statutes to control research, and the growing trend toward using science to further political and ideological agendas is as disturbing to many as using political and ideological viewpoints to control science. That said, regulations to protect human subjects in research are now widespread and likely to become the norm around the world. Every clinical and translational investigator must, of course, know and comply with all applicable regulations, but one's real responsibility is to appreciate and internalize the ethical foundations upon which these regulatory frameworks are built.

Research and Regulations in the United States

In the United States, which probably has the longest-standing and most comprehensive framework for regulation and oversight of research in the world, laws have been passed and policies adopted to protect research integrity from scientific misconduct, ensure proper management of grants and contracts, to protect animals from cruelty and abuse, and protect human research subjects and their privacy.

Responsibility for protection of human subjects is a shared responsibility among all members of the clinical research team, and all can and should be held accountable for meeting those responsibilities. The responsibilities of the investigators and sponsors have been discussed earlier. To ensure that investigators and sponsors are fulfilling their responsibilities in this regard, a system of review and oversight was created by laws that vested specific responsibilities in review committees dubbed Institutional Review Boards (IRBs) in the regulations, but more appropriately designated Research Ethics Review Boards (RERBs). Another generic term commonly used outside the United States is "ethics committee," and in Canada, the preferred term is Research Ethics Board (REB).

Whatever they are called, their charge is the same: to prospectively and on an ongoing basis review proposed research involving human subjects; to insure that the science is sound; that the risks are reasonable in relation to the potential benefits; that the provisions for protecting the safety, interests, and well-being of participants are adequate and appropriate; that the work is conducted in compliance with all applicable ethical principles and regulatory requirements; and that informed consent is properly obtained and documented. RERBs are granted authority under the regulations in the

United States to approve, disapprove, or require modification in proposed research as a condition of approval. They are also granted very broad authority by institutions to take whatever actions are deemed necessary to minimize risks and ensure the safety and well-being of human research subjects.

In the United States, these RERBs are subjected to two basic sets of regulations, the first promulgated under the Food, Drug, and Cosmetic Act known as Title 21, Parts 50 and 56 (21 CFR 50, 56). These are applicable to all research involving human participants conducted under the authority of the Food and Drug Administration. These regulations apply to all studies associated with the testing and development of new drugs, devices, and biological agents subjected to the rules of interstate commerce (see Chapter 23 for additional details).

The FDA regulations are largely identical to a second set of regulations promulgated under an entirely separate authorizing statute, the Public Health Service Act. These regulations are officially known as the Federal Policy for Protection of Human Subjects in Research, more commonly called the Common Rule. The Common Rule resulted from adoption by 17 Federal agencies of the core of the regulations originally established by the Department of Health, Education, and Welfare (HEW), now the Department of Health and Human Services (HHS). Each agency encoded Subpart A of the HHS regulations (45 CFR 46) into its own agency-specific regulations where it resides today. HHS went on to create additional sections pertaining to special protections for vulnerable populations including pregnant women and fetuses (Subpart B), prisoners (Subpart C), and children (Subpart D). A few of the federal agencies have adopted in whole or in part some, or all of the subparts of the HHS regulations, creating an interesting and confusing environment rife with various interpretations and applications of their provisions. Several agencies, such as the Department of Defense, the Department of Energy, and the Department of Education maintain their own office to oversee the functions of IRBs, but much of the oversight falls to the HHS Office for Human Research Protections (OHRP) and the FDA.

Under the prevailing regulations, federally supported research is subject to the Common Rule and any specific provisions adopted by the individual funding agencies. Most nonfederally supported human research is subject to the FDA regulations except for privately supported research that is not intended to create products for interstate commerce, and therefore, is subject to no oversight at all.

Some research falls under both sets of regulations, importantly, most investigator-initiated, federally supported research of new drugs, biologics, or medical devices, all frequent targets of translational research.

Of note, concerns about the adequacy and appropriateness of the regulatory framework for human research in the United States have intensified in recent years. A strong sense that the current approach has become excessively bureaucratic and compliance-focused is now pervasive, even at the level of the federal agencies charged with oversight and enforcement of the regulations. These concerns provoked calls for reform that were heard in Washington, resulting in issuance of an Advanced Notice of Proposed Rulemaking (ANPRM) in 2011 (http://www.gpo.gov/fdsys/pkg/FR-2011-07-26/pdf/2011-18792.pdf) by the OHRP and the Office of Science and Technology Policy.

The tentativeness of the proposed reforms (seven specific actions were proposed, accompanied by more than 70 questions seeking public guidance on their possible implementation) reflected the many challenges facing any meaningful reform of the existing regulatory framework. Perhaps because the proposals focused more on reducing perceived regulatory and administrative burdens on scientists than on strengthening protections for human subjects, they generated great controversy and even a degree of polarization within the scientific and ethics communities which are yet to be resolved. Some, including this author, argued that the protectionist approach embodied in the existing framework nearly a half-century ago is no longer appropriate or effective in the changing world of science today, and that consideration be given to wholesale revision of the current approach, replacing it with a model based more upon professionalism. The ANPRM evoked significant public comment and discussion, and as of this writing, no further action has been taken. Concerns and frustrations among members of the scientific, ethics, and regulatory communities continue to grow while meaningful reform is debated.

The Role of the Sponsors

Generally speaking, the legal responsibilities of sponsors are detailed in regulations related to product development, such as FDA regulations and the European Clinical Trials Directives. Virtually every corporation engaged in development and marketing of biomedical products is well aware of these legal requirements, as failure to adhere to them rigorously will undoubtedly impede the approval and launch of a new product, and the associated costs are enormous. When clinical and translational investigators are working with or for corporate sponsors, most of these responsibilities are assumed by the company, although investigators are required to make several commitments and disclosures. It is critical that every investigator participating in a clinical trial subjects to FDA oversight carefully read and fully understand the terms of agreement set forth in the FDA Form 1572 (see Box 24.1).

BOX 24.1 Commitments and Responsibilities of a Clinical Trial Investigator as Detailed in Food and Drug Administration Form 1572

- I agree to conduct the study (or studies) in accordance with the relevant, current protocol(s) and will only make changes in a protocol after notifying the sponsor, except when necessary to protect the safety, rights, or welfare of subjects.
- I agree to personally conduct or supervise the described investigation(s).
- I agree to inform any patients, or any persons used as controls, that the drugs are being used for investigational purposes and I will ensure that the requirements relating to obtaining informed consent in 21 CFR Part 50 and Institutional Review Board (IRB) review and approval in 21 CFR Part 56 are met.
- I agree to report to the sponsor adverse experiences that occur in the course of the investigation(s) in accordance with 21 CFR 312.64.

- I have read and understood the information in the investigator's brochure, including the potential risks and side effects of the drug.
- I agree to ensure that all associates, colleagues, and employees assisting in the conduct of the study (or studies) are informed about their obligations in meeting the above commitments.
- I agree to maintain adequate and accurate records in accordance with 21 CFR 312.62 and to make those records available for inspection in accordance with 21 CFR 312.68.
- I will ensure that an IRB that complies with the requirements of 21 CFR Part 56 will be responsible for the initial and continuing review and approval of the clinical investigation...
- I agree to comply with all other requirements regarding the obligations of clinical investigators and all other pertinent requirements in 21 CFR Part 312.

Even more critical for the translational investigator working without a corporate sponsor is the need to understand and adhere to the requirements for special exemptions for investigational new drugs, biologics, and devices, commonly known as INDs and IDEs. No investigational agents or devices are permitted to be administered to human subjects or entered into interstate commerce (shipped across state lines) with an exemption from these restrictions. What many independent investigators do not fully appreciate is the magnitude and seriousness of the responsibilities they assume when they step into the role of sponsor and holder of an IND or IDE (see Chapter 23 for additional details).

Any investigator who ventures into this area should study the relevant portions of the FDA regulations, set forth in Title 21 of the Code of Federal Regulations, specifically Parts 312 (drugs) and 812 (devices). Institutions employing investigators engaged in such work should require those investigators to undertake special training to ensure that they fully understand their responsibilities before the work begins. Requiring all translational and clinical investigators to demonstrate their preparedness for conducting this type of research by successfully passing a rigorous certification examination is one way to ensure that the qualifications stipulated by the FDA are met and to ensure the quality of the research and well-being of research participants.

Nongovernmental Agencies and Associations

Dozens of nongovernmental, nonprofit organizations exist around the world that conduct, promote, and oversee responsible clinical research, such as the World Health Organization, International Conference on Harmonization (ICH), and the International Clinical Epidemiology Network. There are also many professional organizations that offer guidance to investigators engaged in clinical research, such as the DoH and the Council of International Organizations in Medical Sciences and the World Medical Association. Some of these agencies and organizations have issued guidelines that are internationally recognized, including the DoH and the ICH Good Clinical Practice (GCP) guidelines. Every clinical investigator should be well-versed in the content of these guidelines, as they are considered international standards for responsible conduct of clinical research. So should investigators fully understand and comply with the uniform standards for preparation and submission of manuscripts established by the International Committee of Medical Journal Editors (ICMJE, 2007). These guidelines require that all clinical research studies be duly registered in a recognized clinical trial registry.

Some investigators complain that the array of regulations and guidelines that have multiplied in recent years have become a serious impediment to their research particularly when regulations are not consistent from one agency to another as illustrated earlier. While this may even be the case, we must recognize that the proliferation of guidelines and requirements is, at least in part, a reaction to events that have occurred in the past, events indicating that some scientists cannot, or at least will not, properly conduct and report their research when left to their own devices.

Some of the guidelines, such as the ICH-GCP guidelines (available at http://www.pdacortex.com/ICH_GCP_Download.htm), are actually proactive efforts to harmonize and standardize procedures for conducting research on new drugs and devices to facilitate their approval by oversight agencies, in Europe referred to as competent authorities. Scientists and sponsors of their work would be well advised to invest more time and effort in faithfully conducting their activities in concert with all applicable guidelines requirements, lest they find themselves confronted with even more. However, it is also the responsibility of various regulatory agencies within a government and across governments to commit to harmonizing and streamlining the regulations and guidelines related to human research and to actually accomplish this goal.

Two caveats are important to stress. First, when scientists and sponsors demonstrate a commitment to conducting their activities responsibly, pressure for further regulation diminishes. Second, bureaucracies, particularly government bureaucracies, are not very adaptable to change and are very protective of their presumed area of authority. Thus it is difficult to modify them when the purpose for which they were established has diminished. The old adage "an ounce of prevention is worth a pound of cure" certainly applies.

Public Perception of Clinical Research

Not surprisingly, the public's perception of clinical research is probably as varied as the many populations and cultures of the world. In the United States, surveys indicate strong public support for research of all kinds, and particularly for translational and clinical research. This support, however, is qualified by concerns about misconduct, and in view of the dramatic drop in public confidence in the pharmaceutical industry and the FDA in recent years, public support for clinical research declines when associated with corporate sponsorship.

In other parts of the world, perceptions and attitudes may differ. For example, in Japan, the citizenry has not been interested in participating in clinical research. Thus, clinical studies and trials involving Japanese subjects are far fewer than those involving North American or European subjects. In most of Europe and Asia, support for clinical research is strong among those who see the endeavor as a profitable business venture. In recent years, several US-based international companies have identified opportunities to conduct research in developing countries with "ascending markets," and the practice of "offshoring" clinical trials to other countries has led to dramatic growth in clinical trials outside the United States, particularly in Eastern Europe, Asia, and South America.

There has also been a dramatic surge of clinical research in Africa, where tropical diseases and HIV infection are both endemic and epidemic. Vast differences in culture and economics between non-African sponsors and investigators, and the people of many poor African nations give rise to concerns of exploitation and concerns substantiated by several actual cases.

Still, despite the problems that have occurred, research offers hope for better health and better lives, and it indeed seems that such hope springs eternal. As discussed earlier, scientists conduct their work within a social context and are subject to public scrutiny and skepticism unless a foundation of trust and respect is well established. Toward this end, scientists and their sponsors should heed the advice of Alexander Pope, "What conscience dictates to be done, or warns me not to do, this teach me more than hell to shun, that more than heaven pursue." To retain public confidence, the responsible scientist must of course comply with regulations, but more importantly, be driven by his conscience.

Protection of Human Subjects

Responsible investigators view protections for the safety and well-being of their subjects as more of a responsibility, rather than a regulatory requirement. Accordingly, the policies and procedures that have been implemented to ensure that the interests of human subjects are properly protected should be respected as a valuable component of a culture of safety across the research domain.

Review and Oversight

The review and approval process for proposed research should be one of reaffirmation that the investigator and sponsor have fulfilled their responsibilities to their subjects. Well-designed studies, proposed by well-qualified and responsible investigators, with appropriate consideration of ethical principles, risk mitigation strategies, and respectful approaches to recruitment and enrollment of subjects are rarely challenged. Good science done well is everyone's goal.

Oversight of clinical research should not be viewed as a police function although often it is both by the "regulator" and the investigator. Effective oversight is a mechanism to ensure safety, objectivity, and integrity of the research. The policies and procedures that have been adopted almost universally for protection of human subjects are part of the critical infrastructure for conducting human research safely, much like the international air transportation safety system. However, in contrast to the air transportation system where uniform and consistent rules have been established across the entire world, this is still a goal to be achieved both among countries and regulatory agencies within a country.

Institutional Review Boards and Ethics Committees

Recognized by various names around the world, these committees are responsible for ensuring that the interests and well-being of research participants are properly attended before research is initiated. The cardinal rule for all human research, whether stated in regulations or guidelines, is that all proposed studies must be reviewed and approved by a duly constitute ethical review committee prior to initiation of proposed studies, and that all changes to proposed research be reviewed and approved prior to implementation. Failure to adhere faithfully to these two fundamental practices demonstrates that an investigator either does not understand or is unwilling to accept his responsibilities to his subjects.

IRBs and ethics committees (ECs), as they are commonly referred to in much of the world, should be comprised of individuals with varied backgrounds and expertise appropriate to the research under consideration so that they may exercise collective wisdom in their reviews and judgments. Committees should call upon experts in specific fields for consultation when necessary expertise is lacking among the regular membership of the committee. Review committees must conduct their activities in a manner that engenders respect from the research community, just as investigators must conduct their activities responsibly. When ECs and IRBs are idiosyncratic, arbitrary, or irresponsible, they lose the respect of the scientists who depend upon them. IRBs and ECs also have a responsibility to conduct their activities with appropriate efficiency and rigor, and institutions have a responsibility to ensure that they have the resources and support to do so. The review and approval process is not an administrative process—it is part of the foundation of responsibly conducted human research. It is not unusual for different IRBs to have different concepts of what is "ethical." Sometimes this is based on lack of knowledge concerning the particular disease entity. However, sometimes it represents a true difference of deeply held moral principles that differ from one culture to another. Because the IRB is supposed to represent the ethical opinion of the community in which the research is being conducted, varying approval standards are likely to be present. This fact is particularly relevant for international studies. For example, in some areas of the world, genetic data cannot be collected no matter the safeguards. Thus the investigator embarking on a multicenter study needs to consider these potential obstacles when designing the study.

Data and Safety Monitoring Boards, Medical Monitors, and Subject Advocates

IRBs and ECs appropriately engage in prospective review and approval of human research, but they generally lack the ability and resources to provide ongoing, real-time monitoring of safety during the actual conduct of research, and many argue that it is not appropriate for them to do so, as they may also lack the necessary expertise. Still, the need for this type of monitoring exists, and it must be objective. Toward this end, a well-designed study will include provisions for boards to monitor study data as they accumulate, medical monitors to observe actual procedures and research tests to ensure safety of participants, and subject advocates, when appropriate, to offer participants support and a stronger voice to insure that their interests are fairly and appropriately considered throughout a study, and particularly to ensure the objectivity of the voluntary decision-making and consent process.

Many investigators and sponsors overlook the need for these additional safety procedures and personnel during the design phase of their research. Were they to build them into their study designs before submitting protocols for review, the approval process would likely become far less onerous and more efficient. Guidelines for establishing data monitoring committees and safety monitoring boards are available to investigators and sponsors. It is their responsibility to know these guidelines and to use them in designing and conducting their studies.

Medicolegal Framework, Liability, and Risk Management

The often heard retort that "Anyone can sue for anything" has been the prevailing approach to medicolegal and liability issues in clinical and translational research. Tragically, there have been serious injuries and deaths, sometimes easily preventable ones that are directly attributable to poorly designed and improperly conducted research. Not surprisingly, such events often result in lawsuits, and such lawsuits are generally not favorable to the public's perception of the scientific endeavor, not to ignore the research participants who were harmed in the process.

It is trite but true that the best way to avoid lawsuits in this setting is to do the work responsibly in the first place. Here, application of good risk management strategies is very useful, and IRB review, data monitoring committees, medical monitors, and subject advocates should be part of this process, but no one is more critical than the investigator and members of his/her clinical research team.

Properly designed human research should include provisions for care and compensation in the event of research-related injury. In most of the world, there is no requirement for such coverage or compensation, and concern about this issue is growing. Translational research by its very nature may carry an inherently higher level of risk than other types of clinical studies, and investigators and sponsors bear responsibility not only to minimize the likelihood of injury, but also to properly care for and compensate subjects when they are harmed as a consequence of participation in research.

Most commonly, assignment and acceptance of this responsibility has been vague at best. Federal regulations require only that potential subjects be informed whether or not care and/or compensation for injury will be provided, not who will provide it, how much will be provided, or for how long. Frequently, investigators, government and foundation sponsors, and even corporate sponsors reserve the right to bill a subject's own medical insurance company for the cost of care required after a research-related injury. This approach seems to ignore the moral obligation imputed to those who conduct research to accept responsibility for the consequences of their actions. Corporations, but not governments or foundations, generally provide insurance coverage to protect them from all sorts of liabilities, and providing such insurance in the case of research-related injury would seem to be an ethically responsible and reasonable cost of doing business. Calls for greater clarity and responsibility are becoming louder and more widespread. This is an issue that needs and deserves urgent attention and action from the research community, government, and industry. It is likely that if governments establish a policy in this area where they are the funding entity, a similar approach would be followed by other sponsors.

INDIVIDUALS AND THE CLINICAL RESEARCH PROCESS

Clinical and translational research take place in a very complex environment in which a cast of players have very specific roles, each requiring varying levels of training and expertise. These individuals also have their own personal motivation for engaging in clinical research and the entire endeavor is dependent upon the willingness of patients and normal healthy individuals to volunteer as research subjects.

Motivation and Expectations of the Volunteer Subject

Surveys of research participants reveal several motivations for participation: these include altruism, a desire to do something to help others; a desire to improve diagnosis, treatment, or prevention of a condition affecting oneself or a loved one; access to potential new treatments, even if yet unproven to be safe and effective, but offering hope when little remains; or simply as a way to earn a bit of money through direct compensation for participation. There are likely others.

Whatever one's motivation for volunteering to participate in a translational or clinical research study might be, volunteers enter the relationship with certain expectations. Among these are an expectation that their safety will be considered a priority, that they will be treated with respect, including respect for their privacy; that discomforts, risks, and inconvenience will be minimized; that they will be properly informed before, during, and after the study about the research objectives and its outcome; and that they will be appropriately compensated or reimbursed for their contributions to the research. None of these are unreasonable expectations, and the research team and sponsors should strive to meet them.

"Patient centricity" is currently in vogue within the clinical research community. Presumably, this means that those engaged in designing and conducting clinical and translational research will do so in a manner that focuses more upon meeting the needs and serving the interests of patient-subject than those of the investigators and sponsors, In many instances, however, the focus of current "patient-centric" initiatives is on what can be done to encourage participation in clinical trials. Better communications, greater sensitivity to the inconveniences experienced by subjects, realistic scheduling of study visits, and other efforts to improve the relationship between subjects and the research team are no doubt warranted and can be very helpful, but to be truly patient centric, their motivation must be more than to provide incentives to induce the subjects to serve the needs of the investigator.

Achieving true "patient centricity" in clinical research requires that those who intend to do research involving human subjects engage patients and volunteers as true partners. The value of such partnership was amply demonstrated decades ago when AIDS activists and patient groups cooperated to become a community force behind the design and execution of clinical trials of potential HIV therapies. Another manifestation of patient-focused research is commonly referred to as community-based participatory research. Investigators who have embraced this approach have recognized that community members and subjects can provide valuable input into study design and execution to facilitate research that might otherwise be difficult or even impossible.

Motivation of the Scientist and Conflicts of Interest

Surely, translational and clinical scientists are motivated by a desire to better understand physiology and disease processes, as already discussed, but those are not the sole motivations. Just as their research is conducted in a societal context, so too are individual scientists motivated by considerations other than a mere quest for knowledge and understanding. It is only natural that scientists would seek some measure of recognition for their contributions and appropriate compensation for their efforts. These are very legitimate interests, as are their interests in being promoted academically, being respected and even honored by their colleagues, or reaping the financial rewards that sometimes accompany discovery and innovation.

Immersed in this environment with multiple motivating factors, competition among them is inevitable. At times, two interests can be so divergent as to be in direct conflict. A scientist engaging in translational and clinical research has a primary responsibility to the well-being of research participants and integrity of the science itself; these should never be in conflict. These primary responsibilities may be in competition with secondary interests, and an individual who finds that his or her secondary interests may compromise their primary interest or responsibility is conflicted.

Conflicts of interest can pose both legal and ethical dilemmas, and when an investigator's conflicted interests could contribute to harm of a research subject or the integrity of one's science, great care must be taken to properly disclose and manage that conflict. Whenever possible, conflicts of interest should be eliminated. When they cannot be eliminated without compromising the primary interest, they must be managed in a manner that minimizes the likelihood that the conflict could do harm, and this will generally involve independent and external oversight of the investigator's activities.

In clinical and translational research, the operational guideline that has been recommended (Association of American Medical Colleges, 2001, 2002) is that of a "rebuttable presumption" that conflicts of interest should be eliminated and that they should be tolerated only when there is a compelling reason, such as improving safety in an early device trial. Institutions have moved aggressively in the last decade to formulate and implement policies and procedures for identifying, disclosing, reporting, and managing conflicts of interest in science.

Not just investigators, but IRBs, institutions and sponsors are all subject to conflicts of interests. For example, a competitor evaluating a research application likely is conflicted and should excuse himself from the review process. Whenever a party to the clinical research process fails to responsibly address potential conflicts openly, there is a great potential for harm to the entire endeavor. To preserve the integrity of science and public respect for the clinical research process, there is probably no more important step to be taken that to insulate to the fullest extent possible the research process from other motivating factors, including financial incentives (Cohen, 2001). However, doing so calls for a level of individual integrity and sometimes sacrifice that can be difficult to realize in a world so fiercely competitive and economically driven.

An important development in the United States is the recent passage and implementation of legislation (the so-called "Sunshine Act") that requires the pharmaceutical and medical device industry to publically report all gifts, contributions, and funds given to physicians, including those involved in research. It is too early to know the full impact of this legislation, but some warn of a chilling effect on relationships between academia and industry. While some argue that disclosure of legitimate financial relationships ought not to pose problems for either party, others express concern that the public is not likely to understand and may misinterpret such financial relationships that will certainly be scrutinized more closely than ever before.

PROFESSIONALISM IN CLINICAL RESEARCH

Increasingly the concept of professionalism is entering the realm of clinical research. The concept of professionalism is hardly novel, but what it really means in relation to research has not been as clearly articulated as one might expect. To paraphrase Justice Louis Brandeis, a profession is characterized by demonstrated mastery of a specific body of knowledge, as opposed to mere skills; engagement in the endeavor primarily for the benefit of others, rather than oneself; a willingness to engage in self-regulation according to standards established by the profession itself; and the measure of success is not financial.

Professionalism, in the context of clinical research, is an emerging paradigm. For far too long, individuals have been permitted to participate as investigators and members of the clinical research team without formal training and without demonstrated mastery of the knowledge base specific to the endeavor. Over the past decade, an expectation has emerged that all investigators and other individuals engaged in human subjects research have a working knowledge of the principles underlying ethical conduct of human subjects research and the regulatory requirements adopted to ensure adequacy of protections for human subjects. Certainly this is not an unreasonable expectation, and it is now a requirement at virtually every academic center in the United States, but not uniformly around the world. Every trainee in basic and translational research is similarly expected and generally required to undertake training is the principle of responsible conduct of research. Today the traditional apprentice model for training young clinical investigators, a model common to skilled trades rather than professional endeavors, is being replaced by rigorous training programs for translational investigators in all aspects of what might be considered an emerging discipline known as *pharmaceutical medicine.*

Over the past several years, examination-based certification of clinical investigators has been introduced by several professional organizations to complement existing professional certification programs for other members of the *patient-oriented research* team, including clinical research coordinators, research monitors, IRB professionals, and research administrators. Accreditation, another validation tool of professionalism, is now available for human research protection programs and research sites, and such accreditation is rapidly becoming the gold standard for such programs not only in the United States, but also around the world.

Recognition of the value and importance of the professional paradigm seems to be rapidly accelerating. A multi-stakeholder working group recently published its recommendations for competency-based education, training, and certification of clinical research professionals. Concurrently, standards and processes for accreditation of clinical research sites have been recommended by a broad-based working group convened by the Institute of Medicine in the United States, and these are currently being developed as part of an effort to build a global system for clinical research with the support of industry and academia. In the United States, the National Board of Registration in Medicine, which administers the globally recognized United States Medical Licensing Examination to evaluate the knowledge base of physicians prior to mandatory licensing, recently initiated a program to similarly assess the education and training of physician investigators and clinical research monitors. Whether professional recognition of clinical and translational investigators by medical specialty boards will ultimately emerge remains to be seen, but there is certainly a trend in that direction. Opportunities and even requirements for rigorous education and training of investigators are not limited to the United States; interest in certification and accreditation for clinical research is growing, particularly in Japan, China, Korea, Canada, India, and elsewhere, where government, academic, and industry leaders acknowledge the benefits these tools can provide.

This emergence of the professional paradigm may bode well for the future of clinical research, as it offers not only an opportunity to raise the standards of training and performance across the board, but also provides firm footing for a more desirable and viable career pathway in clinical and translational research, a pathway that is greatly needed if we are to build a strong workforce to carry on this critically important endeavor in the future.

REFERENCES

Association of American Medical Colleges, 2001. Protecting Subjects, Preserving Trust, Promoting Progress: Policy and Guidelines for the Oversight of Individual Financial Interests in Human Subjects Research. http://www.aamc.org/members/coitf/firstreport.pdf.

Association of American Medical Colleges, 2002. Protecting Subjects, Preserving Trust, Promoting Progress II: Principles and Recommendations for Oversight of an Institution's Financial Interests in Human Subjects Research. http://www.aamc.org/members/coitf/2002coireport.pdf.

Batty, D., 2006. Drug Trials Q & A. http://www.guardian.co.uk/society/2006/aug/02/health.medicineandhealth.

Berg, P., Baltimore, D., Brenner, S., Roblin III, R.O., Singer, M.F., 1975. Summary statement of the Asilomar conference on recombinant DNA molecules. Proc. Natl. Acad. Sci. 72 (6), 1981–1984.

Broad, W.J., March 8, 2005. Hans Bethe, Prober of Sunlight and Atomic Energy, Dies at 98. The New York Times. http://www.nytimes.com/2005/03/08/science/08bethe.html?pagewanted5print&position5.

Citizens for Responsible Care and Research. TeGenero AG TGN1412 Clinical Trial. http://www.circare.org/foia5/tgn1412.htm.

Cohen, J.J., 2001. Trust us to make a difference: ensuring public confidence in the integrity of clinical research. Acad. Med. 76, 209–214.

Department of Health, 2006. Final Report of the Expert Scientific Group on Phase One Clinical Trials (Chairman: Professor Gordon W. Duff). http://www.dh.gov.uk/en/Publicationsandstatistics/Publications/PublicationsPolicyAndGuidance/DH_063117.

Gross, C.P., 2001. Financial conflict of interest and medical research: beware the medical-industrial complex. J. Phil. Sci. Law. http://www6.miami.edu/ethics/jpsl/archives/newsedit/gross.html.

Hwang, W.S., Roh, S.I., Lee, B.C., Kang, S.K., Kwon, D.K., Kim, S., Kim, S.J., Park, S.W., Kwon, H.S., Lee, C.K., Lee, J.B., Kim, J.M., Ahn, C., Paek, S.H., Chang, S.S., Koo, J.J., Yoon, H.S., Hwang, J.H., Hwang, Y.Y., Park, Y.S., Oh, S.K., Kim, H.S., Park, J.H., Moon, S.Y., Schatten, G., 2005. Retraction of article originally published in *Science Express* on 19 May 2005. Science 308 (5729), 1777–1783.

ICMJE (International Committee of Medical Journal Editors), 2007. Uniform Requirements for Manuscripts Submitted to Biomedical Journals: Writing and Editing for Biomedical Publication. http://www.icmje.org/index.html.

Medewar, P.B., 1982. Pluto's Republic. Oxford University Press, Oxford.

Nabel, E., 2008. Testimony before the House Subcommittee on Labor-HHS-Education Appropriations. http://www.nhlbi.nih.gov/directorspage/index.php?page5congressionaltestimony.

National Commission of Protection of Human Subjects in Biomedical and Behavioral Research, 1979. The Belmont Report: Ethical Principles and Guidelines for Protection of Human Subjects of Research. http://www.hhs.gov/ohrp/humansubjects/guidance/belmont.htm.

National Human Genome Research Institute, 2008. President Bush Signs Genetic Information Nondiscrimination Act of 2008. http://www.genome.gov/24519851.

Noddings, N., 1984. Caring, a Feminine Approach to Ethics and Moral Education. University of California Press, Berkeley, CA.

Chapter 25

Clinical Research in the Public Eye

Mary Woolley

Research!America, Alexandria, VA, United States

Chapter Outline

Key Points

- Appreciating the value of research does not come naturally to the nonscientist and that means it does not come naturally to most decision-makers. In order to bridge this gap, scientists have an obligation to engage.
- Patients are taking stronger and stronger roles in driving attention to and support for research and are increasingly vocal about perceived lack of urgency on the part of researchers and institutions.
- Patient-centered research can only be worthy of the description if patients are part of the design and evaluation, as well as being volunteer participants.
- Though Americans generally support clinical research and recognize its importance, they are typically not well informed about the process, and many say that lack of transparency and trust are reasons they do not participate.

INTRODUCTION

This chapter examines the public and political contexts in which clinical research takes place, and the role the science community plays in shaping public and policymaker discourse and decision-making. Gaining an understanding of the links between science and the body politic, including the increasing demands for transparency and accountability, is fundamental to the long-term success of science.

Public opinion surveys consistently demonstrate that health is among Americans' top domestic concerns, ranking among the top 10 and often ranking above issues such as education, taxes, immigration, terrorism, and national security (Brown, 2013; Bloomberg, 2014; Kaiser, 2014). This should come as no surprise; health, for its own sake and as a necessary component of educational and economic success, has long been a leading concern for people everywhere. Intense debate over cost and access to health-care delivery has only intensified public concern. Indeed, public opinion surveys demonstrate that health-care costs rank as the nation's top health concern (Research!America, 2013a). In the same study, about half of respondents said research is a way to help control health spending, with a considerable additional percentage saying they are "not sure". This is just a sampling of the public and political context of research, and especially clinical research, examined in this chapter.

Thanks to research-based advances in public health and medicine, death rates in the United States steadily declined in the 20th century (Centers for Disease Control and Prevention, 1999), but researchers now warn that continuation of this trend is not assured. For example, absent effective intervention to reverse current obesity trends, the high prevalence of obesity observed among younger people today is likely to be exceeded in future generations (Olshansky et al., 2005). "Despite differences of opinion on how the high prevalence of obesity and other adverse health conditions among U.S. children will affect longevity, scientists generally agree that these liabilities will increase the burden of various chronic diseases that children will face in the future. Public health interventions could attenuate some or all of the concern, or the problem could grow worse" (Reither et al., 2011). Increasing Americans' life expectancy and indeed global healthy life expectancy will demand not only a continuous stream of breakthroughs in medical and health research that can cure or treat diseases of aging (Olshansky et al., 2001) but in addition will require putting what we know to work more effectively in the health-care delivery setting. "The increased complexity of health care requires a sustainable system that gets the right care to the right people when they need it, and then captures the results for improvement. The nation needs a health-care system that learns" (Institute of Medicine, 2011).

In the 20th century, concerns about health translated into a supportive policy and funding environment for medical research and public health advances in the United States. Public demand and expectations were met in many ways—consider the polio vaccine, the ability to manage diabetes and heart disease, and later HIV/AIDS, as well as the defeat of many childhood cancers. Indeed, a new era in human longevity was ushered in; average life expectancy is now 77 years, an increase of nearly 17 years since 1930 (Sierra and Kohanski, 2013). Thanks in large part to a sustained commitment to federally funded research; many formerly devastating scourges have been contained or eliminated (NIH, 2014) and have fueled economic growth. The Human Genome Project is among the noteworthy examples of the return on investment in basic research. Between 1988 and 2012, the human genome sequencing projects, associated research, and industry activity, which cost the taxpayer $3.8 billion, generated an economic impact of $796 billion from 1988 to 2003 with personal income exceeding $244 billion (Battelle Memorial Institute, 2011).

But times have changed. The early years of the 21st century have demonstrated that we can no longer take for granted that our elected representatives will consistently invest dollars in research. They have a long list of domestic public priorities to address in the face of fiscal constraints not known since the 1930s, compounded by numerous international challenges, and modern communications expose policymakers to the same news—often confusing, sometimes disturbing—about the conduct and outcomes of research as for everyone else. Watching research in almost "real time" exposes us all—much more often than in past decades—to the three steps forward, one step back, nonlinear, often contradictory scientific process. Appreciating the value of research does not come naturally to the nonscientist and that means it does not come naturally to most decision-makers, and thanks to that "real-time exposure," there are many more questions raised about research than in the past. In order to bridge this gap, scientists have an obligation to engage.

So much that is valuable and good in our nation's medical and scientific research is underappreciated by the public and its elected representatives. Some of this is due to the very nature of research, which can take an indeterminate length of time and involve many setbacks as competing hypotheses are explored. In addition, many developments take place outside public view, in the prepublication demesnes of academia or in the proprietarily protected laboratories of the private sector. It is difficult for the nonscientist, including public decision-makers, to appreciate the years and even decades of work, and the resources that work requires to bring a new medication to approval, and even harder to appreciate why a new discovery takes on average 17 years to reach the patient (Morris et al., 2011).

The United States provides the best training in the sciences and remains, for now, dominant in the global pharmaceutical and device industries. But, neither is an immutable fact of life. While the United States leads the world in research and development spending, at current rates China is expected to assume the top spot in or before 2022 (Battelle Memorial Institute, 2013). Research and development (R&D) spending in the life sciences in the United States was projected to grow slightly in 2014, a rebound over 2013 levels with the growth primarily from smaller biopharmaceutical innovators and medical device manufacturers driving medical progress (Battelle Memorial Institute, 2013), but 45% of researchers are pessimistic about the longer-term future of their budgets, compared with 33% who were more optimistic (Battelle Memorial Institute, 2013). Much of the uncertainty surrounding the pace of private sector R&D investments has to do with the translation of R&D into new or improved products that reach the market, an "uneven endeavor throughout industry" (Battelle Memorial Institute, 2013). Many industry executives report having only some or very limited success in moving a new discovery to the bedside (Battelle Memorial Institute, 2013).

Total federal R&D funding has declined by $24 billion in nominal terms since 2010, partly as a consequence of budget sequestration, which forced major spending cuts in 2013 (Wren, 2014). If current trends continue, it will be difficult for the United States to maintain its accustomed leadership "edge," since for the first time in more than half a century the United States is faced with determined and increasingly viable competition for international leadership in higher education, research, and innovation. Several other countries are mounting an explicit challenge taking a page from the US playbook in assigning an increasing percentage of gross domestic product (GDP) to R&D. Among those nations on the move are China, Germany, the United Kingdom, Japan, and South Korea (Battelle Memorial Institute, 2013).

More and more clinical research is being conducted in other nations as the cost—in time and money—of its conduct in the United States becomes more unwieldy and other nations offer all the quality of high-impact research without an onerous regulatory burden. There is a great deal of "disturbance in the force" of clinical research in the United States today.

US leadership at the global level will continue only if the public and its policymakers demand it, and even if the demand surfaces and response follows, it will be difficult for the United States to maintain its accustomed leadership "edge," since for the first time in several generations, America faces explicit and viable competition for international leadership in higher education and innovation, with many countries mounting a robust challenge.

THE CULTURAL CONTEXT OF RESEARCH

The Lens of Health and Health Care

Although health care in the United States has made progress in recent generations, the need for further improvement is evident. The 2013 annual report from the Agency for Healthcare Research and Quality (AHRQ) revealed stagnant or worsening perceptions by patients regarding access and disparities within health care (Agency for Healthcare Research and Quality, 2014). Increasing concerns about the cost of health care, as previously noted, are an important additional contextual concern. In addition to its main focus on accessibility of health insurance, the much-publicized Affordable Care Act (ACA), health care reform law, planted the seeds for change in how the system operates. Among these new initiatives is an increased focus on comparative effectiveness research, the study of what treatment options work best in a given circumstance. Funding under the ACA established the Patient-Centered Outcomes Research Institute (PCORI), which is designed to support comparative effectiveness research and similar avenues of discovery (Kaiser, 2013). A portion of ACA funding is allocated to the Department of Health and Human Services (DHSS) and AHRQ to support in the dissemination of PCORI's research findings (Government Accountability Office, 2011).

The process of making the health-care system more transparent is underway, with hospitals now required to publish data on certain outcomes (Centers for Medicare & Medicaid Services, 2014), and the Affordable Care Act making hospitals more accountable for quality through measures such as the prevention of avoidable readmissions. Hospital leaders now seem to hold favorable attitudes toward efforts to make these kinds of data public, with about 75% agreeing that public reporting of such data is a good idea (Lindendauer et al., 2014). Increased pressure to make all clinical research data public is palpable; the era of transparency is upon us.

Shifting Power to the Patient

Patients—all of us at some point—view clinical research, if we think of it at all, through the lens of our own health and health-care experience.

Patients are taking stronger and stronger roles in driving attention to and support for research and are increasingly vocal about perceived lack of urgency on the part of researchers and institutions. Patients and other public activists are having increased impact (Michaels et al., 2012; Shattuck et al., 2011). Perhaps the greatest success story of recent generations is advocacy for AIDS research, which in the 1980s and early 1990s led to significantly increased research conducted through the auspices of the National Institutes of Health (NIH) and to the Food and Drug Administration (FDA) fast-tracking treatments that ultimately saved lives and forged true partnerships between patients, advocates, researchers, and regulators in designing, conducting, and evaluating research. The fact that HIV/AIDS is now a manageable chronic illness in developed nations is testimony to this partnership.

Another landmark example of advocacy affecting policy change is the women's health movement, initiated in the late 1980s and carried forward by such organizations as the Society for Women's Health Research. Before advocates spoke out, the differences between men and women were largely ignored in research, leading to missed and misdiagnosed heart disease and other illnesses that develop and progress differently in women than in men

(Women's Health Resource Center, 2001). Advocacy by patients and advocacy organizations played a role in a decision by the FDA in 2014 to display information on the percentage of female and minority participation in clinical trials for medications intended for all populations. The Society said the data validate the group's long-standing concerns that the number of women and minorities included in research is not sufficient for clinical relevance (FDA, 2014a,b).

Over the objections of many in the science community, but with increasingly strong support from Congress, members of the public are now more involved than ever in funding decisions about research. Once the way was cleared by demand and then statute, patients became more powerful and more involved in federally supported research funding and policy decisions. This is a welcome development in several ways. The adage "nothing about me without me" is attributed to the disability community, but it is an apt description of the importance of designing and implementing clinical research protocols with the end user in mind. Failure to engage the patient throughout the research process is not only increasingly viewed as patronizing and elitist, but also slows down research since protocols that are tone-deaf to the patient experience are not likely to succeed. Compliance will drop off; volunteers will be harder to recruit; researchers will be discouraged, and resources will be ill spent. New cures, treatments, and preventions will be longer in the making. Patient-centered research can only be worthy of the description if patients are part of the design and evaluation, as well as being volunteer subjects.

Public Input and National Institutes of Health

In 1997, Congress asked the Institute of Medicine to review the research priority-setting process at the NIH, the largest single funder of health research in the United States, specifically evaluating:

- Factors and criteria used by NIH to make funding allocations;
- The process by which funding decisions are made;
- Mechanisms for public input; and
- The impact of congressional statutory directives on funding decisions.

Following the release of *Scientific Opportunities and Public Needs* (Committee on the NIH Research Priority-Setting Process, 1998) the NIH Director's Council of Public Representatives (COPR) was established by then NIH Director Harold Varmus, MD, "to facilitate interactions between NIH and the general public" (Director's Council of Public Representatives, n.d.). Among COPR's influential reports, *Report and Recommendations on Public trust in Clinical Research* notes a public perception that clinical trial researchers "tend to disregard the perspective of the community and the public at large." It recommended "change in the culture of the scientific community to ensure that medical research is viewed in the context of a long-term commitment to the community, not a one-time research study" (Director's Council of Public Representatives, 2005).

An earlier COPR report, *Enhancing Public Input and Transparency in the National Institutes of Health Research Priority-Setting Process*, begins by quoting Dr. Elias Zerhouni, NIH Director from 2002 to 2008, on the importance of public involvement in the nation's research agenda: "Engaging the public is a major priority, it is a national priority, it is not an option" (Director's Council of Public Representatives, 2004). Commitment to public engagement is an ongoing feature of the NIH research environment. "We need to be tireless in making people aware of why this is a very important investment in our nation," said NIH Director Francis Collins in a 2014 lecture at Massachusetts Institute of Technology (Trafton, 2014).

Public Perception of Research

Research!America regularly commissions surveys of the public to learn more about how Americans perceive various aspects of research, including clinical research. Though Americans generally support clinical research and recognize its importance, they are typically not well informed about the process and many say that lack of transparency and trust are reasons they do not participate (Research!America, 2013b).

- Fifty percent say they would recommend clinical trial participation to their family and friends, and more than two-thirds believe taking part in a trial is as valuable to our health-care system as giving blood (Figs. 25.1 and 25.2).
- Only 16% of those surveyed say that they or a family member has participated in a clinical trial (Fig. 25.3; Research!America, 2013b).
- More than two-thirds (72%) of Americans say it is likely they would participate in a clinical trial if recommended by their doctor (Fig. 25.4), but only 22% say a doctor or other health-care professional has ever talked to them about medical research (Fig. 25.5; Research!America, 2013b).

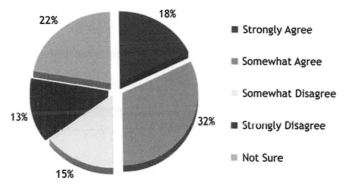

Level of agreement - I would recommend taking part in a clinical trial to my friends and family.

■ Strongly Agree

■ Somewhat Agree

▪ Somewhat Disagree

■ Strongly Disagree

▪ Not Sure

FIGURE 25.1 Clinical trial recommendations. *Research!America, 2013b. National Poll: Clinical Research. Conducted in partnership with Zogby Analytics. Available from: www.researchamerica.org/uploads/June2013clinicaltrials.pdf.*

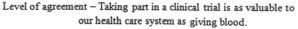

Level of agreement – Taking part in a clinical trial is as valuable to our health care system as giving blood.

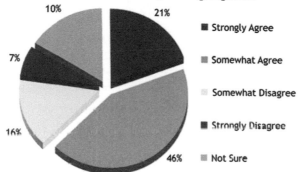

■ Strongly Agree

■ Somewhat Agree

▪ Somewhat Disagree

■ Strongly Disagree

▪ Not Sure

FIGURE 25.2 Value of clinical trials to health-care system. *Research!America, 2013b. National Poll: Clinical Research. Conducted in partnership with Zogby Analytics. Available from: www.researchamerica.org/uploads/June2013clinicaltrials.pdf.*

Have you or anyone in your family ever participated in clinical trials?

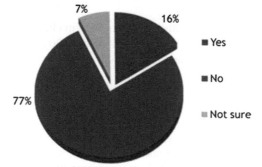

▪ Yes

▪ No

▪ Not sure

FIGURE 25.3 Level of participation in clinical research. *Research!America, 2013b. National Poll: Clinical Research. Conducted in partnership with Zogby Analytics. Available from: www.researchamerica.org/uploads/June2013clinicaltrials.pdf.*

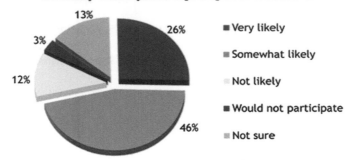

FIGURE 25.4 Doctor recommendations influence clinical trial participation. *Research!America, 2013b. National Poll: Clinical Research. Conducted in partnership with Zogby Analytics. Available from: www.researchamerica.org/uploads/June2013clinicaltrials.pdf.*

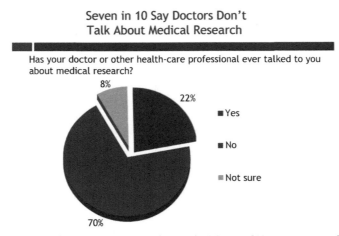

FIGURE 25.5 7 in 10 Americans say doctors don't talk about medical research. *A Research!America survey of U.S. adults conducted in partnership with Zogby Analytics in May 2013.*

- Fifty three percent have heard about clinical trials online, and thirty nine percent have heard of them through the media. Only 24% had heard about a clinical trial from their health-care provider (Fig. 25.6; Research!America, 2013b), despite a doctor being the most preferred source of clinical trial information among 60% of respondents (Fig. 25.7; Research!America, 2013b).
- It is concerning that 31% believe patients are sometimes included in clinical trials when they are receiving medical treatment without their consent, and another 28% are not sure (Fig. 25.8; Research!America, 2013b).
- Thirty seven percent say that they admire "a great deal" those who took part in clinical trials, compared with 69% for organ donors and 61% for blood donors (Fig. 25.9; Research!America, 2013b).
- Also of concern is that only 45% say clinical trial participants are treated fairly and with respect "all the time" by investigators; another 30% believe participants are treated that way "occasionally" (Fig. 25.10; Research!America, 2013b).
- Respondents are about evenly divided as to whether those enrolled in clinical trials receive better medical care, do not receive better care, or do not know (Fig. 25.11; Research!America, 2013b).
- More than half (53%) say lack of information and lack of trust are reasons why individuals do not participate in clinical trials (Fig. 25.12; Research!America, 2013b).

Privacy Issues

Health and research are not exempt from the nation's growing concern over privacy issues (Boulos et al., 2009). Prompted in part by consumer worries over reports of employer and insurer discrimination, Congress passed the Health Insurance Portability and Accountability Act (HIPAA) in 1996. This act increases the security of patient information by providing federal protections for individually identifiable health information and gives patients rights with respect to their own

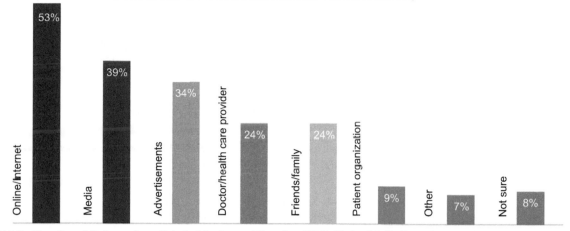

FIGURE 25.6   How the public hears about clinical trials. *Research!America, 2013b. National Poll: Clinical Research. Conducted in partnership with Zogby Analytics. Available from: www.researchamerica.org/uploads/June2013clinicaltrials.pdf.*

Physicians, Internet are Preferred Sources of Info on Clinical Trials

Where would you go to get information about clinical trials? (multiple responses allowed)

My doctor	60%
Online	57%
Hospitals	27%
Pharmaceutical companies	25%
Patient organizations	18%
Government institutions	16%
Foundations/charities	16%
Friends or family	14%
Other	>1%
Not sure	15%

FIGURE 25.7   Majority of Americans say they would go to their doctor or to the internet for clinical trial information. *A Research!America survey of US adults conducted in partnership with Zogby Analytics in May 2013.*

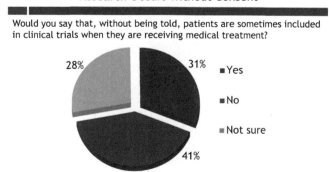

Opinions Split on Whether Clinical Research Occurs without Consent

Would you say that, without being told, patients are sometimes included in clinical trials when they are receiving medical treatment?

28%   31%   ▪Yes

▪No

▪Not sure

41%

FIGURE 25.8   Opinions of Americans are split on whether patients are included in clinical trials without their consent. *A Research!America survey of US adults conducted in partnership with Zogby Analytics in May 2013.*

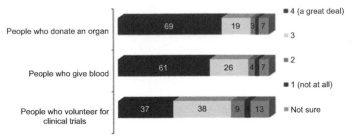

FIGURE 25.9 Admiration for clinical trial participants. *Research!America, 2013b. National Poll: Clinical Research. Conducted in partnership with Zogby Analytics. Available from: www.researchamerica.org/uploads/June2013clinicaltrials.pdf.*

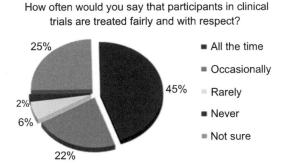

FIGURE 25.10 Treatment of clinical trial participants. *Research!America, 2013b. National Poll: Clinical Research. Conducted in partnership with Zogby Analytics. Available from: www.researchamerica.org/uploads/June2013clinicaltrials.pdf.*

information (Department of Health and Human Services, 2014). While the Department of Health and Human Services emphasizes balance within this law that allows for disclosure of information when appropriate for patient treatment and other purposes, HIPAA has drawn criticism from academic medical centers, which claim it imposes unnecessary barriers to research at their facilities—the nation's major research institutions—thus slowing the pace of research, increasing costs, and significantly hindering participation of individuals in important research studies (Steinberg et al., 2009). As use of digital patient records increases in the future, this issue will demand greater attention in order to strike a balance between privacy and confidential sharing of information to support medical progress.

Advocates backed proactive legislation that became the Genetic Information Nondiscrimination Act (GINA) (Genetics Alliance et al., 2010; HR Res 493, 2008). GINA, signed into law in 2008, is intended to protect individuals from discrimination based on their genetic information as it relates to employment and health insurance, while paving the way for more personalized health care based on advances in genetics and genomic medicine, which can be used to better detect, diagnose, and treat disease. The law allows people to speak honestly about their family history and empowers them to undergo genetic tests that may reveal a health risk or problem. Because of GINA's consumer protections, health insurance

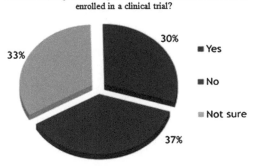

FIGURE 25.11 Level of medical care received in clinical trial. *Research!America, 2013b. National Poll: Clinical Research. Conducted in partnership with Zogby Analytics. Available from: www.researchamerica.org/uploads/June2013clinicaltrials.pdf.*

companies cannot deny or alter coverage because of test results, nor can employers use genetic information to adversely affect an employee's standing or career growth (US National Library of Medicine, 2014). It is worth noting that it took 13 years to passage from the time this legislation was first introduced in the Congress (Asmonga, 2008).

During an age when mobile devices drive productivity but unauthorized access to personal data erodes public trust, there is reason to believe that privacy can be compromised. People are concerned about this, and yet a majority say they are willing to share their health information for reasons of better understanding diseases, improving medical research, improving patient care, and better tracking of public health concerns (Fig. 25.13; Research!America, 2013b).

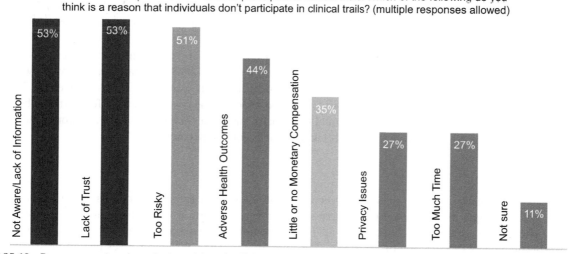

FIGURE 25.12   Reasons more Americans don't participate in clinical trials. *Research!America, 2013b. National Poll: Clinical Research. Conducted in partnership with Zogby Analytics. Available from: www.researchamerica.org/uploads/June2013clinicaltrials.pdf.*

The Internet and an Empowered Public

Nothing has facilitated consumer participation in health decision-making—including access to and awareness of clinical trials—like the Internet.

In 2014, 87% of Americans reported that they use the Internet (Pew, 2014), and 72% of the users say they have looked online for health information in the past year (Pew, 2013). As previously noted, 53% of Americans said they have heard about clinical trials from the Internet—14% higher than the second-highest source, which is traditional media. Only 24% reported hearing about such trials from their health-care providers (Research!America, 2013b). So-called "peer-to-peer" or "social health care" is becoming more prevalent, with 8% of Internet users posting a health-related question or story online over the course of a year, and 16% of online health information seekers attempting to connect with those who share their condition or concerns (Pew, 2013). In particular, in the case of rare disorders with no established patient support system, social media can provide the most effective avenue to connect with others suffering from the same disease around the

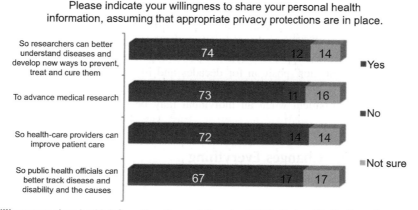

FIGURE 25.13   Patient willingness to share health information. *Research!America, 2013b. National Poll: Clinical Research. Conducted in partnership with Zogby Analytics. Available from: www.researchamerica.org/uploads/June2013clinicaltrials.pdf.*

world. In 2012, the parents of a child suffering from a rare disease, glycosylation disorder caused by NGLY1 mutations, utilized the power of social media, creating a "Google dragnet" to help identify more cases and to publish online essays that went viral on other social media platforms including Twitter. The posts brought together parents whose children had the same condition, as well as experts from academia. After initially being rejected for study by academic researchers for lack of sufficient patients, the condition is now being studied by the NIH (Mnookin, 2014).

Consumers who are thirsty for health information can find an overwhelming amount of information about health and research for health at their fingertips every minute of the day. While it has always been difficult for nonexperts to judge the quality of health information, the Internet with its wide variations in both content and content providers intensifies that reality. With the rise of the Web and tools like blogging and social media, it is easier to publish content for the world to see and present it as a reasonably legitimate source, sometimes with little to no burden of accuracy or attribution. Still, websites like WebMD.com, MayoClinic.org, the NIH-sponsored Medlineplus.gov, and many others provide users with trusted and high-quality health information.

It is fairly well understood today that there are positive and negative effects of patients' use of the Internet depending on the nature of the complaint, the information source, the patient, and the source of treatment if any. However, most Americans still rely primarily on health professionals for dealing with health or medical issues (Fox et al., 2009). This reinforces Research!America-commissioned survey findings that doctors are the preferred source for learning about clinical trials; nearly three-quarters of Americans say they would likely participate in a clinical trial if it was recommended by their physician (Research!America, 2013b), with the Web taking a more complementary role.

Continuing Challenges

Sadly, the history of clinical research includes the reprehensible Tuskegee experiments on African-American men, the deaths of two young people during flawed gene therapy experiments (Thompson, 2000; Food and Drug Administration, 2007), and, more recently, complaints from patient advocates about the mishandling of informed consent for a federally funded study evaluating oxygen saturation targets in premature infants. The subsequent investigation of the prematurity research protocol by the Office for Human Research Protections of the US DHSS "has had the effect of damaging the reputation of the investigators and, even worse, casting a pall over the conduct of clinical research to answer important questions in daily practice," (Drazen et al., 2013).

Untoward events have rekindled long-standing issues of trust between the American people and the research community, particularly among minorities: 61% of African-Americans now cite "lack of trust" as the top reason why they do not participate in clinical trials, with the percentage exceeding 50% among other races (Hispanics, non-Hispanic whites, and Asian-Americans; Research!America, 2013a). Also concerning is that more than half of Americans believe health-care services are not based on recent research, while 31% believe patients are sometimes included in clinical trials when they are receiving medical treatment without being told (Research!America, 2013a).

Increased scrutiny of clinical research investigators' financial relationships with industry has led policymakers to demand increased transparency and accountability in the use of federal funds. These concerns stem from the essential but complex relationships that exist among government, academia, and industry, particularly the partnerships between NIH-funded researchers and industry, necessary to take discoveries from "bench to bedside," but increasingly difficult to regulate. In 2011, the DHSS issued a final rule revising the 1995 regulations on financial conflicts of interest of extramural investigators. "The revised regulations institute a more rigorous approach to the management of investigators', significant financial interests, and resulting financial conflicts of interest to enhance the objectivity and integrity of the research process. These include changes to address investigator disclosure, institutional management of financial conflicts of interest, and federal oversight" (Rockey, 2011).

Meanwhile, the FDA under leadership appointed by both political parties over several administrations has been in a "damned if you do, damned if you do not" position for decades and is woefully underresourced given its mandate to protect public health and safety. The public, in turn, is of two minds about regulation: low tolerance for perceived risk, no matter how remote, but loud and impatient when an individual may be dying while awaiting an as yet unapproved experimental treatment.

Electronic Media Technology Changes Everything

It is no secret that technology increases capacity to conduct research, disseminate information, and build partnerships. From social media to health IT, technology is pervasive; it is in our lives 24/7. How we leverage technology to speed research for health can make a difference.

Social media, a now ubiquitous example of technological innovation, is making medical information and research more accessible to the public, creating an environment in which, instead of relying on providers to manage the flow of information, patients are empowered to seek out pertinent information from other sources (Thompson, 2014). Increasingly, technology exists for patients, advocates and providers to share information, educate, and engage with one another (Thompson, 2014).

Social media has a host of applications in research for health. In addition to empowering information gathering and sharing, social media facilitates research by helping to optimize efforts of investigators and trial participants alike. Studies show that social media can be a successful recruitment tool for research (O'Connor et al., 2014). To improve trial participation, social media outreach works in conjunction with existing platforms that support accrual. The **I'm In** campaign, sponsored by the Pharmaceutical Research and Manufacturer of America (PhRMA) and the National Minority Quality Forum, aims to boost clinical trials participation among minority populations. The campaign has leveraged social media to help build an online "engagement network" that connects patients, clinical trial sponsors, clinical investigators, health-care professionals, their institutions, and advocacy organizations across the nation (I'm In Clinical Trials for Better Health, 2014).

Perhaps the earliest electronic platform is "ClinicalTrials.gov" set up by the National Library of Medicine of the NIH functions as a centralized database of clinical trials, both public and private, taking place around the world (NIH, 2000). Recent efforts to increase inclusion of clinical trials in the ClinicalTrial.gov Website are driven by new regulations and policies by HHS and NIH that more clearly define the standards for clinical trial inclusion and expand the required information about the study that must be published on the Website (Rockey, 2014).

Across the continuum of health care, there are other technological advances that are streamlining clinical research. For example, PCORI's National Patient-Centered Clinical Research Network, PCORnet, partners patients, providers, and health systems to improve health care and advance medical knowledge (PCORnet, 2014). PCORI has invested more than $100 million in the network, collecting patient-generated information to facilitate research into care outcomes—and, by extension, improvement of care processes—across 29 different health networks (Institute of Medicine Roundtable on Value & Science-Driven Health Care, 2014). The goal is ultimately to join together the clinical information of as many as 30 million patients for research purposes.

The changing world of electronic medical records is also paving the way for many new efforts and is a complimentary component of PCORnet. Digital information reduces redundancies among trials and communicates information more effectively and efficiently. Further improvements, often driven by patient demand to "free the data" will drive "interoperability" (Healthcare Information and Management Systems Society, 2013) and speed clinical research.

Research Information and the News Media: Scientists' Role

The difficulty faced by the public in knowing how to understand and value research also affects the news media, as nonscience trained journalists must sift through seemingly contradictory peer-reviewed studies issued by the hundreds each month. The resulting confusion—about the benefits or harms of drinking milk or running for exercise or of the possible risks versus benefits of the human papillomavirus (HPV) vaccine—can make it easy for the public to scoff at science and ignore its findings. The 2014 Ebola outbreak is an example of conflicting views regarding the quarantine of individuals who were potentially exposed to the virus, particularly health-care workers. As state and local officials instituted mandatory enforced quarantines for health-care workers returning to the United States from African countries hardest hit by the outbreak, scientists insisted the policies were unnecessary and did not constitute evidence-based solutions (Flegenheimer et al., 2014).

Oftentimes, media coverage omits the nuances of scientific reports for the sake of brevity, clarity, or, in some cases, even editorial bias. The research community must do much more to assist health and science journalists, as well as general reporting journalists, in interpreting scientific literature in ways that their editors, producers, and audiences will find valuable and interesting. This need not mean misrepresentation or oversimplification, but rather asks scientists to learn new ways to communicate about research to nonscience trained people. When the medium permits and the learning opportunity exists, researchers who have established a relationship with a journalist can work with that person to more fully inform the public about the research process, including, for example, its nonlinear, exploratory nature. In addition, the scientific community should exercise more responsibility in countering the ill effects of inaccurate or hurried reporting, speaking out when journalistic, not to mention scientific standards of accuracy, have not been met.

With the rise of blogs and social media tools like Twitter as information sources, more traditional media including newspapers and broadcast outlets are seeking new ways to deliver research news and make it meaningful. Most major newspapers devote a section or weekly page, both print and online, to consumer health and research news, with plain-language reports on medical research findings, related graphics and important caveats. Increasingly, blogs, Twitter accounts, and other social vehicles, both those affiliated with news outlets and independent, are pushing out content in easily digestible bits of information. In 2013, 40% of journalists said social media networks are "very important" to their work (Willnat and Weaver, 2014). Changing practices in journalism present an opportunity for the research community to connect the dots for nonscientists and benefit their career at the same time. Xuan Liang et al. studied the role of public communications in boosting a researcher's profile and academic career. The researchers found that scientists whose work was mentioned on Twitter had higher *h*-index scores than scientists whose work was not mentioned. Similarly, researchers who had more interactions with reporters also had higher *h*-indices.[1]

Increasingly, prominent peer-reviewed journals like the *Annals of Internal Medicine, Nature Communications, Scientific Reports, eLife, PLOS Pathogens*, the *Journal of the American Cancer Society* and *BioMed Central*, an open access publisher, are directly accessed by journalists and the general public and some require that authors provide some degree of information for the nonresearcher. According to a study by the Research Information Network, open access articles in *Nature Communications*, had been viewed more than twice as often as subscription-only articles and cited more frequently (Nature Communications, 2014). "Our job is to accelerate discovery, not rest on our laurels because something is working. I wanted to put a flag in the ground which said we are committed to open access," said Sam Burridge, managing director for open research at Nature Publishing Group, in a 2014 interview with *The Times Higher Education*. Increasingly, philanthropic funders are joining the open access movement. In 2014, the Bill & Melinda Gates Foundation announced that grant recipients must make their research publicly available online, a policy that went into effect on January 1, 2015.

The Celebrity Factor

A sometimes-disturbing cultural trend involves the public's fascination with celebrities, including the mental health or addictions of young celebrities. Yet, movie stars, musicians, athletes, and famous politicians can be effective and thoughtful advocates for health research. Examples over the years include Mary Tyler Moore for Type 1 diabetes, Betty Ford for addiction, Michael J. Fox for Parkinson disease, Nancy Reagan and Seth Rogen for Alzheimer disease, Magic Johnson for AIDS research, Glenn Close for mental health, and Gwyneth Paltrow, Christina Applegate, and several others for the Stand Up to Cancer campaign (Stand Up to Cancer, 2014). In addition, Fox and Reagan have been effective advocates for embryonic stem-cell research. Joe Perry, lead guitarist of Aerosmith, lent his support for the Geoffrey Beene Foundation's Rock Stars of Science (Geoffrey Beene Foundation, 2012) initiative that called attention to innovative cancer research.

On any given day, a celebrity may be seen in the halls of Congress—and on television and mobile device screens—lobbying and testifying on behalf of research into a certain health condition or treatment option. Such advocacy and media campaigns can offer useful public information or call attention to much-needed information. Sometimes these appearances or campaigns are bankrolled by pharmaceutical companies and designed as marketing efforts for a certain medication. Recent examples include comedian "Larry the Cable Guy" for Prilosec and golfer Phil Mickelson for Enbrel. Though less common than it was in years past, marketing may not make clear whether financial compensation is involved. News media have begun to more actively and consistently question whether such relationships exist and disclose them to their audiences (Petersen, 2002).

Certainly, a compensated arrangement does not mean that a spokesperson, celebrity, or otherwise, is insincere in taking a certain position. The point is that it is in the public's interest to be advised of any financial interest before drawing a conclusion. The same is true in regard to disclosure by researchers and clinicians of their commercial relationships and interests. Not long ago, such relationships were entirely hidden from public view. That mentality has essentially vanished. Unfortunately, it is replaced on occasion by naive notions of decision-making that is completely free of any potential conflicts of interest. The public's expectations are well served when conflicts of interest are disclosed, not when it is supposed that none exist at all.

Unrealistic Expectations?

A challenge for investigators in today's highly transparent society—with the public watching the "3-steps-forward, 1-step-back" mechanisms of research almost in real time—is that external forces can influence the science agenda and even raise

---

1. A high value in *h*-index indicates a high scientific impact of the researcher. "The *h*-index (Hirsch, 2005) is a bibliometric indicator that quantifies the scientific impact of a given researcher and embodies a figure of merit." According to Hirsch, a researcher has "index *h* if *h* of his or her *Np* papers have at least *h* citations each and the other ($Np - h$) papers have $\leq h$ citations each," so that a high value in *h*-index indicates a high scientific impact of the researcher (Liang et al., 2014).

unrealistic expectations for discoveries or their timelines. This can make many in the science community uneasy, even though history has shown that progress can be stimulated in this fashion.

Opinions differ in the research community regarding the value of very ambitious, very public challenges with attendant infusions of dollars—for example, "The Decade of the Brain" from the 1990s and the "War on Cancer" initiated by President Richard Nixon in 1971. In the late 1990s and early 2000s, the doubling of the NIH budget over five years significantly raised congressional, if not public, expectations, giving rise after the subsequent slowdown to researcher lamentations about the ill-advised nature of the effort. There is, however, no evidence, nor any reason to believe, that the significant funds appropriated to NIH in these undertakings would have been forthcoming at all, much less spread over a longer period or distributed in any other fashion than they were as a result of a concerted advocacy campaign. The bolus of funds appropriated to the research-based agencies by the American Recovery and Reinvestment Act (ARRA) of 2009, also known as the "stimulus" bill, provided much-needed resources to science-based institutions across the nation during the depths of the Great Recession. Disgruntlement about the one-time nature of this funding was expressed not only by some members of Congress but also by some in the science community. There are no data demonstrating public disenchantment with any of these programs, although anecdotes to that effect are repeated frequently. Some in Congress have expressed disenchantment with the "absence of progress" as an outcome of the NIH doubling period and as a result of the ARRA funding. Context is called for: the timeframe needed for discoveries to show benefits to patients is long and fluctuations in the NIH budget since the doubling have disrupted promising studies. Both argue for increased engagement by the research community in dialog with elected officials.

SCIENCE AND POLITICS

The Small Voice of Science: Stepping Up to Strengthen It

Researchers are often surprised by the limited amount of influence the science community exerts in public and policy discourse. For instance, in 2005 and 2006, amid reports of rising autism rates among young children, a parents' campaign, "Autism Speaks," backed by a major corporate leader (General Electric Vice Chairman Bob Wright), a media personality (radio host Don Imus), and a political figure (Robert F. Kennedy, Jr.), questioned whether the preservative thimerosal in infant vaccines caused autism. This group's efforts to link autism to vaccines gained significant momentum despite the fact that numerous, robust studies found no link between thimerosal and autism (Miles and Takahashi, 2007; Stehr-Green et al., 2003; Thompson et al., 2007; Institute of Medicine, 2004) This debate has calmed somewhat in recent years as the medical community has pushed back against nonscientific claims and as outbreaks of previously rare vaccine-preventable diseases like whooping cough occurred in communities where parents had eschewed vaccinations. But science has not fully overcome the antivaccine movement, and there will be other major challenges to scientific evidence in the future.

There are several reasons why members of the public, and likewise their elected representatives, do not necessarily consider scientific findings as the final word. When members of the public express skepticism, they are, after all, acting like a well-trained scientist! One obvious reason for public skepticism is that, over the decades, scientists, very much including physicians, trained the public to accept their word without question, more often than not missing the opportunity to explain the way that science works or even offering the caveat "here is what we know at this time." As science evolves, conclusions shift; this can be all too easily lampooned when it comes to advice on lifestyle choices: A glass of red wine is good for your health one day and bad for it the next. For a decade, eggs were verboten because of their cholesterol content, and then new research concluded that eggs are a healthy addition to one's diet. Simply acknowledging the reality of learning more and altering advice, as we do so ("you will know as soon as we do"), rather than apologizing for change, would take the science community a long way forward in earning and maintaining public support and the political potency that accompanies it. In the words of Abraham Lincoln, "With public sentiment, nothing can fail. Without it, nothing can succeed."

For the nonscientist—in other words, the vast majority of our citizenry—it is hard to understand the special meaning of common words when used in scientific context, for example, the word "theory." To the nonscientist, and for that matter the scientist in a general conversation, "theory" implies possibility, not the virtual, but never final certainty it conveys to scientists. Then too, scientific findings periodically conflict with religious beliefs, with outcomes that harken back to the days of Galileo. It behooves members of the science community to become more comfortable in conversations that draw on values and backgrounds that are different from their own. Listening and finding ways to connect to those with whom disagreement is the starting point is not just important, it is critical if science is to continue to receive public support.

To demonstrate the power of smart public advocacy by the science community, consider that many science-based organizations have been instrumental in driving public and policymaker support for embryonic stem-cell research.

More than 60% of Americans now believe embryonic stem-cell research has merit and either "strongly" or "somewhat" favors expanding federal funding for this research (Research!America, 2011).

In addition to the synergies between patient and science communities, science and technology is an innovation driver, and that is not lost on the business community. Opportunities exist for far stronger partnerships between industry and science to push for pro-research policies. In fact, such leading business coalitions as the National Association of Manufacturers, and the Biotechnology Industry Organization actively participate in efforts to increase federal research funding, and a 2011 Research!America survey found that 70% of small business leaders perceive federally funded research as important to private sector innovation (Fig. 25.14; Research!America, 2013c).

One key reason that science does not necessarily receive the "respect" it deserves is that science and scientists simply do not engage as scientists with the broader community. Let us face it: How many scientists make a point of explaining what they do and why they do it? How many scientists engage in advocacy or public outreach? According to recent polls, scientists and research institutions remain largely invisible to the general public. Seventy percent of those surveyed nationwide could not name a living scientist and more than half could not name any institution, company, or organization where medical or health research is conducted (Research!America, 2013a,b,c,d). People visit their doctor, but they do not visit the researcher who discovered the treatment their doctor prescribes. Their physician may even be a researcher, but unless he or she identifies himself or herself as such, how would the patient know? People recognize their doctor in a range of community settings; they often have no idea there are scientists there as well.

Former US congressman and Research!America Chairman John Edward Porter issued a call to action for the science community in an editorial in Science in 2014, "To put it bluntly, scientists remain largely invisible to the public. Yes, this says a great deal about the nation's people, but it says even more about scientists and their lack of engagement with the public... Scientists must take off their lab coats and engage the people of their communities and states" (Porter, 2014).

In another 2014 editorial in *Science*, scientists were urged to proactively engage with policymakers to elevate research in legislative and regulatory agendas. "If scientists do not act now to cultivate new, well-informed champions for science, other interests will quickly eclipse science in vying for attention and loyalty. This cultivation is particularly important now, at a time of intense scrutiny of the federal government's investment in all programs" (Woolley and Leshner, 2014).

The Role and Influence of the White House

The White House can and often does take a leadership role to support research. For example, President Obama launched a "grand challenges" initiative in 2013, establishing "ambitious but achievable goals that require advances in science and technology" (Office of the Press Secretary, 2013). Among those challenges is his *Brain Research through Advancing Innovative Neurotechnologies* (BRAIN) Initiative, which is intended to dramatically advance efforts to treat, cure, and prevent Alzheimer disease, epilepsy, traumatic brain injury, and other brain disorders.

Previously, George W. Bush pledged during his campaign to complete the NIH budget doubling in his first years in office, and followed through, despite the fact that the clarion call of 9/11 caused the nation to alter priorities in short order. President Bush also signed into law the America COMPETES Act, which was intended to double the budget of the National Science Foundation (NSF) and in other ways advance scientific discovery. There is in fact a long history of the

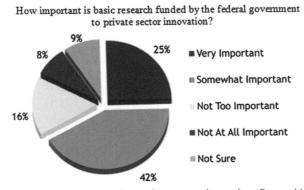

FIGURE 25.14 Importance of federal funding of basic research to private sector innovation. *Research!America, 2013d. Small Business Leaders Nationwide Survey. Conducted in partnership with Zogby Analytics. Available from: http://www.researchamerica.org/uploads/Feb2013smallbizsurvey.pdf.*

role of Presidents in advancing research via their bully pulpit and sometimes their personal experience. For example, Eisenhower outspokenly supported heart disease research following his heart attack; John F. Kennedy helped advance the cause of mental health research; and Nixon's "War on Cancer" begun in the 1970s, did not "cure" cancer as once hoped, but many lives subsequently have been saved, especially those of children coping with the many forms of pediatric cancers. The White House Office of Science and Technology Policy can also prove influential as analyst and advisor to the president. This office works to broaden science, technology, education, and math (STEM) education and promote open data, among other initiatives.

But there is a very important, though much less public way that the White House influences science. Even in the most rancorous political environment, the President's annual budget proposal tends to set the broad parameters for discretionary spending priorities such as funding for the NIH. The President typically establishes funding levels for research; Congress rarely appropriates funding much higher or lower than what the president proposes. There have been exceptions, however, and in addition to appropriating funds, Congress plays a pivotal role in determining the nature and pace of medical progress.

How Congress Influences Medical Research

Congress, with the power of the federal purse and the charge to enact policy and hold agencies accountable, exerts enormous influence on medical research. Congress shares responsibility with the executive branch for establishing budget priorities and addressing such research-relevant issues as US tax policy. Further, Congress authorizes and reauthorizes federal agencies, including the NIH. "Authorizing" legislation not only establishes maximum expenditure levels, or "caps" on federal spending for each agency, it guides the general policies and priorities under which each agency operates. While federal agencies are given modest discretionary authority to establish regulations and allocate funding, most of what they do is guided by authorizing legislation.

And while the rule may be that the funding prioritization put forward in the President's budget guides congressional action, there certainly have been exceptions. Congress, not the Executive Branch, initiated the doubling of the NIH budget that occurred between fiscal year 1999 and fiscal year 2003. In the 113th Congress (2013—14), several members of Congress introduced ambitious legislation that would compensate for a decade of stagnant NIH funding by significantly boosting annual appropriations for NIH over the next 10 years. The prospects for these efforts are unclear, but they demonstrate that it is important to engage both the White House and Congress in seeking to support more robust federal support of medical and health research.

Congress has also played, and will continue to play, an important role in establishing, and from time to time modifying and updating, such relevant policies as research subject protections, privacy rules, conflict of interest rules, reimbursement guidelines, and the variables the FDA uses to assess the clinical proof of safety and efficacy when deciding whether to approve a new drug or medical device. In the case of FDA, changes like these are typically tied to reauthorization of "user fees." Under current law, the bioscience industry pays user fees to supplement the resources FDA receives through appropriations to review the safety and efficacy of new medical products. In exchange, FDA agrees to performance goals establishing timeframes for completing the review of these new products. When the legislation reauthorizing user fees is negotiated, there are often additional provisions included in the legislation relevant to the review process and other factors affecting the nature and pace of medical progress. In the 2012 reauthorization, for example, there were provisions that require FDA to reassess the criteria it uses to evaluate the relative risk and benefit of new medical products. There were also provisions to incentivize the development of treatments for childhood diseases and other unmet medical needs.

For the research community, the point is that Congress is more than willing to go beyond "top line" funding and basic structural issues, if advocates make a strong enough case that congressional intervention is necessary. This reality can work to the benefit of research, or restrict its conduct; all the more reason for researchers to engage.

There are additional reasons for scientists to engage. The two bodies of Congress often directly query specific areas of science, sometimes to alarming effect. In the 113th Congress, for example, legislation was introduced in the House of Representatives that would end the NSF's ability to allocate funding across its various directorates; restrict the length of time any individual researcher could be funded by NSF; markedly reduce funding for geoscience and social sciences research; and prohibit funding for research that is in any way duplicative of any other federally funded research. These provisions were decried by the research community as counterproductive to scientific advancement and human progress. As of this writing, concern runs deep that other voices and other priorities may trump that of science. Legislative attempts like these are not new. Over the last decade, similar attempts have been made to defund specific grants at NIH and NSF. In 2003, an amendment was offered to the Fiscal Year 2004 Labor, Health and Human Services and Education appropriations bill in the House to defund five NIH grants studying sexual behavior. Republicans and Democrats alike spoke out against this threat to research and the peer-review process but the amendment was defeated by a narrow two vote margin (Congressional Record, 2003).

Many elements impact Congressional decision-making, including the fiscal climate, broader ideological considerations regarding the size of government and the tax burden, the force of patient demand and public opinion, the dictum that "all politics is local," high-profile news events, and members' personal perspectives as shaped by their individual experience, as well as the impact of advocates. All too often, scientists themselves do not weigh in as the process unfolds, or even at all.

Making the Case for Research to Congress

Appropriations for the NIH—the principal source of taxpayer-supported research funding—was virtually flat over the period 2003—12, losing more than 20% of its purchasing power (amfAR, 2013).

The science community bears some responsibility for this reality. Researchers, both inside and outside the health and medical fields, continue to eschew public and political engagement even when such engagement is nonpartisan in nature. Scientists' general attitude of avoidance makes it easier for officials to neglect science funding.

Very few members of Congress have any science background or training, which means that talking about and acting on behalf of science does not come naturally to most. Of the 535 members of the Senate and House of Representatives in the 113th Congress, 9 were scientists and 24 were health-care professionals (Manning, 2014). In the 114th Congress as elected, only 4 House members are scientists (none in the Senate) and 33 members from both the House and Senate are health-care professionals (Bellantoni, 2014).

Because it is impractical to become an expert in every area of public policy, members of Congress tend to develop deep expertise in only a few policy topics. These usually include a subset of the issues under the jurisdiction of the committees on which they serve, as well as those of particular importance to their constituents (e.g., a member who represents a district with a major defense contractor becomes well versed in the issues surrounding defense contracting). It is the responsibility of the legislative director and legislative assistants who serve members of the House and Senate to advise them on the full array of issues that come before Congress. This is one of several reasons why it is so important for the science community to forge bonds with congressional staff as well as members themselves. It is also a fact that advocates who make a compelling enough case to a member of Congress can convince them to champion an issue that had not been on their radar previously.

What does this mean for scientists? More scientists need to get to know their senators and representative in Congress, and get to know their legislative directors and legislative assistants. If a researcher's representatives are not on committees with direct jurisdiction over issues related to clinical research or even science more broadly, that scientist may have to work hard to cultivate interest in advancing medical research, but it is eminently doable. When relationships are formed with policymakers and their staff, scientists are viewed as a resource for guidance on research-related issues. These bonds can form the foundation of enduring support for science. Policymakers rarely or never come to scientists; scientists must take the initiative to connect.

Individual research institutions can and frequently have convinced their representatives in Congress to become champions of public research funding. Local universities can in fact create congressional champions for science funding. The fact that more than 80% of NIH funding is allocated through grants to research institutions throughout the nation is one of the agency's most important selling points. In fiscal 2014 alone, 50,000 NIH grants—representing more than 80% of the institutes' $30.1 billion budget for the year—went to 2500 institutions in municipalities around the country. In any congressional district, the more science that is conducted, the more jobs will be created and the greater the benefits to the local constituency. Statistics like these are not lost on federal policymakers.

The Role and Influence of Patient Advocacy Organizations

There is no doubt that grassroots initiatives can have a dramatic effect on the federal role in medical progress. As previously discussed, in the 1980s and 1990s, HIV/AIDS activists overcame prejudice, complacency, and fear to propel research forward; their activism is in no small measure responsible for the development of antiretrovirals and affordable access to them in the United States and abroad. For those who obtain access, this medical breakthrough can transform HIV/AIDS from a death sentence to a manageable chronic condition (Epstein, 1995). The breast cancer movement is another compelling example of patients and their loved ones not taking "no" for an answer and elevating the priority of research into this all too common form of cancer (Lerner, 2001).

There are numerous examples of parents single-handedly attracting congressional attention and federal resources to childhood diseases such as Duchenne muscular dystrophy and childhood cancer (Parent Project Muscular Dystrophy and

Children's Cancer Recovery Foundation). Other patient advocacy organizations, large and small, have an enormous influence on the federal funding and policy climate. The Alzheimers Association, the American Cancer Association, the American Heart Association, and the American Diabetes Association are some of the many iconic grassroots patient organizations with significant influence on Capitol Hill.

Researchers who visibly work with, instead of ignoring, patient advocacy organizations, not only increase the impact of research advocacy on Capitol Hill, they also stand to develop relationships that can advance science more rapidly as patient insights can be better understood and incorporated into research design and execution.

The Role and Influence of Philanthropy

Individual philanthropists and philanthropic foundations also contribute in meaningful ways to advancing medical progress. Many of these organizations focus on funding clinical research. It is common that philanthropy, as well as patient organizations, do not fund all of the costs associated with the conduct of research at a university or independent research institution, frequently limiting indirect cost reimbursement to a small fraction of actual costs. This can and frequently does generate tension between funders and research administrators, a situation that calls out to be eased, so that all the partners in advancing research can work together effectively.

Fund-raising efforts for research spearheaded by patient groups are not a new phenomenon but achieved increased recognition in 2014 with the ALS Foundation's "Ice Bucket Challenge," which took hold on social media, raising over $115 million dollars for amyotrophic lateral sclerosis. Also other examples include the Susan G. Komen Race for the Cure, a fundraising effort that has helped generate over $610 million dollars in breast cancer research over three decades, $58 million in the fiscal year 2012 (Komen, 2012). Leukemia and Lymphoma Society's (LLS) "Team in Training" is a fundraising effort that trains advocates for athletic events while raising awareness for blood cancers. Since 1988, LLS has celebrated the participation of over 500,000 athletes, raising $1.3 billion dollars used for investment in research into these diseases (Leukemia and Lymphoma Society, 2014). The earliest modern-day success of patient and concerned community fundraising for research was the March of Dimes whose storied history began in 1938 in response to polio (Rose, 2010).

While patient advocacy and philanthropic organizations provide meaningful resources to medical research, there are too many nonscientists who view their giving as a possible replacement for federal and industry financing. This is true not only because these organizations do not pay full costs of research, but also because, in total, their resources are dwarfed by those of the federal government. While information on total research spending by patient advocacy and philanthropic organizations is difficult to access, the best estimates to date place philanthropic spending at approximately $1.3 billion and patient advocacy spending in the range of $1 billion per annum. That is about 1.7% of total US spending on medical R&D (Research!America, 2013d). The bottom line is that even though advocacy for federally funded research can be an uphill battle, the effort is more than worthwhile, it is an imperative if US medical progress is to continue.

(Note: Generally, philanthropic organization steer clear of political advocacy, focusing instead on identifying promising research and helping to finance it. That said, major philanthropic organizations like the Bill and Melinda Gates Foundation and the Milken Institute directly and indirectly support advocacy, the former with a focus on addressing global infectious disease, and the latter with a focus on speeding US medical progress.)

Public-Private Collaboration

In many areas of science, collaborations across academic disciplines and between industry and academia are increasingly important tools for taking advantage of intellectual and creative capital while husbanding monetary capital in these uncertain times for federal research dollars.

The previously mentioned BRAIN Initiative, launched in 2013, encapsulates this new sensibility. The initiative places NIH and several other federal agencies (both inside and outside health and medicine) in partnership with companies like GlaxoSmithKline, universities, and philanthropic foundations to ultimately gain a deeper understanding of the human brain.

Another example is the Foundation for the National Institutes of Health Biomarkers Consortium, a research partnership with broad participation from stakeholders across the health field, including government, industry, academia, patient advocacy, and other nonprofit private sector organizations. In addition to the Foundation for NIH, founding members include the National Institutes of Health, Food and Drug Administration and the Pharmaceutical Research and Manufacturers of America. The consortium seeks to identify, develop, and qualify biomarkers to achieve improvements in drug development, clinical care, and regulatory decision-making, often opening exciting new opportunities for clinical research.

Among its many projects, the consortium successfully completed clinical research that affirmed adiponectin, a hormone derived from fat cells, as a robust biomarker predictive of the efficacy of certain medicines for type 2 diabetes

(The Biomarkers Consortium, 2010). Given the growing prevalence and disabling side effects associated with this disease, a biomarker that may speed the results of clinical trials for new treatments is an accomplishment indeed.

A different example of public-private partnerships is provided by Pfizer, which has teamed up with various not-for-profit organizations like the American Academy of Pediatrics, the AMA Foundation Minority Scholars Award, and the NIH Medical Research Scholars Program to establish grants and other programs to improve medical training and, ultimately, health (Pfizer, 2014). The popularity of these kinds of partnerships often spreads as each institution finds needed synergies with one another.

In the global health arena, product development partnerships, which bring together philanthropy, government, and business to address such unmet global health needs as vaccines to prevent HIV/AIDSs, new antimalarial medicines, and rapid diagnostics for tuberculosis are working hard to address the commercial barriers and speed the development process in order to save millions of lives across the globe.

What is the contextual link between these partnerships, politics, and policy? The unique power of public-private partnerships is that they cross ideological lines and engage a diversity of politically active stakeholders. For policymakers who believe the federal government is too large, the fact that the government is not "doing it all" in these partnerships is appealing, and for those who believe the government has a significant role to play in advancing research, the fact that government involvement is integral to the success of these partnerships resonates. Public-private partnerships are not only popular in their own right, they tend to spread good will where it is needed, helping to counter mistrust of government and mistrust of industry, ease the impact of competing ideologies on public policy related to medical research, and finally, because public and private interests both have a stake in these partnerships, they stand up well when fiscal considerations drive major budget cuts. These partnerships will no doubt play an increasingly important role in the medical and health research arenas, in no small measure because they enjoy a strong base of stakeholder support.

CONCLUSION

NIH Director Francis Collins has said he fears that while the current era of knowledge holds vast promise, America and its leaders may lack the political will to tap into it. Which unfunded grant may have surfaced a major new insight into Alzheimer disease, cancer, or diabetes? Which of these unfunded grants caused the layoff of a promising young student scientist who might have been on the path to win the Lasker or Nobel someday? "What keeps me awake at night is the impact of these financial strains on our most critical resource—the talented scientists in our nation's finest universities who have been the source of phenomenal innovative breakthroughs over the decades, making America the envy of the whole world" (Driving Innovations, 2014).

Without a healthy policy environment and strong federal funding for clinical and, indeed, all forms of research, Collins' fears may come to pass. We may see fewer potential drugs, diagnostic tools, and prevention and intervention strategies make it into the development pipeline, let alone clinical practice.

Former NIH Director Elias Zerhouni outlined a compelling vision for the future of medical and health research:

We are in a revolutionary period of medicine that I call the four Ps: predictive, personalized, preemptive and participatory. This requires patient involvement well before disease strikes. As opposed to the doctor-centric, curative model of the past, the future is going to be patient-centric and proactive. It must be based on education and communication.

Zerhouni (2007)

What is needed to make the vision a reality? Given American expectations for ever-enhanced health and quality of life and sustained world leadership, we should:

- Reinvigorate US policy to strengthen investment in scientific research and innovation to foster health, improve national security, and drive economic prosperity.
- Facilitate research at the convergence between biological, behavioral, and social sciences and the physical, chemical, mathematical, and computational sciences.
- Minimize the barriers to collaborative partnerships among researchers who work in the government, academic, and industrial sectors.
- Encourage a sustainable long-term vision of innovation and high-risk, high-reward research.
- Revamp our educational system to reemphasize the STEM fields.

These actions, deemed urgent necessities by many health and science leaders, are *not* well understood by the public, the politicians, private-sector policymakers, or the press. Too often, scientists appear to misunderstand or simply pay no heed to the many new realities of public awareness and concerns—essentially the public context of their research. As previously quoted, Abraham Lincoln captured the riskiness of such an attitude many years ago, "With public sentiment, nothing can fail. Without it, nothing can succeed."

In so many important ways, the old ways—the old public contexts—are not the new ways or the new public contexts of clinical research. In further shaping the new ways, more and more public influence will assuredly be felt.

Progress depends on whether clinical researchers join with others in bringing their knowledge and their influence to the councils of public and private decision-making, keeping in mind that many of those councils are informal in nature. Researchers who do not take every opportunity to say and convey "I work for you; I am proud to serve the public's interest" will do a disservice to the research community at large. Researchers who fail to understand that they serve the public's interest—and thus bear a responsibility to engage with, listen to, and prove responsive to the public—will impede the future progress of research.

White House science adviser John Holdren noted the importance of continually making the case for science and research, "Everybody in the science and technology community who cares about the future of the world should be tithing 10 percent of his or her time to interacting with the public in the policy process" (Holdren, 2008).

Fortunately, it is easier to engage the public, the media, and lawmakers than many researchers anticipate, given public willingness to support greater government and private-sector investment in science for the benefit of our nation's health and economic growth. What may seem awkward and self-serving at first will in time become as much a habit of mind as wearing a seat belt and every bit as life enhancing.

Elected officials will continue to need researchers to reach out to them and their constituents to help everyone understand the full impact and excitement of research being conducted in their district, state, and nation. It is to be expected that the public will continue to press for greater input in the research process. News media will continue to struggle to report science news clearly and accurately in a limited space or timeframe. Social media, informed or otherwise, will proliferate and fiscal, ideological, and political challenges to science will continue to arise.

The research community has the opportunity every day to shape this multifaceted context. Individual researchers have the power to help research become an even more vital force for the health of their nation and world.

ACKNOWLEDGMENT

With thanks to Suzanne Ffolkes, Eleanor Dehoney, Anna Briscno, Ashley Kilgore, Glorymar Rivera, Caitlin Leach, Adam Katz, Tim Tassa, Scott Harris.

REFERENCES

Agency for Healthcare Research and Quality, 2014. National Healthcare Quality & Disparities Reports, 2013. AHRQ Pub. No. 14-0005-1.

amfAR, 2013. The Costs of Flat Funding for Biomedical Research. Available from: http://www.amfar.org/uploadedFiles/_amfarorg/On_the_Hill/Issue%20Brief%20-%20The%20Costs%20of%20Flat%20Funding%20NIH%20Research_080513.pdf.

Asmonga, D., 2008. Getting to know GINA: an overview of the genetic information Nondiscrimination Act. J. Am. Health Inf. Manag. Assoc. 79 (7), 18, 20, 22. Available from: http://www.ncbi.nlm.nih.gov/pubmed/18672722.

Battelle Memorial Institute, 2011. Economic Impact of the Human Genome Project. Available from: http://www.battelle.org/docs/default-document-library/economic_impact_of_the_human_genome_project.pdf?sfvrsn=2.

Battelle Memorial Institute, 2013. 2014 Global R&D Funding Forecast. Available from: http://www.battelle.org/docs/tpp/2014_global_rd_funding_forecast.pdf?sfvrsn=4.

Bellantoni, C. (Ed.), 2014. Guide to the New Congress. CQ Roll Call, 58. Available from: http://info.cqrollcall.com/rs/cqrc/images/CQNews_NewMemberGuide2014.pdf?mkt_tok=3RkMMJWWfF9wsRoiua%2BZKXonjHpfsX57OkkW6K2lMI%2F0ER3fOvrPUfGjI4CSMBqI%2BSLDwEYGJlv6SgFTbbBMbhh2bgPUxM%3D.

Bloomberg, 2014. Bloomberg US National Poll. Available from: http://www.bloomberg.com/infographics/2014-06-11/national-poll.html#poll.

Boulos, M.N.K., Curtis, A.J., AbdelMalik, P., 2009. Musings on privacy issues in health research involving disaggregate geographic data about individuals. Int. J. Health Geogr. 8 (46).

Brown, A., 2013. Americans Say Economy Is Top Worry for Nation's Future. Gallup, Washington, DC. Available from: http://www.gallup.com/poll/163298/americans-say-economy-top-worry-nation-future.aspx.

Centers for Disease Control and Prevention, 1999. Control of infectious diseases. MMWR Morb. Mortal. Wkly. Rep. 48 (29), 621—629.

Centers for Medicare & Medicaid Services, 2014. Hospital Compare. Available from: http://www.medicare.gov/hospitalcompare/search.html.

Children's Cancer Recovery Foundation, 2014. Children's Cancer Recovery Foundation. Available from: www.childrenscancerrecovery.org.

Committee on the NIH Research Priority-Setting Process, Health Sciences Policy Program, Health Sciences Section, Institute of Medicine, 1998. Scientific Opportunities and Public Needs: Improve Priority Setting and Public Input at the National Institutes of Health. National Academy Press, Washington, DC.

Congressional Record, 2003. US House. Available from: https://www.congress.gov/crec/2003/07/10/CREC-2003-07-10-pt1-PgH6565.pdf.

Director's Council of Public Representatives, National Institutes of Health, 2004. Enhancing Public Input and Transparency in the National Institutes of Health Research Priority-Setting Process. Available from: http://copr.nih.gov/reports/enhacning.pdf.

Director's Council of Public Representatives, National Institutes of Health, 2005. Report and Recommendations on Public Trust in Clinical Research. Available from: http://copr.nih.gov/reports/public_trust_clinical_research.pdf.

Director's Council of Public Representatives, National Institutes of Health, n.d. About COPR (Fact sheet). Available from: http://copr.nih.gov/factsheet.asp.

Drazen, J.M., Solomon, C.G., Greene, M.F., 2013. Informed consent and SUPPORT. N. Engl. J. Med. 368 (20), 1923−1931.

Driving innovations through federal investment. In: Hearing Before the Full Appropriations Committee, Senate, 113th Cong., 2, 2014 (testimony of Dr. Francis Collins).

Epstein, S., 1995. The construction of lay expertise: AIDS activism and the forging of credibility in the reform of clinical trials. Sci. Technol. Hum. Values 20 (4), 408−437.

Flegenheimer, M., Shear, M.D., Barbaro, M., October 26, 2014. Under Pressure, Como Says Ebola Quarantines Can Be Spent at Home. The New York Times. Available from: http://www.nytimes.com/2014/10/27/nyregion/ebola-quarantine.html?_r=0.

Food and Drug Administration, 2007. FDA Statement on Gene Therapy Clinical Trial. Available from: http://www.fda.gov/NewsEvents/Newsroom/PressAnnouncements/2007/ucm108955.htm.

Food and Drug Administration, 2014a. Women's Health Research. Available from: http://www.fda.gov/scienceresearch/specialtopics/womenshealthresearch/default.htm.

Food and Drug Administration, 2014b. Drug Trials Snapshot. Available from: http://www.fda.gov/Drugs/InformationOnDrugs/ucm412998.htm.

Fox, S., Jones, S., Pew Research Center Internet Project, 2009. The Social Life of Health Information. Available from: http://www.pewinternet.org/files/old-media/Files/Reports/2009/PIP_Health_2009.pdf.

Genetics Alliance, et al., 2010. Genetic Information Nondiscrimination Act. Available from: http://www.ginahelp.org/#.

Geoffrey Beene Foundation, 2012. Geoffrey Beene Rock Stars of Science. Available from: http://www.geoffreybeene.com/rockstars/.

HR Res 493, 110 Cong., 154 Cong. Rec. 2956, 2008 (enacted).

Healthcare Information and Management Systems Society, 2013. Definition of Interoperability. Available from: http://www.himss.org/files/FileDownloads/HIMSS%20Interoperability%20Definition%20FINAL.pdf.

Hirsch, J.E., 2005. An index to quantify an individual's scientific research output. PNAS 102 (46), 16569−16572.

Holdren, J.P., 2008. Science and technology for sustainable well-being. Science 319 (5862), 424−434.

I'm In Clinical Trials for Better Health, 2014. What's I'm In? Available from: https://www.joinimin.org/SitePages/whats-imin.aspx.

Institute of Medicine Roundtable on Value & Science-Driven Health Care, 2014. Integrating Research and Practice: Health System Leaders Working Toward High-Value Care. National Academy Press, Washington, DC.

Institute of Medicine, 2004. Immunization Safety Review: Vaccines and Autism. National Academy Press, Washington, DC.

Institute of Medicine, 2011. The Learning Health System and Its Innovation Collaboratives. National Academy Press, Washington, DC.

Henry J. Kaiser Family Foundation, 2013. Summary of the Affordable Care Act. Available from: http://kff.org/health-reform/fact-sheet/summary-of-the-affordable-care-act/.

Henry J. Kaiser Family Foundation, 2014. Kaiser Health Tracking Poll: May 2014. Available from: http://kff.org/report-section/kaiser-health-tracking-poll-may-2014-findings/.

Susan G. Komen Foundation, 2012. Annual Report 2011−2012. Available from: http://ww5.komen.org/uploadedFiles/Content/AboutUs/Financial/SGK002_AR_2011_2012.pdf.

Lerner, B.H., 2001. The Breast Cancer Wars: Hope, Fear, and the Pursuit of a Cure in Twentieth-Century America. Oxford University Press, New York.

Leukemia and Lymphoma Society, 2014. Team in Training. Available from: http://www.teamintraining.org/tellmemore.

Liang, X., et al., 2014. Building buzz: (Scientists) communicating science in new media environments. J. Mass Commun. Q. 91 (4), 772−791.

Lindenauer, P.K., et al., 2014. Attitudes of hospital leaders toward publicly reported measures of health care quality. JAMA Intern. Med. 174 (12), 1904−1911.

Manning, J.E., 2014. Membership of the 113th Congress: A Profile. CRS Publication No. R42964.

Michaels, M., et al., 2012. The promise of community-based advocacy and education efforts for increasing cancer clinical trials accrual. J. Cancer Educ. 174 (12), 67−74.

Miles, J.H., Takahashi, T.N., 2007. Lack of association between Rh status, Rh immune globulin in pregnancy and autism. Am. J. Med. Genet. A 143 (13), 1397−1407.

Mnookin, S., 2014. One of a Kind: What Do You Do If Your Child Has a Condition That Is New to Science? The New Yorker. Available from: http://www.newyorker.com/magazine/2014/07/21/one-of-a-kind-2.

Morris, Z., Wooding, S., Grant, J., 2011. The answer is 17 years, what is the question: understanding time lags in translational research. J. R. Soc. Med. 104 (12), 510−520.

National Institutes of Health, 2014. Impact of NIH Research. Available from: http://www.nih.gov/about/impact/index.htm.

National Institutes of Health, 2000. ClinicalTrials.gov. Available from: https://www.clinicaltrials.gov.

National Library of Medicine, 2014. The Genetic Information Nondiscrimination Act (GINA). Available from: http://ghr.nlm.nih.gov/spotlight=thegeneticinformationnondiscriminationactgina.

Nature Communications, 2014. Citation Analysis. Available from: http://www.nature.com/press_releases/ncomms-report2014.pdf.

O'Connor, A., Jackson, L., Goldsmith, L., Skirton, H., 2014. Can I get a retweet please? Health research recruitment and the Twittersphere. J. Adv. Nurs. 70 (3), 599−609.

Office of the Press Secretary, The White House, 2013. BRAIN Initiative. Available from: http://www.whitehouse.gov/the-press-office/2013/04/02/fact-sheet-brain-initiative.

Olshansky, S.J., Carnes, B.A., Desesquelles, A., 2001. Prospects for human longevity. Science 291 (5508), 1491−1492.

Olshansky, S.J., et al., 2005. A potential decline in life expectancy in the United States in the 21st century. N. Engl. J. Med. 352 (11), 1138−1145.

Parent Project Muscular Dystrophy, 2014. Parent Project Muscular Dystrophy. Available from: http://www.parentprojectmd.org/.

Petersen, M., August 23, 2002. CNN to Reveal When Guests Promote Drugs for Companies. The New York Times. Available from: http://www.nytimes.com/2002/08/23/business/cnn-to-reveal-when-guests-promote-drugs-for-companies.html.

Pew Research Center Internet Project, 2013. Health Online 2013. Available from: http://www.pewinternet.org/fact-sheets/health-fact-sheet/.

Pew Research Center Internet Project, 2014. Internet User Demographics. Available from: http://www.pewinternet.org/data-trend/internet-use/latest-stats/.

Pfizer, 2014. Public-Private Partnerships. Available from: http://www.pfizer.com/responsibility/grants_contributions/medical_and_academic_partnerships/public_private_partnerships.

Porter, J.E., 2014. Time to speak up for research. Science 344 (6189), 1207.

Reither, E.N., Olshansky, S.J., Yang, J., 2011. New forecasting methodology indicates more disease and earlier mortality ahead for today's American children. Health Aff. 30 (8), 1562−1568.

Research!America, 2011. Your Candidates-Your Health Survey, October 2011. Available from: http://www.researchamerica.org/uploads/YourCandidatesPoll2011.pdf.

Research!America, 2013a. National Poll. Available from: http://www.researchamerica.org/uploads/Nov13nationalpollwithASH.pdf.

Research!America, 2013b. National Poll: Clinical Research. Available from: http://www.researchamerica.org/uploads/June2013clinicaltrials.pdf.

Research!America, 2013c. Clinical Trials: Poll Data of Minority Populations. Available from: http://www.researchamerica.org/uploads/clinicaltrialsminorities.pdf.

Research!America, 2013d. Small Business Leaders Nationwide Survey. Available from: http://www.researchamerica.org/uploads/Feb2013smallbizsurvey.pdf.

Rockey, S.J., 2011. Charting a Renewed Course for Managing Financial Conflicts of Interest. National Institutes of Health. Available from: http://nexus.od.nih.gov/all/2011/08/23/charting-a-renewed-course-for-managing-financial-conflicts-of-interest/.

Rockey, S., 2014. A Proposed HHS Regulation and NIH Policy to Further the Impact of Clinical Trials Research. National Institutes of Health. Available from: http://nexus.od.nih.gov/all/2014/11/19/clinical-trials/.

Rose, D., 2010. A History of the March of Dimes. March of Dimes. Available from: http://www.marchofdimes.org/mission/a-history-of-the-march-of-dimes.aspx.

Shattuck, P.T., Wagner, M., Narendorf, S., Sterzing, P., Hensley, M., 2011. Post−high school service use among young adults with an autism spectrum disorder. Arch. Pediatr. Adolesc. Med. 165 (2), 141−146.

Sierra, F., Kohanski, R.A., American Federation for Aging Research, 2013. Geroscience offers a new model for investigating the links between aging biology and susceptibility to aging-related chronic diseases. Public Policy Aging Rep. 23 (4), 7−9.

Stand Up to Cancer, 2014. Hollywood Stars Align for SU2C Telecast & New "We're Calling You" Campaign. Available from: http://su2c.standup2cancer.org/press_release/view/hollywood_stars_align_for_su2c_telecast_new_were_calling_you_campaign.

Stehr-Green, P., Tull, P., Stellfeld, M., Mortenson, P.B., Simpson, D., 2003. Autism and thimerosal-containing vaccines. Am. J. Prev. Med. 25 (2), 101−106.

Steinberg, M., Rubin, E.R., Association of Academic Health Centers, 2009. The HIPAA Privacy Rule: Lacks Patient Benefit, Impedes Research Growth. Available from: http://www.aahcdc.org/Portals/0/pdf/AAHC_HIPAA_Privacy_Rule_Impedes_Research_Growth.pdf.

The Biomarkers Consortium, 2010. The Biomarkers Consortium Completes First Project to Show That Adiponectin Is a Predictive Biomarker for Type 2 Diabetes. Available from: http://www.biomarkersconsortium.org/press_release_adiponectin_predictive_biomarker.php.

The National Patient-Centered Clinical Research Network, 2014. PCORnet. Available from: http://www.pcornet.org/.

Thompson, W.W., et al., 2007. Early thimerosal exposure and neuropsychological outcomes at 7 to 10 years. N. Engl. J. Med. 357, 281−1292.

Thompson, L., 2000. Human gene therapy. Harsh lessons, high hopes. FDA Consum. 34 (5), 19−24. Available from: http://www.ncbi.nlm.nih.gov/pubmed/11590787.

Thompson, M.A., American Society of Clinical Oncology, 2014. Social Media in Clinical Trials. Available from: http://meetinglibrary.asco.org/content/11400101-144.

Trafton, A., October 24, 2014. Francis Collins: Scientists Must Engage With the World. MIT News. Available from: http://newsoffice.mit.edu/2014/francis-collins-compton-lecture-1029.

US Department of Health and Human Services, 2014. Health Information Privacy. Available from: http://www.hhs.gov/ocr/privacy/.

US Government Accountability Office, 2011. HHS Research Awards: Use of Recovery Act and Patient Protection and Affordable Care Act Funds for Comparative Effectiveness Research. Report No. GAO-11−712R.

Willnat, L., Weaver, D.H., 2014. The American Journalist in the Digital Age. Available from: http://news.indiana.edu/releases/iu/2014/05/2013-american-journalist-key-findings.pdf.

Women's Health Resource Center, The University of California − San Francisco, 2001. Women's Health Matters: A Brief History of Women's Health in the US. Available from: http://www.whrc.ucsf.edu/whrc/healthed/nwlswinter2001.pdf.

Woolley, M., Leshner, A.L., 2014. Out of sight, out of mind. Science 346 (6211), 791.

Wren, K., American Association for the Advancement of Science, 2014. Science Policy and Education Leaders Push for Progress in Climate Change, STEM Education. Available from: http://www.aaas.org/news/science-policy-and-education-leaders-push-progress-climate-change-stem-education.

Zerhouni, E.A., 2007. The promise of personalized medicine. NIH Medline Plus 2 (1), 2−3. Available from: http://www.nlm.nih.gov/medlineplus/magazine/issues/pdf/winter2007.pdf.

Research in Special Populations

Chapter 26

Research in Special Populations: Acute Illnesses; Critical Care; and Surgical Patients

Todd W. Rice and Gordon R. Bernard

Vanderbilt University School of Medicine, Nashville, TN, United States

Chapter Outline

Key Points

- Pragmatic trials have gained popularity in the design of studies in critically ill patients and may allow for integration of clinical studies within routine medical care of these patients.
- Utilizing nonprotocolized "usual care" as a control group may contribute inefficiencies in studies enrolling critically ill patients.
- Appropriate controls for research in surgical patients can be difficult. While not strictly forbidden, the ethical and regulatory details of using sham surgical procedures as controls should be considered carefully.
- Since the acute illnesses found in critically ill and surgical patients are dynamic and potentially rapidly fatal, informed consent of the patient is often not feasible and consent must be obtained from surrogates.
- Enriching enrollment with patients whose risk of the outcome of interest is attributable to the disease being studied may make the study more efficient. However, it also will limit enrollment and make for more difficult inclusion and exclusion criteria.
- Endpoints and outcomes for critical illness surgery have become more patient-oriented, focusing on longer-term morbidity and mortality and less on short-term surrogate outcomes, such as biomarkers.
- Adverse event reporting in trials enrolling critically ill and surgical patients can be complicated, due to the dynamic course of their care. Prospectively defining and collecting outcomes of interest, such as organ failure, death, or infections, will standardize the collection of these events and potentially make improved reporting.

Clinical and Translational Science. http://dx.doi.org/10.1016/B978-0-12-802101-9.00026-0

INTRODUCTION

As the number of clinical studies has dramatically increased, many difficulties in conducting research in populations of acutely ill patients have been encountered. Although these barriers may initially seem cumbersome and prohibitive, the research community has traversed these encumbrances remarkably well, often leading the way in innovative study design, new clinically relevant outcomes, and novel statistical analyses. This chapter will discuss a few of the unique nuances encountered in conducting clinical research in acutely ill patients, focusing on those with surgical and critical illnesses. Since many of these issues, such as study design, informed consent, randomization, and clinically relevant, yet attainable endpoints, are common to these research populations, they will serve as the sections for discussion in lieu of specific patient populations. However, many acutely ill populations have nuances that are unique to them. These special research conundrums, such as waiver of consent for emergency research, sham surgery, or determining adverse events in critically ill patients, will be discussed within the appropriate research section.

TRIAL DESIGN

Type of Study

The most appropriate study design varies for each clinical trial, depending on numerous factors, such as the patient population being studied, nature of the disease, treatment, prophylactic or diagnostic modality under investigation, availability of other accepted treatments, and known safety information (Friedman et al., 1998a,b). The relatively slowly changing nature and prolonged disease course of many chronic diseases make them ideal candidates for randomized, crossover designs (Friedman et al., 1998a,b), which represent the most efficient clinical study design. Although the order of the interventions is randomly assigned to ensure adequate blinding and avoid any treatment–period interaction, all patients enrolled receive both treatment arms of the study separated by a designated washout period, allowing comparisons to be made in the same patient, which eliminates interpatient variability. This significant reduction in between-patient noise results in a considerable increase in power. Often, these studies only require 10–20 total patients to achieve adequate statistical power. Unfortunately, the dynamic nature of acute illnesses usually limits the use of a crossover design in studying interventions in these populations.

Retrospective cohort studies are commonly used for investigating rare outcomes in relatively common diseases or in illnesses that are uncommon or rare. Furthermore, retrospective cohort studies are also utilized to find associations between risk factors and disease occurrences or outcomes. The retrospective nature allows a significantly large cohort populated with the disease or outcome of interest to be reasonably identified. The incidence, prevalence, and natural history of the disease or outcome can then be analyzed. For example, retrospective cohort design is often used to identify very rare side effects of already approved treatments. These studies can often be conducted with waiver of consent, since the information of interest has been collected prior to undertaking the study. The retrospective study design, however, merely allows determination of associations between findings and does not allow for the determination of cause-and-effect relationships. Statistical techniques, designed to abrogate potential biases in retrospective studies, may strengthen the association and increase the confidence in the extrapolation to a cause-and-effect relationship. Two examples of statistical methods that may result in more confidence in the resultant associations include dose–effect analyses and analyses adjusted for propensity scores. Analyses that demonstrate a dose–effect relationship (i.e., the risk ratio or odds ratio of the outcome increases as the dose of the exposure increases) between the exposure and the outcome strengthen the association (Raghunathan et al., 2014). Propensity score is a statistical method used in an attempt to try to adjust for potential confounders, especially confounding by indication, in cohort studies. A propensity score is developed from multivariable analysis of factors thought to contribute to the likelihood of receiving a specific treatment or intervention. Multiple factors from the multivariable analysis are then combined in a propensity score, which informs on the likelihood of each specific patient receiving that intervention. The propensity score can be used in two ways in adjusted analyses. Propensity score matching, which matches control patients to cases, is based on their individual propensity scores. The matched control cohort is then compared to the case cohort. The disadvantage of this technique is that it might exclude some cases or controls if a match is not available. The second method is to include the propensity score in a multivariable model, thus providing an outcome that is adjusted for (and independent of) the likelihood of receiving the intervention (Connors et al., 1996). This does not exclude patients in the cohort without a matching propensity score, but results in an output that is more difficult to interpret (i.e., the odds ratio for the propensity score is not very informative). Another potential negative of using propensity score adjustment is that, depending on the number of variables included in the propensity scoring, patients can obtain the same propensity score despite being widely different in the individual components making up the score. For example, a patient may get the same propensity for having a myocardial infarction by being diabetic, over the age of 55 years old, male, or hypertensive.

Prospective studies in acute illnesses employ either a sequential or parallel design. Utilizing historical controls as the comparator group represents one of the most common sequential study designs in clinical research of patients with acute illnesses (Friedman et al., 1998a,b). The historical control design, also known as a pre/post, or sequential analysis design, is commonly used to evaluate the effect of new interventions that have been, or will be, widely implemented. The study describes preimplementation outcomes and compares them to postimplementation outcomes. These are often used to report the effects of quality improvement projects. For example, an ICU begins to utilize a protocol to deliver nutrition. In describing the effect of this protocol on patient care, clinical outcomes from patients cared for in that ICU immediately prior to implementation of the protocol can be compared to those from a similar duration of time shortly after implementation of the protocol (Rice et al., 2005; Spain et al., 1999). In addition to quality improvement projects, this design is also commonly used for studies of interventions which have been so widely adopted as standard of care that randomizing patients not receiving them would receive criticism or in situations where implementing the intervention would also "carryover" into patients that might serve as controls. However, since patients are not randomized to care with or without the protocol, bias may exist in the study populations. In addition, the continual evolution of care over time may also contribute to differences in the outcomes (i.e., temporal bias), making the actual effect of the new intervention sometimes difficult to ascertain.

In parallel design studies, the different arms of the study occur concurrently, reducing or eliminating any differential effect of evolving care practices. Parallel design studies can be randomized or nonrandomized. Although group demographics can be manipulated in nonrandomized studies to effectively match characteristics that are known to significantly impact the outcomes of interest, some bias will likely continue to exist as not all characteristics can be matched and some characteristics which influence outcome may not even be known a priori. Propensity score methodology can also be used in prospective, nonrandomized studies, in an attempt to reduce bias. The purpose of randomization is to optimize similarity between study groups with regard to both known and unknown characteristics. The randomized, double-blind, placebo-controlled trial is considered the gold standard for determining the efficacy or effectiveness of a new therapeutic modality (Friedman et al., 1998a,b). Unfortunately, these trials are difficult to conduct in acutely ill patients. They often require a large number of patients. Since acute illnesses are unplanned and limited in both duration and prevalence, finding an appropriate number of eligible patients in a reasonable time period for completing the study is often challenging. As such, enrollment may require multiple centers, often with varying standards of care, which introduces additional variability and increases difficulty in detecting differences in the outcome of interest. Furthermore, coordinating the conduct of the trial across multiple centers takes enormous resources, including time, effort, and money.

Recently, some studies have utilized a group allocation design, or cluster randomization, where the study population is an entire community, hospital, or intensive care unit instead of individual patients (Donner and Klar, 2004; Hahn et al., 2005; Puffer et al., 2003; Noto et al., 2015). These study designs are frequently used for analyzing the effect of feedback or educational tools on changing clinical practice (Beck et al., 2005; Jain et al., 2006). The process of providing feedback or educating the staff may potentially carry over to all patients such that care of patients randomized to the control group in the same hospital or unit would be altered by the feedback or education intended solely for the intervention group. In addition, cluster randomization has been utilized to investigate treatment algorithms for patients otherwise unable to consent. For example, a study evaluating the effects of emergency medical teams in patients with cardiorespiratory collapse utilized cluster randomization of hospitals (Hillman et al., 2005). Individual hospitals were randomized to providing care via an emergency medical team versus their standard emergency response. Due to variation in care among centers, group allocation design requires a large number of centers participating in the study to minimize bias (Connelly, 2003; Campbell et al., 2004). Furthermore, the results from the study apply to the unit of study, namely the group being studied, and may not extrapolate well to individual patients as outcomes for individuals within clusters may be correlated (Rice et al., 2005).

An adaptation of the cluster randomized design is one that uses multiple crossovers between the arms of the study in each unit (Noto et al., 2015; Reddy et al., 2014). The crossovers serve at least two roles—they reduce intercluster variability as each cluster contributes data to both arms of the study. In addition, they provide more potential "units" for randomization. When an individual unit is randomized to the intervention and control group multiple times, it is akin to having additional "units" in the cluster randomization.

Selecting a Control Group

The choice of an appropriate control group represents one of the single most important decisions in clinical research, as an inappropriate comparator may invalidate the results of the study. Poor matching of control cohorts to case cohorts in epidemiologic studies predisposes the study to bias and complicates the interpretation of its results. Choosing an appropriate control for a prospective, interventional clinical trial carries similar importance. For investigations of new drugs or biologics, comparison to placebo, preferably in a blinded manner, represents the optimal control group (World Medical Association Declaration of Helsinki, 2000). The placebo-controlled study design is both preferable and ethical in acute diseases where no widely accepted pharmacologic therapy has been shown to improve outcomes, such as Acute Respiratory Distress Syndrome (ARDS). However, using placebo as the lone treatment for the control group can be problematic in diseases with readily accepted or proven therapies. In more chronic diseases, withholding a beneficial treatment for a short period of time to investigate a potential new treatment may be reasonable if the duration of time that patients spend off the proven treatment does not drastically alter the course of the disease. In addition, the use of "rescue" medication can be allowed (and even collected as an outcome), and safety parameters can be enacted which if met would result in the patient being removed from the study and returned to the widely accepted, standard treatment. Studies of new treatments for asthma have removed patients from some or all of their asthma medications with a safety net in place that if the patient has a certain number of exacerbations during the study, they are removed from the study and restarted on their standard asthma medications. The enactment of safety parameters in patients who deteriorate, however, becomes very difficult in critically ill patients as deterioration may represent the natural progression of the disease. Furthermore, using placebo as the control becomes difficult and possibly unethical if an accepted drug therapy already exists for the acute illness, especially if such a therapy has been shown to reduce mortality or significantly reduce morbidity. For example, assume drug X has already been shown to reduce bleeding and mortality in patients with severe trauma. A novel drug, Y, is proposed as a therapy for bleeding in trauma patients. A randomized, blinded trial comparing drug Y versus placebo would represent the ideal study design to evaluate the isolated effect of drug Y on bleeding. However, this would require withholding a treatment with known benefit from patients with a potentially lethal condition, which would violate one of the basic principles of conducting clinical research combined with medical care. The Declaration of Helsinki states that "The benefits, risks, and effectiveness of a new method should be tested against the current best prophylactic, diagnostic, or therapeutic methods (World Medical Association Declaration of Helsinki, 2000)."

Alternatively, drug Y could be compared directly to the known treatment, drug X. Unfortunately, this does not obviate the ethical issue, as half of the patients in the study (i.e., those randomized to the drug Y group) would still be deprived of the best known therapy. One possible solution to this dilemma is to utilize the proven therapy as a "rescue treatment," similar to reinstituting the asthma medications if a patient experiences a predefined number of exacerbations. Patients would be randomized to drug Y versus drug X. All patients who continue to experience significant bleeding after administration of the randomized therapy would receive drug X as a "rescue" treatment. Administration of open-label drug X as rescue therapy would then represent one of the study endpoints. Unfortunately, this type of design is only possible in situations where timing of the already accepted therapy is not vital. In many critical care diseases, such as sepsis, acute myocardial infarction, stroke, ARDS, or trauma, time to initiate treatment is prognostic, and delays in initiating effective treatment have been shown to result in worse outcomes (Moore et al., 1997; Garnacho-Montero et al., 2003, 2006; Levy et al., 2005).

In instances where the new drug therapy and the approved therapy do not negatively interact, one possible study design is to compare the investigational treatment versus placebo as an "add-on" therapy to the already accepted treatment. For example, in the case of bleeding trauma patients, the already approved drug X would be given per primary team either in a protocolized or unprotocolized manner as part of usual routine care. Patients would then be randomized to receive drug Y or placebo, in conjunction with drug X in those patients whose primary medical team chose to utilize drug X as part of routine care. If the already approved therapy is not universally employed, this design may also allow some evaluation of the isolated effect of drug Y versus placebo on bleeding in the subset of patients who were not treated with drug X. Unfortunately, many times, patients who are not given the widely accepted therapy are disparate from those who are making the results less generalizable and limiting the usefulness in the overall disease population.

Placebo-Controlled Studies in Surgical Patients or Nonsurgical Invasive Procedures

The Declaration of Helsinki declares "The benefits, risks, burdens and effectiveness of a new method should be tested against those of the best current prophylactic, diagnostic, and therapeutic methods (World Medical Association Declaration of Helsinki, 2000)." This does not preclude the use of placebo-controlled studies, but implies that they should be limited to conditions where no proven prophylactic, diagnostic, or therapeutic method exists. Although relatively rare in studies evaluating surgical and nonsurgical invasive procedures, placebo-controlled trials are commonly employed to test the risks, benefits, and effectiveness of new pharmacologic therapies. The placebo is usually an inert substance with identical appearance to the active medicine being studied. Subjects randomized to the placebo arm forgo the hoped benefits (and possible risks) of the experimental treatment, but are presumed to not incur any additional risks from the placebo treatment. The ethics of placebo surgery, or sham controls in the case of nonsurgical invasive interventions, however, have been widely debated (Albin, 2005; Horng and Miller, 2002, 2003; Miller, 2005; Miller and Kaptchuk, 2004; Miller, 2003; Weijer, 2002; Macklin, 1999; Flum, 2006; Angelos, 2007). Like patients receiving placebo medications, patients randomized to sham surgery or procedures forego the hoped benefits, but unlike their pharmacologic counterparts, may incur significant risks of the placebo procedure itself. In addition to the rare risks of receiving anesthetics, these patients also may incur the risk of damage to nearby structures, bleeding, and wound infection from the placebo surgical incision or the sham procedure. Critics argue that this represents an unacceptable risk for no possible benefit. As such, it violates one of the essential standards of research, namely to minimize the risk of harm to participants, and constitutes an infringement on the rights of research subjects (Weijer, 2002; Macklin, 1999).

Without the use of sham surgical controls, however, participants are aware of whether or not they received the surgical intervention. Unblinded patients who have undergone a large surgical procedure tend to be biased toward believing that they received benefit from that procedure. This bias may lead to false-positive results and a perception that the surgical intervention is beneficial. Since these biases have a larger effect on subjective outcomes, such as pain, function, or quality of life scores, their influence is greater in surgical interventions for chronic conditions. One recent randomized, placebo-controlled, blinded trial of arthroscopic lavage versus arthroscopic debridement versus sham knee surgery in patients with osteoarthritis demonstrated no benefit to either arthroscopic lavage or debridement when compared to sham surgery (Moseley et al., 2002). Patients in all three arms reported similar improvements in knee pain and function over the 2 year follow-up for the study. Unfortunately, simply blinding patients may not alleviate these biases. The surgeon performing the procedure cannot reasonably be blinded and unless they continue to perpetuate the sham, they may influence the post-intervention perceptions of the patient. Therefore, sham surgical controls are best suited for studies where the surgeon does not have an ongoing relationship with the patient after the procedure is completed (Angelos, 2003).

Using sham surgery as a control may be unethical, however, when studying surgical interventions in acute surgical conditions, especially those which result in significant morbidity or mortality and are devoid of other treatment options. For example, few would consider a study comparing appendectomy versus sham surgery for patients with appendicitis or a perforated viscus ethical. Studies in these conditions must compare the risks and benefits of surgery versus medical management and not employ placebo surgery (Salminen et al., 2015). When conditions do not allow sham interventions to be used as a control, the effect of treatment bias can be minimized by utilizing objective outcomes which are less likely to be affected by patient or physician perception, such as mortality, infection rates (if surveillance is standardized among all patients), rate of complications, need for surgery (or second operation), or development of new organ failures (Salminen et al., 2015).

Sham procedures are not limited to surgical interventions (Kaptchuk et al., 2000). As more disease states are being treated with less-invasive interventions, the issue of sham controls has expanded to other situations, such as medical devices and nonsurgical interventions (Schwartz et al., 2007; Flachskampf et al., 2007). Many medical interventions and devices pose similar problems assessing outcomes in the face of treatment bias. For example, percutaneous coronary stenting has long been considered the standard of care for patients with noncritical coronary artery disease. However, this intervention was readily adopted as a standard practice without ever being compared to a sham intervention accessing the arterial circulation without actually deploying stents. In fact, the benefit of percutaneous stenting has been questioned. A trial comparing percutaneous coronary stenting demonstrated no additional benefit to best medical care with combined pharmacologic treatments in patients with noncritical coronary artery disease (Boden et al., 2007). Although this study did not utilize a sham intervention as the control, the bias of receiving the intervention was minimized by utilizing the objective outcomes of combined mortality or nonfatal myocardial infarction as the primary outcomes.

Despite the controversy, sham controls in studies of surgical or other invasive interventions or medical devices are often necessary for rigorous experimental design and may be instrumental in excluding false-positive trial results (Albin, 2005; Horng and Miller, 2002; Miller, 2003). Guidelines have been proposed under which these trials can be conducted. A placebo-controlled trial of an invasive procedure or medical device can be ethically justified if: (1) the placebo control is methodologically necessary to test the clinically relevant hypothesis, (2) the valuable information to be gained justifies the risk of the placebo control, (3) the risk of the placebo control itself has been minimized and does not exceed the threshold of acceptable research risk, and (4) the administration and risks of the placebo control are adequately disclosed to the participants and authorization is obtained via informed consent (Horng and Miller, 2003).

USUAL CARE IN CRITICALLY ILL PATIENTS

Studies Comparing Usual Care Components

A number of the prospective, randomized trials conducted in patients with critical illness over the last 15 years have investigated components of usual care and not pharmaceutical agents. Numerous investigations have explored the risks and benefits of common aspects of usual care such as patient positioning (Guerin et al., 2013; Drakulovic et al., 1999), ventilator procedures (Ventilation, 2000; Brower et al., 2004a; Stewart et al., 1998), fluid management (ProCESS Investigators et al., 2014; Rivers et al., 2001; Wiedemann et al., 2006), nutrition (Rice et al., 2005, 2011, 2012), or use of the pulmonary artery catheter (Wheeler et al., 2006; Richard et al., 2003). Many of these studies have utilized pragmatic trial designs, with limited "research-only" interventions, and cluster randomization. Selecting appropriate control arms in these nonpharmacologic intervention studies has proven challenging (see Table 26.1). Ideally, any intervention under investigation should be compared to some measure of usual care. However, determining which measure, or defining what constitutes usual care, remains a topic of considerable debate (Silverman and Miller, 2004; Deans et al., 2004, 2007; Eichacker et al., 2002; Brower et al., 2004a,b). The choice of control group intervention may be relatively simple if the spectrum of usual care for that intervention is limited. For example, supine positioning represents a logical control for studies investigating the effect of different body positions in the care of critically ill patients. Unfortunately, variation in most aspects of usual care is large, unexplained, and unsupported by evidence. Selection of a control group in these cases represents a significant challenge, which becomes even more arduous when the intervention being studied is encompassed within the spectrum of usual care. Many completed studies in patients with critical illness have focused on comparing two controlled interventions which both lie within the spectrum of usual care. For example, prospective trials have compared methods of ventilator weaning (Ely et al., 1996), tidal volumes of 6–12 ml/kg of predicted body weight (Ventilation, 2000), a "wet" versus "dry" fluid management strategy (Wheeler et al., 2006), trophic versus full-calorie enteral feedings (Rice et al., 2012), and two commonly utilized transfusion thresholds (Hebert et al., 1999). Critics have argued, however, that these trials should include one arm representing uncontrolled usual care (Deans et al., 2004, 2007; Eichacker et al., 2002). Inclusion of this "wild-type" group would allow for comparison to "unrestricted" physician practices in clinical care. Unfortunately, comparing one intervention within the spectrum of usual care with unrestricted physician practice often encompasses significant overlap between the intervention and usual care arms, which results in the loss of statistical power and the subsequent requirement to enroll additional patients to adequately address the question of interest. Alternatively, the study could compare two different interventions within the spectrum of usual care and a third study arm of unrestricted usual care. This design also requires 50% more patients due to the additional study arm. Further complicating the study interpretation would be an inability to describe or even ascertain what care was actually provided in the unrestricted physician practice group given its uncontrolled nature and likelihood of change over the course of the study. For example, studies of pulmonary artery catheters in guiding treatment of many critical illnesses have resulted in results that are difficult to interpret due to the inability to describe the treatment received in one or both of the study populations (Richard et al., 2003; Sandham et al., 2003; Harvey et al., 2005; Binanay et al., 2005). On the other hand, proposing guidelines for the usual practice arm would likely render it no longer representative of unrestricted physician practice. Likewise, usual practice in clinical trials is likely to suffer from a Hawthorne effect or a change in practice simply because it is being studied as part of a research project.

Protocolized Nonstudy Treatment

Protocol-driven care has reduced unnecessary variation and improved patient outcomes in critically ill patients (Holcomb et al., 2001; Meade and Ely, 2002). Consequently, the use of protocols has dramatically increased over the last decade. Protocols are now routinely utilized for multiple aspects of clinical care in the ICU, including ventilator management and weaning (Ely et al., 1996; Kollef et al., 1997; Brochard et al., 1994), sedation(Kress et al., 2000; Girard et al., 2008; Mehta et al., 2012), deep venous thrombosis and peptic ulcer disease prophylaxis, nutrition (Rice et al., 2005; Spain et al., 1999; Barr et al., 2004), and tight glucose control (Van den Berghe et al., 2001; NICE-SUGAR Study Investigators et al., 2009). In addition to being used to direct care in both the intervention and control arms, protocols can also be employed in clinical trials to standardize procedures outside of those being directly investigated in the study. This protocolization of routine care reduces noise in the study by decreasing variability in the care provided to patients in the study. Standardizing non–study-related procedures assumes increased importance in studies where the primary endpoint is directly affected by using protocols or where the research team and/or patients are not blinded to treatment allocation. For example, standardizing ventilator and weaning procedures may be very useful if duration of ventilation represents one of the primary outcomes. Simply implementing protocols for patient care, however, is only part of the solution. Standardization of treatment can only be achieved if compliance with the protocols occurs in the vast majority of patients. Recently, studies have utilized random checks of important aspects of the protocolized care to document compliance rates to the protocols (Wiedemann et al., 2006).

Standardizing important aspects of usual care through protocols does have some drawbacks. For example, the simple act of standardizing part of the care may render it different than usual care. In addition, the requirement to use protocols in

TABLE 26.1 Different Study Design Options for Investigating Nonpharmacologic Interventions Within the Usual Care Spectrum

Treatment Arm	Example	Advantages	Disadvantages
Two Arm Trial			
Protocolized intervention #1 within spectrum of usual care versus uncontrolled usual care (care left entirely to treating physician's discretion)	Protocol versus uncontrolled physician practice for sedation of mechanically ventilated patients	• Allows comparison to usual care practices (i.e., effectiveness trials) if usual care is uniform or evidence based • Investigates benefit/harm of individual customization of care by physicians versus controlled or protocolized intervention	• Uncontrolled usual care difficult to define or describe • Usual care subject to change over time • Usual care subject to the influence of research environment (Hawthorne effect) • Some patients in usual care arm may not receive "best current care" • Overlap of treatment arms may result in the loss of statistical power and/or require enrollment of additional patients • Differences in outcome may arise from protocolization and not the actual intervention under study
Two Arm Trial			
Protocolized intervention #1 within usual care spectrum versus protocolized intervention #2 within usual care spectrum	Mechanical ventilation with upper limit of usual tidal volume versus lower limit of usual tidal volume Trial comparing low transfusion threshold versus high transfusion threshold (both within usual care spectrum)	• Easy to describe/define both arms • Good separation of treatment arms, resulting in increased statistical power	• Does not allow individual titration of treatment to unique patient needs (e.g., customization of airway pressures for each patient) • Often tests two extremes within the usual care spectrum • If one arm demonstrates benefit, it still may not be superior to usual care (i.e., treatment effects may be U-shaped curve) • May limit enrollment if physicians have strong preference for which arm they think is best
Three Arm Trial			
Protocolized intervention #1 within usual care spectrum versus protocolized intervention #2 within usual care spectrum versus uncontrolled usual care		• Able to compare two interventions to usual care • Easy to describe/define the two controlled interventions • Investigates benefit/harm of customization of care versus two controlled interventions • May help define whether treatment effects fit U-shaped curve • May answer both efficacy and effectiveness in one study • Improves safety if usual care is superior to both controlled arms	• Need more enrolled patients than two arm trial resulting in: • Increased expense • Increased number of adverse outcomes • Uncontrolled usual care difficult to define or describe • Usual care subject to change over time • Usual care subject to the influence of research environment (Hawthorne effect) • Usual care that is controled or protocolized may not represent actual usual care • Some patients in usual care arm may not receive "best current care"

the nonintervention aspects of the study may inhibit enrollment, especially from physicians or institutions where protocols, or the principles used in the protocols, are not routinely used or accepted. For example, a significant number of eligible patients were excluded from the ARDS network study comparing conservative and liberal fluid management strategies because patient characteristics prevented, or the primary team refused to allow, the use of the required ventilator management strategy (Wheeler et al., 2006). Further, the overall improvement in outcomes derived from the use of protocols may result in reduced statistical power and require larger sample sizes to detect significant differences. Although this poses some challenge in designing future studies, most agree that improving outcomes is part of minimizing risks to patients and should be inherent to all clinical research.

INFORMED CONSENT

The Belmont Report outlines respect for persons, beneficence, and justice as the three fundamental principles in the ethical conduct of human research. Informed consent is required by the moral principle of respect for persons and is essential to the principle of autonomy. One of the basic principles of the Declaration of Helsinki states "In any research on human beings, each potential subject must be adequately informed of the aims, methods, anticipated benefits and potential hazards of the study and the discomfort it may entail. He or she should be informed that he or she is at liberty to abstain from participation in the study and that he or she is free to withdraw his or her consent to participation at any time. The physician should then obtain the subject's freely-given informed consent, preferably in writing (World Medical Association Declaration of Helsinki, 2000)." Legally effective informed consent must therefore contain three elements: information, comprehension, and freedom from undue coercion. In other words, participants must be offered information about the research, possess the capability of understanding the information, and voluntarily agree to participate in the study without coercion and with the freedom to retract their approval without the fear of penalty or repercussions.

Ensuring that acutely ill patients consent to participate in research without coercion can be challenging. Ideally, adequate time should be allotted during the consent process for the participant to have any questions answered to his or her satisfaction. In many studies of chronic diseases, where the disease course is more prolonged, the potential participant is encouraged to take the consent form home and discuss his or her possible participation in the study with relatives and friends. However, the fact that patients with acute illnesses often need urgent medical care may preclude any delay in obtaining consent. This is especially true when the question being studied involves aspects of the treatment for the acute illness and when the illness under study requires life-saving treatment emergently, such as in patients with life-threatening illnesses such as major trauma, acute myocardial infarction, stroke, septic shock, cardiovascular collapse, or respiratory failure. In these cases, special care must be taken to ensure that consent is not hurried and that the participant understands the risks, benefits, and alternatives to participating in the study, while also ensuring that the consent does not delay potentially life-saving care. If these conditions cannot be simultaneously met, providing timely medical care must take precedence and the patient should not be enrolled in the study.

Informed consent for research involving patients undergoing surgery may also be prone to coercion. Some ethicists have argued that consent for research should not occur on the day of surgery in the preoperative holding area because of increased patient anxiety prior to surgery and of the possibility that the participant will feel coerced into agreeing to participate to either obtain their surgery or please their surgeon. Peer-reviewed literature on the subject, however, suggests that this may not be the case. A large study of patients consented on the day of surgery found that all participants reported being capable of making the decision to participate (Mingus et al., 1996). In addition, a retrospective survey of patients enrolled on the day of surgery for trials studying clinical anesthesia found that most patients reported understanding the consent form and the purpose of the trial and did not feel obligated to participate (Brull et al., 2004).

In addition, when the person obtaining consent is also the treating physician or a member of the primary team caring for the patients, the risk of therapeutic misconception must be minimized (Appelbaum et al., 1987; Miller and Brody, 2003; Miller and Rosenstein, 2003; Appelbaum et al., 2004; Chen et al., 2003). In other words, special precautions should be undertaken to ensure that the person providing consent, whether the participant or their legally authorized representative, understands that the procedures are for the purpose of developing new knowledge (i.e., research) and does not mistakenly believe that they are being performed with the primary goal of providing benefit to the participant (i.e., treatment). Although an individual subject may benefit as a result of participating in a clinical trial, the primary goal of the research is to develop new knowledge about the condition. If protocols are being compared, as often occurs in critical care research, potential subjects should be made aware of the experimental nature of these protocols and whether they would be utilized in patients not enrolled in the study (Silverman et al., 2005). Ascribing therapeutic intent to research protocols represents another means of therapeutic misconception. One way of avoiding the appearance of therapeutic misconception is to have a well-informed member of the research team, who is completely independent of the patient—doctor relationship, to obtain informed consent for the study. This may be more difficult in acute illnesses where obtaining consent expeditiously is required.

Surrogate Informed Consent

Informed consent should be prospectively obtained from the participant prior to the initiation of any research procedures. In the past, investigators utilized a mechanism of deferred consent to conduct emergency research (Abramson et al., 1986; Levine, 1995; Miller, 1988). Criticism of this practice arouse on the grounds that participants could not truly consent for procedures that had already occurred and resulted in the practice being abandoned. Today, the preferred mechanism for obtaining consent involves having the participant sign a written informed consent form explaining the research. Unfortunately, obtaining consent directly from the participant is often not possible in many acute disease processes, such as trauma, stroke, shock, or respiratory failure, due to the prevalence of cognitive impairment in these populations. Although only temporary in many instances, the acquired cognitive dysfunction which impairs the capacity of these patients to consent for themselves often coincides temporally with the time the patient is being considered for enrollment in the study. For example, in the PAC-MAN study investigating the effects of pulmonary artery catheter utilization in critically ill patients in the United Kingdom, only 2.6% of the initial 500 patients were able to consent for themselves (Harvey et al., 2006). According to the research ethics established by the Declaration of Helsinki, such cognitively impaired patients should not be included in the research unless the aims of the research are to promote the health of the population represented and the research cannot instead be performed on patients able to legally consent for themselves (World Medical Association Declaration of Helsinki, 2000).

Three types of legally authorized representatives are identified for granting permission for individuals who are unable to consent for themselves to participate in research. Health-care agents, or durable powers of attorney executed while the individual had decision-making capacity, possess the greatest power to make such decisions. Next in line are legal guardians or conservators, who are appointed by a judicial body to make decisions for an individual determined to be incompetent. Unfortunately, none of these legally authorized representatives are established in the vast majority of cases, owing to both the significant expense and prior planning which is required for their enactment. In most instances, where a health-care agent, durable power of attorney, legal guardian, or conservator has not been previously appointed, a surrogate designated to make health-care decisions should be approached for consent. The same person identified to make health-care decisions should also be the surrogate for making decisions regarding research participation. In the previously discussed pulmonary artery catheter study, 81% of patients were enrolled using surrogate consent from a relative (Harvey et al., 2006). Although the surrogate is usually the next of kin, the hierarchy of who can function in this role (i.e., spouse, adult child, parent, sibling, other relative, other interested party) is determined by local legislation. With laws varying from location to location, identifying the legally authorized surrogate is often quite arduous, especially when it must be done as quickly as possible. In addition, telling a patient's closest mate that they do not have the legal authority to consent for the patient even though they have clearly been the patient's primary caretaker can also cause problems, sometimes culminating in family disagreements or even feuds. Most researchers who routinely obtain surrogate consent for investigations can relate at least one story of proceeding through the consent process and near the end (or sometimes even after consent is signed), finding out that the person who consented is "not really the patient's wife, but...."

Surrogate consent is not applicable in all research situations. Some local regulations may restrict the type of research that can be conducted under surrogate consent. For example, research that does not hold out any possibility of direct benefit to the participant, such as genetic research, may be prohibited under certain governing statutes. Furthermore, consent from a proxy often does not accurately reflect the patient's wishes despite that being the commission of the proxy. Multiple studies have found that the decision of the proxy differs from that of the patient in 20—35% of cases (Shalowitz et al., 2006; Ciroldi et al., 2007; Coppolino and Ackerson, 2001).

Despite some concerns over the possible loss of the use of legally collected research data (Nichol et al., 2006), the ethical principle of autonomy and self-determination dictates that the acutely cognitively impaired participant who regains the capacity to make health-care decisions should be "reconsented" with the full informed consent document, even if this occurs after research procedures have been completed. Although this introduces the possibility that a participant could refuse to allow his or her legally collected data to be utilized in the research analyses, this seems especially important given the relatively high rate of patient—surrogate consent disparity. This is tantamount to a participant who initially consents, but subsequently discontinues participation in a research study prior to completion of the study. In fact, one of the basic principles of conducting medical research is that participants remain free to withdraw their consent to participate at any time without reprisal (World Medical Association Declaration of Helsinki, 2000). Despite studies showing disparity between patients and surrogates in up to one-third of cases, data suggest that cognitively impaired, acutely ill patients enrolled using proxy consent usually agree to allow the use of previously collected data and to remain in the study after they regain the ability to consent. When examined prospectively, only 3.2% of patients who regained mental competency refused to provide retrospective consent to participate in the study for which their proxy previously consented them (Harvey et al., 2006).

Waiver of Consent in Emergency Research

Traditionally, waiver of consent had been restricted to noninterventional, minimal risk studies. These studies employed research procedures that would not require consent outside of the research context, such as collection of data acquired during regular care. In addition, the nature of the studies was such that the informed consent document represented the only link between the patient and their identifiable information. Furthermore, obtaining consent was often not practicable, such as that encountered when conducting a cohort or retrospective observational study with thousands of patients. In these cases, loss of confidentiality was the major risk of the research, and protecting privacy became the main concern for oversight. Protected health information could be collected as long as the identifiers were adequately protected from improper disclosure and destroyed at the earliest opportunity consistent with the research.

This traditional utilization of waiving consent, however, only applied to studies which represented no more than minimal risk. Novel or previously unused methods, devices, drugs, or other biological agents intended for emergency medical care clearly represented greater than minimal risk interventions, and thus, trials studying these interventions did not qualify for the waiver. As such, these investigations could only be conducted in patients who were able to consent for themselves or who had a legally authorized representative or appropriate surrogate immediately accessible and willing to consent on their behalf. Most recognize, however, that in certain time-sensitive situations, such as acute cardiac conditions, trauma, emergent surgical conditions, acute brain injury, or other emergent critical care diagnoses, obtaining informed consent from either the patient or their legally authorized representative is impractical, if not impossible. This requirement for informed consent represented a prohibitive impediment (Levine, 1995; Lotjonen, 2002), which rendered most research in these emergency medical conditions virtually impossible to conduct. However, the limited data supporting most diagnostic, therapeutic, and clinical decisions used in these populations made researching them particularly important. In 1995, emergency and critical care physicians joined with cardiologists and neurologists to develop and publish a consensus statement on emergency research (Biros et al., 1995). This prompted both the United States Food and Drug Administration (FDA) and the Office of Human Research Protections (OHRP) to issue regulations in 1996 entitled, "Emergency Research Consent Waiver," which permitted a waiver of the general requirements for informed consent in certain research settings involving greater than minimal risk to participants. Although potentially applicable to emergent surgical or critically ill patients, use of this exception from informed consent (EFIC) has largely been used in emergency department and pre-hospital research. As such, further description of the methods and regulations for this exception are detailed in chapter 27.

OUTCOMES

Associated Versus Attributable Outcomes

Associated, or absolute, outcomes include all outcomes of interest occurring in patients with the disease or exposure under investigation, regardless of their causal relationship to the disease or exposure (Rubenfeld, 2003; Wood et al., 2003). Attributable outcomes are defined as only those outcomes resulting directly from the disease or exposure. Associated outcomes encompass both those resulting directly from the disease being studied (i.e., attributable outcomes) and those not caused by the disease but occurring in patients who have the disease (i.e., unattributable outcomes). For example, some deaths may result from factors unrelated to the actual disease being studied, such as comorbidities or withdrawal of care, but happen to occur when the patient also has the disease or exposure being studied. Since these deaths do not directly result from the disease, they are considered unattributable deaths. They are unlikely to be affected by treatments designed specifically for the condition being studied, because they are not directly related to that condition. However, these deaths are still counted as associated outcomes, because they occurred in patients with the disease or exposure. Although calculating associated outcomes is relatively straightforward, as it merely represents the total number of events in people with the disease, assigning attribution is often more challenging. For example, a patient with atherosclerotic coronary disease and dilated cardiomyopathy develops ARDS after being involved in an automobile collision. On hospital day 5, he develops ventilator-associated pneumonia and renal failure with hyperkalemia. That evening, the patient develops ventricular tachycardia and suffers a cardiac arrest. Resuscitation is unsuccessful. This death is clearly associated with the trauma as it occurred in the patient's peritraumatic period. However, determining the actual cause of this death is much more difficult. Was it attributable to the traumatic event, the patient's underlying cardiomyopathy, the hyperkalemia secondary to the renal failure, ARDS, or some combination of these?

Ideally, attributable risk should be utilized in power calculations to calculate sample sizes. Outcomes which occur in patients with the disease being studied but are not directly attributable to the disease have little chance of being altered by interventions targeting the disease. Inclusion of these unattributable events in the outcome calculations reduces the

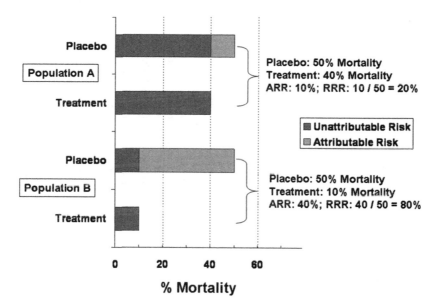

FIGURE 26.1 Attributable versus Unattributable Risk. Differences in risk reduction for varying amounts of absolute mortality that is attributable to the disease being studied. In population A, only 20% (10% out of a total of 50%) of the total mortality is attributable to the disease compared to 80% (40% out of a total of 50%) of the same absolute total mortality in population B. For both populations, the treatment eliminates mortality from the disease (attributable mortality), but does not affect unattributable mortality. This results in significantly larger reductions in associated mortality in population B where the attributable mortality is higher in the population. *ARR*, absolute risk reduction; *RRR*, relative risk reduction.

statistical power to detect an effect. Unfortunately, attributable risk is often difficult to determine precisely, resulting in most studies being powered for sample size based on absolute, or associated, risk. For example, suppose the overall mortality rate in a sample population (Population A) from septic shock is 50%, yet 80% of that mortality is secondary to comorbidities or withdrawal of care (factors not directly related to septic shock). The mortality rate that can be attributed to septic shock is then only 10% (i.e., 20% of 50%). Even if a highly effective, new intervention eliminates all deaths directly attributable to septic shock, the overall mortality rate will still be 40%, because the unattributable deaths would not be affected by the treatment (Fig. 26.1). In other words, the highly effective intervention would only produce a 10% absolute (and 20% relative) improvement in overall mortality, since it eliminates the 10% risk of dying from septic shock, but does not affect the 40% residual unattributable mortality from other causes. Compare that to a population (Population B) with the same overall mortality rate (50%), but in which 80% of all deaths are directly caused by septic shock. The same highly effective intervention, which eliminates deaths attributable to septic shock, would result in a 10% overall mortality rate, or a 40% absolute (or 80% relative) reduction in mortality. Although the highly effective treatment has the identical effect of eliminating all deaths caused by septic shock, the first scenario would require 400 patients in each arm of the study to detect the relatively small absolute difference of 10% mortality with 80% power and a two-sided alpha of 0.05, while the second would require only 20 participants in each arm because the effect in the entire group is fourfold higher (40% vs. 10% absolute reduction in mortality). Similarly, enrolling patients where death is unpreventable, regardless of whether or not it is attributable to the disease of interest, reduces the power of the study to detect the efficacy of an intervention. Exclusion criteria (i.e., moribund status or imminent death) are used in most studies in an attempt to enrich the cohort of study patients with those who are at high risk of death from, and likely to be responsive to, the therapy, but this can only be effective if the high likelihood of death is recognized or can be accurately predicted.

Endpoints

Many different outcomes have been utilized as endpoints in clinical trials of patients with acute illnesses, especially those with critical illness. These endpoints have evolved over time with new statistical approaches, changing disease courses, and more effective treatments. A detailed discussion of the advantages and disadvantages of these endpoints is beyond the scope of this chapter, but a brief overview is outlined in Table 26.2. Initially, most patient-oriented investigations in acute illnesses were conducted at individual research centers and enrolled up to a few hundred patients. As the limited number of patients available for enrollment did not provide adequate statistical power to look at endpoints such as mortality, these single-center studies utilized important clinical or physiologic markers that were easy to measure and presumed to correlate

TABLE 26.2 Advantages and Disadvantages of Different Clinical Endpoints for Clinical Research Studies

Endpoint	Examples	Advantages	Disadvantages
Clinical surrogate (i.e., physiologic or biochemical surrogate)	• Oxygenation • Pulmonary compliance • Change in biological marker, such as cytokine levels or laboratory values	• Objective measure that can be done at the bedside • Adequate statistical power with only a few hundred patients • Demonstrate desired physiologic effect (i.e., "proof of concept") • Easily understood by clinicians • Can be conducted at a single center or relatively few sites	• Often does not correlate with clinical outcome (e.g., oxygenation and mortality) • Not accepted as "definitive" trial for regulatory approval • Does not usually result in high acceptance into practice
Morbidity measures	• Length of ventilation • Length of hospital stay • Length of ICU stay	• Measure of disease morbidity • Usually fairly easy to measure • Continuous nature of endpoint may increase statistical power • Easily understood by clinicians	• May be discordant with mortality (i.e., early deaths decrease lengths of stay) • Some subjectivity in the measurements • May be influenced by physician bias (in unblinded studies)
Short-term Mortality	• 28-day all-cause mortality • Hospital all-cause mortality • ICU all-cause mortality	• Allows for reasonably short follow-up • No need for long-term patient tracking • Objective outcomes • Compares important clinical outcomes • Easily understood by clinicians	• Requires larger sample sizes • Misses differences in mortality after end of measurement period (i.e., after ICU discharge or after 28 days) • Dichotomous endpoint results in reduced statistical power • Requires collaboration of multiple sites
Combined endpoints	• Ventilator-free days • ICU-free days	• Combined measure of morbidity and mortality • Continuous nature of endpoint may increase statistical power	• May be discordant with mortality (i.e., more ventilator-free days, but higher mortality) • Not routinely measured or considered as outcomes by clinicians
Long-term morbidity	• Neurocognitive measures • Neuromuscular weakness • Quality-of-life (QOL) measures	• Important clinical outcome (especially as more patients survive) • Allows measure of long-term effect of disease • Functional measure (provide more information than survival measures)	• Often subjective measurements • Must follow patients postdischarge (difficult to find, do not return for follow-up visits, etc.) • More expensive • Longer follow-up results in longer time to complete studies and risk for loss to follow-up • Some measures (i.e., QOL) still being validated

| Long-term mortality | • 60-day mortality
• 180-day mortality | • Better representation of disease mortality
• Renders information on posthospital survival | • Must follow patients postdischarge (difficult to find, do not return for follow-up visits, etc.)
• More expensive
- Longer follow-up results in longer time to complete studies and risk for loss to follow-up
• Requires collaboration of multiple sites to enroll adequate number of patients |
| QOL measures | • Health-related QOL
• Short form health survey (SF-36, SF-12, SF-8)
• Stroke-Specific Quality of Life Scale
• Quality-adjusted life years | • Very patient centered
• Represents morbidity in survivors
• Can incorporate cost | • Must follow patients postdischarge (difficult to find, do not return for follow-up visits, etc.)
• Requires patients (or caregivers) to fill out surveys
• Longer follow-up results in longer time to complete studies and risk for loss to follow-up
• Requires collaboration of multiple sites to enroll adequate number of patients
• Relevance may be less if differences in mortality in groups |

with prognosis as surrogate endpoints. For example, one surrogate marker frequently utilized as a primary endpoint for studies of patients with respiratory failure requiring mechanical ventilation was oxygenation. Hypoxia in these patients was often dramatic, easily measured via arterial blood gases or pulse oximetry, and presumed to correlate with disease severity and overall clinical course. However, epidemiologic data emerged demonstrating that most patients with acute respiratory failure or trauma died from multisystem organ dysfunction rather than refractory hypoxia or respiratory failure (Stapleton et al., 2005; Montgomery et al., 1985; Estenssoro et al., 2002; Ferring and Vincent, 1997). Larger trials were subsequently designed to evaluate hospital or 28-day all-cause mortality, as many believed that survival to the earlier of 28 days or hospital discharge represented successful treatment of patients with these critical illnesses. In fact, some of these larger trials confirmed that not only did improvement in oxygenation not translate into reduced mortality (Brower et al., 2004a,b; Gattinoni et al., 2001; Taylor et al., 2004), but also sometimes an intervention might improve survival despite worsening hypoxia (Ventilation, 2000). Although improvement in many surrogate outcomes or biomarkers do not translate into improved clinical outcomes, these easy-to-measure physiologic variables continue to represent useful endpoints for phase II "proof-of-concept" investigations to ensure a physiologic effect prior to investing lots of time and resources examining clinical outcomes which require enrollment of large numbers of patients in subsequent phase III clinical trials (Marshall et al., 2005).

As care of the acutely ill patient has improved, short-term mortality rates have decreased. These decreasing event rates have made demonstrating a statistically significant improvement in short-term mortality through additional interventions increasingly difficult. As the overall event rate decreases, more patients are needed to be enrolled in a study to encompass the number of events needed for adequate power to detect a significant difference. In other words, the same relative reduction corresponds to a lower absolute reduction when the overall rate is lower. For example, patients with ARDS experienced 50% mortality rates 20 years ago. Today, the mortality rates are 20% (Guerin et al., 2013; Rice et al., 2011, 2012; Papazian et al., 2010; National Heart Lung and Blood Institute ARDS Clinical Trials Network et al., 2014), meaning that a 10% relative reduction in mortality today represents only a 2% overall reduction compared to 5% two decades ago. In addition, the larger number of survivors has highlighted the significant effect that morbidity has played in the disease course. Duration of ventilator support or length of ICU or hospital stay has become increasingly popular outcomes to study. Unfortunately, differences in mortality complicate interpretation of these early morbidity measures, as patients who die early in the disease course have fewer days on the ventilator. This resulted in the derivation of composite endpoints, such as ventilator-free days (Marshall, 1999; Schoenfeld and Bernard, 2002). These outcomes have become known as "failure-free" outcomes, and they encompass measures of both morbidity and mortality. For example, ventilator-free days measure the number of days a patient is both alive and off the ventilator. The continuous nature of these data results in increased statistical power and the need for fewer participants to detect a significant difference compared to dichotomous outcome variables such as mortality. As such, the failure-free methodology has been expanded to evaluate other organ failures (i.e., CNS failure—free, cardiovascular failure—free, or any organ failure—free days). Furthermore, this technique has also been utilized in place of lengths of stay, which can also be complicated by mortality. ICU-free or hospital-free days combine measures of both lengths of stay and mortality as patients must be both out of the ICU and alive to accumulate ICU-free days. However, failure-free data should not be interpreted in isolation, but instead should be interpreted in conjunction with mortality data, especially if the two are disparate. For example, an intervention, which increases the number of ventilator-free days, but demonstrates a trend toward worse mortality, may appear effective if only ventilator-free days are examined. In reality, the trend toward worse mortality should raise concern about the real effect of the intervention. In addition, some have questioned whether organ failure—free days are really patient-centered outcomes, as many patients do not understand what having an organ failure means, and different organ failures may mean more to patients than others.

More recently, many studies have demonstrated that patients surviving acute illnesses often have significant morbidity even after hospital discharge, probably lasting months or even years (Davidson et al., 1999; Hopkins et al., 1999, 2005; Angus et al., 2001; Needham et al., 2013). These data suggest that measures of long-term morbidity, such as cognitive function, ability to perform activities of daily living, or hospital readmissions, may represent clinical outcomes as important as short-term survival data, especially in these times of improving survival. Neuromuscular weakness and/or neuropsychiatric problems frequently delay return to normal living or work and sometimes never completely resolve. Recognition of this, along with the increasing importance on quality of life, has resulted in a renewed emphasis on long-term outcomes, including mortality, neurocognitive functioning, and quality of life or quality-adjusted life years (QALY) measurements months or years after hospital discharge. Unfortunately, many of these long-term outcomes are difficult to measure. Patients are easily lost to follow-up after discharge from the hospital making even 6- or 12-month survival assessments challenging. Finding and convincing participants who have resumed life after their critical illness, traumatic event, or surgery to return for long-term follow-up visits can also be difficult. Although neurocognitive testing and quality of life questionnaires are slowly becoming validated outcome measures (Black et al., 2001), they take considerable time and effort to complete. Some can be conducted by telephone, but many still require face-to-face visits making the data burdensome to collect and the studies expensive to complete (Jackson et al., 2004). Despite these hindrances, most researchers agree that reducing long-term morbidity and enhancing outcomes such as long-term neurocognitive functioning and quality of life will continue to rise in importance in the care of acutely ill patients over the next decade.

ADVERSE EVENTS

An integral part of clinical research is adverse event reporting, because it represents one of the major mechanisms of evaluating the risk of an investigational intervention and allows these risks to be compared to any possible benefits. Adverse events in clinical trials are usually classified according to three criteria: seriousness, expectedness, and relatedness to the research study procedures (Silverman et al., 2006). Serious and unexpected adverse events that are thought to be related to the research should be reported in a timely manner to the appropriate regulatory agencies, including funding sources, institutional review board or ethics committees, and data safety monitoring boards. In studies evaluating new pharmacologic agents, biologics, or medical devices, the serious adverse event should also be reported to the overseeing agency, such as the FDA, by either the individual investigator or study sponsor. Unfortunately, determining if an adverse event is serious, unexpected, and related to the research can be challenging, especially in acutely ill patients who have dynamic, and often deteriorating, disease courses.

Determining whether an event qualifies as serious is usually defined by the study protocol prior to initiating the study, although the criteria do vary slightly from study to study. For example, most studies consider an adverse event serious if it is fatal or immediately life threatening, permanently disabling, severely incapacitating, or requires or prolongs inpatient hospitalization. In addition, events which require medical or surgical intervention to prevent one of the above from occurring may also be defined as serious. Unfortunately, determining whether an adverse event is unexpected or related to the research is often less clear. This is especially true for research involving acutely ill patients, such as those with trauma, or requiring critical care. Due to the nature and severity of the acute illness, adverse events occur in the natural disease course for these patients frequently, even when they are not enrolled in research protocols. For example, patients with trauma often develop acute lung injury or renal failure secondary to systemic inflammation and altered coagulation homeostasis. When these patients develop new lung injury or renal failure while enrolled in a study of a new investigational treatment, it is difficult, if not impossible, to ascertain whether these adverse events are related to the study agent or simply secondary to the underlying disease state. Complications after procedures represent another challenging area for discerning adverse events. Deciding whether postoperative infections are more severe or frequent than expected in patients enrolled in a study investigating a new surgical procedure is an ambiguous determination which is often left to the sole discretion of the local investigator.

One definition utilized for unexpected events is any event differing from the type, severity, or frequency listed in the current study protocol or investigator's brochure or that is unexpected in the course of treatment for the acute illness. However, based on their individual experience in caring for patients with the acute illness, investigators may differ in their perceptions of whether the events are more severe or frequent than normally seen or whether the event was unexpected in the course of caring for patients with that disease (Raisch et al., 2001). For example, thrombocytopenia is commonly seen in patients with severe sepsis. However, investigators' perceptions of the frequency and severity of thrombocytopenia in these patients differ. Consequently, one investigator may report every episode of thrombocytopenia as an unexpected adverse event, while a second may only classify episodes that pass a certain threshold, such as platelet counts lower than $30,000/mm^3$, and a third may not consider any level of thrombocytopenia as unexpected for the disease state.

Determining the relationship of an adverse event to investigational procedures can also be challenging. Obviously, any event that occurs after randomization but prior to administration of the investigational agent is highly unlikely to be related to the intervention. However, any event that follows a reasonable temporal sequence after a study intervention could reasonably be related to the intervention. One way to increase the certainty that an event is related to a study intervention is to rechallenge the participant after the event has resolved. Unfortunately, this is rarely possible in studies of patients with acute illness because by the time the adverse event has resolved, the patient either no longer has the acute illness or no longer meets criteria for needing the investigational agent.

Adverse event reports are utilized to help determine the overall risk to benefit ratio of the study and/or investigational procedures. The inconsistent adverse event reporting makes determining the actual incidence of adverse events difficult (Hayashi and Walker, 1996; Chou et al., 2007). This is especially true for studies conducted in multiple centers across numerous countries where adverse event reporting regulations vary. Furthermore, in many multicenter studies, local ethics boards are not aware of many of the adverse events occurring at other sites, and even when they are aware, the actual incidence is difficult to determine (Maloney, 2005; DeMets et al., 2006). Not knowing true incidences of adverse events may hinder evaluation of the overall risks of the investigational intervention, both during the conduct of the study and in the interpretation of the final results. In these situations, an independent, central group responsible for oversight of the safety of the study, such as a data safety monitoring board, is vital to summarizing and interpreting the adverse events collectively (Champion et al., 2007; Califf et al., 2003; Emanuel et al., 2004).

Fortunately, many randomized controlled studies in acute illnesses remove the onus for deciding whether to report these events from the local investigators by systematically collecting many of the adverse outcomes. The studies are designed to prospectively collect a number of these "expected" events that occur in the natural course of the disease, such as organ failures, as part of routine data collection for the study. This standardizes the reporting throughout all sites participating in the study and relieves the local investigator of the decision of whether or not the outcome is truly serious, unexpected, and/ or related to the research procedures. This is especially important in conducting multicenter and/or international studies, where the reporting regulations and perception of adverse events are likely to vary widely among sites. The nature of the events that are systematically collected can vary from study to study, but range from nonserious abnormal lab values to serious new organ failures. Some studies in acute illnesses with high underlying mortality even prospectively collect all deaths regardless of perceived relationship to the study. In addition, many studies prospectively define and collect known adversities related to the study intervention, to obtain more robust safety data for the intervention. For example, a study investigating different feeding techniques would prospectively define and collect gastrointestinal intolerances, which are expected to occur in a number of patients, to more objectively classify the incidence of these events and better define the safety of the intervention (Rice et al., 2011, 2012). Recently, some studies have begun prospectively defining and collecting serious and life-threatening adverse events, without expedited reporting to the regulatory agencies. However, this requires an alteration in the regulatory agency guidelines for adverse event reporting. As such, approval for this should be obtained directly from the regulatory agency prior to initiating the study. In a large, multicenter study of a new pharmaceutical agent for treating severe sepsis, the anticoagulant properties of the agent increased the risk of bleeding in patients. With this knowledge, the investigators who designed the study prospectively defined all bleeding and serious bleeding. With regulatory approval, these events were collected via the case report form throughout the conduct of the study instead of being reported through the expedited adverse event reporting system (Bernard et al., 2001; Ranieri et al., 2012). This allowed for more accurate and objective reporting of the bleeding risk and a better assessment of the overall risk to benefit ratio for the agent. Unfortunately, events which are truly unexpected for the disease process or previously unknown for the intervention must continue to be reported through impromptu adverse event reporting from local investigators. However, the prospective collection of commonly seen adverse outcomes as part of data amassment for the study reduces the variability of local investigator judgment and reporting. Overall, this standardizes the reporting of these events, resulting in more objective classification of their incidence and a better overall understanding of the risks of the intervention.

CONCLUSION

Patient-oriented, clinical research in acutely ill patients has encountered numerous challenges over the last two decades, including issues with consent, study design, appropriate endpoints, protocolized care, and adverse event reporting. Fortunately, the experience, intellect, and dedication of the researchers in these fields have adapted to overcome these barriers. Undoubtedly, as the field continues to expand and advance, additional obstacles will be encountered. However, similar to previous hindrances, the entire research community will work together to utilize these difficulties as opportunities to advance the entire research field in patients with acute surgical or critical illnesses.

REFERENCES

Abramson, N.S., Meisel, A., Safar, P., 1986. Deferred consent. A new approach for resuscitation research on comatose patients. JAMA 255 (18), 2466–2471.

Albin, R.L., 2005. Sham surgery controls are mitigated trolleys. J. Med. Ethics 31 (3), 149–152.

Angelos, P., 2003. Sham surgery in research: a surgeon's view. Am. J. Bioeth. 3 (4), 65–66.

Angelos, P., 2007. Sham surgery in clinical trials. JAMA 297 (14), 1545–1546.

Angus, D.C., Musthafa, A.A., Clermont, G., et al., 2001. Quality-adjusted survival in the first year after the acute respiratory distress syndrome. Am. J. Respir. Crit. Care Med. 163 (6), 1389–1394.

Appelbaum, P.S., Roth, L.H., Lidz, C.W., et al., 1987. False hopes and best data: consent to research and the therapeutic misconception. Hastings Cent. Rep. 17 (2), 20–24.

Appelbaum, P.S., Lidz, C.W., Grisso, T., 2004. Therapeutic misconception in clinical research: frequency and risk factors. IRB 26 (2), 1—8.

Barr, J., Hecht, M., Flavin, K.E., et al., 2004. Outcomes in critically ill patients before and after the implementation of an evidence-based nutritional management protocol. Chest 125 (4), 1446—1457.

Beck, C.A., Richard, H., Tu, J.V., et al., 2005. Administrative Data Feedback for Effective Cardiac Treatment: AFFECT, a cluster randomized trial. JAMA 294 (3), 309—317.

Bernard, G.R., Vincent, J.L., Laterre, P.F., et al., 2001. Efficacy and safety of recombinant human activated protein C for severe sepsis. N. Engl. J. Med. 344 (10), 699—709.

Binanay, C., Califf, R.M., Hasselblad, V., et al., 2005. Evaluation study of congestive heart failure and pulmonary artery catheterization effectiveness: the ESCAPE trial. JAMA 294 (13), 1625—1633.

Biros, M.H., Lewis, R.J., Olson, C.M., et al., 1995. Informed consent in emergency research. Consensus statement from the coalition conference of acute resuscitation and critical care researchers. JAMA 273 (16), 1283—1287.

Black, N.A., Jenkinson, C., Hayes, J.A., et al., 2001. Review of outcome measures used in adult critical care. Crit. Care Med. 29 (11), 2119—2124.

Boden, W.E., O'Rourke, R.A., Teo, K.K., et al., 2007. Optimal medical therapy with or without PCI for stable coronary disease. N. Engl. J. Med. 356 (15), 1503—1516.

Brochard, L., Rauss, A., Benito, S., et al., 1994. Comparison of three methods of gradual withdrawal from ventilatory support during weaning from mechanical ventilation. Am. J. Respir. Crit. Care Med. 150 (4), 896—903.

Brower, R.G., Bernard, G., Morris, A., 2004a. Ethics and standard of care in clinical trials. Am. J. Respir. Crit. Care Med. 170 (2), 198—199.

Brower, R.G., Lanken, P.N., MacIntyre, N., et al., 2004b. Higher versus lower positive end-expiratory pressures in patients with the acute respiratory distress syndrome. N. Engl. J. Med. 351 (4), 327—336.

Brull, R., McCartney, C.J., Chan, V.W., et al., 2004. Are patients comfortable consenting to clinical anesthesia research trials on the day of surgery? Anesth. Analg. 98 (4), 1106—1110 (table).

Califf, R.M., Morse, M.A., Wittes, J., et al., 2003. Toward protecting the safety of participants in clinical trials. Control Clin. Trials 24 (3), 256—271.

Campbell, M.K., Thomson, S., Ramsay, C.R., et al., 2004. Sample size calculator for cluster randomized trials. Comput. Biol. Med. 34 (2), 113—125.

Champion, H.R., Fingerhut, A., Escobar, M.A., et al., 2007. The role of data and safety monitoring in acute trauma resuscitation research. J. Am. Coll. Surg. 204 (1), 73—83.

Chen, D.T., Miller, F.G., Rosenstein, D.L., 2003. Clinical research and the physician-patient relationship. Ann. Intern. Med. 138 (8), 669—672.

Chou, R., Fu, R., Carson, S., et al., 2007. Methodological shortcomings predicted lower harm estimates in one of two sets of studies of clinical interventions. J. Clin. Epidemiol. 60 (1), 18—28.

Ciroldi, M., Cariou, A., Adrie, C., et al., 2007. Ability of family members to predict patient's consent to critical care research. Intensive Care Med. 33 (5), 807—813.

Connelly, L.B., 2003. Balancing the number and size of sites: an economic approach to the optimal design of cluster samples. Control. Clin. Trials 24 (5), 544—559.

Connors Jr., A.F., Speroff, T., Dawson, N.V., Thomas, C., Harrell Jr., F.E., Wagner, D., et al., 1996. The effectiveness of right heart catheterization in the initial care of critically ill patients. SUPPORT investigators. JAMA 276 (11), 889—897.

Coppolino, M., Ackerson, L., 2001. Do surrogate decision makers provide accurate consent for intensive care research? Chest 119 (2), 603—612.

Davidson, T.A., Caldwell, E.S., Curtis, J.R., et al., 1999. Reduced quality of life in survivors of acute respiratory distress syndrome compared with critically ill control patients. JAMA 281 (4), 354—360.

Deans, K.J., Minneci, P.C., Eichacker, P.Q., et al., 2004. Defining the standard of care in randomized controlled trials of titrated therapies. Curr. Opin. Crit. Care 10 (6), 579—582.

Deans, K.J., Minneci, P.C., Suffredini, A.F., et al., 2007. Randomization in clinical trials of titrated therapies: unintended consequences of using fixed treatment protocols. Crit. Care Med. 35 (6), 1509—1516.

DeMets, D.L., Fost, N., Powers, M., 2006. An Institutional Review Board dilemma: responsible for safety monitoring but not in control. Clin. Trials 3 (2), 142—148.

Donner, A., Klar, N., 2004. Pitfalls of and controversies in cluster randomization trials. Am. J. Public Health 94 (3), 416—422.

Drakulovic, M.B., Torres, A., Bauer, T.T., et al., 1999. Supine body position as a risk factor for nosocomial pneumonia in mechanically ventilated patients: a randomised trial. Lancet 354 (9193), 1851—1858.

Eichacker, P.Q., Gerstenberger, E.P., Banks, S.M., et al., 2002. Meta-analysis of acute lung injury and acute respiratory distress syndrome trials testing low tidal volumes. Am. J. Respir. Crit. Care Med. 166 (11), 1510—1514.

Ely, E.W., Baker, A.M., Dunagan, D.P., et al., 1996. Effect on the duration of mechanical ventilation of identifying patients capable of breathing spontaneously. N. Engl. J. Med. 335 (25), 1864—1869.

Emanuel, E.J., Wood, A., Fleischman, A., et al., 2004. Oversight of human participants research: identifying problems to evaluate reform proposals. Ann. Intern Med. 141 (4), 282—291.

Estenssoro, E., Dubin, A., Laffaire, E., et al., 2002. Incidence, clinical course, and outcome in 217 patients with acute respiratory distress syndrome. Crit. Care Med. 30 (11), 2450—2456.

Ferring, M., Vincent, J.L., 1997. Is outcome from ARDS related to the severity of respiratory failure? Eur. Respir. J. 10 (6), 1297—1300.

Flachskampf, F.A., Gallasch, J., Gefeller, O., et al., 2007. Randomized trial of acupuncture to lower blood pressure. Circulation 115 (24), 3121—3129.

Flum, D.R., 2006. Interpreting surgical trials with subjective outcomes: avoiding UnSPORTsmanlike conduct. JAMA 296 (20), 2483—2485.

Friedman, L.M., Furberg, C.D., DeMets, D.L., 1998a. Basic Study Design. Fundamentals of Clinical Trials. Springer Verlag, New York, NY, 41—60.

Friedman, L.M., Furberg, C.D., DeMets, D.L., 1998b. Fundamentals of Clinical Trials. Springer-Verlag Publishing, New York, NY.

Garnacho-Montero, J., Garcia-Garmendia, J.L., Barrero-Almodovar, A., et al., 2003. Impact of adequate empirical antibiotic therapy on the outcome of patients admitted to the intensive care unit with sepsis. Crit. Care Med. 31 (12), 2742–2751.

Garnacho-Montero, J., Aldabo-Pallas, T., Garnacho-Montero, C., et al., 2006. Timing of adequate antibiotic therapy is a greater determinant of outcome than are TNF and IL-10 polymorphisms in patients with sepsis. Crit. Care 10 (4), R111.

Gattinoni, L., Tognoni, G., Pesenti, A., et al., 2001. Effect of prone positioning on the survival of patients with acute respiratory failure. N. Engl. J. Med. 345 (8), 568–573.

Girard, T.D., Kress, J.P., Fuchs, B.D., Thomason, J.W., Schweickert, W.D., Pun, B.T., et al., 2008. Efficacy and safety of a paired sedation and ventilator weaning protocol for mechanically ventilated patients in the intensive care (Awakening and Breathing Controlled trial): a randomised controlled trial. Lancet 371 (9607), 126–134.

Guerin, C., Reignier, J., Richard, J.C., Beuret, P., Gacouin, A., Boulain, T., et al., 2013. Prone positioning in severe acute respiratory distress syndrome. N. Engl. J. Med. 368 (23), 2159–2168.

Hahn, S., Puffer, S., Torgerson, D.J., et al., 2005. Methodological bias in cluster randomised trials. BMC Med. Res. Methodol. 5 (1), 10.

Harvey, S., Harrison, D.A., Singer, M., et al., 2005. Assessment of the clinical effectiveness of pulmonary artery catheters in management of patients in intensive care (PAC-Man): a randomised controlled trial. Lancet 366 (9484), 472–477.

Harvey, S.E., Elbourne, D., Ashcroft, J., et al., 2006. Informed consent in clinical trials in critical care: experience from the PAC-Man Study. Intensive Care Med. 32 (12), 2020–2025.

Hayashi, K., Walker, A.M., 1996. Japanese and American reports of randomized trials: differences in the reporting of adverse effects. Control. Clin. Trials 17 (2), 99–110.

Hebert, P.C., Wells, G., Blajchman, M.A., et al., 1999. A multicenter, randomized, controlled clinical trial of transfusion requirements in critical care. Transfusion Requirements in Critical Care Investigators, Canadian Critical Care Trials Group. N. Engl. J. Med. 340 (6), 409–417.

Hillman, K., Chen, J., Cretikos, M., et al., 2005. Introduction of the medical emergency team (MET) system: a cluster-randomised controlled trial. Lancet 365 (9477), 2091–2097.

Holcomb, B.W., Wheeler, A.P., Ely, E.W., 2001. New ways to reduce unnecessary variation and improve outcomes in the intensive care unit. Curr. Opin. Crit. Care 7 (4), 304–311.

Hopkins, R.O., Weaver, L.K., Pope, D., et al., 1999. Neuropsychological sequelae and impaired health status in survivors of severe acute respiratory distress syndrome. Am. J. Respir. Crit. Care Med. 160 (1), 50–56.

Hopkins, R.O., Weaver, L.K., Collingridge, D., et al., 2005. Two-year cognitive, emotional, and quality-of-life outcomes in acute respiratory distress syndrome. Am. J. Respir. Crit. Care Med. 171 (4), 340–347.

Horng, S., Miller, F.G., 2002. Is placebo surgery unethical? N. Engl. J. Med. 347 (2), 137–139.

Horng, S., Miller, F.G., 2003. Ethical framework for the use of sham procedures in clinical trials. Crit. Care Med. 31 (3 Suppl.), S126–S130.

Jackson, J.C., Gordon, S.M., Ely, E.W., et al., 2004. Research issues in the evaluation of cognitive impairment in intensive care unit survivors. Intensive Care Med. 30 (11), 2009–2016.

Jain, M.K., Heyland, D., Dhaliwal, R., et al., 2006. Dissemination of the Canadian clinical practice guidelines for nutrition support: results of a cluster randomized controlled trial. Crit. Care Med. 34 (9), 2362–2369.

Kaptchuk, T.J., Goldman, P., Stone, D.A., et al., 2000. Do medical devices have enhanced placebo effects? J. Clin. Epidemiol. 53 (8), 786–792.

Kollef, M.H., Shapiro, S.D., Silver, P., et al., 1997. A randomized, controlled trial of protocol-directed versus physician-directed weaning from mechanical ventilation. Crit. Care Med. 25 (4), 567–574.

Kress, J.P., Pohlman, A.S., O'Connor, M.F., et al., 2000. Daily interruption of sedative infusions in critically ill patients undergoing mechanical ventilation. N. Engl. J. Med. 342 (20), 1471–1477.

Levine, R.J., 1995. Research in emergency situations. The role of deferred consent. JAMA 273 (16), 1300–1302.

Levy, M.M., Macias, W.L., Vincent, J.L., et al., 2005. Early changes in organ function predict eventual survival in severe sepsis. Crit. Care Med. 33 (10), 2194–2201.

Lotjonen, S., 2002. Medical research in clinical emergency settings in Europe. J. Med. Ethics 28 (3), 183–187.

Macklin, R., 1999. The ethical problems with sham surgery in clinical research. N. Engl. J. Med. 341 (13), 992–996.

Maloney, D.M., 2005. Institutional review boards (IRBs) and working with adverse event reports. Hum. Res. Rep. 20 (12), 1–3.

Marshall, J.C., Vincent, J.L., Guyatt, G., et al., 2005. Outcome measures for clinical research in sepsis: a report of the 2nd Cambridge Colloquium of the International Sepsis Forum. Crit. Care Med. 33 (8), 1708–1716.

Marshall, J.C., 1999. Organ dysfunction as an outcome measure in clinical trials. Eur. J. Surg. Suppl. 584, 62–67.

Meade, M.O., Ely, E.W., 2002. Protocols to improve the care of critically ill pediatric and adult patients. JAMA 288 (20), 2601–2603.

Mehta, S., Burry, L., Cook, D., Fergusson, D., Steinberg, M., Granton, J., et al., 2012. Daily sedation interruption in mechanically ventilated critically ill patients cared for with a sedation protocol: a randomized controlled trial. JAMA 308 (19), 1985–1992.

Miller, F.G., Brody, H., 2003. A critique of clinical equipoise. Therapeutic misconception in the ethics of clinical trials. Hastings Cent. Rep. 33 (3), 19–28.

Miller, F.G., Kaptchuk, T.J., 2004. Sham procedures and the ethics of clinical trials. J. R. Soc. Med. 97 (12), 576–578.

Miller, F.G., Rosenstein, D.L., 2003. The therapeutic orientation to clinical trials. N. Engl. J. Med. 348 (14), 1383–1386.

Miller, B.L., 1988. Philosophical, ethical, and legal aspects of resuscitation medicine. I. Deferred consent and justification of resuscitation research. Crit. Care Med. 16 (10), 1059–1062.

Miller, F.G., 2003. Sham surgery: an ethical analysis. Am. J. Bioeth. 3 (4), 41–48.

Miller, F.G., 2005. Ethical issues in surgical research. Thorac. Surg. Clin. 15 (4), 543–554.

Mingus, M.L., Levitan, S.A., Bradford, C.N., et al., 1996. Surgical patients' attitudes regarding participation in clinical anesthesia research. Anesth. Analg. 82 (2), 332–337.

Montgomery, A.B., Stager, M.A., Carrico, C.J., et al., 1985. Causes of mortality in patients with the adult respiratory distress syndrome. Am. Rev. Respir. Dis. 132 (3), 485–489.

Moore, F.A., Moore, E.E., Sauaia, A., 1997. Blood transfusion. An independent risk factor for postinjury multiple organ failure. Arch. Surg. 132 (6), 620–624.

Moseley, J.B., O'Malley, K., Petersen, N.J., et al., 2002. A controlled trial of arthroscopic surgery for osteoarthritis of the knee. N. Engl. J. Med. 347 (2), 81–88.

National Heart, Lung, and Blood Institute ARDS Clinical Trials Network, Truwit, J.D., Bernard, G.R., Steingrub, J., Matthay, M.A., Liu, K.D., et al., 2014. Rosuvastatin for sepsis-associated acute respiratory distress syndrome. N. Engl. J. Med. 370 (23), 2191–2200.

Needham, D.M., Dinglas, V.D., Morris, P.E., Jackson, J.C., Hough, C.L., Mendez-Tellez, P.A., et al., 2013. Physical and cognitive performance of acute lung injury patients one year after initial trophic vs full enteral feeding: EDEN trial follow-up. Am. J. Respir. Crit. Care Med. 188, 567–576.

NICE-SUGAR Study Investigators, Finfer, S., Chittock, D.R., Su, S.Y., Blair, D., Foster, D., et al., 2009. Intensive versus conventional glucose control in critically ill patients. N. Engl. J. Med. 360 (13), 1283–1297.

Nichol, G., Powell, J., van Ottingham, L., et al., 2006. Consent in resuscitation trials: benefit or harm for patients and society? Resuscitation 70 (3), 360–368.

Noto, M.J., Domenico, H.J., Byrne, D.W., Talbot, T.S., Rice, T.W., Bernard, G.R., Wheeler, A.P., 2015. Chlorhexidine bathing and healthcare-associated infections: a randomized clinical trial. JAMA 313 (4), 369–378.

Papazian, L., Forel, J.M., Gacouin, A., Penot-Ragon, C., Perrin, G., Loundou, A., et al., 2010. Neuromuscular blockers in early acute respiratory distress syndrome. N. Engl. J. Med. 363 (12), 1107–1116.

ProCESS Investigators, Yealy, D.M., Kellum, J.A., Huang, D.T., Barnatro, A.E., Weissfeld, L.A., et al., 2014. A randomized trial of protocol-based care for early septic shock. N. Engl. J. Med. 370 (18), 1683–1693.

Puffer, S., Torgerson, D., Watson, J., 2003. Evidence for risk of bias in cluster randomised trials: review of recent trials published in three general medical journals. BMJ 327 (7418), 785–789.

Raghunathan, K., Shaw, A., Nathanson, B., Stürmer, T., Brookhart, A., Stefan, M.S., Setoguchi, S., Beadles, C., Lindenauer, P.K., 2014. Association between the choice of IV crystalloid and in-hospital mortality among critically ill adults with sepsis. Crit. Care Med. 42 (7), 1585–1591.

Raisch, D.W., Troutman, W.G., Sather, M.R., et al., 2001. Variability in the assessment of adverse events in a multicenter clinical trial. Clin. Ther. 23 (12), 2011–2020.

Ranieri, V.M., Thompson, B.T., Barie, P.S., Dhainaut, J.F., Douglas, I.S., Finfer, S., et al., 2012. Drotrecogin alfa (activated) in adults with septic shock N. Engl. J. Med. 366 (22), 2055–2064.

Reddy, S.K., Bailey, M.J., Beasley, R.W., Bellomo, R., Henderson, S.J., Mackle, D.M., et al., 2014. A protocol for the 0.9% saline versus Plasma-Lyte 148 for intensive care fluid therapy (SPLIT) study. Crit. Care Resusc. 16 (4), 274–279.

Rice, T.W., Swope, T., Bozeman, S., et al., 2005. Variation in enteral nutrition delivery in mechanically ventilated patients. Nutrition 21 (7–8), 786–792.

Rice, T.W., Wheeler, A.P., Thompson, B.T., deBoisblanc, B.P., Steingrub, J., Rock, P., for the NHLBI ARDS Network, 2011. Enteral omega-3 fatty acid, gamma-linolenic acid, and anti-oxidant supplementation in acute lung injury (ALI). JAMA 306 (14), 1574–1581.

Rice, T.W., Wheeler, A.P., Thompson, B.T., Steingrub, J., Hite, R.D., Moss, M., , et al.for the NHLBI ARDS Network, 2012. Randomized trial of initial trophic vs full enteral feeding in patients with acute lung injury (ALI) (EDEN). JAMA 307 (8), 795–803.

Richard, C., Warszawski, J., Anguel, N., et al., 2003. Early use of the pulmonary artery catheter and outcomes in patients with shock and acute respiratory distress syndrome: a randomized controlled trial. JAMA 290 (20), 2713–2720.

Rivers, E., Nguyen, B., Havstad, S., et al., 2001. Early goal-directed therapy in the treatment of severe sepsis and septic shock. N. Engl. J. Med. 345 (19), 1368–1377.

Rubenfeld, G.D., 2003. Epidemiology of acute lung injury. Crit. Care Med. 31 (4 Suppl.), S276–S284.

Sandham, J.D., Hull, R.D., Brant, R.F., et al., 2003. A randomized, controlled trial of the use of pulmonary-artery catheters in high-risk surgical patients. N. Engl. J. Med. 348 (1), 5–14.

Schoenfeld, D.A., Bernard, G.R., 2002. Statistical evaluation of ventilator-free days as an efficacy measure in clinical trials of treatments for acute respiratory distress syndrome. Crit. Care Med. 30 (8), 1772–1777.

Schwartz, M.P., Wellink, H., Gooszen, H.G., et al., 2007. Endoscopic gastroplication for the treatment of gastro-oesophageal reflux disease: a randomised, sham-controlled trial. Gut 56 (1), 20–28.

Shalowitz, D.I., Garrett-Mayer, E., Wendler, D., 2006. The accuracy of surrogate decision makers: a systematic review. Arch. Intern Med. 166 (5), 493–497.

Silverman, H.J., Miller, F.G., 2004. Control group selection in critical care randomized controlled trials evaluating interventional strategies: an ethical assessment. Crit. Care Med. 32 (3), 852–857.

Silverman, H.J., Luce, J.M., Lanken, P.N., et al., 2005. Recommendations for informed consent forms for critical care clinical trials. Crit. Care Med. 33 (4), 867–882.

Silverman, D.I., Cirullo, L., DeMartinis, N.A., et al., 2006. Systematic identification and classification of adverse events in human research. Contemp. Clin. Trials 27 (3), 295–303.

Spain, D.A., McClave, S.A., Sexton, L.K., et al., 1999. Infusion protocol improves delivery of enteral tube feeding in the critical care unit. J. Parenter. Enter. Nutr. 23 (5), 288–292.

Stapleton, R.D., Wang, B.M., Hudson, L.D., et al., 2005. Causes and timing of death in patients with ARDS. Chest 128 (2), 525–532.

Stewart, T.E., Meade, M.O., Cook, D.J., et al., 1998. Evaluation of a ventilation strategy to prevent barotrauma in patients at high risk for acute respiratory distress syndrome. Pressure- and Volume-Limited Ventilation Strategy Group. N. Engl. J. Med. 338 (6), 355–361.

Salminen, P., Paajanen, H., Rautio, T., et al., 2015. Antibiotic therapy vs appendectomy for treatment of uncomplicated acute appendicitis: the APPAC randomized clinical trial. JAMA 313 (23), 2340–2348.

Taylor, R.W., Zimmerman, J.L., Dellinger, R.P., et al., 2004. Low-dose inhaled nitric oxide in patients with acute lung injury: a randomized controlled trial. JAMA 291 (13), 1603–1609.

Van den Berghe, G., Wouters, P., Weekers, F., et al., 2001. Intensive insulin therapy in the critically ill patients. N. Engl. J. Med. 345 (19), 1359–1367.

Ventilation with lower tidal volumes as compared with traditional tidal volumes for acute lung injury and the acute respiratory distress syndrome. The Acute Respiratory Distress Syndrome Network. N. Engl. J. Med. 342 (18), 2000, 1301–1308.

Weijer, C., 2002. I need a placebo like I need a hole in the head. J. Law Med. Ethics 30 (1), 69–72.

Wheeler, A.P., Bernard, G.R., Thompson, B.T., et al., 2006. Pulmonary-artery versus central venous catheter to guide treatment of acute lung injury. N. Engl. J. Med. 354 (21), 2213–2224.

Wiedemann, H.P., Wheeler, A.P., Bernard, G.R., et al., 2006. Comparison of two fluid-management strategies in acute lung injury. N. Engl. J. Med. 354 (24), 2564–2575.

Wood, K.A., Huang, D., Angus, D.C., 2003. Improving clinical trial design in acute lung injury. Crit. Care Med. 31 (4 Suppl.), S305–S311.

World Medical Association Declaration of Helsinki, 2000. Ethical principles for medical research involving human subjects. JAMA 284 (23), 3043–3045.

Chapter 27

Research in the Emergency Care Environment

James Quinn[1] and Daniel J. Pallin[2]

[1]Stanford University, Stanford, CA, United States; [2]Harvard Medical School, Boston, MA, United States

Chapter Outline

Key Points

- Special considerations apply to emergency care research.
- Challenges include identification of eligible subjects, timely enrollment and interventions, and the consenting process including the use of exception from informed consent process.
- Understanding challenges in the emergency care environment will help identify specific strategies to help implement clinical trials in this environment.

INTRODUCTION

The emergency care environment comprises three distinct settings: prehospital, emergency department (ED), and critical care areas within the hospital. Furthermore, the ED patient population can be divided into those who are acutely ill and those who are not.

THE ENVIRONMENT AND UNIQUE CHALLENGES OF EMERGENCY CARE RESEARCH

The most recent public data show that the volume of ED patients has increased by 36% over the past decade to approximately 130 million visits in 2010 (Centers for Disease Control and Prevention, 2010; The High Concentration, 2006; Owens et al., 2010). These visits include critically ill patients as well as the worried well and those who use the ED for its convenience in the absence of severe illness. It is estimated that such visits comprise 20% of ED volume and may grow as more patients gain insurance through political reform while at the same time there remains a shortage of primary care providers (Weber et al., 2008; Pallin et al., 2013a; Schuur and Venkatesh, 2012). Increasingly, EDs are the principal gateway to the hospital, accounting for nearly half of all admissions by 2006 (Schuur and Venkatesh, 2012). With hospital closures and bed shortages, ED overcrowding has become a crisis (Fisher et al., 2009).

The sheer volume and diversity of patients make the emergency care environment a fertile place for research, but there are unique challenges. Overcrowding and boarding of patients in the ED has been shown to negatively impact patient outcomes as resources are stressed (Fisher et al., 2009). As a result physicians, nurses, and all ED staff are focused on ED patient care and have little time for research efforts that are often seen as too burdensome. Unlike the controlled environments of a specialty clinic or General Clinical Research Center (GCRC), the ED is a chaotic overcrowded environment of patients with ill-defined symptoms, unclear diagnoses, and varying acuity, presenting special challenges for research.

Aside from the contextual challenges, the inherent nature of emergency care makes research difficult. Many patients are unable to give informed consent due to altered mental status. Many crucial hypotheses can only be tested if research procedures are initiated within minutes of initial contact, but this time pressure flies in the face of institutional review board (IRB) requirements that patients have time to decide whether to participate in research, and also requires huge investments in preparation and rehearsal, because the moments of enrollment and initial data collection are hectic.

When patients with chronic diseases are screened for research in office and community settings, their eligibility is already well defined. But screening emergency care patients for eligibility is like searching for a needle in a haystack, because their diagnosis is unknown at the time of screening. As an example of the huge challenges presented by minutiae in planning research in the emergency care environment, this fact (diagnosis unknown) often causes researchers to overestimate eligible patients in their client population when they only look post hoc at discharge diagnosis. At discharge patients may have had the disease process or condition of interest but it may not had been recognized in time nor have met eligibility criteria for intervention, especially in the case of many studies which mandate timely therapeutic intervention. In reality some of these patients of interest would never be identified during ED presentation and not been included for trial enrollment.

Further challenges beyond trying to identify the eligible participants include obtaining consent and implementing specific protocols in a timely fashion. In addition to presenting with ill-defined symptoms patients may require treatment before anyone considers a research question or attempts to activate any research infrastructure. Patients are also often acutely ill lacking capacity for consent and if they have a legally authorized representative (LAR) they are often not present making it difficult to enroll patients with diseases that require immediate treatment and/or intervention.

Finally, it is not uncommon to have multiple competing investigators and studies at any one institution. Patients in emergency care environments are often cared for by multiple physicians and consultants with different teams being in charge from ED arrival through the hospital course. A plan to deal with and manage multiple care and investigator teams, as well as potentially competing protocols, is necessary and necessitates a huge investment, relative to research in traditional settings; and this huge investment is often invisible after a publication has been released or when funds are sought for a planned project.

EXAMPLES OF EARLY SUCCESS

Research in the emergency care environment has grown over the past 30 years, as Emergency Medicine and Critical Care Medicine have become recognized medical disciplines. Not surprisingly, with the growth of these specialties, physicians in these environments have developed leadership roles and learned how to navigate the challenges of the environment and become successful researchers. Emergency care investigators are now experts in clinical trials, observational cohort studies, and large epidemiological studies as well as basic scientists in the area of animal and bioengineering that leads naturally to the area of translational research. They have become emergency research experts in the area of trauma, injury prevention, cardiovascular, neurological, pain management, ultrasound, pediatrics, geriatrics infectious disease, psychiatry, simulation/education, information technology, health policy as well as providers and researchers in the area of international medicine, global disaster/emergency response, and the various effectiveness of these programs. It is impossible to highlight all areas of emergency medicine research, but a few examples will be provided.

Researchers in emergency medicine have become experts in the field of clinical decision rules and prehospital care research. They have also collaborated and lead important trials with critical care physicians in the management of critically ill patients in the area of resuscitation and sepsis research.

The developments of clinical decision rules is an example of how emergency physicians can use a large population of nonemergent patients and try to develop the more efficient management of patients through improved disposition and utilization of resources. As physicians who constantly deal with the uncertainty of diagnosis and the demand for correct diagnosis, it is no surprise that emergency medicine is a place where risk ratification in the development of clinical decision rules was truly born. Over 20 years ago, the methodological standards for clinical decision rules were developed and implemented by Dr. Ian Stiell and his group at the University of Ottawa (Stiell and Wells, 1999). They developed, validated, and implemented decision rules providing the evidence base for use of diagnostic studies relying on ionizing radiation, for injuries to the ankle, knee, cervical spine, and head (Stiell et al., 1992a, 1995, 2001a, 2003). Other investigators from emergency medicine have subsequently used these standards for the development of clinical decision rules and risk stratification for a variety of ED presentations including syncope, subarachnoid hemorrhage, and cardiac resuscitation (Perry et al., 2013; Quinn et al., 2004; van Walraven et al., 2001). This group also demonstrated the ability to do resuscitation research in inpatient and prehospital environments, involving not only emergency physicians but also critical care physicians and prehospital providers. This includes a landmark trial of high-dose epinephrine in cardiac arrest, and several other successful cardiac arrest trials (Stiell et al., 1992a,b, 1996, 2001a,b). They then studied the use of prehospital interventions by demonstrating the most efficient tiered response when implementing a paramedic system and determining which prehospital interventions are effective (Stiell et al., 2004, 2008).

Other noteworthy areas in which emergency care research has made great progress include the following: the roll-out of public-access defibrillation via education and installation of automatic external defibrillators; systematic investigation of sepsis resuscitation concepts and protocols; definition and tracking of the epidemic of community-associated methicillin-resistant *Staphylococcus aureus*; use of wound adhesives; management of psychiatric and substance abuse−related emergencies; management of infants with fever; and child abuse detection (Valenzuela et al., 1993, 2000; Rivers et al., 2001; Filbin et al., 2014; Pro et al., 2014; Talan et al., 1999; Quinn et al., 1997; Baraff, 1994, 2000; Pallin et al., 2008; Talan, 2008, 2010; D'Onofrio et al., 2010; Lindberg et al., 2012; Chambers, 2013; Pallin et al., 2013a,b).

Emergency care investigators also have been resourceful in developing collaborations and national networks. Initially with little infrastructure academic departments of emergency medicine invested internal resources to develop their research infrastructure to improve research productivity and participate in collaborative networks. Examples of such networks include the National Center for Infectious Disease and Center for Disease Control's EMERGEncy ID NET for infections and the Multicenter Airway Research Collaboration (MARC) originally for emergency asthma care which is now the Emergency Medicine Network (EMNet), that includes MARC the National ED Inventories (NEDI), National ED Safety Study (NEDSS), Emergency Medicine Shock Research Network (EMSHOCKNET), and ED 24 research network (ED24) (Pro et al., 2014; Cairns et al., 2010; Cearnal, 2006; Kaji et al., 2010; Koroshetz and Brown, 2013; Koroshetz, 2010; Mitka, 2012). This initial investment in research infrastructure with internal resources along with early successes of emergency medicine investigators led to the growth of emergency care research and investment from external sources.

BUILDING AN EMERGENCY CARE RESEARCH SITE

Having the infrastructure and support to hire and train appropriately qualified research coordinators and project managers is important. Trials in the emergency care department are complicated because of the environment and the clinical time pressure constraints. They require careful and often unique regulatory oversight and compliance particularly in the area of human subjects protection including the handling of protected health information. Research coordinators working in the emergency care environment are required to respond quickly, implement protocols in a timely fashion, and be available 24 h a day to do so. They require special motivation to work in the environment and need to be supported with excellent training and oversight. One also needs to consider adequate staffing needs. This involves working with human resources at one's institution to ensure compliance when hiring and staffing. It is important that researchers consider this infrastructure and ensure they have a critical number of trials and funding sources (including internal support) to ensure that they can build and fund the essential infrastructure for this environment. When recruiting for clinical trials the goal is to be inclusive, consider all eligible patients and to be operational 24 h a day, 7 days a week since the emergency care areas never close.

Critically ill patients who present to the ED receive care from multiple providers and hospital services (e.g. emergency medicine, surgery, and critical care), in multiple settings (e.g. ED, OR, and ICU). Engaging these providers, whether or not they are researchers, is important, and difficult. It is important to consider all potential providers and services that could be impacted by a particular research protocol. All such providers and services need to be educated and be generally supportive of the research to ensure the protocol will be followed and implemented efficiently. Those providers who need to be investigators and perform specific roles in the research should be compensated for time and effort. Likewise services such as radiology, laboratory, and pharmacy need to be supportive and compensated for any research-related services. They also need to ensure they have the capacity, time, and commitment to perform those services since these protocols need to be done in an expedited manner and often during nonbusiness hours. It is recommended that before the trial begins, potential patients are identified at their entry point into the emergency care environment then followed through their hospital care to determine all the treatment teams, providers, and services that may come in contact with those patients to ensure that they are all considered when implementing the protocol.

Finally, in this day and age, a successful researcher in the emergency care environment has to be as much a good leader and manager as an effective researcher. The ED research team usually is large with multiple investigators and personnel to manage.

It is important that the researcher be engaged at the administrative level to ensure that adequate regulatory and administrative compliance is followed. This includes a good working knowledge and relationship with their IRB, working relationships with their grants and management office including contacting and financial administration. Understanding the policies and rules of one's institution will help not only at running studies at one's own institution but also contracting with other institutions and determining the most appropriate structure for collaborative activities.

Some of these arrangements can be through consortium agreements with collaborating institutions where each will provide specific infrastructure and research personnel. At other institutions, the approach may differ, consisting of collaboration (business associates agreements), service agreements, and consulting agreements for the researchers. This structure is dependent on the relationship and abilities of the institution and those researchers. Thus, multicenter studies can have a substantial impact on the structure of the research team. For example, an experienced research institution may be able to provide infrastructure to enroll into an emergency care trial. In this case they would be investing research knowledge, infrastructure and intellectual property, and a consortium agreement where they would be totally responsible for running the trial at their institution which would be appropriate. In other nonresearch institutions such arrangements are likely not to be feasible. Thus, different institutions will have varying rules and regulations and how they will contract with outside institutions; therefore, it is important when building a research team to work closely with one's institution and understand the mechanisms for contracting the people involved so that one may optimize and build their research teams accordingly.

FUNDING OF INFRASTRUCTURE IN THE EMERGENCY CARE ENVIRONMENT

Large successful emergency care research networks have demonstrated the importance of investing in research infrastructure. The Neurological Emergencies Treatment Trials (NETT), the Pediatric Care Research Network (PECARN), and the Resuscitation Outcomes Consortium (ROC) have all received support from the National Institutes of Health (NIH) that has invested in substantial infrastructure that has allowed these networks to be successful (Table 27.1).

These networks all have slightly different goals and missions; however, they all have expertise in building research teams, identifying subjects, providing prompt recruitment, and dealing with important ethical issues involved in the consenting processes. Large networks also allow for feedback and trial design considerations that will allow the experience of various sites to give feedback into how particular study may be most successfully implemented.

During the past 15 years, the challenges and difficulty in funding and performing emergency care research are starting to be addressed. Not only with the large research networks like NETT, PECARN, and ROC but also by several roundtables held and supported by NIH that led to the formation of the NIH Office of Emergency Care Research (OECR). Established in 2012 and housed in NIH's National Institute of General Medical Sciences, OECR is a focal point for basic, clinical and translational emergency care research, and training across NIH. It coordinates, catalyzes, and communicates about NIH funding opportunities in emergency care research and fosters the training of future researchers in emergency care. In addition to a K-12 program to train a new generation of emergency care researchers; NIH has supported innovation with new adaptive design research lead by emergency care researchers (Meurer et al., 2012; Teng et al., 2009).

As a result, funding emergency care research continues to grow not only for clinical trials but also in the area of research training and comparative effectiveness research (see Fig. 27.1). There are new mandates by America's Affordable

TABLE 27.1 NIH Sponsored Networks With Emergency Care Research

Network	Focus	Year Started	Website
Neurological Emergencies Treatment Trial (NETT)	Large simple clinical trials focused on very acute injuries of the brain, spinal cord, and peripheral nervous system with the goal of improving patient outcomes through innovative research	2006	www.nett.umich.edu/nett/welcome
Resuscitation Outcomes Consortium (ROC)	Clinical trial network focusing on research in pre-hospital cardiopulmonary arrest and severe traumatic injury	2006	https://roc.uwctc.org
Pediatric Emergency Care Applied Research Network (PECARN)	Multiinstitute research into the prevention and management of acute illnesses and injuries in children	2001	www.pecarn.org

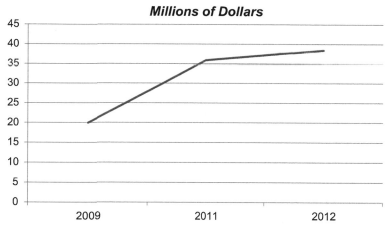

FIGURE 27.1 NIH funding in emergency medicine. *Adapted from Blue Ridge Institute for Medical Research. http://www.brimr.org/NIH_Awards/2013/ NIH_Awards_2013.htm.*

Health Choices Act of 2009, which puts the focus on cost effectiveness in all areas of health care. One of the top priorities of this group is comparative effectiveness research investigating the value of neurological and orthopedic imaging modalities when ordered by emergency physicians (Medicine Io, 2009).

THE ROLE OF INDUSTRY

Historically, public investment (i.e., from the NIH) in research has focused on laboratory research, and clinical trials were viewed as the purview of pharmaceutical companies. This is no longer the case, as the federal government and multiple charitable foundations sponsor clinical research. Nevertheless, understanding the potential role of for-profit corporations in research is valuable. From the perspective of the practicing academic clinical researcher, work that is sponsored by industry can be viewed as falling into three simple groups: "my research," "our research," and "their research."

"My research" refers to research planned and executed by an academic with no input from industry other than money. Most pharmaceutical and biotech companies offer grants to support this type of research. These funding programs are typically known as "IIR" programs, standing for "investigator-initiated research" programs. IIR grants are typically modest in size, from about $5000 to $50,000, but can be very helpful launching an academic career and supporting new ideas for experienced researchers.

"Our research" refers to research that is planned and executed jointly by academics and corporations. Such work is typically financed by the corporation, and there is typically a give-and-take between the academic investigator and the corporation regarding every aspect of the research, from goal setting to interpretation of results. Conflict of interest is a real possibility here, and the key to avoid it is to start with clear expectations. Expectations are embodied in a contract or agreement between the sponsor and the investigator's institution. A good contract gives the investigator full freedom to publish any results that emerge without influence from the sponsor. It is important to remember that the sponsor's first duty is to its shareholders, not to patients or the annals of science. Nevertheless, an enormous amount of very important research is published every week, in top journals, with input and sponsorship from corporations and intellectual contribution from academics. Engaging in shared research with corporations is a major career endeavor, which takes decades to develop. Huge trials can be executed with budgets running into the tens or hundreds of millions of dollars. As with any other aspect of research, the key to doing this well is to do it with mentorship from someone who has extensive experience and who is able to engage in frank discussions about career development, conflict of interest, and bias.

"Their research" refers to research that is designed, executed, and published entirely by a corporation. The role of the academic investigator is simply to perform "work for hire," by following the sponsor's protocol and enrolling patients. These trials are extremely valuable for the academic researcher for several reasons. First, the protocols are typically of very high quality, having been developed by a well-financed team of experts and having passed FDA scrutiny. Thus they provide wonderful learning opportunities for junior faculty and research coordinators, and set high standards for academic-initiated research. Second, these studies typically pay the investigator's institution quite a lot of money for each enrolled participant. This money is used to hire research coordinators, and those research coordinators can be made available for unsponsored research, or research that has sponsorship that is "shoestring" or inadequate. Thus, conducting "their" research can launch and/or provide ongoing support for academicians and their departments.

Dealing with industry in any of the above contexts requires a nuanced understanding of federal law. Several complex concepts must be mastered, including conflict of interest, fraud and abuse, fair market value, and medicare coverage analysis. More detailed discussion is beyond the scope of this chapter, but the reader is urged to devote enough time to these topics to understand them well. Academic institutions typically have rich resources to help investigators navigate these waters, variously known as "Compliance Office," "Office for Interactions with Industry," or by other names.

IMPLEMENTATION OF EMERGENCY CARE RESEARCH

Recruitment

Providing evidence that one has enough patients of interest and can identify and enroll them in a timely manner is crucial to determine whether it is possible to do a specific trial and is often required before receiving funding. Pilot studies provide the most precise determination of potential subjects as well as a preview of potential implementation problems. However they require a significant investment of resources for a small number of patients and cannot be done practically without funding or support. Prospective and retrospective screening through chart review is more practical and can be helpful to estimate recruitment capabilities. These reviews can be problematic, however, as they often fail to identify problems such as real-time identification, consent issues, and other complexities of implementing a specific protocol. As a result they usually overestimate the number of patients that a site can enroll. Once investigators can demonstrate, the participants are available for enrollment, and they should focus on specific strategies to enhance recruitment. Recruitment into emergency care trials has three important elements identification of subjects, consent, and implementation.

Identification of Subjects

More than any other setting the emergency care environment requires a timely and efficient method for identifying patients who meet inclusion into perspective studies. This is particularly important when the time from presentation and the need for treatment is short (e.g., tPA for stroke, administration of fluids for penetrating injuries, blood substitutes for shock, cardiac catheterization, therapeutic hypothermia, and acute brain injury studies). Even for relatively nonemergent conditions patients who present to the ED have a limited time for enrollment. EDs are continuously trying to expedite the care of nonemergent patients and improving patient flow to decrease length of stay in order to meet the demands of increasing volume and overcrowding. For example, patients who suffer a laceration may have a length of stay of only a few of hours and often have treatment started within 15–20 min of arrival leaving a narrow window for identification and enrollment into a potential trial.

Researchers have traditionally tried ED staff education and the placement of posters to remind staff of potential trials. ED staff education is important but trials are unlikely to be successful if they depend on treating practitioners and staff to identify eligible participants. The use of research coordinators or a specific research staff member who is either dedicated to reviewing current patients in the ED or alerted to incoming patients is essential to capture all the patients. The use of dedicated research staff to screen and identify eligible patients can be accomplished in several ways. Some EDs and institutions have incorporated research coordinators to work closely with clinical teams such as stroke and trauma teams. As part of the team they can respond to specific clinical alerts. Some of these positions are funded through research grants other are funded through programs within the hospital. For other conditions institutions have engaged student volunteers to be physically present in the ED to screen prospective patients (Hollander and Singer, 2002). However, the skill and experience of these individuals to enroll patients into complex clinical trials is generally unproven, and also federal labor laws require nuanced interpretation when volunteers provide real assistance (dol). These programs are more for prospective observational studies and to do the initial screening for clinical trials (Hollander et al., 2004). In order to have trained coordinators continually present in the ED there has to be a sufficient volume of patients to justify the effort and cost of this full-time presence. The most successful situation is a hybrid approach. In this approach a network is formed to manage multiple studies within the same clinical center or multiple centers. This allows a more efficient use of dedicated, experienced personnel to screen and identify remotely and respond to a needed care area to enroll patients in a timely manner. Having a large number of trials going at once allows one to use economy of scales to implement skilled coordinators in this fashion. This does require some innovative methods using remote electronic medical records access and clinical informatics to generate clinical alerts in addition to established paging systems.

Using clinical staff to alert study staff or having study staff manually screen ED admission logs or screen paper medical records for potential patients is not only inefficient and untimely but the manual screen also lacks the optimal privacy intended in the Health Insurance Portability and Accountability Act (HIPAA). Many researchers and institutions have turned to other methods to better identify and screen patients that are more timely and compliant with their privacy standards than manual screening. These include electronic screening of patients and being incorporated into paging alerts for clinical conditions such as stroke codes, trauma alerts, and arrest codes within the hospital.

Some conditions are managed with teams that are activated by group pager messages (e.g., stroke, cardiac arrest, trauma, or sepsis response teams). Research teams can be included on these group notifications. With the move of hospitals to electronic health records, a stream of continuous and instantaneously updated clinical information flows throughout the hospital. It is now possible to generate clinical alert systems that are similar to clinical decision support systems that are immediate and alert investigators and their teams of potential patients (Quinn and Durski, 2004). These types of programs have been used by several institutions and investigators and are available on a wide variety of presenting complaints, vital signs, orders, and results that will instantaneously be recognized by screening systems and appropriate alerts sent to staff in a HIPAA-compliant manner (Embi et al., 2005a,b). For example, in a sepsis study the research staff could be alerted any time a person presented with a low systolic blood pressure, fever, or elevated lactate (Weber et al., 2010). That patient could then be screened through the electronic health record on-site or remotely for potential eligibility. This could happen quickly and efficiently and allow patients to be rapidly enrolled into a trial. When such a system is implemented, it has been shown to greatly improve the enrollment of such patients and allowing for an intervention being undertaken before treatment had begun. Similarly patients with TIA now routinely get MRI instead of CT within 3 h of arrival in our ED. Screening MRI of the head orders from our ED has identified all possible patients for our TIA studies (see Fig. 27.2). These messages are immediate, secure, HIPAA compliant and only go to eligible researchers who have IRB protocols approved to screen in this manner. These alert systems have and will continue to evolve over the next several years and be available at more institutions as they solve concerns with data sharing and management of these large networks.

Consent and Enrollment

Once patients are screened and identified, the time frame to consent and enroll them is far shorter in emergency care trials. The unique time pressures that affect the consent and enrollment process are further stressed by the fact that most patients may not be able to consent due to an existing neurological impairment or acute disease process. Consent by a LAR of the patient is an option in these settings and is governed by various state laws and corresponding institutional review boards. However, emergency care trials are still faced with problems seeking consent with an LAR or power of attorney that may not be available to consider participation in a clinical trial. When the patient, power of attorney or LAR is available the acuteness of an unsuspecting illness often takes place in an emotional setting. It can be a difficult time to approach anyone for consent. This is especially true for those seeking consent that often have no established relationship or built in trust. Those obtaining consent to emergency care situations can be aided by aligning themselves with members of the treatment team for support; this includes getting treating physicians involved in updating the patient and family early and then approaching patients only after they have been updated by the treatment team. Working with specially trained social workers or trauma counselors can be helpful. Social workers and counselors have been helpful in the area of organ donations; however, their availability and cost will vary with size and type of institution (Elding and Scholes, 2005). If time permits, involving primary care givers with existing medical relationship and including them in discussions with the patient can be helpful in having patients comfortable in enrolling and also can be helpful with compliance and follow-up (Bland, 2008; Smith et al., 2004a,b).

Studies have successfully used other techniques to improve consent and enrollment times. In the field administration of stroke treatment with magnesium (FAST-MAG) was a pilot trial where the EMTs carried a dedicated study cell phones for

From: nobody@stanford.edu
Subject: MR ANGIO HEAD secure: Alert 189313
Date: May 3. 2016 at
To:

Order: MR ANGIO HEAD Tue May 3, 16,
Reason: MRA head and neck stroke protocol
Attending:

This alert brought to you by the Stanford Center for Clinical Informatics https://clinicalinformatics.stanford.edu

This message was encrypted In transit via the Stanford Secure Email service.

FIGURE 27.2   Real-time email alert for an MRI order used to identify subjects for TIA trials.

investigators to elicit consent in the out-of-hospital setting (Saver et al., 2006). The study found that the use of the cell phone decreased prehospital treatment and allowed for initiation of study procedures over 100 min sooner than prior trials where consent was obtained after hospital arrival. It is unclear whether verbal consent for such studies would be approved by most IRBs. Most require at least faxed consent for out-of-hospital enrollments including patients eligible for trials who are being transferred. Unfortunately fax machines are rather outdated technology and often not available to power of attorneys or LARs. Further work using electronic signature and secure web-based consents that could be signed using a smart phone or tablet appears to be most promising.

When informed consent is not possible exception from informed consent (EFIC) has been used successfully to enroll patients into studies in the prehospital and emergency care setting. Federal regulations (FDA Final Rule 21 CFR 50.24) allow for enrollment of participants into studies without prior consent when a number of very specific conditions are met (Herlitz, 2002). Notification must be done and consent obtained from the patient or LAR as soon as possible after the intervention. Such studies are required to follow additional guidelines including the requirements for community consultation and public disclosure before and after the study. The types of community consultation and public disclosure are at the discretion of local IRBs. However, targeting the community consultation to groups with an interest and perspective on the condition or disease of study is most valuable (Ernst et al., 2005; Ernst and Fish, 2005). For example, using a focus group of patients with epilepsy would be appropriate for community consultation for epilepsy studies as would a brain injury support group or stroke support group when studying acute brain injury or stroke. The community consultation focuses on getting feedback on the study itself but also often on whether they would have agreed to participate or authorized a loved one to participate in the proposed study when they had their emergency presentation. The individuals in these groups often have an interest and perspective missing from other community groups. Once feedback is received from the community consultation, one should focus on public disclosure with attempts at being as transparent as possible to the public and providing a method for opting out of the research in advance of the study (Silbergleit et al., 2012a,b).

While initially somewhat controversial, experience with EFIC trials has resulted in a more standardized approach, and their successful implementation and acceptance from the public and as a result, greater acceptance by patients, IRBs and the FDA (Dickert et al., 2013; Biros et al., 2009). Certain trials in the emergency care setting cannot be done without EFIC. The consenting process may be unfeasible in an unconscious or altered patient and/or result in drugs or treatments being given outside their therapeutic window. Investigators testing therapies with narrow therapeutic windows in the emergency care environment should consider whether the EFIC process should be considered rather than risking coming up with a failed therapy because it was not administered quickly enough in the trial (Fig. 27.3—Process For EFIC Implementation).

Trial Design Considerations

When implementing trials in the emergency care environment investigators need to make sure the trial and intervention can be started quickly and implemented efficiently. It is helpful if the intervention can be completed in one care area or at least started quickly and easily continued if the patient is transferred between units (i.e., ED to ICU or operating room).

Complex trials with prolonged and complex screening protocols can be difficult to implement even if they are well funded and amply staffed. Investigators should realize that if a therapeutic trial is overly complex not only will it be hard to complete but also will affect the trial's ability to be reproduced and therefore, the generalizability of the results. The impact and time required for what appears to be a simple or routine intervention should not be underestimated. Accordingly, most successful large trials in EDs have been prospective cohort studies and relatively simple clinical trials (Stiell et al., 2008, 2004, 1999; Hoffman et al., 1998; Silbergleit et al., 2012a,b; Alldredge et al., 2001). The successful clinical trials done in the emergency care area have some features in common. First, they avoid requiring special environments and specific investigators to be present during implementation. In general, physicians investigators need to be on call and readily available to discuss and approve enrollment and intervene in special circumstances. However, the trials need to be able to be implemented in a timely manner by on-site and prehospital personnel and coordinators with clear inclusion and exclusion criteria. It is important to learn from relatively simple trials successfully completed in the ED. Cardiac and stroke thrombolysis trials are examples of how studies with clear inclusion/exclusion criteria and efforts to streamline drug administration can rapidly recruit and be implemented in the emergency care environment. Similarly seizure trials such as PHTSE (Prehospital Treatment of Status Epilepticus) and RAMPART (Rapid Administration of Medication Prior to Arrival) where intervention was started and completed in the prehospital environment are examples of simple large clinical trials done successfully in the emergency care environment (Silbergleit et al., 2012a,b; Alldredge et al., 2001). More complex emergency care trials like PROTECT III (Progesterone for the Treatment of Traumatic Brain Injury) that required prolonged drug administration and adherence to Brain Trauma Foundation treatment guidelines was successful because it

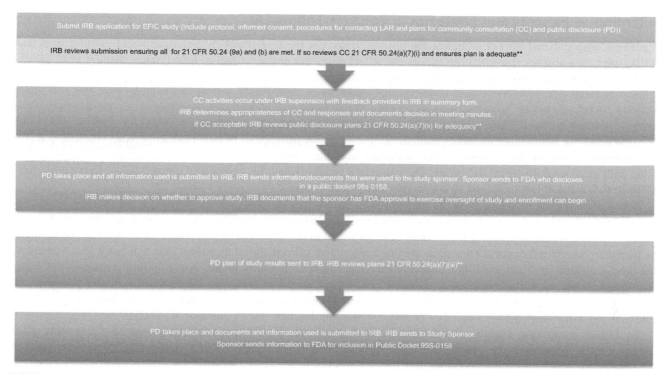

FIGURE 27.3 Suggested Flow Chart for EFIC Implementation. **The CC and PD plans may need to be revised one or more times based on IRB review.

was done under EFIC and the time sensitive intervention in the ED was limited to getting study drug administered in a timely fashion.

As clinical trials become more complex and require more interventions in the emergency care environment they become more difficult to implement and complete. In addition to the increased time and cost of personnel training, the likelihood of protocol violations increases with this complexity. That risk can be minimized with strict oversight by specific study coordinators. However, this is not without significant cost and requires 24/7 presence for enrollment and ongoing trial implementation. Trials that are complicated require significant, dedicated infrastructure including dedicated coordinators and collaborative teams primarily administered through central coordinating centers with central data center collection and management. Without this infrastructure even relatively simple studies will have difficulties being done in the emergency care environment.

It is recommended that investigators meet early on with clinical operational leaders to ensure that trial designs are harmonious with real-world processes in emergency care. Specifically this includes working with emergency care clinical staff (MD and nursing) and when necessary emergency medical service (EMS) staff early on in protocol development. This will ensure that proposed trial is optimally designed for efficiency in the emergency care environment. For example, the NETT has an executive steering committee that helps design and determines the suitability for the emergency care environment, thus, guiding investigators during the trial design and funding stages in an effort to maximize study efficiency and effectiveness. Such a committee also can help develop solutions for accurate and complex data acquisition. For example in the RAMPART study, medication study boxes were also voice box recorders with GPS and time stamps to document medication administration and study procedures. This was done as an alternative to placing a further documentation burden with its inherent risks of incompletion on the paramedics.

It is also important to engage and stimulate nonresearch staff. Providing some sense of ownership over the trial is an important way to improve successful implementation. This can be done through regular updates on enrollment and patient outcomes. Furthermore, providing nonresearch patient care staff with feedback and expressions of gratitude is often overlooked. Having a system and mechanism in place is important. Simple "thank you" cards to give thanks and feedback to helpful patient care staff a few days after enrollment is a relatively easy way to develop awareness and goodwill. It also provides opportunities to further alert and educate patient care staff about a trial. Many patient care staff may have missed or forgotten about specific trials. Thus constant awareness for the need to continually engage and educate during enrollment is imperative for a successful trial.

Retention of Patients and Strategies to Enhance Follow-Up

If patients drop out and do not complete the primary outcome, the investment in recruitment and enrollment is lost. Furthermore, attrition and bias become factors that affect study interpretation and eventual value (Britton et al., 1995). Attrition is defined as a loss of numbers due to resignation or death and is problematic in all clinical trials. Bias in terms of medical research refers to a systematic condition that could not be remedied by repeating the study over and over again. Attrition contributes to bias when participants are not lost randomly and reflect a bias of certain characteristics that sustain better or worse outcomes. While randomization and intention to treat analysis attempt to control for random effects they cannot account for nonrandom treatment termination. For example, participants of one treatment could consistently feel so ill as a result of drug adverse effects that they withdraw from the study at a higher rate than another treatment group. One group with more disability may also be less likely to return and this is another example of potential bias. Every effort must be made to minimize lost to follow-up and curtail this potential bias.

A key issue impacting retention is the time to follow up for participants. Clearly, the longer the follow-up the more participants will be lost. Selection of short-term outcomes can alleviate some attrition but concise short-term outcomes are often not the focus of many clinical trials. Many outcomes tend to involve a one or two hour examination and/or testing of the patient to ascertain the primary and secondary outcomes. This often requires the patient to return to the site. Sometimes this is unavoidable to detect small but important outcomes such as disability or neuropsychiatric outcomes. If possible, clear concise outcomes like termination of seizure on hospital arrival or death should be used as it can accurately be ascertained in most patients. Using tools like the National Death Index and Social Security Death Index is possible in following up participants who appeared to be lost to follow-up but have died (Quinn et al., 2008).

To further complicate matters, research follow-up is complicated when potential participants are transients either because they have had their emergency while traveling or because they are homeless. If a study is multicenter, there may be opportunity for patients to follow up at another site. If participants need to travel from a distance, every effort should be made to make study visits coordinate with clinical care. Consideration should also be made for those who have physical disabilities and logistical difficulties. The time and cost associate with return visits are key reasons for attrition or failure to recruit. If patients are compensated for transportation costs and their time and their visits are at no cost, they are more likely to participate and remain in the study.

Other predictors of attrition include older age, male sex, lower educational attainment, functional impairment, poor cognitive or verbal intelligence, and significant medical comorbidities (Sharma et al., 1986). Despite attempts to minimize attrition, it will occur. Among large population-based epidemiological studies of patients who only require follow-up interviews, 20% are frequently lost to follow-up. Factors that can help decrease attrition include informed consent that clearly conveys the full commitment required for participation, a strong relationship between the study coordinator, care providers and the participants, and consistency among the research staff in maintaining contact with participants who have been recruited. Research staff should be in close contact with study participants, record telephone follow-ups in a log and keep track of the best time to contact patients. While it is generally ideal to have follow-up conducted in a controlled environment, if necessary participants may be seen by local physicians and have standard-of-care tests (including lab tests) and assessments done remotely especially if patients have disabilities and are bed bound in a facility. Doing follow-up in these facilities has been made easier with the guidance of the Office of Human Research Protections (OHRP) which allows the centers to cooperate with follow-up requirements and not become technically engaged in research and bound by other IRBs and institutional subcontracts (Services, USDoHaH). Investigators should also be aware that patients enrolled in EFIC trials must be followed as mandated by federal law even if they drop out. All efforts must be made to account for all patients who participate in EFIC trials. This includes the follow-up of participants who later become prisoners and are incarcerated. In such cases, special approval to enroll and follow up such patients is required from a local IRB and OHRP.

CONCLUSION

The emergency care environment is a unique care area and thus requires special consideration as a unique research area. Efforts to overcome the challenges of the setting can be accomplished with special attention to several factors. These include building collaborative research teams, leveraging effective management and contracting resources, developing effective strategies for immediately identifying, recruiting and enrolling patients, and effectively adapting efficient trials (including EFIC trials). Special attentions to these aspects are fundamental to performing clinical research in this frequently chaotic environment. Finally leveraging relationships between members of the treatment team, primary care givers, outside facilities, and the research team can develop collaborative relationships to enhance patient care and research. This will ensure an environment that will maximize patient follow-up.

REFERENCES

Alldredge, B.K., Gelb, A.M., Isaacs, S.M., Corry, M.D., Allen, F., Ulrich, S., et al., 2001. A comparison of lorazepam, diazepam, and placebo for the treatment of out-of-hospital status epilepticus. N. Engl. J. Med. 345 (9), 631–637.

Baraff, L.J., 2000. Management of fever without source in infants and children. Ann. Emerg. Med. 36 (6), 602–614.

Baraff, L.J., 1994. Outpatient management of fever in selected infants. N. Engl. J. Med. 330 (13), 938–939 author reply 9–40.

Bland, P., 2008. Consent guidance will strengthen doctor-patient relationship. Practitioner 252 (1707), 4.

Biros, M.H., Sargent, C., Miller, K., 2009. Community attitudes towards emergency research and exception from informed consent. Resuscitation 80 (12), 1382–1387.

Britton, A., Murray, D., Bulstrode, C., McPherson, K., Denham, R., 1995. Loss to follow-up: does it matter? Lancet 345 (8963), 1511–1512.

Centers for Disease Control and Prevention, 2010. Emergency Department Visits. Available from: http://www.cdc.gov/nchs/fastats/emergency-department.htm.

Chambers, H.F., 2013. Cellulitis, by any other name. Clin. Infect. Dis. 56 (12), 1763–1764.

Cairns, C.B., Maier, R.V., Adeoye, O., Baptiste, D., Barsan, W.G., Blackbourne, L., et al., 2010. NIH roundtable on emergency trauma research. Ann. Emerg. Med. 56 (5), 538–550.

Cearnal, L., 2006. The birth of the NETT: NIH-funded network will launch emergency neurological trials. Ann. Emerg. Med. 48 (6), 726–728.

D'Onofrio, G., Jauch, E., Jagoda, A., Allen, M.H., Anglin, D., Barsan, W.G., et al., 2010. NIH roundtable on opportunities to advance research on neurologic and psychiatric emergencies. Ann. Emerg. Med. 56 (5), 551–564.

http://www.dol.gov/whd/regs/compliance/whdfs71.pdf.

Dickert, N.W., Mah, V.A., Baren, J.M., Biros, M.H., Govindarajan, P., Pancioli, A., et al., 2013. Enrollment in research under exception from informed consent: the Patients' Experiences in Emergency Research (PEER) study. Resuscitation 84 (10), 1416–1421.

Embi, P.J., Jain, A., Clark, J., Harris, C.M., 2005a. Development of an electronic health record-based Clinical Trial Alert system to enhance recruitment at the point of care. AMIA Annu. Symp. Proc. 231–235.

Embi, P.J., Jain, A., Clark, J., Bizjack, S., Hornung, R., Harris, C.M., 2005b. Effect of a clinical trial alert system on physician participation in trial recruitment. Archives Intern. Med. 165 (19), 2272–2277.

Elding, C., Scholes, J., 2005. Organ and tissue donation: a trustwide perspective or critical care concern? Nurs. Crit. care 10 (3), 129–135.

Ernst, A.A., Weiss, S.J., Nick, T.G., Iserson, K., Biros, M.H., 2005. Minimal-risk waiver of informed consent and exception from informed consent (Final Rule) studies at institutional review boards nationwide. Acad. Emerg. Med. official J. Soc. Acad. Emerg. Med. 12 (11), 1134–1137.

Ernst, A.A., Fish, S., 2005. Exception from informed consent: viewpoint of institutional review boards—balancing risks to subjects, community consultation, and future directions. Acad. Emerg. Med. 12 (11), 1050–1055.

Fisher, J., Sokolove, P.E., Kelly, S.P., 2009. Overcrowding: harming the patients of tomorrow? Acad. Emerg. Med. 16 (1), 56–60.

Filbin, M.R., Arias, S.A., Camargo Jr., C.A., Barche, A., Pallin, D.J., 2014. Sepsis visits and antibiotic utilization in U.S. emergency departments*. Crit. Care Med. 42 (3), 528–535.

Hollander, J.E., Singer, A.J., 2002. An innovative strategy for conducting clinical research: the academic associate program. Acad. Emerg. Med. 9 (2), 134–137.

Hollander, J.E., Sparano, D.M., Karounos, M., Sites, F.D., Shofer, F.S., 2004. Studies in emergency department data collection: shared versus split responsibility for patient enrollment. Acad. Emerg. Med. 11 (2), 200–203.

Herlitz, J., 2002. Consent for research in emergency situations. Resuscitation 53 (3), 239.

Hoffman, J.R., Wolfson, A.B., Todd, K., Mower, W.R., 1998. Selective cervical spine radiography in blunt trauma: methodology of the National Emergency X-Radiography Utilization Study (NEXUS). Ann. Emerg. Med. 32 (4), 461–469.

Kaji, A.H., Lewis, R.J., Beavers-May, T., Berg, R., Bulger, E., Cairns, C., et al., 2010. Summary of NIH medical-surgical emergency research roundtable held on April 30 to May 1, 2009. Ann. Emerg. Med. 56 (5), 522–537.

Koroshetz, W.J., Brown, J., 2013. NIH: developing and funding research in emergency care and training the next generation of emergency care researchers. Health Aff. 32 (12), 2186–2192.

Koroshetz, W.J., 2010. Setting NIHTFoRiE. NIH and research in the emergency setting: progress, promise, and process. Ann. Emerg. Med. 56 (5), 565–567.

Lindberg, D.M., Shapiro, R.A., Laskey, A.L., Pallin, D.J., Blood, E.A., Berger, R.P., et al., 2012. Prevalence of abusive injuries in siblings and household contacts of physically abused children. Pediatrics 130 (2), 193–201.

Mitka, M., 2012. NIH signals intent to boost funding of emergency care research and training. JAMA 308 (12), 1193–1194.

Meurer, W.J., Lewis, R.J., Tagle, D., Fetters, M.D., Legocki, L., Berry, S., et al., 2012. An overview of the adaptive designs accelerating promising trials into treatments (ADAPT-IT) project. Ann. Emerg. Med. 60 (4), 451–457.

Medicine Io, 2009. Initial Priorities for Comparative Effectiveness Research. The National Academies Press.

Owens, P.L., Barrett, M.L., Gibson, T.B., Andrews, R.M., Weinick, R.M., Mutter, R.L., 2010. Emergency department care in the United States: a profile of national data sources. Ann. Emerg. Med. 56 (2), 150–165.

Pallin, D.J., Egan, D.J., Pelletier, A.J., Espinola, J.A., Hooper, D.C., Camargo, Jr., C.A., 2008. Increased US emergency department visits for skin and softtissue infections, and changes in antibiotic choices, during the emergence of community-associated methicillin-resistant Staphylococcus aureus. Ann.Emerg. Med. 51 (3), 291–298.

Pallin, D.J., Allen, M.B., Espinola, J.A., Camargo Jr., C.A., Bohan, J.S., 2013a. Population aging and emergency departments: visits will not increase, lengths-of-stay and hospitalizations will. Health Aff. (Millwood) 32 (7), 1306–1312.

Pallin, D.J., Binder, W.D., Allen, M.B., Lederman, M., Parmar, S., Filbin, M.R., et al., 2013b. Clinical trial: comparative effectiveness of cephalexin plustrimethoprim-sulfamethoxazole versus cephalexin alone for treatment of uncomplicated cellulitis: a randomized controlled trial. Clin. Infect. Dis. 56 (12), 1754−1762.

Perry, J.J., Stiell, I.G., Sivilotti, M.L., Bullard, M.J., Hohl, C.M., Sutherland, J., et al., 2013. Clinical decision rules to rule out subarachnoid hemorrhage for acute headache. JAMA 310 (12), 1248−1255.

Pro, C.I., Yealy, D.M., Kellum, J.A., Huang, D.T., Barnato, A.E., Weissfeld, L.A., et al., 2014. A randomized trial of protocol-based care for early septic shock. N. Engl. J. Med. 370 (18), 1683−1693.

Quinn, J.V., Stiell, I.G., McDermott, D.A., Sellers, K.L., Kohn, M.A., Wells, G.A., 2004. Derivation of the San Francisco Syncope Rule to predict patients with short-term serious outcomes. Ann. Emerg. Med. 43 (2), 224−232.

Quinn, J., Wells, G., Sutcliffe, T., Jarmuske, M., Maw, J., Stiell, I., et al., 1997. A randomized trial comparing octylcyanoacrylate tissue adhesive and sutures in the management of lacerations. JAMA 277 (19), 1527−1530.

Quinn, J., Durski, K., 2004. A real-time tracking, notification, and web-based enrollment system for emergency department research. Acad. Emerg. Med. 11 (11), 1245−1248.

Quinn, J., Kramer, N., McDermott, D., 2008. Validation of the social security death index (SSDI): an important readily-available outcomes database for researchers. West. J. Emerg. Med. 9 (1), 6−8.

Rivers, E., Nguyen, B., Havstad, S., Ressler, J., Muzzin, A., Knoblich, B., et al., 2001. Early goal-directed therapy in the treatment of severe sepsis and septic shock. N. Engl. J. Med. 345 (19), 1368−1377.

Schuur, J.D., Venkatesh, A.K., 2012. The growing role of emergency departments in hospital admissions. N. Engl. J. Med. 367 (5), 391−393.

Stiell, I.G., Wells, G.A., 1999. Methodologic standards for the development of clinical decision rules in emergency medicine. Ann. Emerg. Med. 33 (4), 437−447.

Stiell, I.G., Clement, C.M., McKnight, R.D., Brison, R., Schull, M.J., Rowe, B.H., et al., 2003. The Canadian C-spine rule versus the NEXUS low-risk criteria in patients with trauma. N. Engl. J. Med. 349 (26), 2510−2518.

Stiell, I.G., Wells, G.A., Vandemheen, K.L., Clement, C.M., Lesiuk, H., De Maio, V.J., et al., 2001a. The Canadian C-spine rule for radiography in alert and stable trauma patients. JAMA 286 (15), 1841−1848.

Stiell, I.G., Wells, G.A., McDowell, I., Greenberg, G.H., McKnight, R.D., Cwinn, A.A., et al., 1995. Use of radiography in acute knee injuries: need for clinical decision rules. Acad. Emerg. Med. 2 (11), 966−973.

Stiell, I.G., Greenberg, G.H., McKnight, R.D., Nair, R.C., McDowell, I., Worthington, J.R., 1992a. A study to develop clinical decision rules for the use of radiography in acute ankle injuries. Ann. Emerg. Med. 21 (4), 384−390.

Stiell, I.G., Hebert, P.C., Wells, G.A., Vandemheen, K.L., Tang, A.S., Higginson, L.A., et al., 2001b. Vasopressin versus epinephrine for in hospital cardiac arrest: a randomised controlled trial. Lancet 358 (9276), 105−109.

Stiell, I.G., Hebert, P.C., Wells, G.A., Laupacis, A., Vandemheen, K., Dreyer, J.F., et al., 1996. The Ontario trial of active compression-decompression cardiopulmonary resuscitation for in-hospital and prehospital cardiac arrest. JAMA 275 (18), 1417−1423.

Stiell, I.G., Hebert, P.C., Weitzman, B.N., Wells, G.A., Raman, S., Stark, R.M., et al., 1992b. High-dose epinephrine in adult cardiac arrest. N. Engl. J. Med. 327 (15), 1045−1050.

Stiell, I.G., Nesbitt, L.P., Pickett, W., Munkley, D., Spaite, D.W., Banek, J., et al., 2008. The OPALS Major Trauma Study: impact of advanced life-support on survival and morbidity. CMAJ 178 (9), 1141−1152.

Stiell, I.G., Wells, G.A., Field, B., Spaite, D.W., Nesbitt, L.P., De Maio, V.J., et al., 2004. Advanced cardiac life support in out-of-hospital cardiac arrest. N. Engl. J. Med. 351 (7), 647−656.

Smith, S.R., Jaffe, D.M., Fisher Jr., E.B., Trinkaus, K.M., Highstein, G., Strunk, R.C., 2004a. Improving follow-up for children with asthma after an acute Emergency Department visit. J. Pediatr. 145 (6), 772−777.

Smith, S.R., Jaffe, D.M., Petty, M., Worthy, V., Banks, P., Strunk, R.C., 2004b. Recruitment into a long-term pediatric asthma study during emergency department visits. J. Asthma 41 (4), 477−484.

Saver, J.L., Kidwell, C., Eckstein, M., Ovbiagele, B., Starkman, S., 2006. Physician-investigator phone elicitation of consent in the field: a novel method to obtain explicit informed consent for prehospital clinical research. Prehosp. Emerg. Care 10 (2), 182−185.

Silbergleit, R., Biros, M.H., Harney, D., Dickert, N., Baren, J., 2012a. Implementation of the exception from informed consent regulations in a large multicenter emergency clinical trials network: the RAMPART experience. Acad. Emerg. Med. 19 (4), 448−454.

Stiell, I.G., Wells, G.A., Field, B.J., Spaite, D.W., De Maio, V.J., Ward, R., et al., 1999. Improved out-of-hospital cardiac arrest survival through the inexpensive optimization of an existing defibrillation program: OPALS study phase II. Ontario Prehospital Advanced Life Support. JAMA 281 (13), 1175−1181.

Silbergleit, R., Durkalski, V., Lowenstein, D., Conwit, R., Pancioli, A., Palesch, Y., et al., 2012b. Intramuscular versus intravenous therapy for prehospital status epilepticus. N. Engl. J. Med. 366 (7), 591−600.

Sharma, S.K., Tobin, J.D., Brant, L.J., 1986. Factors affecting attrition in the Baltimore longitudinal study of aging. Exp. Gerontol. 21 (4−5), 329−340.

Services USDoHaH. Guidance on Engagement of Institutions in Human Subjects Research. Available from: http://www.dhhs.gov/ohrp/humansubjects/guidance/engage08.html.

The High Concentration of US Health Care Expenditures, 2006. Available from: archive.ahrq.gov/research/findings/factsheets/costs/expriach/expendria.pdf.

Talan, D.A., Citron, D.M., Abrahamian, F.M., Moran, G.J., Goldstein, E.J., 1999. Bacteriologic analysis of infected dog and cat bites. Emergency medicine animal bite infection study group. N. Engl. J. Med. 340 (2), 85−92.

Talan, D.A., 2008. MRSA: deadly super bug or just another staph? Ann. Emerg. Med. 51 (3), 299−302.

Talan, D.A., 2010. Lack of antibiotic efficacy for simple abscesses: have matters come to a head? Ann. Emerg. Med. 55 (5), 412−414.

Teng, C.S., Lo, W.T., Wang, S.R., Tseng, M.H., Chu, M.L., Wang, C.C., 2009. The role of antimicrobial therapy for treatment of uncomplicated skin and soft tissue infections from community-acquired methicillin-resistant *Staphylococcus aureus* in children. J. Microbiol. Immunol. Infect. 42 (4), 324–328.

Valenzuela, T.D., Roe, D.J., Nichol, G., Clark, L.L., Spaite, D.W., Hardman, R.G., 2000. Outcomes of rapid defibrillation by security officers after cardiac arrest in casinos. N. Engl. J. Med. 343 (17), 1206–1209.

Valenzuela, T.D., Spaite, D.W., Meislin, H.W., Clark, L.L., Wright, A.L., Ewy, G.A., 1993. Emergency vehicle intervals versus collapse-to-CPR and collapse-to-defibrillation intervals: monitoring emergency medical services system performance in sudden cardiac arrest. Ann. Emerg. Med. 22 (11), 1678–1683.

van Walraven, C., Forster, A.J., Parish, D.C., Dane, F.C., Chandra, K.M., Durham, M.D., et al., 2001. Validation of a clinical decision aid to discontinue in-hospital cardiac arrest resuscitations. JAMA 285 (12), 1602–1606.

Weber, E.J., Showstack, J.A., Hunt, K.A., Colby, D.C., Grimes, B., Bacchetti, P., et al., 2008. Are the uninsured responsible for the increase in emergency department visits in the United States? Ann. Emerg. Med. 52 (2), 108–115.

Weber, S., Lowe, H.J., Malunjkar, S., Quinn, J., 2010. Implementing a real-time complex event stream processing system to help identify potential participants in clinical and translational research studies. AMIA Annu. Symp. Proc. 2010, 472–476.

Chapter 28

Psychiatric Disorders

Alan F. Schatzberg
Stanford University, Stanford, CA, United States

Chapter Outline

INTRODUCTION

Psychiatric research has been helped by active collaboration with the field of psychology where research methods have historically been a major focus. Still there are considerable methodological issues that pose hurdles for those studying psychiatric disorders. These include the difficulty in making a truly valid diagnosis; reliance on rating scales to assess change; effects of medications on various biological and clinical measures; inability to biopsy or directly visualize the brain, etc. Herein we discuss key issues that arise in clinical psychiatric research.

DIAGNOSTIC ISSUES

Adults

Mental disorders do not have clear pathological or physical stigmata in contrast to many medical disorders. There is no clinical anatomical pathological specimen or biopsy nor is there a culture and sensitivity test or other laboratory measure to help diagnosis. Instead psychiatry has no way to validate clinical signs and symptoms to make a diagnosis. Thus from the very start, clinical population definition and characterization are challenges that must be met by both researchers and clinicians. To meet this challenge the American Psychiatric Association (2015) on a regular basis promulgates a classification system—the *Diagnostic and Statistical Manual of Mental Disorders*, which is now in its fifth edition (DSM-5). This text provides a categorical classification system for psychiatric and related behavioral disorders in both adults and children. Specific criteria are provided for each syndrome. Syndromes are clustered under types of disorders—mood disorders, anxiety disorders, psychotic disorders, personality disorders, etc. The system is highly reliable but suffers from a lack of face validity. It can be used to assure that investigators at various sites or clinicians in disparate offices are similarly diagnosing subjects. It does not, however, have true face validity since we have no pathology tests to confirm the diagnosis. Thus from the outset investigators always must be concerned that down the road laboratory tests, including genetics, could bring into question the validity of the very diagnostic group they have studied.

Historically, researchers in classification of psychiatric disorders were divided into two camps—lumpers or splitters. Lumpers saw many types of disorders as being variants of a larger category and then characterized patients as being more or less depressed or more or less anxious. Splitters devised categories to discriminate among disorders. DSM-III, DSM-IV, and DSM-5 are largely categorical in nature. DSM-5 has added dimensional measures of anxiety, depression, etc. to a classification system that still uses broad categories (Kraemer et al., 2004).

While there are similarities between the DSM- and the International Classification of Diseases (ICD)-based classification systems, the two are not exactly the same (Sorensen et al., 2005). Most researchers employ DSM categories for their studies in part because of a long track record of research using them. Applying the DSM criteria can be performed via a checklist; however, the conduct of the interview may vary considerably between sites, resulting in a lack of agreement regarding primary and secondary diagnoses. To that end, researchers developed structured interviews to help in arriving at a diagnosis. Most notably the Structured Clinical Interview for Diagnosis (SCID) is the Cadillac of such approaches for Axis I disorders (Spitzer et al., 1992; First et al., 1995). There are briefer tools—e.g., the Mini International Neuropsychiatric Interview, that can also be employed for rapid screening (Sheehan et al., 1998). Personality disorders are assessed via another SCID module (First et al., 1997). Structured interviews for assessing and diagnosing children are also available—e.g., Kiddie Schedule for Affective Disorders and Schizophrenia (Schaffer et al., 1993; Kaufman et al., 1997).

For some disorders greater detail on lifetime course is needed, and structured interviews for specific types of disorders have also been developed and these appear reliable. One such example is the LIFE (Keller et al., 1987) that can be used to determine total lifetime duration of depression in a patient, number of previous episodes experienced, duration of each, etc. Other tools for assessing other features—e.g., childhood abuse—can be added depending on the population or issues being assessed.

Recently, the National Institute of Mental Health (NIMH) has initiated research into the application of research domain criteria (RDoC) to define behavioral constructs such as fear that can then be studied from biological and pharmacological perspectives (Insel et al., 2010). These are discussed in a later subsection of this chapter. To date a limited number of constructs are being studied.

Special Age- and Gender-Based Populations

Childhood disorders are increasingly being studied in psychiatry. For many years, it was thought that many of the major psychiatric disorders were seen only in adults. As researchers began to look more closely at children, a number of maladaptive behavioral and mood features were noted, and investigators began to argue that these features signified that a specific syndrome—e.g., bipolar disorder—was manifesting itself already in young children. This has become a very controversial area since many children receive medication for specific symptoms with, at times, very limited data supporting use. Until we have long-term follow-up data to point to what happens to children with specific symptoms, we are at a loss to really know whether they do have a specific disorder, are merely going through a developmental phase, etc. One example is the rates of adolescent females who meet criteria for major depression, which are not only very high (as high as 50%) but also much higher than seen in adults, suggesting this is indeed a common phase young girls may go through and not depression in the adult sense. In DSM-5, a new syndrome was added to address the probable increase in misdiagnosed childhood bipolar disorder. Disruptive mood dysregulation disorder has been added as an alternative diagnosis to help avoid mislabeling children as having bipolar disorder. There is a real danger in merely applying a symptom checklist to determine diagnosis. Another example is an overlap one sees in hyperactive children with putative symptoms of hypomania. Are these the same disorder? To assist in this area, there are structured instruments but without face validity and follow-up data this area remains a difficult one to study.

At the other end of the age spectrum, geriatrics poses other problems. Often the depressive symptoms seen in this age group are different than in younger subjects. They may be more commonly agitated or delusional than are younger subjects. At one time the term *involutional melancholia* was used, but this was not incorporated in later DSMs since such patients frequently had previous episodes earlier in life. In contrast, as subjects age some may also develop depression because of cerebrovascular changes (Krishnan et al., 2004). These are generally not given a separate DSM designation if there is no clear-cut evidence of cardiovascular disease. These syndromes can be less responsive to treatment.

Patients can be classified as having a depression that is secondary to a medical disorder and this is common. Comorbid medical diseases can confuse the presenting picture—e.g., Is the patient slowed down because of depression or fatigued from a comorbid illness? These issues are discussed in further detail later. The Geriatric Depression Scale was developed specifically for this age group and has, at times, been more sensitive in this population than the more commonly used scales in adults—e.g., the Hamilton Depression Rating Scale (HDRS) (Yesavage et al., 1982; Murphy et al., 2004).

Women's reproductive cycles are often associated with specific symptoms—e.g., premenstrual dysphoria—and syndromes such as postpartum depression. Specific syndromes for premenstrual dysphoria and postpartum depression are included in DSM-5 while others such as menopausal-related depressions are not. The latter are often seen in women with previous histories of major depression and are thought to be largely not of new onset. Specific rating instruments for premenstrual dysphoria and postpartum depression have been commonly used in research on these conditions.

TYPES OF STUDIES

As in other areas of research in clinical medicine, studies range the gamut from observational through pathophysiology and treatment. The basic components of each of these are well described in other chapters in this volume. Herein we provide some brief descriptions and examples of the types of studies in psychiatry.

Epidemiological

Epidemiology studies have been a key component of psychiatric research for more than 40 years. A major boost to the field was noted in the early 1980s with the NIMH-funded national Epidemiologic Catchment Area study (Regier et al., 1984). This study aimed to establish prevalence rates for the major psychiatric disorders in several geographically dispersed regions around the United States. The study was made feasible by the development of structured diagnostic instruments that could be applied to assess the presence of a host of DSM-III disorders whose presence could be defined according to DSM-III criteria. The study told us much about mental disorders in this country. Of particular note were the high lifetime rates of major depression, phobic disorders, substance abuse, and antisocial personality. Many wondered about some of the prevalence rates but these have often proven to be if anything conservative, with subsequent large-scale replication studies—such as the National Comorbidity Survey (Kessler et al., 2005) that used DSM-IV criteria. These studies have been very useful platforms for related studies in health outcomes, services utilization, genetics, etc.

Observational

Since there are little in the way of diagnostic tests in psychiatry, observational studies have remained important in the field. These can be both cross-sectional as well as longitudinal in nature. In cross-sectional studies, investigators are frequently attempting to define the core features as well as the correlates of specific disorders—e.g., negative versus positive symptoms in schizophrenia; the neuropsychological deficits in major depression or schizophrenia; nature of compulsive thoughts or acts in obsessive compulsive disorders, etc. Longitudinally, studies have commonly explored decline in function in time in disorders such as schizophrenia, relationship of depression with pseudodementia in elderly patients to the development of Alzheimer disease, etc.

Mechanism and Physiology

Considerable psychiatric research is devoted to defining normative functioning of key systems and then comparing a paradigm in a patient population with healthy controls. For example, there has been considerable work in understanding hormonal responses to peptides—e.g., adrenocorticotropin and cortisol responses to corticotropin-releasing hormone (CRH); thyroid hormone response to thyroid releasing hormone; etc. Many of these are based on simulating central peptide stimulation of pituitary responses. Understanding normative and abnormal responses has led to a number of interesting hypotheses regarding alterations in central CRH or thyrotropin-releasing hormone in depression. As described later under biologic studies, there has been considerable effort expended into understanding how central catecholamine, monoamines, and other neurotransmitters may be involved in the pathogenesis or drug response in depression.

Over a decade ago, investigators began to explore allelic variations for specific genes and how they may play a role in normal and abnormal function. For example, variants of the transporter promoter for serotonin were reported to be related to disordered emotional processing and risk for depression (Caspi et al., 2003). Another example was haplotypes for the CRH receptor as associated with the development of depression or posttraumatic stress disorder in subjects exposed to major psychological trauma (Bradley et al., 2008). These studies opened up methods for assessing gene—environment interactions and risk for developing specific disorders under specific types of untoward circumstances. One advance has been to explore a number of genes that control a specific system such as the hypothalamic—pituitary—adrenal (HPA) axis (Schatzberg et al., 2014).

Disease Mechanisms

Genetics offers some real opportunities for mapping the biology of a disease and its progression. Postmortem tissue has been used to assess messenger RNA levels for specific genes activated or turned down in specific disorders. These studies have pointed to a number of candidate genes involving kinases and excitatory toxins—e.g., glutamate—potentially playing roles in depression and schizophrenia, and these could tell us something regarding the risk for brain atrophy in several disorders. Excessive glucocorticoid activity in the brains of lower animals has been associated with hippocampal tissue loss either reflecting atrophy or suppressed neurogenesis, and investigators have explored variants of glucocorticoid receptors and chaperone proteins as conferring protection or risk for developing depression.

Related to these studies is epigenetic research where early handling of rodent pups were reported to result in changes in methylation or acetylation status of specific genes, and these can have enduring genetic and behavioral effects (Weaver et al., 2004). Such changes in humans are only beginning to be explored using modern genetic screening tools and this could tell us much regarding gene—environment interactions in psychiatric patients. One example is the effect of early child abuse on methylation of FKBP-5 cochaperone to the glucocorticoid receptor (Klengal et al., 2013). Genes that control histone processes have been reported to be associated with schizophrenia (Kiebir et al., 2014).

A number of other genes have been reported to be associated with schizophrenia, and many of these appear to have something to do with synaptic formation, branching, etc. Related to this potentially is the much replicated observation that fathers over the age of 40 are significantly more likely to father schizophrenic offspring, a risk perhaps due to fraying of DNA in chromosomes with aging (Perrin et al., 2007). This may relate to the role telomere length plays in conferring risk for psychopathology (Gotlib et al., 2015). Again, understanding how key genes may place individuals at risk can provide necessary clues regarding disease progression.

Treatment

Somatic—drug or device administered—treatment and psychotherapy are key areas of research in the field. Here, well-established methods have been developed, although the nonspecific aspects of response in general and to placebo in particular are thorny issues that are discussed in detail later. These studies usually utilize standard double-blind, random assignment designs but also can involve biologic markers and imaging. For example, positron emission tomography (PET) can be used to assess receptor occupancy to drug versus placebo. Generally, these latter studies are smaller in scope and are more focused because of the difficulty in generating data on a number of biological measures—e.g., imaging or neuroendocrine studies. One exception is that a number of recent large-scale, multicenter trials have utilized pharmacogenetics, and these are telling us much about risk of adverse events and to a lesser extent treatment response.

Translational Research

In clinical medicine, including psychiatry, there is frequently a blurring between clinical and translational research. In psychiatry, a number of potential treatments are being developed based on biologic observations. Examples include deep brain stimulation for refractory major depression, glucocorticoid antagonists for mood disorders, CRH antagonists for anxiety and depressive disorders, intravenous ketamine for depression, m-Glu 2/3 agents in schizophrenia, etc. Each of these has been in the clinic. However, to date other than ketamine, the novel treatments have encountered high placebo responses and difficulty separating treatment specific from placebo responses (e.g., Binneman et al., 2008).

TOOLS

Behavioral Measures

Diagnostic interview tools are designed to allow for a qualitative assessment and are not necessarily designed to assess change. Instead, clinical rating scales that are quantitative are employed to assess severity at baseline and then change with treatment. Unfortunately, there are no biological endpoints that are commonly used—no fasting lipid level—to judge efficacy. The rating scales need to be evaluated for validity (i.e., do they measure what they say they will?), as well as whether they are reliable—do two disparate raters come up with the same total and individual item scores when assessing one patient? Generally, new scales are tested against other known scales where test properties have been assessed. The Brief Psychiatric Rating Scale (BPRS) was one of the first developed to assess a variety of patients (Overall and Gorham, 1976). It includes items and subscales to assess depression, anxiety, and psychosis. It has been used in literally thousands of studies and over time has been modified into other scales, e.g., the Positive and Negative Symptom Scale (PANSS) to focus on a specific syndrome such as schizophrenia (Kay et al., 1987). One issue that does arise in evaluating instruments has to do with whether the test can sensitively measure change. The BPRS and PANSS were developed for treatment assessment. Some scales that are used routinely for such purpose—e.g., the HDRS (Hamilton, 1960)—were in fact not developed for longitudinal assessment but became commonly used for it in the absence of other tools at that time. The HDRS was designed to help in making a diagnosis of depression. There are now a host of rating instruments for all sorts of purposes and a sample of some of the common ones and their major applications are summarized in Table 28.1. For further description of instruments, the reader is referred to the second edition of the *Handbook of Psychiatric Rating Scales*, published by the American Psychiatric Publishing Company (Rush et al., 2008) and for copies of many of the instruments to *Rating Scales in Mental Health*, third edition (Sajatoric and Ramirez, 2012) published by Johns Hopkins Press.

In a single site study, raters must be trained to administer specific instruments. Training often involves rating live patients or taped interviews. Some tests that measure particular symptoms can be more difficult to master (e.g., the Scale for the Assessment of Negative Symptoms or the Scale for the Assessment of Positive Symptoms—Andreasen, 1984a,b) than are others (e.g., PANSS), and this may affect the choice of instrument (Andreasen, 2008), particularly in larger-scale studies with multiple raters. Specific raters can be assessed in comparison to others using the same clinical material, thus establishing interrater reliability. Over the course of a study that may last more than 6 months, reassessing raters is important to insure consistency over time. Booster sessions may be needed to insure consistency over time. In studies that use more than one rater, interrater reliability within and across sites is important. Having a preset method for training and assessment, including frequency, minimum standards, etc., is essential. At times, quality control for raters' assessments may be conducted by consultants who are off-site and who review.

Not all psychometric instruments are rater-administered. Rather, some are self-report, and these can be quite sensitive and at times a better reflection of how someone is faring, e.g., responding to a particular therapy. Generally, the Federal Food and Drug Administration (FDA) has required an objective, blinded measure for primary outcomes. However, voice-activated telephone-based self-assessments are at times used for assessing patients at baseline to avoid rater-based inflation of baseline scores to insure inclusion in a particular study (see later discussion).

While physically not invasive, rating scale sessions can become time-consuming and physically even taxing when too many scales are administered in a single session. Distributing the assessment of subjects at baseline over more than one session is often helpful as is careful consideration of the number of tests administered, their frequency, their length, etc. Moreover, having too many scales or tests can affect the statistical analysis in terms of power, significance level, etc. Thus researchers should prioritize what dimensions or symptoms are most important and begin with them and determine what can be eliminated based on the time to be spent. Some scales will allow for assessing multiple dimensions, e.g., anxiety and depression, and may be time-economical in some studies.

Biological Measures

Trying to understand the biological basis of psychiatric disorders has been a major focus of research for decades. A great deal of the theoretical underpinning of early work was based on pharmacological observations. For example, monoamines were thought to be involved in the pathogenesis or pathophysiology of mood disorders because of a number of observations on drug effects (Schildkraut, 1965). Reserpine, an early antihypertensive, was reported to cause depression in a small percentage of patients and then was shown to deplete monoamine stores. Tricyclic antidepressants were discovered serendipitously to be effective in depression and then were shown to block reuptake of monoamines. These led to

TABLE 28.1 Representative Diagnostic Instruments and Clinical Rating Scales

Name	Population	Application	Comments
Diagnostic Instruments			
Structured Clinical Interview for DSM-IV Axis I Disorders (SCID-I)	Adults	Over 50 DSM-IV Axis I diagnoses. Structured interview	Very detailed. Can be used to rule in or rule out Axis I disorders. Can take 21 h to administer
Structured Clinical Interview for DSM-IV Axis II Personality Disorders (SCID-II)	Adults	All DSM-IV Axis II diagnoses. Structured interview	Assessment of possible personality disorder diagnoses
Mini International Neuropsychiatric Interview (MINI)	Adults	17 Axis I DSM-IV diagnoses. Structured interview	Rapid assessment instrument (15 min)
Diagnostic Interview Schedule (DIS)	Adults	Multiple Axis I disorders. Structured interview	Developed for DSM-III, updated for DSM-IV. Takes 2 h to administer to patients
Schedule for Affective Disorders and Schizophrenia for School-Age Children; Present and Lifetime Version (K-SADS-PL)	Children and adolescents	Semistructured interview of child and parent. DSM-IV Axis I disorders	Duration of 150 min for patients
National Institute of Mental Health Diagnostic Interview Schedule for Children (NIMH DISC)	Children and adolescents	Interview-based. DSM-IV and ICD-10. Over 30 disorders. Versions for parents, youths, and teachers	Over 2 h in duration
General Symptom Measures			
Brief Psychiatric Rating Scale (BPRS)	Adults	A variety of symptoms and symptom clusters. Objective rater	Often thought of as the original schizophrenia scale but can be used for other disorders
Symptom Checklist-90 Revised (SCL-90-R)	Adults	Clusters and subscales for anxiety, depression, psychosis, etc.	Earlier versions had fewer items. Can assess anxiety and depression simultaneously
Minnesota Multiphasic Personality Inventory (MMPI-2)	Adults	Self-report of measures of temperament and psychological disturbances	Classic instrument. Large control samples. Extensive assessment. Not DSM-IV- or ICD-10-oriented
General Health Questionnaire (GHQ)	Adults	Assesses psychiatric caseness in adults with medical illnesses. Self-report	Brief screener. Leads to more formal assessment if needed
Anchored Brief Psychiatric Rating Scale for Children (BPRS-C)	Children and adolescents	Interview of child and parent. Broad symptom assessment	Brief assessment of symptoms. Not diagnostic. Duration 20–30 min
Behavior Assessment System for Children—Second Edition (BASC-2)	Children and adolescents	Broad range of symptoms and behaviors. Multiple scales for parents, teachers, and children. Objective raters	Relatively rapid. Useful in various settings
Conners' Rating Scales—Revised (CRS-R)	Children and adolescents	Psychiatric and behavioral disorders screen. Various forms for parents, teachers, and adolescent self-report	Particularly useful for attention deficit disorder

Syndrome-Specific Scales

Mood Disorders

Scale	Population	Assessment	Comments
Hamilton Rating Depression Scale for Depression (HAM-D)	Adults and adolescents	Depressive symptoms. Objective rater	Classic scale originally developed for baseline assessment but used longitudinally. Multiple forms. Has many anxiety symptoms
Montgomery–Asberg Depression Rating Scale (MADRS)	Adults	Depressive symptoms. Objective rater	More focused on depressive symptoms. Briefer than HAM-D
Beck Depression Inventory (BDI)	Adults	Depressive symptoms. Self-report	Has more in the way of psychological and cognitive items
Center for Epidemiologic Studies Depression Scale (CES-D)	Adults	Self-report. Screen for depressive symptoms in community sample	Brief (5 min) assessment—not totally inclusive of all depressive symptoms
Geriatric Depression Scale (GDS)	Geriatric subjects	Depressive symptoms	Sensitive tool for geriatric depressives. Can be used longitudinally
Children's Depression Rating Scale (CDRS)	Children	Objective rater	Derived from the HAM-D
Young Mania Rating Scale (YMRS)	Adults	Manic symptoms. Objective rater	First specific scale for hypomanic and manic symptoms. Can be used longitudinally. Brief scale

Anxiety Disorders

Scale	Population	Assessment	Comments
Hamilton Anxiety Scale (HAM-A)	Adults	Anxiety symptoms. Objective rater	Used in depression and generalized anxiety disorders. Less overlap with HAM-D than vice versa
Beck Anxiety Inventory (BAI)	Adults	Anxiety symptoms. Self-report	Used in generalized anxiety disorder studies. Developed to separate anxiety and depressive disorders
Panic Disorder Severity Scale (PDSS)	Adults	Overall severity of panic disorder. Objective rater	Assesses panic attack and disorder symptoms. Brief scale—5 to 10 min
Brief Social Phobia Scale (BSPS)	Social phobia, adults	Objective rater. Social phobia symptoms	Brief instrument. Three subscale scores
Liebowitz Social Anxiety Scale (LSAS)	Social phobia, adults	Objective rater of social phobia symptoms	Can be used to assess symptom change longitudinally
Yale–Brown Obsessive Compulsive Scale (Y-BOCS)	Obsessive compulsive disorder (OCD) adults	Objective rater of overall OCD symptoms	Can assess change in obsessive compulsive symptoms longitudinally

Psychotic Disorders

Scale	Population	Assessment	Comments
Positive and Negative Syndrome Scale (PANSS)	Schizophrenia, adults	Assesses positive and negative symptoms of schizophrenia	Expansion of the BPRS
Scale for the Assessment of Positive Symptoms (SAPS)	Schizophrenia, adults	Specific tool for positive symptoms	More complex to administer than PANSS
Scale for the Assessment of Negative Symptoms (SANS)	Schizophrenia, adults	Specific tool for negative symptoms	More complex to administer than PANSS

Continued

TABLE 28.1 Representative Diagnostic Instruments and Clinical Rating Scales—cont'd

Name	Population	Application	Comments
Global and Quality-of-Life Measures			
Global Assessment Scale (GAS)	Adults	Overall psychosocial function. Objective rater—generally clinician	Tied to DSM-IV diagnoses. Overall function—no subscales
Clinical Global Impression (CGI) scales—Severity of Illness and Improvement Subscales	Adults	Objective rater generally. Assesses global status and improvement over time. Patient self-report version available	7-Point Likert scale, used for drug studies
Sheehan Disability Scale	Anxious or depressive adults	Self-rated. Assesses disability related to anxiety or depressive symptoms	Can be used longitudinally to assess treatment outcome
Mini Mental Status Evaluation (MMSE)	Adults—generally geriatric and potentially demented	Screens for organic and cognitive deficits. Rater-administered	Classic screen for central organic symptoms
Quality-of-Life Scale (QLS)	Adults with schizophrenia	Objective rater. Semistructured interview	Multiple domains of function. 45 min to administer
Social Adjustment Scale (SAS)	Adults	Objective rater and self-report versions	Assesses multiple domains of social function. Has excellent normative comparison date
Columbia Suicide Severity Rating Scale (CC-SSRS)	Adults and adolescents	Objective rater; semistructured assessment	Assesses multiple aspects of suicide ideation and behavior
Adverse Events—Clinical Trials			
Udvalg for Kliniske Undersogelser (UKU) Side Effect Rating Scale	Adults in drug studies; can be used clinically as well	Objective clinician rating used to assess general adverse events with treatment	Detailed measure that can be used longitudinally
Barnes Akathisia Rating Scale (BARS)	Adults often with schizophrenia	Objective rater. Specific measure of akathisia—used in drug trials	Assesses akathitic symptoms in antipsychotic studies. Had particular use in typical antipsychotic studies
Abnormal Involuntary Movement Scale (AIMS)	Adults often	Objective rater. Assess drug-induced dyskinesias	Can be used to assess tardive dyskinesia
Rating Scale for Extrapyramidal Side Effects (Simpson–Angus EPS Scale)	Antipsychotic drug study participants	Objective ratings	Assesses various extrapyramidal symptoms in clinical trials
Arizona Sexual Experience Scale (ASEX)	Adults—men and women	Self-administered. Screens for sexual function. Used in antidepressant trials. Gender-based versions	Brief scale

hypotheses that dysregulation of monoamine neurotransmitters was involved in depression. Early studies involved peripheral measures, e.g., 24-h urinary catecholamines, to theoretically assess central activity. While some metabolites of norepinephrine, e.g., 3-methoxy-4-hydroxyphenyglycol (MHPG), may derive preferentially from central stores, they still have major peripheral components. Even cerebrospinal fluid monoamine metabolites may be derived mainly from sources outside the brain. Some studies employed responses to challenges with pharmacological or hormonal agents; growth hormone response to clonidine, α_2, often a presynaptic noradrenergic agent, was used to test norepinephrine responsivity. Suppression in cortisol in response to dexamethasone, a synthetic steroid, was used to assess overactivity of the HPA axis with an emphasis on central CRH drive. But all of these were several steps away from what should be the area of focus. Another consideration was that some of these measures were largely affected by drug status such that they often required that patients be off their medications for a week or longer to be useful. A third issue was whether the measure was a so-called trait or state marker. Did they change with clinical state? Were they related to the underlying trait? For all of these reasons, these approaches to looking at the brain were not optimal. At the time and still today, however, there were few alternatives, although brain imaging and genetics are proving useful approaches for understanding activity of brain regions and delineating key neurocircuitry.

Imaging

In the past decade, development of genetic and brain imaging tools have revolutionized research in this area. Computer-enhanced techniques to image both the brain structure and activity are opening up ways of studying the underlying basis of psychiatric disorders. Initially, computer-assisted tomography (CT) allowed for study of ventricular size and more global pathology. Magnetic resonance imaging (MRI) is generally more sensitive and has allowed for assessing surface as well as deeper structures. These have allowed for determining size of key regions—amygdala, hippocampus, prefrontal cortex, etc.—involved in neuropsychiatric disorders. MRI can be used to assess blood flow in response to emotional or cognitive challenges, and this technique has led to studies that have identified which regions are activated during specific tasks. Moreover, these studies have demonstrated increased or decreased activity in response to specific stimuli as well as even at rest in patients versus controls. Similarly, PET measures of fluoridated glucose have been used to assess activity, although here there is less in the way of second-to-second dynamic response than of activity over a consolidated period of minutes. PET has been successfully used to assess binding of serotonin and dopamine receptors and transporters in specific brain regions—giving researchers a direct look at activity. It has also been used to validate pharmacological targets and for determining optimal drug dosages (Hargreaves, 2002; Meyer et al., 2009).

Interpretations of functional activity findings can at times be problematic. If an increase in regional activity on fMRI is seen in a patient group over a control group during a challenge, what should one infer? Is there a defect in function in a region that then requires greater activation to perform a task? Is energy preferentially switched from a less key to a more vital area? Frequently, the interpretations involve a template of theoretical percepts rather than clear certitude. Again, in the absence of clear organic pathology, inferences can be wrong.

Genetics

Genetics offers a method for assessing trait markers of psychiatric and medical disorders. Here psychiatric disorders are generally complex such that simple Mendelian approaches are not very informative. Their complexity may be a reflection of a group of disorders that have been given the same name, e.g., schizophrenia, and have been treated as a disease when these disorders are actually syndromes. Unfortunately, complex genetic disorders require large populations to identify key genes underlying specific disorders. Until recently, studies had been too small (on the order of 1000 subjects with the disorder) to demonstrate clear and replicable findings, and this is the current state of affairs. Whether this would be remedied by larger-scale studies was a hope that now appears to being fulfilled with a recent report of a number of genes being involved in risk for schizophrenia in a study that included 37,000 patients and 113,000 controls (Schizophrenia Working Group of the Psychiatric Genomics Consortium, 2014). Such studies may provide entrée to understanding the individual pathways (proximate or intermediate phenotypes) that then lead to the distant phenotype—schizophrenia. If correct, then identifying the intermediate phenotype with association with variance in a specific gene may allow the use of a Mendelian approach. As in other common chronic disorders, e.g., hypertension and diabetes, there is likely to be a substantial gene–environment interaction that will need to be controlled to understand fully the role of genetic variation in the development of psychiatric disorders.

Thus, one approach has been to explore genetic variation not as a differentiator of a diseased population from a healthy control group but rather to look at how a variant may have a clinically relevant effect. This may then employ environmental effects that can interact with genes to result in or affect an illness. This has been exemplified in the interaction between significant untoward life events and the short/short (s/s) form of the serotonin transporter where four or more such stressors were associated with a significantly increased risk for developing depression only in s/s serotonin transporter subjects (Caspi et al., 2003). In individuals with fewer stressors, significant increases in risk for depression were not observed. Similar data have been demonstrated for haplotypes for CRH receptor genes (Bradley et al., 2008).

Genetic variations have been studied in conjunction with both functional and structural imaging. For example, alleles for neurotrophins have been associated with reduced hippocampal volumes in depressives (Frodl et al., 2007). S/s alleles for the serotonin transporter promoter have been associated with difficulty in processing emotional stimuli as evidenced by fMRI responses in depressives. In schizophrenia, alleles for the catechol-O-methyltransferase have been associated with difficulty in prefrontal cortex tasks (Egan et al., 2001). A CRH allele has been reported to be associated with increased amygdala activation in response to "negative faces" on fMRI (Hsu et al., 2012). These are opening ways for understanding aspects of a specific syndrome but are not necessarily identifying the entire process underlying the disorder. This is in part in keeping with these disorders being complex—either involving multiple genes in a group of patients with a disorder or perhaps a specific gene being associated with a disorder in one family but not in another.

STATISTICAL AND DESIGN ISSUES

Power

There are a number of issues that enter into why many studies in medicine are underpowered for establishing statistical significance to support making clear conclusions about specific hypotheses. For one, the study may not have clear hypotheses to be tested, leaving the investigator without a base to calculate the probability of demonstrating significance. This is understandable often in a pilot study in an innovative area but is less easily justified in larger-scale studies. In larger-scale studies, the problem is generally an overestimate of the effect size resulting in an underestimate of the sample size needed to demonstrate statistical significance. In some instances, recruitment of subjects may lag projections; at times, this is because subjects who are medication-free for some biological studies may be particularly difficult to recruit today. A host of different providers—primary care physicians, gynecologists, pediatricians—now prescribe most of the psychotropics being used. Another issue is that protocols may be so time-consuming to discourage subjects participating. Obviously a simple pen-and-pencil questionnaire for subjects is simpler than a more arduous battery of tests and interviews.

At times, it is clear that the effect size of a biological difference or of a drug effect may be modest requiring numbers well beyond pilot or one-center studies. But it may not be feasible to conduct a large study. How then is innovative work to be undertaken if large samples are required? Indeed, some have argued that smaller N studies should emphasize effect sizes rather than testing statistical significance that can be employed in larger-scale trials (Schatzberg and Kraemer, 2000; Kraemer and Kupfer, 2006). If statistical significance needs to be tested, highly focused questions tested in more uniform populations can reduce the noise in the system and allow for testing major differences. Generally, significant interactions are difficult to assess in small N studies.

In genetic studies, power has also been a thorny issue, particularly in so-called complex disorders. Power issues have, if anything, been intensified by the great advances in the so-called SNP chips that can array up to one million variations on a single chip. Understandably, this generates a huge amount of data, and investigators call for high probability values such as 10^{20} for confidence. This is sure to create some distortions in findings with real findings being overlooked or vice versa. Using split samples to detect and retest differences can be employed, as can focusing on candidate genes defined on an a priori basis. For further discussion of these issues, see Chapters 16 and 17.

Predictors and Moderators

There has been increasing attention to defining variables that may have an effect on drug response. These variables are called predictors or moderators depending on whether a main effect for an active treatment is observed in a study. Rationale for and statistical methods for assessing these moderators or predictors have been developed by Kraemer et al. (2006). Such approaches do require sufficiently large samples. They can be highly clinically relevant and ultimately lead to more tailored treatments. Variables can be demographic descriptors (age, gender, etc.) or clinical measures (symptoms, duration of illness, etc.), biological measures (e.g., catecholamine metabolite levels), or genetic tests (so-called pharmacogenetics). Usually, these types of data analyses are used to provide information for hypothesis-generating rather than for hypothesis-proving.

SPECIAL ISSUES

Treatment Studies

Introduction

Much of clinical research in psychiatry explores optimal treatments for specific disorders. Here, as with medical disorders, there are a number of common issues blinding, random assignment, etc.—and these are discussed in detail in other chapters. For psychiatry, a number of somewhat unique issues do arise. These include competency to signed informed consent; use of patients versus symptomatic volunteers; high nonspecific or placebo responses to participating in trials; small effect sizes of specific treatments; no hard, biologic endpoints to judge endpoints or effects; etc. A detailed discussion of key design issues, statistics, and placebo response in psychopharmacology by this author can be found elsewhere (Kraemer and Schatzberg, 2009).

Informed Consent

Informed consent is a requirement for all clinical research. Obviously, competency to give informed consent is key for all research but may become particularly thorny in psychiatry where some patients—particularly those with psychotic disorders—may not be competent to care for themselves. At times, this worry has been extended to include all patients with psychiatric disorders; however, the vast majority of patients with mental disorders are fully competent to sign informed consent. Competency to sign for a psychometric pen-and-pencil test is less of a concern often than it is for being a subject in a clinical trial. For those studies where competency issues are confronted (e.g., schizophrenia, dementia, etc.), detailed assessment is often necessary. In rare instances, some trials may require the so-called substituted judgment, e.g., legally appointing a guardian to sign for the patient. There are helpful guides for dealing with these issues (Berg and Appelbaum, 2005). In the case of minors, parents or guardians serve to sign the consent and in some jurisdictions the child must also indicate their agreement to participate.

Placebo Response

Treatment in psychiatry is often associated with high placebo response rates. The reason for this has often been studied and is highly debated but has never been fully clarified. There is clearly a nonspecific positive benefit from participating in a clinical trial. Being supportive to a subject can have tangible benefits that can result in improvement in mood or lessening of anxiety. A relatively recent metaanalysis of antidepressant studies pointed to a consistent growth of placebo response rates over the past three decades (Walsh et al., 2002; Papakostas and Fava, 2009). Similarly high rates of placebo response are also seen in generalized anxiety disorder and panic disorder but often lower in psychotic or obsessive compulsive disorders (Khan et al., 2005b). Even in such disorders as Alzheimer disease, study participants can improve transiently (or not demonstrate the expected worsening) on placebo. Not uncommonly this response has been thought not to be enduring but the effect may last for months and confuse results (Khan et al., 2008). High placebo response rates limit one's ability to demonstrate efficacy, resulting in potential type II errors regarding an agent not being effective. To combat this, some studies have attempted to limit the number of visits over the course of the trial or the length of individual sessions; however, others have used independent off-site raters via video conferencing—enhanced educations of raters, etc. All of these have not been consistently effective.

One hotly debated area has been whether we should eliminate using placebo comparisons (Schatzberg and Kraemer, 2000). Some would argue that for more severely ill patients, placebos are unethical. For a long time this was the prevailing view in Europe; however, even there the field has moved to placebo-controlled trials. Eliminating them has been perhaps most cogently argued in patients with psychotic disorders where worsening without active treatment is a real concern. However, even in schizophrenic patients, one sees relatively common improvement on placebo and only infrequently does one see worsening on placebo, suggesting the argument about undue risk with placebo may be overstated. The FDA would accept superiority of an investigational agent over an improved one for a specific disorder as an alternative to superiority over placebo; however, such superiority over an active drug is rarely demonstrated, limiting the practicality of this approach.

The so-called placebo responses probably contain several components in psychiatric studies (Walsh et al., 2002; Schatzberg and Kraemer, 2000; Khan et al., 2003; Khan et al., 2005a; Kraemer and Schatzberg, 2009). Drug—placebo differences are more robust in more severely depressed outpatients (Khan et al., 2005a). However, requiring minimal severity levels for entry into studies may result in inflation of baseline scores that often drop in parallel in both active and control groups at initiation of dosing. This then can obscure actual drug—placebo differences. Some have incorporated variable and blinded start dates of drug to overcome this; others have used totally independent raters—sometimes over video or audio telecommunication devices. These may help, but they are not foolproof solutions. Another has been to use one rating scale for entry and blindly using another for assessing effects of treatment. Raters, however, who fill out two scales will often allow one to affect scoring on the other, defeating the purpose. In a study by O'Reardon et al. (2007), there was also an imbalance between the two treatment groups on the scale preselected for use to assess primary efficacy and that had an effect on outcome. Last, some studies enroll all subjects who have a minimal score on a rating instrument but only analyze those with scores above a higher threshold.

A number of other methodological modifications have been used to decrease placebo responses, particularly in depression studies, but here too, results have often not been dramatic. For one, a single-blind placebo run-in has often been employed, but it does not appear to help separate drug from placebo (Trivedi and Rush, 1994). Another approach that has helped is to limit the number of active cells to decrease expectation bias. More than four cells in a study have been associated with higher placebo responses and less in the way of separation from active drug (Khan et al., 2004).

One possible explanation for small drug—placebo differences has to do with the relatively smaller effects of treatment in this area and the difficulty in generating large enough samples to guarantee adequate power. In psychiatry, efficacy studies tend to be smaller than in cardiovascular medicine in part because exclusion criteria often slow enrollment. If large, multicenter studies are conducted, small differences may be statistically significant but not dramatic—limiting the enthusiasm for the approach. In medicine, this has been less of an issue as best seen in the use of aspirin as a prophylactic. Indeed, one must, at times, make a decision whether one should emphasize effect size or statistical significance in psychiatric research. Effect sizes may be far more meaningful for clinicians regardless of whether statistical significance is attained. Some have argued that statistical significance should not be emphasized in early trials, rather advocating looking at the effect size and its clinical relevance (Kraemer and Kupfer, 2006; Schatzberg and Kraemer, 2000).

Special populations may more commonly demonstrate responses to placebo. This is seen particularly in children and geriatric patients with major depression. In the former, suggestibility regarding the effects of medications may be important. In the latter, providing the patient with contact and support and overcoming social isolation may have nonspecific effects. Moreover, elderly patients may want to express their appreciation by overestimating their responses in the trial.

Defining Outcome

Traditionally drug trials in psychiatry have defined the response based on degree of change in rating scale scores. For example, a common metric to assess response in antidepressant trials is a 50% reduction in HDRS score over 6—8 weeks. The degree and trajectory of response do suggest the patient is improving. It is not intended to mean that the patient has responded fully. Generally, studies have used ANOVA to assess differences between treatments on intent to treat samples—often defined as receiving at least one dose. Many studies have used a criterion of a 50% reduction in score to assess efficacy. This categorical approach uses up power and is generally used for secondary analyses. In cases where there is a dramatic or robust effect for a particular agent, categorical approaches actually can prove more sensitive. To address the problem of dropouts, some investigators have argued for applying the so-called mixed effects models that project out the slope of improvement of a particular patient beyond the dropout point to the expected conclusion (Goldstein et al., 2004; DeBattista et al., 2006). These have now become more commonly used in psychiatry.

A stronger categorical definition of outcome is remission. This is frequently more clinically relevant and signifies full improvement or recovery. This definition produces lower "response" rates than do traditional response criteria, but applying remission as an outcome can demonstrate a more robust response. Remission is becoming the goal of treatment in depression, in part because remission is associated with better social adjustment and lower likelihood of recurrence. Recent efforts have been expended to define key issues to be addressed in the design of trials that wish to claim superiority in terms of remission. One key issue has to do with the durability of the so-called remission. At least 3 weeks of maintaining remission has been advised to support the claim rather than one endpoint rating, indicating improvement in scores (Rush et al., 2006).

Blinding

Blinding of somatic treatment studies can be at times difficult if a particular therapy has major untoward adverse events. This is a reasonable concern although not uncommonly when asked if a patient was on one or another treatment, the clinicians' responses or guesses have not indicated the blind was violated. The issue comes up more commonly if the patient is not blinded to the treatment but the rater is. Here, geographical separation of treating and rating staff can help, as can instructing the patient to avoid indicating treatment being received. This can be monitored by recording of sessions and using blind assessors. In such trials, all subjects should be assessed at the same frequency to maintain the blinding of the rater. One recent notable area where this remains a problem is with ketamine, a potent anesthetic agent that is a drug of abuse but which at lower doses can produce rapid antidepressant effects. Unfortunately, it produces dissociation that is a predictor of antidepressant response and that makes maintaining a blind difficult (Sanacora and Schatzberg, 2015).

Psychotherapy Studies

Psychotherapy is a commonly used and effective treatment of many psychiatric disorders and as such is a focus of considerable research. In these types of studies, subjects cannot be easily blinded, although, as indicated above, using an independent rater can insure the blind. At times, studies attempt to compare efficacy of drug and psychotherapy in specific populations. These comparisons can be problematic. For one, patients who agree to be randomized to one or the other treatment may not be the same as other depressed patients, particularly those who require medication—i.e., more severely ill patients. Indeed, psychotherapy is often as effective as medication in less severely ill patients. When exploring more recurrent or more severely ill patients, medication is often more effective. In some instances, designs have focused on comparing drug versus placebo in subjects receiving psychotherapy. This provides for assessing the so-called combination strategies but can be misleading regarding conclusions about drug responses if the open psychotherapy trial was associated with a too-vigorous response, leaving little room for further improvement. Another strategy has been to compare an intensive psychological therapy with a more modest one, often added to a medication given to all subjects. This is used in more severely ill patients. Here, open label drug treatment may provide great effect.

In multicenter psychotherapy trials, not only must interrater reliability of raters be established, but investigators must insure the reliability and similarity of administration of the therapy across sites. This need led many years ago to the development of manualized treatments—such as interpersonal therapy—that could be taught to treaters across sites and could be reliably applied. Therapists' adherence to the manual can be assessed by audiotaping the sessions and using independent assessors to determine adherence using prespecified criteria. Another issue in psychotherapy studies is sometimes subtle, but important. Interpersonal issues such as empathy, concern, affect regulation, etc. can have an effect on outcome independent of the specific treatment being evaluated. These qualities can be assessed psychometrically and used in the analysis as a covariate. Last, psychotherapy studies generally specify the length and frequency of sessions and provide for maximum levels of contact.

Behavioral Research in Medical Disorders

Psychiatric complications of medical disorders—coronary disease, HIV infection, etc.—are common foci of research. These studies may pose unique problems. For one, the symptoms of the medical illnesses can mimic the psychiatric disorder, e.g., psychomotor retardation of depression and fatigue seen in cancer or a metabolic disorder. Teasing these apart can be difficult and for specific disorders, researchers have developed rating instruments to separate out the medical and psychiatric components of symptoms.

At times, these approaches may overlook common underlying biology. For example, while interferon can produce fatigue in hepatitis C patients, it may similarly cause the fatigue of depression through a common biologic mechanism increasing adrenocorticotropic hormone release. Thus, the two may be more difficult to separate than one might think. Similarly, cardiovascular disease or diabetes may be commonly linked to depression, and at first glance one might intuit that a postmyocardial infarction patient is understandably depressed rather than the two disorders may share a common endothelial or vascular component. Thus, they could be more intimately related. It is possible that these various disorders share a host of genetic risks. Interestingly, some of the genetic variants seen in studies on patients with diabetes or hypertension also appeared in patients with psychiatric disorders such as bipolar disorder (Wellcome Trust Case Control Consortium, 2007). Indeed, immune-related genes are emerging in larger-scale genetic studies of schizophrenia and other psychiatric disorders (Schizophrenia Working Group, 2014).

In summary, behavioral disorders represent a challenge for researchers. Application of reliable methods—diagnostic assessments, clinical ratings, etc.—can help; however, without clear validators of diagnosis or outcome there are inherent problems that such strategies may not fully overcome. Ultimately understanding the underlying pathophysiology of major disorders will help shape future research. In the interim, careful attention to methodological detail can be helpful.

A PRACTICAL SCHEMATIC APPROACH (FIG. 28.1)

Any research in psychiatric disorders must pay attention to a number of issues in initiating and carrying through a research project. Some do obviously overlap with research in other specialties; others are unique. Obviously research can take one of the two major approaches—hypothesis-generating and hypothesis-testing. This fork requires a decision regarding whether a specific testable question can be framed and a determination that there will be sufficient power to assess

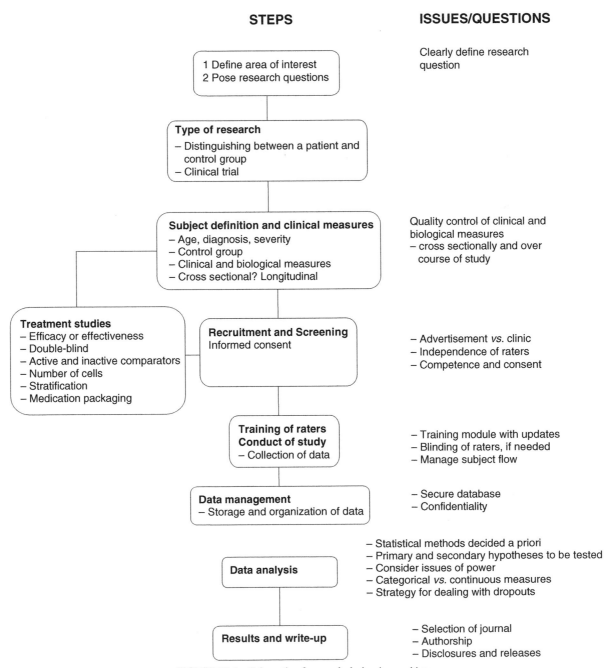

FIGURE 28.1 Schematic of research design in psychiatry.

statistical significance. In hypothesis testing, smaller samples are required to generate new leads, but there will be less certainty regarding the results. At times, hypothesis generation is looked down upon as being less scientific or less certain, but in clinical research where generating large enough samples is time-consuming and difficult and where we are always struggling to learn more about diseases, hypothesis generation is important for advancing knowledge. Once a question is framed—e.g., biologic differentiators between patients with a disorder or relative efficacy—the investigator must define the patient population to be studied. One would first start with what is the diagnosis or disorder being studied? How will it be defined? Here ICD or DSM criteria have generally been used. How will ascertainment of the diagnosis be done? While one can utilize a clinical interview and a checklist from DSM, it is better to apply a structured interview to determine diagnosis. Next, one needs to determine age, gender, and even ethnicity of subjects. These may be of varying importance. For example, if one is exploring reproductive hormones in depression, one is likely to study women who are premenopausal. Ethnicity plays a role in genetics, where ethnic background can exert a major effect on allelic distribution. Having once determined the basic demographics, clinical characterization of the sample is needed. Here rating scales are generally used to determine how depressed or anxious or disorganized a patient may be. At times, minimal scores on such scales are also used to insure a minimal level of severity or to attempt to insure less heterogeneity in the sample.

Another approach that is emerging is applying RDoC criteria to define samples of patients or controls (Insel et al., 2010). It can also be used to assess an RDoC dimension across patients and controls. The approach builds on core behaviors (e.g., fear or anhedonia) that can be assessed by rating instruments or biological measures. The hope is to develop ultimately a classification system based on specific behaviors. How this approach will perform in describing patients seeking treatment is in great debate and is not clear.

In biologic studies the investigator needs to determine whether the drug or medication status of the patient is important. This will depend on whether the test is affected by state and by how much. For some measures, such as structural genetics and structural imaging, drug status generally matters little. For catecholamines, medication effects can be important. In some cases drug may have an effect, but it can be small and insignificant in a large enough sample, and a large drug-free sample may not be feasible.

Once one makes those decisions, selection of a control group is key. Generally healthy controls are selected. But here too there are different considerations in health. Some studies include controls with no first-degree relatives with psychiatric disorders. But since these disorders are so common is this not a super-healthy group? Generally, one should make an effort to match the overall sample on gender distribution and mean age. One-to-one matching is difficult to achieve and not easily done. Some will match by gender and decade, but again this can be difficult to define. A 39-year-old patient and a 41-year-old control can be more similar than are a 39-year-old patient and a 31-year-old control. Statistical tools such as covarying out age can be used to control for imbalances, but these may require more power than is available, particularly in a smaller pilot study.

Having defined the characteristics of the populations, the investigator must set the numbers of subjects to be studied. This will generally be based on an assessment of the effect size or degree of difference between two groups and the resultant number needed to attain a statistically significant separation. Here the statistical test to be applied to the question will have an effect on the sample size needed with some tests requiring greater power.

Once the population is selected, researchers must then develop a plan for recruitment of patients and controls. This will first require approval of the protocol and consent forms by an independent Institutional Review Board (IRB). For severely and cognitively impaired patients substituted judgment may be needed. Once approved, the investigator can next begin to recruit patients and control subjects. Here affiliation with a clinical program can be helpful, but, at times, it will not yield a sufficient flow to allow for completion of the studies. Many studies advertise for patients in news media, through flyers and on the Web. Here the investigator must be careful to determine the severity of the patient. Some patients may not have a severe enough disorder to assess a particular therapy. Rating scales are essential, with minimal scores for inclusion defined a priori. To insure against score inflation in the rating process, some large-scale industry-funded trials have utilized remote assessment by an independent rater to determine severity. This can miss important but subtle findings. For any studies that employ multiple raters, a plan for training raters and assessing competence is required. Videotapes can be used to assess interrater reliability at a given point in time as well over the course of the study. This is essential if one is to be sure that a clinical rating done in January 2009 yields a similar result to December 2010. To that end, booster sessions and periodic assessment of reliability are needed.

On a biologic study, assay test characteristics must be established at first use and quality control methods instituted to insure similar values of a test specimen over time. These will include the percentage variance on repeated tests of a sample—degree or coefficient of association. Laboratories generally use samples with known concentrations—at times, artificially "spiked" with a known concentration of the test compound and measure the concentration at repeated intervals to help adjust the test performance. This is essential in biological research where samples are assayed over a longer course of study or where concentrations in one study are to be compared with previous study results. Such quality control is essential for virtually all biologic research, including brain imaging.

On a drug treatment study a number of decisions need to be made, particularly the blinding, comparison group(s), stratification, etc. For an initial pilot study, an open label design can be used, but this may not be optimal. Instead, blinding the study and using a placebo comparison are better but may not be feasible initially. For one, too few subjects may be available at a given site at a point in time. If the N is sufficiently large, an active drug can be compared to placebo. This provides more power than does comparison with both an active and inactive comparator, but it may be difficult without an active cell to determine whether the failure to demonstrate significant differences truly represents a lack of efficacy (i.e., a negative study where an active comparator did separate from placebo) rather than a failed study where neither cell separated and the study is inherently uninformative.

In drug studies, dosing of active medications should be specified a priori whenever possible. This includes the minimum and maximum doses, protocols for increasing and decreasing doses, etc. Flexible dosing in antidepressant trials has yielded greater drug—placebo separation than has a fixed dose approach (Khan et al., 2003). Time of day of drug administration is the key, as are the number of daily doses. Related to the dosing, duration of the trial and the frequency of ratings are important considerations. Generally, drug trials in psychiatry are 3—8 weeks in duration, with shorter trials used for more severe disorders such as mania. Weekly ratings are generally used in studies of 6 weeks or shorter. In studies of 8—12 weeks patients are often rated weekly until 4 weeks and then biweekly for the remaining months.

Having collected all the data, the investigator generally comes to a point where the data set is fixed or locked in the case of a double-blind study. Then data analysis can begin. Generally, the analytic plan is preset with the first priority being the testing of a particular question with secondary and even fully exploratory analyses coming later. Some of these—in the case of a drug study—can be prioritized and pursued sequentially, particularly if the primary endpoint has been met. In biologic studies these are often more open-ended depending on the questions that may have arisen after the study was begun, particularly if other research has yielded new information and questions to be pursued.

Once the analyses are completed, the investigator team begins to write up the data for publication. It is best when possible to have decided prior to data analysis what papers will be written and in what order. This is particularly important in larger studies where individual papers can interrelate but can be faced with different publication schedules in independent journals. Having a plan decided early on can decrease confusion for the field and maintain group cooperation among the collaborators.

Clinical and translational research in psychiatry faces not only the common problems seen in biomedical research but also unique ones encountered in this specialty. Careful attention to detail helps to overcome many of these, but inherently the diagnostic schema in psychiatry is based on clinically descriptive information and not physical data, and this obstacle requires investigators to remain modest in their view of the validity and generalizability of the findings, both for today and for tomorrow where other data, e.g., genetics, could result in a redefinition of the specific disorder studied. Still, the field's challenges and the power of the brain in human existence make this an exciting area of inquiry.

SUMMARY

Psychiatric research provides unique challenges in medicine, largely because studying the brain in man in vivo is generally indirect. In this chapter we have provided an overview of the common challenges encountered and discussed specific issues that need to be considered in designing clinical and translational research studies. Methods for assessing and solving these issues are provided as is a schematic model for approaching research in this area and a table of representative diagnostic instruments and clinical rating scales.

REFERENCES

American Psychiatric Association, 2015. Diagnostic and Statistical Manual of Mental Disorders, Fifth Edition (DSM-5). American Psychiatric Association, Washington, DC.

Andreasen, N.C., 1984a. Scale for the Assessment of Negative Symptoms (SANS). University of Iowa, Iowa City, IA.

Andreasen, N.C., 1984b. Scale for the Assessment of Positive Symptoms (SAPS). University of Iowa, Iowa City, IA.

Andreasen, N.C., 2008. Scale for the assessment of positive symptoms (SAPS); scale for the assessment of negative symptoms (SANS). In: Rush, A.J., First, M.B., Blacker, D. (Eds.), Handbook of Psychiatric Measures, second ed. American Psychiatric Press, Washington, DC, pp. 483—487.

Berg, J.W., Appelbaum, P.S., 2005. Subjects' capacity to consent to neurobiological research. In: Pincus, H.A., Liberman, J.A., Ferris, S. (Eds.), Ethics in Psychiatric Research: A Resource Manual for Human Subjects Protection. American Psychiatric Association, Washington, DC, pp. 81—106.

Binneman, B., Feltner, D., Kolluri, S., et al., 2008. A 6-week randomized, placebo controlled trial of CP-313, 318 (a selective CRHR1 antagonist) in the treatment of major depression. Am. J. Psychiatry 165, 617–620.

Bradley, R.G., Binder, E.B., Epstein, M.P., et al., 2008. Influence of child abuse on adult depression: moderation by the corticotropin-releasing hormone receptor gene. Arch. Gen. Psychiatry 65 (2), 190–200.

Caspi, A., Sugden, K., Moffitt, T.E., et al., 2003. Influence of life stress on depression: moderation by a polymorphism in the 5-HTT gene. Science 301 (5631), 386–389.

DeBattista, C., Belanoff, J., Glass, S., et al., 2006. Mifepristone versus placebo in the treatment of psychosis in patients with psychotic major depression. Biol. Psychiatry 60 (12), 1343–1349.

Egan, M.F., Goldberg, T.E., Kolachana, B.S., et al., 2001. Effect of COMT Val108/158 Met genotype on frontal lobe function and risk for schizophrenia. Proc. Natl. Acad. Sci. U.S.A. 98 (12), 6917–6922.

First, M.B., Spitzker, R.L., Wililams, J.B.W., et al., 1995. Structural Clinical Interview for DSM-IV (SCID-I) Research Version (User's Guide and Interview). Biometrics Research, New York State Psychiatric Institute, New York.

First, M.B., Gibbon, M., Spitzker, R.L., et al., 1997. Structural Clinical Interview for DSM-IV Axis II Personality Disorders (SCID-II). American Psychiatric Association, Washington, DC.

Frodl, T., Schule, C., Schmitt, G., et al., 2007. Association of the brain-derived neurotrophic factor Val66Met polymorphism with reduced hippocampal volumes in major depression. Arch. Gen. Psychiatry 64 (4), 410–416.

Goldstein, D.J., Lu, Y., Detke, M.J., Wiltse, C., Mallinckrodt, C., Demitrack, M.A., 2004. Duloxetine in the treatment of depression: a double-blind placebo-controlled comparison with paroxetine. J. Clin. Psychopharmacol. 24 (4), 389–399.

Gotlib, I.H., Lemoult, J., Colich, N.L., et al., 2015. Telomere length and cortisol reactivity in children of depressed mothers. Mol. Psychiatry 20, 615–620.

Hamilton, M., 1960. A rating scale for depression. J. Neurol. Neurosurg. Psychiatry 23, 56–62.

Hargreaves, R., 2002. Imaging substance receptors (NR1) in the living human brain using position emission tomography. J. Clin. Psychiatry 63 (Suppl. 11), 18–24.

Hsu, D.T., Mickey, B.J., Langenecker, S.A., et al., 2012. Variation in the corticotropin-releasing hormone receptor 1 (CRHR1) gene influences fMRI signal responses during emotional stimulus processing. J. Neurosc. 32, 3253–3260.

Insel, T., Cuthbert, B., Garvey, M., et al., 2010. Research domain criteria (RDoC): toward a new classification framework for research on mental disorders. Am. J. Psychiatry 167, 748–751.

Kaufman, J., Birmaher, B., Brent, D., et al., 1977. Schedule for affective disorders and schizophrenia for school-age children: present and lifetime version (K-SADS-PL): initial reliability and validity data. J. Am. Acad. Child. Adolesc. Psychiatry 36, 980–989.

Kay, S.R., Fizbein, A., Opler, L.A., 1987. The positive and negative syndrome scale (PANSS) for schizophrenia. Schizophr. Bull. 13, 261–276.

Keller, M.B., Lavori, P.W., Friedman, B., et al., 1987. The longitudinal interval follow-up evaluation. A comprehensive method for assessing outcome in prospective longitudinal studies. Arch. Gen. Psychiatry 44 (6), 540–548.

Kessler, R.C., Chiu, W.T., Demler, O., Merikangas, K.R., Walters, E.E., 2005. Prevalence, severity, and comorbidity of 12-month DSM IV disorders in the National Comorbidity Survey Replication. Arch. Gen. Psychiatry 62 (6), 617–627.

Khan, A., Khan, S.R., Walens, G., Kolts, R., Giller, E.L., 2003. Frequency of positive studies among fixed and flexible dose antidepressant clinical trials: an analysis of the food and drug administration summary basis of approval reports. Neuropsychopharmacology 28 (3), 552–557.

Khan, A., Brodhead, A.E., Kolts, R.L., Brown, W.A., 2005a. Severity of depressive symptoms and response to antidepressants and placebo in antidepressant trials. J. Psychiatr. Res. 39 (2), 145–150.

Khan, A., Kolts, R.L., Rapaport, M.H., Krishnan, K.R., Brodhead, A.E., Browns, W.A., 2005b. Magnitude of placebo response and drug-placebo differences across psychiatric disorders. Psychol. Med. 35 (5), 743–749.

Khan, A., Kolts, R.L., Thase, M.E., Krishnan, K.R., Brown, W., 2004. Research design features and patient characteristics associated with the outcome of antidepressant clinical trials. Am. J. Psychiatry 161 (11), 2045–2049.

Khan, A., Redding, N., Brown, W.A., 2008. The persistence of the placebo response in antidepressant clinical trials. J. Psychiatr. Res. 42 (10), 791–796.

Kiebir, O., Chaumetter, B., Fatgo-Vilas, M., et al., 2014. Family based associated study of common variants, rare mutation study and epistatic interaction detection in HDAC genes in schizophrenia. Schizophr. Res. 160, 97–103.

Klengel, T., Mehta, D., Anacker, C., et al., 2013. Allele-specific FKBP5 DNA demethylation mediates gene-childhood trauma interactions. Nat. Neurosci. 16, 33–41.

Kraemer, H.C., Kupfer, D.J., 2006. Size of treatment effects and their importance to clinical research and practice. Biol. Psychiatry 59 (11), 990–996.

Kraemer, H.C., Frank, E., Kupfer, D.J., 2006. Moderators of treatment outcomes: clinical, research, and policy importance. JAMA 296 (10), 1286–1289.

Kraemer, H.C., Noda, A., O'Hara, R., 2004. Categorical versus dimensional approaches to diagnosis: methodological challenges. J. Psychiatr. Res. 38 (1), 17–25.

Kraemer, H.C., Schatzberg, A.F., 2009. Statistics, placebo response and clinical trial design in psychopharmacology. In: Schatzberg, A.F., Nemeroff, C.B. (Eds.), Textbook of Psychopharmacology. American Psychiatric Association, , Washington, DC.

Krishnan, K.R., Taylor, W.D., McQuoid, D.R., et al., 2004. Clinical characteristics of magnetic resonance imaging-defined subcortical ischemic depression. Biol. Psychiatry 55 (4), 390–397.

Meyer, J.H., Wilson, A.A., Sagrati, S., et al., 2009. Brain monoamine oxidase A binding in major depressive disorder; relationship to selective serotonin reuptake inhibitor treatment, recovery, and recurrence. Arch. Gen. Psychiatry 66, 1304–1312.

Murphy Jr., G.M., Hollander, S.B., Rodrigues, H.E., Kremer, C., Schatzberg, A.F., 2004. Effects of the serotonin transporter gene promoter polymorphism on mirtazapine and paroxetine efficacy and adverse events in geriatric major depression. Arch. Gen. Psychiatry 61 (11), 1163–1169.

O'Reardon, J.P., Solvason, H.B., Janicak, P.G., et al., 2007. Efficacy and safety of transcranial magnetic stimulation in the acute treatment of major depression: a multisite randomized controlled trial. Biol. Psychiatry 62 (11), 1208−1216.

Overall, J.R., Gorham, D.R., 1976. The brief psychiatric rating scale. In: Guy, W. (Ed.), ECDEU Assessment Manual for Psychopharmacology. National Institute of Mental Health, Rockville, MD, pp. 157−169.

Papakostas, G.I., Fava, M., 2009. Does the probability of receiving placebo influence clinical trial outcome? A meta-regression of double-blind, randomized clinical trials in MDD. Eur. Neuroopsychopharmacol. 19, 34−40.

Perrin, M.C., Brown, A.S., Malaspina, D., 2007. Aberrant epigenetic regulation could explain the relationship of paternal age to schizophrenia. Schizophr. Bull. 33 (6), 1270−1273.

Regier, D.A., Myers, J.K., Kramer, M., et al., 1984. The NIMH Epidemiologic Catchment Area program. Historical context, major objectives, and study population characteristics. Arch. Gen. Psychiatry 41 (10), 934−941.

Rush, A.J., Kraemer, H.C., Sackeim, H.A., et al., 2006. Report by the ACNP task force on response and remission in major depressive disorder. Neuropsychopharmacology 31 (9), 1841−1853.

Rush, A.J., First, M.B., Blacker, D., 2008. Handbook of Psychiatric Measures, second ed. American Psychiatric Press, Washington, DC.

Sagatovic, M., Raminez, L.F., 2012. Rating Scales in Mental Health, third ed. Johns Hopkins Press, , Baltimore, MD.

Sanacora, G., Schatzberg, A.F., 2015. Ketamine: promising path or false prophecy in the development of novel therapeutics for mood disorders. Neuropsychopharmacology 40, 259−267.

Schaffer, D., Schwab-Stone, M., Fisher, P., et al., 1993. The diagnostic interview for children − revised version (DISC-R), I: preparation, field testing, interrater reliability, and acceptability. J. Am. Acad. Child. Adolesc. Psychiatry 32, 643−650.

Schatzberg, A.F., Kraemer, H.C., 2000. Use of placebo control groups in evaluating efficacy of treatment of unipolar major depression. Biol. Psychiatry 47 (8), 736−744.

Schatzberg, A.F., Keller, J., Tennakoon, L., et al., 2014. HPA axis genetic variation, cortisol, and psychosis in major depression. Mol. Psychiatry 259−267.

Schildkraut, J.J., 1965. The catecholamine hypothesis of affective disorders; a review of supporting evidence. Am. J. Psychiatry 122, 509−522.

Schizophrenia Working Group of the Psychiatric Genomics Consortium, 2014. Biological insights from 108 schizophrenia-associated genetic loci. Nature 511, 421−427.

Sheehan, D.V., Lecrubier, Y., Sheehan, K.H., et al., 1998. The Mini-International Neuropsychiatric Interview (M.I.N.I.): the development and validation of a structured diagnostic psychiatric interview for DSM-IV and ICD-10. J. Clin. Psychiatry 59 (Suppl. 20), 22−33.

Sorensen, M.J., Mors, O., Thomsen, P.H., 2005. DSM-IV or ICD-10-DCR diagnoses in child and adolescent psychiatry: does it matter? Eur. Child. Adolesc. Psychiatry 14 (6), 335−340.

Spitzer, R.L., Williams, J.B.W., Gibbon, M., et al., 1992. The Structural Clinical Interview for DSM-III-R (SCID), I: history, rationale, and description. Arch. Gen. Psychiatry 49, 624−629.

Trivedi, M.H., Rush, H., 1994. Does a placebo run-in or a placebo treatment cell affect the efficacy of antidepressant medications? Neuropsychopharmacology 11 (1), 33−43.

Walsh, B.T., Seidman, S.N., Sysko, R., Gould, M., 2002. Placebo response in studies of major depression: variable, substantial, and growing. JAMA 287 (14), 1840−1847.

Weaver, I.C., Cervoni, N., Champagne, F.A., et al., 2004. Epigenetic programming by maternal behavior. Nat. Neurosci. 7 (8), 847−854.

Wellcome Trust Case Control Consortium, 2007. Genome-wide association study of 14,000 cases of seven common diseases and 3,000 shared controls. Nature 447, 661−678.

Yesavage, J.A., Brink, T.L., Rose, T.L., et al., 1982. Development and validation of a geriatric depression screening scale: a preliminary report. J. Psychiatr. Res. 17 (1), 37−49.

Chapter 29

Research in Special Populations: Geriatrics

Stephanie Studenski and Luigi Ferrucci

NIA, Baltimore, MD, United States

Chapter Outline

Key Points

- While older adults have the highest rates of disease and health-care use, they are underrepresented in all aspects of research including clinical and translational studies of etiology, natural history, and treatment.
- Including more older animals and humans in research will promote greater generalizability, but age-related multiple coexisting conditions and functional limitations lead to challenges with implementation and interpretation of results.
- Study designs that anticipate and address multiple morbidity and functional limitations can promote greater inclusion of the aged in research.
- The emerging field of geroscience is based on shared models of molecular and cellular etiology across multiple diseases and aging.

INTRODUCTION

Older adults have higher rates of co-occurring diseases and conditions than any other age group and are most likely to receive many types of costly interventions, including medications, procedures, and therapies. The evidence base for care of older adults is often derived from basic, translational and clinical research carried out on single diseases in younger, less complex populations. To create an accurate evidence base for health care of older adults, modern aging research builds on a unique set of concepts and practical approaches. While there are numerous barriers to implementing changes in clinical and translational studies of aging, many feasible strategies exist. The goals of this chapter are to present a conceptual framework to approach health problems of aging in the context of new knowledge about the shared molecular and cellular changes that lead to both aging and multiple chronic conditions, to provide examples of how research studies of aging build on this conceptual framework, and to address aspects of research design that can be modified to promote more effective translation, broader age participation, and to specifically address unique problems of aging.

WHAT IS DIFFERENT ABOUT AGING RESEARCH?

Disease and Aging

Most chronic conditions, including heart disease, lung disease, diabetes, hypertension, kidney disease, cancer, and arthritis, are 2–10 times more common in persons of age 65 and over compared to younger adults (Fig. 29.1). Consequently, multiple coexisting conditions are much more likely in older than in younger adults (Fig. 29.2). Similarly, both the prevalence of medication use and the likelihood of multiple medication use increases with age (Fig. 29.3). Not unsurprisingly, older adults use more health-care services including hospitalization, most surgical procedures, diagnostic tests, and ambulatory services (Fig. 29.4). In the context of clinical research, the high prevalence of disease in older adults means that the majority of individuals with almost any target condition will be older adults, but most of these older adults will have concurrent conditions and treatments that could modify underlying biological and physiological processes, affect response to intervention, and alter the course of disease, and thus potentially confound research findings.

A Conceptual Model to Account for the Causes and Consequences of Altered System Structure and Function With Aging

An approach to inquiry that is based on individual diseases provides valuable scientific insights, but fails to address many critical issues. Not only the prevalence of diagnosed disease is higher among older adults, also aging brings interindividual variability in susceptibility to abnormal, often subclinical and unrecognized pathophysiologic processes. This individual variability in susceptibility produces tremendous heterogeneity in health status among older adults. There are several possible causes for unrecognized health problems in older adults. First, some symptoms or findings may be attributed to "normal aging." Classic examples of problems attributed to normal aging include forgetfulness, insomnia, or musculo-skeletal pain. Second, some conditions become apparent only when a system is stressed. Examples might include renal insufficiency or diastolic cardiomyopathy. Third, some conditions involve multiple interacting organ systems, are largely recognized as geriatric syndromes, and may lack clear guidelines for diagnosis. Typical geriatric syndromes include delirium, frailty and failure to thrive, falls, and urinary incontinence. Fourth, some conditions are not routinely

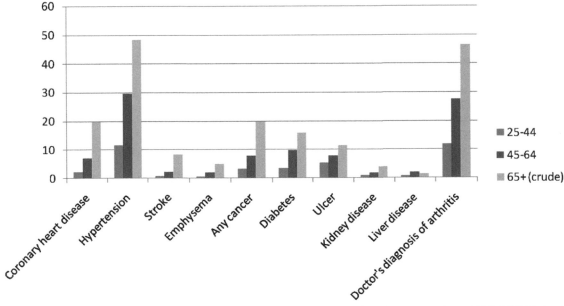

FIGURE 29.1 Prevalence of chronic conditions by age group. *Data derived from the National Health Interview Survey 1997–2005, National Center for Health Statistics, Trends in Health and Aging. http://www.cdc.gov/nchs/agingact.htm.*

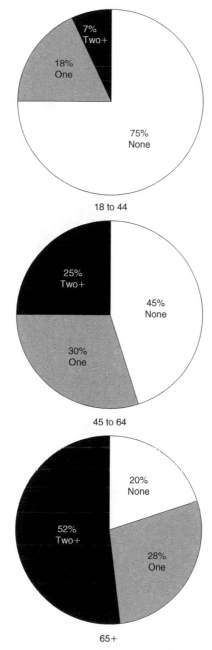

FIGURE 29.2   Prevalence of multiple conditions by age group.

acknowledged in usual medical care and may simply be overlooked; classic examples are vision and hearing deficits, mood disorders, and mobility disorders.

The combined effects of multiple interacting diseases, unrecognized health problems, and altered physiology of aging influence how research should be approached in the older adult. An alternate conceptual framework accounts for the effect of age on the development and manifestation of illness (Fig. 29.5A–C) (Ferrucci et al., 2005). Homeostatic and regulatory mechanisms underlying system function and system interactions help to protect generally healthy organisms from disease (Fig. 29.5A). When a robust living system develops a disease, it is likely that a powerful pathologic process has overridden all available homeostatic and regulatory mechanisms (Fig. 29.5B). Disease is then diagnosed and treated by identifying and correcting the pathologic process. This disease model applies to many conditions, underlies most of the clinical research, and is the basis for much of medical practice, but does not work well in the aged. With aging, variable loss of organ system capacity to adapt and respond to challenges results in reduced tolerance to stress and increased vulnerability to system

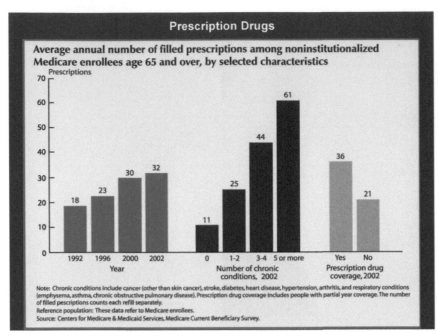

FIGURE 29.3 Prescription drug use among Medicare enrollees.

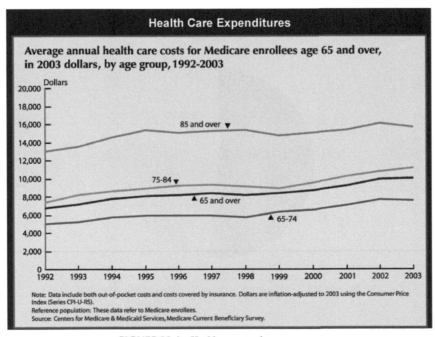

FIGURE 29.4 Health-care use by age group.

failure. The age-related loss of self-correcting mechanisms manifests clinically in unique ways (Fig. 29.5C). First, since organ system capacity is already reduced or because compensatory processes are less available, milder underlying pathologic processes can precipitate disease. Second, multiple modest pathologic processes might together contribute to a condition, so that a single unifying pathologic process no longer exists. Third, failure of a widely shared compensatory process could cause multiple diseases and conditions to present simultaneously. An interactive, multiple system framework to explain the onset of age-related conditions is congruent with emerging concepts and methods of systems biology and may yield new methodological approaches to aid aging research. Modular systems, nodes, and intersystem communication with positive and negative feedback are concepts that can be usefully applied to complex problems of aging at every level of organization, from genes and gene products to cellular machinery, tissues, organs, whole organisms, or even societies (Barabasi and Oltvai, 2004; Kitano, 2004; Loscalzo et al., 2007).

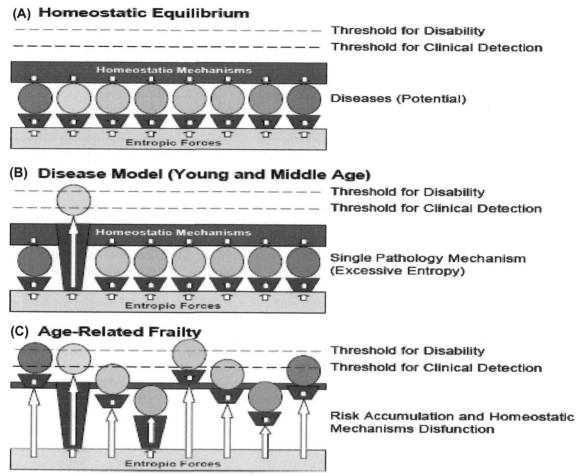

FIGURE 29.5 Diseases in young and old age. Homeostatic equilibrium is maintained by robust function and interconnections between multiple physiologic systems (A). When one system is dysregulated or impaired, as in a specific disease, disability can result in the specific areas of function affected by the disease (B). The derangement of general homeostatic mechanisms characteristic of frailty induces a multiple systems impairment that emerges clinically as frailty as well as development of multiple diseases and complex patterns of disability (C). *With permission from Ferrucci L., Windham, G., Fried, L.P., 2005. Frailty in older persons. GENUS LXI, 39–53.*

More recently, the term, "geroscience" has been coined to refer to an integrative model of the shared molecular and cellular alterations that predispose to both aging and multiple diseases (Burch et al., 2014; Kennedy et al., 2014). Mitochondrial dysfunction is an examples of such a process that has been shown to contribute to both aging and a wide ranges of conditions, such as Alzheimer disease, cardiomyopathy, diabetes, hypertension, renal failure, Parkinson disease, and cancer (Wallace, 2011). Age-related impairment of autophagy, a process of clearing out damaged molecules and cellular elements, has similarly been proposed as an underlying contributor to primary aging as well as to Alzheimer and Parkinson disease, amyotrophic lateral sclerosis, Paget disease, and several forms of cancer (Choi et al., 2013; Schneider and Cuervo, 2014). There is also recent evidence that the number of copies of mitochondrial DNA in the circulating blood is inversely associated with the risk of developing frailty, independent of age and sex (Ashar et al., 2014). A shared manifestation of these types of molecular and cellular alterations is loss of function, whether accumulating across individual physiological systems or whole organism attributes such as physical and cognitive function. Thus advocates of geroscience speak of "healthspan" in addition to "life span" and promote attention to measures across multiple organ systems and whole organism capacities.

The Effect of Aging on Practical Aspects of Clinical Research

The consequences of altered regulation in aging create challenges for the design and implementation of clinical research. As stated earlier, many clinically important phenomena of aging do not fit the traditional model of disease causation (Inouye et al., 2007). They are often not organ-specific and represent clinical manifestations of dysregulated systems. Examples include not only classic geriatric syndromes but also altered body composition, failures of temperature regulation, and systemic effects of chronic low-level inflammation. The presence of multiple conditions and altered physiology modify not only the approach to causation, but also the selection of important outcomes and the goals of medical care and thus the goals of clinical research. Treatment goals for younger adults may encompass cure, prolonged survival, and prevention of major health events. While such treatment goals do sometimes make sense for the older adult, treatment decisions are more often an overt trade-off between benefits such as symptom relief, improved function, or quality of life versus the risks and burdens of diagnosis and treatment. Thus when developing or testing a novel intervention for a problem of aging, it is important to assess beneficial and harmful effects not only in one target system, but also in others. It is also essential to assess impact on the organism as a whole. Therefore, important research topics include causes, recognition, and treatment of symptoms; ways to maintain or improve function; or modifiable factors that affect quality of life. To account for these important health states, reliable and valid indicators of functional status and life expectancy become critical.

Research into such age-related phenomena often leads to fundamental challenges in research design and implementation. As relatively novel areas of inquiry, there is often a paucity of well-established research terms or practices (Inouye et al., 2007). Investigators often must build novel multisystem frameworks to study causation, design novel eligibility criteria, define new forms of valid and reproducible measures, account for innumerable cofactors, consider new types of interventions, and even apply novel analytic techniques (Ferrucci et al., 2004). Pragmatically, studies may need to screen for important unrecognized conditions, anticipate problems with tolerance for study procedures and treatments, and incorporate multisystem indicators of function and well-being.

Even within more traditional research that focuses on a single disease, multiple barriers have limited participation of older adults. While contemporary clinical trials no longer have an explicit upper age limit, there is still widespread underrepresentation of older adults compared to the age distribution of the condition under study. For most common conditions, the mean age of the trial sample is up to 20 years younger than the mean age of the population with the condition (Heiat et al., 2002). This phenomenon occurs in studies of coronary heart disease (Lee et al., 2001), congestive heart failure (Heiat et al., 2002), cancer (Hutchins et al., 1999), Parkinson disease (Mitchell et al., 1997), and even dementia (Gill et al., 2004). In consequence, the evidence base for the treatment of conditions that are common among older adults is based on data from a population significantly younger and less likely to have coexisting diseases or altered physiology that might modify risks and benefits. This disparity in age between the sources of evidence for clinical practice and the population to whom the interventions are applied in clinical practice is the most compelling rationale for increasing participation of older adults in traditional disease-based clinical trials.

HOW AN AGING PERSPECTIVE AFFECTS RESEARCH TOPICS AND APPROACHES

Overview

Whether the research question involves genetics, cells, systems, or whole persons, the core concepts of gerontological science apply. While the term, "translational" is relevant to a wide range of inquiry, in the context of this chapter, we restrict the term to mean the "first translation" or bidirectional translation between basic and clinical science. The following sections will provide examples from contemporary aging research to illustrate applications from translational, interventional, and population-based research.

Translational Research

The advent of geroscience has invigorated translational research in the field of aging and begun to engage investigators from disparate fields that more traditionally focused on a single organ system. The potential applications of geroscience are based on the integrating concept of "healthspan" or "active life expectancy." Healthspan represents the duration of life while the organism is functioning well. As opposed to a focus on specific diseases or longevity alone, healthspan captures the capacity and resilience of multiple systems and their impact on the organism as a whole. According to a recent Commentary, "the lynchpin is that aging itself is the predominant risk factor for most diseases and conditions that limit healthspan. Moreover, interventions that extend life span in model organisms often delay or prevent many chronic diseases. This knowledge launched the era of geroscience, which strives to understand how aging enables chronic disease and seeks to develop novel multi-disease preventative and therapeutic approaches.... By understanding how aging enables pathology, new therapeutics will arise for multiple chronic diseases" (Kennedy et al., 2014).

Geroscience posits a set of fundamental processes that contribute to both aging and multiple conditions. While different authors cite varying numbers of such processes, they generally include genomic instability, epigenetic alterations, disrupted stem cell growth and regeneration, macromolecular damage, proteostasis, dysregulated mitochondrial energetics, altered cell signaling, and inflammation (Burch et al., 2014; Kennedy et al., 2014; Seals and Melov, 2014). These processes, of course, interact and cumulatively impact healthspan and disease. The geroscience approach has major conceptual and pragmatic implications. From a scientific conceptual perspective, the existence of molecules that are somewhat similar and perform similar functions across species (ortholog families) allowed for the identification of evolutionarily conserved pathways associated with longevity and healthspan. Most prominent among them are insulin/insulin-like growth factor signaling, target of rapamycin (TOR), the protein kinase A pathway, and the protein deacetylase silent information regulator 2 (SIR2) (Kennedy et al., 2014). Knowledge about these fundamental processes has led to novel therapeutic strategies. For example, rapamycin, originally used in organ transplantation and cancer, was found to suppress the TOR pathway and lead to longer-lived invertebrates. Experimental testing in rodent models supported longevity effects (Ehninger et al., 2014). However, rapamycin is far from being the perfect molecule for healthspan extension and despite its positive effects, it also has serious side effects, including immune suppression, predisposition to arthritis, and accelerated cataracts (Lamming et al., 2013). Rapamycin is active in a signaling pathway that could extend healthspan if the side effects could be eliminated or minimized. This line of thought has led to the ongoing search for related molecules deemed "rapalogs" (Lamming et al., 2013). Various aspects of the TOR pathway are known to be sensitive to nutrient levels, so that TOR-related pathways are felt to have an influence on longevity through some mechanism that is shared by caloric restriction (Ingram and Roth, 2014). The SIR2 pathway influences life span and led to the discovery of sirtuins, including resveratrol, which have been shown in invertebrates but not in mice to prolong life span. Interestingly, most of the positive effects of resveratrol that have been demonstrated in mammals were restricted to groups on a high-fat diet, although some positive effects at the biochemical or tissue levels have been detected also in animal models on a normal diet. Other existing molecules which have been shown to influence key aging-related pathways and have potential to alter not only life span but also healthspan and multiple chronic conditions include metformin and statins, although these effects have not been demonstrated in humans (Kennedy and Pennypacker, 2014).

As part of the Geroscience initiatives, recommendations have been made to refine pragmatic aspects of translational preclinical aging research. Especially in the case of animal models such as rodents, major efforts are needed to study more diverse genetic populations, to study aged animals in addition to genetically altered young animals, and to develop a standardized set of indicators of healthspan and physiological function. Such indicators can be developed in parallel to similar measures in humans such as physical and cognitive function, exercise capacity, body temperature, energy expenditure, resistance to stress, body composition, inflammatory markers, and multiple "omics" (Seals and Melov, 2014).

The need to obtain information on multiple physiological systems in human studies of aging has been a great incentive to research on biomarkers (Butler et al., 2004). A unique approach to the development of biomarkers of aging is to perform broad discovery studies that scan the genome, epigenome, and proteome to identify patterns associated with aging, and in particular to healthy aging and expanded healthspan (Burch et al., 2014; O'Connell et al., 2007; Spindler and Mote, 2007). This new approach is possible because of the availability of different types of microarrays relatively at a low cost and because of the considerable improvement in sensitivity and specificity of mass spectrometry. In particular, the study of gene expression has moved to clinical research, mostly by measuring patterns of gene expression rather than single genes. By looking for a pattern of gene expression changes with aging across multiple tissues, scientists have identified aging-specific gene expression patterns and demonstrated that certain interventions can reverse the gene pattern from "old" to "young" (Boehm and Slack, 2006; Kato and Slack, 2008). It has been proposed that aging can probably manifest by a finite number of these patterns, similar to quantum mechanics, because the homeostatic network can compensate for damage by evolving through different discrete states of equilibrium, selecting from time to time the least energetically expensive and most stable (Carlberg and Dunlop, 2006). In the near future, most studies of frail older persons will include patterns of gene expression, analogous to recent developments in oncology (Croce, 2008).

Clinical Trials

Clinical trials for age-related conditions are based on the usual research designs and methods but are modified to address special issues of aging. Some trials focus on specific diseases, some target multisystem geriatric problems, and some examine ways to provide care to complex aging populations. Conceptually, clinical trials targeting prevention and treatment of chronic diseases are not specific to the elderly. It was once thought that treatment efficacy would be independent of age for common conditions such as diabetes and hypertension, but that thinking has been disproved. For example, age influences the efficacy of lifestyle and metformin interventions on diabetes control (Knowler et al., 2002).

In contrast, the scientific rationale for an intervention for a complex problem of aging is often based on a multisystem mechanism of action. As outlined in the introduction to this chapter, changes in health that occur in younger and older individuals differ mostly because of differences in susceptibility and ability to respond to stress. Hence, at least in theory, only interventions that have effects across multiple physiological systems and that are aimed at strengthening resistance to disease or improving homeostatic capacity to deal with stress should be considered unique to clinical trials in the elderly. Randomized controlled trials aimed at preventing the so-called "geriatric syndromes," such as disability, depression, falls, or sarcopenia, exemplify this concept. The increased susceptibility of frail older individuals for these conditions is often explained by multiple parallel pathways and alternative physiopathologic mechanisms, none of which substantially prevail over the others. The traditional approach to intervention, based on a single specific target mechanism, is unlikely to show robust effectiveness because, even in the optimal scenario, only one of the many parallel branches in the causal pathway can be positively affected. For example, sarcopenia (low muscle mass) of aging is caused by a combination of mechanisms, including poor nutrition and a procatabolic state (Roubenoff and Hughes, 2000). At least in theory, interventions targeting only nutrition or only hormones will have limited effectiveness, while interventions combining both mechanisms may act synergistically.

The existence of multiple causal pathways introduces a novel layer of complexity in the design and conduction of these trials. Such complexity is exemplified in Fig. 29.6. On the left is the schematic representation of a traditional clinical trial, in this case a prevention trial. The intervention is designed to modify a specific risk factor (LDL cholesterol) to reduce the underlying pathology (atherosclerosis) and, after potential confounding is taken into account, reduce the risk of outcome development (myocardial infarction or stroke). This is straightforward. Assuming that the randomization is successful and unbiased, theoretically we only need to measure the outcome, and having some information on all the intermediate steps may be useful but not mandatory. In contrast, as illustrated in Fig. 29.6 on the right, when dealing with a trial for the prevention of a "geriatric" outcome, the level of complexity grows exponentially. Since multiple pathways are potentially involved and may affect the outcome, we must collect adequate information on all of them. Since regulatory agencies expect potential interventions to have a clear target and the mechanism of action, the best option is to map within the trial design multiple potential mechanisms and assess all of them. However, taking this approach also permits information to be collected that may shed light on the mechanism of the specific pathologic condition targeted and on the mechanism and side effects of the intervention. Effectiveness is demonstrated against a composite outcome. In Fig. 29.6 on the right, a clinical trial tests the effectiveness of testosterone administration on the prevention of mobility loss in frail older persons. Mobility loss is clearly multifactorial. Some potential pathways in the figure are potentially affected by the intervention, including changes in executive cognitive function, mood, inflammation, and body composition. Older persons with more severe mobility limitations tend to have more than one of these problems, but the pattern of multiple mechanisms varies widely across individuals. Different patterns may explain (or not) variability of effectiveness between individuals. Thus,

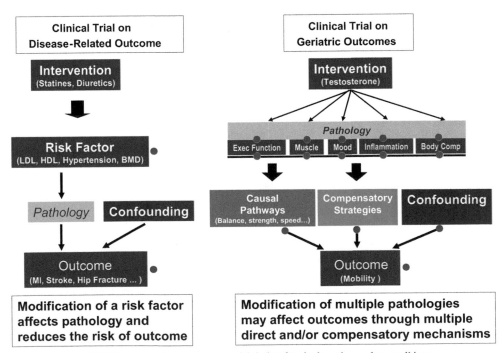

FIGURE 29.6 Contrasting clinical trial design for single and complex conditions.

such a design not only provides solid information on the drug effectiveness, but also sheds light on the mechanism of effectiveness and provides information useful for targeting individuals most likely to be responsive. In this way, even if the intervention tested in the trial is not effective, new experience and knowledge is accumulated that may help in the development of new potentially more effective treatments. This dual purpose for clinical trials has been advocated, for example, in the area of dementia in response to the multiple trails that failed to demonstrate effectiveness of interventions that targeted beta-amyloid (Becker et al., 2014).

Clinical trials of complex problems of aging should be designed with a unique perspective on selection of appropriate primary outcomes, eligibility criteria, and approach to the intervention. Primary outcomes may focus on combined morbid events, prevention of disability, or health-care utilization rather than control of a specific disease process. Eligibility criteria are often more inclusive and represent a population with high rates of comorbid conditions. Interventions are often more flexible and protocolized to deal with interruptions, modifications for coexisting limitations, and constraints on ability to participate. Given the tendency to involve multiple contributing factors and effect modifiers, it is often important to identify and track multiple indicators of system functions. Given the similar tendency toward multiple potential mechanisms of action, it is again useful to plan for multiple potential indicators. The following paragraphs provide examples of how aging influences the research questions and approach to clinical trials.

Exercise has been an attractive intervention for problems of aging because of its pleiotropic effects on numerous organ systems, potential benefits for function, and quality of life, and because exercise clearly induces changes even in very old people (Fiatarone et al., 1994; Nelson et al., 2007). Intrinsically pleiotropic effects such as those from exercise may be effective through multiple mechanisms. Not surprisingly, there is overwhelming evidence that physical activity may improve health and prevent disability through a number of different, potentially synergistic mechanisms, including increased muscle strength, higher aerobic capacity, lower levels of proinflammatory cytokines and higher levels of anti-inflammatory cytokines, increased antioxidant capacity and upregulation of antioxidant enzymes, improved endothelial reactivity, and many others. There have now been hundreds and perhaps even thousands of clinical trials of various types of exercise in a wide range of older populations (Nelson et al., 2007; Baker et al., 2007; Huang et al., 2005). The field has advanced to the point that large phase 3 clinical trials are possible. For example, LIFE-M was the largest multisite randomized clinical trial of physical exercise in sedentary older adults and required extensive preliminary work accomplished in LIFE-P (Espeland et al., 2007; Pahor et al., 2006). This study reflects contemporary thinking about the causes, detection, and prevention of mobility disability in late life. Because, like many areas in aging, consensus definitions and procedures are still in evolution, the preliminary LIFE-P trial was needed to establish characteristics of the primary outcome, clarify screening procedures, and refine the intervention protocol. For the main trial, LIFE-M, the primary outcome measure was the onset of major mobility disability, defined as the inability to walk 400 m. The investigators argue that mobility disability is an appropriate primary outcome measure because it is a major source of loss of autonomy and reduced quality of life and because mobility disability predicts inability to remain in the community, increased subsequent morbidity and mortality, and higher hospitalization rates. Eligibility for both LIFE trials included ages 70–89, sedentary lifestyle, evidence of increased risk of future mobility disability as reflected in a Short Physical Performance Battery (SPPB) score of 9 or less and baseline absence of mobility disability as reflected in the ability to complete a 400-m walk. The SPPB score has been shown to be a powerful predictor of future disability, health-care utilization, and mortality (Guralnik et al., 1994, 1995). While potential participants with active uncontrolled medical conditions or recent cardiac events were excluded, the study allowed participants to have multiple chronic conditions, including hypertension, diabetes, congestive heart failure, arthritis, and atrial fibrillation. While participants underwent a physical examination and 12-lead electrocardiogram, they did not have formal exercise testing for ischemia prior to intervention. The rationale for this choice was that the intervention was limited to moderate-intensity exercise and the study excluded persons with active symptoms or recent events (Gill et al., 2000). The LIFE exercise intervention was based primarily on a progressive, initially supervised program of brisk walking combined with functional strength training and balance exercises. Multiple adaptations were required to protocolize accommodations to patient symptoms and limitations, to develop protocols for monitoring health status during exercise, and to refine plans for restarting exercise after an interval health event. Running such a large complex, multicenter trial in the frail population was an important learning experience for the investigators and for the entire field. While feasibility was clearly demonstrated, a substantial amount of expertise and human and financial resources were required to finalize this project.

The Shingles Prevention Study is representative of a clinical trial targeted at a common condition of aging. It was a multisite placebo-controlled clinical trial of a single dose of zoster vaccine that required novel approaches to design and implementation (Oxman et al., 2005). The primary outcome, developed for the study, was the herpes zoster burden of illness score. The score reflected the integrated accumulated pain and discomfort attributable to zoster over time as measured by the Zoster Brief Pain Inventory, which had itself been shown to be associated with function and health-related quality of life (Coplan et al., 2004). An extensive telephone- and site-based monitoring system was created to detect and ascertain episodes of zoster and monitor pain over time. Eligibility criteria were generally liberal and excluded subjects primarily for reasons of safety or probability of surviving to the end of the study.

Multifactorial interventions for the prevention and treatment of geriatric syndromes have been somewhat successful even in situations where single intervention had been ineffective (Fiatarone et al., 1994). For example, Fiatarone et al. (1994) demonstrated that high-intensity resistance exercise training maximally counteracts muscle weakness and physical frailty in very elderly people when associated with nutritional supplementation, while nutritional supplementation by itself is not effective. Perhaps, the place where multifactorial interventions have been most successful is in falls prevention. There is trial-based evidence to support the efficacy of a combined assessment and multifactorial intervention to reduce falls (Tinetti et al., 1994). Accordingly, current guidelines support multicomponent fall prevention programs, although a recent metaanalysis challenges this view (Campbell and Robertson, 2007). Surprisingly, this type of multifactorial approach has not been fully explored in other geriatric syndromes and, most surprising of all, has still not permeated pharmacological trials. On the surface, such an approach seems entirely logical. If the increased susceptibility to geriatric syndromes and diseases is due to a multisystem dysregulation of the signaling network that maintains a stable homeostasis, then increasing the effectiveness and stability of this network through multiple simultaneous interventions makes sense.

Finally, some intervention studies target complex care management. Since older adults have multiple coexisting and often unrecognized conditions, they might benefit from highly coordinated care. There is a great need to examine the effectiveness of various approaches to care coordination using reproducible interventions. Over the last decade, there have been multiple clinical trials of special care coordination for complex elders, often termed "Geriatric Evaluation and Management." Results of these trials have been mixed but encouraging (Ellis and Langhorne, 2004; Stuck et al., 2002). One classic example is the Veterans Affairs Geriatric Evaluation and Management study, a multisite trial using a two-by-two factorial design trial of inpatient and outpatient geriatric evaluation and management in frail older veterans (Cohen et al., 2002). The rationale for the study was that acutely ill older adults with indicators of risk for functional decline require a structured multidisciplinary assessment and coordinated implementation of a wide range of therapeutic maneuvers to promote recovery and prevent disability. Given this conceptual framework, this study required innovations in definitions of subject eligibility and in intervention design. The primary outcomes were survival and health-related quality of life at 1 year. Eligibility for the study included hospitalization, age 65 or older, and the presence of frailty as indicated by two or more of the following indicators: limitations in basic activities of daily living, stroke in the last 3 months, history of falls, difficulty walking, malnutrition, dementia, depression, prior unplanned hospital admissions, prolonged bed rest, and incontinence. The intervention was a team-based, protocolized assessment that included screening for geriatric syndromes, function, nutrition, mood, cognition, caregiver capacity, and social situation. A structured plan of care was developed and implemented. Controls received their usual Veterans Affairs health-care services.

Population-Based Research

Population-based research on problems of aging is usually prospective a cohort study of representative groups of older adults. After over 20 years of experience with aging populations, these studies have achieved high levels of methodological sophistication and have produced landmark findings about innumerable aging phenomena. In addition to a focus on the natural history of common chronic conditions, such studies monitor age-relevant outcomes such as falls and fractures (Ensrud et al., 2007), disability (Newman et al., 2006), frailty (Fried et al., 2001), cognitive function (Lopez et al., 2007), and body composition (Newman et al., 2005). These longitudinal studies have mapped the development, progression, and consequences of early or preclinical states of many age-related conditions, including cardiovascular disease (Newman et al., 2001, 2003a), peripheral arterial disease (McDermott et al., 2004), renal function (Sarnak et al., 2008), and peripheral nerve dysfunction (Baldereschi et al., 2007). Innovations in use of portable technology have fostered community-based aging studies of physical activity (McDermott et al., 2007) and sleep (Iber et al., 2004). Large epidemiological studies are now engaging in collaborations to examine complex genetic contributions to aging and are linking to important emerging translational issues as described in the chapter section earlier (Kahn, 2007).

Community-based studies have created extensive and well-developed structures for studying functional status in aging. Functional status and disability are central to aging research because these states can be considered summary indicators of multiple conditions and physiologic impairments. Large community-based studies have explicated the natural history of age-related disability. They have examined contributors to disability, including the influence of demographic factors, innumerable biomarkers, and psychological and social factors (Guralnik et al., 1996; Ostir et al., 1999). There are thought to be two main pathways to disability: catastrophic and progressive. Younger older adults are more likely to experience the former whereas the most aged are more likely to experience the latter (Ferrucci et al., 1996). Disability states are not as fixed as once thought; rather disability is dynamic with periods of exacerbation and recovery that are influenced by health and social factors (Gill and Kurland, 2003; Gill et al., 2001a; Hardy and Gill, 2004). Performance-based measures of physical function, such as the short physical performance battery or gait speed, are powerful predictors of future survival, function, and health-care use (Guralnik et al., 1989, 1995; Cesari et al., 2005; Studenski et al., 2003). Short-term improvements or declines in walking speed have been found to influence 5- and 10-year survival, independent of multiple medical and functional factors (Hardy et al., 2007; Perera et al., 2005).

Large community-based studies have also served to identify previously unrecognized clinical conditions related to aging. Among those of great current interest, sarcopenia, or age-related low muscle mass, is emerging as a potentially diagnosable and treatable condition with serious and potentially preventable sequelae. As a rapidly evolving area of inquiry, there remains controversy within the aging research community about the role of muscle mass itself versus muscle composition and quality, neuromuscular influences, and the effects of obesity on muscle composition and function in late life (Solomon and Bouloux, 2006; Schrager et al., 2007; Roubenoff, 2003; Roth et al., 2006; Newman et al., 2003b). This area of inquiry combines clinically relevant outcomes such as functional status with a multiple systems approach encompassing muscle, nerve, bone, endocrine, and vascular systems, as well as cellular influences from satellite and muscle progenitor cells as well as subcellular effects of mitochondrial changes with age (Solomon and Bouloux, 2006; Schrager et al., 2007; Roubenoff, 2003; Roth et al., 2006; Newman et al., 2003b; Marzetti and Leeuwenburgh, 2006; Ehrhardt and Morgan, 2005).

THE EFFECT OF AGING ON THE PRAGMATICS OF RESEARCH

Overview

One could consider any human research study to be a respectful mutual partnership between investigators, participants, funding agencies, regulators, institutions, providers, significant others, and communities, in which success is based on a balanced solution to the needs and concerns of every party. The challenge and opportunity with every study, but perhaps, especially with older adults, is to creatively incorporate a user-friendly research structure into rigorous science. Strategies that promote success can be organized on two dimensions (Table 29.1). The first dimension represents general research goals such as (1) achieve and retain a representative sample of adequate size; (2) measure the right things; (3) if the study involves a treatment, employ an effective, feasible, and reproducible intervention; and (4) anticipate analytic challenges. The second dimension represents phases of research planning and implementation and includes (1) initial information gathering; (2) study design; (3) pilot studies; and (4) study implementation. Various strategies apply to goals and phases. Many examples are provided in the table, but some influence design so pervasively that they are highlighted here. These globally important strategies include (1) create inclusive research questions; (2) use advisory boards; (3) maximize the benefit/burden ratio; and (4) monitor processes continuously.

Primary aims of research studies focus on key or most important effects. Some studies aim for specificity of effect and prioritize precision by excluding study candidates with potentially effect-modifying coexisting conditions. Conversely, study designers could select an important main effect that is age-relevant and present despite the coexistence of modifying factors. Even studies of common conditions of aging such as diabetes, hypertension, or arthritis could target main effects that are robust in the presence of other common coexisting conditions. Thus studies could create liberal inclusion criteria and exclude potential participants based only on factors related to safety.

Advisory boards can help assure that concerns of participants, primary care providers, or other key constituencies are anticipated and addressed. Boards should be composed of constituency representatives such as target participants; significant other participants; and a range of gatekeepers who control access to participants, use of space, or organizational support for the study. Boards can benefit investigators by suggesting sources and anticipating needs of participants, as well as concerns of providers, significant others, and institutions. Boards can help investigators communicate with larger community groups and frontline providers. Boards can also review study materials and recommend user-friendly practices. Boards can help solve problems such as alternative strategies when recruitment is lagging. Representative participants, either on boards or not, can inform investigators in real time about problems with the study.

The benefit/burden ratio is a powerful influence on success in clinical research with older participants. While altruism alone may support enrollment and retention in some studies, it is often insufficient for older, more vulnerable populations. Research studies for older adults should prioritize benefits of participation for all. Rewards for participation are especially critical for studies of interventions that are hard to blind, such as exercise or surgical procedures. Control groups without tangible benefit can lead to negative expectations and higher dropout rates. Other ways to increase the benefit/burden ratio are to design the study so that the most burdensome assessment activities are not essential for all participants. Within a

TABLE 29.1 Overview of Strategies to Promote Participation of Older Adults in Clinical Trials

Phase of Research	Main Strategies
Research question	Explore options to enhance generalizability
	Maximize benefit-to-burden ratio
	Maximize use of primary data through ancillary pilots and substudies
	Incorporate less burdensome alternatives to invasive gold standard tests
Design	Identify opportunities for benefit for all participants including controls
	Consider frequency, site, and duration of participation
	Plan for flexibility in schedules, sites, and protocols
	Build in and budget for retention activities
Participant sample	Minimize exclusion criteria
	Consider the needs of nonparticipants involved in the study, such as family members and health-care providers
Outcome measures	Select easily accessible primary outcome measures
	Prespecify alternate data collection strategies to use when the primary strategy fails
	Consider alternatives to a single fixed time point for outcome assessment
Independent or predictor measures	Prioritize order of collection
	Plan for participant inability to perform tests, and code reasons why
Intervention	For all study arms, particularly control groups, identify opportunities for participants to benefit from the research activities
	Minimize the burdens on participants: travel, time, effort, risk, and cost
Pilot studies	Use pilot studies to assess the potential magnitude of missing data
	Use this experience to modify design and plans
Implementation	Promote a sense of belonging with stable staff, personal attention, and rewards for participation that promote study membership
	Provide feedback when possible to enhance participation and retention
	Provide transportation if needed
	Maximize convenience and flexibility of scheduling
	Have protocols for identifying participants at risk of missing data
	Have protocols for backup data collection alternatives
	Operationalize complex interventions through pilot studies
Data tracking	Plan for ongoing tracking and reporting of retention and missing data
	Be prepared to implement enhanced efforts if problems arise
Missing data assessment	Quantify amount of missing data (problems minor when $<5\%$)
	Characterize missing data rates by items, waves, and participants
	Understand the reasons and mechanisms for missing data
Analysis	Know the analytic problems that result from missing data or nonparametric outcome measures
	Know the appropriate use, advantages, and disadvantages of various analytic strategies for missing data
	Plan and implement analytic strategies for multicomponent interventions

research study, research aims that are mechanistic and use invasive or uncomfortable procedures could be defined as secondary and require only a subset of participants. Investigators should be aware that many types of barriers are hard to anticipate without the user perspective. Barriers identified by older adults and their significant others include travel costs, need for help from significant others to participate, discomfort, and fatigue. Similarly, serious organizational barriers may be hard to recognize. Health-care providers may be concerned about study time demands or the impact of the study on usual operations due to interference with space, schedules, noise, or other problems. Finally, a major source of motivation and interest in participation derives from interactions with study staff. All study staff need to have special training in issues of aging, including communications in the face of hearing or vision problems, dealing with concerned significant others, and adapting study procedures due to problems with mobility, vision, hearing, dyspnea, or other limitations. While many research budgets are tight, it is dangerous to reduce costs by cutting out activities that maximize benefit and reduce burden.

Finally, successful research studies in older adults should plan to measure and track aspects of recruitment and satisfaction. At a relatively low cost, the investigator should estimate screening costs and the screen-to-recruit yield; have backup plan in case of slow accrual; keep track of participant concerns and be prepared to address them; have an early detection system for risk of dropouts, such as persons who repeatedly reschedule or who have had a recent serious illness; and have plans to increase attention or adapt study procedures to assure that continued participation remains attractive.

Samples

It is important to recognize that increasing proportional enrollment of older patients does not necessarily improve the external validity of a study. If the older adults who are recruited are chronologically old, but biologically healthy and functional, they may still not represent the aging population of interest. For example, the Systolic Hypertension in the Elderly Program (SHEP) was a double-blind, randomized, placebo-controlled trial in men and women aged 60 years and older with isolated systolic hypertension that tested the efficacy of antihypertensive drug therapy to prevent stroke. Given the tight inclusion criteria, 16 centers screened 450,000 individuals to enroll the 4736 participants who were finally randomized, thus a screen-to-recruit yield of 1% (Prevention, 1991). While it can be argued that SHEP results have altered the clinical management of isolated systolic hypertension in older people, the study cannot inform treatment in the large majority of older individuals with complex comorbidity and disability that would been ineligible for the SHEP. The solution to this problem is to relax the inclusion criteria while at the same time implementing strategies that minimize the risk of harm to more frail older individuals.

Clearly, there is a fine equilibrium in the trade-off between inclusiveness and excess risk, and the search for this pivotal point of equilibrium is probably the most difficult choice to make in the design of modern clinical trials for chronic disease treatment. Thus, the keyword should be "enabling participation," even though this approach requires more frequent and in-depth tests for discovering early toxicity and side effects, problems in the consenting process and IRB approval, need to provide transportation and to interact with proxies. In general, the implementation of these strategies will increase substantially the cost and complexity of these trials.

As with any age group, the best source of potential participants depends on who is most likely to know about or have access to people with the target condition. Certainly specialty clinics often have the highest concentrations of individuals with a target condition, but often, the age distribution of patients in a specialty clinic is decidedly shifted toward midlife rather than late life. Primary care clinics are likely to have greater total numbers of older adults with common chronic conditions, but providers can be busy and too distracted to help with recruitment. Targeted electronic reminder systems accompanied by easy one-click notification of study personnel can reduce barriers to primary care referral for research (Belnap et al., 2006). Other sources of participants can include community interest groups, direct targeting, or referral from significant others. For example, Parkinson disease and prostate cancer support groups have been used for recruitment. Direct targeting through mailings or advertisements can work well for conditions that individuals can recognize themselves. These conditions, especially geriatric syndromes and symptoms, may be more recognized by older adults than by their providers. For example, studies of incontinence, falls, or sleep disorders might be able to find more candidates with direct recruitment in the community than from clinics. For some conditions, such as cognitive impairment or other disabling conditions, the key initiator of interest in a trial is a significant other such as a spouse or an adult child. For most recruitment plans, it is wise to become familiar with the potential sources and to estimate success rates through early contact and pilot studies. Some trials have incorporated formal aspects of social marketing to promote recruitment (Nichols et al., 2004).

Efficient screening for eligibility is important for reducing both study costs and burden on potential participants. In general, screening over several phases of increasing time, effort, and cost is most likely to yield an eligible and willing population at the lowest cost and burden (Ferrucci et al., 2004). Depending on the initial source, a first screening can be of medical records (electronic or not), other health information sources, or by telephone interview. This first screening is most useful for excluding persons who clearly do not have the condition of interest or who have major unmodifiable exclusions. Most institutional review boards (IRBs) have defined processes for obtaining a waiver of informed consent and HIPAA regulations for the purpose of screening potential candidates for clinical trials.

Successful recruitment and retention of older adults depends on attention to some special needs (McMurdo et al., 2005; Witham and McMurdo, 2007). Be sure to consider factors related to access and transportation, including travel support, ease of access to the study site from parking locations, and even use of home visits for data collection. Be sure to plan to accompany participants with limitations in mobility, cognition, or sensory disorders if they are expected to visit multiple testing sites.

While most experts in research ethics recommend a general assumption of competence to consent, there are some special issues that arise in older adults. First, the prevalence of cognitive impairment and dementia is higher than in the general population. Second, several conditions, such as stroke or Parkinson disease, may be associated with impaired decisional capacity. Third, special settings that provide care to older adults, such as nursing homes, can have high concentrations of persons with decisional impairment. In general, it is not necessary or appropriate to perform formal tests of cognition and memory on all prospective study candidates to determine competence to consent. Rather, it is reasonable to define a process for all prospective study participants that assures that the candidate understands what he or she is agreeing to do. Many studies have developed brief checklists of key information that can be used to determine if the candidate has understood and can recall the major aspects of the study. Research can be performed on persons with limited capacity to consent, but often the IRB will have a higher expectation of direct benefit of participation for the individual. Processes for appropriate identification and use of proxies for consent can vary between states and even institutions, but all also include procedures for assessing participant "assent." For assent, it is expected that the participant is initially provided a description of the study, in terms he or she might understand and is also continuously assessed for willingness to cooperate with aspects of study procedures.

Recruiting minority older adults compounds the challenges of recruitment of older adults (Stahl and Vasquez, 2004). All of the issues that exist in minority recruitment in general, including distrust, socioeconomic constraints, and cultural barriers, are applicable. Further complicating the successful recruitment of older minorities can be very high degrees of protectiveness by significant others and generally higher rates of disability in older minorities than in Caucasians. Solutions to challenges of recruiting older minorities tend to reflect the same approaches used in other age groups, including building long-term relationships with minority communities and trusted leaders, using culturally and linguistically appropriate study materials, accommodating to the needs of the participant, and prioritizing the benefits of participation (Stahl and Vasquez, 2004).

Measures

Since the older population is heterogeneous, it is important to characterize study participants along several dimensions, in addition to measures of disease severity and outcomes. For some special problems of aging, the most important measures of interest may not even be related to a specific disease. This section will first address general principles of measurement related to aging and then review commonly used instruments to characterize important dimensions related to aging.

In general, because older adults are heterogeneous, it is important to use measures that capture the range of expected values. The main concern is to avoid large proportions of older participants with values at the floor or ceiling of the measure at baseline. Similarly, there is an increased risk of missing data in older adults, due to issues such as inability to perform the measure, contraindications to testing, missed sessions due to illness, or other reasons. The fact of missing data itself can be informative, since it might indicate worsening health or function. It is wise to learn about possible sources of missing data during pilot studies and to develop a set of missing data codes that can be used later to disentangle effects. Other general aspects of measurement to be considered include adaptations for visual or hearing problems and planning for possibly longer data collection sessions.

Inability to perform a test is a common challenge in aging research. For example, a participant who cannot complete a gait speed test does not just have missing data, the missing data is informative. A task that has progressively difficult levels may generate more and more missing data as a participant's abilities decrease. Some performance measures have incorporated "can't do" directly into the metric. For example, the SPPB, a measure of lower-extremity physical abilities, gives a score of 0 for persons who cannot perform a task (Guralnik et al., 1994, 1995). Some timed tests, such as the Digit Symbol Substitution Test of cognition (Wechsler, 1987), count the number of items completed in a fixed amount of time. Such tests naturally accommodate inability, since the number of completed items will be 0. Performance tests that measure distance, such as the 6-min walk test (Troosters et al., 2002), also accommodate inability because persons who cover no distance over 6 min can be scored. Another strategy is to count the number of missing items, such as the number of missed words or tones in hearing tests (Lichtenstein et al., 1988; Macphee et al., 1988).

Important Measures for Aging Populations

See Table 29.2 for an overview of brief measures that are commonly used in aging research. It is a good idea to plan to collect data on these important domains in most studies that include older adults. A core data set that includes summary measures of physical, cognitive, and emotional function can probably be collected using the shortest valid measures in about 20 min. For questionnaires with response options, consider interviewer—rather than self-administration. Be sure to use props such as single pages that present each set of response options in large type to allow the participant to review and select his/her choice.

Interventions

While much of the literature about clinical trials tends to focus on pharmacologic interventions, there are many other types of interventions that have great potential benefit for older adults. Interventions for problems of aging include exercise, counseling, therapeutic procedures and devices, environmental adaptations, as well as educational or organizational maneuvers targeted at health-care providers and systems. Each type of intervention brings its own opportunities and challenges. Many of these potential interventions do not have easily defined placebo forms. This lack of convincing placebo results in participants who are not blind to treatment arm assignment. Such participants may be especially vulnerable to expectation biases, and results may be distorted, especially if self-report of function is a primary outcome. Novel placebo forms of exercise or counseling can include sham or diluted interventions that are considered unlikely to affect the outcome. Since a general principle of research with older adults is to maximize benefit—burden ratio, trials should try to limit the use of placebos. Some trials can compare two or more forms of an intervention or use a wait-list control design that provides eventual access to the intervention. Another challenge with nonpharmacologic interventions is standardization of procedures, dose, and intensity. Many such interventions require that the "treatment" evolves over time depending on response to initial efforts. Thus clinical trials of counseling, exercise, or care provision can take months or even years of effort and pilot studies to operationalize and develop training procedures for interventionists.

Since complex problems of aging can involve multiple systems and have multiple contributors, many geriatric interventions, such as fall prevention or geriatric assessment, are multifactorial. In fact, single interventions for complex problems often do not make clinical sense, because many contributors to the problem would remain unaddressed. These complex multifactorial interventions present special challenges. First, clear decision rules must be developed to define who gets which part of the interventions or whether everyone gets all of them. Second, systems that track the elements of the intervention must be developed and implemented. Third, results of complex multifactorial interventions can be especially difficult to interpret, since it is not possible to determine what the "active ingredients" were. Alternative designs for multifactorial trials might include (1) a combined intervention versus a single intervention; (2) a combined intervention versus another combined intervention, with the difference based on the lack of one key element, to test its contribution; and (3) stepped or sequential interventions that allow interval assessment of effect.

Analysis

Research studies of human aging are at increased risk of missing data since older adults can develop new serious health problems or otherwise develop barriers to participation. While exclusion of subjects at risk of missing data is one strategy, such an approach will reduce participation of older adults. Other strategies to maximize key data are available. Options include passively available information, alternate plans for data acquisition, flexibility in timing follow-up data collection, and the use of combined outcomes. Passively acquired information about survival can be obtained from resources such as the National Death Index, and in some cases, there are similar resources to determine health-care utilization or living situation. While important outcomes such as functional ability or symptom status can rarely be obtained passively, some practical adaptations can reduce missing data. Investigators can plan for data collection in the home for frail elders, use proxy respondents, and develop protocols for assigning outcomes in the face of missing data. Proxy responses cannot be used for self-perceived states such as symptoms, but are helpful for observable phenomena, such as behaviors and activities (Magaziner et al., 1997; Neumann et al., 2000). Formal adjudicated decision processes can be developed to assign status in an unbiased way. For example, in the LIFE-P trial (Pahor et al., 2006), the main outcome is observed inability to walk

TABLE 29.2 Brief Measures That are Frequently Used to Assess Older Adults in Clinical Research

Domain	Instrument Name (References)	Key Characteristics
Disability	Katz Activities of Daily Living (Katz et al., 1963)	6 items, usually by self-report
	Late Life Function and Disability Index (Haley et al., 2002; Jette et al., 2002)	Function scale has 32 items, scored 0–100
		Disability scale has 16 items, scored 0–100
	Functional Independence Measure (FIM) (Linacre et al., 1994; Stineman et al., 1996)	Widely used in rehabilitation, professional rating
		18 items (8 basic activities, 5 mobility, 5 cognition), each scored 0–7
		Total FIM 0–126
		Motor FIM (activities plus mobility) 0–91
	National Health Interview Survey (Fitti and Kovar, 1987)	16 basic and instrumental items, with detail about level of difficulty, use of assistive devices
		No total scoring protocol
Physical function	SF-36 physical function	10 items score 0–100
	Short Physical Performance Battery (SPPB) (Guralnik et al., 1995; Guralnik et al., 1994)	3 items: timed walk, chair rises, tandem stands
		Score 0–12
	Gait speed (Studenski et al., 2003)	Timed usual or fast pace walk of a fixed distance
		Healthy older adult usual speed is >1.0 m/s
	6-minute walk test (Enright et al., 2003)	Distance covered in 6 min of walking
		Healthy older adult mean >500 m
		Older adults in assisted living <300 m
	Expanded SPPB (Simonsick et al., 2001)	Longer tandem stands, one-foot stand, narrow walk time, added to usual SPPB
		Especially useful for persons with scores of 10–12 on SPPB
		Score range 0–4
	Get up and go (Podsiadlo and Richardson, 1991)	Time to rise from a chair, walk 10 ft, turn, walk back, and sit
		Healthy elders complete in <10 s

Domain	Test	Description	Scoring	Notes
Cognitive function	Mini Mental Status Examination (Folstein et al., 1975)	Widely used multidimensional screening test	Score 0–30, cognitive impairment <24	
	Montreal Cognitive Assessment (Nasreddine et al., 2005)	Screening test with expanded capacity to detect mild cognitive impairment	Score 0–30	Score <26 suggests mild cognitive impairment
	Teng Modified Mini Mental State (3MS) (Teng and Chui, 1987)	Screening test with expanded capacity	Score 0–100	80 or less suggests cognitive impairment
	Digit Symbol Substitution Test (Wechsler, 1987)	Number of symbols translated into minutes		
Mood	Center for Epidemiologic Studies Depression Scale (CES-D) (Andresen et al., 1994)	Depression screening test	20 items, each scored 0–3, total score 0–60	Abnormal is 16 or higher
	Geriatric Depression Scale (Yesavage et al., 1982)	Depression screening test	15 and 30 item forms, scores 0–15, 0–30	For the 15-item scale, abnormal is 6 or higher; For the 30-item scale, abnormal is 10 or higher
Sleep	Pittsburgh Sleep Quality Index (Buysse et al., 1989)	Score 0–21, 5 or higher is indicative of a sleep problem		

400 m. In participants with missing data for this variable, a decision process was developed that defined inability to walk 400 m as having occurred in a participant who is observed to be bedridden or unable to walk 10 ft.

When outcome assessments are scheduled at fixed time points, a missed visit results in missing data. If the analysis plan is structured to examine time to event, then all available data can be used, even if some time points are missing. Some outcomes can be based on "time in state"; this approach has been used to assess reports of restricted activity days over time (Gill et al., 2001b). A novel approach to outcome monitoring is "triggered sampling." Participants receive frequent low-burden assessments, such as telephone calls. A change in status precipitates an in-person interview (Fried et al., 2006). Competing events, such as death, before a primary outcome event such as stroke, pneumonia, or disability can lead to bias (Satagopan et al., 2004). One approach is to predefine combined outcomes such as "death or primary outcome" in addition to the primary outcome alone (Ferrucci et al., 2004). A variety of analytic strategies exist to analyze data accounting for competing risks (Satagopan et al., 2004).

Multifactorial problems of aging require multiple independent or predictor measures. Participants with health limitations may be unable to complete all assessments, increasing the risk of missing data (Chatfield et al., 2005). One solution is to prioritize the set of independent variables so that the most essential are collected first (Berkman et al., 2001). The number of variables to be collected can be reduced if measures that are not expected to change are captured only at baseline (Cornoni-Huntley et al., 1986). Long sessions for data collection can lead to fatigue; sessions can include time for rest breaks, data collection can be broken down into separate shorter encounters, and information can be collected using multiple methods including telephone, in home, and on-site encounters (Berkman et al., 2001). Travel can be reduced by delivering equipment, such as activity or heart rhythm monitors or diaries, to the participant's residence.

CONCLUSIONS AND RECOMMENDATIONS

Since older adults have high rates of health problems and are the most frequent users of health care, clinical research to build the knowledge base for care of the aged is essential. Without efforts to change the way we perform research, we will continue to mistakenly extrapolate information from studies in younger adults to decisions about care of the aged. Investigators can incorporate general concepts about aging processes into research questions and study design. Many practical solutions exist that can foster increased participation of older adults in research studies. Better knowledge about health and aging can result in important new insights into fundamental biological processes, better care and more rationale use of health care resources, and ultimately into improved quality of life and function for older adults.

REFERENCES

Andresen, E.M., Malmgren, J.A., Carter, W.B., Patrick, D.L., 1994. Screening for depression in well older adults: evaluation of a short form of the CES-D (Center for Epidemiologic Studies Depression Scale). Am. J. Prev. Med. 10, 77—84.

Ashar, F.N., Moes, A., Moore, A.Z., et al., 2014. Association of mitochondrial DNA levels with frailty and all-cause mortality. J. Mol. Med. 93 (2).

Baker, M.K., Atlantis, E., Fiatarone Singh, M.A., 2007. Multi-modal exercise programs for older adults. Age Ageing 36, 375—381.

Baldereschi, M., Inzitari, M., Di Carlo, A., Farchi, G., Scafato, E., Inzitari, D., 2007. Epidemiology of distal symmetrical neuropathies in the Italian elderly. Neurology 68, 1460—1467.

Barabasi, A.L., Oltvai, Z.N., 2004. Network biology: understanding the cell's functional organization. Nat. Rev. Genet. 5, 101—113.

Becker, R.E., Greig, N.H., Giacobini, E., Schneider, L.S., Ferrucci, L., 2014. A new roadmap for drug development for Alzheimer's disease. Nat. Rev. Drug Discov. 13, 156.

Belnap, B.H., Kuebler, J., Upshur, C., et al., 2006. Challenges of implementing depression care management in the primary care setting. Adm. Policy Ment. Health 33, 65—75.

Berkman, C.S., Leipzig, R.M., Greenberg, S.A., Inouye, S.K., 2001. Methodologic issues in conducting research on hospitalized older people. J. Am. Geriatrics Soc. 49, 172—178.

Boehm, M., Slack, F.J., 2006. MicroRNA control of lifespan and metabolism. Cell Cycle (Georgetown, Tex.) 5, 837—840.

Burch, J.B., Augustine, A.D., Frieden, L.A., et al., 2014. Advances in geroscience: impact on healthspan and chronic disease. J. Gerontol. 69 (Suppl. 1), S1—S3.

Butler, R.N., Sprott, R., Warner, H., et al., 2004. Biomarkers of aging: from primitive organisms to humans. J. Gerontol. 59, B560—B567.

Buysse, D.J., Reynolds 3rd, C.F., Monk, T.H., Berman, S.R., Kupfer, D.J., 1989. The Pittsburgh Sleep Quality Index: a new instrument for psychiatric practice and research. Psychiatry Res. 28, 193—213.

Campbell, A.J., Robertson, M.C., 2007. Rethinking individual and community fall prevention strategies: a meta-regression comparing single and multifactorial interventions. Age Ageing 36, 656—662.

Carlberg, C., Dunlop, T.W., 2006. An integrated biological approach to nuclear receptor signaling in physiological control and disease. Crit. Rev. Eukaryot. Gene Expr. 16, 1—22.

Cesari, M., Kritchevsky, S.B., Penninx, B.W., et al., 2005. Prognostic value of usual gait speed in well-functioning older people—results from the Health, Aging and Body Composition Study. J. Am. Geriatr. Soc. 53, 1675—1680.

Chatfield, M.D., Brayne, C.E., Matthews, F.E., 2005. A systematic literature review of attrition between waves in longitudinal studies in the elderly shows a consistent pattern of dropout between differing studies. J. Clin. Epidemiol. 58, 13—19.

Choi, A.M., Ryter, S.W., Levine, B., 2013. Autophagy in human health and disease. N. Engl. J. Med. 368, 1845—1846.

Cohen, H.J., Feussner, J.R., Weinberger, M., et al., 2002. A controlled trial of inpatient and outpatient geriatric evaluation and management. N. Engl. J. Med. 346, 905—912.

Coplan, P.M., Schmader, K., Nikas, A., et al., 2004. Development of a measure of the burden of pain due to herpes zoster and postherpetic neuralgia for prevention trials: adaptation of the brief pain inventory. J. Pain 5, 344—356.

Cornoni-Huntley, J., Brock, D.B., Ostfeld, A.M., Taylor, J.O., Wallace, R.B., Lafferty, M.E. (Eds.), 1986. Established Populations for Epidemiologic Studies of the Elderly: Resource Data Book. National Institute on Aging, Bethesda, MD.

Croce, C.M., 2008. Oncogenes and cancer. N. Engl. J. Med. 358, 502—511.

Ehninger, D., Neff, F., Xie, K., 2014. Longevity, aging and rapamycin. Cell Mol. Life Sci. 71, 4325—4346.

Ehrhardt, J., Morgan, J., 2005. Regenerative capacity of skeletal muscle. Curr. Opin. Neurol. 18, 548—553.

Ellis, G., Langhorne, P., 2004. Comprehensive geriatric assessment for older hospital patients. Br. Med. Bull. 71, 45—59.

Enright, P.L., McBurnie, M.A., Bittner, V., et al., 2003. The 6-min walk test: a quick measure of functional status in elderly adults. Chest 123, 387—398.

Ensrud, K.E., Ewing, S.K., Taylor, B.C., et al., 2007. Frailty and risk of falls, fracture, and mortality in older women: the study of osteoporotic fractures. J. Gerontol. 62, 744—751.

Espeland, M.A., Gill, T.M., Guralnik, J., et al., 2007. Designing clinical trials of interventions for mobility disability: results from the lifestyle interventions and independence for elders pilot (LIFE-P) trial. J. Gerontol. 62, 1237—1243.

Ferrucci, L., Guralnik, J.M., Simonsick, E., Salive, M.E., Corti, C., Langlois, J., 1996. Progressive versus catastrophic disability: a longitudinal view of the disablement process. J. Gerontol. 51, M123—M130.

Ferrucci, L., Guralnik, J.M., Studenski, S., Fried, L.P., Cutler Jr., G.B., Walston, J.D., 2004. Designing randomized, controlled trials aimed at preventing or delaying functional decline and disability in frail, older persons: a consensus report. J. Am. Geriatr. Soc. 52, 625—634.

Ferrucci, L., Windham, G., Fried, L.P., 2005. Frailty in older persons. GENUS LXI, 39—53.

Fiatarone, M.A., O'Neill, E.F., Ryan, N.D., et al., 1994. Exercise training and nutritional supplementation for physical frailty in very elderly people. N. Engl. J. Med. 330, 1769—1775.

Fitti, J.K., Kovar, M.G., June 1987. The supplement on aging to the 1984 national health interview survey. Vital Health Stat. 1—115.

Folstein, M.F., Folstein, S.E., McHugh, P.R., 1975. "Mini-mental state". A practical method for grading the cognitive state of patients for the clinician. J. Psychiatr. Res. 12, 189—198.

Fried, L.P., Tangen, C.M., Walston, J., et al., 2001. Frailty in older adults: evidence for a phenotype. J. Gerontol. 56, M146—M156.

Fried, T.R., Byers, A.L., Gallo, W.T., et al., 2006. Prospective study of health status preferences and changes in preferences over time in older adults. Archiv. Intern. Med. 166, 890—895.

Gill, T.M., Kurland, B., 2003. The burden and patterns of disability in activities of daily living among community-living older persons. J. Gerontol. 58, 70—75.

Gill, T.M., DiPietro, L., Krumholz, H.M., 2000. Role of exercise stress testing and safety monitoring for older persons starting an exercise program. JAMA 284, 342—349.

Gill, T.M., McGloin, J.M., Gahbauer, E.A., Shepard, D.M., Bianco, L.M., 2001a. Two recruitment strategies for a clinical trial of physically frail community-living older persons. J. Am. Geriatr. Soc. 49, 1039—1045.

Gill, T.M., Desai, M.M., Gahbauer, E.A., Holford, T.R., Williams, C.S., 2001b. Restricted activity among community-living older persons: incidence, precipitants and health care utilization. Ann. Intern. Med. 135, 313—321.

Gill, S.S., Bronskill, S.E., Mamdani, M., et al., 2004. Representation of patients with dementia in clinical trials of donepezil. Can. J. Clin. Pharmacol. 11, e274—85.

Guralnik, J.M., Branch, L.G., Cummings, S.R., Curb, J.D., 1989. Physical performance measures in aging research. J. Gerontol. 44, M141—M146.

Guralnik, J.M., Simonsick, E.M., Ferrucci, L., et al., 1994. A short physical performance battery assessing lower extremity function: association with self-reported disability and prediction of mortality and nursing home admission. J. Gerontol. 49, M85—M94.

Guralnik, J.M., Ferrucci, L., Simonsick, E.M., Salive, M.E., Wallace, R.B., 1995. Lower-extremity function in persons over the age of 70 years as a predictor of subsequent disability. N. Engl. J. Med. 332, 556—561.

Guralnik, J.M., Fried, L.P., Salive, M.E., 1996. Disability as a public health outcome in the aging population. Annu. Rev. Public Health 17, 25—46.

Haley, S.M., Jette, A.M., Coster, W.J., et al., 2002. Late life function and disability Instrument: II. Development and evaluation of the function component. J. Gerontol. 57, M217—M222.

Hardy, S.E., Gill, T.M., 2004. Recovery from disability among community-dwelling older persons. JAMA 291, 1596—1602.

Hardy, S.E., Perera, S., Roumani, Y.F., Chandler, J.M., Studenski, S.A., 2007. Improvement in usual gait speed predicts better survival in older adults. J. Am. Geriatr. Soc. 55, 1727—1734.

Heiat, A., Gross, C.P., Krumholz, H.M., 2002. Representation of the elderly, women, and minorities in heart failure clinical trials. Arch. Intern Med. 162, 1682—1688.

Huang, G., Gibson, C.A., Tran, Z.V., Osness, W.H., 2005. Controlled endurance exercise training and VO_{2max} changes in older adults: a meta-analysis. Prev. Cardiol. 8, 217—225.

Hutchins, L.F., Unger, J.M., Crowley, J.J., Coltman Jr., C.A., Albain, K.S., 1999. Underrepresentation of patients 65 years of age or older in cancer-treatment trials. N. Engl. J. Med. 341, 2061−2067.

Iber, C., Redline, S., Kaplan Gilpin, A.M., et al., 2004. Polysomnography performed in the unattended home versus the attended laboratory setting−Sleep Heart Health Study methodology. Sleep 27, 536−540.

Ingram, D.K., Roth, G.S., 2014. Calorie restriction mimetics: can you have your cake and eat it, too? Ageing Res. Rev. 20.

Inouye, S.K., Studenski, S., Tinetti, M.E., Kuchel, G.A., 2007. Geriatric syndromes: clinical, research, and policy implications of a core geriatric concept. J. Am. Geriatr. Soc. 55, 780−791.

Jette, A.M., Haley, S.M., Coster, W.J., et al., 2002. Late life function and disability instrument: I. Development and evaluation of the disability component. J. Gerontol. 57, M209−M216.

Kahn, A.J., 2007. Epigenetic and mitochondrial mechanisms in aging and longevity: a report of a meeting of the Longevity Consortium, Napa, California, October 25−27, 2006. J. Gerontol. 62, 577−582.

Kato, M., Slack, F.J., 2008. microRNAs: small molecules with big roles - *C. elegans* to human cancer. Biol. Cell/Under Auspices Eur. Cell Biol. Organ. 100, 71−81.

Katz, S., Ford, A.B., Moskowitz, R.W., Jackson, B.A., Jaffe, M.W., 1963. Studies of illness in the aged. The index of Adl: a standardized measure of biological and psychosocial function. JAMA 185, 914−919.

Kennedy, B.K., Pennypacker, J.K., 2014. Drugs that modulate aging: the promising yet difficult path ahead. Transl. Res. 163, 456−465.

Kennedy, B.K., Berger, S.L., Brunet, A., et al., 2014. Geroscience: linking aging to chronic disease. Cell 159, 709−713.

Kitano, H., 2004. Biological robustness. Nat. Rev. Genet. 5, 826−837.

Knowler, W.C., Barrett-Connor, E., Fowler, S.E., et al., 2002. Reduction in the incidence of type 2 diabetes with lifestyle intervention or metformin. N. Engl. J. Med. 346, 393−403.

Lamming, D.W., Ye, L., Sabatini, D.M., Baur, J.A., 2013. Rapalogs and mTOR inhibitors as anti-aging therapeutics. J. Clin. Invest. 123, 980−989.

Lee, P.Y., Alexander, K.P., Hammill, B.G., Pasquali, S.K., Peterson, E.D., 2001. Representation of elderly persons and women in published randomized trials of acute coronary syndromes. JAMA 286, 708−713.

Lichtenstein, M.J., Bess, F.H., Logan, S.A., 1988. Validation of screening tools for identifying hearing-impaired elderly in primary care (erratum appears in JAMA July 4, 1990, 264 (1), 38). JAMA 259, 2875−2878.

Linacre, J.M., Heinemann, A.W., Wright, B.D., Granger, C.V., Hamilton, B.B., 1994. The structure and stability of the functional independence measure. Arch. Phys. Med. Rehabil. 75, 127−132.

Lopez, O.L., Kuller, L.H., Becker, J.T., et al., 2007. Incidence of dementia in mild cognitive impairment in the cardiovascular health study cognition study. Arch. Neurol. 64, 416−420.

Loscalzo, J., Kohane, I., Barabasi, A.L., 2007. Human disease classification in the postgenomic era: a complex systems approach to human pathobiology. Mol. Syst. Biol. 3, 124.

Macphee, G.J., Crowther, J.A., McAlpine, C.H., 1988. A simple screening test for hearing impairment in elderly patients. Age Ageing 17, 347−351.

Magaziner, J., Zimmerman, S.I., Gruber-Baldini, A.L., Hebel, J.R., Fox, K.M., 1997. Proxy reporting in five areas of functional status. Comparison with self-reports and observations of performance. Am. J. Epidemiol. 146, 418−428.

Marzetti, E., Leeuwenburgh, C., 2006. Skeletal muscle apoptosis, sarcopenia and frailty at old age. Exp. Gerontol. 41, 1234−1238.

McDermott, M.M., Liu, K., Greenland, P., et al., 2004. Functional decline in peripheral arterial disease: associations with the ankle brachial index and leg symptoms. JAMA 292, 453−461.

McDermott, M.M., Hoff, F., Ferrucci, L., et al., 2007. Lower extremity ischemia, calf skeletal muscle characteristics, and functional impairment in peripheral arterial disease. J. Am. Geriatr. Soc. 55, 400−406.

McMurdo, M.E., Witham, M.D., Gillespie, N.D., 2005. Including older people in clinical research. BMJ 331, 1036−1037.

Mitchell, S.L., Sullivan, E.A., Lipsitz, L.A., 1997. Exclusion of elderly subjects from clinical trials for Parkinson disease. Arch. Neurol. 54, 1393−1398.

Nasreddine, Z.S., Phillips, N.A., Bedirian, V., et al., 2005. The Montreal Cognitive Assessment, MoCA: a brief screening tool for mild cognitive impairment. J. Am. Geriatr. Soc. 53, 695−699.

Nelson, M.E., Rejeski, W.J., Blair, S.N., et al., 2007. Physical activity and public health in older adults: recommendation from the American College of Sports Medicine and the American Heart Association. Circulation 116, 1094−1105.

Neumann, P.J., Araki, S.S., Gutterman, E.M., 2000. The use of proxy respondents in studies of older adults: lessons, challenges, and opportunities. J. Am. Geriatrics Soc. 48, 1646−1654.

Newman, A.B., Gottdiener, J.S., McBurnie, M.A., et al., 2001. Associations of subclinical cardiovascular disease with frailty. J. Gerontol. 56, M158−M166.

Newman, A.B., Arnold, A.M., Naydeck, B.L., et al., 2003a. "Successful aging": effect of subclinical cardiovascular disease. Arch. Intern Med. 163, 2315−2322.

Newman, A.B., Kupelian, V., Visser, M., et al., 2003b. Sarcopenia: alternative definitions and associations with lower extremity function. J. Am. Geriatr. Soc. 51, 1602−1609.

Newman, A.B., Lee, J.S., Visser, M., et al., 2005. Weight change and the conservation of lean mass in old age: the Health, Aging and Body Composition Study. Am. J. Clin. Nutr. 82, 872−878 quiz 915−6.

Newman, A.B., Simonsick, E.M., Naydeck, B.L., et al., 2006. Association of long-distance corridor walk performance with mortality, cardiovascular disease, mobility limitation, and disability. JAMA 295, 2018−2026.

Nichols, L., Martindale-Adams, J., Burns, R., et al., 2004. Social marketing as a framework for recruitment: illustrations from the REACH study. J. Aging Health 16, 157S–176S.

O'Connell, K., Gannon, J., Doran, P., Ohlendieck, K., 2007. Proteomic profiling reveals a severely perturbed protein expression pattern in aged skeletal muscle. Int. J. Mol. Med. 20, 145–153.

Ostir, G.V., Carlson, J.E., Black, S.A., Rudkin, L., Goodwin, J.S., Markides, K.S., 1999. Disability in older adults. 1: prevalence, causes, and consequences. Behav. Med. 24, 147–156.

Oxman, M.N., Levin, M.J., Johnson, G.R., et al., 2005. A vaccine to prevent herpes zoster and postherpetic neuralgia in older adults. N. Engl. J. Med. 352, 2271–2284.

Pahor, M., Blair, S.N., Espeland, M., et al., 2006. Effects of a physical activity intervention on measures of physical performance: results of the lifestyle interventions and independence for Elders Pilot (LIFE-P) study. J. Gerontol. 61, 1157–1165.

Perera, S., Studenski, S., Chandler, J.M., Guralnik, J.M., 2005. Magnitude and patterns of decline in health and function in 1 year affect subsequent 5-year survival. J. Gerontol. 60, 894–900.

Podsiadlo, D., Richardson, S., 1991. The timed "Up & Go": a test of basic functional mobility for frail elderly persons. J. Am. Geriatr. Soc. 39, 142–148.

Prevention of stroke by antihypertensive drug treatment in older persons with isolated systolic hypertension. Final results of the Systolic Hypertension in the Elderly Program (SHEP). SHEP Cooperative Research Group. JAMA 265, 1991, 3255–3264.

Roth, S.M., Metter, E.J., Ling, S., Ferrucci, L., 2006. Inflammatory factors in age-related muscle wasting. Curr. Opin. Rheumatol. 18, 625–630.

Roubenoff, R., Hughes, V.A., 2000. Sarcopenia: current concepts. J. Gerontol. 55, M716–M724.

Roubenoff, R., 2003. Sarcopenia: effects on body composition and function. J. Gerontol. 58, 1012–1017.

Sarnak, M.J., Katz, R., Fried, L.F., et al., 2008. Cystatin C and aging success. Arch. Intern Med. 168, 147–153.

Satagopan, J.M., Ben-Porat, L., Berwick, M., Robson, M., Kutler, D., Auerbach, A.D., 2004. A note on competing risks in survival data analysis. Br. J. Cancer 91, 1229–1235.

Schneider, J.L., Cuervo, A.M., 2014. Autophagy and human disease: emerging themes. Curr. Opin. Genet. Dev. 26, 16–23.

Schrager, M.A., Metter, E.J., Simonsick, E., et al., 2007. Sarcopenic obesity and inflammation in the InCHIANTI study. J. Appl. Physiol. 102, 919–925.

Seals, D.R., Melov, S., 2014. Translational geroscience: emphasizing function to achieve optimal longevity. Aging 6, 718–730.

Simonsick, E.M., Newman, A.B., Nevitt, M.C., et al., 2001. Measuring higher level physical function in well-functioning older adults: expanding familiar approaches in the Health ABC study. J. Gerontol. 56, M644–M649.

Solomon, A.M., Bouloux, P.M., 2006. Modifying muscle mass - the endocrine perspective. J. Endocrinol. 191, 349–360.

Spindler, S.R., Mote, P.L., 2007. Screening candidate longevity therapeutics using gene-expression arrays. Gerontology 53, 306–321.

Stahl, S.M., Vasquez, L., 2004. Approaches to improving recruitment and retention of minority elders participating in research: examples from selected research groups including the National Institute on Aging's Resource Centers for Minority Aging Research. J. Aging Health 16, 9S–17S.

Stineman, M.G., Shea, J.A., Jette, A., et al., 1996. The Functional Independence Measure: tests of scaling assumptions, structure, and reliability across 20 diverse impairment categories. Arch. Phys. Med. Rehabil. 77, 1101–1108.

Stuck, A.E., Egger, M., Hammer, A., Minder, C.E., Beck, J.C., 2002. Home visits to prevent nursing home admission and functional decline in elderly people: systematic review and meta-regression analysis. JAMA 287, 1022–1028.

Studenski, S., Perera, S., Wallace, D., et al., 2003. Physical performance measures in the clinical setting. J. Am. Geriatr. Soc. 51, 314–322.

Teng, E.L., Chui, H.C., 1987. The Modified Mini-Mental State (3MS) examination. J. Clin. Psychiatry 48, 314–318.

Tinetti, M.E., Baker, D.I., McAvay, G., et al., 1994. A multifactorial intervention to reduce the risk of falling among elderly people living in the community. N. Engl. J. Med. 331, 821–827.

Troosters, T., Gosselink, R., Decramer, M., 2002. Six-minute walk test: a valuable test, when properly standardized. Phys. Ther. 82, 826–827 author reply 7–8.

Wallace, D.C., 2011. Bioenergetic origins of complexity and disease. Cold Spring Harb. Symp. Quant. Biol. 76, 1–16.

Wechsler, D., 1987. Wechsler Memory Scale-Revised Manual. Psychological Corporation, New York.

Witham, M.D., McMurdo, M.E., 2007. How to get older people included in clinical studies. Drugs Aging 24, 187–196.

Yesavage, J.A., Brink, T.L., Rose, T.L., et al., 1982. Development and validation of a geriatric depression screening scale: a preliminary report. J. Psychiatr. Res. 17, 37–49.

Chapter 30

Clinical Research in Neurology

Keren Regev and Howard L. Weiner

Harvard Medical School, Boston, MA, United States

Chapter Outline

Key Points

- Increase in life expectancy has led to substantial increases in the number of individuals with neurological conditions, a trend that is expected to accelerate during this century.
- CNS drugs take 35% longer to complete clinical trials and gain regulatory approval compared to other new prescription medicines.
- The blood—brain barrier, formed by the endothelial cells of the brain capillaries, restricts access to cells and blood-borne compounds and facilitates nutrients essential for normal metabolism to reach brain cells. This very tight regulation of brain homeostasis results in the inability of some small and large therapeutic compounds to cross the blood—brain barrier.
- Current FDA-approved MS drugs target the inflammatory component of the disease and are effective in reducing relapse rate and MRI activity for relapsing forms of disease.
- Neuroprotective strategies include increasing axon stability, altering processes that damage axons, remyelination by promoting differentiation of oligodendrocyte precursors into myelin-producing cells, and the use of mesenchymal or bone marrow—derived stem cells.
- Alzheimer disease research revolves around the two hallmarks of disease pathology, the neuritic plaque, and the neurofibrillary tangle (NFT). Neuritic plaques are largely composed of the fibrillogenic 42-amino acid length form of the beta amyloid protein (Aβ42). NFTs are composed primarily of hyperphosphorylated aggregations of the microtubule-associated protein tau.
- In amyotrophic lateral sclerosis (ALS), the identification of mutations in genes for TDP-43, FUS, and C9ORF72 has led to the concept that aberrant RNA metabolism plays an important role in ALS pathogenesis.
- Immunotherapy for glioma includes developing personalized vaccines based on an individual patient's tumor, by utilizing them as antigen sources for potent immunostimulatory cells such as dendritic cells or by harvesting immunostimulatory chaperone protein/tumor peptide complexes.

INTRODUCTION

The goal of clinical science is to identify therapeutics to relieve human suffering. Brain disorders represent an enormous disease burden both in terms of human suffering and economic cost. Many brain disorders are chronic and incurable conditions whose disabling effects continue for decades. Thus, the overall disease burden from disorders of the nervous system is greater than suggested by mortality figures alone. The extent of this burden is underlined by data from the World Health Organization (2006).

Therapy for neurological diseases such as cerebrovascular diseases, amyotrophic lateral sclerosis (ALS), multiple sclerosis (MS), Parkinson disease (PD), Alzheimer disease (AD), neuropathic pain, epilepsy, and headaches are in many instances incomplete as they aim to arrest or slow the progression of disease, the worsening of disability, the flair of an episodic disease, or relieve symptoms.

To compare the impact of different diseases, the WHO estimates for each disease category the "disability-adjusted life years" or DALYs—years of healthy life that are lost to disability as well as death. Each disease is assigned a weighting factor according to the severity of the resulting disability. Using this measure, brain disorders emerge as a leading contributor to global disease burden.

The economic cost of brain disorders is correspondingly large. These include not only the cost of treatment, but also the lost productivity both of patients and their caregivers. In Europe, for example, Andlin-Sobocki et al. (2005) estimated the total cost of brain disorders to be €386 billion per year in 2004 prices (including direct costs of treatment and care plus indirect cost of lost workdays and lost productivity)—twice the estimated cost of cancer.

The increase in life expectancy that occurred in the 20th century has led to substantial increases in the number of individuals with neurological conditions, a trend that is expected to accelerate during this century. Cerebrovascular disease currently accounts for the majority of global disability from neurological disorders as measured in DALYs and will account for 4% of total DALYs globally by 2030. Other conditions, such as Alzheimer disease and Parkinson disease prevalence, will increase, and that increase will be greatest in developing countries (Ferri et al., 2005; Dorsey et al., 2007).

The growth in the burden of neurological disease increases the global demand for effective therapy; however, several challenges confront the development of new therapies. Some of these challenges are generic and some are specific for neurological diseases.

FEATURES UNIQUE TO NEUROLOGIC DISEASES

The uniqueness and complexity of development of treatment for neurologic diseases is reflected by the relative poor success rate of FDA approval for central nervous system (CNS) drugs ($\sim$8%) as compared to other therapeutic areas ($\sim$11%) and the fact that CNS drugs take 35% longer to complete clinical trials and gain regulatory approval compared to other new prescription medicines (Kola and Landis, 2004).

The Challenges

Target Selection

Identifying targets is always difficult in the drug development process. It is particularly challenging for neurological indications due to the complexity of the human CNS, the paucity of validated targets, and the "orphan" nature of many neurological diseases.

CNS targets that have been clinically validated in human CNS disease include the dopamine, serotonin (5HT1b), gamma aminobutyric acid-A, N-methyl-D-aspartate (NMDA) receptors, norepinephrine and 5HT transporters, and monoamine oxidase and catechol-O-methyltransferase. MS, which is one of the most successful examples of disease modification in neurologic disease, has several validated targets that include the very late antigen-4 (VLA-4) integrin, type 1 interferon receptors, CD20, CD52, CD25, and the S1P receptor.

Animal Models

Animal CNS disease models are important tools that have led to significant breakthroughs in both understanding the underlying disease processes and discovering new therapeutic targets. They have, however, often proven disappointing in predicting therapeutic efficacy (Rothstein, 2009; Geerts, 2009; Akhtar et al., 2008). This has led to considerable debate about the use of animal models in drug discovery. Most animal models fail to successfully replicate human CNS disease, probably owing to the enormous complexity of the human cerebral cortex required to support language, self-awareness, and comprehension. Other factors that may contribute to this "translational disconnect" include differences in pharmacological effects of the same compound on human versus rodent targets, differences in neurotransmitter wiring in human versus rodent brains, difficulties in modeling the presymptomatic state, differences in placebo effects, and the fundamentally different readouts in cognitive states.

Blood—Brain Barrier

The CNS is protected by barriers which control the entry of compounds into the brain, thereby regulating brain homeostasis. The blood—brain barrier (BBB), formed by the endothelial cells of the brain capillaries, restricts access to cells and blood-borne compounds and facilitates nutrients essential for normal metabolism to reach brain cells. This very tight regulation of brain homeostasis results in the inability of some small and large therapeutic compounds to cross the BBB (Ghersi-Egea et al., 1995).

The BBB is the bottleneck in brain drug development and is probably the single most important factor limiting the development of neurotherapeutics.

Small molecules with appropriate lipophilicity, molecular weight (mw), and charge will diffuse from blood into the CNS. However, the majority of small molecules (mw > 500 daltons, Da), proteins, and peptides do not cross the BBB. It has been reported that ∼98% of the small molecules and nearly all large molecules (mw > 1 kD, kilodaltons) do not cross or only minimally cross the BBB. This includes recombinant proteins and monoclonal antibodies, which are being increasingly used to target diseases such as AD (Siemers et al., 2010; De Mattos et al., 2001).

Current data suggest that <0.1% of systemic concentrations of a monoclonal antibody will enter the brain, as assessed by the analysis of cerebrospinal fluid (CSF), being used as an imperfect surrogate for brain (Rubenstein et al., 2003).

Strategies under investigation to overcome the problem of the BBB can be roughly divided into two broad categories. The first category comprises techniques that allow or facilitate the crossing of drugs through the BBB (e.g., molecular Trojan horses, proton-coupled oligopeptide transporters, exosomes, liposomes, nanoparticles, chimeric peptides, prodrugs), while the second category consists of techniques that bypass the BBB altogether via direct delivery to the CNS. In the second category, several techniques have been investigated, including BBB disruption, and intrathecal (IT), intracerebroventricular, and intranasal delivery.

Intrathecal Administration

New evidence demonstrates that IT delivery may be a promising route of administration for large biomolecules. Numerous studies in small and large animal models of stroke and the neurodegenerative lysosomal storage diseases (Tsai et al., 2007) show that IT delivery of protein therapeutics can result in widespread penetration into the brain parenchyma, improvement of neuropathology, and improved clinical and behavioral outcomes. Tolerability to treatment was generally good in these studies. Similarly, case reports and studies in human patients with a neurodegenerative lysosomal storage disease or breast cancer metastases to the CNS have suggested generally acceptable tolerability and therapeutic effectiveness with few adverse events after the IT delivery (Calias et al., 2014).

Biomarkers

Biological markers (biomarkers) are objectively measured features that can be evaluated to describe normal biological processes, pathogenic processes, or pharmacologic responses to a therapeutic intervention. A biomarker can be defined as a biological variable that has a statistically significant relationship with parameters of disease states or the activity of a drug. Biomarkers are becoming an essential part of clinical development, largely because they offer the prospect of more homogenous patient populations in clinical trials through patient selection and stratification and provide markers of disease modification. Because biomarkers can increase the statistical power of clinical trials and predict drug efficacy more quickly than conventional clinical endpoints, they hold the potential to substantially accelerate product development and increase confidence of demonstrating therapeutic efficacy in phase III trials. For most disorders, biomarker measurements derive from the determination of a biological variable in blood samples, but, because of the existence of the BBB, this is not true for CNS disorders.

A CNS disorder where biomarkers play a critical role in patient selection and stratification is MS. Magnetic resonance imaging (MRI) is essential to this process, particularly when used in combination with contrast agents, which permits new MS plaques (areas of demyelination) to be quantified (Palmer and Alavijeh, 2012). MRI also plays a key role in the identification of prodromal MS, termed clinically isolated syndrome (CIS). First-line therapies (beta interferon drugs and glatiramer acetate) delay the conversion from CIS to clinically definite MS, which makes CIS an attractive target for disease-modifying medicines (Palmer, 2012). Another example of selecting patients most likely to respond to pharmacotherapy is seen with the use of perfusion-weighted (PWI) and diffusion-weighted MRI (DWI) in clinical trials of stroke. A mismatch of PWI and DWI lesion volumes indicates those patients most likely to respond to treatment with thrombolytic or neuroprotective drugs (Lansberg et al., 2011). Patient stratification on this basis markedly improves the statistical power of stroke studies assessing the therapeutic efficacy of drug candidates. Positron emission tomography (PET) is crucial for beta-amyloid (Aβ) imaging to distinguish between AD, which is characterized by Aβ deposition, and frontotemporal dementia, which is not associated with Aβ deposition.

Biochemical measurements in CSF can also be used to establish target engagement. For example, changes in CSF concentrations of Aβ following administration of drug to candidates that act to reduce Aβ concentration in the brains of AD patients (Rinne et al., 2010).

Endpoints

In addition to the challenges described, difficulties arise because of the complexity of the CNS and the hurdle of developing clinical endpoints that successfully capture therapeutic effects on multiple functional domains controlled by complex brain circuits. Some of these problems arise because historically the clinical scales that were developed to describe subgroups of domains affected by CNS diseases are not precise or linear. One example is the (expanded disability status scale) EDSS in MS, which is largely influenced by lower extremity walking function, but less so by cognition or upper extremity function. Furthermore, it has been difficult to determine and agree on what degree of change represents a clinically meaningful outcome.

DISEASE EXAMPLES

Multiple Sclerosis

MS is classified as an organ-specific autoimmune disease. Genomewide association studies have linked HLA and immune system genes to MS, confirming that immunological factors contribute to disease pathogenesis (Sormani et al., 2009). Fig. 30.1 summarizes the immunological cascade in MS.

In the early stages of MS, scattered foci of inflammation occur in the CNS causing clinical symptoms reflecting the localization of the inflammatory focus. With resolution of inflammation, patients enter a clinical remission. Studies using MRI have revealed that the frequency of new lesions seen on MRI exceeds that of clinical relapses by approximately 10−1.

Approximately 50% of patients with relapsing remitting MS transition to secondary-progressive MS (SPMS) within a decade of the initial diagnosis characterized by slowly advancing neurological disability in the absence of relapses. Mechanisms underlying the transition from relapsing-remitting MS (RRMS) to SPMS are not entirely understood. Some argue that there is a transition from a primarily inflammatory pathology to one that is neurodegenerative and no longer dependent on inflammation. Others believe that the inflammation becomes compartmentalized within the CNS and is no longer initiated in the periphery in an episodic manner. The inflammation in the CNS involves the innate immune system including microglia and astrocytes.

A subtype of MS, primary progressive MS (PPMS), occurs in ~15% of patients and is characterized by continuous progression of neurological disability in the absence of relapses from the onset. Mechanisms underlying PPMS are presumed to be similar to those underlying SPMS.

For nearly a century, experimental autoimmune encephalomyelitis (EAE) models have provided a great insight into the mechanism of immune-initiated inflammation within the CNS. EAE can be induced in a variety of animal species and strains by immunization with CNS constituents or passive transfer of T cells or antibodies from immunized animals, resulting in immunologically mediated inflammatory injury to the CNS. The EAE model has yielded important information about immune-mediated CNS tissue injury, and although the EAE model has led to new therapies for relapsing forms of MS, not all treatments that work in EAE have been successfully translated to patients. No single EAE model reliably mimics all aspects of MS, importantly; no generally accepted models exhibit the marked neurodegeneration observed in later stages of MS although recently, models that may be useful to investigate the axonopathy seen in SPMS have been developed.

MS drugs are divided into disease-modifying drugs and symptomatic treatments; the goals of disease-modifying therapy in MS, reducing relapse rate and disease progression, do not necessarily improve quality of life in the short term. An entirely separate approach targets MS symptoms; symptomatic therapies may significantly benefit patients by reducing morbidity or improving quality of life.

Currently all FDA-approved MS drugs that are defined as disease-modifying drugs target the inflammatory component of the disease and are generally effective in reducing relapse rate and MRI activity (measured by gadolinium-enhancing or new T2 lesions).

Antiinflammatory therapy may indirectly lead to neuroprotection by preventing axonal injury and transection because axons are transected at sites of acute inflammation; however, none of the currently available drugs have proved effective in

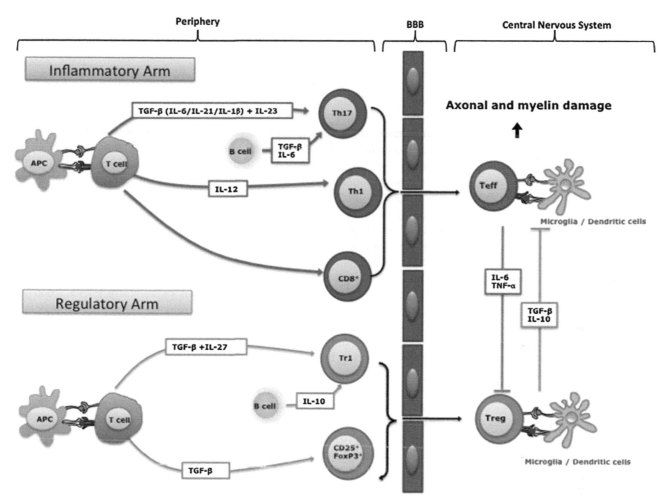

FIGURE 30.1 Immune mechanisms in multiple sclerosis. Autoreactive T cells in the periphery are activated by antigen presenting cells in the context of antigen presenting cells in their cytokine microenvironment and differentiate into either inflammatory (Th1, Th17, CD8 T cells) or regulatory T cells. Guided by adhesion molecules, T cells migrate through the blood—brain barrier (BBB) and into the central nervous system (CNS). In the CNS, autoreactive T cells recognize myelin antigens are reactivated, proliferate, and initiate myelin damage. This inflammatory arm of the autoimmune process is counterbalanced by a regulatory arm. In the CNS, inflammatory T cells lead to activation of microglia, macrophages, and astrocytes, which contribute and perpetuate CNS damage. Autoreactive B cells and autoantibodies can also be detected in blood and cerebrospinal fluid. They can pass through a damaged BBB and, upon activating the complement system with ensuing formation of the membrane attack complex, damage myelin. Antibodies can also bind to macrophages and stimulate them to antibody-dependent cytotoxicity.

the later stages of MS, presumably because peripheral inflammation is not essential to the ongoing neurodegeneration in the later stages of the disease. Different potential neuroprotective strategies are under investigation: some aim to increase axon stability or alter processes that damage axons, others include remyelination strategies that promote differentiation of oligodendrocyte precursors into myelin-producing cells or the use of mesenchymal or bone marrow—derived stem cells (Akhtar et al., 2008). To date, no phase III trial of neuroprotective therapy in MS has been positive.

Recombinant interferon-beta (IFNβ) was first approved for RRMS in 1993 (IFNβ-1b, Betaseron/Betaferon), representing a long-awaited breakthrough in the management of MS (Bermel and Rudick, 2007). Subsequently, two IFNβ-1a preparations (Avonex and Rebif) were approved. The pivotal studies for the different types of IFNβ consisted of placebo-controlled phase III trials showing efficacy in reducing the annualized relapse rate by 30—40%. Secondary outcomes included different MRI measures (gadolinium-enhancing lesions, new or enlarging T2 lesions). In addition to proving the feasibility of therapeutic intervention in RRMS, these trials set an example for future MS drug trials using MRI-based outcome measures. The second drug to be approved for RRMS was glatiramer acetate (Copaxone), a mixture of random oligomers of amino acids enriched in myelin basic protein (Johnson, 2012). Both IFNβ and GA are delivered by self-administered injections, varying in frequency from once daily to once weekly. Subsequent studies have shown that IFNβ and GA are also effective in reducing the risk of conversion from CIS to definite MS (Kappos et al., 2006; Pichl, 2014).

According to current guidelines, both IFNβ and GA are indicated as first-line drugs for the treatment of RRMS and for patients with CIS with a high risk of conversion to MS. Recent head-to-head studies of IFNβ and GA have demonstrated a similar efficacy for reducing relapse frequency in RRMS (O'Connor et al., 2009; Mikol et al., 2008). Both IFNβ and GA have a favorable long-term safety and tolerability profile.

Until recently there have been two main options for patients displaying breakthrough disease on first-line drugs: chemotherapy drugs such as cyclophosphamide and natalizumab. Mitoxantrone is a cytostatic agent used in cancer therapy that also displays some degree of immunosuppressive activity. The largest randomized trial in MS included 194 patients with worsening of RRMS or SPMS and demonstrated a moderate effect on relapses and disease progression (Hartung et al., 2002; Piehl, 2014). Cyclophosphamide, a commonly used anticancer agent and immunosuppressive drug, has been shown to be effective in patients with active inflammation who have not responded to first-line therapy (Stankiewicz et al., 2013).

The management of breakthrough disease was revolutionized with the approval of natalizumab for highly active RRMS in 2006. Natalizumab is a humanized monoclonal antibody that binds to the α4-integrin of VLA-4, a surface marker present on immune cells, and is given by monthly IV infusion. Antibody binding impedes the transmigration of leukocytes across the BBB. The marketing approval for natalizumab was based on two pivotal phase III trials: AFFIRM in which natalizumab monotherapy was compared to placebo and SENTINEL in which the antibody was combined with once-weekly IFNβ (Piehl, 2014; Polman et al., 2006; Rudick et al., 2006). Trial data demonstrated a 67% decrease in annual relapse rate and a reduction in gadolinium-enhancing lesions by >90%. The proportion of patients displaying progression of disability was reduced by 40%; this finding represented an important step in terms of clinically relevant outcome measures compared to previous disease-modifying drugs. Long-term follow-up in the postmarketing setting demonstrated that natalizumab is generally well tolerated with beneficial effects on other treatment-related outcome measures (Holmen et al., 2011). Unfortunately, natalizumab increases the risk of progressive multifocal leukoencephalopathy (PML), a serious and potentially lethal opportunistic brain infection caused by the JC virus. For this reason, natalizumab is approved only as monotherapy, mainly in MS patients displaying breakthrough disease on first-line drugs. The risk of PML increases with >2 years of treatment and is also affected by prior use of immunosuppressant. Therefore, natalizumab provides a good example of how the risk-benefit profile must be determined at an individual level, with the drug representing an effective option with a good safety profile in JC virus serology-negative patients with high disease activity (Piehl, 2014).

Fingolimod (Gilenya) was the first oral treatment approved for RRMS and is also the first in class for drugs targeting sphingosine-1-phosphate receptors (S1PRs). This fungal derivative and structural analog of sphingosine acts as an antagonist on S1PRs. The primary mechanism of action of fingolimod in MS is thought to be the sequestration of lymphocytes in lymph nodes due to blocking of S1PR1, which is present at variable levels on different types of immune cells (Matloubian et al., 2004). Fingolimod was approved based on two pivotal phase III studies: FREEDOMS to assess fingolimod versus placebo over 2 years (Kappos et al., 2010) and TRANSFORMS with the same doses of fingolimod versus weekly IFNβ over 1 year (Cohen et al., 2010). These two studies showed a reduction in relapse rate of 60% compared to placebo and of 52% compared to IFNβ. Secondary outcome measures in the FREEDOMS study were also significant, and close to 90% of treated patients were free of gadolinium-enhancing lesions compared to 65% in the placebo group. The deaths of two study subjects in the TRANSFORMS study, one due to herpes encephalitis and the other due to a primary varicella zoster virus (VZV) infection, caused concerns over the risk of opportunistic infections. It is therefore recommended that VZV immunity status should be determined before initiating therapy. Another safety concern is bradycardia and heart block; therefore, care should be taken in patients with known heart conditions, especially as several cases of sudden death in MS patients treated with fingolimod have been reported (Pelletier and Hafler, 2012; Piehl, 2014).

Teriflunomide is a selective immunosuppressant with antiinflammatory properties that exerts its therapeutic effect by blocking the mitochondrial enzyme dihydroorotate dehydrogenase (DHO-DH), a necessary enzyme for de novo pyrimidine synthesis in proliferating lymphocytes. In the TEMSO trial, once-daily doses of 7 and 14 mg teriflunomide were compared to placebo over 2 years in patients with RRMS (O'Connor et al., 2011). The results showed a 31% reduction in annual relapse rate with both doses compared to placebo, while an effect on disability progression was only observed with the higher dose (26% relative risk reduction). The primary MRI outcome measure, increase in total lesion volume, was reduced by 39% and 67% in the 7- and 14-mg dose groups, respectively, versus placebo (Miller et al., 2013). The proportion of patients free from gadolinium-enhancing lesions was 51% in the 7-mg group and 64% in the 14-mg group, compared with 39% in the placebo group. A similar design was used in the TOWER study, in which teriflunomide at 14 mg daily reduced annual relapse rate by 36% and disability progression by 31% compared with placebo (Wolinsky et al., 2013). A smaller phase III study, TENERE, comparing teriflunomide with three times weekly IFNβ1a, failed to meet its primary outcome measure the proportion of patients experiencing treatment failure (Piehl, 2014; Vermersch et al., 2014).

Dimethyl fumarate (also known as BG-12) is the methyl ester of fumaric acid, an intermediate in the citric acid cycle. Different combinations of fumaric acid esters, including DMF, have been used for topical and/or oral use for many years, mainly for the treatment of psoriasis. Two different DMF regimens were examined in phase III trials, 240 mg twice or three times daily. In the phase III DEFINE study, annual relapse rate was reduced to 0.17 and 0.19 in the twice- and three times-daily DMF groups, compared with 0.36 in the placebo group, corresponding to relative reductions of 53% and 48%, respectively, in the active arms (Gold et al., 2012). The phase III CONFIRM trial also included an active, open-label comparator arm with GA, although the study was not powered to compare the two active treatments directly (Fox et al., 2012). The relative relapse rate reduction compared to placebo was 44% for twice-daily DMF, 51% for three times-daily DMF, and 29% for GA. Several MRI measures were also significantly affected by treatment, including a reduction in new or enlarging T2 hyperintense lesions by 85% and 74% with DMF twice and three times daily, respectively.

Alemtuzumab is a humanized monoclonal antibody directed against CD52, a surface antigen expressed on various subtypes of lymphocyte and originally approved for the treatment of hematological malignancies. In two parallel phase III studies, CARE-MS I and II, the effect of alemtuzumab compared to IFNβ was examined over 2 years, with relapse rate and time to 6-month sustained accumulation of disability as coprimary endpoints (Cohen et al., 2012; Coles et al., 2012). Treatment-naïve patients with a disease duration of <5 years were eligible for enrollment in CARE-MS I, whereas previously treated patients with <10 years disease duration were also included in CARE-MS II (Piehl, 2014). In CARE-MS I, alemtuzumab demonstrated a 55% reduction in relapses and no significant reduction in risk of sustained progression compared to IFNβ. A similar reduction in relapses of 49% and a 42% reduced risk of sustained disability were observed in CARE-MS II. Both studies confirmed a relatively high rate of autoimmune side effects and an increased risk of herpes infections. Taken together, alemtuzumab demonstrates very good efficacy, but a high rate of adverse events requiring close monthly monitoring for at least 4 years after the last infusion.

The current FDA-approved disease-modifying drugs and a description of their clinical trials are summarized in Table 30.1. Several outcome measures have been used in recent RRMS clinical trials to quantify clinical disease activity including relapses, MRI disease burden (lesion counts, lesion, and brain volumes), and neurodegeneration (optical coherence tomography). The selection of the outcome measures and the results of these trials should be interpreted with an understanding of the following. (1) MS is a heterogeneous disease, and inclusion criteria should be carefully reviewed when results are applied to individual patients. (2) MS evolves over time from a predominantly inflammatory disease to a secondary progressive or neurogenerative process. As such, a clinical outcome measure may have different efficacy in the early stages of the disease compared to the advanced stages. (3) A surrogate may seem to indicate a positive correlation with the intervention, but the relationship could be attributed to other mechanisms. If the measure is not something that is in the causal pathway for MS, it may mislead researchers about the efficacy of the treatment or progression of the disease (Lavery et al., 2014).

Alzheimer Disease

AD is a progressive neurodegenerative disorder first described over 100 years ago but still without disease-modifying therapy. Memory loss is the most notable symptom at the early stage, but as the disorder advances, patients experience difficulties with language and perception, followed by neuropsychiatric and behavioral abnormality, muscle mass loss, and mobility deterioration. The global prevalence of dementia for people over the age of 60 is estimated as high as 46.8 million in 2015, and the figure is forecasted to double every 20 years (Reitz et al., 2011), indicating that AD has become a modern epidemic.

Definitive diagnosis of AD is usually reached only upon postmortem examination or brain biopsy, demonstrating the presence of amyloid and tangles, although PET imaging of amyloid and tau plus CSF examination allow for diagnosis in life. The clinical diagnosis of AD is based on symptom presentation and is often supported by biological testing which can aid in diagnosis, especially differentiating AD from other dementias. Among the biological changes that can be used in the diagnosis of AD, brain atrophy in the hippocampus and entorhinal cortex is well described in the earliest stages. Similarly, bilateral hypometabolism in the temporal lobe, parietal cortex, and posterior cingulate cortex are consistently observed in AD with FDG PET. Recently, disease-specific ligands for use with PET, such as the amyloid-specific Pittsburgh compound B (PIB), florbetaben, and AV-45, have become available for use in AD research and appear to have good specificity and sensitivity for AD diagnosis. Analysis of CSF proteins can be used in diagnostic assessment, although consistency in cutoff points and protein level measures across laboratories is still lacking. Decreased levels (<190 pg/mL) of CSF Aβ are expected in AD, hypothetically due to the accumulation of Aβ into plaques in the brain. Concomitant increases in CSF tau and hyperphosphorylated tau occur and are good predictors of AD diagnosis.

TABLE 30.1 FDA-Approved Disease-Modifying Drugs and Their Clinical Trials

Drug treatment	Study name/group	Publication year	Primary outcome
Interferon beta-1a (Avonex)	Multiple sclerosis collaborative research group	1990	Worsening in disability defined as deterioration by 1.0 point on EDSS
Interferon beta-1b (Betaseron)	Interferon beta-1b multiple sclerosis study group	1993	(1) Annualized relapse rate; (2) proportion of relapse-free patients
Copolymer 1 (Glatiramer acetate, Copaxone)	Copolymer 1 multiple sclerosis study group	1995	Average number of relapses at 2 years
Interferon beta-1a (Rebif)	PRISMS	1998	Number of relapses
Interferon beta-1a (Avonex)	CHAMPS	2000	Conversion to CDMS
Interferon beta-1a (Rebif)	ETOMS	2001	Conversion to CDMS
Interferon beta-1a (Rebif)	EVIDENCE	2005	(1) Annualized relapse rate; (2) proportion remaining relapse free at 2 years
Interferon beta-1b (Betaseron)	BENEFIT	2006	(1) Time to conversion to CDMS; (2) time to McDonald-defined multiple sclerosis
Interferon beta-1a (Avonex)	CHAMPIONS	2006	Conversion to CDMS
Natalizumab (Tysabri)	AFFIRM	2006	(1) Annualized relapse rate; (2) progression of disability as measured by sustained worsening of EDSS at 12 weeks
Natalizumab and interferon beta-1a (Tysabri and Avonex)	SENTINEL	2006	(1) Annualized relapse rate; (2) progression of disability as measured by sustained worsening of EDSS at 12 weeks
Interferon beta-1a and glatiramer acetate (Rebif and Copaxone)	REGARD	2008	Time to relapse at 96 weeks
Interferon beta-1b and glatiramer acetate (Betaseron and Copaxone)	BECOME	2009	Number of combined active lesions in the first year
Fingolimod (Gilenya)	TRANSFORMS	2010	Annualized relapse rate
Fingolimod (Gilenya)	FREEDOMS	2010	Annualized relapse rate accompanied by change in EDSS
Teriflunomide (Aubagio)	TEMSO	2011	Annualized relapse rate
BG-12 (Dimethyl fumarate, Tecfidera)	DEFINE	2012	Proportion of patients who had a relapse at 2 years
BG-12 and glatiramer acetate (Tecfidera, Copaxone)	CONFIRM	2012	Annualized relapse rate at 2 years
Alemtuzumab and interferon beta-1a (Lemtrada and Rebif)	CARE MS I	2012	(1) Annualized relapse rate; (2) progression of disability as measured by sustained worsening of EDSS at 6 months
Alemtuzumab and interferon beta-1a (Lemtrada and Rebif)	CARE MS II	2012	(1) Annualized relapse rate; (2) progression of disability as measured by sustained worsening of EDSS at 6 months

To date, the US FDA has approved five drugs for the treatment of AD. Most studies suggest that these agents provide only symptomatic improvement in AD, and the effort to discover treatments capable of altering the natural history of AD is rigorous. Therefore, clinical trials of new therapies in AD in the coming years will continue to be a mainstay of AD research.

In 1974 Drachman and Leavitt suggested that memory was related to the cholinergic system and was age dependent. Around the same time it was demonstrated that the severity of dementia in AD was correlated with the extent of cholinergic loss in the nucleus basalis of Meynert (Davis et al., 1995). AD was conceptualized as a cholinergic disease, similar to the way that PD is considered a dopaminergic disease. The cholinergic hypothesis drove drug development and trials throughout the 1980s and 1990s. From animal and human studies, administration of cholinesterase inhibitors stimulated memory and learning process. Agents as donepezil, galantamine, and rivastigmine inhibit the catabolic enzyme AChE to maintain or at least to delay the decrease of acetylcholine levels in synaptic clefts and are FDA approved for patients with mild-to-moderate AD. Donepezil is a highly selective and reversible central AChE inhibitor, rivastigmine inhibits both AChE and butyrylcholinesterase, and galantamine is a reversible and competitive inhibitor of AChE and is also an allosteric modulator by binding to cholinergic nicotinic receptors at an additional site. Randomized trials of cholinesterase inhibitors were parallel group 3- and 6-month studies. Subjects were assigned to therapy or placebo, and cognitive performance was assessed with the AD Assessment Scale, which included the cognitive subscale (ADAS-cog) and the noncognitive (ADAS-noncog) portion. Since these initial trials, the ADAS-cog has remained the cognitive scale used in most trials conducted in mild-to-moderate AD. Despite being widely prescribed, the magnitude of clinical efficacy of the three compounds remains modest and symptomatic, and usually for a limited period, without evidence of substantial impact on disease course (Stella et al., 2015).

The neurodegenerative process of AD comprises several neurotoxic mechanisms including abnormalities of excitatory amino acids such as glutamate, the most widespread excitatory neurotransmitter. Its excitotoxicity is closely linked to chronic calcium (Ca^{2+}) influx through Ca^{2+} channels of NMDA receptors and ultimately leads to neuronal Ca^{2+} overload, a phenomenon involved in cell damage and death. Memantine is an uncompetitive glutamatergic receptor antagonist that engenders a blockage or regulation of NMDA receptors, providing a more suitable neurotransmission required for synapses functioning (Bullock, 2006).

Much of the current focus in AD research revolves around the two hallmarks of disease pathology, the neuritic plaque and the neurofibrillary tangle (NFT). Neuritic plaques are largely composed of the fibrillogenic 42-amino acid length form of the beta amyloid protein (Aβ-42). NFTs are composed primarily of hyperphosphorylated aggregations of the microtubule-associated protein tau. Most attempts at developing disease-modifying drugs have focused on the Aβ cascade (Fig. 30.2), assuming that deregulation in amyloid precursor protein (APP) leads to accumulation of Aβ1-42 peptide in the brain parenchyma. Cleavage mechanisms of APP implicate two pathways: the nonamyloidogenic (or secretory pathway) and the amyloidogenic pathway. In the nonamyloidogenic pathway, APP is first cleaved by α-secretase, releasing a soluble *N*-terminal (sAPPα) and a *C*-terminal fragment, which is cleaved by γ-secretase to originate a smaller *C*-terminal fragment. In the amyloidogenic pathway, APP is alternatively cleaved by β-secretase releasing a smaller *N*-terminal fragment (sAPPβ) and a longer *C*-terminal fragment that contains the full amyloidogenic sequence of amino acids. A further cleavage of APP by γ-secretase yields the Aβ peptide. Aβ1-42 is prone to aggregation, being the most neurotoxic Aβ peptide in the AD pathogenesis (Recuero et al., 2004). Aβ oligomers are considered the most toxic forms of the amyloid derivatives; they interact with neurons and glial cells leading to several pivotal mechanisms that ultimately lead to cell death.

Over the past years, passive and active immunotherapies have been developed with the main purpose of changing the course of AD pathology through the efflux of Aβ deposits from the brain to the peripheral circulation, or the reduction of parenchymal Aβ peptide aggregation. With respect to passive immunization, the monoclonal antibodies solanezumab, bapineuzumab, gantenerumab, crenezumab, and ponezumab are currently being tested with the purpose of inducing the clearance of brain Aβ burden and to improve clinical outcomes with the expectation of favorable impact on cognition and functionality (Moreth et al., 2013). Preliminary results from randomized clinical trials involving different compounds have achieved these goals with acceptable safety, while others brought disappointing neurobiological and clinical results or were related to serious adverse occurrences.

Strongly encouraging, neurobiological data from the first generation of active immunotherapy were tested in experimental models, reducing amyloid burden from brain parenchyma and providing the rationale for subsequent investigations in AD patients (Winblad et al., 2014). AN1792 was the first active Aβ immunotherapy in clinical trials with patients at mild-to-moderate AD disease stages. Although this compound was effective in reducing Aβ from brain parenchyma and in providing some cognitive benefits, the trial was interrupted at phase II because only 19.7% of patients presented antibody response to this treatment, and 6% developed serious cytotoxic T cell reactions and acute meningoencephalitis (Gilman et al., 2005).

Approaches based on other disease-modifying strategies are ongoing. For instance, there is a growing knowledge body on compounds involving inhibition of γ-secretase and β-secretase to preclude amyloidogenic cleavage of APP.

Tau is an endogenous microtubule-associated protein critical to axonal transport and neuronal health and function. The hyperphosphorylation of tau can occur through activity of a variety of kinases, but glycogen synthase kinase 3β appears to be a primary mechanism for tau hyperphosphorylation. Active and passive tau-directed immunotherapies involving

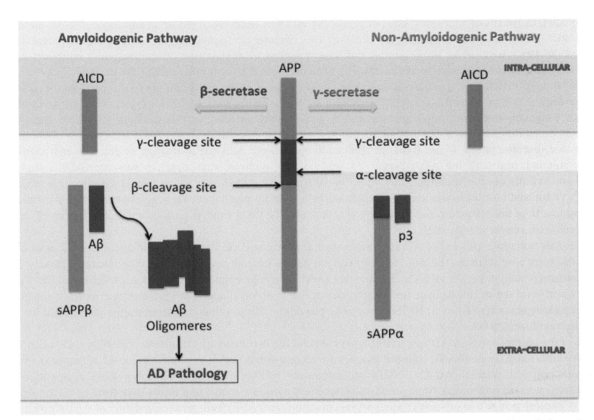

FIGURE 30.2 The Amyloid cascade. Cleavage mechanisms of amyloid precursor protein (APP) implicate two and mutually exclusive pathways: the nonamyloidogenic (or secretory pathway) and the amyloidogenic pathway. In the nonamyloidogenic pathway, APP is first cleaved by α-secretase, releasing a soluble *N*-terminal (sAPPα) and a *C*-terminal fragment, which is cleaved by γ-secretase to originate a smaller *C*-terminal fragment. In the amyloidogenic pathway, APP is alternatively cleaved by β-secretase releasing a smaller *N*-terminal fragment (sAPPβ) and a longer *C*-terminal fragment that contains the full amyloidogenic sequence of amino acids. A further cleavage of APP by γ-secretase yields the Aβ peptide. Aβ1-42 is prone to aggregation, being the most neurotoxic Aβ peptide in the Alzheimer disease (AD) pathogenesis.

experimental models are in progress with promising results. Reducing the levels of tau oligomers, preventing tau aggregation, and blocking hyperphosphorylation or microtubule destabilization could preclude the neurofibrillary pathologies (Herrmann and Spires-Jones, 2015).

In addition to the two hallmarks and neurotransmitter system impairment, there are several other features found in AD, including inflammation, oxidative stress, mitochondrial dysfunction, and neurotrophin deficiency. These aspects provide new perspectives in developing AD treatments. The innate immune system may play an important role in AD, and therapy aimed at microglia or the peripheral innate immune response has shown promise in animal models (Frenkel et al., 2008).

Amyotrophic Lateral Sclerosis

ALS is an adult-onset neurodegenerative disorder characterized by selective motor neuron death. Both upper motor neurons in the motor cortex and lower motor neurons in the brain stem and in the ventral horn of the spinal cord are affected. Patients develop a progressive muscle phenotype characterized by spasticity, hyperreflexia, fasciculations, muscle atrophy, and paralysis. ALS is usually fatal within 3−5 years after the diagnosis. No treatment prevents, halts, or reverses the disease, although a small delay in mortality occurs with the drug riluzole (Bensimon et al., 1994).

The incidence of ALS is ~2/100 000/year, in most cases (~90%), there is no family history of ALS, and these patients are classified as sporadic ALS patients. However, a clear family history is present in ~10% of ALS patients, and these patients suffer from familial ALS (FALS). Mutations in more than 10 different genes are known to cause FALS. The most common ones are found in the genes encoding superoxide dismutase 1 (SOD1; ~20%), fused in sarcoma (FUS; 1−5%), and TDP-43 (TARDBP; 1−5%). Recently, a hexanucleotide repeat expansion (GGGGCC)n in the C9ORF72 gene was identified as the most frequent cause of FALS (~40%) in the Western population (Renton et al., 2011).

The precise cause of selective motor neuron death in ALS is unknown. Many pathogenic mechanisms have been proposed such as excitotoxicity, oxidative damage, mutant proteins, immune dysregulation, mitochondrial dysfunction, and growth factors. Advances in understanding the biology of motor neuron death in ALS has led to more than 32 compounds being tested in phase II/III clinical trials in ALS during the past two decades. Discovery of new genes associated with ALS led to the hypothesis that aberrant RNA metabolism has a role in the pathogenesis of both familial and sporadic ALS.

A wide range of in vitro and in vivo models are available to both study disease biology and screen therapeutic compounds. Since the generation of the hSOD1G93A transgenic ALS mouse model, which successfully replicated many aspects of the disease pathology observed in patients with ALS, including progressive degeneration of both spinal and corticospinal motor neurons and an increase in astrogliosis, it has become one of the major in vivo tools for drug screening. However, the validity of life-span extension in mice as a readout for success in clinical trials is strongly debated as a result of ineffective translation from animal models to clinical efficacy. Although important information was gained from using disease models, their utility as a tool to inform the potential efficacy of compounds for patient survival has not been proven.

Recently, skin fibroblasts from people with ALS were used to produce induced pluripotent stem cells (iPS) that are capable of differentiating into motor neurons and glia (Poppe et al., 2014). These advances in stem cell technology offer potential motor neuron models of sporadic ALS that can help understand disease pathogenesis and screen new drugs.

The primary goals of most clinical trials in ALS are to slow disease progression as measured by either function or survival. Functional scales include measures of strength, pulmonary function, and a questionnaire called the ALS functional rating scale-revised (ALSFRS-R). In addition to muscle weakness, people with ALS suffer from many other ALS-related symptoms. Improving ALS symptomatic management is highly important to ALS patients; however, there have been very few studies to determine best approaches to manage most of the symptoms of ALS.

Clinical trials in ALS require a degree of certainty about the diagnosis based on the El Escorial diagnostic criteria. There are currently no reliable blood, CSF, or imaging biomarkers for either disease diagnosis or progression. To assess efficacy of an intervention, phase II or III clinical trials often enroll people early in the disease course. This would include people classified as possible or probable or definite by El Escorial criteria. The reasons for these criteria include wanting to treat people as early as possible in their disease course, minimizing patient to patient variability in rate of disease progression, and for studies whose primary outcome measure is function, ensuring that people can complete the study.

Even though ALS is highly heterogeneous among patients, most phase II or III clinical trials in ALS have not addressed phenotypic or genotypic heterogeneity.

Recent discoveries in the ALS field resulted in a paradigm shift, which will have an important impact on the development of new therapeutic strategies. It becomes more and more clear that many genes linked to ALS encode for RNA/DNA-binding proteins and/or influence RNA metabolism. The identification of mutations in genes for TDP-43, FUS, and C9ORF72 led to the idea that aberrant RNA metabolism plays a crucial role in ALS pathogenesis (reviewed in Poppe et al., 2014). These insights shifted the focus from "ALS as a proteinopathy" to "ALS as a ribonucleopathy."

A great deal of attention is focused on TDP-43, as the protein is present in ubiquitinated cytoplasmic neuronal inclusions in most patients with ALS (Lemmens et al., 2010). TDP-43 is mainly present in the nucleus where it is involved in transcription, RNA splicing, and micro-RNA processing (Neumann et al., 2006). The protein can shuttle between the nucleus and the cytoplasm. In the cytoplasm, TDP-43 is associated with transport granules where mRNA is transported to the sites of translation. In conditions of cellular stress, TDP-43 is incorporated in cytoplasmic stress granules until stress is resolved. ALS-associated mutations seem to induce a shift of TDP-43 from the nucleus to the cytoplasm, where aggregates are formed. Like TDP-43, FUS is an RNA-binding protein involved in different processes of RNA metabolism. It is mainly localized in the nucleus, but it is also present in the cytoplasm. Most ALS-linked mutations disrupt nuclear import, inducing the mislocalization of FUS in stress granules and in inclusions. These FUS-positive inclusions are observed in brain and spinal cord of patients with FUS mutations (Lee et al., 2011).

Both TDP-43 and FUS target more than 5000 mRNAs emphasizing their important role in normal cell functioning. However, it is still a matter of debate whether the depletion of these proteins from the nucleus (loss of function), the increased cytoplasmic aggregation propensity (gain of function), or both are involved in disease pathogenesis.

In 2012, a major breakthrough in ALS genetics was the discovery of hexanucleotide repeat expansions in the C9ORF72 gene as a common cause of ALS. The physiological function of the gene product is still unknown, but the protein is structurally related to DENN proteins, which are GDP/GTP exchange factors implicated in membrane trafficking (Vance et al., 2009). The mechanism by which these repeat expansions can cause disease evolution is intriguing the ALS research field (Poppe et al., 2014).

Hopefully, these new insights in TDP-43, FUS, and C9ORF72 pathology will result in the development of a completely new class of therapeutic strategies; however, better insights into the exact pathogenic mechanisms are essential before these findings can be successfully translated into new therapies. We have recently identified that the innate immune system may play an important noncell autonomous role in ALS related to increased mir-155 expression in microglia and in peripheral monocytes (Butovsky et al., 2012).

In summary, ALS is an orphan neurodegenerative disorder that has one available treatment with minimal impact on survival. In addition to the experience in clinical management and clinical trial design and conduct, an enormous amount of new and promising information about ALS genetics, pathophysiology, and biomarkers has become available.

Glioma

Gliomas are the most frequent and most devastating brain malignancies. More than half of all gliomas are glioblastomas (GBMs), the most aggressive and infiltrating type. Despite aggressive use of surgery, radiotherapy (RT), and chemotherapy, GBMs are invariably associated with tumor recurrence and overall poor prognosis. Historically, the overall survival of patients with GBMs is 12—15 months and has not significantly changed in the past decades.

Extensive clinical research conducted to determine optimal therapeutic approaches highlights the significance of phase III trials in creating the standard treatment for GBM patients.

The phase III EORTC trial created standard of care for GBMs by demonstrating that concurrent and adjuvant chemoradiation with temozolomide (TMZ) conferred a 2.5-month survival benefit as compared to RT alone (Levine et al., 2013). This led to the current standard of care for newly diagnosed disease that includes maximal safe resection, followed by 6 weeks of radiation and concurrent daily-TMZ chemotherapy, followed by at least 6 months of adjuvant TMZ.

However, it should be borne in mind that the improvement of combined RT/TMZ is quite modest and the median survival is still limited.

Immunotherapy for glioma immunology is a rapidly developing and expanding area for therapeutics in cancer treatment. Harnessing the immune system to attack tumors holds the promise to specifically eradicate cancer cells while leaving normal tissues unharmed.

Immunotherapy for gliomas has been intensively investigated in the past decade. Many different approaches have been investigated, including developing personalized vaccines based on individual patients' tumors, either by utilizing them as antigen sources for potent immunostimulatory cells such as dendritic cells (Stupp et al., 2005) or by harvesting immunostimulatory chaperone protein/tumor peptide complexes (Prins et al., 2011). An alternative strategy is a vaccine directed against specific tumor antigens such as EGFRvIII (See et al., 2011). This has the advantage of not being limited by the availability of fresh tumor; nevertheless, clinical trials did not show promising results presumably because not all GBM patients express the antigens in question, and this approach is vulnerable to immune-editing by the tumor. To date, a number of small but promising clinical trials of these various forms of GBM immunotherapy have been conducted, and several larger clinical trials are currently underway.

Oncolytic viruses are another novel option for patients with high-grade gliomas. Viruses are chosen for their ability (either innate or genetically engineered) to infect and kill cancer cells while leaving normal cells unharmed. Oncolytic viruses proposed for glioma therapy include HSV-1, adenovirus, reovirus, measles, and Newcastle virus. To date, oncolytic viruses have been largely delivered locally to gliomas at the time of surgery (Sampson et al., 2010; Forsyth et al., 2008). Oncolytic measles virus therapy takes advantage of the potential for certain measles virus strains to infect and kill tumor cells while leaving normal cells relatively spared. Phase I trials have confirmed safety in Rhesus macaques, with no toxicity seen 36 months from treatment (Chiocca et al., 2004). This approach is currently in phase I trial in recurrent glioma patients. Another viral approach has been the use of adenovirus as a substrate for targeted gene therapy in the setting of high-grade glioma. This represents a gene therapy approach more than a true oncolytic virus, but many of the principles are similar. An adenovirus encoding a prodrug-converting enzyme is locally delivered at surgery and patients subsequently receive systemic IV ganciclovir therapy, then the nontoxic prodrug is converted by adenovirally delivered HSV-thymidine kinase into an active cytotoxic nucleoside analog within tumor cells. In an international multicenter, randomized, open-label, phase III study featuring 150 newly diagnosed GBM patients randomized to resection with intraoperative injection of the agent of study versus resection and standard of care, no significant difference in overall survival was seen between the groups (Myers et al., 2008). The study demonstrated that adenovirus-mediated gene therapy was clinically safe and large-scale virus production was feasible. It remains uncertain whether this relative lack of impact on overall survival reflects poor delivery of the virus to the tumor or limited efficacy of the gene therapy approach itself.

Although advances in technology have led to the introduction of many promising new therapeutic agents and modalities for brain tumor treatment, these agents' efficacy has often been limited by our ability to deliver these agents across the BBB. Combining novel agents or therapeutic strategies with advances in surgical technique to deliver these agents locally within brain tumors has provided a new avenue for treatment in neurooncology.

Biodegradable wafers impregnated with chemotherapy agents and placed within resection cavities at the time of surgery have been tested in the treatment of gliomas. Randomized controlled trials comparing postresection implantation of carmustine (BCNU)-impregnated polymer wafers with resection alone for recurrent malignant glioma revealed a modest increase in survival in the absence of TMZ. Median survival was prolonged by 8 weeks in patients receiving the chemo wafer therapy compared to 23 weeks in patients undergoing resection alone (Westphal et al., 2013).

One limitation of chemo wafers is the fact that they rely on diffusion through tissue, which is limited by the degree to which an agent is able to permeate the surrounding parenchyma. To address this, several investigators pioneered a technology termed convection-enhanced delivery (CED) (Brem et al., 1995). The technology consists of stereotactically inserting one to three small-diameter silastic tubes in neoplastic tissue or surrounding brain. Agents are then delivered via slow continuous infusions for several days. The continuous infusion creates convection currents within brain parenchyma that ultimately results in a large volume of distribution. Ideally, a single catheter can perfuse the volume of an entire lobe. Clinical trials evaluating CED to deliver other agents and using improved catheter placement technologies are ongoing.

Acute Ischemic Stroke

Stroke is one of the leading causes of death in the United States and the leading cause of adult disability. Cerebral infarction is the result of severe enough ischemia for a sufficient time to result in cell death. The progression toward infarction includes protein synthesis failure, anaerobic metabolism, release of neurotransmitters, energy failure, and ultimately anoxic depolarization. At the initiation of ischemia, there is a core of infarction surrounded by viable but impaired tissue (penumbra). Expeditious reperfusion can save the impaired tissue and return it to normal function. Failure to save this tissue results in permanent structural changes and ultimately necrosis of neuronal cells. The goal of acute ischemic stroke treatment is to improve patient outcome months and years after the event; this objective can be achieved by early recanalization of an occluded artery with use of a thrombolytic drug or neurothrombectomy device to salvage ischemic tissue at risk of infarction and reduce the size of the infarct, which leads to a better functional outcome.

Volume of infarction does relate to ultimate outcome, but the relationship is nonlinear, and depends on location, age, and other factors.

An alternative or complementary strategy to recanalization is neuroprotection aimed at interruption of the ischemic cascade for tissue preservation. Numerous aspects such as energy supply failure, membrane depolarization, excitatory amino acid release, intracellular calcium accumulation, free radical elaboration, and cellular edema can be targeted (Bobo et al., 1994). Despite successes in animal models, no neuroprotective agent has yet been successful in humans.

Because ischemia causes time-dependent tissue injury, time is critical in the initiation of acute stroke therapies. Many variables influence individual response to duration of ischemia including collateral flow and age. Functional imaging, originally with PET, but now more commonly with CT perfusion and MR perfusion imaging, can be used to study the ischemic penumbra. When mismatch of infarct and perfusion deficit is identified, salvageable penumbra may exist.

Until recently, the only therapy with proven efficacy was intravenous alteplase (Hossmann, 1994). This drug was initially shown to be beneficial when therapy was initiated within 3 h after stroke onset in the two landmark National Institute of Neurological Disorders and Stroke (NINDS) trials (Fisher and Saver, 2015). Efficacy within 3 h was also supported by the subgroup results of five additional trials (Wardlaw et al., 2012) and the largest trial, the Third International Stroke Trial (Sandercock et al., 2012). Although the European Cooperative Acute Stroke Study III (ECASS-3) trial results supported the efficacy of alteplase when initiated within 3–4.5 h of stroke onset (Hacke et al., 2008), the extent of reduction in disability was slight and less robust than the effect when given between 0 h and 3 h.

Another widespread approach to revascularization therapy for acute ischemic stroke is the use of endovascular devices to remove clots (neurothrombectomy). A growing number of such devices have been tested in patients with acute ischemic stroke: aspiration devices initially, and more recently the stent retrievers and a direct aspiration first pass technique (Turk et al., 2014). Most recently, four concurrently conducted randomized trials [MR CLEAN (Berkhemer et al., 2015), ESCAPE (Goyal et al., 2015), EXTEND IA (Campbell et al., 2015), and SWIFT PRIME (Saver et al., 2015)] that tested neurothrombectomy reported them to be superior to medical treatment alone. In all of these superiority trials, a vascular occlusion had to be documented by CT or magnetic resonance angiography before enrollment and, except in the MR CLEAN trial, a small baseline region of ischemic injury should be seen on plain noncontrast CT or CT perfusion. In all four trials, patients were randomly assigned to best medical treatment versus endovascular therapy. Symptomatic intracranial hemorrhage is a highly feared complication of acute stroke recanalization therapies because it carries with it a high risk of poor outcome including death. Pooled analysis of these trials is needed to analyze benefits and risks in patient subgroups, but collectively they suggest that neurothrombectomy has become a proven intervention for the treatment of acute ischemic stroke.

Recanalization, the reestablishment of arterial patency can be assessed by angiography and indirectly by transcranial Doppler ultrasound (Alexandrov et al., 2004), and graded.

A variety of measures can be used to assess recovery post stroke, and there is no consensus about which measure or what cutoffs to use. The most common efficacy endpoint measures in acute stroke therapy trials are the modified Rankin Scale and the NIHSS (Fisher et al., 2007), most often performed at 3 months post stroke. Cognitive outcomes are also gaining recognition as important poststroke measures. No single measure of cognitive function post stroke is accepted, and measurements are complicated by aphasia.

As mentioned previously, neuroprotective agents targeting different pathways involved in ischemic cell injury including excitotoxicity, extracellular matrix modulation, free radical generation, downstream oxidative damage, and prevention of apoptosis (Grupke et al., 2015) have been investigated, with discouraging clinical results thus far. There are some explanations to the failure of model translation. In most of the animal studies, the animals are either pretreated with the neuroprotective agent, or treatment is started within a very short time frame from a surgical vessel occlusion. The time of infarct and the time to treat are distinct and consistent. This is not so in a heterogeneous human population where the best measurement of the time of infarct is designated "last known normal."

While animal studies typically involve infarction of one vascular territory through a controlled ligation or occlusion of a parent vessel, strokes in human subjects vary widely and will lead to intrinsically widely differing outcomes.

One other striking mechanistic difference between animal studies and human trials is the course of the affected parent vessel. In animal studies, the parent vessel is mechanically occluded for some length by a form of physical obstruction. After a predetermined and uniform period of time, this obstruction is removed and the infarcted bed is reperfused. In these subjects, administered medicines may have more bioavailability to the ischemic bed than they would in a human stroke patient who continues to have a permanent occlusion.

Finally, animal studies typically use healthy animals of similar ages. The human patients are of widely varying ages and have a nonuniform medical comorbidity profile. This nonuniformity among patients complicates clinical studies, introducing confounding variables. Furthermore, small animal models may have higher tolerance for ischemia than the human brain.

CONCLUSION

Neurologic diseases are a major cause of worldwide disability, and an expected increase in life expectancy will lead to a significant increase in the burden of neurologic diseases, hence the need for effective therapy is growing. Unfortunately, clinical trials in CNS diseases have a relatively poor success rate (apart from relapsing forms of MS), probably owing to the complexity of the CNS, the translational barrier from animal model to humans, the challenge in selecting endpoints that capture the multiple domains involved in neurological disease, the BBB, and the paucity of established biomarkers.

Recent discoveries have led both to the development of novel therapeutic approaches that are currently under clinical investigation and to advances in technology that allows effective drug delivery to CNS. The combination of the two gives hope to both relieve of symptoms and developing effective disease-modifying therapy.

REFERENCES

Akhtar, A.Z., Pippin, J.J., Sandusky, C.B., 2008. Animal models in spinal cord injury: a review. Rev. Neurosci. 19, 47–60.

Alexandrov, A.V., Molina, C.A., Grotta, J.C., et al., 2004. Ultrasound-enhanced systemic thrombolysis for acute ischemic stroke. N. Engl. J. Med. 351, 2170–2178.

Andlin-Sobocki, P., Jönsson, B., Wittchen, H.U., Olesen, J., June 2005. Cost of disorders of the brain in Europe. Eur. J. Neurol. 12 (Suppl. 1), 1–27.

Davis, R.E., Doyle, P.D., Carroll, R.T., Emmerling, M.R., Jaen, J., March 1995. Cholinergic therapies for Alzheimer's disease. Palliative or disease altering? Arzneimittelforschung 45 (3A), 425–431.

Bensimon, G., Lacomblez, L., Meininger, V., 1994. The ALSRSG. A controlled trial of riluzole in amyotrophic lateral sclerosis. N. Engl. J. Med. 330, 585–591.

Berkhemer, O.A., Fransen, P.S., Beumer, D., et al., 2015. A randomized trial of intraarterial treatment for acute ischemic stroke. N. Engl. J. Med. 372, 11–20.

Bermel, R.A., Rudick, R.A., 2007. Interferon-beta treatment for multiple sclerosis. Neurotherapeutics 4, 633–646.

Bobo, R.H., Laske, D.W., Akbasak, A., et al., March 15, 1994. Convection-enhanced delivery of macromolecules in the brain. Proc. Natl. Acad. Sci. U.S.A. 91 (6), 2076–2080.

Brem, H., Piantadosi, S., Burger, P.C., et al., April 22, 1995. Placebo-controlled trial of safety and efficacy of intraoperative controlled delivery by biodegradable polymers of chemotherapy for recurrent gliomas. The Polymer-brain tumor treatment group. Lancet 345 (8956), 1008–1012.

Bullock, R., January—March 2006. Efficacy and safety of memantine in moderate-to-severe Alzheimer disease: the evidence to date. Alzheimer Dis. Assoc. Disord. 20 (1), 23—29.

Butovsky, O., Siddiqui, S., Gabriely, G., et al., 2012. Modulating inflammatory monocytes with a unique microRNA gene signature ameliorates murine ALS. J. Clin. Invest. 122 (9), 3063—3087.

Calias, P., Banks, W.A., Begley, D., Scarpa, M., Dickson, P., November 2014. Intrathecal delivery of protein therapeutics to the brain: a critical reassessment. Pharmacol. Ther. 144 (2), 114—122.

Campbell, B.C.V., Mitchell, P.J., Kleinig, T.J., et al., 2015. Endovascular therapy for ischemic stroke with perfusion-imaging selection. N. Engl. J. Med. 372, 1009—1018.

Chiocca, E.A., Abbed, K.M., Tatter, S., et al., 2004. A phase I open-label, dose-escalation, multi-institutional trial of injection with an E1B-Attenuated adenovirus, ONYX- 015, into the peritumoral region of recurrent malignant gliomas, in the adjuvant setting. Mol. Ther. 10, 958—966.

Cohen, J.A., Barkhof, F., Comi, G., et al., 2010. Oral fingolimod or intramuscular interferon for relapsing multiple sclerosis. N. Engl. J. Med. 362, 402—415.

Cohen, J.A., Coles, A.J., Arnold, D.L., et al., 2012. Alemtuzumab versus interferon beta 1a as first-line treatment for patients with relapsing-remitting multiple sclerosis: a randomised controlled phase 3 trial. Lancet 380, 1819—1828.

Coles, A.J., Twyman, C.L., Arnold, D.L., et al., 2012. Alemtuzumab for patients with relapsing multiple sclerosis after disease-modifying therapy: a randomised controlled phase 3 trial. Lancet 380, 1829—1839.

De Mattos, R.B., Bales, K.R., Cummins, D.J., et al., July 17, 2001. Peripheral anti-A beta antibody alters CNS and plasma A beta clearance and decreases brain A beta burden in a mouse model of Alzheimer's disease. Proc. Natl. Acad. Sci. U.S.A. 98 (15), 8850—8855.

Dorsey, E.R., Constantinescu, R., hompson, J.P., et al., 2007. Projected number of people with Parkinson disease in the most populous nations, 2005 through 2030. Neurology 68, 384—386.

Drachman, D.A., Leavitt, J., February 1974. Human memory and the cholinergic system. A relationship to aging? Arch. Neurol. 30 (2), 113—121.

Ferri, C.P., Prince, M., Brayne, C., et al., 2005. Global prevalence of dementia: a Delphi consensus study. Lancet 366, 2112—2117.

Fisher, M., Saver, J.L., July 2015. Future directions of acute ischaemic stroke therapy. Lancet Neurol. 14 (7), 758—767.

Fisher, M., Hanley, D.F., Howard, G., et al., 2007. Recommendations from the STAIR V meeting on acute stroke trials, technology and outcomes. Stroke 38, 245—248.

Forsyth, P., Roldán, G., George, D., et al., March 2008. A phase I trial of intratumoral administration of reovirus in patients with histologically confirmed recurrent malignant gliomas. Mol. Ther. 16 (3), 627—632.

Fox, R.J., Miller, D.H., Phillips, J.T., et al., 2012. Placebo-controlled phase 3 study of oral BG-12 or glatiramer in multiple sclerosis. N. Engl. J. Med. 367, 1087—1097.

Frenkel, D., Puckett, L., Petrovic, S., et al., 2008. A nasal Proteosome adjuvant Activates microglia and prevents amyloid deposition Ann Neurol. 63 (5), 591—601.

Geerts, H., 2009. Of mice and men: bridging the translational disconnect in CNS drug discovery. CNS Drugs 23, 915—926.

Ghersi-Egea, J.F., Leininger-Muller, B., Cecchelli, R., et al., 1995. Blood—brain interfaces: relevance to cerebral drug metabolism. Toxicol. Lett. 82—83, 645—653.

Gilman, S., Koller, M., Black, R.S., et al., May 10, 2005. AN1792(QS-21)-201 Study Team. Clinical effects of Abeta immunization (AN1792) in patients with AD in an interrupted trial. Neurology 64 (9), 1553—1562.

Gold, R., Kappos, L., Arnold, D.L., et al., 2012. Placebo-controlled phase 3 study of oral BG-12 for relapsing multiple sclerosis. N. Engl. J. Med. 367, 1098—1107.

Goyal, M., Demchuk, A.M., Menon, B.K., et al., 2015. Randomized assessment of rapid endovascular treatment of ischemic stroke. N. Engl. J. Med. 372, 1019—1030.

Grupke, S., Hall, J., Dobbs, M., Bix, G.J., Fraser, J.F., February 2015. Understanding history, and not repeating it. Neuroprotection for acute ischemic stroke: from review to preview. Clin. Neurol. Neurosurg. 129, 1—9.

Hacke, W., Kaste, M., Bluhmki, E., et al., 2008. Thrombolysis with alteplase 3 to 4.5 hours after acute ischemic stroke. N. Engl. J. Med. 359, 1317—1329.

Hartung, H.P., Gonsette, R., Konig, N., et al., 2002. Mitoxantrone in progressive multiple sclerosis: a placebo-controlled, double-blind, randomised, multicentre trial. Lancet 360, 2018—2025.

Herrmann, A., Spires-Jones, T.J., January 2015. Clearing the way for tau immunotherapy in Alzheimer's disease. Neurochem. 132 (1), 1—4.

Holmen, C., Piehl, F., Hillert, J., et al., 2011. A Swedish national post-marketing surveillance study of natalizumab treatment in multiple sclerosis. Mult. Scler. 17, 708—719.

Hossmann, K.A., 1994. Viability thresholds and the penumbra of focal ischemia. Ann. Neurol. 36, 557—565.

Johnson, K.P., 2012. Glatiramer acetate for treatment of relapsing-remitting multiple sclerosis. Expert Rev. Neurother. 12, 371—384.

Kappos, L., Polman, C.H., Freedman, M.S., et al., 2006. Treatment with interferon beta-1b delays conversion to clinically definite and McDonald MS in patients with clinically isolated syndromes. Neurology 67, 1242—1249.

Kappos, L., Radue, E.W., O'Connor, P., et al., 2010. A placebo-controlled trial of oral fingolimod in relapsing multiple sclerosis. N. Engl. J. Med. 362, 387—401.

Kola, I., Landis, J., 2004. Can the pharmaceutical industry reduce attrition rates? Nat. Rev. 3, 711—715.

Lansberg, M.G., Lee, J., Christensen, S., Straka, M., De Silva, D.A., et al., June 2011. RAPID automated patient selection for reperfusion therapy: a pooled analysis of the Echoplanar Imaging Thrombolytic Evaluation Trial (EPITHET) and the Diffusion And Perfusion Imaging Evaluation for Understanding Stroke Evolution (DEFUSE) Study. Stroke 42 (6), 1608—1614.

Lavery, A.M., Verhey, L.H., Waldman, A.T., 2014. Outcome measures in relapsing-remitting multiple sclerosis: capturing disability and disease progression in clinical trials. Mult. Scler. Int. 2014, 262350.

Lee, E.B., Lee, V.M., Trojanowski, J.Q., November 30, 2011. Gains or losses: molecular mechanisms of TDP43-mediated neurodegeneration. Nat. Rev. Neurosci. 13 (1), 38−50.

Lemmens, R., Moore, M.J., Al-Chalabi, A., Brown Jr., R.H., Robberecht, W., May 2010. RNA metabolism and the pathogenesis of motor neuron diseases. Trends Neurosci. 33 (5), 249−258.

Levine, T.P., Daniels, R.D., Gatta, A.T., et al., February 15, 2013. The product of C9orf72, a gene strongly implicated in neurodegeneration, is structurally related to DENN Rab-GEFs. Bioinformatics 29 (4), 499−503.

Matloubian, M., Lo, C.G., Cinamon, G., et al., 2004. Lymphocyte egress from thymus and peripheral lymphoid organs is dependent on S1P receptor. Nature 427, 355−360.

Mikol, D.D., Barkhof, F., Chang, P., et al., October 2008. REGARD study group. Comparison of subcutaneous interferon beta-1a with glatiramer acetate in patients with relapsing multiple sclerosis (the REbif vs Glatiramer Acetate in Relapsing MS Disease [REGARD] study): a multicentre, randomised, parallel, open-label trial. Lancet Neurol. 7 (10), 903−914.

Miller, A., Comi, G., Confavreux, C., et al., 2013. Teriflunomide efficacy and safety in patients with relapsing multiple sclerosis: results from TOWER, a second, pivotal, phase 3 placebo-controlled study. Neurology 80, S01−S004.

Moreth, J., Mavoungou, C., Schindowski, K., May 11, 2013. Passive anti-amyloid immunotherapy in Alzheimer's disease: what are the most promising targets? Immun. Ageing 10 (1), 18.

Myers, R., Harvey, M., Kaufmann, T.J., et al., July 2008. Toxicology study of repeat intracerebral administration of a measles virus derivative producing carcinoembryonic antigen in rhesus macaques in support of a phase I/II clinical trial for patients with recurrent gliomas. Hum. Gene Ther. 19 (7), 690−698.

Neumann, M., Sampathu, D.M., Kwong, L.K., et al., October 6, 2006. Ubiquitinated TDP-43 in frontotemporal lobar degeneration and amyotrophic lateral sclerosis. Science 314 (5796), 130−133.

O'Connor, P., Filippi, M., Arnason, B., et al., 2009. 250 microg or 500 microg interferon beta-1b versus 20 mg glatiramer acetate in relapsing-remitting multiple sclerosis: a prospective, randomised, multicentre study. Lancet Neurol. 8, 889−897.

O'Connor, P., Wolinsky, J.S., Confavreux, C., et al., 2011. Randomized trial of oral teriflunomide for relapsing multiple sclerosis. N. Engl. J. Med. 365, 1293−1303.

Palmer, A.M., Alavijeh, M.S., October 2012. Translational CNS medicines research. Drug Discov. Today 17 (19−20), 1068−1078.

Palmer, A., 2012. Pharmacotherapeutic options for the treatment of multiple sclerosis. Clin. Med. Insights Ther. 4, 1−24.

Pelletier, D., Hafler, D.A., 2012. Fingolimod for multiple sclerosis. N. Engl. J. Med. 366, 339−347.

Piehl, F., 2014. A changing treatment landscape for multiple sclerosis: challenges and opportunities. J. Intern. Med. 275 (4), 364−381.

Polman, C.H., O'Connor, P.W., Havrdova, E., et al., 2006. A randomized, placebo-controlled trial of natalizumab for relapsing multiple sclerosis. N. Engl. J. Med. 354, 899−910.

Poppe, L., Rué, L., Robberecht, W., Van Den Bosch, L., December 2014. Translating biological findings into new treatment strategies for amyotrophic lateral sclerosis (ALS). Exp. Neurol. 262 (Pt B), 138−151.

Prins, R.M., Soto, H., Konkankit, V., et al., March 15, 2011. Gene expression profile correlates with T-cell infiltration and relative survival in glioblastoma patients vaccinated with dendritic cell immunotherapy. Clin. Cancer Res. 17 (6), 1603−1615.

Recuero, M., Serrano, E., Bullido, M.J., et al., July 16, 2004. Abeta production as consequence of cellular death of a human neuroblastoma overexpressing APP. FEBS Lett. 570 (1−3), 114−118.

Reitz, C., Brayne, C., Mayeux, R., 2011. Epidemiology of Alzheimer disease. Nat. Rev. Neurol. 7 (3), 137−152.

Renton, A.E., Majounie, E., Waite, A., et al., 2011. A hexanucleotide repeat expansion in C9ORF72 is the cause of chromosome 9p21-linked ALS-F70. Neuron 72 (2), 257−268.

Rinne, J.O., Brooks, D.J., Rossor, M.N., et al., April 2010. 11C-PiB PET assessment of change in fibrillar amyloid-beta load in patients with Alzheimer's disease treated with bapineuzumab: a phase 2, double-blind, placebo-controlled, ascending-dose study. Lancet Neurol. 9 (4), 363−372.

Rothstein, J.D., 2009. Current hypotheses for the underlying biology of amyotrophic lateral sclerosis. Ann. Neurol. 65 (Suppl. 1), S3−S9, 8.

Rubenstein, J.L., Combs, D., Rosenberg, J., et al., 2003. Rituximab therapy for CNS lymphomas: targeting the leptomeningeal compartment. Blood 101, 466−468.

Rudick, R.A., Stuart, W.H., Calabresi, P.A., et al., 2006. Natalizumab plus interferon beta-1a for relapsing multiple sclerosis. N. Engl. J. Med. 354, 911−923.

Sampson, J.H., Heimberger, A.B., Archer, G.E., et al., November 1, 2010. Immunologic escape after prolonged progression-free survival with epidermal growth factor receptor variant III peptide vaccination in patients with newly diagnosed glioblastoma. J. Clin. Oncol. 28 (31), 4722−4729.

Sandercock, P., Wardlaw, J.M., Lindley, R.I., et al., 2012. The benefits and harms of intravenous thrombolysis with recombinant tissue plasminogen activator within 6 h of acute ischaemic stroke (the third international stroke trial [IST-3]): a randomised controlled trial. Lancet 379, 2352−2363.

Saver, J.L., Goyal, M., Bonafe, A., et al., 2015. Stent-retriever thrombectomy after intravenous t-PA vs t-PA alone in stroke. N. Engl. J. Med. 372 (24), 2285−2295 undefined.

See, A.P., Pradilla, G., Yang, I., et al., June 2011. Heat shock protein-peptide complex in the treatment of glioblastoma. Expert Rev. Vaccines 10 (6), 721−731.

Siemers, E.R., Friedrich, S., Dean, R.A., et al., 2010. Safety and changes in plasma and cerebrospinal fluid amyloid beta after a single administration of an amyloid beta monoclonal antibody in subjects with Alzheimer disease. Clin. Neuropharmacol. 33, 67−73, 37.

Sormani, M.P., Bonzano, L., Roccatagliata, L., et al., 2009. Magnetic resonance imaging as a potential surrogate for relapses in multiple sclerosis: a meta-analytic approach. Ann. Neurol. 65, 268–275.

Stankiewicz, J.M., Kolb, H., Karni, A., Weiner, H.L., 2013. Role of immunosuppressive therapy for the treatment of multiple sclerosis. Neurotherapeutics 10 (1), 77–88. http://dx.doi.org/10.1007/s13311-012-0172-3.

Stella, F., Radanovic, M., Canineu, P.R., et al., August 2015. Anti-dementia medications: current prescriptions in clinical practice and new agents in progress. Ther. Adv. Drug Saf. 6 (4), 151–165.

Stupp, R., Mason, W.P., van den Bent, M.J., et al., 2005. Radiotherapy plus concomitant and adjuvant temozolomide for glioblastoma. N. Engl. J. Med. 352, 987–996.

Tsai, S.Y., Markus, T.M., Andrews, E.M., et al., 2007. Intrathecal treatment with anti-Nogo-A antibody improves functional recovery in adult rats after stroke. Exp. Brain Res. 182, 261–266.

Turk, A.S., Frei, D., Fiorella, D., et al., 2014. ADAPT FAST study: a direct aspiration first pass technique for acute stroke thrombectomy. J. Neurointerv Surg. 6, 260–264.

Vance, C., Rogelj, B., Hortobágyi, T., et al., February 27, 2009. Mutations in FUS, an RNA processing protein, cause familial amyotrophic lateral sclerosis type 6. Science 323 (5918), 1208–1211.

Vermersch, P., Czlonkowska, A., Grimaldi, L.M., et al., May 2014. Teriflunomide versus subcutaneous interferon beta-1a in patients with relapsing multiple sclerosis: a randomised, controlled phase 3 trial. Mult. Scler. 20 (6), 705–716.

Wardlaw, J.M., Murray, V., Berge, E., et al., 2012. Recombinant tissue plasminogen activator for acute ischaemic stroke: an updated systematic review and meta-analysis. Lancet 379, 2364–2372.

Westphal, M., Ylä-Herttuala, S., Martin, J., et al., August 2013. ASPECT Study Group. Adenovirus-mediated gene therapy with sitimagene ceradenovec followed by intravenous ganciclovir for patients with operable high-grade glioma (ASPECT): a randomised, open-label, phase 3 trial. Lancet Oncol. 14 (9), 823–833.

Winblad, B., Graf, A., Riviere, M.E., Andreasen, N., Ryan, J.M., 2014. Active immunotherapy options for Alzheimer's disease. Alzheimers Res. Ther. 6 (1), 7.

Wolinsky, J.S., Narayana, P.A., Nelson, F., et al., 2013. Magnetic resonance imaging outcomes from a phase III trial of teriflunomide. Mult. Scler. 19, 1310–1319.

World Health Organization, 2006. Global Burden of Neurological Disorders Estimates and Projections. http://www.who.int/mental_health/neurology/.

Chapter 31

Research in Pediatrics

Lisa Bomgaars[1], Stacey Berg[1] and Ann R. Stark[2]

[1]Texas Children's Hospital, Houston, TX, United States; [2]Monroe Carell Jr. Children's Hospital, Nashville, TN, United States

Chapter Outline

Key Points

- Special considerations apply to translational research in children.
- Challenges include developmental changes in physiology and metabolism, small size, limited number of available subjects, and need for long-term follow-up.
- Specific regulatory and ethical issues must be addressed.

INTRODUCTION

Children comprise a large percentage of the medical patient population, but often have been underrepresented in clinical research studies. Many specific considerations must be taken into account when performing translational and clinical research in the pediatric population. These considerations include pediatric-specific disease processes, developmental differences in physiology and metabolism, practical limitations related to physical size, small numbers of subjects available for inclusion in studies, and the need for long-term follow-up to determine study outcomes. In addition, specific regulatory and ethical issues must be considered when performing clinical research in children. All of these issues make translational research in children both different from and sometimes more difficult than research in adults. If outcomes are to improve for children with medical conditions for which prevention or treatment are not yet optimal, it is critical that pediatric studies be conducted by researchers who are knowledgeable in how to deal with these difficulties.

Clinical and Translational Science. http://dx.doi.org/10.1016/B978-0-12-802101-9.00031-4

WHAT IS DIFFERENT ABOUT PEDIATRIC RESEARCH?

Developmental Physiology

Normal growth and development result in changes in physiologic processes that influence disease processes as well as drug disposition. For instance, physiologic changes such as hormonal variability in adolescents may influence disease states such as insulin resistance in diabetes mellitus or migraine headache exacerbations and are important to consider in development of clinical research studies (Sillanpaa, 1983; Amiel et al., 1986). Developmental changes also influence many practical issues relevant to the conduct of research in children, such as evaluation or monitoring techniques. The most dramatic changes affecting these physiologic processes occur during the first 12 months of life. However, an understanding of developmental changes occurring throughout childhood is critical for the appropriate design and evaluation of clinical studies in this age group.

Maturational differences across the pediatric age spectrum may have a significant impact on drug disposition. Developmental changes influencing the absorption, distribution, metabolism, and elimination of drugs should be considered in the design of pediatric studies involving medications. Other pharmacologic variables that may be unique to specific age groups include maturational differences in receptor number, binding, affinity, and coupling of those receptors with intracellular effects, although the full impact of these processes requires further elucidation (Rakhmanina and van den Anker, 2006; Ward et al., 2006). A detailed review of developmental pharmacology is beyond the scope of this chapter; however, a brief review of developmental factors influencing drug absorption, distribution, metabolism, and elimination is provided in the following sections.

Absorption

Administration of drugs by a nonvascular route requires that the agent overcomes biologic, chemical, mechanical, and physical barriers to be absorbed (Kearns et al., 2003). Changes in gastric acid secretion, intestinal motility, and bile acid excretion contribute to the differences in absorption following oral drug administration seen in neonates and children. Neonates and young infants have reduced intragastric hydrochloric acid production compared to older children and adults, resulting in a relatively elevated gastric pH during the neonatal period (Agunod et al., 1969; Euler and Ament, 1977; Euler et al., 1977; Kearns et al., 2003). This may result in enhanced absorption of basic drugs and reduced absorption of acidic drugs during infancy and early childhood (Huang and High, 1953; Morselli et al., 1980). Lipophilic medications that require solubilization in the intestine may also be influenced by age-related changes in biliary function, with neonates having limited capacity absorption due to lower intraluminal bile salt concentration (Poley et al., 1964; Boehm et al., 1997; Abdel-Rahman et al., 2007). Drug absorption is also influenced by gastric emptying and intestinal motility. These processes are decreased in neonates and infants, reaching adult levels around the age of 4−6 months (Gupta and Brans, 1978; Berseth, 1989; Di Lorenzo et al., 1995), and may result in a delay in time to attain peak levels of orally administered agents in children less than 6 months of age.

The percutaneous absorption of drugs is influenced by skin hydration, perfusion, and structure. Preterm infants and neonates have decreased skin stratum corneum thickness that may lead to more efficient absorption (Nachman and Esterly, 1971; Barker et al., 1987). Furthermore, children have greater cutaneous perfusion and epidermal hydration compared to adults, potentially resulting in further enhancement of transdermal absorption (Rutter, 1987; Fluhr et al., 2000; Nikolovski et al., 2008). The absorption and dispersion of intramuscularly administered medications is related to blood flow to the muscle and muscular contractions (Greenblatt and Koch-Weser, 1976). In children with reduced muscle blood flow, skeletal muscle mass, or muscular contractions, absorption of intramuscularly administered medications may be slower and less predictable.

Infants and young children also have a greater body surface area to body mass ratio as compared to adults. This is an important consideration for studies of topical agents as it may result in greater relative systemic exposure for a given dose, and therefore possible increases in adverse effects (Goutieres and Aicardi, 1977; West et al., 1981; Peleg et al., 1998).

Distribution

Distribution of drugs to the site of action and throughout the body influences both therapeutic efficacy and potential toxicity and must be taken into account when evaluating these parameters. Developmental changes in body composition influence the distribution of medications, with the greatest changes in body composition occurring during infancy and adolescence.

Preterm infants and neonates have greater total body water spaces and extracellular fluid as compared to older children and adults, leading to larger apparent volumes of distribution for drugs distributing into these spaces and reduced plasma concentrations of drugs administered at the same weight-based dose (Friis-Hansen, 1961; Siber et al., 1975; Hartnoll et al., 2000). The greater proportion of water in neonatal adipose tissue as compared to adults may also contribute to the greater apparent volume of distribution in this age range (Hartnoll et al., 2000). The amount and types of circulating proteins are also different in neonates and young infants, leading to altered plasma protein binding (Ehrnebo et al., 1971). This has specific implications for drugs that are highly protein-bound. In neonates and young infants who have decreased plasma proteins this may result in increased free drug concentrations and therefore increased drug effect. Significant physiologic changes also occur during puberty, including a rapid increase in body size, hormonal fluctuations, and marked gender-specific changes in body composition. These body composition changes include alterations in the distribution and an increase in the amount of adipose tissue, increases in lean body mass and bone mineral content, and a decrease in the percentage of total body water to fat-free mass (Siervogel et al., 2003; Veldhuis et al., 2005).

Metabolism

The ontogeny of drug-metabolizing enzymes should also be considered when performing pediatric drug studies. Maturational changes in both phase I and phase II enzymes have implications for both drug efficacy and toxicity in pediatric patients and may result in a need for alterations in dose or dosing interval in some age groups. The activity of phase I enzymes, primarily cytochrome P450 enzymes responsible for the biotransformation of endogenous compounds and drugs, may change significantly during fetal development and infancy. A common general developmental profile appears to exist for most P450 enzymes and is characterized by low to absent activity at birth with a gradual increase over the first 3 months of life (Blake et al., 2005). Notable exceptions to this are CYP2D7, which is the fetal form of CYP3A4 and disappears during the first month of postnatal life (Lacroix et al., 1997), and CYP2D6, where activity is concordant with genotype by 2 weeks of age (Blake et al., 2007). Phase II enzymes, resulting in inactivation or detoxification of drugs primarily by conjugation, also vary with age. The ontogeny of phase II enzymes has not been evaluated to the same extent as that of phase I enzymes. However, pharmacokinetic evaluation for drugs that are substrates for these enzymes suggests a pattern of acquisition of function during postnatal life (Blake et al., 2005).

Elimination

Developmental changes in renal function may also contribute to differences in drug elimination and necessitate dosing adjustments of renally cleared medications. Glomerular filtration and active tubular secretion both undergo development following birth, with slower development occurring in preterm infants (Kelly and Seri, 2008). Glomerular filtration rate increases dramatically in the 2 weeks following birth (in infants with birth weight more than 1.5 kg) secondary to changes in renal blood flow and recruitment of nephrons (Arant, 1978; Guignard et al., 1975) and occurs more slowly in preterm infants (Aperia et al., 1981; Vanpee et al., 1988). Adult levels are attained by 8–12 months of age, with renal tubular secretion increasing more slowly, reaching adult values during the first year of life (Kearns et al., 2003).

An appreciation of the potential impact of these developmental changes, in part, led to increased awareness regarding the need to evaluate drugs in the pediatric population, and subsequently to legislation in the United States that encourages industry sponsorship of pediatric clinical trials (Food and Drug Administration Modernization Act, 1997; Best Pharmaceuticals for Children Act, 2002; Pediatric Research Equity Act, 2003; Abdel-Rahman et al., 2007). Since the inception of these programs, many agents evaluated have been found to have substantive differences in dosing, safety, or efficacy as compared to adults with studies in infants and premature infants particularly lacking (Benjamin et al., 2006; Laughon et al., 2011). This highlights the concept that unique pediatric dosing reflecting growth and maturational stages is often necessary and provides evidence that pediatric dosing should not be determined by simply applying weight-based calculations to adult doses (Rodriguez et al., 2008). The NICHD has supported pediatric clinical trials for the Best Pharmaceuticals for Children Act drug development program through the Pediatric Trials Network, an alliance of clinical research sites in the United States (NICHD, 2014; PTN, 2014).

In addition to dosing requirements, consideration of maturational differences is also critical to the appropriate design and evaluation of pediatric studies. Benjamin et al. (2008) evaluated the reasons for the failure of six dose-ranging antihypertensive efficacy studies that were performed on drugs already approved for use in adults. They showed that successful studies evaluated broad dose ranges with little overlap, utilized pediatric formulations, and evaluated appropriate pediatric endpoints. Failing studies showed a lack of acknowledgment of the differences between pediatric and adult populations. For example, the failing studies did not develop pediatric dosing formulations, resulting in a wide range of exposure within each weight stratum, and did not evaluate dose ranges higher than the corresponding adult dose. Both of these deficits could contribute to a failure to demonstrate a significant dose response. Maturational differences may also make it difficult to study drugs because there is no appropriate formulation for young children. For example, ethanol is a common excipient in many oral medications, yet pharmacokinetic and pharmacodynamic factors make it undesirable for use in children (Marek and Kraft, 2014). As part of the US Pediatric Formulations Initiative, the NICHD has sponsored pediatric formulation workshops to address the knowledge gaps that limit the availability of pediatric formulations (Giacoia et al., 2012; Abdel-Rahman et al., 2012).

In summary, an appreciation of the developmental differences across the pediatric age spectrum is critical for the conduct and assessment of pediatric studies. The physiologic differences occurring in this population are complex and influence multiple factors including the pharmacology and response to medications, age-appropriate monitoring techniques, measurements, and study endpoints.

Limitations Related to Body Size

An important limitation in research in children is body size, especially as it relates to sampling from a total blood volume that is substantially smaller than that of an adult. In term or premature newborn infants, the blood volume is approximately 80—90 mL/kg. Thus, a premature infant with a birth weight of 1200 g might have a blood volume of approximately 110 mL, and substantial losses would necessitate transfusion to maintain vascular volume and oxygen-carrying capacity. In addition, premature or older critically ill infants require frequent blood sampling for clinical monitoring, leaving little room for additional sampling for research purposes. Furthermore, additional venipunctures, capillary blood draws from a heel puncture, or catheter placement for blood sampling for research are not always acceptable. As a result, blood sampling for research purposes in infants is frequently timed to coincide with blood drawn for clinical monitoring or relies on excess blood remaining from clinical samples. Translational research studies must be designed to take these challenges into consideration. One approach is to use archived blood samples collected for the mandated newborn screens for metabolic disorders for correlation with clinical outcomes in childhood. For example, this technique has been used to identify the association of elevated concentrations of neonatal cytokines and coagulation factors with cerebral palsy in children (Nelson et al., 1998). Similarly, sampling of cord blood may be an alternative to obtaining blood samples directly from the infant in the first few hours after birth (Goepfert et al., 2004).

When performing pharmacokinetic studies, the development of analytical techniques for measuring drug and metabolites using microvolumes of blood can minimize the amount of blood required. Sample collection using dried blood sampling (DBS) allows for the use of very small sample volumes ($\sim 30\ \mu L$) of whole blood. This methodology is also advantageous as it does not require sample processing or special storage. In a recent study evaluating the pharmacokinetics of metronidazole in premature infants, use of DBS was shown to provide results similar to standard plasma sampling methods (Cohen-Wolkowiez et al., 2013). Modeling strategies can also be used to reduce the amount of samples drawn in a study. Simulation can be used to predict optimal sampling times, and population-based pharmacokinetic modeling can be used to analyze data sets with small numbers of samples per patient. The use of these techniques can make pharmacokinetic studies in the neonatal population more feasible.

ORPHAN (RARE) DISEASES

Population Available for Study Is Smaller in Children Than in Adults

By definition, children occupy a life span for a shorter period of time than an adult. Children less than 18 years of age represent a minority (25%) of the population in the United States (Hirschfeld, 2005). Furthermore, many diseases relevant to the pediatric population are rare or infrequent, with a small number of children requiring treatment with medications as compared to adults (Mathis and Iyasu, 2007). This fact results in limited numbers of potential pediatric patients to participate in clinical research studies as compared to the adult population. The Office of Rare Diseases of the National Institutes of Health (NIH) defines a rare disease as one that has a prevalence of less than 200,000 affected individuals in the US population (http://rarediseases.info.nih.gov). Many diseases in pediatrics are considered rare or "orphan diseases," both because of the low incidence of many pediatric disorders and the fact that children encompass only a quarter of the population.

Small Numbers Mean Multicenter Trials Are Usually Required

In contrast to adults in whom death is primarily disease-related, the leading cause of death in children 0—17 years of age is unintentional injuries followed by birth defects in children 1—4 years of age and cancer in children 5—14 years of age

(Heron, 2007). The incidence of new cancer cases in children is estimated to be 17.9 per 100,000, with an estimated 14,102 total number of American children less than 20 years diagnosed with cancer of all types in 2011 (Howlander et al., 2014). When considering acute lymphoblastic leukemia, the most common type of cancer in children, the estimated number of patients diagnosed in 2011 dropped to 2914 (Howlander et al., 2014). Given that not all of them will be eligible for or consent to participation in clinical trials, the number of subjects available is small and research must be very carefully designed to identify the most critical questions and to answer as many questions as is feasible in any given trial. Even then, detection of small differences in the outcome between interventions is not always practical. In contrast, for heart disease, the leading cause of death in adults in the United States (Heron, 2007), large numbers of patients are available for enrollment in multiple studies that may identify small differences between arms. The landmark ALLHAT study (Antihypertensive and Lipid-Lowering Treatment to prevent Heart Attack Trial) evaluated the effect of antihypertensive and cholesterol-lowering treatment on the incidence of coronary heart disease and enrolled 42,418 patients at ~600 sites over a 4-year period (Davis et al., 1996; ALLHAT, 2002). Although this trial was a large US—Canadian multisite trial, the study enrollment numbers exceed those reasonably attainable in any pediatric treatment study.

The small number of patients with any particular pediatric diagnosis has necessitated the development of cooperative study groups. Pediatric oncology was among the first areas where a cooperative group approach to translational research in children was broadly applied. The Children's Oncology Group (formerly the Pediatric Oncology Group and Children's Cancer Group) and the Pediatric Brain Tumor Consortium currently perform numerous multicenter clinical trials to provide large enough study populations for adequate statistical power to detect clinically important differences in therapy and disease parameters for children with cancer.

Similar cooperative group approaches to translational research have been applied to other childhood conditions. Approximately 11.4% of births in the United States are preterm and approximately 56,000 infants are born annually with birth weight less than 1500 g (Hamilton et al., 2014). The NICHD-sponsored Neonatal Research Network (https://neonatal. rti.org/) was established in 1986 to conduct multicenter trials and observational studies to improve the evidence for clinical management decisions in this population. Other NIH-sponsored multicenter pediatric networks include the National Collaborative Pediatric Critical Care Research Network (http://www.cpccrn.org) and the Pediatric Heart Network (http://www.pediatricheartnetwork.org). Additional consortia may be funded under the National Pediatric Research Network Act of 2013 (https://www.govtrack.us/congress/bills/113/hr225/text).

Lack of Pediatric Guidelines for Most Drugs

Historically, more than 70% of drugs have dosing, safety, or efficacy information that is insufficient for pediatric labeling, leading to the use of the term "therapeutic orphans" to describe this situation in children (Shirkey, 1968; Wilson, 1999). Lack of appropriate pediatric dosing information places children at risk for both ineffective treatment as well as potential toxicity. Furthermore, while pediatric indications frequently parallel those in adults, medical conditions requiring treatment may be unique to children, limiting available data to guide the pediatrician (Blumer, 1999). A review of pediatric patients less than 18 years of age treated at 31 tertiary care pediatric hospitals in the United States shows that most (78.7%) received at least one drug administered outside of the Food and Drug Administration (FDA)-approved age range for any indication of the drug (Shah et al., 2007). Because the study evaluated only 90 drugs and off-label use was defined solely on age criteria, this figure may underestimate the magnitude of "off-label" use when all drugs and indications for administration are considered.

In response to the paucity of pediatric data available for drug labeling, legislation has been implemented in the United States, and recently in Europe, to assist with the acquisition of pediatric dosing, safety, and efficacy data. In 1997 the United States Congress passed the Food and Drug Administration Modernization Act (FDAMA). This Act provided the industry an incentive of 6 months of marketing exclusivity for all products containing the active moiety in return for the performance of studies conforming to the details of a written request issued by the FDA. The Best Pharmaceuticals for Children Act (BPCA) succeeded FDAMA in 2002, continuing the incentive and establishing an office of pediatric therapeutics at the FDA. This Act also established a process for studying on-patent and off-patent drugs in children. The Pediatric Research Equity Act (PREA) was signed into law in 2003. PREA codified many elements of the Pediatric Rule that was enjoined in 2002 and required all applications for a new active ingredient, indication, dosage form, regimen, or route of administration contain a pediatric assessment unless the applicant has obtained a waiver or deferral. Subsequent amendments include the institution of an FDA internal review committee to evaluate all drugs that apply to pediatrics and a requirement to make more data from pediatric submissions publicly available. BPCA and PREA legislation were made permanent in 2012 under the Food and Drug Administration Safety and Innovation Act.

The European Union Regulation on Medicinal Products for Pediatric Use was entered into force in 2007 and closely resembles the BPCA legislation. This legislation requires that any company applying to the European Medicines Agency (EMEA) to market a new drug must include a pediatric investigation plan or obtain a waiver if the drug is not suitable for children [Regulation (EC) No. 1902/2006; Regulation (EC) No. 1901/2006]. The incentive for compliance is a 6-month extension of exclusivity for studies that incorporate data from the pediatric studies into the Summary of Product Characteristics. The legislation also provides funding to study off-patent drugs, mandates that data from all trials in children be publicly available, and establishes a Pediatric Committee within the EMEA.

These legislative changes have proven to be a successful stimulus for the performance of pediatric studies in the United States and have resulted in an improvement in pediatric information available for drug labeling (Rodriguez et al., 2008). As of October 2014, the FDA has issued 386 written requests for pediatric studies, resulting in 204 drugs granted exclusivity, with 538 pediatric label changes since 2007 (FDA, 2014a,b,c). Despite this success, off-label use of drug in children remains common, resulting in a policy statement from the American Academy of Pediatrics outlining recommendations for practitioners (Frattarelli et al., 2014).

Need for Very Long-Term Follow-up to Determine Outcomes

An important difference of research in children compared to adults is that adverse effects or toxicities of treatments may not appear for a prolonged period after exposure. The following examples demonstrate that the follow-up period for children participating in research, in contrast to adults, may need to extend through childhood or well beyond to fully characterize the outcomes of certain interventions.

Late Effects of Childhood Cancer Treatment

Many adverse effects of childhood cancer therapy are not apparent for many years after treatment (reviewed in Landier et al., 2015). Forty percent of survivors of childhood cancer treatment have severe, disabling, or life-threatening chronic health conditions 30 years after a cancer diagnosis (Oeffinger et al., 2006). Many of these conditions are related to toxicities of the therapy. Cognitive impairment often worsens gradually after cranial radiation and may occur in some children after chemotherapy, even without radiation (Moleski, 2000; Mulhern et al., 2004; Mulhern et al., 1998; Peterson et al., 2008). Total body irradiation and cranial radiation are associated with growth failure and obesity in survivors, especially in patients who were treated at early ages (Gurney et al., 2003; Oeffinger et al., 2003; Sanders, 2008). Young women are at risk for premature ovarian failure years after treatment with alkylating agents (Green et al., 2009). Patients who received anthracyclines in infancy or early childhood sometimes develop congestive heart failure when they "outgrow" their remaining healthy myocardium, or when their cardiovascular systems are stressed, as by pregnancy (Steinherz and Steinherz, 1991; Steinherz et al., 1991; Hudson, 2007; Hudson et al., 2007). Women who received chest wall radiation in childhood have a high risk of developing breast cancer by age 50 (Moskowitz et al., 2014). Second malignancies and cardiovascular events are the greatest causes of mortality, other than recurrent disease, in childhood cancer survivors (Mertens et al., 2008). Without studies specifically designed for prolonged follow-up, much of this important information would be lacking.

Outcome of Premature Infants

Survival has improved among premature infants born at extremely low birth weight (ELBW, less than 1000 g), or extremely low gestational age (less than 27 or 28 completed weeks), and is strongly related to gestational age. The proportion of survivors with severe disability is greatest in the most immature infants and has changed little between the mid-1990s and mid-2000s (Moore et al., 2012; Jonsdottir et al., 2012). However, long-term follow-up is essential to test the effect of an intervention in the neonatal period. Although severe disability identified at 2−3 years of age generally persists at school age (Wood et al., 2000), severe abnormalities are often identified later in childhood in children who did not appear abnormal in early infancy. For example, in a follow-up study of ELBW infants born between 1992 and 1995, one-third of those with no neurosensory abnormalities at discharge had a low IQ, limited academic skills, or poor motor skills, and two-thirds had poor adaptive functioning when they were evaluated at 8−9 years of age (Hack et al., 2005).

Studies that have followed very low birth weight (VLBW) infants into adulthood have identified a higher risk of medical, functional, and neurodevelopmental problems than in controls born at term (Saigal, 2014; Doyle and Anderson, 2010). Considerable knowledge about these outcomes has resulted from longitudinal studies including cohorts of infants born in the late 1970s in central-west Ontario (Saigal et al., 2007), Cleveland (Hack et al., 2002), and Helsinki (Hovi et al., 2007). Epidemiologic studies of national registries in Sweden (Lindstrom et al., 2007) and Norway (Moster et al., 2008) have also contributed to our knowledge of adult outcomes. These studies have identified the association of very low birth weight with later development of metabolic conditions including insulin resistance and impaired glucose tolerance, hypertension, sleep-disordered breathing (Paavonen et al., 2007), poorer neurocognitive performance and neurosensory impairments, increased autism spectrum traits (Pyhälä et al., 2014) and psychopathology, and delay in leaving the parental home and beginning sexual partnerships (Kajantie et al., 2008). The Victorian Infant Collaborative Study Group, a geographic cohort of children born extremely preterm or ELBW in 1991−92 in Victoria, Australia, has also documented higher ambulatory blood pressure (Roberts et al., 2014) and increased visual morbidity in adolescents (Molloy et al., 2013) compared to term controls, although self-reported quality of life was similar (Roberts et al., 2013).

Strategies Used During Neonatal Cardiac Surgery

In a trial in infants undergoing an arterial switch operation for transposition of the great vessels, low-flow cardiopulmonary bypass was associated with less likelihood of clinical or electroencephalographic seizures and a shorter time to recover normal EEG activity after surgery than hypothermic circulatory arrest (Newburger et al., 1993). In follow-up at 1 year of age, infants who underwent circulatory arrest had a greater risk of neurologic abnormalities and scored lower on the Bayley Psychomotor Development Index, but not on the Mental Development Index or a test of visual recognition memory (Bellinger et al., 1995). When children were reevaluated at 8 years of age, circulatory arrest continued to be associated with more motor morbidity and low-flow cardiopulmonary bypass was associated with more behavioral problems, but the groups did not differ in most of the outcomes that were assessed (Bellinger et al., 2003). Furthermore, shorter durations of circulatory arrest had little influence on later neurodevelopmental outcomes (Wypij et al., 2003).

These results demonstrate that it is critical to include long-term follow-up in many types of pediatric studies. Important challenges include tracking of subjects, the need for multiple interim visits to maintain long-term contact with patients and families, the use of appropriate controls, adequate sample size to address questions of safety, transportation to study site and incentives or compensation for the time to collect study data, and sufficient funding to support the research infrastructure needed to overcome these challenges (Vohr et al., 2004b).

PEDIATRIC CONDITIONS AS FOCUS OF INQUIRY

Prematurity

Prematurity is obviously a condition unique to children that has no direct correlation to adult diseases and therefore requires special approaches to translational research. Although ~85% of VLBW infants (birth weight less than 1500 g) survive to hospital discharge, a substantial proportion suffer complications, including bronchopulmonary dysplasia (chronic lung disease of prematurity), necrotizing enterocolitis, and neurosensory abnormalities including intracranial hemorrhage, white matter injury, and retinopathy of prematurity. Trials of management strategies to reduce these complications are urgently needed, yet are difficult to perform.

The trial of early dexamethasone treatment to reduce the risk of chronic lung disease in ELBW infants performed by the NICHD Neonatal Research Network (Stark et al., 2001) illustrates some of the challenges involved in performing clinical trials in this population, including the need for a large sample size, competing outcomes, and design issues. The primary outcome selected was a combination of death before 36 weeks postmenstrual age or chronic lung disease, because these are competing outcomes, with death precluding the development of chronic lung disease. Secondary outcomes included complications of corticosteroid therapy as well as complications of prematurity. A clinically meaningful change was considered to be a reduction in the primary outcome from 55% to 44%, a 20% reduction, which required a sample of 532 infants in each group. However, because of concern about possible unintended consequences, the projected enrollment was increased to 600 per group to be able to exclude a substantial increase in cerebral palsy at follow-up projected for 18−22 months corrected age. A trial of this size was anticipated to continue enrollment for at least 2 years and precluded another intervention trial in this population. Thus, to maximize the results of this undertaking, the dexamethasone trial was combined in a two-by-two factorial design with a trial of minimal ventilatory support (Carlo et al., 2002). Furthermore, stratification by birth weight (or gestational age) is essential in studies of this type due to marked differences in outcome with increasing maturity. Stratification by center is typically necessary as well, due to marked differences in populations and in short-term and long-term outcomes among sites (Vohr et al., 2004a). This particular trial was ultimately terminated early due to an increased rate of spontaneous intestinal perforation in infants who had received dexamethasone and who had also been treated with indomethacin.

Investigation of other important and highly controversial management strategies in neonatal intensive care poses additional challenges, including potential opposing effects on the components of a combined outcome, as illustrated by the NICHD Neonatal Research Network trial of aggressive or conservative phototherapy to reduce serum bilirubin concentration in ELBW infants (Morris et al., 2008). The combined primary outcome of death or neurodevelopmental impairment at 18–22 months corrected age was not different between groups. Among secondary outcomes, although the rate of death was similar between groups, aggressive phototherapy reduced the rate of neurodevelopmental impairment. Subjects had been stratified by birth weight (501–750 and 751–1000 g), and in the higher birth weight stratum, aggressive phototherapy was associated with a marginally reduced combined outcome and a reduction in severe hearing loss. However, in the lower birth weight group, aggressive phototherapy was associated with a 5% decrease in the rate of neurodevelopmental impairment that was offset by a 5% increase in the rate of death. Similarly, the NICHD Neonatal Research Network SUPPORT trial compared target ranges of oxygen saturation of 85–89% or 91–95% in infants between 24 and 27 weeks gestational age (SUPPORT Study Group, 2010). Although the combined primary outcome of severe retinopathy of prematurity or death was not different between groups, lower saturation was associated with a reduction of severe retinopathy in survivors that was offset by an increased mortality rate. These opposing and unanticipated findings must be considered as study results are applied.

Investigation of some interventions, such as the timing of initiation of enteral feedings and the rate of advancement to full-volume feedings, pose other methodologic challenges. These include bias due to the inability to mask a trial of feeding, the lack of an objective definition of feeding intolerance, and the possibility that clinicians might obtain more radiographs in a more aggressive feeding group and thus be more likely to detect necrotizing enterocolitis, an important complication that occurs mostly in infants who have been fed (Tyson et al., 2007). These challenges may require additional strategies such as the use of Bayesian methods to calculate probabilities rather than reliance on conventional confidence limits (Lilford et al., 1995; Arnold et al., 2014).

Childhood Cancers That Do Not Occur in Adults

Although cancer occurs in adults as well as children, many pediatric cancers are distinctly different from adult tumors and require a different approach to research. For example, Wilms tumor is the most common malignant renal tumor in children, with an incidence of 7.6 cases per million children less than 15 years of age. However, this tumor very rarely occurs in adults (Dome et al., 2006). In contrast, the most common type of malignant renal tumor occurring in adults is renal cell carcinoma, with less than 1% of all renal cell carcinomas occurring in children (Rakozy, 2008). Treatment advances for Wilms tumor would not be possible using data from adult studies of renal malignancies as the physiology is very different; indeed, therapy for Wilms tumor may include treatment with actinomycin D, vincristine, doxorubicin, and radiation, none of which are utilized in the treatment of renal cell carcinoma.

Other diseases such as hypertension may occur in both children and adults, yet the manifestations and etiology of the disease may be very different in children, requiring research to be conducted specifically in children rather than attempting to use adult data to address pediatric questions. While the cause of hypertension in adults is generally unknown and therefore termed essential, hypertension occurring in preadolescents is commonly associated with underlying renal disease (Wyszynska et al., 1992; Bartosh and Aronson, 1999). Such differences in the disease and physiology are critical to consider when developing studies of hypertensive agents in children. A review of pediatric studies of antihypertensive agents known to be effective in adults noted that evaluation of a traditional adult endpoint may contribute to a negative study in children (Benjamin et al., 2008). While adult hypertension studies commonly evaluate both diastolic and systolic blood pressure as an efficacy endpoint, this review noted that successful pediatric studies evaluated change in diastolic blood pressure as the primary endpoint. This finding may be related to the fact that in children, diastolic blood pressure has less physiologic variability between observations than does systolic blood pressure (Benjamin et al., 2008).

REGULATORY AND ETHICAL ENVIRONMENT FOR PEDIATRIC RESEARCH (SEE ALSO CHAPTERS 23 AND 24)

The Nuremberg Code (1949) and the Belmont Report (National Commission on Protection of Human Subjects in Biomedical and Behavioral Research, 1979) provide touchstones for the ethical conduct of human subject research, emphasizing that "the voluntary consent of the human subject is absolutely essential" for permissible medical experiments, and grounding this requirement in the principle of respect for persons, which requires that individual autonomy be respected. Furthermore, the subject must have "sufficient knowledge and comprehension of the elements of the subject matter involved as to enable him/her to make an understanding and enlightened decision" for his/her consent to be valid. In addition, both ethical and regulatory constraints require that human subject research be reviewed for reasonableness and appropriate balance of the risks presented by the research and the potential benefits to be garnered by the individual or by society, for provisions to protect the safety and privacy of the subjects, and for equitable selection of subjects (see Chapter 24). A large body of regulation and guidance and an equivalent volume of research attest to the difficulty of achieving, or even measuring, the quality of institutional review board (IRB) review of and informed consent for human subject research (Joffe et al., 2001a,b; Tait et al., 2003, 2005; Flory and Emanuel, 2004).

The ethical and regulatory environment for research in children is more complex than that in adults because children (individuals under the age of legal majority) often lack the cognitive capacity and almost always lack the legal competence to provide informed consent for their own participation. Thus a parent or other representative is almost always required to provide permission for a child's participation, while the child provides age-appropriate assent (see later discussion). In addition, children receive additional protections in terms of the risks they are permitted to be exposed to as research participants. For fetuses and neonates of uncertain viability, there are even further protections in terms of permissions and limitations of research risks.

Risk Categories

The Code of Federal Regulations at 45 CFR 46 subpart C divides research involving children into four categories, each of which requires special considerations regarding additional protections, parental permission, and the child's assent (Table 31.1). In all categories, adequate provision for soliciting permission and assent are required (see the next section). For category 404 research, no additional protections are required. For category 405 research, to approve the study, the IRB must find that the risk is justified by the anticipated benefit to the subjects and that the relation of the anticipated benefit to the risk is at least as favorable to the subjects as that are presented by available alternative approaches. For category 406 research, IRB must find that the risk represents a minor increase over minimal risk; the intervention or procedure presents experiences to subjects that are reasonably commensurate with those inherent in their actual or expected medical, dental, psychological, social, or educational situations; and the intervention or procedure is likely to yield generalizable knowledge about the subjects' disorder or condition which is of vital importance for the understanding or amelioration of the subjects' disorder or condition. Category 407 research does not fit into any of the above categories and cannot be approved by a local IRB until the Secretary of Health and Human Services has consulted with a panel of experts and determined that the research presents a reasonable opportunity to further the understanding, prevention, or alleviation of a serious problem affecting the health or welfare of children and that it will be conducted in accordance with sound ethical principles.

Assent/Permission Procedures

Parents (or guardians) provide permission for children to participate in research in a way analogous to adults consenting to their own participation. Thus it is the parent who must have a good understanding of the purpose of the research, its potential risks and benefits, the voluntary nature of participation, etc. The IRB may find that the permission of one parent is sufficient for research in categories 404 or 405. For research in categories 406 and 407, both parents must give their permission unless one parent is deceased, unknown, incompetent, or not reasonably available, or when only one parent has legal responsibility for the care and custody of the child. There are some circumstances, such as research in abused or neglected children, where parental permission is replaced by some other, equally rigorous, mechanism for protecting child subjects (§46.408(c)).

TABLE 31.1 Ethical Requirements Before Enrolling Pediatric Subjects in Research

CFR 45: Part 46	Requirement
§46.404	Research not involving greater than minimal risk
§46.405	Research involving greater than minimal risk but presenting the prospect of direct benefit to the individual
§46.406	Research involving greater than minimal risk and no prospect of direct benefit to individual subjects, but likely to yield generalizable knowledge about the subject's disorder or condition
§46.407	Research not otherwise approvable which presents an opportunity to understand, prevent, or alleviate a serious problem affecting the health or welfare of children

As children develop the capacity to consent to research along a developmental continuum that begins early in childhood and reaches near-adult capacity by the teenage years (reviewed in Committee on Clinical Research Involving Children, 2004), their assent (not consent) is also sought. The age of 7 years is often used as a guideline for when investigators should begin to seek assent from children for research participation (American Academy of Pediatrics, 1995), although this practice is highly variable (Mammel and Kaplan, 1995; Whittle et al., 2004; Kimberly et al., 2006). According to the regulations "assent means a child's affirmative agreement to participate in research. Mere failure to object should not, absent affirmative agreement, be construed as assent" (§46.402(b)). The IRB is to "take into account the ages, maturity, and psychological state of the children involved" (§46.408(a)) in determining whether assent is required for participation in a particular study. In addition, if the research presents the prospect of an important direct benefit to the child that is only available in the context of the research, the federal regulations permit waiver of the requirement for assent (§46.408(a)).

Controversies continue over whether parents may give permission for their children to take on all the same kind of risks that a competent adult may consent to on his own behalf. In *T.D. v. New York*, a New York court found "unacceptable the provisions that allow for consent to be obtained on behalf of minors for participation in greater than minimal risk nontherapeutic research from the minor's parent or legal guardian." In *Grimes v. Kennedy-Krieger*, a court found that "in Maryland a parent…cannot consent to the participation of a child or other person under legal disability in nontherapeutic research or studies in which there is any risk of injury or damage to the health of the subject." (The court later clarified that "any risk" meant any more than minimal risk.) It is unclear whether these cases imply that additional limits may be imposed on category 406 or 407 research (Kopelman, 2006).

45 CFR 46 subpart B contains regulations outlining "additional protections for pregnant women, human fetuses, and neonates involved in research." In general, the regulations require that such research only be carried out if there is no other means to acquire important biomedical knowledge, and then only after any appropriate studies in preclinical models and less vulnerable subjects have assessed the potential risks. The members of the research team may not be involved in decisions about termination of pregnancy or the viability of a neonate (§46.204(i), (j)). In the case of nonviable neonates, the neonate's life may not be artificially maintained in the course of participation, and conversely, the research may not terminate heartbeat or respiration (§46.205(c)).

Consent and permission for research in neonates is complicated and is based both on the viability of the neonate and, during pregnancy, on the ascertainment of whom research participation might benefit. If the research holds out the prospect of direct benefit to the pregnant woman, or to both the pregnant woman and the fetus, or if there is no prospect of benefit for the woman or the fetus but the risk to the fetus is minimal, the woman's consent is obtained. However, if the research holds out the prospect of direct benefit solely to the fetus, then the consent of both the pregnant woman and the father is obtained (§46.204(d),(e)).

Viable neonates fall under the rules for research in children. For neonates of uncertain viability informed consent may be obtained from either parent of the neonate, or if neither parent is available, then informed consent may be obtained from either parent's legally authorized representative (§46.205(b)). For nonviable neonates, both parents must consent to any research participation (§46.205(c)).

There are additional ethical considerations in obtaining consent for research in pregnant women and neonates and for testing of fetal interventions. These include assessment of the balance of a potentially life-saving treatment of the fetus with risks of maternal mortality or morbidity in the present pregnancy or future pregnancies (Chervenak and McCullough, 2007). Examples of such studies include fetal surgery for congenital diaphragmatic hernia (Harrison et al., 2003), comparison of endoscopic laser surgery and serial amnioreduction to treat severe twin-to-twin transfusion syndrome (Senat et al., 2004), and prenatal versus postnatal repair of meningomyelocele (Adzick et al., 2011).

CONCLUSION

Translational research in children differs from that in adults because the disorders studied, the techniques and study designs required, the outcomes to be measured, and the regulatory environment in which the research takes place are different. A thorough understanding of these differences and expertise in the conduct of pediatric-specific studies are required for the safe, ethical, and successful completion of translational studies in children.

STATUTES AND REGULATIONS

Best Pharmaceuticals for Children Act 2002. Public L. 107-109. 115 Stat. 1408, 2002. Code of Federal Regulations Title 45, Part 46: Health and Welfare, Protection of Human Subjects. June 2005. http://www.hhs.gov/ohrp/humansubjects/guidance/45cfr46.htm.

Food and Drug Administration Amendments Act 2007. Public L. 110-85, 2007.
Food and Drug Administration Modernization Act 1997. Public L. 115-105. 111 Stat. 2296, 1997.
Pediatric Research Equity Act 2003. Public L. 108-155. 117 Stat. 1936, 2003, S650.
Regulation (EC) No. 1902/2006 of the European Parliament and the Council of 20 December 2006 amending Regulation 1901/2006 on Medical Products for Pediatric Use. Regulation (EC) No. 1901/2006 of the European Parliament and the Council of 12 December 2006 on Medical Products for Pediatric Use and Amending Regulation (EEC) No. 1768/92, Directive 2001/20/EC, Directive 2001/83/EC and regulation (EC) No. 726/2004. European Parliament, 2006.
FDA 2014a (accessed December 8, 2014) Written Requests Issued. http://www.fda.gov/Drugs/Development ApprovalProcess/DevelopmentResources/ucm050002.htm
FDA 2014b (accessed December 8, 2014) Pediatric Exclusivity Granted. http://www.fda.gov/Drugs/Development ApprovalProcess/DevelopmentResources/ucm050005.htm
FDA (2014c) New Pediatric Labeling Information Database as of December 8, 2014. http://www.accessdata.fda.gov/ scripts/sda/sdNavigation.cfm?sd=labelingdatabase
NICHD (accessed 2014) http://bpca.nichd.nih.gov/clinical/network/Pages/index.aspx.
PTN (accessed 2014) https://www.pediatrictrials.org/.

CASES

Grimes v. Kennedy-Krieger Institute, Inc. 366 Md 29, 782 A2d 807 (2001), reconsideration denied (October 11, 2001), 2001.
T.D. et al. *v. New York State Office of Mental Health* et al. December 22, 1997. 91 N.Y. 2d 860, 690 N E. 2d 1259, 668 N.SY.S.2 d 153 [T.D. v. N.Y. State of Mental Health, 228 AD.2 d 95 (Court, 1996).], 1996.

REFERENCES

Abdel-Rahman, S.M., Reed, M.D., Wells, T.G., et al., 2007. Considerations in the rational design and conduct of phase I/II pediatric clinical trials: avoiding the problems and pitfalls. Clin. Pharmacol. Ther. 81, 483−494.
Abdel-Rahman, S., Amidon, G., Kaul, A., et al., 2012. Summary of the NICHD-BPCA pediatric formulation initiatives Workshop-Pediatric Biopharmaceutics Classification System (PBCS) working group. Clin. Ther. 34, 311−324.
Adzick, N.S., Thom, E.A., Spong, C.Y., et al., 2011. A randomized trial of prenatal versus postnatal repair of myelomeningocele. N. Engl. J. Med. 364, 993−1004.
Agunod, M., Yamaguchi, N., Lopez, R., et al., 1969. Correlative study of hydrochloric acid, pepsin, and intrinsic factor secretion in newborns and infants. Am. J. Dig. Dis. 14, 400−414.
ALLHAT Officers and Coordinators for the ALLHAT Collaborative Research Group, 2002. Major outcomes in high-risk hypertensive patients randomized to angiotensin-converting enzyme inhibitor or calcium channel blocker vs diuretic: the Antihypertensive and Lipid-Lowering Treatment to Prevent Heart Attack Trial (ALLHAT). JAMA 288, 2981−2997.
American Academy of Pediatrics, 1995. Guidelines for the ethical conduct of studies to evaluate drugs in pediatric populations. Pediatrics 95, 286−294.
Amiel, S.A., Sherwin, R.S., Simonson, D.C., et al., 1986. Impaired insulin action in puberty. A contributing factor to poor glycemic control in adolescents with diabetes. N. Engl. J. Med. 315, 215−219.
Aperia, A., Broberger, O., Elinder, G., et al., 1981. Postnatal development of renal function in pre-term and full-term infants. Acta Paediatr. Scand. 70, 183−187.
Arant Jr., B.S., 1978. Developmental patterns of renal functional maturation compared in the human neonate. J. Pediatr. 92, 705−712.
Arnold, C., Pedroza, C., Tyson, J.E., 2014. Phototherapy in ELBW newborns: does it work? Is it safe? The evidence from randomized clinical trials. Semin. Perinatol. 38, 452−464.
Barker, N., Hadgraft, J., Rutter, N., 1987. Skin permeability in the newborn. J. Invest. Dermatol. 88, 409−411.
Bartosh, S.M., Aronson, A.J., 1999. Childhood hypertension. An update on etiology, diagnosis, and treatment. Pediatr. Clin. North Am. 46, 235−252.
Bellinger, D.C., Jonas, R.A., Rappaport, L.A., et al., 1995. Developmental and neurologic status of children after heart surgery with hypothermic circulatory arrest or low-flow cardiopulmonary bypass. N. Engl. J. Med. 332, 549−555.
Bellinger, D.C., Wypij, D., duPlessis, A.J., et al., 2003. Neurodevelopmental status at eight years in children with dextro-transposition of the great arteries: the Boston Circulatory Arrest Trial. J. Thorac. Cardiovasc. Surg. 126, 1385−1396.
Benjamin Jr., D.K., Smith, P.B., Jadhav, P., et al., 2008. Pediatric antihypertensive trial failures: analysis of end points and dose range. Hypertension 51, 834−840.
Benjamin Jr., D.K., Smith, P.B., Murphy, M.D., et al., 2006. Peer-reviewed publication of clinical trials completed for pediatric exclusivity. JAMA 296, 1266−1273.
Berseth, C.L., 1989. Gestational evolution of small intestine motility in preterm and term infants. J. Pediatr. 115, 646−651.
Blake, M.J., Castro, L., Leeder, J.S., et al., 2005. Ontogeny of drug metabolizing enzymes in the neonate. Semin. Fetal Neonatal Med. 10, 123−138.

Blake, M.J., Gaedigk, A., Pearce, R.E., et al., 2007. Ontogeny of dextromethorphan O- and N-demethylation in the first year of life. Clin. Pharmacol. Ther. 81, 510−516.

Blumer, J.L., 1999. Off-label uses of drugs in children. Pediatrics 104, 598−602.

Boehm, G., Braun, W., Moro, G., et al., 1997. Bile acid concentrations in serum and duodenal aspirates of healthy preterm infants: effects of gestational and postnatal age. Biol. Neonate 71, 207−214.

Carlo, W.A., Stark, A.R., Wright, L.L., et al., 2002. Minimal ventilation to prevent bronchopulmonary dysplasia in extremely low birth weight infants. J. Pediatr. 141, 370−374.

Chervenak, F.A., McCullough, L.B., 2007. Ethics of maternal-fetal surgery. Semin. Fetal Neonatal Med. 12, 426−431.

Cohen-Wolkowiez, M., Sampson, M., Bloom, B., et al., 2013. Determining population and developmental pharmacokinetics of metronidazole using plasma and dried blood spot samples from premature infants. Pediatr. Infect. Dis. J. 32, 956−961.

Committee on Clinical Research Involving Children, 2004. Board on health sciences policy. In: Field, M.J., Behrman, R.E. (Eds.), Ethical Conduct of Clinical Research Involving Children. National Academies Press, Washington, DC, pp. 146−211.

Davis, B.R., Cutler, J.A., Gordon, D.J., et al., 1996. Rationale and design for the Antihypertensive and Lipid Lowering Treatment to Prevent Heart Attack Trial (ALLHAT). ALLHAT Research Group. Am. J. Hypertens. 9, 342−360.

Di Lorenzo, C., Flores, A.F., Hyman, P.E., 1995. Age-related changes in colon motility. J. Pediatr. 127, 593−596.

Dome, J., Perlman, E.J., Ritchey, M.L., Coppes, M.J., Kalapurakal, J., Grundy, P.E., 2006. Renal tumors. In: Pizzo, P., Poplack, D. (Eds.), Principles and Practice of Pediatric Oncology, fifth ed. Lippincott Williams & Wilkins, Philadelphia, PA, pp. 905−932.

Doyle, L.W., Anderson, P.J., 2010. Adult outcome of extremely preterm infants. Pediatrics 126, 342−351.

Ehrnebo, M., Agurell, S., Jalling, B., et al., 1971. Age differences in drug binding by plasma proteins: studies on human foetuses, neonates and adults. Eur. J. Clin. Pharmacol. 3, 189−193.

Euler, A.R., Ament, M.E., 1977. Gastrin concentration in the neonate. Pediatrics 60, 791.

Euler, A.R., Byrne, W.J., Cousins, L.M., et al., 1977. Increased serum gastrin concentrations and gastric acid hyposecretion in the immediate newborn period. Gastroenterology 72, 1271−1273.

Flory, J., Emanuel, E., 2004. Interventions to improve research participants' understanding in informed consent for research: a systematic review. JAMA 292, 1593−1601.

Fluhr, J.W., Pfisterer, S., Gloor, M., 2000. Direct comparison of skin physiology in children and adults with bioengineering methods. Pediatr. Dermatol. 17, 436−439.

Frattarelli, D., Galinkin, J., Green, T., et al., 2014. Off-label use of drugs in children. Pediatrics 133, 563−567.

Friis-Hansen, B., 1961. Body water compartments in children: changes during growth and related changes in body composition. Pediatrics 28, 169−181.

Giacoia, G., Taylor-Zapata, P., Zajicek, A., 2012. Eunice Kennedy Shriver National Institute of Child Health and Human Development Pediatrics Formulation Initiative: proceedings from the Second Workshop on Pediatric Formulations. Clin. Ther. 34, S1−S10.

Goepfert, A.R., Andrews, W.W., Carlo, W., et al., 2004. Umbilical cord plasma interleukin-6 concentrations in preterm infants and risk of neonatal morbidity. Am. J. Obstet. Gynecol. 191, 1375−1381.

Goutieres, F., Aicardi, J., 1977. Accidental percutaneous hexachlorophene intoxication in children. BMJ 2, 663−665.

Green, D.M., Sklar, C.A., Boice Jr., J.D., Mulvihill, J.J., et al., 2009. Ovarian failure and reproductive outcomes after childhood cancer treatment: results from the Childhood Cancer Survivor Study. J. Clin. Oncol. 27, 2374−2381.

Greenblatt, D.J., Koch-Weser, J., 1976. Intramuscular injection of drugs. N. Engl. J. Med. 295, 542−546.

Guignard, J.P., Torrado, A., Da Cunha, O., et al., 1975. Glomerular filtration rate in the first three weeks of life. J. Pediatr. 87, 268−272.

Gupta, M., Brans, Y.W., 1978. Gastric retention in neonates. Pediatrics 62, 26−29.

Gurney, J.G., Ness, K.K., Stovall, M., et al., 2003. Final height and body mass index among adult survivors of childhood brain cancer: childhood cancer survivor study. J. Clin. Endocrinol. Metab. 88, 4731−4739.

Hack, M., Flannery, D.J., Schluchter, M., et al., 2002. Outcomes in young adulthood for very-low-birth-weight infants. N. Engl. J. Med. 346, 149−157.

Hack, M., Taylor, M., Drotar, D., et al., 2005. Chronic conditions, functional limitations, and special health care needs of school-aged children born with extremely low-birth-weight in the 1990s. JAMA 294, 318−325.

Hamilton, B.E., Martin, J.A., Osterman, M.J.K., et al., 2014. Births: preliminary data for 2013. Natl. Vital Stat. Rep. 63 (2).

Harrison, M.R., Keller, R.L., Hawgood, S.B., et al., 2003. A randomized trial of fetal endoscopic tracheal occlusion for severe fetal congenital diaphragmatic hernia. N. Engl. J. Med. 349, 1916−1924.

Hartnoll, G., Betremieux, P., Modi, N., 2000. Body water content of extremely preterm infants at birth. Arch. Dis. Child. Fetal Neonatal 83, F56−F59.

Heron, M., 2007. Deaths: leading causes for 2004. Natl. Vital Stat. Rep. 56, 1−95.

Hirschfeld, S., 2005. Pediatric patients and drug safety. J. Pediatr. Hematol. Oncol. 27, 122−124.

Hovi, P., Andersson, S., Eriksson, J.G., et al., 2007. Glucose regulation in young adults with very low birth weight. N. Engl. J. Med. 356, 2053−2063.

Howlander, N., Noone, A.M., Krapcho, M., et al. (Eds.), April 2014. SEER Cancer Statistics Review, 1975−2011. National Cancer Institute, Bethesda, MD. http://seer.cancer.gov/csr/1975_2011/. Based on November 2013 SEER data submission, posted to the SEER Website.

Huang, N.N., High, R.H., 1953. Comparison of serum levels following the administration of oral and parenteral preparations of penicillin to infants and children of various age groups. J. Pediatr. 42, 657−658.

Hudson, M.M., 2007. Anthracycline cardiotoxicity in long-term survivors of childhood cancer: the light is not at the end of the tunnel. Pediatr. Blood Cancer 48, 649−650.

Hudson, M.M., Rai, S.N., Nunez, C., et al., 2007. Noninvasive evaluation of late anthracycline cardiac toxicity in childhood cancer survivors. J. Clin. Oncol. 25, 3635–3643.

Joffe, S., Cook, E.F., Cleary, P.D., et al., 2001a. Quality of informed consent: a new measure of understanding among research subjects. J. Natl. Cancer Inst. 93, 139–147.

Joffe, S., Cook, E.F., Cleary, P.D., et al., 2001b. Quality of informed consent in cancer clinical trials: a cross-sectional survey. Lancet 358, 1772–1777.

Jonsdottir, G.M., Georgsdottir, I., Haraldsson, A., Hardardottir, H., Thorkelsson, T., Dagbjartsson, A., 2012. Survival and neurodevelopmental outcome of ELBW children at 5 years of age: comparison of two cohorts born 10 years apart. Acta Paediatr. 101, 714–718.

Kajantie, E., Hovi, P., Raikkonen, K., et al., 2008. Young adults with very low birth weight: leaving the parental home and sexual relationships — Helsinki Study of Very Low Birth Weight Adults. Pediatrics 122, e62–e72.

Kearns, G.L., Abdel-Rahman, S.M., Alander, S.W., et al., 2003. Develop-mental pharmacology — drug disposition, action, and therapy in infants and children. N. Engl. J. Med. 349, 1157–1167.

Kelly, L., Seri, I., 2008. Renal development physiology: relevance to clinical care. NeoReviews 9.

Kimberly, M.B., Hoehn, K.S., Feudtner, C., et al., 2006. Variation in standards of research compensation and child assent practices: a comparison of 69 institutional review board-approved informed permission and assent forms for 3 multicenter pediatric clinical trials. Pediatrics 117, 1706–1711.

Kopelman, L.M., 2006. Children as research subjects: moral disputes, regulatory guidance, and recent court decisions. Mt. Sinai J. Med. 73, 596–604.

Lacroix, D., Sonnier, M., Moncion, A., et al., 1997. Expression of CYP3A in the human liver — evidence that the shift between CYP3A7 and CYP3A4 occurs immediately after birth. Eur. J. Biochem. 247, 625–634.

Landier, W., Armenian, S., Bhatia, S., 2015. Late effects of childhood cancer and its treatment. Pediatr. Clin. North Am. 62 (1), 275–300.

Laughon, M., Benjamin, D., Capparelli, E., et al., 2011. Innovative clinical trial design for pediatric therapeutics. Expert Rev. Clin. Pharmacol. 4, 643–652.

Lilford, R.J., Thornton, J.G., Braunholtz, D., 1995. Clinical trials and rare diseases: a way out of a conundrum. BMJ 311, 1621–1625.

Lindstrom, K., Winbladh, B., Haglund, B., et al., 2007. Preterm infants as young adults: a Swedish national cohort study. Pediatrics 120, 70–77.

Mammel, K.A., Kaplan, D.W., 1995. Research consent by adolescent minors and institutional review boards. J. Adolesc. Health 17, 323–330.

Marek, E., Kraft, W.K., 2014. Ethanol pharmacokinetics in neonates and infants. Curr. Ther. Res. Clin. Exp. 76, 90–97.

Mathis, L.L., Iyasu, S., 2007. Safety monitoring of drugs granted exclusivity under the Best Pharmaceuticals for Children Act: what the FDA has learned. Clin. Pharmacol. Ther. 82, 133–134.

Mertens, A.C., Liu, Q., Neglia, J.P., Wasilewski, K., et al., 2008. Cause-specific late mortality among 5-year survivors of childhood cancer: the Childhood Cancer Survivor Study. J. Natl. Cancer Inst. 100, 1368–1379.

Moleski, M., 2000. Neuropsychological, neuroanatomical, and neurophysiological consequences of CNS chemotherapy for acute lymphoblastic leukemia. Arch. Clin. Neuropsychol. 15, 603–630.

Molloy, C.S., Wilson-Ching, M., Anderson, V.A., et al., 2013. Visual processing in adolescents born extremely low birth weight and/or extremely preterm. Pediatrics 132, e704–e712.

Moore, T., Hennessy, E.M., Myles, J., Johnson, S.J., Draper, E.S., Costeloe, K.L., Marlow, N., 2012. Neurological and developmental outcome in extremely preterm children born in England in 1995 and 2006: the EPICure studies. BMJ 345, e7961.

Morris, B.H., Oh, W., Tyson, J.E., et al., 2008. Aggressive vs. conservative phototherapy for infants with extremely low birth weight. N. Engl. J. Med. 359, 1885–1896.

Morselli, P.L., Franco-Morselli, R., Bossi, L., 1980. Clinical pharmacokinetics in newborns and infants. Age-related differences and therapeutic implications. Clin. Pharmacokinet. 5, 485–527.

Moskowitz, C.S., Chou, J.F., Wolden, S.L., Bernstein, J.L., et al., 2014. Breast cancer after chest radiation therapy for childhood cancer. J. Clin. Oncol. 32, 2217–2223.

Moster, D., Lie, R.T., Markestad, T., 2008. Long-term medical and social consequences of preterm birth. N. Engl. J. Med. 359, 262–273.

Mulhern, R.K., Kepner, J.L., Thomas, P.R., et al., 1998. Neuropsychologic functioning of survivors of childhood medulloblastoma randomized to receive conventional or reduced-dose craniospinal irradiation: a Pediatric Oncology Group study. J. Clin. Oncol. 16, 1723–1728.

Mulhern, R.K., Merchant, T.E., Gajjar, A., et al., 2004. Late neurocognitive sequelae in survivors of brain tumours in childhood. Lancet Oncol. 5, 399–408.

Nachman, R.L., Esterly, N.B., 1971. Increased skin permeability in preterm infants. J. Pediatr. 79, 628–632.

National Commission on Protection of Human Subjects in Biomedical and Behavioral Research, 1979. The Belmont Report: Ethical Principles and Guidelines for Protection of Human Subjects of Research. http://www.hhs.gov/ohrp/humansubjects/guidance/belmont.htm (accessed 11.06.2008.).

Nelson, K.B., Dambrosia, J.M., Grether, J.K., Phillips, T.M., 1998. Neonatal cytokines and coagulation factors in children with cerebral palsy. Ann. Neurol. 44, 665–675.

Newburger, J.W., Jonas, R.A., Wernovsky, G., et al., 1993. A comparison of the perioperative neurologic effects of hypothermic circulatory arrest versus low-flow cardiopulmonary bypass in infant heart surgery. N. Engl. J. Med. 329, 1057–1064.

Nikolovski, J., Stamatas, G.N., Kollias, N., et al., 2008. Barrier function and water-holding and transport properties of infant stratum corneum are different from adult and continue to develop through the first year of life. J. Invest. Dermatol. 128, 1728–1736.

Nuremberg Code, 1949. Directives for Human Experimentation. Office of Human Subjects Research. http://ohsr.od.nih.gov/guidelines/nuremberg.html (accessed 30.08.2008).

Oeffinger, K.C., Mertens, A.C., Sklar, C.A., et al., 2006. Chronic health conditions in adult survivors of childhood cancer. N. Engl. J. Med. 355 (15), 1572–1582.

Oeffinger, K.C., Mertens, A.C., Sklar, C.A., et al., 2003. Obesity in adult survivors of childhood acute lymphoblastic leukemia: a report from the Childhood Cancer Survivor Study. J. Clin. Oncol. 21, 1359—1365.

Paavonen, E.J., Strang-Karlsson, S., Raikkonen, K., et al., 2007. Very low birth weight increases risk for sleep-disordered breathing in young adulthood: the Helsinki Study of Very Low Birth Weight Adults. Pediatrics 120, 778—784.

Peleg, O., Bar-Oz, B., Arad, I., 1998. Coma in a premature infant associated with the transdermal absorption of propylene glycol. Acta Paediatr. 87, 1195—1196.

Peterson, C.C., Johnson, C.E., Ramirez, L.Y., et al., 2008. A meta-analysis of the neuropsychological sequelae of chemotherapy-only treatment for pediatric acute lymphoblastic leukemia. Pediatr. Blood Cancer 51, 99—104.

Poley, J.R., Dower, J.C., Owen, C.A., et al., 1964. Bile acids in infants and children. J. Lab. Clin. Med. 63, 838—846.

Pyhälä, R., Hovi, P., Lahti, M., Sammallahti, S., Lahti, J., Heinonen, K., Pesonen, A.K., Strang-Karlsson, S., Eriksson, J.G., Andersson, S., Järvenpää, A.L., Kajantie, E., Räikkönen, K., 2014. Very low birth weight, infant growth, and autism-spectrum traits in adulthood. Pediatrics 134, 1075—1083.

Rakhmanina, N.Y., van den Anker, J.N., 2006. Pharmacological research in pediatrics: from neonates to adolescents. Adv. Drug Deliv. Rev. 58, 4—14.

Rakozy, C., 2008. Kidney-tumor. Pathol. Com. http://www.pathologyoutlines.com/kidneytumour.html (revised 8.07.2008.).

Roberts, G., Burnett, A.C., Lee, K.J., et al., 2013. Quality of life at age 18 years after extremely preterm birth in the post-surfactant era. J. Pediatr. 163, 1008—1013.e1.

Roberts, G., Lee, K.J., Cheong, J.L., et al., 2014. Higher ambulatory blood pressure at 18 years in adolescents born less than 28 weeks' gestation in the 1990s compared with term controls. J. Hypertens. 32, 620—626.

Rodriguez, W., Selen, A., Avant, D., et al., 2008. Improving pediatric dosing through pediatric initiatives: what we have learned. Pediatrics 121, 530—539.

Rutter, N., 1987. Percutaneous drug absorption in the newborn: hazards and uses. Clin. Perinatol. 14, 911—930.

Saigal, S., 2014. Functional outcomes of very premature infants into adulthood. Semin. Fetal Neonatal Med. 19, 125—130.

Saigal, S., Stoskopf, B., Boyle, M., Paneth, N., Pinelli, J., Streiner, D., Goddeeris, J., 2007. Comparison of current health, functional limitations, and health care use of young adults who were born with extremely low birth weight and normal birth weight. Pediatrics 119, e562—e573.

Sanders, J.E., 2008. Growth and development after hematopoietic cell transplant in children. Bone Marrow Transplant. 41 (2), 223—227.

Senat, M.V., Deprest, J., Boulvain, M., et al., 2004. Endoscopic laser surgery versus serial amnioreduction for severe twin-to-twin transfusion syndrome. N. Engl. J. Med. 351, 136—144.

Shah, S.S., Hall, M., Goodman, D.M., et al., 2007. Off-label drug use in hospitalized children. Arch. Pediatr. Adolesc. Med. 161, 282—290.

Shirkey, H., 1968. Therapeutic orphans. J. Pediatr. 72, 119—120.

Siber, G.R., Echeverria, P., Smith, A.L., et al., 1975. Pharmacokinetics of gentamicin in children and adults. J. Infect. Dis. 132, 637—651.

Siervogel, R.M., Demerath, E.W., Schubert, C., et al., 2003. Puberty and body composition. Horm. Res. 60, 36—45.

Sillanpaa, M., 1983. Changes in the prevalence of migraine and other headaches during the first seven school years. Headache 23, 15—19.

Stark, A.R., Carlo, W.A., Tyson, J.E., et al., 2001. Adverse effects of early dexamethasone treatment in extremely low-birth-weight infants. N. Engl. J. Med. 344, 95—101.

Steinherz, L., Steinherz, P., 1991. Delayed cardiac toxicity from anthracycline therapy. Pediatrician 18, 49—52.

Steinherz, L.J., Steinherz, P.G., Tan, C.T., et al., 1991. Cardiac toxicity 4 to 20 years after completing anthracycline therapy. JAMA 266, 1672—1677.

SUPPORT Study Group of the Eunice Kennedy Shriver NICHD Neonatal Research Network, Carlo, W.A., Finer, N.N., et al., 2010. Target ranges of oxygen saturation in extremely preterm infants. N. Engl. J. Med. 362, 1959—1969.

Tait, A.R., Voepel-Lewis, T., Malviya, S., 2003. Do they understand? (part I): parental consent for children participating in clinical anesthesia and surgery research. Anesthesiology 98, 603—608.

Tait, A.R., Voepel-Lewis, T., Malviya, S., et al., 2005. Improving the readability and processability of a pediatric informed consent document: effects on parents' understanding. Arch. Pediatr. Adolesc. Med. 159, 347—352.

Tyson, J.E., Kennedy, K.A., Lucke, J.F., Pedroza, C., 2007. Dilemmas initiating enteral feedings in high risk infants: how can they be resolved? Semin. Perinatol. 31, 61—73.

Vanpee, M., Herin, P., Zetterstrom, R., et al., 1988. Postnatal development of renal function in very low birthweight infants. Acta Paediatr. Scand. 77, 191—197.

Veldhuis, J.D., Roemmich, J.N., Richmond, E.J., et al., 2005. Endocrine control of body composition in infancy, childhood, and puberty. Endocrinol. Rev. 26, 114—146.

Vohr, B.R., Wright, L.L., Dusick, A.M., et al., 2004a. Center differences and outcomes of extremely low birth weight infants. Pediatrics 113, 781—789.

Follow-up care of high-risk infants. In: Vohr, B., Wright, L.L., Hack, M., et al. (Eds.), 2004b. Pediatrics, vol. 114, pp. S1377—S1397.

Ward, R.M., Lane, R.H., Albertine, K.H., 2006. Basic and translational research in neonatal pharmacology. J. Perinatol. 26 (Suppl. 2), S8—S12.

West, D.P., Worobec, S., Solomon, L.M., 1981. Pharmacology and toxicology of infant skin. J. Invest. Dermatol. 76, 147—150.

Whittle, A., Shah, S., Wilfond, B., et al., 2004. Institutional review board practices regarding assent in pediatric research. Pediatrics 113, 1747—1752.

Wilson, J.T., 1999. An update on the therapeutic orphan. Pediatrics 104, 585—590.

Wood, N.S., Marlow, N., Costeloe, K., et al., 2000. Neurologic and developmental disability after extremely preterm birth. N. Engl. J. Med. 343, 378—384.

Wypij, D., Newburger, J.W., Rappaport, L.A., et al., 2003. The effect of duration of deep hypothermic circulatory arrest in infant heart surgery on late neurodevelopment: the Boston Circulatory Arrest Trial. J. Thorac. Cardiovasc. Surg. 126, 1397—1403.

Wyszynska, T., Cichocka, E., Wieteska-Klimczak, A., et al., 1992. A single pediatric center experience with 1025 children with hypertension. Acta Paediatr. 81, 244—246.

Chapter 32

Cancer as a Paradigm for Translational and Clinical Biomedical Research

César Serrano[1] and George D. Demetri[2]

[1]Vall d'Hebron University Hospital, Barcelona, Spain; [2]Dana-Farber Cancer Institute and Ludwig Center at Harvard, Harvard Medical School, Boston, MA, United States

Chapter Outline

Key Points

- Cancer research has progressed from empiric to mechanistic: this has allowed great progress in translational and clinical research which has improved medical practice and changed patient outcomes for the better.
- The mechanisms by which genetic, genomic, epigenetic, and immunobiologic alterations are key to cancer have contributed to more precise definition of cancer types and cancer vulnerabilities for targeted therapies.
- Translational cancer research promotes a view of the science as a continuum from preclinical research to biomarker and treatment testing in patients with cancer of many types.
- A comprehensive understanding of recurrent genomic and nongenomic alterations across a variety of human cancer types and its connection with critical pathways can be linked with clinical data to gain important insights and stimulate rational translational research.
- International cooperative consortia and collaborations are essential for a cancer research model based on both genomewide, large-scale studies, but also on smaller-scale hypothesis-driven discovery science, which eventually should converge to bring real advances in cancer patients.

INTRODUCTION

The explosion of basic science resulting from public funding of cancer research has led to extraordinary advances in the understanding of critical mechanisms which drive cancer initiation and progression, together with a parallel and successful development of a wide range of therapies that directly target these crucial oncogenic drivers and other antineoplastic mechanisms in cancer patients. The combination of all these processes and discoveries resulting from this continuous flow between laboratory research, therapeutic discovery and development, and clinical research in cancer patients makes cancer research a paradigm of the translational and clinical research process.

The recent advances in translational cancer research are not random scientific events. Cancer is a major public health problem in the Unites States, as in many other parts of the world. Data collected by the US National Cancer Institute (NCI) and the Centers for Disease Control and Prevention (CDC) place cancer as the second leading cause of death in the United States, exceeded only by heart disease. The estimated number of deaths in the United States in 2014 was 585,720 people, about 1600 people a day (Siegel et al., 2014); this serves as an example of the relevance of this disease in the overall population, and the urgent need to generate meaningful data impacting on patients in the short- and medium term. The public commitment to improving cancer outcomes is notable, as anticancer research ranks among the top-funded disease areas by the US NCI and the National Institutes of Health (NIH), the single largest funder of biomedical research in the world.

Several advances in translational oncology over the past two decades have enabled the positive inflection to the curve of progress. First, a technology revolution has allowed the scientific community to obtain high-resolution, large-scale data from different fields, such as genomics, proteomics, or metabolomics. Whereas initial "high-throughput" reports from 2010 were confined to single samples (Pleasance et al., 2010), studies of hundreds of samples are currently the norm. Second, there has been a remarkable shift from cancer diagnosis based exclusively on histopathology, to the identification of tumor types and subtypes based on defined genomic and molecular biologic characteristics, with great progress in the field of prognostic and predictive biomarkers. For instance, the human epidermal growth factor receptor-2 (HER2) was noted to be overexpressed in ~15% of patients with breast cancer (Slamon et al., 1987). This molecular feature not only served to define a subgroup of breast tumors associated with aggressive clinical behavior and decreased survival, but also identified a population that could be molecularly targeted with new therapeutics: the patients selected for this specific protein over-expression greatly benefited from trastuzumab, a monoclonal antibody that directly binds to, and inhibits, HER2 oncogenic signaling (Baselga et al., 1996). Another key factor is the substantial effort which has been made to develop new models for fundamental discovery research in cancer. Particularly, in vivo models that are more relevant to human cancer pathophysiology than older immortal cancer cell lines have enabled a more accurate and meaningful assessment of mechanisms of human tumorigenesis and neoplastic disease states. Together, all these aspects of cancer research have contributed to reduce the time from the discovery of a new basic science concept to the clinical proof of concept, and subsequently to the application of a safe and effective new therapy in clinical practice. Thus, the US Food and Drug Administration (FDA) approved trastuzumab for the treatment of metastatic breast cancer in 1998, a full 13 years after HER2 was identified as a critical oncogenic driver in breast cancer (Slamon et al., 1987, 2001). In contrast, more recent discoveries, such as chromosomal rearrangements dysregulating activity of the anaplastic lymphoma kinase (ALK) in non−small-cell lung cancer (NSCLC) (Soda et al., 2007), rapidly translated into proof-of-concept trials (phase 1 and 2) and pivotal regulatory studies with the ALK inhibitors crizotinib which led to FDA approval in the record time of 4 years after the discovery of ALK fusions in NSCLC (Shaw et al., 2013), as well as other kinase inhibitors (e.g., ceritinib or alectinib) to overcome resistance to crizotinib.

In this chapter, we will explore a number of strategies related to cancer translational research that are hallmarks of what makes this work a model for effective biomedical research. The goal here is not to survey the full range of current research in these areas. Rather than being encyclopedic, we will concentrate here on landmark developments in translational cancer discovery, as well as applied research in diagnostics and therapeutics, that have accelerated the pace of new developments to improve clinical outcomes for cancer patients over the past two decades. By understanding the underlying principles of these research and development processes, we will better be able to advance from basic, preclinical and clinical research in other fields and suggest prospects to improve the success of translational medical research overall.

CANCER: FROM THE EDWIN SMITH PAPYRUS TO THE MOLECULAR GENETIC ERA

The importance of an accurate diagnosis in cancer is undeniable, particularly given the newest therapeutic agents which selectively target molecules and pathways that are specific to small subsets of this complex and heterogenous disease. Historically, "cancer" was categorized mainly according to the anatomic site and presumed tissue of origin. Thus, the

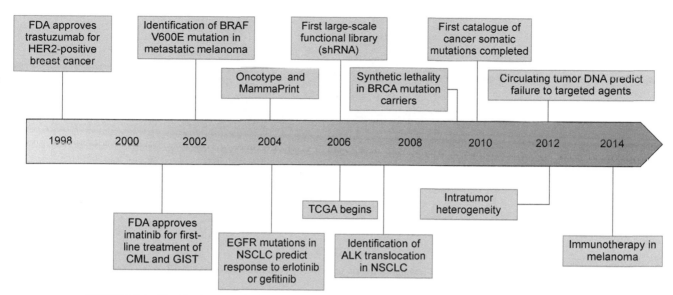

FIGURE 32.1 Timeline including landmark achievements in translational cancer research in the last two decades.

Edwin Smith Papyrus (3000 BC)—the earliest written description of cancer—depicts a bulging tumor of the breast as a grave disease without curative treatment. One thousand five hundred years later, the Ebers Papyrus (1500 BC) likely describes a soft-tissue sarcoma and includes reference to other possible cancers arising from the skin, uterus, stomach, and rectum (Hajdu, 2011). It was not until the 19th century when Dr. Johannes Müller (1801–58) provided some granularity among different types of cancer beyond the tissue of origin, as he first distinguished microscopically epithelial and connectivity tissue tumors. With improved technology of biochemical staining, tissue micromorphology and histopathology became the gold standard of diagnostics in cancer, despite the wide interobserver differences between the most expert pathologists. The most recent advances in tumor biology have shown that "cancer" is a complex web of hundreds, if not thousands, of different diseases which can be parsed and linked in novel subsets across anatomic or tissue-based definitions—and these definitional groups have important clinical implications for therapeutics and patient outcomes (Fig. 32.1).

Defining Subgroups in Cancer

Early approaches to distinguish and classify tumors beyond the limits of tissue morphology and histopathology were based on technologic advances in the field of immunohistochemistry, a process of detecting antigens (e.g., proteins) in cells of a tissue section by antibodies that specifically bind to those antigens. Several immunohistochemistry markers are currently in use to assist in the diagnosis of cancer. Of particular interest are those immunomarkers that either (1) are pathognomonic and unequivocally define the diagnosis of a specific type of cancer or (2) have prognostic and/or therapeutic implications for a type of cancer. Hematologic malignancies led the way in the immune characterization of human liquid tumors to separate different subtypes of leukemias and lymphomas. In solid tumors, the HER2 overexpression in aggressive breast carcinomas was another example.

Immunostaining also allowed the identification of previously obscure subtypes of neoplasms, exemplified by the expression of the CD117 antigen (a component of the protein product of the *KIT* proto-oncogene). Gastrointestinal stromal tumors (GISTs) were a poorly understood group of lesions identified purely on tissue morphology in the early 1980s. In a seminal discovery from 1998, Hirota et al. demonstrated that human GIST cells express the KIT tyrosine kinase which is commonly dysregulated due to oncogenic gain-of-function mutations in the *KIT* gene (Hirota et al., 1998). It is now well established that ~90% of metastatic GISTs express CD117 by immunohistochemical (IHC) staining, although this does not predict the underlying mutational status of the *KIT* gene, even though different IHC patterns can be described (e.g., diffuse cytoplasmic, membranous, or perinuclear patterns of staining) (Fletcher et al., 2013). Consequently, CD117 by IHC became a diagnostic marker for routine clinical use and is helpful to make a critically important diagnosis in gastrointestinal neoplasms.

The widespread use of immunohistochemistry coupled with advances in technology to sequence and analyze human tissue samples has contributed substantially to parsing into separate categories many cancers previously considered to be

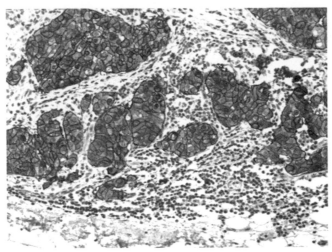

FIGURE 32.2 Representative immunohistochemical staining of human epidermal growth factor receptor-2 (HER2) in HER2-positive breast cancer. Uniform, intense membrane stain can be observed.

homogeneous. Thus, breast cancer has been changed in the past two decades after the identification of several molecular subgroups by DNA microarray and RNA expression technologies (Perou et al., 2000). These tools have clarified that breast cancer is a heterogeneous family of diseases, divisible into several clinical subtypes, which has been facilitated by the use in routine practice of standard IHC antibodies to assay steroid hormone receptors [estrogen receptor (ER) and progesterone receptor (PR)] and HER2. Based on ER, PR, and HER2 immunoexpression, at least four major disease subtypes have been established so far in breast cancer: two types of predominantly ER-positive tumors (luminal A and luminal B), and two predominantly ER-negative tumors (basal-like and HER2+/ER− subtypes) (Blows et al., 2010). Luminal A breast cancers express higher levels of ER compared to luminal B tumors. Additionally, HER2-positive tumors (Fig. 32.2) can be subdivided into HER2+/ER+ and HER2+/ER−. Importantly, intrinsic breast cancer subtypes carry different prognosis and therapeutic features. Thus, HER2-positive and basal-like tumors have less favorable outcomes than luminal subtypes, which likely reflect these subtypes' higher proliferative activity and predisposition to early and frequent relapse (Carey et al., 2006). The important translational and clinical impact is that these subtypes differ in the utility of therapeutic agents directed against hormone receptors and HER2.

It is important to bear in mind that immune-detectable protein expression is generally not exclusive for any single tumor type, and the same immunomarkers may be found in multiple neoplasms and inflammatory conditions. Thus, besides breast cancer, HER2 overexpression is also seen in ~20% of gastric cancers (Ruschoff et al., 2012). Although its role in prognosis remains controversial, this finding correlates with distinctive clinicopathologic features (intestinal-type and gastroesophageal junction predominance) and response to HER2-targeted therapies (Bang et al., 2010). It is important for researchers and clinicians to recognize the technical limitations of protein expression analyses by IHC, since the simplest interpretations may not necessarily apply to all tumor contexts. For instance, expression of CD117 expression might be found in other mesenchymal neoplasms resembling GIST, such as leiomyosarcomas (Raspollini et al., 2004), which have no activating mutation in the *KIT* gene and therefore do not use KIT as an oncogenic driver (Serrano et al., 2005). Alternatively, *KIT* oncogenic mutations have been found in a small subset of melanomas. As in HER2-positive gastric cancers, *KIT*-mutant melanomas show a singular clinical behavior (not related to sun exposure, and only distributed in acral and mucosal regions) (Curtin et al., 2006) and benefit from KIT inhibitory therapies (Hodi et al., 2013), which clearly show that mutationally activated KIT drives tumor progression in this subset of melanomas. However, there is no correlation between KIT mutation or amplification and CD117 expression (Beadling et al., 2008).

In summary, translational cancer research in different tumor models and clinical settings has defined a relevant biomarker as one that improves accuracy in the diagnosis of cancer or that helps to identify a clinically meaningful therapeutic target in the biology of a particular tumor type, which may vary depending on each cellular context.

A Genetic Definition of Cancer

Histopathology and immunohistochemistry have assisted in the diagnosis of cancer for several decades, but the basic and translational science advances in cancer molecular genetics now enable clinicians and investigators to define distinctive disease subcategories. Indeed, all the paradigmatic examples of human cancer categorization described in the previous section have an underlying genetic event eventually leading to the phenotypic features that would otherwise be difficult to separate.

DNA emerged as the main character in cancer research after the discovery of its structure (Watson and Crick, 1953), the determination that DNA is the molecular substrate of inheritance (Avery et al., 1944), and that damage to the primary structure through mutations leads to cancer (Loeb and Harris, 2008). DNA from cancer and normal cells within an individual derive from the same progenitor fertilized egg, and therefore the first attempts in cancer research aimed to decipher those sequences of DNA that were different in the cancer genome. Some of the genetic events were recurrent in certain cancer types [i.e., *KIT* mutations in GIST, or *BCR—ABL* translocation in chronic myeloid leukemia (CML)] (Hirota et al., 1998; Gorre et al., 2001), or further contributed to define a subset of tumors with different biology and/or clinical behavior within a particular cancer type [i.e., *HER2* amplification in breast cancer, or mutations in the gene encoding epidermal growth factor receptor (EGFR) in NSCLC] (Slamon et al., 1987; Paez et al., 2004). By contrast, other genes were found mutated across diverse types of cancer, such as p53 (Efeyan and Serrano, 2007) and *RAS* oncogenes (Malumbres and Barbacid, 2003). As we will see later, the emergence of rapid, inexpensive, and large-scale DNA sequencing technologies has allowed researchers to obtain a detailed and comprehensive view of changes in cancer genome (Stratton et al., 2009). Table 32.1 shows a sampling of genes relevant to specific diseases distributed by cancer type.

In all, translational research in cancer genetics has been a key driving force leading the change in paradigm from diagnosis based on morphology to a more comprehensive view of cancer as a highly diverse set of diseases that are linked by certain hallmarks and clinical behaviors (Fig. 32.1). The exponentially improving tools and technologies for genetics and genomics have contributed greatly to the advances in cancer-relevant translational research which have led to clinical advances. Thus, we have learned from these early studies that somatically mutated genes in cancer may act as oncogenes or as suppressors of tumorigenesis. Positively acting oncogenes are most often genes normally involved in signal transduction, for which mutation of just one allele may be sufficient to contribute to cancer development. Examples of such oncogenes include receptor tyrosine kinases (RTKs such as the products of genes including *EGFR, HER2, KIT*), signaling amplifiers and intermediates (e.g., *RAS or BRAF*), and transcription factors or other transcription-regulating proteins (*MITF, myc, cyclin D1*). By contrast, "tumor suppressor genes" require aberrant function of both alleles, which often result in abrogation of protein function. Tumor suppressor genes are often involved in the control of the cell cycle (*p53, RB1, CDKN2A*), prevention of apoptosis and proliferation (*PTEN, NF1*), or DNA repair (*BRCA1, BRCA2*). Extensive work has characterized many forms of cancer by these criteria, briefly summarized in Table 32.2.

Basic research has described cancer development as a multistep process consisting of progressive acquisition of heritable genetic variations that are naturally selected during tumorigenesis (Vogelstein and Kinzler, 1993) that ultimately

TABLE 32.1 Representative Targets and Kinase Inhibitors

Genetic Alteration	Therapeutic Agent	Tumor Type
Tyrosine Kinases		
ALK	Crizotinib	NSCLC
BCR/ABL	Imatinib	CML
EGFR	Erlotinib, gefitinib	NSCLC
	Cetuximab, panitumumab	mCRC, head and neck cancer
HER2	Trastuzumab	Breast and gastric cancers
KIT	Imatinib	GIST
PDGFRB	Imatinib	DFSP
Serine—Threonine Kinases		
BRAF V600E	Vemurafenib, dabrafenib	Melanoma
MEK1/2	Trametinib	Melanoma
mTOR	Everolimus	Breast and kidney cancers

ALK, anaplastic lymphoma kinase; *CML*, chronic myeloid leukemia; *DFSP*, dermatofibrosarcoma protuberans; *EGFR*, epidermal growth factor receptor; *GIST*, gastrointestinal stromal tumor; *HER2*, human epidermal growth factor receptor-2; *mCRC*, metastatic colorectal cancer; *NSCLC*, non—small-cell lung cancer.

TABLE 32.2 Technologies for Characterizing Tumors

Invasive Techniques
Mutation analysis
Sanger sequencing
Exome sequencing
Whole genome sequencing
DNA copy-number assessment
Comparative genome hybridization to DNA microarrays
Gene expression profiling
cDNA microarray
RNA sequencing
Large-scale functional studies
shRNA libraries
ORF libraries
CRISPR/Cas9 libraries
Drug-screen libraries
Noninvasive Techniques
Serum biomarkers
Circulating cell tumor DNA
Circulating tumor cells
PET-scan

results in distinctive and complementary capabilities that enable abnormal cell growth, cell accumulation, aberrant differentiation, and metastatic dissemination (Hanahan and Weinberg, 2011). These processes include several common "hallmark" mechanisms summarized by Hanahan and Weinberg in their landmark reviews.

Translation of Cancer Gene Expression Profiling to the Clinic

The hypothesis that phenotypic diversity of human cancer might be accompanied by a corresponding diversity in gene expression patterns eventually took shape in 2000. In their seminal study, Perou et al. (2000) proved that systematic investigation of gene expression patterns captured with complementary DNA (cDNA) microarrays led to an improved molecular taxonomy of human breast cancer (Perou et al., 2000). Briefly, cDNA microarray technology consists of labeling RNA samples obtained from patients and control subjects with distinguishable fluorescent dyes and hybridized to gene-specific probes composed of single strands of cDNA (Fodor et al., 1993). Relative levels of gene expression are estimated by measuring the fluorescence intensity of each probe. A hierarchical clustering method is used to group experimental samples on the basis of similarity in their patterns of expression. This technology was first used in a set of 65 surgical specimens of human breast tumors from 42 different individuals. In this study, two broad subgroups of breast cancer could be defined based on the lineage of the two types of cells present in the human mammary gland: basal cells and luminal cells. The gene expression cluster defining basal epithelial cells included keratin 5, keratin 17, integrin-β4, and laminin, whereas ER and ER-associated transcription factors clustered in a subgroup with characteristics of luminal cells. Moreover, overexpression of HER2 oncogene (Fig. 32.2) was associated with higher expression of a specific subset of genes and lower levels of ER and ER-associated genes (Perou et al., 2000).

This technology was rapidly applied to cancer across a diverse set of studies in multiple tumor types to investigate whether patterns of expression could be used to classify tumors and predict outcome and response to chemotherapy. Following with the example of breast cancer, a number of studies identified gene sets with prognostic and/or predictive value (Sotiriou and Piccart, 2007). Although early findings were initially taken with skepticism—because the list of differentially expressed genes from different studies apparently evaluating the same problem barely showed any overlap—a later study nicely demonstrated a high degree of concordance in predicting the outcome using five different gene expression signatures with little gene overlap (Fan et al., 2006). Of note, these prognosis signatures predicted outcome of disease more accurately than standard systems based on clinical and histologic criteria (Sotiriou and Piccart, 2007).

A 70-gene expression signature (Amsterdam signature, MammaPrint) (van de Vijver et al., 2002) and a 21-gene expression signature (Recurrence Score, Oncotype DX) (Paik et al., 2004) are both currently been tested in each prospective randomized clinical trials (MINDACT and TAILORx, respectively), which examine whether these signatures associated with risk of recurrence for women with early-stage breast cancer can be used to assign patients to the most appropriate and effective treatment. For instance, in the MINDACT trial node-negative and node-positive breast cancer patients have their risk assessed through both traditional clinical—pathological factors and the 70-gene profile. If both methods classify the patient's risk of relapse as low, adjuvant chemotherapy is withheld; if both methods classify the patient's risk of relapse as high, then chemotherapy is proposed; if the methods give discordant results, patients are randomized to follow the clinical—pathological method or to follow the genomic results, therefore aiming to identify the best type of therapy for the individual patient (Cardoso et al., 2007). Results from these two clinical trials are eagerly awaited by the entire scientific community, but it seems that we have begun a new era of personalized medicine guided by molecular biomarkers predicting prognosis and disease response to therapeutic agents.

Massive Parallel DNA Sequencing: Deciphering the Complexity of Cancer

Early Beginnings of Cancer Genomics

Cancer genomics aims to understand recurrent DNA alterations in specific cancer types through systematic, large-scale, high-resolution analyses of some or all of the genome. Application of emerging technologies in the field has been instrumental to boost the development that we have witnessed in the past decade (Metzker, 2010). Particularly, the daunting effort of analyzing the entire human genome—the Human Genome Project (HGP)—at the turn of the millennium paved the way for next steps in cancer research by means of a twofold contribution: on the one hand, the HGP sparked the field of wide-genome analyses, on the other hand, the reference human sequence was available for cancer research purposes (International Human Genome Sequencing Consortium, 2004).

First unbiased, genomewide, sequencing screenings were launched in early 2000s and examined tens of tumor samples with Sanger sequencing. These early studies narrowed down the number of genes analyzed to only those already known to be mutated in cancer encoding actionable RTKs and RTK-dependent signaling pathways (Tables 32.1 and 32.3). This strategy rapidly harvested successful results. In one of the first landmark reports, gain-of-function BRAF mutations were found in two-thirds of human melanomas and at lower frequency in a wide range of other human cancers (Davies et al., 2002). All BRAF mutations clustered in two hot spots within the somatic kinase domain of the *BRAF* gene (exons 11 and 15), and the single

TABLE 32.3 First-Line Food and Drug Administration—Approved Targeted Therapies as Monotherapy

Drug	Target	Tumor	RR (%)	mPFS (mo)	mOS (mo)
Crizotinib	ALK	NSCLC	74	10.9	N.R.
Denosumab	RANK	GCTB	41—58	N.R.	N.R.
Erlotinib	EGFR	NSCLC	61	9.7	19.3
Imatinib	BCR/ABL	CML	82	>60	>60
Imatinib	KIT	GIST	45	18	51
Imatinib	PDGFRA	DFSP	29.2	20.4	N.R.
Letrozole	ER/PR	Breast cancer	32	9.4m	34m
Vemurafenib	BRAF	Melanoma	53	6.8	15.9

ALK, anaplastic lymphoma kinase; *CML*, chronic myeloid leukemia; *DFSP*, dermatofibrosarcoma protuberans; *EGFR*, epidermal growth factor receptor; *ER*, estrogen receptor; *GCTB*, giant cell tumor of the bone; *GIST*, gastrointestinal stromal tumor; *m*, median; *mOS*, median overall survival; *mPFS*, median progression-free survival; *N.R.*, not reported; *NSCLC*, non—small-cell lung cancer; *PR*, progesterone receptor; *RR*, response rate.

substitution in exon 15, V600E, accounted for up to 80% of all mutations. Oncogenic activation of BRAF led to elevated kinase activity and constitutive activation of downstream signaling intermediates MEK1/2 and ERK1/2 (Davies et al., 2002). This landmark report shed light on melanomas and enlightened a decade of translational progress based on the relevance of RAS/MAPK pathway in melanoma biology. RAS/MAPK signaling turned out to be a critical lineage-dependent survival pathway starting in melanocytes (Garraway et al., 2005; Johannessen et al., 2013; Michaloglou et al., 2005; Pollock et al., 2003). Furthermore, inhibition of RAS/MAPK pathway with BRAF and/or MEK inhibitors significantly impacted on outcome in a disease formerly deficient in active therapeutic agents (Chapman et al., 2011; Flaherty et al., 2012).

Another significant study focused on mutations activating the other crucial signaling cascade in human cancer together with RAS/MAPK: the PI3K/AKT/mTOR signaling pathway. Similarly, PIK3CA mutations were screened in over 100 samples from patients with various cancer types (Samuels et al., 2004). A total of 92 somatic PIK3CA mutations were found in colorectal (32%), glioblastoma (27%), gastric (25%), breast (8%), and lung (4%) cancers. PIK3CA mutations demonstrated to increase kinase activity and therefore are regarded as oncogenes.

Besides BRAF and PIK3CA, several other screenings provided further insights predominantly in genes known to be involved in cancer, such as EGFR in NSCLC (Paez et al., 2004), fibroblast growth factor receptor 2 (FGFR2) in endometrial carcinoma (Dutt et al., 2008), or Janus kinase 2 (JAK2) in myeloproliferative disorders (Levine et al., 2005). Conversely, other studies discovered novel genetic alterations apparently surprising a priori. For instance, recurrent isocitrate dehydrogenase 1 (IDH1) somatic mutations were found in 12% of glioblastoma multiforme patients using next-generation sequencing (NGS) technologies (Parsons et al., 2008). Interestingly, identification of IDH1 mutations contributed to define a subset of IDH1-mutated patients—World Health Organization (WHO) grade II and III astrocytomas and oligodendrogliomas, and in glioblastomas that developed from lower-grade lesions—and established an unrecognized link between cell metabolism and cancer (Yan et al., 2009).

The Cancer Genome Atlas

It soon became obvious that it was necessary to increase the numbers of tumor sample data sets to enhance the power to detect recurrent genomic events. The emergence of massive parallel sequencing (MPS)—also called NGS—revolutionized the field because it had important advantages compared to prior methods: first, MPS is capable of sequencing large numbers of different DNA sequences in a single reaction (up to 1000 gigabases in a single run); second, it allows for the use of a single technology platform for all categories of genome analysis (mutations, copy number alterations, translocations, transcript levels, alternative splicing, chromatin structure); and third, it is possible to produce large number of reads at low cost in the so-called desktop sequencers (Metzker, 2010).

With both the improvement in genomic technologies and the aforementioned initial successes, cancer characterization using a genomic approach was prioritized by the US NCI and the International Cancer Genome Consortium (ICGC). The Cancer Genome Atlas (TCGA) project began in 2006 and was fully launched in 2009, involving researchers in more than 15 countries (Hudson et al., 2010). TCGA initial focus consisted of detailed evaluation of the mutational spectrum of various cancers, aiming to create a comprehensive catalog of cancer somatic mutations. By January 2015, TCGA had analyzed the whole genome in more than 10,000 tumors across more than 25 tumor types. Additionally, TCGA involves the creation of sophisticated mathematical methods and bioinformatics platforms for data analyses, and free release of data in raw and published formats. Together, it is undeniable that TCGA has become a milestone, as it provides the entire scientific community with a comprehensive catalog of cancer genes to inform basic biomedical research in cancer as well as to contribute to therapeutic development through the identification of potentially meaningful actionable targets (Fig. 32.1).

TCGA achievements and innovations are above and beyond the cancer genome and particular genes themselves. The more that we have studied cancer genes with deeper genomewide, large-scale analyses, the more complex the genomic landscape of cancer has turned out to be. In one study, analysis of 27 cancer types revealed that genomic events are quite heterogenous across different forms of cancer: the median frequency of nonsynonymous mutations vary by more than 1000-fold across cancer types (Lawrence et al., 2013). Pediatric tumors, such as Ewing sarcoma or malignant rhabdoid tumor, are associated with key driver genetic events (EWS translocations and SMARCB1 inactivation, respectively) despite having a very low mutation frequency [e.g., 0.1 per megabase (Mb)], whereas the highest mutation rates (100/Mb) are found in tumor linked to extensive exposure to well-known carcinogens, such as tobacco smoke (NSCLC) and ultraviolet radiation (melanoma). In the same study, two more types of heterogeneity were found: one involved natural groupings concerning to mutational spectrum. For instance, lung cancers were dominated by a mutational spectrum of C→A mutations, consistent with tobacco exposure, while C→T mutations were predominant in melanomas due to ultraviolet-induced covalent bonds. The so-called regional heterogeneity is the second type and corresponds to the variation of mutation rates across the genome. Mutation rates are very heterogeneous across the genome and correlates with DNA replication time and gene expression levels (Lawrence et al., 2013).

Several studies have looked in depth into various tumor types, and mutational landscapes have provided enormous amounts of data about biology, target candidates, or mechanism of resistance to anticancer agents. For instance, samples from 50 heavily pretreated metastatic castration-resistant prostate cancer (CRPC) patients were compared with 11 treatment-naïve, high-grade localized prostate cancers. Among many relevant findings, this study identified, as a common mechanism of hormone resistance, multiple driver mutations and copy number alterations in genes associated with androgen receptor (AR) signaling, particularly AR-interacting cofactors, such as FOXA1, MLL, MLL2, and ASH2L (Grasso et al., 2012). A different study in 32 primary head and neck squamous cell carcinomas unexpectedly identified inactivating mutations in NOTCH (Agrawal et al., 2011), at a time when NOTCH was regarded purely as an oncogene. Subsequent studies demonstrated that NOTCH and NOTCH-related genes are involved in regulation of squamous differentiation. The translational message is that the finding of the mutation informed the design of the experiments to seek alterative explanations about the protein function in cancer and normal biology.

TCGA has further contributed to a better understanding and categorization of cancer types. Indeed, all TCGA studies have shown that virtually all single-tissue cancer types can be further divided into several molecular subtypes. The most relevant subclassification is perhaps that from glioblastoma multiforme, due to the presence of actionable/druggable genes depending on the subtype. Four distinct categories were described based on gene expression and mutational analysis, each subtype showing distinctive profiles: proneural (PDGFRA, IDH1, p53), neural (neuron biomarkers), classical (EGFR, CDKN2A), and mesenchymal (NF1, PTEN, CHI3L1, MET) (Verhaak et al., 2010).

TCGA studies have not only provided relevant data regarding mutations and their consequences in each tumor type, but also are informing new concepts in biology. Cancer is considered a multistep process consisting of continuous acquisition of heritable genetic variations that are naturally selected during cancer progression (Vogelstein and Kinzler, 1993). However, some cancers might acquire, in a determined moment of their evolution, intense clustering of double-strand DNA breaks in well-circumscribed genomic regions, which differs substantially from the classical pattern of step-by-step acquisition of novel, unique events (Stephens et al., 2011). This phenomenon, termed "chromothripsis," occurs in a one-off cellular crisis, although key drivers of this new cell state still remain unknown. Nevertheless, if cancer cells are able to survive this genomic catastrophe, the resultant clones may have taken a considerable evolutionary leap forward with respect to the original normal cells or even the primary steps of neoplastic transformation.

TCGA consortium has recently taken a step forward and launched the Pan-Cancer initiative, which aims to inform about the relevance of biological processes through a combined analysis of multiple different types of data (Weinstein et al., 2013). This ambitious project includes molecular profiles at the DNA, RNA, protein, and epigenetic levels in 12 cancer types. Exome sequencing, copy number variation, DNA methylation, mRNA expression and sequence, microRNA expression, and transcript splice variation are the projected assessments within this initiative. Nonetheless, and despite all the landmark successes achieved by large-scale studies, several challenges are yet to be solved. Distinguishing between driver and passenger mutations has emerged as a matter of debate in the scientific community, and it is based on the fact that not all acquired somatic changes in the genome of cancer cells are involved in cancer development and/or progression (Stratton et al., 2009). A driver mutation confers a growth advantage on the cancer cell and is positively selected following Darwinian principles, and therefore it is considered casually implicated in the process of oncogenesis. Conversely, so-called passenger mutations are present in the cancer genome and accumulate during the subsequent cell division, but do not confer any clonal growth advantage and thus do not contribute to cancer development. Therefore, evolving computational approaches are warranted to better discriminate true cancer drivers and avoid false positives.

A second challenge concerns the impact of TCGA or high-throughput-driven science on patients' outcome. The ultimate test of cancer genomic is its ability to improve diagnostics and therapeutics. However, most of the targeted therapies recently approved or currently being tested in clinical trials were devised before TCGA was fully launched, because initial large-scale research focused on protein kinases—which are considered prime targets for cancer therapeutics due to their druggability and central role in cancer growth and survival. Thus, novel therapeutic approaches against crucial survival genes discovered within TCGA are urgently needed. Nonetheless, progress is underway. For instance, transcription factors were considered undruggable until an oncogene Myc inhibitor came on the scene. Small molecule JQ1 binds competitively to acetyl-lysine recognition motifs, also known as bromo and extraterminal (BET) family of bromodomains. BET bromodomain proteins are regulatory factors of c-Myc, and their inhibition substantially downregulates Myc transcription and profoundly impairs proliferation in vivo and in vitro (Filippakopoulos et al., 2010).

Finally, we should not dismiss some critical voices from renowned investigators in the field questioning the current cancer research model that prioritizes funding and resources to genomewide, large-scale studies, to the detriment of hypothesis-driven research (Weinberg, 2010). Indeed, both models have obtained undeniable achievements in the past decades, and their coexistence is essential to keep progress in basic and applied biomedical research on the right track. The cancer research processes described earlier represent a massive social and scientific experiment to test how to accelerate progress in medicine when new technology becomes widely available.

CANCER DRIVERS AND PERSONALIZED MEDICINE

The history of anticancer pharmacotherapy has been dominated by agents with very limited selectivity for cancer cells versus normal or healthy cells. Conversely, current therapeutic development lies in the rational discovery of treatments that selectively interfere with molecules considered to be crucial in specific cellular processes—neoplastic transformation, cell proliferation, invasion, metastases, or angiogenesis—across several tumor types and within a determined type of cancer alike. This is what we call personalized medicine (Tables 32.1 and 32.3).

Targets in Cancer Therapy

What Makes a Molecule a Good Target Candidate?

It is necessary to test between 5000 and 10,000 compounds in the most preliminary phase of drug discovery to eventually achieve that one single drug obtain approval of national or international authorities for its use in cancer patients. As a first step, chemical or biological agents are screened aiming to identify the subset of molecules that effectively and specifically hit their targets. Correct identification of cancer targets is nontrivial, since it certainly improves and speeds up the process of drug development. However, which properties best suit the definition of a good target candidate?

Any quest for anticancer agents must have deep roots in searching for each cancer's Achilles heel. In other words, the first indispensable requisite to consider any molecule as a target candidate for anticancer therapies is that the molecule should participate in a cellular function critical for cancer survival, and complete abolition of this function is expected to derail the entire oncogenic process. This is the case of CML. Reciprocal translocation between chromosomes 9 and 22 generates a chimeric gene encoding BCR—ABL fusion tyrosine kinase and constitutes the essential initiating and driver event in CML (Kurzrock et al., 1988). Imatinib is a specific inhibitor of the BCR-ABL tyrosine kinase and has efficacy in CML. Following preclinical studies, a randomized trial demonstrated significant superiority of imatinib over first-line standard treatment and therefore was approved for the treatment of CML (O'Brien et al., 2003).

The large majority of targeted agents currently being tested or approved in cancer are commonly directed against initiating/driver processes. However, novel insights in cancer biology constitute a source of emerging mechanisms against which it is possible to design innovative therapeutic strategies. For instance, it has not become apparent until recently that cancer homeostasis is a frail environment that depends upon the critical activity of heat shock proteins (HSPs). HSPs are chaperone proteins involved in the response to stress through the maintenance of protein stability and guidance of the normal folding of proteins, among several other functions (Whitesell and Lindquist, 2005). There is an increased expression of HSPs in cancer cells as an adaptive response to the numerous proteotoxic stressors to which cancer cells are subjected (i.e., hypoxia, acidosis, or unfolded proteins). Over the past decade, several small-molecule drugs targeting the molecular chaperone HSP90 have been identified as potential anticancer agents aiming to disrupt the proteome of cancer cells through accumulation and aggregation of unfolded proteins (Whitesell and Lindquist, 2005).

Any molecule might also be regarded as a good target candidate for cancer therapeutics if it is differentially and uniformly present in either a single tumor type or well-defined subsets of various tumor types. To this purpose, a target candidate, to be considered as such, should both represent a key oncogenic event and define a group of tumors. Therefore, cancer diagnosis accuracy, supported not only by morphological appearance but also by molecular features, is decisive in modern anticancer therapeutics. In this sense, we have mentioned earlier some paradigmatic examples of dual targets and diagnosis markers, such as HER2 in 10—15% of breast cancer (Slamon et al., 1987)—targeted by trastuzumab, and KIT (CD117) in up to 90% of GIST (Hirota et al., 1998)—targeted by imatinib. Obviously, target detection by clinical routine laboratory techniques increases the likelihood of implementing targeted treatments and recruiting patients for clinical trials.

A target has to be druggable if we want to consider it for further development in the preclinical and the clinical settings. Current genomewide association studies and deep sequencing screenings are identifying a myriad of molecules, pathways, and mechanisms relevant in cancer biology. However, the great majority of modern targets challenge drug discovery efforts because it involves a shift from traditional pharmacology based on enzymatic targets to protein—protein interaction. More insights about druggable and nondruggable targets are detailed later.

There is a long way from science to medicine, but the first step always consists in systematic preclinical characterization. Thus, translational cancer research first validates the activity and relevance of target inhibition by a given drug compound. As a consequence, translational research further contributes to a nuanced understanding of a determined area of biology. To this end, biological models are indispensable for functional target characterization in terms of activity, range of efficacy, and mechanisms, but also pharmacokinetic, pharmacodynamics and toxicity. Therefore, functional studies in appropriate preclinical models are necessary to culminate the effort of target candidates' evaluations.

In vitro and in vivo models are both extensively used in cancer research (Table 32.4). In vitro systems (i.e., human tumor cell lines with clinically representative genetic background) are well suited for first-pass validation aiming to (1) sift out less significant target candidates, (2) obtain preliminary functional data, and (3) prioritize candidates for more intensive in vivo validation. Conversely, in vivo models (i.e., patient-derived xenografts) are used as the last step prior to drug development in cancer patients and often involve complex and costly systems that provide highly informative and clinically relevant data. Therefore, the simplicity and immediacy of in vitro technology and results, as well as the quality of the information obtained throughout in vivo assays have to be weighted and placed correctly along the experimental process. In general, every functional assay has a limited range of information, and target discovery and validation require data from different sources to determine the biological weight of evidence for a genetic alteration (Chin and Gray, 2008).

Oncogene Addiction

So far it has been made clear that cancer is a multistep process in which a variable number of genetic alterations are gradually accumulated in less or greater extent, which in turn defines cancer capabilities (Vogelstein and Kinzler, 1993; Hanahan and Weinberg, 2011). Some of these oncogenic events are truly responsible for the growth advantage acquired by cancer cells (driver mutations), whereas other alterations might also contribute to tumor phenotype, but are dispensable. This observation is based on the heightened state of addiction (equal to dependency on) of cancer cells to oncogenes and tumor suppressor genes that enable tumor initiation and/or maintenance (Weinstein, 2002). Additionally, oncogene addiction can be exploited as a vulnerability of cancer cells with therapeutic purposes. This explains why modern cancer drug discovery focuses on the rational development of therapeutic strategies directed against driver events (Stratton et al., 2009). This concept, initially presented as oncogene addiction, has been later widened with other types of addiction to various hallmarks of cancer, particularly to several stresses that cancer cells need to handle for survival (i.e., mitotic stress, proteotoxic stress, etc.) (Luo et al., 2009).

Oncogenic addiction has been validated in cancer patients, as pharmacological suppression of a given driver molecule within a particular tumor type results in tumor shrinkage (response) and delay of tumor regrowth (progression-free survival). Notably, anticancer agents designed to directly target critical drivers in various tumor types show intriguing differences in terms of clinical efficacy. We would expect that any targeted therapy would hold back cancer progression, as is the case with imatinib in CML: after 5 years of treatment, 90% of the patients are still progression free (Druker et al., 2006). Likewise, inhibition of BRAF V600E kinase activity with vemurafenib, a specific BRAF inhibitor, achieves dramatic and unprecedented responses in up to 60% of melanoma patients with BRAF V600E mutation. However, and unlike imatinib in CML, half of the patients will have progressed within the first 6 months on therapy (Sosman et al., 2012). Leaving aside pharmacologic issues, it is possible that certain cancer types might rely more on oncogenic drivers than

TABLE 32.4 Functional Models for Cancer Research

In Vitro

Human cell line panels

Matched pairs of cell lines

3D cell cultures

Mouse-models explants

In Vivo

Cell line xenograft

Patient-derived xenograft

Genetically engineered mouse model

Phase 0–1 clinical trial

others, particularly if the overall genomic landscape tends to be simple and therefore, redundancy and adaptation mechanisms are limited (Hanahan and Weinberg, 2011). Indeed, melanoma is the tumor type with the highest rate of somatic mutations (Lawrence et al., 2013).

Druggable Versus Nondruggable Targets

Despite the impressive number of relevant molecules recently characterized in the biology of cancer by high-throughput efforts, the spectrum of druggable targets is considerably more narrow. On the one hand, drugs inhibit biochemical function (antagonist) rather than enhance it (agonist). Although some strategies attempt to restore the function of tumor suppressor genes (see in the next section how p53-MDM2 interaction can be targeted), logic leaves oncoproteins as the most suitable targets in cancer drug development. On the other hand, classic oncoproteins with enzymatic activity have well-defined active or allosteric sites, which constitute attractive binding regions for drug inhibition. Thus, the majority of current drug targets are G-protein-coupled receptors, nuclear receptors, or enzymes.

Oncogenically active kinases are actionable master regulators of downstream signal transduction in several cancer types. This explains why tyrosine kinases (i.e., KIT, EGFR, ALK) and serine/threonine kinases (i.e., BRAF V600E, mTOR) have been the main focus of drug development during most of the past two decades (Tables 32.1 and 32.3). By contrast, targeting nonenzymes is challenging because most of them lack obvious binding pockets or exert their function through protein–protein interactions, which usually involve larger and more diffuse contact areas (Arkin and Whitty, 2009).

FROM THE BENCH TO THE BEDSIDE

Numerous changes have shaken the daily routine practice in clinical oncology in just a little more than a decade. There is no better illustrative example of this fact than the current shift from cytotoxic chemotherapeutic agents to the rational development of compounds with better defined mechanisms of action. Although cytotoxic chemotherapies continue to be the backbone of cancer therapy, in the past 10 years over 40 molecularly targeted agents have been approved for the treatment of cancer by the US FDA, whereas fewer than 10 new cytotoxic chemotherapeutic agents were approved in the same period of time (Huang et al., 2014). Unlike cytotoxic regimens, normal cells are usually unaffected by targeted therapies, and therefore the tolerance is better. This greater target selectivity leads to a wider therapeutic window, which in turn enables the use of active drug doses against the intended cancer cell population (Fig. 32.3).

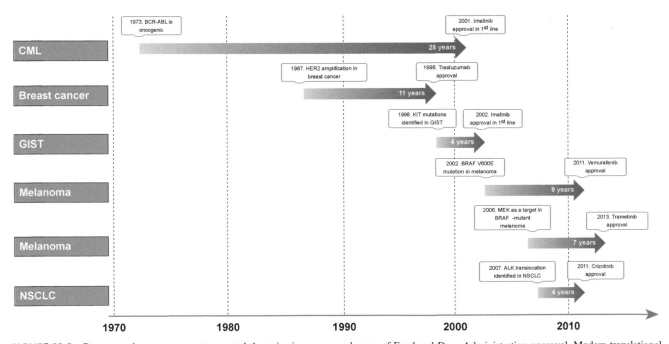

FIGURE 32.3 Representative cancer targets, targeted therapies in cancer, and years of Food and Drug Administration approval. Modern translational cancer research is accelerating the time from target discovery to clinical testing and drug approval. The number of targeted therapies approved in the last decade is higher than all the molecular agents previously approved.

The current state of the art is also exemplified by the increasingly extensive use of biomarkers, which contribute to an improved selection of patients based on molecular predictors. Various clinical programs around the world either already use or are implementing evolving gene panels containing mutational variants that guide decision-making in the daily clinical practice for both standard treatments and drug testing in clinical trials.

Together, these achievements are gradually changing the management of cancer patients toward a more personalized type of oncology.

Biomarkers and Clinical Trials

The concept of clinical trials has been reshaped—particularly those in earlier phases of drug development—due to the emergence of molecular targeted therapies. Indeed, the current *mantra* encourages that clinical investigation should involve drug development and biomarker validation in parallel. Ideally, robust and reproducible biomarkers should be developed and validated in a preclinical stage for later implementation in clinical studies.

Biomarkers

Biomarkers have assumed an increasingly relevant role in cancer therapeutics because targeted agents can maximize their activity when applied in the appropriate cellular context, and some of them have reached the daily routine use in clinical laboratories. The intended use of biomarkers can be academically simplified with the aim of responding to three different questions: should I treat this patient?; which drug should I use?; and is this dose effective in this patient? (Fig. 32.4).

Prognostic biomarkers address the first question because they allow clinicians to distinguish—within a determined cancer type—between tumors with favorable outcome and tumors with poor outcome. This certainly serves to guide health professionals in clinical decision-making and therapeutic approach. We can refer here to the earlier discussions about multigene expression tests such as MammaPrint and Oncotype DX, which are being validated to determine if their use can identify patients at low risk of recurrence and avoid often harmful adjuvant chemotherapeutic treatments. Also, there are numerous prospective and retrospective studies that link several intrinsic tumor aggressiveness properties with impaired prognosis, such as expression of growth factor receptors, signaling intermediates, or cell cycle components. A classic example would be the proliferation marker Ki-67 (Jonat and Arnold, 2011).

By contrast, predictive biomarkers, when present, define an increased likelihood of benefit from a particular treatment. Usually, these molecules are both biomarker and targets of the drugs for which they predict activity. Thus, in HER2-driven breast cancers HER2 amplification predicts activity for trastuzumab (Slamon et al., 1987, 2001); similarly, in ALK-translocated NSCLC identification of rearrangements involving ALK ensures a higher rate of success with crizotinib (Soda et al., 2007), and so on. In addition, there is a subset of predictive biomarkers that, if present, determine low or null chances of benefit from a determined target agent. Oncogenic KRAS mutation in metastatic colorectal cancer (mCRC) is perhaps the best well-known and standardized example of negative predictive biomarkers. Although EGFR is expressed in nearly 85% of mCRC and was an eligibility criterion for many of the clinical trials, it became clear after few retrospective studies using samples from pivotal, phase III clinical trials that EGFR-positivity was not a determinant of clinical efficacy. Approximately 30–40% of mCRCs carry KRAS oncogenic mutations, which in turn activates downstream RAS/MAPK

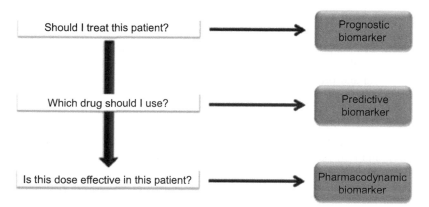

FIGURE 32.4   Cancer biomarkers can be used to determine prognosis, predict the activity of an anticancer agent, and determine the on-target effect of a given drug.

pathway independently of upstream EGFR control. Hence, it was found that the clinical activity of cetuximab and pan-itumumab anti-EGFR monoclonal antibodies is restricted to KRAS wild-type mCRC (Amado et al., 2008; Karapetis et al., 2008). Importantly, a take-home-message of KRAS and mCRC is the necessity that all biomarkers should be thoroughly evaluated in the preclinical stage and incorporated up front before large and costly clinical trials, which would minimize as well the number of patients receiving ineffective treatments, often with added toxic effects. However, in other cases the evidence is not that straightforward and more level of granularity in certain biological particularities is required to tune-up many other biomarkers. For instance, cell-cycle regulator cyclin-dependent kinase 4 (CDK4) is a key driver event in well-differentiated and dedifferentiated liposarcomas, where it is found amplified in >90% of the cases. However, a recent phase II clinical trial failed to show significant activity with a CDK4 inhibitor (palbociclib) despite the fact that a target population was molecularly selected before inclusion (Dickson et al., 2013). Conversely, palbociclib achieved a remarkable response rate in combination with antihormone treatment in a hormone receptor—positive breast cancer population using similar selection criteria as that of the previous study (Finn et al., 2015). Reasons for, respectively, failure and success remain unknown.

Finally, pharmacodynamic or surrogate biomarkers accurately reflect on-target effect of a certain drug and thus can be used to predict clinical benefit and guide dose selection of new anticancer agents at earlier time points. Pharmacodynamic biomarkers are closely related to phase I clinical trials and will be developed later.

Clinical Cancer Research

Early-stage clinical trials aim to define the therapeutic window of a given drug, selecting the most effective dose while staying within the safety range. In cytotoxic chemotherapy, optimal dosing was established in phase I studies by incremental dose escalation until maximum tolerated dose (MTD) was reached. However, the current framework for defining optimal dosing and treatment efficacy of molecularly targeted therapies is rapidly evolving. Traditional designs are less applicable because some drugs never reach the MTD, biomarkers facilitate when drugs are effectively hitting their targets, and disease stabilization is a common outcome—meaningless for traditional response criteria—such as Response Evaluation Criteria in Solid Tumors (RECIST) (Weber et al., 2015). Modern phase I trial design often includes a dose-escalation scheme based not only on toxicity evaluation, but also on sufficient pharmacokinetic, pharmacodynamic, and biomarker data. Once the activity range is established, an expansion cohort of patients is analyzed aiming to obtain more insights on toxicity, data on preliminary efficacy, and the recommended phase II dose (RP2D). Design usually allows cohort enrichment with patients whose tumors have molecular aberrations that might increase the likelihood of responses. Likewise, pathway inhibition assessment in patients' normal and/or tumor tissue at both baseline and other time points is currently pursued by an increasing number of clinical trials thanks to the introduction of pharmacodynamic biomarkers. All these data are considered collectively to better define drug dosing and schedule, thus playing a pivotal role in the clinical development of drugs subsequently granted worldwide approval after larger phase III, placebo-controlled studies such as KIT-inhibitor sunitinib in GIST (Demetri et al., 2009).

The American Society of Clinical Oncology (ASCO) recently published a policy statement update given the significant advances in this field. In this report, the ASCO working group reaffirmed the critical importance of phase I trials and emphasized their therapeutic intent (Weber et al., 2015). Thus, and irrespective of the undeniable successes achieved in the past years, it is important to be aware that the vast majority (more than 95%) of agents selected in preclinical studies to be clinically tested fail to show robust therapeutic effects and/or safe toxicity profile in phase I/II clinical trials and are subsequently dismissed (Ocana et al., 2011). Therefore, there are clear biological, ethical, and financial imperatives to increase the odds of the successful approval of cancer therapies. In this sense, the field is rapidly evolving and some drugs have obtained accelerated US FDA approval (i.e., pembrolizumab in melanoma) after remarkable results shown in phase I/II clinical studies. Interestingly, these trials accrue hundreds of patients to better assess efficacy, especially if there is a strong biological rational behind and promising activity is seen in early dose-finding, phase I cohorts.

Current Strategies in Molecular Targeted Therapy

Treatments based on molecular targeted therapies aim to exploit cancer vulnerabilities following the principle of oncogene addiction (Weinstein, 2002), and remarkable successes are achieved with single-agent strategies (Table 32.3). Regarding small molecule inhibitors (SMIs), we already cited the example of imatinib, which targets ABL and KIT kinases, obtaining spectacular results in CML (Kurzrock et al., 1988) and GIST (Demetri et al., 2002), respectively. As for monoclonal antibodies (mAbs), it is worth to mention monotherapy treatment with anti-RANK ligand (RANKL) human mAb deno-sumab in giant cell tumor of the bone (GCTB), an aggressive osteolytic tumor for which no standard medical treatment is

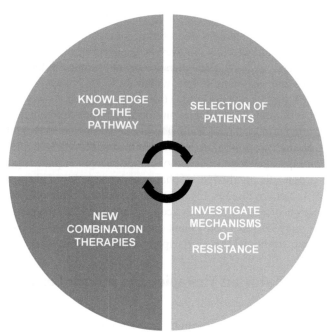

FIGURE 32.5 Schematic view of current progress in translational cancer research. The continuous flowing between lab and clinical research is instrumental for the recent advances in the cancer field.

available. Denosumab recently gained the FDA approval as first-line treatment for GCTB after a phase II study showed that it was associated with tumor responses and reduction of morbid surgeries (Chawla et al., 2013). Several other approaches targeting the main capabilities of cancer are in progress (Hanahan and Weinberg, 2011). Of particular interest is the elucidation of the role of immunity in cancer. Cancer cells express PD-L1, which interacts with PD-1 on activated T lymphocytes, resulting in the abolition of T cell—mediated antitumor response. Subsequent PD-1 blockade in melanoma induces tumor regression by inhibiting adaptive immune resistance (Tumeh et al., 2014).

However, the current trend in preclinical and clinical cancer research is drug combination (Fig. 32.5). In general, driver mechanisms of cancer oncogenicity are redundant rather than unique, and dual or triple therapies might maximize the likelihood of tumor eradication (Stommel et al., 2007). Thus, the first intuitive attempts investigated combinations of targeted agents and cytotoxic chemotherapeutic agents. Examples of this may include anti-EGFR mAbs cetuximab or panitumumab in mCRC (Serrano et al., 2010), or trastuzumab in HER2-positive breast cancer (Slamon et al., 2001). Overall, mAbs are much better tolerated than SMIs in combination with chemotherapy. A different approach evaluated the combination of two active SMI in advanced melanoma, dabrafenib (BRAF inhibitor) and trametinib (MEK inhibitor), which doubled the mPFS (median progression-free survival) of monotherapy with a BRAF inhibitor (Robert et al., 2015). Furthermore, the combination of concomitant BRAF and MEK inhibition underscored the relevance of targeting the same signaling pathway by two different mechanisms. This same principle also applies to the combination of two anti-HER2 mAbs, trastuzumab—inhibition of HER2 kinase activation—and pertuzumab—inhibition of HER2 dimerization. The addition of pertuzumab to the standard trastuzumab and docetaxel in HER2-positive metastatic breast cancer extended the overall survival with the median duration of response being 15.7 and 7.7 months, respectively (Swain et al., 2015). Nonetheless, we should be aware that the therapeutic window of combinations of kinase inhibitors is expected to be considerably narrow and therefore it will challenge their clinical development (Garraway and Janne, 2012).

THE CRUSADE TO OVERCOME DRUG RESISTANCE IN CANCER

Tumor Heterogeneity

Despite the outstanding benefits obtained by the aforementioned molecular strategies, cancer in metastatic stage is considered incurable because most of the tumors will eventually progress on anticancer therapies. Recent molecular characterization of tumors reveals a challenging and complex landscape both prior and after treatment with targeted therapies and directly points at inter- and intratumor heterogeneity as the leading cause of treatment failure.

In a landmark study from Dr. Swanton's group (Gerlinger et al., 2012), multiregion genetic analysis of four renal tumors showed spatially separated somatic mutations and chromosomal imbalances. Remarkably, two-thirds of all the mutations were heterogeneous and thus not detectable in every sequenced region. Study of paired samples from primary tumor and their corresponding metastases allowed researchers to dig into tumor phylogeny. Ubiquitous somatic mutations in all analyzed regions from primary and metastatic lesions within the same patient were deemed initiating, while shared (by several regions and/or metastases, but not all) and unique (to one site) mutations were considered to have emerged later in the multistep process of tumor evolution. Interestingly, and despite tumor heterogeneity, convergent phenotypic evolution was observed, as several confluent mechanisms leading to loss of PTEN—a PI3K pathway-negative regulator—were discovered. Together, tumor heterogeneity sets the start point for future cancer research and drug development based on the following facts: (1) presence of multiple, heterogeneous tumor clones in a given patient prevents tumor eradication with a single drug; (2) likewise, use of prognostic, predictive, or pharmacodynamic biomarkers from a single tumor-biopsy will unlikely give an actual portray of the entire disease; (3) targeted agents put selective pressure on a certain type of tumor clones, but allow Darwinian selection of cell clones with acquired advantageous characteristics; and (4) parallel genetic evolution of separate tumor clones may lead to a convergent phenotype that creates resistance to selective inhibition with molecular agents. Other studies have confirmed these principles in several other neoplasms (Yap et al., 2012).

Mechanisms of Resistance to Targeted Therapies and Future Perspectives in Drug Development

The greatest part of the recent progress in cancer research has been conditioned by the relevance of RTKs and RTK-dependent signaling pathways. Thus, the most solid models of progression on anticancer treatments are commonly related to resistance to targeted therapies. And resistance to molecular therapies may refer to de novo/primary resistance—primary refractoriness—and acquired/secondary resistance—tumor progression in the face of an ongoing treatment to which the tumor was initially sensitive.

Three broad mechanisms of acquired resistance to kinase inhibitors have been proposed: (1) target reactivation, (2) activation of upstream and/or downstream pathways, and (3) activation of bypass effectors (Garraway and Janne, 2012).

In relation to target reactivation, additional somatic mutations within the target gene itself are one of the most common drug resistance mechanisms. Secondary mutations impair kinase inhibitor efficacy because they perturb the conformational state of the kinase. This results in drug binding inhibition, reactivation of kinase activity, and tumor cell proliferation. Many SMIs exploit a conserved threonine residue within the ATP-binding site for binding specificity. Therefore, acquired substitution of the gatekeeper threonine with bulkier aminoacids constitutes a shared resistance mechanism to targeted therapies across several RTK-driven tumor types, including GIST (imatinib), CML (imatinib), or NSCLC (erlotinib) (Azam et al., 2008).

The second mechanism of resistance is frequently observed in BRAF-driven melanoma and concerns activation of upstream and/or downstream effectors of the pathway. Thus, feedback activation of upstream RTKs such as EGFR or PDGFR after treatment with BRAF inhibitors contributes to acquired resistance in the absence of secondary mutation in the target (Sun et al., 2014). This same mechanism, although with some biological particularities, also explains why BRAF-mutant colorectal cancers fail to respond to BRAF inhibition (Prahallad et al., 2012).

Tumors addicted to a single genetic event usually seek to reopen the crucial, driver-inhibited pathway. However, target inhibition might be bypassed by activation of alternative pathways, which in turn restore tumor cell proliferation. Tumors with lower rate of secondary mutations in the target are more prone to this mechanism, such as EGFR-driving neoplasms (NSCLC, mCRC, or head and neck cancer) (Chong and Janne, 2013). Thus, acquired resistance to erlotinib in EGFR-mutant NSCLC mediated by focal amplification of MET proto-oncogene perfectly exemplifies this situation (Engelman et al., 2007).

Mechanisms of resistance are multiple and heterogeneous in cancer, and thus it has been proposed that selective cotargeting of multiple core and emerging cancer capabilities will result in effective and durable therapies for cancer patients (Fig. 32.5). A major challenge in molecular medicine will be to target secondary events early enough to avoid treatment resistance. Accordingly, the cancer research field is moving in this direction, and many of the current strategies rely on the concept of drug combination to achieve synthetic lethality (Fig. 32.6). Two genes are synthetically lethal if mutation of either alone is compatible with viability but mutation of both leads to death. In applying synthetic lethality to drug discovery, target screenings attempt to find vulnerable genes that, when chemically inhibited, kills cells that harbor the cancer-related alteration, but spares normal cells (Kaelin, 2005). Subsequent studies are specifically looking into cancer vulnerabilities using large-scale, high-throughput in vitro screens comparing different conditions in the absence or presence

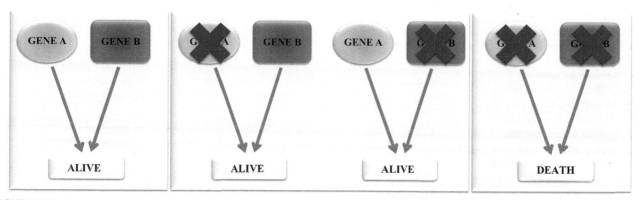

FIGURE 32.6   Schematic concept of synthetic lethality. It can be applied to cancer treatment if either of one of the two genes is targeted while the second gene is nonfunctional due to a genetic event.

of drug inhibitors. These studies include large-scale loss-of-function screens with RNA interference (Root et al., 2006), gain-of-function screens using open reading frames (Yang et al., 2011), or knockout screens with the CRISPR—Cas9 system (Shalem et al., 2014), to cite some relevant examples (Table 32.2).

More into translational and clinical fields, a number of strategies aiming to counteract tumor heterogeneity are under consideration. For instance, combination of targeted therapies and immunotherapy may improve the long-term efficacy of treatment in selected patients with low levels of genetic instability (Andre et al., 2013). Also, clinical trials combining inhibitors of the two main critical signaling pathways in cancer—RAS/MAPK and PI3K/AKT/mTOR—are undergoing (Tolcher et al., 2015). Additionally, new approaches aiming to monitor the disease are required, since spatially limited tumor biopsies are unlikely to represent the complexity of the cancer burden in any individual patient. "Liquid biopsies" based on amplification and detection of circulating tumor DNA from patients' blood samples have the promise to provide high resolution to detect the cumulative genomic abnormalities, including emerging resistant clones to targeted therapies. It was recently shown that this technology is suitable to detect emerging KRAS clones in mCRC patients weeks before the patient progressed on anti-EGFR treatment (Misale et al., 2012). Finally, clinical trials are evolving accordingly. Thus, new adaptive dose-finding designs are being developed to allow better guidance in the dose escalation, thereby permitting the study of an increased number of doses and combinations (Iasonos and O'Quigley, 2014). In summary, cancer research is moving from defining the mechanisms that are responsible for keeping cancer cells alive to translating this knowledge systematically into orthogonal approaches to target the unique vulnerabilities of different cancers. The complexity of cancer is mind-boggling, but the technology and research methods have promise to overcome many prior challenges and to allow practice-changing advances that were inconceivable in 1971 when the first "War on Cancer" was begun. "From mechanism to treatment, then to proof, then to practice" is at the heart of translational research, and cancer research embodies this as a paradigm applicable to all other fields of biomedical research.

LIST OF ACRONYMS AND ABBREVIATIONS

ALK Anaplastic lymphoma kinase
AR Androgen receptor
ASCO American Society of Clinical Oncology
BET Bromo and extraterminal
CDC Center for Disease Control and Prevention
CDK4 Cyclin-dependent kinase 4
CML Chronic myeloid leukemia
CRPC Castration-resistant prostate cancer
DFSP Dermatofibrosarcoma protuberans
ER Estrogen receptor
FDA Food and Drug Administration
FGFR2 Fibroblast growth factor receptor 2
GCTB Giant cell tumor of the bone
GIST Gastrointestinal stromal tumor
HER2 Human epidermal growth factor 2
HGP Human Genome Project

HSP Heat shock protein
ICGC International Cancer Genome Consortium
IDH1 Isocitrate dehydrogenase 1
JAK2 Janus kinase 2
mAb Monoclonal antibody
mCRC Metastatic colorectal cancer
MPS Massive parallel sequencing
MTD Maximum tolerated dose
NCI National Cancer Institute
NGS Next-generation sequencing
NIH National Institutes of Health
NSCLC Non−small-cell lung cancer
PR Progesterone receptor
RECIST Response Evaluation Criteria in Solid Tumors
RP2D Recommended phase II dose
RTK Receptor tyrosine kinase
SMI Small molecule inhibitor
TCGA The Cancer Genome Atlas
WHO World Health Organization

REFERENCES

Avery, O.T., Macleod, C.M., McCarty, M., 1944. Studies on the chemical nature of the substance inducing transformation of pneumococcal types: induction of transformation by a desoxyribonucleic acid fraction isolated from *Pneumococcus* type III. J. Exp. Med. 79 (2), 137−158.

Agrawal, N., Frederick, M.J., Pickering, C.R., Bettegowda, C., Chang, K., Li, R.J., et al., 2011. Exome sequencing of head and neck squamous cell carcinoma reveals inactivating mutations in NOTCH1. Science 333 (6046), 1154−1157.

Arkin, M.R., Whitty, A., 2009. The road less traveled: modulating signal transduction enzymes by inhibiting their protein−protein interactions. Curr. Opin. Chem. Biol. 13 (3), 284−290.

Amado, R.G., Wolf, M., Peeters, M., Van Cutsem, E., Siena, S., Freeman, D.J., et al., 2008. Wild-type KRAS is required for panitumumab efficacy in patients with metastatic colorectal cancer. J. Clin. Oncol. 26 (10), 1626−1634.

Azam, M., Seeliger, M.A., Gray, N.S., Kuriyan, J., Daley, G.Q., 2008. Activation of tyrosine kinases by mutation of the gatekeeper threonine. Nat. Struct. Mol. Biol. 15 (10), 1109−1118.

Andre, F., Dieci, M.V., Dubsky, P., Sotiriou, C., Curigliano, G., Denkert, C., et al., 2013. Molecular pathways: involvement of immune pathways in the therapeutic response and outcome in breast cancer. Clin. Cancer Res. 19 (1), 28−33.

Baselga, J., Tripathy, D., Mendelsohn, J., Baughman, S., Benz, C.C., Dantis, L., et al., 1996. Phase II study of weekly intravenous recombinant humanized anti-p185HER2 monoclonal antibody in patients with HER2/neu-overexpressing metastatic breast cancer. J. Clin. Oncol. 14 (3), 737−744.

Blows, F.M., Driver, K.E., Schmidt, M.K., Broeks, A., van Leeuwen, F.E., Wesseling, J., et al., 2010. Subtyping of breast cancer by immunohisto-chemistry to investigate a relationship between subtype and short and long term survival: a collaborative analysis of data for 10,159 cases from 12 studies. PLoS Med. 7 (5), e1000279.

Bang, Y.J., Van Cutsem, E., Feyereislova, A., Chung, H.C., Shen, L., Sawaki, A., et al., 2010. Trastuzumab in combination with chemotherapy versus chemotherapy alone for treatment of HER2-positive advanced gastric or gastro-oesophageal junction cancer (ToGA): a phase 3, open-label, rand-omised controlled trial. Lancet 376 (9742), 687−697.

Beadling, C., Jacobson-Dunlop, E., Hodi, F.S., Le, C., Warrick, A., Patterson, J., et al., 2008. KIT gene mutations and copy number in melanoma subtypes. Clin. Cancer Res. 14 (21), 6821−6828.

Carey, L.A., Perou, C.M., Livasy, C.A., Dressler, L.G., Cowan, D., Conway, K., et al., 2006. Race, breast cancer subtypes, and survival in the Carolina Breast Cancer Study. JAMA 295 (21), 2492−2502.

Curtin, J.A., Busam, K., Pinkel, D., Bastian, B.C., 2006. Somatic activation of KIT in distinct subtypes of melanoma. J. Clin. Oncol. 24 (26), 4340−4346.

Cardoso, F., Piccart-Gebhart, M., Van't Veer, L., Rutgers, E., 2007. The MINDACT trial: the first prospective clinical validation of a genomic tool. Mol. Oncol. 1 (3), 246−251.

Chapman, P.B., Hauschild, A., Robert, C., Haanen, J.B., Ascierto, P., Larkin, J., et al., 2011. Improved survival with vemurafenib in melanoma with BRAF V600E mutation. N. Engl. J. Med. 364 (26), 2507−2516.

Chin, L., Gray, J.W., 2008. Translating insights from the cancer genome into clinical practice. Nature 452 (7187), 553−563.

Chawla, S., Henshaw, R., Seeger, L., Choy, E., Blay, J.Y., Ferrari, S., et al., 2013. Safety and efficacy of denosumab for adults and skeletally mature adolescents with giant cell tumour of bone: interim analysis of an open-label, parallel-group, phase 2 study. Lancet Oncol. 14 (9), 901−908.

Chong, C.R., Janne, P.A., 2013. The quest to overcome resistance to EGFR-targeted therapies in cancer. Nat. Med. 19 (11), 1389−1400.

Davies, H., Bignell, G.R., Cox, C., Stephens, P., Edkins, S., Clegg, S., et al., 2002. Mutations of the BRAF gene in human cancer. Nature 417 (6892), 949−954.

Dutt, A., Salvesen, H.B., Chen, T.H., Ramos, A.H., Onofrio, R.C., Hatton, C., et al., 2008. Drug-sensitive FGFR2 mutations in endometrial carcinoma. Proc. Natl. Acad. Sci. U.S.A. 105 (25), 8713–8717.

Druker, B.J., Guilhot, F., O'Brien, S.G., Gathmann, I., Kantarjian, H., Gattermann, N., et al., 2006. Five-year follow-up of patients receiving imatinib for chronic myeloid leukemia. N. Engl. J. Med. 355 (23), 2408–2417.

Dickson, M.A., Tap, W.D., Keohan, M.L., D'Angelo, S.P., Gounder, M.M., Antonescu, C.R., et al., 2013. Phase II trial of the CDK4 inhibitor PD0332991 in patients with advanced CDK4-amplified well-differentiated or dedifferentiated liposarcoma. J. Clin. Oncol. 31 (16), 2024–2028.

Demetri, G.D., Heinrich, M.C., Fletcher, J.A., Fletcher, C.D., Van den Abbeele, A.D., Corless, C.L., et al., 2009. Molecular target modulation, imaging, and clinical evaluation of gastrointestinal stromal tumor patients treated with sunitinib malate after imatinib failure. Clin. Cancer Res. 15 (18), 5902–5909.

Demetri, G.D., von Mehren, M., Blanke, C.D., Van den Abbeele, A.D., Eisenberg, B., Roberts, P.J., et al., 2002. Efficacy and safety of imatinib mesylate in advanced gastrointestinal stromal tumors. N. Engl. J. Med. 347 (7), 472–480.

Efeyan, A., Serrano, M., 2007. p53: guardian of the genome and policeman of the oncogenes. Cell Cycle 6 (9), 1006–1010.

International Human Genome Sequencing Consortium, 2004. Finishing the euchromatic sequence of the human genome. Nature 431 (7011), 931–945.

Engelman, J.A., Zejnullahu, K., Mitsudomi, T., Song, Y., Hyland, C., Park, J.O., et al., 2007. MET amplification leads to gefitinib resistance in lung cancer by activating ERBB3 signaling. Science 316 (5827), 1039–1043.

Fletcher, C., Bridge, J., Hogendoorn, P., Mertens, F., 2013. World Health Organization Classification of Tumours of Soft Tissue and Bone, fourth ed. IARC, Lyon.

Fodor, S.P., Rava, R.P., Huang, X.C., Pease, A.C., Holmes, C.P., Adams, C.L., 1993. Multiplexed biochemical assays with biological chips. Nature 364 (6437), 555–556.

Fan, C., Oh, D.S., Wessels, L., Weigelt, B., Nuyten, D.S., Nobel, A.B., et al., 2006. Concordance among gene-expression-based predictors for breast cancer. N. Engl. J. Med. 355 (6), 560–569.

Flaherty, K.T., Robert, C., Hersey, P., Nathan, P., Garbe, C., Milhem, M., et al., 2012. Improved survival with MEK inhibition in BRAF-mutated melanoma. N. Engl. J. Med. 367 (2), 107–114.

Filippakopoulos, P., Qi, J., Picaud, S., Shen, Y., Smith, W.B., Fedorov, O., et al., 2010. Selective inhibition of BET bromodomains. Nature 468 (7327), 1067–1073.

Finn, R.S., Crown, J.P., Lang, I., Boer, K., Bondarenko, I.M., Kulyk, S.O., et al., 2015. The cyclin-dependent kinase 4/6 inhibitor palbociclib in combination with letrozole versus letrozole alone as first-line treatment of oestrogen receptor-positive, HER2-negative, advanced breast cancer (PALOMA-1/TRIO-18): a randomised phase 2 study. Lancet Oncol. 16 (1), 25–35.

Gorre, M.E., Mohammed, M., Ellwood, K., Hsu, N., Paquette, R., Rao, P.N., et al., 2001. Clinical resistance to STI-571 cancer therapy caused by BCR-ABL gene mutation or amplification. Science 293 (5531), 876–880.

Garraway, L.A., Widlund, H.R., Rubin, M.A., Getz, G., Berger, A.J., Ramaswamy, S., et al., 2005. Integrative genomic analyses identify MITF as a lineage survival oncogene amplified in malignant melanoma. Nature 436 (7047), 117–122.

Grasso, C.S., Wu, Y.M., Robinson, D.R., Cao, X., Dhanasekaran, S.M., Khan, A.P., et al., 2012. The mutational landscape of lethal castration-resistant prostate cancer. Nature 487 (7406), 239–243.

Garraway, L.A., Janne, P.A., 2012. Circumventing cancer drug resistance in the era of personalized medicine. Cancer Discov. 2 (3), 214–226.

Gerlinger, M., Rowan, A.J., Horswell, S., Larkin, J., Endesfelder, D., Gronroos, E., et al., 2012. Intratumor heterogeneity and branched evolution revealed by multiregion sequencing. N. Engl. J. Med. 366 (10), 883–892.

Hajdu, S.I., 2011. A note from history: landmarks in history of cancer, part 1. Cancer 117 (5), 1097–1102.

Hirota, S., Isozaki, K., Moriyama, Y., Hashimoto, K., Nishida, T., Ishiguro, S., et al., 1998. Gain-of-function mutations of c-kit in human gastrointestinal stromal tumors. Science 279 (5350), 577–580.

Hodi, F.S., Corless, C.L., Giobbie-Hurder, A., Fletcher, J.A., Zhu, M., Marino-Enriquez, A., et al., 2013. Imatinib for melanomas harboring mutationally activated or amplified KIT arising on mucosal, acral, and chronically sun-damaged skin. J. Clin. Oncol. 31 (26), 3182–3190.

Hanahan, D., Weinberg, R.A., 2011. Hallmarks of cancer: the next generation. Cell 144 (5), 646–674.

Hudson, T.J., Anderson, W., Artez, A., Barker, A.D., Bell, C., Bernabe, R.R., et al., 2010. International network of cancer genome projects. Nature 464 (7291), 993–998.

Huang, M., Shen, A., Ding, J., Geng, M., 2014. Molecularly targeted cancer therapy: some lessons from the past decade. Trends Pharmacol. Sci. 35 (1), 41–50.

Iasonos, A., O'Quigley, J., 2014. Adaptive dose-finding studies: a review of model-guided phase I clinical trials. J. Clin. Oncol. 32 (23), 2505–2511.

Johannessen, C.M., Johnson, L.A., Piccioni, F., Townes, A., Frederick, D.T., Donahue, M.K., et al., 2013. A melanocyte lineage program confers resistance to MAP kinase pathway inhibition. Nature 504 (7478), 138–142.

Jonat, W., Arnold, N., 2011. Is the Ki-67 labelling index ready for clinical use? Ann. Oncol. 22 (3), 500–502.

Kurzrock, R., Gutterman, J.U., Talpaz, M., 1988. The molecular genetics of Philadelphia chromosome-positive leukemias. N. Engl. J. Med. 319 (15), 990–998.

Karapetis, C.S., Khambata-Ford, S., Jonker, D.J., O'Callaghan, C.J., Tu, D., Tebbutt, N.C., et al., 2008. K-ras mutations and benefit from cetuximab in advanced colorectal cancer. N. Engl. J. Med. 359 (17), 1757–1765.

Kaelin Jr., W.G., 2005. The concept of synthetic lethality in the context of anticancer therapy. Nat. Rev. Cancer 5 (9), 689–698.

Loeb, L.A., Harris, C.C., 2008. Advances in chemical carcinogenesis: a historical review and prospective. Cancer Res. 68 (17), 6863–6872.

Levine, R.L., Wadleigh, M., Cools, J., Ebert, B.L., Wernig, G., Huntly, B.J., et al., 2005. Activating mutation in the tyrosine kinase JAK2 in polycythemia vera, essential thrombocythemia, and myeloid metaplasia with myelofibrosis. Cancer Cell 7 (4), 387–397.

Lawrence, M.S., Stojanov, P., Polak, P., Kryukov, G.V., Cibulskis, K., Sivachenko, A., et al., 2013. Mutational heterogeneity in cancer and the search for new cancer-associated genes. Nature 499 (7457), 214–218.

Luo, J., Solimini, N.L., Elledge, S.J., 2009. Principles of cancer therapy: oncogene and non-oncogene addiction. Cell 136 (5), 823–837.

Malumbres, M., Barbacid, M., 2003. RAS oncogenes: the first 30 years. Nat. Rev. Cancer 3 (6), 459–465.

Metzker, M.L., 2010. Sequencing technologies—the next generation. Nat. Rev. Genet. 11 (1), 31–46.

Michaloglou, C., Vredeveld, L.C., Soengas, M.S., Denoyelle, C., Kuilman, T., van der Horst, C.M., et al., 2005. BRAFE600-associated senescence-like cell cycle arrest of human naevi. Nature 436 (7051), 720–724.

Misale, S., Yaeger, R., Hobor, S., Scala, E., Janakiraman, M., Liska, D., et al., 2012. Emergence of KRAS mutations and acquired resistance to anti-EGFR therapy in colorectal cancer. Nature 486 (7404), 532–536.

O'Brien, S.G., Guilhot, F., Larson, R.A., Gathmann, I., Baccarani, M., Cervantes, F., et al., 2003. Imatinib compared with interferon and low-dose cytarabine for newly diagnosed chronic-phase chronic myeloid leukemia. N. Engl. J. Med. 348 (11), 994–1004.

Ocana, A., Pandiella, A., Siu, L.L., Tannock, I.F., 2011. Preclinical development of molecular-targeted agents for cancer. Nat. Rev. Clin. Oncol. 8 (4), 200–209.

Pleasance, E.D., Cheetham, R.K., Stephens, P.J., McBride, D.J., Humphray, S.J., Greenman, C.D., et al., 2010. A comprehensive catalogue of somatic mutations from a human cancer genome. Nature 463 (7278), 191–196.

Perou, C.M., Sorlie, T., Eisen, M.B., van de Rijn, M., Jeffrey, S.S., Rees, C.A., et al., 2000. Molecular portraits of human breast tumours. Nature 406 (6797), 747–752.

Paez, J.G., Janne, P.A., Lee, J.C., Tracy, S., Greulich, H., Gabriel, S., et al., 2004. EGFR mutations in lung cancer: correlation with clinical response to gefitinib therapy. Science 304 (5676), 1497–1500.

Paik, S., Shak, S., Tang, G., Kim, C., Baker, J., Cronin, M., et al., 2004. A multigene assay to predict recurrence of tamoxifen-treated, node-negative breast cancer. N. Engl. J. Med. 351 (27), 2817–2826.

Pollock, P.M., Harper, U.L., Hansen, K.S., Yudt, L.M., Stark, M., Robbins, C.M., et al., 2003. High frequency of BRAF mutations in nevi. Nat. Genet. 33 (1), 19–20.

Parsons, D.W., Jones, S., Zhang, X., Lin, J.C., Leary, R.J., Angenendt, P., et al., 2008. An integrated genomic analysis of human glioblastoma multiforme. Science 321 (5897), 1807–1812.

Prahallad, A., Sun, C., Huang, S., Di Nicolantonio, F., Salazar, R., Zecchin, D., et al., 2012. Unresponsiveness of colon cancer to BRAF(V600E) inhibition through feedback activation of EGFR. Nature 483 (7387), 100–103.

Ruschoff, J., Hanna, W., Bilous, M., Hofmann, M., Osamura, R.Y., Penault-Llorca, F., et al., 2012. HER2 testing in gastric cancer: a practical approach. Mod. Pathol. 25 (5), 637–650.

Raspollini, M.R., Amunni, G., Villanucci, A., Pinzani, P., Simi, L., Paglierani, M., et al., 2004. c-Kit expression in patients with uterine leiomyosarcomas: a potential alternative therapeutic treatment. Clin. Cancer Res. 10 (10), 3500–3503.

Robert, C., Karaszewska, B., Schachter, J., Rutkowski, P., Mackiewicz, A., Stroiakovski, D., et al., 2015. Improved overall survival in melanoma with combined dabrafenib and trametinib. N. Engl. J. Med. 372 (1), 30–39.

Root, D.E., Hacohen, N., Hahn, W.C., Lander, E.S., Sabatini, D.M., 2006. Genome-scale loss-of-function screening with a lentiviral RNAi library. Nat. Methods 3 (9), 715–719.

Siegel, R., Ma, J., Zou, Z., Jemal, A., 2014. Cancer statistics, 2014. CA Cancer J. Clin. 64 (1), 9–29.

Slamon, D.J., Clark, G.M., Wong, S.G., Levin, W.J., Ullrich, A., McGuire, W.L., 1987. Human breast cancer: correlation of relapse and survival with amplification of the HER-2/neu oncogene. Science 235 (4785), 177–182.

Slamon, D.J., Leyland-Jones, B., Shak, S., Fuchs, H., Paton, V., Bajamonde, A., et al., 2001. Use of chemotherapy plus a monoclonal antibody against HER2 for metastatic breast cancer that overexpresses HER2. N. Engl. J. Med. 344 (11), 783–792.

Soda, M., Choi, Y.L., Enomoto, M., Takada, S., Yamashita, Y., Ishikawa, S., et al., 2007. Identification of the transforming EML4-ALK fusion gene in non-small-cell lung cancer. Nature 448 (7153), 561–566.

Shaw, A.T., Kim, D.W., Nakagawa, K., Seto, T., Crino, L., Ahn, M.J., et al., 2013. Crizotinib versus chemotherapy in advanced ALK-positive lung cancer. N. Engl. J. Med. 368 (25), 2385–2394.

Serrano, C., Mackintosh, C., Herrero, D., Martins, A.S., de Alava, E., Hernandez, T., et al., 2005. Imatinib is not a potential alternative treatment for uterine leiomyosarcoma. Clin. Cancer Res. 11 (13), 4977–4979 author reply 9–80.

Stratton, M.R., Campbell, P.J., Futreal, P.A., 2009. The cancer genome. Nature 458 (7239), 719–724.

Sotiriou, C., Piccart, M.J., 2007. Taking gene-expression profiling to the clinic: when will molecular signatures become relevant to patient care? Nat. Rev. Cancer 7 (7), 545–553.

Samuels, Y., Wang, Z., Bardelli, A., Silliman, N., Ptak, J., Szabo, S., et al., 2004. High frequency of mutations of the PIK3CA gene in human cancers. Science 304 (5670), 554.

Stephens, P.J., Greenman, C.D., Fu, B., Yang, F., Bignell, G.R., Mudie, L.J., et al., 2011. Massive genomic rearrangement acquired in a single catastrophic event during cancer development. Cell 144 (1), 27–40.

Sosman, J.A., Kim, K.B., Schuchter, L., Gonzalez, R., Pavlick, A.C., Weber, J.S., et al., 2012. Survival in BRAF V600-mutant advanced melanoma treated with vemurafenib. N. Engl. J. Med. 366 (8), 707–714.

Stommel, J.M., Kimmelman, A.C., Ying, H., Nabioullin, R., Ponugoti, A.H., Wiedemeyer, R., et al., 2007. Coactivation of receptor tyrosine kinases affects the response of tumor cells to targeted therapies. Science 318 (5848), 287−290.

Serrano, C., Markman, B., Tabernero, J., 2010. Integration of anti-epidermal growth factor receptor therapies with cytotoxic chemotherapy. Cancer J. 16 (3), 226−234.

Swain, S.M., Baselga, J., Kim, S.B., Ro, J., Semiglazov, V., Campone, M., et al., 2015. Pertuzumab, trastuzumab, and docetaxel in HER2-positive metastatic breast cancer. N. Engl. J. Med. 372 (8), 724−734.

Sun, C., Wang, L., Huang, S., Heynen, G.J., Prahallad, A., Robert, C., et al., 2014. Reversible and adaptive resistance to BRAF(V600E) inhibition in melanoma. Nature 508 (7494), 118−122.

Shalem, O., Sanjana, N.E., Hartenian, E., Shi, X., Scott, D.A., Mikkelsen, T.S., et al., 2014. Genome-scale CRISPR-Cas9 knockout screening in human cells. Science 343 (6166), 84−87.

Tumeh, P.C., Harview, C.L., Yearley, J.H., Shintaku, I.P., Taylor, E.J., Robert, L., et al., 2014. PD-1 blockade induces responses by inhibiting adaptive immune resistance. Nature 515 (7528), 568−571.

Tolcher, A.W., Khan, K., Ong, M., Banerji, U., Papadimitrakopoulou, V., Gandara, D.R., et al., 2015. Antitumor activity in RAS-driven tumors by blocking AKT and MEK. Clin. Cancer Res. 21 (4), 739−748.

Vogelstein, B., Kinzler, K.W., 1993. The multistep nature of cancer. Trends Genet. 9 (4), 138−141.

van de Vijver, M.J., He, Y.D., van't Veer, L.J., Dai, H., Hart, A.A., Voskuil, D.W., et al., 2002. A gene-expression signature as a predictor of survival in breast cancer. N. Engl. J. Med. 347 (25), 1999−2009.

Verhaak, R.G., Hoadley, K.A., Purdom, E., Wang, V., Qi, Y., Wilkerson, M.D., et al., 2010. Integrated genomic analysis identifies clinically relevant subtypes of glioblastoma characterized by abnormalities in PDGFRA, IDH1, EGFR, and NF1. Cancer Cell 17 (1), 98−110.

Watson, J.D., Crick, F.H., 1953. Molecular structure of nucleic acids; a structure for deoxyribose nucleic acid. Nature 171 (4356), 737−738.

Weinstein, J.N., Collisson, E.A., Mills, G.B., Shaw, K.R., Ozenberger, B.A., Ellrott, K., et al., 2013. The Cancer Genome Atlas Pan-Cancer analysis project. Nat. Genet. 45 (10), 1113−1120.

Weinberg, R., 2010. Point: hypotheses first. Nature 464 (7289), 678.

Whitesell, L., Lindquist, S.L., 2005. HSP90 and the chaperoning of cancer. Nat. Rev. Cancer 5 (10), 761−772.

Weinstein, I.B., 2002. Cancer. Addiction to oncogenes—the Achilles heal of cancer. Science 297 (5578), 63−64.

Weber, J.S., Levit, L.A., Adamson, P.C., Bruinooge, S., Burris, H.A., Carducci, M.A., et al., 2015. American Society of Clinical Oncology policy statement update: the critical role of phase I trials in cancer research and treatment. J. Clin. Oncol. 33 (3), 278−284.

Yan, H., Parsons, D.W., Jin, G., McLendon, R., Rasheed, B.A., Yuan, W., et al., 2009. IDH1 and IDH2 mutations in gliomas. N. Engl. J. Med. 360 (8), 765−773.

Yap, T.A., Gerlinger, M., Futreal, P.A., Pusztai, L., Swanton, C., 2012. Intratumor heterogeneity: seeing the wood for the trees. Sci. Transl. Med. 4 (127), 127ps10.

Yang, X., Boehm, J.S., Salehi-Ashtiani, K., Hao, T., Shen, Y., Lubonja, R., et al., 2011. A public genome-scale lentiviral expression library of human ORFs. Nat. Methods 8 (8), 659−661.

Chapter 33

Maintaining an Emphasis on Rare Diseases With Research Initiatives and Resources at the National Center for Advancing Translational Sciences

Stephen C. Groft and Rashmi Gopal-Srivastava

National Institutes of Health, Bethesda, MD, United States

Chapter Outline

Key Points

- Innovative rare diseases research includes the importance of genomewide association studies (GWAS), epidemiological studies, natural history studies, and a team approach of various medical specialists within a collaborative rare diseases community.
- Conducting natural history studies is an important approach to understanding the development, the progression, and the appropriate treatment of rare diseases.
- The National Center for Advancing Translational Sciences (NCATS) develops new approaches, demonstrates their use, and disseminates findings which improve the treatment and health of patients living with a rare disease.
- Team Science provides a collaborative community comprised of various medical specialists who are critical to innovative rare disease research, creative interventions, and insightful care of those living with rare diseases.
- Research networks create a model for research collaboration.
- The importance of medical devices research and development communities requires the training of research investigators with an emphasis on rare diseases, international classification of rare diseases, and orphan products development.

INTRODUCTION

Rare diseases translation is the process of turning observations in the laboratory, clinic, and community into interventions that improve the health of individuals and the public—from diagnostics and therapeutics to medical procedures and

Clinical and Translational Science. http://dx.doi.org/10.1016/B978-0-12-802101-9.00033-8

behavioral changes. Translational science is the field of investigation focused on understanding the scientific and operational principles underlying each step of the translational process. Patient involvement is a critical feature of all stages in translation. Development of many orphan products for rare diseases requires adhering to translational science principles, utilization of novel clinical trial designs for data collection and analyses, and flexibility in the regulatory review process during investigational research studies and postmarketing uses of products by clinicians and patients.

Rare diseases occur infrequently. A rare disease becomes a common disease when the diagnosis is received. It is no longer an obscure, difficult-to-pronounce disease that most people have never heard about across their life span. The definition of a rare disease, established by amendments to the Orphan Drug Act of 1983 and the Rare Diseases Act of 2002, was set as a disease or condition occurring in less than 200,000 people in the United States. The European Union considers diseases to be rare when they affect no more than 5 in 10,000 persons. An estimated 6—8% of the population has a rare disorder. Many patients are living without a diagnosis.

The increased emphasis on research of rare diseases is due to a number of interrelated factors. The most significant has been the expanding role of patients, families, and patient advocacy groups (PAGs) as research partners. Despite the commitments of many organizations to meet the needs of patients and families, very few diseases currently have treatments available to them. Many rare disorders involve multiple organs and systems requiring the care of numerous medical specialists to provide adequate care and increase the understanding of the disease by the patient, their families, and caregivers. Similarly, the development of an intervention for the prevention and treatment of rare diseases with an orphan product requires a team approach with very close collaboration among patients and PAGs, the biopharmaceutical industry, the academic research investigators, philanthropic foundations, sequencing centers, government regulatory and research agencies, health-care providers and services, and reimbursement organizations.

Many translational research programs and resources are now available from several of the National Institutes of Health (NIH) research institutes and centers to facilitate the development of orphan products. The Rare Diseases Clinical Research Network (RDCRN) program has been supported by the Office of Rare Diseases Research (ORDR) and other institutes and centers at NIH including the National Center for Advancing Translational Sciences (NCATS).

Even with the numerous barriers and difficulties developing orphan products, optimism prevails. Translational research efforts and increasing public—private partnerships continue to expand. The emphasis on personalized medicine with targeted therapies directed to specific molecular pathways or specific genetic mutations will add to additional products for very small and well-identified patient populations with rare diseases. We look forward to correct interpretation of results from sequencing procedures to improve diagnostic capabilities of the community for the millions of patients with rare diseases around the globe, and especially those living with a disease without a diagnosis.

TRANSLATIONAL SCIENCE SPECTRUM

The translational science spectrum represents each stage of research along the path from the biological basis of health and disease to interventions that improve the health of individuals and the public. The spectrum is not linear or unidirectional; each stage builds upon and informs the others. At all stages of the spectrum, the NCATS develops new approaches, demonstrates their usefulness, and disseminates the findings. Patient involvement is a critical feature of all stages in translation (Fig. 33.1).

Translational science is the field of investigation focused on understanding the scientific and operational principles underlying each step of the translational process. NCATS studies translation on a systemwide level as a scientific and operational problem. The Center convenes expert teams from diverse scientific disciplines—including efficacy, toxicity, data sharing, biomarkers, and clinical application—to reduce, remove, or bypass significant bottlenecks across the entire continuum of translation and to train the future translational science workforce (National Center for Advancing Translational Sciences). The spectrum of translational research is not linear or unidirectional; each stage builds upon and informs the others (National Center for Advancing Translational Sciences).

Increasingly, development of many orphan products for rare diseases requires adhering to translational science principles, utilization of novel clinical trial designs for data collection and analyses, and flexibility in the regulatory review process during investigational research studies and postmarketing.

DEFINITION OF RARE DISEASES

Rare diseases individually may occur infrequently. Collectively they may become very common when viewed from a global perspective. A rare disease becomes a common disease to us and to our families when we receive the diagnosis. It is no longer an obscure, difficult-to-pronounce disease that most people have never heard about across their life span.

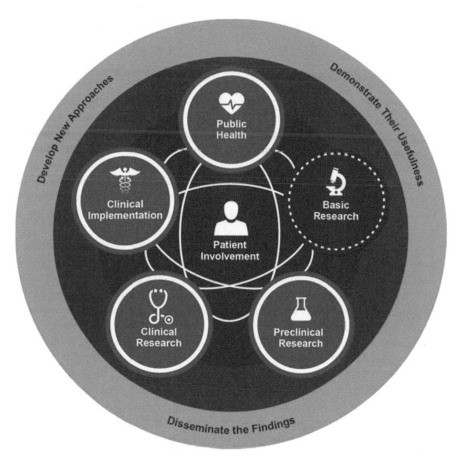

FIGURE 33.1 Translational science spectrum.

Unknown to us, there may be an affected biochemical pathway, missing or excessive enzymes, or an environmental exposure resulting in changes to our comfortable and more familiar well-being. Was that strange neurological disease that changed the life of my uncle the same disease that I have? Is the cancer that I have different or the same as my mother's? None of these are rare disorders in our eyes. These and many other questions about the heritability of diseases, genetic heterogeneity, recessive or dominant inheritance cast a cloud over our lives. All common and rare diseases are not at the same level of public understanding despite the different prevalence of these diseases. Most of the more than 6500 rare diseases do not have appropriate diagnostic and treatment interventions available to the patients, families, clinicians, and all health-care providers who provide needed services for the millions of patients with rare diseases. Phenotypic expression and genetic variability is causing us to rethink the common diseases. Occurring at a rapid rate is the development of an understanding of even more specific molecular differentiation of the common diseases into multiple and related rare diseases. The definition of a rare disease, established by amendments to the Orphan Drug Act of 1983 and the Rare Diseases Act of 2002 was set as a disease or condition occurring in less than 200,000 people in the United States. The European Union considers diseases to be rare when they affect no more than 5 in 10,000 persons.

INTERNATIONAL CLASSIFICATION OF DISEASES

Rare and common diseases are receiving additional emphasis by multidisciplinary review teams in the expanding classification by the World Health Organization (WHO). The revised and updated classification of diseases will identify new diseases and variants of existing diseases based on more sophisticated genomic analyses. Approximately 5400 rare diseases are proposed to be included in the 11th version of the International Classification of Diseases (ICD 11); refinements to diagnostic criteria and variants of existing diseases to the WHO's International Classification of Diseases are now available for public review and comment at their website (International Classification of Diseases; Aymé et al., 2015).

Providing access to proposed changes will help gain a global consensus on diseases from a description of the symptoms and the molecular differentiation of disorders now possible by improving genomic sequencing capabilities.

With acceptance of these refinements in differentiation of diseases and their treatments, greater expansion of personalized or precision medicine can be expected.

EVOLVING APPROACHES TO RARE DISEASES RESEARCH

Chance encounters with patients with rare diseases generate research interest even though research of rare diseases is often not viewed as career track enhancers. Perceptions persist related to the difficulties encountered recruiting patients, the inability to establish multidisciplinary research teams, and the failure of funding agencies, academic institutions, and foundations to provide adequate personnel and financial resources to conduct research on rare diseases. Fortunately, significant changes in emphasis enable investigators to move forward in research portfolios for several rare diseases. Research with these disorders offer unique challenges in study design and analyses but provide for the rapid expansion of the knowledge base for a particular rare disease or a group of related rare diseases. Breakthrough discoveries with rare diseases are quickly viewed for opportunities to increase the understanding of the more common diseases and possible common pathways.

The increased emphasis on research of rare diseases is due to a number of interrelated factors. The most significant has been the expanding role of patients, families, and PAGs as research partners. Increasing emphasis by our Congress has directed the public sector organizations such as the NIH and the Food and Drug Administration (FDA) to expand their current interest and support for rare diseases, orphan products, and translational research programs. Increased participation of academic research centers has occurred concurrently with the expanding emphasis by federal funding sources, and the biopharmaceutical industry increased their interest in research and development of potential diagnostic and treatment interventions. The voices of the rare disease community including those from diseases with very small patient populations are being heard and responses to their needs are occurring. Despite the commitments of many organizations to meet the needs of the patients and families, very few diseases currently have treatments available to them. More efficient and more effective procedures are needed to meet these needs more quickly and at less expense than the estimated 1 billion dollars to deliver a new molecular entity to the marketplace. In recent years, progress is being made as interconnected pathways for multiple diseases in a common pathway are now being used as a research option leading to treatments for many different diseases. Breakthroughs in developing treatments of diseases in the common pathways will most likely lead to the development of the traditional blockbuster products generating billions of dollars in revenue, which hopefully will be used to expand research efforts even more into more pathways for the investigation of rare and common diseases.

Importance of Genomewide Association Studies

Large-scale genomic studies and bioinformatics analyses suggest that discoveries from genomewide association studies (GWAS) may pinpoint specific pathways for disease development or different phenotypic expression. Knowledge gained from these studies is applicable to common and rare diseases. Thus, there is a need for epidemiological studies and natural history studies that are important sources of this valuable and accurate information with applications leading to a correct diagnosis and appropriate treatments. Knowledge of different phenotype traits of rare and common diseases increases from GWAS studies from larger cohorts, and the results can be verified with data available from other cohorts with different phenotype and genotype expression. We are discovering common diseases frequently include subpopulations of rare diseases such as the number of different diseases under the name of anemia and lymphomas. Identifying different subtypes and pathways are opening the door to the development of molecular-targeted interventions. These differences are also important to compare similarities and differences in diseases and conditions in the pediatric and geriatric populations as well as implications for specific treatments.

Team Approach

Many rare disorders involve multiple organs and systems requiring numerous medical specialists to provide adequate care and increase the understanding of the disease by the patient, their families, and caregivers. Similarly, the development of an intervention for the prevention and treatment of rare diseases with an orphan product requires a team approach with very close collaboration among patients and PAGs, the biopharmaceutical industry, the academic research investigators, philanthropic foundations, sequencing centers, government regulatory and research agencies, health-care providers and services, and reimbursement organizations—frequently referred to as the "rare diseases community."

TRANSLATIONAL RESEARCH EFFORTS AND RESOURCES AT NCATS AND OTHER NIH INSTITUTES

Many translational research programs and resources are now available from several of the NIH research institutes and centers. These include novel programs at NCATS such as the Therapeutics for Rare and Neglected Diseases (TRND) program, the Bridging Interventional Development Gaps (BrIDGs) program, the RDCRN program, Tissue Chip for Drug Screening program, and the Repurposing Program of Approved and Investigational Products (Colvis and Austin; Colvis et al., 2013). TRND projects focus on rare and neglected diseases and are collaboration between NCATS intramural laboratories with preclinical drug development expertise and extramural labs with disease-area/target expertise. These projects enter at various stages of development and are taken to a stage needed to attract external organization to adopt for final clinical development (McKew and Pilon, 2013). These projects serve to develop new generally applicable platform technologies and paradigms. BrIDGs program provides in-kind, government contract-based services to overcome obstacles in later-stage preclinical development. BrIDGs projects address any disease or disorder, regardless of prevalence or incidence (Website). The Tissue Chip for Drug Screening program is an interagency collaborative effort between NCATS, Defense Advanced Research Projects Agency (DARPA), and the FDA. It aims to develop tissue chips that mimic human physiology to screen for safe, effective drugs using best ideas in engineering, biology, and toxicology (Fabre et al., 2014).

THE RARE DISEASES CLINICAL RESEARCH NETWORK PROGRAM—A MODEL FOR COLLABORATION

To facilitate multisite natural history studies and clinical trials for rare diseases, the ORDR at the NCATS established the RDCRN program, in collaboration with six NIH institutes/centers. The RDCRN is a successful and innovative international clinical studies network of 22 distinct clinical research consortia, a central Data Management and Coordinating Center (DMCC), and 130 PAGs. The research conducted in this network explores the natural history, epidemiology, diagnosis, and treatment of more than 200 rare diseases. Each consortium is required to conduct two multisite clinical studies on a minimum of three related rare diseases, develop a training program for new investigators, involve PAGs as research partners, have a clinical trials infrastructure in place, and provide information about rare diseases to investigators, patients, and general public. The RDCRN—DMCC supports the consortia by supplying infrastructure; user-friendly resources for the public and Web-based recruitment and referral tools; logistical and administrative assistance; and data coordination, management, and sharing. The goal of RDCRN is to contribute to the research and treatment of rare diseases by working together as a team to identify biomarkers for disease risk, disease severity and activity, and clinical outcome, while encouraging the development of new approaches to diagnosis, prevention, and treatment.

The direct involvement of PAGs in RDCRN is a major feature of this program. This program is successful due to PAGs' engagement to facilitate clinical research in rare diseases. PAG representatives are included as research partners in RDCRN consortia and are involved in all consortium activities including study design, recruitment, education, and administrative support. Additionally, some PAG groups invest financially in consortium research activities. Collectively, all PAGs from each consortium have created a Coalition of Patient Advocacy Groups (RDCRN-CPAG) and chairpersons of the CPAG are engaged as partners in RDCRN oversight activities as voting members of the RDCRN Steering Committee.

RDCRN productivity is evidence of the effectiveness of the collaborative model using the team approach. The Consortia of the RDCRN have enrolled more than 31,000 patients with rare disorders in 90 active protocols including longitudinal studies and clinical trials studies and trained 208 investigators in rare disease research since its establishment in 2003 (Fig. 33.2; Groft et al., 2013; Krischer et al., 2014).

RDCRN, an initiative of NCATS, is comprised of 22 consortia, a DMCC, and PAGs as partners. These novel programs complement several existing programs from the National Cancer Institute (NCI) and their Experimental Therapeutics (NExT) program, the NeuroNEXT program from the National Institute of Neurological Disorders and Stroke (NINDS), the Best Pharmaceuticals for Children Act and the Newborn Screening Translational Research Network from the National Institute of Child Health and Human Development (NICHD), and the Centers for Accelerated Innovation from the National Heart, Lung, and Blood Institute (NHLBI).

RARE DISEASES RESEARCH AND ORPHAN PRODUCT DEVELOPMENT

Research of rare diseases and development of orphan products continues to grow around the world. In the United States, more than 3876 active orphan product designations have been made and 262 additional designations have been provided

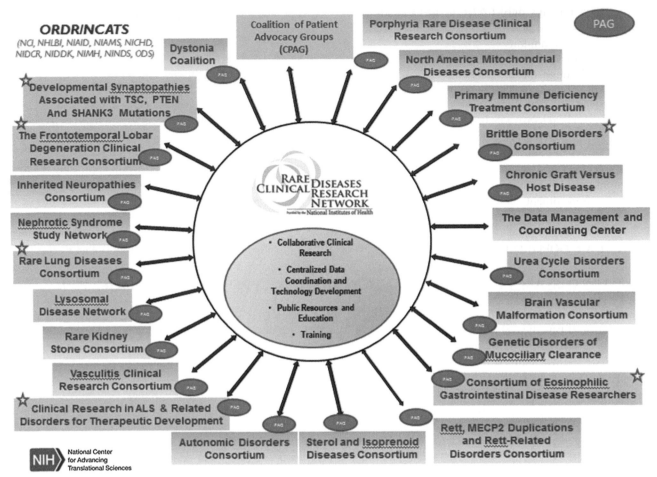

FIGURE 33.2   Pictorial representation of the Rare Diseases Clinical Research Network.

thus far in 2016. There were 354 orphan product designations in 2015. Obtaining the orphan product designation from the Office of Orphan Products Development (OOPD) at the FDA provides incentives such as 7-year marketing exclusivity, eligibility for research grants, and possible exemption from an estimated 1 million dollars in filing fees for an Investigational New Drug (IND) application or a New Drug Application (NDA). The FDA in the United States has approved products for 29 indications thus far in 2016 and more than 579 Approved Orphan Products/Indications by the FDA since 1983. The Pharmaceutical Research and Manufacturers of America (PhRMA) reported more than 650 compounds in development for rare diseases. Current levels of activities indicate a continued emphasis on rare diseases. The NIH Clinical Center Hospital research portfolio contains more than 948 research protocols for ~584 rare diseases. NIH continues to provide information on the status of research of rare diseases and orphan products through the Research, Condition, and Disease Categorization (RCDC) process. IN FY 2015 NIH provided $3.639 billion for nearly 9400 rare diseases research projects and included ~$785 million for ~1650 orphan product research projects (Estimates of Funding for Various Research; RCDC Estimates of Funding for Rare Diseases Website; RCDC Estimates of Funding for Orphan Drugs Website).

The combined collaborative efforts of the multiple partners included under the International Rare Diseases Research Consortium umbrella reflects the attainable IRDiRC goal of 200 approved orphan product indications by 2020. Since 2010, there have been 195 approvals reported by the FDA and EMA (International Rare Diseases Research Consortium Website).

RDCRN AND CLINICAL AND TRANSLATIONAL SCIENCE AWARDS PROGRAMS ON EDUCATION AND TRAINING RESOURCES

Providing the resources to train, cultivate, and sustain future leaders of the translational research workforce is a key NCATS goal. The Center offers translational education and training to support a pipeline of investigators who can move basic research findings into the clinic to improve human health. The examples of translational education and training research resources are organized by intended user and highlight the wide variety of education and training opportunities available to the community. These resources are designed to prepare clinician-scientists to better address the unique challenges of translational research (Education and Training Resources). Each consortium of RDCRN also has a training component. Through these RDCRN training components more than 200 young investigators have been trained in rare diseases clinical research and several of these trainees are now leading individual sites. The RDCRN website provides educational information and resources for health-care providers, scientists, patients, and their families.

ASSESSMENT OF UNMET MEDICAL DEVICE NEEDS FOR RARE DISEASES

The Institute of Medicine (IOM) Rare Diseases and Orphan Products (2010) report recommended that the FDA and NIH perform an assessment of unmet device needs, identify impediments to meeting these needs, and examine options for overcoming impediments. Similarly, the 2011 Section 740 Rare and Neglected Diseases Report to Congress suggested that the FDA perform an assessment of the barriers and incentives associated with the development of devices for rare diseases (Rare Diseases and Orphan Products, 2010).

The FDA's OOPD, Office of Policy and Planning, and Center for Devices and Radiological Health (CDRH), in collaboration with ORDR at NCATS, NIH, will conduct a needs assessment for medical devices for rare diseases and publish the findings. The results of this inquiry will be a publication for all audiences.

This project will seek to advance medical device development for rare disease patients by addressing the recurrent call to document the needs of medical device for these patients. This assessment will evaluate and document examples of compelling unmet medical device needs for rare diseases across sections of various medical specialties with a report and publication as deliverables. Rare diseases of focus will be those with no approved devices for the treatment or diagnosis of a rare disease and when an improvement of existing devices is needed. This project will be completed in four phases, some of which will occur simultaneously, by performing a needs assessment of key stakeholders. The phases are as follows:

- Phase I: Clinicians
- Phase II: Researchers
- Phase III: Patients (including caregivers and advocates)
- Phase IV: Industry

Approaches to obtaining information will be Web-based survey of clinicians from CDRH advisory committee, pediatric device consortia, American Medical Association (AMA), and American Association of Physicians (AAP) Request for Information (RFI) from NIH to academic researchers including RDCRN and electronic bulletin board or focus groups to gather information from patients and industry. This effort will document and highlight rare disease device needs and provide detailed information so that stakeholders may understand the demand and potential for device innovation for rare diseases patient.

AN EVEN BRIGHTER PATH FORWARD

Even with the numerous barriers and difficulties developing orphan products, optimism prevails. The promises of newer therapies resulting from years of supporting research and development and the commitment of the biomedical, the diagnostics, and the devices industries provide that hope. Translational research efforts and increasing public—private partnerships continue to develop small molecules, enzyme replacement therapies, monoclonal antibodies, therapeutic proteins, gene therapy interventions, and RNA interference (RNAi) treatments. Added expertise in regenerative medicine and nanotechnology will result in personalized medicine treatments and companion diagnostics. The repurposing of existing approved and investigational products will provide valuable treatments to the rare diseases community.

Emphasis on rare diseases is increasing as we gain more information from clinical trials, clinical studies, natural history studies, and patient registries. We are experiencing a need to address transitional care from pediatric to adult populations in different clinical settings. With limited access to knowledgeable and skilled physicians at local institutions or even in every country, we must seek better ways to improve access to key health-care providers through telehealth or telemedicine technologies when travel is difficult or not possible to the treatment sites. The question has been raised—are we reaching the tipping point on the costs of orphan drugs and reimbursement? At the present time, with 270 orphan products approved for ~500 rare diseases, the health-care systems are able to provide resources to ensure access to approved treatments. When the number of approved products increase by a factor of three to four, uncertainties and limitation of funding resources surrounding reimbursement will occur more frequently. Many pharmaceutical companies have established programs to make available the needed orphan products for rare diseases regardless of the ability of the patients' ability to pay for the product.

We anticipate access to more approved drugs with labeling using pharmacogenomics to inform the best use of the drug in smaller and more select patient populations of rare and common diseases. The emphasis on personalized medicine with targeted therapies directed to specific molecular pathways or specific genetic mutations will add to additional products for very small and well-identified patient populations with rare diseases. We look forward to correct interpretation of results from sequencing procedures to improve diagnostic capabilities of the community for the millions of patients with rare diseases around the globe, and especially those living with a disease without a diagnosis.

REFERENCES

Aymé, S., Bellet, B., Rath, A., March 26, 2015. Rare diseases in ICD11: making rare diseases visible in health information systems through appropriate coding. Orphanet J. Rare Dis. 10, 35. http://dx.doi.org/10.1186/s13023-015-0251-8.

Colvis, C.M., Austin, C.P., January 2014. Innovation in therapeutics development at the NCATS. Neuropsychopharmacology 39 (1), 230—232. http://dx.doi.org/10.1038/npp.2013.247. No abstract available. PMID:24317308.

Colvis, C.M., Devaney, S., Brady, L.S., Hudson, K.L., January 2013. Partnering for therapeutics discovery. Clin. Pharmacol. Ther. 93 (1), 24—25. http://dx.doi.org/10.1038/clpt.2012.188. Epub 2012 Nov 28. Review. No abstract available. PMID:23187876.

Education and Training Resources for (a.) Clinicians, Researchers and Health/Science Professionals, (b.) Undergraduate and Graduate Students, and (c.) High School Students. https://ncats.nih.gov/translation/resources.

Estimates of Funding for Various Research, Condition, and Disease Categories (RCDC). http://report.nih.gov/categorical_spending.aspx.

Fabre, K.M., Livingston, C., Tagle, D.A., September 2014. Organs-on-chips (microphysiological systems): tools to expedite efficacy and toxicity testing in human tissue. Exp. Biol. Med. 239 (9), 1073—1077. http://dx.doi.org/10.1177/1535370214538916. Epub 2014 Jun 24. PMID:24962171, Website: Tissue Chip for Drug Screening Program. http://www.ncats.nih.gov/tissuechip/about.

Field, M.J., Boat, T.F. (Eds.), 2010. Rare diseases and orphan products: Accelerating Research and Development. The Institute of Medicine (IOM) of the National Academies, Washington, DC. ISBN-13:978-0-309-15806-0 and ISBN-10:0-309-15806-0.

Groft, S.C., Pharm, D., Gopal-Srivastava, R., 2013. A model for collaborative clinical research in rare diseases: experience from the rare diseases clinical research network program. Clin. Investig. 3 (11), 1—7.0.

International Classification of Diseases World Health Organization Beta Website. http://apps.who.int/classifications/icd11/browse/f/en#/.

International Rare Diseases Research Consortium Website. http://www.irdirc.org/.

Krischer, J.P., Gopal-Srivastava, R., Groft, S.C., Eckstein, D.J., August 2014. The Rare Diseases Clinical Research Network's organization and approach to observational research and health outcomes research. J. Gen. Intern Med. 29 (Suppl. 3), S739—S744. http://dx.doi.org/10.1007/s11606-014-2894-x. PMID:25029976, Website: Rare Diseases Clinical Research Network (RDCRN). http://www.rarediseasesnetwork.org/.

McKew, J.C., Pilon, A.M., February 2013. NIH TRND program: successes in preclinical therapeutic development. Trends Pharmacol. Sci. 34 (2), 87—89. http://dx.doi.org/10.1016/j.tips.2012.10.001. Epub 2012 Nov 19. Website: Therapeutics for Rare and Neglected Diseases Program. https://ncats.nih.gov/trnd.

National Center for Advancing Translational Sciences. https://ncats.nih.gov/translation/spectrum.

National Center for Advancing Translational Sciences. https://ncats.nih.gov/files/translation-factsheet.pdf.

RCDC Estimates of Funding for Orphan Drugs Website. http://report.nih.gov/categorical_spending_project_listing.aspx?FY=2014&ARRA=N&DCat=Orphan Drug.

RCDC Estimates of Funding for Rare Diseases Website. http://report.nih.gov/categorical_spending_project_listing.aspx?FY=2014&ARRA=N&DCat=Rare Diseases.

Website: Bridging Interventional Development Gaps Program. https://ncats.nih.gov/bridgs.

Infrastructure

Chapter 34

Clinical and Translational Science Infrastructure

David Robertson[1] and Gordon H. Williams[2]

[1]Clinical & Translational Research Center, Elton Yates Professor of Medicine, Pharmacology and Neurology, Vanderbilt University, Nashville, TN, United States; [2]Hormonal Mechanisms of Cardiovascular Injury Laboratory, Brigham and Women's Hospital, and Professor of Medicine, Harvard Medical School, Boston, MA, United States

Chapter Outline

Key Points

- *General Clinical Research Center (GCRC)* or *clinical research center (CRC)* is a medical facility usually in an academic environment which conducts patient-centered clinical research.
- *National Center for Advancing Translational Sciences (NCATS)* is one of 27 institutes and centers (ICs) at the *National Institutes of Health* which was signed into creation by President Obama on December 23, 2011 under the *Consolidated Appropriations Act of 2012*.
- *NCATS* was established to transform and accelerate the translational research process to expedite the delivery of treatments to patients more quickly.
- The *Clinical and Translational Science Award (CTSA)* provides innovative methods, cutting-edge informatics, and a nationwide network of medical research institutions—*hubs*—that collaborate locally, regionally, and nationally to foster innovation across the broad spectrum of clinical translational science.
- CTSA Accrual to Clinical Trials (CTSA ACT) is the developed nationwide network that shares electronic health records (EHRs) across the country.
- *Repurposing* is the process of using existing therapies to treat new diseases based on commonalities in molecular causation.
- *Translation* is the process of turning observations in the laboratory or clinic into interventions that improve the health of individuals and the public.
- *Translational Science* is the field of investigation focused on understanding the scientific principles and underlying each step of the translational process.
- A *studio* is a different kind of committee focused on issues relevant to a broad range of public health disciplines and issues.
- NCATS' *Rare Diseases Clinical Research Network (RDCRN)* consists of physician-scientists and their multidisciplinary teams who work together (often at multiple sites) to study rare diseases which have no known effective treatments or cures.
- The *National Institutes of Health* and the *US Department of Energy* collaborated on the international *Human Genome Project* from 1990 to 2003 to understand the similarities, differences, and functions of the human genome (3 billion DNA base pairs) which are critical for a healthy life.

INTRODUCTION

For 50 years, the General Clinical Research Center's (GCRC) programs and their antecedents provided an academic infrastructure which enabled the extraordinary achievements of 20th-century patient-oriented research (Ahrens, 1995). This program aimed "to make available to medical scientists the resources which are necessary for the conduct of clinical research" and "to provide the clinical research infrastructure for investigators who receive their primary research funding from the other components of the NIH." In addressing these goals the GCRCs were one of the great success stories of the American biomedical research enterprise (Newman and Greathouse, 1963). Serving in the 1960s and 1970s as well-appointed laboratories accommodating dozens of clinical investigators at each site, the GCRCs were great economies of scale. They were also in a sense the institutional commons, where the academic physicians in both the basic and clinical departments encountered each other daily and where proximity provided fertile ground for intellectual creativity which stimulated collaborative research (Luft, 1997). And finally, in the last two decades of the 20th century, when NIH study sections were losing confidence in the patient-oriented research paradigm as an efficient means to discover new knowledge (Williams et al., 1997), the GCRCs kept clinical research alive at its time of greatest threat. Surely this is a unique legacy in the history of the NIH.

These over 70 GCRCs were dispersed throughout the nation's research-intensive universities and medical centers. At any given time they typically hosted ~6000 clinical research projects being carried out by nearly 9000 clinical investigators. A GCRC might include outpatient and inpatient facilities, core laboratories to facilitate clinical assessment, and sometimes a metabolic unit for the provision of dietary interventions and energy balance. There was also support for personnel such as nurses, laboratory technicians, biostatisticians, research subject advocates, and a director of informatics. In some respects a larger GCRC resembled a small hospital, given the diversity and integration of its resources, and it tended to function as an organic whole with its own character and sense of mission. In the frenetic hospital environment of managed care, the GCRC often seemed like an oasis of tranquility where patients and normal subjects might enjoy an unhurried communication with physicians and nurses reminiscent of an earlier era and a place where a controlled environment was essential to unravel pathophysiologic processes.

At most institutions, the GCRC had a characteristic organizational structure. The GCRC's principal investigator was a Dean or Associate Dean for clinical research at an institution, whereas the GCRC Program Director was a clinical investigator and user of the GCRC, who was charged with the day-to-day management of the GCRC with the assistance of the GCRC Scientific Advisory Committee (GAC). The GAC was composed of both basic and clinical scientists and functioned somewhat like an NIH study section. It met monthly or twice-monthly to oversee the management of center resources and to consider applications for the use of center facilities for research projects. Clinical investigators from all segments of the medical and nursing school and in some cases other schools of the university could apply for the use of GCRC resources by preparing a proposal for the GAC to consider. If the GAC approved the study, it was given a priority score reflecting enthusiasm for its science, value for human health, and need for GCRC resources.

REINVENTING THE CLINICAL RESEARCH INFRASTRUCTURE

Any research community must evolve in response to changes in scientific opportunity, driven by advances in scientific knowledge and the methods and technology available to engage the discovery process. Such advances impinge on the research process.

All these developments had important implications for the clinical research infrastructure. As the new millennium arrived, there was increasing discussion of the dissociation between opportunities and capabilities of the clinical research enterprise in Academic Health Centers (Nathan, 1998; Williams, 1999; Robertson, 2000; Nathan and Nathan, 2016).). The GCRCs had served the community well, but by 2000 their budget constraints (in constant dollars as a fraction of the NIH budget, almost all GCRCs had less funding in 1999 than they had had 30 years before, in 1969) had seriously compromised their capabilities (Vaitukaitis, 2000). Moreover, education of young clinical investigators had faltered, and there was a major decline in the number of young and midcareer clinical investigators (Ahrens, 1994). There was concern that their success rate in competing for NIH funding was declining. An analysis of the GCRC program for the AAMC in 2000 made several recommendations, including more emphasis on genomics and proteomics, more emphasis on research off-site (a "virtual GCRC"), more emphasis on informatics to facilitate research, increased application of imaging technologies, core facilities with a national user catchment, improved trainee education, and discarding the name General Clinical Research Centers in favor of a more exciting and evocative name (Robertson, 2000).

In addition to these concerns, there were societal and economic issues that also changed the clinical research enterprise. Some people wondered why all the support going into the National Institutes of Health (NIH) was not improving

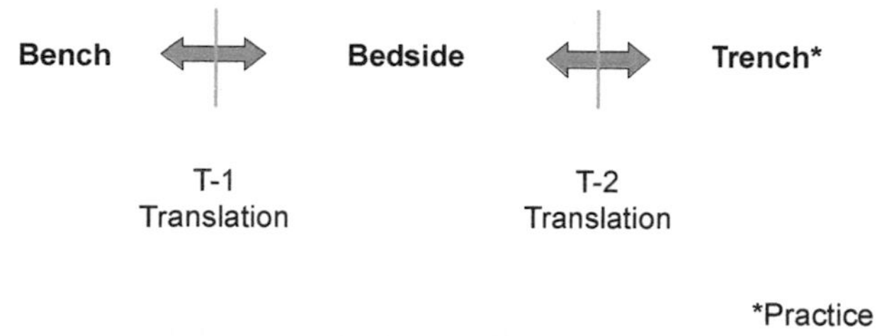

Bench ⬌ Bedside ⬌ Trench*

T-1
Translation

T-2
Translation

*Practice

FIGURE 34.1 Impediments in clinical and translational research. The translation block 1 (T-1) block is the barrier between discovery and its elaboration in individual human subjects. While this is often considered to be between bench and clinical science, it also could be between population and patient-oriented science. The translation block 2 (T-2) block occurs in moving the discovery into medical practice.

health in a more dramatically apparent fashion. One paradigm (Fig. 34.1) of analysis focused on failure of smooth *translation* of scientific knowledge along the way from scientific concept to improved human health. This focused on two impediments that were limiting greater progress: one when a discovery is needed to move from its point of demonstration, either in the laboratory or at the bedside, to the next level of clinical investigation (translation block 1 or T-1). A second impediment (translation block 2 or T-2) was defined as taking a discovery from its demonstration in human research to clinical practice. The latter is sometimes subdivided further, for example, into T-3 and other potential blocks. In this context, T-3 is usually defined as dissemination and implementation of research translation into practice/community/large populations (see Chapter 7).

Novel methods of clinical discovery were emerging, for example, the combined use of deidentified medical records together with linked DNA (Roden et al., 2008), which came to fruition nearly a decade ago, and the extraordinary developments in imaging technology that began to permit observation of human physiology in real time and were providing fruitful new paradigms for research.

At the same time the debate over the problems of clinical research in the universities was emerging, there were also changes in the pharmaceutical industry and biotechnology companies. In spite of many opportunities such as the flourishing of high-throughput screening, combinatorial chemistry, improved drug design, and the information emanating from pharmacogenetics (Giacomini et al., 2007), the economic outlook of the pharmaceutical industry seemed bleak (Fitzgerald, 2008). While there were many ways this faltering could be described, few are more telling than the change in the number of new molecular entities approved by the US Food and Drug Administration. In 1996 there were 53 such approvals. In 2007, this had fallen to only 17. There are many reasons for this change in the fate of new drugs and in the efficiency of the pharmaceutical industry in bringing them to market, but some believed that more interaction between scientists in the AHCs and the private sector might yield increased benefits for the public. One hope was that through the proposed *Clinical and Translational Research Institutes,* a more robust collaboration between academic and private sectors could work to the advantage of both (Fig. 34.2).

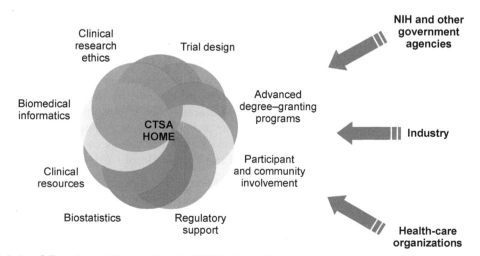

FIGURE 34.2 Clinical and Translational Science Award (CTSA): internal and external components. The CTSA provides an organizational structure for clinical research and improves its optimized linkage to entities outside the academic health center. *Courtesy NCRR Office of the Director.*

As defined by NCATS, *translational science* is the field of investigation focused on understanding the scientific and operational principles underlying each step of the translational process (National Institutes of Health, 2015b). At the heart of NCATS' Clinical and Translational Research movement, *translation* is the process of turning observations in the laboratory, clinic, and community into interventions that improve the health of individuals and the public (National Institutes of Health, 2015b).

While such relationships must be managed with great care, a failure of interaction might be contributing to the dearth of new drugs and devices for public health. If fostered in an open and cooperative framework, such public and private partnerships entailing better scientific and intellectual exchange might be fruitful and advantageous for both constituencies.

CLINICAL AND TRANSLATIONAL SCIENCE INSTITUTES

Events both within and outside, the AHCs were driving a debate about a viable strategy to overcome problems clinical research and its practitioners and trainees were facing (Schechter et al., 2004). From this debate, the Clinical and Translational Science Award (CTSA) program emerged in 2005 (Zerhouni, 2005, 2006). There was a broad consensus that conceptually it was an important and transformative step for clinical research in the 21st century.

In responding to the CTSA initiative, most institutions generally strove to create an overarching entity, based on a reconsideration of all aspects of their clinical research infrastructure, its strengths, and its limitations. The entire spectrum of clinical research was analyzed, and the availability and utilization of current resources and expected needs of the local and regional research community were evaluated. This often entailed an intense but thoughtful process of deconstruction and subsequent reconstruction of existing research components into the emerging local CTSA. As has been documented in many other activities of this type, there has been considerable debate and experimentation as to how to accomplish the overarching goal of this program. The initial challenges included restructuring without damaging what is beneficial; expanding educational and infrastructure opportunities without creating fiscal inefficiencies; developing a new administrative structure without adding bureaucratic impediments that limit adaptability, flexibility, and responsiveness.

By January 2009, the emerging CTSA consortium comprised 38 AHCs in 23 states and aimed to grow to 60 by 2012 (National Center for Research Resources, 2008a). The central driver of this program is the conviction that there is a distinct discipline of clinical and translational research and that AHCs need to establish the critical linkage between clinical research resources and the provision of sustained interdisciplinary training in a supportive and dedicated intellectual environment beyond that provided by traditional academic departments. The discipline of clinical and translational science is needed to ensure that rapidly emerging and fundamental advances in biomedical and behavioral sciences will animate patient-oriented research. The discipline requires the development of well-structured and well-recognized career development pathways that are intertwined with original and fundamental research addressing the methods and spirit of clinical research. This goal will be easiest to achieve in an academic home with a dedicated faculty and staff of multiple disciplines who have a transformative vision, mission, and strategy.

In 2014, CTSA developed a network of sites that shared *electronic health records (EHRs)* to help locate potential research participants more effectively and efficiently, and IRBrely—an initiative to develop a reliance agreement template, agreed upon workflows and procedures, and a mutually used institutional review board (IRB) for investigators conducting multisite studies (National Institutes of Health, 2015a).

Within a wider framework, the mission statement proposed for the CTSA program was *to transform the local, regional, and national environment to facilitate clinical and translational science, thereby increasing the efficiency, quality, and speed of clinical research (National Institutes of Health, 2016d)*. This transformation was to be achieved in part by the creativity emanating from the individual new academic homes. There is usually an institute (comprised of faculty and programs) that integrates clinical research across multiple departments, schools, institutions, and hospitals. The CTSA includes faculty who conduct original research, develop graduate and postgraduate training curricula, and lead programs that integrate clinical and translational science across the landscape of the clinical research enterprise in the broadest sense. Traditionally, a given CTSA had the following aims (National Institutes of Health, 2016d):

1. Provide resources for original research on novel methods and approaches to translational and clinical science;
2. Promote the enabling technologies and knowledge base that will facilitate the broad spectrum of clinical and translational science, including all clinical disciplines and all types and sizes of studies;
3. Integrate translational and clinical science by fostering collaboration between departments and schools of an institution and between institutions and industry;
4. Create a point of contact for partnerships with industry, foundations, and community physicians as appropriate;

5. Foster research education, training, and career development leading to an advanced degree (master's or PhD) for the next generation of clinical investigators (including physicians, nurses, dentists, pharmacists, and other allied health professionals);
6. Conduct self-evaluation activities and participate in a national evaluation of the CTSA program.

As is readily apparent, the challenge of implementing this program was formidable and the scope of any single CTSA by necessity would be limited by structures of a given time, adaptability, and available funds. Indeed, most individuals involved in this process believed that there would not be a single CTSA that would overcome the challenges but rather, the consortium of CTSAs that would achieve this end.

Under NCATS' (National Center for Advancing Translational Sciences) leadership since 2011, CTSA supports a nationwide network of medical research institutions, referred to as *hubs,* which work together to improve the translational research process and the delivery of treatments to patients in a more efficient manner. In 2016, CTSA's current program goals include:

1. Training and cultivating the translational science workforce;
2. Engaging patients and communities in all phases of the translational process;
3. Promoting the integration of underserved or special populations in translational research throughout the human life span;
4. Innovating processes to increase the efficiency and quality of translational research;
5. Advancing the utilization of cutting-edge informatics (National Institutes of Health, 2016f).

At the heart of CTSA, collaborations are fostered to improve quality, efficiency, speed, and safety throughout the clinical and translational research processes. In essence, the CTSA program aims to expedite the translation of research discoveries into health benefits for patients (National Institutes of Health, 2015c).

NATIONAL CENTER FOR ADVANCING TRANSLATIONAL SCIENCES

NCATS, CTSA, and Clinical Trial Recruitment Innovation

One of the more spectacular failures related to cancer trials. In an analysis of more than 7500 registered Phase II and Phase III cancer trials on ClinicalTrials.gov from 2005 to 2011, 20% were never completed due to the inability to recruit participants.

On December 23, 2011, President Obama signed the *Consolidated Appropriations Act, 2012* (P.L. 112-74) into law which enabled the *NIH* to establish the *NCATS (National Institutes of Health, 2016b).* The CTSA program was repositioned under the leadership of NCATS to *build upon its existing strengths and to ensure its further development in alignment with the NCATS mission to enhance the development, testing, and implementation of diagnostics and therapeutics across a wide range of human diseases and conditions (National Institutes of Health, 2014).*

NCATS was also encouraged to govern CTSA's direction by formalizing evaluation processes, advancing innovation in training, education programs, and community engagement in research initiatives.

Recently, to implement ways to improve the success of recruitment, NCATS' CTSA program launched the *CTSA Accrual to Clinical Trials (CTSA ACT)* to develop a nationwide network that shares EHRs across the country by devising a "federated" network with common standards, data terminology, and shared resources. *CTSA ACT* investigators focus on the following:

1. Data harmonization across the EHR platform;
2. Technical needs assessment and implementation;
3. Regulatory approaches to ensure compliance with protocols for data access and participant contact;
4. Governance development to establish proper agreements among institutions (National Institutes of Health, 2015a).

By linking EHRs, potential participants who meet study criteria can be matched to clinical trials, enrolled more efficiently, and investigators will be enabled to complete more studies. In turn, by recruiting participants in clinical trials, which is a critical factor, more treatments will be tested and expedited more quickly to patients with challenging health issues.

CTSA ACT's technology platform includes two major components: *Informatics for Integrating Biology and the Bedside (i2b2)* deidentifies patient records, and *Shared Health Research Informatics Network (SHRINE)* is a search engine for i2b2 for patient criteria. SHRINE queries all network institutions and matches participants to study criteria. By deidentifying patient information, privacy is protected while advancing research is ensured. In 2015, the CTSA ACT technology was utilized at 13 hubs (National Institutes of Health, 2015a).

In 2012, CTSA investigators aimed at improving trial accrual by developing the patient registry *Frontiers: The Heartland Institute for Clinical and Translational Research* which has a CTSA-federated hub called *HERON* (*Healthcare Enterprise Repository for Ontological Narration*) (National Institutes of Health, 2015a). *HERON* has a patient registry of 30,000 participants who were asked during office visits if they would be interested in clinical trials, they signed a consent form to authorize research-related contact and the inclusion of their medical records within the registry (National Institutes of Health, 2015a). Initially, HERON was utilized at University of Kansas and recently expanded to include 10 hospitals and 14 community partners in Kansas and Missouri.

University of California created a similar registry system, UC ReX, which demonstrated an efficient way to find and enroll qualified participants in studies at UC (National Institutes of Health, 2015a).

In 2016, NCATS plans to fund two *Recruitment Innovation Centers (RICs)*, which will focus on improving accrual at multisite clinical research centers through innovative informatics and technology-driven approaches, through the CTSA program. The RICs will assess the feasibility of recruitment, select research sites for multisite studies, ensure privacy safeguards, and provide permissions for data access. In turn, participants will be linked with research opportunities, trial participant recruitment will be ensured, and evidence-based understanding of health issues and effective treatments will follow (National Institutes of Health, 2015a).

NCATS' Major Programs and Mission

NCATS' mission (2016) includes the following directives:

1. Develop and demonstrate the usefulness of innovative methods, new approaches, technologies, resources and collaborative models that enhance the development, testing and implementation of diagnostics and therapeutics across a wide range of human diseases and conditions;
2. Expedite the delivery of treatments to patients more effectively;
3. Disseminate data, analyses and methodologies to the community (National Institutes of Health, 2016b).

In essence, NCATS develops new technologies, resources, and collaborative research models; demonstrates their usefulness; and then disseminates the data, analysis, and methodologies for use by the worldwide research community.

NCATS' most prominent programs include the following:

1. *NCATS' CTSA* program provides a national network of medical research institutions—hubs—who collaborate locally, regionally, and nationally to foster innovation across the spectrum of translational science. CTSA's innovative methodologies and cutting-edge informatics are utilized by the federated network of hubs, and CTSA investigators are enabled to train the research workforce of multisite clinical trials. CTSA is considered the flagship of NCATS.
2. *Rare Diseases Clinical Research Network (RDCRN)* enables the study of rare diseases and the development of treatments for rare diseases. RDCRN supports studies of over 200 rare diseases and facilitates multidisciplinary team collaborations and data sharing.
3. *Extracellular RNA Communication (ExRNA)* enables the further investigation of the newly discovered cell-to-cell signaling process via ExRNA communication which will deepen understanding of various diseases.
4. *RNA Interference (RNAi)* promotes the development and improvement of RNAi screening to better understand gene function and identify treatment targets.
5. *Illuminating the Druggable Genome (IDG)* aims to broaden scientific comprehension of understudied protein families.
6. *NCATS Chemical Genomics Center (NCGC)* investigators advance small molecule therapeutic development through assay (test) design, high-throughput screening, and medicinal chemistry.
7. *Therapeutics for Rare and Neglected Diseases (TRND)* supports preclinical development of candidates who treat rare or neglected diseases and enables investigational new drug application.
8. *Bridging Interventional Development Gaps (BrIDGs)* enables investigators to advance promising therapeutic agents from the late-stage of preclinical development to the investigational new drug application and clinical testing phases.
9. *Discovering New Therapeutic Uses for Existing Molecules (New Therapeutic Uses)* program focuses on drug repurposing—finding effective new treatments and cures for diseases by using drugs developed and approved for other disease treatment.

10. *Pfizer's Centers for Therapeutic Innovation (CTI)* fosters early scientific discovery and the translation into new medications via public—private resource sharing.
11. *Matrix Combination Screening* enables the screening of a long list of potential drugs to find the most effective drug combinations for patients.
12. *Toxicology in the 21st Century (Tox21)* is an initiative designed to improve toxicity testing methods and provide a more efficient evaluation of the effects of chemicals on human health.
13. *Tissue Chip for Drug Screening* program aims to develop bioengineered devices for more effective prediction of the safety or toxicity of drugs in humans.
14. *Stem Cell Translation Laboratory* aims to advance translational methods in stem cell research by transforming induced pluripotent stem cells into many other cell types for use in clinical research and therapies (National Institutes of Health, 2016g).

NCATS' Innovative Technological Programs

In addition to CTSA ACT's recruitment technology and registry programs, NCATS also fosters the following technologically advanced research innovation programs:

1. *Assay Development and Screening Technology (ADST)* enables the advancement of therapeutic development through research, innovative assay (test) designs, and chemical library screening methods.
2. *Chemistry Technology* enables experts at NCATS to develop screening approaches which can be used by scientists in the pursuit of innovative therapeutic development.
3. *Small Business Innovation Research (SBIR) and Small Business Technology Transfer (STTR)* programs support NCATS' mission to transform the translational science process by helping small businesses to develop and commercialize new technologies.
4. *Genetic and Rare Diseases Information Center (GARD)* collaborates with the National Human Genome Research Institute and offers comprehensive information on rare genetic diseases to patients, health-care providers, families, and the public.
5. *Global Rare Diseases Patient Registry Data Repository (GRDR)* aims to evolve into a Web-based resource that stores deidentified patient information from many different registries for rare diseases in one database (National Institutes of Health, 2016g).

NCATS' Rare Diseases Clinical Research Network

Even though thousands of diseases affect humans, only about 500 have approved treatments. Developing a drug, intervention or device can take up to 14 years and cost ~$2 billion or more, and only about 5% ever make it out of clinical trials. Many years may pass before patients who can benefit from a new drug are identified and treated. NCATS aims to quicken the translation from laboratory discoveries to beneficial patient treatment, as well as find cures for more than 200 rare diseases (National Institutes of Health, 2016c).

Through NCATS' RDCRN consortia, physician-scientists and their multidisciplinary teams work together (often at multiple sites) to study rare diseases which have no known effective treatments or cures. There are a number of rare diseases which have no Food and Drug Administration (FDA)-approved treatment, and there is an increasing interest in understanding rare diseases and the process of discovering, repurposing, and deploying effective treatments.

Dr. Christopher Austin, Director of NCATS, defines the success of *repurposing* as *the process of using existing therapies to treat new diseases based on commonalities in molecular causation of clinically disparate conditions (National Institutes of Health, 2016h).*

As proven by NCATS' RDCRN investigators, repurposing the drug sirolimus (Rapamune), which was originally approved for organ rejection in patients receiving implants, proved effective for a patient with the lung disease, lymphangioleiomyomatosis, and moved the disease into the treatable category for the first time.

NCATS' focus on the identification of rare disease commonalities and shared underlying molecular causation may in fact accelerate the development of treatments for multiple diseases at once as in the case of repurposing drugs (National Institutes of Health, 2016a).

HUMAN GENOME PROJECT

From 1990 to 2003, the NIH and the US Department of Energy cocreated the international Human Genome Project which determined the sequence of the human genome and identified the genes it contains. In humans, a copy of the entire genome equates to 3 billion DNA base pairs. As researchers understand the blueprint of a person, and the similarities, differences, and functions of genes, investigators can better understand what is critical for a healthy life (National Institutes of Health, 2016e).

CLINICAL AND TRANSLATIONAL SCIENCE AWARD

Structural Components of CTSA

One major difference in the organizational framework of the CTSA is that, unlike the GCRC program, which was largely under local control, the CTSA is a consortium governed by a cooperative agreement mechanism. This mechanism, a U54 award from the NIH, entails substantial joint ownership with the NIH, with active participation of NIH staff in consortium decision-making (see Box 34.1). This occurs in the context of substantial layers of involvement among individuals throughout the CTSA who have special interests in intellectual or disciplinary components of the CTSAs. The interaction of these components operates largely through teleconferences and the CTSA Wiki, which allows rapid communication and sharing of ideas among individuals with common interests and concerns (National Center for Research Resources, 2008b). At the top of the governance is the CTSA consortium Oversight Committee, which includes the CTSA principal investigators together with many NIH staff. A subgroup of this group acts as the CTSA Executive Committee. In addition to this committee, other major CTSA components are organized into working groups in the following way:

- **Education and career development.** A key component of a CTSA is one or more graduate degree–granting and postgraduate programs in clinical and translational science. The CTSA will train investigators from diverse disciplines, such as medicine, pediatrics, surgery, dentistry, nursing, and pharmacology, as well as study coordinators, project managers, and other key clinical research personnel relevant to clinical and translational sciences. Topics include clinical research design, epidemiology, biostatistics, pharmacology, biomedical informatics, ethics, behavioral science, engineering, and law. In many respects this is an expansion of a previous funded NIH program termed K-30 awards.

- **Clinical Research Center.** Under the CTSA framework, there is expansion of the mandate of the traditional clinical research center (CRC) far beyond its walls, in both a physical and metaphorical sense. The CRC and CTSA resources provide an environment to promote participation in clinical and translational research and help determine the most efficient and effective ways to interact with participants in clinical trials. Examples of resources identified at the CTSA website as appropriate include inpatient beds, outpatient facilities, or community-based exam rooms; mobile clinical research units that assist in community research; the recruitment of research participants; temporary research participant recruitment/enrollment sites; research nurses; research coordinators; phlebotomists; scheduling services; and services for research specimen collection and shipping. These resources are made available to the broadest possible cadre of investigators, especially those early in their professional careers and those at a crucial point in their career development. Issues addressed are the common and differing needs of a clinical study, a clinical trial, and a population study. The former likely needs a facility that intensively studies a small number of subjects in a tightly controlled environment to differentiate between the inherent pathophysiologic and environmental characteristics of a disease (see Chapter 2). Critical for drawing the correct conclusions are the study of normal subjects under similar protocols. Clinical trialists, on the other hand, need access to a large number of subjects with some definition of the individual subject's characteristics. By definition, they do not need or want a controlled environment because the answers to their questions have to be applicable to a free-living community (see Chapter 3). Finally, the epidemiological, outcome or health services study (see Chapter 4 and Chapter 7) does not require interaction with any person/subject but access to some data from a large population. All these areas of research have strong advocates for the strength and importance of their field. The challenge for the CTSA program will be how to balance these competing demands for resources.

- **Pilot clinical and translational studies.** One of the most successful CTSA programs has been the provision of support for pilot clinical and translational studies. These have especially been beneficial to young investigators and have also been a mechanism to extend new support to outcomes and community-based research. But more broadly they have allowed clinical and translational trainees and researchers to generate preliminary data for submission of research grant applications. They have also been utilized to improve clinical study design, biostatistics, clinical research ethics, informatics, and regulatory pathways, to develop new technologies.

- **Biomedical informatics.** The effective implementation of a CTSA depends crucially on communication within the CTSA itself and with all collaborating organizations. Internal, intrainstitution, and external interoperability is imperative to allow for communication among CTSAs and their research partners in government, clinical research networks, pharmaceutical companies, and research laboratories.

BOX 34.1 Draft conceptual framework of the National Clinical and Translational Science Award (CTSA) consortium

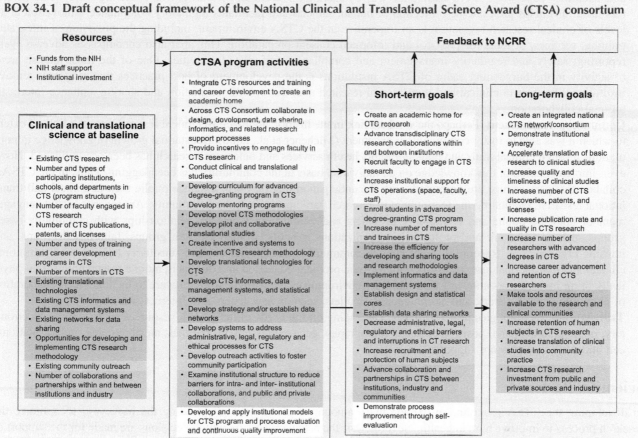

This document provides operational information about the governance and strategic plan of the CTSA consortium. This working document was prepared by the Trans-NIH CTSA Evaluation Subcommittee. This framework was made to provide a high-level starting point to outline short- and long-term objectives, serving as a guide for CTSA evaluation. With time, this structure is evolving based on experience.

Color code: dark yellow, CTS research; green, CTS training and mentoring; blue, CTS methods and technologies; tan, CTS community liaison and human subjects protection; lilac, CTS partnerships; light yellow, evaluation and quality improvement.

Courtesy NCRR Office of the Director.

- **Research design, biostatistics, and clinical research ethics.** Research in these diverse areas is increasingly important to the clinical investigation process, but in some respects the published research on these processes is more circumscribed than generally recognized. The CTSA encourages the development of innovative and creative research programs that bridge these functions with other CTSA activities. Issues particularly relevant include limiting risk to participants; preventing bias; improving enrollment; capturing appropriate data; developing design and analysis plans for studies of unique populations or very small numbers of subjects; and issues related to diseases with limited treatment options. Systematic study of informed consent and all aspects of the bioethical background of clinical and translational research is an appropriate concern.
- **Regulatory knowledge and support.** One area universally acknowledged to be problematic in the American clinical research enterprise in 2005 was the flood of regulatory paperwork which clinical investigators were compelled to navigate. This task was made more difficult by a lack of coordination and integration. A lack of administrative interoperability greatly slowed the process of clinical research at almost every level. With the tools of informatics, some of this regulatory burden will be able to be lifted. Indeed, the CTSAs with advanced informatics underpinning their regulatory

and scientific "paperwork" were leading the way by sharing this technology with others. Fostering teams of investigators itself also helps promote the protection of human subjects and facilitates regulatory compliance. Innovation at all levels of clinical research regulation is an imperative in the CTSA environment, including the provision of integrated training, services, or tools for protocol and informed consent preparation. This area also encompasses adverse event reporting; safety and regulatory management and compliance; and related activities. One of the most intense areas of activity in the burgeoning cadre of CTSA institutions is the development of best practices that reduce or remove institutional barriers to clinical and translational research and, through dissemination and sharing, enhance interinstitutional collaboration.

- **Community engagement.** Perhaps the weakest link in the historic approach to clinical research was the limited extent to which it was able to foster community engagement (Michener et al., 2008). Substantial growth of knowledge in populations through clinical trials and epidemiologic, health services and outcomes research has been accomplished. However, implementation of this knowledge on an individual basis in the community has been lagging. Thus, in the CTSAs, collaborative partnerships are promoted, with the anticipation that they will enhance public trust in clinical and translational research, facilitating the recruitment of research participants from the community. Approaches include engagement of both the public and community providers and establishing long-term relationships with community-based groups such as voluntary and professional organizations, schools, women's health groups, faith-based groups, and housing organizations. Resources might include community outreach and cultural sensitivity training for institutional clinical and translational researchers; community and provider education and outreach; mobile research units that physically bring investigators and staff to community environments; development of software to facilitate the collaboration of community practitioners; and communication outlets, such as newsletters and Internet sites.
- **Translational technologies and resources.** Depending on the needs of the CTSA, resources such as mass spectrometry, imaging, ultrasound, positron emission tomography, gene expression, proteomics, metabolomics, and translational cell and gene therapies are supported.

Scientific Review

In developing the CTSA program, an extraordinary amount of effort went into finding the best ways to optimize the research process to improve human health. Nevertheless, the actual process by which decisions are made for the support of the research projects of individual clinical investigators on the CTSAs is so crucial to CTSA programmatic success that, even if everything else succeeds, but the identification of the best and most productive scientific ideas and projects for funding does not occur, the CTSA cannot succeed.

To understand how a CTSA within a given institution operates in practice, it is valuable to consider the review process. It touches on all other activities of the CTSA and depends on the successful operation of a wise and diverse scientific review committee (SRC). In this section we will consider the steps by which applications from clinical investigators for award of CTSA resources are evaluated and awarded. This process varies considerably across institutions, reflecting local philosophical and scientific cultural patterns of research activity. However, with time it is likely that there will be some consensus toward optimized methods. This trend will be driven by the strong emphasis within the CTSA program on detailed evaluative processes of all aspects of methodology to find what works well and is cost-effective and what works less well. Because of the current variability, we will review one approach not as a model but as an example (Fig. 34.3).

At Vanderbilt University's CTSA, a voucher system is used for small investigator requests up to $2000. This voucher system was introduced to deal with relatively inexpensive but important and time-sensitive requests made by clinical investigators to the CTSA. Such vouchers are usable for specialized research assays, research consultation, or any other items needed in the translational research process. Vouchers are commonly used to cover costs of requested core assays as taken from the menu of options and prices in our extensive core laboratory system. The vouchers are easily available and require only a few minutes at the StarBRITE portal of the Vanderbilt CTSA, with the voucher request typically reviewed and awarded within 4–24 h. Few innovations in Vanderbilt's CTSA have been as flexible, efficient, and economical as this voucher system.

For requests for funding greater than $2000, a fuller and more detailed application is required. The initial step takes place at the time of receipt of the application by the IRB and the SRC scientific review analyst. There is a coordinated application process whereby concurrent IRB and scientific components are prepared within the same parent document. This represents an economy of effort for the investigator since many aspects of our previously separate forms for IRB and scientific review required entry of duplicate information. The IRB review is conducted by the appropriate IRB committee. The scientific review is initiated when the relevant portion of the application goes to a scientific review administrator, who ascertains if the application is complete and contains the kinds of information the SRC will require to judge its merit.

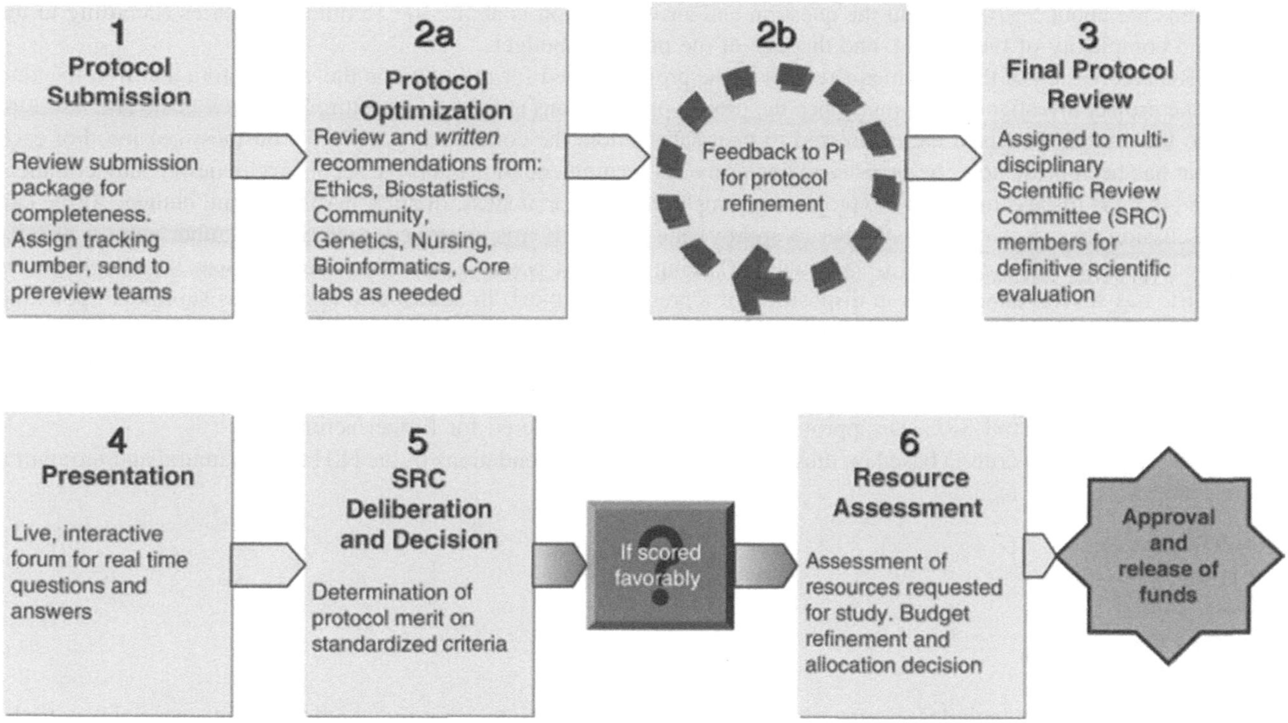

FIGURE 34.3 **Scientific review process on a Clinical and Translational Science Award (CTSA).** This depicts the steps and manner of scientific review in the Vanderbilt/Meharry CTSA. *Courtesy of Jill Pulley and Lynda Lane.*

The next step is concurrent administrative prereview by the biostatistician, the nurse administrator, the informatics specialist, and the research safety advocate. Both the conduct of the study and its budgetary aspects are evaluated for clarity and appropriateness. Occasionally other expert reviewers from the faculty will be called in if the project is deemed to require such specialized expertise because of its nature or its methodology. Any of the above prereviewers might judge that the project is not yet ready for formal review in its current form and might therefore recommend that the investigator participate in a clinic or studio. These exercises are among the most successful innovations of Vanderbilt's CTSA. These individuals review the science and, if appropriate, also the IRB material and then convene at a time when the investigator can present the project briefly, after which there is discussion of ways the project can be improved. Specific questions from the investigator or from the consultants convened are dealt with in this 60- to 90-min session.

The most important innovation at Vanderbilt has been the **studio**. A *studio* is a different kind of committee convened to focus on issues relevant to the broad range of public health disciplines. This committee systematically includes individuals with expertise in special aspects of the processes and disciplines of T-2 research. As in the clinic, the project is discussed and the advice of the studio participants emerges from these conversations with the investigator. At the conclusion of the studio or clinic session the investigator revises his/her project to reflect the new ideas and advice from the committee. It is then brought back into the review process in a substantially improved form. It is noteworthy that as the value of these clinic and studio experiences has permeated the environment at Vanderbilt, there has been increasing utilization of these "think tanks" before submission of a proposal to the CTSA.

The input of the various staff involved in the prereview of the project is then passed back to the investigator so that the project can be improved. If there are study design concerns, the biostatistician may consult separately with the investigator. If significant changes in the research project have emerged during the clinic or studio, or in response to the initial prereview input, an IRB re-review of the project may be necessary at this time. Usually, however, the initial IRB review occurs at a time when the protocol is optimized so that this committee's time is optimally utilized.

At this point the protocol is analyzed for budgetary items, with careful costing out of the various requested components. The protocol and budget are then sent to the primary and secondary reviewers on the SRC. For applications requiring substantial ($10 000) resources, the principal investigator is invited to attend the SRC meeting to make a brief presentation of the project and to address any questions the reviewers or the committee have concerning the project. In practice, most presentations are about 5—10 min and the question and answer session is about 10—15 min. This varies according to the quality and complexity of the project, and the size of the proposed budget.

The SRC receives all of the documents related to the projects slated for presentation the week before the SRC meeting at which the project investigator presents. Once the project presentations at the SRC meeting have been made and issues of concern to the committee raised and addressed to their satisfaction, the committee goes into a business session. For each project that has been presented, the primary reviewer makes recommendations and the secondary reviewer may comment briefly, and there is discussion about the project in terms of its appropriateness, quality, feasibility, and budget. At the end of this discussion there is a vote on whether to approve the project. If it is approved, each SRC member votes a priority score based upon the NIH scale of 1.0—5.0, with 1.0 being the most positive possible review.

The SRC has several alternatives in disposition of a presented project. In addition to approval as submitted, potential committee actions may include conditional approval, deferral, and disapproval. With conditional approval, the investigator must respond to a specific concern, and this does not require representation to the SRC. With deferral, there is usually a more serious concern and if the investigator chooses to go forward with the study, there will need to be a protocol revision and representation to the full SRC. Disapproved studies are not considered for further action.

Projects are judged by criteria based on those from the 2004 recommendations of the NIH for extramural and intramural research proposals. They include:

- Significance
- Approach
- Innovation
- Investigators
- Environment

In addition, Vanderbilt includes several components in a composite sixth category which relates in a special way to the CTSA and its goals. These include the following questions: (1) Will the study promote the training/career of a young investigator? (2) Will the study advance the development of preliminary data for an important grant submission, e.g., an R01, P01, K23, etc.? (3) Will the study likely result in a publication? (4) Will the study improve the translational research infrastructure in some way? (5) Does the project enhance community outreach?

Collective Initiatives of the CTSA Consortium

While local issues were the foremost preoccupation of most CTSAs in the early phase of the program, increasingly the emerging consortium of CTSAs sought ways to work cooperatively on issues that are overarching for all individual institutions (Fig. 34.4). In early deliberations, a number of these strategic concepts were developed, and it is anticipated that these will constitute the major tasks that will be undertaken in the consortium's first 5 years. Most of these include concepts that originated from the time of the program initiation, but, with the experience over the first few years of its existence, many have been refined and embraced in a more systematic way. The following are some of the most important ones:

- **Clinical research management system:** The basis of this strategic concept is the inefficiencies in the current informatics underpinning the clinical research management in many institutions and the potential improvements that may accrue with fuller integration of bioinformatics and improved management principles. These include interactions with the IRB and especially the efficiency and speed of contract evaluation for clinical research projects and collaborations. There is also a need for more systematic subject recruitment for studies, and in some cases, especially in rare disorders, patient recruitment at a national and international level needs to be facilitated (Watson, 2008). Other issues in this area are monitoring the conduct of trials to completion and reporting, careful balance of financial issues, and ensuring that clinical investigators always maintain public trust.
- **Career support ladder:** This is the training component and aims at the efficient education of emerging investigators with the best possible curriculum and broad exposure to multiinstitutional research, including the international dimension. This educational program can emphasize team-building and study management and may benefit from inclusion of business education aspects, perhaps in collaboration with individuals associated with business schools. Local successes in new training structures and better oversight of young investigators as they transition through promotions and benchmarks may be able to be adopted across the consortium. An additional aspect is to enhance relationships between basic researchers and clinical investigators.
- **National research inventory:** Not all CTSAs can do all things equally well. A goal for the program is to capitalize on individual capabilities and strengths. Presently, it is not easy to know exactly what these are, at least from the point of

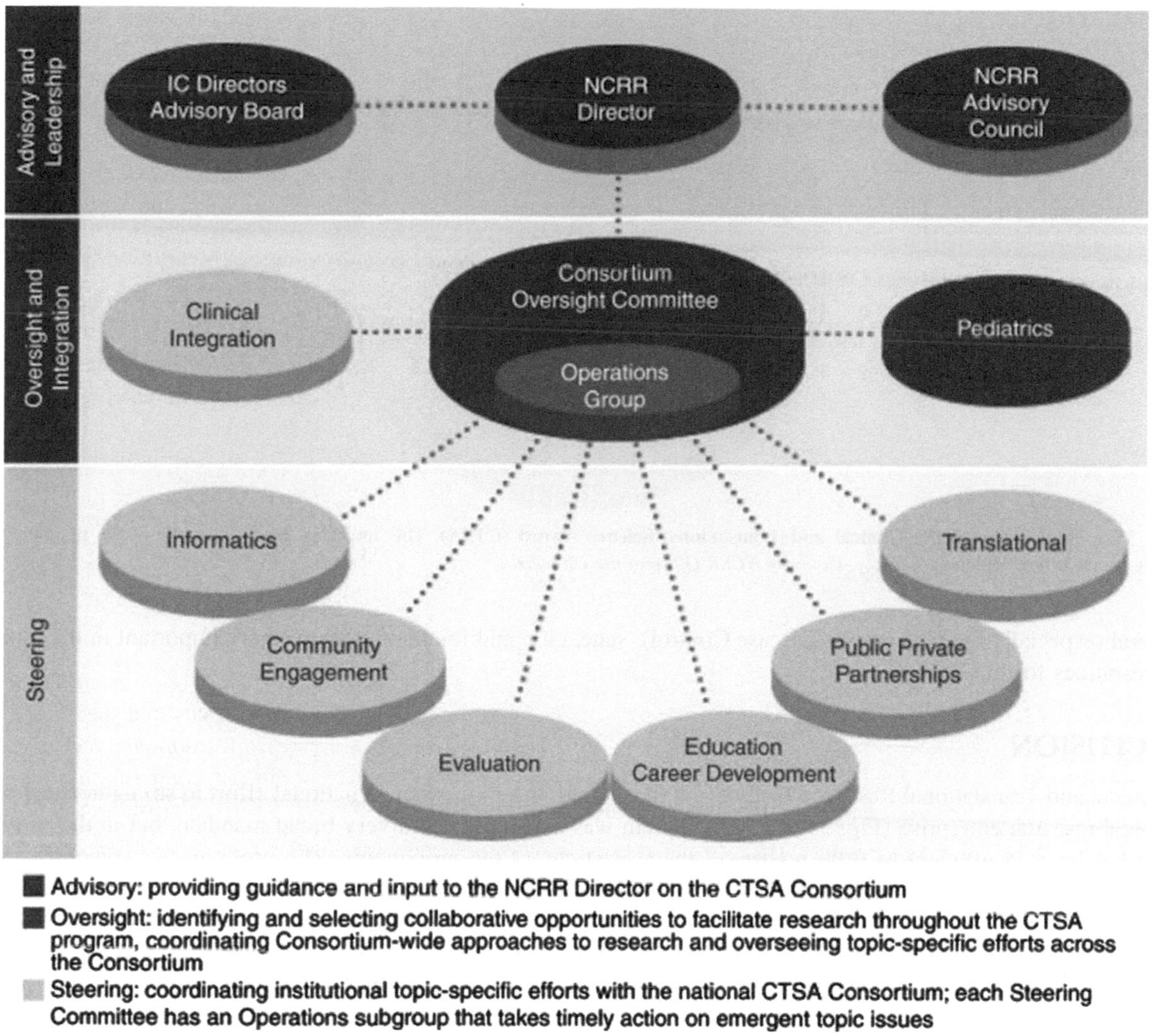

■ Advisory: providing guidance and input to the NCRR Director on the CTSA Consortium

■ Oversight: identifying and selecting collaborative opportunities to facilitate research throughout the CTSA program, coordinating Consortium-wide approaches to research and overseeing topic-specific efforts across the Consortium

Steering: coordinating institutional topic-specific efforts with the national CTSA Consortium; each Steering Committee has an Operations subgroup that takes timely action on emergent topic issues

FIGURE 34.4 Clinical and Translational Science Award (CTSA) consortium interactions and governance. The figure depicts components in the national consortium and their relationship to each other. *Courtesy NCRR Office of the Director.*

view of a given investigator seeking to identify such expertise. Establishment of a national research inventory would be able to provide a structure that would enable the identification of such nuclei of strength. This would make possible, in some cases, outsourcing of certain tests or sometimes even the research itself to the most capable sites. This could best be achieved through the establishment of a national searchable interface which would address technical cores such as bioinformatics, biostatistics, and data coordinating centers, as well as the establishment of a national bio bank. This might include systematic collection of biospecimens, tissues, and fluids. It would be IT-enabled with standardization of quality, usability, annotation, and consent. The value of tapping into the national comparative animal model core at NCRR would enhance this aspect of consortium activities.

- **National clinical data system:** A problem in clinical and translational research is the incapability of clinical data criteria used in different institutions and in different projects. This limits the utilization that can be made of data gathered from diverse centers. With national standards, some of this problem could be overcome. This would entail solutions involving entry criteria being defined with informatics solutions being brought to the table, but considerable discussion about what basic health information would be required for systematic inclusion.
- **National model for community engagement:** Community participatory research is an important aspect of the CTSA. This is for many of the CTSA institutions a relatively underserved need. Health disparities, their identification, and delineation of ways to address them are an important aspect of community engagement. The support of these practices exceeds the capabilities of any one source of funding, and creative financial sources for community research including

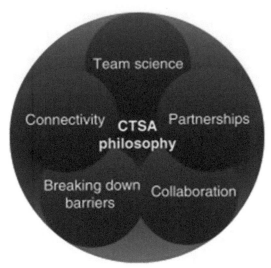

FIGURE 34.5 Philosophy of the Clinical and Translational Science Award (CTSA). This indicates the five themes of the program aimed at improving research function and efficiency. *Courtesy NCRR Office of the Director.*

federal (especially the Centers for Disease Control), state, city, and foundation sources are important in the leveraging of resources for this activity.

CONCLUSION

The Clinical and Translational Research Institutes in the United States constitute a national effort to strengthen and broaden the clinical research enterprise (Fig. 34.5). This program was launched with a very broad mandate, but in the early years, funding has been insufficient to fully realize all the aspirations of the proponents. The program has sometimes seemed administratively cumbersome and time-consuming during the start-up period, and it has received thoughtful criticism (Morrison, 2008). Nevertheless, such problems are common in new entities, and some were noted in the rollout of the GCRCs many years ago (Newman and Greathouse, 1963). What is clear is that the CTSAs are introducing new ways of organizing clinical research, enlarging the community of clinical and translational investigators, and providing a level of evaluation of research processes that seems almost certain to produce rapid improvement in efficiency and quality of the organization and practice of clinical research. While funds, education, and infrastructure are important elements of the transformation of the clinical research enterprise, the most critical component will rest on attracting bright, imaginative young people to occupy this home for clinical research. If that goal is achieved, the CTSA program will be declared a success.

REFERENCES

Ahrens Jr., E.H., 1994. The lamentable state of basic patient-oriented research: a call for action. J. Intern. Med. 235, 293–295.

Ahrens Jr., E.H., 1995. The birth of patient-oriented research as a science (1911). Perspect. Biol. Med. 38, 548–553.

Fitzgerald, G.A., 2008. Drugs, industry, and academia. Science 320, 1563.

Giacomini, K.M., Brett, C.M., Altman, R.B., Benowitz, N.L., Dolan, M.E., Flockhart, D.A., Johnson, J.A., Hayes, D.F., Klein, T., Krauss, R.M., Kroetz, D.L., McLeod, H.L., Nguyen, A.T., Ratain, M.J., Relling, M.V., Reus, V., Roden, D.M., Schaefer, C.A., Shuldiner, A.R., Skaar, T., Tantisira, K., Tyndale, R.F., Wang, L., Weinshilboum, R.M., Weiss, S.T., Zineh, I., 2007. The pharmacogenetics research network: from SNP discovery to clinical drug response. Clin. Pharmacol. Ther. 81, 328–345.

Luft, F.C., 1997. The role of the general clinical research center in promoting patient-oriented research into the mechanisms of disease. J. Mol. Med. 75, 545–550.

Michener, J.L., Yaggy, S., Lyn, M., Warburton, S., Champagne, M., Black, M., Cuffe, M., Califf, R., Gilliss, C., Williams, R.S., Dzau, V.J., 2008. Improving the health of the community: Duke's experience with community engagement. Acad. Med. 83, 408–413.

Morrison, L., 2008. The CTSAs, the congress, and the scientific method. J. Invest. Med. 56, 7–10.

Nathan, D.G., 1998. Clinical research: perceptions, reality, and proposed solutions. National institutes of health Director's Panel on clinical research. JAMA 280, 1427–1431.

Nathan, D.G., Nathan, D.M., 2016. Eulogy for the clinical research center. J. Clin. Invest. 126 (7), 2388–2391.

National Center for Research Resources, 2008a. Clinical and Translational Science Awards. http://www.ncrr.nih.gov/publications/clinicaldiscipline/CTSA_FactSheet.pdf.

National Center for Research Resources, 2008b. CTSAs: Translating Discoveries to Medical Practice [Online]. http://www.ctsaweb.org/.

National Institutes of Health [Internet], updated May 16, 2014. Proposed CTSA Initiatives. Bethesda (MD), Available from: http://ncats.nih.gov/advisory/concepts/council.

National Institutes of Health [Internet], updated February 2015a. CTSA consortium tackling clinical trial recruitment roadblocks. Bethesda (MD). Available from: https://ncats.nih.gov/pubs/features/ctsa-act.

National Institutes of Health [Internet], updated Winter 2015b. NCATS Fact Sheet: Transforming Translational Research & Spanning the Full Spectrum of Translation. Bethesda (MD), Available from: https://ncats.nih.gov/files/NCATS-factsheet.pdf.

National Institutes of Health [Internet], updated April 23, 2015c. Communities & Research: CTSA Program. Bethesda (MD), Available from: http://ncats.nih.gov/ctsa/community.

National Institutes of Health [Internet], updated February 2016a. Celebrating Progress at Rare Disease Day 2016. Bethesda (MD). Available from: http://ncats.nih.gov/director/message2016#feb2016.

National Institutes of Health [Internet]. updated February 18, 2016b National Center for Translational Sciences (NCATS): Mission, Important Events, Legislative Chronology, Director, Programs. Bethesda (MD). Available from: http://www.nih.gov/about-nih/what-we-do/nih-almanac/national-center-advancing-translational-sciences-ncats .

National Institutes of Health [Internet], updated March 28, 2016c. Rare Diseases Clinical Research Network (RDCRN). Bethesda (MD), Available from: http://ncats.nih.gov/rdcrn.

National Institutes of Health [Internet], updated March 16, 2016d. CTSA Program Funding Information. Bethesda (MD). Available from: http://ncats.nih.gov/ctsa/funding.

National Institutes of Health [Internet], updated April 12, 2016e. Human Genome Project. Bethesda (MD), National Library of Medicine (US), Available at: https://ghr.nlm.nih.gov/primer/hgp/description.

National Institutes of Health [Internet], updated March 29, 2016f. About the CTSA Program. Bethesda (MD), Available from: http://ncats.nih.gov/ctsa/about.

National Institutes of Health [Internet], updated April 13, 2016g. NCATS Programs & Initiatives. Bethesda (MD), Available from:https://ncats.nih.gov/programs.

National Institutes of Health [Internet], updated March 21, 2016h. Director's Corner: A New Horizon for Discovering Rare Disease Therapies. Bethesda (MD), Available from: http://ncats.nih.gov/directorNational.

Newman, E.V., Greathouse Jr., J.S., 1963. The relationship of the medical school and hospital to the clinical research center. J. Med. Educ. 38, 514–517.

Robertson, D., 2000. Reinventing the general clinical research centers for the post-genome era. In: Task Force on Clinical Research (Ed.), For the Health of the Public: Ensuring the Future of Clinical Research, vol. 2. Association of American Medical Colleges, Washington, DC, pp. 37–46.

Roden, D., Pulley, J., Basford, M., Bernard, G., Clayton, E., Balser, J., Masys, D., 2008. Development of a large-scale de-identified DNA Biobank to enable Personalized medicine. Clin. Pharmacol. Ther.

Schechter, A.N., Perlman, R.L., Rettig, R.A., 2004. Why is revitalizing clinical research so important, yet so difficult? Perspect. Biol. Med. 47, 476–486.

Vaitukaitis, J.L., 2000. Reviving patient-oriented research. Acad. Med. 75, 683–685.

Watson, M.S., Epstein, C., Howell, R.R., Jones, M.C., Korf, B.R., McCabe, E.R., Simpson, J.L., 2008. Developing a national collaborative study system for rare genetic diseases. Genet. Med. 10, 325–329.

Williams, G.H., 1999. The conundrum of clinical research: bridges, linchpins, and keystones. Am. J. Med. 107, 522–524.

Williams, G.H., Wara, D.W., Carbone, P., 1997. Funding for patient-oriented research. Critical strain on a fundamental linchpin. JAMA 278, 227–231.

Zerhouni, E.A., 2005. Translational and clinical science—time for a new vision. N. Engl. J. Med. 353, 1621–1623.

Zerhouni, E.A., 2006. Clinical research at a crossroads: the NIH roadmap. J. Invest. Med. 54, 171–173.

Education, Training and Career Choices

Chapter 35

Education, Training and Career Choices: Training Basic, Clinical, and Translational Investigators

Katherine E. Hartmann, Elizabeth Heitman and Nancy J. Brown

Vanderbilt University School of Medicine, Nashville, TN, United States

Chapter Outline

Key Points

- The optimal training of clinical and translational scientists shares many characteristics with the training of other biomedical scientists.
- Immersion in a scientific field through a mentored research apprenticeship remains the lynchpin of training.
- A didactic curriculum should be designed to increase knowledge in an area of core scientific content, as well as competencies and methodologies relevant to translational research.
- Participation in formal career development activities enhances skills needed for leadership, collaboration, and project management.

INTRODUCTION

Advances in molecular biology and human genetics, in informatics and technology, in tailored interventions, and means to reach into communities provide us with an unprecedented opportunity to understand the pathogenesis of human disease, to develop new diagnostic and therapeutic strategies, to enhance care, and to impact human health.

Clinical and Translational Science. http://dx.doi.org/10.1016/B978-0-12-802101-9.00035-1

637

Fundamentally, the translation of scientific and technological advances into better human health requires the training of health-care professionals equipped to assimilate rapidly evolving knowledge, to evaluate advances critically, to observe the effect of interventions in their own patients, to communicate their observations effectively, and to advocate for improvements. To this end, in 2006 the Association of American Medical Colleges (AAMC)'s Task Force II on Clinical Research recommended that every future physician "should be expected to understand translational and clinical research." Since then, both the Liaison Committee on Medical Education (LCME) and the Accreditation Council for Graduate Medical Education (ACGME) have promulgated standards for providing medical students and residents with core competencies in understanding clinical and translational research (Table 35.1).

Translating recent advances into better health requires the training of scientists equipped to make discoveries in humans, to develop and test new diagnostic tools and treatments safely in people, to evaluate the effectiveness of clinical interventions, and to implement the most effective approaches in traditional and novel settings to improve health. This chapter focuses on advanced training for clinical and translational investigators, including physician-scientists and other doctorally trained clinical and translational researchers.

OVERVIEW

The optimal training of clinical and translational scientists shares many characteristics with the training of other biomedical scientists. Successful clinical and translational investigators must be fluent in the language of science. Like their counterparts in the basic sciences, they must know how to design and conduct hypothesis-driven research. Developing these skills requires mentoring and protected time as well as resources such as money and space. Like other scientists, clinical and translational investigators require passion, commitment, and perseverance in the pursuit of their careers.

Clinical and translational researchers, however, also require a unique knowledge base and skill set. Whether dealing with individual human subjects or populations, clinical and translational investigators must understand how to study complex and integrated systems, in which there is considerable variability. Conducting research in individual human beings or communities requires an awareness of the ethical principles that govern such research. Clinical and translational research also requires facility with the regulatory systems and processes that have evolved to ensure the ethical conduct of research. Finally, clinical and translational investigators now and in the future must function as members of cross-disciplinary research teams.

Throughout the latter half of the 20th century, trainees in clinical and translational investigation acquired this knowledge base and skill set by immersing themselves in a research question under the tutelage of an experienced clinical and translational researcher. Today, the mentored research apprenticeship remains the lynchpin in the training of clinical and translational investigators. However, the increasing density of our knowledge about the genetic and molecular bases of disease; the sophistication of technological advances in biomarker development, imaging, informatics, and analysis techniques; as well as the growing complexity of regulatory requirements calls for a breadth of expertise beyond that which can be attained solely through the mentored research apprenticeship. Likewise, as noted earlier, these advances increasingly require the clinical and translational investigator to function within and lead multidisciplinary teams. Thus, the training of successful clinical and translational investigators best includes three components:

1. participation in a core didactic curriculum;
2. immersion in a scientific field through a mentored research apprenticeship; and
3. participation in formal career development and leadership development activities.

We address first the core didactic curriculum.

TABLE 35.1 Requirements for Medical Student and Resident Education in Clinical and Translational Research

Liaison Committee on Medical Education (Effective 1 July 2016)

"7.3 Scientific Method/Clinical/Translational Research.
The faculty of a medical school ensure that the medical curriculum includes instruction in the scientific method (including hands-on or simulated exercises in which medical students collect or use data to test and/or verify hypotheses or address questions about biomedical phenomena) and in the basic scientific and ethical principles of clinical and translational research (including the ways in which such research is conducted, evaluated, explained to patients, and applied to patient care)."

Accreditation Council for Graduate Medical Education (Effective 1 July 2016)

"The curriculum must advance residents' knowledge of the basic principles of research, including how research is conducted, evaluated, explained to patients, and applied to patient care."

Liaison Committee on Medical Education, Standards for Accreditation of Medical Education Programs Leading to the M.D. Degree, Published 2015. Standards and Elements Effective July 1, 2016 Accreditation Council for Graduate Medical Education, Common Program Requirements, approved 2007 with approved focus revision effective July 1, 2016.

DIDACTIC CURRICULUM

Clinical and translational research spans a wide spectrum from first-in-human studies conducted in a small number of subjects to health outcomes research conducted in large populations. The skills required to conduct these types of research differ, as does the infrastructural support. In describing the core competencies that should be addressed by a didactic curriculum, we will take a reductionist approach, describing those core competencies common to all types of research, basic or clinical and translational, those competencies required by patient-oriented or "T-1" translational investigators and those competencies required by "T-2" or higher translational investigators (Table 2). For the purposes of this schema, we define "T-1" research as that which translates discovery from "bench-to-bedside" or from the laboratory to human and back to the laboratory, whereas "T-2 through T-4" research encompasses the translation of research from the bedside into clinical evidence (T-2), from evidence to practice (T-3) and to populations (T-4).

Core Competencies Common to All Types of Research

Strength in a Core Content Area

Regardless of where an investigator falls on the spectrum from basic to T-4 translational research, he or she must develop and maintain a substantive focus in a particular biological, disease, clinical, or prevention area. A PhD-trained scientist or a physician-scientist working in a new area must acquire and maintain a depth of knowledge encompassing normal biology, physiology, pathophysiology, and clinical issues relevant to his or her content focus. This knowledge should be adequate to identify important areas that warrant further study, to design appropriate research with plausible etiologic pathways, and to

TABLE 35.2 Recommended Knowledge Base for Patient-Oriented or T-1 Translational Research

Core competencies

Biostatistics
- Sample size and power calculation
- Simple parametric and nonparametric methods
- Methods for controlling for confounders and effect modifiers

Epidemiology
- Case-control and cohort designs

Study design
- Randomized clinical trials
- Crossover, nested, factorial, and group allocation designs

Research ethics

Pharmacology and drug development human genetics

Molecular biology

Human physiology and measurement techniques

Writing and communication

Exposure to translational science

Genomics

Proteomics

Imaging technology

Biomarker development

Biomedical informatics

Data set acquisition and management

Human biology (required in area of research)

Molecular biology

Pathophysiology

Animal models

Laboratory techniques

interpret the research in a way that is useful to medicine and health. Although much of this knowledge can be acquired through immersion in a specific research question, trainees may benefit from directed, individualized didactic work. This individualization of the didactic program is critical to the training of a clinical and translational investigator. By way of example, the patient-oriented investigator involved in vaccine development may choose to pursue didactic work in virology or immunology, whereas the epidemiologist studying Alzheimer's may require in-depth knowledge of neurobiology or imaging techniques. In this regard, clinical and translational investigation is not an academic discipline per se, and training programs must be tailored to meet the needs of the individual.

Critical Review

Closely related to the need to develop depth in a core content area is the need to master skills in critical review of the literature. Critical review skills include the ability to assemble systematically all relevant literature, to critically read and synthesize the literature, to assess appropriateness and weaknesses of study design and analytic approaches, and to identify appropriate and feasible approaches that offer comparable or superior rigor. Students should emerge from training confident in their ability to compile supporting information to provide historical context and biological plausibility. Critical review is also essential during and at the conclusion of research to solidify the potential impact of the findings.

Communication, Manuscript, and Grant Writing

The ability to communicate one's ideas and scientific findings may be the single most important skill required by any investigator. Teaching abstract writing, presentation skills, and manuscript writing usually falls to the mentor. Courses or workshops in effective presentation skills, however, may allow for increased exposure, opportunities for videotaping, and access to new methods for preparing visual aids. Likewise, courses in manuscript writing can hone an investigator's ability to organize and focus a manuscript, to write effectively for his or her audience, to target papers and interpret letters from editors, and to respond effectively to reviewers.

Beyond requisite skills in the preparation of conventional academic products (e.g., abstracts, manuscripts, and presentations), trainees should also gain experience in developing grants in the style and approach required to obtain foundation and federal funding. They should learn to develop concise and compelling background sections; to specify the research hypothesis; to identify the appropriate study population; to describe measurement tools, analysis strategies, and human subjects' concerns; to discuss anticipated findings and alternative approaches, to develop a budget; and to elucidate project management.

Research Ethics and Responsible Conduct of Research

Clinical and translational scientists, such as bench scientists, pursue careers in research out of a motivation to improve human health. It follows, then, that clinical and translational scientists must know how to conduct research in a way that does not compromise the health or welfare of the individuals and populations they study. In addition, because scientific progress rests on the integrity of communicated discovery, investigators must understand how to maximize the integrity of their work and to ensure the integrity of the research record.

Trainees and faculty understand ethics differently and require distinct kinds of instruction and reflection at different stages in their careers. Nevertheless, as the US Public Health Service's Ryan Commission concluded in 1995, all those engaged in biomedical research or research education need to know essential concepts and standards of responsible conduct of research (RCR). Every investigator must be able to identify and address ethical issues that may arise in his or her career and interpret regulations and policies that affect his or her work.

Historically, formal instruction in research ethics and RCR in biomedical research training programs evolved in response to concern over instances of research misconduct or mistreatment of human subjects. As a result, certain aspects of education in research ethics and RCR are federally mandated. Since 1989, the National Institutes of Health (NIH) has required institutional research training programs to include instruction in research ethics and RCR. In 2000, the Office of Research Integrity (ORI) of the Department of Health and Human Services defined the core areas of RCR as data acquisition, management, ownership, and sharing; conflict of interest and conflict of commitment; protection of human research subjects; animal welfare; research misconduct; publication practices and responsible authorship; mentor and trainee responsibilities; peer review; and collaborative science. Also in 2000, the NIH mandated that all key personnel involved in NIH-funded research involving human subjects should complete formal instruction in the protection of human subjects. Typically such training addresses historical perspectives and addresses ethical principles and federal regulations related to the conduct of research with human participants, including the requirements and application of informed consent, and the protection of privacy and confidentiality.

Ideally, instruction in ethics and RCR integrates an understanding of the principles and regulations governing the nine core areas into a framework of professional responsibility and behavior expected of every competent investigator. Despite the intuitive link between education in research ethics and ethical behavior, the impact of formal instruction on research practice is not known. Educators in research ethics typically reject the idea that training in RCR will affect the morality or character of trainees. Education in research ethics can, nonetheless, increase trainees' sensitivity to and awareness of ethical issues in research, increase their understanding of ethical standards and regulations, and improve their moral reasoning skills. These effects, in turn, should enhance both the quality of their research and the trustworthiness of the research enterprise.

Study Design and Basic Epidemiology

Skill in design allows investigators to maximize the likelihood of observing an etiologic effect and to minimize bias. Trainees in either T-1 or T-2 clinical and translational research should be able to describe traditional and emerging research designs, compare and contrast strengths and weaknesses, and describe key aspects of "real-world" implementation, including specification of approaches to calculation of sample size, field logistics, human subjects' considerations, and data analysis, as it relates to the design. Researchers must understand advantages and limitations of each design for addressing specific problems, including practical aspects of their use, and the implications for subsequent data analysis. Researchers should be able to write a study protocol, to develop study instruments, to create randomization schedules, to enroll subjects, to validate measurements, and to collect data including means to assess and assure data quality.

Epidemiologists or T-2 translational investigators require in-depth training in epidemiological theory and methods as specified in the section on "Measurement." Nevertheless, although patient-oriented or T-1 translational investigators typically design interventional studies, they require sufficient familiarity with the design of observational studies to avoid common errors in study design and data interpretation. They should understand the rationale for selecting a specific study design, such as a cohort study versus a case–control study. In the case of cohort studies, they should understand the pragmatic differences between prospective and retrospective cohort studies, issues involved in the assembly and follow-up of a cohort, concepts related to measuring exposure, and ascertaining outcome ascertainment, as well as methods for calculating rates of occurrence and effect. Investigators need to understand conditions requisite to ensure the validity of a case–control study, considerations in the appropriate selection of controls, and potential sources of bias. Emphasis is also needed on the liabilities of analyses conducted in very large genetic and electronic data which can include poorly operationalized measures, inability to adequately control confounding that follows from the nature of the data, and false discovery.

Core Competencies for T-1 Translational Research

Biostatistics and Data Management

All basic, clinical, and translational investigators require a depth of training in biostatistics; among clinical and translational investigators, because T-1 and T-2 translational researchers deal with different types of data, the educational needs in biostatistics differ. T-1 translational investigators must have sufficient grounding in biostatistics to collaborate with biostatisticians. Patient-oriented researchers who conduct studies in relatively small sample populations can become sufficiently proficient in statistics to design studies and tackle straightforward analyses. Investigators must be familiar with sample size and power calculation. They should understand the appropriate use of the Student's t test, one-way analysis of variance (ANOVA), two-way ANOVA, and nonparametric tests. They should be competent in using common statistical software packages and presenting data in a sophisticated manner. They should understand the fundamental biostatistical concepts related to multivariable analyses in the presence of confounders and effect modifiers. T-1 clinical investigators should be familiar with regression models including linear, binary logistic, proportional odds logistic, conditional logistic, and Cox proportional-hazard analysis. They must understand the basic concepts related to repeated measures analysis, including mixed-effect regression models.

Beyond understanding study design and biostatistics, T-1 translational investigators must know how to collect and manage or supervise the collection and management of data. They should be proficient in data security, management, quality control, and documentation methods. They should know how to design and write a data management plan and should be familiar with the principles of database design. They should be familiar with the cognitive science underlying the design of paper and computerized forms. Translational investigators must understand privacy and security requirements for physical and electronic research records, including principles of deidentification. They should become sufficiently familiar with new information technologies for acquiring and managing research data to collaborate with biomedical informatics experts.

Pharmacology and Drug Development

The development of therapeutic or diagnostic modalities often involves the administration of new pharmacological agents to humans. Thus, the T-1 clinical and translational investigator requires a basic understanding of pharmacokinetics, drug metabolism and elimination, mechanisms underlying variation in drug responses, drug—drug interactions, and pharmacogenetics. To advance therapeutics, investigators must become knowledgeable about the steps in drug development, including the identification of targets and drug discovery, preclinical testing, and phase I through phase IV testing in humans. Investigators should understand the role and responsibilities of the sponsor or an investigator under an investigational new drug application. In addition to becoming familiar with the process involved in the development of new molecule entities, T-1 clinical and translational investigators should appreciate the special considerations in the development of devices, vaccines, or biological entities.

Genetics/Genomics

The description of the human genome provides an extraordinary opportunity to study the role of genetic variants in human diseases. The development of high-throughput genotyping mechanisms affords investigators the ability to acquire large amounts of genetic data quickly from well-phenotyped populations of interest. Therefore, clinical investigators must understand basic concepts of human genetics and heritability, as well as methods of polymorphism and mutation discovery, and genotyping. They should be familiar with the strengths and weaknesses of association studies and linkage analysis. Investigators must understand ethical issues regarding genetic studies and the use of DNA samples. They should appreciate the unique statistical methodologies used to calculate sample size in genetic studies, as well as informatics methods for detecting patterns in clinical and high-dimensionality genetic and molecular data, and understand the importance of replication.

Molecular Medicine

Few investigators conduct bench research and patient-oriented research concurrently, although many may conduct these types of research sequentially. Indeed, the highest impact patient-oriented or T-1 translational research is driven by a clear appreciation of the molecular basis of a disease. In many cases, investigators can achieve this breadth of knowledge by leveraging collaboration. Goldstein and Brown, in their wonderful commentary on the clinical investigator, expound on the importance of "an intimate collaboration between two individuals that allows them jointly to cover a range that neither could cover alone." They write further: "Such collaborations work best when each of the partners has some training and experience in the discipline of the other so that they can readily exchange ideas and insights" (1997: 2808). In short, either as an individual investigator or a collaborator, the patient-oriented investigator must have some grounding in current molecular biology. Because many trainees enter their research training after a substantial hiatus from medical school, they often require formal coursework to become acquainted with the most recent advances in molecular biology.

Core Competencies for Epidemiologic, Health Outcomes and T-2 Translational Research

Epidemiology increasingly embodies the multidisciplinary nature of clinical and translational research. On the one hand, epidemiology encompasses the science of public health devoted to the study of pathological conditions—such as cancer, cardiovascular disease, communicable diseases—that account for a substantial burden of disease, present opportunities for prevention, or require population-based approaches to understanding causes or interventions. On the other, epidemiology is a methodological discipline, dedicated to advancing the sophistication and integration of study design and data analysis methods to achieve more rigorous and accurate measures of complex associations.

Examples of the public health science of epidemiology include the Framingham Study, a community cohort that shaped understanding of the role of lipids in myocardial infarction or stroke, and observational research that identified the viral etiology of cervical cancer. Such research sets the stage for further basic science, as well as for treatment and prevention studies. The quantitative science of epidemiology has fostered the development of tools such as probability-based methods to predict individual cardiovascular disease risk and methods for analyzing the complex interactive effects of genetic variation and environmental factors on the expression of disease. This combined focus on disease processes and analytic methods prepares the trainee in epidemiology to contribute across the spectrum of translational research. Training in epidemiology provides skills relevant from the bench to the bedside, from research evidence to practice, from practice to populations, *and* from population-level observations back to the laboratory.

Within the increasingly multidisciplinary field of epidemiology, some traditional subdisciplines merit definition. **Clinical epidemiology** has its roots in the development of gold standard methods for conduct of clinical trials. It encompasses research focused on the study of patients and individual-level data that originates in clinical care settings. Clinical epidemiology includes the evaluation of diagnostic and screening test characteristics and measures how elements of the self-reported history of illness, the physical examination, and diagnostic tools contribute to reliable, reproducible definitions of disease entities as well as to accurate clinical diagnoses. **Pharmacoepidemiology** addresses the efficacy, effectiveness, safety, and costs of pharmacologic treatments, alternative medicine, medical devices, and surgical interventions. Health services, health-care outcomes, and evidence-based practice research may also be grouped as **health-care epidemiology**, which focuses on the organization, delivery, costs, and quality of care as they influence health outcomes. This encompasses the emerging areas of research on patient-centered outcomes research that extends consideration of clinical relevance of intervention effects to include patient values and perception about desired results, acceptability of treatments, understanding of anticipated outcomes and risks, and engagement in decision-making. Health-care epidemiology also includes the conduct of research and the synthesis of research evidence for the purposes of informing medical care recommendations, guidelines, and policy. It covers the explicit evaluation of systems for implementing best practices based on evidence.

Within these specific fields, epidemiologists commonly apply tools such as economic analysis, quality of life measures, and randomized intervention trials. Likewise, investigators engaged in utilization and outcomes research routinely employ community-based sampling of participants, estimation of attributable risk, and statistical methods such as survival models, propensity scores, and sensitivity models. Beyond this focus on health outcomes and components of health care, clinician epidemiologists increasingly bring clinical expertise, training in pathophysiology, training in human genetics, and understanding of tools from the laboratory and diagnostic sciences into fieldwork in epidemiology. A deep understanding of the proposed mechanisms and correlates of disease can enhance large-scale epidemiologic research, leading to such advances as the more efficient use of specimen banks, the refinement of inclusion and exclusion criteria to reduce risk of confounding, the improved implementation and validation of outcomes, and the novel use of imaging or diagnostic tests. In other words, a translational researcher working in epidemiology has the potential to carry his or her work to the population and back again to the preclinical and clinical environment to inform laboratory developments, advances in research methods, and improvements in clinical care.

For this reason, trainees in epidemiology would do well to pursue explicitly a strategy that capitalizes on the dual nature of the discipline, by developing both content expertise and specialized skills in developing and utilizing analytic tools that have cross-cutting relevance. This strategy promotes expertise in a content area of research and methodological breadth that enhances collaborations with colleagues in other content areas. For instance, a reproductive health researcher whose own work focuses on determinants of preterm birth might develop a methodological expertise in the design and conduct of studies using time-to-event data. These skills could lead to collaborations with colleagues who study topics as diverse as outcomes of ovarian cancer treatment or contraceptive uptake and continuance. Likewise, a researcher focused on pharmacotherapy in asthma care, who develops expertise in cost-effectiveness methods, might also apply that expertise in collaborations regarding the cost-effectiveness of new childhood immunization programs or other areas of pediatric chronic disease.

Theory and Methods

Trainees of even strong epidemiology programs may acquire little more than a primer on study design and statistical tools for calculation of measures of effect in logistic and proportional hazard models. To achieve depth, trainees should seek preparation that unites (1) exposure to causal logic and probability, (2) deep understanding of the constructs of confounding and effect measure modification, (3) rigorous analysis of the ways in which current tools fall short of the ideal goals of estimation, and (4) facility with the assumptions of the mathematical models used to guide study design, analysis, and proper interpretation of results. Such rigor will prepare them to advance the quality of research in their discipline.

Familiar examples of advances in methods that have permanently changed quality of research in the past decades include the now widespread use of power calculations to describe the precision of a study as it relates to power to reach valid conclusions and emphasis on adjusted models for mitigating the influence of confounding factors and assessing the potential influence of bias. Attention to theories of causal inference led to exposition of the flaws of hospital-based and clinical controls in case—control studies. Interest in webs of causality, rather than a focus on a single risk factor, has pushed development of new approaches such as multilevel modeling and applications of game theory to better capture the complexity of the interdependence of factors such as stress, nutrition, community characteristics, psychological traits, and genetic predisposition.

Epidemiology has also increasingly concerned itself with the accuracy and reliability of measurement and contributed to creation and validation of tools ranging from quality-of-life instruments to mathematical approaches like sensitivity analysis to assess how dependent the results of research are on the accuracy of the measured (and unmeasured) covariates. Expertise in the development and validation of methods for research is another potential focus of methods training to achieve specialized expertise. Researchers who strive to apply and advance the best available methods, and who remain students of the quantitative aspects of the discipline, will be leaders in both content and methods translation.

Measurement

The ability to define operational constructs (i.e., base population, inclusion and exclusion criteria, sampling frame) and to define exposures and outcomes for research requires familiarity with approaches to the design and validation of measures. Researchers should be able to specify definitions for outcomes, exposures, and covariates and should also be able to evaluate and develop tools for specialized measures, for instance, quality of life, substance dependence, functional status, and comorbidity. They should learn to identify the limitations of selected measures and to apply approaches for estimating the error introduced by chosen measures. Further, researchers should be able to produce the descriptive epidemiology of a given condition, including case definition, to identify critical candidate confounders, calculate primary measures of disease morbidity and mortality, and make appropriate comparisons by person, place, and time.

Analysis

The analysis of data, from studies or from large, administrative data sets, is perhaps the most common activity of epidemiologists. Such work requires a level of comfort with a range of data types—including dichotomous and multilevel outcomes. Analyses are often focused on understanding events over time and therefore require, at a minimum, a basic working understanding of time-to-event and repeated measures data. Interest in diagnostic or screening tools should be matched with skills in assessment of test properties and with the statistical implications of predictive modeling. Regardless of the methodological areas of greatest relevance to the new investigator, understanding the conceptual and statistical implications, as well as strengths and limitations of other modeling methods for data analysis, is essential.

Breadth in Exposure and Depth in Skills

Epidemiologic researchers in academic medical settings should cultivate familiarity with the range of topics that are common to both health-care research and epidemiology (Table 35.3). Exposure, even in areas that will not personally apply, creates common ground for collaborations and provides a basis from which to determine whether the skills of others working in related methodological areas could bring additional value to their own content-focused research. Investigators should develop independent areas of expertise in design and data analysis. To become and remain current in a quantitative domain requires a focus on a limited number of skill areas and methods. Target areas for developing specialized skills should be carefully chosen to mesh with content interest.

DEGREE-GRANTING PROGRAMS IN CLINICAL OR TRANSLATIONAL RESEARCH

Investigators engaged in epidemiologic or outcomes research generally complete a Master of Public Health or a PhD in epidemiology. The Council on Education for Public Health (CEPH) accredits schools of public health and some other public health programs offered in settings other than schools of public health. For a listing of CEPH-accredited schools and programs of public health see http://www.ceph.org/accredited/.

For clinical investigators engaged in patient-oriented research, a number of academic medical centers have developed training programs in clinical and early translational research. Many of these programs are targeted at MDs engaged in postdoctoral training. Some programs grant a master's degree, while others grant certification. Among degree-granting programs, some provide an MPH with special emphasis on clinical investigation or patient-oriented research, while others provide a master's in Clinical Investigation. As yet, no accrediting body exists for these programs and curricula vary widely across programs.

TABLE 35.3 Recommended Knowledge Base for Epidemiology and T-2 to T-4 Translational Research

Familiarity with

Clinical decision-making
Medical informatics
Comparative effectiveness research
Organization of health care (national, international, health systems)
Disparities in access and/or outcomes
Medical errors, patient safety, quality of care
Research ethics (data privacy, research in clinical settings)
Risk communication
Health-care data sources
Health-care financing, economics, and policy
Patient-centered research
Community-engaged research
Advocacy issues
Rare event and/or postmarketing surveillance
Professional development

Core health-care epidemiology methods[a]

Study design and analysis:
• Randomized trials
• Prospective intervention studies
• Evaluation of rare events
Diagnostic/screening test evaluation
Validation of measures
Scale development
Evidence-based practice
• Cost-effectiveness
• Decision analysis
Large data set acquisition and management
• Transfer and translation
• Linkage
• Data protection
• Risk of false discovery
• Approaches to replication
Obtaining and use of medical records data
Risk adjustment
Qualitative methods
Systematic evidence review
• Obtaining and abstracting "evidence"
• Metaanalysis; metaregression
• Measures of quality of the literature

Exposure to other core epidemiology methods[a]

Study design and analysis
• Large-scale cohorts
• Nested cohort study designs
• Case—control comparisons
Secondary data analysis
Administrative and vital records data
• Identification of relevant sources
• Assessment of quality and validity
Agnostic approaches to discovery in large data
Sampling and population-based recruitment
Multilevel/cluster analysis
Sensitivity analysis/simulations
Propensity models
Probability-based models
Survey design
Time-to-event data and adjusted hazard models
Repeated measures

Human biology (required in area of core content)

Genetics/molecular biology
Natural history of disease
Physiology and pathophysiology
Age/gender influences

[a]Recommend independent level of expertise in two methodological areas.

Trainees who wish to delve even deeper into clinical and translational investigation may wish to pursue a PhD. While a few programs offer PhDs in clinical and translational research per se, trainees may also opt to pursue PhDs in either their core scientific area or in methodologies such as epidemiology, biostatistics, informatics, biomedical engineering, and genetics, to name a few examples. In traditional MD/PhD programs, students complete their graduate studies and thesis between the preclinical years and clinical years of medical school. This timing can preclude participation in patient-oriented research, as students may not have adequate clinical knowledge or skills to conduct research in humans. In addition, because students typically enter extended clinical training after completion of their graduate work, they spend years away from science before returning to their postdoctoral training. The net effect is that many trainees are sidetracked during this hiatus. A simple innovation of moving the PhD to after the completion of clinical training permits students to engage in clinical and translational research during their PhD and to transition directly from the training phase into the career development phase of their research career.

THE MENTORED RESEARCH EXPERIENCE

Qualities of a Good Mentor

A good mentor serves as a role model; therefore it goes without saying that the mentor must exemplify the high-impact, well-funded scientist the trainee wishes to become. While every good mentor must be a successful scientist, however, not every successful scientist will be a good mentor. Good mentors value their trainees and make time for them. Good mentors are fair and honest and are willing to tell their trainees when they are making mistakes. Good mentors are compassionate and provide direction and hope when trainees face obstacles. Good mentors provide opportunities to their trainees and prepare them to become independent.

Using a Mentor Effectively

Finding a good mentor is only the first step to developing a successful mentor–trainee relationship. The trainee must also commit to the relationship. The trainee must be serious about the research endeavor and work hard. The trainee must take initiative. The trainee should honor deadlines. A trainee must be willing to take advice and must be honest and give feedback.

A trainee and mentor should meet early to establish goals and expectations. The trainee should provide his or her curriculum vitae (CV) prior to the meeting. Together the trainee and mentor should draft a time line and establish goals for the research project, the trainee's first (or next) presentation, first (or next) publication, and first (or next) grant. The trainee and mentor should discuss the allocation of the trainee's time among research, clinical activities, teaching, and administrative activities.

The Mentorship Committee

The mentorship committee may be the most underutilized resource available to trainees, whether in basic, translational, or clinical research. Many physician-scientists include a mentorship committee in the career development plan of their proposal for career development funding. Likewise, many academic departments assign mentorship committees for young PhD- and physician-scientists at the time of their appointment to the faculty.

When properly configured, a mentorship committee can provide promising young investigators with access to scientific expertise, resources, and wisdom beyond what is available through their primary mentor. The trainee and his or her primary mentor should identify individuals within the institution whose expertise would enhance the trainee's scientific development. This may include individuals from other departments. The inclusion of senior scientists in both basic and clinical or translational research can bring important depth to the committee and may promote multidisciplinary collaborations.

Scheduling a meeting of several busy, senior investigators can be a daunting task. Nevertheless, it is important to convene the mentorship committee every 6 months, even if the trainee meets with individual committee members more frequently. As a group, a committee of senior investigators will often generate ideas and may see issues that each individual member had missed or had been reticent to raise. Without the impetus of a mentorship committee meeting to focus them, the trainee or even the mentor may lose track of goals and time lines. Here, the mentorship committee holds the mentor, as much as the trainee, to task. The trainee should provide the members of the committee with his or her curriculum vitae and any recent publications, should present a short synopsis of recent progress, and, together with the committee, should set or revise goals for the next 1, 3, and 5 years.

CAREER DEVELOPMENT RESOURCES

The best training in clinical and translational science cannot provide an individual with all of the skills necessary to develop an independent research career, to manage complex projects, or to lead multidisciplinary groups. The mentor plays a critical role in guiding a trainee along his or her career path. Increasingly, however, educational programs for basic and clinical and translational investigators strive to expose their trainees to the skills they will need to succeed in a career. These include knowledge of or skills in negotiation, management, financial management, conflict resolution, organizational behavior, communication, time management, networking, and promotion and tenure.

FUNDING FOR TRAINING CLINICAL AND TRANSLATIONAL INVESTIGATORS

Predoctoral trainees in the basic sciences typically receive tuition support, as well as a stipend. Trainees in clinical or translational research may also pursue funding support through the National Institutes of Health (NIH), Ruth L. Kirschstein National Research Service Award (NRSA), research training grants, and fellowships (http://grants.nih.gov/training/nrsa. htm). This program funds both predoctoral and postdoctoral trainees and provides both individual and institutional grants and fellowships. Both predoctoral and postdoctoral training grants provide tuition support; however, the post-doctoral grants provide less money for tuition than do the predoctoral grants. Since many PhD or MD scientists pursue training in clinical and translational research after obtaining their initial doctorate, this can place a burden on them or their institution. The Research Career Development (K-series) Award Programs are designed to fund the transition from mentored postdoctoral trainee to independent investigator. To this end, they provide salary support for protected research time and typically require the trainee to dedicate 75% of his or her time to research. Typically, the K-series awards also provide money for tuition. The "K-Kiosk" at NIH (http://grants.nih.gov/training/careerdevelopmentawards.htm) provides current information about career development awards. A number of private foundations also offer funding for training and career development.

FURTHER READING

Altman, D.G., Machin, D., Bryant, T.N., Gardner, M.J., 2000. Statistics with Confidence: Confidence Intervals and Statistical Guidelines. BMJ Publishing, London.

Association of American Medical Colleges Task Force II on Clinical Research, 2006. Promoting Translational and Clinical Science: The Critical Role of Medical Schools and Teaching Hospitals. Available at: www.amc.org.

Bulger, R.E., Heitman, E., Reiser, S.J., 2002. The Ethical Dimensions of the Biological and Health Sciences. Cambridge University Press, New York.

Council of State and Territorial Epidemiologists, 2007. Competencies for Epidemiologists. http://www.cste.org/competencies.asp (accessed November 2008).

Feltcher, R.K., Fletcher, S.W., Wagner, E.H., 1996. Clinical Epidemiology: The Essentials. Lippincott Williams and Wilkins, Philadelphia.

Friedman, L.M., Furberg, C.D., DeMets, D.L., 1998. Fundamentals of Clinical Trials. Springer-Verlag, New York.

Goldstein, J.K., Brown, M.S., 1997. The clinical investigator: bewitched, bothered and bewildered—but still beloved. J. Clin. Invest. 99, 2803—2812.

Hennekens, C., 1999. Epidemiology in Medicine. Lippincott Williams and Wilkins, Philadelphia.

Inouye, S.K., Fiellin, D.A., 2005. An evidence-based guide to writing grant proposals for clinical research. Ann. Intern. Med. 142, 274—282.

Institute of Medicine of the National Academies, 2013. The CTSA Program at NIH: Opportunities for Advancing Clinical and Translational Research, ISBN 978-0-309-28474-5.

Katz, M., 2006. Study Design and Statistical Analysis—a Practical Guide for Clinicians. Cambridge University Press, New York.

National Institutes of Health, 2014. Physician-Scientist Workforce Group Report. http://acd.od.nih.gov/reports/psw_report_acd_06042014.pdf.

Ramani, S., Gruppen, L., Kachur, E.K., 2006. Twelve tips for developing effective mentors. Med. Teach. 28, 404—408.

Rosner, B., 1999. Fundamentals of Biostatistics. Duxbury Press, Boston, MA.

Steneck, N.H., Bulger, R.E., 2007. The history and purpose of instruction in the responsible conduct of research (RCR). Acad. Med. 82, 829—834.

Strachan, T., Read, A., 1999. Human Molecular Genetics. John Wiley & Sons, New York.

Tobin, M.J., 2004. Mentoring: seven roles and some specifics. Am. J. Resp. Crit. Care Med. 170, 14—117.

Whitcomb, M.E., 2007. The need to restructure MD-PhD training. Acad. Med. 82, 623—624.

Chapter 36

A Stepwise Approach to a Career in Translational Research

William F. Crowley, Jr.

Massachusetts General Hospital, Boston, MA, United States

Chapter Outline

Key Points

- Translational research thus now encompasses an ever-broadening spectrum of human physiology, genetics, pathophysiology, phenotyping, pharmacology, natural history, and/or proof of concept studies using interventional drugs or devices in appropriately selected disease models.
- In contemporary clinical research, this increasing ability to use patients with specific diseases as human models can define specific pathways, proteins, or genes underlying these disorders in a fashion not previously attributable.
- For most clinical investigators, the launching point for a career in clinical research starts with their clinical interactions that occur when faced with a patient with a specific disease.
- Young investigator should select the most severe examples of the disorder they can find.
- Normative data and control populations are needed.
- Engage relevant basic researchers and their technologies.
- Identify appropriate mentors across one's career.
- Search widely to obtain funding for one's research.
- The perils of senior research leadership.

DEFINITIONAL ISSUES

In the past, translational research has been variably defined. For the purposes of this chapter, a relatively restricted definition will apply that refers specifically to that subset of human investigations that address the "first translational block." This term refers to that spectrum of clinical research activities that focus on the increasingly dynamic interface between bedside and bench (as portrayed in Fig. 36.1). This type of clinical investigation contrasts with the "second translational block," a term coined by the Institute of Medicine's Clinical Research Roundtable (Sung et al., 2003), referring to those activities and difficulties encountered in achieving a more widespread implementation of new treatments into everyday medical practice that have been previously determined to be safe and efficacious in randomized clinical trials.

Clinical and Translational Science. http://dx.doi.org/10.1016/B978-0-12-802101-9.00036-3

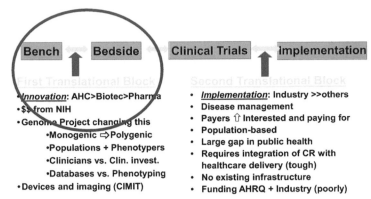

FIGURE 36.1   The two translational blocks in clinical research as defined by the Institute of Medicine's Clinical Research Roundtable. *AHC*, Academic Health Center; *NIH*, National Institutes of Health. *(Reproduced with permission from Sung, N.S., Crowley, W.F. Jr., Genel, M., Salber, P., Sandy, L., Sherwood, L.M., Johnson, S.B., Catanese, V., Tilson, H., Getz, K., Larson, E.L., Scheinberg, D., Reece, E.A., Slavkin, H., Dobs, A., Grebb, J., Martinez, R.A., Korn, A., Rimoin, D., 2003. Central challenges facing the national clinical research enterprise. JAMA 289, 1278–1287.)*

Historically, this first step of clinical solely with translating new findings discovered by basic researchers into the human. More recently, however, the opportunity to use phenotypic information derived from the study of patients, families, tissues, DNA, and/or other bodily fluids to drive the direction of basic research is increasingly at hand. This new bidirectional flow of information is the most exciting element of translational research driven largely by the new "omic" tools derived from the Human Genome Project. Additionally, novel devices and pharmaceutical "therapeutic probes" can serve as the empowering first steps of translational research as they both unearth an intriguing variability of patient responses to these interventions that often redefines a disease spectrum. What was previously thought of to be a relatively homogenous group of patients with a disease is now divisible into discrete subsets of "responders" and "nonreponders" by these new probes. Basically, any of these tools (genetic, devices, or pharmacologic agents) are now able to provide information that can initiate this first step of translating new insights in diseases into new treatments. Traditionally, most of these translational activities have occurred in Academic Health Centers (AHCs). They are typically supported by funding from the NIH or disease-oriented foundations and often occur in the milieu created by the NIH's Clinical and Translational Science Awards.

HISTORICAL PERSPECTIVE

Between the 1930s and 1960s (see Fig. 36.2), clinical observations were the most powerful drivers of the discovery of novel biology as new therapies were few in number. This situation was true because of the combination of two features. The first was the severity of diseases that, due to the lack of effective therapies at the time, typically presented themselves to their physician-investigators in the later stages of their natural history. The second was the relative weakness of the then contemporary measurement tools of biomedical investigation that were cumbersome, imprecise, and frequently insensitive often involving expensive biologic assays. In addition, poor fundamental understandings of normal physiology and a dearth of specific therapies that might have helped to sort patients into various response states were the norm. Finally, there

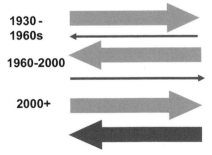

FIGURE 36.2   Trends in bench-to-bedside research in the past century.

was a lack of synergistic animal models. Collectively, these environmental features weakened the ability of basic science to support biomedical research. Despite these limitations, this period was a relative "golden era" of clinical investigation as typified by pioneering clinical investigators such as Fuller Albright, the father of modern endocrinology. Based upon keen bedside observations and a few general laboratory tests, he was able to deduce and identify several new clinical syndromes that still bear his name.

Beginning in the 1960s, this situation changed dramatically. The power, speed, and precision of new scientific disciplines such as biochemistry and molecular biology and their related diagnostic tools increased dramatically relative to the existing measurements. Consequently, a dramatic shift of emphasis occurred in medical research. Basic science suddenly offered substantial advantages over bedside investigation as a career path. The result was several decades in which basic research became the preferred career pathway for talented young physician scientists interested in biomedical research. Space allocations, resources, and promotions within academic centers were optimized for those who pursued a career in laboratory research. Clinical research, on the other hand, was relegated to less talented people who were supported largely by industrial funding.

With the advent of the Human Genome Project in the 1990s and the development of its new derivative "omic" tools, this situation began to change once again. Dramatic examples began to emerge whereby the study of patients and families with genetic conditions such as Huntington's disease, amyotrophic lateral sclerosis, and Alzheimer's disease employed these new genetic tools of the Human Genome Project to isolate new genes, identify novel pathways, and redefine biochemical targets for subsequent therapeutic interventions that had been largely overlooked by the basic science community. In each of these conditions, the genes and pathways determined to be pathophysiologic by these relatively "unbiased" genetic approaches were novel. These discoveries demonstrated clearly to basic researchers that their previous research efforts had not been appropriately targeting defects that were truly pathogenic. With these striking examples in hand, it became clear that refocusing the bedside-to-bench component of biomedical research was now not only possible but really mandatory. This remains the current state of the evolution in the field today with only the addition of devices and pharmacologic probes that have added considerable power to sort disease states in dramatic fashions. Collectively then all of these new approaches provide novel information regarding underlying pathophysiologic disease mechanisms and clearly target new potential therapeutics.

Most recently, the more complex genetic disorders, i.e., those common disease states that consume most of the healthcare resources globally today and are polygenic in nature, have also begun to fall prey to a new set of genetic tools. These include genome-wide association studies (GWAS), the Haplotype Map, and cross-species bioinformatic approaches. During the past 20 years, these tools and their derivatives have provided powerful resources for translational investigators (see Chapters 16 and 17 for details). The result of this last decade of tool generation from genetic research is that aspiring young clinical investigators are now empowered an ensemble of new enabling techniques available that collectively make this translational interface quite dynamic and appealing. However, specific training programs and mentoring are required for these investigators to take maximal advantage of these tools. However, this impressive ensemble of new tools offers young clinical researchers unique opportunities to initiate their biologic investigations in patients and families with diseases making the human the species of choice for study. The current challenge remains to blend these powerful contemporary tools with fundamental human phenotyping studies to forge an exciting new era of opportunities for translational researchers.

Translational research thus now encompasses an ever-broadening spectrum of human physiology, genetics, pathophysiology, phenotyping, pharmacology, natural history, and/or proof of concept studies using interventional drugs or devices in appropriately selected disease models. In contemporary clinical research, this increasing ability to use patients with specific diseases as human models can define specific pathways, proteins, or genes underlying these disorders in a fashion not previously attributable. These tools are also increasingly used to modify and reorder the future priorities and directions of basic researchers. As a result of this important biomedical research trend, the first translational block arena has become an increasingly exciting and bidirectional activity potentially enormous impact. It is thus becoming more attractive to younger clinical investigators seeking to be "where the action is." This chapter will outline a stepwise approach to building clinical research careers in this arena of "bedside-to-bench and back" form of clinical investigation.

STEP 1: THE STARTING POINT

For most clinical investigators, the launching point for a career in clinical research start with their clinical interactions that occur when faced with a patient with a specific disease. Fuller Albright first pointed out that this presence of a disease process in a patient identifies the system affected as an important basic biological process gone awry for the astute physician scientist. Using our patients afflicted with a disease as the starting point of investigations provides excellent assurance to young investigators that they will be devoting their research efforts to the study of something essential as attested to by the clinical consequences of its going awry. This advantage stands in sharp contrast to the approaches typically employed by basic investigators who commence their investigation within a scientific discipline—specific manner (e.g., biochemistry, molecular biology, or cell biology) where individual proteins or pathways emerge as novel findings seeking further study. Only after considerable work (if they are lucky) can they then focus upon broad biological processes or families of genes/proteins as the initiation point of their inquiries and then attempt to seek disease relevance for them. Thus, translational investigators theoretically start with a potential advantage over their basic science colleagues. They are assured of the biologic relevance of their quest since it is guaranteed by the abnormalities existing in the patients standing before them at the initiation of their investigation.

Such an advantaged starting point, however, does not guarantee success. Rather, young investigators must next make three transformative leaps that mark their transition from clinician to clinical investigator. The first is that they must synthesize the diverse clinical, pathological, and pathophysiologic observations that have been previously made on this disease into a cohesive schema that enables them to formulate specific hypotheses as to the underlying pathophysiologic mechanisms. Second, once empowered by these potential hypotheses, they must assemble the relevant toolkit of contemporary measurements to test them with both accuracy and precision. Finally, they must be able to test these hypotheses in a sufficient number of patients to make their conclusions statistically valid to overcome the heterogeneity that exists in most disease states, even monogenic disorders. On occasion, this testing can even occur in a single patient if the findings are dramatic enough, just as the finding of a single black swan was all that was required to dash the European myth that all swans were white. All of these transformative events—hypotheses generation, selection of appropriately powerful and precise measurements, and testing these hypotheses in sufficient numbers—have been the traditional "weakest links" in human investigation.

STEP 2: THE NEED FOR NORMATIVE DATA AND CONTROL POPULATIONS

Once a young investigator contemplates initiating such a human study, he or she should select the most severe examples of the disorder they can find. These patients represent the most egregious and hence most easily identifiable errors in any gene, protein, or pathway. Whether studying monogenic disorders where linkage and linkage disequilibrium are powerful tools that enable single family studies or polygenic disorders where GWAS are required, starting with the most glaring phenotypes is helpful. Eventually, however, as such investigations progress, they typically lead to the study of ever-more subtle abnormalities of these aberrant pathways that can often overlap the normal population. For example, patients and families with homozygous hypercholesterolemia revealed defects in the (low-density lipoprotein) LDL receptor as the abnormal gene responsible for this disorder that ultimately proved to be a target for intervention by the statins (Goldstein and Brown, 1974). However, these investigations soon blossomed to identify a series of other ancillary genetic defects with milder abnormalities that overlapped the cholesterol levels observed in the normal population and eventually required assembly of large population-based normative data (Hobbs and Cohen, 2006) to define. Of particular importance for these milder abnormalities is an understanding of and control of the relevant environmental factors capable of modifying the phenotype(s) for the particular disease such as diet, etc. These environmental factors are likely responsible both for permitting a relatively modest change in gene function and/or its expression to cause a severe disease and often confound the investigator's search for the genetic underpinnings of a disease since many genes can have their phenotypes modified similarly and dramatically by these environment changes. Such a typical progression requires a firm understanding of the normal range of the processes under study, whether be it a quantitative serum measurements of marker proteins (HDL (high-density lipoprotein) and triglyceride levels), intracellular levels of proteins, or subtle polymorphisms in gene sequences. In each case, assembling a large and well-phenotyped normal population to serve as controls is an essential next step to make their contrasts with disorders both statistically meaningful as well as medically relevant.

STEP 3: ENGAGING RELEVANT BASIC RESEARCHERS AND THEIR TECHNOLOGIES

Emerging tools and technologies continually create new opportunities for clinical investigation. Young investigators must thus be prepared to seek out, acquire, and incorporate them into their evolving research programs as soon as they are discovered and validated. They must be equally prepared to discard older and less precise tools for continued success. To update their repertoire of tools, continuous interactions with a wide range of basic collaborators are critical for the translational researcher to sustain a competitive career at the interface of bench-to-bedside. Consequently, success as a translational investigator requires recognizing and accommodating to the fact that they often must devote at least as much time to basic science lectures and labs as they will with patients.

Ultimately, this requirement to continually master the ever-changing interface of clinical investigation with a wide range of basic research technologies quickly emerges as *the most* substantial challenge to sustaining a long-term career in translational research. The dazzling repertoire of basic techniques, ever-changing arrays of enabling machinery, and increasing burdens of complex statistical manipulation all become daunting obstacles to building and sustaining successful translational careers for young clinical investigators. These challenges loom even larger as most physician-investigators must assume increasing patient care, teaching, and administrative responsibilities as they become successful faculty members. Juggling the difficulties of maintaining cutting edge scientific information while balancing these other career requirements is challenging and requires a disciplined restriction of the time devoted to teaching and patient care. A young investigator must avoid dilution with too much or too early administration and/or providing general teaching and medical care, until or unless this becomes a critical part of their long-term career game plan. Inevitably, balancing these conflicts requires appropriate mentorship within our complex academic organizations that are subject to so many often conflicting and diverse funding constraints. Some of these institutional pressures are not in the best interest of an emerging investigator's career development, and the young investigators often need skillful guidance in how to handle and often avoid them until a more appropriate time in their career.

STEP 4: IDENTIFYING TRACTABLE PROBLEMS

Assuming that young clinical investigators have appropriately chosen their patients and disease process, they are in the proper setting for career development in a successful AHC where they have access to both patients and basic scientists, and have positioned themselves to capitalize on emerging technologies/collaborations, the next most important career step is to focus on a discrete but previously unsolved part of their disease problem that has now been more tractable by current technology. This step involves a creative leap by the nascent investigators with appropriate guidance from a more senior mentor. Mentors should help young investigators sort through the various options, ever-sharpening their thought processes and assisting them in paring down their plans to focus upon the right problem with the right tools and at the right time. Recognizing that successful long-term careers represent a blending of both short- and long-term projects, developing a diversified portfolio of research projects is also critical. Broad, long-term projects give young investigators a shape and structure for meaningful and coherent careers. However, "early hits" are equally necessary to provide some early successes to demonstrate the positive rewards of a career in research. In addition, such projects are necessary to provide sufficient preliminary data to build independent grant support. Achieving the proper portfolio balance between these two types of projects requires continual refinement and discussion between the young investigators and their mentors.

STEP 5: IDENTIFYING APPROPRIATE MENTORS ACROSS A CAREER

The next step is anticipated by the previous, i.e., finding the correct mentor(s). In contemporary clinical research, just as technologies are ever-changing and represent both opportunities and challenges for young clinical investigators, so too the complex task of obtaining the proper mentorship is an ongoing challenge. Certain mentors are critical for the early development of scientific careers whereas others become crucial for acquisition of new technologies or for longer term leadership opportunities. Thus, young clinical investigators must be ever on the lookout for appropriate mentorship. Good mentors are rare treasures in academic medicine and are usually well known for these traits. Typically, they have achieved a considerable degree of professional success personally and are both comfortable with their own accomplishments and secure in their egos. This maturity factor is essential to providing objective advice quite independent of the particular career that the mentor may have chosen. Good, selfless mentors stand in contrast to more self-interested advisors who often suggest a project that helps the mentor more than the young investigator. To be sure, often an element of self-interest on both parts is appropriate in project selection. However, these potential conflicts have to be discussed openly and frankly with a clear plan for the eventual independence of the younger researcher as soon as appropriate. Thus, young investigators must be ever-vigilant for dated or self-interested advice and be prepared to rapidly resolve apparently conflicting sources of career advice to achieve a true synthesis of how best to pursue success over a long-term career.

Above all, primary mentors must have good scientific vision as proven by their record of publications, grant support, and, most importantly, track record of successful prior trainees. When considering working for an established investigator, a reasonable request for the young investigator is to see the list of *all* prior trainees, what projects they pursued when under the mentor's supervision, and current information on where they are and what they are doing now. Due diligence should involve calling these prior trainees and asking them detailed questions about the mentor and their experiences with that person as a trainee.

Young investigators often become confused about mentorship issues, particularly if they are new to research and just emerging from clinical training. Frequently, they wrongly choose physicians who they admire for their clinical skills. While successful mentorship and clinical acumen often overlap, since research careers involve development of nonclinical skill sets, the most successful clinical investigators are often quite focused in their clinical interests. Often, they are not the leading physicians caring for large patient populations as such clinicians frequently have little time to conduct and stay contemporary in their scientific careers. So selection of a mentor is a tricky career step, and one best pursued with considerable screening, discussion, due diligence, and advice from others will be successful in analogous fields. Good clinical investigators are well known among their peers, and seeking advice of other successful scientists is very helpful.

STEP 6: OBTAINING SUCCESSFUL INDEPENDENT FUNDING

Achieving successful peer-reviewed funding is divided into two distinct phases. The first is obtaining the proper degree and duration of support for the prolonged period of clinical research training required to provide a safe temporal harbor in which to acquire the skill sets necessary to launch an independent career. The second is a progression on to obtaining independent research funding.

The earliest training for young clinical investigator typically involves support for 2–3 years on an NIH-sponsored training grant (T32) or equivalent individual research award (F32) or foundation grant. These competitive awards occur immediately following completion of clinical subspecialty training. This training is most often done under the directorship of a single mentor who is the principle investigator (PI or Co-PI) of a training grant, most typically one funded by the NIH. It is typical that, during this period of time, the young investigator becomes involved in a Masters Program in Clinical Research (K30) funded by the NIH's Clinical and Translational Center Program. This didactic experience acquaints them with the core analytic disciplines of clinical research (biostatistics, epidemiology, bioinformatics, genetics, etc.). However, this didactic experience must be balanced by an early and equally intensive mentored research experience to be optimally successful.

Thereafter, the NIH's K23 Clinical Investigator Awards or equivalent support from subspecialty societies can provide the next 5 years of funding. This next period permits further consolidation and practical use of the learned techniques in clinical research during their mentored research project. This next 5-year block should also involve an intense period of acquiring and publishing important preliminary data, obtaining considerable local and national opportunities to present their data for critique by other investigators, and beginning to assemble the various elements required to submit their own independent research project, typically in the third year of their K23 funding. Obtaining a first independent grant is a challenging task, as is successfully renewing that initial funding. Each of these individual steps in the independent award pathway represents an important career milestone that requires a great deal of thought, planning, and counsel.

Once these initial training grants and a new independent grant support have been obtained, the next step from a funding perspective is for the newly minted, independent, typically R01-funded investigator (see Chapter 39 for details on NIH's R01 grant program) to obtain an NIH K24 Mentorship Grant. This mentoring award can stabilize 50% of their salary funding for up to a decade, as it is renewable once for a second 5-year term in some NIH Institutes.

Once these early and mid-career awards are obtained, the final and more complex task is to decide how broad a program the now mid-career investigator wishes to mount. This choice becomes a critical one. Considerable individuality exists in this final step of career building as an ever-rising tide of administration, teaching, and patient care responsibilities begin to vie for the emerging investigator's time and effort. Assuming mentorship responsibilities for younger investigators becomes yet another important but time-consuming facet of a successful mid-career physician scientist. In addition, the idea of whether to expand their independent program to obtain Center Grants, SCORE Grants, or other specialized programs represents yet other opportunities for young investigators to expand the breadth of their research programs. However, each of these opportunities requires time, organizational abilities, and interpersonal skills in addition to a breadth of science appetite. By the time these options present themselves, the successful clinical investigator is typically mentoring an expanding number of younger individuals.

STEP 7: THE PERILS OF SENIOR LEADERSHIP

Assuming that clinical investigators have successfully hurtled each of these career markers, they soon discover that they have become quite valuable members of the academic scientific community. Because successful clinical investigators have had to be firmly grounded and knowledgeable in a broad range of important areas such as patient care, teaching, administration, and research, they have learned to speak the several "dialects" used to converse within the Towers of Babel that represent our contemporary AHCs. Senior clinical investigators bring a firm understanding of the clinical missions that weigh heavily on the operations of an academic center's core financial and patient care objectives. To have survived to such senior leadership positions, they also must have achieved a deep appreciation of the difficulty of sustaining research funding from the peer review process. Hence, they are often sought to assume valuable roles in research administration. Moreover, they typically have had to become quite successful educators and mentors. Thus, they offer academic centers seasoned hands in the complex task of educating and training younger generations, regulating and supervising the research process, and melding the ever-conflicting patient care and research mandates that bedevil all AHCs. Finally, most clinical investigators are fundamentally focused on battling the human suffering caused by disease and disease-related processes. Hence, they have great value to the biotechnology and pharmaceutical industry both as consultants or recruits. Thus, they not only represent extremely valuable role models for the younger faculty, they are also important assets for their academic centers and the broader biomedical research community whose regulatory burden only continues to grow.

Consequently, a vast playing field of career opportunities now unfolds before the successful clinical investigators as they emerge as senior leaders within AHCs. While such choices are pleasant, all are dilutive to their research efforts due to their time-consuming nature. This feature requires constant refinement and negotiation and is ever-changing in its pattern of opportunities/challenges for career investigators. That said, successful mid- or senior clinical investigators become extremely valuable members of the biomedical research community.

SUMMARY

Careers in translational research are uniquely challenging and rewarding. Continually updating technological tools; reeducating oneself across an entire career; juggling the conflicting mandates of teaching, patient care, and research; and repeatedly competing for peer-reviewed funding is not for the faint of heart. However, research at the translational interface represents one of the most deeply satisfying careers that exist in academic medicine. The leverage of success in these positions is vast and the ability to potentially contribute to the alleviation of human suffering is irresistibly attractive, making all of the challenges seemingly small compared to the opportunity to accomplish a greater good in this pathway.

REFERENCES

Goldstein, J.L., Brown, M.S., 1974. Binding and degradation of low density lipoproteins by cultured human fibroblasts: comparison of cells from a normal subject and from a patient with homozygous familial hypercholesterolemia. J. Biol. Chem. 249, 5153–5162.

Hobbs, H., Cohen, J., 2006. Sequence variations in PCSK9, low LDL, and protection against coronary heart disease. N. Engl. J. Med. 354 (12), 1264–1272.

Sung, N.S., Crowley Jr., W.F., Genel, M., Salber, P., Sandy, L., Sherwood, L.M., Johnson, S.B., Catanese, V., Tilson, H., Getz, K., Larson, E.L., Scheinberg, D., Reece, E.A., Slavkin, H., Dobs, A., Grebb, J., Martinez, R.A., Korn, A., Rimoin, D., 2003. Central challenges facing the national clinical research enterprise. JAMA 289, 1278–1287.

Chapter 37

Physician Careers in the Pharmaceutical Industry

Ronald L. Krall

University of Pittsburgh, Pittsburgh, PA, United States; GlaxoSmithKline

Chapter Outline

Key Points

- Successful drug discovery and development is entirely dependent on a deep understanding of the clinical manifestations of disease.
- Drug development proceeds by the systematic posing of key questions, each of which requires specialized expertise to answer. These areas of expertise define medical-scientific "jobs" in the pharmaceutical industry.
- In Phase 1, translational scientists and clinical pharmacologists attempt to determine whether a drug can be safely administered to humans.
- In Phase 2, translational scientists and experts in clinical investigation aim to prove that the candidate drug delivers the desired biological effect (mechanism) and clinical concept.
- In Phase 3, experts in clinical investigation and drug safety aim to prove that the drug delivers the desired clinical benefit to large groups of patients.
- Once a drug is approved (Phase 4), experts in drug safety, epidemiology, outcomes research, and clinical investigation partner together to confirm that the marketed drug continues to deliver the expected health benefits.
- Physicians with expertise in disease, clinical investigation, and evidence generation contribute to the determination of standards of evidence required for the regulatory approval of medicines.
- Physician careers can be built within any of these disciplines through their application to successive drug development projects.
- Physicians can move beyond their specific discipline training to become leaders of projects or departments.
- There are certain responsibilities that require medical expertise: safe administration to humans, ethical conduct of human subject research, and the accurate communication of medical information.

INTRODUCTION

A chemical compound travels a complex path from its discovery to approved medicine. Whether the compound is a small molecule or a protein biopharmaceutical, specific phases of drug development define that path. These phases, illustrated in Fig. 37.1, build the evidence that a medicine is safe and effective for its intended use.

Each of these phases is characterized by a focused set of research questions. These questions define the experiments that must be carried out. In Phase 1, the first question is "Can this compound be safely administered to humans?" To answer this question the researcher must first develop nonclinical evidence that exposing human subjects to the compound in question can be expected to be safe. Such evidence includes experiments such as determining the tissue distribution of the target in humans, so as to understand the theoretical pharmacologic effects, and pharmacology and toxicology studies in animals to determine the target organs and the serum concentrations associated with toxicity. Satisfied that administration to humans can be safe, the first human experiments confirm safety, define the observable effects, determine the consequences of "too much drug," and define the absorption, distribution, metabolism, and excretion—the pharmacokinetics—of the compound.

These and subsequent experiments go on to test the premise of the compound's discovery: does the compound have the expected pharmacologic effect in humans? For example, Tysabri (natalizumab) was developed on the back of substantial in vitro and in vivo evidence that activated T lymphocytes that passed through the blood—brain barrier contributed to the inflammatory lesions of multiple sclerosis (MS). The first step in testing that hypothesis in humans—the "Proof of Mechanism" experiment—was the confirmation that natalizumab, which binds to the alpha-4 integrin adhesion molecule that enables transit across the blood—brain barrier, reduced the number of MRI gadolinium enhancing (inflammatory) lesions in the brains of patients with MS (Tubridy et al., 1999).

Classical Phase 2 clinical studies are designed to answer two questions: Does the compound have efficacy in a specific disease state and what is the optimal dose and regimen? These questions may be structured differently for some compounds and some experiments. For example, the question may be "Is there a dose or concentration response with regard to some specific efficacy measure?" Alternatively, the question may be about therapeutic margin: "Is there an adequate margin between the doses (concentrations) needed for efficacy and those that produce adverse effects?" This question may be answered by the comparison of dose/concentration response curves for efficacy and specific adverse events. Some describe Phase 2 as the "learning" phase of drug development and describe the purpose of Phase 3 clinical experiments to be the confirmation of the hypothesis derived from Phase 2 (Sheiner, 1997).

Compounds are selected for entry into drug development with an expectation or hypothesis of how they will contribute to the treatment of a disease. This expectation is expressed as the "concept" for the medicine. For example, "a new drug" will be as effective as "the best drug" used to treat disease A, as measured by B, but not have the side effect C known to be seen in 32% of persons treated with "the best drug." Most development projects aim to test whether that concept is

FIGURE 37.1 Phases of medicine development.

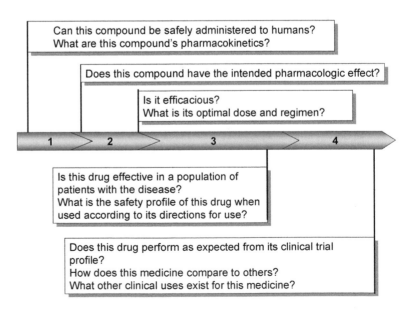

achievable during Phase 2 of drug development and hence incorporate a milestone of "Proof of Concept" into this phase of development.

Phase 3 clinical experiments are meant to demonstrate that the clinical effect seen in Phase 2 is translated into a meaningful clinical benefit, with a safety profile that is acceptable, given that benefit. These experiments form the bulk of the patient experience on which regulatory approval depends. Evidence from these (and all the other clinical) experiments form the basis for product labeling.

Once approved and marketed, medicines continue to be studied (Phase 4). Research aims to confirm that when used in practice a medicine performs as expected from its clinical trial experience. Observational trials, registers, and formal outcome studies may be carried out to determine the long term health benefits presumed to result from shorter-term or surrogate outcomes (effects of statin treatment on cardiovascular events, for example) and to confirm the safety profile found in registration clinical trials. Clinical trials may be conducted to compare a new medicine with other medicines used to treat the same disease or to determine whether the new medicine is effective in a different indication.

The development of a medicine can be thought of as a series of gates or milestones: Safe Administration to humans; Proof of Mechanism; Proof of Concept; Evidence for Regulatory Approval; Confirmation of Benefit and Safety in Marketing. The decisions made at these milestones are fundamentally medical-scientific decisions. Thus it is that physicians have a central role in determining whether compounds continue to travel the "medicine path."

MEDICAL-SCIENTIFIC POSITIONS IN THE PHARMACEUTICAL INDUSTRY

Knowledge of how health and illness are experienced by the patient is a unique perspective that only physicians can bring to a pharmaceutical company's efforts to discover, develop, and market medicines. Within pharmaceutical companies there are a variety of scientific jobs in which physicians can practice their knowledge of medicine. In addition, there are opportunities for physicians to exercise managerial and leadership skills. Finally, in most companies there are responsibilities that fall uniquely to physicians.

Experimental or Translational Medicine

Experimental medicine, discovery medicine, and translational medicine are terms used to describe the application of medicine and clinical investigation to the translation of observations from basic science to humans (Zerhouni, 2005) The importance of translational medicine to advancing health and treatment of disease has been recognized by the Clinical and Translational Science Award grants of the NIH and the funding of translational medicine units by the Medical Research Council in the United Kingdom.

In the pharmaceutical industry, physicians working in translational/discovery medicine work closely with laboratory scientists to determine the relevance of biological targets to human physiology and disease. They design and execute experiments in humans to determine the effect of intervening in specific biological pathways.

Translational medicine is central to medicine discovery and development. Observations in humans about mechanisms of disease need to be brought back into the biology and chemistry laboratory to create a platform for the discovery of new drug candidates. For example, the discovery of a new gene associated with a disease creates the platform for a drug discovery effort. The gene, its products, and pathway all become potential targets for intervention. Chemists and biologists work together to understand the targets in the pathway that are tractable—readily amenable to chemistry. Those tractable targets need to be studied in humans to determine their tissue distribution. The upstream and downstream consequences of modifying the activity of the target need to be studied in cellular preparations and in animals.

The excitement of translational medicine in drug development begins when there is first evidence that a target is tractable—that there is chemistry that can lead to a molecule that may modify the activity of the target. The translational scientist begins the work of determining how to tell whether a compound is affecting the target in humans—they pursue the identification and validation of biomarkers. In what tissues or organs is that pathway active? Is there an upstream or downstream pathway intermediate that can be measured in serum or CSF? Is there an imaging method that will reveal the activity of the target in the absence and presence of the compound? The translational medicine physician is asking the questions that lead to the experiments that underpin the "Proof of Mechanism" milestone decision. The science performed by physicians in these roles is critically important to pharmaceutical R&D because it directs the first experiments conducted in humans and validates the hypotheses that underpin each drug discovery program.

Cystic fibrosis is an autosomal recessive genetic disease characterized primarily by persistent, recurrent pulmonary infections and progressive pulmonary failure. Nearly 2000 mutations have been identified in the responsible protein, the cystic fibrosis transmembrane conductance regulator (CFTR), affecting the folding of the protein, localization of the channel in the cell membrane, or the activity of the channel (Monterisi et al., 2013).

One type of loss of function is a defective channel gate in the CFTR resulting in reduced chloride ion transport. The G551D missense mutation in which aspartate is substituted for glycine is representative of this type of CFTR loss of function. Errors of this type are found in ~4% of cystic fibrosis patients (Bobadilla et al., 2002).

Ivacaftor was found to increase the CFTR channel opening in vitro, leading to an expectation that it might rescue epithelial cell function in the human airway (Van Goor et al., 2009). Subsequent development and clinical trials confirmed dramatic improvements in lung function, weight, lung disease stability, and patient-reported outcomes in cystic fibrosis patients with the G551D mutation, leading to fast-track approval of Kalydeco (ivacaftor).

Clinical Pharmacology

Classical clinical pharmacology—the study of drug action in humans—remains a core discipline in drug development. Clinical pharmacologists focus on four fundamental problems: safe administration of compounds to humans; how the body handles the compound; how other drugs, disease states, food, and other body states affect the disposition of the compound; and how formulations and route of administration affect absorption and distribution.

Clinical pharmacologists design and execute first time in human studies where the safe exposure of humans to experimental compounds is determined. The safe execution of these experiments depends on a thorough understanding of physiology and the pharmacology of the compound.

Clinical pharmacologists work together with clinical pharmacokineticists to determine the absorption, distribution, metabolism, and excretion of drugs. This understanding is fundamental to determining the mode of administration, dose, and regimen of a medicine. An important part of both Phase 2 and 3 clinical studies is the determination of the dose and/or concentration response relationships of drug action. Clinical pharmacologists use the kinetic information from Phase 1 trials to construct the doses tested in those experiments and timing and number of drug concentration and pharmacodynamic measurements. Based on the understanding of a drug's metabolism and its effect on the cytochrome P450 enzyme system, clinical pharmacologists design and carry out drug interaction studies. They also study the effect of food on the absorption and distribution of compounds to determine whether compounds can be taken with food.

Prescribing physicians and patients do not, and probably should not have to, recognize the complexity of the formulation of the final drug product—the tablet or capsule form that patients actually take. Clinical pharmacologists work closely with formulation chemists to test the absorption and distribution characteristics of formulations to choose an optimal final design. The design of more convenient dosage forms such as once daily or sublingual fast-dissolving tablets requires the close cooperation of the clinical pharmacologist and formulation scientist.

Nifedipine is a calcium channel blocker. Its original immediate release formulation was approved for the treatment of angina. This formulation was associated with a 30–40% incidence of reflex tachycardia that prevented its use as an antihypertensive.

Clinical pharmacology studies of intravenous infusion of nifedipine demonstrated that reflex tachycardia was associated with the rate of rise of serum concentration (Kleinbloesem et al., 1987). Since the immediate-release oral formulation was characterized by rapid release of nifedipine into the gastrointestinal tract and a steep rise of nifedipine concentration in plasma, an oral constant delivery formulation was developed. This formulation, which also enabled once-daily administration, was not associated with reflex tachycardia and resulted in the product Procardia XL, which was approved for the treatment of angina and hypertension (Urquhart, 2000).

Clinical Research Physician

"Clinical research physician" is used here to mean the physician responsible for the clinical studies that test the efficacy and safety of medicines and the integration of all the data on a medicine into comprehensive and accurate prescribing information. These physicians represent medicine on the team of scientists responsible for developing a compound into a medicine.

This physician role is usually responsible for the Phase 3 studies of a compound in development and for the Phase 4 studies of medicines. Clinical research physicians design, interpret, and communicate the results of these studies. A core expertise for these positions is knowledge of clinical investigation—the design and conduct of clinical experiments.

Clinical research physicians begin working on a compound long before it enters Phase 3 development. They participate in developing the concept for the medicine, perhaps even before Phase 1, contributing their knowledge of medicine so the concept is grounded in the reality of treating patients. Their contribution ensures that the design of the Phase 2 experiments aligns with the medicine "concept" and paves the road for Phase 3 and beyond.

In the mid-1980s insomnia was generally recognized as a manifestation of psychiatric disease. The primary treatments were with benzodiazepines. The most commonly used benzodiazepine, triazolam, was characterized by rapid onset and short duration of action. While triazolam was quite effective, patients experienced rebound insomnia—increased difficulty sleeping after stopping triazolam—and there were many, and widely publicized, cases of transient amnesia upon awakening.

In 1983, an NIH Consensus Conference created a classification for insomnia according to the duration of the symptoms: transient, short term, and chronic (Consensus Conference, 1984). The Conference characterized these different insomnias. For example, transient insomnia occurred in persons whose usual sleep was normal, but was disrupted for several days, usually by some environmental factors (phase shift, noise, new sleep environment).

Zolpidem was a new, nonbenzodiazepine hypnotic under development for the treatment of insomnia in 1983. The development of zolpidem was specifically designed to demonstrate efficacy in these three newly defined insomnias and to show the absence of rebound insomnia and effects on memory. Landmark trials, especially in transient insomnia, led to the approval of zolpidem (Ambien) for all three types of insomnia and labeling that included the absence of rebound insomnia and next-day memory impairment (Roth et al., 1995; Walsh et al., 1990; Scharf et al., 1994).

Medicine Safety and Pharmacovigilance

The last 20 years experience of medicines has led us to recognize the importance of a systematic approach to studying drug safety and managing drug risk. More than any of the disciplines described in this chapter, pharmacovigilance has evolved within the pharmaceutical industry and the bodies that regulate it. The World Health Organization defines pharmacovigilance as "the science and activities relating to the detection, assessment, understanding and prevention of adverse effects or any other drug-related problem" (Regulation (EC) 542/95, 1995).

Physicians engaged in pharmacovigilance specialize in understanding the mechanisms of drug-related toxicity, early identification of safety signals from clinical trials, and adverse events that occur during marketing. Placing risks in the context of benefits, reaching decisions about appropriate directions for use, and developing plans to minimize risk are areas of special expertise for physicians who focus on drug safety.

Most pharmaceutical companies have a committee or board to independently assess the safety of administering compounds to humans and to set conditions for human exposure. Boards are composed of the most senior and experienced pharmaceutical physicians and toxicologists and may include external consultant physicians. The purview of these boards extends to the entire life cycle of compounds/medicines, from the first exposure to humans to the prescribing information that describes the safety profile, the warnings and precautions that physicians should follow when treating patients.

Studying safety begins with the first exposure of humans to a new test compound. The safety physician carries out epidemiology studies of the disease and of medicines that treat the disease, seeking a deep understanding of the symptoms and events known to occur in patients who will be the subjects of the new compound research. He/she brings together all that is known about the pharmacologic target, pathway, and other therapies that affect the target and pathway and couples it with the observations in animal toxicology studies to create a backdrop for anticipating and interpreting events that may happen with a new compound. These data guide the design of methods for monitoring the safety of subjects in the early human experiments.

Before the first time in human experiment, the safety board meets and reviews the relevant data to approve the exposure of humans and to set the conditions for human experiments. The board approves the type of subjects who will be exposed, the starting dose, the dose increments, and the top dose for single dose exposure, along with the observations and testing to be carried out to assure subject safety. As Phase 1 experiments progress, the safety board will approve longer duration of treatment, amend the monitoring program based on the observations in the completed subjects, and agree to ever-broader population of subjects who may receive the compound (gender, age, different disease stages, etc.).

An adverse event is defined as "any response to a drug that is noxious and unintended, and that occurs at doses used in humans for prophylaxis, diagnosis, or therapy, excluding failure to accomplish the intended purpose" (Karch and Lasagna, 1975). While this definition has not changed materially in almost 30 years, we know that the way in which an investigator seeks information about adverse events influences both the number and type of events collected. An important role of the safety physician is to decide what data will be collected and the methodology used to query for adverse events.

As Phase 1 and then Phase 2 experiments accumulate, the safety physician, together with other physician experts, collates the safety data to identify trends and patterns of adverse events. From these data plans for mitigating or preventing the identified risks are developed and incorporated into ongoing and new trials.

Most companies carry out systematic reviews of the safety experience with a development compound. These are often coupled with the development progression milestones, for example, at the completion of Phase 2 and the Proof of Concept decision. These systematic reviews result in the drafting of safety information about a compound that is incorporated into investigator brochures, informed consent statements, and filings with regulatory agencies. When a compound reaches the end of Phase 3, this core safety information becomes the information that is included in the proposed prescribing information in new drug applications.

There are at least seven adverse events that are classically related to medicines: Stevens–Johnson syndrome, agranulocytosis, prolonged QT interval, hepatitis, renal failure, seizures, and anaphylaxis. A good pharmacovigilance system has an evidence base on these events, training for their recognition, and procedures for studying these events should they occur during clinical trials. Rapid intervention with information and guidance for investigators can prevent poor outcomes from these medical emergencies. Safety physicians guide the collection of data on these kinds of events and carry out the assessment of their relation to the study compound.

Once a medicine is approved for prescribing, the safety physician monitors the safety experience as the population of exposed patients grows and evaluates the accuracy of the safety information and risk mitigation procedures. Treating physicians report adverse events to pharmaceutical manufacturers, the Food and Drug Administration (FDA), and other regulatory agencies around the world. The safety physician must attempt to pursue all possible data on serious adverse events to determine, to the extent possible, the relationship between the administered drug and the event.

All the adverse events reported in association with a drug are regularly reviewed to look for unexpected events, especially the more serious and rare events that may not have been seen during development clinical trials. Since the typical medicine will only have been administered to 3000–6000 patients before marketing, adverse events rarer than 1 in 2000 are unlikely to have been seen. Most serious adverse events that lead to drug withdrawal occur much less frequently and thus are only seen once a medicine is marketed. It is the responsibility of the safety physician to monitor the spontaneously reported adverse events for these rare, serious events.

Medicines for which the safety experience has already identified potential safety concerns should be studied more systematically during their marketing. Patient registries are often established to study possible effects on pregnancy or on the fetus. Epidemiology studies are valuable for assessing retrospectively whether patients are achieving benefit from therapy, or experiencing an adverse event more frequently than with other medicines. Safety physicians develop an expertise in interpreting these kinds of studies as part of their safety monitoring expertise.

In 2001, the European Medicine Evaluation Agency promulgated regulation that required pharmaceutical manufacturers establish a system for pharmacovigilance (EU Directive). The system is expected to (1) monitor medicines as they are used in everyday practice and in clinical research to identify previously unrecognized or changes in the patterns of their adverse effects; (2) assess the risks and benefits of medicines to determine what action, if any, is necessary to improve their safe use; (3) provide information to users to optimize safe and effective use of medicines; and (4) monitor the impact of any actions taken (Regulation (ED) No. 541/95, 1995). The regulation also required a physician employee resident in the European Union be designated the "Qualified Person for Pharmacovigilance." This physician carries a personal responsibility to report to regulatory agencies any safety concern for a marketed medicine, independent of the company decision on that safety matter. The system for pharmacovigilance requires the ability to assure that all persons employed by the company report any adverse event they become aware of, that the company has procedures in place to collect and report those events to regulatory agencies, and that the company has procedures and experts to assess the safety of their experimental compounds and marketed medicines and develop methods to mitigate known safety concerns.

Tysabri (natalizumab) is an antibody to alpha-4 integrin adhesion molecule on T lymphocytes. Tysabri was first shown to reduce the number of new inflammatory lesions in patients with MS (Tubridy et al., 1999) and subsequently to reduce relapses by 68% compared to placebo (Polman et al., 2006), a response far greater than seen with other medicines approved for MS. This evidence led to FDA approval in 2004.

During the first 2 years of marketing, several cases of progressive multifocal leukoencephalopathy (PML) were reported in MS patients treated with Tysabri. PML is known to be caused by reactivation of John Cunningham (JC) virus in the central nervous system, usually in the setting of immune suppression. When these cases were recognized, Tysabri was voluntarily withdrawn from the market. Careful analysis of these adverse events led to recognition of a much higher risk of PML in patients who had previously been treated with immunosuppressants, tested positive for JC virus antibodies, and had more than 24 months exposure to Tysabri.

Subsequently, the manufacturer and the FDA developed a risk management program, the TOUCH Prescribing Program (Tysabri Prescribing Information), for reintroduction of Tysabri to the market. This program required prescribers, pharmacies, and patients to be qualified for the use of Tysabri and to carry out regular assessments of treated patients and report the results to the manufacturer. This program has allowed patients who benefit to receive Tysabri while minimizing the risk of PML.

Medical Affairs

Physicians who work in medical affairs have responsibility for approved and marketed drugs. They are responsible for the communication of accurate information about these products to physicians, patients, regulatory agencies, health-care payers, and providers. They must monitor and interpret newly emerging information from clinical trials, adverse event reports, publications in the scientific literature, regulatory decisions taken elsewhere in the world, and incorporate it into product information.

Product information appears in many forms. Besides the formal prescribing information, information about products is communicated through sales representatives' presentations, print and television advertising, symposia, scientific meetings, etc. All of these presentations of information about a medicine need to be reviewed by the medical affairs physician to be certain it is accurate and complete and consistent with regulatory guidance. In this role, the medical affairs physician works closely with the commercial arm of the company to provide the medical guidance to and oversight of promotional activities, even extending to review of the training of sales representatives.

Often, clinical research physicians (described previously) also carry out a medical affairs role. Working with experts, they develop new ideas for the use of a medicine and translate those ideas into Phase 4 clinical research studies.

Ropinirole is a dopamine agonist originally developed and approved for the treatment of Parkinson's disease. In Parkinson's disease it acts by replacing the dopamine deficit that results from progressive dropout of presynaptic dopaminergic neurons.

Restless legs syndrome (RLS) is characterized by uncontrollable urge to move the legs accompanied by uncomfortable and unpleasant sensations in the legs. RLS leg movements disrupt sleep but also occur in the day, especially in the evenings.

The uncontrolled movements of RLS led to trials of dopaminergic agents levodopa, levodopa plus carbidopa, and bromocriptine with some reported favorable outcomes but substantial side effects. These observations led to the design of well-controlled, double-blind trials of ropinirole (Trenkwalder et al., 2004; Allen et al., 2004). These trials showed reductions of leg movements and symptoms that led to the approval of a new indication for ropinirole for the treatment of RLS.

Regulatory Affairs/Regulatory Science

Each country in the world has its own regulations for medicines. While there has been great progress in harmonizing those regulations, there remain real differences among countries and regions. Knowing those regulations and being able to interpret both the regulation and the actions of agencies are critical to the success of developing and marketing medicines.

Physicians bring knowledge of medicine to the interpretation of regulation and regulatory agency decisions. What stands for evidence of efficacy of a drug for type 2 diabetes? How long should a depressed patient who has responded to therapy be treated? These are the kinds of questions that are answered partly by scientific evidence and partly by the history of regulatory decisions.

Physicians who focus on regulatory affairs become expert in the law, regulation, and guidance that governments around the world create to regulate medicines. They provide guidance to the company on how to conduct research and communicate information about medicines in a way that accords with regulatory requirements and the existing scientific evidence. This expertise is recognized as the discipline of regulatory science, "the science of developing tools, standards, and approaches to assess the safety, efficacy, quality, and performance of regulated products" (Califf and Ostroff, 2015; Woodcock, 2014) and for which there are now multiple degree programs in both European and American universities.

The FDA approves drugs after reviewing a sponsor's New Drug Application (NDA). The NDA includes all the data on the pharmacology, toxicology, clinical safety, and efficacy and manufacturing of the proposed new product. Generally, the FDA is required to accept an NDA for review within 60 days of submission and to act on the application in 1 year.

There are, however, four pathways to accelerated drug approval. Fast-track designation, reserved for drugs intended to treat a serious condition, allows for frequent interaction with the FDA during development and the submission of portions of the NDA ahead of the final application. Priority review status may be granted when an NDA is submitted for a drug that is expected to provide a significant improvement for a serious condition. Priority review status brings the target date for action on an NDA forward by 4 months. Accelerated approval allows for approval of drugs on the basis of a surrogate or intermediate clinical endpoint that is expected to predict clinical benefit. "Breakthrough therapy" designation, intended for serious or life-threatening diseases, can be granted based on preliminary clinical evidence that the drug may demonstrate substantial improvement over existing therapies. The FDA aims to work with sponsors of breakthrough therapy—designated drugs to find the most efficient pathway to approval.

In addition, the Generating Antibiotic Incentives Now (GAIN) Act provides additional incentives to the development and review of antibiotics with particular promise. The Orphan Drug Act, first passed in 1982, has been amended several times to provide incentives and fast-track approval to drugs for rare diseases.

In 2014, two-thirds of novel new drugs approved by the FDA were reviewed and approved through one or more of the accelerated review pathways, four antibiotics were approved under the GAIN Act, and 17 drugs were approved for Orphan Drug indications.

Iressa (gefitinib) is an oral, selective inhibitor of epidermal growth factor receptor (EGFR) tyrosine kinase. In the first Phase 2 trial in non-small-cell lung cancer (NSCLC) completed in 2000, 10% of patients showed tumor shrinkage of 50% or more (Kris et al., 2000). Initial results of a Phase 3 randomized controlled trial in NSCLC patients who had failed two or more chemotherapy regimens showed an objective tumor response rate of 10.6% and a median duration of response of 7 months (Thatcher et al., 2005). On the basis of these results, Iressa was granted accelerated approval in the United States as a third-line treatment for NSCLC.

Survival results of the same trial, however, failed to show a clinical benefit, and in 2005, Iressa was withdrawn from the US market, despite remarkable responses in a small number of patients. Subsequently, Lynch et al. (2004) showed that patients with NSCLC who responded to Iressa had specific somatic mutations of the EGFR tyrosine kinase domain, and after development of a commercial test kit that detected these mutations and a trial in patients with these specific mutations that showed 50% of patients had tumor shrinkage that lasted an average of 6 months (Douillard et al., 2014), FDA granted orphan drug status to Iressa and approved it again in 2015.

Pharmacoepidemiology and Health Outcomes Research

With the growth of large health-care information databases has come the ability to learn about drug utilization, benefits, risks, and costs. Physicians skilled in epidemiology and data mining techniques provide valuable insight into the effects of drugs in real-world use.

This kind of information greatly informs the decision to enter into research in a disease. It can define the unmet needs of current therapy, the aspects of disease that if treated would make a meaningful difference to patients and health-care systems. From knowledge of the disease, as seen from health-care system data, the pharmacoepidemiologist can assure relevant characteristics of disease are assessed during clinical trials, and even guide the development of new methodologies to measure those characteristics.

Increasingly health-care systems want to be assured they are getting value for their health expenditures. By developing studies of benefit and cost, physicians construct models to predict the effect of the introduction of new medicines into a specific health care environment.

Project Leadership

With experience in pharmaceutical medicine, the pharmaceutical physician who develops and demonstrates leadership skills can have the opportunity to lead drug development project teams.

The exact makeup, authority, and ways of working of medicine development project teams vary from company to company, but all have one thing in common: the need for leadership. Only some of the leaders of these teams are physicians, but when physicians do lead them they add their understanding of medicine to the leadership they practice. Leadership is the key skill required: the ability to set a vision for the project and to gather people to that vision.

Developing drugs requires bringing together many disparate experts. The kind of leadership required in this setting is one that rallies highly skilled experts around a common goal and concept—it is consultative, inclusive, and inspirational. Physicians who master this kind of leadership can be unusually effective leaders because of the medical background they bring to the task.

Some companies organize medicine development project teams into therapy area—focused units, or even business units. For example, three different teams developing an antipsychotic, an antidepressant, and a drug for behavioral dysfunction associated with Alzheimer's might be brought together to form a psychiatry drug development unit. Physicians with demonstrated leadership talent at the individual medicine level can compete for these very challenging portfolio leadership positions. These kinds of leadership positions are usually at the vice-president level or above and represent some of the most important executive positions in a pharmaceutical company.

Management

Most companies organize their scientific disciplines into departments or group scientific disciplines that work together to deliver a business process into functions. For example, physicians and PhD clinical scientists may be grouped together in a clinical development department. Physicians have the opportunity to manage such departments or functions, selecting and training the people within the department, designing the business policies and procedures for which the department is responsible, and managing the resources (people and money) allocated to the department.

In larger pharmaceutical companies these departments are global. Persons in the department are located on multiple sites in different countries around the world. Leaders of such departments must learn the local cultures and employment laws to adapt global practices to accommodate local differences. Managing large global departments well returns great value to the company because the disciplines benefit from the cultural, medical, and societal diversity.

Managers of departments are usually members of senior management teams, teams that are responsible for major parts of the pharmaceutical business. They participate in setting the strategic direction for the business and translate that direction into year-on-year operational objectives and budgets. These managers may be responsible for the spending of large amounts of money and will get substantial training on how to manage those fiscal responsibilities.

Rodney Harris was appointed Head of Clinical Development for North America after successfully completing the US, European, and International registration of a new drug for migraine. Dr. Harris was now responsible for 750 physicians, clinical scientists, statisticians, clinical research associates, data managers and programmers, and affiliated staff, and he was expected to begin 22 clinical studies in his first year of responsibility.

Nine months into his tenure, Dr. Harris reported to his management that he was 6 months behind schedule in starting his clinical trials. Analysis of his resources told him that his department was severely understaffed for the work that had been assigned to it. Dr. Harris presented two options for remedying the situation: hiring a contract clinical research organization and hiring 100 additional staff. Based on Dr. Harris' recommendation, the company agreed to expand the size of Clinical Development to 850 persons.

Medical Officer: Decisions That Require Medical Input

The safety of medicines, the ethical conduct of clinical research, and the communication of medical information about medicines make up a constellation of responsibilities that require the unique expertise and experience of physicians. These domains are collectively referred to as "medical governance," and senior physicians in the company, usually under the leadership of a Chief Medical Officer, are accountable for defining a system that assures, to the extent possible, that all of the company's activities in these domains are consistent with the highest standards of ethical research and medical practices. Examples of decisions that are part of medical governance and for which physicians are responsible are as follows.

Safe Administration to Humans

What is the evidence that is sufficient to justify the first exposure of humans to an experimental agent? How should the "first time in humans" experiment be carried out? Is the safety profile of a compound or medicine acceptable given the potential (or known) benefits?

Ethical Conduct of Clinical Research

Do the potential benefits of a clinical research experiment justify the possible risks to human subjects? Have subjects been adequately informed of the risks and benefits such that they can give fair and free informed consent? Are the conduct of research and the disclosure of research results consistent with the covenant with the subjects of the research? Is the company's clinical research carried out in accord with the Declaration of Helsinki, the ethical principles of the Belmont Report, and other recognized statements of ethical research principles?

Medical Information

Is the information about a medicine complete and accurate? Is it being communicated in a way that is helpful to physicians and patients? How should prescribing physicians be notified of new safety findings about a medicine?

The Chief Medical Officer is responsible for establishing and maintaining a system and practice for medical governance to assure these kinds of decisions are made robustly, consistently, and transparently.

A PHARMACEUTICAL CAREER

Mid-career physicians traditionally enter the pharmaceutical industry into a clinical research, clinical pharmacology, safety physician, or translational physician position. They are assigned to one or more compound development projects or medicines and take on the responsibilities described previously. Because most physicians will not have training or experience in true pharmaceutical medicine, companies offer induction courses, mentoring or coaching, a community of pharmaceutical physicians to learn from, and in some cases formal didactic education in pharmaceutical medicine.

Learning pharmaceutical medicine is an important part of the early career of pharmaceutical physicians because it prepares physicians for taking on broader responsibility for medicines. In the United Kingdom, Ireland, Switzerland, and Mexico, pharmaceutical medicine is a recognized medical subspecialty. In the United Kingdom, the Faculty of Pharmaceutical Medicine offers a diploma that combines course work with experience.

Having developed a working knowledge of pharmaceutical medicine and performed well in the first position, physicians may decide to broaden their knowledge by moving into a different position. For example, a clinical research physician may next take a position in safety/pharmacovigilance or in regulatory affairs. Gaining experience in the different positions provides breadth that can be applied to later jobs.

After the first several years of experience in the industry, most physicians make a choice about remaining close to their area of medical specialization. Keeping current in a medical specialty is invaluable because it is that medical expertise physicians uniquely bring to the medicine development process. However, some physicians choose to pursue broader careers, recognizing that they will not remain an expert in their specialty.

In either case the pharmaceutical physician must make another choice: is he/she interested in leadership and management positions? If the answer is "yes," it is wise to seek leadership and management training to give the first leadership or management opportunity the best chance of success. The first position may be as head of a subteam of a medicine development team, or as a manager of a unit within a department. Successful in that first role, the opportunities are endless.

Because opportunities for growth may not exist within one's own company, physicians may move to another company. It is not unusual to find that senior physicians in the industry have worked in several different companies over a career. Different companies have different cultures and practices, and there is benefit to the experience of working in different companies just as there is in working at different universities.

SUMMARY

Making medicines—which is what a career in pharmaceutical medicine is all about—is a rewarding endeavor. It offers the potential to change the lives of hundreds, thousands, even millions of people. For the physician-scientist, pharmaceutical medicine is a career path that deserves consideration.

REFERENCES

Allen, R., Becker, P.M., Bogan, R., Schmidt, M., Kushida, C.A., Fry, J.M., Poceta, S., Winslow, D., 2004. Ropinirole decreases periodic leg movements and improves sleep parameters in patients with restless legs syndrome. Sleep 27 (5), 907–914.

Bobadilla, J.L., Macek, M., Fine, J.P., Farrell, P.M., June 2002. Cystic fibrosis: a worldwide analysis of CFTR mutations—correlation with incidence data and application to screening. Hum. Mutat. 19 (6), 575–606.

Califf, R.M., Ostroff, S., 2015. FDA as a catalyst for translation. Sci. Transl. Med. 7 (296), 296–297.

Consensus Conference, 1984. Drugs and Insomnia. The use of medications to promote sleep. JAMA 251, 2410–2414.

Douillard, J.Y., et al., January 7, 2014. First-line gefitinib in Caucasian EGFR mutation-positive NSCLC patients: a phase-IV, open-label, single-arm study. Br. J. Cancer 110, 55–62.

EU Directive 2001/83/EEC and Regulation (EC) No. 2309/93.

Karch, F.E., Lasagna, L., 1975. Adverse drug reactions — a critical review. JAMA 234, 1236–1241.

Kleinbloesem, C.H., van Brummelen, P., Danhof, M., Faber, H., Urquhart, J., Breimer, D.D., 1987. Rate of increase in the plasma concentration of nifedipine as a major determinant of its hemodynamic effects in humans. Clin. Pharmacol. Ther. 41, 26–30.

Kris, M.G., Natale, R.B., Herst, R.S., 2000. Efficacy of gefitinib, an inhibitor of the epidermal growth factor tyrosine kinase, in symptomatic patients with platinum-based chemotherapy. J. Clin. Oncol. 18, 2095–2103.

Lynch, T.J., Bell, D.W., Sordella, R., Gurubhagavatula, S., Okimoto, R.A., Brannigan, B.W., Harris, P.L., Haserlat, S.M., Supko, J.G., Haluska, F.G., Louis, D.N., Christiani, D.C., Settleman, J., Haber, D.A., 2004. Activating mutations in the epidermal growth factor receptor underlying responsiveness of non-small-cell lung Cancer to gefitinib. NEJM 350 (21), 2129−2139.

Monterisi, S., Casavola, V., Zaccolo, M., May 2013. Review Local modulation of cystic fibrosis conductance regulator: cytoskeleton and compartmentalized cAMP signalling. Br. J. Pharmacol. 169 (1), 1−9.

Polman, C.H., O'Connor, P.W., Havrdova, E., et al., 2006. A randomized, placebo-controlled trial of natalizumab for relapsing forms of multiple sclerosis. N. Engl. J. Med. 354 (9), 899−910.

Regulation (EC) 542/95, 1995. Official Journal of the European Communities, No. L55/15.

Regulation (ED) No 541/95, 1995. Official Journal of the European Communities, No. L55/7.

Roth, T., Roehrs, T., Vogel, G., May 1995. Zolpidem in the treatment of transient insomnia: a double-blind, randomized comparison with placebo. Sleep 18 (4), 246−251.

Scharf, M.B., Roth, T., Vogel, G.W., Walsh, J.K., May 1994. A multicenter, placebo-controlled study evaluating zolpidem in the treatment of chronic insomnia. J. Clin. Psychiatry 55 (5), 192−199.

Sheiner, L.B., 1997. Learning versus confirming in clinical drug development. Clin. Pharmacol. Ther. 61, 275−291.

Thatcher, N., Chang, A., Parikh, P., et al., 2005. Gefitinib plus best supportive care in previously treated patients with refractory advanced non-small-cell lung cancer: results from a randomised, placebo-controlled, multicentre study (Iressa Survival Evaluation in Lung Cancer). Lancet 366 (9496), 1527−1537.

Trenkwalder, C., Garcia-Borreguero, D., Montagna, P., Lainey, E., de Weerd, A.W., Tidswell, P., Saletu-Zyhlarz, G., Telstad, W., Ferini-Strambi, L., on behalf of the TREAT RLS 1 Study Group, 2004. Ropinirole in the treatment of restless legs syndrome: results from the TREAT RLS 1 study, a 12 week, randomised, placebo controlled study in 10 European countries. J. Neurol. Neurosurg. Psychiatry 75, 92−97.

Tubridy, N., Behan, P.O., Capildeo, R., Chaudhuri, A., Forbes, R., Hawkins, C.P., Hughes, R.A.C., Palace, J., Sharrack, B., Swingler, R., Young, C., Moseley, I.F., MacManus, D.G., Donoghue, S., Miller, D.H., The UK Antegren Study Group, 1999. The effect of anti-alpha4 integrin antibody on brain lesion activity in MS. Neurology 53 (3), 466−472.

Tysabri Prescribing Information. http://www.accessdata.fda.gov/drugsatfda_docs/label/2012/125104s0576lbl.pdf.

Urquhart, J., 2000. Controlled drug delivery, pharmacologic and therapeutic aspects. J. Intern. Med. 248, 357−376.

Van Goor, F., Hadida, S., Grootenhuis, P.D., Burton, B., Cao, D., Neuberger, T., Turnbull, A., Singh, A., Joubran, J., Hazlewood, A., Zhou, J., McCartney, J., Arumugam, V., Decker, C., Yang, J., Young, C., Olson, E.R., Wine, J.J., Frizzell, R.A., Ashlock, M., Negulescu, P., November 2009. Rescue of CF airway epithelial cell function in vitro by a CFTR potentiator, VX-770. Proc. Natl. Acad. Sci. U.S.A. 106 (44), 18825−18830.

Walsh, J.K., Schweitzer, P.K., Sugerman, J.L., Muehlbach, M.J., June 1990. Transient insomnia associated with a 3-hour phase advance of sleep time and treatment with zolpidem. J. Clin. Psychopharmacol. 10 (3), 184−189.

Woodcock, J., 2014. Paving the critical path of drug development: the CDER perspective. Nat. Rev. Drug Discov. 13, 783−784.

Zerhouni, E.A., 2005. Translational and clinical science − time for a new vision. N. Engl. J. Med. 353, 1621−1623.

Research in Academia

Chapter 38

Industry-Sponsored Clinical Research in Academia

Italo Biaggioni

Vanderbilt University, Nashville, TN, United States

Chapter Outline

Key Points

- Industry-sponsored clinical research is integral to the development of new therapies for our patients and, as such, should be a component of the research mission of academic medical centers.
- From the industry perspective, including academia in their clinical trials results in greater regulatory, legal, and financial costs, often without an equivalent payout in research performance.
- Increased regulations, the need for dedicated administrative and research personnel and lack of incentives are barriers to academic investigators in participation in industry-sponsored research.
- Self-supported clinical trials centers offer a model to remove these barriers, by acting as a single point of contact with industry, facilitating contracts and budget negotiations, ensuring compliance with regulations, and providing dedicated research personnel to improve the efficiency and quality of clinical research.

INTRODUCTION

The participation of Academic Health Centers (AHCs) in industry-sponsored clinical trials is not without its challenges and controversies. Industry often judges AHCs to be overpriced and inefficient. The leadership of AHCs often perceives industry-sponsored clinical research (ISCR) to be second tier in scientific importance compared to basic research, and are concerned that institutional resources might end up subsidizing ISCR. Academic investigators perceive a lack of institutional support for their efforts and do not understand why so much of the study budget is taken up by the institution's indirect cost recovery, while regulatory mandates are increasingly complex and expensive for academic institutions to maintain, particularly compared to private practice. The media, on the other hand, have focused on unfortunate, and sometimes negligent, medical mishaps and have implied that the main motivation for the participation of AHCs in ISCR is a financial reward. Finally, even though patients remain supportive of clinical research, only a small percentage actually participates in clinical research.

Clinical and Translational Science. http://dx.doi.org/10.1016/B978-0-12-802101-9.00038-7

There is probably some truth in each of the perceptions and complaints from the different players: industry, AHC leaders, AHC investigators, and the public. We will attempt to discuss and clarify these various issues, and will focus on models of institutional support for ISCR that would hopefully offer a win–win situation for all involved and at the same time advance science with the ultimate goal of developing novel therapies to improve patient care.

THE PUBLIC PERSPECTIVE

Patients are the reason behind clinical research. Their participation makes clinical research possible. Furthermore, patients are the ones who, sooner or later, directly or indirectly, foot the bill for drug development. Despite a few highly publicized tragedies in clinical trials, the general public remains supportive of clinical research (Comis et al., 2003; Harris Interactive, 2006). Losing their trust would be a devastating, and perhaps irreparable, loss for clinical research. The media have understandably covered cases involving adverse outcomes in clinical trials, and have tended to focus on how the investigators' financial incentives may contribute to these mishaps. No doubt this is a valid point when financial conflicts of interests are involved (Couzin and Kaiser, 2005). Whether conflict of interest actually contributed to an adverse outcome is irrelevant; the public's perception that financial interests can introduce a bias is what we must avoid. The response from academia ranges from simply informing study participants about the conflicts of interest in the consent form, to prohibiting investigators with significant financial conflicts of interest from leading or participating in the research. The media have played a role in bringing these issues to the forefront, but it is not helpful when they imply that the per-patient payment received by an AHC is pure profit to the investigator, without clarifying that this amount includes the actual cost of doing the study and that the residuals stay in the institution and can only be used for academic pursuits.

It has been argued that participants in clinical trials have better health outcomes than those of nonparticipants (Braunholtz et al., 2001). Such comparisons, however, are difficult to make given the many possible confounding factors. When these are taken into account the differences in outcomes are not as obvious (Peppercorn et al., 2004). Peppercorn et al. reviewed 26 published studies that compared the outcomes of cancer patients enrolled in a clinical trial with those not enrolled. Fourteen studies showed some evidence that trial participants had better outcomes, but only nine of the trials were designed to compare the outcomes of the participants with those nonparticipants who would have been eligible for the trials (to avoid selection bias). Of these, three studies suggested better outcomes among trial participants than those among nonparticipants. No studies showed that participants had worse outcomes than nonparticipants. Therefore, whereas the evidence is not conclusive, it seems likely that participating in clinical trials can have a positive effect on health outcomes. It has been suggested that this is not related to the experimental intervention itself, but to the better adherence to a standard therapy applied to study participants (Braunholtz et al., 2001). In any case, we cannot, and should not, imply a beneficial effect of study participation when consenting patients.

THE ACADEMIC HEALTH CENTER PERSPECTIVE

ISCR offers the possibility of advancing the three main missions of an AHC: patient care, research, and training. Patients remain supportive of clinical research, and have a positive attitude toward AHCs in part because their physicians are involved in research (Gabriel et al., 2004). If we believe that the main purpose of biomedical research is to improve health, clinical trials are the last step in the biomedical research continuum that ultimately is aimed at improving patient care (Kaitin et al., 1993). The biomedical journals with the highest impact are, in effect, clinical trials journals. Academic investigators, therefore, perceive their participation in ISCR as critical to their being at the forefront of patient care, and this improves faculty satisfaction and retention. ISCR also opens the possibility of training faculty and fellows in organized clinical research, and exposes them to approaches, techniques, and regulations that they can then apply to investigator-initiated research. Finally, it provides another source of financial support in the face of uncertainties about the stability of federal research funding or private donations.

No doubt that there are drawbacks in academic participation in ISCR that the leadership has to take into account (Bodenheimer, 2000). In most cases investigators are presented with a finished protocol, for which they have had limited or nil intellectual input. Only a few "opinion leaders" may be consulted at the onset to contribute to study design. Furthermore, the "scientific return" of ISCR can vary widely, from breakthrough novel therapeutic approaches with the potential to change clinical practice, to the study of "me-too" drugs or phase IV marketing studies that appear designed to increase market share rather than advance science. Also important are ethical issues that can arise from the conflict between science and profit inherent to ISCR. In some cases the inclusion/exclusion criteria of a given trial may not reflect the actual patient population the drug will be given to if approved. There are understandable ethical reasons why this may be the case; we would not enroll in a cardiovascular study a patient with terminal cancer. However, we need to be careful that inclusion criteria are not chosen to increase the likelihood of a beneficial clinical effect and decrease the occurrence of adverse events.

At the fringe of ethical research are perhaps phase IV marketing trials, funded by the marketing side of industry rather than its research and development arm, that seem aimed more at increasing use patterns of key prescribers, or as a way to enroll patients into a chronic and often expensive therapy. Finally, academic participation in ISCR creates numerous conflict of interest issues that the institution needs to manage (Friedberg et al., 1999), but for which there are rarely clear solutions. Academic investigators who consult for industry, for example, may find themselves not being able to participate in the study they helped to design, or in preparing their final report. Even if the previous emphasis on the negative aspects of industry—physician relations are being reexamined in favor of a more balanced view (Rosenbaum, 2015), AHC will still need to dedicate effort in managing potential conflicts of interest.

In summary, ISCR presents significant challenges to AHCs that require significant effort and resources to manage. We cannot, however, turn our backs on the type of research that most often advances clinical practice. Whatever we think about our current system, the fact remains that novel drug development relies on a profit incentive. As long as it is conducted ethically, we should participate. The challenge is to provide adequate safeguards that ensure ethical conduct of research without unduly burdening investigators.

THE INDUSTRY PERSPECTIVE

Drug development continues to be an expensive and risky proposition (Mattison et al., 1988; Bienz-Tadmor et al., 1992; DiMasi et al., 2003). Many would argue that it is increasingly so. Despite the discovery of a variety of new therapeutic targets from the genome project and other advances in molecular biology (Bumol and Watanabe, 2001), bringing these ideas to market has proven challenging. On an average, every new approved drug requires $1.24 billion in research investment (including the costs of failures) (Fordyce et al., 2015) and 10 years of development. The reason for this high cost is that still only 16% of potential medicines that reaches clinical trials are ultimately approved (Owens, 2014). The US drug industry invested approximately $51 billion in biomedical research in 2013, the same as the US government (Owens, 2014). Despite an increase in investment in drug development, fewer new drugs are being approved by the FDA (Fordyce et al., 2015), from 35.5 new molecules approved in 1994, to 23.3 in 2003 and 19 in 2007 (Eapen et al., 2013). The reasons for this phenomenon are several fold. One reason is that the low-hanging fruit may be already picked; i.e., we currently have satisfactory treatments for most common diseases, making it increasingly more difficult and more expensive, to develop novel therapies. Nonetheless, the pharmaceutical industry remains one of the most profitable (Henry Kaiser Foundation, 2007) and would likely be willing to increase their investment in drug development if this was associated with an improvement in the likelihood of success of clinical trials. This, however, has not been the case. The number of studies needed to obtain FDA approval and the number of patients needed per study are increasing. Studies are more complex in design, the regulatory burden is greater, and the time to completion of studies and approval remains long.

In the face of these challenges, industry's response has been to streamline operations and restrain cost increases. One of the results of this has been to outsource the management of their clinical studies to contract research organizations (CROs) (Shuchman, 2007). These entities can indeed be more efficient, but their mission is to complete the clinical trial within budget and on time, and not to advance science or provide support to investigators. They add a cost structure and bureaucratic layer between sponsors and investigators.

Of greater impact has been the shift of site selection for clinical trials away from AHCs and into private practices, or site management organizations (SMOs). SMOs are for-profit businesses that organize networks of community physicians or treat patients with the main goal of including them in research studies (Vincent-Gattis et al., 2000). In 1991, 80% of industry money for clinical trials was invested in AHCs, but by 1998 that percentage was reduced to 40% (Bodenheimer, 2000). The pharmaceutical industry has also followed the globalization trend seen in other industries, and now routinely enrolls sites outside the US. Indeed, fewer US investigators are participating in large-scale clinical trials and this translates in FDA approval based on studies where the US patients constitute the minority of subject (Eapen et al., 2013). This is driven by financial and regulatory issues, and by the observation that, at least in some studies, the US sites performed more poorly than non-US sites (Eapen et al., 2013). In some cases, this has opened questions about the ethics of enrolling research subjects in countries where drug approval is not being sought, implying absence of the critical ethical concept of beneficence because the population will not benefit from their participation in the research.

Compared to private and for-profit practices, AHCs are seen as inefficient. Contracts between industry and academia are complicated by the need to ensure institutional safeguards, including issues of the tax-exempt nature of the most AHCs, rights to publication, indemnification, and liability. These issues are of less concern to private practitioners who, not having lawyers on their payroll or an endowment to protect as AHCs do, are more likely to agree to the contractual outlines suggested by industry.

Another burden from the viewpoint of industry is the reliance of AHC on institutional review boards (IRBs), even though the use of a central IRB is far more efficient and is supported by the FDA and the Office for Human Research Protection for multicentre clinical trials (Eapen et al., 2013). Few AHCs are willing to cede control for fear of liability. Ironically, there is little evidence that IRBs improve the ethics or science of multicentre clinical trials (Infect Dis Soc Amer, 2009). To the sponsor, the idea of having a site-specific consent form and the risk of IRB-suggested changes in the protocol introduces an administrative burden they do not have to bear if they select private practices. Many AHCs have invested heavily in their IRB so that they can now handle ISCR effectively. The additional costs are passed on to the industry. Vanderbilt has pioneered a different approach to streamline the approval process, that relies on participating institutions accepting the review process of another institution while maintaining autonomy over local context issues (https://www.irbchoice.org/p/, last accessed September 2016.

Supporting the institutional organizations that oversee clinical research and protect the mission of the AHC requires substantial resources. In addition to the IRB, these include the contracts office, conflict of interest committee, and others. Institutions have to recover these costs through their indirect cost structure. The institutional indirect cost rate is negotiated with the federal government to recover facilities and administrative cost. AHCs cannot offer a lower cost structure to industry, lest they be seen as subsidizing industry. Indirect costs charged to industry are about 30% of total cost. These costs are not borne by private practitioners because they do not need to support contracts offices or IRB. This places academic investigators at a disadvantage because a portion of their study budget is directed toward institutional indirect cost recovery, a portion that is often within the residual margin ("profit") of the study budget. Sponsors may increase the study budget for academic investigators to cover a part, but rarely all, of the indirect costs.

The fact that a growing proportion of ISCR is currently carried out by private or for-profit practitioners creates a self-fulfilling prophecy. Private practices do not have the additional expenses of indirect costs, no administrative burden, less legal restraints to contract signing, and use centralized IRB. The process by which private practices process study initiation, therefore, has become the gold standard for industry that no longer understands why AHCs are more expensive, slower, more contentious, and have unique requirements. This self-perpetuates the perception among industry that AHCs are inefficient and difficult to deal with.

Finally, there is a concern that AHCs may not deliver the patient population required for clinical trials. AHCs often become tertiary referral facilities, and the patient population they see, therefore, may be too sick or complicated to meet the inclusion/exclusion criteria for many phase III trials. In response to this, some AHCs have established community networks of investigators.

THE INVESTIGATORS' PERSPECTIVE

Not surprisingly, investigators feel caught in the middle between the expectations of industry and the requirements of AHCs. The institutional structures required to protect the integrity of an AHC (contracts' offices, IRB, conflict of interest committee) are perceived by investigators, correctly or not, as burdens. There is often a disconnect between what the leadership of an AHC consider costly investments made to maintain the clinical research enterprise, and the lack of support perceived by faculty investigators (Oinonen et al., 2001). Moreover, whereas indirect costs are recovered as a separate line item in federal grants, they are included in the study budget in industry contracts. Unless investigators can negotiate a higher reimbursement, they end up paying for the indirect cost, or at least they perceive it that way.

There is also the perception that the regulatory burden associated with clinical research is increasing so much so that it seems improbable that one can be effective in clinical research without dedicated support personnel. The downside of hiring dedicated research personnel is that it forces investigators to commit to their continuous involvement in clinical research (the "treadmill effect").

MATCHING INDUSTRY NEEDS AND ACADEMIC HEALTH CENTER PRIORITIES

At first sight, industry and AHCs have opposing motives and even incompatible objectives. However, their interests are closer than would appear, particularly in regards to efficient conduct of clinical trials. Both are interested in timely enrollment of study subjects and trial completion. It is estimated that for each day's delay in gaining FDA approval of a drug, the manufacturer loses, on an average, $1.3 million (Bodenheimer, 2000); sales of atorvastatin (Lipitor) were $7.7 billion in 2011, or $21.1 million a day. For industry, shorter study initiation and on-schedule completion of subject enrollment will translate into a faster FDA approval and longer patent-protected marketing.

AHCs and investigators are harmed by unnecessary delays. The financial and scientific return to investigators is related to the number of patients they enroll in the study. High enrollment makes it more likely that investigators will participate in the design and publication of current and future studies. Unless a critical number of patients are enrolled, investigators may not be able to recover the investment they have made in supporting a dedicated personnel infrastructure, and institutions will not recuperate the substantial start-up costs required to initiate studies. Because enrollment is competitive among sites, a delay in study initiation places the academic site at a disadvantage.

Thus, even though much has been said about the disparate goals of industry and academia, they are both interested in an ethical and efficient conduct of clinical trials, while ensuring subject safety and compliance with regulations.

Academic Clinical Trials Centers As a Solution

In response to the decline in the share of ISCR carried out in academic centers, and the perception they are inefficient, the solution adopted by many academic centers has been to centralize institutional resources and streamline processes to improve efficiency and support clinical investigators.

In many cases these are centralized offices, often a part of the contracts office, which act as a single contract point with industry, and act as a matchmaker between industry and investigators (Paller et al., 2002). A few offer more comprehensive service options to investigators that include assistance in subject recruitment and supply of research coordinators. In a few cases, their goal has been to create a network of investigators in the community, or to function as a full-service CRO (e.g., the Duke Clinical Research Institute).

The purpose of full-service Clinical Trial Centers (CTCs) is to promote clinical research by supporting their clinical investigators. Among the services typically provided are cost estimation of studies and budget negotiations, facilitation of the contract approval process, IRB preparation, assistance with patient recruitment, and provision of research coordinators to assist in performing studies. Even though these offerings are considered a service to investigators, savvy administrators will realize that they are in the institution's interest as well. Many investigators may not have enough experience to adequately estimate the real cost of the study. Ensuring that the budget that actually cover the cost of the trial is essential; otherwise we will end up subsidizing industry. It also ensures that we include adequate start-up costs to compensate for institutional investments in IRB and contracts offices. In particular, contract negotiations are becoming more arduous for the reasons stated earlier. It is not in the institution's interest to prolong negotiations unnecessarily. This only increases the cost of the contracts office and leaves investigators at a disadvantage in regards to competitive enrollment. If there are deal-breakers identified early on, it is better to advice investigators against pursuing the trial. Similarly, assisting investigators with patient enrollment will result in greater revenue to the institution because the study reimbursement, and its indirect cost recovery, is tied to patient enrollment.

Most of these services, therefore, can be offered with a relatively small investment that can be justified by the improvement in efficiency. The most difficult service to provide, but the most valuable, is research coordinators to assist with the conduct of the trial. Maintaining a pool of trained and dedicated research coordinator is not only costly, but implies underwriting a financial risk that many institutions are not prepared to undertake. This is, however, the single most effective intervention to recruit new investigators into ISCR. Few sponsors will recruit a site without dedicated research personnel, and without the probability of continuous funding it is difficult for individual investigators to take on this financial commitment. Strict adherence to federal, local, and institutional regulations is a very important added value of having a pool of trained research coordinators working for the institution.

The purpose of these CTCs is usually not to centralize the conduct of all ISCR, but rather to support established investigators in the hope they will grow their research enterprises, and to act as an incubator to bring new investigators into this area of research in the hope they will become independent investigators.

Financing Clinical Trial Centers

Even though most of these initiatives can be started with relatively little investment, creating a stable funding mechanism is an important consideration as part of their planning and establishment. Many institutions consider them cost centers, but with such financial structure the services they are able to provide are limited and usually do not include research coordinator support. In a few cases the center is appropriated with the indirect cost recovery linked to their clinical research activities. This has the advantage of providing a performance-based incentive. In most cases, a fee-for-service structure is created based on reimbursement by the investigators for the support provided; for example, nursing support is charged monthly on an hourly basis. The main limitation of this approach is that payment to the investigator from the sponsor is usually delayed; payment is triggered by milestones (e.g., completion of all queries from case report forms) and is aggregated after a certain number of patients are completed. Thus it could be a few months before the first payment is received by the investigator, leaving them with the liability of paying for services provided by the CTC from other sources.

At Vanderbilt we decided to take a different approach, by partnering with investigators. When a new study is evaluated, a decision is made about what portion of the study would be performed by the center and what by the investigators. The budget is divided accordingly and the study is initiated. When a milestone payment from the sponsor arrives, the reimbursement is divided automatically. This represents a greater financial risk to the institution because if the study underperforms the institution will not recover its investment in personnel. On the other hand, if the study overperforms the center shares in the profits. This arrangement also benefits the investigators because it removes the financial uncertainties and the responsibility of maintaining dedicated research personnel, thus encouraging the participation of investigators in ISCR.

Any of these financial approaches to support the clinical trials center requires an initial investment from the institution. This investment will hopefully be recovered by the increase in ISCR within the institution. However, it is important to emphasize that the institutions should not invest in a clinical trials center hoping for a substantial financial reward. Given the competitive nature of ISCR, its global nature, and the increasing involvement of CROs, the profit margin is usually too thin to expect a significant financial return. The incentive for the creation of these centers, therefore, should be to fulfill the missions of the institution, rather than being mainly a financial one.

If the goal of the CTCs is to promote clinical research, it can accomplish this with a simple strategy that can be summarized as: (1) increasing the number of investigators, (2) increasing the number of trials performed, and (3) increasing the number of patients enrolled in trials. This applies for both industry-funded and federally funded research. We have outlined strategies to increase the number of faculty involved in research, by streamlining the regulatory burdens, supporting their activities, and providing incentives to their participation in ISCR. This, plus acting as a point of contact to industry, will increase the number of trials carried out at the institution. Finally, the CTC should have one of its pivotal goals to improve the efficiency by which investigators identify and enroll patients. As mentioned previously, patient enrollment is a key to the scientific and financial success of AHC involvement in ISCR.

Information Technology Solutions to Improve Patient Enrollment

Even if the main motivation for creating clinical trials centers is not a financial one, it is undoubtedly important to maintain their financial viability. Most of this depends on improving enrollment of patients into clinical trials. Patient enrollment is arguably the most effort-intensive and inefficient process in clinical research (Lovato et al., 1997; Sung et al., 2003). Approximately two out of three patients screened are not ultimately enrolled in clinical trials (Kroll, 2007). Thus anything we do to improve subject enrollment would make a significant advancement in the efficiency by which we perform clinical research. Surprisingly, there is little published in this area, but the most effective solutions derive from information technology.

Several institutions, including Vanderbilt, developed HIPAA-compliant searchable database of subjects that had expressed an interest in participating in clinical research, in some cases by self-registering via Internet interfaces (Harris et al., 2005). This model leads to the establishment of Research Match (https://www.researchmatch.org, last accessed September 2016), a national registry created as a collaborative project for institutions in the Clinical and Translational Science Awards (CTSAs) consortium (Harris et al., 2012). It is possible to automatically link the subjects' diseases of interest to a database of active clinical trials, thus performing an automated match between willing participants and clinical studies actively enrolling. Others have developed automated alert systems, triggered when a patient who meets inclusion criteria enters the system and automatically notifies research personnel (Embi et al., 2005).

CONCLUSION

In summary, we have presented the rationale why AHCs should participate in ISCR, and how they can support clinical research via the establishment of a CTC. The objectives of these centers are similar to those of the General Clinical Research Centers (GCRCs), but have a different organizational and financial structure. Both complement each other and serve not only industry-sponsored research, but also federally funded clinical research.

We also present an approach to promote clinical research through the services provided by the CTCs; the most important of them is the availability of dedicated research personnel. This is a financially demanding and risky investment by the institution, but proven strategies for cost recovery are presented. Finally, it is also important to improve the efficiency of how clinical research is performed at academic institutions, in particular dealing with patient identification and enrollment. For this, IT solutions offer the best alternative.

ACKNOWLEDGMENT

This work was supported in part by CTSA award no. UL1TR000445 from the National Center for Advancing Translational Sciences.

REFERENCES

Bienz-Tadmor, B., Dicerbo, P.A., Tadmor, G., Lasagna, L., 1992. Biopharmaceuticals and conventional drugs: clinical success rates. Biotechnology 10, 521–525.

Bodenheimer, T., 2000. Uneasy alliance – clinical investigators and the pharmaceutical industry. N. Engl. J. Med. 342, 1539–1544.

Braunholtz, D.A., Edwards, S.J., Lilford, R.J., 2001. Are randomized clinical trials good for us (in the short term)? Evidence for a "trial effect". J. Clin. Epidemiol. 54, 217–224.

Bumol, T.F., Watanabe, A.M., 2001. Genetic information, genomic technologies, and the future of drug discovery. JAMA 285, 551–555.

Comis, R.L., Miller, J.D., Aldige, C.R., Krebs, L., Stoval, E., 2003. Public attitudes toward participation in cancer clinical trials. J. Clin. Oncol. 21, 830–835.

Couzin, J., Kaiser, J., 2005. Gene therapy. As Gelsinger case ends, gene therapy suffers another blow. Science 307, 1028.

DiMasi, J.A., Hansen, R.W., Grabowski, H.G., 2003. The price of innovation: new estimates of drug development costs. J. Health Econ. 22, 151–185.

Eapen, Z.J., Vavalle, J.P., Granger, C.B., Harrington, R.A., Peterson, E.D., Califf, R.M., 2013. Rescuing clinical trials in the United States and beyond: a call for action. Am. Heart J. 165, 837–847.

Embi, P.J., Jain, A., Clark, J., Bizjack, S., Hornung, R., Harris, C.M., 2005. Effect of a clinical trial alert system on physician participation in trial recruitment. Arch. Intern. Med. 165, 2272–2277.

Fordyce, C.B., Roe, M.T., Ahmad, T., Libby, P., Borer, J.S., Hiatt, W.R., Bristow, M.R., Packer, M., Wasserman, S.M., Braunstein, N., Pitt, B., DeMets, D.L., Cooper-Arnold, K., Armstrong, P.W., Berkpwitz, S.D., Scott, R., Prats, J., Galis, Z.S., Stockbridge, N., Peterson, E.D., Califf, R.M., 2015. Cardiovascular drug development. Is it dead or just hibernating? J. Am. Coll. Cardiol. 65, 1567–1582.

Friedberg, M., Saffran, B., Stinson, T.J., Nelson, W., Bennett, C.L., 1999. Evaluation of conflict of interest in economic analyses of new drugs used in oncology. JAMA 282, 1453–1457.

Gabriel, S.E., Lymp, J.F., Fallon, M., Maurer, M.S., Kamath, C.C., Nayar, V.R., Seltman, K.D., 2004. Why patients choose academic medical centers for their care: findings and implications from Mayo Clinic. Group Pract. J. 53, 24–31.

Harris, P.A., Lane, L., Biaggioni, I., 2005. Clinical research subject recruitment: the volunteer for Vanderbilt research program. J. Am. Med. Inf. Assoc. 12, 608–613.

Harris, P.A., Scott, K.W., Lebo, L., Hassah, N., Lightner, C., Pulley, J., 2012. ResearchMatch: a national registry to recruit volunteers for clinical research. Acad. Med. 87, 66–73.

Harris Interactive, 2006. Nationwide Survey Reveals Public Support of Clinical Research Studies on the Rise. http://www.harrisinteractive.com/news/printerfriend/index.asp?NewsID5 3 23.

Henry Kaiser Foundation, 2007. Trends and Indicators in the Changing Health Care Marketplace. http://www.kff.org/insurance/7031/ti2004-1-21.cfm.

Infectious Disease Society of American, 2009. Grinding to a halt: the effects of increasing regulatory burden on research and quality improvement efforts. Clin. Infect. Dis. 49, 328–335.

Kaitin, K.I., Bryant, N.R., Lasagna, L., 1993. The role of the research-based pharmaceutical industry in medical progress in the United States. J. Clin. Pharmacol. 33, 412–417.

Kroll, J.A., 2007. An Industry in Evolution. Centerwatch, Boston, MA.

Lovato, L.C., Hill, K., Hertert, S., Hunninghake, D.B., Probstfield, J.L., 1997. Recruitment for controlled clinical trials: literature summary and annotated bibliography. Control Clin. Trials 18, 328–352.

Mattison, N., Trimble, A.G., Lasagna, L., 1988. New drug development in the United States, 1963 through 1984. Clin. Pharmacol. Ther. 43, 290–301.

Oinonen, M.J., Crowley Jr., W.F., Moskowitz, J., Vlasses, P.H., 2001. How do Academic Health Centers value and encourage clinical research? Acad. Med. 76, 700–706.

Owens, B., 2014. Mapping biomedical research in the US. Lancet 384, 11–14.

Paller, M.S., Hostetler, L., Dykhuis, D.A., December 2002. Clinical trials at AHCs: the perspective of an academic clinical trials office. Acad. Med. 77 (12 Pt 1), 1201–1206. PMID: 12480622.

Peppercorn, J.M., Weeks, J.C., Cook, E.F., Joffe, S., 2004. Comparison of outcomes in cancer patients treated within and outside clinical trials: conceptual framework and structured review. Lancet 363, 263–270.

Rosenbaum, L., 2015. Reconnecting the dots – reinterpreting industry-physician relations. N. Engl. J. Med. 372, 1860–1864.

Shuchman, M., 2007. Commercializing clinical trials – risks and benefits of the CRO boom. N. Engl. J. Med. 357, 1365–1368.

Sung, N.S., Crowley Jr., W.F., Genel, M., Salber, P., Sandy, L., Sherwood, L.M., Johnson, S.B., Catanese, V., Tilson, H., Getz, K., Larson, E.L., Scheinberg, D., Reece, E.A., Slavkin, H., Dobs, A., Grebb, J., Martinez, R.A., Korn, A., Rimoin, D., 2003. Central challenges facing the national clinical research enterprise. JAMA 289, 1278–1287.

Vincent-Gattis, M., Webb, C., Foote, M., 2000. Clinical research strategies in biotechnology. Biotechnol. Annu. Rev. 5, 259–267.

Chapter 39

Governmental Support of Research

Sten H. Vermund[1] and Salim Abdool Karim[2,3]

[1]Vanderbilt University, Nashville, TN, United States; [2]University of KwaZulu–Natal, Durban, South Africa; [3]Columbia University, New York, NY, United States

Chapter Outline

Key Points

- At the Federal government level in the United States, clinical and translational research support is provided primarily by two US Department of Health and Human Services (DHHS) agencies: The National Institutes of Health (NIH) and The Centers for Disease Control and Prevention (CDC).
- In the United States, the NIH provides the majority of governmental research funding, and it is the premier extramural funding source for biomedical and biobehavioral research in health sciences.
- NIH research grants are competitive and peer-reviewed.
- The NIH's R series research grants are the principal source of support for clinical and translational research.
- Within the NIH Research Project Grant Program, the R01 is the grant most commonly awarded to investigators, and it provides a multisite epidemiological consortia option which supports expert collaborations, and increases scientific productivity as collected core data are shared across the networks/consortia.
- The CDC is the public health arm of the US Federal government and it predominantly focuses on moving the clinical and other research discoveries to the community.
- Although the CDC does not have the substantial extramural research funds as does the NIH, historically, the CDC research efforts steer toward partner institutions and selected extramural investigators.
- T1 research focuses on discoveries in biology or other potentially health-related areas, and seeks to make them relevant to the clinical setting.
- T2 research translates clinical discoveries into broader medical and public health applications.
- The CDC primarily supports T3 research which is the scaling-up of already proven clinical or public health interventions to make them more effective and available in real-world settings.
- The European Community (EC) represents a historic economic, legal, and social alliance of 28 of Europe's 49 nations.

Clinical and Translational Science. http://dx.doi.org/10.1016/B978-0-12-802101-9.00039-9

- The EC offers competitive grants to science administrators, investigators, and research institutions within its member nations.
- The EC's FP7 awards support HIV/AIDS, tuberculosis (TB), and malaria research.
- The EC also supports research activity in neurosciences, cancer, cardiovascular diseases, poverty-related diseases, and other clinical and translational research topics.
- Canada has a system of research support that is a hybrid of the US and UK systems.
- The South African Medical Research Council (SAMRC) promotes the improvement of the health and the quality of life of the population by supporting research, development, and technology transfer, and their intramural program focuses on the disease burden experienced by South Africans.
- Building a healthy nation through research is the mission of the SAMRC. Similarly, a number of other middle income nations have national research organizations that support research at varying levels.

INTRODUCTION

Biomedical and biobehavioral research funds are available from a wide variety of government sources around the globe. As a consequence of global fiscal dynamics, the more prosperous "Western" nations contribute a disproportionate share of global research funds toward clinical and translational research. Some of these funds are limited to application from selected institutions or citizens from the nation that is providing the fiscal support. Those funds that are available through open competition to persons affiliated with recognized institutions are provided from a relatively limited number of sources that will be highlighted here. Institutions that do not have offices that handle grants and contracts and/or have no approved mechanisms and entities to provide research ethics reviews are typically ineligible to receive funds from government sources. Government funding sources whose focus is on basic science exclusively and/or are not focused on clinical or translational research are not discussed here (e.g., the National Science Foundation of the United States). A number of public–private partnerships between government sources of funds and private sources, especially foundations, have begun to blur distinctions of funding sources, especially for larger scientific endeavors where more money is needed to achieve a "big science" goal. This may or may not become an increasing component of future funding for selected activities; many foundations may opt to retain their autonomy through maintaining grant portfolios independent of government influences, but may choose to partner with governments for specific activities. A recent example is the investment made by the Bill & Melinda Gates Foundation in a large field trial of combination HIV prevention interventions anchored on expanded HIV testing and antiretroviral treatment, the HIV Prevention Trials Network 071 protocol (the "PopART" intervention); a majority of funding is coming from the President's Emergency Plan for AIDS Relief (PEPFAR) and the National Institutes of Health (NIH), yet the Gates Foundation is contributing substantial co-funding support (Hayes et al., 2014; Cori et al., 2014; Vermund et al., 2013).

OVERVIEW

Clinical and translational research requires substantial investment, particularly for large studies that may need to meet regulatory standards. Large observational and experimental research studies are often designed with the assistance of prior exploratory or feasibility studies, costly in and of themselves. Communities must be engaged, educated as to the studies, and consulted. Research subjects must be recruited, enrolled, queried, examined, give needed specimens, followed, and assessed. All this costs money for research staff, community outreach, transport, reimbursements and/or incentives, questionnaire administration, physical examinations, specialized examinations (noninvasive or invasive) of blood, other bodily fluids, biopsies or of specific organs. Research staff can include nurses, counselors, data entry personnel, epidemiologists, biostatisticians, phlebotomists, physicians, laboratory personnel, community workers, ethicists, quality control specialists, and many others. When regulatory authorities such as the US Food and Drug Administration, the European Medicines Agency or the Medicines Control Council of South Africa will be approached for drug, biologics (e.g., vaccine or immunotherapy) or device licensure, the documentary requirements inflate the research costs.

In this chapter we present governmental sources of funding, concentrating on the largest single funding source, the NIH in the United States. We offer definitions of common research funding terms and give examples of the types of grants that are available. We also provide examples of research funding and government research agencies from around the world. These include multinational partnerships, such as the European Community (EC), as well as individual national health research agencies (e.g., the United Kingdom, Sweden, France, Germany, Canada, Australia, New Zealand, Japan, Thailand, India, and China). We describe one national funding agency from a middle-income nation in more detail, the Medical Research Council (MRC) of South Africa, to contrast it with the NIH. To illustrate the diversity of funding options, we also present one example of a regional research institute (from Catalonia in Spain) and an additional example of a multilateral partnership (of the Netherlands, Australia, and Thailand). Nongovernmental sources of funding are discussed in a subsequent chapter in this book (see Chapter 40).

UNITED STATES GOVERNMENT SCIENTIFIC PROGRAMS

In the United States, principal sources of governmental research funding in clinical and translational science are from the 11 operating divisions of the US Department of Health and Human Services (DHHS), including eight agencies in the US Public Health Service and three human services agencies (see Table 39.1). Health research support may also come from such agencies as the Environmental Protection Agency, the Department of Energy, the Department of Defense, or the National Aeronautics and Space Administration (NASA) related to environmental health, radiation exposure, military medicine, or the health of airline workers and astronauts. NIH-type research can be supported by these agencies, as in the case of NASA supporting protein crystallography experiments in the weightlessness of space, for example, during its International Space Station or previous Space Shuttle missions. State governments typically support research through programs targeting their state institutions. One example is the University of California's university-wide AIDS Research Program founded in 1983 through the Legislature of the State of California. This annual augmentation of the University of California's research budget has supported research projects related to AIDS at nonprofit institutions within California. We will not discuss such state or local programs here as they are highly restricted in who may apply (typically institutions within the state or even just within a given public sector university system within a state). About 2% of the US federal budget went to research and development (R&D) in 2013, but only about 10% of this was for medical and life science research per se (http://www.aaas.org/program/rd-budget-and-policy-program).

At the Federal government level, clinical and translational research support is provided primarily by two of the DHHS agencies: the NIH and the Centers for Disease Control and Prevention (CDC). One of the authors (S.H.V.) has worked at the NIH and the other author (S.A.K.) has worked at the South African Medical Research Council (SAMRC), to be described later in this chapter. The other agencies in the DHHS (see Table 39.2) provide grants for service programs, service training, or for program monitoring and evaluation (M&E), but their support for clinical and translational research is comparatively quite limited. The Agency for Healthcare Research and Quality, the Health Resources and Services Administration, and the Substance Abuse and Mental Health Services Administration concentrate on health services research. The Food and Drug Administration may support small, highly targeted research grants in their direct mission of regulating food, drugs, biologics, and devices, though they have an active intramural research program, supporting its own employees to do scientific research of relevance to the agency. The NIH and its counterparts overseas, such as the MRCs and Institut Pasteur laboratories in several nations also have vibrant intramural research programs (see definitions in Box 39.1), in contrast to extramural work in which the agencies support grants for work done outside the agencies themselves, as in universities or research institutes.

TABLE 39.1 Agencies of the United States Department of Health and Human Services

Agency	Abbreviation	2013 Budget in US Dollars (billions)	2013 Employees
Administration for Children and Families	ACF	49.73	1287
Administration for Community Living	ACL	2.07	148
Agency for Healthcare Research and Quality[a]	AHRQ	0.43	314
Agency for Toxic Substances and Disease Registry[a]	ATSDR	Within CDC budget	Included in CDC
Centers for Disease Control and Prevention[a]	CDC	10.24	11,086
Centers for Medicare & Medicaid Services	CMS	763.14	5951
Food and Drug Administration[a]	FDA	4.03	14,829
Health Resources and Services Administration[a]	HRSA	8.10	1965
Indian Health Service[a]	IHS	5.31	15,429
National Institutes of Health[a]	NIH	29.15	18,948
Substance Abuse and Mental Health Services Administration[a]	SAMHSA	3.35	627

[a]Part of the U.S. Public Health Service.
U.S. Department of Health and Human Services, http://www.hhs.gov/about/foa/opdivs/index.html.

TABLE 39.2 Nations Whose Institutions Received at Least One NIH
Grant in Fiscal Years 2007 and/or 2013

Nation	2007 Grants	2013 Grants
Argentina	8	9
Australia	59	22
Austria	2	0
Bangladesh	1	2
Belgium	4	2
Benin	0	1
Bolivia	0	1
Botswana	1	0
Brazil	16	11
Cambodia	1	0
Canada	198	134
Chile	4	1
China[a]	22	14
Colombia	1	5
Costa Rica	0	1
Czech Republic	2	0
Denmark	6	3
Dominican Rep.	1	0
Egypt	2	0
Estonia	1	0
Ethiopia	0	2
Finland	3	2
France	16	10
Gambia	0	1
Germany	18	10
Ghana	0	7
Greece	1	0
Guatemala	0	1
Haiti	5	2
Iceland	5	5
India	16	9
Israel	28	14
Italy	12	0
Jamaica	1	0
Japan	1	0
Kenya	4	10
Korea Rep of	1	1
Lebanon	1	1
Lithuania	0	1
Malawi	4	1

Continued

TABLE 39.2 Nations Whose Institutions Received at Least One NIH Grant in Fiscal Years 2007 and/or 2013—cont'd

Mali	0	2
Mexico	10	0
Mozambique	0	2
Netherlands	14	6
New Zealand	7	2
Nicaragua	0	2
Nigeria	2	8
Norway	0	1
Pakistan	6	2
Peru	9	12
Philippines	0	1
Poland	2	0
Portugal	0	2
Romania	0	1
Russia	3	0
Rwanda	0	1
Senegal	1	0
Singapore	3	0
South Africa	24	45
Spain	6	1
Sri Lanka	0	2
Suriname	0	1
Sweden	16	8
Switzerland	24	11
Tanzania U. Rep.	0	2
Thailand	6	6
Tunisia	1	0
Uganda	6	13
United Kingdom	93	54
Uruguay	3	1
Vietnam	0	1
Zambia	0	4
Zimbabwe	1	6

[a]CHINA includes three grants to Hong Kong in 2007 and one in 2013.
http://report.nih.gov/award/index.cfmot=&fy=2013&state=Foreign&ic=&fm=&orgid=&distr=&rfa=&pid=#tab1.

The Indian Health Service relies on support from the CDC and the NIH for needed research in support of Native Americans. Other Federal health agencies have programs that solicit occasional external research activities, but these focus on service training and program evaluation rather than clinical and translational research. Hence, we will concentrate on the NIH and the CDC.

BOX 39.1 Key Definitions Essential to Understanding Grant-Related Issues

Intramural research supports investigators within a given agency itself, in contrast to extramural research that supports grants and/or contracts outside the agency. If one works for a university or research organization, for example, one would apply for extramural grant or contract support from an agency like the National Institutes of Health (NIH) in the United States or the Medical Research Council (MRC) in the United Kingdom or South Africa, among others. The university or research organization employee would not be eligible to apply for intramural funds, while the employee of the NIH or the MRC itself would be restricted only to those intramural funds.

A research grant is directed by the investigator, not by the funding agency. In a cooperative agreement-type grant, the government representatives will take an active collaborative and/or supervisory role of investigator-driven science. This represents a spectrum; some cooperative agreements resemble contracts in that the government partners are making all the major decisions, though they may be administered as grants. A research and development (R&D) contract or a research support contract represents work whose product is predefined by the funding agency, though an investigator may have influence over exactly how the work will be done.

The National Institutes of Health

The NIH is the premier extramural funding source for biomedical and biobehavioral research in the health sciences. In the 2013−14 fiscal year, about 11% of NIH research expenditures went to support the research of its own intramural scientists in Bethesda, Maryland, or in satellite laboratories and venues elsewhere in Maryland (e.g., Rockville, Frederick, and Baltimore) or elsewhere (e.g., Research Triangle Park, NC, and Hamilton, MT; Fig. 39.1). Thus, it is widely believed that 89% of the research expenditures go for extramural research grants (see Box 39.1). This is not the case. About 8% of NIH funds support administrative work and miscellaneous expenses, 53% fund extramural research project grants, 3% support research training or research nested within career development awards, 8% fund other types of research grants, 9% support research centers, and 10% fund R&D contracts (Fig. 39.1). Recipients of extramural grants and contracts include universities, contract research organizations, research institutes, small businesses, and other eligible organizations that compete for funds.

The context for the NIH share of the US federal research investment is presented in Fig. 39.2. The department of defense (DoD) is the only entity spending more in research dollars, though DoD research funding is very sparse for health.

Research grants are awarded as both direct and indirect costs by the NIH. Direct costs represent the actual funds available to support the research through faculty and staff salary support, equipment, supplies, travel costs, and expenses incurred by subcontractors. Indirect costs (officially called Facilities and Administrative Costs or "F&A") are incurred "for common or joint objectives and cannot be identified specifically with a particular project or program." Institutions negotiate

FIGURE 39.1 Distribution of the US$30.1 billion US National Institutes of Health budget, in the total enacted budget authority Fiscal year 2014, by category of funding. *From an NIH Website: http://report.nih.gov/NIHDatabook/Charts/Default.aspx?showm=Y&chartId=5&catId=1. Data are from the President's Budget Request, p. 29. http://officeofbudget.od.nih.gov/pdfs/FY15/FY2015_Overview.pdf.*

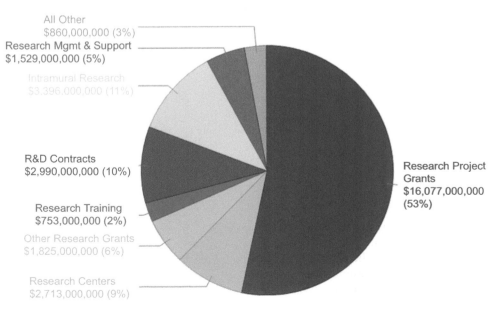

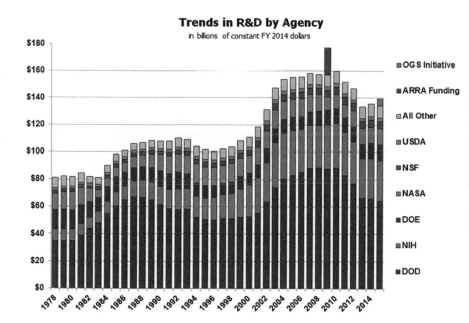

Trends in R&D by Agency

in billions of constant FY 2014 dollars

Legend:
- OGS Initiative
- ARRA Funding
- All Other
- USDA
- NSF
- NASA
- DOE
- NIH
- DOD

FIGURE 39.2 Historical trends in federal Research & Development (R&D) by agency are presented here. Most data come from annual American Association for the Advancement of Science R&D budget reports, with additional data from the National Science Foundation's National Center for Science and Engineering Statistics, and from the historical tables provided by the White House Office of Management and Budget (OMB). Constant-dollar conversions (adjusting for inflation) use OMB's chained price index. *http://www.aaas.org/sites/default/files/Agencies_0.jpg. AAAS Report: Research & Development series and analyses of FY 2015 budget request. 1976-1994 figures are NSF data on obligations in the Federal Funds survey.* © 2014 AAAS.

their indirect cost rates with the Federal government based on a complex array of expenses that they demonstrate to Federal auditors to be research-related. These might include utilities, space rental or maintenance, library book and journal expenses, university-wide information technology costs, salaries of librarians, chairs, or deans, and many other generic university or institute expenses. Putting it simply, direct costs benefit the investigator directly in his/her efforts to answer a given research question, while indirect costs benefit the investigator's institution in its broader effort to support the research and teaching mission.

It is difficult to disaggregate the direct cost support for research versus the indirect costs and fees for such research from readily available public sources, but these data are available from the NIH upon request. It is also difficult to discern why some indirect costs exceed 70% for some universities, while being less than 45% for others whose missions and infrastructures are similar. However, review of audits during indirect cost negotiations reveals why these may differ from institution to institution; cost of doing business is much higher in some geographic venues than others, for example. As well as indirect costs, research or research service contracts may be awarded to for-profit entities that charge a "fee," or profit, beyond the costs of the research itself.

A daunting array of grant types are available to support research or research training through a competitive, peer-reviewed process, including:

1. Research Grants (R series)
2. Career Development Awards (K series)
3. Research Training and Fellowships (T series and F series)
4. Program Project or Center Grants (P series)
5. Resource Grants (various series)
6. Trans-NIH Programs (various initiatives).

NIH announces the availability of potential grant support though a Funding Opportunity Announcement (FOA). An FOA, regardless of the type of grant solicitation, is a public document that is released by a Federal agency to inform the public that it intends to award grants, cooperative agreements (grants with more Federal involvement than with a standard grant), or contracts. There are several types of FOAs:

1. Program announcements that suggest a given topic that the agency (or Institute or Center within the agency) wishes to support, but does not typically guarantee a sum of money to support it.
2. Requests for applications (RFA) that give very specific guidelines as to topic and type of research being solicited, typically with a reserved pool of resources specifically earmarked for the given RFA.
3. Requests for proposals solicit contract proposals for very specific, government-directed tasks or products.
4. Other terms are used to solicit grants by various agencies, including notices of funding availability or solicitations.

Two web sources of FOAs are available of particular relevance for biomedical and biobehavioral researchers:

1. http://www.grants.gov/web/grants/search-grants.html provides information for all agencies, not just the NIH.
2. http://grants.nih.gov/grants/guide/index.html accesses the NIH Guide for Grants and Contracts that can be queried electronically.

Universities and research organizations around the world monitor these sites for opportunities applicable to investigators at their institutions. While only eligible institutions may apply, eligibility varies depending upon what type of award is being applied for. Also, while citizens of any nation may apply for most unsolicited NIH research grant awards, other grants are restricted to defined subgroups. Only US citizens or permanent residents can apply for most career development awards, for example. For a tiny fraction of NIH programs, only institutions in low- and middle-income countries (LMIC) can apply, as with the selected programs of the NIH Fogarty International Center (FIC). In 2014, some nations that are classified by the World Bank as LMIC were deemed ineligible by the FIC for this privileged status by virtue of their membership in the Group of 20 forum for the world's major economies, though they can apply for many other NIH research grants. Only employees of eligible businesses can apply for certain types of grants in the Small Business Innovation Research (SBIR) or Small Business Technology Transfer (STTR) Programs. There are too many specific examples to list, but the NIH helps walk prospective investigators through the maze of Institutes and Centers (ICs) and possible grant awards through its Office of Extramural Research Website (http://www.grants.nih.gov/grants/oer.htm).

Research Grants (R series) are a principal source of support for basic science research, but are also used to support clinical and translational research. The unsolicited, investigator-initiated "R01" represents the grant most familiar to investigators and most commonly awarded within the NIH Research Project Grant Program. The NIH has an online database of all funded grants, which also provides aggregated summary data (http://report.nih.gov/fundingfacts/fundingfacts.aspx) called RePORT (NIH Research Portfolio Online Reporting Tool). Of 50,567 grants awarded by NIH in fiscal year 2013, over half (51.6%) were R01 or R01-equivalent grants. As stated by the Office of Extramural Research at the NIH, the R01 supports "a discrete, specified, circumscribed research project," and it is used by all of the ICs.

In fiscal year 2013, the 49,652 Research Project Grants (including R01 and equivalent) grants that NIH awarded averaged $441,106 in total costs (direct and indirect costs combined). In the face of inflationary and regulatory pressures of medical research in recent years, this is especially challenging for a lot of clinical and translational research content. In theory, there is no specific dollar limit for an R01, but IC directors are reluctant to tie up too much money in one particular grant. Hence they typically require investigators to secure advanced permission if grants are intended for submission with greater or equal to US$500,000 in direct costs per year. The R01 award is granted for 3−5 years in nearly all cases. Much clinical research is impractical for the R01 mechanism as clinical trials or epidemiological studies may be multisite, multiyear, costly, and are beyond the scope of the conventional R01. However, R01 awards may collaboratively nest within clinical trials networks or multisite epidemiological consortia to great mutual advantage; networks and consortia gain added scientific productivity through expert collaborations, while R01 investigators gain the specimens and core data already being collected by the networks/consortia.

Other research grants are less commonly used, but are valuable for specific purposes. The NIH Small Grant Program (R03) provides modest funding (often US$50,000 direct support per year for one to three years) to support projects for which such funding makes sense. As per the Office of Extramural Research, these can include "pilot or feasibility studies, collection of preliminary data, secondary analysis of existing data, small, self-contained research projects, development of new research technology, etc." About half of the 27 ICs that comprise the NIH use the R03 mechanism. This is especially helpful to young investigators as the length of the grant application is half that of an R01 (7 pages of the aims and the research proposal compared to 13 pages) and the competitiveness for the R03 is less severe than for the higher stakes R01.

NIH provides support for conferences and scientific meetings (R13 and U13) when prior approval for a proposal has been gained from the funding IC. These awards, like nearly all NIH awards, are awarded based on a competitive peer-review process. Foreign institutions are not eligible to apply and support may be from 1 to 5 years. One often sees NIH credited for support of research conferences; often this support has been provided by supplementing competitively awarded research or training grants or through the R13 and U13 mechanisms.

The NIH Academic Research Enhancement Award (AREA) supports small research projects (up to three grant years up to US$300,000 aggregate direct costs) in US academic institutions that have not been major recipients of NIH research grant funds. AREA grants are typically received by smaller colleges and universities that are not primarily research-oriented, rather being focused on their teaching missions. These awards can provide smaller research laboratories with support, but are less frequently used for more costly clinical or translational research.

NIH Exploratory/Developmental Research Grant Awards (R21) and NIH Clinical Trial Planning Grant (R34) Programs are intended to enable exploratory and developmental research though the early stages of project development. Pilot and feasibility studies are encouraged in these mechanisms, with work leading to clinical trials focused on the R34. Funds for R21 awards are limited to 1 or 2 years of funding and a combined budget for direct costs for up to a 2-year project period up to US$275,000 aggregate direct funding; most ICs support R21 awards. Only a few ICs support the R34 mechanism and these awards vary considerably, but never more than 3 years, nonrenewable, and up to $450,000 direct cost support in aggregate. Innovation is deemed a priority for R21 and R34 grants. However, since the same peer-review groups that see the R01 proposals also review the R21 grants, it is often the case that R21 awards are already well developed, but are merely "cheaper R01s." Risk in research funding is rarely embraced by the typical NIH peer-review group; since the competition is so fierce, the grants judged most-likely-to-succeed are often favored for funding. Innovative grants may be passed over, particularly if there are feasibility concerns. No preliminary data are required, unlike the R01 mechanism, but it is unlikely that R21 grant proposals are enhanced in a competitive funding environment by a complete absence of preliminary data.

A special category of NIH research grants support small business research, typically in partnership with academic institutions. STTR (R41 and R42) grants seek to stimulate innovative partnerships between small businesses and research institutions through commercialization of innovative technologies. SBIR (R43 and R44) grants seek to spawn techno-logical innovation in the private sector by supporting research or R&D within for-profit institutions for ideas that have potential for commercialization. The two phases that are funded by the NIH for SBIR and STTR are to establish the technical/scientific merit and feasibility of the proposed efforts (Phase I) and to continue the research or R&D efforts initiated in Phase I when the early work shows promise (Phase II). Only US small business concerns can apply and the grant principal investigator (PI) must be employed by a small business partner. Fully 2.8% of the NIH research budget is mandated by the US Congress to be spent on the small business grants in fiscal year 2014, compared to 2.5% in 2006. In 2013, there were 5304 SBIR/STTR grants compared to 1844 in 2006.

From 1998 to 2003, the US Congress and President expanded the NIH budget twofold in a bipartisan effort to enhance biomedical and behavioral science discovery. Since 2004, NIH has been subject to near-zero growth. In billions of dollars the trends are as follows: 2004, $27.2; 2005, $27.5; 2006, $27.3; 2007, $29.1; 2008, $29.6; 2009, $29.5; 2010, $30.2; 2011, $29.9; 2012, $30.9; and 2013, $29.1. This has resulted in an abrupt decline in real purchasing power, particularly since 2007, corresponding to a time of fiscal exigency with the continuing US military engagements in Iraq and Afghanistan, and the global banking and mortgage fiscal crisis of 2008. Many investigators have been troubled by failure to secure research funding with awards that would have been competitive previously; funding thresholds are more stringent given that more grant proposals are chasing fewer real dollars, given inflation and flat funding. The overall success rate for research project grants like R01s was only 17% in 2013, meaning that of submitted grants, about one in every six was funded eventually, i.e., including resubmissions. Given 2015 US Congressional sentiment, it is unlikely that the NIH budget will grow in the near future.

An especially important mechanism for support of clinical research is the Research Project Cooperative Agreement (U01). While Office of Extramural Research sources state that U01 awards are to support "discrete, specified, circum-scribed projects to be performed by investigator(s) in an area representing their specific interests and competencies," this does not at all capture the breadth of U01 usage. NIH uses a U01 mechanism to instill substantial scientific and/or pro-grammatic collaboration between the awarding IC and the investigators. There are many "U-series" types of cooperative agreements that do not fall under preordained dollar limits. Many of the large clinical trials networks are funded via U01 awards, including large oncology and HIV/AIDS cooperative networks that extend far beyond US borders (Box 39.2). In fiscal year 2013, over 10,000 cooperative agreement grants (U series) were made by the NIH, of which 2096 were U01s with an average U01 award size of about US$800,000.

The R29 award was a former, now-discontinued new investigator award to help jump-start the careers of newly trained investigators. A new award at the NIH continues this tradition, the NIH Pathway to Independence Award (K99 and R00 series). Providing up to 5 years of support, this award consists of Phase I, 1–2 years of "mentored support for highly promising, postdoctoral research scientists," and Phase II, up to 3 years of research support for persons who are able to secure an independent research position. Eligible PIs are those postdoctoral candidates with clinical or research doctorates and no more than 5 years of postdoctoral research training. While foreign institutions are not eligible to apply, the PI need not be a US citizen.

Program Project and Center Grants (P series) are large, multiproject activities that include a diversity of research activities. Comprehensive Cancer Centers, Centers for AIDS Research, Specialized Programs of Research Excellence in Human Cancer, and centers for many specific diseases or activities (e.g., arthritis, diabetes, coronary heart disease, TB, vaccine trials, maternal and child health) are but a few such programs. In Fiscal Year 2006, 1488 such awards were made of an average US$1,879,583 per award that year. In Fiscal Year 2014, the investment was similar, or $2.8 million. These programs include Research Program Project Grants (P01) that support multiproject research involving "a number of

---

BOX 39.2 Two Examples of U01 Cooperative Agreement Grants That Support NIH Clinical Trials Networks

National Cancer Institute (NCI) Clinical Trials Cooperative Group Program:
http://dctd.cancer.gov/ProgramPages/ctep/major_ctcgp.htm.

American College of Radiology Imaging Network	http://www.acrin.org
Alliance for Clinical Trials in Oncology/American College of Surgeons Oncology Group	http://www.acosog.org
Cancer and Leukemia Group B	http://www.calgb.org
Children's Oncology Group	http://www.childrensoncologygroup.org
Eastern Cooperative Oncology Group	http://www.ecog.org
European Organisation for Research & Treatment of Cancer	http://www.eortc.org/
Gynecologic Oncology Group	http://www.gog.org
National Cancer Institute of Canada, Clinical Trials Group	http://www.ctg.queensu.ca
National Surgical Adjuvant Breast and Bowel Project	http://www.nsabp.pitt.edu
North Central Cancer Treatment Group	http://ncctg.mayo.edu
Radiation Therapy Oncology Group	http://www.rtog.org
Southwest Oncology Group	http://www.swog.org

National Institute of Allergy and Infectious Diseases (NIAID) HIV/AIDS Clinical Trials Networks:
http://www.niaid.nih.gov/about/organization/daids/networks/pages/daidsnetworks.aspx

AIDS Clinical Trials Group (ACTG)	https://actgnetwork.org/
HIV Prevention Trials Network (HPTN)	http://www.hptn.org/
HIV Vaccine Trials Network (HVTN)	http://www.hvtn.org/
International Maternal Pediatric Adolescent AIDS Clinical Trials (IMPAACT)	http://impaactnetwork.org/
International Network for Strategic Initiatives in Global HIV Trials (INSIGHT)	http://insight.ccbr.umn.edu/index.php
Microbicide Trials Network (MTN)	http://www.mtnstopshiv.org/

---

independent investigators who share knowledge and common resources." There must be a common theme of the overall research endeavor that is directed toward a well-defined research program goal. Other programs and centers include:

1. Exploratory Grants (P20) that support planning activities related to large multiproject program project grants;
2. Center Core Grants (P30) that support shared resources/facilities either for a collaborative research effort by investigators from different disciplines in a multidisciplinary approach or to support investigators from the same discipline who focus on a common research problem through complementary research;
3. Specialized Centers (P50) that support varying parts of the research or R&D spectrum, including supportive activities such as protracted patient care necessary to the research effort.

Program Project and Center Grants are rarely awarded to overseas institutions. A 5-year-old program called the Comprehensive International Program of Research on AIDS was designed to nurture such awards. Before abandoning the program, the National Institute of Allergy and Infectious Diseases made 24 R03, three U01 (to Cambodia, Haiti, and Senegal) and five U19 (to China, Peru, Thailand, and two to South Africa) grants to institutions and investigators in 24 developing countries.

Resource Grants are unusual grants that provide research-related support or access to resources. Resource-Related Research Projects (R24) awards are highly diverse and ICs use them in various ways to enhance research infrastructures or to address problems where diverse expertise is supported in addressing a complex research problem. Equally diverse are the Education Projects (R25) that promote biomedical research by training and dissemination of scientific discovery, including application in public health and community arenas.

Trans-NIH Programs are also supported across the ICs. Examples include:

1. NIH Bioengineering Consortium (BECON)
2. Biomedical Information Science and Technology Initiative (BISTI)

3. The NIH Blueprint for Neuroscience Research
4. Research Supplements to Promote Diversity in Health-Related Research, efforts to nurture researchers from underrepresented racial and ethnic minority groups in the United States, within existing NIH research grants
5. Genome-Wide Association Studies (GWAS)
6. Presidential Early Career Award for Scientists and Engineers (PECASE)
7. NIH Roadmap Initiatives, awards related to the long-term strategy plan (the Roadmap) for the NIH, including NIH Director's Pioneer Award (DP1) and the NIH Director's New Innovator Program (DP2).

Space does not permit us to review these and other programs in detail, such as the recent precision medicine initiative, but all are described in NIH websites.

In addition to research, many grants support research training, the so-called K and T series grants. K series career development awards are deemed enormously valuable for their recipients (US citizens or permanent residents only) for the career start-up provided. K24 and K05 awards are for more senior investigators doing some mid- or late-career shifting of professional emphasis. T32 grants typically support trainees within a given area of laboratory expertise and many such grants have been renewed for multiple cycles. The F series grants are small awards that help both pre- and postdoctoral trainees with research needs.

The NIH is a principal driver of clinical and translational research both in the United States and beyond US borders (Matherlee, 1995). Its role and budget stagnation since 2003 have been the topic of much commentary (Campbell, 2009; Crowley et al., 2004; Cripe et al., 2005; Moses et al., 2005, 2012; Loscalzo, 2006; Owens, 2014).

Awards to Non-US Institutions From the NIH

Foreign institutions are supported routinely by the NIH, but this support has decline substantially in recent years. In fiscal year 2007, 689 grants went to foreign institutions directly from the NIH. In fiscal year 2013, this number was only 480 grants for a total of $232.2 million. In 2007, 55 nations were represented among institutions awarded NIH grants, about the same as the 57 nations in 2013 (Table 39.2). In 2007, Canada (198 grants or 27.9%), the United Kingdom (93 or 13.5%), and Australia (59 or 8.6%) accounted for over half of the grants awarded to non-US institutions. In 2013, more than half the grants were received by four countries: Canada (134 grants or 28.7%), the United Kingdom (54 or 11.3%), South Africa (45 grants or 9.4%), and Australia (22 or 4.9%). That South Africa should have gone from 24 NIH grants in 2007 to 45 in 2013 is a notable commentary on that nation's increasing competitiveness in research in recent years. In 2007, 23.2% of the grants (160 of 689) went to 29 LMIC institutions, while in 2013, 39.6% of the grants (190 of 480) went to 35 LMIC institutions. This comparatively increased competitiveness of investigators in LMIC may be attributed, in part, to efforts by the FIC at NIH as well as institutional and human resource development efforts by many nations to develop indigenous research capacities in LMIC around the globe.

The Centers for Disease Control and Prevention

CDC, the public health arm of the US Federal government, does not have the substantial extramural research dollar pool of the NIH. Nonetheless, some extramural support for research is provided, almost always as a cooperative agreement grant. This is most often focused on state public health departments, but academic partners or organizations with an overseas presence may be supported. Often CDC research support is in the form of a network or consortium making possible studies across multiple venues. The service mandate of the CDC also drives extramural funding, most notably in the PEPFAR that has run since 2003 and continues (in 2015) to have broad bipartisan support in the US Congress. Largely through the CDC and the US Agency for International Development, the US Government has committed over US$52 billion through fiscal year 2013, largely to bilateral HIV/AIDS and TB programs, but also through the multilateral Global Fund to Fight HIV/AIDS, TB, and Malaria. This makes PEPFAR the largest disease-focused international development initiative in world history. While PEPFAR grants are not for research, but are rather for service delivery, there is a large M&E component and opportunities for "implementation science" and quality improvement studies (Nash et al., 2011; Ciampa et al., 2012; Howard et al., 2014). In addition, PEPFAR investments create clinical and public health contexts that investigator-initiated research grants can exploit for nested initiatives (Moon et al., 2013; Brentlinger et al., 2014).

Evaluations of public health and health systems innovations have elements of translational research, namely moving discovery in clinical research ("the bedside") to the community ("scaling-up"). So-called "T1" research is often thought of as a discovery in biology or other potentially health-related area of science and "T2" research is translation of this discovery into a health application (Hait, 2005). The research work supported by CDC and US Federal agencies other than the NIH may often be viewed as "T3" research, namely the scaling-up of already proven clinical or public health interventions to make them more effective and available in real-world settings. The NIH less often embraces T3 research, even though its discoveries and their implementation can have disproportionately large impacts on health indicators, leaving this agenda to CDC and other agencies that have only modest funding to support research.

Historically, CDC research efforts were steered toward partner institutions and selected extramural investigators. Since about 2000, CDC has made efforts to mimic the peer-review approaches and truly competitive processes of its sister institution, the NIH. At present, CDC research solicitations are typically formal RFAs, analogous to the NIH. This has increased interest in the extramural academic and research organization communities in CDC-supported grant opportunities.

Due to the magnitude of US funding and its availability to institutions both outside and within the United States, we have highlighted the types of grants of greatest interest to investigators in the clinical and translational sciences. There are many other funding opportunities from government sources in the EC, Canada, and even in developing countries.

SCIENTIFIC PROGRAMS IN EUROPE, CANADA, AND AUSTRALIA/NEW ZEALAND

The European Community

EC represents a historic economic, legal, and even social alliance of 28 of Europe's 49 nations (as of early 2015). The EC has taken an active role in supporting health research within its member nations (Box 39.3). These nations include a preponderance of Europe's finest research institutions. Competitive grants are available both in response to solicitations from EC science administrators and also for investigator-initiated work. While leadership is limited to PIs and institutions from member nations, collaborations beyond EC borders occur to foster the best scientific partnerships. One of the EC's most valued functions is communication, linking investigators to information in Europe that can support their research endeavors. The EC itself funds research and some of their successes are seen in the STAR Projects Website (http://ec.europa.eu/research/star/home.html).

Joint Technology Initiatives have sought are a good example of what the EC has organized for its member nation scientists (http://ec.europa.eu/research/jti/index_en.cfm). These initiatives have to create a critical mass of expertise and research attention in the areas of emphasis across Europe. The rationale for these investments includes strengthening the competitiveness of European industry and the attractiveness of Europe as a location for research-related investments. Most of these investments are not in health-related research, but those projects that are health-related include networks that cross national boundaries, just as the NIH networks cross state and national lines (e.g., the European HIV/AIDS research activities in the Seventh Framework Programme (FP7) that ran from 2006 to 2013, though many projects are still active as this program in early 2015 (http://ec.europa.eu/research/health/infectious-diseases/poverty-diseases/aids_en.html)).

One example of a funding approach in the EC is the sponsorship of a body of grants and networks in one particular area of emphasis. In Box 39.4, the FP7 awards that supported HIV/AIDS, TB, and malaria research are presented. As with their earlier Joint Technology Initiative predecessors, the FP7 programs permit EC funding to be combined with other funding sources, including private investments (Gaspar et al., 2012; Goldman et al., 2013; Romero et al., 2009; Sautter et al., 2011). There is a wide array of sponsored research activity in neurosciences, cancer, cardiovascular diseases, poverty-related diseases, and many other clinical and translational topics. Research funding from the EU FP7 initiative rose steadily from 2007 to 2013 to 10.84 billion euros in 2013 (about US$14.98 billion as per 30 December 2013 exchange rate), but

---

BOX 39.3 The 28 Member States of the European Community as of January 2015

Austria	Estonia	Italy	Portugal
Belgium	Finland	Latvia	Romania
Bulgaria	France	Lithuania	Slovakia
Croatia	Germany	Luxembourg	Slovenia
Cyprus	Greece	Malta	Spain
Czech Republic	Hungary	Netherlands	Sweden
Denmark	Ireland	Poland	United Kingdom

Note: Several other European nations have treaty relationships that link them closely to the EU, including Norway, Iceland, Lichtenstein, Switzerland and some western Balkan nations.

BOX 39.4 Projects and Networks Funded in the European Community for Research in HIV/AIDS, Tuberculosis (TB) and Malaria, Sorted Alphabetically by Acronym

AnoPopAge	Population age structure and age structure modification via *Wolbachia* in *Anopheles gambiae*
ARTEMIP	The safety pharmacology of artemisinins when used to reverse pathophysiology of malaria in pregnancy
CHAARM	Combined Highly Active Anti-retroviral Microbicides
CHAIN	Collaborative HIV and Anti-HIV Drug Resistance Network
CRIMALDDI	The Coordination, Rationalization and Integration of Antimalarial Drug Discovery and Development Initiatives
CUT'HIVAC	Cutaneous and Mucosal HIV Vaccination
ENAROMaTIC	European Network for Advanced Research on Olfaction for Malaria Transmitting Insect Control
ESI-TBVI	Establishment, Strategy and Initial activities of the TuBerculosis Vaccine Initiative: Coordination of European efforts with global research initiatives
EUCO-Net	European Network for global cooperation in the field of HIV & TB
EuroCoord	European network of HIV/AIDS cohort studies to coordinate at European and International level clinical research on HIV/AIDS
EuroNeut-41	European consortium on NEUTralising antibodies using gp41
Evimalar	Towards the establishment of a permanent European Virtual Institute dedicated to Malaria Research
FAST-XDR-DETECT	Development of a two-approach plate system for the fast and simultaneous detection of multiple drug resistant (MDR) and extensively drug resistant (XDR) strains of *Mycobacterium tuberculosis*
HIV-ACE	Targeting assembly of infectious HIV particles
HIVIND	The antiretroviral roll out for HIV in India — generating evidence to promote adherence and patient follow-up
HOMITB	Host and mycobacterial molecular dissection of immunity and pathogenesis of tuberculosis
iNEF	Inhibiting Nef: a novel drug target for HIV-host interactions
INYVAX	Optimization of the development of Poverty-Related-Diseases (PRD) vaccines by a transversal approach, addressing common gaps and challenges
MALACTRES	Multi-drug resistance in malaria under combination therapy: Assessment of specific markers and development of innovative, rapid and simple diagnostics
MALSIG	Signaling in life cycle stages of malaria parasites
MALVECBLOK	Population biology and molecular genetics of vectorial capacity in *A. gambiae*: targeting reproductive behavior and immunity for transmission-refractory interventions
MEPHITIS	Targeting protein synthesis in the apicoplast and cytoplasm of plasmodium
NANOMAL	Development of a handheld antimalarial drug resistance diagnostic device using nanowire technology
NATT	New Approaches to Target Tuberculosis
NEWTBVAC	Discovery and preclinical testing of new vaccine candidates for tuberculosis
NGIN	Next Generation HIV-1 Immunogens inducing broadly reactive Neutralizing antibodies
NOstress	Unraveling the molecular mechanism of nitrosative stress resistance in tuberculosis
NOVSEC-TB	Novel secretion systems of *M. tuberculosis* and their role in host–pathogen interaction
OPTIMALVAC	Initiative on Optimizing malaria Vaccine lab assays evaluation
PENTA LABNET	Paediatric European Network Treatment AIDS Laboratory Network
PRD College	Poverty Related Diseases-College: International Programme on BioMedicine and Development
PreMalStruct	Structural analysis of the placental chondroitin-4-sulfate (CSA) binding interactions involved during pregnancy associated malaria
PregVax	*Plasmodium vivax* Infection in Pregnancy
REDMAL	Clinical Development of a Pfs48/45 protein-based malaria transmission blocking vaccine
StopLATENT-TB	LATENT TUBERCULOSIS: New tools for the detection and clearance of dormant *M. tuberculosis*
STOPPAM	Strategies TO Prevent Pregnancy-Associated Malaria
TB PAN-NET	TB PAN-NET: Pan-European network for study and clinical management of drug resistant tuberculosis
TBsusgent	Sustaining research momentum over the coming decades: mentoring the next generation of researchers for tuberculosis
TB-EURO-GEN	Genetic analysis of the host–pathogen interaction in tuberculosis
TB-VIR	*M. tuberculosis* W-Beijing genetic diversity and differential virulence and immune responses
THINC	Targeting HIV integration co-factors, targeting cellular proteins during nuclear import or integration of HIV
TransMalariaBloc	Blocking Malaria Transmission by vaccines, drugs and immune mosquitoes: Efficacy assessment and targets
TM-REST	TM-REST: a new platform for fast molecular detection of multiple drug resistant (MDR) and extensively drug resistant (XDR) strains of *M. tuberculosis* and of drug resistant malaria

http://ec.europa.eu/research/health/infectious-diseases/poverty-diseases/projects/l_fp7_en.htm.

health represented only 1.003 billion euros in 2013 (US$1.39 billion), 9.3% of this aggregate research investment (http://ec.europa.eu/research/fp7/index_en.cfm?pg=budget). It is difficult to contrast NIH budgets in health to those of the EC since European nations still have their national and regional research investments, i.e., the EC investments are just a fraction of the aggregate investments made by European nations in health R&D. In addition, national health systems pay for many expenses (investigator salaries, for instance) that US investigators have to pay from their NIH grants. The newest EU collaborative endeavor is under the rubric of HORIZON 2020 (http://ec.europa.eu/programmes/horizon2020/en/what-horizon-2020), continuing and extending the FP7 process of multinational European research support.

National Initiatives

Individual European nations continue their strong domestic research programs through MRC in the United Kingdom, the Karolinska Institute in Sweden, the Institut Pasteur in France, the Robert Koch Institute in Germany, and many others. Typically, there is at least one biomedical research institution per nation and sometimes many such institutions in a single nation. European institutions may represent both Federal research institutions and universities combined. (In contrast in the United States, Federal research institutions and academic institutions are separate entities with rare exceptions such as the Uniformed Services University of the Health Sciences that serves to train physicians, nurses, and public health and other health specialists for military service; even that unique institution must apply competitively for extramural research funds to the NIH, as civilian schools must do.) National support of research in Europe is largely reserved for the nationals of the nation providing the funds, through cross-national collaborations are frequent. Since universities and nationally sponsored research establishments are often integrated, European investigators may have more stable salary support than do US investigators, but typically have less total money for research. This historic stability has been threatened by European fiscal challenges in the early 21st Century, and Government partnerships with industry are more common and increasingly encouraged in Europe, keeping with a global trend in research (Demotes-Mainard et al., 2006; van Ommen et al., 2014).

Research institutes at the regional level are also found in much of Europe. One example is the Center for Epidemiological Studies on HIV/AIDS of Catalonia (Centre d'Estudis Epidemiològics sobre les ITS i Sida de Catalunya or CEESCAT), which is a research arm of the Department of Health and Social Security in Catalonia, one of the 17 autonomous communities and two autonomous cities that constitute the country of Spain. A regional HIV/AIDS research center, CEESCAT has had Catalonia-specific research projects, Spanish and European collaborations, and international partnerships. CEESCAT's research focus is on HIV prevention and risk reduction, including research in substance abuse and sexual risk behaviors. Regional institutions may collaborate with outside experts, but their funds have been focused historically upon the region that is supporting the work with its local tax dollars. However, a CEESCAT-linked foundation worked in Guatemala in an independent service effort, demonstrating that even regionally supported initiatives can open research collaborations beyond the region's own borders. Unfortunately, funding challenges in Spain and Catalonia since 2008 have led to substantial cuts in both local and international investments, including the closure of the aforementioned Guatemalan collaboration.

Canada has a system of research support that is a hybrid of the UK and US systems. In fact, this is symbolized by the change in the name and increase in funding level that occurred in 2000. The former MRC of Canada was disbanded after 40 years, replaced by the Canadian Institutes of Health Research (Charbonneau, 2000). The doubling of the biomedical research budget in that year reflected an evolution of the Canadian system from its UK-influenced roots to one that was more reminiscent of the US NIH. As with Europe and the United States, Canadian research investments include support of clinical and translational research through institutional, network, and individual investigator-supported initiatives. Network activities can be province-wide, Canada-wide, or may involve collaboration with other networks in the US or elsewhere (Perez et al., 2004; Hachinski et al., 2006; Clark et al., 2007; Paige, 2007; Dainty et al., 2011; Kahn et al., 2014).

The Australian government supports research through its National Health and Medical Research Council, which, among many other activities, supports a clinical trials network based at the University of Sydney. Nations with smaller populations may be more likely to work nationwide or across national boundaries to reach the patients needed for the research (Froom et al., 1990; Montaner et al., 1998; Heller et al., 1999; Anon., 2000; Arroll and Beilby, 2004; Darlow et al., 2005; Nichol et al., 2005; Green et al., 2012; Hoyos et al., 2014; Johnson et al., 2014; Oh et al., 2014; Purvis et al., 2014). Special partnerships in Australo-Asian neighbors such as Papua and New Guinea, Australia, and Thailand have enabled partnerships in larger studies of cross-national importance in Asia (Cardiello et al., 2002; Oman et al., 2002; Alpers, 2003; Fulu et al., 2013; Jewkes et al., 2013). International clinical research partnerships with the United States and Europe are increasingly common, as with the INSIGHT network (Achhra et al., 2010; Lynfield et al., 2014).

The Health Research Council of New Zealand supports clinical and translational research in New Zealand. The University of Auckland hosts a Clinical Trials Research Unit active both in the nation and in multicenter international trials, for example. Special attention to the health of the Maori (the indigenous people of New Zealand) is paid in some of the research work supported by the Health Research Council of New Zealand, analogous to concerns in many other nations about minority populations in research. However, investigators from high income, smaller nations can be frustrated about the relatively small magnitude of their science endeavors (Jull et al., 2005).

SCIENTIFIC PROGRAMS IN ASIA, AFRICA AND SOUTH/CENTRAL AMERICA AND THE CARIBBEAN

Remarkably vibrant public sector research support has emerged in upper middle-income nations. Space does not permit a detailed country-by-country listing, but Box 39.5 reviews a sample of government funding agencies that support investigators in those nations, including South Africa, India, China, Thailand, Brazil, Mexico, and Chile, to name but a few. Richer, but smaller nations such as Singapore and Israel support diverse research programs in their nations. Such nations as Japan and South Korea support some clinical and translational research, though they have trended towards emphasizing the laboratory disciplines. Similarly in China, Thailand and India, to name a few, support for clinical and translational research is somewhat more recent than the recognition of laboratory disciplines. Lower income nations rarely support substantial public sector research programs, but there has been rapid development of research talent and infrastructure in resource-limited nations. Some of the successes in building in-country developing country research capacities can be attributed to investments from the French-supported Institut Pasteur laboratories, the US NIH through training and clinical trials investments overseas, such collaborations as the HIVNAT (The HIV Netherlands Australia Thailand Research Collaboration Thai Red Cross AIDS Research Centre), the overseas UK MRC and US military-sponsored laboratories, the NIH/PEPFAR Medical Education Partnership Initiative, the Wellcome Trust Southern Africa Consortium for Research Excellence, the Consortium of New Southern African Medical Schools, and the TREAT Asia network (Ananworanich et al., 2014; Boettiger et al., 2014; Eichbaum et al., 2014; Frehywot et al., 2014; Mandala et al., 2014; Ménard et al., 2005; Sanchez et al., 2011). Many government-co-sponsored programs in developing nations have emerged with international aid. Examples are numerous, such that Box 39.6 gives only a very partial list, focused on infectious disease research.

The South African Medical Research Council

To give an example of one such developing nation research agency, we highlight the SAMRC, where one of the authors (S.S.A.K.) has worked previously. The SAMRC is a statutory biomedical research body that was established in 1969, in terms of the Medical Research Council Acts (19 of 1969 and 58 of 1991). The governance of the SAMRC is determined by the Minister of Health who appoints the SAMRC Board and its chairperson. The organization's most important functions are "to promote the improvement of the health and the quality of life of the population of the Republic" by performing "research, development and technology transfer." The vision of the SAMRC is: "Building a healthy nation through research" through its mission: "To improve the nation's health and quality of life through promoting and conducting relevant and responsive health research." The SAMRC is funded from the national Treasury through the South African

---

BOX 39.5 Examples of Upper-Middle and Lower-Middle Income Nations That Support Biomedical Research Enterprises With Government Support

Country	Examples of Internet-Based Informational Resources for Government
South Africa	http://www.mrc.ac.za/; http://www.mrc.ac.za/cochrane/cochrane.htm; http://www.saavi.org.za/
India	http://www.icmr.nic.in/; http://www.drdo.gov.in/drdo/English/index.jsp?pg=homebody.jsp; http://dbtindia.nic.in/
China	http://www.chinacdc.cn/en/; http://www.cicams.ac.cn/cicams_en/web/index.aspx
Kenya	http://www.kemri.org/
Brazil	http://portal.fiocruz.br/en/content/home-inglês
Mexico	http://www.conacyt.mx/; http://lanic.utexas.edu/la/mexico/
Chile	http://ri.conicyt.cl/575/channel.html
Russia	http://whodc.mednet.ru/eng/index.php; http://www.mvk.ru/en-GB

BOX 39.6 Examples of Research Institutions in Developing Countries That Are Either Supported With Donor Government Aid and/or Are Co-Supported by the Host Government

Organization (Principal International Partner or Donor)	Websites
Tanzanian Training Centre for International Health, Ifakara (Swiss Tropical Institute)	http://www.healthtrainingifakara.org/
International Centre for Diarrhoeal Disease Research, Bangladesh (multiple partners)	http://www.icddrb.org/
Centre for Infectious Disease Research in Zambia (multiple partners)	http://www.cidrz.org/
Tropical Disease Research Centre (World Health Organization and university partners)	http://www.africa.upenn.edu/Org_Institutes/tdrc_zm.html; http://www.tdrc.org.zm/
The Caribbean Public Health Agency (CARPHA, supported by Pan American Health Organization, and regional Ministries of Health, largely from the English-speaking Caribbean)	http://carpha.org/Home
The Africa Centre, University of KwaZulu-Natal in collaboration with the Medical Research Council of South Africa (Wellcome Trust)	http://www.africacentre.ac.za/
Centre for the AIDS Program of Research in South Africa (multiple partners)	http://www.caprisa.org/
El Instituto de Medicina Tropical 'Alexander von Humboldt,' Universidad Peruana 'Cayetano Heredia' (multiple partners)	http://www.upch.edu.pe/TROPICALES/

Department of Health's budget, and also receives funds from other sources such as the NIH, Bill & Melinda Gates Foundation, the South African Department of Science and Technology, Wellcome Trust, and the US CDC (Table 39.3).

Since inception, the SAMRC has played a key role in supporting medical research in the country and has thereby contributed to, among others, (1) developing and sustaining a sound health research infrastructure in South Africa, (2) building a culture of high-quality research in a developing country setting, (3) developing and supporting health systems research. Additionally, the work of the SAMRC has made a significant impact on public health including the establishment and subsequent improvements in public health surveillance, especially for malaria, TB, HIV/AIDS, alcohol and substance abuse, gender-based violence, nutrition, cancer, injuries and measuring the burden of disease. One of the biggest achievements of the SAMRC was its research on smoking that resulted in major policy changes.

A major revitalization of the SAMRC has occurred since 2012. In November 1997, the SAMRC was reviewed by an external panel as part of the national review of the country's science and technology system—the organization was commended on its significant scientific contributions and its postapartheid transformation "in line with the national objectives of the new South Africa." Further, the Review Panel recommended a substantial increase in the SAMRC's budget with an emphasis that it be used to support priority-driven research. However, 13 years later, in 2010, the next

TABLE 39.3 South African Medical Research Council Funding Sources, 2004—08

Source of Income (ZAR '000s)	2004/5	2005/6	2006/7	2007/8
Government grants	159,695	154,388	157,284	180,222
Contracts, grants, and services rendered	128,560	162,429	176,547	237,216
Other income	19,005	17,537	18,233	39,585
Total	304,260	334,354	352,064	457,023
Year by year increment	—	10%	5%	30%

Note: In 2007 the total budget was South African Rand (ZAR) 457 million, with 180 million from the South African government, 237 million of external income from grants and contracts and 40 million from other income. In September 2004, the exchange rate for the South African Rand was ZAR6.6 5 US$1 and in March 2008, the exchange rate was ZAR8.0 5 US$1.

government-funded external review revealed that the reputation of the SAMRC had declined drastically and was no longer viewed as a front-runner within the wider medical research environment, locally and abroad. The review showed that due to budgetary constraints, support for extramural research units had declined dramatically, failing even to ensure adequate funding to cover intramural staff salaries. Further, the review found that the Executive Management had not been rigorously applying the criteria for establishing or renewing SAMRC units resulting in some poorly conceived units. Some disturbing post-1997 trends included a shift away from recognizing and rewarding scientific excellence (e.g., SAMRC stopped awarding medals for scientific excellence after 1999), an increasing proportion of the MRC budget being devoted to intramural research at the expense of extramural funding to medical schools (the intramural research budget was more than twice the extramural budget), and a diminishing fraction of the SAMRC's research portfolio dedicated to clinical research.

Following the 2010 review, a new SAMRC Board was appointed to oversee the organization's turnaround strategy. In 2012, the Board appointed a President (S.S.A.K.) and new Executive Management. An initial assessment by the incoming President of the organizational state of the MRC in 2012 revealed that, although pockets of excellence in research existed in the MRC, there were a number of challenges (Nordling, 2014). These included a lack of a common vision, a lack of leadership, erosion of staff confidence in the SAMRC's management, an unclear vision, and skewed resource allocations with the majority of the funding being directed to intramural units and salaries. In addition, SAMRC was coping with declining baseline grant resources from the Department of Health, outdated organizational and funding approaches, and shortcomings in fulfilling its mandate as custodian of South African medical research.

A process of revitalization and restructuring started in 2012 and is still in progress as of this writing (early 2015). The principles used to guide the process for re-focusing the SAMRC included, prioritizing its research activities, ensuring the organization fulfills its mission to support the full range of medical research in South Africa and modernizing the way it aims to fulfill its mission. During the restructuring more than half of the MRC's 27 intramural units were closed, leaving 11 units that focus on the top 10 causes of death in South Africa, including HIV, TB, Chronic Diseases, Alcohol and Drug Abuse, and Women's Health.

The revitalization led to a successful request by the SAMRC President to the national Treasury for an increase the organization's budget. The South African government's contribution to the SAMRC increased by 65% from R271 million ($27.1 million) in the fiscal year ending on 31 March 2012 to R446 million ($44.6 million) in the fiscal year starting 1 April 2014 (Fig. 39.3). In the 2016 fiscal year, the budget is slated in terms of the government's medium-term expenditure framework to rise to R648 million ($64.8 million), a 140% increase from 2012. Funding was also secured from international research funders and at the end of 2013 the MRC launched a $40 million 5-year joint program with the NIH to fund research in South Africa on TB, HIV, and HIV-related conditions. In January 2014, the SAMRC announced a partnership with the Bill & Melinda Gates Foundation and the South African Department of Science and Technology, with funding of close to R300 million ($30 million) over 3 years starting in 2014 to set up a new initiative known as Strategic Health Innovation Partnerships (SHIP). SHIP is funding several new health technology projects focusing on developing an AIDS vaccine, new treatments for TB, especially drug-resistant TB, and new drug treatments for malaria.

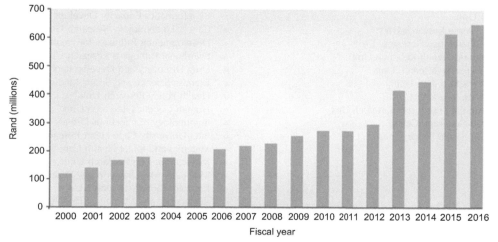

FIGURE 39.3 Historical trends in South African government baseline funding for the Medical Research Council, as published in http://www.nature.com/nm/journal/v20/n2/full/nm0214-113.html?WT.ec_id=NM-201402 (Nordling, 2014). Rand conversion per US$1.00 for February 1 of each year is 6.3 in 2000, 6.0 in 2005, 11.7 in 2010, and 7.7 in 2015.

The SAMRC also established a new high-profile funding opportunity for Medical Schools and Universities to undertake "Flagship Projects" aimed at addressing South Africa's key health problems. A total of R150 million ($15 million), over 3 years, was granted to 12 large multidisciplinary projects at Universities across the country, while 5 MRC intramural projects were granted R40 million ($4 million) over the same period for intramural "Flagship Projects." All 17 of these projects address the established health priorities of the nation and aim to place South African medical research and innovation on a globally competitive platform.

"The SAMRC fulfills its mandate through both intramural and extramural programs, similar to the NIH in the United States. The underlying philosophy for each component differs. Intramural program priorities are based on the disease burden experienced by South Africans, while the extramural investigator-initiated research award grants are based on the quality of science proposed and the track record of the scientists, i.e., investigator-initiated research applications are funded on merit and do not need to be aligned with the country's burden of disease. The extramural program now comprises five components: (1) 3-year awards for University-based "Flagship projects"; (2) 5-year grants for extramural SAMRC units of excellence; (3) 3-year grants for investigator-initiated research; (4) 3—5 year junior or mid-career scholarships and research development awards; and (5) milestone-driven project funding for technology development. The range of funding opportunities provided by the SAMRC is presented below.

Balancing the Needs of Intramural and Extramural Research Units

The key mechanism whereby the SAMRC provides long-term support to high-quality medical research in South Africa is through its funding for research units. In the 2013—14 financial year, 11 intramural and 17 extramural research units were supported, with an additional 6 new extramural units being initiated in 2014 by the SAMRC. In 2014, the budget for extramural research (including Units, Flagship projects and investigator-initiated studies) exceeded the intramural research budget. Intramural units of the SAMRC also apply to international and local research agencies and on average secure almost twice as much funding through external grants than they get from the SAMRC's government funding allocation. The intramural units received an average grant from the SAMRC's government allocation of R10 million ($1 million), while the extramural entities received an average of R2 million ($200,000) in 2014. The extramural units are awarded to research leaders based on track record and potential impact of the proposed research for renewable cycles of 5 years each. Ongoing support for the SAMRC-funded units is contingent upon a successful annual submission detailing their research and operational progress, along with justified budget requests and plans for the coming year. A detailed external review of each unit at the end of its funding cycle is central to determining whether support will be continued for a subsequent cycle. Intramural and extramural research units supported in 2013—14 are listed in Box 39.7.

BOX 39.7 Intramural and Extramural Research Programs, Units and Centers Funded by the South African Medical Research Council in 2013—14

Intramural Research Units	Extramural Research Units
• Alcohol, Tobacco and Other Drug Research Unit	• Anxiety and Stress Disorders Research Unit
• Biostatistics Unit	• Bioinformatics Capacity Development Research Unit
• Burden of Disease Research Unit	• Cancer Epidemiology Research Unit
• Centre for Tuberculosis Research	• Developmental Pathways for Health Research Unit
• Environment and Health Research Unit	• Diarrhoeal Pathogens Research Unit
• Gender and Health Research Unit	• Drug Discovery and Development Research Unit
• Health Systems Research Unit	• Exercise Science and Sports Medicine Research Unit
• HIV Prevention Research Unit	• Health Policy Research Unit
• Non-communicable Diseases Research Unit	• Human Genetics Research Unit
• South African Cochrane Centre	• Immunology of Infectious Disease Research Unit
• Violence, Injury and Peace Research Unit	• Inter-University Cape Heart Research Unit
	• Maternal and Infant Health Care Strategies Research Unit
	• Medical Imaging Research Unit
	• Molecular Mycobacteriology Research Unit
	• Receptor Biology Research Unit
	• Respiratory and Meningeal Pathogens Research Unit
	• Rural Public Health and Health Transition Research Unit

HIV, human immunodeficiency virus.
Note: Local spelling is used throughout.

Supporting Capacity Building Through Grants and Scholarships

The SAMRC aims to building health research capacity by providing and administering scholarships for South Africans studying toward their masters and doctoral degrees in Medical and Health Sciences. Additionally, health research capacity is augmented by providing postdoctoral funding support. There is a particularly strong focus on competitive scholarship funding for clinical research through a new SAMRC Clinician Researcher Programme offering eligible candidates an R500,000 ($50,000) fellowship per annum. In addition, the SAMRC administers a Clinical Scholars Programme for the National Health Research Committee. These new Fellowship programs were specifically initiated to help re-build clinical research and support clinicians wishing to pursue research. All applications for funding are subjected to rigorous review to ensure that the best candidates and or research projects receive funding. The SAMRC also provides funding to host national and international conferences, including research symposia. The maximum amount that may be requested is R200,000 ($20,000) for international conferences and R100,000 ($10,000) for national conferences.

New Initiatives to Recognize and Support Scientific Merit

After several years, the SAMRC has re-initiated calls for applications from the country's foremost scientists to establish extramural research units—six new units were created and funded in 2014. Further, the SAMRC re-instituted recognition of scientific excellence with its annual Scientific Merit Awards. The previous Gold medal award for MRC-employed scientists and Silver medal award for non-MRC scientists have been enhanced by the creation of awards in five categories. New Platinum medals are awarded for a lifetime of exemplary scientific achievement, with seminal contributions that have impacted health, especially for people living in developing countries. Gold medals are awarded for major scientific breakthroughs, while Silver medals are given to recent postdoctoral researchers who have made a major scientific contribution. A new medal award is now also awarded for contributions to building medical research capacity in South Africa. Finally, at the discretion of the MRC President, special awards can also be conferred for exceptional contributions to medical research in South Africa, including contributions to science diplomacy or enhancing the country's status in science. The awards are among South Africa's most prestigious science awards, and reinforce the importance of leadership, innovation, and achievement.

CURRENT SUPPORT FOR CLINICAL AND TRANSLATIONAL RESEARCH

Clinical and translational research has been revitalized as a direct consequence of accountability being demanded by taxpayers, policy-makers, and government officials who want to see how biomedical and behavioral research investments actually improve health indices, with clear "deliverables" from monies invested. The importance of basic research for discovery and understanding of basic biological mechanisms has been recognized and research is active at every level of the organism—organ system, cell, and molecule. Genomics, proteomics, and nanosciences are just a few examples of how basic discovery, thought by some to be arcane and distant from application, is moving into clinical research and application. Hence, it is logical that as basic science succeeds in presenting new tools, clinical and translational research must assess their utility.

The NIH of the United States is a principal global funding source from government resources for research conducted either inside or outside the government itself. NIH initiatives sometimes inspire global action in biomedical research in other parts of the world. By 2013, the NIH had made awards to 62 institutions in 31 states and the District of Columbia to help them improve the translation of basic research into the clinic through the Clinical and Translational Science Awards (CTSAs; Box 39.8). Each award supports the institutions for 5-year periods, and provided $450 million in FY 2014 alone in so-called UL1 type grants to the CTSA network. CTSAs have been established to eliminate barriers between moving basic research discoveries into clinical research and application. Resources are invested into interdisciplinary teams including across disciplines, schools, and universities. Whether for discovering new diagnostic approaches, drugs and other therapeutics, vaccines, or prevention and behavioral strategies, interdisciplinary teams are deemed essential given the complexity of most clinical research and application challenges (Zerhouni, 2006, 2007).

The National Institute for Health Research in the United Kingdom supports 11 comprehensive Biomedical Research Centres (BRCs) as of early 2015 with awards ranging from £9.7 to £113 million each focused on translational research. The British Heart Foundation supports six interdisciplinary Centres of Excellence from 2008 through this writing in early 2015, also focused on translational research (Table 39.4). Even earlier, the Wellcome Trust had invested in translational research such that these aggregate efforts are seeking to move the United Kingdom forward very much in the same spirit as the CTSAs of the NIH. BRCs differ from the CTSAs in that each BRC represents a university and a National Health Service (NHS) trust partnership. The NHS trusts represent regional components of the publicly financed, universal health system in the United Kingdom. Since the United

BOX 39.8 The 61 Medical Research Institutes in the Clinical and Translational Science Award Consortium, From 31 of 50 States and the District of Columbia, Supported by the National Institutes of Health in 2014. None is Supported in a US Territory, e.g., Puerto Rico

Institution and, When Applicable, Principal Entity (U. = University)	Year of Award
Albert Einstein College of Medicine (partnering with Montefiore Medical Center) Albert Einstein-Montefiore Institute for Clinical and Translational Research—New York	2008
Boston U. Clinical and Translational Science (BU-BRIDGE) Institute—Massachusetts	2008
Case Western Reserve UCTSA at Case Western U.—Ohio	2007
Children's National Medical Center Clinical and Translational Science Institute at Children's National—District of Columbia	2010
Columbia U. Irving Institute for Clinical and Translational Research—New York	2006
Dartmouth College The Dartmouth Clinical and Translational Science Institute—New Hampshire	2013
Duke U. Duke Translational Medicine Institute—North Carolina	2006
Emory U. (partnering with Morehouse School of Medicine and Georgia Institute of Technology) Atlanta Clinical & Translational Science Institute (ACTSI)—Georgia	2007
Georgetown U. with Howard U. Georgetown–Howard Universities Center for Clinical and Translational Science—District of Columbia	2010
Harvard U. Harvard Catalyst: The Harvard Clinical and Translational Science Center—Massachusetts	2008
Indiana U. School of Medicine Indiana Clinical and Translational Science Institute—Indiana	2008
Johns Hopkins U. Johns Hopkins Institute for Clinical and Translational Research—Maryland	2007
Mayo Clinic Mayo Center for Translational Science Activities—Minnesota	2006
Medical College of Wisconsin Clinical and Translational Science Institute of Southeast Wisconsin—Wisconsin	2010
Medical U. of South Carolina South Carolina Clinical & Translational Research Institute—South Carolina	2009
Mount Sinai School of Medicine Mount Sinai Institutes for Clinical and Translational Sciences—New York	2009
New York U. School of Medicine (partnering with New York City Health and Hospitals Corporation (HHC)) NYU-HHC Clinical and Translational Science Institute—New York	2009
Northwestern U. Northwestern U. Clinical and Translational Sciences Institute—Illinois	2008
Oregon Health & Science U. Oregon Clinical and Translational Research Institute—Oregon	2006
Penn State Milton S. Hershey Medical Center Penn State Clinical and Translational Science Institute—Pennsylvania	2011
Stanford U. The Stanford Center for Clinical and Translational Education and Research—California	2008
The Ohio State U. The Ohio State U. Center for Clinical and Translational Science—Ohio	2008
The Rockefeller U. Rockefeller U. Center for Clinical and Translational Science—New York	2006
The Scripps Research Institute The Scripps Translational Science Institute—California	2008
The U. of Alabama at Birmingham UAB Center for Clinical and Translational Science—Alabama	2008
The U. of North Carolina at Chapel Hill North Carolina Translational and Clinical Sciences (TraCS) Institute—North Carolina	2008
The U. of Texas Health Science Center at San Antonio The U. of Texas Health Science Center at San Antonio Institute for Integration of Medicine and Science (IIMS)—Texas	2008
The U. of Utah U. of Utah Center for Clinical and Translational Science—Utah	2008
Tufts U. Tufts Clinical and Translational Science Institute—Massachusetts	2008
U. of Arkansas for Medical Sciences Arkansas Translational Research Institute—Arkansas	2009
U. of California Los Angeles UCLA Clinical and Translational Science Institute—California	2011
U. of California, Davis UC Davis Clinical and Translational Science Center—California	2006
U. of California, Irvine Institute for Clinical and Translational Science—California	2010
U. of California, San Diego San Diego Clinical and Translational Research Institute—California	2010
U. of California, San Francisco The UCSF Clinical and Translational Science Institute—California	2006
U. of Chicago The U. of Chicago Institute for Translational Medicine—Illinois	2007
U. of Cincinnati U. of Cincinnati Center for Clinical and Translational Science and Training—Ohio	2009

U. of Colorado Denver Colorado Clinical and Translational Sciences Institute—Colorado	2008
U. of Florida U. of Florida Clinical and Translational Science Institute—Florida	2009
U. of Illinois at Chicago UIC Center for Clinical and Translational Science—Illinois	2009
U. of Iowa U. of Iowa Institute for Clinical and Translational Science—Iowa	2007
U. of Kansas Medical Center Frontiers: The Heartland Institute for Clinical and Translational Research—Kansas	2011
U. of Kentucky Research Foundations Kentucky Center for Clinical and Translational Science—Kentucky	2011
U. of Massachusetts Medical School, Worcester U. of Massachusetts Center for Clinical and Translational Science—Massachusetts	2010
U. of Miami Miami Clinical and Translational Science Institute—Florida	2012
U. of Michigan Michigan Institute for Clinical and Health Research—Michigan	2007
U. of Minnesota Twin Cities U. of Minnesota Clinical and Translational Science Institute—Minnesota	2011
U. of New Mexico Health Sciences Center U. of New Mexico Clinical and Translational Science Center—New Mexico	2010
U. of Pennsylvania U. of Pennsylvania Institute for Translational Medicine and Therapeutics—Pennsylvania	2006
U. of Pittsburgh U. of Pittsburgh Clinical and Translational Science Institute—Pennsylvania	2006
U. of Rochester School of Medicine and Dentistry U. of Rochester Clinical and Translational Sciences Institute—New York	2006
U. of Southern California Southern California Clinical and Translational Science Institute—California	2010
U. of Texas Health Science Center at Houston U. of Texas Houston Center for Clinical and Translational Sciences—Texas	2006
U. of Texas Medical Branch UTMB Institute for Translational Sciences—Texas	2009
U. of Texas Southwestern Medical Center at Dallas UT Southwestern Center for Translational Medicine—Texas	2007
U. of Washington The Institute of Translational Health Sciences at the U. of Washington—Washington	2007
U. of Wisconsin—Madison U. Of Wisconsin Madison Institute for Clinical and Translational Research—Wisconsin	2007
Vanderbilt U. (partnering with Meharry Medical College) Vanderbilt Institute for Clinical and Translational Research—Tennessee	2007
Virginia Commonwealth U. Center for Clinical and Translational Research—Virginia	2010
Washington U. Washington U. Institute of Clinical and Translational Sciences—Missouri	2007
Weill Cornell Medical College (partnering with Hunter College) CTSA at Weill Cornell Medical College—New York	2007
Yale U. Yale Center for Clinical Investigation—Connecticut	2006

Note: No CTSA consortium award for any institution in these 19 states: Alaska, Arizona, Delaware, Hawaii, Idaho, Louisiana, Maine, Mississippi, Montana, Nebraska, Nevada, New Jersey, North Dakota, Oklahoma, Rhode Island, South Dakota, Vermont, West Virginia, Wyoming.

https://www.ctsacentral.org/institutions.

States has no such public and universal health-care system (with the exception of the systems for members of the military, veterans, or native Americans living on reservations), the CTSAs must rely on their own consortia of clinical entities, potentially limiting the generalizability of some of their scientific insights (Honey, 2007).

Investigators in Britain have sought to catalog their government's investments into biomedical research, including clinical and translational support. For example, immunology, vaccinology, and antimicrobial resistance investments have been summarized in peer-reviewed publications (Fitchett et al., 2013; Head et al., 2014).

CONCLUSION

The public sector investments into biomedical and biobehavioral research are small when weighed against global investments into arms, war, and defense. After the September 11, 2001 attack on the World Trade Center, world military spending in 2003 increased by about 11% in real terms (Sköns et al., 2004). This was preceded by an increase of 6.5% in 2002 such that world military spending was an estimated US$956 billion (in 2004 dollars) in 2003. Global military expenditure estimates are typically 2 years old when published; in 2012, over $1.7 trillion in annual expenditures were

TABLE 39.4 Biomedical Research Centres Funded by the UK National Health Service's National Institute for Health Research (NIHR) 2012–17 5-Year Cycle

Name	Name Including Host Organization and Academic Partner	Research Themes	5-Year Funding (2012–17)
NIHR Cambridge Biomedical Research Centre	NIHR Biomedical Research Centre at Cambridge University Hospitals NHS Foundation Trust and the University of Cambridge	Brain Injury and Repair; Cancer; Cardiovascular Disease; Dementia and Neurodegenerative Disorders; Genomics; Immunity, Infection and Inflammation; Mental Health; Metabolism, Endocrinology and Bone; Population Health Science; Transplantation and Regenerative Medicine; Women's Health	£110,073,288
NIHR Great Ormond Street Biomedical Research Centre	NIHR Biomedical Research Centre at Great Ormond Street Hospital for Children NHS Trust and University College London	Molecular Basis of Childhood Disease; Diagnostics and Imaging in Childhood Disease; Gene, Stem and Cellular Therapies; Novel Therapies for Childhood Diseases	£35,660,795
NIHR Guy's and St Thomas' Biomedical Research Centre	NIHR Biomedical Research Centre at Guy's and St Thomas' NHS Foundation Trust and King's College London	Cancer; Cardiovascular Disease; Cutaneous Medicine; Environment, Respiratory Health and Allergy; Imaging and Bioengineering; Infection and Immunity; Transplantation; Translational Genetics	£58,724,775
NIHR Imperial Biomedical Research Centre	NIHR Biomedical Research Centre at Imperial College Healthcare NHS Trust and Imperial College London	Biobanking; Cancer; Cardiovascular; Gastroenterology and Hepatology; Genetics and Genomics; Imaging; Hematology; Infection; Interventional Public Health; Neonatal Medicine; Neuroscience; Obesity, Diabetes, Endocrinology and Metabolism; Pediatrics; Renal Medicine and Transplantation; Respiratory; Rheumatology; Stratified Medicine; Surgery and Technology; Women's Health	£112,992,915
NIHR Moorfields Biomedical Research Centre	NIHR Biomedical Research Centre at Moorfields Eye Hospital NHS Foundation Trust and UCL Institute of Ophthalmology	Gene Therapy; Genotyping, Phenotyping and Informatics; Inflammation and Immunotherapy; New Technologies and Devices; Regenerative Medicine and Pharmaceutics; Visual Assessment and Imaging	£26,548,596
NIHR Newcastle Biomedical Research Centre	NIHR Biomedical Research Centre at Newcastle upon Tyne Hospitals NHS Foundation Trust and Newcastle University	The Ageing Body; The Ageing Brain; The Ageing Limbs; Genomics; Fibrosis & Repair; Mitochondria & Neuromuscular Disease	£16,636,125
NIHR Oxford Biomedical Research Centre	NIHR Biomedical Research Centre at Oxford University Hospitals NHS Trust and the University of Oxford	Biomedical Informatics and Technology; Blood; Cancer; Cardiovascular; Dementia and Cardiovascular Disease; Diabetes; Functional Neuroscience and Imaging; Genomic Medicine; Immunity and Inflammation; Infection; Prevention and Population Care; Surgical Innovation and Evaluation; Translational Physiology; Vaccines	£97,988,771
NIHR Royal Marsden Biomedical Research Centre	NIHR Biomedical Research Centre at The Royal Marsden NHS Foundation Trust and The Institute of Cancer Research	Breast Cancer; Cancer Genetics; Cancer Imaging; Cancer Therapeutics; Clinical Studies; Molecular Pathology; Prostate Cancer; Radiotherapy	£61,543,735

Continued

TABLE 39.4 Biomedical Research Centres Funded by the UK National Health Service's National Institute for Health Research (NIHR) 2012–17 5-Year Cycle—cont'd

Name	Name Including Host Organization and Academic Partner	Research Themes	5-Year Funding (2012–17)
NIHR Southampton Biomedical Research Centre	NIHR Biomedical Research Centre at University Hospitals Southampton NHS Foundation Trust and the University of Southampton	Nutrition, Growth and Development; Nutrition, Lifestyle and Healthy Ageing	£9,677,372
NIHR Maudsley Biomedical Research Centre	NIHR Biomedical Research Centre at the South London and Maudsley NHS Foundation Trust and King's College London	Bioinformatics and Statistics; BioResource, Biomarkers and Genomics; Clinical and Population Informatics; Developmental Disorders; Disorders of Affect and Addiction and Their Interface with Medicine; Neuroimaging; Neuropsychiatric Disorders; Patient and Carer Participation; Clinical Trials — Innovation and Implementation	£48,873,298
NIHR University College London Hospitals Biomedical Research Centre	NIHR Biomedical Research Centre at University College London Hospitals NHS Foundation Trust and University College London	Cancer; Cardiometabolic; Infection, Immunity and Inflammation; Neuroscience	£98,204,505

http://www.nihr.ac.uk/about/biomedical-research-centres.htm.

documented, having declined a tiny 0.5% compared to 2011, the first decline since 1998 (http://www.globalissues.org/article/75/world-military-spending#USMilitarySpending). The principal investor in global spending in defense and war was the United States, accounting for over $600 billion in military spending in 2014, about 35% of global military spending for only 4.4% of the world's population. Along with the near-collapse of the global marketplace in 2008 and subsequent austerity measures in many societies, the lack of growth in the budgets of the NIH, the MRC, and many comparable global research support agencies has caused clinical and translational research to lose ground, putting in peril young investigators competing for a shrinking research funding pool.

The diminishing magnitude of short-term discretionary spending by the world's leading biomedical research investors, be they governments or foundations, has led to a rise in the average age of grantees. As documented by the president of Johns Hopkins University for grants from the NIH, "The average age at which an investigator with a medical degree receives her first R01 or equivalent grant has risen from less than 38 y in 1980 to more than 45 y as of 2013. The number of principal investigators for R01s who are 36 y of age or younger has declined from 18% in 1983—3% in 2010. Today, more than twice as many R01s are awarded to principal investigators who are over 65 y as are under 36 y, a reversal from only 15 y ago. A similar decline can be seen if one looks beyond R01 s at all NIH research grants: the percent of all grant funding awarded to scientists under the age of 36 has dropped from 5.6% in 1980—1.3% in 2012 (Daniels, 2015)." The global research community welcomes the day that taxpayers all over the world can increase their support for biomedical and biobehavioral research, reaping a harvest from reduced defense spending, global financial reform, and improved global priorities for government investments (Box 39.9).

BOX 39.9 Problem Set

The following problem set reinforces key concepts, with answers provided.

1. The National Institutes of Health (NIH) in the United States provides the most clinical and translational research funding of any single entity worldwide. Describe briefly what research funding is available for non-US citizens.

   Answer: Nearly all unsolicited research grants (a so-called "R01") are openly competed and may be awarded to anyone deemed qualified with a highly competitive proposal from any legitimate organization that can support the work successfully. Some specifically reserved research support (identified in a so-called request for applications) may be reserved for US institutions, but even then, the investigator need not be a US citizen or permanent resident. However, a number of categories of grants are typically off-limits to non-US citizens or permanent residents, especially T32-type training grants and institutional capacity-building awards. The NIH has liberalized its pool of eligible investigators in past decades such that more support goes abroad than used to be the case.

2. The South African Medical Research Council is both a grant-giving and a grant-receiving institution in that it awards South African government funds but can also receive funds awarded by other entities such as the NIH or the Bill & Melinda Gates Foundation. Which source of funding was most prevalent in the 1990s? Is this true in 2015?

   Answer: The SAMRC budget was largely funded by the South African government in the 1990s, but now extramural funds from abroad represent a substantial majority of the overall funds used by the SAMRC today. This suggests the competitiveness of South African investigators and the research opportunities and challenges facing South Africa today.

3. The HIVNAT is an acronym referring to an HIV-related research partnership involving three nations. Comment on this program.

   Answer: The Netherlands and Australia have partnered with Thailand (hence the acronym's derivation) in HIV research based at the Thai Red Cross, adjacent to the Chulalongkorn University School of Medicine in Bangkok. Overseas research support and expertise combines with Thai talent, infrastructure, and more highly endemic HIV rates to make research possible that is of relevance to developing countries.

4. The National Institutes of Health doubled in its budget in the 1990s with broad support from both the principal US political parties. From 2003 to 2015, there was an abrupt halt in this rise in research support. Speculate why this might be and how it has affected biomedical research efforts supported by the NIH.

   Answer: Inflation continues to erode what the research dollar can buy, such that there has been a net decline in what the NIH supported in 2014 compared to what it supported in 2003. This has been devastating for career development of many junior investigators who were trained in the late 20th/early 21st century to fill the acknowledged gap in clinical and translational research. It is speculated by many that the Iraq War, also begun in 2003, the Afghan war (longest war in US history) along with historically high US budget deficits, tax cuts for higher income Americans, and the threat of recession have contributed to the loss of bipartisan support for raising the budget of the NIH to meet or exceed the rate of inflation.

5. Define the difference between intramural and extramural funding.

   Answer: An agency's intramural funding is available to the employees of that agency alone, whereas its extramural funding is available to persons soliciting these funds from an outside institution.

REFERENCES

Achhra, A.C., Amin, J., Law, M.G., et al., 2010. Immunodeficiency and the risk of serious clinical endpoints in a well studied cohort of treated HIV-infected patients. AIDS 24, 1877–1886.

Alpers, M.P., 2003. Hospital twinning between Australia and Papua New Guinea. P. N. G. Med. J. 46, 81–86.

Ananworanich, J., Puthanakit, T., Suntarattiwong, P., et al., 2014. Reduced markers of HIV persistence and restricted HIV-specific immune responses after early antiretroviral therapy in children. AIDS 28, 1015–1020.

Anon., 2000. Risk adjusted and population-based studies of the outcome for high risk infants in Scotland and Australia. International Neonatal Network, Scottish Neonatal Consultants, Nurses Collaborative Study Group. Arch. Dis. Child. Fetal Neonatal Ed. 82 (2), F118–F123.

Arroll, B., Beilby, J., 2004. Gathering meaningful outcomes in interventional trials in general practice. Primary Care Alliance for Clinical Trials (PACT) network. Aust. Fam. Physician 33, 659, 662.

Boettiger, D.C., Kerr, S., Ditangco, R., et al., 2014. Trends in first line antiretroviral therapy in Asia: results from the TREAT Asia HIV observational database. PLoS One 9, e106525.

Brentlinger, P.E., Silva, W.P., Buene, M., et al., 2014. Management of fever in ambulatory HIV-infected adults in resource-limited settings: prospective observational evaluation of a new Mozambican guideline. J. Acquir. Immune Defic. Syndr. 67, 304–309.

Campbell, E.G., 2009. The future of research funding in academic medicine. N. Engl. J. Med. 360, 1482–1483.

Cardiello, P.G., van Heeswijk, R.P., Hassink, E.A., et al., 2002. Simplifying protease inhibitor therapy with once-daily dosing of saquinavir soft-gelatin capsules/ritonavir (1600/100 mg): HIVNAT 001.3 study. J. Acquir. Immune Defic. Syndr. 29, 464–470.

Charbonneau, L., 2000. MRC changes its name, doubles its budget. CMAJ 162 (7), 1029.

Ciampa, P.J., Tique, J.A., Jumá, N., et al., 2012. Addressing poor retention of infants exposed to HIV: a quality improvement study in rural Mozambique. J. Acquir. Immune Defic. Syndr. 60, e46–e52.

Clark, D.R., McGrath, P.J., MacDonald, N., 2007. Members' of Parliament knowledge of and attitudes toward health research and funding. CMAJ 177, 1045–1051.

Cori, A., Ayles, H., Beyers, N., et al., 2014. HPTN 071 (PopART): a cluster-randomized trial of the population impact of an HIV combination prevention intervention including universal testing and treatment: mathematical model. PLoS One 9 (1), e84511.

Cripe, T.P., Thomson, B., Boat, T.F., Williams, D.A., 2005. Promoting translational research in Academic Health Centers: navigating the 'roadmap'. Acad. Med. 80, 1012–1018.

Crowley Jr., W.F., Sherwood, L., Salber, P., et al., 2004. Clinical research in the United States at a crossroads: proposal for a novel public–private partnership to establish a national clinical research enterprise. JAMA 291, 1120–1126.

Dainty, K.N., Scales, D.C., Brooks, S.C., et al., 2011. A knowledge translation collaborative to improve the use of therapeutic hypothermia in post-cardiac arrest patients: protocol for a stepped wedge randomized trial. Implement. Sci. 6, 4.

Daniels, R.J., 2015. A generation at risk: young investigators and the future of the biomedical workforce. Proc. Natl. Acad. Sci. U.S.A. 112, 313–318.

Darlow, B.A., Hutchinson, J.L., Simpson, J.M., Henderson-Smart, D.J., Donoghue, D.A., Evans, N.J., 2005. Variation in rates of severe retinopathy of prematurity among neonatal intensive care units in the Australian and New Zealand Neonatal Network. Br. J. Ophthalmol. 89, 1592–1596.

Demotes-Mainard, J., Canet, E., Segard, L., 2006. Public–private partnership models in France and in Europe. Therapie 61 (325–334), 313–323.

Eichbaum, Q., Bowa, K., Pires, P., et al., 2014. Challenges and opportunities for new medical schools in Africa: the Consortium of New Southern African Medical Schools (CONSAMS). Acad. Med. 89 (8 Suppl.), S108–S109.

Fitchett, J.R., Head, M.G., Atun, R., 2013. Infectious disease research investments: systematic analysis of immunology and vaccine research funding in the UK. Vaccine 31, 5930–5933.

Frehywot, S., Mullan, F., Vovides, Y., et al., 2014. Building communities of practice: MEPI creates a commons. Acad. Med. 89 (8 Suppl.), S45–S49.

Froom, J., Culpepper, L., Grob, P., et al., 1990. Diagnosis and antibiotic treatment of acute otitis media: report from International Primary Care Network. BMJ 300, 582–586.

Fulu, E., Jewkes, R., Roselli, T., et al., 2013. Prevalence of and factors associated with male perpetration of intimate partner violence: findings from the UN Multi-country Cross-sectional Study on Men and Violence in Asia and the Pacific. Lancet Glob. Health 1, e187–207.

Gaspar, R., Aksu, B., Cuine, A., 2012. Towards a European strategy for medicines research (2014–2020): the EUFEPS position paper on Horizon 2020. Eur. J. Pharm. Sci. 47, 979–987.

Goldman, M., Compton, C., Mittleman, B.B., 2013. Public–private partnerships as driving forces in the quest for innovative medicines. Clin. Transl. Med. 2, 2.

Green, S.E., Bosch, M., McKenzie, J.E., et al., 2012. Improving the care of people with traumatic brain injury through the Neurotrauma Evidence Translation (NET) program: protocol for a program of research. Implement. Sci. 7, 74.

Hachinski, V., Iadecola, C., Petersen, R.C., et al., 2006. National Institute of Neurological Disorders and Stroke–Canadian Stroke Network vascular cognitive impairment harmonization standards. Stroke 37, 2220–2241.

Hait, W.N., 2005. Translating research into clinical practice: deliberations from the American Association for Cancer Research. Clin. Cancer Res. 11, 4275–4277.

Hayes, R., Ayles, H., Beyers, N., et al., 2014. HPTN 071 (PopART): rationale and design of a cluster randomised trial of the population impact of an HIV combination prevention intervention including universal testing and treatment—a study protocol for a cluster randomized trial. Trials 15, 57.

Head, M.G., Fitchett, J.R., Cooke, M.K., et al., 2014. Systematic analysis of funding awarded for antimicrobial resistance research to institutions in the UK, 1997–2010. J. Antimicrob. Chemother. 69, 548–554.

Heller, R., O'Connell, R., Lim, L., et al., 1999. Variation in in-patient stroke management in ten centres in different countries: the INCLEN multicentre stroke collaboration. J. Neurol. Sci. 167, 11—15.

Honey, K., 2007. Translating medical science around the world. J. Clin. Invest. 117, 2737.

Howard, L.M., Tique, J.A., Gaveta, S., et al., 2014. Health literacy predicts pediatric dosing accuracy for liquid zidovudine. AIDS 28, 1041—1048.

Hoyos, C.M., Killick, R., Keenan, D.M., et al., 2014. Continuous positive airway pressure increases pulsatile growth hormone secretion and circulating insulin-like growth factor-1 in a time-dependent manner in men with obstructive sleep apnea: a randomized sham-controlled study. Sleep 37, 733—741.

Jewkes, R., Fulu, E., Roselli, T., et al., 2013. Prevalence of and factors associated with non-partner rape perpetration: findings from the UN multi-country cross-sectional study on men and violence in Asia and the Pacific. Lancet Glob. Health 1, e208—218.

Johnson, D.W., Badve, S.V., Pascoe, E.M., et al., 2014. Antibacterial honey for the prevention of peritoneal-dialysis-related infections (HONEYPOT): a randomized trial. Lancet Infect. Dis. 14, 23—30.

Jull, A., Wills, M., Scoggins, B., Rodgers, A., 2005. Clinical trials in New Zealand — treading water in the knowledge wave? N. Z. Med. J. 118 (1221), U1638.

Kahn, S.R., Shapiro, S., Wells, P.S., et al., 2014. Compression stockings to prevent post-thrombotic syndrome: a randomised placebo-controlled trial. Lancet 383, 880—888.

Loscalzo, J., 2006. The NIH budget and the future of biomedical research. N. Engl. J. Med. 354, 1665—1667.

Lynfield, R., Davey, R., Dwyer, D.E., et al., 2014. Outcomes of influenza A(H1N1)pdm09 virus infection: results from two international cohort studies. PLoS One 9, e101785.

Mandala, W.L., Cowan, F.M., Lalloo, D.G., et al., 2014. Southern Africa consortium for research excellence (SACORE): successes and challenges. Lancet Glob. Health 2, e691—e692.

Matherlee, K.R., 1995. The outlook for clinical research: impacts of federal funding restraint and private sector reconfiguration. Acad. Med. 70, 1065—1072.

Ménard, D., Maïro, A., Mandeng, M.J., et al., 2005. Evaluation of rapid HIV testing strategies in under equipped laboratories in the Central African Republic. J. Virol. Methods 126, 75—80.

Montaner, J.S., Reiss, P., Cooper, D., et al., 1998. A randomized, double-blind trial comparing combinations of nevirapine, didanosine, and zidovudine for HIV-infected patients: the INCAS Trial. Italy, The Netherlands, Canada and Australia Study. JAMA 279, 930—937.

Moon, T.D., Silva, W.P., Buene, M., et al., 2013. Bacteremia as a cause of fever in ambulatory, HIV-infected Mozambican adults: results and policy implications from a prospective observational study. PLoS One 8, e83591.

Moses IIIrd, H., Dorsey, E.R., Matheson, D.H., Their, S.O., 2005. Financial anatomy of biomedical research. JAMA 294, 1333—1342.

Moses 3rd, H., Dorsey, E.R., 2012. Biomedical research in an age of austerity. JAMA 308, 2341—2342.

Nash, D., Wu, Y., Elul, B., Hoos, D., El Sadr, W., International Center for AIDS Care and Treatment Programs, 2011. Program-level and contextual-level determinants of low-median CD4$^+$ cell count in cohorts of persons initiating ART in eight sub-Saharan African countries. AIDS 25, 1523—1533.

Nichol, G., Steen, P., Herlitz, J., et al., 2005. The International Resuscitation Network Investigators. International Resuscitation Network Registry: design, rationale and preliminary results. Resuscitation 65, 265—277.

Nordling, L., 2014. Changing Council. Nat. Med. 20, 113—116.

Oh, T.G., Bailey, P., Dray, E., et al., 2014. PRMT2 and RORγ expression are associated with breast cancer survival outcomes. Mol. Endocrinol. 28, 1166—1185.

Oman, K.M., Baravilala, W., Sapuri, M., Hays, R., 2002. The tropical triangle: a health education alliance for the Southwest Pacific. Educ. Health 15, 346—352.

Owens, B., 2014. Mapping biomedical research in the USA. Lancet 384, 11—14.

Paige, C.J., 2007. The future of health research is hanging in the balance. CMAJ 177, 1057—1058.

Perez, G., MacArthur, R.D., Walmsley, S., Baxter, J.A., Mullin, C., Neaton, J.D., Terry Beirn Community Programs for Clinical Research on AIDS; Canadian Trials Network, 2004. A randomized clinical trial comparing nelfinavir and ritonavir in patients with advanced HIV disease (CPCRA 042/CTN 102). HIV Clin. Trials 5, 7—18.

Purvis, T., Moss, K., Denisenko, S., et al., 2014. Implementation of evidence-based stroke care: enablers, barriers, and the role of facilitators. J. Multidiscip. Healthc. 7, 389—400.

Romero, K., de Mars, M., Frank, D., et al., 2009. The Coalition against Major Diseases: developing tools for an integrated drug development process for Alzheimer's and Parkinson's diseases. Clin. Pharmacol. Ther. 86, 365—367.

Sanchez, J.L., Johns, M.C., Burke, R.L., et al., 2011. Capacity-building efforts by the AFHSC-GEIS program. BMC Public Health 11 (Suppl. 2), S4.

Sautter, J., Olesen, O.F., Bray, J., Draghia-Akli, R., 2011. European Union vaccine research—an overview. Vaccine 29, 6723—6727.

Sköns, E., Perdomo, C., Perlo-Freeman, S., Stålenheim, P., 2004. Military expenditure. Chapter 10. In: SIPRI Yearbook 2004: Armaments, Disarmament and International Security (Stockholm International Peace Research Institute). Oxford University Press, London.

van Ommen, G.J., Törnwall, O., Bréchot, C., et al., 2014. BBMRI-ERIC as a resource for pharmaceutical and life science industries: the development of biobank-based Expert Centres. Eur. J. Hum. Genet. http://dx.doi.org/10.1038/ejhg.2014.235 (19 November 2014 Epub ahead of print).

Vermund, S.H., Fidler, S.J., Ayles, H., Beyers, N., Hayes, R.J., 2013. Can combination prevention strategies reduce HIV transmission in generalized epidemic settings in Africa? the HPTN 071 (PopART) study plan in South Africa and Zambia. J. Acquir. Immune Defic. Syndr. 63 (Suppl. 2), S221—S227.

Zerhouni, E.A., 2006. Clinical research at a crossroads: the NIH roadmap. J. Invest. Med. 54 (4), 171—173.

Zerhouni, E.A., 2007. Translational research: moving discovery to practice. Clin. Pharmacol. Ther. 81 (1), 126—128.

BIBLIOGRAPHY

Internet Resources

California HIV/AIDS research program (University of California Research Grants Program Office) supporting AIDS research at nonprofit institutions within California. http://www.californiaaidsresearch.org/.

Center for Epidemiological Studies on HIV/AIDS of Catalonia (CEESCAT): http://www.ceeiscat.cat/.

Centers for Disease Control and Prevention: http://www.cdc.gov.

Comprehensive International Program of Research on AIDS (CIPRA): http://archive.hhs.gov/news/press/2004pres/20041118.html.

European Commission Research: http://ec.europa.eu/research/index.cfm.

European Commission Research: STAR Projects: http://ec.europa.eu/research/star/home.html.

Foreign Institution Grantees of NIH in Fiscal Year 2015: http://report.nih.gov/award/index.cfm?ot=&fy=2015&state=Foreign&ic=&fm=&orgid=&distr=&rfa=&om=n&pid=&view=statedetail.

Health Research Council of New Zealand: About Maori Health Research: http://www.hrc.govt.nz/news-and-publications/publications/maori.

Health Research Council of New Zealand: http://www.hrc.govt.nz/.

HORIZON 2020 — The EU Framework Programme for Research and Innovation: http://ec.europa.eu/programmes/horizon2020/.

INSIGHT (International Network for Strategic Initiatives in Global HIV Trials): http://insight.ccbr.umn.edu/index.php.

Institut Pasteur (France): http://www.pasteur.fr/en.

Karolinska Institute (Sweden): http://ki.se/en/ki/startpage-kise.

Medical Research Council (MRC) in the UK: http://www.mrc.ac.uk/index.htm.

Microbicide Trials Network: MTN-003: Vaginal and Oral Interventions to Control the Epidemic [VOICE]: http://www.mtnstopshiv.org/node/70.

National Institute for Health Research: Biomedical Research Centres: http://www.nihr.ac.uk/about/biomedical-research-centres.htm.

National Institutes of Health: http://www.nih.gov.

National Priorities Project: http://www.nationalpriorities.org/costofwar_ home.

NHMRC Clinical Trials Centre, University of Sydney: http://www.ctc.usyd.edu.au/.

Robert Koch Institute (Germany): http://www.rki.de/EN/Home/homepage_node.html.

South African Medical Research Council: Publications: http://www.mrc.ac.za/annualreport/annual.html (2007).

Uniformed Services University of the Health Sciences. http://www.usuhs.mil/.

University of Auckland Clinical Trials Research Unit: http://nihi.auckland.ac.nz/.

Chapter 40

The Role of Nonprofit, Nongovernmental Funding in Support of Biomedical Research

Rose Marie Robertson and Suzie Upton

American Heart Association, Dallas, TX, United States

Chapter Outline

Key Points

- The dependence of Federal research funding on unpredictable governmental budgets and the perceived conflicts of interest created by industry funding can create problems for investigators.
- Funding from nonprofit philanthropic sources avoid these problems and should become familiar to all investigators early in their careers.
- Working with philanthropic organizations or individuals requires understanding their specific goals and a decision about their fit with the aims of your research.
- Fostering continuing research funding by philanthropic sources depends on appropriate stewardship by investigators, an essential research skill.

INTRODUCTION

The preponderance of funding for biomedical research in the United States is provided by the Federal government, including the NIH, CDC-P, HRS, AHRQ, FDA, and the sometimes overlooked DOD, as covered in Chapter 39, or by the pharmaceutical, device, and biotech industry, covered in Chapter 42. Obtaining Federal funding is considered an important marker of academic success, but at the present time, the investment of the United States government in biomedical research continues to be subject to the annual decisions of Congress and can fluctuate considerably in amount and availability (with periods of enthusiasm for investment in research alternating with increasingly frequent periods of government shutdown or budget sequestration). There is thus both safety and flexibility in being knowledgeable about the alternative sources such as industry. However, this should not be the only alternative focus of your efforts to seek support for your research, since relationships with industry can limit your participation in other professional activities, e.g., serving on writing groups for

Clinical and Translational Science. http://dx.doi.org/10.1016/B978-0-12-802101-9.00040-5

some clinical practice guidelines. An increasingly important additional resource you should be sure to become familiar with lies in the private, nonprofit funding of research, based on the support of independent foundations (whether individual or family, corporate or government-related), voluntary health organizations (VHOs), and gifts from individuals. These resources can be an important supplement even to careers primarily supported by Federal funding. They also provide advantages that will be outlined in this chapter, both for the individual investigator and for their institution. In this chapter, we will review the various forms of private philanthropic funding of research and discuss the advantages they offer as well as some of the specific initiatives they have undertaken. Suggestions for interacting with these organizations are included.

OVERVIEW OF PHILANTHROPIC FUNDING ON BIOMEDICAL RESEARCH

Philanthropy has contributed more to the biomedical research enterprise in the United States than most these days would guess. In fact, it has been crucial to both infrastructure and to medical advances that we now take for granted. Hospitals that care for those without the means to pay have been founded by religious orders and by individuals, often businessmen, since the 1700s, and many of our leading medical schools and research institutes bear the names of individuals or families that provided initial or ongoing support. Seminal philanthropic funding has catalyzed the development of therapies such as penicillin, insulin, the polio vaccine, heart transplants, renal dialysis, and much of today's success against cancer. Donors have endowed professorships and created laboratories, and philanthropy has also been essential for the translation of improved health measures out into communities—from the Rockefeller Foundation's critical campaign against yellow fever, through the substantial funding provided by the Robert Wood Johnson Foundation ($700 million over 19 years) to reduce smoking in the United States (http://www.rwjf.org/content/dam/web-assets/ 2011/04/tobacco-control-work–1991-2010-doi 3/26/2016; RWJF Retrospective Series: Tobacco-Control Work, 1991–2010), to the Clinton Foundation work against HIV/AIDS in Africa and against childhood obesity in the United States and the Gates Foundation's battles against malaria, leprosy, polio, and other diseases insufficiently addressed worldwide. Despite these important contributions, the growth of research both within and supported externally by industry, as well as government funding, has led in the United States to an environment in which private philanthropy now comprises only a small portion of funding for medical research and public health. However, this funding can be flexible, risk-tolerant, nimble, and offered in most cases with far less bureaucracy than government grants. Because of this, philanthropic funding is highly sought after by biomedical researchers today and continues to have an important impact.

In recent years, this impact has been increasing because of external forces threatening the contributions of the other sectors of biomedical research funding in the United States. As recently as 2015, an analysis of medical research funding in the United States raised cautions. While the research environment in the United States continues to be envied in many other parts of the world, leading to a steady stream of young investigators seeking positions in US labs and clinical research enterprises, other nations such as Germany, Japan, and China have demonstrated a strong commitment to research, and their scientists, both basic and clinical, are increasingly represented in international forums. Moses et al. pointed out that in terms of global biomedical research, the US share of provided funding has fallen from 57% to 44% over just the past decade. And while medical research spending in China is still proportionally much less than that in the United States, it has increased rapidly over the last 7 years for which records are available (2004–11), growing at 17% per year, while in the United States in the same time period it has increased at 1% per year, less than the rate of biomedical inflation (Fig. 40.1).

Giving USA has chronicled philanthropy in American society over the past 60 years, noting the impact of economic downturns on giving, but also pointing out especially the environment that has existed since the late 1960s, when it was recognized that foundations had the potential to direct significant resources at multiple social problems, including education and health. And while the concept of leaving a legacy via bequests was not new, bequests began to rise in the 1970s and remain important. Over the last half-century, the public's appreciation of medical research has grown, as has the understanding that individual philanthropy can have a real impact on discovery. Movies such as *Lorenzo's Oil* in 1992 have raised expectations, and the courageous families of children with rare disorders continue to push for attention, via support groups, individual fund-raising, or through advocacy. The recent New York Times coverage of the Hempel family and their struggle with Niemann-Pick I is an educational example.

While health care currently is responsible for ~17% of the US economy, and the system is generally regarded as expensive, inefficient, and fraught with inequities, the funding for health services research has only recently begun to grow. And while many areas or research funding increased from 1996 to 2006, the situation between then and 2012 has been dismal, not keeping up with biomedical inflation. The overall reduction in the economy following 2008 had an important

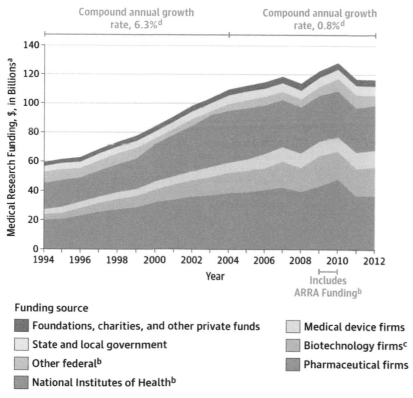

Funding source

- ◼ Foundations, charities, and other private funds
- ◻ State and local government
- ◻ Other federal[b]
- ◼ National Institutes of Health[b]
- ◻ Medical device firms
- ◼ Biotechnology firms[c]
- ◼ Pharmaceutical firms

Data were calculated according to methods outlined in eTable 1 in the Supplement.
ARRA indicates American Recovery and Reinvestment Act.

a -data adjusted to 2012 dollars using the Biomedical Research and Development Price Index
b -NIH and Other Federal sources include funds from ARRA in 2008 and 2009. ARRA indicates American Recovery and Reinvestment Act
c -data from 1994-2002 and 2011-2012 were estimated based on linear regression analysis of industry market share
d -compound annual growth rate (CAGR) supposing that year A is x and year B is y. CAGR =(y/x)(1/(B-A))-1. The CAGR was calculated separately for two different periods with a single overlapping year: 1994-2004 and 2004-2012. The cut point was chosen at 2004 given the changes seen in funding from the NIH in that year.

FIGURE 40.1 US funding for medical research by source 1994−2012. *ARRA*, American Recovery and Reinvestment Act. Data for the NIH and other Federal sources includes ARRA funds in 2009 and 2010. *Moses, H., Matheson, D.H.M., Cairns-Smith, S., George, B.P., Palisch, C., Dorsey, E.R, 2015. Scientific discovery and the future of medicine: the Anatomy of Medical Research − US and International Comparisons. JAMA 313 (2), 174−189.*

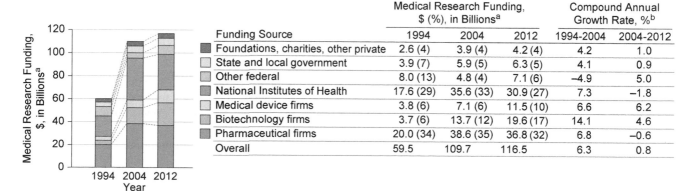

Funding Source	Medical Research Funding, $ (%), in Billions[a]			Compound Annual Growth Rate, %[b]	
	1994	2004	2012	1994-2004	2004-2012
◼ Foundations, charities, other private	2.6 (4)	3.9 (4)	4.2 (4)	4.2	1.0
◻ State and local government	3.9 (7)	5.9 (5)	6.3 (5)	4.1	0.9
◻ Other federal	8.0 (13)	4.8 (4)	7.1 (6)	−4.9	5.0
◼ National Institutes of Health	17.6 (29)	35.6 (33)	30.9 (27)	7.3	−1.8
◻ Medical device firms	3.8 (6)	7.1 (6)	11.5 (10)	6.6	6.2
◻ Biotechnology firms	3.7 (6)	13.7 (12)	19.6 (17)	14.1	4.6
◻ Pharmaceutical firms	20.0 (34)	38.6 (35)	36.8 (32)	6.8	−0.6
Overall	59.5	109.7	116.5	6.3	0.8

FIGURE 40.2 Growth in US funding for medical research by source 1994−2012. American Medical Association.

negative impact, and recovery has been slow for science, despite the modest impact of American Recovery and Reinvestment Act funding. While nonprofit, nongovernmental sources cannot make up for a fall in Federal sources, they have contributed to the recovery of scientific funding, as have increased funding from biotechnology and the medical device industry, at least in their specific areas (Fig. 40.2).

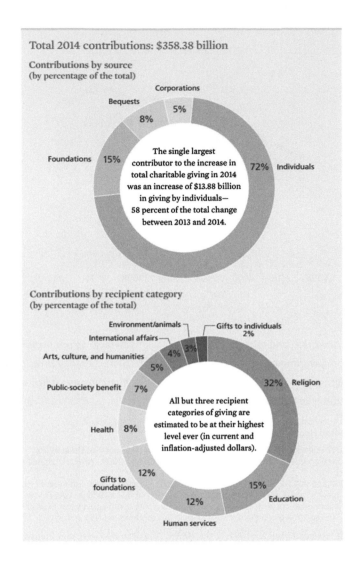

DISTINCTIONS BETWEEN DIFFERENT TYPES OF PHILANTHROPIC FUNDERS

There are five main categories of nongovernmental, nonprofit funders of biomedical research, differing in their sources of funding, the basis for their funding decisions, and their other activities.

- Voluntary health agency/public charity—Funds are derived from public contributions; research is aimed at diagnosing, preventing, treating, or curing specific disease entities; may engage in advocacy to increase Federal funding for research and to encourage policies that improve access to health care. Examples: American Cancer Society, American Heart Association, Multiple Myeloma Research Foundation
- Independent foundation—Endowed by individual(s), family, or company; has governing donor or board of trust; cannot lobby. Examples: Bloomberg Family Foundation; Burroughs Wellcome Fund, Henrietta B. and Frederick H. Bugher Foundation, Robert Wood Johnson Foundation.
- Individual philanthropist—Donates funds privately to institution, investigator, or other agency.
- Operating foundation—Endowed; conducts its own operations rather than providing outside grants/awards. Example: Howard Hughes Medical Institute.
- Community foundation—Supported by multiple local donors; usually funds research in a specific geographic area. Example: Silicon Valley Community Foundation.
- Corporate foundation—Independent from the corporation, but the mission is often closely related to the corporation or corporation's founders. Examples: Avon Foundation, The Merck Company Foundation, the Medtronic Foundation.

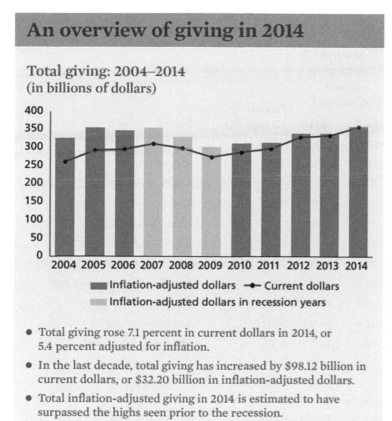

An overview of giving in 2014

Giving USA, 2015. Tables researched and written by Lilly's Family School of Philanthropy Indiana University.

Total giving: 2004–2014
(in billions of dollars)

Inflation-adjusted dollars — Current dollars
Inflation-adjusted dollars in recession years

- Total giving rose 7.1 percent in current dollars in 2014, or 5.4 percent adjusted for inflation.

- In the last decade, total giving has increased by $98.12 billion in current dollars, or $32.20 billion in inflation-adjusted dollars.

- Total inflation-adjusted giving in 2014 is estimated to have surpassed the highs seen prior to the recession.

Voluntary Health Agencies

There are hundreds of voluntary health agencies (VHAs), generally focused on a specific disease or syndrome (e.g., the Multiple Myeloma Research Foundation, the Marfan's Foundation) or an area of medicine (e.g., the American Cancer Society, the American Heart Association, the American Lung Association). Many of these have been continuously active for many decades, and some for more than 100 years, but new organizations continue to be founded by dedicated physicians and other health-care providers, dedicated laypeople, or a combination of both. The National Health Council (http://www.nationalhealthcouncil.org) provides a home for 40–50 of these organizations at any given time, along with other types of health-related organizations, to share ideas about research funding and to sometimes join together on advocacy issues. VHAs are dependent on the charitable contributions of the public for their funding, and a number of them enlist tens of millions of donors/supporters annually and have supported more than $100 million of research grants annually for many years. Even with their broad source of support, of course, there is considerable dependence on the overall state of the economy, and giving per se may not rebound as quickly as other aspects of an individual's or family's finances.

VHAs are committed to the service of those with specific disorders and are generally interested in a broad range of activities, including raising patient and health-care provider awareness, supporting, advocating for policies that are of benefit for prevention activities for the disorders they focus on, implementing quality improvement programs in health care, and in some cases providing direct patient services. Many VHAs are strongly convinced that research is the most effective use of the funds they raise, or at least very high on their list, and are often interested in supporting relevant research in all areas, including basic, clinical, and population levels, and all translational types as well. They may support studies of all types of therapies, from lifestyle through pharmaceutical, surgical, or other interventional approaches, as well as recovery or rehabilitation. While large clinical trials are beyond the financial scope of most, several have devoted considerable attention to supporting early phase trials and can be of great help in recruiting patients with uncommon disorders.

Importantly, funding from these organizations does not constitute a "relationship with industry," nor is it considered to produce a "conflict of interest." Therefore, receiving this funding does not limit your participation in leadership positions or guideline or scientific statement writing groups. Table 40.1 lists the research expenditures for 12 of the largest VHOs and

TABLE 40.1 The Research Grant Expenditures of the Largest Voluntary Health Organizations

Organizations	Research Grant Expenditures (In Millions)		Percent Change (%)	Source
	2011	2014		
American Heart Association (Dallas, TX)	120.7	133.3	+10	http://www.heart.org/
Juvenile Diabetes Research Foundation (New York, NY)	116[a]	119.6	+3	http://jdrf.org/
American Cancer Society (Atlanta, GA)	108.7	99	−8.9	http://www.cancer.org
Leukemia and Lymphoma Society (White Plains, NY)	71.9	79.8	+11	https://www.lls.org
Cystic Fibrosis Foundation (Bethesda, MD)	73.2	45.08	−38[b]	https://www.cff.org
American Diabetes Association (Alexandria, VA)	33.5	29.2	−13	http://main.diabetes.org/
Natural Multiple Sclerosis Society (New York, NY)	40.2	48.4	+20	http://www.nationalmssociety.org/
March of Dimes (White Plains, NY)	30.5	31.9	+5	http://www.marchofdimes.org/
Muscular Dystrophy Association (Chicago, IL)	38.1	18.4	−51	https://www.mda.org/
Alzheimer's Association (Chicago, IL)	24.9	29.1	+16	http://www.alz.org/
Arthritis Foundation (Atlanta, GA)	14.2	10	−30	http://www.arthritis.org/
American Lung Association (New York, NY)	5.8	9[c]	+55	http://www.lung.org/

[a]JDRF foundation figure from 2010.
[b]Specific figures on medical research unknown as figures contain expenditures on patient care programs.
[c]2013–14 figure.

the increases or decreases over the past 3–4 years (the most recent accessible data). In general, funds available to this area of philanthropy, primarily from their large numbers of individual donors, have continued to increase as the economy has gradually recovered.

These organizations generally have great freedom to develop novel approaches in research structures and can often be nimble in shifting focus to new areas or to undertaking radically different aims. For example, the Leukemia and Lymphoma Society shifted its focus in the past from simply funding the best of all investigator-initiated proposals that were submitted to it, to inviting Translational Research Program proposals for potential drug targets that could have a specific beneficial impact on disease as well as inviting New Idea proposals for potentially transformative concepts for which there are no preliminary data available yet. The American Heart Association, in addition to its long-standing portfolio of investigator-initiated grants, has added Strategically Focused Research Networks in a number of areas relevant to cardiovascular diseases, including stroke, such as Prevention, Hypertension, Women and Heart Disease and Disparities, with others to follow, and has also mounted a one-time, paradigm-shifting effort to recruit and fund a leader and a research team (not limited to type of investigator, belonging to the usual type of research institution, or type of research) to "cure coronary heart disease," that is, to prevent and/or reverse coronary heart disease and its consequences. Branded as One Brave Idea, this approach is a fundamentally different kind of model for funding innovation—it is supported by a collaborative effort of the American Heart Association, Verily (formerly Google Life Sciences), and AstraZeneca for a total $75 million over at least 5 years, and it is designed to remove the barriers and silos that plague traditional research projects.

There are additional benefits to working with a VHO that focuses on diseases of interest to you. Many of these organizations are gathering places for investigators whose research interests may complement your own and who are often willing to serve as mentors or collaborators from outside your own institution. The organizations may also undertake direct or indirect services to patients, programs that screen for disease, guide prevention efforts, or even assist with disease management. Some write clinical practice guidelines and derive and publish performance measures, and some directly guide or provide quality improvement activities for practice in hospitals or practices. VHOs can and do lobby and/or advocate efforts in support of patients and in support of Federal research and can provide an opportunity to learn more about the policy implications of your research or the policies that need to be changed to benefit patients who suffer with the disorder. Local state, city, and even school district advocacy is also undertaken.

Many VHOs have a strong commitment to the development and/or maintenance of an investigative workforce in their disease area(s). For some, especially those with a long view of the time horizon in which they can impact the diseases they "own," the support of early-career individuals is an important internal value and a major proportion of their funds may be directed to this population.

Because VHOs often include laypersons who either suffer with or are close to those who suffer from a disease or disorder, they offer you the opportunity to connect with and even be a champion for those with either common or rare diseases. This can bring access to information that allows you to be truly patient-focused and to study those aspects that are most central to the burden of the disease. Commitment to such organizations over time may allow your exposure to patient groups willing to provide essential samples and to participate in clinical trials.

Independent Foundations

Independent foundations are the largest funders in terms of assets. They include family foundations, such as the Bill & Melinda Gates Foundation, the Robert Wood Johnson Foundation, the Milken Family Foundation, and the Eli & Edythe Broad Foundation (and others such as those listed in Table 40.2), established by bequests or by endowments from living donors and run by family members or designees, as well as foundations that can be formed when nonprofit entities such as health plans or hospitals are converted to for-profit status. While some founders of foundations such as the Gates Foundation and Milken Foundation have derived their wealth from corporate activities in other than the health or medical industry, the founders' interest in improving health has led them to make important contributions to research in the United States and worldwide.

Community (e.g., The Cleveland Foundation), corporate, and operating foundations (such as the Howard Hughes Medical Institute) comprise a much smaller proportion of this category.

TABLE 40.2 Grant Expenditures for the 15 Largest Foundation Funders of Medical Research, 2011

Agency	Research Grants Expenditure (In Millions)[a]		Percent Change (%)	Website Address
	2009	2011		
Bill & Melinda Gates Foundation	544.8	838.7	35.0	www.gatesfoundation.org
The Starr Foundation	50.2	58.7	14.5	www.starrfoundation.org
The Simons Foundation	13.8	48.2	71.4	www.simonsfoundation.org
The Leona M. and Harry B. Helmsley Charitable Trust	No data	40.4		helmsleytrust.org
Eli & Edythe Broad Foundation	35.1	31.2	−12.5	www.broadfoundation.org
Gary and Mary West Foundation	No data	21.0		www.gmwf.org
Burroughs Wellcome Fund	21.0	17.7	−18.6	www.bwfund.org
Wallace H. Coulter Foundation	20.0	14.8	−35.1	www.whcf.org
Doris Duke Charitable Foundation	8.3	14.7	43.5	www.ddcf.org
W. M. Keck Foundation	9.4	13.3	29.3	www.wmkeck.org
The G. Harold & Leila Y. Mathers Charitable Foundation	10.4	12.0	13.3	www.mathersfoundation.org
The William K. Warren Foundation	8.8	11.6	24.1	www.williamkwarrenfoundation.org
The Charles A. Dana Foundation, Inc.	7.8	10.2	23.5	www.dana.org
Richard M. Schulze Family Foundation	No data	10.1		www.schulzefamilyfoundation.org
The Hsieh Family Foundation	No data	10.0		N/A
Total:	729.6	1909.8[b]	61.8	

[a]© The Foundation Center's Statistical Information Service, 2011. Table includes all grants to recipient organizations classified in this topic area, and grants to other recipient types for activities classified in this topic area. Used by permission.
[b]Data for 2009 could not be found for The Leona M. and Harry B. Helmsley Charitable Trust, Gary and Mary West Foundation, Richard M. Schulze Family Foundation, and The Hsieh Family Foundation, thus they are not included in the total for 2011, nor in the net change from 2009 to 2011.

In terms of their sources of support, independent foundations are generally not dependent on multiple individual donors in the manner that voluntary health agencies are. Their endowments are, of course, subject to their investment vehicles, but in general their assets will often rise and fall in a manner approximately equivalent to the stock market. They are accountable to their trustees or boards of directors, but these groups are often quite flexible and not as risk-averse as other funders of research. Thus they may be willing to provide funds for fundamental research on novel ideas or on proof-of-principle studies. Their nonprofit status depends on their expending a specific percentage of their assets for charitable purposes each year, and while these purposes do not have to include medical research, it often does, as you will see in Table 40.2.

Individual Philanthropists—Major Gifts

Millions of individuals in the United States donate time, expertise, or money to research organizations or voluntary health agencies, and all can be considered philanthropists. However, some wealthy individuals may devote considerable time and effort to deciding how best to use their wealth to the greatest effect. A number of these individuals who choose to support research with funds from their personal wealth will do so, not via the family foundations discussed earlier, but to a VHO or to an individual researcher or academic organization. These donations are often referred to as major gifts, meaning those that are of special significance to the organization receiving them and to the giver. In either case, it is important that the gift be appropriately recognized, and the relationship valued.

As an investigator, you may be involved in discussions with major donors (or foundations) on behalf of your own institution or for a voluntary health agency for which you volunteer. You will often be working with an organizational professional who can help guide the donor through the sometimes-complex mechanisms by which gifts can be made, and they can help by outlining the benefits that these mechanisms may have on tax burden, estate, etc. However, you will be a critical part of the conversation and need to understand several issues.

Major gifts most often arise in the setting of a real and ongoing relationship with the entity being honored with a gift. This relationship, whether with an investigator who has also been a treating physician for the donor or a family member, or with an organization committed to a disease or disorder of interest to the donor, is based on trust. Thus, you must avoid promises or suggestions of what can be achieved with the funds in question that are other than factual and as accurate as possible. Enthusiasm for your science and the ability to describe its potential in lay terms is important. But hype and false promises destroy trust.

Major gifts are often the embodiment of a wish, a hope, or a dream. Whether prompted by the desire to produce in some way a healthier future for their families, themselves, or for the rest of humanity, they represent an essential part of an individual's legacy, and most often a selfless part. Donors generally recognize that science most often moves in steps and incrementally and that progress can usually not outstrip the pace of the biology needed to test even the most paradigm-shifting discoveries, and most donors are quite aware of that. However, they are, appropriately enough, impatient with the pace of progress that they see as being hampered by a lack of collaboration, openness of access to new discoveries, and more concern with individual career progress than with progress for the patient. Recent progress with open science, open access, and collaborative or team science are seen as positive by most donors.

In addition to representing science in an accurate manner, it behooves you to provide appropriate follow-up if a request is successful. Recognition of the donor and acknowledgment of their contribution should always follow, but it is also essential that stewardship of gifts also include letting the donor know what their gift has produced. Regular contact to describe the progress of the science or the career progress of individuals through their careers is very much appreciated and will help to build long-lasting mutually respectful and sincere relationships. Neither the donor nor the investigator/organization can achieve their goals alone, and clarifying throughout the process how they can create more by working together is essential.

Corporate Foundations

The concept of corporations as givers was legally established in the United States in the second half of the last century, and this concept expanded rapidly. Some nonmedical corporations develop foundations to support health issues, either through supporting patients or funding research, with the Avon Foundation and Geoffrey Beene Foundation having been examples of these. In addition, corporations that are involved in the biomedical industry have also formed foundations, some of which are designed to provide patient assistance. A number of pharmaceutical companies have formed foundations, some of which rank among the largest foundations in the United States. Their grants support many different kinds of activities, ranging from infrastructure support at academic medical centers to support for educational activities to fellowship support of early-career investigators, with examples such as those listed below for the Pfizer Foundation (see Section Individual Philanthropists—Major Gifts).

FUNDING TYPE

Charitable contributions—Donations to 501(c)(3) not-for-profit organizations, used to support patient education and advocacy, improvements in patient care or general operating support, among other things.

Fellowships—Support provided to US medical schools, teaching hospitals, and other institutions intended to aid junior faculty and talented researchers early in their careers.

Independent Grants for Learning & Change (IGL&C)—Medical education grants from the office of IGL&C in support of health-care quality improvement initiatives and independent education for health-care professionals, including but not limited to, continuing medical education.

Nonmedical education grants—Funding to medical, scientific, and patient organizations for a variety of nonmedical education purposes, such as general operating support and support of meetings.

Scholarships—Funding to augment specialty training and encourage development of senior faculty scientists, similar to fellowships.

Visiting professorships—Funding that enables schools and teaching hospitals to recruit an expert of their choice to lead a 3-day information exchange on predetermined topics with students, faculty, nurses, and others.

Funding requests for health-care activities—Funding in support of a wide array of initiatives and charitable causes including patient advocacy and education.

Collaborations—Funding a health care–related activity or project undertaken by Pfizer and one or more organizations to advance specific and discrete shared objectives. The Organization conducts activities in their own name; however, Pfizer may be involved with the creation of output and/or receives the right to use materials being created.

Coalitions—Funding a health care–related activity where two or more separate and distinct organizations, in addition to Pfizer, seek to formally and publicly join forces over a period of time to achieve a common objective.

RECIPIENT TYPE

Academic medical centers—Accredited medical schools or academic hospitals in which a majority of the physicians are faculty members; may include other health professional schools, such as public health, nursing, or graduate schools in the life sciences.

Other medical centers—A single entity or loose conglomerate not directly affiliated with a medical school or university, usually focused on patient care and residency training.

Medical associations—Professional membership organizations for physicians.

Professional associations—Professional membership organizations for health-care professionals such as nurses, nurse practitioners, physician assistants, and pharmacists.

Patient organizations—Nonprofit organizations that represent the interests of patients by providing information and services, support for increased medical research, and/or access to health care.

Scientific associations—Organizations that exist to promote an academic or scientific discipline through activities including regular meetings or conferences for the presentation and discussion of new research results, and publishing or sponsoring academic journals.

Civic organizations—Nonprofit organizations that promote social welfare.

FOUNDATIONS ATTACHED TO GOVERNMENT AGENCIES

Since the mid-1990s, federal agencies have also engaged in or encouraged the creation of independent, nonprofit foundations, which provide them options for accessing charitable donations to support their missions, in much the same way as the private foundation arms of public or state universities. For the latter, which are dependent on the appropriations of their state governments, 501(c)(3) philanthropic arms allow them to raise funds to supplement public moneys. Several federal agencies, including the CDC, the NIH, the FDA, and NASA, are using a similar strategy to allow interested philanthropic sources to support activities they value.

The National Foundation for the Centers for Disease Control and Prevention, commonly referred to as the CDC Foundation, was established in 1995 and serves such a role for national and international health issues that have broad support, but requires additional capital, particularly in times when the funds appropriated by Congress are meager.

For example, the Foundation had sufficient assets in FY2010 to support programs such as the Bloomberg Initiative to Reduce Tobacco Use ($7.3 million), the Meta-Leadership Summits for Preparedness ($1.7 million), and a program of disease surveillance and response in Cameroon, the Democratic Republic of the Congo, and the Central African Republic ($2.8 million).

Major foundation sources for the CDC Foundation included the Bloomberg Family Foundation ($10.2 million in 2009), the Doris Duke Charitable Foundation ($5.8 million between 2005 and 2010), the Robert Wood Johnson Foundation ($24.2 million from 2004 to 2011), as well as seven-figure grant totals from the Woodruff Foundation, the Marcus Foundation, the John S. and James L. Knight Foundation, the Ellison Medical Foundation, and the Avon Foundation for Women. Not unexpectedly, the Foundation has also received $28.1 million between 2003 and 2010 from the Bill & Melinda Gates Foundation (see Section Corporate Foundations).

The Foundation for the NIH (FNIH), established a year later in 1996, was authorized by Congress to support the research mission of the NIH and to foster public health through scientific discovery, translational research and the dissemination of research results through specific high-impact public partnerships. It has been very successful in raising funds to organize and administer research programs, to support the education of early-career researchers, to organize educational events, and to support a wide range of health challenges. Some examples of completed efforts have been the 2013 Geoffrey Beene Global NeuroDiscovery Challenge with the goal to identify male/female differences in the pathogenesis and presentation of Alzheimer disease, the completion in 2007 of the Genetic Association Information Network (GAIN), an ambitious program to genotype existing research studies in six major common diseases, and combine the results with clinical data to create a significant new resource for genetic researchers. The FNIH manages the Biomarkers Consortium and the Bill & Melinda Gates Foundation's Grand Challenges in Global Health initiative, among many other projects.

CONCEPTS IN NONPROFIT FUNDING

While philanthropic funding does support research along the entire continuum, from basic to clinical and population research, including the translational phases in between, nonprofit organizations vary in the disease area of biomedical research that they focus on. Many foundations and large voluntary health agencies have primarily funded basic discovery aimed at the diseases of their focus and have also felt that providing a scientific workforce for the future by funding early-career investigators is an appropriate investment that would ultimately lead to the benefit of patients in many ways, including new therapies. The Burroughs Wellcome Fund, the Robert Wood Johnson Foundation, and the Howard Hughes Institute are only a few of the groups that make such investments in "human capital" as well as in research itself. Many of the VHOs also invest in human capital, with awards ranging from the medical student level to those aimed at midcareer investigators.

The traditional focus of philanthropic funders on basic research has changed substantially over the past several decades. While large clinical trials are generally beyond the cost ceiling of most such funders, patient-oriented mechanistic clinical studies and proof-of-principle clinical trials have been encouraged by some, and several have used their close relationships with patients to provide an infrastructure that can support drug discovery and later phase clinical trials. The Multiple Myeloma Research Foundation has been especially successful in this regard. Novel concepts, including the idea that such investment in the development of therapeutic agents might provide funds coming back to a foundation to support subsequent research, have gained interest. Certainly, clinical investigators should not ignore VHOs in their search for funding.

Similarly, the recognition that much clinical research founders in a gulf between basic discovery of a druggable target or even of a drug itself and the point at which industry or venture capitalists might invest, has led to the funding of science accelerators, designed to propel potentially useful and even commercializable science across this gulf. While accelerator funding has been the function of research institutes or some academic institutions (Johns Hopkins is an example), it has now been undertaken at the NIH and also by some VHOs, such as the American Heart Association.

CONCLUSION

The contributions of philanthropists, whether as individuals or in foundations, whether deriving from families or from corporate entities, and whether supporting the fight against common or rare diseases, have had and continue to have an important impact on the support of scientific discovery and the career development of investigators. Knowledge of the nonprofit, nongovernmental funding resources in your area of research can not only provide useful financial support, but can help you become part of a community of lay and professional colleagues working together to combat disease. Be sure not to overlook the opportunities they provide.

PROBLEM SET

The following problem set reinforces key concepts, with answers provided:

1. A donor whose child has a rare disorder is devoted to the physician-researcher who has been the child's long-term health-care provider. The donor wishes to support research on this disorder and proposes a donation of a million dollars to a VHO to be granted to this researcher's laboratory.

   What should the VHO consider? If the VHO accepts the gift with the understanding that it may only be granted to this specific researcher at this specific institution, the VHO has relinquished its variance power (the right and responsibility to grant the funds in accordance with the criteria it determines for its research program) and will be acting simply as a pass-through. The funds cannot be considered to be a donation. In addition, the VHO's reputation for funding the most meritorious research available may be damaged. The VHO may still decide to accept the gift, but it does pose problems. What should the donor consider? The donor should carefully consider their goals. If their primary goal is to reward this physician-researcher for the care they have provided and to demonstrate their faith in the investigator's research, they may wish to donate the funds directly to the investigator's institution for their use. If the goal is to support the best and most promising research on this problem, wherever it is done, they should consider whether to have the VHO act for them to survey applications from across the country and award the funds to the researcher(s) with the best ideas. Their level of trust in the researcher, the researcher's institution, and the VHO's capabilities will be major factors in this decision.

2. A postdoctoral MD-PhD investigator with a productive track record of early basic research experience, with several publications. She is finishing the clinical portion of her fellowship training and wants to get solid training in clinical research, where she thinks her future may lie. She has already begun to work with an experienced senior investigator on aspects of an ongoing clinical trial and has developed a novel idea that she would like to test within the trial. Her mentor suggests that she apply for an early-career grant that could support her through her fellow to faculty transition. A national VHO with a local branch in her city has such grant opportunities for clinical investigators. She could also occupy an industry-funded fellowship at her institution.

   What are this early-career investigator's chances of getting this VHO grant award? How can she improve her chances? Since the funding environment has been difficult in recent years, it is hard to be optimistic about more than a 10–15% success rate for a single application. To improve her chances, she should work with the Grants Management Office to identify all applicable funding programs and apply to more than one. In addition, she should learn all she can about the goals of the VHO, so that her project can, if it is possible and truthful, be described in terms of how it will help them meet their aims and goals. She should also examine the publicly available information about the review processes of the organization. Information about who, or at least what type of reviewer, will be reviewing her application can be invaluable. While it might seem that taking the industry-funded fellowship would be easier, it would not provide as valuable an addition to her cv as a personal award. It also would create a "relationship with industry" which would limit some professional activities she might wish to engage in.

ACKNOWLEDGMENTS

The authors gratefully acknowledge Ms. Luisa Tam for her expert assistance with research and Nancy Brown and Paul Kalil for their very helpful review of the manuscript.

REFERENCE

Moses III, H., Matheson, D.H.M., Cairns-Smith, S., George, B.P., Palisch, C., Dorsey, E.R., 2015. The Anatomy of Medical Research: US and International Comparisons. JAMA 313 (2), 174–189.

Chapter 41

Modern Drug Discovery and Development

Daniel E. Salazar[1] and Glenn Gormley[2]

[1]EMS Pharma, Hortolàndia, SP, Brazil; [2]Daiichi Sankyo Inc., Edison, NJ, United States

Chapter Outline

Key Points

- This chapter presents an overview of the drug discovery and development process from the laboratory bench through postmarketing approval.
- Aspects of drug discovery include discovering drug targets, identifying and optimizing lead compounds, preclinical studies, and phase 1 through phase 4 clinical studies.
- New paradigms such as personalized medicine are also discussed. Numerous examples are included to illustrate various processes and principles.

INTRODUCTION

The Irony of Innovation

In 2012, global pharmaceutical sales were estimated to be $965 billion, with the majority of sales in North America, Japan, and China, according to Intercontinental Marketing Services (IMS Health, 2013). The IMS Institute for Healthcare Informatics predicts that the pharmaceutical market will reach nearly $1200 billion by 2016 (IMS Health, 2013). The irony of this investment in innovation is that although there are numerous targets, greater investment, and better technology, the number of compound approvals has been declining. In the period from 1995 to 1999, the average annual number of new molecular entities (NMEs) approved in the United States was 37.6; from 2004 to 2012 this number decreased to 26 (Kaitin and DiMasi, 2011).

Clinical and Translational Science. http://dx.doi.org/10.1016/B978-0-12-802101-9.00041-7

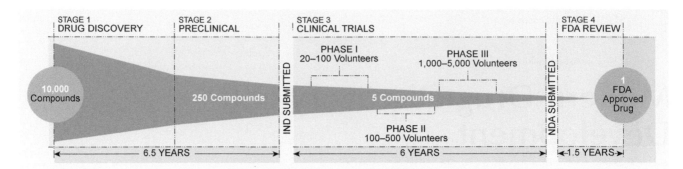

FIGURE 41.1 Process of drug development. *(Reproduced from CMR; Centre for Medicines Research International Ltd, 2006).*

The research and development of new medicines is a lengthy process, taking over 11.5 years from the identification of a drug candidate to its introduction to the market (Fig. 41.1). The development times are surprisingly similar between the United States, Europe, and Japan and also have not changed much over the past decades (CMR; Centre for Medicines Research International Ltd, 2006). Not only is time for new drug development very long but attrition is also severe. According to the Pharmaceutical Research and Manufacturers of America (2014), for every 10,000 compounds synthesized, only one will be approved by the Food and Drug Administration (FDA). Of these approvals, only one in five will cover the cost of development and the estimated the cost of the development of one compound reaching the market was over $800 million (Vernon et al., 2010).

This chapter presents an overview of the drug discovery and development process from the laboratory bench through postmarketing approval. A brief history of drug discovery through the ages is included to provide perspective on the current situation and anticipated changes in the near future.

THE ORIGINS OF DRUG DISCOVERY

Discovery of Small-Molecule Drugs

Small-molecule drug discovery began with observations that various natural products produced a desirable response in the patient e.g., the use of the foxglove plant in the mid-1700s for the treatment of heart failure (Goldman, 2001). Beginning in the 19th century, advances in chemistry enabled isolation of the active substance and de novo synthesis of active compounds such as aspirin by Bayer (Vane, 2000). Subsequently, chemical modifications of key functionalities to enhance efficacy were achieved. For example, the discovery of penicillin in 1928 and its introduction into clinical use in the 1940s was followed by the introduction of semisynthetic penicillins over the next 40 years (Rolinson, 1998).

Importantly, these same advances in chemistry also allowed for the discernment of distinct enzymes and cellular receptors by pharmacologists and biochemists, thereby providing the basis for rational drug discovery and development that evolved during the mid-20th century and is still in use today (Fig. 41.2). Once a target is identified, a screening method is developed based on knowledge of the disease and validated. Libraries of natural product extracts or pure chemicals are "screened" against a cellular, tissue, or even whole animal model to discover a "hit," that is, a chemical or mixture with recognizable activity. When a hit is discovered, analog synthesis proceeded to improve activity in the screen and hopefully develop a structure–activity relationship. This is known as the lead identification stage. Once a lead or several lead compounds are identified that have sufficient pharmacological activity, preliminary animal toxicology and pharmacokinetic studies as well as studies characterizing the pharmaceutical properties of the compounds are conducted to support lead compound optimization. Usually, additional analogs have to be synthesized to obtain a compound deemed sufficient in all these aspects and thus suitable for clinical development. The clinical candidate then undergoes the studies required by regulatory authorities to proceed to clinical testing and hopefully make it to the market. Most of the time, the initial clinical candidate fails due to poor pharmacokinetics, safety, or efficacy.

An example of the 20th-century small-molecule drug discovery process was the discovery of the histamine 2 (H_2) receptor antagonists. In 1956, Code discovered that histamine could stimulate gastric acid secretion (Code, 1956). In addition, several different investigators postulated that there were subtypes of histamine receptors (Folkow et al., 1948;

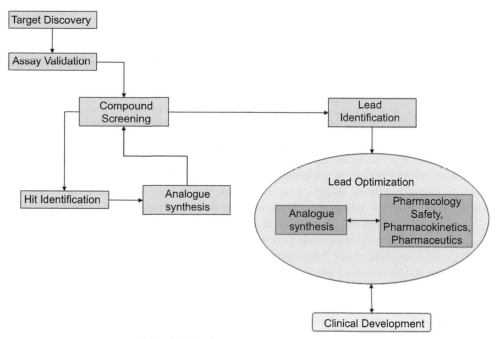

FIGURE 41.2   General process of drug discovery.

Ash and Schild, 1966). Based on this foundation, James Black and his colleagues began studying the ability of histamine analogs to block histamine-induced gastric acid secretion and identified several lead compounds (Molinder, 1994). As Molinder describes, the initial lead H_2 receptor antagonists had no oral bioavailability, and several had very poor pharmaceutical characteristics (Molinder, 1994). In addition, the first H_2 receptor antagonist tested clinically, metiamide, produced severe agranulocytosis in several patients and was discontinued from development. Eventually, improvements in pharmacokinetics and drug safety were made through exploration of additional molecules. This research led to the discovery, development, and marketing of cimetidine, the first selective H_2 receptor antagonist. Following clinical studies demonstrating the efficacy of cimetidine in gastric and duodenal ulcer, it would go on to revolutionize the treatment of these ailments, becoming the first drug to achieve "blockbuster" status as defined by more than $1 billion in global annual sales. It is available today over the counter without a prescription. Finally, while cimetidine is a very good drug, it has enough affinity for histamine 1 (H_1) receptors that, at high concentrations, can produce side effects (Freston, 1987); it has a fairly short plasma half-life and thus must be dosed at least twice per day; and it can significantly inhibit the metabolism of other medications (Ostro, 1987). Subsequently discovered and marketed H_2 receptor antagonists were more selective for the H_2 receptor, had somewhat longer half-lives, and were less susceptible to metabolic drug interactions (Ostro, 1987), illustrating that clinical feedback to drug discovery is essential for new drugs to provide meaningful advances in therapeutics.

Discovery of Biological Therapies

The first biologic therapy, taka-diastase, was discovered by Dr. Jokichi Takamine, a Japanese chemist, in 1894. Diastase is a digestive enzyme whose name comes from the term "diastase," which means "enzyme," and "Taka," which means "best" or "excellent" in Greek and which is also the first half or Dr. Takamine's name. In 1894, Takamine was granted the first patent on a microbial enzyme in the United States. He licensed his preparation to Parke–Davis and Company under the brand name Taka-diastase, and the company marketed the product as a digestive aid for the treatment of dyspepsia. The product was enormously successful (Bennett, 2001).

The discovery and introduction of insulin in 1921 by Banting and Best (1922), another early biologic therapy, is an excellent example of translational medicine. These investigators went from demonstrating that pancreatic extracts were effective in a dog model of diabetes to proof of concept in a 14-year-old boy in about 1 year (Macleod, 1922). Between 1922 and 1982 insulin was extracted from bovine or pig pancreas with attendant side effects of frequent immune response. Commercial production of human insulin began in 1982 (Gualandi-Signorini and Giorgi, 2001). Semisynthetic human insulin was obtained by an enzymatic method, in which trypsin catalyzes the substitution of alanine, in position B30 of pig insulin, with threonine. The insulin thus produced has the same amino acid sequence as that of the human one, and it was completely free of other pancreatic and gastroenteric hormones. Today human insulin is produced by recombinant techniques. The yeast *Saccharomyces cerevisiae* acts as a living laboratory after insertion of purified DNA that dictates the synthesis of a single chain precursor different from proinsulin in one of the yeast's plasmids. The protein, produced by yeast continuously during fermentation, is harvested from the culture medium, isolated by centrifugation and crystallization, converted to insulin ester by trypsin transpeptidization, and finally hydrolyzed (Gualandi-Signorini and Giorgi, 2001).

The development of recombinant technology has resulted in the approval of over 80 biological products in the United States (US Food and Drug Administration Center for Drug Evaluation and Research, 2014). Over 30 of these, such as insulin, are considered as replacement therapies, either hormones or enzymes that are administered for insufficient or defective protein production. For example, Epogen, a recombinantly produced form of erythropoietin, was initially developed for treatment of anemia in patients undergoing hemodialysis (Winearls et al., 1986) and was approved in 1989.

Among the most important biological therapies are antibodies. It was long recognized in the 20th century that antibodies could be useful in a variety of indications. The pioneering work of Kohler and Milstein (1975) demonstrated that murine antibodies could be produced in vitro from fused B and tumor cells. Early attempts to use these murine monoclonal antibodies (mAbs) in humans met with failure due to development of human antimouse antibodies (HAMA) sometimes accompanied by allergic reactions and shock (Mountain and Adair, 1992). Using recombinant technology, eventually the immunogenicity of murine mAbs could be reduced by producing chimeric forms (murine variable region with human constant domain) and eventually "humanized" mAbs wherein most of the variable region is replaced with human sequences (Mountain and Adair, 1992).

The first approved therapeutic murine mAb was launched in 1986 under the trade name Orthoclone OKT3, indicated for use during organ transplantations (Weidle et al., 2013). A number of other antibodies followed, some of which were chimeric wherein the constant domains of an antibody were human and the variable domains were murine under the trade names Remicade and Erbitux, which are being used for the treatment of rheumatoid arthritis and colorectal cancer, respectively. Later, antibodies were completely humanized, as in the case of Herceptin, which is being used for the treatment of breast cancer. In 2002, the first fully human antibody under the trade name Humira has been isolated using antibody phage display, which is directed against TNF-α (Weidle et al., 2013).

DRUG DISCOVERY IN THE 21ST CENTURY

In the 21st century, the overall process of drug discovery is largely unchanged from the previous century; however, the biotechnology and informatics revolutions of the late 1980s and 1990s have changed the specifics of the stages of drug discovery in manifold ways. Due to these advances, many activities that were formerly conducted as part of lead optimization, such as absorption, pharmacokinetic, and toxicology profiling, are now conducted as high-throughput screening (HTS) assays closely associated with lead identification. Drug target discovery is being radically changed by the greatly increased ability to identify mutations associated with disease, and advances in recombinant technology have opened the door to new types of biological therapies that are entering clinical development.

Drug Target Discovery

Contemporary drug target discovery employs techniques such as genomics, proteomics, genetic association, and reverse genetics as well as clinical and in vivo studies to identify potential targets (Fig. 41.3) (Lindsay, 2003). During validation, the demonstration of the role of the target in disease phenotype, modulation of gene expression, and/or protein function in both cell and animal models is used to support the target selection before using the target in the drug discovery process.

The elucidation of the human genome is having a fundamental impact on drug target discovery. For example, researchers now have characterized the molecular targets of all FDA-approved drugs showing that there are only 324 different molecular targets for approved drugs and that most of these targets belong to only 10 gene families (Overington et al., 2006; Imming et al., 2006). Furthermore, while an exact determination of the potential number of drug targets is not possible, estimates based on the human genome suggest there may be somewhere between 600 and 1500 drug targets (Hopkins and Groom, 2002).

Although the effects of highly penetrant genes on diseases has long enabled drug discovery, e.g., familial hypercholesterolemia and drugs for the treatment of atherosclerosis (Lusis et al., 2004), there are now newer techniques that allow for the discovery of the association of low-penetrance genes with human disease. A recent study linked 108 regions of the genome to schizophrenia (Schizophrenia Working Group of the Psychiatric Genomics Consortium, 2014). Already drug discovery researchers are combing through the results to validate new drug targets for schizophrenia (Dolgin, 2014).

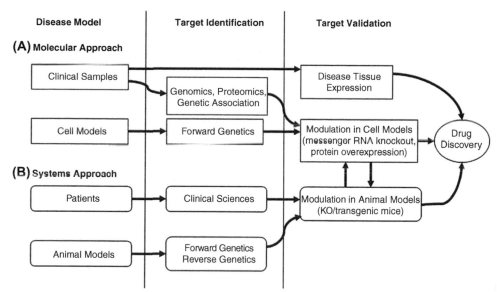

FIGURE 41.3 An overview of molecular- and system-based approaches to target discovery. Target discovery is composed of three steps: the provision of disease models/tissues, target identification, and target validation. The "molecular" approach (A) uses techniques such as genomics, proteomics, genetic association, and reverse genetics, whereas the "systems" approach (B) uses clinical and in vivo studies to identify potential targets. During validation, modulation of gene expression and/or protein function in both cell and animal models is used to confirm the role of the target prior to passing into the discovery pipeline. *Adapted from Lindsay, M.A., 2003. Target discovery. Nat. Rev. Drug Discov. 2, 831–838 with permission.*

Clearly the areas where genetics has had the biggest impact on drug discovery are in rare diseases and oncology. Fig. 41.4 shows the increase in discovery of genes responsible for rare diseases using whole exome sequencing (Boycott et al., 2013). It is estimated that by 2020 nearly all rare disease–causing mutations will have been identified (Boycott et al., 2013). Fig. 41.5 shows a historical view of the role of cancer genetics in drug discovery (Chin et al., 2012). Clearly, the time from the discovery of a driver mutation in cancer to emergence of a new therapy is decreasing substantially. Importantly, along with the genetic mutation being the basis for drug target discovery, an in vitro test for the mutation can be developed that, when applied to the patient's tumor sample, can determine likelihood to respond to the drug, i.e., personalized medicine. The discovery and development of vemurafenib is an excellent example of the impact of cancer genomics and personalized medicine.

Lead Identification

Once a drug target has been selected, the type of modality to modulate the target has to be decided. Given the advances in recombinant technology, the options for different biological modalities are emerging rapidly; however, technology has greatly improved the ability to identify small-molecule leads as well. Accompanying the advancements in the ability to discover new targets for drugs, the fields of HTS and combinatorial chemistry have dramatically altered the small-molecule lead identification process. HTS is currently conducted with protein, cell, and even organism-based assays (Nicholson et al., 2007). Many drug discovery laboratories can now perform over 100,000 screening assays per week (Fox et al., 2006). In addition, it was estimated that in 2004, the 54 HTS laboratories surveyed generated over 700 lead compounds and 100 compounds tested in humans.

The ability to rapidly screen compounds was accompanied in parallel by the ability to synthesize new molecules. During the 1980s, it was recognized that creation of new molecules for screening against targets and for lead optimization was rate limiting in the discovery of new drugs. In the late 1980s and 1990s, many pharmaceutical companies built on the work of Merrifield in solid-phase peptide synthesis (Krieger et al., 1976) and formed groups devoted to combinatorial synthesis, the simultaneous preparation of all possible combinations of two or more mutually reactive sets of chemical monomers, either as individual compounds or mixtures (Rabinowitz and Shankley, 2006). These groups were able to increase the output of compounds per chemist and thereby create large diverse chemical libraries for screening against many targets and, most importantly, rapidly generate libraries based on a "hit" to obtain a lead or for subsequent lead optimization. Depending on the methods used, the libraries produced by combinatorial chemistry can contain mixtures of compounds and sometimes impurities that compromised the ability to interpret screening results. However, using the technique of parallel synthesis, libraries of 10–1000 compounds of good purity can be prepared at one time (Rabinowitz

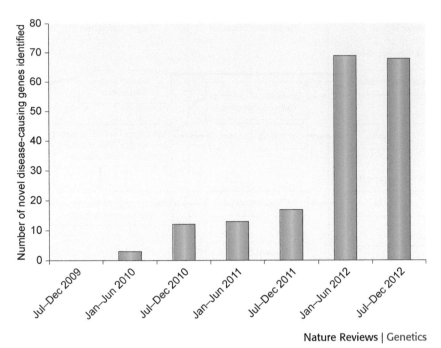

Nature Reviews | Genetics

FIGURE 41.4 Number of identified novel disease-causing genes. *Reproduced from Boycott, K.M., Vanstone, M.R., Bulman, D.E., MacKenzie, A.E., 2013. Rare-disease genetics in the era of next-generation sequencing: discovery to translation. Nature Reviews Genetics 14, 681–691 with permission.*

and Shankley, 2006). These so-called focused libraries are now widely employed to go from hit-to-lead and lead optimization. While certainly combinatorial chemistry is being used to discover drugs, natural product extracts are still yielding drug candidates (Newman and Cragg, 2007) and combinatorial biosynthesis is an area which may further improve the ability to generate novel therapeutic compounds (Floss, 2006).

Although, in general, more than one criterion is considered while assembling a chemical library, the design process is typically driven by a core guiding principle (Medina-Franco et al., 2014). For example, "the library size comes first" was the guiding principle of combinatorial libraries that heavily influenced drug discovery in the 1990s; the goal was to synthesize large libraries, producing millions of molecules to be screened using approaches such as HTS or mixture-based methods. "Chemical diversity comes first" has been the guiding principle to design large libraries or select subsets of molecules from existing libraries for the purpose of identifying novel hits or demonstrating chemical tractability of unexplored targets. More recently, it has been proposed that library design should be focused on maximizing biological diversity instead of chemical diversity. As drug discovery evolves and faces novel challenges, chemical libraries are designed and selected considering "quality comes first" as the guiding principle where "quality" involves a balance of multiple criteria including structural diversity, bioavailability, permeability, drug-likeness, and target focus, among others (Medina-Franco et al., 2014) (Table 41.1).

One type of chemically constrained library, small-molecule fragments, has been combined with high-throughput X-ray crystallography to produce what is termed fragment-based drug discovery (Jhoti et al., 2013). Owing to its ability to detect low-affinity binding, X-ray crystallography has been shown to be a very effective technique to identify the initial fragment hit as well as to guide the fragment hit-to-lead phase given the 3D structural information provided by the protein-fragment crystal structure. These new structure-dependent approaches have not only yielded novel lead compounds but also uncovered new allosteric pockets that can be exploited for drug development. This drug discovery approach has led to the discovery of highly selective compounds such as vemurafenib, which is a potent inhibitor of BRAF with the V600E mutation (Tsai et al., 2008).

Finally, virtual screening, the process of using computer-based methods to discover new ligands on the basis of biological structures, is becoming more widely used in the pharmaceutical industry (Shoichet, 2004); recently, Pang (2007) has demonstrated that virtual screening was able to identify chemicals that penetrate and rescue cells from viral infection. Importantly, in this process, the target structure was predicted by computers solely from the protein's gene sequence and thus may be termed virtual "genome-to-lead" discovery.

As mentioned previously, the options for employing a biological therapy to modulate a target have increased substantially. Currently most therapeutic antibodies and other protein-based scaffolds are produced via phage-display

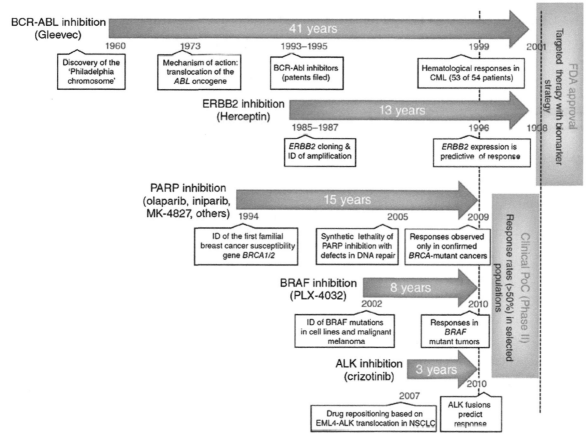

FIGURE 41.5 Cancer genetics is accelerating the time from "driver mutation discovery" to "clinical proof-of-concept" and the approval of new drugs. Historical timelines for developing targeted therapies are highlighted as examples. GLEEVEC received Food and Drug Administration (FDA) approval long after the discovery of the Philadelphia chromosome mutation and hyperactive BCR-ABL protein in chronic myelogenous leukemia (CML). By contrast, the more recent discovery of the chromosomal rearrangements (translocations) of *anaplastic lymphoma kinase* (ALK) in non-small cell lung cancer (NSCLC) has rapidly translated into registration trials for Crizotinib, a "C-MET turned ALK" inhibitor based on tantalizing response rates in ALK-fusion-positive tumors (phase 1 and 2 trial results). Likewise, the development paradigm for selective BRAF inhibitors, as exemplified by PLX4032, underlines the much faster pace of translation (8 years, compared with GLEEVEC or Herceptin) once the driver status (in this case BRAF mutations) has been established for an indication (malignant melanoma). Such accelerated development times are enabled by the broader body of knowledge of cancer biology and mechanisms of actions that have been generated in the cancer field. The FDA approval of Herceptin and the accompanying diagnostic test for HER2 expression (HercepTest) proved the value of biomarker-driven that are informed by mechanistic insights gained from cancer genetics. In a similar vein, it is the functional understanding of DNA repair mechanisms and the *BRCA1* and *BRCA2* mutations in sensitizing tumors to PARP inhibition that inform current registration trials of PARP inhibitors in BRCA-associated cancer types and patients who carry the BRCA mutation. *Reproduced from Chin, L., Anderson, J.N., Futreal P.A., 2011. Cancer genomics: from discovery science to personalized medicine. Nature Medicine 17, 297–303. http://dx.doi. org/10.1038/nm.2323.with permission.*

libraries (Shukra et al., 2014). A phage-display library is constructed from diverse variable regions of the immunoglobulin genes [i.e., single chain of joined heavy (VH) and light variable (VL) fragments (scFv), Fab fragments, or single VH or VL domains]. Each library is constructed from cDNA derived from immune or naive B cells. The DNA library is ligated into a surface protein gene of a phage and will be displayed on their surface as antibody constructs fused with the surface protein. Phages expressing the required specificities are readily isolated and enriched, using antigen-conjugated affinity-binding columns. Eluted phages, similarly reselected, will be used to infect *Escherichia coli* to produce the mAb construct.

Certain limitations regarding antibody-derived therapeutics have emerged. For example, due to their large size, tumor tissue penetration is a potential issue, and due to a planar binding interface, binding to grooves and catalytic sites of enzymes is difficult to achieve. Fragment crystallizable region (Fc)-mediated complement-dependent cytotoxicity and antibody-dependent cellular cytotoxicity can give rise to adverse effects. mAbs have to be produced in mammalian cells; in some cases posttranslational modifications, such as specific glycosylation patterns, are required. High costs of goods, production and purification plants, and often complex intellectual property issues are notable facts.

TABLE 41.1 Examples of Type of Chemical Libraries Confined in Areas of Chemical Space

Type of Confinement	Criterion for Confinement and Examples
Properties	Metric-based rules ADME complaint
Structure	High-density libraries Combinatorial libraries Libraries centered on bioactive and/or privileged scaffolds Libraries enriched with complex structures
Biological relevance	Target family-oriented (e.g., GPCRs, kinases, epigentic targets) Pathway-oriented focused libraries Disease-oriented collections
Molecular function	Protein—protein modulators Peptides and peptidomimetics (including cyclic, polycyclic and stapled peptides, conopeptides, etc.) Covalent drugs Macrocycles Metal-based chemical compounds Synthetic carbohydrates Synthetic lipids

ADME, Absorption, distribution, metabolism and excretion; *GPCR,* G protein—coupled receptor.

To overcome limitations of mAbs, new protein-based scaffolds have been designed and evaluated preclinically, and some of them are being used in clinical studies for the treatment of cancer (Weidle et al., 2013). These entities can be placed into two categories: scaffolds which bind ligands via amino acids in exposed loops and those in which ligand binding is mediated by amino acids in secondary structures, such as β-sheet modules (Fig. 41.6). Adnectins, lipocalins, Kunitz domain-based binders, avimers, knottins, fynomers, atrimers, and cytotoxic T-lymphocyte—associated protein-4 (CTLA4)-based binders, fall into the first category, while darpins, affibodies, affilins, and armadillo repeat protein-based scaffolds are members of the second category (Weidle et al., 2013).

A central problem with antibody-based therapies is that despite the high-affinity antigen binding and highly specific recognition possible in the antigen-combining site of antibody variable domains, the presence of the target antigen in healthy tissues can limit the ability of the antibody to achieve therapeutic effects without inducing unacceptable "on-target" toxicity. The Probody platform leverages the upregulation of protease activity, widely recognized as a hallmark of diseased tissue, particularly in cancer and in inflammatory conditions, to achieve disease-tissue-specific therapeutic activity (Polu and Lowman, 2014).

The approach is based on the use of an IgG antibody, or fragment thereof, such as a variable region, which has been modified to include a masking peptide linked to the N-terminus of the light chain of the antibody through a protease-cleavable linker peptide. The entire construct is fully biosynthetic: the masking peptide, cleavable linker, and antibody light chain comprise one polypeptide chain. In the intact form, the Probody is effectively blocked from binding to the target antigen in healthy tissues; however, once activated by appropriate proteases in the diseased environment, the masking peptide is released, revealing a fully active antibody capable of binding to its target (Polu and Lowman, 2014).

Lead Optimization

The lead optimization stage typically includes further study in various pharmacological models intended to predict the compound's efficacy in humans. Since these are typically specific to the intended indication for the compound, they are beyond the scope of this chapter. Instead, the common elements of the lead optimization phase of drug discovery are discussed in this section. First, in order for a lead compound to progress to the clinic, it must be able to be formulated for delivery by the intended route. If a drug is to be given orally, it should have adequate solubility and permeability, as well as negligible luminal transport by P-glycoprotein for absorption to take place. Second, once absorbed it should have pharmacokinetics consistent with the intended use. Finally, the candidate compound should have a safety profile such that risk does not outweigh the projected benefit of the compound. Lead candidates that do not pass criteria set for these lead optimization elements by drug discovery teams are rejected, resulting in additional synthetic work to find an optimized lead.

The solubility and permeability of small-molecule lead candidates were traditionally determined empirically with a good deal of effort. By examining the computed physical chemical properties of drugs that entered phase 2 clinical

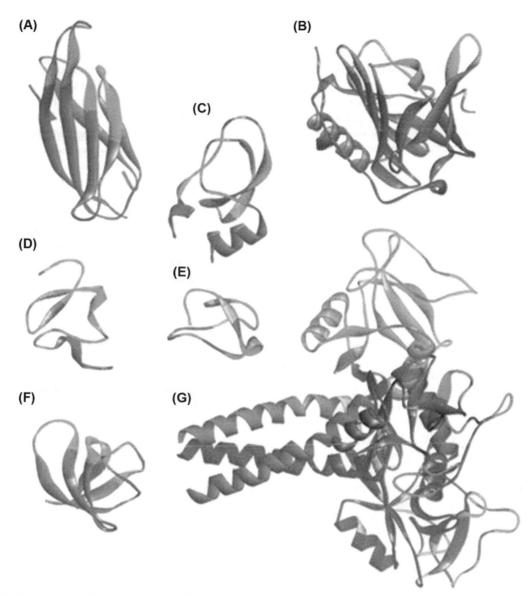

FIGURE 41.6 Loop-based scaffold structures. Scaffold structures are extracted from the Protein DataBank (PDB, November 2011) (when required, models are generated based on available structure data of domains or entities of similar proteins). The scaffolds are displayed using Discovery Studio 31 as ribbon representations to facilitate the domain type identification with the N-terminus placed either on the left or on the top for each structure. The loops that recognize the antigen are colored in green, whereas the framework is indicated in gray, and disulphide bridges are indicated in yellow. A, Adenectin scaffold; B, Anticalin scaffold; C, Kunitz domain scaffold; D, Avimer scaffold; E, Knottin scaffold; F, Fynomer scaffold; G, Atrimer scaffold. The trimeric assembly is indicated in gray, blue, and brown. *Reproduced from Weidle, U.H., Auer, J., Brinkmann, U., Georges, G., Tiefenthaler, G., July–August 2013. The emerging role of new protein scaffold-based agents for treatment of cancer. Cancer Genomics Proteomics 10 (4), 155–168 with permission.*

development, and thus assumed to have adequate pharmaceutical properties, Lipinski et al. (2001) were able to derive the "rule of 5" used to identify compounds that may have absorption or permeability issues. The rule of 5 states that poor absorption or permeation is more likely if one or more of the following conditions are met:

- There are more than 5 H-bond donors (expressed as the sum of OHs and NHs);
- The molecular weight (MWT) is over 500;
- The Log P is over 5 (or MLogP is over 4.15);
- There are more than 10 H-bond acceptors (expressed as the sum of Ns and Os).

Importantly, compound classes that are substrates for biological transporters are exceptions to the rule. To determine if a potential lead compound may be a substrate of intestinal efflux transporters, many companies now assess compound permeability across Caco-2 cell monolayers (Balimane et al., 2006). Caco-2 cells are cultured human adenocarcinoma cells that undergo spontaneous enterocytic differentiation, and it has been demonstrated that the extent of drug absorption correlates well with permeability across Caco-2 monolayers. Between the rule of 5 and Caco-2 permeability, most lead compounds now have good absorption capability.

In similar fashion to rapid assessment of pharmaceutical properties, the advancements in biotechnology and informatics have led to greatly improved ability to predict human drug pharmacokinetics. One of the key foundations for this improved capability was the development of models for animal scale-up by Dedric (1973). He proposed using physiologically based pharmacokinetic models to simulate human pharmacokinetics based on animal data (Fig. 41.7). For certain parameters, he proposed the use of allometric plots, log of parameter versus log of species body weight, to estimate the human parameter. This approach with various permutations is still in use today (Tang and Mayersohn, 2005). The second key foundation was the discovery, in the 1960s, that cytochrome P450 enzymes (CYP450) catalyzed the oxidation of many drugs (Estabrook, 2003). By combining the physiological pharmacokinetic approach with in vitro determination of the rate of drug metabolism, Obach et al. (1997) were able to improve the prediction of human pharmacokinetics for drugs of disparate structures and physicochemical properties.

Today, most large pharmaceutical companies have high-throughput and well-validated in vitro assays for determining compound metabolism by recombinant CYP450 enzymes. High-throughput assays are now used at the lead identification stage to eliminate compounds that might have too rapid a metabolism or too great a potential for drug interactions (Zlokarnik et al., 2005). Highly validated in vitro assays are used at the lead optimization stage to provide more accurate predictions of human pharmacokinetics and to better predict clinical drug−drug interactions (Obach et al., 1997; Walsky and Obach, 2004).

The impact of greater ability to predict the human pharmacokinetics of new compounds is highlighted by data showing that the attrition rate for new drugs due to poor pharmacokinetics or bioavailability appears to have dropped from about 40% in 1991 to about 10% in 2000 (Kola and Landis, 2004). Importantly, pharmaceutical manufacturers have worked together to encourage standardization of the in vitro CYP450 drug interaction studies to allow for better assessment and comparison of different drugs (Bjornsson et al., 2003). This has helped the FDA to formulate guidances for industry that not only support the design of in vitro and clinical drug interaction studies but also allow for drug interaction labeling based on in vitro CYP450 studies (US Dept. of Health and Human Services Food and Drug Administration Center for Drug Evaluation and Research (CDER) Center for Biologics Evaluation and Research (CBER), 1997; US Dept. of Health and Human Services Food and Drug Administration Center for Drug Evaluation and Research (CDER) Center for Biologics Evaluation and Research (CBER), 2006).

An essential element of modern lead compound optimization for safety includes profiling of compounds in vitro for "off target" activity in assays selected to predict major adverse drug events (Whitebread et al., 2005). Many of these assays are intended to predict cardiovascular risk, such as activity in the hERG assay as predictive of the risk of the potentially lethal torsade de pointe arrhythmia, although they can also be predictive of adverse impact on the endocrine, central nervous, or other systems. While these assays are useful to gauge risk for pharmacologically induced toxicity, compounds may also induce toxicity through direct or indirect interactions with DNA or through reactive metabolites that form covalent adducts with proteins or DNA (Liebler and Guengerich, 2005). During the lead optimization stage, companies will conduct in vitro and at times in vivo genotoxicity studies to eliminate compounds with carcinogenic potential as per the FDA guidance (US Dept. of Health and Human Services Food and Drug Administration Center for Drug Evaluation and Research (CDER) Center for Biologics Evaluation and Research (CBER), 1997). In addition, several companies are conducting in vitro and in vivo studies that determine the covalent binding of compounds to microsomal or hepatic proteins (Evans et al., 2004), with the aim of assessing the potential for hepatotoxicity and idiosyncratic toxicities. While these genotoxicity studies and covalent binding studies are not completely predictive of human risk, in many cases any potential risk can outweigh the benefit of the candidate compound especially if there is already a marketed pharmacotherapy for the indication of interest.

Toxicogenomics, the integration of omic technologies, bioinformatics, and toxicology, has seen significant investment in the pharmaceutical industry for both predictive and mechanism-based toxicology in an effort to identify candidate drugs more quickly and economically (Boverhof and Zacharewski, 2006). While this approach may hold future promise for supporting lead optimization efforts and generating new biomarkers for clinical safety evaluation, the key proof-of-principle experiments are yet to emerge. Finally, as part of the lead optimization stage, compounds will undergo short-term in vivo toxicology and safety pharmacology studies. The duration of study, species selected, and doses chosen for study vary from compound to compound largely depending upon the intended indication, the compound's metabolic profile, and the standards of the organization developing the compound. These studies are usually the last tests that a compound must pass before becoming a clinical candidate and proceeding on to the studies required by regulatory guidance before clinical testing.

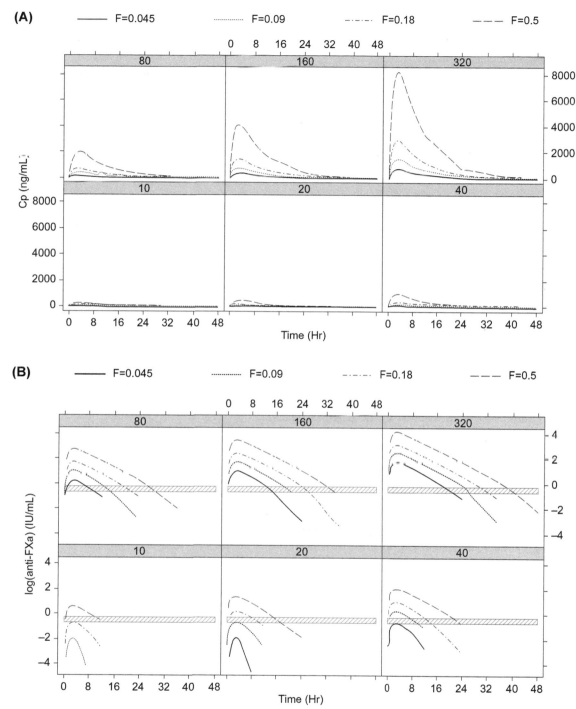

FIGURE 41.7 An example of the simulation of (A) human pharmacokinetics and (B) pharmacodynamics for a range of doses and bioavailabilities for a factor Xa inhibitor from preclinical data. The dose in milligrams is shown above each panel in the simulation. Cp is the plasma concentration of the drug, F is the absolute oral bioavailability, and FXa is the antifactor Xa activity in plasma.

PRECLINICAL DEVELOPMENT

Once a clinical candidate has been identified, preclinical safety, pharmacokinetic, pharmacodynamic, and pharmaceutical studies are conducted to support initiation of clinical trials. While there are some differences in the specific preclinical studies required for initiating clinical trials in various regions, e.g., between the United States and the European Union, the International Conference on Harmonisation (ICH) has produced guidelines that are relatively

TABLE 41.2 Recommended Duration of Repeated-Dose Toxicity Studies to Support the Conduct of Clinical Trials

Maximum Duration of Clinical Trial	Recommended Minimum Duration of Repeated-Dose Toxicity Studies to Support Clinical Trials	
	Rodents	Nonrodents
Up to 2 weeks	2 weeks[a]	2 weeks[a]
Between 2 weeks and 6 months	Same as clinical trial[b]	Same as clinical trial[b]
>6 months	6 months[b, c]	9 months[b, c, d]

[a]In the United States, as an alternative to 2-week studies, extended single-dose toxicity studies can support single-dose human trials. Clinical studies of <14 days can be supported with toxicity studies of the same duration as the proposed clinical study.

[b]In some circumstances, clinical trials of longer duration than 3 months can be initiated, provided that the data are available from a 3-month rodent and a 3-month nonrodent study and that complete data from the chronic rodent and nonrodent study are made available, consistent with local clinical trial regulatory procedures, before extending dosing beyond 3 months in the clinical trial. For serious or life-threatening indications or on a case-by-case basis, this extension can be supported by complete chronic rodent data and in-life and necropsy data for the nonrodent study. Complete histopathology data from the nonrodent should be available within an additional 3 months.

[c]There can be cases where a pediatric population is the primary population, and existing animal studies (toxicology or pharmacology) have identified potential developmental concerns for target organs. In these cases, long-term toxicity testing starting in juvenile animals can be appropriate in some circumstances.

[d]In the European Union, studies of 6 months' duration in nonrodents are considered acceptable. However, where studies with a longer duration have been conducted, it is not appropriate to conduct an additional study of 6 months.

similar between the regions. The ICH guidelines for the United States can be found on the Internet at http://www.fda.gov/cder/guidance/index.htm.

The elements of preclinical development are driven by the clinical plans to evaluate the compound and the regulatory requirements to support those clinical trials. For example, Table 41.2 shows the duration of repeated-dose toxicity studies in rodent and nonrodent studies required by the ICH M3 guideline for phase 1, 2, and 3 studies of various durations in various regions of the world [US Dept. of Health and Human Services Food and Drug Administration Center for Drug Evaluation and Research (CDER) Center for Biologics Evaluation and Research (CBER), 2010]. Typically, most phase 1 studies have treatment durations of no longer than 2 weeks; thus, toxicology studies to support an investigational new drug (IND) filing usually have treatment durations of 2 weeks. Toxicology studies with longer treatment durations are subsequently conducted to support clinical trials of longer duration. The species selection and design for these nonclinical safety studies should be tailored to the compound and the indication that is being pursued (Greaves et al., 2004). It is also required to measure the blood or plasma concentrations as part of these studies or in satellite groups of animals to enhance the value of the toxicological data generated, both in terms of understanding the toxicity tests and eventually in comparison with clinical data as part of the assessment of risk and safety in humans. These toxicology and toxicokinetic studies must be conducted under "Good Laboratory Practice" (GLP) conditions to ensure the quality of the data.

The results of the toxicology studies are used to determine the maximum recommended safe starting dose for initial clinical trials [US Dept. of Health and Human Services Food and Drug Administration Center for Drug Evaluation and Research (CDER), 2005]. The approach is to use a fraction of the no observable adverse effect dose (NOAEL), as determined by the toxicology studies, and convert this to the human equivalent dose by using body surface area (Fig. 41.8). Alternatively, one can perform simulations of the human pharmacokinetics and then select doses for initial studies based on comparison with the pharmacokinetics in animals at doses that produce pharmacological effects and concentrations observed in the toxicology study at the NOAEL. In this approach, one should consider that there may be large differences in plasma protein binding between species. This alternative approach is particularly useful if animal pharmacokinetic—pharmacodynamic (PK/PD) studies have been performed. PK/PD models describe the relationship between drug plasma concentrations and pharmacologic effects by mathematical equations. PK/PD models were first described in the 1960s (Levy, 1964a,b) and are now available for a wide diversity of pharmacological responses (Mager et al., 2003). By combining pharmacokinetic models that predict human pharmacokinetics and PK/PD models developed in animals, it is possible to simulate the dose and time course of pharmacological activity before conducting a clinical study.

With sufficient clinical data linking the pharmacological activity to the clinical endpoint, it is possible to simulate a dose response as well. Enoxaparin, a low MWT heparin whose principle pharmacological activity is inhibition of FXa, is indicated for the prophylaxis of deep vein thromboembolism in several postoperative situations (Lovenox, 2005). Using published data on the safety and efficacy of enoxaparin (Colwell et al., 1995; US Dept. of Health and Human Services Food and Drug Administration Center for Drug Evaluation and Research (CDER), 2006), its pharmacokinetics, and the

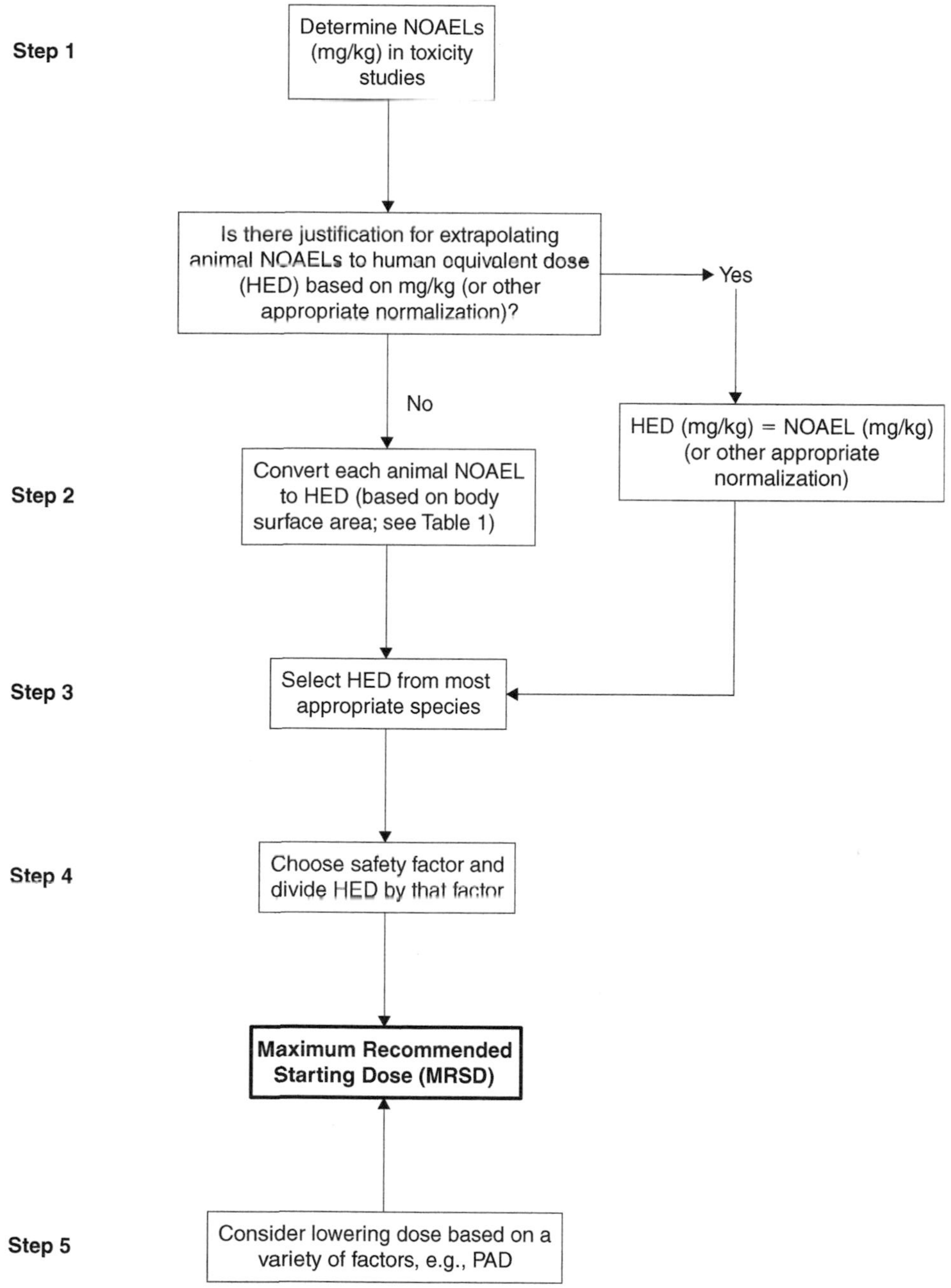

FIGURE 41.8 Algorithm for determining the maximum recommended starting dose for drugs administered systemically to normal volunteers.

relative potency for antifactor Xa activity, we were able to simulate the dose response for the probability of venous thromboembolism and major bleeding for our FXa inhibitor (Fig. 41.9) without conducting a clinical trial.

While clearly these simulations relied on many assumptions, they are helpful when attempting to optimize the design of early clinical trials. For example, if there is a great deal of uncertainty in the predicted human pharmacokinetics or pharmacodynamics, one might consider phase 1 studies of a more limited nature to determine these properties at a limited dose and/or duration and thereby save time and money. This approach is now supported by a recent guideline US FDA Guidance on Exploratory IND studies (Lappin et al., 2006). This guidance allows for single-dose toxicology studies to support single doses less than 1/100th of the estimated pharmacologically active dose up to ≤100 µg. This type of human "microdose" study permits limited characterization of the human pharmacokinetics with modern analytical methodology or imaging studies via positron emission tomography (Boyd and Lalonde, 2007). Whether the pharmacokinetics at these very low,

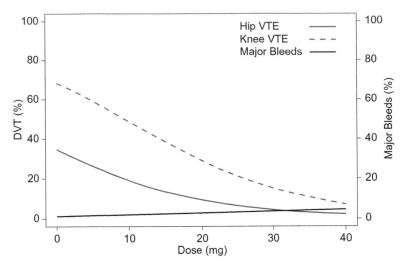

FIGURE 41.9 Simulated dose response for the probability of venous thromboembolism (VTE) postoperatively in patients undergoing hip replacement (Hip VTE) and knee replacement (Knee VTE). The probability of experiencing a major bleeding event (Major Bleeds) was also simulated.

nonpharmacological doses will predict those at higher doses remains uncertain (Fowler et al., 1999), the value for use of such low doses in positron emission tomography imaging studies is clear (LeTorneau et al., 2009). The exploratory IND guidance also provides information on the preclinical studies needed to support clinical studies up to 7 days duration, with the objective of determining the pharmacologically active human dose. The preclinical toxicology-testing strategy from the guidance is displayed in Fig. 41.10. This strategy reduces the amount of preclinical toxicology required to initiate a clinical study; however, as can be seen in Fig. 41.10, the criteria for stopping dose escalation in the study does not permit determination of the human maximum tolerated dose (MTD). The MTD can be important in subsequent clinical development, so the full possible dose range can be explored for safety and efficacy and ultimately the risk benefit of the compound.

CLINICAL DEVELOPMENT

Traditionally, clinical development has been divided into different phases from 1 to 4. The process is described in Fig. 41.1. The development of new drugs is very highly regulated and the development requirements in the United States are defined specifically in the Code of Federal Regulations (21 CFR §312.20). Over the past decade, there has been a profound shift in the design of clinical trials.

Phase 1 Studies

Most drugs that undergo preclinical testing never advance to human testing. However, if extensive preclinical toxicology testing in several animal species has determined that a compound is safe, "first-in-human studies" are initiated in a small number of healthy human volunteers. Phase 1 studies represent the first time that a drug is administered to humans and form the foundation for a successful clinical drug development plan (Hay et al., 2014). The primary goal of phase 1 studies is to investigate the safety and tolerability and, if possible, the MTD, pharmacokinetics, and pharmacodynamics of a product. Since there can be no therapeutic benefit for healthy volunteers, the risk must be very low. After safety is established with the initial dose of the product, the dose may be gradually increased and given more frequently.

A recent study that looked at more than 4000 drugs in phase 1 development from 2004 to 2010 found that the overall success rate of drugs moving from phase 1 clinical trials to FDA approval is 1 in 10, down from 1 in 5 to 1 in 6 seen in previous years (Stein, 2003).

While the safety experience in "first-in-human" studies has been excellent (Silverman, 2013) almost all of the agents previously tested had been small molecules with known pharmacologic mechanisms that were quickly reversible. More recently, drug development has begun to shift to biological molecules targeted mainly for oncology and immunology. In the past decade, worldwide growth in biotech sales rose 353%—from $36 billion in 2001 to $163 billion in 2012 (Kenter and Cohen, 2006). These agents often have novel mechanisms of action, a high degree of species specificity, and/or new immune system targets. Because of these changes, pharmacologic and safety testing in animals may be less predictive than in the past (Ivy et al., 2010).

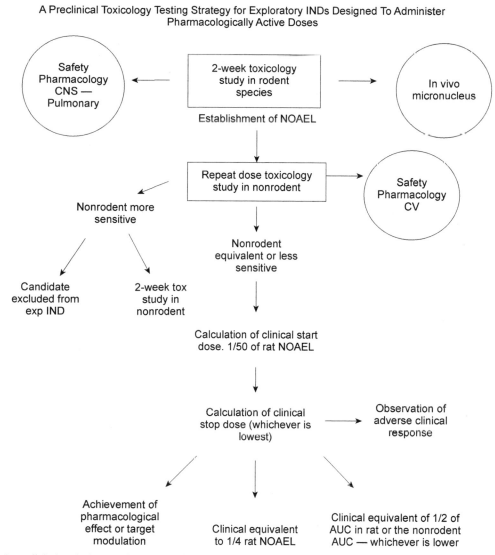

A Preclinical Toxicology Testing Strategy for Exploratory INDs Designed To Administer
Pharmacologically Active Doses

FIGURE 41.10 A preclinical toxicology testing strategy for exploratory investigational new drug applications (eINDs) designed to administer pharmacologically active doses.

A phase 1 study is governed by three principles: safety, ethical conduct, and efficiency. Patient safety is achieved by minimizing the number of participants exposed to serious or life-threatening adverse events. When phase 1 trials involve patients with life-threatening diseases, it is because patients and their physicians have exhausted all other reasonable and standard therapeutic options. This is especially true in oncology trials, where therapeutic options are sometimes limited. The dose and dosing schedule is determined so that the least number of patients are exposed to subtherapeutic doses and the starting dose is determined to minimize the risk to. Efficiency requires that the study will move forward into phase 2 where therapeutic intent is more clear-cut (Suntharalingam et al., 2006).

Although the activity of a drug in animals often correlates to its activity in humans, this is not always the case. Suntharalingam et al. (Expert Scientific Group, 2006) reported on a "first-in-human" study where six volunteers became seriously ill after receiving injections of TGN1412, a novel humanized mAb targeted as a CD28-specific T-lymphocyte receptor agonist for autoimmune diseases or B-cell chronic lymphocytic leukemia. In preclinical studies in rhesus monkeys, large doses were well tolerated. While 100% homology was thought to exist between the binding sites in humans and monkeys, this has now been disputed (Ivy et al., 2010). This was the first human trial with this novel antibody, but important information was available from trials with antibodies for similar targets such as CTLA4. This study, which was unfortunate because of the illness endured by the enrolled subjects, has been extensively studied by European regulatory authorities, and a final report and recommendation is now available from the UK Medicine and Healthcare Products

Regulatory Agency (MHRA) report (Committee for Medicinal Products for Human Use (CHMP), 2007). Also the Committee for Medicinal Products for Human Use (CHMP) of the European Medicines Agency (EMEA) has issued new draft guidelines for first-in-man clinical trials for potentially high-risk medicinal products (Steinbrook, 2002a,b). Many additional efforts are underway (Steinbrook, 2002a,b; Lutterotti et al., 2013) to further protect research subjects after deaths at the University of Rochester and Johns Hopkins, as well as after the death of 18-year-old Jesse Gelsinger in a gene-transfer trial at the University of Pennsylvania. More innovative adaptive trial designs are also being implemented to mitigate patient risk. These designs are discussed later in this chapter.

Not only safety and pharmacology but also potential efficacy can sometimes be indicated from phase 1 studies. For example, a phase 1 trial designed to assess the feasibility, safety, and tolerability of a regimen in multiple sclerosis (MS) patients that used a single infusion of autologous peripheral blood mononuclear cells chemically coupled with seven myelin peptides showed that patients receiving the higher doses ($>1 \times 10^9$) of peptide-coupled cells showed a decrease in antigen-specific T cell responses after peptide-coupled cell therapy. This first-in-human study set the stage for a phase 2 trial to determine if the new treatment can prevent the progression of MS in humans (Wagner et al., 2007).

Proof of Concept (Phase 2)

In cases where safety is established in phase 1 studies of single and multiple doses of the new compound in up to 100 volunteers, the next step is to test the new medicine in patients with the targeted disease. The results of phase 2 clinical trials help determine whether a case can be made, scientifically and commercially, for progressing to phase 3 trials. Different dosage regimens are employed over varying periods of time to obtain a suggestion of therapeutic benefit and the optimal dosage. These data are quite predictive for diseases such as hypertension and diabetes where the blood pressure or fasting blood glucose levels are accepted as evidence of efficacy. For other conditions, biomarkers are now actively being identified to guide discovery, early development, and possibly earlier approval for serious conditions (Williams, 2014).

At the present time, phase 2 success rates are lower than at any other phase of development. Since phase 2 studies are exploratory, using data generated from these trials in clinical practice can be misguided or even dangerous. Phase 2 trials may not be powered to provide a definitive answer regarding efficacy. They may also be subject to selection bias (narrower inclusion criteria than later trials), surveillance bias, and publication bias. However, these issues must be weighed against the reality that, in certain diseases or therapies, a definitive phase 3 study is either unlikely to be done, or cannot be done due to limited numbers of patients with a particular disease. In these instances, clinicians are left with only phase 2 data. As long as these limitations are properly understood and acknowledged, phase 2 data can be meaningfully applied to clinical decision-making.

Many drugs are dropped at each phase of the clinical trial process; the reason for discontinuation in development during phase 1 is often unspecified or strategic, while the reason for termination during phase 2 or 3 is usually related to lack of efficacy. In 2013, 40 drugs were dropped from the oncology development pipeline: 19 in phase 1, 9 in phase 2, and 12 in phase 3 (Eba et al., 2015).

A recent example of a drug whose development was discontinued during phase 2 is tovetumab, a mAb targeting the platelet-derived growth factor (PDGF) receptor. Following completion of a phase 2 study in patients with recurrent glioblastoma multiforme with sepantronium bromide, an inhibitor of survivin expression, patients showed no improvement when the new compound was combined with docetaxel for first-line treatment of HER2-negative metastatic breast cancer. Similarly, parsatuzumab, an anti-EGF-like domain-containing protein 7 (EGFL7) mAb was discontinued following investigation in combination with bevacizumab and FOLFOX in patients with previously untreated metastatic colorectal cancer in combination with paclitaxel and carboplatin (Eba et al., 2015).

In some instances, phases 2 and 3 are combined into a single trial. These combination trials are conducted in several stages; in the early stages, a determination is made as to whether the compound is promising enough for further study. In the later stages, the study drug is compared with a control. Under appropriate conditions, there are advantages to using integrated phase 2/3 designs, including smaller sample size, time and resource savings, and shorter study duration.

Since phase 2/3 trials save time, they are often used in cases where immediate treatment is important, particularly for potentially fatal conditions. One such example involves a randomized phase 2/3 trial that began in Japan in December 2013 for elderly patients with extensive-disease small-cell lung cancer. The purpose of this study was to confirm the superiority of carboplatin plus irinotecan in terms of overall survival over carboplatin plus etoposide. In phase 2, the primary endpoint was the response rate of the carboplatin plus irinotecan arm, and the secondary endpoint was adverse events. In phase 3, the primary endpoint was overall survival, and the secondary endpoints were progression-free survival, response rate, adverse events, serious adverse events, and symptom score. By using this design, more patients who might benefit from treatment received this drug combination (ICH, 1994).

Phase 3 Trials

The final stage before product registration is phase 3. During this phase, larger numbers of patients are treated with one or more doses for longer periods of time in controlled studies versus placebo or active agents. The goal is to confirm efficacy in proof-of-concept studies, but more importantly to evaluate safety during long-term treatment. It is recognized that the statistical power is limited to detect rare adverse reactions with exposure of usually only months with limited numbers treated for longer than 6 months and 1 year. The International Conference on Harmonisation of Technical Requirements for Registration of Pharmaceuticals for Human Use (ICH) is a group that brings together the regulatory authorities and pharmaceutical industry of Europe, Japan, and the United States to discuss scientific and technical aspects of drug registration. This group has produced guidelines for drugs intended for long-term treatment of non-life-threatening conditions (Büller, 2014). It is suggested that the numbers of volunteers and patients exposed to the new agent before registration be about 1500, with 300–600 treated for 6 months and 100 treated for a minimum of 1 year. Depending on the safety observed in animal studies, earlier human studies, and information from related compounds, much greater exposure may be necessary for an adequate assessment of risk and benefit. In most cases, at least two well-designed randomized phase 3 trials must be completed to submit a new drug application to the FDA.

Two examples of modern phase 3 trials are the ENGAGE AF-TIMI 48 and Hokusai-VTE studies. These trials studied edoxaban, a novel, orally available, highly specific direct inhibitor of factor Xa in various stages of development for the treatment and prevention of venous thromboembolism (Giugliano et al., 2013) and prevention of stroke and systemic embolism in patients with nonvalvular atrial fibrillation (NVAF) (Fuji et al., 2010). In addition to dose-ranging phase 2 trials in the target patient populations (Raskob et al., 2010; Yamashita et al., 2012; Weitz et al., 2010; Rohatagi et al., 2012), biosimulation was used to evaluate the potential dose regimens to be studied in phase 3 and thereby reduce the risk of failure (Salazar et al., 2012; Ferrara et al., 2001). Population pharmacokinetics analysis of data from 1281 edoxaban-dosed subjects with intrinsic factors such as renal impairment or NVAF and extrinsic factors such as concomitant medications revealed potentially significant effects of renal impairment and concomitant P-glycoprotein (P-gp) inhibitors on the pharmacokinetics of edoxaban. Exposure–response analysis found that in patients with NVAF, the incidence of bleeding events increased with increasing edoxaban exposure, with steady-state minimum concentration $[(C_{min,ss})]$ showing the strongest association (Ferrara et al., 2001). Clinical trial simulations of bleeding incidence were used to support selection of 30 mg and 60 mg once-daily doses of edoxaban, with 50% dose reductions for patients with moderate renal impairment (CrCl 30–50 mL/min), those receiving concomitant P-gp inhibitors, and patients whose body weight was <60 kg in the ENGAGE AF-TIMI 48 trial.

Both once-daily regimens of edoxaban were noninferior to warfarin with respect to prevention of stroke or systemic embolism in patients with NVAF and were associated with significantly lower rates of bleeding and death from cardiovascular causes (Fuji et al., 2010). In addition, edoxaban administered once daily after initial treatment with heparin was noninferior to high-quality standard therapy and caused significantly less bleeding in a broad spectrum of patients with venous thromboembolism, including those with severe pulmonary embolism (Giugliano et al., 2013).

Phase 4 Trials

The final postmarketing phase of drug testing is a time to explore the safety in larger number of patients after longer-term treatment and also to confirm the efficacy in terms of clinical endpoints. In many instances the clinical team looks at data in different ways; for example, they might look at data from a completed trial to determine if the drug was more effective in a certain subgroup; if there is evidence that this might be the case, a phase 4 trial in that group will be initiated.

In July of 2012, the FDA initiated an accelerated approval program that allows for earlier approval of drugs that treat serious conditions and that fills an unmet medical need based on a surrogate endpoint. A surrogate endpoint is a marker, such as a laboratory measurement, radiographic image, physical sign, or other measure that is thought to predict clinical benefit, but is not itself a measure of clinical benefit. The use of a surrogate endpoint can considerably shorten the time required prior to receiving FDA approval. Drug companies are still required to conduct phase 4 confirmatory trials to evaluate efficacy and safety of a compound. If the confirmatory trial shows that the drug provides a clinical benefit, the FDA grants traditional approval for the drug. If the confirmatory trial does not show that the drug provides clinical benefit, the FDA has regulatory procedures in place that could lead to removing the drug from the market.

Two examples of drugs that received accelerated approval are bevacizumab (Avastin, Genentech), a humanized variant of this anti-VEGF antibody, and gefitinib (Iressa, AstraZeneca). Bevacizumab is a recombinant humanized mAb that blocks angiogenesis by inhibiting vascular endothelial growth factor A (VEGF-A). VEGF-A is a chemical signal that stimulates angiogenesis in a variety of diseases, especially in cancer. Bevacizumab was the first clinically available angiogenesis inhibitor in the United States. Using the accelerated approval process, bevacizumab was approved by the FDA for certain metastatic cancers. It received its first approval in 2004, for combination use with standard chemotherapy for metastatic colon cancer (Cohen et al., 2003). It has since been approved for use in certain lung cancers, renal cancers, ovarian cancers, and glioblastoma multiforme of the brain. It had been approved for breast cancer, but that approval was withdrawn when later studies showed no evidence of effectiveness. This was the first time that the FDA withdrew approval for a drug since the accelerated approval process was developed, and its decision was unpopular with patients and physicians who cited the lack of therapeutic options for patients with metastatic breast cancer.

In May of 2003, gefitinib received accelerated approval by the FDA as monotherapy for patients with locally advanced or metastatic non—small-cell lung cancer (NSCLC) after failure of both platinum-based and docetaxel chemotherapies (US Food and Drug Administration Center for Drug Evaluation and Research, 2005). (The drug received regulatory clearance in the United States via the accelerated approval process.) However, in December 2004, AstraZeneca notified the FDA that they had been unable to demonstrate effectiveness in a confirmatory trial (King). Therefore, in 2005, the FDA recommended that Iressa be withdrawn from the market. This recommendation delayed the launch of Iressa in Europe until 2009 and prompted AstraZeneca to withdraw a second marketing application with the FDA (Mok et al., 2009). To secure approval in Europe, AstraZeneca reconfigured its development program in an attempt to correct mistakes made the first time (Mok et al., 2009). From the beginning of the process, the company had identified the EGFR mutation as a marker, but they did not screen patients for this mutation in earlier trials. It was later determined that only patients with activating mutations in EGFR-TK respond to Iressa (Sequist et al., 2008; Adam and Brantner, 2010). The company developed companion diagnostic tests that allow physicians to identify patients with EGFR mutations. This helped secure European approval of Iressa in July 2009 for the treatment of locally advanced or metastatic NSCLC in patients who have mutations in the genes that control EGFR production. If AstraZeneca acted earlier in screening patients for EGFR mutations, and if they had sought approval with a companion diagnostic, then Iressa would already be available in the United States. In targeted therapy, the diagnostic is often as important as the drug itself (Mok et al., 2009).

NEW DRUG DISCOVERY PARADIGMS

The 21st century has seen increasing costs, expanding timelines, and a higher failure rate for clinical trials. The cost of development for a drug is estimated at about $868 million, and estimates run as high as 15 years to get a drug from target identification to market (US Dept. of Health and Human Services Food and Drug Administration, 2004). Given these high costs as well as the high attrition rate for drug in development, in 2004 the US FDA proposed a "critical path" initiative outlining new ways and opportunities to improve drug development (US Dept. of Health and Human Services Food and Drug Administration, 2006). This "critical path" report was updated in 2006 (Mauro et al., 2002) with the lessons learned since 2004, including a clear consensus that biomarker development, streamlining of clinical trials, and use of modeling and simulation were the most important areas for improving medical product development. A summary and status of these initiatives can be found at www.fda.gov/oc/initiatives/criticalpath/.

An example of such a new paradigm for development of rationally targeted oncology drugs is the development of imatinib (GLEEVEC Novartis) a protein—tyrosine kinase inhibitor originally approved for chronic myelogenous leukemia (CML), a rare form of cancer that affects certain types of white blood cells (Pray, 2008). Based on decades of research, it was determined that imatinib was effective in patients with particular genetic biomarkers, including BCR-ABL (the Philadelphia chromosome), KIT, and PDGFR gene rearrangements. Because of its incredible success, the FDA allowed for the streamlining of clinical trials in other indications. Today imatinib is approved in the United States for the treatment of a wide range of conditions related to these particular biomarkers. The imatinib drug discovery and development program incorporated the use of biomarker development and streamlined clinical trials as promoted by the critical path initiative. It also created one of the drugs at the forefront of personalized medicine (US Food and Drug Administration, 2013).

The Move Toward Personalized Medicine

In 2013, the US FDA published a report (US Food and Drug Administration, 2012) describing the progress that has been made in the streamlining of clinical trials since the initiative began in 2004. They describe the move to "personalized" or "precision medicine," defined as "the use of genomic, epigenomic, exposure, and other data" to define individual disease patterns that could potentially lead to better individual treatment. Personalized medicine generally involves the use of two medical products—typically, a diagnostic device and a therapeutic product—to improve patient outcomes. Diagnostic devices can be in vitro tests that look at genetic factors or in vivo tests, such as electroencephalography, electrocardiography, or diagnostic imaging. These tests or devices are used to identify groups that will benefit from a particular medication. Enrichment is the terms used to describe the prospective use of a patient characteristic (demographic, pathophysiologic, historical, genetic, or other) to select a study population in which detection of a drug effect is more likely than it would be in an unselected population (US Food and Drug Administration, 2013; US Food and Drug Administration, 2012). Diseases with high placebo response rate are particularly good candidates for enrichment designs; a placebo lead-in phase can be used to identify responders to placebo and eliminate them from the second phase of the study.

One form of enrichment strategy involves enrolling high-risk patients who are more likely to respond to treatment. This is known as predictive enrichment and can be based on an understanding of the disease (pathophysiology, tumor receptors) or it can be empiric (e.g., based on history, early response, or response of a biomarker). Ridker et al. (2008) used an enrichment design for the JUPITER trial by enrolling 17,802 apparently healthy men and women with LDL levels of <130 mg/dL and high-sensitivity C-reactive protein (CRP) levels ($\geq$2.0 mg/L) and randomizing them to rosuvastatin 20 mg daily or placebo. The combined primary endpoint was the occurrence of myocardial infarction, stroke, arterial revascularization, hospitalization for unstable angina, or death from cardiovascular causes during the study period. Rosuvastatin reduced the incidence of major cardiovascular events in this group and led to a better understanding of the importance of measuring CRP levels.

Adaptive design is another method for increasing clinical trial efficiency. Adaptive designs can be classified into three categories—prospective, concurrent, and retrospective. An adaptive design allows modifications to be made to the trial and/or statistical analysis of ongoing trials. Prospective adaptations include, but are not limited to, adaptive randomization; stopping a trial early due to safety, futility, or efficacy at interim analysis; dropping the losers (or inferior treatment group); and sample size reestimation (Chow and Chang, 2008). Table 41.3 describes a number of adaptive designs that are currently in use.

Collaborations between pharmaceutical companies, academia, government agencies, and other groups represent a growing trend in drug development as a means of combining resources to increase knowledge and understanding in many therapeutic areas. I-SPY (investigation of serial studies to predict your therapeutic response with imaging and molecular analysis 2) is an example of one such collaboration. This study employed an adaptive phase 2 clinical trial design in a neoadjuvant setting for women with locally advanced breast cancer. I-SPY 2 is a collaborative effort among academic investigators, the National Cancer Institute, the US FDA, and the pharmaceutical and biotechnology industries under the auspices of the Foundation for the National Institutes of Health Biomarkers Consortium. In I-SPY 2, which was launched in March 2010, all enrolled patients had chemotherapy and invasive breast cancers and received the standard of care. Most also received an investigational agent from Abbott Laboratories, Amgen, and Pfizer. The drugs included veliparib, a small-molecule inhibitor of PARP1; conatumumab, a human mAb against APO/TRAIL (tumor necrosis factor-related

TABLE 41.3 Most Common Forms of Adaptive Randomization Designs

Type of Adaptive Design	Description
Adaptive randomization design	Modification of randomization schedules based on varied and/or unequal probabilities of treatment assignment to increase the probability of success
Group sequential design	Prematurely stopping a trial due to safety, futility/efficacy, or both, with the option of additional adaptations based on results of an interim analysis
Sample size reestimation design	Sample size adjustment or reestimation based on interim data
Drop-the-losers design	Dropping the inferior treatment group as well as adding additional arms
Adaptive dose-finding design	Identifying the minimum effective dose and/or the maximum tolerable dose, which is used to determine the dose level for the next phase clinical trials
Biomarker adaptive design	Adaptation based on the response of biomarkers such as genomic markers
Adaptive treatment-switching design	Allows the investigator to switch a patient's treatment from an initial assignment to an alternative treatment if there is evidence of lack of efficacy or safety of the initial treatment
Adaptive-hypotheses design	Allows modifications or changes in hypotheses based on interim analysis results.
Adaptive seamless phase 2/3 design	Combining phases 2b and 3 of clinical development into one trial
Multiple-adaptive design	Combining any of the aforementioned aspects of adaptive design

apoptosis-inducing ligand); AMG 386, a peptibody Fc fragment linked to a peptide that inhibits the proangiogenic factors angiopoietin-1 (Tie-2) and angiopoietin-2; figitumumab, a fully human IgG2 mAb against insulin-like growth factor receptor; and neratinib, a pan-ErbB (HER2 kinase) inhibitor. Over the entire course of the trial, 12 drugs were evaluated. Under an unusual master IND approval from FDA, the investigators were able to add and drop new compounds to the trial without writing new protocols, saving several years and hundreds of millions of dollars. Other coalitions [Personalized

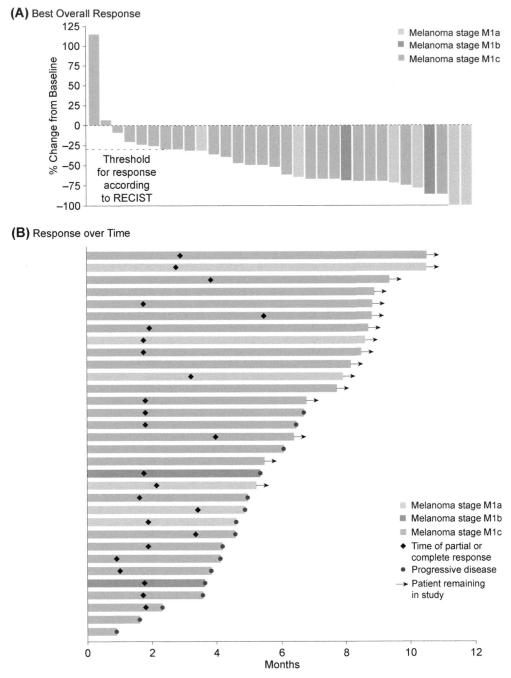

FIGURE 41.11 Antitumor Response in Each of the 32 Patients in the Extension Cohort. All 32 patients had melanoma tumors and were treated at the recommended phase 2 dose of 960 mg twice daily. Panel A shows the best overall response for each of the 32 patients, measured as change from baseline in the sum of the largest diameter of each target lesion. Negative values indicate tumor shrinkage, and the *dashed line* indicates the threshold for a partial response according to response evaluation criteria in solid tumors (RECIST) (i.e., shrinkage by 30%). Two patients had a complete response. Panel B shows the duration and characteristics of the responses in each patient. *Reproduced from McArthur, G.A., Chapman, P.B., Robert, C., Larkin, J., Haanen, J.B., Dummer, R., Ribas, A., et al., March 2014. Safety and efficacy of vemurafenib in BRAF(V600E) and BRAF(V600K) mutation-positive melanoma (BRIM-3): extended follow-up of a phase 3, randomised, open-label study. Lancet Oncol. 15 (3), 323–332 with permission.*

Medicine Coalition (PMC), Partners HealthCare Personalized Medicine (Brigham and Women's Hospital, Mass General Hospital, Harvard University Medical School)] are also joining together and working toward better-designed trials.

Example of the Personalized Medicine Paradigm in Drug Development

Almost 50% of metastatic melanoma patients have a $BRAF^{V600}$ mutation, and the introduction of BRAF inhibitors (BRAFi) has greatly improved their treatment options. Vemurafenib (ZELBORAF), a BRAF enzyme inhibitor, is an example of a drug that fits the criteria for personalized medicine. The compound was first synthesized in 2005 using cocrystallography, a scaffold-based approach to drug discovery. With scaffold-based lead discovery, an initially very small molecule (MWT 150−350) that binds to a protein of interest is identified. Next, it is elaborated based on detailed structural data to generate a hit, which is subsequently optimized to produce a lead, and (in some cases) a drug.

Initial biochemical characterization of vemurafenib revealed that it has mild selectivity for $BRAF^{V600E}$ over the wild-type enzyme. Although this selectivity was modest, in melanoma or colorectal cancer cell lines it was remarkable (Bollag et al., 2012). The substantial increase in cellular selectivity compared to biochemical selectivity is attributed to preferential binding of vemurafenib to the monomeric form of the BRAF enzyme; the $BRAF^{V600E}$ mutation favors the monomeric form.

The first phase 1 trial of vemurafenib included a 55 patient dose escalation followed by a dose expansion in 32 patients (Flaherty et al., 2010). All advanced solid tumor patients were eligible for participation in the dose escalation; however, most patients had a $BRAF^{V600E}$ mutation. Patients received daily drug dosing until the development of toxicity or progression of disease. Vemurafenib was initially developed in a crystalline formulation and administered at a starting dose of 200 mg daily. After 26 patients were treated, accrual was stopped due to poor pharmacokinetics. The drug was then reformulated into a more highly bioavailable powder, and accrual was restarted.

The phase 1 dose expansion enrolled only melanoma patients with $BRAF^{V600E}$ mutations and showed a striking response rate. Twenty-six of thirty-two patients (81%) exhibited a response, and some experienced marked improvement in quality of life, as indicated by decreased narcotic use (Fig. 41.11). Because of these results, vemurafenib was rapidly moved to phase 2 and phase 3 trials in melanoma.

The phase 2 study (BRIM-2) enrolled only patients with a confirmed $BRAF^{V600E}$ mutation (Young et al., 2012). One hundred and thirty-two patients were treated with vemurafenib. The primary endpoint of the study was met with a 52% response rate.

BRIM-3 is a phase 3 study of vemurafenib in $BRAF^{V600E}$ mutant melanoma (McArthur et al., 2014). This two-arm randomized study compared vemurafenib, 960 mg oral twice daily, to dacarbazine chemotherapy, 1000 mg/m² administered every 3 weeks. The initial primary endpoint was overall survival; progression-free survival was later added as a

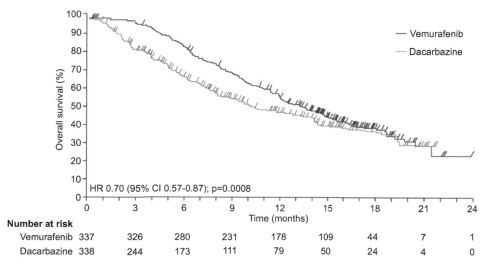

FIGURE 41.12 Overall survival (randomized population, censored at crossover) for patients assigned to vemurafenib or dacarbazine. *Reproduced from McArthur, G.A., Chapman, P.B., Robert, C., Larkin, J., Haanen, J.B., Dummer, R., Ribas, A., et al., March 2014. Safety and efficacy of vemurafenib in BRAF(V600E) and BRAF(V600K) mutation-positive melanoma (BRIM-3): extended follow-up of a phase 3, randomised, open-label study. Lancet Oncol. 15 (3), 323−332 with permission.*

coprimary endpoint. Overall survival at 6 months was 84% in the vemurafenib arm and 64% in the dacarbazine arm. PFS was evaluated in 81% (549/675) of patients with a median value of 5.3 months for the vemurafenib arm versus 1.6 months for the dacarbazine arm. Fig. 41.12 depicts overall survival for both arms in this phase 3 trial.

Vemurafenib is now indicated for the treatment of patients with unresectable or metastatic melanoma bearing the $BRAF^{V600E}$ mutation. The product is combined with the cobas 4800 BRAF V600 Mutation Test, an in vitro diagnostic device intended for the qualitative detection of the $BRAF^{V600E}$ mutation in DNA extracted from formalin-fixed, paraffin-embedded human melanoma tissue. ZELBORAF binds to some forms of mutated BRAF, including V600E, and renders the oncogenic protein inactive. As a result, downstream signaling is inhibited, blocking proliferation and potentially leading to apoptosis and immune-mediated tumor regression. Using this personalized strategy, clinicians are able to select their patients who are most likely to benefit from treatment with vemurafenib.

CONCLUSION

Today, the need for effective translational science in the form of drug discovery and development is as great as ever. For example, resistance to conventional antibiotics has led to the challenge of superinfections, and increased globalization has led to the spread of viruses beyond their areas of origin. At the time of publication of this book, NIH's National Institute of Allergy and Infectious Diseases and GlaxoSmithKline are collaborating to develop a vaccine to protect patients against the spread of Ebola from West Africa.

However, as the landscape changes, so do the opportunities. As we move from a focus on single-receptor diseases to more complex conditions such as cancers with multiple genotypes and long-term neurologic diseases, enhanced cooperation and collaboration between the biopharmaceutical industry and academic research facilities is more important than ever. Developing these innovative partnership models will help to control costs and increase efficiencies. In the same way, the move to more cost-effective and flexible adaptive designs for clinical trials will help patients and investigators to focus on patient health and safety as well as key findings. The focus on more personalized approaches to patient care, which becomes possible due to our expanding knowledge of the human genome, promises to expand exponentially in the 21st century. In fact, the company NantHealth is developing strategies for comprehensive cancer care around their ability to provide full exome sequences in near real time (http://nanthealth.com/science-of-managing-cancer-care). As our population ages, the need for innovative solutions to patient health is increasing. Success in the future lies in our ability to continue to develop and carry out creative translational science solutions in an ever-changing landscape.

REFERENCES

Adam, C.P., Brantner, V.V., 2010. Spending on new drug development. Health Econ. 19, 130–141.

Ash, A.S.F., Schild, H.O., 1966. Receptors mediating some actions of histamine. Br. J. Pharmacol. Chemother. 27, 427–439.

Balimane, P.V., Han, Y.-H., Chong, S., 2006. Current industrial practices of assessing permeability and P-glycoprotein interaction. AAPS J. 8, E1–E13.

Banting, F.G., Best, C., 1922. The internal secretion of the pancreas. J. Lab. Clin. Med. 7, 251–266.

Bennett, J.W., 2001. Adrenalin and cherry trees. Mod. Drug Discov. 4 (12), 47–48, 51.

Bjornsson, T.D., Callaghan, J.T., Einolf, H.J., Fischer, V., Gan, L., Grimm, S., et al., Pharmaceutical Research and Manufacturers of America (PhRMA) Drug Metabolism/Clinical Pharmacology Technical Working Group; FDA Center for Drug Evaluation and Research (CDER), July 2003. The conduct of in vitro and in vivo drug-drug interaction studies: a Pharmaceutical Research and Manufacturers of America (PhRMA) Perspective. Drug Metab. Dispos. 31 (7), 815–832.

Bollag, G., Tsai, J., Zhang, J., Zhang, C., Ibrahim, P., Nolop, K., Hirth, P., November 2012. Vemurafenib: the first drug approved for *BRAF*-mutant cancer. Nat. Rev. Drug Discov. 11 (11), 873–886.

Boverhof, D.R., Zacharewski, T.R., 2006. Toxicogenomics in risk assessment: applications and needs. Toxicol. Sci. 8, 352–360.

Boycott, K.M., Vanstone, M.R., Bulman, D.E., MacKenzie, A.E., 2013. Rare-disease genetics in the era of next-generation sequencing: discovery to translation. Nat. Rev. Genet. 14, 681–691.

Boyd, R.A., Lalonde, R.L., 2007. Nontraditional approaches to first-in-human studies to increase efficiency of drug development: will microdose studies make a significant impact? Clin. Pharmacol. Ther. 81, 24–26.

Büller, H.R., January 2, 2014. Edoxaban versus warfarin for venous thromboembolism. N. Engl. J. Med. 370 (1), 80–81. http://dx.doi.org/10.1056/NEJMc1313883.

Chin, L., Anderson, J.N., Futreal, P.A., 2011. Cancer genomics: from discovery science to personalized medicine. Nature Medicine 17, 297–303. http://dx.doi.org/10.1038/nm.2323.

Chin, L., Hahn, W.C., Getz, G., Meyerson, M., May 1, 2012. Making sense of cancer genomic data. Genes Dev. 26 (9), 1003.

Chow, S.-C., Chang, M., 2008. Adaptive design methods in clinical trials—a review. Orphanet J. Rare Dis. 3, 11.

CMR (Centre for Medicines Research International Ltd, 2006. CMR International. 2006/2007 Pharmaceutical R&D Factbook. CMR, Surrey, UK.

Code, C.F., 1956. Histamine and gastric secretion. In: O'Connor, C.M. (Ed.), Wolstenholme GEW. Churchill, London, pp. 189–219.

Cohen, M.H., Johnson, J.R., Chattopadhyay, S., Tang, S., Justice, R., Sridhara, R., et al., 2003. FDA drug approval summary: gefitinib (ZD1839) (Iressa®) tablets. Oncologist 15 (12), 1344–1351.

Colwell Jr., C.W., Spiro, T.E., Trowbridge, A.A., Stephens, J.W., Gardiner Jr., G.A., Ritter, M.A., December 1995. Efficacy and safety of enoxaparin versus unfractionated heparin for prevention of deep venous thrombosis after elective knee arthroplasty. Enoxaparin Clinical Trial Group. Clin. Orthop. Relat. Res. 321, 19–27.

Committee for Medicinal Products for Human Use (CHMP), March 22, 2007. Guideline on Requirements for First-In-Man Clinical Trials for Potential High-Risk Medicinal Products. European Medicines Agency, London, UK, 1–11. Doc. Ref.EMEA/CHMP/SWP/28367/2007 Corr.

Dedric, R.L., 1973. Animal scale-up. J. Pharmacokin Biopharm. 1, 435–461.

Dolgin, E., 2014. Massive schizophrenia genomics study offers new drug directions. Nat. Rev. Drug Discov. 13, 641–642.

Eba, J., Shimokawa, T., Nakamura, K., Shibata, T., Misumi, Y., Okamoto, H., , et al.Lung Cancer Study Group of the Japan Clinical Oncology Group, January 2015. A phase II/III study comparing carboplatin and irinotecan with carboplatin and etoposide for the treatment of elderly patients with extensive-disease small-cell lung cancer (JCOG1201). Jpn. J. Clin. Oncol. 45 (1), 115–187.

Estabrook, R.W., 2003. A passion for P450s (remembrances of the early history of research on cytochrome P450). Drug Metab. Dispos. 31, 1461–1473.

Evans, D.C., Watt, A.P., Nicoll-Griffith, D.A., Baillie, T.A., 2004. Drug-protein adducts: an industry perspective on minimizing the potential for drug bioactivation in drug discovery and development. Chem Res Toxicol. 17, 3–16.

Expert Scientific Group, 2006. Expert Scientific Group on Phase One Clinical Trials, Final Report. The Stationery Office, Norwich, UK.

Ferrara, N., Hillan, K.J., Gerber, H.P., Novotny, W., 2001. Discovery and development of bevacizumab and anti-VEGF antibody for treating cancer. Nat. Rev. 3, 391–400.

Flaherty, K.T., Puzanov, I., Kim, K.B., Ribas, A., McArthur, G.A., Sosman, J.A., et al., August 26, 2010. Inhibition of mutated, activated BRAF in metastatic melanoma. N. Engl. J. Med. 363 (9), 809–819.

Floss, H.G., 2006. Combinatorial biosynthesis—potential and problems. J. Biotechnol. 124, 242–257.

Folkow, B., Haeger, K., Kahlson, G., 1948. Observations on reactive hyperaemia as related to histamine on drugs antagonizing vasodilatation induced by histamine and on vasodilator properties of adenosine triphosphate. Acta Physiol. Scand. 15, 264–278.

Fowler, J.S., Volkow, N.D., Wang, G.J., Ding, Y.S., Dewey, S.L., July 1999. PET and drug research and development. J. Nucl. Med. 40 (7), 1154–1163.

Fox, S., Farr-Jones, S., Sopchak, L., et al., 2006. High-throughput screening: update on practices and success. J Biomol Screening 11, 864–869.

Freston, J.W., 1987. Safety perspectives on parenteral H_2-receptor antagonists. Am. J. Med. 83, 58–67.

Fuji, T., Fujita, S., Tachibana, S., Kawai, Y., 2010. A dose-ranging study evaluating the oral factor Xa inhibitor edoxaban for the prevention of venous thromboembolism in patients undergoing total knee arthroplasty. J. Thromb. Haemost. 8, 2458–2468.

Giugliano, R.P., Ruff, C.T., Braunwald, E., Murphy, S.A., Wiviott, S.D., Halperin, J.L., et al., November 28, 2013. Edoxaban versus warfarin in patients with atrial fibrillation. N. Engl. J. Med. 369 (22), 2093–2104.

Goldman, P., 2001. Herbal medicines today and the roots of modern pharmacology. Ann. Intern. Med. 135, 594–600.

Greaves, P., Williams, A., Eve, M., 2004. First dose of potential new medicines to humans: how animals help. Nat. Rev. Drug Discov. 3, 226–236.

Gualandi-Signorini, A.M., Giorgi, G., 2001. Insulin formulations—a review. Eur. Rev. Med. Pharmacol. Sci. 5 (3), 73–83.

Hay, M., Thomas, D.W., Craighead, J.L., Economides, C., Rosenthal, J., January 2014. Clinical development success rates for investigational drugs. Nat. Biotechnol. 32 (1), 40–51.

Hopkins, A.L., Groom, C.R., 2002. The druggable genome. Nat. Rev. Drug Discov. 1, 727–730.

ICH, October 27, 1994. ICH Harmonised Tripartite Guideline, The Extent of Population Exposure to Assess Clinical Safety for Drugs Intended for Long-Term Treatment of Non-Life-Threatening Conditions E1. ICH, 1–3.

Imming, P., Sinning, C., Meyer, A., 2006. Drugs, their targets and the nature and number of drug targets. Nat. Rev. Drug Discov. 5, 821–834.

IMS Health, May 2013. IMS World Review 2013. IMS Health.

Ivy, S.P., Siu, L.L., Garrett-Mayer, E., Rubinstein, L., 2010. Approaches to phase 1 clinical trial design focused on safety, efficiency, and selected patient populations: a report from the clinical trial design task force of the National Cancer Institute Investigational Drug Steering Committee. Clin. Cancer Res. 16, 1726–1736.

Jhoti, H., Williams, G., Rees, D.C., Murray, C.W., 2013. The 'rule of three' for fragment-based drug discovery: where are we now? Nat. Rev. Drug Discov. 12, 644.

Kaitin, K.I., DiMasi, J.A., 2011. Pharmaceutical innovation in the 21st century: new drug approvals in the first decade, 2000-2009. Clin. Pharmacol. Ther. 89 (2), 183–188.

Kenter, M.J.H., Cohen, A.F., 2006. Establishing risk of human experimentation with drugs: lessons from TGM1412. Lancet 368, 1387–1391.

King, S., Spotlight On: Can AstraZeneca secure a fitting conclusion to its Iressa-themed history lesson? Available at: http://www.firstwordpharma.com/node/1249737#axzz3LmWpJ2P4.

Kohler, G., Milstein, C., 1975. Continuous cultures of fused cells secreting antibody of predefined specificity. Nature 256 (5517), 495–497.

Kola, I., Landis, J., 2004. Can the pharmaceutical industry reduce attrition rates? Nat. Rev. Drug Discov. 3, 711–715.

Krieger, D.E., Erickson, B.W., Merrifield, R.B., 1976. Affinity purification of synthetic peptides. Proc. Nat. Acad. Sci. U.S.A. 73, 3160–3164.

Lappin, G., Kuhnz, W., Jochemsen, R., Kneer, J., Chaudhary, A., Oosterhuis, B., et al., September 2006. Use of microdosing to predict pharmacokinetics at the therapeutic dose: experience with 5 drugs. Clin. Pharmacol. Ther. 80 (3), 203–215.

LeTorneau, C., Lee, J.J., Siu, L.L., 2009. Dose escalation methods in phase 1 cancer clinical trials. J. Natl. Cancer Inst. 101, 708–720.

Levy, G., 1964a. Relationship between elimination rate of drugs and rate of decline of their pharmacologic effects. J. Pharm. Sci. 53, 342–343.

Levy, G., 1964b. Relationship between rate of elimination of tubocurarine and rate of decline of its pharmacological activity. Br. J. Anaesth. 36, 694−695.

Liebler, D.C., Guengerich, F.P., 2005. Elucidating mechanisms of drug-induced toxicity. Nat. Rev. Drug Discov. 4, 410−420.

Lindsay, M.A., 2003. Target discovery. Nat. Rev. Drug Discov. 2, 831−838.

Lipinski, C.A., Lombardo, F., Dominy, B.W., Feeney, P.J., 2001. Experimental and computational approaches to estimate solubility and permeability in drug discovery and development settings. Adv. Drug Deliv. Rev. 46, 3−26.

Lovenox [package Insert], 2005. Aventis Pharmaceuticals Inc, Bridgewater, NJ.

Lusis, A.J., Fogelman, A.M., Fonarow, G.C., 2004. Genetic basis of atherosclerosis: P\part I: new genes and pathways. Circulation 110, 1868−1873.

Lutterotti, A., Yousef, S., Sputtek, A., Stürner, K.H., Stellmann, J.P., Breiden, P., et al., June 5, 2013. Antigen-specific tolerance by autologous myelin peptide-coupled cells: a phase 1 trial in multiple sclerosis. Sci. Transl. Med. 5 (188), 188ra75.

Macleod, J.J., 1922. Pancreatic extract and diabetes. Can. Med. Assoc. J. 12 (6), 423−425.

Mager, D.E., Wyska, E., Jusko, W.J., 2003. Diversity of mechanism−based pharmacodynamic models. Drug Metab. Dispos. 31, 510−519.

Mauro, M.J., O'Dwyer, M., Heinrich, M.C., Druker, B.J., 2002. STI571: a paradigm of new agents for cancer therapeutics. J. Clin. Oncol. 20, 325−334.

McArthur, G.A., Chapman, P.B., Robert, C., Larkin, J., Haanen, J.B., Dummer, R., Ribas, A., et al., March 2014. Safety and efficacy of vemurafenib in BRAF(V600E) and BRAF(V600K) mutation-positive melanoma (BRIM-3): extended follow-up of a phase 3, randomised, open-label study. Lancet Oncol. 15 (3), 323−332.

Medina-Franco, J.L., Méndez-Lucio, O., Martinez-Mayorga, K., 2014. The interplay between molecular modeling and chemoinformatics to characterize protein-ligand and protein-protein interactions landscapes for drug discovery. Adv. Protein Chem. Struct. Biol. 96, 1−37.

Mok, T.S., Wu, Y.L., Thongprasert, S., Yang, C.H., Chu, D.T., Saijo, N., et al., September 3, 2009. Gefitinib or carboplatin-paclitaxel in pulmonary adenocarcinoma. N. Engl. J. Med. 361 (10), 947−957.

Molinder, H.K.M., 1994. The development of cimetidine: 1964-1976. J. Clin. Gastroenterol. 19, 248−254.

Mountain, A., Adair, J.R., 1992. Engineering antibodies for therapy. Biotechnol. Genet. Eng. Rev. 10, 1−142. http://www.ncbi.nlm.nih.gov/pubmed/1301737.

Newman, D.J., Cragg, G.M., 2007. Natural products as sources of new drugs over the last 25 years. J. Nat. Prod. 70, 461−477.

Nicholson, R.L., Welch, M., Ladlow, M., Spring, D.R., 2007. Small-molecule screening: advances in microarraying and cell-imaging technologies. ACS Chemical Biology 2, 24−30.

Obach, R.S., Baxter, J.G., Liston, T.E., Silber, B.M., Jones, B.C., MacIntyre, F., et al., October 1997. The prediction of human pharmacokinetic parameters from preclinical and in vitro metabolism data. J. Pharmacol. Exp. Ther. 283 (1), 46−58.

Ostro, M.J., 1987. Pharmacodynamics and pharmacokinetics of parenteral histamine (H_2)-receptor antagonists. Am. J. Med. 83, 15−22.

Overington, J.P., Al-Lazikani, B., Hopkins, A.L., 2006. How many drug targets are there? Nat. Rev. Drug Discov. 5, 993−996.

Pang, Y.P., 2007. *In Silico* drug discovery: solving the "target-rich and lead-poor" imbalance using the genome-to-drug-lead paradigm. Clin. Pharmacol. Ther. 81, 30−34.

Pharmaceutical Research and Manufacturers of America (PhRMA), April 2014. 2014 Profile, Pharmaceutical Research and Manufacturers of America (PhRMA), Washington, DC.

Polu, K.R., Lowman, H.B., 2014. Probody therapeutics for targeting antibodies to diseased tissue. Expert Opin. Biol. Ther. 14 (8), 1049−1053.

Pray, L.A., 2008. Gleevec: the breakthrough in Cancer Treatment. Nat. Educ. 1 (1), 37.

Rabinowitz, M., Shankley, N., 2006. The impact of combinatorial chemistry on drug discovery. In: Smith, C.G., O'Donnell, J.T. (Eds.), The Process of New Drug Discovery and Development, second ed. Informa Healthcare, New York, pp. 56−77.

Raskob, G., Cohen, A.T., Eriksson, B.I., et al., 2010. Oral direct factor Xa inhibition with edoxaban for thromboprophylaxis after elective total hip replacement: a randomized double-blind dose-response study. Thromb. Haemost. 104, 642−649.

Ridker, P.M., Danielson, E., Fonseca, F.A., Genest, J., Gotto Jr., A.M., Kastelein, J.J., et al., November 20, 2008. Rosuvastatin to prevent vascular events in men and women with elevated C-reactive protein. N. Engl. J. Med. 359 (21), 2195−2207.

Rohatagi, S., Mendell, J., Kastrissios, H., Green, M., Shi, M., Patel, I., et al., November 2012. Characterisation of exposure versus response of edoxaban in patients undergoing total hip replacement surgery. Thromb. Haemost. 108 (5), 887−895.

Rolinson, G.N., 1998. Forty years of β-lactam research. J. Antimicrob. Chemother. 2641, 589−603.

Salazar, D.E., Mendell, J., Kastrissios, H., Green, M., Carrothers, T.J., Song, S., et al., May 2012. Modelling and simulation of edoxaban exposure and response relationships in patients with atrial fibrillation. Thromb. Haemost. 107 (5), 925−936.

Schizophrenia Working Group of the Psychiatric Genomics Consortium, 2014. Biological insights from 108 schizophrenia-associated genetic loci. Nature 511, 421−427.

Sequist, L.V., et al., 2008. First-line gefitinib in patients with advanced non-small cell lung cancer. Harbouring somatic EGFR mutations. J. Clin. Oncol. 26, 2442−2449.

Shoichet, B.K., 2004. Virtual screening of chemical libraries. Nature 432, 862−865.

Shukra, A.M., Sridevi, N.V., Chandran, D., Maithal, K., 2014. Production of recombinant antibodies using bacteriophages. Eur. J. Microbiol. Immunol. (Bp) 4 (2), 91−98. http://www.ncbi.nlm.nih.gov/pubmed/24883194.

Silverman, E., November 18, 2013. Biotech meds are swelling those pharma pipelines. Forbes.

Stein, C.M., 2003. Managing risk in healthy subjects participating in clinical research. Clin. Pharmacol. Ther. 74, 511−512.

Steinbrook, R., 2002a. Improving protection for research subjects. New Engl. J. Med. 346, 1425–1430 [published correction appears in New Engl J Med. 2002;346:1838-a].

Steinbrook, R., February 28, 2002. Protecting research subjects—the crisis at Johns Hopkins. N. Engl. J. Med. 346 (9), 716–720 [published correction appears in New Engl J Med. 2002;346:1678-a].

Suntharalingam, G., Perry, M.R., Ward, S., Brett, S.J., Castello-Cortes, A., Brunner, M.D., et al., September 7, 2006. Cytokine storm in a phase 1 trial of the anti-CD28 monoclonal antibody TGN1412. N. Engl. J. Med. 355 (10), 1018–1028.

Tang, H., Mayersohn, M., 2005. A novel model for prediction of human drug clearance by allometric scaling. Drug Metab. Dispos. 33, 1297–1303.

Tsai, J., Lee, J.T., Wang, W., Zhang, J., Cho, H., Mamo, S., et al., February 26, 2008. Discovery of a selective inhibitor of oncogenic B-Raf kinase with potent antimelanoma activity. Proc. Natl. Acad. Sci. U.S.A. 105 (8), 3041–3046.

US Dept of Health and Human Services, Food and Drug Administration, Center for Drug Evaluation and Research (CDER), Center for Biologics Evaluation and Research (CBER), July 1997. Guidance for Industry; S2B Genotoxicity: A Standard Battery for Genotoxicity Testing of Pharmaceuticals. US Department of Health and Human Services, Rockville, MD.

US Dept of Health and Human Services, Food and Drug Administration, Center for Drug Evaluation and Research (CDER), Center for Biologics Evaluation and Research (CBER), September 2006. Guidance for Industry; Drug Interaction Studies—Study Design, Data Analysis, and Implications for Dosing and Labeling; Draft Guidance. US Department of Health and Human Services, Rockville, MD.

US Dept of Health and Human Services, Food and Drug Administration, Center for Drug Evaluation and Research (CDER), Center for Biologics Evaluation and Research (CBER), July 1997. Guidance for Industry; M3 Nonclinical Safety Studies for the Conduct of Human Clinical Trials for Pharmaceuticals. US Department of Health and Human Services, Rockville, MD.

US Dept of Health and Human Services, Food and Drug Administration, Center for Drug Evaluation and Research (CDER), Center for Biologics Evaluation and Research (CBER), January 2010. Guidance for Industry; M3(R2) Nonclinical Safety Studies for the Conduct of Human Clinical Trials and Marketing Authorization for Pharmaceuticals. US Department of Health and Human Services, Rockville, MD.

US Dept of Health and Human Services, Food and Drug Administration, Center for Drug Evaluation and Research (CDER), July 2005. Guidance for Industry; Estimating the Maximum Safe Starting Dose in Initial Clinical Trials for Therapeutics in Adult Healthy Volunteers. US Department of Health and Human Services, Rockville, MD.

US Dept of Health and Human Services, Food and Drug Administration, Center for Drug Evaluation and Research (CDER), January 2006. Guidance for Industry, Investigators, and Reviewers; Exploratory IND Studies. US Department of Health and Human Services, Rockville, MD.

US Dept of Health and Human Services, Food and Drug Administration, March 2004. Challenge and Opportunity on the Critical Path to New Medical Products.

US Dept of Health and Human Services, Food and Drug Administration, March 2006. Critical Path Opportunities Report.

US Food and Drug Administration Center for Drug Evaluation and Research, January 2014. Novel New Drugs 2013 Summary. US Food and Drug Administration, Rockville, MD.

US Food and Drug Administration Center for Drug Evaluation and Research, March 4, 2005. Oncologic Drugs Advisory Committee Meeting.

US Food and Drug Administration, December 2012. FDA Draft Guidance for Industry: Enrichment Strategies for Clinical Trials. US Department of Health and Human Services, Rockville, MD.

US Food and Drug Administration, October 2013. Paving the Way for Personalized Medicine: FDA's Role in a New Era of Medical Product Development. US Department of Health and Human Services, Rockville, MD.

Vane, J., 2000. Aspirin and other anti-inflammatory drugs. Thorax 55, S3–S9.

Vernon, J.A., Golec, J.H., Steven, J.S., 2010. Comparative effectiveness regulations and pharmaceutical innovation. Pharmacoeconomics 28 (10), 877–887.

Wagner, J.A., Williams, S.A., Webster, C.J., 2007. Biomarkers and surrogate end points for fit-for purpose development and regulatory evaluation of new drugs. Clin. Pharmacol. Ther. 81, 104–107.

Walsky, R.L., Obach, R.S., 2004. Validated assays for human cytochrome P450 activities. Drug Metab. Dispos. 32, 647–660.

Weidle, U.H., Auer, J., Brinkmann, U., Georges, G., Tiefenthaler, G., July-August 2013. The emerging role of new protein scaffold-based agents for treatment of cancer. Cancer Genomics Proteomics 10 (4), 155–168.

Weitz, J.I., Connolly, S.J., Patel, I., Salazar, D., Rohatagi, S., Mendell, J., et al., September 2010. Randomized, parallel group, multicenter, multinational study phase 2 comparing edoxaban, an oral factor Xa inhibitor, with warfarin for stroke prevention in patients with atrial fibrillation. Thromb. Haemost. 104 (3), 633–641.

Whitebread, S., Hamon, J., Bojanic, D., Urban, L., 2005. In vitro safety pharmacology profiling: an essential tool for successful drug development. Drug Discov. Today 10, 1421–1433.

Williams, R., 2014. Discontinued in 2013: oncology drugs. Expert Opin. Investig. Drugs 24 (1), 1–16.

Winearls, C.G., Oliver, D.O., Pippard, M.J., Reid, C., Downing, M.R., Cotes, P.M., November 22, 1986. Effect of human erythropoietin derived from recombinant DNA on the anaemia of patients maintained by chronic haemodialysis. Lancet 2 (8517), 1175–1178.

Yamashita, T., Koretsune, Y., Yasaka, M., Inoue, H., Kawai, Y., Yamaguchi, T., et al., 2012. Randomized, multicenter, warfarin-controlled phase II study of edoxaban in Japanese patients with non-valvular atrial fibrillation. Circ. J. 76, 1840–1847.

Young, K., Minchom, A., Larkin, J., 2012. BRIM-1, -2 and -3 trials: improved survival with vemurafenib in metastatic melanoma patients with a BRAF(V600E) mutation. Future Oncol. 8 (5), 499–507.

Zlokarnik, G., Grootenhuis, P.D.J., Watson, J.B., 2005. High throughput P450 inhibition screens in early drug discovery. Drug Discov. Today 10, 1443–1450.

Chapter 42

Pharmaceutical and Biotechnology Sector Support of Research

Joann Data

Data Consulting, Sparta, TN, United States

Chapter Outline

Key Points

- The drug development process is complex, lengthy, and involves individuals from many different backgrounds and disciplines (Fig. 42.1).
- Basic science within the pharmaceutical and biotechnology sectors:
    The contrast between the basic research processes in larger pharmaceutical companies versus the many biotechnology organizations is discussed. Large pharma is more internally focused utilizing internal funding, while the biotech industry relies on academia and government and depends almost exclusively on external funding. Both have a common need, working with regulators to bring a safe product to market in the fastest time feasible for patient benefit.
- Developmental research in the pharmaceutical arena has grown in complexity over the decades. The various physiological, pharmacological, pharmacokinetic, and toxicological findings along with stringent chemistry and manufacturing constraints support safe evaluation of a molecule or class of molecules in normal volunteers or patients with a condition for which the molecule has been targeted.
- Clinical research and development is the process by which a molecule eventually becomes a drug assuming no safety concerns and an adequate demonstration of efficacy. The regulatory requirements have regional differences as well as marked differences in formulary acceptance and pricing.
- Marketed product research is very varied and dependent on product/regulatory demands. Examples include a label extension for diseases or populations, combination therapies, different dosage forms, and a change in the manufacturing site. Often comparator studies with similar products or database expansion studies are requested in a specified time frame. In addition, patient and healthcare provider education is refined for thorough product understanding and safe use.

INTRODUCTION

The pharmaceutical and biotechnology sector supports much of the research base for drugs, medical devices, drug—device combination products, diagnostics, nutraceuticals, and over-the-counter agents. In doing so, there are numerous interactions with governmental research groups, such as the National Institutes of Health (NIH) and the National Cancer Institute (NCI), the academic world in both the basic sciences and clinical sciences, and with various regulatory bodies

Clinical and Translational Science. http://dx.doi.org/10.1016/B978-0-12-802101-9.00042-9

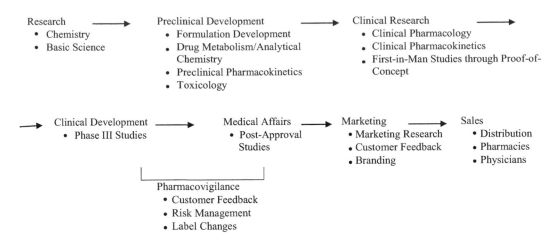

FIGURE 42.1   Organizational responsibilities in the drug development process.

throughout the world. Basic research, clinical research, epidemiology, and consumer/marketing research all play a role in the overall development process. Scientists from many different backgrounds provide vital input to understanding the mechanism of action of the molecular entity, identifying diagnostic criteria for defining patient populations, providing protocol design and evaluation plans for clinical and epidemiology studies, and assisting in device engineering, to name just a few disciplines involved. My background has been in the pharmaceutical and biotechnology sector that has been primarily drug oriented; thus this chapter will have drug development research as its primary focus. For familiarization with the drug development process and nomenclatures to be used, see Fig. 42.1. The first section of this chapter will compare and contrast the similarities and subtle differences between the subsectors of the overall industry. In subsequent sections the steps of the process as outlined in Fig. 42.1 will be elaborated upon, making mention of how larger pharmaceutical companies and biotechnology groups approach the various steps in the process.

THE DRUG DEVELOPMENT PROCESS

In large pharmaceutical companies, most of the companies select 4—6 areas of basic research to focus on while the companies in the biotechnology group usually are involved in only one therapeutic area. The larger companies have a sizable infrastructure for basic research with medicinal chemists, protein chemists, biologists, pharmacologists, and molecular biologists. Some of the personnel work across therapeutic areas and others are focused on a given approach to therapy. These scientists work with academic scientists to validate what they do and to enhance their own capabilities. This initiates interest as to a potential approach to a novel therapy. In the biotechnology arena, often the scientists must rely on the scientists at the NIH, NCI, or academia to support what little they can afford to do internally. They may have been able to perform a limited number of basic experiments but to have the science well defined they are highly reliant on others.

Large companies tend to optimize a lead drug candidate—finding a molecular agent with enhanced activity (potency) or greater "drugability," such as solubility or improved metabolic fate. They may take their lead through proof-of-concept but simultaneously try to optimize the lead with one or more backup compounds. Those in the biotech group may take the first active moiety forward into first-in-man studies. There is rarely a backup strategy.

The biotechnology companies in their beginning years at least have limited funding—private or public. This limited funding drives the decision-making. One must carefully select the studies to do preclinically and clinically to supply the appropriate safety information and demonstrate the compound's activity while using the least amount of dollars. Positive results on both fronts will allow for additional funding and/or interest the larger pharmaceutical companies to purchase the molecular entity or codevelop it with the biotechnology firm.

Another factor that differs between the larger pharmaceutical and the biotechnology companies is the importance of market size. While both organizations wish to have their future products have a large market potential that by definition is 2—3 billion dollars or more for the larger pharmaceutical companies while the smaller companies with a small infrastructure can tolerate a market that is in orders of less magnitude and/or consider development of drugs for orphan indications.

Large companies put much emphasis on time-to-market while the biotechnology group, out of necessity, focuses incremental milestones: submission of an Investigational New Drug (IND) application, first patient dosed, proof-of-concept, etc. Once funding is available through New Drug Application (NDA) submission, time becomes more relevant. The larger companies have the internal infrastructure to do much of this work on the components of the NDA while those in the biotech arena rely on consultants and contract research organizations (CROs) to provide much if not all of that support.

The actual regulatory steps are the same for the two types of drug development sectors, but the approach to the process differs considerably. Larger companies do work in parallel, sometimes with several molecules, while smaller companies do work sequentially as finances allow.

BASIC SCIENCE WITHIN THE PHARMACEUTICAL AND BIOTECHNOLOGY SECTORS

Within the larger pharmaceutical companies, basic science in 4–6 therapeutic areas is extremely well "funded." Those funds are used to bring in the personnel and technology to be state of the art in a given therapeutic area.

The scientists do not need to apply for external funds; however, they are under similar pressures at budget time to demonstrate progress toward compound selection around chosen therapeutic targets. Progress brings funds. The scientists are often thought leaders and are usually well networked. Some companies allow the scientists to have postdoctoral fellows. Internal publications are expected and often there are awards, including financial incentives that are given. Depending on the company, external publications are forbidden, limited, or allowed. Early on these scientists learn the importance of patent protection. They work closely with the patent lawyers to protect the operational space. Projects that can be enhanced with outside help are encouraged. Academic personnel are given grants to perform certain experiments. Early in the development cycle, the lead molecule will be used but as development progresses into the clinical arena, compounds with nearly identical biological properties are provided, so that should a negative and/or unexpected finding be learned, it can be further explored. This allows the finding to be understood before determining if the clinical lead has this same response. Should the finding be a safety concern, the regulators and investigators are notified, and further development may be halted, or patient populations to be studied are modified.

Within the biotechnology group, basic research may be provided solely by the academic institute from which the molecular entity was discovered. Research is minimally supported and the depth and breadth of internal research is narrow. The lead compound may be provided to external bodies with minimal to no funding of dollars and publications allowed. Patent coverage is important but may not be as robust as it occurs in the larger pharmaceutical companies.

Many of the biotechnology companies have a single platform on which the whole company is focused. To have these platforms further evaluated, they need to network with governmental agencies and academia that might evaluate the molecules further. They need to educate the venture capital world about the potential utility of this technology, so as to get additional funds. Governmental agencies wishing to study these molecules are considered an outside validation. Another funding source comes from Small Business Innovation Research (SBIR) grants, or grants from foundations or societies interested in enhancing the knowledge around a given disease.

Both pharmaceutical and biotechnology organizations tend to have scientific advisory boards where thought leaders, primarily from academia, review the science providing feedback to the basic scientists as well as research management within the company. These meetings help to keep the basic scientists, current and networked to the academic world. For the biotech organizations, the chair of the scientific advisory board may present to the board of directors of the company.

In both organizations, the basic researchers are responsible for internal education and the translation of the science into regulated documents to be reviewed by regulators and investigators. They have a role in educating nonscience management and the financial world about the potentials for a therapeutic platform and to contextualize safety concerns.

Another role the scientists, particularly in large companies, play is interfacing with business development personnel. They provide suggestions on technologies that might be available to enhance the internal platform. This may come in the form of new molecular entities, diagnostic tools, including genetic markers or new instrumentation. These tools may be brought into the company and further developed. Depending on how the "contract" is constructed, the originator may get royalties should the technology become successful, or the technology may be returned to the originator for future development with the new data. Relationships can be formed between pharmaceutical companies, between biotech companies, or more commonly between pharmaceutical companies and biotechnology companies. The roles and responsibilities of the two organizations are defined by contract. Funding is usually provided by the larger organization but may be shared in some way as well. These arrangements may be only for the basic science or be continued throughout the drug development process.

In larger companies, the basic scientists tend to be insulated from the marketing/sales side of the organization, until they have identified a drug candidate. With continued success they become more involved in preparing the internal and external environments for the potential of a new therapeutic. In the biotech sector, the company's success is dependent on the ability of the basic researchers to educate the financial world. They must be both savvy business persons as well as excellent researchers.

DEVELOPMENTAL RESEARCH—A CONTRAST TO ACADEMIA AND GOVERNMENT

Academic and governmental organizations historically have been focused on "pure" research in contrast to "applied" research. It is not that the science discovered did not have applications but the focus was more mechanistic in nature or used to explain physiological processes. The pharmaceutical and/or biotechnology companies learn of these findings and internalize them or contract with the discoverers to evaluate potential drug candidates in their systems. Some findings become so predictive of a given activity (positive or negative) that they become standardized, to be performed internally or at specialized contract research organizations on a routine screening basis. The more predictable an experiment the greater the expectation by the regulators and/or investigators of seeing these findings detailed in IND documentation in support of further development.

Pharmaceutical companies frequently internalize these procedures and use them to assist in identifying compounds in their portfolio with a given activity. Larger companies have compound or chemical libraries that are composed of all molecules ever made at the company as well as libraries of compounds that they may have purchased from other sources. For a given approach to therapy, a schema is developed to evaluate the probability that a given entity may have therapeutic potential. An attempt is made to find a receptor or receptors to target. Equally important are defining undesired activities that may be screened via a receptor-based assay. The compound library is put through both positive and negative screens and "hits" identified. Some companies have developed high throughput screening that records activity on many receptors and catalogs this information for future use. Once the hits have been identified, medicinal chemists and molecular biologists study the molecules and their spatial orientation. The molecules are further studied on other receptors and/or in animal models by the biologists and pharmacologists to determine potential therapeutic activity while the chemists work to enhance the activity of the molecules with various structural alterations. In parallel, particularly in the larger companies, the formulation chemists begin evaluation to determine the compound's solubility. They work with animal pharmacokinetic personnel to determine their absorption, distribution, metabolic, and excretion profile. Once a class of molecules is identified with "reasonable" characteristics for its intended use, the "best" entity is put forth as a drug candidate. If the research administration group supports further development, funds are provided for evaluation in mandated toxicology studies as well as safety pharmacology. The pharmaceutical chemists are responsible for supplying adequate quantities of drug substance and drug product for animal and human studies. Chemical engineers work to define the various approaches to the manufacturing process. They work with internal personnel or consider contracting the process. The smaller the organization the more outsourcing is done. Once the drug substance is made, its characteristics must be defined. A certificate of analysis is created and all future manufacturing runs must produce a compound with these characteristics. A standard is produced, so all future lots of drug substance can be compared to this standard. The stability of the standard as well as all future drug lots must be evaluated. Different storage conditions (refrigerated, room temperature and elevated temperature, and various humidity conditions) are evaluated and from these data, shelf life is determined. Multiple formulations are created and evaluated. Once the formulation is optimized, a standard of the drug product is prepared and fully characterized. All future lots of drug product must have the same characteristics (Certificate of Analysis) and with carefully maintained stability under defined conditions. The characteristics of different lots of drug substance and drug product are studied in the various toxicology studies. These results are tabulated so as to define different levels of impurities that may be allowed in clinical supplies. Animal studies (toxicology or safety) may be performed with any lot of drug substance or drug product. Clinical studies may only be done with drug substance and drug product manufactured under current Good Manufacturing Practices. The certificates of analysis must meet those levels previously qualified in animal studies.

Findings in the safety and/or toxicology studies are evaluated by internal scientists as well as external experts. Larger companies may do some or all of their preclinical studies internally while biotech companies do not have the appropriate infrastructure. They must contract out in most cases both the design of the study and preclinical protocols with consultants and the conduct of the studies with preclinical contract research organizations. A "standard" of operations has evolved over the years as to necessary receptor screens and safety pharmacology studies that need to be completed. These may differ slightly from one therapeutic area to the next. Consultants help the biotech organizations in assuring the work is completed in a timely manner while larger companies know what is needed from prior experience. Occasionally, the regulator requests certain studies to be done. This is usually because they have seen findings from other molecular entities, from other companies, and need to determine whether the finding is a class effect. The more requests from regulators the more likely they have a concern that a finding may be associated with a safety concern in man.

Toxicology studies are usually required in two species—one rodent and one nonrodent—but the choice of species is determined by the company based on tolerability and bioavailability by species. Animal findings are evaluated by internal and external scientists, including those in veterinary medicine, clinical medicine, drug metabolism, and pharmacokinetics. Sampling of blood and tissue is done to evaluate exposure and metabolic fate as a part of other animal studies. Unexpected findings require further evaluation. If the margin of safety is not great enough or the finding has been problematic with other molecular entities, the compound may be dropped from development. Larger companies may be able to bring backups forward that do not have these characteristics but smaller biotechnology companies may have no backups or funding to do more. It may mean the end of the company depending on its financial state and breadth of its research platforms.

If internal scientists and external advisors agree that the compound can be given to man, the clinician scientists become more involved, and the necessary documentation are organized to allow human testing. In the United States, this is in the form of an IND that includes an investigational drug brochure and a protocol while other countries have different yet similar regulatory requirements. Once regulatory and local approval is granted the translation of preclinical findings to man begins.

CLINICAL RESEARCH AND DEVELOPMENT

Once a molecular entity makes it to clinical research, its probability of becoming a marketed product increases many fold and as a consequence, the interest in it from both the internal corporate infrastructure as well as the outside financial community intensifies. For the biotech companies, public announcements are common around the first-in-man studies that focus primarily on safety. Larger pharmaceutical companies may not acknowledge a drug candidate until it enters phase 3 or clinical development. The probability of success (becoming a commercial product) for a molecule that enters phase 1 is about 10% while for a phase 3 entry, it is about 60–70%. For a smaller company that may only have the means to take a compound through proof-of-concept, the importance of learning the most possible about potential therapeutic utility is critical. While this is important to the larger pharmaceutical organizations as well, in "big pharma" more attention is paid to safety in the targeted patient population, evaluating clinical pharmacogenetics and pharmacokinetic characteristics, modeling around efficacy and safety, evaluating food effects and drug interactions, and optimizing the formulation. The larger companies may learn much from a lead candidate but plan to market a backup with greater potency, longer half-life, or other properties that may enhance therapeutic acceptance. Often phase 2 is used to not only determine the dose or doses to take into phase 3 but also to evaluate one or more tools that might be used to show therapeutic utility. These tools (diagnostic tests and/or questionnaires) need to be validated as reproducible and predictive. Phase 2 is an excellent time to do this in several small studies.

Another group that is affected by progress of the drug candidate through the drug development process is the biostatistical group. Biostatisticians are the personnel who establish sample size and prepare the analysis plan for the various variables. They produce tables and listings of blinded data to share with data safety monitoring boards and if necessary break the blind. In collaboration with the clinicians and clinical scientists, they determine group size, number of groups to be evaluated, what the primary endpoint should be, how many secondary end points might be useful and are vital to the study's success as well as to that of the overall program. Larger companies have data managers, programmers, and statisticians as part of the clinical organizations while the biotech sector must use outside consultants to help with protocol design and the statistical analysis plan.

Once a compound enters the clinic, physicians both internally and externally work with the clinical research and development groups to position the product. They define what the drug might do, how it might be used, and then how to go about proving it can deliver the desired outcome safely. These interfaces with specialists may be in the form of individual encounters or advisory boards. There are information sharing and suggestions of additional clinical or preclinical studies that might be done to refine the product's use. Ultimately, some of these specialists will serve as investigators and/or product advocates and be the lead or senior author on publications about the product. They may work closely with the marketing organization on speaker programs as the product is launched.

The role of pharmacokinetic modelers has become more prominent in recent years. In the United States they work to refine the product's profile based on its activity in preclinical models and/or clinical research models. The approach enhances the predictability of information about the compound even if the entity has not been evaluated in that setting. If the drug candidate is studied in that setting, data are reviewed and the model is refined as needed. This approach can minimize the number of subjects exposed early to the product and help define parameters to evaluate carefully in future studies. The approach can be used to predict dose or dose modifications for special populations or with concomitant medications.

As the first-in-man studies are completed, there are more interactions with internal regulatory scientists and eventually with external regulatory bodies. In the United States, there are specific meetings that may be requested between the company and the Food and Drug Administration (FDA). There are requirements by companies and investigators to evaluate the potential drugs under Good Clinical Practice guidelines. Safety must be reported annually at a minimum and for severe unexpected life-threatening events within 48 h. If the drug is being considered for the European market, scientific advice might be sought early in the development program. In Europe and Japan, the safety reporting is similar but in some cases more stringent. It is the company's responsibility to devise a development program that incorporates the needs of the various markets. The United States prefers placebo-controlled studies, while in Europe active controls are preferred. There has been an attempt over the last decade to harmonize the regulatory path forward across various regions of the globe and to standardize the approach to the regulatory submissions. The use of electronic submissions from INDs to NDAs has assisted both sponsors and regulators alike in communicating the known information about the drug candidate. The overall process has been improved immensely but country-specific requests are still present.

The scientific advisory boards mentioned previously may change in makeup (more clinical and biostatistical input) and may provide different recommendations for different regions as the drug progresses in the clinic. The larger companies take a global approach to drug development while the smaller biotech companies take a regional approach. If a biotech product is successful in one region, then the company may add a different region or partner the drug for development in different regions. In this setting, scientists from two or more companies may serve as the advisory panel for creating the overall drug development plan.

The role of marketing varies from company to company but tends to have a greater presence and impact in larger companies with bigger economic demands. They "research" the market and profile the market need. They usually want to know early in development that the therapeutic entity can be given safely once daily. If it cannot, even though it is safe and effective twice or more times a day, they may recommend it to be dropped from the portfolio because the product will not be competitive. Larger companies want to dominate a given market while smaller companies are willing to carry a small segment of the market. The marketing organization may recommend certain studies to enhance the product's profile. Many companies have created a targeted product profile and some have even drafted an ideal package insert. As studies are completed that support the desired profile or insert, interest is enhanced. If completed studies do not support the ideal product, the compound may be dropped from development, have development delayed, or allow the business development group an opportunity to out-license the product. The marketing group often recommends certain investigators to evaluate the drug candidate in different regions. They may also suggest evaluation using different tools or instruments to support product utility once the product gets to market.

Once the phase 3 studies have been completed and the drug has the properties of safety and efficacy needed to commercialize the product in the eyes of the company, a regulatory submission is created. Depending on the size of the regulatory group and the completeness of the write-up of previously completed preclinical and clinical studies, the process of integrating information may take anywhere from several to many months. While there is greater overall regulatory harmony between the United States, Europe, and Japan, within a given therapeutic area, there may be differences in regional approaches to approvals. The impact of the inconsistencies between divisions within an agency adds uncertainty as to the timing of the product approval. Once submitted for review the probability of approval increases 8–90%. With this comes greater interest especially from the financial community, both internal and external to the company. Product commercialization is an expensive proposition. Larger companies have established patterns while a biotech company with its first product is again in a position to cost-justify its spending. Not only does the biotech company need to create the market for its future product but it must educate the payer community on both the company and the product capabilities. The process involves no longer only patient and prescribers but pharmacies, formularies, wholesalers, distributors, and many other groups. There is a need to create awareness of a "product-to-be" with no certainty as to when the product may be available. A product trade name needs to be created and branding created. Educational material needs to be created as well as advertising and promotional pieces. These are done in mock form until the final label is approved. There is a tension created between the development scientists and the marketing group. The importance of clearly establishing a product's limitations as well as its potential utility is the focus of this tension. The risk/benefit profile is different in different groups of patients. It is important that the development group works closely with their commercial counterparts to establish how best to use the product. This guidance will need to be refined as more is learned about the product once commercialized. This aspect is discussed further in the next section. Working with regulatory bodies, labels can be enhanced and/or become more restrictive with time. The input of the patient, the providers, and payers shape the product going forward. Epidemiologists, patient advocates, and the legal community all play a role in the commercial success of the product. The more the research and development scientists can learn about the new data being generated in the market place and assisting in the interpretation in light of what was known from preclinical and clinical studies, the more likely the drug will remain on the market. The use may be restricted and the warnings about safety may be sterner but product characteristics rule. A company may choose to remove a product from the market because of a safety concern or limit its market penetration. Alternatively, a regulatory body may restrict or prevent use. Occasionally, a finding is discovered in one member of a class of therapeutics. Class labeling may be required until such time as this new member of a class can be shown not to have the characteristic. This often involves discussions with basic and developmental scientists from academia, government, and the company. Such teamwork enhances the understanding of the product—its potential and its liability.

MARKETED PRODUCT RESEARCH

The research on a molecular entity does not stop with product approval. Frequently, the approval comes with commitments for additional studies to evaluate safety. These studies may be structured as in the developmental phase or they may be "in-use" studies to get patient and physician feedback on the product. Assessing the comparative effectiveness of newly marketed medications has become important in the cost conscious health-care environment of the current decade. There is mandatory postmarketing surveillance where adverse events reported by patient, health-care providers, or interested parties reported to the company are sent quarterly to regulatory bodies. The regulatory agency also may receive these reports directly. The company is responsible for evaluating these reports and determining if there are new adverse events being reported or an increase in the frequency of adverse events stated in the label. These findings may require further evaluation and in almost every incidence leads the company to a better understanding of the disease and any drug—disease interaction. Large databases can be accessed, and epidemiology studies conducted to research whether a finding is disease related, or is a signal or just noise. Tens of thousands of patients exposed to the drug may be necessary to ascertain the drug—disease relationship. Safety findings may require label changes to include special labels. To insure the prescribing communities are aware of safety concerns, "Dear Doctor" letters may be sent with information provided to explain the need for a label change.

In addition, as a condition for approval, phase 4 commitments may be necessary. These may take the form of an evaluation of a drug in the pediatric population, or the creation of a patient registry. Reports of these studies must be provided to the regulatory agencies in certain preagreed times unless extensions are granted.

Other phase 4 activities include studies called "line extensions." These studies are ones which are designed much like phase 3 studies to allow use with previously excluded drugs, use with a new class of agents or in a different medical condition. The dose(s) may be the same or different. All known data in this population are submitted as a supplement NDA (sNDA). The data are evaluated much as the original application was reviewed though the time frame tends to be shorter. If acceptable, the label and the package insert are changed to accommodate this new information.

At other times an sNDA will involve a new product presentation such as an extended delivery product, a liquid formulation instead of a solid, etc. These supplements, as long as the disease state is the same, are less extensive than line extensions and often are based on bioequivalency. Some products also require demonstration of similar activity but the patient exposure is less demanding.

Changes in the site or process of manufacturing of drug substance or drug product require supplements. The extent of the submission and review is dependent on the nature of the change. For biological sites, changes may require additional clinical work to show that the new manufacturing process or site of manufacturing does not alter the molecular configuration and thus the activity of the product.

As additional information is learned about the product, the label is refined. Each label change requires regulatory approval. Some changes are grammatical only and others quite substantial. Some products require patient information leaflets or instructions. These pieces are evaluated by patients to assure that the material presented is clear and instructions easily followed. The regulators evaluate the instructions as well to make sure that the material is consistent with the product's package insert.

CONCLUSION

This chapter has outlined the multiple types of research that are done by or supported by the pharmaceutical and biotechnology industry. These organizations, depending on their size, do nearly all to nearly none of the research with internal scientists. Scientists from many different disciplines are involved in the drug development process. Companies may have large internal basic research groups that add to the general state of knowledge in a given area. Smaller companies may have minimal internal research capabilities or rely heavily on external groups to assist them with the basic experimentation. As the molecules move from the basic research areas into preclinical development, other types of scientists become involved with a more applied focus. Clinical research and development is mainly done externally, thus involving nonpharmaceutical clinical scientists in the clinical research process. The pharmaceutical and biotechnology sector spends much for research. These dollars help support the research base of academia and governmental agencies as well as certain private practitioners. They provide jobs for many internally, and research dollars and consultantships for scientists external to the industry. The focus of research differs in the pharmaceutical industry and accountability is greater. While industrial scientists may not rely solely on grants, they must deliver certain information in a timely manner. Failure of the research team to deliver a substance with the desired activity and safety margin may terminate the project, the company, and/or the scientists' jobs. Science for science sake is not tolerated in the industry. However, excellent science is expected to assure that the decisions made about an entity to be commercialized are sound and can support further development with a high probability of commercial success. Life in the pharmaceutical and biotechnology sector is demanding. Currency of information around a science and/or therapeutic area is essential. There are many opportunities, and flexibility is an important attribute of the scientists. The stronger the person's science base, the more opportunities he/she may encounter. Interest in scientific communications, business, and management provide many and varied employment opportunities in both the pharmaceutical and biotechnology sector.

FURTHER READING

Several good sources for further reading are available for those who wish to delve more deeply into issues addressed in the chapter.

Caudle, K.E., Rettie, A.E., Whirl-Carrello, M., Smith, L.H., Mintzer, S., Lee, M.T.M., Klein, T.E., Calaghan, J.T., 2014. Clinical pharmacogenetics implementation guidelines for CYP2C9 and HLA-B genetypes and phenytoin dosing. Clin. Pharmacol. Ther. 96 (5), 542–548.

Clancy, J.P., Johnson, S.G., Yee, S.W., McDonagh, E.M., Caudle, K.E., Klein, T.E., Cannavo, M., Giacomini, K.M., 2014. Clinical pharmacogenetics Implementation Consortium (CPIC) guidelines for ivacaftor therapy in the context of CFTR genotype. Clin. Pharmacol. Ther. 95 (6), 592–597.

FitzGerald, G.A., 2011. Regulatory science: what it is and why we need it. Clin. Pharmacol. Ther. 89 (2), 291–294.

Kaitin, K.I., 2008. Obstacles and opportunities in new drug development. Clin. Pharmacol. Ther. 83 (2), 210–212.

Kola, I., 2008. The state of innovation in drug development. Clin. Pharmacol. Ther. 83 (2), 227–230.

Mackintosh, D.R., Molloy, V.J., Mathieu, M.P., 2003. Good Clinical Practice: A Question & Answer Reference Guide. Barnett International, a Subsidiary of PAREXEL International Corporation, Waltham, MA.

Milne, C.-P., Kaitin, K.I., 2012. FDA review divisions: performance levels and the impact on drug sponsors. Clin. Pharmacol. Ther. 91 (3), 393–404.

Schneeweis, S., Gagne, J.J., Glynn, R.J., Ruhl, M., Rassen, J.A., 2011. Assessing the comparative effectiveness of newly marketed medications: methodological challenges and implications for drug development. Clin. Pharmacol. Ther. 90 (6), 777–790.

Spilker, B., 1991. Guide to Clinical Trials. Raven Press, New York.

Wagner, J.A., 2008. Back to the future: driving innovation in drug development. Clin. Pharmacol. Ther. 83 (2), 199–202.

Yacobi, A., Skelly, J.P., Shah, V.P., Benet, L.Z., 1993. Integration of Pharmacokinetics, Pharmacodynamics, and Toxicokinetics in Rational Drug Development. Plenum Press, New York.

Section X

Prospectus

Chapter 43

The Future of Clinical Research

Gordon H. Williams[1] and David Robertson[2]

[1]*Hormonal Mechanisms of Cardiovascular Injury Laboratory, Brigham and Women's Hospital, and Professor of Medicine, Harvard Medical School, Boston, MA, United States;* [2]*Clinical & Translational Research Center, Elton Yates Professor of Medicine, Pharmacology and Neurology, Vanderbilt University, Nashville, TN, United States*

Chapter Outline

Key Points

- The definition of translational research is ambiguous but usually divided into four subgroups: (1) clinical trials, (2) human pharmacology, (3) human physiology, pathophysiology and genetics, and (4) gene and cell therapy.
- The medical scientist training program has been a very successful MD/PhD training program, but most graduates are performing bench research.
- During the past decade, the NIH created a new "home" for the clinical investigator called a Clinical and Translational Science Award (CTSA) program, but the program has been modified twice and has largely been replaced by the National Center for Advancing Translational Sciences (NCAT) that is more focused on drug discovery and clinical trials rather than translational human research.
- Most chronic illnesses (hypertension, diabetes, obesity, anxiety, asthma, cancer, autoimmunity, etc.) are not diseases but syndromes.
- Big Data approach is to perform association study between genotype and phenotype. Its critical components are sophisticated genetic analyses; large scale genotyping—potentially metabolomics, lipidomics, etc.; and large amounts (2−20 million patients) of currently available clinical data from clinical records and clinical trial results.
- N-of-One precision approach performs association studies between variants of candidate genes and intermediate phenotypes based on specific disease mechanisms. Its critical components are sophisticated candidate genetic analyses primarily rather than whole genome analyses; translational research studies involving human, animal, and cell studies to identify specific mechanisms responsible for individual diseases in the syndrome; human phenotyping facility with control of environmental factors to study intensively relevant humans—normal and diseased; associated (linked) animal/cell facility to assess mechanisms and animal intermediate phenotypes linked to human ones.

DEFINITION OF TRANSLATIONAL HUMAN RESEARCH

It was six men of Indostan
 To learning much inclined,
 Who went to see the Elephant
 (Though all of them were blind),
 That each by observation
 Might satisfy his mind.
 The First ... [falling against his side, said: 'He]

Clinical and Translational Science. http://dx.doi.org/10.1016/B978-0-12-802101-9.00043-0

... is very much like a wall!'
The Second [feeling his tusk said: 'He]
... is very like a spear!'
The third [feeling his trunk said: 'He
... is very like a snake!'
... And so these men of Indostan
Disputed loud and long,
Each in his own opinion
Exceeding stiff and strong,
Though each was partly in the right,
And all were in the wrong!

Moral

So oft in [clinical research] wars,
The disputants, I ween,
Rail on in utter ignorance
Of what each other mean,
And prate about an Elephant
Not one of them has seen!

The Blind Men and the Elephant (by John Godfrey Saxe)

As John Saxe concluded in his last stanza concerning "theological wars," so the debate concerning the future of clinical/translational research depends on the definition used. Many studies at the cell or gene level may "translate" new information that increases our understanding of how biologic systems function and/or are regulated. Such research is clearly bench-type. Much useful information can come from the study of large databases that may translate new information concerning outcomes and cost benefits of certain diseases and even treatments. Such research is certainly included under the platform of clinical research. Another subset of clinical research is *patient-oriented research* (POR), recently defined as being T-1 research. It has as its core tenet research involved directly with individual human subjects (Williams, 1999) (Fig. 43.1).

Dr. Lee Nadler contends that "A translational researcher is someone who takes something from basic research to a patient and measures an endpoint in a patient" (Nadler, 2007) (Fig. 43.2). One could add to this definition that it is someone who takes information gathered or hypotheses generated from population-based studies and tests these hypotheses in individual subjects to determine to what individual the information (treatment or prevention strategy) applies. Thus, in many respects from the perspective of POR investigators both population-derived and bench-derived data serve the same purpose—hypothesis generation. Neither discipline can test these hypotheses directly because to do so requires studies in well-characterized individual subjects: the province of the POR investigator. Indeed, when some bench or population investigators use their experimental data to extrapolate outcomes in individual subjects, inconsistencies often have been reported much to the consternation of the public and the individual patient. The most recent example of a translational investigator is one who translates from discovery or study in humans, i.e., in a POR study, to proving something (like a therapy) works not only when a POR investigator gives it to a small number of well-characterized and intensely studied patients, but also in the whole community as delivered by the health-care system. This process has been defined as T-2 research.

FIGURE 43.1 Structure of the biomedical research enterprise. *(Adapted with permission from Williams, G.H., 1999. The conundrum of clinical research: bridges, linchpins, and keystones. Am. J. Med. 107, 523.)*

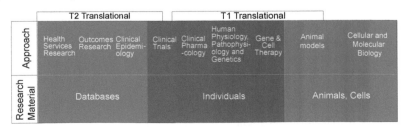

Structure of a Biomedical Research Platform

Definition of a T1 Translational Investigator

- **Who is not a translational investigator**
  - Cloned a gene from a human cell line or tissue
  - Studied human expression profile on chips
  - Immunoperoxidase typing of breast cancers

- **Who is a translational investigator—an individual whose own work attempts to:**
  - Improve diagnosis or prognosis in patients
  - Improve prevention in patients
  - Conceive and execute a new treatment in patients

FIGURE 43.2 Definition of a translational researcher. *(Adapted with permission from Nadler, L., 2007. American Association of Cancer Research Annual Meeting, San Francisco, 8 April 2007.)*

Investigators in each of these groups are translational investigators from their individual perspectives just like the six blind men of Indostan from their individual perspective believed they were defining an elephant. From our perspective, each group of investigators is in the process of translating and therefore all are correct in using this term. While there are many similarities between each group, each uses somewhat different approaches, requires different infrastructure, has somewhat different educational requirements, uses different starting "material" for their studies, and produces different results. However, all are essential to achieve the goal of identifying specific therapy to treat or prevent disease in individual patients. Therefore, the challenge to future progress in human research is to insure that "translational" research does not create new silos but rather builds bridges between the various translating groups.

SUBGROUPING OF BIOLOGICAL SCIENTISTS

Traditionally, POR has been divided into four major subgroups: (1) clinical trials, (2) human pharmacology, (3) human physiology, pathophysiology and genetics, and (4) gene and cell therapy. Often within the POR category the latter two are grouped as "T-1 translational research," implying that both consist of bidirectional bench to bedside research. However, it is important to note that the other two POR subgroups also can have bench research components and all could have population research components. Indeed, often "T-2 translational research" is that at the junction of patient-oriented and population-oriented research with some of the tools and techniques of each being used to study ways to treat or prevent disease in the general community.

POR differs from the other major group comprising the clinical research category by the nature of the research material used. *Population-oriented researchers* use databases, while *patient-oriented researchers* use individuals.

Population-oriented research can be subgrouped into at least three major categories: clinical epidemiology, outcomes, and health-care service (see Fig. 43.1). Bench research differs from the two subgroups of clinical research in using animals, cells, genes, and subcellular components as its research material.

POR investigators use many of the same structural and educational resources as those used by the bench or population scientists as exemplified by the potential overlap of the tools described in this textbook and those found in textbooks of bench or population sciences. For example, both patient and population scientists rely heavily on statistical support and informatics (Fig. 43.3). However, POR investigators need a physical facility in which to interact with their subjects; population scientists do not (Fig. 43.4). Both patient and population scientists need training for coordinators and research nurses. While not essential, both groups of scientists will be most productive if the individual has received formal didactic training, i.e., for the population scientist an MPH degree and for the patient scientist an MCS degree (Fig. 43.5). The change in the training of the clinical investigator during the past 20 years mimics what was a recognized educational need (PhD degree) nearly a half century ago.

It is also important to recognize that these major groupings have adapted many of the tools and techniques they use in their research from each other as described in the individual chapters in this book. For example, many of the advanced statistical techniques used by the patient-oriented and bench researcher are adaptations from those developed for population analyses. The importance of the social and economic environment, addressed in detail by the health services and outcomes researcher, often have been added to the physical environment by patient-oriented researchers in their approach to control environmental factors in their studies.

The other major concept that sometimes distinguishes the patient-oriented investigator from the other two classes of biological scientists is the necessity and complexity of the "team" (Fig. 43.6). While increasingly all biologic scientific endeavors are dependent on some type of teamwork, for the POR investigator it is an absolute requirement.

FIGURE 43.3   Infrastructure needs for the clinical
research enterprise. *IRB*, Institutional review board.

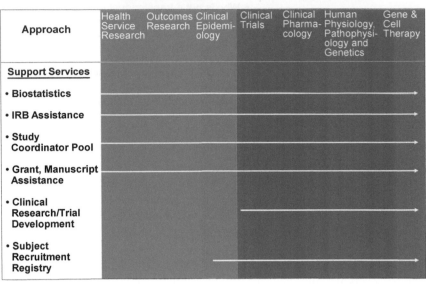

FIGURE 43.3   Infrastructure needs for the clinical
research enterprise. *IRB*, Institutional review board.

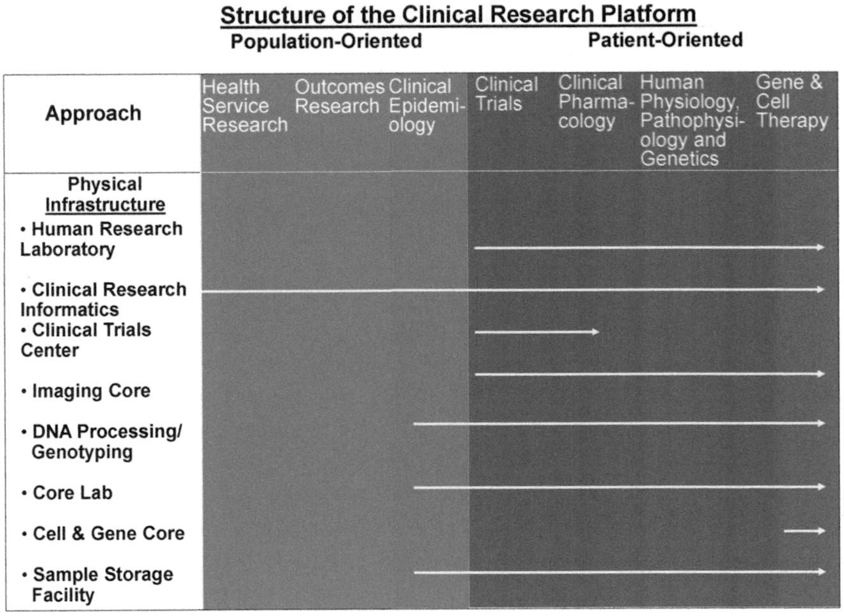

FIGURE 43.4   Support services for the clinical
research enterprise.

This requirement not only changes the cost of doing the research, but also requires POR investigators to have excellent interpersonal skills (and a subdued ego) to effectively engage their team in the research activity.

THE PATIENT-ORIENTED SCIENTIST AT THE BEGINNING OF THE 21ST CENTURY

Several scientists in the latter two decades of the 20th century lamented on the potential demise of POR investigators (Wyngaarden, 1979; Kelly, 1985; Ahrens, 1992; Clinical Research Study Group, 1994; Kelly and Randolph, 1994; Williams et al., 1997; Feinstein, 1999). Infrastructure to support these individuals was crumbling and/or stagnant. There was no specific training program for them in contrast to the many degree programs for the bench scientist and the substantial growth of Masters and PhD degree programs offered by the increasing number of Schools of Public Health. There were no specific start-up packages provided by academia, government, or foundations in contrast to what was

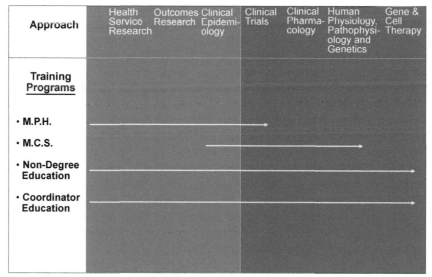

FIGURE 43.5 Educational needs for the clinical research enterprise.

Translational Research Team

- Laboratory-based investigators
  - Preclinical target validation
  - Assessment of biologic markers *in situ*
  - Measurement of surrogate endpoints
- Clinical investigators
  - Plan and investigate the experiment
  - Physicians, radiologists, pathologists
- Statisticians
- Research nurses and coordinators
- Data managers
- Research pharmacists

FIGURE 43.6 Translational investigation is performed by a team not an individual.

occurring for the bench, and in some cases population, scientists. Promotion of the POR investigator was more uncertain and laborious than for the bench and population scientist. Specific funding for their research projects by governmental agencies (e.g., the National Institutes of Health in the United States) and by foundations was declining to less than 10% of total projects in contrast to the greater than 30% of projects two decades earlier (Williams et al., 1997). While support continued from industry, it usually was directed toward phase III or IV clinical trials rather than T-1 translational research.

The Medical Scientist Training Program (MSTP) started by the United States National Institutes of Health in the early 1960s to ensure a continuous stream of well-qualified investigators to engage in the T-1 translational research enterprise, ultimately developed along different lines, more commonly providing basic investigators in clinical departments. While a successful program, its mission did not educate a workforce to perform patient-oriented translational research. While there was substantial growth in potential tools to assist T-1 patient-oriented investigators in performing their research, particularly in imaging, genetics, and biomarkers, these tools usually were not available to most investigators because of cost or restricted access.

THE 21ST CENTURY AND THE FUTURE

At the turn of the century, it became clear in many quarters that substantial changes in the clinical research enterprise world-wide would be required if full advantage was to be taken of the emerging discoveries and hypotheses emanating from bench and population studies (Robertson, 2000). Indeed, the critical lynchpin in this enterprise—the T-1 translational investigator—was "gradually becoming extinct." In some quarters, the intense study of the individual/subject/patient (so-called "N-of-One" studies) was considered nonessential. These groups advocated that with the human genome/proteome efforts, the available large clinical databases and the sophisticated new statistical programs and mathematical models, the need to study the individual subject was both unnecessary and expensive. Thus, the "Big Data" area was born. Some scientists suggested that there not only were impediments to the appropriate testing of the hypothesis generated by bench and/or population researchers, but also inhibitions to the translation of the results from POR studies into routine clinical practice—a new area of clinical research at the transition between patient- and population-oriented research that likely will use some of the tools in each of the former research areas as discussed earlier, so-called T-2 translational clinical research (Sung et al., 2003; Woolf, 2008).

The United States government took the first step to initiate appropriate changes to strengthen the N-of-One approach by implementing the Clinical Research Enhancement Act of 2000. This act codified the needs of the human translational investigator in six areas. First, it established specifically funded educational programs to train the clinical investigator. Second, it provided 5 years of grant funding to support much of the salary of junior translational investigators during the critical early years of their career development. Third, the government recognized that to prevent a future "meltdown" of the human translational investigator (both T-1 and T-2 types), it needed to provide salary support for mid-career investigators to be available to mentor more effectively new entrants into the field. Next, it codified the General Clinical Research Center (GCRC) program—the laboratory and infrastructure for the patient-oriented investigator. Fifth, it provided a school loan forgiveness program for these individuals. Finally, it instructed the NIH to ensure that there is a level playing field for the review of investigator-initiated grant applications by POR investigators. Since then other governments, foundations, and disease-oriented societies have developed parallel programs or ones to fill the gaps left after implementation of the Clinical Research Enhancement Act of 2000.

The success of this experiment is still pending, and indeed the answers remain elusive as in less than 15 years dramatic changes have occurred. By the end of the first decade of the 21st century, the GCRC and K-30 programs were merged into a new expanded entity termed CTSAs (National Institutes of Health, 2005) (see Chapter 34). These linked educational, early support of clinical scientists and infrastructure into a single entity at 60 sites across the United States. A little more than half decade later NIH further modified the implementation of the Clinical Research Enhancement Act of 2000. Support for the human research laboratories (GCRC) was terminated and NIH's direct financial support for human research infrastructure was substantially reduced. It is unclear whether clinical research infrastructure support has been reduced in other countries. Furthermore, both in the United States and elsewhere an increasing emphasis on the Big Data approach has occurred. This approach does not require most of the programs implemented in the Clinical Research Enhancement Act of 2000. Thus, it is clear that the support for human translational investigation is changing rapidly and substantially. This support is exemplified by a number of facts: the changes described above implemented by the United States government; the substantial advances in human genetics; the increasing availability of new imaging modalities at a reasonable cost and appropriate access; and the development of a variety of biomarkers for human diseases; more sophisticated statistical and informatics approaches; and the recognition, acceptance, and support of the "team" approach to human translational research. These changes are often contrasted in the approaches termed Big Data and N-of-One.

NOVEL NEW APPROACHES: BIG DATA AND N-OF-ONE

During the latter half of the 20th century, treatment strategies shifted from those used to treat acute illnesses to ones used for chronic illnesses, e.g., hypertension, diabetes, cancer, obesity, autoimmune diseases, heart failure, coronary artery disease, Alzheimer's disease, psychiatric diseases. This shift was accompanied by a change in the assumptions underlying present treatment strategies for common chronic illnesses. For example:

- Each chronic illness was considered a disease.
- The primary goal was to treat the sign/symptom, e.g., reduce the blood pressure or blood glucose.
- Population data suggest that treating the sign/symptom will reduce morbidity and/or mortality associated with the illness.
- Algorithms to treat the illness were based on population and animal data associated with reducing the sign/symptom of the illness.
- Treatment changes were driven by *failure* to achieve goal level of sign/symptom, e.g., blood pressure or glucose. However, it became increasingly evident by the turn of the century that these approaches had substantial drawbacks, including the following:
- Nonspecificity: many strategies but no individual linkage

- High noncompliance rates
- Increased costs
- Increased off-target adverse effects
- High drug development cost because of heterogeneity of the target population
- Poor reliability of biomarkers to achieve specificity of treatment/prevention programs
  Thus, other approaches were proposed that would overcome these deficiencies and lead to personalized medicine. Among these approaches was "Big Data." At the time of the publication of the first edition of this book (2009), there was but a single article in PubMed linked to Big Data. In 2015, there were 157 articles with representative ones cited (Lichtman et al., 2014; Peters and Khan, 2014; Martin and Félix-Bortolotti, 2014; Gray & Thorpe, 2015; J Comp, 2015; Liebeskind et al., 2015; Kim, 2015; Tanaka et al., 2015; Cook and Collins, 2015; Ramos-Casals et al., 2015). In a brief the Big Data approach is as follows:
- Association study between genotype and phenotype
- Critical components are:
  - Sophisticated genetic analyses
  - Large scale genotyping—potentially metabolomics, lipidomics, etc.
  - Large amounts (2−20 million patients) of currently available clinical data from clinical records and ± clinical trial results
- Steps taken
  - Use high-powered statistics
  - Associate variation in gene structure with clinical data
  - Primary outcome is therapeutic response to treatment in individuals by subjects clustered by common genetic structure.
  The Big Data approach has several advantages over the current approach used to treat chronic illnesses, specifically as follows:
- It has a potential specific biomarker (genetic structure), e.g., "culture and sensitivity testing."
- It is relatively inexpensive since the clinical data has already been collected.
- Major progress has been made in understanding the structure of the gene.
- Sophisticated statistical programs have been developed.
- Some studies in rare diseases have documented the success of this approach.
- It offers the possibility of specific, personalized medicine.
  However, there are also disadvantages of the Big Data approach, including the following:
- The critical linchpin is the quality of the clinical data: no studies have assessed it. Anecdotal evidence suggests the quality of the data is variable.
- Statistical programs adjust for a variety of potential confounders, but no way of knowing if all potential confounders have been identified.
- Even if positive, the results are observational with all the limitation of these types of results.
- This approach likely will not identify mechanisms.
- Normal data will be limited since clinical data sets come from patients. Any such data will need to come from cohort studies that may not reflect a similar background of the patients.
- Unlikely possibility of identifying new therapeutic approaches with current drugs or drug development.
- Little if any data to guide specific prevention strategies.
  An alternate proposed strategy to Big Data is a modification in the strategy envisioned in the Clinical Research Enhancement Act of 2000, so-called "N-of-One precision" strategy.
- Considers chronic illnesses, syndromes not diseases.
- Approach: association studies between variants of candidate genes and intermediate phenotypes based on specific disease mechanisms.
- Critical components:
  - Sophisticated candidate genetic analyses primarily rather than whole genome analyses.
  - Translational research studies involving human, animal, and cell studies to identify specific mechanisms responsible for individual diseases in the syndrome.
  - Human phenotyping facility with control of environmental factors to study intensively relevant humans—normal and diseased.
  - Associated (linked) animal/cell facility to assess mechanisms and animal intermediate phenotypes linked to human ones.

- Steps
  - Identify potential normal physiologic mechanisms using previous human and animal studies that could be altered to produce the illness syndrome
  - Assess status of mechanisms in illness syndrome using intermediate phenotypes in humans and animals
  - Associate variation in candidate gene structure related to mechanism pathway
  - Identify specific individual by variants in candidate gene structure
    The advantages of the N-of-One approach over the current approaches include the following:
- Documents that the one-size-fits-all therapeutic/preventative approaches are wrong
- Identifies mechanisms of diseases
- Treatment/prevention targeted to the individual
- Has a potential specific biomarker (genetic structure), e.g., "culture and sensitivity" testing
- Greater compliance
- More precise and cheaper drug development based on mechanisms documented in humans
- Fewer off-target side effects
- Offers specific, personalized medicine with treatment specificity identified by mechanism and specific subject identified by genotype
  The advantages of the N-of-One approach over the Big Data approach include the following:
- Identification of physiological and pathophysiological mechanisms
- Less complicated genetic analyses
- Less reliance on sophisticated statistical techniques
- Requires many fewer subjects, studied more intently
- Less heterogeneous human data
- Not observational data, but randomized data
- Can identify environmental factors and gene environmental interactions
- More readily identifies specific prevention strategies
- More precise and cheaper drug development based on mechanisms documented in humans
  However, there are disadvantages of the N-of-One approach compared to the Big Data approach including the following:
- Requires a human laboratory—clinical research unit
- Requires tight linkage between human, animal, and cell studies: translational research
- Requires precise identification and control of environmental factors in both human and animal studies.
- Could miss identifying a new mechanism associated with an illness syndrome

SUMMARY

In the not too distant future one could envision that the "laboratory" (mentioned above) used by the human translational investigator would have many of the components illustrated in Fig. 43.7. Each program will be supported by a number of

FIGURE 43.7   The 21st century human translational investigator's laboratory.

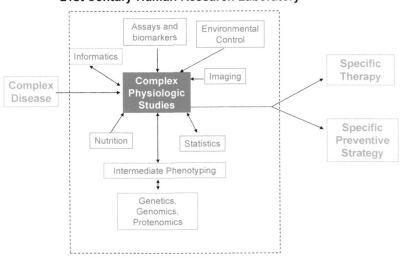

team members (see Fig. 43.6) with varying and complementary skill sets. The investigator will be the team leader rather than the laboratory chief. The investigator and his/her team will have completed substantial, individualized, and specific training programs. A variety of tools will be at the team's disposal, including sophisticated and miniaturized imaging equipment, sophisticated and comprehensive genetic, genomic and proteomic platforms and efficient and knowledgeable nutritional support systems. A variety of laboratory support systems will be in place to help the team identify specific biomarkers for diseases to allow personalized treatment and prevention strategies.

It is important to emphasize that these two new approaches are not necessarily competing ones, but should be viewed as complimentary ones. Finally, the position of human translational investigators in academia, the Academic Health Center and industry will be more in keeping with their role as equal partners with population and bench scientists in the human research enterprise. Indeed, the optimist would say the future for this research field is bright, rewarding, and expanding. Even the pessimist would say this field is certainly looking up. However, all will agree that the fulfillment of these opportunities will largely be dependent on the level and balance of funding provided by the public, through its governments and philanthropic organizations, and by industry—pharmaceutical, biotech, and insurance (Moses et al., 2005). Thus, the next few years will be challenging as these new models evolve to provide infrastructure support for the translational/clinical investigator.

REFERENCES

Ahrens, E.H., 1992. The Crisis in Clinical Research: Overcoming Institutional Obstacles. Oxford University Press, New York and Oxford.

Clinical Research Study Group, 1994. An Analysis of the Review of Patient-Oriented Research (POR) Grant Applications by the Division of Research Grants. National Institutes of Health, Bethesda, MD.

Cook, J.A., Collins, G.S., 2015. The rise of big clinical databases. Br. J. Surg. 102, e93–e101.

Feinstein, A.R., 1999. Basic biomedical science and the destruction of the pathophysiologic bridge from bench to bedside. Am. J. Med. 107, 461–467.

Gray, E.A., Thorpe, J.H., 2015. Comparative Effectiveness Research and Big Data: Balancing Potential with Legal and Ethical Considerations. J. Comp. Eff. Res. 4, 2015, 61–74.

Kelly, W.N., Randolph, M.A. (Eds.), 1994. Careers in Clinical Research: Obstacles and Opportunities. National Academy Press, Washington, DC.

Kelly, W.N., 1985. Personnel needed for clinical research: role of the clinical investigator. Clin. Res. 33, 100–104.

Kim, E.S., 2015. The future of molecular medicine: biomarkers, BATTLEs, and big data. Am. Soc. Clin. Oncol. Educ. Book 22–27.

Lichtman, J.W., Pfister, H., Shavit, N., 2014. The big data challenges of connectomics. Nat. Neurosci. 17, 1448–1454.

Liebeskind, D.S., Albers, G.W., Crawford, K., Derdeyn, C.P., George, M.S., Palesch, Y.Y., Toga, A.W., Warach, S., Zhao, W., Brott, T.G., Sacco, R.L., Khatri, P., Saver, J.L., Cramer, S.C., Wolf, S.L., Broderick, J.P., Wintermark, M., 2015. Imaging in StrokeNet: realizing the potential of big data. Stroke 46, 2000–2006.

Martin, C.M., Félix-Bortolotti, M., 2014. Person-centred health care: a critical assessment of current and emerging research approaches. J. Eval. Clin. Pract. 20, 1056–1064.

Moses, H., Dorsey, E.R., Matheson, D.H., Their, S.O., 2005. Financial anatomy of biomedical research. JAMA 294, 1333–1342.

Nadler, L., 2007. In: American Association of Cancer Research Annual Meeting, San Francisco, 8 April 2007.

National Institutes of Health, 2005. Re-engineering the Clinical Research Enterprise: Translational Research, 2005. http://www.nihroadmap.nih.gov/clinicalresearch/overview-translational.asp.

Peters, S.G., Khan, M.A., 2014. Electronic health records: current and future use. J. Comp. Eff. Res. 3, 515–522.

Ramos-Casals, M., Brito-Zerón, P., Kostov, B., Sisó-Almirall, A., Bosch, X., Buss, D., Trilla, A., Stone, J.H., Khamashta, M.A., Shoenfeld, Y., 2015. Google-driven search for big data in autoimmune geoepidemiology: analysis of 394,827 patients with systemic autoimmune diseases. Autoimmun. Rev. 14, 670–679.

Robertson, D., 2000. Reinventing the General Clinical Research Centers for the Post-genome Era, vol. 2. Association of American Medical College (AAMC) Chapter in for the Health of the Public: Ensuring the Future of Clinical Research, pp. 37–46. http://www.aamc.org/publications.

Sung, M.S., Crowley Jr., W.F., Genel, M., Salber, P., Sandy, L., Sherwood, L.M., Johnson, S.B., Catanese, V., Tilson, H., Getz, K., Larson, E.L., Scheinberg, D., Reece, E.A., Slavkin, H., Dobs, A., Grebb, J., Martinez, R.A., Korn, A., Rimoin, D., 2003. Central challenges facing the national clinical research enterprise. JAMA 289, 1278–1287.

Tanaka, S., Tanaka, S., Kawakami, K., 2015. Methodological issues in observational studies and non-randomized controlled trials in oncology in the era of big data. Jpn. J. Clin. Oncol. 45, 323–327.

Williams, G.H., Wara, D.W., Carbone, P., 1997. Funding for patient-oriented research. Critical strain on a fundamental linchpin. JAMA 278, 227–231.

Williams, G.H., 1999. The conundrum of clinical research: bridges, linchpins, and keystones. Am. J. Med. 107, 522–524.

Woolf, S.H., 2008. The meaning of translational research and why it matters. JAMA 299 (2), 211–213.

Wyngaarden, J.B., 1979. The clinical investigator as an endangered species. N. Engl. J. Med. 301, 1254–1259.

Index